AF776912

Polymers Synthesis and Characterization

Polymers Synthesis and Characterization

Editor

Edina Rusen

MDPI • Basel • Beijing • Wuhan • Barcelona • Belgrade • Manchester • Tokyo • Cluj • Tianjin

Editor
Edina Rusen
Bioresources and Polymer
Science
University Politehnica
Bucharest
Bucharest
Romania

Editorial Office
MDPI
St. Alban-Anlage 66
4052 Basel, Switzerland

This is a reprint of articles from the Special Issue published online in the open access journal *Polymers* (ISSN 2073-4360) (available at: www.mdpi.com/journal/polymers/special_issues/Polym_Synth_Charact).

For citation purposes, cite each article independently as indicated on the article page online and as indicated below:

LastName, A.A.; LastName, B.B.; LastName, C.C. Article Title. *Journal Name* **Year**, *Volume Number*, Page Range.

ISBN 978-3-0365-7271-0 (Hbk)
ISBN 978-3-0365-7270-3 (PDF)

© 2023 by the authors. Articles in this book are Open Access and distributed under the Creative Commons Attribution (CC BY) license, which allows users to download, copy and build upon published articles, as long as the author and publisher are properly credited, which ensures maximum dissemination and a wider impact of our publications.
The book as a whole is distributed by MDPI under the terms and conditions of the Creative Commons license CC BY-NC-ND.

Contents

Preface to "Polymers Synthesis and Characterization"

Polymer science represents a domain of great interest due to the possible applications of polymers in areas that range from the most common to those that are high-tech. To achieve this, synthesis and characterization techniques, as well as the correlation of the chemical structure and morphology with their properties, is critical. From the perspective of synthesis, there are two strategies for obtaining polymers: step-growth (polyaddition, polycondensation, and reversible-deactivation radical polymerization) and chain polymerization (radical polymerization, coordinative polymerization, and cationic and anionic polymerization). Polymer analogous reactions aimed at the control of the hydrophilic/hydrophobic characteristics of the surface or the introduction of responsive moieties onto a polymer backbone are also interesting related topics. Surface modification can also be performed using surface-initiated controlled radical polymerization reactions such as atom transfer radical polymerization (ATRP), reversible addition-fragmentation chain transfer (RAFT), nitroxide-mediated living free-radical polymerization (NMP), and iodine transfer polymerization (ITP). Through these methods, polymer grafts and block copolymers are obtained, displaying different properties than random or alternate polymers obtained by radical copolymerization, which follows the well-known Mayo–Lewis equation. Thus, the physical properties of the polymers can be tailored depending on the structures that are synthesized and their morphology. The right design that encompasses structure, composition, and morphology is the key to the final applications of polymeric structures.

The advancement of polymer applications continues to lead to the expansion of the synthesis and characterization techniques that can be facilitated by novel, smart, multifunctional polymers. There is a correlation between structure, composition, morphology, and properties of applications that leads to substantial benefits for specific applications.

This Special Issue established a collection of articles and reviews that follow the latest developments in polymer synthesis, their characterization techniques, and the correlation between structure and properties.

Edina Rusen
Editor

polymers

MDPI

Article

Green Synthesis of Fe–Cu Bimetallic Supported on Alginate-Limestone Nanocomposite for the Removal of Drugs from Contaminated Water

Inas A. Ahmed [1,*], Hala. S. Hussein [2], Zeid A. ALOthman [3], Abdullah G. ALanazi [3], Norah Salem Alsaiari [4] and Awais Khalid [5]

1 Department of Chemistry, Faculty of Science, King Khalid University, Abha 62224, Saudi Arabia
2 Chemical Engineering & Pilot Plant Department, Engineering Research and Renewable Energy Institute, National Research Centre, Cairo 11865, Egypt
3 Chemistry Department, College of Science, King Saud University, Riyadh 11451, Saudi Arabia
4 Department of Chemistry, College of Science, Princess Nourah bint Abdulrahman University, P.O. Box 84428, Riyadh 11671, Saudi Arabia
5 Department of Physics, Hazara University, Mansehra 21300, Khyber Pakhtunkhwa, Pakistan
* Correspondence: eaahmed@kku.edu.sa

Abstract: In this study Fe–Cu supported on Alginate-limestone (Fe–Cu/Alg–LS) was prepared. The increase in surface area was the main motivation for the synthesis of ternary composites. Scanning electronic microscopy (SEM), energy-dispersive X-ray spectroscopy (EDX), and transmission electron microscopy (TEM) were used to examine the surface morphology, particle size, percentage of crystallinity, and elemental content of the resultant composite. Fe–Cu/Alg–LS was used as an adsorbent for the removal of drugs such as ciprofloxacin (CIP) and levofloxacin (LEV)from contaminated medium. The adsorption parameters were computed using kinetic and isotherm models. The maximum removal efficiency of CIP (20 ppm) and LEV (10 ppm) was found to be 97.3% and 100%, respectively. The optimal conditions were pH 6 and 7 for CIP and LEV, optimum contact time 45, 40 min for CIP and LEV, and temperature of 303 K. The pseudo-second-order model, which confirmed the chemisorption properties of the process, was the most appropriate kinetic model among the ones used, and the Langmuir model, which was the most appropriate isotherm model. Moreover, the parameters of thermodynamics were also assessed. The results imply that the synthesized nanocomposites can be used to remove hazard materials from aqueous solutions.

Keywords: ciprofloxacin (CIP); levofloxacin (LEV); bimetallic; limestone; alginate; adsorption

Citation: Ahmed, I.A.; Hussein, H.S.; ALOthman, Z.A.; ALanazi, A.G.; Alsaiari, N.S.; Khalid, A. Green Synthesis of Fe–Cu Bimetallic Supported on Alginate-Limestone Nanocomposite for the Removal of Drugs from Contaminated Water. *Polymers* **2023**, *15*, 1221. https://doi.org/10.3390/polym15051221

Academic Editor: Edina Rusen

Received: 15 November 2022
Revised: 24 January 2023
Accepted: 30 January 2023
Published: 28 February 2023

Copyright: © 2023 by the authors. Licensee MDPI, Basel, Switzerland. This article is an open access article distributed under the terms and conditions of the Creative Commons Attribution (CC BY) license (https://creativecommons.org/licenses/by/4.0/).

1. Introduction

In the last decade, pharmaceutical residues, typically at low concentrations, have been detected in the water cycle, including surface waters, wastewater, groundwater and, to a lesser extent, drinking-water. Their presence in water, even at very low concentrations is a potential risk not only to human health from exposure to traces of pharmaceuticals via drinking-water, but also to aquatic organisms and other components of the environment [1,2]. Accordingly, there is a significant need to propose cost-effective and feasible alternatives to eliminate these toxic products from the environment. In this investigation, the fluoroquinolone antibiotics ciprofloxacin (CIP) and levofloxacin (LEV) are used as model antibiotics. In river water, CIP and LEV antibiotics were found in amounts ranging from 0.2 to 18.8 ng/L and 0.3 to 6.0 ng/L, whereas industrial wastewater had concentrations of 0.91 to 99.3 ng/L and 0.5 to 19,981 ng/L, respectively [3]. These antibiotics were detected in drinking water after conventional water treatment techniques such as flocculation, sedimentation, and chlorination. As a result, alternative technologies for treating effluents containing antibiotics, such as electrocoagulation [4], membrane filtration [5], advanced oxidative process [6], or adsorption, have received a lot of attention [7–9]. The adsorption process has been suggested as a more viable approach than alternative treatment methods due to a number of advantages, including cheap cost, reusability, accessibility and flexibility, ease of operation, and nonsusceptibility to contaminants and dangerous compounds [10]. Recently, there is a tremendous increase in research for using green nanomaterials as an adsorbent, with an emphasis on their possible use in environmental remediation. Metal-based nanoparticles (for instance, nCu or nFe) are one type of nanomaterial that is frequently suggested for wastewater treatment because of their advantages of higher intrinsic reactivity on their surface sites due to their small particle size and large specific surface area, which is frequently suitable for removal of various organic and inorganic contaminants from surface or ground waters [10–14].

Alginate is a naturally occurring polysaccharide that is commercially harvested from brown algae. It has a highly functionalized backbone (mostly hydroxyl and carboxylic groups) and can be used to create materials that are ionically crosslinked with divalent cations. It has additional properties such as biodegradability, renewability, and nontoxicity which encouraged its contribution with other adsorbents to water treatment [15]. Alginate can be combined with other materials, such as chitosan, hydroxyapatite, or activated carbon to create materials with a variety of properties used in several fields, such as medicine, pharmacy, and environmental protection. Hence, Wasilewska and Deryo-Marczewska used alginate–carbon composites as an effective adsorbent for the removal of non-steroidal anti-inflammatory drugs [16]. Moreover, limestone is a common adsorbent due to its reasonable price and widespread availability in nature. It has been shown that limestone can remove various contaminants, including heavy metals, dyes, and pharmaceuticals. In addition to its heterogeneous surface, the buffering capability, secondary binding site, and repurposing properties are particularly helpful. As a result, limestone can be used as an inexpensive adsorbent in the treatment of water [17].

Green synthesis provides several advantages over conventional synthesis technologies, including being more freely diffused and using less chemicals that are naturally harmful and dangerous. For the creation of metal-based nanoparticles, several natural plants, including green tea, have been employed [18]. According to Zhu et al.'s study [19] on the green synthesis of nano zero-valent iron/Cu using green tea, this approach was shown as more affordable and environmentally friendly than traditional ones. Polyphenols, flavonoids, and other reducing agents found in plant extracts can effectively reduce ferric or ferrous ions and prevent the agglomeration of nanoparticles. Hu et al., [20] investigated the removal of ciprofloxacin with aluminum-pillared kaolin sodium alginate beads with maximum adsorption capacity 68.36 mg/g at pH = 4. Moreover, levofloxacin (LEV) was examined by Al-Jabari et al. [21] for its ability to bind to the surface of superparamagnetic iron oxide nanoparticles (Fe_3O_4) and Fe_3O_4 & SiO_2 nanocomposites. At pH 6.5 and 240 min contact time, the greatest removal effectiveness of 80.2% of LEV was attained.

In this study, zero valence Fe–Cu nanoparticles supported on alginate-limestone (ZVFe–Cu/Alg–LS) were prepared as a successful nanocomposite material for eliminating antibiotics from water. The nanocomposite (ZVFe–Cu/Alg–LS) is presented as a group of adsorbents (zero valence Fe–Cu nanoparticles, alginate polymer and limestone) exhibting high ability to adsorb and degrade the fluoroquinolone antibiotics. It is an environmentally friendly nanocomposite that includes several active sites synthesized for removal of CIP and LEV antibiotics. The efficient removal of CIP and LEV was predicted to benefit from the synergistic effect of the (ZVFe–Cu/Alg–LS) nanocomposites. The operating conditions such as the solution pH, drug concentration, contact time and temperature were investigated beside Langmuir and Freundlich isotherm models. In addition, the kinetic models, were also applied to analyze the experimental equilibrium data.

2. Experimental

2.1. Materials

The materials utilized included limestone from Al-Gomhoria Company (Al-Mansoura, Egypt), calcium chloride ($CaCl_2$), sodium alginate, potassium dichromate ($K_2Cr_2O_7$), copper (II) sulphate pentahydrate ($CuSO_4{\cdot}5H_2O$), and ferrous sulfate ($FeSO_4.7H_2O$) acquired from Sigma-Aldrich. Green tea leaves obtained from a local market (Egypt). Amon Company, Cairo, Egypt, provided the ciprofloxacin (20 ppm) and levofloxacin (10 ppm). Table 1 lists the physicochemical properties of ciprofloxacin and levofloxacin. The sample's pH was accustomed using sodium hydroxide (NaOH) and hydrochloric acid (HCl). None of the chemicals had been refined and were all of commercial purity.

Table 1. Physicochemical properties of ciprofloxacin and levofloxacin.

	Ciprofloxacin (CIP) [22]	**Levofloxacin (LEV)**
Structure	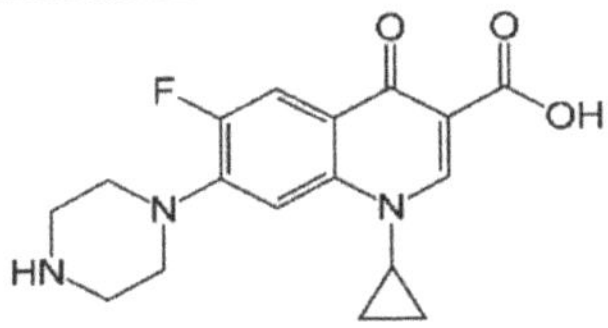	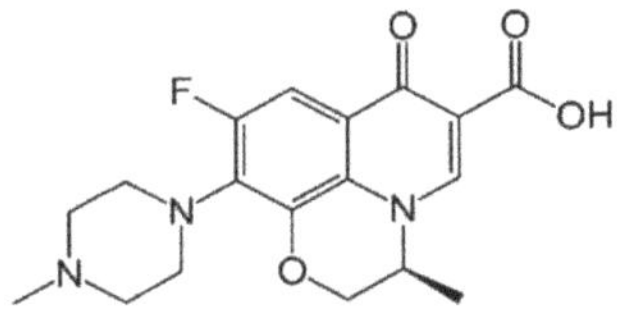
Molecular formula	$C_{17}H_{18}FN_3O_3$	$C_{18}H_{20}FN_3O_4$
Usage	Antibiotic	Antibiotic
Molecular weight (g mol^{-1})	331.346	361.373
pKa	6.09	5.59 (acid) and 7.94 (base) [23]
Water solubility (g L^{-1}) at 20 °C	36	1.44 [24]

2.2. Synthesis of Copper Nanoparticles

The production of green copper nanoparticles was performed using green tea leaf extracts, according to the technique explained by Asghar [25]. Green tea leaves were purchased from a local market in Cairo (Egypt). The leaf extracts of each tea were separated by exactly weighing 10 g of tea leaves and transferring into a 250 mL conical flask already containing 100 mL of DI-H_2O. The mixtures were then heated at 80 °C for 10 min, cooled and filtered. Copper nanoparticles from black tea were synthesized using a $CuSO_4$ solution with the corresponding tea leaves extract. Briefly, $CuSO_4$ (1 mmol/L) and tea leaf extracts were used in a 4:1 ratio by volume, and the solution was subjected to continuous stirring at 80 °C for 10 min. The resultant suspensions were settled at room temperature for 24 h to complete the reaction, then filtered and washed three times with DI-H_2O to remove any unbound molecules. Lastly, Cu-NPs were dried at 65 °C for 3 h.

2.3. Synthesis of Fe^- Nanoparticles

Black tea leaves were used to prepare Fe^- nanoparticles. Black tea leaves was washed many times with tap water and then with double distilled water to eliminate impurities. The washed tea was dried at 105 °C for 24 h in a hot air oven and then powder form obtained after grinding of dried tea. Extraction procedure was carried out according to Anamika et al. [14]. An extraction of tea sample was prepared with a 1:10 ratio of tea powder to water, and the extraction was carried out at 80 °C for 30 min. Then the extract was filtered after cooling at room temperature and then 0.1 M of $FeSO_4$ solution was added to the tea extract at a ratio of 1:3 (1-part $FeSO_4$ and 3-part extract) [26,27]. Then sodium hydroxide solution was added, dropwise. During this process, the ferrous ions were converted into nanometallic particles rapidly. Separation of prepared black nanoparticles was performed by filtration and then washed with double distilled water. Separated nanoparticles were subsequently dried in a vacuum at 50 °C for 24 h.

2.4. Preparation of ZV Fe–Cu/Alg–LS Nanocomposites

In order to formulate ZV Fe–Cu/Alg–LS nanocomposites, 2% (*w*/*v*) adsorbents of sodium alginate and 7 g of limestone were mixed in 100 mL of distilled water. The combination was stirred with a mechanical stirrer and heated on a hot plate to 80 °C. When the mixture of limestone and alginate attained a homogeneous condition, a solution of mixed zero valance Fe–Cu (0.5 Fe–0.5 Cu in 100 mL) was added. Then 0.3 M of calcium chloride was dripped through a syringe injector into to form beads. In order to obtain hardened beads, the beads were submerged in a calcium chloride ($CaCl_2$) solution for 12 h. Washing the adsorbents with distilled water several times removed the excess unbounded calcium chloride from the adsorbent surface.

2.5. *Surface Characterization of the Nanocomposites*

2.5.1. Instruments

The investigation of the adsorbent was carried out using a Genesis-II FTIR spectrometer (ALT, San Diego, CA, USA) (using potassium bromide). Additionally, (SEM) was performed using an Inspect S (FEI Company, Eindhoven, the Netherlands) equipped with an energy-dispersive X-ray analyzer (EDX, Quanta 200, FEL, Eindhoven, the Netherlands).The mineralogical structure of the powdered materials was determined using X-ray diffraction (XRD) and logged on a Philips PW 1050/70 diffractometer (Philips, Am-sterdam, the Netherlands) using a Cu–Kα source with a post-sample Kα filterant, a scanning speed of 1 s/step, a range of 5 to 50 (2θ°) and a resolution of 0.05°/step. The surface area was determined using BEL SORB max (Made in Japan). TEM analysis was performed using the JEM-HR-2001 model (JEOL, Akishima, Japan) with an accelerating voltage of 200 kV to assess the particle sizes of the material. CIP and LEV were detected in aqueous solutions using an Agilent HPLC 1200 Infinity apparatus equipped with a photodiode array detector (Agilent Technologies, Waldbronn, Germany). At 280 nm, the chromatograms were captured. At a temperature of 25 C, an Agilent Zorbax Eclipse Plus C18 column (3.5 mm, 150 mm, 4.6 mm) (Agilent, Newport, CA, USA) was used. A total of 40% water (mobile phase A) and 60% acetonitrile made up (mobile phase C). The flow rate was set at 1.0 mL/min. BEL and SORB max (Made in Japan) assisted in measuring the surface's area, and an OHAUS STARTER 2100 pH meter (Pine Brook, NJ, USA) was used for pH adjustment [28].

2.5.2. Adsorption Process

The adsorption was carried out by mixing an identified quantity of the sorbent with an aqueous solution of CIP and LEV at the necessary concentration in a 100 mL capped flask using a mechanical stirrer. First, a certain quantity of the sorbent and 25 mL of the sorbate solution were mixed and agitated for long enough to allow for sorption equilibrium. After filtering the mixtures using filter paper, HPLC was used to calculate the concentration of the antibiotic and medication in the solution. By adjusting the contact duration t (10–90 min), pharmaceutical solution starting concentration (10–100 ppm), and drug solution beginning pH (2–10) using 0.1 M NaOH and HCl, researchers were able to study the effects of a variety of parameters on sorption. The proportion of antibiotics sorption was assessed by the incoming Equations (1) and (2) [29].

$$\text{Adsorption capacity } q_e = \frac{(C_0 - C_e)V}{W} \quad (1)$$

$$\text{Removal efficiency } \% = \frac{(C_0 - C_e)}{C_0} \times 100 \quad (2)$$

where V (L) and W (g) stand for the volume of the solution and the weight of the adsorbent, respectively, C_0 and C_e also stand for the initial and equilibrium concentrations (mg/L) of CIP and LEV ions, respectively.

3. Results and Discussion

3.1. *Zero Valant Fe–Cu/Alg–LS Nanocomposites Characteristics*

3.1.1. FTIR Study

Figure 1 shows the Fe–Cu/Alg–LS nanocomposites' Fourier-transform infrared spectra before and after CIP and LEV adsorption. A specific band that was associated with O-H (hydroxyl) groups appeared at about 3440 cm^{-1}. The peaks observed at 1334 and 1081 cm^{-1} suggest the existence of OH bending and C-O stretching vibrations [30]. At 1774 cm^{-1}, two additional peaks can be seen that are associated to the stretching vibration of C=O seen in carboxylic and/or carbonyl moiety groups [31]. Additionally, a peak at 700 cm^{-1} that was correspond to C-H out-of-plane bending in benzene derivatives [32] was observed. As demonstrated in Figure 1B, in Fe–Cu/Alg–LS loaded CIP, After the adsorption, numerous functional groups adjusted to different bands, it was noticed that the bands at 3430, 1776, 881, 700 and 416 cm^{-1} shifted to 3428,1774, 879, 696 and 420 cm^{-1}, respectively. Moreover, after adsorption of levofloxacin (LEV) the bands at 3430, 1776, 1384, 881, 700 and 416 cm^{-1} shifted to at 3482, 1774, 1382, 856 and 422 cm^{-1}, respectively. Shifting the bands can explain the presence of H-bonded OH in the adsorption of CIP and LEV on Fe–Cu/Alg–LS nanocomposites.

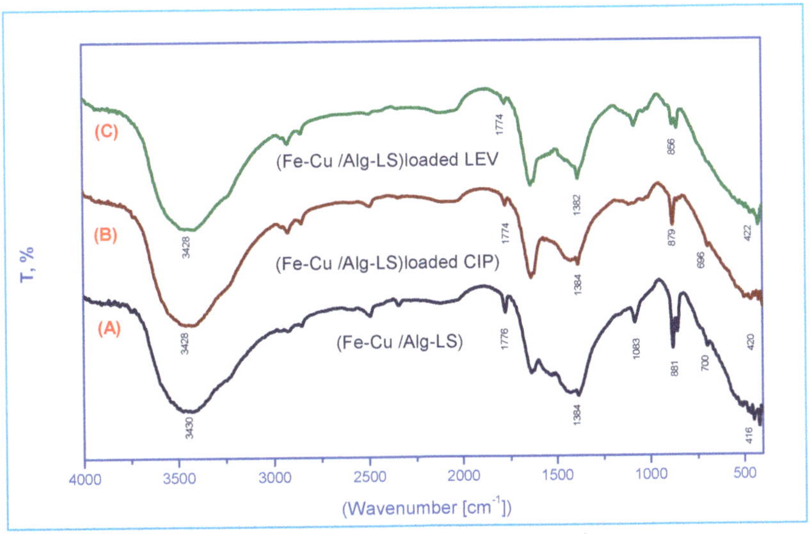

Figure 1. FTIR spectra of (**A**) the Fe–Cu/Alg–LS nanocomposite, (**B**) Fe–Cu/Alg–LS—loaded CIP, (**C**) Fe–Cu/Alg–LS—loaded LEV.

3.1.2. XRD Study

Spectra were analyzed for the phase purity of the Fe–Cu/Alg–LS nanocomposite as shown in Figure 2. The characteristic diffraction peaks located at 43.7° (111) indicated formation of Cu nanocrystals [33]. The diffraction peak observed at 2θ = 44.77° are indexed to (101) denoted the crystalline phase for Fe nanoparticles [34]. The peaks at 29.4 and 47.1° indicated the presence of calcite [35]. The results reveal the formation of copper, ferric and calcite that guarantees the good synthesis of the Fe–Cu/Alg–LS nanocomposites.

3.1.3. SEM and EDX Study

Figure 3 shows the SEM analysis of the generated. Fe–Cu/Alg–LS nanocomposites both before and after the adsorption of ciprofloxacin (CIP) and levofloxacin (LEV). Figure 3 provides SEM images (A–D). The micrographs display rough surfaces, many holes, and nanoparticles dispersed throughout the sample. The surface exhibits an uneven surface overall for the adsorption of the designated antibiotics. Figure 3A's surface morphologies showed more pores than Figure 3B,C, indicating that the nanocomposite has enough space for the adsorption process to take place. According to Figure 3B,C, the nanocomposite had less pores due to the CIP and LEV that cover the composite.

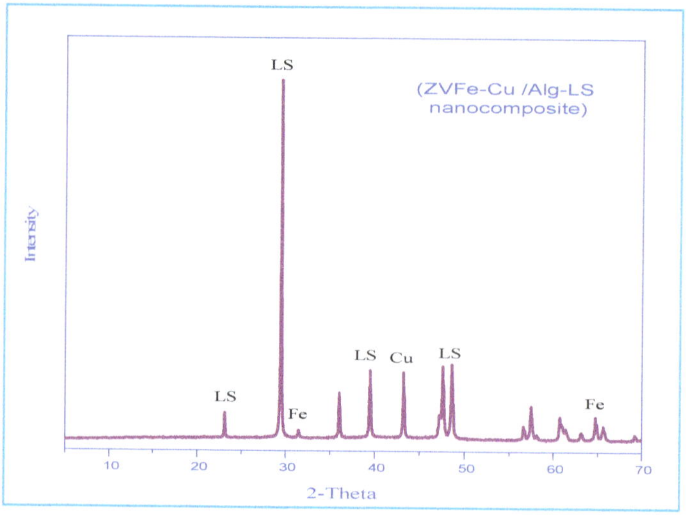

Figure 2. XRD of the Fe–Cu/Alg–LS nanocomposite.

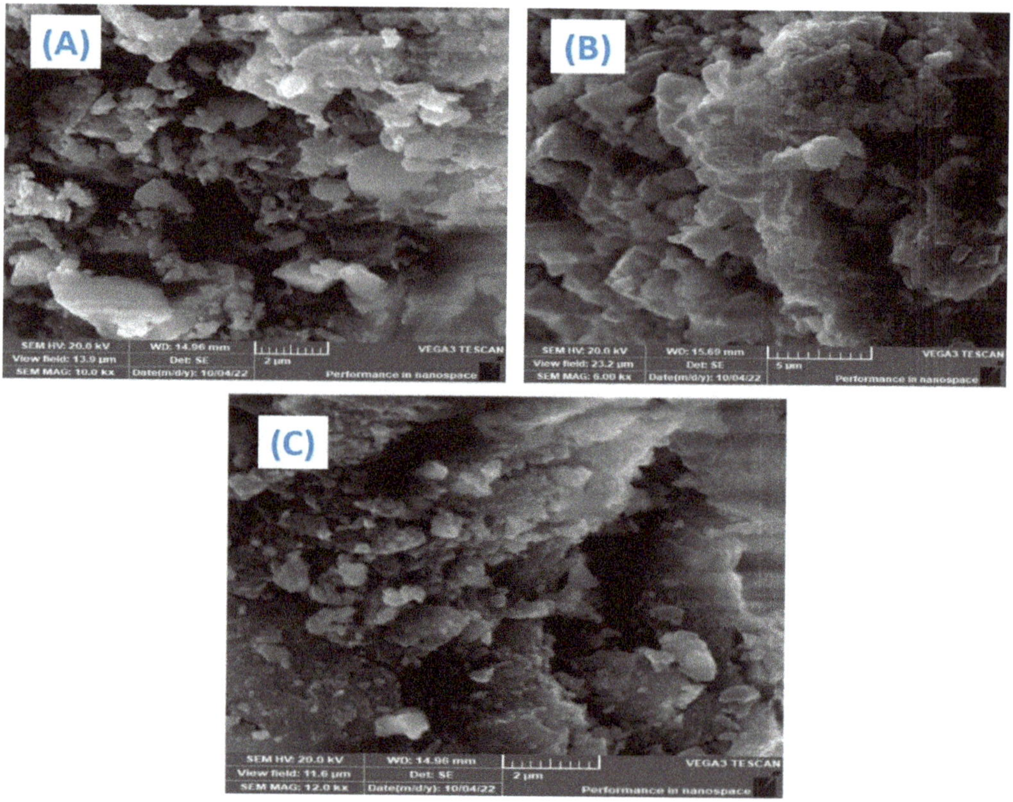

Figure 3. SEM of (**A**) Fe–Cu/Alg–LS nanocomposite, (**B**) Fe–Cu/Alg–LS—loaded CIP, (**C**) Fe–Cu/Alg–LS—loaded LEV.

The EDX analysis of Fe–Cu/Alg–LS nanocomposites are shown in Figure 4. The EDX analysis of the Fe–Cu/Alg–LS nanocomposites before adsorption reveals the peaks corresponding to oxygen, carbon, cupper, ferric and calcium elements. Thus, EDX guarantees the good synthesis of the Fe–Cu/Alg–LS nanocomposites.

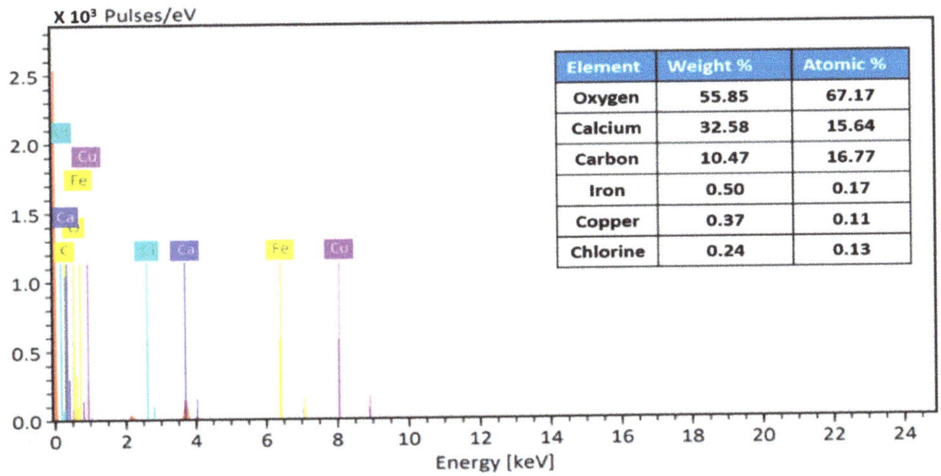

Element	Weight %	Atomic %
Oxygen	55.85	67.17
Calcium	32.58	15.64
Carbon	10.47	16.77
Iron	0.50	0.17
Copper	0.37	0.11
Chlorine	0.24	0.13

Figure 4. EDX of Fe–Cu/Alg–LS nanocomposite.

3.1.4. Transmission Electron Microscopy Study

The TEM and particular area electron diffraction images of the Fe–Cu/Alg–LS nanocomposite is provided in Figure 5. From TEM micrographs, it is clear that the constructed Fe–Cu/Alg–LS nanocomposite exhibited a multilayer structure. This rough surface indicates that nucleation occurred. As seen in Figure 5A, the sample's microstructure and porosity are well suited for enhanced absorption. The wide range in particle size was shown in a histogram of the particle size distribution generated from the TEM images. The particles have an average diameter of 45.54 nm and range in size from 40 to 50 nm.

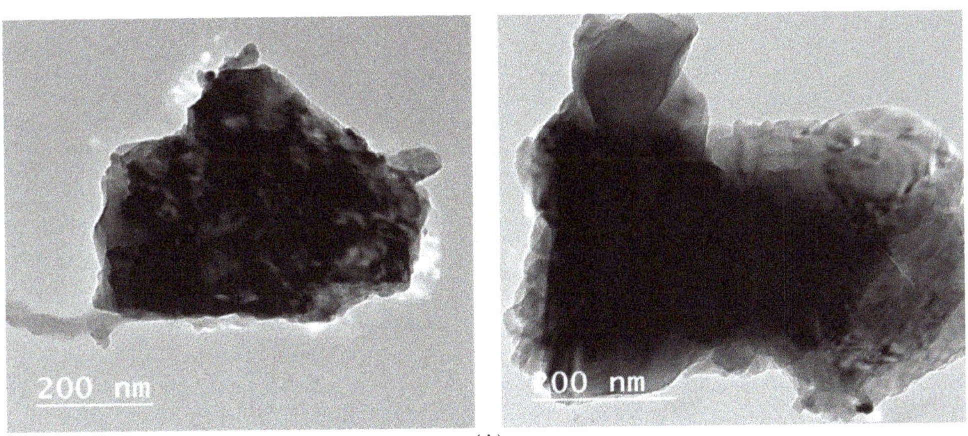

(A)

Figure 5. *Cont.*

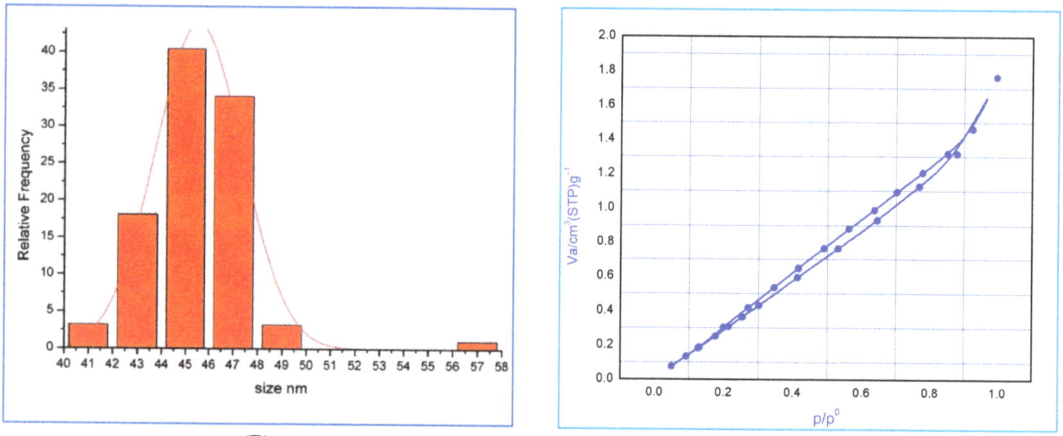

(B) **(C)**

Figure 5. (**A**) TEM analysis of the ZVFe–Cu/Alg–LS nanocomposite (**B**) particle size distribution for ZVFe–Cu/Alg–LS nanocomposite (**C**) Adsorption–desorption nitrogen isotherms.

3.1.5. BET Adsorption—Desorption Measurements

The surface area and porous structure were examined with N_2 adsorption–desorption tests. The N_2 adsorption–desorption isotherms curve of the as-prepared aerogels obtained at 77 K are shown in Figure 5C. Figure 5C reveals a type-IV isotherm for the as-prepared samples calcined at different temperatures, indicating the existence of a mesoporous structure. The surface area data showed that the pore volume and surface area of the Fe–Cu/Alg–LS nanocomposite were 0.04432 $cm^3\ g^{-1}$ and 21.05 $m^2\ g^{-1}$, respectively, as listed in Table 2.

Table 2. The pore size distribution adsorption results (surface area, pore volume and average pore radius).

Sample	Surface Area ($m^2\ g^{-1}$)	Pore Volume ($cm^3\ g^{-1}$)	Pore Radius (nm)
Fe–Cu/Alg–LS	21.0506	0.04432	1.92176

3.2. Performance of the Fe–Cu/Alg–LS Nanocomposite

3.2.1. Effect of pH

An extremely significant factor that affects the removal efficiency of an adsorbent in wastewater treatment is the pH of the solution since the adsorption efficiency is influenced by the pH of the medium. CIP and LEV adsorption was adjusted to make the solution acidic, neutral, and alkaline (2–10). According to Figure 6A, the maximum CIP and LEV removal was obtained at pH 6 and 7, respectively. It is well recognized that the solubility of CIP and LEV is a function of pH, which is explained by the presence of different CIP and LEV chemical species at the different pH values. At low pH values, a highly soluble CIP^+ and LEV^+ species occurs and its fraction value decreases as pH values move from 2 to 7, where the pKa constant value (carboxylic acid group) is located [36]. Finally, as the pH value continues to increase to higher than 7, CIP^+ and LEV^+ becomes more soluble because of the appearance of the CIP^- and LEV- species [37,38]. This behavior can be described through the relationship between CIP and LEV total charge and the surface charge of the Fe–Cu/Alg–LS nanocomposite. As the pH increases up to 6 and 7, the cationic form (CIP^+ and LEV^+) is present, the negative Fe–Cu/Alg–LS nanocomposite surface will perform a significant adsorption of the pollutant. Moreover, the high efficiency may be attributed to the increase of adsorbent surface area and greater availability of adsorption active sites. The removal efficiency decreases significantly after the initial pH value reaches 7. This performance can be associated with the presence of the anionic

CIP and LEV form (CIP^- and LEV^-), which can produce repulsive interactions with the Fe–Cu/Alg–LS nanocomposite negative surface [39].

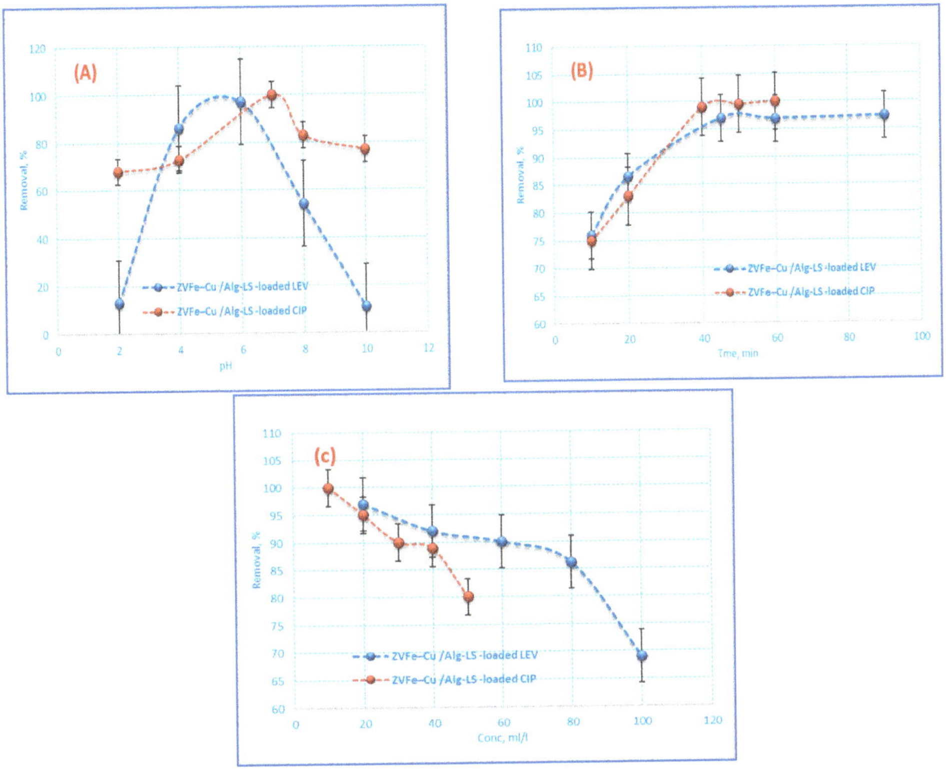

Figure 6. Influences of (**A**) pH, (**B**) contact time and (**C**) initial (CIP and LEV) concentration on the adsorption of CIP and LEV by 0.2 g/25 mL of the nanocomposite at pH 6 (CIP) 7 (LEV) and a contact time of (40 min (CIP) 45 min (LEV)).

3.2.2. Contact Time Effect

Contact time is one of the important influences in the adsorption of the CIP and LEV onto the Fe–Cu/Alg–LS. The effect of contact time on CIP and LEV adsorption on the Fe–Cu/Alg–LS nanocomposite at concentrations of 20 ppm (CIP),10 ppm (LEV) using 0.2 g/25 mL of the nanocomposite and pH 6 (CIP) and 7 (LEV) is presented in Figure 6B. The results illustrate that in CIP and LEV removal efficiency increased to 97.3% and 100% with time. Furthermore, Figure 6B exhibited that the adsorption rate was quick in the first period of time and moderate after 40 min. This may be qualified to the accessibility of abundant free active sites on Fe–Cu/Alg–LS at the initial adsorption stage for CIP and LEV sorption. The rate became very slow after 40 min, and no appreciable CIP and LEV removal was achieved. Hence, equilibrium was reached at about 40 min for CIP and 45 min for LEV. The number of existing active sites reduced with time, and eventually, the adsorbent becomes saturated [40,41]. Consequently, beyond the equilibrium time, no significant uptake of CIP and LEV took place as depicted in Figure 6B. This result could be ascribed to an increase in electrostatic interactions between the surfaces of adsorbents and adsorbates.

3.2.3. Effect of the CIP and LEV Concentrations

Using 0.2 g/25 mL of the nanocomposite, the impact of CIP and LEV concentrations on the adsorption process was investigated at concentrations ranging from 10 to 100 ppm.

In addition, the applied pH was (6 for CIP and 7 For LEV), and a contact time of (45 min for CIP and 40 min for LEV), as shown in Figure 6C. As expected, the increase in the concentration of CIP and LEV has a negative effect on the removal efficiency. Moreover, at high concentrations of antibiotics, the adsorbent surface was saturated with pollutants which decreased the adsorption uptake. Consequently, the removal efficiency was found to be decreased from 97% to 69% for CIP and from 100% to 80% for LEV.

3.3. *Kinetic Models*

The pseudo-first-order, pseudo-second-order, and intraparticle diffusion models were engaged to evaluate the adsorption kinetics of the Fe–Cu/Alg–LS nanocomposite. [37]. The optimum situations were conventional as pH 6 for (CIP) 7 for (LEV), Fe–Cu/Alg–LS G nanocomposite mass of 0.2 g/25 mL, a contact time of (40 min (CIP), 45 min (LEV)), and 20 ppm (CIP), 10 ppm (LEV) as the initial concentrations.

3.3.1. Pseudo-First-Order Reaction Kinetics

Figure 7A [42] provides a description of the PFOR reaction kinetics equation. The following equation is used to represent the current starting phase:

$$\mathrm{Log}\,(q_e - q_t) - \log q_e = -\,K_{ads}\,t/2.303 \qquad (3)$$

where q_t (mg/g) represents the adsorption capacity at time t. K_{ads} (min^{-1}) stands for the rate constant of PFOR adsorption.

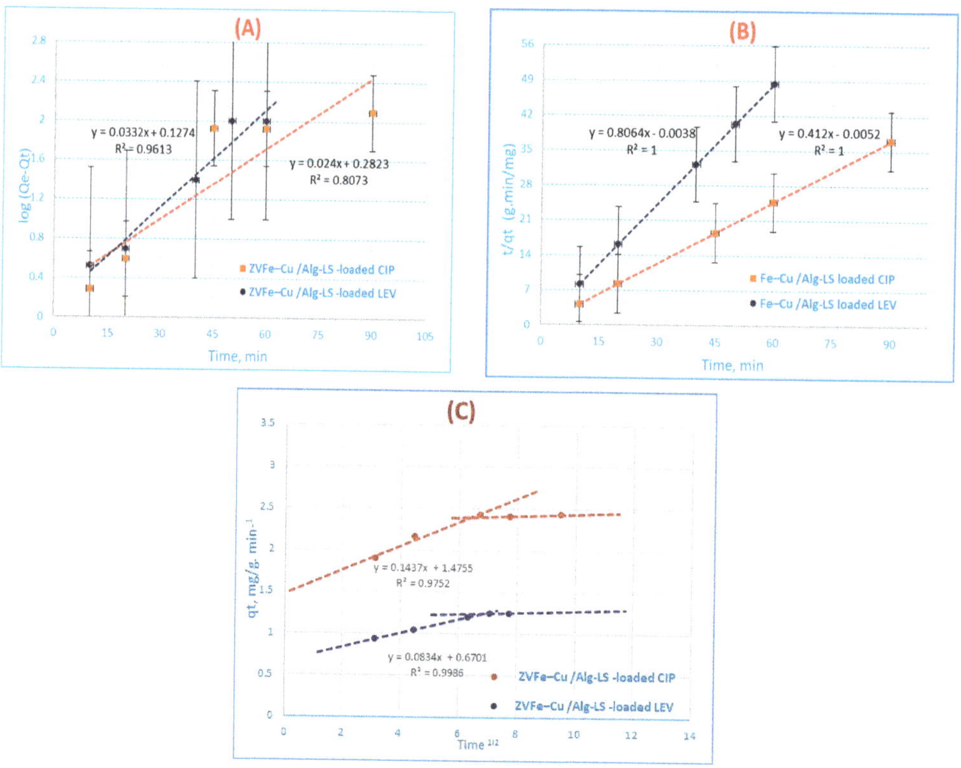

Figure 7. Adsorption kinetics: (**A**) pseudo-first-order reaction (PFORE), (**B**) pseudo-second-order reaction (PSORE), (**C**) Morris–Weber equation for CIP and LEV on the Fe–Cu/Alg–LS nanocomposite (sorption time, (40 min (CIP) 45 min. (LEV) by 0.2 g/25 mL of the nanocomposite at pH 6 (CIP) and 7 (LEV).

In this study, a linear relationship was recognized for the adsorption of CIP and LEV ions onto the Fe–Cu/Alg–LS nanocomposite. The values of q_e and k_{ads} were measured from the slope and intercept by plotting log ($q_e - q_t$) versus t. The PFOR kinetics are illustrated in Figure 7A. The outcomes exhibited correlation coefficients (R^2 = 0.8073, 0.9613) for CIP and LEV. The collected data show that the pseudo-first-order model has a poor fit for the adsorption of CIP and LEV onto the Fe–Cu/Alg–LS nanocomposite.

3.3.2. Pseudo-Second-Order Reaction

The PSOR kinetic model [43] is illustrated in the following equation:

$$t/q = 1/K_2 q_{e2} + t/q_e \tag{4}$$

The PSOR rate constant, denoted by K_2 (g/mg/min), is shown in Figure 7B. When t/qt is plotted versus t, the slopes and intercepts determine the values of the rate constant K_2, equilibrium adsorption capacity q_e, and correlation coefficient (R^2).

PSOR correlation coefficients (R^2) for the Fe–Cu/Alg–LS nanocomposite in Table 3 maintained high values. The outcomes exhibited high correlation coefficients (R^2 = 1) for both CIP and LEV. The statistics imply that the CIP and LEV adsorption suitable for the pseudo-second-order kinetics.

Table 3. Kinetic modeling with the PFOR, PSOR and Morris–Weber equations.

Kinetic Models	Parameter	Ciprofloxacin (CIP)	Levofloxacin (LEV)
PFOR	q_e, exp (mg g^{-1})	2.425	1.24
	q_e, cal (mg g^{-1})	1.915	1.34
	K_{ads} (min^{-1})	0.4120	0.8064
PSOR	R^2	0.8073	0.9613
	q_e, cal (mg g^{-1})	2.427	1.24
	K_2 (g mg^{-1} min^{-1})	0.00007	1.157
Morris–Weber	R^2	1	1
	K_d (mg g^{-1} $min^{0.5}$)	0.0831	0.0727
	R^2	0.1884	0.9783

3.3.3. Morris–Weber Kinetic Equation

The Morris–Weber Equation (5) [44] can be used to represent the intraparticle mass transfer diffusion, as shown in Figure 7C.

$$q = K_d\,(t)^{1/2} \tag{5}$$

where the uptake of metal ions is denoted by the symbol q (g/g), the intraparticle mass transfer diffusion rate constant is denoted by K_d, and the square root of time is denoted by the symbol $t^{1/2}$. Only in the shorter stage, if the intraparticle diffusion and adsorption data overlapped, would it occur. The first part is linear, which is related to the boundary layer effect, as shown by the Morris–Weber equation in Figure 7C. However, the intraparticle diffusion effect may be related to the second component [45]. The fact that practically all sorption occurs within the first 40 min for CIP and 45 min for LEV, with a clear linear trend, indicates that the porosity of nanocomposites exceeds the resistance influencing intraparticle diffusion [46]. For CIP and LEV adsorption, the intraparticle diffusion rate constant value K_d was estimated to be 0.0831 and 0.0727 (g/g min^{-1}), respectively, onto the Fe–Cu/Alg–LG nanocomposite, suggesting CIP and LEV ions move to the composite. The values of K_d for both antibiotics represent the rate of diffusion of pollutants towards the pore of nanocomposite, accordingly the rate of diffusion of CIP is higher than LEV onto nanocomposite. The kinetic modeling with the PFOR, PSOR and Morris–Weber equations are detailed in Table 3.

3.4. Isotherm Model

To adequately understand the adsorption process, isotherm studies are required [47]. The Langmuir, Freundlich, and Dubinin–Radushkevich isotherm models were used to study the adsorption process. The Fe–Cu/Alg–LS nanocomposite was 0.2 g/25 mL in mass, with contact times of 40 min for CIP (20 ppm) and 45 min for LEV (10 ppm) according to the optimized experimental conditions.

The Langmuir isotherm was used to explain the adsorption of any substance on a homogeneous surface with minimal interaction between the molecules that had been adsorbed [48]. The model assumes a homogeneous uptake in accordance with the saturation level of the monolayer on the surface with the highest adsorption. The following gives an illustration of the Langmuir linear equation model [49]:

$$C_e/q_e = 1/K_L\, q_{max} + (1/q_{max})\cdot C_e \tag{6}$$

where K_L ($L\cdot mg^{-1}$) denotes the monolayer's maximum adsorption capacity and q_{max} ($mg.g^{-1}$) denotes its maximum capacity for sorption heat. Figure 8A,B illustrate the Langmuir adsorption isotherm that was constructed on the basis of monolayer adsorption through the adsorption process. The equilibrium absorption of the homogeneous surface of the adsorbents is explained by the Langmuir model.

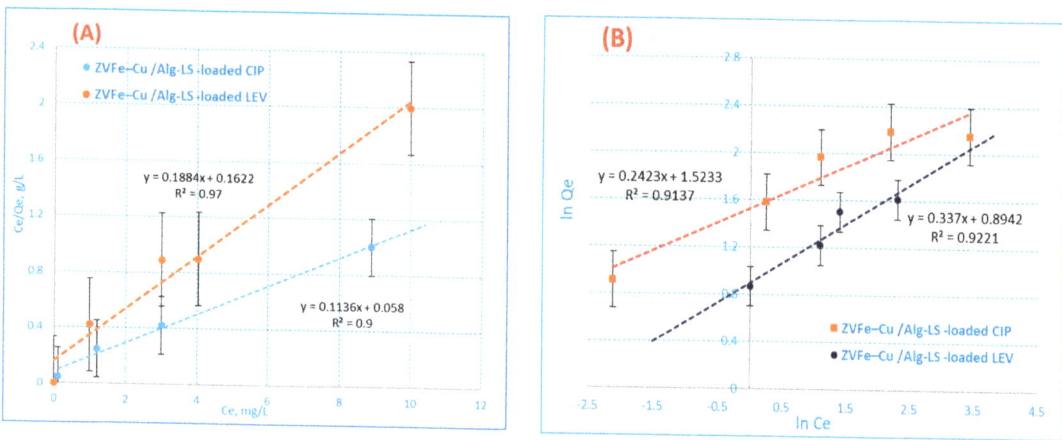

Figure 8. (**A**) Langmuir (**B**) Freundlich adsorption isotherm.

The Freundlich model, particularly for heterogeneous surfaces [50,51], is one of the first empirical equations compatible with the exponential distribution of active centers as follows:

$$\ln q_e = \ln K_f + 1/n \ln C_e \tag{7}$$

If K_f denotes adsorption capacity, n denotes intensity, and K_f is a crucial and relative indicator of adsorption capacity; it denotes a beneficial adsorption extent. Adsorption is considered suitable when n is greater than 1 [52]. The results demonstrate that the Langmuir model performed better than the Freundlich model in describing the experimental data of the Fe–Cu/Alg–LS nanocomposites. The values for the correlation coefficient (R^2) are described in Table 3. For both CIP and LEV adsorption, the R^2 values from the Langmuir model data were 0.9731 and 0.9990, above those of the Freundlich isotherm. The displacement of CIP and LEV ions appears to be a monolayer covered on the surface of the Fe–Cu/Alg–LS nanocomposite, according to the adsorption results. As a result, the outcomes closely matched the Langmuir model.

Dubinin–Radushkevich Isotherm

This model fits exceedingly well with the Gaussian energy distribution and adsorption techniques that were used on a heterogeneous surface. The D-R equation is as follows [53]:

$$\ln q = \ln q_{(D\text{-}R)} - ßε^2 \quad (8)$$

$$ε = RT \ln(1 + 1/C_e) \quad (9)$$

When the ideal gas constant, R, is taken into account, the theoretical adsorption capacity, q(D-R) ($mg{\cdot}g^{-1}$), the activity coefficient, ß ($mol^2\ kJ^{-2}$), the Polanyi potential(ε), T (absolute temperature in K), and E ($kJ\ mol^{-1}$), represented as the free energy change, are as follows:

$$E = 1/(2ß)^{1/2} \quad (10)$$

The E value can be used to identify the type of reaction. Physical forces are predicted to have an impact on the adsorption process if E < 8 $kJ\ mol^{-1}$. If E is between 8 and 16 $kJ\ mol^{-1}$, chemical ion exchange takes place during the sorption process. Particle diffusion may also be used to determine the sorption process if E is more than 16 $kJ\ mol^{-1}$ [54]. Table 4 provides a list of the D-R model simulation data. E values for the absorption of CIP and LEV ions onto the Fe–Cu/Alg–LS nanocomposite were 0.7624 and 0.7446 $kJ\ mol^{-1}$. As a result, if E is less than 8 $kJ\ mol^{-1}$, physical adsorption will affect the sorption [55].

Table 4. Sorption isotherms.

Kinetic isotherm	Parameter	Ciprofloxacin (CIP)	Levofloxacin (LEV)
Langmuir	q_e, cal. ($mg\ g^{-1}$)	8.84	5.3
	K_L ($L\ mg^{-1}$)	0.0012	0.0011
	R^2	0.9731	0.9990
Freundlich	K_F ($mol^{n-1}\ L^n\ g^{-1}$)	4.5850	2.7150
	n	4.12	3.60
	R^2	0.9137	0.9221
D–R model	E ($kJ\ mol^{-1}$)	0.7624	0.7446
	$q_{(D\text{-}R)}$ ($mg\ g^{-1}$)	8.4350	4.4260
	R^2	0.9402	0.8088

3.5. Sorption Thermodynamics

To estimate the thermodynamic action of the CIP and LEV adsorbed onto the Fe–Cu/Alg–LS nanocomposite, the thermodynamic factors were assessed in order to determine the thermodynamic viability and spontaneous nature of the process of adsorption. At varied temperatures (30, 40, and 60 °C), the findings were recorded. The formulae listed below were used to calculate the thermodynamic factors [56,57]:

$$ΔG° = -RT \ln K_d \quad (11)$$

$$ΔG° = ΔH° - TΔS° \quad (12)$$

$$\ln K_d = -ΔH°/RT + ΔS°/R \quad (13)$$

where T is the absolute temperature (K), K_d is the distribution coefficient, and R is the gas constant (8.314 $J\ mol^{-1}\ K^{-1}$). Using Equation (11), the Gibbs free energy was calculated. Furthermore, using Equation (12), G° might be calculated from H. Using Equation (13), the thermodynamic variables S° and H° were calculated (from the intercept and slope). The data showed that the amount of CIP and LEV ions taken up by nanocomposites slightly decreased in direct proportion to the temperature increase. In contrast to the adsorbent particles, the rise in degree of temperature increased the pollutants' solubility in a bulk solution to a larger extent [58]. Table 5 illustrates the Fe–Cu/Alg–LS nanocomposite components' thermodynamic sorption response to CIP and LEV ions.

Table 5. Thermodynamic conditions using 0.2 g/25 mL of the nanocomposite at pH 6 for CIP (10 ppm) and 7 for LEV (20 ppm), with contact times of 40 min and 45 min for CIP and LEV, respectively.

Parameter	T (K)	A%	ln K_L	ΔH° ($KJ \cdot mol^{-1}$)	ΔS° ($J \cdot mol^{-1} \cdot K^{-1}$)	ΔG° ($kJ \cdot mol^{-1}$)	R^2
Ciprofloxacin (CIP)	303 313 333	97.0 97.2 97.6	1.25 1.46 1.65	10.75	−46.12	−3.148 −3.812 −4.568	0.9621
levofloxacin (LEV)	303 313 333	99.0 98.6 98.4	2.50 2.20 2.03	−12.39	−20.49	−6.26 −5.72 −5.45	0.8929

Negative G° show that the adsorption process is feasible and spontaneous. Additionally, negative findings of H° also suggest that LEV was adsorbed onto the Fe–Cu/Alg–LS nanocomposite in an exothermic manner. Positive readings of H° show that CIP was adsorbed endothermically. Given how CIP and LEV adsorbed onto the surfaces of the adsorbents, negative S° for the Fe–Cu/Alg–LS adsorbent demonstrated that randomness declined at the solid–liquid interfaces, demonstrating that the adsorption was energetically stable [59]. For G° values under 80 kJ mol^{-1}, the sorption was of a physical origin. However, if G° was between 80 and 400 kJ mol^{-1}, it might have been chemical [60]. Table 4 shows the "G" values, which show that CIP and LEV sorption were of a physical origin. These results support the D-R isotherm.

3.6. A Comparison Study

A comparison study between the sorption capacities for CIP and LEV with other sorbents in literature are listed in Table 6.

Table 6. A comparison between the sorption capacities for CIP and LEV with other sorbents in previous work.

The Sorbent	Adsorption Capacity, mg/g	Conditions	References
	Ciprofloxacin		
Fe–Cu Bimetallic Supported on Alginate–Limestone Nanocomposite	8.8	20 ppm, 45 min.	Current research
Chemically prepared carbon from date palm leaflets	44.6	Ci = 200 ppm, 2880 min.	[9]
Pillared Clays	122.1	Ci = 18–500 ppm, 1440 min.	[61]
A chemically modified bamboo biochar was prepared from bamboo sawdust	78.43	Ci = 20 ppm, 46 min.	[42]
Protein-modified nanosilica (ProMNS)	85	Ci = 10 ppm, 90 min.	[62]
Activated carbon from Mangosteen Peel	29.76	Ci = 300 ppm, 60 min.	[63]
Fe clay cellulose-acrylamide beads	57.84	Ci= 0.01, ppm	[3]
	Levofloxacin		
Fe–Cu Bimetallic Supported on Alginate–Limestone Nanocomposite	8.8	10 ppm, 40 min.	Current research
Clay nanotubes	442	Ci =10 ppm,1800 min.	[64]
Fe clay cellulose-acrylamide beads	38.01	Ci= 0.01 ppm	[3]
Magnetite (Fe_3O_4—gINPs) nanoparticles from Moringa olifera	22.47	Ci= 4 ppm, 1440 min.	[65]
Hydroxyapatite nanopowder	157.09 (uncalcined Nanohydroxyapatite)	Ci= 25 ppm, 80 min.	[66]

4. Conclusions

Through the use of FTIR, SEM, EDX, and TEM, it was demonstrated that the Fe–Cu/Alg–LS nanocomposite was successfully designed and used to remove the CIP and LEV ions from aquatic solutions. The elimination of CIP and LEV ions was successfully accomplished by the Fe–Cu/Alg–LS nanocomposite. It was also fairly stable at high temperatures. The pharmaceutical concentration and pH level of the solution have a significant impact on sorption capacity. The ideal pH values of 6 and 7 for the adsorption of CIP and LEV ions from contaminated solutions, respectively, were carefully selected. The kinetic models of CIP and LEV ions onto the Fe–Cu/Alg–LS nanocomposite were fitted using the Langmuir adsorption and pseudo-second-order rate equation. After calculating the thermodynamic variables, it was determined that the reaction was spontaneous, exothermic for LEV and endothermic for CIP. The Fe–Cu/Alg–LS nanocomposite sorption was physical. The developed composites demonstrated the potential to be used as an adsorbent in water treatment.

Author Contributions: I.A.A., H.S.H., Z.A.A., A.G.A., N.S.A. and A.K. contributed to methodology and writing of the paper. All authors have read and agreed to the published version of the manuscript.

Funding: This work was assisted financially by the Dean of Science and Research at King Khalid University via the General Research Project (grant number. R.G.P.1/28/43). Also, funded by the Deanship of Scientific Research, King Saud University through Research Group no RG-1441-043 and by Princess Nourah bint Abdulrahman University Researchers Supporting Project number (PNURSP2023R19), Princess Nourah bint Abdulrahman University, Riyadh, Saudi Arabia.

Institutional Review Board Statement: Not applicable.

Informed Consent Statement: Not applicable.

Data Availability Statement: Data on the compounds are available from the authors.

Acknowledgments: The authors are grateful to the Dean of Science and Research at King Khalid University for making financial support available. The authors also extend their appreciation to the Deanship of Scientific Research, King Saud University for funding this work through Research Group no RG-1441-043. This research was also funded by Princess Nourah bint Abdulrahman University Researchers Supporting Project Number (PNURSP2023R19), Princess Nourah bint Abdulrahman University, Riyadh, Saudi Arabia.

Conflicts of Interest: There are no conflicts of interest for the authors to declare.

Sample Availability: Samples of the compounds are available from the authors.

References

1. Ilavský, J.; Barloková, D.; Marton, M. Removal of Specific Pharmaceuticals from Water using Activated Carbon. *IOP Conf. Ser. Earth Environ. Sci.* **2021**, *906*, 012065. [CrossRef]
2. Karungamye, P. Methods used for removal of pharmaceuticals from waste water: Review. *Appl. J. Environ. Eng. Sci.* **2020**, *6*, 412–428.
3. Geetha, G.; Chandrasekaran, N.; Amitava, M. Adsorptive removal of fluoroquinolone antibiotics using green synthesized and highly efficient Fe clay cellulose-acrylamide beads. *Environ. Technol. Innov.* **2022**, *28*, 102783.
4. Oladipo, A.; Mustafa, F.; Ezugwu, O.; Gazi, M. Efficient removal of antibiotic in single and binary mixture of nickel by electrocoagulation process: Hydrogen generation and cost analysis. *Chemosphere* **2022**, *300*, 134532. [CrossRef] [PubMed]
5. Nasrollahi, N.; Vatanpour, V.; Khataee, A. Removal of antibiotics from wastewaters by membrane technology: Limitations, successes, and future improvements. *Sci. Total Environ.* **2022**, *838*, 156010. [CrossRef]
6. Wang, X.; Li, F.; Hu, X.; Hua, T. Electrochemical advanced oxidation processes coupled with membrane filtration for degrading antibiotic residues: A review on its potential applications, advances, and challenges. *Sci. Total Environ.* **2021**, *784*, 146912. [CrossRef]
7. Jie, X.; Xuehai, Y.; Meng, L.; Ye, T.; Xiaodong, L.; Fazhi, Z. Iron-containing palygorskite clay as Fenton reagent for the catalytic degradation of phenol in water. *RSC Adv.* **2021**, *11*, 29537–29542.
8. Mezni, M.; Saied, T.; Horri, N.; Srasra, E. Removal of enrofloxacin from aqueous solutions using illite and synthetic zeolite X. *Surf. Eng. Appl. Electrochem.* **2017**, *53*, 89–97. [CrossRef]
9. El-Shafey, E.-S.I.; Al-Lawati, H.; Al-Sumri, A.S. Ciprofloxacin adsorption from aqueous solution onto chemically prepared carbon from date palm leaflets. *J. Environ. Sci.* **2012**, *24*, 1579–1586. [CrossRef]

10. El-Baz, A.A.; Hendy, I.A.; Dohdoh, A.M.; Srour, M.I. Adsorption technique for pollutants removal; current new trends future challenges—A Review. *EIJEST* **2020**, *33*, 1–24. [CrossRef]
11. Hao, R.; Li, D.; Zhang, J.; Jiao, T. Green Synthesis of Iron Nanoparticles Using Green Tea and Its Removal of Hexavalent Chromium. *Nanomaterials* **2021**, *11*, 650. [CrossRef]
12. Zha, S.; Cheng, Y.; Gao, Y.; Chen, Z.; Megharaj, M.; Naidu, R. Nanoscale zero-valent iron as a catalyst for heterogeneous Fenton oxidation of amoxicillin. *Chem. Eng. J.* **2014**, *255*, 141–148. [CrossRef]
13. Inas, A.A.; Hala, S.H.; Ahmed, H.R.; Najla, A.; Ayman, A.G. Investigation the Effects of Green-Synthesized Copper Nanoparticles on the Performance of Activated Carbon-Chitosan-Alginate for the Removal of Cr(VI) from Aqueous Solution. *Molecules* **2021**, *26*, 2617.
14. Anamika, G.; Shalu, R.; Lata, V.; Jiwan, S.; Samiksha, S.; Yadav, B.C.; Ajay, S.K. Green synthesis of iron nanoparticle from extract of waste tea: An application for phenol red removal from aqueous solution. *Environ. Nanotechnol. Monit. Manag.* **2018**, *10*, 377–387.
15. Silva, E.C.; Soares, V.R.; Fajardo, A.R. Removal of pharmaceuticals from aqueous medium by alginate/polypyrrole/$ZnFe_2O_4$ beads via magnetic field enhanced adsorption. *Chemosphere* **2023**, *316*, 137734. [CrossRef] [PubMed]
16. Wasilewska, M.; Deryło-Marczewska, A. Adsorption of Non-Steroidal Anti-Inflammatory Drugs on Alginate-Carbon Composites—Equilibrium and Kinetics. *Materials* **2022**, *15*, 6049. [CrossRef]
17. Ragab, A.H.; Hussein, H.S.; Ahmed, I.A.; Abualnaja, K.M.; AlMasoud, N. An Efficient Strategy for Enhancing the Adsorption of Antibiotics and Drugs from Aqueous Solutions Using an Effective Limestone-Activated Carbon–Alginate Nanocomposite. *Molecules* **2021**, *26*, 5180. [CrossRef]
18. Jibran, I.; Noor, S.S.; Murtaza, S.; Muhammad, I.; Nawshad, M.; Fares, M.H.; Sara, A.A.; Javed, A.K.; Zia, H.K.; Amit, B.; et al. Synergistic effects of activated carbon and nano-zerovalent copper on the performance of hydroxyapatite-alginate beads for the removal of As^{3+} from aqueous solution. *J. Clean. Prod.* **2019**, *235*, 875–886.
19. Zhu, F.; Ma, S.; Liu, T.; Deng, X. Green synthesis of nano zero-valent iron/Cu by green tea to remove hexavalent chromium from groundwater. *J. Clean. Prod.* **2018**, *174*, 184–190. [CrossRef]
20. Hu, Y.; Pan, C.; Zheng, X.; Liu, S.; Hu, F.; Xu, L.; Xu, G.; Peng, X. Removal of Ciprofloxacin with Aluminum-Pillared Kaolin Sodium Alginate Beads (CA-Al-KABs): Kinetics, Isotherms, and BBD Model. *Water* **2020**, *12*, 905. [CrossRef]
21. Al-Jabari, M.H.; Saleh, S.; Shahid, A.; Reem, B.; Asem, M.; Safyan, A.K. Adsorption study of levofloxacin on reusable magnetic nanoparticles: Kinetics and antibacterial activity. *J. Mol. Liq.* **2019**, *291*, 111249. [CrossRef]
22. Bilgehan, N.; Taylan, D.; Serdar, K. Behavior and Removal of Ciprofloxacin and Sulfamethoxazole Antibiotics in Three Different Types of Full-Scale Wastewater Treatment Plants: Comparative Study. *Water Air Soil Pollut.* **2021**, *232*, 127.
23. Iqbal, A.; Raheela, B.; Muhammad, A.; Sofia, A.; Tania, M.; Shakeel, A.A. Photodegradation of levofloxacin in aqueous and organic solvents: A kinetic study. *Acta Pharm.* **2013**, *63*, 223–229.
24. HMDB. Metabocard for Levofloxacin (HMDB0001929). Available online: https://hmdb.ca/metabolites/HMDB0001929 (accessed on 7 March 2022).
25. Asghar, M.A.; Zahir, E.; Shahid, S.M.; Khan, M.N.; Iqbal, J.; Walker, G. Iron, copper and silver nanoparticles: Green synthesis using green and black tea leaves extracts and evaluation of antibacterial, antifungal and aflatoxin B1 adsorption activity. *LWT* **2018**, *90*, 98–107. [CrossRef]
26. Lingamdinne, L.P.; Kim, I.S.; Ha, J.H.; Chang, Y.Y.; Koduru, J.R.; Yang, J.K. Enhanced Adsorption Removal of Pb(II) and Cr(III) by Using Nickel Ferrite-Reduced Graphene Oxide Nanocomposite. *Metals* **2017**, *7*, 225. [CrossRef]
27. Sofija, S.P.; Dejan, M.K.; Snežana, P.M.; Zoltán, K.; Dragana, D.T.P.; Djurdja, V.K.; Srdjan, D.R. Removal of As(III) and Cr(VI) from aqueous solutions using "green" zero-valent iron nanoparticles produced by oak, mulberry and cherry leaf extracts. *Ecol. Eng.* **2016**, *90*, 42–49.
28. Chaba, J.M.; Philiswa, N.N. Effective adsorptive removal of amoxicillin from aqueous solutions and wastewater samples using zinc oxide coated carbon nanofiber composite. *Emerg. Contam.* **2019**, *5*, 143–149. [CrossRef]
29. Jinsheng, W.; Weihai, C.; Rui, Z.; Can, D. A Study of Sr Sorption Behavior in Claystone from a Candidate High-Level Radioactive Waste Geological Disposal Site under the Action of FeOOH Colloidsice. *Int. J. Environ. Res. Public Health* **2022**, *19*, 9970.
30. Michael, S.C.; Doug, J.H. FT-IR Examination of the Development of Secondary Cell Wall in Cotton Fibers. *Fibers* **2015**, *3*, 30–40.
31. Hong, T.; Yin, J.; Nie, S.P.; Xie, M.-Y. Applications of infrared spectroscopy in polysaccharide structural analysis: Progress, challenge and perspective. *Food Chem.* **2021**, *12*, 100168. [CrossRef] [PubMed]
32. Belaidi, O.; Bouchaour, T.; Maschke, U. Structural Preferences and Vibrational Analysis of 2-Hydroxy-2-methyl-1-phenylpropan-1-one: A Computational and Infrared Spectroscopic Research. *J. Struct.* **2013**, *2013*, 942302. [CrossRef]
33. Betancourt-Galindo, R.; Reyes-Rodriguez, P.Y.; Puente-Urbina, B.A.; Avila-Orta, C.A.; Rodríguez-Fernández, O.S.; Cadenas-Pliego, G.; Lira-Saldivar, R.H.; García-Cerda, L.A. Synthesis, Synthesis of Copper Nanoparticles by Thermal Decomposition and Their Antimicrobial Properties. *J. Nanomater.* **2014**, *2014*, 980545. [CrossRef]
34. Solimana, T.S.; Vshivkova, S.A. Effect of Fe nanoparticles on the structure and optical properties of polyvinyl alcohol nanocomposite films. *J. Non-Cryst. Solids* **2019**, *519*, 119452. [CrossRef]
35. Al-Thawadi, S.; Cord-Ruwisch, R.; Bououdina, M. Consolidation of Sand Particles by Nanoparticles of Calcite after Concentrating Ureolytic Bacteria In Situ. *Int. J. Green Nanotechnol. Biomed.* **2012**, *4*, 28–36. [CrossRef]
36. Ahmed, M.J. Adsorption of quinolone, tetracycline, and penicillin antibiotics from aqueous solution using activated carbons: Review. *Environ. Toxicol. Pharmacol.* **2017**, *50*, 1–10. [CrossRef] [PubMed]

37. Yoosefian, M.; Ahmadzadeh, S.; Aghasi, M.; Dolatabadi, M. Optimization of electrocoagulation process for efficient removal of ciprofloxacin antibiotic using iron electrode; kinetic and isotherm studies of adsorption. *J. Mol. Liq.* **2017**, *225*, 544–553. [CrossRef]
38. Ahmadzadeh, S.; Asadipour, A.; Pournamdari, M.; Behnam, B.; Rahimi, H.R.; Dolatabadi, M. Removal of ciprofloxacin from hospital wastewater using electrocoagulation technique by aluminum electrode: Optimization and modelling through response surface methodology. *Process Saf. Environ. Prot.* **2017**, *109*, 538–547. [CrossRef]
39. Wang, C.J.; Li, Z.H.; Jiang, W.-T. Adsorption of ciprofloxacin on 2:1 dioctahedral clay minerals. *Appl. Clay Sci.* **2011**, *54*, 723–728. [CrossRef]
40. Inchaurrondo, N.; Font, J.; Ramos, C.P.; Haure, P. Natural diatomites: Efficient green catalyst for Fenton-like oxidation of Orange II. *Appl. Catal. B* **2016**, *181*, 481–494. [CrossRef]
41. El-Bendary, N.; El-Etriby, H.K.; Mahanna, H. Reuse of adsorption residuals for enhancing removal of ciprofloxacin from wastewater. *Environ. Technol.* **2021**, *42*, 4438–4454. [CrossRef]
42. Wondimu, K.W.; Beteley, T.M.; Joon, W.K.; Yonas, C. Enhanced Ciprofloxacin Removal from Aqueous Solution Using a Chemically Modified Biochar Derived from Bamboo Sawdust: Adsorption Process Optimization with Response Surface Methodology. *Adsorp. Sci. Technol.* **2022**, *2022*, 2699530.
43. Khana, Z.A.; Mekkya, A.E.M.; Bin Mahfouzc, A.S.; Saleha, T.S.; Mohy Eldin, M.S. Separation of nickel(II) ions from synthetic aqueous solutions with novel dimethylglyoxime-modified Amberlite IRA-420: Kinetic and equilibrium studies. *Desalin. Water Treat.* **2017**, *81*, 123–132. [CrossRef]
44. Margaritis, K.; Thodoris, D.K. Why Is the Linearized Form of Pseudo-Second Order Adsorption Kinetic Model So Successful in Fitting Batch Adsorption Experimental Data? *Colloids Interfaces* **2022**, *6*, 55.
45. Aljeboree, M.; Alshirifi, N.; Alkaim, F. Kinetics and equilibrium study for the adsorption of textile dyes on coconut shell activated carbon. *Arab. J. Chem.* **2017**, *10*, 3381. [CrossRef]
46. Elwakeel, K.; El-Bindary, A.; Kouta, E.; Guibal, E. Functionalization of polyacrylonitrile/Na-Y-zeolite composite with amidoxime groups for the sorption of Cu(II), Cd(II) and Pb(II) metal ions. *Chem. Eng. J.* **2018**, *332*, 727–736. [CrossRef]
47. Ho, Y.S.; McKay, G. The Kinetics of Sorption of Divalent Metal Ions onto Sphagnum Moss Peat. *Water Res.* **2000**, *34*, 735–742. [CrossRef]
48. Meenakshi, S.; Viswanathan, N. Identification of selective ion-exchange resin for fluoride sorption. *J. Colloid. Interface Sci.* **2007**, *308*, 438–450. [CrossRef] [PubMed]
49. Israa, I.N.; Hilal, W.S. Adsorption of Eriochrom Black T Azo Dye onto Nanosized Anatase TiO_2. *J. Environ. Eng. Sci.* **2015**, *2*, 86–92.
50. Ho, Y.-S. Effect of pH on lead removal from water using tree fern as the sorbent. *Bioresour. Technol.* **2005**, *96*, 1292–1296. [CrossRef]
51. Ahmed, H.R.; Inas, A.A.; Dina, M.D.B. The Removal of Brilliant Green Dye from Aqueous Solution Using Nano Hydroxyapatite/Chitosan Composite as a Sorbent. *Molecules* **2019**, *24*, 847.
52. Sebastiano, C.; Claudio, E.; Georgeta, P.; Antonella, G. Combination of interfacial reduction of hexavalent chromium and trivalent chromium immobilization on tin-functionalized hydroxyapatite materials. *Appl. Surf. Sci.* **2021**, *539*, 148227.
53. Inas, A.A.; Najlaa, S.A.; Hussein, H.S.; Ahmed, H.R. Environmentally Friendly Mesoporous Nanocomposite Prepared from Al-Dross Waste with Remarkable Adsorption Ability for Toxic Anionic Dye. *J. Chem.* **2019**, *2019*, 7685204.
54. Inas, A.A.; Hussein, H.S.; Ahmed, H.R.; Najlaa, S.A. Synthesis and Characterization of Silica-Coated Oxyhydroxide Aluminum/Doped Polymer Nanocomposites: A Comparative Study and Its Application as a Sorbent. *Molecules* **2020**, *25*, 1520.
55. Fozia, B.; Jamshed, A.; Shahid, I.; Sobia, N.; Syed, N.; Abbas, B. Study of Isothermal, Kinetic, and Thermodynamic Parameters for Adsorption of Cadmium: An Overview of Linear and Nonlinear Approach and Error Analysis. *Bioinorg. Chem. Appl.* **2018**, *2018*, 3463724.
56. Elgarahy, A.M.; Elwakeel, K.Z.; Elshoubaky, G.A.; Mohammad, S.H. Microwave-accelerated sorption of cationic dyes onto green marine algal biomass. *Environ. Sci. Pollut. Res.* **2019**, *26*, 22704–22722. [CrossRef]
57. Salvestrini, S.; Leone, V.; Iovino, P.; Canzano, S.; Capasso, S. Considerations about the correct evaluation of sorption thermodynamic parameters from equilibrium isotherms. *J. Chem. Thermodyn.* **2014**, *68*, 310–316. [CrossRef]
58. Zhao, W.; Zhu, J.; Wei, W.; Ma, L.; Zhua, J.; Xie, J. Comparative study of modified/non-modified aluminum and silica aerogels for anionic dye adsorption performance. *RSC Adv.* **2018**, *8*, 29129. [CrossRef] [PubMed]
59. Haleemat, I.; Folahan, A.; Olalekan, S.F.; Bhekumusa, J.X. Adsorption of Cr (VI) on synthetic hematite (α-Fe_2O_3) nanopar-ticles of different morphologies. *Korean J. Chem. Eng.* **2014**, *31*, 142–154.
60. Shujauddin, K.; Zhang, L.; Aimin, L.; Muhammad, I.; Zhang, X. Microwave-assisted hydrothermal carbonization of furfural residue for adsorption of Cr(VI): Adsorption and kinetic study. *Pol. J. Environ. Stud.* **2020**, *29*, 1671–1681.
61. Maria, E.R.; Miria, B.; Karim, S. Removal of Ciprofloxacin from Aqueous Solutions Using Pillared Clays. *Materials* **2017**, *10*, 1345.
62. Pham, T.; Vu, T.; Nguyen, H.; Le, P.; Hoang, T. Adsorptive Removal of Antibiotic Ciprofloxacin from Aqueous Solution Using Protein-Modified Nanosilica. *Polymers* **2020**, *12*, 57. [CrossRef] [PubMed]
63. Tran, Q.T.; Do, T.H.; Ha, X.L.; Nguyen, H.P.; Nguyen, A.T.; Ngo, T.C.Q.; Chau, H.D. Study of the Ciprofloxacin Adsorption of Activated Carbon Prepared from Mangosteen Peel. *Appl. Sci.* **2022**, *12*, 8770. [CrossRef]
64. Abukhadra, M.; Mohamed, A.; El-Sherbeeny, A.; Soliman, A. adsorption of toxic and biologically active levofloxacin residuals from wastewater using clay nanotubes as a novel fixed bed; column performance and optimization. *ACS Omega* **2020**, *5*, 26195–26205. [CrossRef] [PubMed]

65. Altaf, S.; Zafar, R.; Zaman, W.Q.; Ahmad, S.; Yaqoob, K.; Syed, A.; Khan, A.J.; Bilal, M.; Arshad, M. Removal of levofloxacin from aqueous solution by green synthesized magnetite (Fe_3O_4) nanoparticles using Moringa olifera: Kinetics and reaction mechanism analysis. *Ecotoxicol. Environ. Saf.* **2021**, *226*, 112826. [CrossRef] [PubMed]
66. Ciobanu, G.; Harja, M. Studies on the Sorption of Levofloxacin from Aqueous Solutions onto Nanohydroxyapatite. *Rev. Roum. Chim.* **2018**, *63*, 593–601.

Disclaimer/Publisher's Note: The statements, opinions and data contained in all publications are solely those of the individual author(s) and contributor(s) and not of MDPI and/or the editor(s). MDPI and/or the editor(s) disclaim responsibility for any injury to people or property resulting from any ideas, methods, instructions or products referred to in the content.

polymers

Article

Study of Formulation and Process Variables for Optimization of Piroxicam Nanosuspension Using 3^2 Factorial Design to Improve Solubility and In Vitro Bioavailability

Yahya Alhamhoom [1], Sandip M. Honmane [2,*], Umme Hani [1,*], Riyaz Ali M. Osmani [3], Geetha Kandasamy [4], Rajalakshimi Vasudevan [5], Sharanya Paramshetti [3], Ravindra R. Dudhal [2], Namrata K. Kengar [2] and Manoj S. Charde [6]

1 Department of Pharmaceutics, College of Pharmacy, King Khalid University, Abha 62529, Saudi Arabia
2 Department of Pharmaceutics, Annasaheb Dange College of B. Pharmacy, Ashta, Shivaji University, Kolhapur 416301, Maharashtra, India
3 Department of Pharmaceutics, JSS College of Pharmacy, JSS Academy of Higher Education and Research (JSS AHER), Mysuru 570015, Karnataka, India
4 Department of Clinical Pharmacy, College of Pharmacy, King Khalid University, Abha 62529, Saudi Arabia
5 Department of Pharmacology, College of Pharmacy, King Khalid University, Abha 62529, Saudi Arabia
6 Department of Pharmaceutical Chemistry, Government College of Pharmacy, Karad, Shivaji University, Kolhapur 415124, Maharashtra, India
* Correspondence: sandiphonmane@gmail.com (S.M.H.); uahmed@kku.edu.sa (U.H.); Tel.: +91-8600392878 (S.M.H.); +965-98712387 (U.H.)

Abstract: Piroxicam is a Biopharmaceutical Classification System (BCS) Class II drug having poor aqueous solubility and a short half-life. The rationale behind the present research was to develop a Piroxicam nanosuspension to enhance the solubility and thereby the in vitro bioavailability of the drug. Piroxicam nanosuspension (PRX NS) was prepared by an anti-solvent precipitation technique and optimized using a full-factorial design. Herein, the nanosuspension was prepared using polymer polyvinylpyrrolidone (PVP) K30® and Poloxamer 188® as a stabilizer to improve the solubility and in vitro bioavailability of the drug. Nine formulations were prepared based on 3^2 full-factorial experimental designs to study the effect of the formulation variables such as concentration of poloxamer 188 (%) (X_1) and stirring speed (rpm) (X_2) as a process variable on the response of particle size (nm) and solubility (μg/mL). The prepared NS was characterized by phase solubility, Fourier-transform infrared (FT-IR), differential scanning calorimetry (DSC), X-ray powder diffraction (XRPD), transmission electron microscopy (TEM), particle size, zeta potential, entrapment efficiency, and percent drug release. DSC and XRPD analysis of freeze-dried NS formulation showed conversion of PRX into a less crystalline form. NS formulations showed a reduction in the size from 443 nm to 228 nm with -22.5 to -30.5 mV zeta potential and % drug entrapment of 89.76 ± 0.76. TEM analysis confirmed the size reduction at the nano level. The solubility was increased from 44 μg/mL to 87 μg/mL by altering the independent variables. The solubility of PRX NS in water was augmented by 14- to 15-fold (87.28 μg/mL) than pure PRX (6.6 μg/mL). The optimized formulation (NS9) at drug-to-stabilizer concentration exhibited a greater drug release of approximately 96.07% after 120 min as compared to the other NS formulations and pure PRX (36.78%). Thus, all these results revealed that the prepared NS formulations have improved the solubility and in vitro dissolution compared to the pure drug. Furthermore, an increase in the drug release was observed from the NS than that of the pure PRX. All these outcomes signified that the prepared PRX NS showed an increase in solubility and in vitro dissolution behavior; which subsequently would aid in attainment of enhanced bioavailability.

Keywords: polymers; nanosuspension; solubility; bioavailability; anti-solvent precipitation; in vitro drug dissolution

Citation: Alhamhoom, Y.; Honmane, S.M.; Hani, U.; Osmani, R.A.M.; Kandasamy, G.; Vasudevan, R.; Paramshetti, S.; R. Dudhal, R.; K. Kengar, N.; Charde, M.S. Study of Formulation and Process Variables for Optimization of Piroxicam Nanosuspension Using 3^2 Factorial Design to Improve Solubility and In Vitro Bioavailability. *Polymers* **2023**, *15*, 483. https://doi.org/10.3390/polym15030483

Academic Editor: Edina Rusen

Received: 14 December 2022
Revised: 8 January 2023
Accepted: 16 January 2023
Published: 17 January 2023

Copyright: © 2023 by the authors. Licensee MDPI, Basel, Switzerland. This article is an open access article distributed under the terms and conditions of the Creative Commons Attribution (CC BY) license (https://creativecommons.org/licenses/by/4.0/).

1. Introduction

There has been a growing need for alternative drug delivery systems and techniques owing to the increasing number of newly investigated drugs, increased sensitivity to clinical findings, and escalating healthcare costs. The current drug delivery system is rapidly evolving and has increased productivity. There has been significant advancement and research in delivery systems for the optimization of therapy and its cost-effectiveness. Newer pharmaceutical and biopharmaceutical product categories are accelerating the development of drug delivery technology. Often, conventional methods are unable to deliver these novel entities effectively. Thus, advanced delivery systems have therefore become increasingly important in today's world. Most of the recently developed drug molecules have poor solubility, which can lead to significant formulation issues and poor bioavailability.

Solubility is one of the key parameters to achieve the desired concentration of drug in the systemic circulation necessary for achieving the required pharmacological response. Following oral administration, poorly water-soluble drugs require high dosages to attain therapeutic plasma concentration. Water insolubility is one of the central elements which restrict the usage of many potential drug moieties and other active compounds. Because of this, the bioavailability of the drug is less and it fails to reach the site of action [1]. Solubility is closely associated with bioavailability. It greatly increases the bioavailability of the dosage form. Only highly soluble drug molecules can cross the cell membrane and show their desired therapeutic effect by reaching the site of action [2,3]. In vitro dissolution of the drug is related to its in vivo bioavailability [2]. The primary strategy in this study domain is enhancing the solubility of BCS class II drugs. Solubility of the drugs can be improved by a variety of techniques, such as particle size reduction, micro-emulsion, micellar solubilization, solvent deposition, Super Critical Fluid (SCF) process, solid dispersion, nanosuspension (NS), cryogenic techniques, inclusion complex formation-based techniques, hydrotropy, co-crystallization, complexation, liquid-solid system, etc. [2,4].

The majority of the aforementioned solubility enhancement methods, including NS, may be utilized to make drugs more soluble. In developing an ideal formulation, several factors, such as stability at various temperatures, solubility, compatibility of the solvent and excipients, and photostability, are essential. Thus, the present study aimed in developing NS to resolve issues relating to low solubility and poor bioavailability [5]. A drug that is weakly water soluble and free of any matrix material can be utilized to develop NS [6]. NS improves medication safety and effectiveness by resolving the challenges of low solubility and bioavailability as well as by changing the pharmacokinetics of the drug [7]. NSs are colloidal dispersions having surfactant-stabilized drug particles that are nanoscale in size [4,8]. Pure drug particles are dispersed in the aqueous medium to create a biphasic system known as an NS. The suspended particle has a diameter of less than 1 μm. The increase in surface area and saturation solubility of the drug particles results from the reduction in drug particle size, which accelerates the dissolution. The increased vapor pressure of the particles leads to increased saturation solubility and solution velocity of the nanoparticles. Because of these properties, NS is the best technique for enhancing the water solubility and dissolution rate of drugs [9,10]. The particle size distribution, surface charge, crystalline state, dissolution rate, and saturation solubility are the main characteristics of oral NSs. A zeta potential of at least ± 30 mV is needed for an electrostatically stabilized NS to be physically stable. A general guideline line of ± 20 mV will be sufficient in the case of a combined steric and electrostatic stabilization. For pharmaceuticals that exist in several polymorphs, the crystallinity of the NS is crucial [11].

The drug Piroxicam(PRX) [4-Hydroxy-2-methyl-N-(2-pyridinyl)-2H-1, 2-benzothiazine-3-carboxamide 1,1-dioxide] belongs to the class of anti-inflammatory drugs. PRX demonstrates prolonged and delayed oral absorption [12]. It is a highly protein-bound medication that is slowly removed from the body, increasing the half-life to up to 36 to 86 h. It has various side effects such as diarrhea, constipation, headache, dizziness, and ringing in the ears. Although it has a variety of side effects, PRX has a stronger pharmacological efficacy because it is a potent anti-inflammatory drug [12]. In the US, PRX is approved for the treatment of rheumatoid arthritis and osteoarthritis [13]. PRX is a BCS class II medication that has high permeability and low solubility [2]. It exhibits a slow and gradual absorption when taken orally and is proven to be ulcerogenic. Therefore, the need of the hour is to develop novel formulations that would accelerate its absorption in the GI tract and might give quick relief from rheumatoid arthritis and osteoarthritis with a reduction in its dose and dose-dependent side effects. The NS development approach has viable potential for enhancing solubility of poorly soluble drugs and is also cost effective, simple and robust. PRX in the form of NS is a practically executable and promising way to mitigate this problem. Hence, the objective of the current study was to investigate the effects of a formulation variable (polymer concentration) and process parameters (stirring speed and time) on the NS formulation of PRX with the intent of attainment of improved solubility and in vitro bioavailability. In the present study, NS has been chosen as an approach to enhance the solubility, dissolution release and in vitro bioavailability of PRX. Differential scanning calorimetry (DSC), X-ray powder diffraction (XRPD), and transmission electron microscopy (TEM) analysis were performed to determine the solid-state properties of the drug in physical mixtures and NSs in comparison with the free PRX.

2. Materials and Methods

2.1. Materials

PRX was procured from Zydus Cadila Healthcare Ltd., Goa, India. The following chemicals were obtained commercially: PVP K30®, Poloxamer 188®, Methanol, Hydrochloric Acid and Dichloromethane (Loba Chemie, Mumbai, India). All the other chemicals, solvents and reagents used were of analytical grade and were stored and used as per the supplier's instructions.

2.2. Methods

2.2.1. Pre-Formulation Study

Pre-formulation studies were carried out to determine the characteristics of the drug and excipients, particularly on the physicochemical, physicomechanical, and biopharmaceutical aspects [14].

2.2.2. Organoleptic Evaluation

The drug sample was evaluated for organoleptic properties. The organoleptic evaluation was conducted by observing the appearance, color, and odor of the drug sample.

2.2.3. Melting Point

Using a micro-controlled based melting point apparatus (SMP10/1, Stuart, UK), the melting point of the drug was determined. The drug sample was inserted into a capillary tube with one end closed. The capillary tube was inserted into a silicone oil bath, which was heated with the help of an electrical heating coil in a controlled manner. The temperature at which the drug sample started melting was noted as the melting point temperature. The average of triplicate readings was noted and compared with the literature value.

2.2.4. Ultraviolet-Visible (UV-Visible) Spectrophotometry

The absorbance maxima (λ_{max}) of PRX were determined in various solvents such as methanol, methanolic HCl, phosphate buffer solution (PBS) pH 6.8, and PBS pH 7.4 in the range of 200–400 nm by using a UV-Visible double beam spectrophotometer (Shimadzu 1800, Tokyo, Japan) [14].

2.2.5. Determination of Calibration Curve

The calibration curve of PRX has been investigated in different solvents such as methanol, methanolic HCl, PBS pH 6.8, and PBS pH 7.4 at specific wavelengths [14].

2.2.6. Solubility of Drug

Drug solubility was assessed in a variety of solvents, including distilled water, PBS pH 7.4, and methanol. In order to create saturated solutions, the excess drug was added to the vehicles, which were then shaken continuously for 48 h at a temperature of 25 ± 0.5 °C. The solutions were filtered, diluted and analyzed using UV spectrophotometry (Shimadzu 1800, Japan) [14].

2.2.7. Fourier-Transform Infrared (FT-IR) Spectroscopy

Drug-excipient compatibility was confirmed by FT-IR spectroscopy (Bruker Alpha II). The spectra were obtained using the KBr pellet method within the range of 4000 cm^{-1} to 400 cm^{-1}. Briefly, the pellets were prepared with KBr in a ratio of 1:100 and force was applied for several minutes to obtain uniform thin pellets. These pellets were placed in between two plates in a sample holder and scanned. The absorbance was plotted against their corresponding wavenumber [12].

2.2.8. Preparation of NS

Preparation of drug solution: the required amount of the drug was dissolved in 4 mL dichloromethane to obtain a clear solution [15].

Preparation of polymer solution: the required amounts of Poloxamer 188 and PVP K30 were dissolved in water. Using a mechanical stirrer (Remi RQT 124 AD, Mumbai, India), the polymer solution was homogenized at 1000–1200 rpm. Then, the drug solution was added dropwise into the polymer solution using a syringe and stirred continuously followed by sonication for 20 min. From the preliminary study, based on complete mixing and the optimal particle size, stirring time was optimized at 15 min [15,16].

2.3. Optimization of Formulation

A randomized, 3^2 full factorial design with two factors and three levels was employed to systematically study the nanosuspension formulation (Table 1). A total of nine experimental trials were performed at all possible combinations. The concentration of stabilizer and stirring speed were identified as the independent variables based on the experiments conducted during the optimization, which were altered at three different levels, i.e., low, medium, and high. Solubility (μg/mL) and particle size (nm) were considered as dependent variables (responses) The response variables used were solubility and particle size (nm). The development and evaluation of the statistical experimental design were accomplished by utilizing the Design-Expert 8.0 software (Stat-Ease Inc., Minneapolis, USA). The effect of two independent variables, stabilizer concentration (X_1) and stirring speed (X_2), on the response (Y) was studied.

Table 1. 3^2 full factorial design with composition and independent variables.

Formulation Code	Composition			Speed (rpm)	Total Volume (mL)
	PRX (mg)	PVP K30® (mg)	Poloxamer 188® (%w/v)		
NS1	50	50	0.1 (−1)	1000 (−1)	100
NS2	50	50	0.1 (−1)	1800 (0)	100
NS3	50	50	0.1 (−1)	2600 (+1)	100
NS4	50	50	0.3 (0)	1000 (−1)	100
NS5	50	50	0.3 (0)	1800 (0)	100
NS6	50	50	0.3 (0)	2600 (+1)	100
NS7	50	50	0.5 (+1)	1000 (−1)	100
NS8	50	50	0.5 (+1)	1800 (0)	100
NS9	50	50	0.5 (+1)	2600 (+1)	100

−1: low level, 0: medium level, and +1: high level of independent variables.

Formulation optimization using the desirability function: using Design-Expert 8.0 software (Stat-Ease Inc., USA), various response surface methodology (RSM) computations were carried out for the current optimization research. All of the response variables were developed in a linear model with quadratic terms. Additionally, using the Design-Expert software output files, linear plots and 3D graphs were developed. Analysis of variance (ANOVA) was used to determine the importance of these characteristics on the variables.

The optimization process employed the desirability function once the mathematical model had been fitted. The results were combined to find a product with the desired properties during formulation optimization. The desirability function predicts the ideal values for the independent variables by combining all the results into one variable. The least desirable value for the replies is represented by a desirability value of 0, while the most desirable value is represented by a desirability value of 1.

2.3.1. Freeze-Drying of Optimized Formulation

The optimized batch (NS9) was lyophilized using mannitol as a cryoprotectant for 24 h under controlled conditions to obtain dry powder using a laboratory-scale lyophilizer (Alpha 1–2 L Dplus, Martin Christ Gefriertrocknungsanlagen GmbH, Germany). The product was stored in an airtight container until further characterization [17].

2.3.2. Characterization of NS

Determination of phase solubility of NS: for the determination of the phase solubility of PRX, different concentrations of Poloxamer 188 were prepared. Briefly, an excess amount of the drug (1 g) was added to each of the 250 mL flasks containing 25 mL of stabilizer poloxamer 188 having three distinct concentrations (0.1%, 0.3%, and 0.5%) and at different speeds. The flasks were properly sealed and agitated for 48 h at 37 °C at 100 rpm in an orbital shaker cum incubator (Orbit™ 1000 multipurpose digital shaker). For the establishment of the equilibrium, they were kept in the incubator for a further 24 h. Five (5) mL of the supernatant was filtered and appropriately diluted, and the amount of drug in the filtrate was evaluated photometrically using a UV-Visible spectrophotometer (Shimadzu 1800, Tokyo, Japan) at 354 nm [18].

Particle size, polydispersity index and zeta potential analysis: by using the dynamic light scattering (DLS) technique, the mean particle size, polydispersity index and zeta potential were determined using a particle size analyzer (Horiba Scientific SZ-100). Freshly prepared NS was diluted 100 times with distilled water and analyzed [17,19].

Determination of Entrapment Efficiency (EE): entrapment efficiency (EE) of the optimized formulation was determined by quantitatively estimating the amount of drug loaded into the NS. The NS formulation was ultracentrifuged (Optima KE-90-IVD, Beckman Coulter, Pasadena, CA, USA) for 20 min; the resulting supernatant was diluted sufficiently with methanol for subsequent UV-spectrophotometric analysis at 334 nm [20].

$$\%\ \text{EE} = \frac{\text{Total amount of drug} - \text{Free drug}}{\text{Total amount of drug}} \times 100 \quad (1)$$

Transmission Electron Microscopy (TEM): the size and shape of the optimized NS were evaluated by means of TEM (H-7500, Hitachi, Tokyo, Japan).

Differential Scanning Calorimetry (DSC) analysis: a differential scanning calorimeter (Mettler-Toledo DSC 821e, Columbus, OH, USA), was used to carry out the thermal analysis for confirming the compatibility between the drug and the excipient. The thermal behavior of the drug and the optimized formulation was investigated via DSC analysis. Briefly, the samples were weighed accurately and sealed hermetically in aluminum pans and crimped and were heated from 25 to 250 °C at a heating rate of 10 °C/min. Throughout the measurement, nitrogen gas was purged over the sample cell with a flow rate of 50 mL/min [21].

X-ray Powder Diffraction (XRPD): XRPD is a crucial method used to determine the crystalline or amorphous nature of the sample. Using a powder X-ray diffractometer (AXS

D8 Advances, Bruker Ltd., Germany), diffractograms of the pure drug and NS formulation were obtained with tube anode Cr over the interval of 10–70°/2θ using copper as an X-ray target, with 1.54 Å wavelength [21,22].

In vitro dissolution study: a USP Type II dissolution testing paddle apparatus was used for determining the in vitro dissolution. An amount of sample which is equivalent to 10 mg of PRX was added to the glass jar of apparatus containing 900 mL of PBS pH 6.8 maintained at a temperature of 37 °C [18,23]. Then, for 2 h, the paddle was rotated at 75 rpm. Three (3) mL of sample was taken out at predetermined intervals, filtered, and properly diluted. The concentration of the drug dissolved in the medium was then determined with the help of a UV spectrophotometer (Shimadzu 1800, Tokyo, Japan). To maintain a constant volume of a medium during the dissolution, 3 mL of fresh medium was always replaced in the glass jar.

3. Result and Discussion

3.1. Pre-formulation Study

3.1.1. Organoleptic Properties

For identifying the compound, the organoleptic features of the drug sample, such as appearance, color, and odor were examined. Table 2 displays outcomes that adhere to the standards set out in the existing literature.

Table 2. Organoleptic properties of PRX.

Sr. No.	Identification Test	Observation	Standard
1	Appearance	Off-white powder	Off-white to light tan or light-yellow powder
2	Odor	Odorless	Odorless
3	Color	White	Off-white

3.1.2. Melting Point

The melting point of PRX was determined using the capillary method. It has been noted in the range of 198–200 °C. The experimental values are in good agreement with the criteria in the published literature [13].

3.1.3. Determination of Maximum Wavelength (λ_{max})

The absorption maxima (λ_{max}) of PRX in the methanol, methanolic HCl, PBS pH 7.4, and PBS pH 6.8 is represented in Figure 1 and the same has been depicted in Table 3.

Table 3. λ_{max} of PRX in different solvents.

Solvent	λ_{max} (nm)	
	Observed	Standard [12]
Methanol	334	334
Methanolic HCl	334.8	334
PBS pH 7.4	354	354
PBS pH 6.8	354	354

3.1.4. Preparation of Calibration Curve

The calibration curve of PRX was plotted in methanol, methanolic HCl, PBS pH 6.8, and PBS pH 7.4. It showed the linear relationship in the concentration range of 2-10 μg/mL. The R^2 values were obtained above 0.99.

3.1.5. Solubility of Drug

In contrast to being poorly soluble in water, PRX readily dissolves in methanol, ethanol, and dichloromethane, among other organic solvents. PRX solubility was investigated in a variety of solvents including water, PBS pH 7.4, and methanol. The investigation shows that the drug was freely soluble in methanol, soluble in PBS pH 7.4, and very slightly soluble in water (Table 4).

Table 4. Solubility of the drug in different solvents.

Sr. No.	Solvent	Reference Solubility (mg/mL)	Observed Solubility (mg/mL)
1	Methanol	5.0	4.44
2	PBS pH 7.4	0.592	0.6606
3	Water	0.0076	0.0066

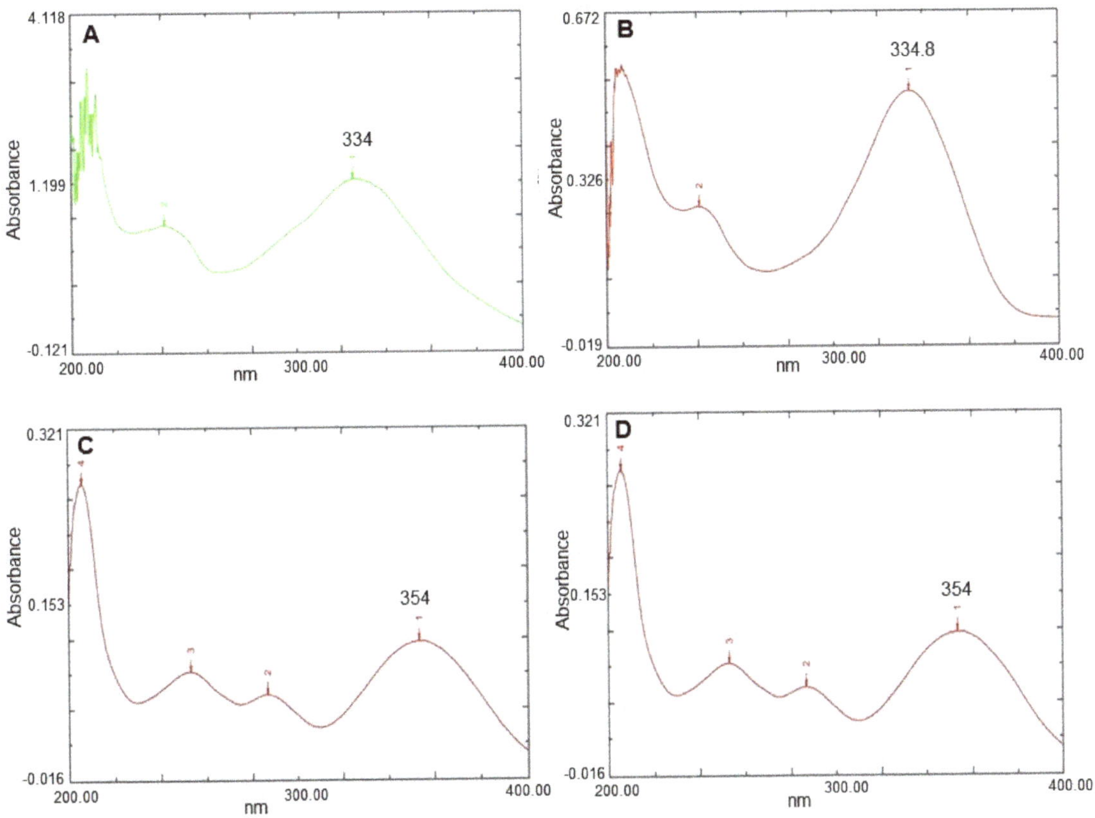

Figure 1. UV spectra depicting λ_{max} of PRX in methanol (**A**), methanolic HCl (**B**), PBS pH 6.8 (**C**) and PBS pH 7.4 (**D**).

3.1.6. FT-IR Studies

FT-IR studies predict the possible interaction between the drug and the excipients. FT-IR spectra (Figure 2) of pure PRX, its physical mixture and optimized formulation depicted all the characteristic IR peaks corresponding to their functional groups as reported in the literature. (Table 5).

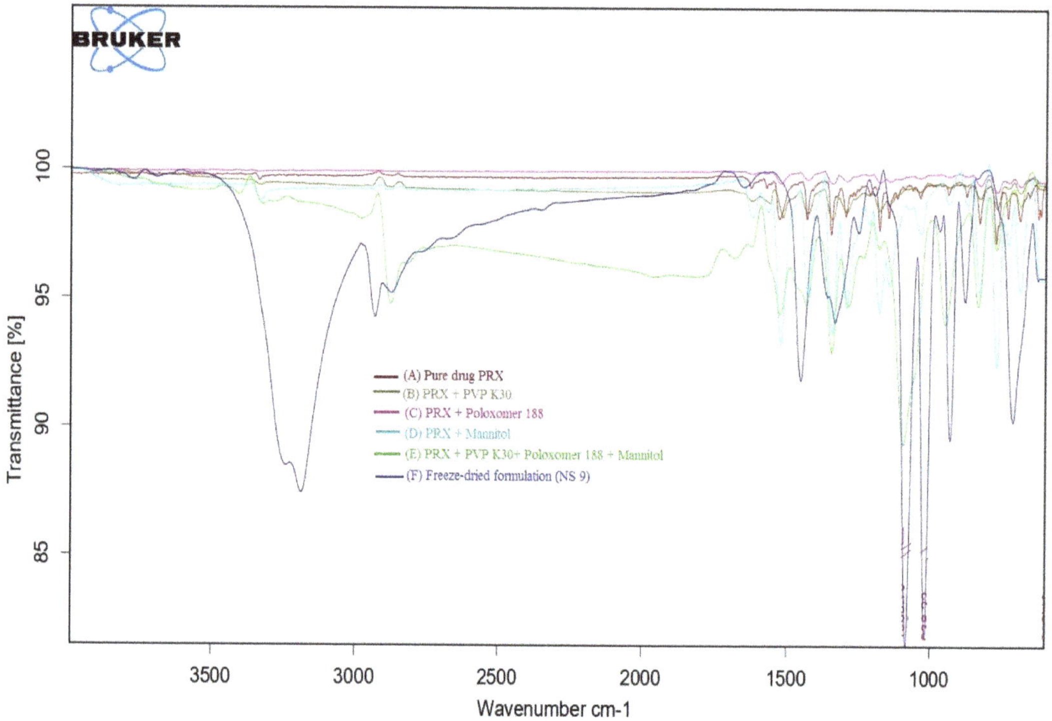

Figure 2. Overlain FT-IR spectra of PRX (A), physical mixture of PRX and PVP K30® (B), physical mixture of PRX and Poloxamer 188® (C), physical mixture of PRX and mannitol (D), physical mixture of PRX, PVP K30®, Poloxamer 188® and mannitol (E), and freeze-dried optimized formulation (NS9) (F).

Table 5. Interpretation data of the FT-IR spectral analysis.

Characteristics Functional Groups.	Standard Range (cm^{-1})	Observed Peak for Pure PRX (cm^{-1})	Observed Peak for NS Physical Mixture (cm^{-1})
–OH and –NH Stretching	3650–3300	3337.34	3337.72
Aromatic –C=C–H	3300–2700	2374.30	2878.27
C=O Stretching	1850–1680	1628.71	1628.93
Aromatic –C=C–	1680–1450	1575.84	1575.95
Ar–NH	1360–1250	1349.93	1342.49
N-CH_3 Stretching	1220–1050	1148.62	1147.59
–SO_2-N= Group	1070–1050	1065.53	1059.18
o-disubstituted phenyl	750	772.14	771.90

All the characteristic peaks were observed in the IR spectra of pure PRX and in the physical mixture of PRX, poloxamer 188®, PVP K 30®, and mannitol. This indicates that there is no interaction between the PRX and excipients.

3.2. Preparation of NS

PRX NS was prepared by the antisolvent-precipitation technique. The aqueous phase containing suitable polymer and stabilizer (PVP K30® and Poloxamer 188®) was used as the antisolvent and dichloromethane was used as a solvent.

3.2.1. Full Factorial Design

The experimental design shown in Table 1 is a factorial design for two variables at three different levels using −1, 0, and +1 corresponding to a 3^2 full factorial design. It is possible to evaluate the impact of distinct variables (primary effects) and their second-order effects using this efficient second-order experimental design with a small number of tests. Further, this design has the advantage of calculating the quadratic response surface, which is not estimable using a factorial design at two levels. A full factorial design was used to carefully evaluate the variables.

Data from the experimental runs were subjected to regression and graphical analysis, which produced the equations in Table 6 with F ratios that were statistically significant ($p < 0.05$) and Adj-R^2 values in the range of 0.7–1. The data were well-fit by these model equations.

Table 6. Results of statistical analysis of the 3^2 full factorial experimental design.

Response	Sources			
	Model *p* Value	**R^2**	**Adjusted R^2**	**Predicted R^2**
Particle Size	0.0053	0.8255	0.7673	0.5975
Solubility	0.0005	0.9217	0.8956	0.8126

The significance of the effect on particle size (Y_1) was established using ANOVA and the linear regression equation fit to the data was:

$$Y_1 = 326.56 - 70.50 \times X_1 - 41.67 \times X_2 \quad (2)$$

As the stabilizer concentration and stirring speed rise, respectively, the negative sign at coefficients X_1 and X_2 indicate a reduction in particle size. Because super-saturation occurs with increasing drug concentration in the aqueous phase and leads to fast precipitation on diffusion, particle size was reduced. In order to prevent agglomeration, the drug particles were reduced in size up to nanosized ranges and well-protected by a stabilizer. Moreover, considerable stabilizer upholds Oswald's ripening while too little stabilizer causes agglomeration or aggregation and increases particle size (a phenomenon in which smaller particles get smaller due to more solubility, and larger particles become larger through re-precipitation of small particles on it). The stabilizer concentration was found to be ideal between 0.2 and 0.6 percent. The same observation has been reported by Ahire et al. [24]. The increase in the stirring speed also results in the reduction in the particle size.

The effect on solubility (Y_2) was established to be meaningful by ANOVA and the linear regression equation fit to the data was:

$$Y_2 = 67.23 + 11.80 \times X_1 + 6.36 \times X_2 \quad (3)$$

Poor wetting of the drug surface is a sign of hydrophobicity, which contributes to low solubility. As a result of this, the particles agglomerate instead of dispersing. Because of the increased surface area brought on by the reduction in particle size, the drug's solubility in the NS was enhanced shown in Figure 3. The increase in solubility (Y_2) is predicted by the positive coefficient of X_1 and X_2 as the stabilizer concentration and stirring speed, respectively.

3.2.2. Formulation Optimization Using the Desirability Function

The pharmaceutical formulation is optimized by studying the effects of different levels of variables on the responses. During the optimization process, all measured responses that can have an impact on the product's quality were taken into account. Considering the effect of stabilizer concentration (0.5%) and stirring speed of 2600 rpm and manipulating data from Table 6 it was observed that the optimized formulation (NS9) showed a decrease in particle size (228 nm) and high solubility (87.28 μg/mL).

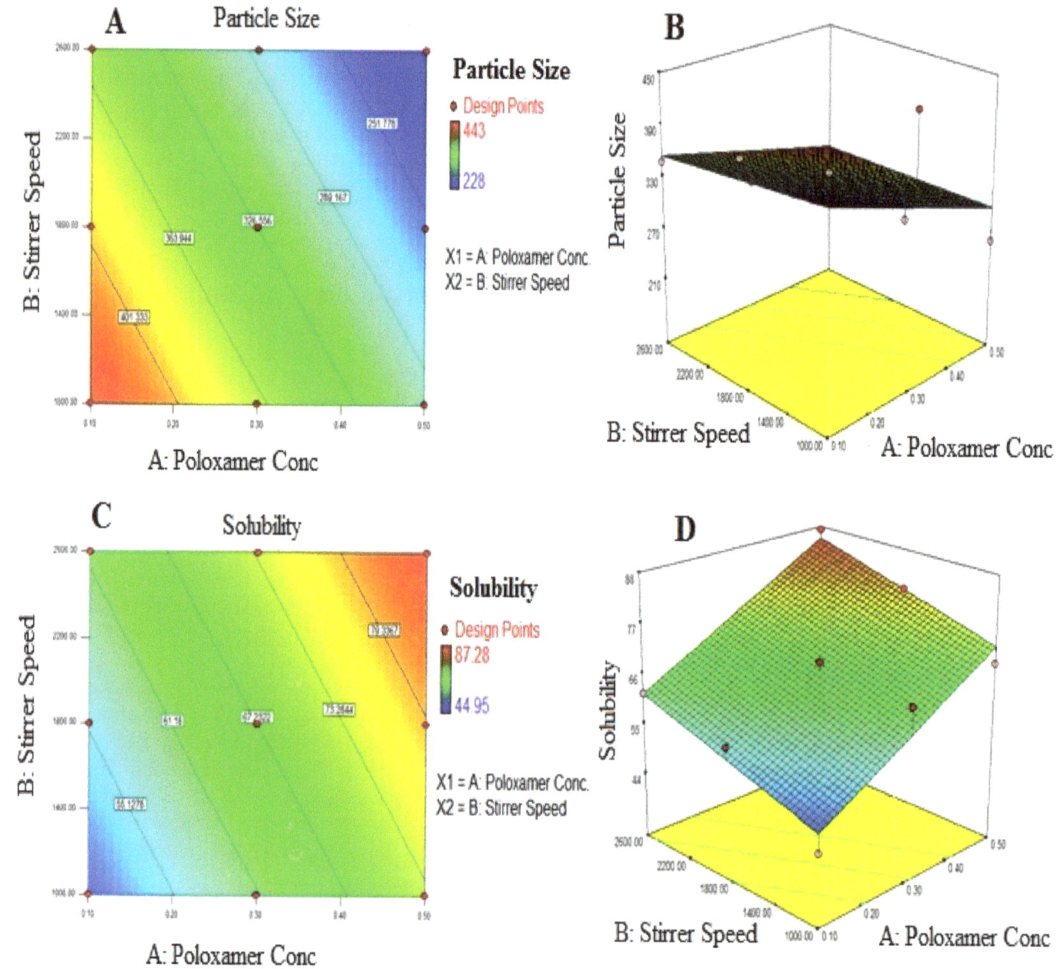

Figure 3. Linear and 3D response surface plot for independent variable's effects on the response particle size (**A**,**B**), respectively, and on response solubility (**C**,**D**), respectively.

3.3. Characterization of NS

3.3.1. Phase Solubility of NS

The initial solubility of PRX in distilled water was found to be 6.6 μg/mL. The solubility study of NS was performed at different concentrations of stabilizer, i.e., poloxamer 188® and at a different speed. The results showed an increase in the solubility of PRX in the NS formulations (Table 7).

3.3.2. Particle Size and Zeta Potential

The particle size in the NS formulation was found to be decreasing with an increase in the poloxamer 188® concentrations and an increase in the speed as shown in Table 7. The particle size and zeta potential of the optimized formulation are shown in Figure 4.

Table 7. Particle size, PI, zeta potential, PDE, and solubility of the NSs.

Formulation Code	Particle Size (nm)	PI	Zeta Potential (mV)	* PDE ± SD	* Solubility ± SD (µg/mL)
NS 1	412	0.706	−22.5	89 ± 0.14	44.95 ± 0.63
NS 2	389	0.713	−24.7	89.48 ± 1.02	57.86 ± 1.21
NS 3	342	0.794	−25.6	89.4 ± 0.21	61.74 ± 0.63
NS 4	443	0.705	−29	88.42 ± 0.32	66.6 ± 1.04
NS 5	334	0.615	−35.9	90.54 ± 0.80	69.65 ± 1.96
NS 6	287.7	0.722	−34.2	88.87 ± 0.95	68.96 ± 1.24
NS 7	258	0.703	−26.4	88.44 ± 1.05	68.96 ± 0.41
NS 8	240	0.566	−28.2	89.41 ± 0.75	79.09 ± 0.24
NS 9	228	0.458	−30.5	89.76 ± 0.76	87.28 ± 0.24

* All values are expressed as mean ± SD (n = 3).

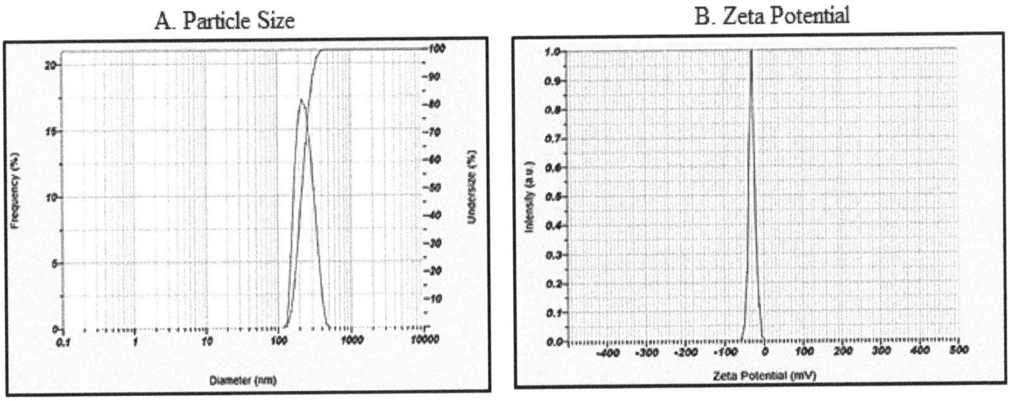

Figure 4. Particle size distribution curve (**A**) and zeta potential peak (**B**) of the optimized formulation NS9.

Entrapment efficiency.

The amount of non-capsulated PRX was determined by an indirect method. The entrapment efficiency of NSs is shown in Table 7. The optimized formulation (NS9) exhibited the highest drug entrapment, i.e., 89.76 ± 0.76%.

3.3.3. TEM Study

TEM analysis was used to depict the morphology of suspended PRX nanoparticles, and the resulting TEM micrograph of the optimized NS is shown in Figure 5. TEM analysis revealed that the suspended nanoparticles were roughly spherical or irregular in shape with uniform distribution; these findings were in good agreement with the particle size analysis outcomes of the optimized formulation NS9 suggesting an average particle size of 228 nm.

3.3.4. DSC STUDY

DSC was used to examine the thermal behavior of the drug and nanoparticles. The pure drug shows a sharp endothermic peak, which corresponds to its melting point, which was observed at 203.15 °C. However, a significant shift in the melting peak of the pure drug (PRX) in the nanoformulation was observed at 198.05 °C, indicating a reduction in crystallinity compared to pure PRX (Figure 6). This indicated the change in the crystalline nature of PRX during the preparation of nanosuspension. The shift also may be due to the presence of stabilizers (Poloxamer 188® at 51.05 °C and PVP K30® at 152.15 °C) in the formulation when compared with the pure drug. A sharp endothermic peak at 164.05 °C represents the melting of mannitol used as a cryoprotectant in the formulation.

The conclusions drawn from the DSC analysis were in good agreement with the noted FT-IR outcomes and were further validated using the results of the XRPD analysis.

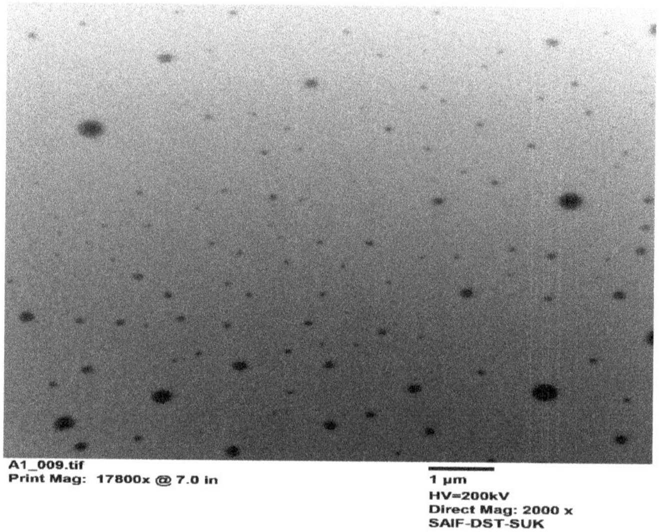

Figure 5. TEM micrograph of optimized PRX NS formulation (NS9).

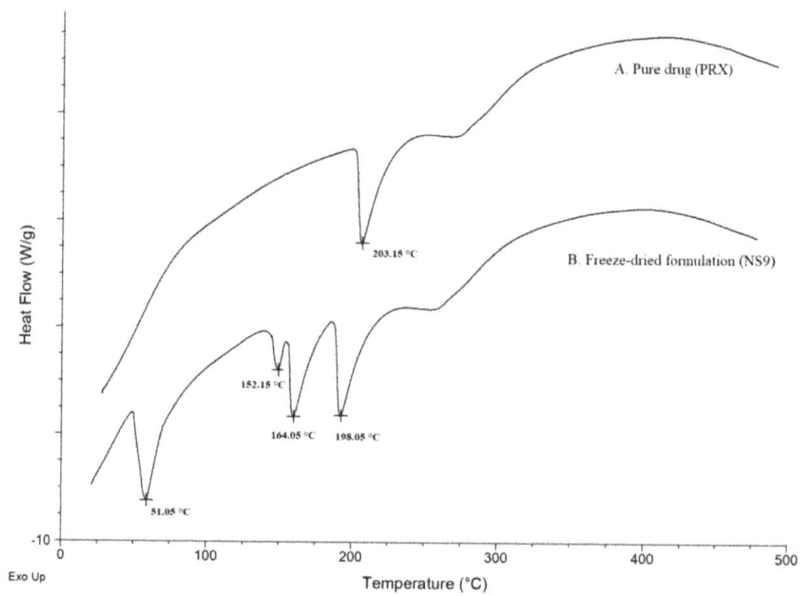

Figure 6. Overlain DSC thermograms of the pure drug (PRX) (A) and freeze-dried formulation (NS9) (B).

3.3.5. XRPD Analysis

To characterize the crystalline nature of PRX within the NS and to evaluate the mode of interaction between PRX and NS, XRPD data of pure PRX and freeze-dried formulation (NS9) samples were acquired (Figure 7). Typical diffraction peaks at 8.62°, 11.65°, 12.49°, 13.28°, 14.51°, 15.86°, 16.70°, 17.71°, 18.85°, 21.74°, and 27.80° were used to identify the crystalline nature of PRX. The pure PRX exhibits an intense crystalline peak between 10°

and 30°. However, the peaks in the NS9 formulation were less intense; indicating a decrease in crystallinity. Additionally, the peaks at 9.10°, 20.22°, 36.42°, 40.80°, and 45.10° are the peaks of D-mannitol used in freeze drying which was absent in pure drug (PRX). Similar kinds of results were reported in previous research studies in the literature [25,26].

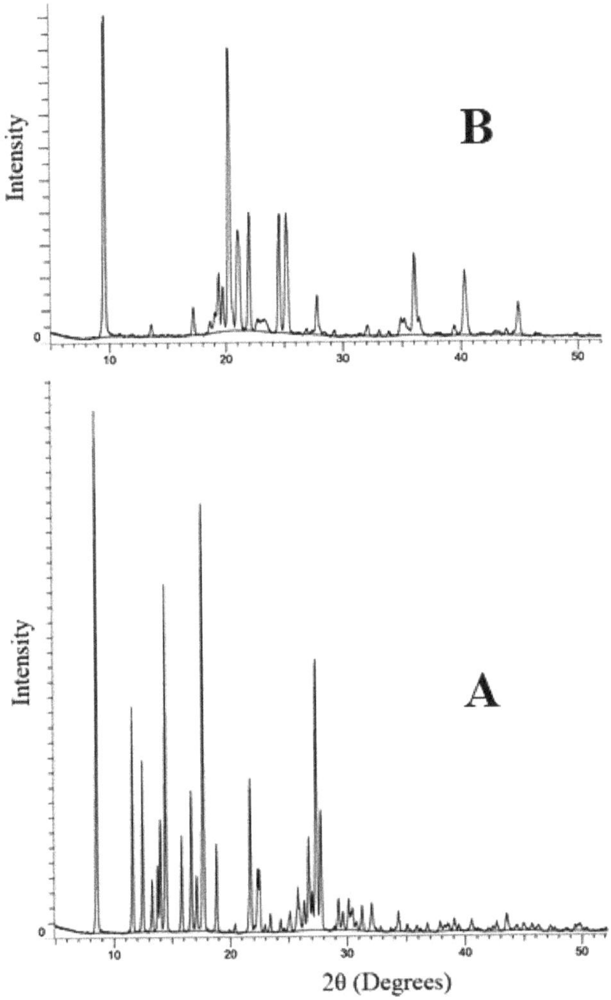

Figure 7. XRPD diffractograms of (**A**) pure drug PRX and (**B**) freeze-dried optimized formulation NS9.

3.3.6. In Vitro Dissolution Study

The in vitro dissolution profile of all the NS formulations (Figure 8) revealed that the PRX was released significantly faster than that of the pure drug. The optimized formulation (NS9) exhibits a greater drug release of approximately 96.07% after 120 min as compared to the other NS formulations and pure PRX (36.78%). According to the dissolution profile, all formulations exhibit burst release, which may be caused by the solubilized drug present in the NS. PRX was more easily dissolved by the formulation when the stabilizer, Poloxamer 188®, was used in higher concentrations and stirred at higher rpm. This improvement can be due to the ability of PVP K30® and stabilizer Poloxamer 188® to form a complex with PRX altering its crystalline nature, which leads to an increase in the solubility and dissolution rate of the PRX in NS than pure PRX. Dissolution efficiency is the area under the dissolution curve within a given range of time. The amount of drug dissolved and the time of solution in contact with the region of absorption, i.e., the GI tract, directly correlate with drug absorption, which increases drug bioavailability.

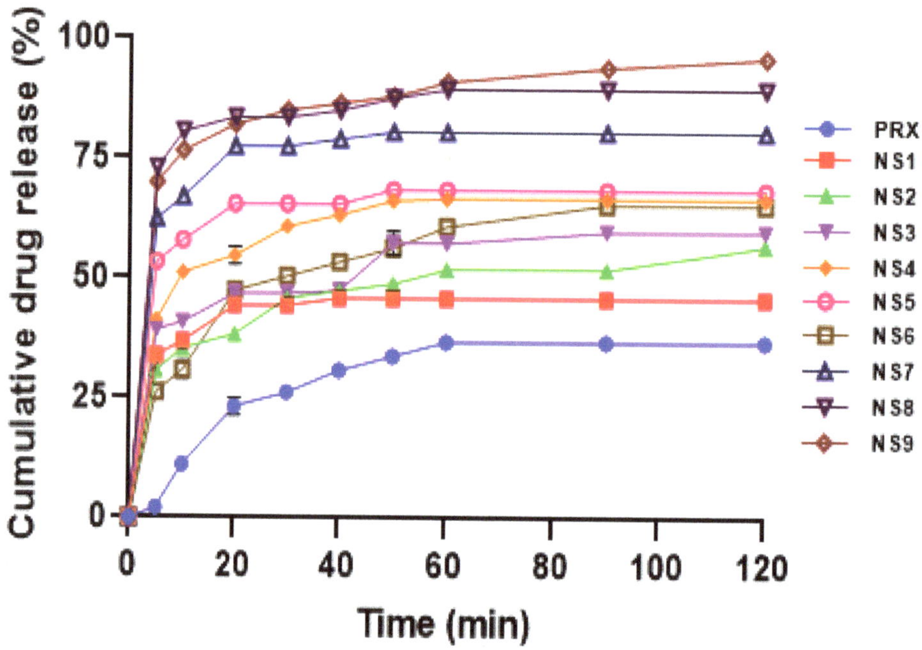

Figure 8. In vitro drug release profiles of prepared PRX NS formulations.

3.3.7. Mathematical Modeling Studies

The R^2 values from fitting the experimental model's in vitro dissolution data into the various release kinetic models are shown in Table 8. Pure PRX drug release data were extremely well-described by the Higuchi model, whereas the Korsmeyer-Peppas model was found to best fit for NS. The Korsmeyer-Peppas model was the most accurate model for describing the release mechanism for all formulations, while all other models were the least accurate.

Table 8. Kinetic profiles of in vitro drug release of optimized PRX NS (NS9).

Dissolution Medium	Zero Order	First Order	Higuchi Model	Hixson-Crowell	Korsmeyer–Peppas		Diffusion Mechanism
	R^2				R^2	Release Exponent (n)	
PBS pH6.8	0.82	0.9797	0.9413	0.9432	0.9929	0.098	Quasi Fickian Diffusion

The highest R^2 value noted for the Korsmeyer-Peppas model (0.9929) established it to be the best-fitting model and the release exponent (n) value of NS was below 0.5 which indicated that NS had followed Quasi-Fickian release kinetics [27]. This Fickian diffusion corroborates to the transport of water via the diffusion process which was driven by the concentration gradient, i.e., drug release from high concentration to low concentration.

4. Conclusions and Future Scope

Nano-sizing is the finest method to improve the solubility and dissolution rate of poorly water-soluble drugs. The anti-solvent precipitation method implied in current research vocation is a cost-effective and simple approach for formulating NS. This research work was aimed at optimizing the NS formulation of PRX by varying concentrations of stabilizers and stirring speed. Optimization was successfully achieved implying 3^2 full factorial design as a Quality by Design (QbD) approach. The obtained 3D surface response analysis results and plots revealed that increasing the stabilizer concentration and stirring speed had resulted in the reduction in particle size and increase in solubility. Different NS batches have depicted a reduction in the particle size from 443 nm to 228 nm and the solubility has increased from 44 µg/mL to 87 µg/mL over changing the independent variables. From all the noted results, it was concluded that the solubility of PRX in water was enhanced by 14- to 15-fold (87.28 µg/mL) than that of the pure PRX (6.6 µg/mL). The optimized NS formulation (NS9) exhibited a greater in vitro drug release of approximately 96.07% post-120 min with respect to the other NS formulations and pure PRX (36.78%). Moreover, the in vitro drug release data from the PRX NS was fitted into diverse mathematical models to predict the drug release mechanism. The release kinetics outcomes established that release profile best fits into the Korsmeyer-Peppas model suggesting the Fickian diffusion as drug release mechanism. Furthermore, from the DSC and XRPD analysis data, it was evident that the PRX was converted into a less crystalline form when formulated into NS, which could be a valid justification for the enhanced PRX solubility. In a nutshell, from all the noted results, it was evident that the NS formulations had improved the solubility and in vitro dissolution of PRX quite significantly in comparison with the pure drug. Hence, from the in vitro-in vivo correlation (IVIVC) point of view, we conclude that prepared and optimized NS formulation would also lead to an increase in the bioavailability of PRX. However, to validate the in vivo performance and to establish the in vivo pharmacokinetic parameters in detail, a multicentric, randomized in vivo pharmacokinetic study should be undertaken in near future. To conclude, the present research outcomes have clearly indicated that NS formulation approach could be a potential alternative to the conventional methods of solubility as well as bioavailability enhancement and need to be explored much for the diverse promising existing and newly developed BCS Class II drugs.

Author Contributions: Conceptualization, S.M.H., R.A.M.O. and U.H.; methodology, S.M.H., R.A.M.O., Y.A. and N.K.K.; software, R.R.D. and M.S.C.; validation, R.R.D. and S.M.H.; formal analysis, R.A.M.O., S.P., G.K. and R.V.; investigation, S.M.H., R.A.M.O. and S.P.; resources, S.P., R.R.D., G.K. and R.V.; data curation, M.S.C., R.A.M.O. and U.H.; writing—original draft preparation, Y.A., S.M.H., R.R.D. and S.P.; writing—review and editing, Y.A, G.K., R.V. and R.A.M.O.; visualization, S.M.H., N.K.K. and R.V.; supervision, M.S.C. and Y.A..; project administration, G.K., R.V., R.A.M.O. and U.H.; funding acquisition, Y.A., U.H., G.K. and R.V. All authors have read and agreed to the published version of the manuscript.

Funding: This research work is funded by Deanship of Scientific Research at King Khalid University, Saudi Arabia, through Large Program under grant number (RGP-2/121/43) 1443.

Institutional Review Board Statement: Not applicable.

Informed Consent Statement: Not applicable.

Data Availability Statement: The data in support of the findings of this study will be available from the corresponding author upon request.

Acknowledgments: The authors extend their appreciation to the Deanship of Scientific Research at King Khalid University, Saudi Arabia, for funding this work through Large Program under grant number (RGP-2/121/43) 1443. The authors are also thankful to the Principal and Management of the Annasaheb Dange College of B. Pharmacy, Ashta, Maharashtra, India for providing the obligatory support and facility to complete this research.

Conflicts of Interest: The authors declare no conflict of interest.

References

1. Gajdziok, J.; Vranikova, B.; Kostelanska, K.; Vetchy, D.; Muselik, J. *Drug Solubility and Bioavailability Improvement: Possible Methods with Emphasis on Liquisolid Systems Formulation*, 1st ed.; GringVerlag Publication: Munich, Germany, 2018; Volume 8.
2. Raj, A.; Harindran, J. Enhancement of Bioavailability using different Solubility Enhancement Techniques: A Review. *Fire J. Med. Pharm. Sci.* **2014**, *1*, 27–42.
3. Savjani, K.T.; Gajjar, A.K.; Savjani, J.K. Drug Solubility: Importance and Enhancement Techniques. *ISRN Pharm.* **2012**, *2*, 28–36. [CrossRef]
4. Geetha, G.; Poojitha, U.; Khan, A. Various Techniques for Preparation of Nanosuspension—A Review. *Int. J. Pharma Res. Rev.* **2014**, *3*, 30–37.
5. Jacob, S.; Nair, A.B.; Shah, J. Emerging role of Nanosuspensions in drug delivery systems. *Biomater. Res.* **2020**, *24*, 2–16. [CrossRef]
6. Sutradhar, K.B.; Khatun, S.; Luna, I.P. Increasing Possibilities of Nanosuspension. *J. Nanotechnol.* **2013**, *2013*, 346581. [CrossRef]
7. Patel, V.R.; Agrawal, Y.K. NS: An approach to enhance solubility of drugs. *J. Adv. Pharm. Technol.* **2011**, *2*, 81–87.
8. Arunkumar, N.; Deecaraman, M.; Rani, C. Nanosuspension technology and its applications in drug delivery. *Asian J. Pharm.* **2009**, *3*, 168–173. [CrossRef]
9. Lakshmi, P.; Kumar, G.A. Nanosuspension Technology: A Review. *Int. J. Pharm. Pharm. Sci.* **2010**, *2*, 35–40.
10. Yadollahi, R.; Vasilev, K.; Simovic, S. Nanosuspension Technologies for Delivery of Poorly Soluble Drugs. *J. Nanomater.* **2015**, *2015*, 1–13. [CrossRef]
11. Singare, D.S.; Marella, S.; Gowthamrajan, K.; Kulkarni, G.T.; Vooturi, R.; Rao, P.S. Optimization of formulation and process variable of nanosuspension: An industrial perspective. *Int. J. Pharm.* **2010**, *402*, 213–220. [CrossRef]
12. Brodgen, R.N.; Heel, R.C.; Speight, T.M.; Avery, G.S. Piroxicam: A Review of its Pharmacological Properties and Therapeutic Efficacy. *Drugs* **1981**, *22*, 165–187.
13. *USP 30/NF 25*; The United States Pharmacopoeia 30/National Formulary 25. United States Pharmacopoeial Convention, Inc.: Rockville, MD, USA, 2007.
14. Bachhav, A.A.; Ahire, S.A.; Jadhav, A.G. Preformulation Study of Piroxicam. *Int. J. Pharm. Sci. Res.* **2019**, *10*, 811–818.
15. Suhesti, T.S.; Fudholi, A.; Martien, R.; Martono, S. Pharmaceutical Nanoparticle Technologies: An approach to improve drug solubility and dissolution rate of Piroxicam. *Res. J. Pharm. Tech.* **2017**, *10*, 968–974. [CrossRef]
16. Pal, S.L.; Jana, U.; Manna, P.K.; Mohanta, G.P.; Manavalan, R. Nanoparticle: An overview of preparation and characterization. *J. Appl. Pharm. Sci.* **2011**, *1*, 228–234.
17. Jadhav, P.; Yadav, A. Formulation, optimization, and in vitro evaluation of polymeric nanosuspension of flurbiprofen. *Asian J. Pharm. Clin. Res.* **2019**, *12*, 183–191. [CrossRef]
18. Patnaik, S.; Chunduri, L.A.; Akilesh, M.S.; Bhagavatham, S.S.; Kamisetti, V. Enhanced dissolution characteristics of piroxicam–Soluplus®nanosuspensions. *J. Exp. Nanosci.* **2016**, *11*, 916–929. [CrossRef]
19. Yang, H.; Teng, F.; Wang, P.; Tian, B.; Lin, X.; Hu, X. Investigation of a nanosuspension stabilized by Soluplus®to improve bioavailability. *Int. J. Pharm.* **2014**, *477*, 88–95. [CrossRef]
20. Bakhshayesh, A.R.; Akbarzadeh, A.; Alihemmati, A.; Nasrabadi, H.T.; Montaseri, A.; Davaran, S.; Abedelahi, A. Preparation and characterization of novel anti-inflammatory biological agents based on piroxicam-loaded poly-ε-caprolactone nano-particles for sustained NSAID delivery. *Drug Deliv.* **2020**, *27*, 269–282. [CrossRef]
21. Honmane, S.M.; Chimane, S.M.; Bandgar, S.A.; Patil, S.S. Development and Optimization of Capecitabine loaded Nanoliposomal System for Cancer Delivery. *Indian J. Pharm. Edu. Res.* **2020**, *54*, 376–384. [CrossRef]
22. Honmane, S.; Kadam, A.; Choudhari, S.; Patil, R.; Ansari, S.A.; Gaikwad, V. Effect of polymers and process parameters in augmenting the compactability and dissolution behaviour of oxcarbazepine spherical agglomerates. *Drug Deliv. Sci. Technol.* **2021**, *64*, 102578. [CrossRef]
23. Maulvi, F.A.; Dalwadi, S.J.; Thakkar, V.T. Improvement of dissolution rate of aceclofenac by solid dispersion technique. *Powder Technol.* **2011**, *207*, 4754. [CrossRef]
24. Ahire, E.; Thakkar, S.; Wad, M.D.; Misran, M. Parenteral Nanosuspensions: A brief review from solubility enhancement to more novel and specific applications. *Acta Pharm. Sin. B* **2018**, *8*, 733–755. [CrossRef] [PubMed]
25. Ammanage, A.; Rodriques, P.; Kempwade, A.; Hiremath, R. Formulation and evaluation of buccal films of Piroxicam co-crystals. *Future J. Pharm. Sci.* **2020**, *6*, 1–11. [CrossRef]
26. Altay Benetti, A.; Bianchera, A.; Buttini, F.; Bertocchi, L.; Bettini, R. Mannitol Polymorphs as Carrier in DPIs Formulations: Isolation Characterization and Performance. *Pharmaceutics* **2021**, *13*, 1113. [CrossRef]
27. Sahoo, S.; Chakraborti, C.K.; Behera, P.K. Development and Evaluation of Gastroretentive Controlled Release Polymeric Suspensions Containing Ciprofloxacin and Carbopol Polymers. *J. Chem. Pharm. Res.* **2012**, *4*, 2268–2284.

Disclaimer/Publisher's Note: The statements, opinions and data contained in all publications are solely those of the individual author(s) and contributor(s) and not of MDPI and/or the editor(s). MDPI and/or the editor(s) disclaim responsibility for any injury to people or property resulting from any ideas, methods, instructions or products referred to in the content.

polymers

MDPI

Article

Dosimetric Characterization of DSF/NaOH/IA-PAE/*R.* spp. Phantom Material for Radiation Therapy

Damilola Oluwafemi Samson [1,2,*], Ahmad Shukri [1], Nurul Ab. Aziz Hashikin [1], Siti Hajar Zuber [1], Mohd Zahri Abdul Aziz [3,*], Rokiah Hashim [4], Mohd Fahmi Mohd Yusof [5], Nor Ain Rabaiee [6] and Sylvester Jande Gemanam [7]

1 School of Physics, University Sains Malaysia, Penang 11800, Malaysia
2 Department of Physics, Faculty of Science, University of Abuja, Abuja 900211, Nigeria
3 Advanced Medical and Dental Institute, University Sains Malaysia, Penang 13200, Malaysia
4 School of Industrial Technology, University Sains Malaysia, Penang 11800, Malaysia
5 School of Health Sciences, University Sains Malaysia, Kota Bharu 16150, Malaysia
6 Department of Radiology, Kulliyyah of Medicine, International Islamic University, Kuantan 25200, Malaysia
7 Department of Physics, Benue State University, Makurdi 102119, Nigeria
* Correspondence: damilola.samson@uniabuja.edu.ng (D.O.S.); mohdzahri@usm.my (M.Z.A.A.)

Abstract: Background: Different compositions of DSF/NaOH/IA-PAE/*R.* spp. composite particleboard phantoms were constructed. **Methods:** Photon attenuation characteristics were ascertained using gamma rays from ^{137}Cs and ^{60}Co. Absorbed doses at the location of an ionization chamber and Gafchromic EBT3 radiochromic films were calculated for high-energy photons (6 and 10 MV) and electrons (6, 9, 12, and 15 MeV). **Results:** The calculated $TPR_{20,10}$ values indicate that the percentage discrepancy for 6 and 10 MV was in the range of 0.29–0.72% and 0.26–0.65%. It was also found that the relative difference in the d_{max} to water and solid water phantoms was between 1.08–1.28% and 5.42–6.70%. The discrepancies in the determination of PDD curves with 6, 9, 12, and 15 MeV, and those of water and solid water phantoms, ranged from 2.40–4.84%. Comparable results were found using the EBT3 films with variations of 2.0–7.0% for 6 and 10 MV photons. Likewise, the discrepancies for 6, 9, 12, and 15 MeV electrons were within an acceptable range of 2.0–4.5%. **Conclusions:** On the basis of these findings, the DSF/NaOH/IA-PAE/*R.* spp. particleboard phantoms with 15 wt% IA-PAE addition level can be effectively used as alternative tissue-equivalent phantom material for radiation therapy applications.

Keywords: dosimetric properties; tissue-equivalent phantom; absorbed dose; IA-PAE; radiation therapy

Citation: Samson, D.O.; Shukri, A.; Hashikin, N.A.A.; Zuber, S.H.; Aziz, M.Z.A.; Hashim, R.; Yusof, M.F.M.; Rabaiee, N.A.; Gemanam, S.J. Dosimetric Characterization of DSF/NaOH/IA-PAE/*R.* spp. Phantom Material for Radiation Therapy. *Polymers* **2023**, *15*, 244. https://doi.org/10.3390/polym15010244

Academic Editor: Edina Rusen

Received: 17 October 2022
Revised: 15 November 2022
Accepted: 17 November 2022
Published: 3 January 2023

Copyright: © 2023 by the authors. Licensee MDPI, Basel, Switzerland. This article is an open access article distributed under the terms and conditions of the Creative Commons Attribution (CC BY) license (https://creativecommons.org/licenses/by/4.0/).

1. Introduction

Phantoms have become essential for quality assurance (QA) and quality control (QC) in a variety of medical procedures involving radiation. The earliest phantoms consisted of water or wax, but wax phantoms had a number of issues. Wax formulations differed greatly depending on the type of wax used and, at low energies, deviated from tissue equivalence [1]. On the other hand, water has been described as the standard and most universal phantom material for dosimetry measurements of photon and electron beams. As the use of liquid water can prove to be challenging and inconvenient in certain situations, because of its surface tension and the uncertainty in positioning the detector near the surface, solid homogeneous phantom materials have achieved substantial recognition [2]. The benefit of these phantoms allows the measurement of the interaction of ionizing radiation in the human body, which enables the range of doses in various organs and tissues to be measured according to their sensitivity. The most widely used tissue-equivalent material (TEM) are those that are both easy to work with and relatively inexpensive. The usage of natural, readily available, and cheap phantom material, such as wood (*Rhizophora* spp.), is often of interest.

Rhizophora spp. (*R.* spp.) has received increasing attention for industrial applications due to its fast-growing nature, high productivity, quick maturity, and high strength, with advancement in processing technology and increased market demand. The chemical composition of *R.* spp. is very similar to those of TEM normally utilized as phantoms for radiation therapy when compared with other wood species [3–8]. Moreover, *R.* spp. possesses convenient morphological characteristics and physiological adaptations, with moisture content ranging from 5–10% and basic physical density within 0.90–1.04 gcm^{-3} [9]. Various researchers have shown that *R.* spp. is a highly attractive material for use as an effective TEM for a wide range of benefits, including high-energy photon and electron radiation therapy, as well as X-ray imaging [5–7,10–12]. However, due to its shortcomings, such as the tendency to be warped, cracked, degraded, and weakened over time, the usage of appropriate resins with unique characteristics in the development of *R.* spp. particleboards has been reported [5–8,11,12]. The type of these curing resins and their chemical properties are also crucial criteria that should be well-decided for particleboard phantom formation and structure.

Among the various forms of modifying resins, synthetically-based ones are the most commonly adopted [13], but prolonged human exposure to non-renewable resources has been shown to cause chronic toxicity, myeloid leukemia mortality, and lymphohematopoietic malignancies [14–16]. In relation to TEM studies, synthetically based resins were also found not to be compatible with the intended density and radiation attenuation properties (RAPs) of *R.* spp. particleboards as compared to water [17]. On the other hand, bio-based materials, such as soy protein (DSF—defatted soy flour) developed in wood resin, have been validated through specific independent studies because of their ready availability and low cost, coupled with the fact that they are biodegradable, biocompatible, and eco-friendly [6,7,15,16,18–20].

DSF is a highly oxygenated carbon compound, which makes it attractive for use in the development of phantom materials equivalent to tissue and water, and it can be appropriate either as an uncured or a cured bio-based adhesive [18]. Uncured DSF was, however, identified as a weak adhesive, and a chemical change is needed to break the internal bonds and disperse the polar protein molecules [6,16,18–24]. The most commonly used cross-linking agents for DSF are itaconic acid polyamidoamine-epichlorohydrin (IA-PAE), epoxy, formaldehyde, glutaraldehyde, and glyoxal. Since some of these curing agents also have a deleterious environmental impact, as well as being non-renewable, IA-PAE has been considered as an alternative cross-linker for DSF-based adhesives [6,18–20,24]. The cross-linking reaction of DSF with NaOH/IA-PAE resins is highly regarded for their incomparable multifunctionality, enhanced physical and mechanical characteristics, stable water resistance, and good wood-bonding ability [6].

The current study aims to construct and examine the dosimetric characterization of bio-based particleboard phantoms for radiation therapy by integrating DSF-based resins—*R.* spp. particles of size ≤74 μm, NaOH (10 wt%)—and four different treatment levels of IA-PAE (0, 5, 10, and 15 wt%). High energy photon attenuation measurements were ascertained using a Ludlum setup with ^{137}Cs and ^{60}Co sources with effective photon energies of 0.662 and 1.250 MeV. A linear accelerator (LINAC) was utilized to determine the dosimetric characteristics of the DSF/NaOH/IA-PAE/*R.* spp. particleboard phantoms. This is done using a cylindrical Farmer-type ionization chamber (IC) (NE 2581/334) and Gafchromic EBT3 radiochromic films to evaluate the tissue–phantom ratio ($TPR_{20,10}$), percentage depth dose (PDD) and beam profile of the samples for high energy photon (6 and 10 MV) and electron (6, 9, 12 and 15 MeV) beams. The findings were compared with those of appropriate standard phantom materials (water and solid water) utilized in radiation therapy.

2. Materials and Methods

2.1. Preparation of Bio-Based Adhesives

As previously reported [6,7], the synthesized IA-PAE solution had a solid content of 55.96 ± 0.01 wt%, a pH of 6.68 at 27.58 °C, and an apparent viscosity of 100.40 ± 0.25 mPa.s, comparable to commercial PAE-soy protein (C-PAE-SF) and IA-PAE reported by Gui et al. [19]. The DSF-based bio-adhesives were prepared at room temperature by dissolving 35 g of DSF under steady mechanical stirring in distilled water (65, 50, 45, and 40 g) for 0.5 h following the procedure described by Samson et al. [6]. Various concentrations of the prepared IA-PAE (0, 5, 10 and 15 wt%) were then applied to the uniform mixtures and moderately stirred for 0.5 h. The cured DSF/IA-PAE slurry mixture was maintained at pH 11.0 with 2N of NaOH (10 wt%) solution, since pH 11.0 is the optimum condition for cross-linking reactions [25].

2.2. Fabrication of DSF/NaOH/IA-PAE/*R. spp.* Particleboard Phantoms

At the start of the process, all sample formulations with particle size $\leq$ 74 µm were thoroughly mixed by hand as they were applied for 0.5 h to the DSF/NaOH/*R.* spp. mixture with different IA-PAE content. Thereafter, they were mixed evenly for another 10 min using a rotary mixer machine to ensure the uniformity of the samples. By using a mould of dimension (30 cm × 30 cm × 1.0 cm), the mixtures were subsequently cold-compressed for 10 min using a hydraulic press machine (0.49 MPa, 5 min, and 0 °C) fitted with stops to achieve a target density of 1.0 gcm^{-3} at room temperature and relative humidity of 55%. The stacked mats were then constructed using a hydraulic hot press machine at 170 °C for 20 min with 20 MPa [6]. A total of 150 units of DSF/NaOH/IA-PAE/*R.* spp. particleboard phantoms were developed. Tables 1 and 2 have been tabulated to examine the physico-mechanical and dimensional stability properties (PMDSP) (MC—moisture content, SC—solid content, IB—internal bonding, MOR—modulus of rupture, MOE—modulus of elasticity, TS—thickness swelling, and WA—water absorption), elemental compositions, effective atomic numbers (Z_{eff}), and electron densities (N_{el}) of the particleboards and standard phantom materials. As can be seen, Table 1 shows that sample A_{15} with 15 wt% IA-PAE concentrations provides the ascribable parameters and meets the minimum requirements of Type 8, Type 13, and Type 18, according to JIS A-5908 [26]. The Z_{eff} and N_{el} of these phantoms, which were the exclusive parameters used to characterize various types of materials, were found to be comparable to those of water and other commercial phantom materials (Table 2).

Table 1. PMDSP of particleboard phantoms.

Sample	Physico-Mechanical Property					Dimensional Stability Property			
	MC (%)	SC (%)	IB (MPa)	MOR (MPa)	MOE (GPa)	TS (%)		WA (%)	
						2 h	24 h	2 h	24 h
A_0	**7.44 ± 0.21**	**33.18 ± 0.40**	**0.11 ± 0.09**	**5.39 ± 0.57**	2.04 ± 0.10	34.62 ± 1.8	50.3 ± 1.7	60.2 ± 1.1	66.1 ± 0.9
A_5	8.08 ± 0.30	34.10 ± 1.08	0.65 ± 0.02	14.18 ± 0.31	4.46 ± 0.51	21.11 ± 1.0	25.2 ± 1.3	31.4 ± 1.6	36.3 ± 1.0
A_{10}	7.11 ± 0.27	35.06 ± 0.56	0.69 ± 0.06	17.60 ± 0.45	7.30 ± 0.13	10.20 ± 1.2	11.0 ± 1.8	24.2 ± 0.8	30.6 ± 1.2
A_{15}	7.05 ± 0.19	37.31 ± 1.01	0.72 ± 0.01	18.97 ± 0.22	7.89 ± 0.11	10.01 ± 0.9	10.5 ± 1.2	20.7 ± 1.0	23.9 ± 0.5

Remark: A_0 = Uncured *R.* spp., A_5 = DSF/NaOH/IA-PAE/*R.* spp., A_{10} = DSF/NaOH/IA-PAE/*R.* spp., and A_{15} = DSF/NaOH/IA-PAE/*R.* spp. indicate 0, 5, 10, and 15 wt% IA-PAE. Data are expressed as an average ± standard deviation (SD).

Table 2. Elemental compositions, Z_{eff}, and N_{el} of phantom samples.

Sample	Weight Fraction of Elements in Each Sample (%)													Z_{eff}	$N_{el} \times 10^{23}$
	H	C	O	N	Na	Mg	*p*	S	Cl	K	Ca	Fe	Zn		
A_0	-	51.01	46.24	2.64	-	-	-	0.11	-	-	-	-		7.18 [a]	3.39
A_5	-	51.48	42.10	4.71	0.55	0.20	-	-	0.15	0.21	0.60	-	-	7.45 [a]	3.26
A_{10}	-	51.07	43.92	4.02	0.28	-	-	-	0.11	0.24	0.31	0.05		7.51 [a]	3.33
A_{15}	-	53.11	41.28	2.56	0.43	0.04	0.12	0.05	0.32	0.18	1.75	0.07	0.09	7.72 [a]	3.34
Solid water	8.10	67.20	19.90	2.40	-	-	-	-	0.10	-	2.30	-		7.29 [b]	3.32
Water	11.20	-	88.80	-	-	-	-	-	-	-	-	-		7.42 [b]	3.34
Virtual water	7.70	68.70	18.90	2.30	-	-	-	-	0.10	-	2.30	-	-	6.12 [c]	3.38
PMMA	8.00	60.00	31.96	-	-	-	-	-	-	-	-	-	-	5.85 [c]	3.87
Polystyrene	7.74	92.26	-	-	-	-	-	-	-	-	-	-	-	5.29 [c]	3.43

[a] Curent study, [b] Sahoo et al. [27], [c] Schoenfeld et al. [28].

Using the gravimetric technique (Equation (1)), the average particleboard densities (ρ) were assessed, and the propagation of uncertainty was deduced based on the external dimensions using Equation (2):

$$\rho = \frac{m}{l \times w \times h} \tag{1}$$

$$d\rho = \left(\frac{dm}{m} + \frac{dl}{l} + \frac{dw}{w} + \frac{dh}{h}\right)\rho \tag{2}$$

where, m, l, w, and h denote the respective particleboard mass, length, width, and thickness; dm, dl, dw, and dh are the uncertainties in m, l, w, and h, respectively.

The computed tomography (CT) image modality was achieved based on a previous technique detailed by Samson et al. [7]. The parameters of the various standard phantom materials, compared with DSF/NaOH/IA-PAE/*R.* spp., are listed in Table 3. The results of the average density show that the density of A_{15} is within the range found for water and other commercial phantom materials. According to the results, the mean HU values and ED of the A_{15} were near to those acceptable standard reference equivalent materials [7], while a significant variation is observed with A_o, A_5, and A_{10}, respectively. This might explain the fact that better attenuation abilities were observed when X-ray beams passed through the corresponding sample (A_{15}). Therefore, the A_{15} sample formulation showed the potential to replicate human tissue because it has a comparable dynamic and is higher in terms of stability as a medical phantom.

Table 3. Parameters of various standard phantom materials compared with DSF/NaOH/IA-PAE/*R.* spp.

Phantom Materials	Manufacturer	Density (g/cm^3)	Mean HU Value	ED × 10^{23} (Electrons/cm^3)
A_0	Current study	1.07	−89.71	3.39 [a]
A_5	Current study	0.96	−55.55	3.26 [a]
A_{10}	Current study	0.99	−33.01	3.33 [a]
A_{15}	Current study	1.01	−12.79	3.34 [a]
Solid water	Gammex, Middleton, WI, USA	1.04	5.30	3.32 [b]
Water	-	1.00	−9.01	3.34 [b]
Virtual water	Med-Cal, Middleton, WI, USA	1.04	-7 ± 7	3.38 [c]
Polymethyl methacrylate	-	1.19	133	3.87 [c]
Polystyrene	-	1.06	140.5	3.43 [c]

[a] Current study, [b] Sahoo et al. [27], [c] Schoenfeld et al. [28].

2.3. Measurement of RAPs

The attenuation properties were determined using a Ludlum lead equivalent setup, as depicted in Figure 1. ^{137}Cs and ^{60}Co sealed sources with effective gamma energies of 0.662 and 1.250 MeV were used to provide the incident photons. The sources and the Ludlum NaI(TI) detector, with diameters of 2.5 cm and 6.5 cm, were encapsulated in a lead container with collimation of diameter 0.5 cm and thickness of 2 cm to simulate the line source projection and avoid leakage. An aluminum (Al) plate of dimension 7 cm × 7 cm, with an approximate thickness of 0.1 cm, was used as an attenuator to produce the scattered photons. The optimum distance between the source compartment and the Al plate and between the Al plate and the detector compartment was 30 cm, whereas the distance between the phantom samples and the detector compartment was 6.2 cm. The transmitted photons from the source were collected and detected using the Ludlum scintillation detector connected to a single channel analyzer (SCA).

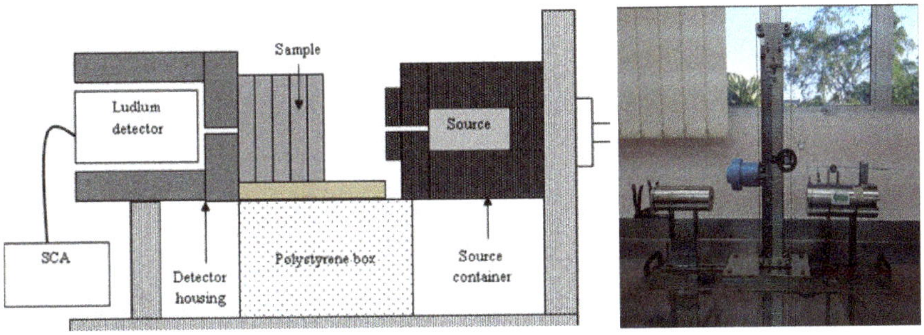

(a) (b)

Figure 1. Ludlum setup: (**a**) schematic diagram and (**b**) actual experimental setup.

The linear attenuation coefficient (LAC) and mass attenuation coefficient (MAC) are the fundamental parameters to evaluate the dosimetric and radiation shielding performance of any composite material. These commonly used parameters provide some information on the possibility of photon interaction processes with matter per unit thickness. As a photon beam propagates through a homogeneous medium, the beam intensity at depth t is assigned as I_t, whereas the beam intensity at a reference point in the absorbing material ($t = 0$), is assigned as I_o. This can be described by the familiar Beer-Lambert's law (Equation (3)) [6,7].

$$\mu = \frac{1}{x}\ln\left(\frac{I_o}{I_t}\right) \tag{3}$$

$$\mu_m = \frac{\mu}{\rho} = \frac{1}{\rho t}\ln\left(\frac{I_o}{I_t}\right) = \frac{A}{M}\ln\left(\frac{1}{T}\right) \tag{4}$$

where μ (cm^{-1}) denotes the LAC, ρ ($g\cdot cm^{-3}$) is the density, x (cm) and t ($g\cdot cm^{-2}$) are the physical thickness and mass thickness (mass per unit area), T is the transmittance, M (g) is the mass of the sample material, A is the cross-sectional area (cm^2), and μ_m (cm^2g^{-1}) indicates the total MAC. The total μ_m values were calculated on the basis of the mixture rule by using the weight fraction (ω_i) for each element i of the particleboard materials, as expressed in Equations (5) and (6):

$$\mu_m = \left(\frac{\mu}{\rho}\right)_{DSF-based\ particleboard} = \omega_1\left(\frac{\mu}{\rho}\right)_1 + \omega_2\left(\frac{\mu}{\rho}\right)_2 + \ldots = \sum_{i=1}^{N}\omega_i\left(\frac{\mu}{\rho}\right)_i \tag{5}$$

$$\omega_i = \frac{n_i A_i}{\sum_i n_i A_i} = \frac{\widetilde{\rho}_i}{\rho} \tag{6}$$

where n_i denote the number of atoms of the ith individual element, A_i is the atomic weight, and $\widetilde{\rho}_i$ is the actual mass density. The related cumulative discrepancies in the experimental MAC were obtained by using the propagation of error relationship from ambiguities in $I_o, I_t,$ x and areal density (ρ) [7]:

$$\Delta\left(\frac{\mu}{\rho}\right) = \frac{1}{\rho}\sqrt{\left[\left(\frac{\Delta I_t}{I_t}\right)^2 + \left(\frac{\Delta I_o}{I_o}\right)^2 + \left(\ln\frac{\Delta I_o}{I_t}\right)^2 + \left(\frac{\Delta x}{x}\right)^2\right]} \tag{7}$$

where ΔI_t, ΔI_o, and Δx are the errors in the intensities I_t, I_o, and thickness x of the sample material, respectively. Paired t-test using SPSS (V22.0) was used to calculate any variations in μ_m values, as compared with the value of water ascertained via the photon cross-section database (XCOM) [29]. The half-value layer (HVL–$X_{1/2}$) is used to assess how far X-ray penetrates the particleboard samples, which were used to verify the performance of the patient's radiation exposure. It can be defined, as given in Equation (8), whereas Equation (9) is the relationship between the mean free path (MFP $-\ \lambda$) and $X_{1/2}$ [7,8].

$$\text{HVL, } X_{1/2} = \frac{0.693}{\mu_m \times \rho} \quad (8)$$

$$\text{MFP, } \lambda = \frac{X_{1/2}}{0.693} \quad (9)$$

2.4. Dosimetric Evaluation of DSF/NaOH/IA-PAE/R. spp. Particleboard Phantoms

Samples with up to 15 wt% IA-PAE addition were selected because of their optimum characteristics, and a total of 34 units of DSF/NaOH/IA-PAE/*R.* spp. particleboard phantoms of sizes 30 cm × 30 cm × 1.0 cm and 30 cm × 30 cm × 0.5 cm, simulating the dimensions of widely used solid water phantom slabs (CIRS Inc., Norfolk, VA, USA), were fabricated. Additionally, two of these slabs were designed with slots to accommodate the cylindrical IC. The Farmer-type IC was used due to its unique features, such as high precision, stability, dose rate independence, excellent linearity, little to no fading, and equivalency to soft tissue nature. All experimental measurements with both photon and electron beams were carried out on the medical Elekta Synergy PRIMUS LINAC at the Department of Oncological and Radiological Science, Advanced Medical and Dental Institute, Universiti Sains Malaysia (USM).

2.5. Determination of Photon Beam Quality Index ($TPR_{20,10}$—Tissue Phantom Ratio)

Samples of DSF-based particleboard and solid water phantoms were mounted and aligned on the central axis of the beam, followed by the insertion of the IC into an electrometer (Model PTW-Unidos ET10008/081134) at depths z = 20 cm and z = 10 cm below the water surface at 10 cm × 10 cm field size and 100 cm SSD, as depicted in Figure 2. Before taking any reading, the IC and electrometer were warmed up for 10 min. For each of the phantom samples, three exposures were made at the two depths, and the average charge collected was evaluated. The expression related to the charge collected at the two depths can be expressed as:

$$TPR_{20,10} = \frac{Q_{20}}{Q_{10}} \quad (10)$$

where Q_{20} and Q_{10} are the respective charge (nC) collected at depths z = 20 cm and z = 10 cm for DSF-based particleboards, water, and solid water phantoms, respectively.

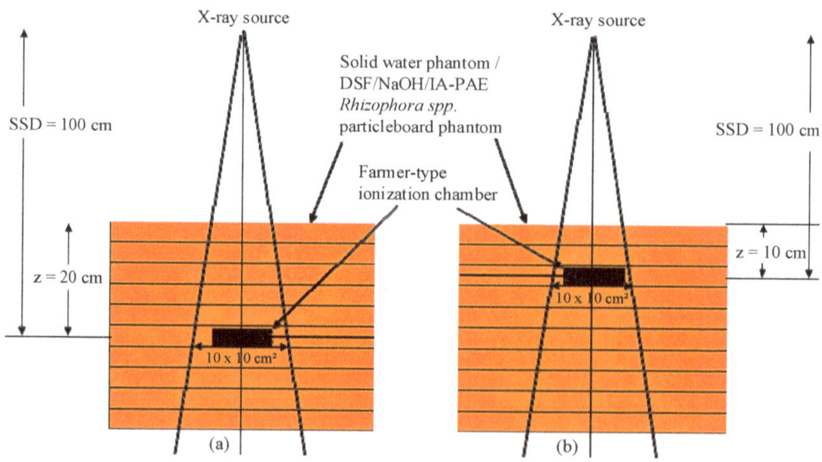

Figure 2. Measurement setup for $TPR_{20,10}$: (**a**) depths z = 20 cm and (**b**) depth z = 10 cm below the water surface at field size (10 cm × 10 cm) and SSD (100 cm).

2.6. PDD Evaluation Using IC

Samples of 15 cm thickness were placed on the LINAC couch to establish the photon and electron beams with backscatter. The calibrated IC with an inner volume of 0.6 cm^3, connected to the electrometer, was placed in the chamber slot to acquire PDDs in the DSF-based particleboards, water, and solid water phantoms, as displayed in Figure 2. The slabs of solid water were selected for the phantom material, as it was found to be appropriate for dosimetry of high energy photon and electron beams [30]. The IC and phantoms were positioned in an isocentre distance of the LINAC at an SSD of 100 cm using the front pointer device, and the field size was set at 10 cm x 10 cm on the surface in accordance with the calibration parameters in the dosimetry protocol of IAEA TRS-398:2000 [31]. Exposures were rendered using photon beams of 6 and 10 MV and electron beams of 6, 9, 12, and 15 MeV with 100 monitor units (MU). Particleboard slabs were added above the IC to assess the ionization at depth below the surface of the phantoms, and the SSD and field size were subsequently readjusted. During these measurements, both the gantry and collimator angles were set to zero degrees. The PDDs were measured from the phantom surface at 0 cm until a depth of approximately 20 cm was reached along the central axis, with a measurement interval of 1 mm from the surface to 2 cm depth followed by 2.5 cm and then 3 cm up to 25 cm. The PDD determination for each depth took 6 s. Exposure using electron beams was achieved by adopting an applicator to the LINAC treatment head. After each exposure, a time delay of 120 s was applied before the next phantom slab was inserted in order to take proton production into account. The PDD values were expressed as a percentage of the absorbed dose at a given depth D′ to the absorbed dose at a specified reference depth (maximum depth) D″ along the central axis of the phantom samples (Equation (11)). The discrepancy in the calculated PDD was estimated as a percentage (D%), as given in Equation (12):

$$PDD = \frac{D'}{D''} \times 100\% \quad (11)$$

$$D(\%) = \frac{PDD_{(DSF)} - PDD_{(water/solid\ water)}}{PDD_{(water/solid\ water)}} \times 100 \quad (12)$$

where $PDD_{(DSF)}$ is the PDD for the constructed DSF-based *R.* spp. particleboard phantom samples and $PDD_{(water/solid\ water)}$ is the PDD for the water and solid water phantoms. The PDD curves were plotted for 6 and 10 MV photons, as well as for 6, 9, 12, and 15 MeV electrons.

2.7. PDD Evaluation Using Gafchromic EBT3 Radiochromic Films

The Gafchromic EBT3 radiochromic film sheets (Lot #: 05161903), with dimensions of 20.3 cm × 25.4 cm, were inserted between the DSF-based particleboard phantoms and solid water phantoms in a portrait orientation due to their near tissue- and water-equivalent characteristics, and 10 cm of phantom material was placed under the film to ensure sufficient backscatter. Phantom slabs were inserted above the film, and the SSD and field size were readjusted afterward. The measurements were repeated until a depth of almost 20 cm, and the results were compared with that of water and solid water phantoms. All films were marked with reference points to indicate the film orientations relative to the gantry. Irradiation was made parallel to the beam for a static 10 × 10 cm^2 field size at 100 cm SSD and with a dose ranging from 0 to 700 cGy. Three films were exposed for each photon and electron energy. The irradiated films were kept at room temperature for 24 hrs post-irradiation to allow time for the polymerization reactions in the film to stabilize and produce a stable optical density measurement [32]. The films were then processed with an EPSON Expression 10,000 XL flatbed scanner. To acquire images, a desktop computer was interfaced with the scanner, and VeriSoft® software 5.1 was used for image scanning and capture. The experimental setup for the PDD evaluation is highlighted in Figure 3.

The PDD data were normalized to the maximum dose, expressed as a percentage, and the percentage variation was measured, as indicated in Equations (11) and (12).

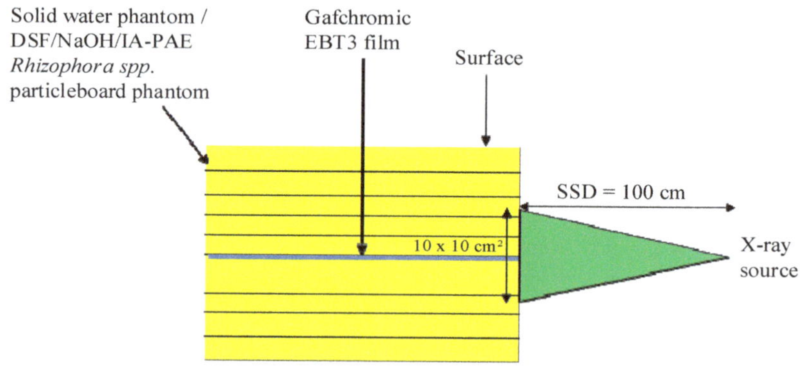

Figure 3. Experimental setup used for PDD evaluation.

3. Results and Discussion

3.1. Density Measurement of DSF-Based R. spp. Particleboard Phantoms

Figure 4 displays the variation of average densities with point distribution of DSF-based *R.* spp. composite particleboard phantom slabs. It is seen that the constructed particleboard phantoms exhibit acceptable quality values of average densities to those of water (1.00 gcm^{-3}) and solid water phantom (1.04 gcm^{-3}) in the range between 0.99 ± 0.01 –1.04 ± 0.03 gcm^{-3}, making them potentially suitable for use in the fabrication of tissue-equivalent phantom materials. This is attributed to better adhesive-coated particles that provide intimate contact with the mat's wood particles and, thus, increase the bonding capabilities of the particleboards. This revealed that the combination of DSF/NaOH/IA-PAE with an increased percentage concentration of IA-PAE up to 15 wt% leads to an improvement in the average mass density of the particleboards approaching the value of water. These findings are in good agreement with previous studies of the average density of particleboard phantoms for dosimetric applications at high photon and electron energies utilized in radiation therapy [5,10].

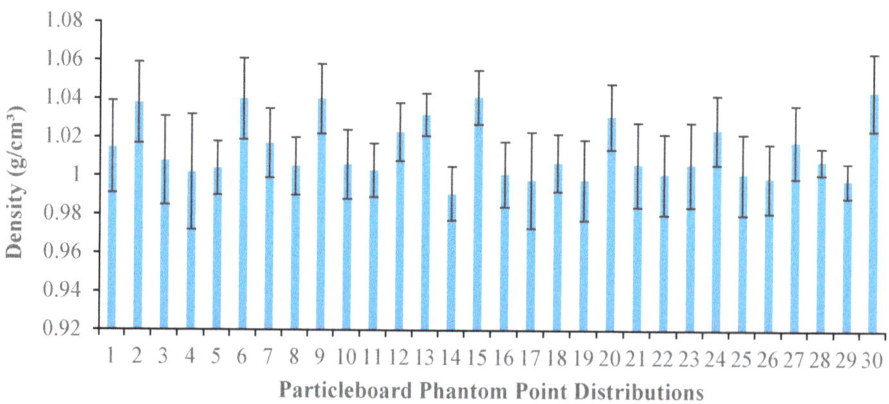

Figure 4. Average density with point distributions of DSF/NaOH/IA-PAE/*R.* spp. composite particleboard phantoms.

3.2. Evaluation of RAPs

The experimental computation has been performed in order to obtain the total LAC and MAC values for photon energies of 0.662 and 1.250 MeV and compared with those of solid water phantom and theoretical values of water using cross-section data (XCOM), as displayed in Table 4. The errors in density, thickness, incident, and transmitted gamma-ray intensities were used to evaluate the uncertainties in experimental MAC. The dependency of MAC values on photon energies can be explained by the dominance of partial photon interactions (e.g., photoelectric absorption, coherent scattering, incoherent scattering, and pair production) with the samples. As we know, the photoelectric effect dominates below, and pair production dominates above 1 MeV, whilst Compton scattering dominates at around 1 MeV [8,33]. The calculated values of LAC ranged from 0.059–0.083 cm^{-1} for 0.662 MeV photon energy, while for 1.250 MeV, the observed LAC values were within 0.043–0.056 cm^{-1}. Additionally, the observed total MAC values for these photon energies ranged between 0.059–0.082 cm^2g^{-1} at 0.662 MeV, whereas for 1.250 MeV, the MAC values were found to range between 0.041–0.056 cm^2g^{-1}. The estimated errors in experimental total MAC values for all the samples were less than 0.028%. As observed from Figure 5a, the difference of total MAC values with the incident photon energy for all composite particleboard samples and those of solid water phantom and water (XCOM) is almost identical as IA-PAE concentration increases with A_{15}, depicting higher MAC values for both photon energies, potentially providing a useful approximation of tissue-equivalent phantom materials. As expected, by increasing the incident photon energy, the total MAC values in all samples decreased slightly. This behavior may be due to the incoherent scattering process, which becomes the dominant mechanism in this region [34]. This can be ascribed to the fact that the Compton scattering cross-section is inversely proportional to the incoming photon energy (E^{-1}) and varies linearly with atomic number. In all the investigated samples, by increasing the incident photon energy, the highest HVL and MFP values were found for samples containing A_0 and A_5, while the lowest values were found for water (XCOM) and solid water phantom (Figure 5b,c). It was also observed that, in all samples and for all energies, A_{15} has the lowest HVL and MFP values with approximately no noticeable difference relative to those of solid water phantom and water (XCOM), which implies a higher radiation absorption ability. A comparison between the calculated values shows reasonable agreement with 15 wt% IA-PAE, solid water phantom, and theoretical values of water (XCOM), as depicted by the χ^2 values (Table 5). It can be seen that, among the selected samples, A_{15} provided the least values of χ^2 (0.044). This revealed, with an insignificant difference, the closest value of RAPs to those of solid water phantom and the theoretical value for water (XCOM).

Table 4. LAC and MAC values of DSF/NaOH/IA-PAE/*R.* spp. particleboards and solid water phantoms in comparison with water (XCOM).

Sample	Average ρ (gcm^{-3})	^{137}Cs (0.662 MeV)			^{60}Co (1.250 MeV)		
		μ (cm^{-1})	μ/ρ (cm^2/g)	$\sigma_{\mu/\rho}$ ± (%)	μ (cm^{-1})	μ/ρ (cm^2/g)	$\sigma_{\mu/\rho}$ ± (%)
A_0	1.040	0.063	0.061	0.028	0.043	0.041	0.023
A_5	1.038	0.067	0.064	0.021	0.045	0.044	0.017
A_{10}	1.002	0.070	0.070	0.019	0.054	0.054	0.015
A_{15}	1.006	0.083	0.082	0.009	0.056	0.056	0.011
Solid water	1.040	0.085	0.082	0.013	0.058	0.056	0.017
Water (XCOM)	1.000		0.086			0.059	

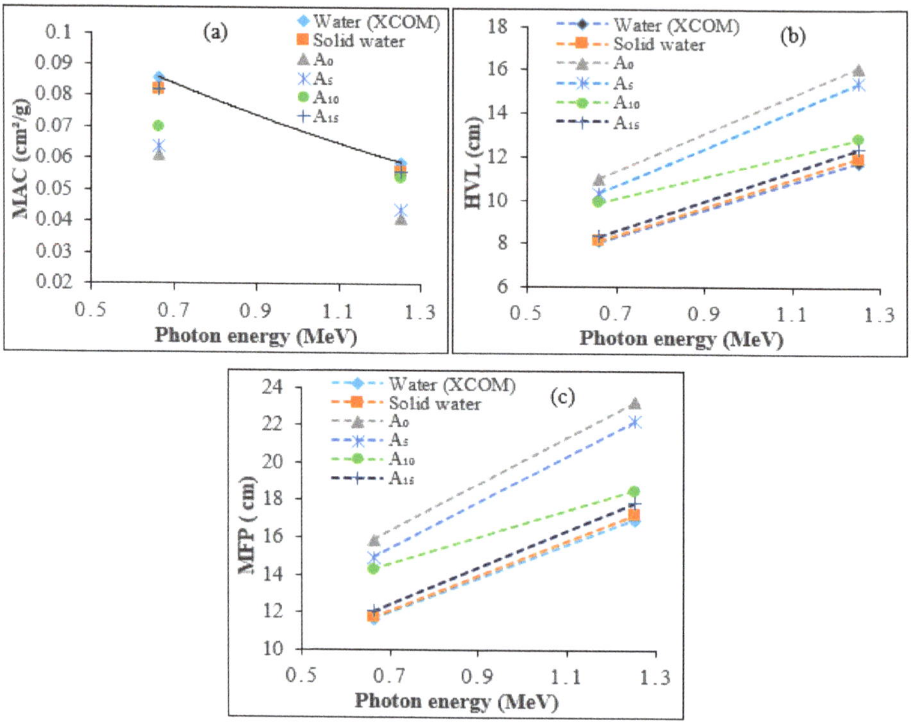

Figure 5. RAPs of DSF-based composite phantoms, water (XCOM), and solid water phantom against gamma energies: (**a**) MAC, (**b**) HVL, and (**c**) MFP.

Table 5. χ^2 values for MAC of DSF-based *R.* spp. particleboards and solid water phantoms.

Sample	χ^2 Water (XCOM)	
	^{137}Cs (0.662 MeV)	^{60}Co (1.250 MeV)
A_0	0.797	0.612
A_5	1.098	0.779
A_{10}	0.709	0.111
A_{15}	0.198	0.044
Solid water	0.095	0.031

3.3. *Dosimetric Characteristics of DSF-Based R. spp. Particleboard Phantoms*

3.3.1. Measurement of Photon Beam Quality Index

The tissue-phantom ratio ($TPR_{20,10}$) remains the most appropriate parameter for ascertaining the beam quality of a clinical photon beam. It is believed that material with near $TPR_{20,10}$ to water has similar RAPs to those of water and soft tissue [35]. The measured $TPR_{20,10}$ values of DSF-based *R.* spp. particleboards (sample A_{15}), solid water, and water phantoms for 6 and 10 MV photon beams with the use of IC are presented in Tables 6 and 7. The result indicates that the percentage discrepancies of sample A_{15} in comparison to those of solid water and water phantoms are in the range between 0.29–0.72% for 6 MV photons. Likewise, the discrepancies for the 10 MV photon beam are within the acceptable range of 0.26–0.65%. These results are in good agreement with previous work on the $TPR_{20,10}$ of renewable resources in the respective photon energy ranges [5,10].

Table 6. $TPR_{20,10}$ measurement for DSF/NaOH/IA-PAE/*R.* spp. particleboards (sample A_{15}), water, and solid water phantoms for 6 MV photons.

Phantom	Depth (cm)	Charge Collected (nC)			Mean	Ratio	Discrepancy (%)		
		1	2	3			W/S	W/*R*	S/*R*
Water	20 10	9.786 14.04	9.791 14.01	9.792 14.03	9.789 14.027	0.698	-	-	-
Solid water	20 10	9.511 13.69	9.509 13.69	9.512 13.69	9.511 13.690	0.695	0.43	-	-
DSF-based	20 10	9.845 14.22	9.836 14.20	9.841 14.20	9.841 14.207	0.693	-	0.72	0.29

Note: W, S, and *R* depict the water, solid water, and DSF-based (DSF/NaOH/IA-PAE/*R.* spp.) phantoms.

Table 7. $TPR_{20,10}$ evaluation for DSF/NaOH/IA-PAE/*R.* spp. particleboards (sample A_{15}), water, and solid water phantoms for 10 MV photons.

Phantom	Depth (cm)	Charge Collected (nC)			Mean	Ratio	Discrepancy (%)		
		1	2	3			W/S	W/*R*	S/*R*
Water	20 10	12.23 15.86	12.20 15.86	12.22 15.86	12.217 15.860	0.770	-	-	-
Solid water	20 10	12.05 15.74	12.08 15.73	12.08 15.73	12.070 15.733	0.767	0.39	-	-
DSF-based	20 10	12.32 16.10	12.31 16.10	12.31 16.09	12.313 16.097	0.765	-	0.65	0.26

Sample A_{15} with *p*-values of 0.071 and 0.069 for 6 and 10 MV photons showed no significant difference to those of water and solid water phantoms in the photon beam radiation quality, as presented in Tables 8 and 9. These findings demonstrated that DSF/NaOH-based *R.* spp. particleboard phantoms with 15 wt% IA-PAE (sample A_{15}) provide the ascribable characteristics that are proper as appropriate tissue-equivalent phantom materials.

Table 8. Paired *t*-test of the $TPR_{20,10}$ measurement for 6 MV photons.

Phantom	Paired Differences					*t*	df	Sig. (2-Tailed)
	Mean (d)	Std. Dev. (σ_d)	Std. Error Mean	95% Confidence Interval Difference				
				Lower	Upper			
Water	0.698	0.0082	0.0004	0.6967	0.6993	1709.7	20	0.081
Solid water	0.695	0.0016	0.0082	0.6924	0.6976	851.2	20	0.073
DSF-based	0.693	0.0022	0.0011	0.6896	0.6964	641.59	20	0.071

Table 9. Paired *t*-test of the $TPR_{20,10}$ measurement for 10 MV photons.

Phantom	Paired Differences					*t*	df	Sig. (2-Tailed)
	Mean (d)	Std. Dev. (σ_d)	Std. Error Mean	95% Confidence Interval Difference				
				Lower	Upper			
Water	0.770	0.0008	0.0004	0.7690	0.7715	1942.2	20	0.086
Solid water	0.767	0.0011	0.0005	0.7654	0.7689	1414.4	20	0.077
DSF-based	0.765	0.0010	0.0005	0.7638	0.7670	1486.8	20	0.069

3.3.2. Determination of PDD Photon Beams Using IC

The measured PDD values of 6 and 10 MV photon beams with the use of IC for sample A_{15}, water, and solid water phantoms are shown in Figure 6. The computed profiles were normalized to the maximum dose in the depth-dose profile positioned symmetrically opposite the IC within the photon beam to ensure that the profiles being compared were identical to those of water and solid water phantoms. The dose first increases steadily below the surface dose (d_s), reaches a maximum value (d_{max}) at z_{max}, then decreases almost gradually until it reaches d_{ext} at the patient's exit position. The discrepancies in the d_max in comparison to those of water and solid water phantoms were at most 1.08% and 1.28% for 6 MV photons (Figure 6a). On the other hand, for 10 MV photons (Figure 6b), the observed percentage differences in the d_{max} were found to be 5.42% and 6.70% at the dose build-up region, which is the region from the phantom surface to the depth at d_max and the equilibrium region. The greatest difference was recorded for 10 MV photons for all the phantoms, which is similar to previous observations by Yusof et al. [5] and Banjade et al. [10]. The observed values of the surface dose were found to range between 2.29% and 2.34% for 6 MV photons. Similarly, the surface dose values for 10 MV photons were found to be 4.69% and 5.29%, respectively. The PDD values for the examined particleboard phantoms at a depth beyond d_{max} indicates no significant difference with percentage difference within the limit of 0.09–0.16% for 6 MV, while for 10 MV photon, the variations in the depth beyond d_{max} were found to range between 0.37–0.70%, which is consistent with those of water and solid water phantoms.

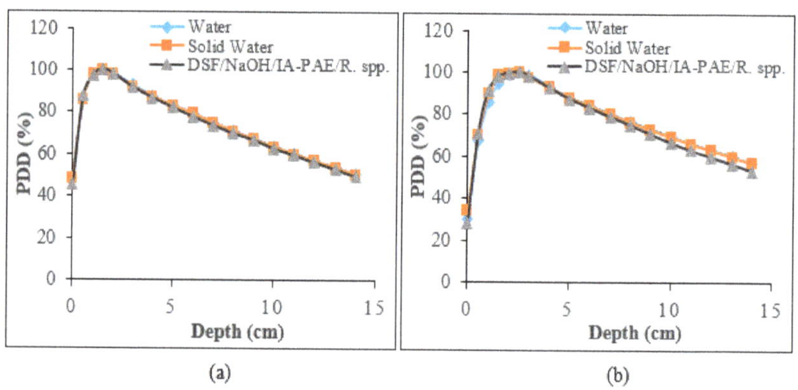

Figure 6. PDD curves for water, solid water, and DSF/NaOH/IA-PAE/*R.* spp. particleboard phantoms using IC for: (**a**) 6 MV and (**b**) 10 MV photons.

3.3.3. PDD Curves for Gafchromic EBT3 Radiochromic Films for Photon Beams

Figure 7 shows a comparison between the estimated PDD profiles for sample A15, water, and solid water phantoms for 6 MV and 10 MV photon beams using Gafchromic EBT3 radiochromic films. As shown in the Figures, the percentage variations in the d_{max} of the particleboard phantom, relative to those of water and solid water phantoms, were found to be 1.03% and 1.68% for 6 MV photons (Figure 7a), whereas the contrast between the measured PDD at all depths for 10 MV photon beams indicates good consistency with a difference within the range of 5.42% and 5.92% (Figure 7b). Overall, the results depict agreement with those of water and solid water phantoms in the build-up region for 6 MV photons with lower variations in the PDD values at d_{max}, whereas the variations were marginally higher for 10 MV photons with discrepancies found within 5% and 7%. This can be ascribed to the fact that the dominant free electron population originating in the build-up region continues to cause further interactions as a result of pair production, Compton scattering, and the photoelectric effect. High energy electrons are emitted as high energy photons (10 MV) interact with the phantoms and are absorbed by their interaction with the phantom. The resulting electrons will reduce with depth inside the phantoms owing to the continuously reduced energy fluence of the photons. The corresponding results for the surface dose of fabricated particleboards with water and solid water phantoms were found to be within 2.23–2.44% and 4.48–4.84%. With regards to the depth beyond d_{max}, the PDD values showed agreement to those of water and solid water phantoms, with percentage deviation in the interval of 0.01–0.02% and 0.06–0.07%, respectively. These trends are similar to what was reported for IC performance.

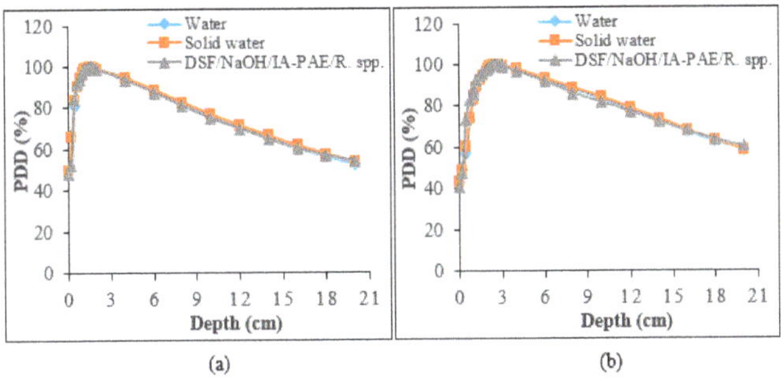

Figure 7. PDD plots with the use of EBT3 film for water, solid water, and DSF/NaOH/IA-PAE/*R.* spp. particleboard phantoms ascertained for: (**a**) 6 MV and (**b**) 10 MV photons.

3.3.4. Beam Profile Comparison at Reference Dose (d_{ref}) and Maximum Dose (d_{max})

The comparison of the relative dose plots against distance from the central axis of sample A_{15} and solid water phantoms for the beam profile curves for 6 and 10 MV photons are presented in Figures 8 and 9. As can be seen from the figures, the DSF-based particleboards reveal remarkable beam profiles with good dose homogeneity and beam symmetry in comparison to those of solid water phantoms. There was a consistency between the constructed particleboard and solid water phantom plots in both the dose plateau and the penumbra regions. Table 10 addressed the variation of flatness of the beam profiles at d_{ref} and d_{max} between the DSF-based particleboards and solid water phantoms for both photon energies. Overall, the beam uniformity enhanced as the photon energy increased, with 10 MV photons having a reduced percentage discrepancy of beam flatness values at d_{ref} and d_{max} compared with that of 6 MV photons relative to solid water phantom. This has demonstrated the appropriateness of DSF/NaOH/IA-PAE/*R.* spp. particleboards to be utilized as phantom material for high-energy photons in medical health applications.

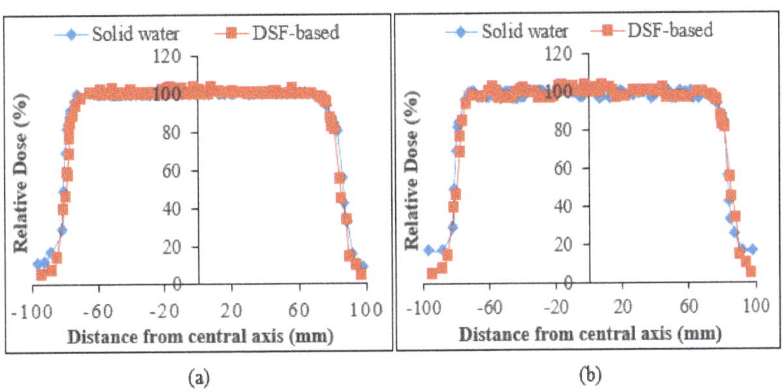

Figure 8. Beam profile for 6 MV photons evaluated at: (**a**) d_{max} and (**b**) d_{ref}.

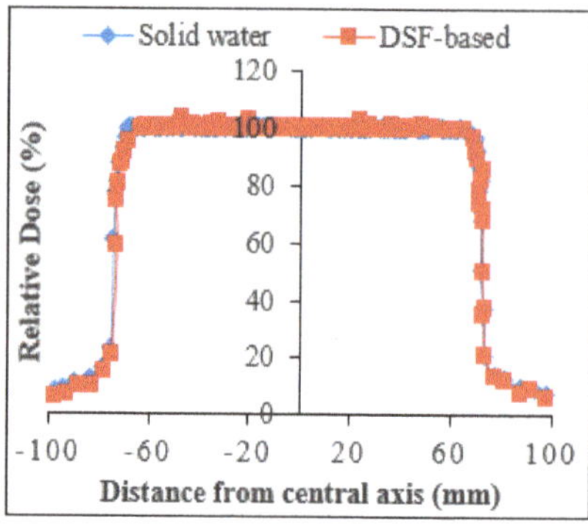

(a)

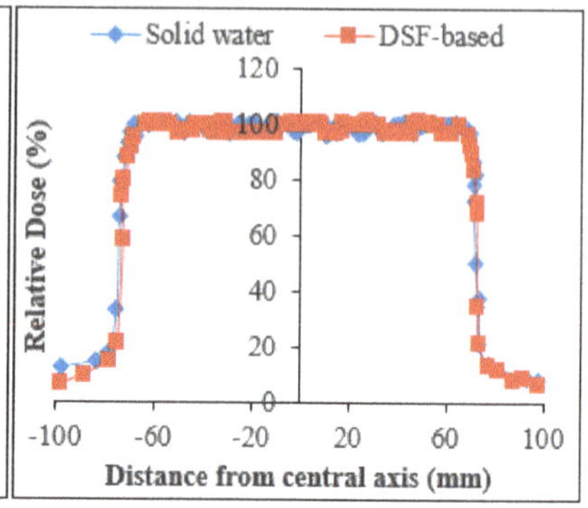

(b)

Figure 9. Beam profile for 10 MV photons measured at: (**a**) d_{max} and (**b**) d_{ref}.

Table 10. Beam profile flatness for DSF/NaOH/IA-PAE/*R.* spp. particleboards (sample A_{15}) compared to that of solid water phantom for 6 MV and 10 MV photons.

Phantom	Beam Flatness				Discrepancy (%)			
	6 MV		10 MV		6 MV		10 MV	
	5 cm	z_{max}	5 cm	z_{max}	5 cm	z_{max}	5 cm	z_{max}
Solid water	2.582	2.416	1.933	1.816	-	-	-	-
DSF-based	2.761	2.596	2.010	1.923	6.93	7.45	3.98	5.89

3.3.5. Determination of PDD for Electron Beams Using IC

The PDD curves of the electron beams for the particleboard phantoms showed an improved surface dose when compared with that of water and solid water phantoms, as displayed in Figure 10a–d. DSF-based phantom delivers a reasonably homogeneous dose from the surface to a specific depth, after which the dose falls off rapidly with increasing depth, eventually to near zero values. As can be seen from the figures, the percentage dose variations in d_{max} between the DSF-based *R.* spp. Particleboards, with respect to water and solid water phantoms for the four electron beam energies, were within 2.40–3.87%, 3.52–3.59%, 4.36–4.55%, and 2.82–4.63%, respectively. In addition, the percentage difference at which the electron PDD beyond the depth of z_{max} drops off sharply as a result of the scattering and continuous energy loss by the incident electrons. The therapeutic range (R90 and R80) and half-value depth range (R50) were found to be within the limit and similar to those of water and solid water phantoms for 6, 9, 12, and 15 MeV electrons (Table 11).

Table 11. Comparison of PDD curves between DSF/NaOH/IA-PAE/*R.* spp., water and solid water phantoms for different electron beams.

Depth	Percentage Difference of PDD (%)							
	Water				Solid Water			
	6 MeV	9 MeV	12 MeV	15 MeV	6 MeV	9 MeV	12 MeV	15 MeV
z_{max}	2.06	1.33	1.27	1.81	1.51	0.79	1.01	1.30
d_{50}	2.34	2.75	2.73	2.22	1.69	1.56	1.09	2.02
d_{80}	2.69	2.54	2.84	2.05	1.71	2.33	1.34	2.67
d_{90}	2.51	2.78	2.64	2.40	1.60	2.89	1.78	2.44

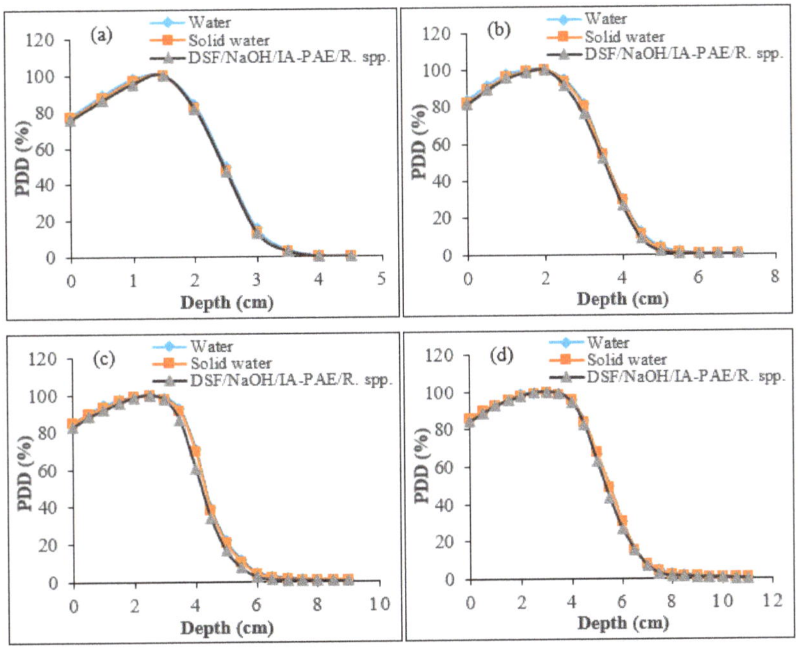

Figure 10. PDD curves for water, solid water, and DSF/NaOH/IA-PAE/*R.* spp. particleboard phantoms using IC for: (**a**) 6 MeV, (**b**) 9 MeV, (**c**) 12 MeV, and (**d**) 15 MeV electrons.

3.3.6. Evaluation of PDD for Electron Beams Using Gafchromic EBT3 Radiochromic Film

Figure 11a–d depicts the PDD profiles between DSF-based particleboard phantom, water, and solid water phantom evaluated from their surfaces for 6, 9, 12, and 15 MeV electron beams using Gafchromic EBT3 radiochromic films. As shown in the figures, comparable results were found in the constructed particleboards at the selected electron beams range to those of water and solid water phantoms. In this case, for 6, 9, 12, 15 MeV electrons discrepancies found were within 1.49–1.90%, 1.89–3.01%, 1.74–3.53%, and 2.38–3.84%, respectively. These findings indicate that, at 6 MeV, DSF-based particleboards depicted good agreement to those of water and solid water phantom with minimum discrepancies, whereas 9, 12, and 15 MeV give maximum values of percentage of discrepancies. Additionally, it can be observed that variations in percentage between the examined phantoms were lower at a depth beyond z_{max} in comparison to that in the build-up region. According to the obtained results, the dissimilarities in the discrepancy of the surface dose values were found to improve in the range between 1.45–1.63%, 1.51–1.79%, 1.53–2.17%, and 1.98–2.70%, which showed good agreement with the results of the IC. The observed reduction in surface dose can be assigned to a slight reduction in backscatter. This confirms that EBT3 radiochromic film is suitable and provided surface dosimetry measurements in 6, 9, 12, and 15 MeV electrons beam fields.

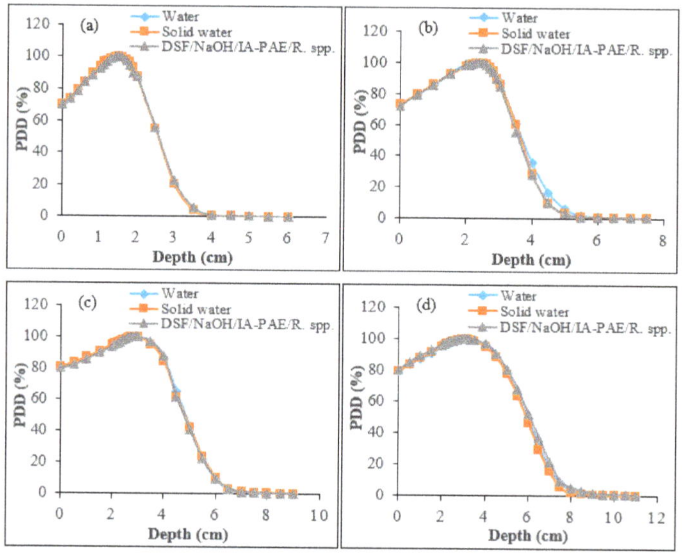

Figure 11. PDD plots for water, solid water, and DSF/NaOH/IA-PAE/*R.* spp. particleboard phantoms with the use of Gafchromic EBT3 radiochromic film for: (**a**) 6 MeV, (**b**) 9 MeV, (**c**) 12 MeV, and (**d**) 15 MeV electrons.

4. Conclusions

The RAPs and dosimetric characterization of DSF/NaOH/*R.* spp. particleboard phantoms as a tissue-equivalent phantom material with different amounts of IA-PAE (0, 5, 10, and 15 wt%) have been demonstrated. The ascertained average mass density exhibited acceptable quality values to those of water and solid water phantom in the range between 0.99 ± 0.01 gcm^{-3}–1.04 ± 0.03 gcm^{-3}. The PMDSP, z_{eff}, and N_{el} values were found to be satisfactory. Comparison between the calculated RAPs values shows a reasonable agreement with 15 wt% IA-PAE, solid water phantom, and theoretical values of water (XCOM), as indicated by the χ^2 values (0.044). The dosimetric computation results of DSF/NaOH/IA-PAE/R. spp. particleboard phantoms from IC showed good agreement with Gafchromic EBT3 radiochromic films, and they were benchmarked with those of water and solid water phantoms for the selected high energy photons and electrons, demonstrating the possibility to use these dosimeters under extremely intense radiation fields and confirming the effectiveness of the DSF/NaOH/IA-PAE/*R.* spp. particleboards. The fabricated particleboard phantom (sample A_{15}) was shown to be ideal for use in radiation therapy dosimetry as tissue-equivalent phantom material within the range of 1% variations to those of water and solid water phantoms.

Author Contributions: Conceptualization, D.O.S., M.Z.A.A., A.S., N.A.A.H. and R.H.; Data curation, D.O.S., M.Z.A.A. and S.H.Z.; Formal analysis, D.O.S., M.Z.A.A., M.F.M.Y. and N.A.R.; Funding acquisition, M.Z.A.A., A.S., R.H. and M.F.M.Y.; Investigation, D.O.S., R.H. and S.J.G.; Methodology, D.O.S., M.Z.A.A., N.A.A.H., S.H.Z. and M.F.M.Y.; Project administration, A.S.; Resources, A.S. and S.J.G.; Software, D.O.S., S.H.Z. and N.A.R.; Supervision, A.S. and M.F.M.Y.; Validation, D.O.S., M.Z.A.A., N.A.A.H. and M.F.M.Y.; Visualization, R.H., N.A.R. and S.J.G.; Writing—original draft, D.O.S.; Writing—review and editing, D.O.S., M.Z.A.A., A.S. and R.H. All authors have read and agreed to the published version of the manuscript.

Funding: This research was funded by the Universiti Sains Malaysia under Fundamental Grant Research Scheme, Ministry of Higher Education [Grant No. FRGS/1/2022/STG07/USM/02/2].

Institutional Review Board Statement: Not applicable.

Informed Consent Statement: Not applicable.

Data Availability Statement: Not applicable.

Acknowledgments: We express our appreciation and gratitude to the Advanced Medical and Dental Institute, Universiti Sains Malaysia and the Malaysian Nuclear Agency for allowing this research to be conducted using their facilities.

Conflicts of Interest: The authors declare no conflict of interest. The funders had no role in the design of the study; in the collection, analyses, or interpretation of data; in the writing of the manuscript; or in the decision to publish the results.

References

1. DeWerd, L.A. *The Phantoms of Medical and Health Physics: Devices for Research and Development*; Kissick, M., Ed.; Springer: Berlin/Heidelberg, Germany, 2014. [CrossRef]
2. Khan, F.M. *The Physics of Radiation Therapy*, 4th ed.; Lippincott Williams & Wilkins: Philadelphia, PA, USA, 2010.
3. Bradley, D.A.; Tajuddin, A.A.; Sudin, C.W.A.C.W.; Bauk, S. Photon attenuation studies on tropical hardwoods. *Int. J. Radiat. Appl. Instrum. Part A Appl. Radiat. Isot.* **1991**, *42*, 771–773. [CrossRef]
4. Tajuddin, A.A.; Sudin, C.W.A.C.W.; Bradley, D.A. Radiographic and scattering investigation on the suitability of *Rhizophora* spp. as tissue-equivalent medium for dosimetric study. *Radiat. Phys. Chem.* **1996**, *47*, 739–740. [CrossRef]
5. Yusof, M.F.M.; Abd Hamid, P.N.K.; Tajuddin, A.A.; Hashim, R.; Bauk, S.; Isa, N.M.; Isa, M.J.M. Mass attenuation coefficient of tannin-added *Rhizophora* spp. particleboards at 16.59–25.56 keV photons, and ^{137}Cs and ^{60}Co gamma energies. *Radiol. Phys. Technol.* **2017**, *10*, 331–339. [CrossRef]
6. Samson, D.O.; Mat Jafri, M.Z.; Hashim, R.; Sulaiman, O.; Aziz, M.Z.A.; Yusof, M.F.M.; Shukri, A. *Rhizophora* spp. particleboards incorporating defatted soy flour bonded with NaOH/IA-PAE: Towards a water equivalent phantom material. *Radiat. Phys. Chem.* **2020**, *176*, 109057. [CrossRef]
7. Samson, D.O.; Shukri, A.; Mat Jafri, M.Z.; Hashim, R.; Sulaiman, O.; Aziz, M.Z.A.; Yusof, M.F.M. Characterization of *Rhizophora* spp. particleboards with soy protein isolate modified with NaOH/IA-PAE adhesive for use as phantom material at photon energies of 16.59–25.26 keV. *Nucl. Eng. Technol.* **2021**, *53*, 216–233. [CrossRef]
8. Anugrah, M.A.; Suryani, S.; Ilyas, S.; Mutmainna, I.; Fahri, A.N.; Tahir, D. Composite gelatin/*Rhizophora* spp. particleboards/PVA for soft tissue phantom applications. *Radiat. Phys. Chem.* **2020**, *173*, 108878. [CrossRef]
9. Duke, N.C.; Allen, J.A. Rhizophora mangle, *R. samoensis*, *R. racemosa*, *R.* × *harrisonii* (Atlantic-East Pacific red mangrove). In *Species Profiles for Pacific Island Agroforestry*; Agroforestry Net, Inc.: Holualoa, HI, USA, 2006; Volume 10, pp. 1–18.
10. Banjade, D.P.; Tajuddin, A.A.; Shukri, A. A study of *Rhizophora* spp. wood phantom for dosimetric purposes using high-energy photon and electron beams. *Appl. Radiat. Isot.* **2001**, *55*, 297–302. [CrossRef]
11. Alshipli, M.; Kabira, N.A.; Hashim, R.; Marashdeh, M.W.; Tajuddin, A.A. Measurement of attenuation coefficients and CT numbers of epoxy resin and epoxy-based *Rhizophora* spp. particleboards in computed tomography energy range. *Radiat. Phys. Chem.* **2018**, *149*, 41–48. [CrossRef]
12. Hamid, P.N.K.A.; Yusof, M.F.M.; Hashim, R.; Tajuddin, A.A. Characterization and attenuation properties of corn starch-bonded *Rhizophora* spp. particleboards as water equivalent phantom material at 16.59-25.26 XRF photons and 99mTc gamma energies. *Int. J. Environ. Eng.* **2018**, *9*, 254–270. [CrossRef]
13. Xu, X.; Ren, S.; Li, L.; Zhou, Y.; Peng, W.; Xu, Y. Biodegradable engineered fiber scaffolds fabricated by electrospinning for periodontal tissue regeneration. *J. Biomater. Appl.* **2020**, *36*, 55–75. [CrossRef]
14. Kariuki, S.W.; Wachira, J.; Kawira, M.; Murith, G. Formaldehyde use and alternative biobased binders for particleboard formulation: A review. *J. Chem.* **2019**, *2019*, 5256897. [CrossRef]
15. Huang, X.; Chen, Y.; Li, J.; Li, J.; Gao, Q.; Mao, A. Development of a strong soy protein-based adhesive with excellent antibacterial and antimildew properties via biomineralized silver nanoparticles. *Ind. Crops Prod. Part A* **2022**, *188*, 115567. [CrossRef]
16. Liu, Z.; Chen, M.; Xu, Y.; Zhang, J.; Huang, X.; Luo, J.; Li, J.; Shi, S.Q.; Gao, Q. Preparation of a strong and multiple-function soybean flour adhesive via the construction of tannin microspheres with a core–shell structure. *Compos. Part B Eng.* **2022**, *242*, 110114. [CrossRef]
17. Ngu, K.T.; Bauk, S.; Hashim, R.; Tajuddin, A.A.; Shukri, A. Fabrication of formaldehyde-based *Rhizophora* spp. particleboards and their mass attenuation coefficients at 15.77, 17.48, 21.18 and 25.27 keV photon energies. *J. Phys. Sci.* **2015**, *26*, 27–33.
18. Gao, D.; Fan, B.; Zhang, B.; Mi, Y.; Zhang, Y.; Gao, Z. Storage stability of polyamidoamine-epichlorohydrin resin and its effect on the properties of defatted soybean flour-based adhesives. *Int. J. Adhes. Adhes.* **2019**, *91*, 92–101. [CrossRef]
19. Gui, C.; Wang, G.; Wu, D.; Zhu, J.; Liu, X. Synthesis of a bio-based polyamidoamine-epichlorohydrin resin and its application for soy-based adhesives. *Int. J. Adhes. Adhes.* **2013**, *44*, 237–242. [CrossRef]
20. Zhang, X.; Zhu, Y.; Yu, Y.; Song, J. Improve performance of soy flour-based adhesive with a lignin-based resin. *Polymers* **2017**, *9*, 261. [CrossRef]
21. Gupta, P.; Nayak, K.K. Characteristics of protein-based biopolymer and its application. *Polym. Eng. Sci.* **2015**, *55*, 485–498. [CrossRef]

22. Frihart, C.R.; Satori, H. Soy flour dispersibility and performance as wood adhesive. *J. Adhes. Sci. Technol.* **2013**, *27*, 2043–2052. [CrossRef]
23. Li, Y.; Chen, H.; Dong, Y.; Li, K.; Li, L.; Li, J. Carbon nanoparticles/soy protein isolate bio-films with excellent mechanical and water barrier properties. *Ind. Crops Prod.* **2016**, *82*, 133–140. [CrossRef]
24. Xia, C.; Wang, L.; Dong, Y.; Zhang, S.; Shi, S.Q.; Cai, L.; Li, J. Soy protein isolate-based films cross-linked by epoxidized soybean oil. *RSC Adv.* **2015**, *5*, 82765–82771. [CrossRef]
25. Schmitz, J.F., Jr. Enzyme Modified Soy Flour Adhesives. Ph.D. Thesis, Iowa State University, Ames, IA, USA, 2009.
26. *JIS A-5908*; Japanese Industrial Standard Particleboards. Standardization Promotion Department. Japanese Standards Association: Tokyo, Japan, 2015.
27. Sahoo, S.; Selvam, T.P.; Vishwakarma, S.R.; Chourasiya, G. Monte Carlo modeling of ^{60}Co HDR brachytherapy source in water and in different solid water phantom materials. *J. Med. Phys.* **2010**, *35*, 15–22. [CrossRef] [PubMed]
28. Schoenfeld, A.A.; Harder, D.; Poppe, B.; Chofor, N. Water equivalent phantom materials for ^{192}Ir brachytherapy. *Phys. Med. Biol.* **2015**, *60*, 9403–9420. [CrossRef] [PubMed]
29. Berger, M.J.; Hubbell, J.H. *XCOM: Photon Cross Sections on a Personal Computer*; U.S. Department of Commerce: Washington, DC, USA, 1987; pp. 1–10.
30. Feye, A.T. Percentage depth dose and beam profile measurements for electron and photon beam in reference field size for different energies. *Int. J. Sci. Eng. Res.* **2018**, *9*, 1460–1464.
31. Mather, S.J.; Mansi, L. IAEA Technical Report Series. *Eur. J. Nucl. Med. Mol. Imaging* **2008**, *35*, 1030–1031. [CrossRef] [PubMed]
32. Richley, L.; John, A.C.; Coomber, H.; Fletcher, S. Evaluation and optimization of the new EBT2 radiochromic film dosimetry system for patient dose verification in radiotherapy. *Phys. Med. Biol.* **2010**, *55*, 2601–2617. [CrossRef] [PubMed]
33. Mann, K.S.; Rani, A.; Heer, M.S. Shielding behaviors of some polymer and plastic materials for gamma-rays. *Radiat. Phys. Chem.* **2015**, *106*, 247–254. [CrossRef]
34. Aygün, B. High alloyed new stainless steel shielding material for gamma and fast neutron radiation. *Nucl. Eng. Technol.* **2020**, *52*, 647–653. [CrossRef]
35. Thomadsen, B.; Constantinou, C.; Ho, A. Evaluation of water-equivalent plastics as phantom material for electron-beam dosimetry. *Med. Phys.* **1995**, *22*, 291–296. [CrossRef]

Disclaimer/Publisher's Note: The statements, opinions and data contained in all publications are solely those of the individual author(s) and contributor(s) and not of MDPI and/or the editor(s). MDPI and/or the editor(s) disclaim responsibility for any injury to people or property resulting from any ideas, methods, instructions or products referred to in the content.

polymers

Article

Synthesis and Characterization of Eco-Friendly Bio-Composite from Fenugreek as a Natural Resource

Nayem Hossain [1], Mohammad Asaduzzaman Chowdhury [2,*], Tauhidul Islam Noman [2], Md. Masud Rana [2], Md. Hasan Ali [2], Raja Saad Alruwais [3], Md. Shafiul Alam [2], Khalid A. Alamry [4], Mahmood D. Aljabri [5,*] and Mohammed M. Rahman [4,6,*]

1 Department of Mechanical Engineering, IUBAT-International University of Business Agriculture and Technology, Dhaka 1230, Bangladesh
2 Department of Mechanical Engineering, Dhaka University of Engineering and Technology (DUET), Gazipur 1700, Bangladesh
3 Chemistry Department, Faculty of Science and Humanities, Shaqra University, Dawadmi 11912, Saudi Arabia
4 Chemistry Department, King Abdulaziz University, P.O. Box 80203, Jeddah 21589, Saudi Arabia
5 Department of Chemistry, University College in Al-Jamoum, Umm Al-Qura University, Makkah 21955, Saudi Arabia
6 Center of Excellent for Advanced Materials Research (CEAMR) & Department of Chemistry, King Abdulaziz University, P.O. Box 80203, Jeddah 21589, Saudi Arabia
* Correspondence: asad@duet.ac.bd (M.A.C.); mdjabri@uqu.edu.sa (M.D.A.); mmrahman@kau.edu.sa (M.M.R.)

Abstract: The present study show the usability of starch (tamarind) based-bio-composite film reinforced by fenugreek by various percentages to replace the traditional petrochemical plastics. The prepared bio-composite films were systematically characterized using the universal testing machine (UTM), soil degradation, scanning electron microscope (SEM), X-ray diffraction (XRD), thermogravimetric analyzer (TGA), and antibacterial tests. The experiments showed that a lower percentage of fenugreek improves biodegradation and mechanical strength. More than 60% of biodegradation occurred in only 30 days. Almost 3 N/mm^2 tensile strength and 6.5% tensile strain were obtained. The presence of micropores confirmed by SEM images may accelerate the biodegradation process. Antibacterial activity was observed with two samples of synthesized bio-composite, due to photoactive compounds confirmed by FTIR spectra. The glass transition temperature was shown to be higher than the room temperature, with the help of thermal analysis. The prepared bio-composite containing 5% and 10% fenugreek showed antibacterial activities.

Keywords: starch bio-composite; environmental concern; natural resource; biodegradability; antimicrobial activity

Citation: Hossain, N.; Chowdhury, M.A.; Noman, T.I.; Rana, M.M.; Ali, M.H.; Alruwais, R.S.; Alam, M.S.; Alamry, K.A.; Aljabri, M.D.; Rahman, M.M. Synthesis and Characterization of Eco-Friendly Bio-Composite from Fenugreek as a Natural Resource. *Polymers* **2022**, *14*, 5141. https://doi.org/10.3390/polym14235141

Academic Editors: Edina Rusen and Sergio Torres-Giner

Received: 19 September 2022
Accepted: 10 November 2022
Published: 25 November 2022

Publisher's Note: MDPI stays neutral with regard to jurisdictional claims in published maps and institutional affiliations.

Copyright: © 2022 by the authors. Licensee MDPI, Basel, Switzerland. This article is an open access article distributed under the terms and conditions of the Creative Commons Attribution (CC BY) license (https://creativecommons.org/licenses/by/4.0/).

1. Introduction

Our everyday human life has been transformed by plastic playing an important part in many aspects of life. Global plastic production from petrochemical sources is increasing rapidly, owing to its extraordinary mechanical, versatility, and barrier properties [1]. As a raw material, around 4% of the total extracted fossil fuels are consumed for the production of these bio-composites. Increasing future demand suggests that by the year 2050, 20% of total fossil fuel extracted internationally may be consumed to produce plastic [2]. The production of these plastics is creating a big challenge, as very few of them are recycled or reused [3,4]. These plastics can remain in the environment for a long period of time, even 1000 years. Moreover, the harmful effect of this waste on the environment is very high. A significant amount of carbon dioxide and other toxic gases are released from this waste, which is harmful to human health and nature [5]. Because of these effects on human health and the environment, finding an alternative has become inevitable.

Bio-composite can be a suitable alternative to petrochemical plastic. Significant development is being made in bio-composite, to make it usable. The production of these plastics has also significantly increased, although the production of bio-composites is only 0.59% of total plastic production [6,7]. The data suggest that more research is necessary to replace harmful plastics completely in any kind of application. One big drawback of bio-composite is its high price compared to conventional plastic, which makes it uncompetitive in the market and limits its use. One alternative to reduce the cost can be the use of waste or by-products of the agri-food industry, which are produced every day in huge quantities [8]. This low-cost waste and by-products are rich in protein, making them competitive candidates to be used as raw material to synthesize bio-composite [9].

The materials of bio-composite that need to biodegrade quickly should have antibacterial properties to kill harmful viruses and bacteria produced from food, but should not have harmful effects on human health and the environment, in order to be used as biodegradable food packaging. Many researchers have synthesized composite bio-composite to improve physio-chemical properties, but such a type of bio-composite contains harmful additives, such as sulfuric acid or titanium dioxide [10–14]. Thus, it is necessary to focus on the research of synthesizing biodegradable bio-secured plastic that is not harmful to human beings and the environment, is not expensive, has antibacterial properties, is available in nature, and represents properties allowing it to be used as a true alternative to petrochemical plastic.

Starch is an available, biodegradable, and low-cost material used as a renewable polymer in wide number of applications. The production of biodegradable films using starch shows promising results, although shortcomings are still there in mechanical properties, dimensional stability, hydrophilicity, and light permissibility. A nonstructured and plasticized version of starch is known as thermoplastic starch, prepared by adding plasticizers to the mixture of starch. Plasticizers can penetrate starch molecules and form hydrogen bonds, which are necessary to increase the durability of the bio-composite [14]. Thermoplastic starch materials are cost-effective, biodegradable, abundant in nature, and renewable. However, humidity causes recrystallization problems in these materials, which drastically decreases their mechanical properties [15]. The other shortcomings of these materials are their hydrophilic character and lower thermal stability [16,17].

Throughout human history, it has been known that Fenugreek is consumed as a food and used as medicine. Its seeds are used in spices, to increase the taste of food. Numerous medicinal properties such as hypocholesterolemic, gastric stimulant, antidiabetic agent, hepatoprotective effect, lactation aid, anticancer, galactagogue, and antibacterial, are available in the seeds of fenugreek. Some of these effects are attributed to the intrinsic dietary fiber constituent. The texture of food is changed by the dietary fiber, which is almost 25% of the seeds. It is also used as a food stabilizer, emulsifying agent, and adhesive, because of its high fiber, gum, and protein content. The protein is more soluble in an alkaline solution. It helps us with digestion, and it can modify food [18,19]. Fenugreek contains up to 60% starch. Different industrial products such as polysaccharides, kernel powder, gum, starch, and oil are produced (Table 1) from fenugreek [20–22].

Table 1. Percentages of chemical compounds present in fenugreek seed.

Constituent	Moisture	Ash	Fat	Protein	Fiber	NFE
Percentages	10.17 ± 0.06	2.63 ± 0.15	7.47 ± 0.31	27.55 ± 1.56	3.79 ± 0.31	48.39 ± 1.33

Tamarind is a commercially valuable plant that grows in different parts of Asia, Africa, and America. It is an evergreen plant belonging to the Fabaceae family and Caesalpinioideae subfamily. Different parts of the tree, including leaves, seed, shell, and fiber are used in pharmaceutical, food, electrochemical, biofuel, composite, water, and textile industries [23]. Around 55% pulp, 34% seed, and 11% shell and fiber are available in a typical tamarind pod. Tartaric acid, reducing sugar, and minerals including calcium, phosphorus, and potassium are available in the pulp of tamarind seeds [24]. The pulp has antimicrobial properties and can be used as a preservative. Tamarind seeds contain Zn, Fe, Mg, P, Na,

K and Ca as minerals [25]. Tamarind seeds contain polysaccharides that are naturally biodegradable and biocompatible. A typical tamarind fruit fiber contains cellulose, hemicelluloses, lignin, wax, and moisture. The tamarind shell which covers the pulp contains carbohydrates, free tartaric acid, and protein [26]. The leaves of tamarind are composed of lipids, vitamins, fatty acids, and flavonoids [27]. Table 2 shows the percentages of chemical compounds present in the tamarind seed [28].

Table 2. Percentages of chemical compounds present in tamarind seed.

Constituent	Moisture	Protein	Fat/Oil	Crude Fiber	Carbohydrates	Total Ash	Total Sugar
Percentages	9.4–11.3	13.3–26.9	4.5–16.2	7.4–8.8	50.0–57.0	1.6–4.2	11.3–25.3

In the current situation of increased pollution worldwide because of synthetic petrochemical plastic, biodegradable composite can be a good source that will help to minimize environmental pollution. Both tamarind and fenugreek are abundant in the local area, and can be grown in vast quantities because of the good quality of the soil. Therefore, it can be said that both tamarind and fenugreek can be used as raw materials to manufacture bio-composite as an alternative source.

The current research work shows the synthesis and characterization of biodegradability properties of bio-composites synthesized from naturally available and cheap sources, which can be used as an alternative to synthetic plastic. The purpose of this work is to show the usability of naturally cheap sources of a biodegradable composite material that can kill bacteria, so that the material can be used for food-packaging applications. The bio-composite in this research work was synthesized from naturally available fenugreek seeds. The synthesized bio-composites were characterized by biodegradable, mechanical, FTIR, SEM, XRD, thermal and antimicrobial tests.

2. Materials and Methodology

2.1. Materials

The bio-composites were synthesized by tamarind, fenugreek, distilled water, vinegar, and glycerin. After collecting the tamarind seeds from the local market of the Gazipur district, Bangladesh, they were washed properly with deionized water, boiled for 30 min, and blended to make the extract of starch. Fenugreek was also collected from the local market of the Gazipur district of Bangladesh. The collected fenugreek was also washed properly with deionized water, boiled for 30 min, and blended, and thus starch was obtained. The environmental lab of IUBAT within the department of civil engineering supplied the necessary distilled water for the experiments. Glycerin and white vinegar were also collected from the local market of the Gazipur District of Bangladesh.

2.2. Production of Bio-Composite

Table 3 shows the synthesized bio-composites at different percentages of fenugreek. Initially, with the help of a precise electronic scale, all the ingredients were carefully and precisely weighted. After measuring, the mixing of the ingredients was performed using a magnetic stirrer shown in Figure 1, followed by blending. Clumping was avoided by stirring for seven minutes. Heat was applied to the process at 100 °C temperature. A thick and translucent mixture was obtained after some time. Aluminum foil was used for pouring the mixture, and bubbles were removed if found. The desired bio-composites were obtained after six hours of natural cooling. The thickness of the obtained bio-composite was 1 mm, and it was opaque and chocolate-colored. The obtained bio-composite samples were then taken for characterization. All the samples were made in a dry environment.

Table 3. The used constituents with their percentages.

Sample	Distilled Water	White Vinegar	Glycerol	Tamarind Seed Starch	Fenugreek Powder
S1	69% (360 mL)	8% (40 mL)	6% (30 mL)	12% (60 gm)	5% (26 gm)
S2	66% (360 mL)	8% (40 mL)	6% (30 mL)	11% (60 gm)	10% (54 gm)
S3	62% (360 mL)	7% (40 mL)	5% (30 mL)	11% (60 gm)	15% (85 gm)

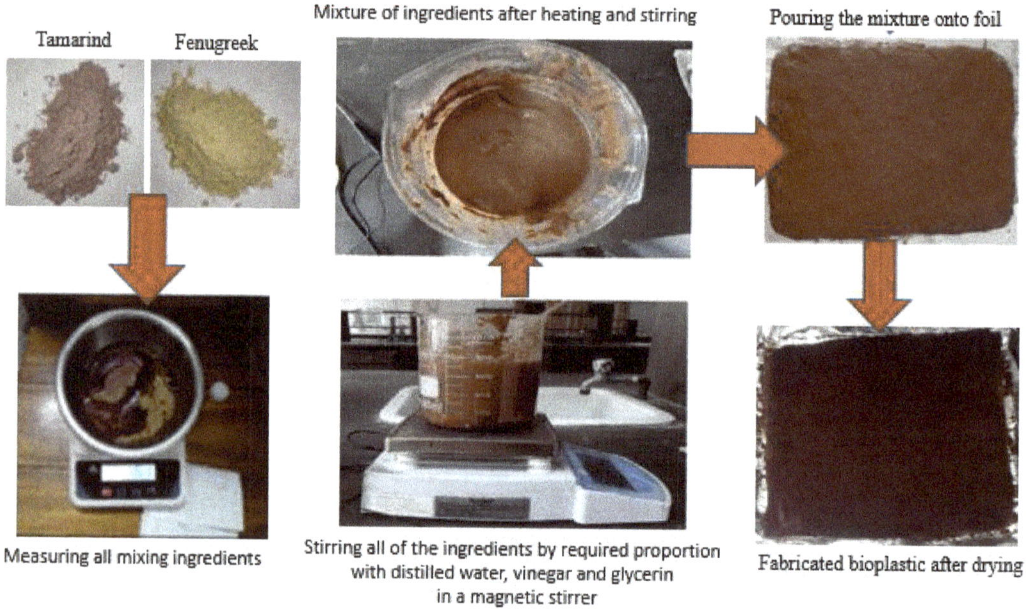

Figure 1. Bio-composite preparation from the natural sources.

2.3. Characterization

2.3.1. Biodegradation Test

The synthesized bio-composites were subjected to different characterization processes. For the biodegradability test, each sample was cut to a size of 50 mm × 20 mm × 1 mm. The average weight of each sample was 10 gm. The samples were buried at 2 cm depth. The pH value of the soil was 6. Weight loss was measured by burying the bio-composite samples in soil for 7, 15, and 30 days, in aerobic conditions. Before burying the bio-composite samples under the soil, the weight of each sample was measured carefully, using a precise electronic balance. After the test, each sample was removed from the soil, cleaned with water, dried, and the weight was taken again (Table 3). The biodegradability was measured using the following formula:

$$\text{Biodegradability } (\%) = \frac{W1 - W2}{W1} \times 100$$

Here, $W1$ = the weight of the bio-composite sample before the biodegradable test.
$W2$ = the weight of the bio-composite sample after the biodegradable test [29–34].

2.3.2. Mechanical Test

After production, the bio-composite samples were taken for mechanical testing. A universal testing machine controlled by a computer called CMT-10 was employed to evaluate the tensile properties of the bio-composite samples. All the tests were conducted maintaining the ASTM D638-77 standard method. The samples were cut with a dimension of 100 mm × 30 mm, in a dry environment. Then the samples were hung on a ring, using a thread at the bottom part of the samples with an attached hook, to place the loads. Maintaining a 2 mm/min strain rate, the force–distance data was measured at room temperature, and the loads were applied until the samples failed. The total length of the failure samples was measured carefully, and recorded. The total applied loads were recorded as well. Elongation and tensile strength were measured with the help of a stress–strain curve. For the same condition, 5 experiments were done for each sample, and the average value was considered.

2.3.3. Scanning Electron Microscopy Test

The surface microstructure of the synthesized bio-composites was analyzed by a Hitachi brand scanning electron microscope, model number S-4800. For analyzing the surface of these bio-composite samples, the bio-composites were submerged in liquid nitrogen and cut into 0.5 cm^2-sized samples. Then, cryo-fracturing was performed. The cryo-fractured samples were fixed onto the support using adhesive tape and mounted on aluminum stubs. Coating of the bio-composite samples was performed with gold-palladium, to observe the microstructure.

2.3.4. X-ray Diffraction Test

The X-ray diffraction analysis of the synthesized bio-composites was performed using a Rigaku, Tokyo, Japanese-made X-ray diffractometer, where scattering speed was 0.02(θ) s^{-1} ranging 5 to 60° (2θ) angles maintaining 40 kV voltage and 35 mA current, to determine the crystallographic structure.

2.3.5. Thermal Analysis

An SDT 650 TA-Instrument was employed to analyze the TGA of the synthesized bio-composites at a heating rate of 5 °C and weighing 10–25 mg, from room temperature to 500 °C temperature. TGA was performed in a nitrogen environment, with the lid kept hermetically sealed.

2.3.6. Antimicrobial Test

The evaluation of the antimicrobial properties of the developed bio-composite samples was performed using the ASTM-E2149-01 standard. Here, the evaluation of microbial growth and the resistivity of the non-leaching antimicrobial-treated specimens was performed under dynamic conditions. The microbiology department of Dhaka University supplied the necessary microbial for the experiments. The Kirby–Bauer disk-diffusion method was used to examine the antimicrobial performance of the synthesized bio-composites. *Staphylococcus aureus* was the bacteria cell subjected to the microbial test. The bacterial cell and sample concentrations were 1000 CFU/mL and 200 mg/mL, respectively, and the disc size was 6 mm × 6 mm.

3. Results and Discussion

3.1. Biodegradation

The biodegradation process of the synthesized bio-composites is represented by Figures 2a–c, 3a–d, 4a–d, 5a–d and 6. The weight loss of the bio-composite films was estimated after 7, 15, and 30 days. Linear biodegradation was observed in the bio-composite samples, which indicated obedience to pseudo-zero-order kinetics in reaction rates and constant, as well as independent rates of biodegradation [35]. From the figures, it can also be seen that the bio-composite films had a slower degradation initially. However, after 30 days the degradation was enhanced. The results also suggest that the addition of fenugreek slowed the degradation process of the starch-based bio-composite films. A sample with a lower percentage of fenugreek was obtained with maximum degradation among all synthesized bio-composite films. Ecologically-friendly and biodegradable properties were proved by the outcomes demonstrated.

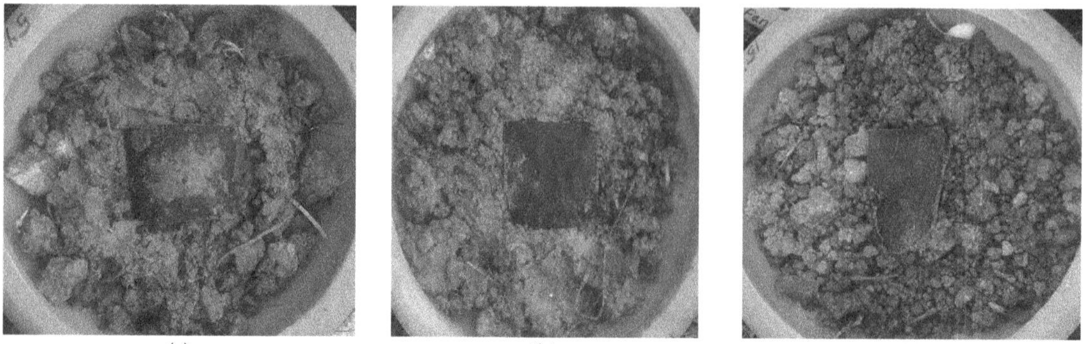

(a) (b) (c)

Figure 2. Prepared samples for soil biodegradation test (**a**) 5% fenugreek, (**b**) 10% fenugreek, and (**c**) 15% fenugreek.

(a) (b) (c) (d)

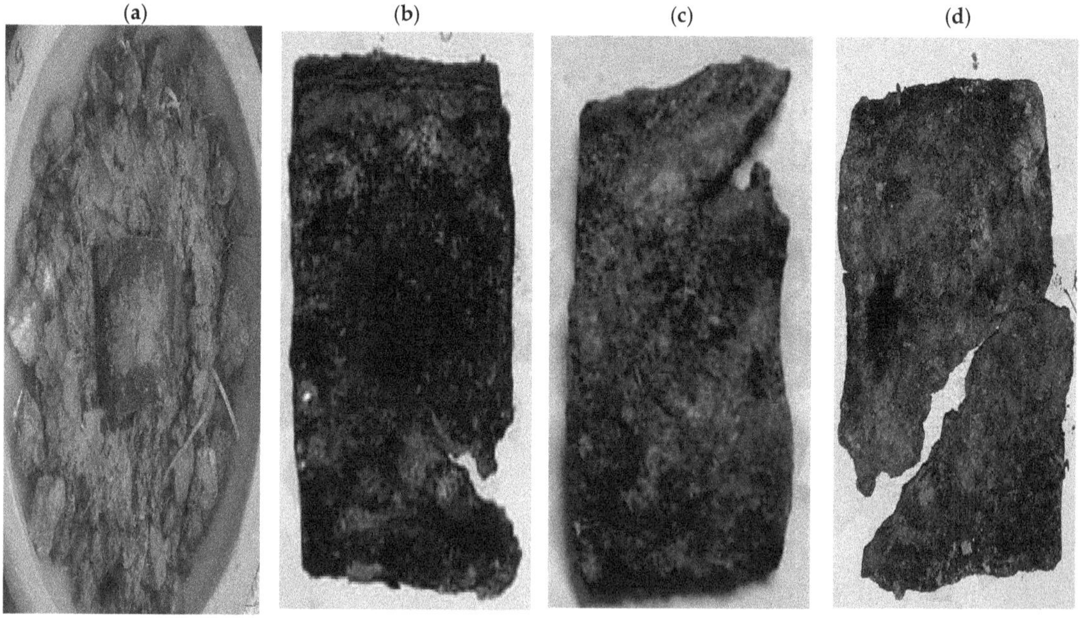

0% 6% 25% 62.5%

Figure 3. 5% fenugreek containing bio-composite degradation sample on (**a**) Day 1, (**b**) Day 7, (**c**) Day 15, and (**d**) Day 30.

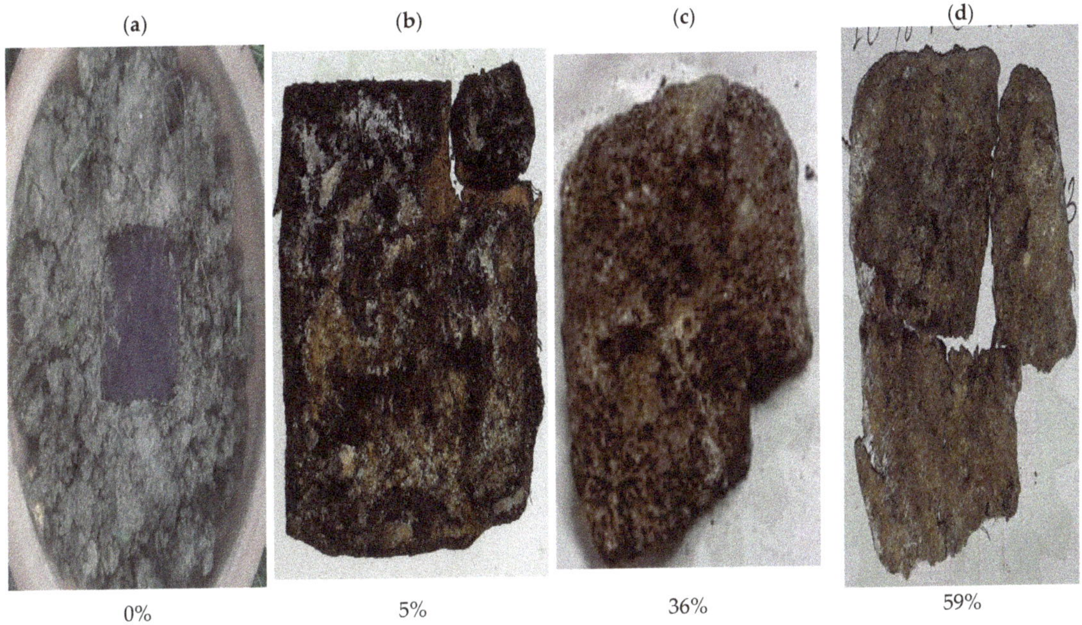

Figure 4. 10% fenugreek containing bio-composite degradation sample on (**a**) Day 1, (**b**) Day 7, (**c**) Day 15, and (**d**) Day 30.

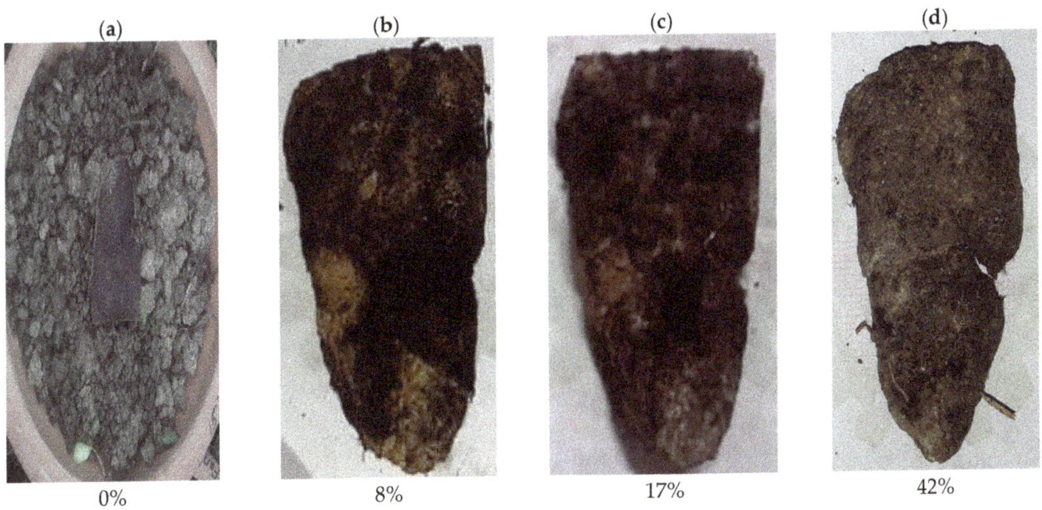

Figure 5. 15% fenugreek containing bio-composite degradation sample on (**a**) Day 1, (**b**) Day 7, (**c**) Day 15, and (**d**) Day 30.

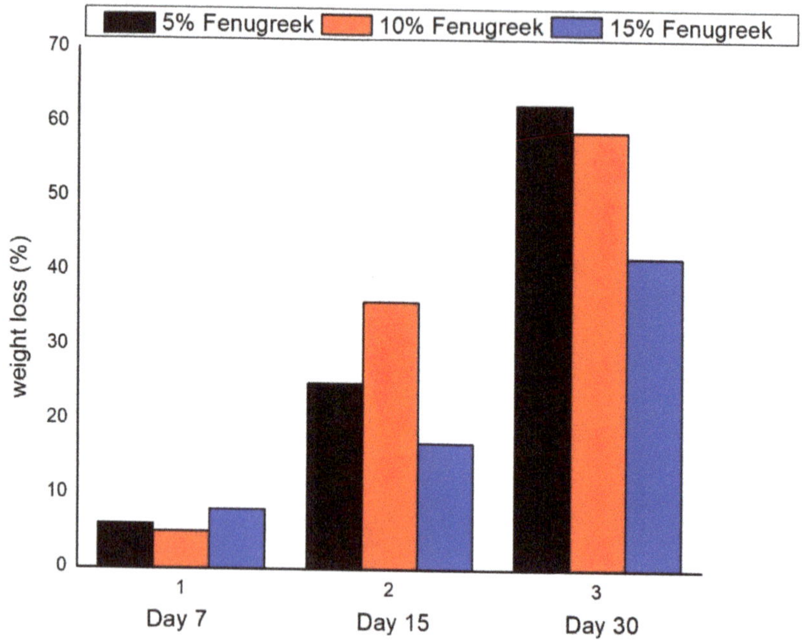

Figure 6. Biodegradibility comparison of the synthesized bio-composites at different percentages of fenugreek.

3.2. *Tensile Strength*

Figure 7a–c shows the tensile behavior of the developed bio-composites. Table 4 shows the comparison of tensile strength, tensile strain, and the Young's modulus of the prepared samples. An increase in the percentage of fenugreek decreased both the tensile strength and the tensile strain of the bio-composites, up to a certain level. Maximum stress development was reflected in the tensile strength of the film during the tensile testing [36]. The higher concentration of fenugreek decreased tensile strength and increased flexibility [37]. The presence of 15% of fenugreek exhibited the lowest tensile strength of 1.357 N/mm^2, due to the increased water presence in the matrix of the film, provided by the hygroscopic nature of fenugreek. The disarrangement of the polymer network was caused by the fenugreek present in the film, which worked to decrease the tensile strength and increase the film's flexibility [38]. Lower concentration percentages of fenugreek were distributed properly in the matrix, resulting in higher mechanical strength. However, minimum tensile strain was obtained after the addition of 10% of fenugreek. Tensile strain slightly increased after that percentage.

Table 4. Comparison of tensile strength, tensile strain and Young's modulus of the prepared samples.

Sample	Tensile Strength (MPa)	Standard Deviation	Tensile Strain (%)	Standard Deviation	Young's Modulus (MPa)
S1	2.905	0.110	6.797	0.353	0.427
S2	1.435	0.071	2.189	0.094	0.655
S3	1.357	0.062	3.029	0.125	0.448

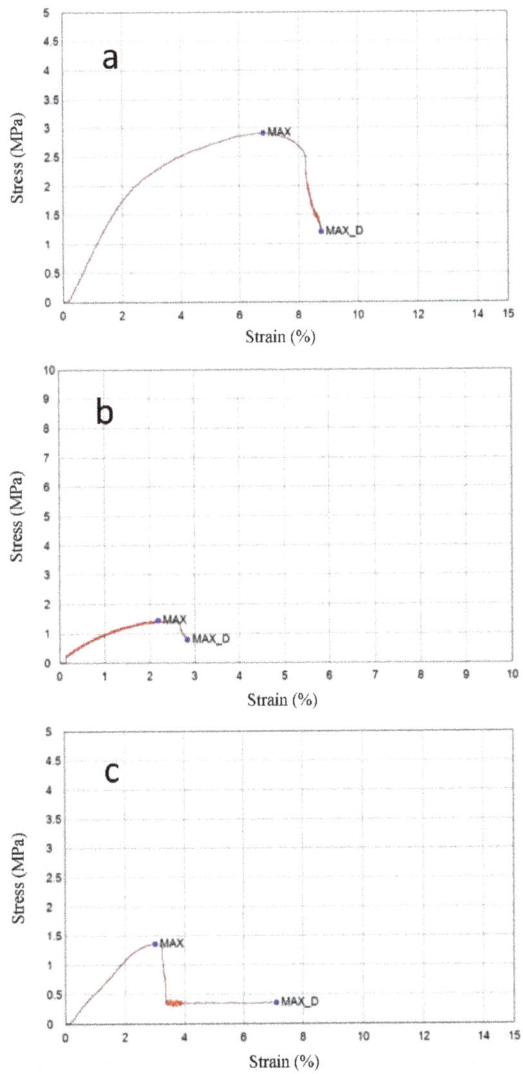

Figure 7. Tensile behavior of the bio-composite samples at various ratios of fenugreek (**a**) 5%, (**b**) 10%, and (**c**) 15%.

3.3. Scanning Electron Microscopy

The microstructure of the developed bio-composites is seen in Figures 8a–d, 9a–d and 10a–d. Defects such as voids, cracks, and micropores are seen in SEM images, and result in more porous and irregular surfaces. The samples containing a higher percentage of fenugreek show more micropores, due to the cohesion between the tamarind starch and fenugreek. Voids and micropores may affect the modulus of elasticity and tensile strength of the bio-composite samples [39,40]. The biodegradation process of the bio-composite samples may be accelerated in the soil, due to the interaction between micropores and microorganisms [41]. Impurities are seen with globular-like fibers on the surface in higher magnification images, because potency is provided to the biomass by the non-cellulosic contained in the natural biomass [42].

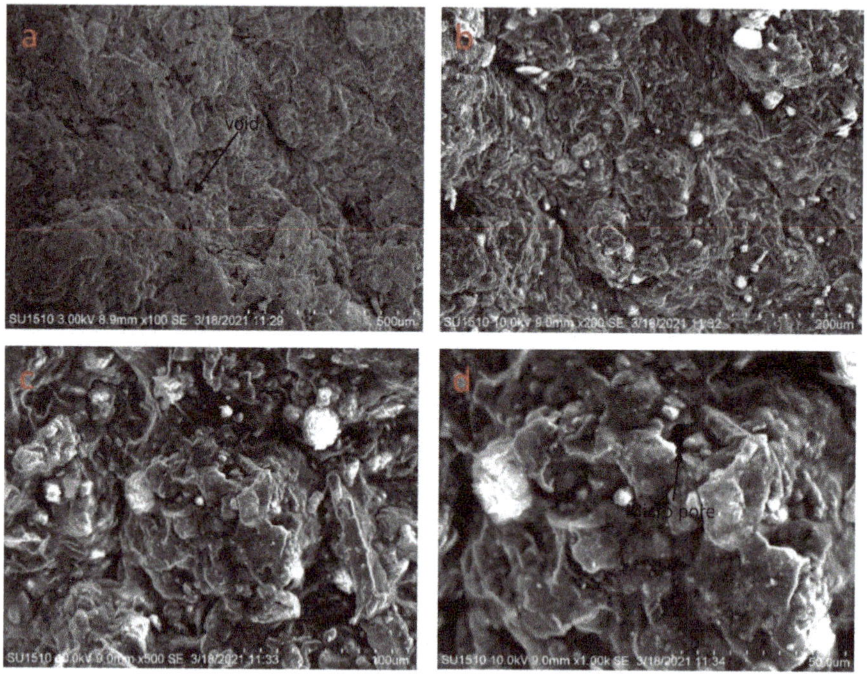

Figure 8. SEM images of 5% fenugreek bio-composite at (**a**) 500 μm (**b**) 200 μm (**c**) 100 μm (**d**) 50 μm.

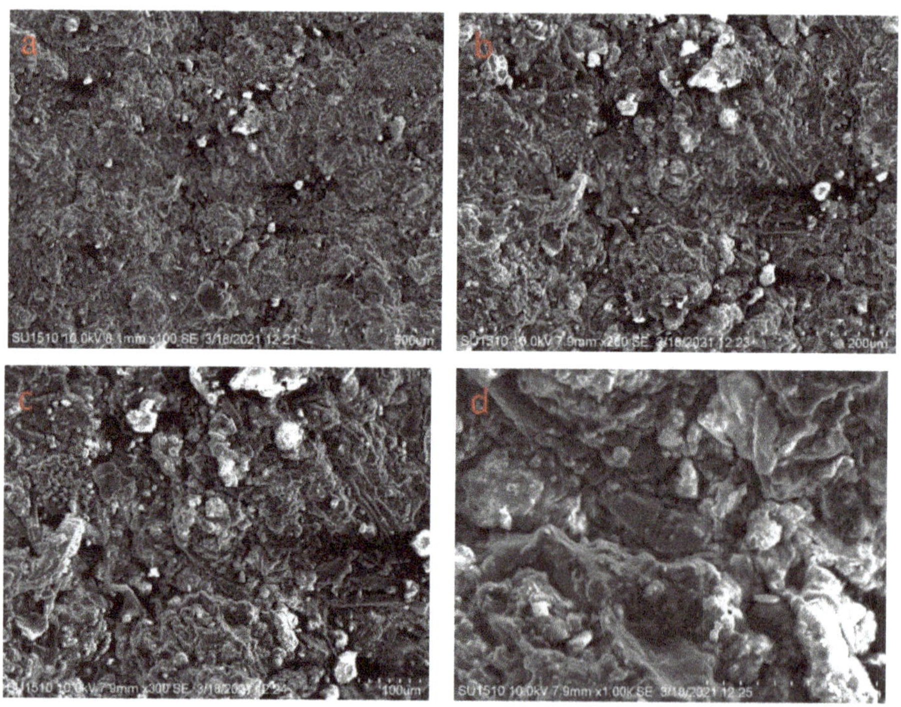

Figure 9. SEM images of 10% fenugreek bio-composite at (**a**) 500 μm (**b**) 200 μm (**c**) 100 μm (**d**) 50 μm.

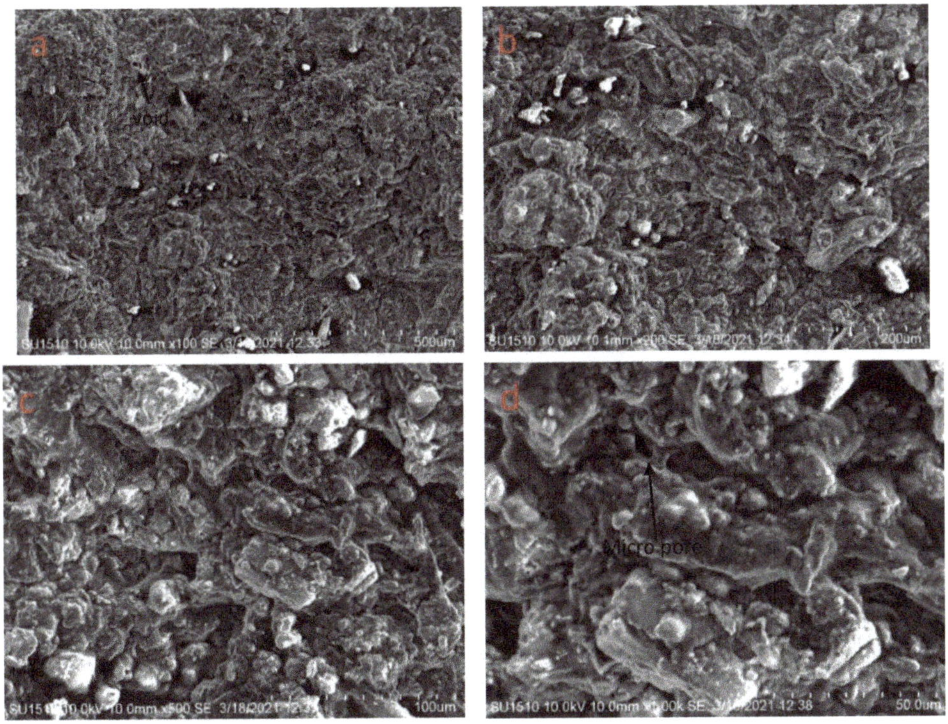

Figure 10. SEM images of 15% fenugreek bio-composite at (**a**) 500 μm (**b**) 200 μm (**c**) 100 μm (**d**) 50 μm.

3.4. X-ray Diffraction Analysis

Figure 11a–c shows the X-ray diffraction patterns of the bio-composite samples with different percentages of fenugreek, where intensity is shown vertically and the 2θ region is shown horizontally. The bio-composite sample with 5% fenugreek shows only two distinct peaks. Small peaks in the region of 2θ are represented in the bio-composite by the XRD pattern. Sharp peaks confirm the crystalline nature of the cellulose [43]. The peak at 2θ = 20° is attributed to the parallel-chain-segment aggregate region, which confirms the bio-composite semi-crystalline nature [44]. The region of 2θ = 28° also relates to the occurred disturbance in the cellulose at the time of the acetylation process, which can cause breakage in the micro-fibrillar structure of the cellulose [45,46]. However, the bio-composite samples with 10% and 15% fenugreek show no peaks, confirming the amorphous structure of the synthesized bio-composite samples.

3.5. Thermal Analysis

Figure 12a–c shows the TGA analysis of the bio-composite samples. Similar trends are observed in all the samples with variable temperatures. However, the graphs indicate that weight loss is reduced, due to the higher fenugreek percentage at elevated temperatures. Initially, 5% of weight loss is observed, because of moisture evaporation as the temperature changes from 25 to 100 °C [47,48]. Moreover, cellulose decomposes rapidly at around 350 °C because of the removal of hemicelluloses. Only about 30% weight is left for the pyrolysis of the cellulose skeleton at 350 °C [49].

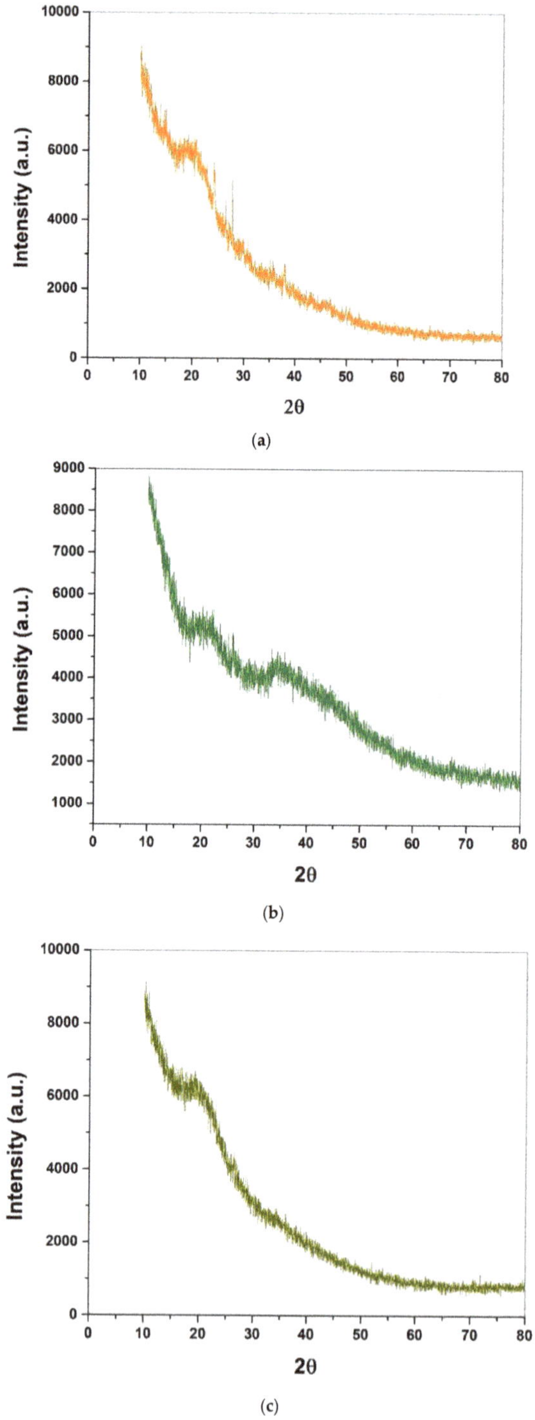

Figure 11. XRD analysis of the bio-composite samples at various ratio of fenugreek (**a**) 5%, (**b**) 10%, and (**c**) 15%.

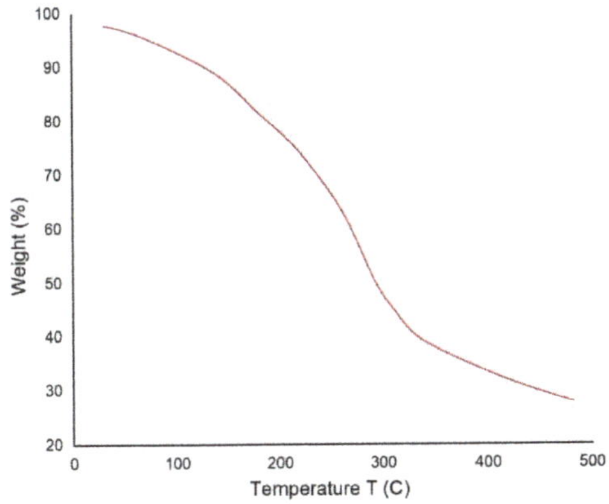

(a)

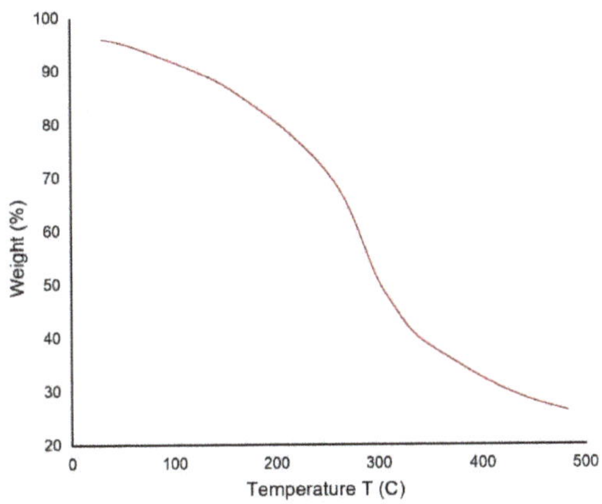

(b)

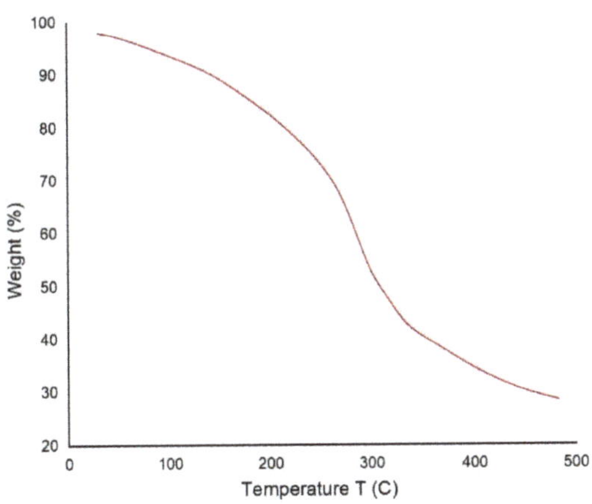

(c)

Figure 12. TGA analysis of the bio-composite samples at various ratios of fenugreek (**a**) 5% (**b**) 10% and (**c**) 15%.

3.6. Antibacterial Activity

The antibacterial analysis of the bio-composite samples against *Staphylococcus aureus* is seen in Figure 13a–c. The samples showed some form of antibacterial activity against *Staphylococcus aureus*, due to the presence of active phytocompounds shown in FTIR analysis. The antibacterial performance was obtained due to the presence of fenugreek. Reducing and/or capping agents containing phytochemicals may be present in the samples, which kill the virus. Moreover, inherent antibacterial activities may also be present in those botanic compounds. In addition, different types of metabolites such as alkaloids, terpenoids, tannins, flavonoids, and glycosides are available in fenugreek. These metabolites have antibacterial properties which make fenugreek active against bacterial strains [50].

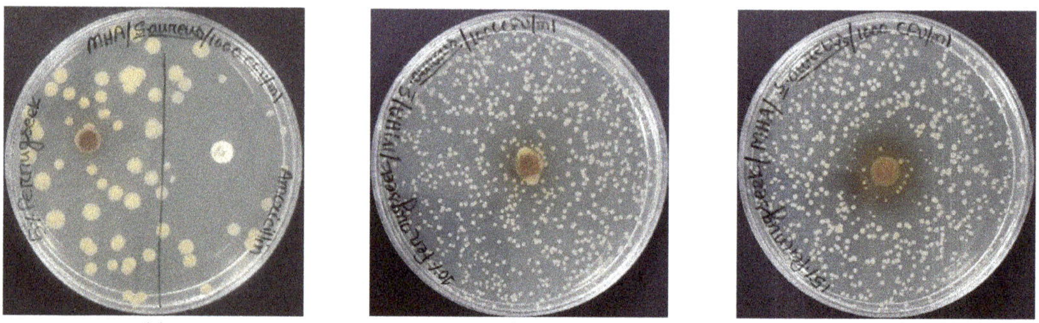

(**a**) (**b**) (**c**)

Figure 13. Antimicrobial analysis of the bio-composite samples at various ratio of fenugreek (**a**) 5%, (**b**) 10%, and (**c**) 15%.

4. Conclusions

Starch-based bio-composite films with varying percentages of fenugreek were successfully developed, using vinegar and glycerin. The properties of the bio-composite films were significantly affected by the concentration percentages of fenugreek. It was observed from the biodegradability test that a maximum of 62.5% of the bio-composite film was biodegraded in only 30 days. Lower percentages of fenugreek distributed properly in the film resulted in higher mechanical strength. Higher percentages of fenugreek led to the formation of an amorphous structure of bio-composite films, confirmed with XRD analysis. Thermal stability was assured for the synthesized bio-composites by thermal analysis. Some samples showed antimicrobial activity against *Staphylococcus aureus* bacteria, which confirmed the usability of the synthesized bio-composites in packaging applications.

Author Contributions: Conceptualization, N.H. and M.A.C.; methodology, T.I.N.; software, M.M.R. (Md. Masud Rana); validation, M.H.A., M.S.A. and N.H.; formal analysis, M.A.C., K.A.A. and M.M.R. (Md. Masud Rana); investigation, M.A.C., K.A.A., M.D.A. and M.M.R. (Md. Masud Rana); resources, M.A.C., K.A.A., R.S.A., M.D.A. and M.M.R. (Mohammed M. Rahman); data curation, K.A.A., M.A.C., R.S.A., M.D.A. and M.M.R. (Mohammed M. Rahman); writing—original draft preparation, N.H.; writing—review and editing, K.A.A., M.A.C., M.D.A. and M.M.R. (Mohammed M. Rahman); visualization, K.A.A., M.A.C., M.D.A. and M.M.R. (Mohammed M. Rahman); supervision, M.A.C.; project administration, M.A.C., K.A.A., M.D.A. and M.M.R. (Mohammed M. Rahman); funding acquisition, K.A.A., R.S.A., M.D.A. and M.M.R. (Mohammed M. Rahman). All authors have read and agreed to the published version of the manuscript.

Funding: Deanship of Scientific Research at Umm Al-Qura University for supporting this work with Grant Code: 22UQU4290372DSR06.

Institutional Review Board Statement: Not applicable.

Informed Consent Statement: Not applicable.

Data Availability Statement: Data will be available upon reasonable request.

Acknowledgments: The authors would like to thank the Deanship of Scientific Research at Umm Al-Qura University for supporting this work with Grant Code: (22UQU4290372DSR06).

Conflicts of Interest: The authors declare no conflict of interest.

References

1. Sanyang, M.; Sapuan, S.; Jawaid, M.; Ishak, M.; Sahari, J. Effect of plasticizertype and concentration on tensile, thermal and barrier properties of biodegradable films based on sugar palm (Arengapinnata) starch. *Polymers* **2015**, *7*, 1106–1124. [CrossRef]
2. Lebreton, L.; Andrady, A. Future scenarios of global plastic waste generation and disposal. *Palgrave Commun.* **2019**, *5*, 6. [CrossRef]
3. Blank, L.M.; Narancic, T.; Mampel, J.; Tiso, T.; O'Connor, K. Biotechnological upcycling of plastic waste and other non-conventional feedstocks in a circulareconomy. *Curr. Opin. Biotechnol.* **2020**, *62*, 212–219. [CrossRef]
4. Jogi, K.; Bhat, R. Valorization of food processing wastes and by-products for bioplastic production. *Sustain. Chem. Pharm.* **2020**, *18*, 100326. [CrossRef]
5. Méité, N.; Konan, L.K.; Tognonvi, M.T.; Doubi, B.I.H.G.; Gomina, M.; Oyetola, S. Properties of hydric and biodegradability of cassava starch-based bio-composites reinforced with thermally modified kaolin. *Carbohydr. Polym.* **2021**, *254*, 117322. [CrossRef]
6. European Bioplastic. 2020. Bioplastic Market Data 2019 [WWW Document]. Available online: https://docs.european-bioplastic.org/publications/market_data/Report_Bio-composites_Market_Data_2019.pdf (accessed on 21 September 2020).
7. Plastics Europe. 2020. Plastics: The Facts 2019 [WWW Document]. Available online: https://www.plasticseurope.org/application/fles/9715/7129/9584/FINAL_web_version_Plastics_the_facts2019_14102019.pdf (accessed on 10 June 2020).
8. Morone, P.; Koutinas, A.; Gathergood, N.; Arshadi, M.; Matharu, A. Food waste: Challenges and opportunities for enhancing the emerging bio-economy. *J. Clean. Prod.* **2019**, *221*, 10–16. [CrossRef]
9. Jiménez-Rosado, M.; Rubio-Valle, J.F.; Perez-Puyana, V.; Guerrero, A.; Romero, A. Use of heat treatment for the development of protein-based bioplastic. *Sustain. Chem. Pharm.* **2020**, *18*, 100341. [CrossRef]
10. Amin, M.R.; Chowdhury, M.A.; Kowser, M.A. Characterization and performance analysis of composite bioplastic synthesized using titanium dioxide nanoparticles with corn starch. *Heliyon* **2019**, *5*, e02009. [CrossRef] [PubMed]
11. Xie, F.; Pollet, E.; Halley, P.J.; Avérous, L. Starch-based nano-bioplastic. *Prog. Polym. Sci.* **2013**, *38*, 1590–1628. [CrossRef]
12. Roy, K.; Thory, R.; Sinhmar, A.; Pathera, A.K.; Nain, V. Development and characterization of nano starch-based composite flms from mung bean (Vignaradiata). *Int. J. Biol. Macromol.* **2020**, *144*, 242–251. [CrossRef]
13. Zoungranan, Y.; Lynda, E.; Dobi-Brice, K.K.; Tchirioua, E.; Bakary, C.; Yannick, D.D. Influence of natural factors on the biodegradation of simple and composite bioplastic based on cassava starch and corn starch. *J. Environ. Chem. Eng.* **2020**, *8*, 104396. [CrossRef]
14. Peidayesh, H.; Ahmadi, Z.; Khonakdar, H.A.; Abdouss, M.; Chodák, I. Fabrication and properties of thermoplastic starch/Montmorillonite composite using dialdehyde starch as a crosslinker. *Polym. Int.* **2020**, *69*, 317–327. [CrossRef]
15. Peidayesh, H.; Heydari, A.; Mosnáčková, K.; Chodak, I. In situ dual crosslinking strategy to improve the physico-chemical properties of thermoplastic starch. *Carbohydr. Polym.* **2021**, *269*, 118250. [CrossRef] [PubMed]
16. Peidayesh, H.; Mosnáčková, K.; Špitalský, Z.; Heydari, A.; Šišková, A.O.; Chodák, I. Thermoplastic Starch–Based Composite Reinforced by Conductive Filler Networks: Physical Properties and Electrical Conductivity Changes during Cyclic Deformation. *Polymers* **2021**, *13*, 3819. [CrossRef] [PubMed]
17. Paluch, M.; Ostrowska, J.; Tyński, P.; Adurski, W.; Konkol, M. Structural and Thermal Properties of Starch Plasticized with Glycerol/Urea Mixture. *J. Polym. Environ.* **2022**, *30*, 728–740. [CrossRef]
18. Meghwal, M.; Goswami, T.K. A review on the functional properties, nutritional content, medicinal utilization and potential application of fenugreek. *J. Food Process. Technol.* **2012**, *3*, 9. [CrossRef]
19. Wani, S.A.; Kumar, P. Fenugreek: A review on its nutraceutical properties and utilization in various food products. *J. Saudi Soc. Agric. Sci.* **2018**, *17*, 97–106. [CrossRef]
20. Nagar, C.K.; Dash, S.K.; Rayaguru, K. Tamarind seed: Composition, applications, and value addition: A comprehensive review. *J. Food Process. Preserv.* **2022**, *46*, e16872. [CrossRef]
21. Doharey, V.; Sharma, N. The permutation role of fenugreek seeds starch and Gunda glue as a binder in Paracetamol tablets. *J. Pharm. Sci. Res.* **2010**, *2*, 64–68.
22. Pasha, I.; AsimShabbir, M.; Haider, M.A.; Afzal, B. Muhammad Farhan Jahangir Chughtai, Shabbir Ahmad and Muhammad SajidManzoor, Biochemical Evaluation of Trigonellafoenumgraecum (Fenugreek) with Special Reference to Phenolic Acids. *Pak. J. Sci. Ind. Res. Ser. B Biol. Sci.* **2017**, *60*, 154–161.
23. Ghaffaripour, S.; van den Bilcke, N.; Samson, R. The importance of seed reserve on performance and breeding of tamarind seedlings. *Sci. Hortic.* **2017**, *222*, 145–152. [CrossRef]
24. Okello, J.; John, B.; Okullo, L.; Eilu, G.; Nyeko, P.; Joseph, O. Physicochemical composition of *Tamarindusindica* L. (Tamarind) in the agro-ecological zones of Uganda. *Food Sci. Nutr.* **2018**, *6*, 1179–1189. [CrossRef] [PubMed]
25. Okello, J.; John, B.; Okullo, L.; Eilu, G.; Nyeko, P.; Joseph, O. Mineral composition of Tamarindusindica LINN (tamarind) pulp and seeds from different agro-ecological zones of Uganda. *Food Sci. Nutr.* **2017**, *5*, 959–966. [CrossRef] [PubMed]

26. Radhakrishnan, P.G.; Varghese, S.P.; Das, B.C. Application of ethylene diamine hydroxypropyl tamarind fruit shell as adsorbent to remove Eriochrome black T from aqueous solutions—Kinetic and equilibrium studies. *Separ. Sci. Technol.* **2018**, *53*, 417–438. [CrossRef]
27. Mansingh, B.B.; Binoj, J.S.; PremSai, N.; Hassan, S.A.; Siengchin, S.; Sanjay, M.R.; Liu, Y.C. Sustainable development in utilization of *Tamarindusindica* L. and its by-products in industries: A review. *Curr. Res. Green Sustain. Chem.* **2021**, *4*, 100207. [CrossRef]
28. Kumar, C.S.; Bhattacharya, S. Tamarind Seed: Properties, Processing and Utilization. *Crit. Rev. Food Sci. Nutr.* **2008**, *48*, 1–20. [CrossRef]
29. Chowdhury, M.A.; Badrudduza, M.D.; Hossain, N.; Rana, M.M. Development and characterization of natural sourced bioplastic synthesized from tamarind seeds, berry seeds and licorice root. *Appl. Surf. Sci. Adv.* **2022**, *11*, 100313. [CrossRef]
30. Shanmathy, M.; Mohanta, M.; Thirugnanam, A. Development of biodegradable bio-composite films from Taro starch reinforced with bentonite. *Carbohydr. Polym. Technol. Appl.* **2021**, *2*, 100173.
31. Marichelvam, M.K.; Manimaran, P.; Sanjay, M.R.; Siengchin, S.; Geetha, M.; Kandakodeeswaran, K.; Boonyasopon, P.; Gorbatyuk, S. Extraction and development of starch-based bio-composites from Prosopis Juliflora Plant: Eco-friendly and sustainability aspects. *Curr. Res. Green Sustain. Chem.* **2022**, *5*, 100296. [CrossRef]
32. Behera, L.; Mohanta, M.; Thirugnanam, A. Intensification of yam-starch based biodegradable bioplastic film with bentonite for food packaging application. *Environ. Technol. Innov.* **2022**, *25*, 102180. [CrossRef]
33. Baidurah, S.; Takada, S.; Shimizu, K.; Ishida, Y.; Yamane, T.; Ohtani, H. Evaluation of biodegradation behavior of poly(butylene succinate-co-butylene adipate) with lowered crystallinity by thermally assisted hydrolysis and methylation-gas chromatography. *J. Anal. Appl. Pyrolysis* **2013**, *103*, 73–77. [CrossRef]
34. Cucina, M.; Carlet, L.; De Nisi, P.; Somensi, C.A.; Giordano, A.; Adani, F. Degradation of bio-composites in organic waste by mesophilic anaerobic digestion, composting and soil incubation. *Waste Manag.* **2021**, *134*, 67–77. [CrossRef] [PubMed]
35. Sachdeva, A.; Vashist, S.; Chopra, R.; Puri, D. Antimicrobial activity of active packaging film to prevent bread spoilage. *Int. J. Food Sci. Nutr.* **2017**, *2*, 29–37.
36. Goncalves, S.M.; dos Santos, D.C.; Motta, J.F.G.; dos Santos, R.R.; Chavez, D.W.H.; de Melo, N.R. Structure and functional properties of cellulose acetate films incorporated with glycerol. *Carbohydr. Polym.* **2019**, *209*, 190–197. [CrossRef] [PubMed]
37. Nigam, S.; Das, A.K.; Patidar, M.K. Synthesis, characterization and biodegradation of bio-composite films produced from Parthenium hysterophorus by incorporating a plasticizer (PEG600). *Environ. Chall.* **2021**, *5*, 100280. [CrossRef]
38. Gaaz, T.S.; Sulong, A.B.; Akhtar, M.N.; Kadhum, A.A.; Mohamad, A.B.; Al-Amiery, A.A. Properties and applications of polyvinyl alcohol. Halloysite Nanotub. *Nanocomposite Mol.* **2015**, *20*, 22833–22847. [CrossRef]
39. Jayakumar, S.; Bhuyar, P.; Pugazhendhi, A.; Rahim, M.H.; Maniam, G.P.; Govindan, N. Effects of lightintensity and nutrients on the lipid contentof marine microalga (diatom) Amphiprora sp. for promising biodiesel production. *Sci. Total Environ.* **2021**, *768*, 145471. [CrossRef]
40. Rohmawati, B.; AtikahNataSya'idah, F.; Rhismayanti, R.; Alighiri, D.; Tirza Eden, W. Synthesis of bio-composite-based renewable cellulose acetate from teak wood (Tectonagrandis) biowaste using glycerol-chitosan plasticizer. *Orient. J. Chem.* **2018**, *34*, 1810–1816. [CrossRef]
41. Kaith, B.S.; Mittal, H.; Jindal, R.; Maiti, M.; Kalia, S. Environment benevolent biodegradable polymers: Synthesis, biodegradability, and applications. In *Cellulose Fibers: Bio- and Nano-Polymer Composites*; Springer: Berlin/Heidelberg, Germany, 2011; pp. 425–451.
42. Montgomery, D.C. *Design and Analysis of Experiments*, 5th ed.; Wiley: New York, NY, USA, 2001; p. 697.
43. Hassan, M.; Berglund, L.; Abou-Zeid, R.; Hassan, E.; Abou-Elseoud, W.; Oksman, K. Nanocomposite film based on cellulose acetate and lignin-rich rice straw nanofibers. *Materials* **2019**, *12*, 595. [CrossRef]
44. Chen, J.; Xu, J.; Wang, K.; Cao, X.; Sun, R. Cellulose acetate fibers prepared from different raw materials with rapid synthesis method. *Carbohydr. Polym.* **2016**, *137*, 685–692. [CrossRef]
45. Chowdhury, M.A.; Hossain, N.; Noman, T.I.; Hasan, A.; Shafiul, A.; Abul, K.M. Biodegradable, physical and microbial analysis of tamarind seed starch infused eco-friendly bioplastic by different percentage of Arjuna powder. *Results Eng.* **2022**, *13*, 100387. [CrossRef]
46. Huang, F.Y. Thermal properties and thermal degradation of cellulose tri-stearate (CTs). *Polymers* **2012**, *4*, 1012–1024. [CrossRef]
47. Fortunati, E.; Pugliaa, D.; Luzia, F.; Santulli, C.; Kenny, J.M.; Torrea, L. Binary PVA bionanocomposites containing cellulose nanocrystals extracted from different natural sources. *Carbohydr. Polym.* **2013**, *97*, 825–836. [CrossRef] [PubMed]
48. Melikoglu, A.Y.; Bilek, S.E.; Cesur, S. Optimum alkaline treatment parameters forthe extraction of cellulose and production of cellulose nanocrystals from apple pomace. *Carbohydr. Polym.* **2019**, *215*, 330–337. [CrossRef] [PubMed]
49. Nigam, S.; Das, A.K.; Patidar, M.K. Valorization of Partheniumhy sterophorus weed for cellulose extraction andits application for bioplastic preparation. *J. Environ. Chem. Eng.* **2021**, *9*, 105424. [CrossRef]
50. Al-Timimi, L.A.N. Antibacterial and Anticancer Activities of Fenugreek Seed Extract. *Asian Pac. J. Cancer Prev.* **2019**, *20*, 3771–3776. [CrossRef]

polymers

MDPI

Article

Development of Polyphenol-Functionalized Gelatin-Poly(vinylpyrrolidone) IPN for Potential Biomedical Applications

Lidia Escutia-Guadarrama [1,2], David Morales [1], Daniel Pérez-Calixto [3] and Guillermina Burillo [1,*]

[1] Departamento de Química de Radiaciones y Radioquímica, Instituto de Ciencias Nucleares, Universidad Nacional Autónoma de México, Circuito Exterior, Ciudad Universitaria, Ciudad de Mexico 04510, Mexico
[2] Departamento de Química Analítica, Facultad de Química, Universidad Nacional Autónoma de México, Circuito Exterior, Ciudad Universitaria, Ciudad de Mexico 04510, Mexico
[3] Subdirección de Genómica Poblacional, Instituto Nacional de Medicina Genómica, Ciudad de Mexico 14610, Mexico
* Correspondence: burillo@nucleares.unam.mx

Abstract: Owing to their suitable physical and chemical properties, hydrogels have been considered a convenient choice for wound dressings because of the advantages that they offer, such as maintaining the moist environment required for wound healing. In this research, interpenetrating hydrogels of polyphenol-functionalized gelatin (GE), a water-soluble protein derived from natural polymer collagen with excellent biocompatibility, no immunogenicity, and hydrophilicity, and polyvinylpyrrolidone (PVP), a hydrophilic, non-toxic, biodegradable, biocompatible polymer that is soluble in many solvents, widely used in biomedical applications, particularly as a basic material for the manufacturing of hydrogel wound dressings, were synthesized. Gallic acid (GA) was selected in this work to study whether the interpenetrating polymer networks (IPNs) synthesized can provide antioxidant properties given that this material is intended to be used as a potential wound dressing. The obtained IPN hydrogels showed improved mechanical properties in comparison with pristine gelatin network (*net*-GE), a porous structure, and good thermal stability for biological applications. The antioxidant capacity of the IPNs functionalized with GA was compared to Trolox standards, obtaining a radical scavenging activity (RSA%) equivalent to a Trolox concentration of 400 μM.

Keywords: gelatin; poly(vinylpyrrolidone); hydrogel; radiation crosslinking; interpenetrating network; antioxidant

Citation: Escutia-Guadarrama, L.; Morales, D.; Pérez-Calixto, D.; Burillo, G. Development of Polyphenol-Functionalized Gelatin-Poly(vinylpyrrolidone) IPN for Potential Biomedical Applications. *Polymers* **2022**, *14*, 4705. https://doi.org/10.3390/polym14214705

Academic Editor: Edina Rusen

Received: 21 September 2022
Accepted: 2 October 2022
Published: 3 November 2022

Publisher's Note: MDPI stays neutral with regard to jurisdictional claims in published maps and institutional affiliations.

Copyright: © 2022 by the authors. Licensee MDPI, Basel, Switzerland. This article is an open access article distributed under the terms and conditions of the Creative Commons Attribution (CC BY) license (https://creativecommons.org/licenses/by/4.0/).

1. Introduction

Skin, the largest organ of the human body, serves as the primary barrier for protection against the surrounding environment and external damage, such as exogenous pathogens and mechanical injuries, [1]. For that reason, skin injuries are a health concern worldwide. A wound can be defined as a defect or a break in the skin, or as a result of the presence of an underlying medical or physiological condition [2]. The degree of damage depends on the depth of the wound, which can be superficial, partial-thickness or full-thickness. Wound repair is a physiological process that includes four overlapping phases, namely, hemostasis, inflammation, tissue formation or proliferation, and tissue remodeling [3]. All of these are required to occur in the proper sequence and time in order to achieve successful wound healing. However, there are several factors that interfere with one or more of these phases, resulting in irregular wound healing or even in extra damage [4]. Deficient or inappropriate care of wounds may threaten individual health, triggering infections, other pathological reactions, and even death [5].

Due to the risk of complications and infections, different strategies are widely used in the treatment of both acute and chronic wounds; these strategies include, but are not restricted to, the use of wound dressings, which are physical barriers that isolate and protect the wound from the harmful factors of the external environment and further damage [6]. Different kinds of wound dressings are applied on damaged tissue depending on the type, depth, and size of the injury. In general, an ideal wound dressing should meet particular requirements: it can effectively prevent the invasion of bacteria, has a strong liquid absorption capacity but also maintains high humidity at the wound site, and promotes the rapid healing of wounds. Moreover, a wound dressing should possess a certain strength to resist pulling during use, proper adhesion performance, so that it does not cause any extra damage when it is removed from the wound.

Hydrogels are polymeric materials with three-dimensional (3D) crosslinked networks that are capable of absorbing large amounts of water or biological fluids, resulting in materials with unique properties, such as softness, flexibility, high water absorption, and water retention capacity [7]. Owing to their suitable physical properties, hydrogels have become very popular biomaterials for use in drug delivery [8,9], tissue engineering [10], pharmaceuticals [11], and sensors [12,13]. Additionally, hydrogels have been considered a convenient choice for wound dressings because of the advantages that they offer, such as maintaining the moist environment required for wound healing, while at the same time removing the excess wound exudate and promoting wound recovery via granulation and re-epithelialization [14].

Because of their outstanding biodegradability and biocompatibility, natural polymers have been used for decades in the biomedical and pharmaceutical fields. These materials have the capacity to retain relevant biochemical cues, and some of them present structures that can mimic the extracellular matrix of native tissues [15]. Gelatin (GE) is a water-soluble protein derived from natural polymer collagen. It offers many advantages for clinical applications, such as excellent biocompatibility, nonimmunogenicity, and hydrophilicity [16,17]. However, gelatin has poor in vivo mechanical strength and elasticity due to its high swelling capacity in water [18], resulting in the quick breakage of hydrogels. Moreover, gelatin has low thermal stability, which restricts its use in biomedicine [19]. This challenge can be overcome by using crosslinking techniques (physical, chemical, and radiation-induced) and reinforcing materials [18,20].

A number of chemical crosslinking molecules, such as formaldehyde and glutaraldehyde, have been used in attempting to crosslink GE and other polymers. However, it has been documented that the use of these agents results in toxicity issues [17]. Genipin is a naturally occurring crosslinker, and it has been considered an alternative to glutaraldehyde, as it is a nontoxic molecule. Nevertheless, its elevated cost restricts its use [21].

Radiation-induced crosslinking is a suitable alternative to crosslinking gelatin molecules without the need for chemical reagents, being a safe and environmentally friendly technique. For that reason, it also ensures that the biocompatibility of the materials is not altered after the crosslinking process [22]. Additionally, it allows the simultaneous sterilization of the irradiated polymer [23,24].

Another method to improve the mechanical strength of hydrogels is to prepare them as an interpenetrating polymer network (IPN); this is defined as a combination of at least two polymer networks in which the polymers have been crosslinked and entangled but not covalently bonded to each other [25]. IPNs obtained from a mixture of synthetic and natural polymers offer the advantage of combining the mechanical properties of the synthetic polymer chain and the biological properties and biocompatibility of the natural component. In the present study, poly (vinylpyrrolidone) (PVP) was chosen as the synthetic polymer to obtain IPN systems. PVP is a hydrophilic, non-toxic, biodegradable, biocompatible linear synthetic polymer that is soluble in many solvents and interacts with many hydrophilic and hydrophobic materials. It has been widely used in biomedical applications, particularly as a basic material for the manufacturing of hydrogel wound dressings [26–28].

Gallic acid (GA) is a flavonoid compound also named 3,4,5-trihydroxybenzoic acid. It is found in almost all plants, including fruits, leaves, and wildflowers [29,30], and it has been reported to possess powerful health benefits, such as antioxidant, anti-inflammatory, analgesic, neuroprotective, anticancer, and anti-diabetic properties [31].

Since antioxidants are considered to help control wound oxidative stress and thereby accelerate wound healing, GA was selected in this work to study whether it can provide antioxidant properties to the synthesized IPNs given that this material is intended to be used as a potential wound dressing.

2. Materials and Methods

2.1. Reagents

Gelatin type A (GE), polyvinylpyrrolidone (PVP), gallic acid (GA), Folin–Ciocalteu phenol reagent (FC), 6-hydroxy-2,5,7,8-tetramethyl-chroman-2-carboxylicacid (Trolox), *N,N,N′,N′*-Tetramethylethylendiamine (TEMED), *N,N′*-Methylenebis(acrylamide) (MBAAm), azobisisobutyronitrile (AIBN), (1-Ethyl-3-(3-dimethylaminopropyl) carbodiimide hydrochloride) (EDC-HCl), *N*-hydroxysuccinimide (NHS), and 2,2-diphenyl-2-picrylhydrazyl (DPPH) were purchased from Sigma Aldrich (Estado de México, México) and used without further purification. Ethanol and methanol were supplied from JT Baker (Ciudad de México, Mexico). Deionized water was obtained from a Milli-Q plus apparatus (Millipore, Molsheim, France). *N*-vinyl-2-pyrrolidone (NVP) and azobisisobutyronitrile (AIBN) were also obtained from Sigma Aldrich. NVP was vacuum-distilled, and AIBN was recrystallized in methanol prior to use.

2.2. Synthesis of Hydrogels

2.2.1. Gamma Radiation Crosslinking of Gelatin Hydrogels

Gelatin hydrogels were prepared from 5% and 10 wt% aqueous gelatin solutions. Gelatin type A was dissolved in deionized water at 40 °C with constant stirring. Then, 6 mL aliquots of the solutions were transferred to glass ampoules and bubbled with argon for 20 min in order to eliminate oxygen.

Gamma-ray irradiation of the ampoules was performed at ambient temperature using a Gammabeam 651PT (Nordion International Inc., Ottawa, ON, Canada) ^{60}Co gamma-ray source with an average dose rate of 6.1 KGy h^{-1} at various total absorbed doses (5, 10, 15, and 30 KGy) at the Institute of Nuclear Sciences of Universidad Nacional Autónoma de México (UNAM).

The crosslinked hydrogels were washed for 48 h with distilled water, changing the solvent 3 or 4 times a day. Once cleaned, the samples were vacuum-dried at 45 °C until reaching a constant weight. The percentage of reticulation was calculated with Equation (1):

$$\text{Gel (\%)} = (W_1/W_0) \times 100 \quad (1)$$

where W_1 and W_0 are the insoluble crosslinked GE after extraction with distilled water and the initial crosslinked GE, respectively.

The initial formation of the gel dose was estimated using the Charlesby–Pinner equation, shown in Equation (2) [32]:

$$S + S^{1/2} = (G_d/2G_c) + (9.65 \times 10^5/M_nG_c) \times (1/D_g) \quad (2)$$

where S is the soluble part present in the product (if one knows the percentage of the original sample and the insoluble sample after crosslinking, the remanent percentage is the soluble part); G_d is the radiochemical yield of scission (the number of chain scissions per 100 eV); G_c is the radiochemical yield of crosslinking (the number of crosslinkings per 100 eV); M_n is the number-average molecular weight at the beginning of the experiment; and D_g is the absorbed dose, converted to kGy from Mrad.

2.2.2. Synthesis of IPN GE/PVP Hydrogels

IPNs were prepared by the free radical polymerization of NVP. MBAAm (2%), TEMED (0.55%), AIBN (3%), and NVP monomer (10% and 15%) were thoroughly dissolved in an ethanol:water mixture (70:30, *v:v*). GE crosslinked hydrogels were immersed in 5 mL of this solution for 48 h. After that time, the solutions were bubbled with Ar for 10 min at room temperature, and ampoules were sealed by heat. The polymerization reaction was performed for 4 h in a water bath at 70 °C.

The gels were subsequently washed with an ethanol:water mixture (70:30, *v:v*) and vacuum-dried at 45 °C until reaching a constant weight.

2.2.3. Coupling of GA

The carbodiimide-mediated coupling reaction was used to covalently modify the IPN GE/PVP hydrogels with the polyphenol GA according to the procedure of Xu et al. [33] with some modifications. In brief, GA (1 mmol) and EDC-HCl (3 mmol) were dissolved in 10 mL of ethanol in an ice bath. Then, 3 mmol of NHS was added under continuous stirring. After the reaction for 1 h at 4 °C, swollen samples of *net*-GE and IPN GE/PVP hydrogels of approximately 0.0500 g were introduced into the reaction mixture. The reaction was further performed at 4 °C for 1 h and 24 h more at room temperature.

Afterward, the samples were washed with distilled water for 48 h with at least 4 changes of solvent each day to remove the free GA, EDC-HCl, and NHS prior to characterization.

2.3. Physicochemical Characterization

2.3.1. FT-IR Characterization

The total attenuated reflectance–Fourier transform infrared (ATR-FTIR) spectra of the different samples were recorded with a Perkin-Elmer spectrum 100 spectrometer (Perkin-Elmer Cetus Instruments, Norwalk, CT, USA) with a universal ATR with a diamond tip, performing 16 scans per sample in the range of 4000–650 cm^{-1}. The spectra were analyzed using PerkinElmer Spectrum Spectroscopy software (Version 6.2.0.0055).

2.3.2. Thermogravimetric Analysis

The thermal stability and decomposition temperatures of the hydrogels were measured using a TGA Q50 (TA Instruments, New Castle, DE, USA). Each sample was pulverized and weighted (15–20 mg) before being heated (temperature ramp rate 10 °C/min) under a nitrogen atmosphere in the range of 25–800 °C. The degradation temperature was calculated as the midpoint of the 1st derivative peak of the TGA thermograms for GE, PVP, and IPN GE/PVP.

2.3.3. Scanning Electron Microscopy (SEM)

The hydrogels were immersed in deionized water at room temperature until equilibrium was reached. The swollen hydrogel was then frozen in liquid nitrogen for 10 min and fractured carefully prior to being subjected to lyophilization (Labconco, Kansas City, MO, USA). The interior morphology of a specimen of the freeze-dried hydrogels was observed using scanning electron microscopy (JEOL, Tokyo, Japan). Before conducting the SEM observation, the specimen was fixed on aluminum stubs and coated with gold for 60 s.

2.3.4. Equilibrium Swelling Time and Water Content Test

Hydrogel cylinders of 2 cm in diameter and 1 cm in height were previously dried and studied to determine their water adsorption capacities. The study was performed by immersing the samples in distilled water at room temperature. The mass gain was measured gravimetrically by intervals of time from 5 min until 100 h of immersion; in each measurement, the sample was taken out from the distilled water, placed on an absorbent tissue paper to remove the excess water, and weighted. This was repeated until the weight remained unchanged.

Equation (3) was used to calculate the swelling percentage:

$$\text{Swelling } (\%) = [(W_s - W_d)/W_d] \times 100 \quad (3)$$

where W_s and W_d represent the weights of the swollen and dry samples, respectively.

The dried hydrogel samples were placed in 5 mL of distilled water until they fully swelled. They were then blotted with tissue paper and weighted. The water content was calculated according to Equation (4):

$$\text{Water content (\%)} = (W_t - W_0)/W_t \times 100 \quad (4)$$

where W_t is the weight of the swollen hydrogel after reaching equilibrium, and W_0 is the weight of the dried hydrogel.

All measurements were carried out with triplicated samples and averaged.

2.3.5. Mechanical Testing

The hydrogel samples were characterized using an mechanical microindentation assay as described in [34]. In short, microindentation tests were performed using a commercial microindenter setup, the FT-MTA03 Micromechanical Testing and Assembly System (FemtoTools AG, Zúrich, Switzerland). The force versus displacement data were obtained with a FT-S200 tip (spherical tip with a diameter of 50 μm), with a measurement range of ±200 μN and a maximum resolution of 0.0005 μN (at a sampling rate of 10 Hz). Since the measurements were performed while the hydrogel was submerged in MilliQ water, attraction forces between the hydrogel surface and the tip were not observed. Thus, the Hertz model [34] was used to fit the experimental data after a preprocessing step. The data analysis was performed using Wolfram Mathematica 12.1 (Wolfram Research, Oxford, UK), assuming a Poisson's ratio of 0.45 [35].

2.3.6. Gallic acid Content

The gallic acid content of the hydrogel–polyphenol conjugates was measured according to the Folin–Ciocalteu method [36,37]. Briefly, triplicated samples of IPNs with gallic acid were immersed in 300 μL of 1N Folin–Ciocalteu phenol reagent solution mixed with 2.4 mL of sodium carbonate solution 0.5 M. The volume was adjusted to 3 mL with 300 μL of distilled water. Samples of *net*-GE without gallic acid were used as controls.

The absorbance of the samples was determined at 750 nm using a UV–Vis spectrophotometer (Specord 200 plus, Analytik Jena, Germany). Solutions of gallic acid with concentrations in the range of 5–50 ppm were used as standards for the calibration curve. The gallic acid content of the hydrogels was expressed as mg gallic acid equivalent per gram of sample (GAE/g sample).

2.3.7. Radical Scavenging Activity

The radical scavenging activity of the hydrogels was determined spectrophotometrically by measuring the decrease in the absorbance of the DPPH radical at 520 nm [38]. In brief, DPPH was dissolved in anhydrous ethanol to prepare a stock solution (100 μM). Samples of polyphenol-modified IPNs were then incubated with 2 mL of DPPH solution at room temperature. The radical scavenging activity (%) of the samples was compared to Trolox, a water-soluble derivative of vitamin E, regularly used as a reference standard for antioxidant capacity. All standards and samples were kept in the dark for 30 min. Absorbance values were measured at 517 nm using a UV–Vis spectrophotometer (Specord 200 plus, Analytik Jena, Germany). The free radical scavenging effect was calculated as follows (Equation (5)):

$$\text{Radical scavenging activity (\%)} = [1 - (A/A_0)] \times 100 \quad (5)$$

where A_0 is the absorbance of the blank (DPPH solution alone), and A is the absorbance of DPPH incubated with the hydrogels. All tests were performed in triplicate and averaged.

3. Results and Discussion

Antioxidant IPN hydrogels were prepared according to the global synthesis process depicted in Figure 1. Gamma radiation was used to permanently crosslink GE via the formation of covalent bonds. The resulting *net*-GE hydrogels were interpenetrated with a second network of PVP, synthesized via free radical polymerization with NVP monomers using N,N′-Methylenebisacrylamide (MBAAm) as a crosslinking agent, with N,N,N′,N′-Tetramethyl ethylenediamine (TEMED) as the activator and azobisisobutyronitrile (AIBN) as the initiator.

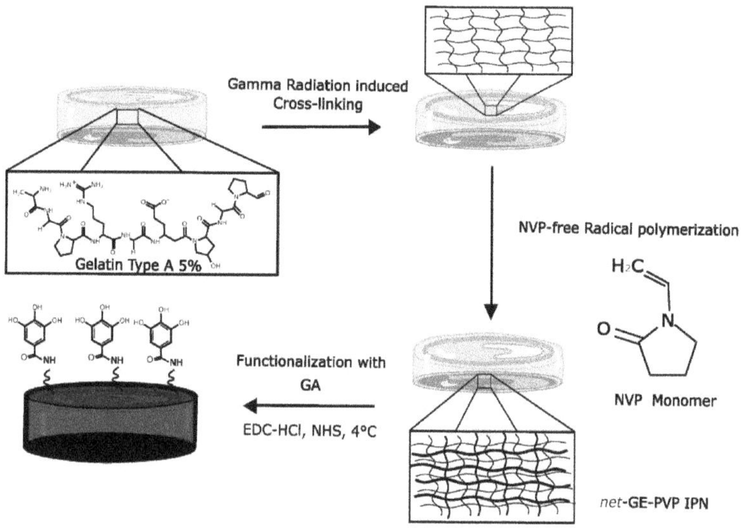

Figure 1. Scheme of the synthesis route to obtain GA-functionalized IPN hydrogels.

3.1. Crosslinking of GE

Net-GE hydrogels were synthesized via gamma-ray crosslinking in aqueous media. Gamma radiation induces water radiolysis, which yields proton radicals and hydroxyl radicals. According to the proposal of Wan Ishak et al. [22], OH radicals remove H in the gelatin chain and induce the formation of gelatin radicals, which, after recombination, form covalent bonds between gelatin chains.

Four total absorbed doses were selected for this study: 5, 10, 15, and 30 kGy. The hydrogels obtained had differences mainly in their mechanical stability. While the 5 and 10 kGy hydrogels were mechanically unstable and their manipulation was complicated, the hydrogels with a total dose of 30 kGy were brittle.

3.2. Radiochemical Yield Ratios of Crosslinking and Scission

The radiochemical yield ratios of crosslinking and scission (degradation) and the dose of incipient gel D_g calculated using the Charlesby–Pinner equation can be observed in Table 1.

Table 1. Radiochemical yield ratios and initial gel dose of *net*-GE at different initial GE concentrations.

Concentration	5%	10%
G_d/G_c	−0.0436	0.1115
D_g (kGy)	0.2829	0.3978

The radiochemical yield of the degradation and crosslinking ratio was almost zero, indicating that only crosslinking takes part in this system. Degradation increased with the initial concentration of GE. Due to these results, a 5% GE concentration was chosen to synthesize IPNs and their subsequent characterization.

3.3. FT-IR Analysis

The FTIR-ATR spectra of GE, PVP, and IPN GE/PVP are presented in Figure 2. GE shows bands at 3285 cm^{-1} corresponding to O-H and N-H stretching, 2937 cm^{-1} corresponding to C-H stretching (methyl groups), 1629 cm^{-1} corresponding to C=O stretching (amide I), and 1520 cm^{-1} corresponding to N-H bending (amide II) [39]. The PVP spectrum shows peaks at 3424 cm^{-1} corresponding to O-H stretching, 2921 cm^{-1} corresponding to C-H stretching (polymer chain), 1654 cm^{-1} corresponding to C=O stretching (pyrrolidone ring), and 1430 cm^{-1} corresponding to C-H bending (polymer chain) [40]. IPN GE/PVP shows the characteristic bands of both GE and PVP, but it is more similar to GE because the quantity of PVP is low, and the two polymers have the same functional groups. Nevertheless, it is possible to identify PVP because this material has a peak at 712 cm^{-1}, which is attributed to the C-H rock vibration of polymer long chains.

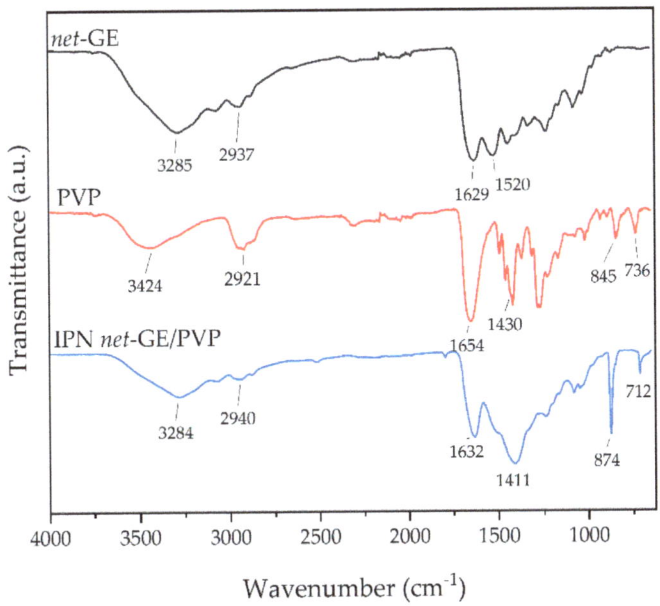

Figure 2. FTIR-ATR stacked spectra.

3.4. Thermal Analysis

In Figure 3, we can see the TGA of *net*-GE, PVP, and IPN *net*-GE/PVP, and it is possible to observe the thermal decomposition steps of each material. GE and PVP show only one step at 330 °C and 442 °C, respectively, while IPN shows two steps at 343 °C and 449 °C, corresponding to starting compounds. This proves that GE and PVP were incorporated into the system in an interpenetrated way; i.e., only physical interactions occurred between both polymers. The thermal stability of IPN is very similar to that of GE.

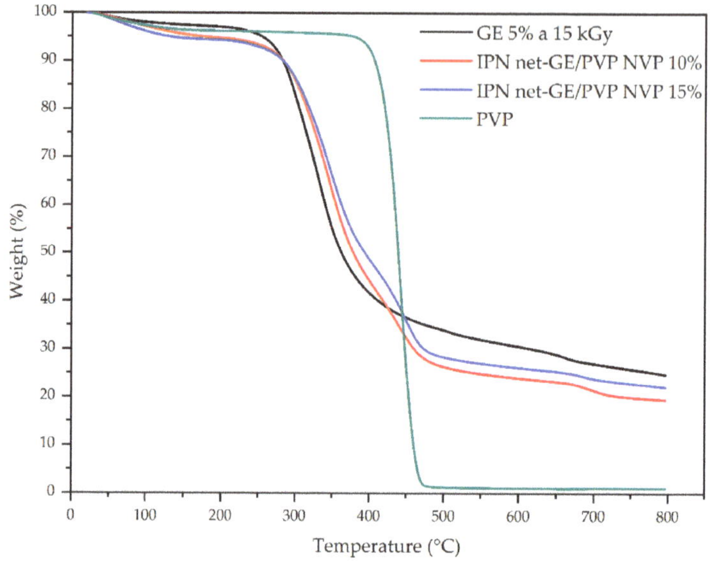

Figure 3. TGA thermograms under N_2 of *net*-GE, PVP, and IPN *net*-GE/PVP.

3.5. Morphology of the Hydrogels

The SEM micrographs of the freeze-dried hydrogels are shown in Figure 4. Figure 4a corresponds to a sample of *net*-GE crosslinked with a total absorbed dose of 15 kGy. It has a porous structure with a homogeneous appearance. We can observe that a second network is appreciable in the cross-section of the sample IPN *net*-GE/PVP, obtained by the free radical polymerization and crosslinking of NVP 10% and 15% solutions (Figure 4b and 4c, respectively). It is notable that an increase in monomer concentration generates a reduction in the total void space of the hydrogel structure, which correlates with the decreased values of the equilibrium swelling and an increase in the Young's modulus value.

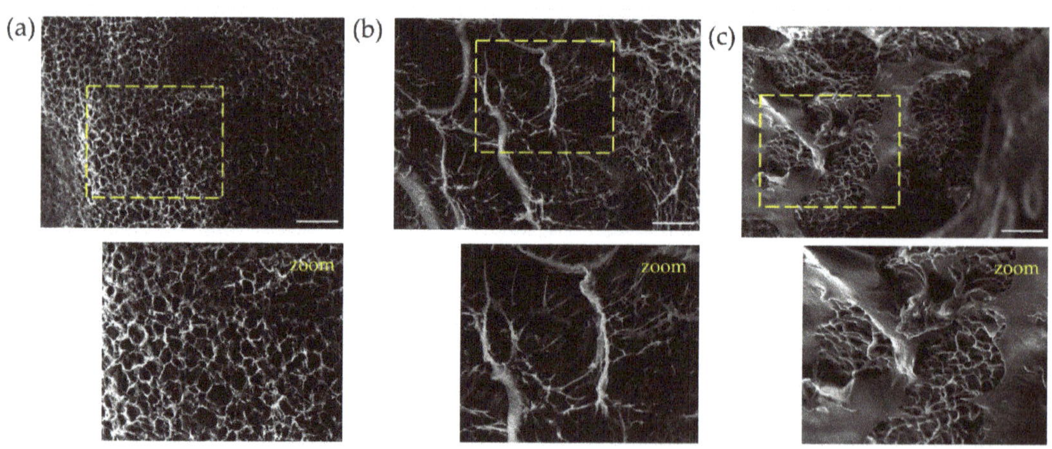

Figure 4. SEM micrographs of cross-section morphology of hydrogels after lyophilization. (**a**) *net*-GE, (**b**) *net*-GE/PVP IPN (10% NVP solution), and (**c**) *net*-GE/PVP IPN (15% NVP solution). Magnification: 1000×. Scale bar: 50 μm.

3.6. Swelling Behavior and Water Content

In the equilibrium swelling studies, it was possible to observe different swelling capacities. In Figure 5, we show the effect of radiation dose on the degree of the equilibrium swelling of the GE gels of 5% and 10% concentrations (Figures 5a and 5b, respectively).

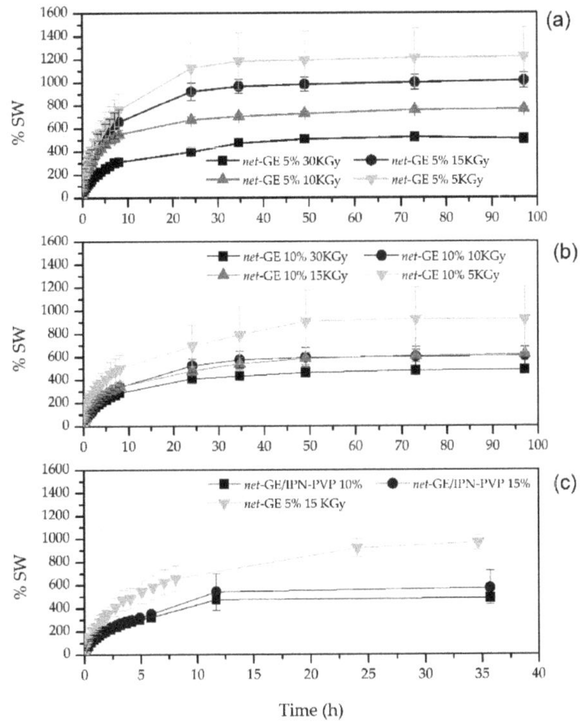

Figure 5. Swelling of *net*-GE synthesized with different GE concentrations: (**a**) 5% and (**b**) 10% in distilled water at room temperature (**c**).

The hydrogels with a total dose of 5 kGy had the highest swelling capacity, around 1220.22 ± 252.43% for GE 5% and 923.86 ± 283.16% for GE 10%; however, their mechanical stability was very poor and decreased with time, resulting in the loss of their integrity after 30 h of being immersed. It is important to clarify that the elevated values of the measurement errors, particularly in the samples of 5 kGy, were due to the loss of material over the course of time. Because of that, the registered values of mass started to differ considerably in the measurements of the samples with more than 20 h of swelling.

The samples of GE 10% irradiated with a total dose of 10 and 15 kGy had very similar behavior.

In general, the hydrogels obtained from the 10% gelatin solutions had a lower swelling capacity than those obtained from gelatin concentrations of 5%.

The samples with 5 and 10% concentrations of GE and a total dose of 30 kGy were discarded because they had to be manipulated very carefully, as they were very fragile and had the lowest swelling capacity (507.83 ± 42.36% and 485.40 ± 16.00%, respectively).

The water content of all analyzed samples and the equilibrium swelling data are reported in Table 2.

Table 2. Water content and equilibrium swelling of hydrogels of 5% and 10% GE with different total absorbed doses, expressed as percentages.

GE Concentration	Total Absorbed Dose (kGy)	Water Content (%) [a]	Equilibrium Swelling (%) [b]
5%	5	92.22 ± 1.64	1220 ± 252.43
	10	88.43 ± 0.25	764.81 ± 18.28
	15	91.01 ± 0.54	1014.79 ± 68.98
	30	83.49 ± 1.19	507 ± 42.36
10%	5	89.76 ± 2.59	923.86 ± 283.16
	10	85.71 ± 1.62	606.12 ± 82.48
	15	86.09 ± 1.22	622.69 ± 61.08
	30	82.01 ± 0.47	485.40 ± 16.00

[a] Data are shown as means ± SD (n = 3); [b] data are means ± SD (n = 3).

3.7. Mechanical Stability Measurements

To determine whether the interpenetrated PVP network impacts the mechanical stability of the hydrogels, three different conditions (*net*-GE, *net*-GE/PVP 10%, and *net*-GE/PVP 15%) were mechanically characterized using a microindentation assay (see the Materials and Methods Section). The results of the mechanical characterization presented as mean ± standard deviation of the Young's modulus (**E**) are reported in Figure 6 and summarized in Table 3. At least two independently manufactured samples per condition were characterized, and at least 10 measurements per sample were performed.

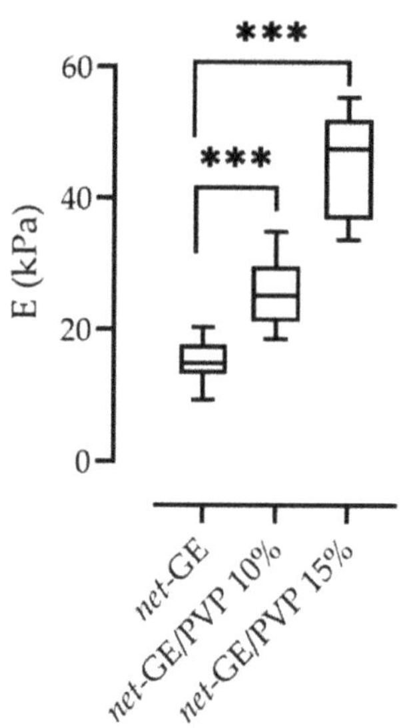

Figure 6. Stiffness of GE-based hydrogels. Young's modulus was measured using a microindentation test and by fitting the Hertz model to the force–distance curves. A one-way analysis of variance (ANOVA) with Tukey correction was employed for multiple comparisons. It was considered statistically significant when $p < 0.05$; *** $p < 0.0001$.

Table 3. Elastic modulus of gelatin-based hydrogels.

	E (kPa)
net-GE	15.82 ± 2.59
net-GE/IPN-PVP 10%	25.71 ± 4.75
net-GE/IPN-PVP 15%	44.85 ± 7.55

As can be seen in Figure 6, Young's modulus increases significantly with the increase in PVP content, suggesting a positive correlation between PVP concentration and **E**. These results show that the addition of PVP increases the stiffness of the gelatin network, providing better mechanical stability. The tunability of the mechanical properties of this material, which depend on the total absorbed dose of gamma radiation and the interpenetration of a second polymeric network, is an advantage that allows the gamma radiation crosslinked gelatin-based hydrogels to be suitable for biomedical applications, such as wound dressings and tissue engineering.

3.8. Gallic Acid Content and Radical Scavenging Activity

The amount of GA attached to the *net*-GE/PVP hydrogels was determined according to the Folin–Ciocalteu method, which is easy to perform, rapid, and low-cost. Polyphenols react with Folin–Ciocalteu reagent in alkaline media to form a blue phosphotungstic–phosphomolybdenum complex that is quantifiable by UV–Vis spectrophotometry at 750 nm. A standard calibration curve of GA, shown in Figure 7a, was used to quantify the amount of GA in the hydrogel samples of known mass employing the Beer–Lambert Law.

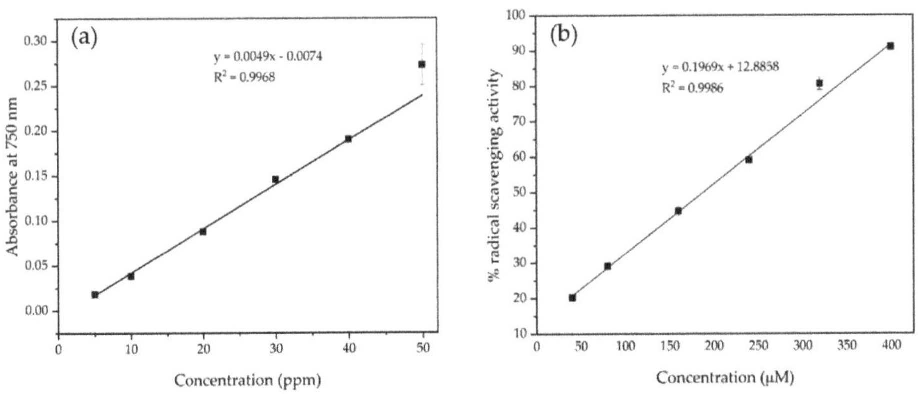

Figure 7. (**a**) Calibration curve of gallic acid; (**b**) calibration curve of Trolox.

Table 3 shows the amounts of GA coupled with the IPN *net*-GE/PVP hydrogels after an amide coupling reaction.

The antioxidant properties of the *net*-GE/PVP IPN hydrogels were investigated using a 2,2-diphenyl-1-picrylhydrazyl (DPPH) radical scavenging assay. Trolox was used as a reference compound to compare the antioxidant efficiency of the prepared hydrogels.

The DPPH radical has an absorption maximum at 517 nm. Once it is reduced in the presence of an antioxidant molecule, its color changes from purple to colorless due to the formation of stable hydrazine, noted as DPPH-H, upon the absorption of hydrogen from an antioxidant molecule. The antioxidant effect of that molecule is proportional to the decrease in the absorbance values at 517 nm.

The radical scavenging activity (RSA%) is related to the antioxidant activity of the analyzed samples. A calibration curve with increasing concentrations of Trolox, shown in

Figure 7b, was used to set a reference of antioxidant capacity and to compare the radical scavenging ability of the hydrogels with that of Trolox standards.

Figure 8 shows that there is a significant difference between the antioxidant capacity of the functionalized hydrogels of *net*-GE and *net*-GE/PVP. Both samples were compared to *net*-GE hydrogels without GA.

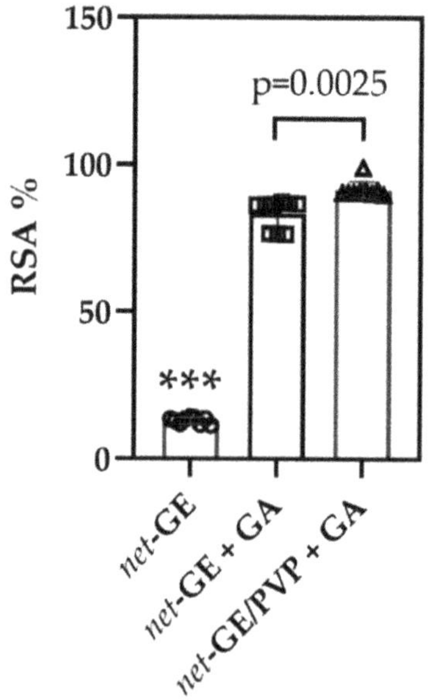

Figure 8. Comparation of radical scavenging activity (RSA%) of *net*-GE/PVP IPN and *net*-GE hydrogels functionalized with GA. *net*-GE was used as a control. A one-way analysis of variance (ANOVA) with Tukey correction was employed for multiple comparisons. It was considered statistically significant when $p < 0.05$; *** $p < 0.0001$.

Table 4 shows that strong antioxidant abilities were observed for net-GE/PVP + GA IPN hydrogels, which inhibited 91.61 ± 1.36% of DPPH radical, while *net*-GE + GA showed an inhibition of 83.28 ± 1.22%. Samples of pristine *net*-GE were used as negative controls.

Table 4. Amounts of coupled GA.

Sample	Gallic Acid Content (mg GA/g Sample) [a]	Antioxidant Activity [b]
net-GE (pristine)	0.20 ± 0.21	13.085 ± 1.22
net-GE + GA	1.39 ± 0.19	83.281± 5.85
IPN *net*-GE/PVP	1.64 ± 0.34	91.61 ± 1.36

[a] Data are shown as means ± SD (n = 3); [b] data are means ± SD (n = 3).

A polymeric material with antioxidant properties is of relevant importance for the development of wound dressings. It has been found that the presence of reactive oxygen species (ROS) in the wound bed promotes a pro-inflammatory environment, which is associated with the oxidative stress of cells and an impaired wound healing process [4].

Previous studies have reported that GA exhibits a potent anti-inflammatory effect due to the reduction of inflammatory cell accumulation and the downregulation of major pro-inflammatory cytokines, namely, interleukins, tumor necrosis factor α (TNF-α), and nuclear factor κB (NF-κB) [41,42]. This kind of bioactive material could improve wound management versus the use of dressings, which only protect the wound surface and moisturizes the wound bed, in standard care. In this context, the material studied in this research could potentially restore an appropriate ROS balance, particularly in chronic wounds.

In the next stage of this research, it is necessary to evaluate the biocompatibility of the material by performing in vitro and in vivo tests.

4. Conclusions

In this study, we developed IPN hydrogels composed of GE and PVP networks, with covalent crosslinking in each network but only physical interactions between them. The crosslinking degree of the *net*-GE hydrogels increased with an increase in the total absorbed dose of gamma radiation, while the equilibrium swelling decreased when the total absorbed dose was augmented. After optimizing the concentration of the initial GE solution and the total absorbed dose of gamma radiation for the GE hydrogels, 15 kGy was selected as the best condition to obtain mechanically stable materials in order to achieve adequate manipulation without breakage. IPN formation was accomplished via the free radical polymerization of NVP inside the crosslinked network of GE, confirmed by thermogravimetric and FT-IR analyses.

The antioxidant capacity of the functionalized IPNs was evaluated by a DPPH analysis, demonstrating that the synthesized materials had a radical scavenging activity comparable to that of a Trolox standard with a concentration of 400 μM. The results suggest that the polyphenol-functionalized IPN *net*-GE/PVP could be considered as a candidate in the use of polymeric biomaterials for improving the wound healing process.

Further research must be conducted to evaluate the biocompatibility of the materials and their performance in vivo to enhance the wound healing process, and to evaluate their potential as a bioactive material for wound dressing.

Author Contributions: L.E.-G., conceptualization; methodology; and formal analysis in the synthesis and characterization, the investigation, and writing of the draft manuscript. D.P.-C., mechanical characterization, analysis, and discussion of the manuscript. D.M., methodology and experimentation. G.B., conceptualization, methodology, investigation direction, resources, writing—review and editing, project administration, and funding acquisition. All authors have read and agreed to the published version of the manuscript.

Funding: This research was funded by Dirección General de Asuntos del Personal Académico (DGAPA), Universidad Nacional Autónoma de México, under Grant IN200322.

Institutional Review Board Statement: Not applicable.

Informed Consent Statement: Not applicable.

Data Availability Statement: The data presented in this study are available on request from the corresponding author.

Acknowledgments: We thank Alejandra Ortega and Benjamín Leal from ICN UNAM for technical support. We also thank the laboratory uNTec at FC/UNAM and Genaro Vazquéz Victorio for the facilities to carry out the mechanical characterization. L.E.-G. thanks CONACYT for the postdoctoral scholarship (CVU 390627).

Conflicts of Interest: The authors declare no conflict of interest.

References

1. Xia, J.; Zhang, H.; Yu, F.; Pei, Y.; Luo, X. Superclear, Porous Cellulose Membranes with Chitosan-Coated Nanofibers for Visualized Cutaneous Wound Healing Dressing. *ACS Appl. Mater. Interfaces* **2020**, *12*, 24370–24379. [CrossRef] [PubMed]
2. Boateng, J.S.; Matthews, K.H.; Stevens, H.N.E.; Eccleston, G.M. Wound Healing Dressings and Drug Delivery Systems: A Review. *J. Pharm. Sci.* **2008**, *97*, 2892–2923. [CrossRef] [PubMed]
3. Gurtner, G.C.; Werner, S.; Barrandon, Y.; Longaker, M.T. Wound Repair and Regeneration. *Nature* **2008**, *453*, 314–321. [CrossRef]
4. Guo, S.; DiPietro, L.A. Critical Review in Oral Biology & Medicine: Factors Affecting Wound Healing. *J. Dent. Res.* **2010**, *89*, 219–229. [CrossRef] [PubMed]

5. Jayakumar, R.; Prabaharan, M.; Sudheesh Kumar, P.T.; Nair, S.V.; Tamura, H. Biomaterials Based on Chitin and Chitosan in Wound Dressing Applications. *Biotechnol. Adv.* **2011**, *29*, 322–337. [CrossRef] [PubMed]
6. Park, H.; Baek, S.; Kang, H. Biomaterials to Prevent Post-Operative Adhesion. *Materials* **2020**, *13*, 3056. [CrossRef]
7. Ahmed, E.M. Hydrogel: Preparation, Characterization, and Applications: A Review. *J. Adv. Res.* **2015**, *6*, 105–121. [CrossRef]
8. Li, J.; Mooney, D.J. Designing Hydrogels for Controlled Drug Delivery. *Nat. Publ. Gr.* **2016**, *1*, 16071. [CrossRef]
9. Dreiss, C.A. Hydrogel Design Strategies for Drug Delivery. *Curr. Opin. Colloid Interface Sci.* **2020**, *48*, 1–17. [CrossRef]
10. Chaudhary, S.; Chakraborty, E. Hydrogel Based Tissue Engineering and Its Future Applications in Personalized Disease Modeling and Regenerative Therapy. *Beni-Suef Univ. J. Basic Appl. Sci.* **2022**, *11*, 3. [CrossRef]
11. Peppas, N.A.; Bures, P.; Leobandung, W.; Ichikawa, H. Hydrogels in Pharmaceutical Formulations. *Eur. J. Pharm. Biopharm.* **2000**, *50*, 27–46. [CrossRef]
12. Sun, X.; Agate, S.; Salem, K.S.; Lucia, L.; Pal, L. Hydrogel-Based Sensor Networks: Compositions, Properties, and Applications—A Review. *ACS Appl. Bio Mater.* **2021**, *4*, 140–162. [CrossRef] [PubMed]
13. Wang, C.; Liu, Y.; Qu, X.; Shi, B.; Zheng, Q.; Lin, X.; Chao, S.; Wang, C.; Zhou, J.; Sun, Y.; et al. Ultra-Stretchable and Fast Self-Healing Ionic Hydrogel in Cryogenic Environments for Artificial Nerve Fiber. *Adv. Mater.* **2022**, *34*, 2105416. [CrossRef] [PubMed]
14. Firlar, I.; Altunbek, M.; Mccarthy, C.; Ramalingam, M. Functional Hydrogels for Treatment of Chronic Wounds. *Gels* **2022**, *8*, 127. [CrossRef] [PubMed]
15. Puertas-Bartolomé, M.; Mora-Boza, A.; García-Fernández, L. Emerging Biofabrication Techniques: A Review on Natural Polymers for Biomedical Applications. *Polymers* **2021**, *13*, 1209. [CrossRef]
16. Kumari, A.; Yadav, S.K.; Yadav, S.C. Biodegradable Polymeric Nanoparticles Based Drug Delivery Systems. *Colloids Surf. B Biointerfaces* **2010**, *75*, 1–18. [CrossRef] [PubMed]
17. Rujitanaroj, P.; Pimpha, N.; Supaphol, P. Wound-Dressing Materials with Antibacterial Activity from Electrospun Gelatin Fiber Mats Containing Silver Nanoparticles. *Polymer* **2008**, *49*, 4723–4732. [CrossRef]
18. Wisotzki, E.I.; Hennes, M.; Schuldt, C.; Engert, F.; Knolle, W.; Decker, U.; Käs, J.A.; Zink, M.; Mayr, S.G. Tailoring the Material Properties of Gelatin Hydrogels by High Energy Electron Irradiation. *J. Mater. Chem. B* **2014**, *2*, 4297–4309. [CrossRef]
19. Biswal, D.; Anupriya, B.; Uvanesh, K.; Anis, A.; Banerjee, I.; Pal, K. Effect of Mechanical and Electrical Behavior of Gelatin Hydrogels on Drug Release and Cell Proliferation. *J. Mech. Behav. Biomed. Mater.* **2016**, *53*, 174–186. [CrossRef]
20. Mondragon, G.; Peña-Rodriguez, C.; González, A.; Eceiza, A.; Arbelaiz, A. Bionanocomposites Based on Gelatin Matrix and Nanocellulose. *Eur. Polym. J.* **2015**, *62*, 1–9. [CrossRef]
21. Zhao, Y.; Sun, Z. Effects of Gelatin-Polyphenol and Gelatin–Genipin Cross-Linking on the Structure of Gelatin Hydrogels. *Int. J. Food Prop.* **2018**, *20*, S2822–S2832. [CrossRef]
22. Ishak, W.H.W.; Ahmad, I.; Ramli, S.; Amin, M.C.I.M. Gamma Irradiation-Assisted Synthesis of Cellulose Nanocrystal-Reinforced Gelatin Hydrogels. *Nanomaterials* **2018**, *8*, 749. [CrossRef] [PubMed]
23. Hafeez, S.; Islam, A.; Gull, N.; Ali, A.; Khan, S.M.; Zia, S.; Anwar, K.; Khan, S.U.; Jamil, T. γ-Irradiated Chitosan Based Injectable Hydrogels for Controlled Release of Drug (Montelukast Sodium). *Int. J. Biol. Macromol.* **2018**, *114*, 890–897. [CrossRef] [PubMed]
24. Fazolin, G.N.; Varca, G.H.C.; de Freitas, L.F.; Rokita, B.; Kadlubowski, S.; Lugão, A.B. Simultaneous Intramolecular Crosslinking and Sterilization of Papain Nanoparticles by Gamma Radiation. *Radiat. Phys. Chem.* **2020**, *171*, 108697. [CrossRef]
25. Kaur, H.; Chatterji, P.R. Interpenetrating Hydrogel Networks. Swelling and Mechanical Properties of the Gelatin-Polyacrylamide Interpenetrating Networks. *Macromolecules* **1990**, *40*, 401–410. [CrossRef]
26. Ajji, Z.; Othman, I.; Rosiak, J.M. Production of Hydrogel Wound Dressings Using Gamma Radiation. *Nucl. Instrum. Methods Phys. Res. Sect. B Beam Interact. Mater. Atoms* **2005**, *229*, 375–380. [CrossRef]
27. Razzak, M.T.; Darwis, D.; Zainuddin; Sukirno. Irradiation of Polyvinyl Alcohol and Polyvinyl Pyrrolidone Blended Hydrogel for Wound Dressing. *Radiat. Phys. Chem.* **2001**, *62*, 107–113. [CrossRef]
28. Rosiak, J.M.; Olejniczak, J. Medical Applications of Radiation Formed Hydrogels. *Radiat. Phys. Chem.* **1993**, *42*, 903–906. [CrossRef]
29. Ng, T.B.; He, J.S.; Niu, S.M.; Pi, Z.F.; Shao, W.; Liu, F.; Zhao, L. A Gallic Acid Derivative and Polysaccharides with Antioxidative Activity from Rose (Rosa Rugosa) Flowers. *J. Pharm. Pharmacol.* **2010**, *56*, 537–545. [CrossRef]
30. Ma, J.; Luo, X.D.; Protiva, P.; Yang, H.; Ma, C.; Basile, M.J.; Weinstein, I.B.; Kennelly, E.J. Bioactive Novel Polyphenols from the Fruit of Manilkara Zapota (Sapodilla). *J. Nat. Prod.* **2003**, *66*, 983–986. [CrossRef]
31. Mansouri, M.T.; Soltani, M.; Naghizadeh, B.; Farbood, Y.; Mashak, A.; Sarkaki, A. A Possible Mechanism for the Anxiolytic-like Effect of Gallic Acid in the Rat Elevated plus Maze. *Pharmacol. Biochem. Behav.* **2014**, *117*, 40–46. [CrossRef] [PubMed]
32. Charlesby, A. *Atomic Radiation and Polymers*; Pergamon Press Ltd.: Oxford, UK; London, UK, 1960; ISBN 0471359246.
33. Xu, Y.; Wei, Z.; Xue, C.; Huang, Q. Assembly of Zein–Polyphenol Conjugates via Carbodiimide Method: Evaluation of Physicochemical and Functional Properties. *Lwt* **2022**, *154*, 112708. [CrossRef]
34. Pérez-Calixto, D.; Amat-Shapiro, S.; Zamarrón-Hernández, D.; Vázquez-Victorio, G.; Puech, P.H.; Hautefeuille, M. Determination by Relaxation Tests of the Mechanical Properties of Soft Polyacrylamide Gels Made for Mechanobiology Studies. *Polymers* **2021**, *13*, 629. [CrossRef] [PubMed]
35. van Otterloo, J.; Cruden, A.R. Rheology of Pig Skin Gelatine: Defining the Elastic Domain and Its Thermal and Mechanical Properties for Geological Analogue Experiment Applications. *Tectonophysics* **2016**, *683*, 86–97. [CrossRef]

36. Sánchez-Rangel, J.C.; Benavides, J.; Heredia, J.B.; Cisneros-Zevallos, L.; Jacobo-Velázquez, D.A. Analytical Methods The Folin–Ciocalteu Assay Revisited: Improvement of Its Speci Fi City for Total Phenolic Content Determination. *Anal. Methods* **2013**, *5*, 5990–5999. [CrossRef]
37. Ikawa, M.; Shaper, T.D.; Dollard, C.A.; Sasner, J.J. Utilization of Folin-Ciocalteau Phenol Reagent for the Detection of Certain Nitrogen Compounds. *J. Agric. Food Chem.* **2003**, *51*, 1811–1815. [CrossRef]
38. Brand-Williams, W.; Cuvelier, M.E.; Berset, C. Use of a Free Radical Method to Evaluate Antioxidant Activity. *LWT-Food Sci. Technol.* **1995**, *28*, 25–30. [CrossRef]
39. Lee, D.H.; Arisaka, Y.; Tonegawa, A.; Kang, T.W.; Tamura, A.; Yui, N. Cellular Orientation on Repeatedly Stretching Gelatin Hydrogels with Supramolecular Crosslinking. *Polymers* **2019**, *11*, 2095. [CrossRef]
40. Rahma, A.; Munir, M.M.; Khairurrijal; Prasetyo, A.; Suendo, V.; Rachmawati, H. Intermolecular Interactions and the Release Pattern of Electrospun Curcumin-Polyvinyl(Pyrrolidone) Fiber. *Biol. Pharm. Bull.* **2016**, *39*, 163–173. [CrossRef]
41. Cheng, Y.; Li, X.; Tse, H.-F.; Rong, J. Gallic Acid-L-Leucine Conjugate Protects Mice against LPS-Induced Inflammation and Sepsis via Correcting Proinflammatory Lipid Mediator Profiles and Oxidative Stress. *Oxid. Med. Cell. Longev.* **2018**, *2018*, 1081287. [CrossRef]
42. Albouchi, F.; Avola, R.; Maria Lo Dico, G.; Calabrese, V.; Graziano, A.C.E.; Abderrabba, M.; Cardile, V. Melaleuca Styphelioides Sm. Polyphenols Modulate Interferon Gamma/Histamine-Induced Inflammation in Human NCTC 2544 Keratinocytes. *Molecules* **2018**, *23*, 2526. [CrossRef] [PubMed]

polymers

Article

Phosphazene-Containing Epoxy Resins Based on Bisphenol F with Enhanced Heat Resistance and Mechanical Properties: Synthesis and Properties

Ilya V. Tarasov [1,*], Anastasiya V. Oboishchikova [1], Roman S. Borisov [1,2], Vyacheslav V. Kireev [1] and Igor S. Sirotin [1,*]

1 Department of Plastics, Mendeleev University of Chemical Technology of Russia, Miusskaya sq. 9, 125047 Moscow, Russia
2 Topchiev Institute of Petrochemical Synthesis, Russian Academy of Sciences, Leninskii pr. 29, 119991 Moscow, Russia
* Correspondence: ilya.v.tarasov@gmail.com or tarasov.i.v@muctr.ru (I.V.T.); sirotin.i.s@muctr.ru (I.S.S.)

Abstract: Organophosphazenes are of interest due to the combination of increased mechanical and thermal properties of polymer materials obtained with their use, however, they are characterized by a complex multi-stage synthesis. Moreover, the high viscosity of phosphazene-containing epoxy resins (PhER) makes their processing difficult. To simplify the synthesis of PhER, a one-step method was developed, and bisphenol F was chosen, which also provided a decrease in viscosity. In the current study, PhER were formed by a one-stage interaction of hexachlorocyclotriphosphazene (HCP) with bisphenol F isomers and epichlorohydrin in the presence of alkali, which was a mixture of epoxycyclophosphazenes (ECPh) with a functionality from 1 to 4 according to the results of MALDI-TOF analysis. Conventional epoxy resins based on bisphenol F, also formed during the process, showed high mechanical properties and glass transition temperature, and the reactivity of the obtained resins is similar to the base epoxy resins based on bisphenols A and F. Cured PhER had higher or the same mechanical properties compared to base epoxy resins based on bisphenol A and F, and a glass transition temperature comparable to base epoxy resins based on bisphenol F: glass transition temperature (Tg) up to 174.5 °C, tensile strength up to 74.5 MPa, tensile modulus up to 2050 MPa, tensile elongation at break up to 6.22%, flexural strength up to 146.6 MPa, flexural modulus up to 3630 MPa, flexural elongation at break up to 9.15%, and Izod impact strength up to 4.01 kJ/m^2. Analysis of the composition of the obtained PhER was carried out by ^1H and ^{31}P NMR spectroscopy, MALDI-TOF mass spectrometry, X-ray fluorescence elemental analysis, and contained up to 3.9% phosphorus and from 1.3% to 4.2% chlorine. The temperature profile of the viscosity of the resulting epoxy resins was determined, and the viscosity at 25 °C ranged from 20,000 to 450,000 Pa·s, depending on the ratio of reagents. The resins studied in this work can be cured with conventional curing agents and, with a low content of the phosphazene fraction, can act as modifiers for traditional epoxy resins, being compatible with them, to increase impact strength and elasticity while maintaining the rest of the main mechanical and processing properties, and can be used as a resin component for composite materials, adhesives, and paints.

Keywords: epoxy resin; epoxy oligomer; phosphazene; bisphenol F; mechanical properties; glass transition temperature

Citation: Tarasov, I.V.; Oboishchikova, A.V.; Borisov, R.S.; Kireev, V.V.; Sirotin, I.S. Phosphazene-Containing Epoxy Resins Based on Bisphenol F with Enhanced Heat Resistance and Mechanical Properties: Synthesis and Properties. *Polymers* **2022**, *14*, 4547. https://doi.org/10.3390/polym14214547

Academic Editor: Edina Rusen

Received: 21 September 2022
Accepted: 24 October 2022
Published: 27 October 2022

Publisher's Note: MDPI stays neutral with regard to jurisdictional claims in published maps and institutional affiliations.

Copyright: © 2022 by the authors. Licensee MDPI, Basel, Switzerland. This article is an open access article distributed under the terms and conditions of the Creative Commons Attribution (CC BY) license (https://creativecommons.org/licenses/by/4.0/).

1. Introduction

Epoxy resins are one of the most common thermosetting polymers, having a wide range of applications, such as: coatings, adhesives, casting compounds, resins, and binders for composite materials in various areas, from household to critical in aircraft, rocket, mechanical engineering, and electronics and electrical engineering [1].

At the same time, a number of thermoplastics and other types of thermosetting polymers are used for the same applications. A new generation of binders is being developed with enhanced mechanical properties, thermal and fire resistance, such as: bismaleimides, polyimides, and cyanate ethers [2]. However, they are inferior to epoxides in cost, manufacturability, and in some cases in mechanical properties and toxicity during processing [2,3].

With their advantages, base epoxy resins also have a number of disadvantages, as they often have low and unstable performance characteristics, in particular, combustibility and relatively low heat resistance [4]. An effective way to increase the heat resistance of epoxy matrices is to modify them with polyfunctional resins that crosslink and form a joint mesh structure with base resins. Examples of such resins are triglycidyl-p-aminophenol, tetraglycidyl-4,4′-methylenedianiline, and epoxidized novolaks [4,5].

The problem of flammability of epoxy resins is more complex and can be solved using various additives and/or components:

- Halogenated epoxy resins
- Additive-type fire retardant
- Halogen-free organoelement compounds

The traditional way to reduce flammability is to use halogen-containing epoxy resins, in particular based on brominated bisphenol A. However, these types of resins release toxic gases when in contact with a flame, which limits their use in areas that involve human contact [6,7].

The modern global trend is the use of halogen-free polymer materials as additives to improve the fire resistance of polymeric materials [6]. A cheap and common option is the use of inorganic additives, such as aluminum or magnesium hydroxides, polyphosphates, and red phosphorus, but they significantly reduce the mechanical properties and transparency of materials and degrade the technological properties important for processing [6,8].

Organophosphates proved to be more functional. They are better combined with the base polymer, but at the same time they can act as a plasticizer and reduce heat resistance [9].

Functional organoelement compounds combine the advantages of non-toxicity, preservation of the properties of the base polymer, and heat resistance, capable of forming a common three-dimensional structure with the base epoxy resin [7,10]. An example of such modifiers are glycidyl esters of phosphorus acids, with a decrease in flammability, which increase the mechanical and adhesive strength of the material [10,11]. However, they are not well-compatible with base epoxy resins [11,12]. Additionally, the disadvantage of phosphorus-containing fire retardants is their lower heat resistance compared to modified epoxy bases [6,7,13,14], which limits their use in engineering plastics and high-temperature binders.

One of the promising classes of compounds that can meet the growing needs of high-tech industries are phosphazenes. The main chain of organophosphazenes consists of alternating phosphorus and nitrogen atoms, and at the phosphorus atom there are organic radicals introduced by substitution of a halogen in halophosphazenes. The nature of organic substituents, usually introduced by the reaction of nucleophilic substitution of chlorine, can vary widely and determines the properties of the final polymer or oligomer.

There are two main synthetic approaches that make it possible to obtain functional phosphazenes capable of forming covalent bonds with epoxy matrices:

(1) Synthesis of organophosphazenes with reactive epoxy groups for addition to the epoxy component [15–40].
(2) Synthesis of organophosphazenes with reactive amino groups for use as a hardener or its component [41–49].

Epoxyphosphazenes are well-compatible with the epoxy matrix and at the same time improve its mechanical properties and heat resistance, probably due to the formation of a special three-dimensional polymer network at the junction nodes, where phosphazene cycles are located [50].

Previously, the synthesis of epoxyphosphazenes was primarily scientific due to the complexity of scaling and the large number of intermediate steps [15–19,21,25,27,28,30–33].

At present, a number of works have appeared that present fairly simple synthesis methods, with a level of manufacturability close to the production processes of basic epoxy resins [20,22–24,26,29,34,37]. Some of them, such as alkoxyphosphazenes, are thermally unstable [51]. Thus, the synthesis of aromatic organophosphazenes is preferred. Examples of such processes are one-stage methods for obtaining: (a) phosphazene-containing epoxy resins based on bisphenol A [52,53] and (b) based on resorcinol [54]. The resulting PhER consisted mainly of tetra- and penta-epoxyphosphazenes and an organic epoxy monomer, which is essentially an active diluent that reduces the viscosity and average functionality to a level acceptable for further processing, and these PhER demonstrated increased mechanical properties and heat resistance. Due to the fact that the viscosity of such resins at ambient temperature is more than 200 Pa·s, which is a value close to the processing limit, studies were carried out to obtain PhER based on bisphenol A and phenol, including the one-stage method, with the purpose of reducing viscosity and increasing the phosphorus content to reduce flammability [50,55].

In order to obtain PhER with a lower viscosity, which ensures the processability of processing, and a higher phosphorus content, for a potential reduction in flammability, in comparison with PhER described in [52–55], while maintaining their advantages in mechanical properties, PhER were synthesized by direct interaction of HCP, bisphenol F, and epichlorohydrin. To assess the impact on the physical properties of the use of bisphenol F and the content of the phosphazene fraction, the analysis of the composition of the obtained phosphazene-containing epoxy resins and tests for mechanical properties, glass transition temperature, and viscosity were carried out.

2. Materials and Methods

In this article, PhER were synthesized by a single-stage interaction of hexachlorocyclotriphosphazene with an isomer mixture of bisphenol F, in the medium of epichlorohydrin, which acts both as a reagent and as a solvent (Figure 1). The epoxy group formed by the catalyst and the HCl acceptor NaOH was used in one case in solid form, and in the other in the form of an aqueous solution.

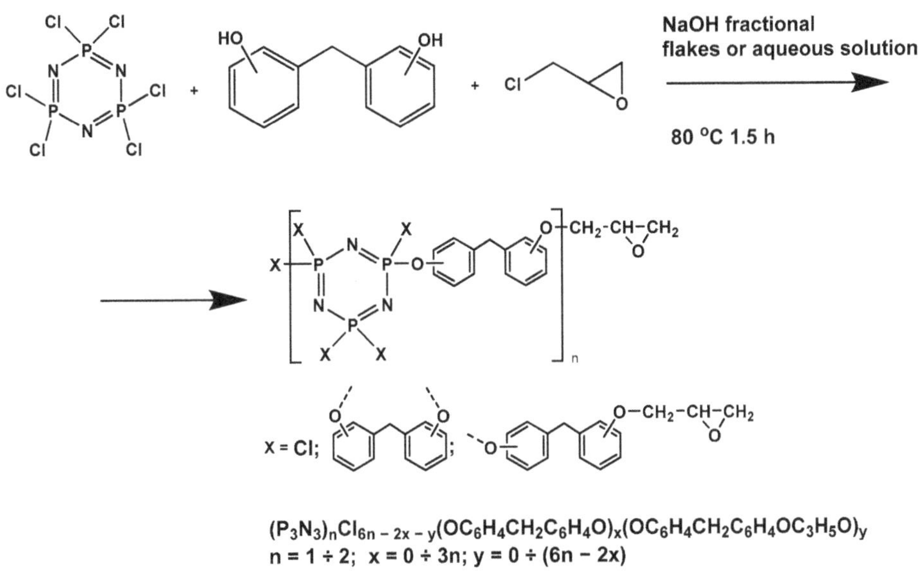

Figure 1. Single-stage synthesis of phosphazene-containing epoxy resins by direct interaction of hexachlorocyclotriphosphazene (HCP) and bisphenol F in epichlorohydrin medium (this work).

2.1. Materials

Hexachlorocyclotriphosphazene (HCP), a white crystalline substance ($T_{m.p.}$ = 113.0 °C; ^{31}P NMR spectrum, singlet with δ_P = 19.9 ppm), was obtained by the method in [56].

Bisphenol F (BPF), a white powder (mixture of isomers 2,2′-, 2,4′-, and 4,4′-dioxydiphenyldimethylmethane, $T_{m.p.}$ = 162–164 °C) (Manufacturer: Hangzhou Sartort Biopharma Co., Ltd., Hangzhou, China).

Epichlorohydrin (ECH) is a colorless liquid, the content of the main substance is 99.8%, and it was used after preliminary distillation, $_{b.p.}$ = 116 °C (Manufacturer: Solvay, Tavaux, France).

Potassium hydroxide, in the form of white flakes (the content of the main substance is 90%), was used without purification. The water content was determined by acid-base titration and was about 10% (Manufacturer: JSC "Caustic" Volgograd, Russia).

Acetone is a colorless liquid, and the content of the main substance is 99.8%. It was used as a solvent during isolation without preliminary purification (Supplier: Component-Reaktiv LLC, Moscow, Russia).

Diglycidyl ether of bisphenol A (DGEBA) is a clear-cloudy viscous liquid at room temperature. Commercial-grade KER 828 was used without pre-treatment (Manufacturer: Kumho P & B Chemicals, Inc., Seoul, South Korea).

Diglycidyl ether of bisphenol F (DGEBF) is a clear-cloudy viscous liquid at room temperature. The commercial brand Ipox er 1016 was used without preliminary purification (Supplier: NORTEX LLC, Moscow, Russia).

Diaminodiphenylsulfone (DDS) (Aradur 9664-1) is a fine powder with a particle size smaller than 64 μm and $_{m.p.}$ = 175 °C (Manufacturer: Huntsman Corporation, Spain).

2.2. Synthetic Methods

2.2.1. Synthesis of Phosphazene-Containing Resins Based on Bisphenol F (PNA-BPF) with Batch Introduction of Solid Alkali NaOH

HCP, BPF, and epichlorohydrin were loaded into a 2 L batch reactor equipped with a mechanical stirrer and a direct condenser cooled by recycled water in HCP:BPF ratios from 1:8 to 1:24 with a constant excess of 16-fold ECH (Table 1).

Table 1. The amounts of initial reagents for the synthesis of PhER.

Sample	Molar Ratio HCP:BPF:ECH	HCPg (Mole)	BPFg (Mole)	ECHmL (Mole)	NaOHg (Mole)
PNA-BPF-S	1:8:16	35 (0.101)	161.11 (0.805)	1387.865 (17.606)	64.363 (1.609)
PNA-BPF-1	1:8:16	35 (0.101)	161.11 (0.805)	1387.865 (17.606)	64.363 (1.609)
PNA-BPF-2	1:9:16	35 (0.086)	181.249 (0.905)	1639.022 (20.917)	72.408 (1.810)
PNA-BPF-3	1:12:16	23 (0.066)	158.809 (0.793)	1574.182 (20.096)	63.444 (1.586)
PNA-BPF-4	1:18:16	15 (0.043)	155.356 (0.776)	1675.045 (21.376)	62.064 (1.552)
PNA-BPF-5	1:24:16	10.5 (0.030)	144.999 (0.724)	1626.414 (20.757)	57.927 (1.448)

The reaction mixture was heated with stirring to a temperature of 45–50 °C and kept at this temperature until complete dissolution of the solid reagents. Then, the mixture was heated to 77 °C, after which the first portion of sodium hydroxide was loaded in the form of a powder, then at intervals of 5 min, but taking into account that the temperature of the reaction mixture should not exceed 85 °C, the rest of the sodium hydroxide was loaded in equal portions over 1/8 of the total. Sodium hydroxide was taken in a stoichiometric amount. After loading the entire amount of sodium hydroxide, the reaction was carried out for 50 min at a temperature of 80 °C, distilling off the epichlorohydrin–water mixture in a vacuum.

At the end of the reaction, from the resulting suspension, ECH was distilled off on a rotary evaporator, and then the resulting mixture of resin and salt was dissolved in an excess of acetone, and the solution was subsequently filtered from the precipitate through a paper filter, after which acetone was distilled off from the filtrate on a rotary evaporator. The product was a yellow viscous liquid. The yield was 91.2–94.3%.

2.2.2. Synthesis of Phosphazene-Containing Resins Based on Bisphenol F (PNA-BPF-S) with Batch Introduction of NaOH Alkali Solution

The synthesis of PNA-BPF-S was carried out in exactly the same way as the synthesis of PNA-BPF, except that the alkali was loaded in portions in the form of a solution (Table 1). The yield was 76%.

2.3. *Methods of Analysis*

2.3.1. Epoxy Group Content

The epoxy group content was determined by back acid-base titration in acetone according to GOST R 56752-2015 (ISO 3001:1999).

2.3.2. Elemental Analysis

Elemental analysis was carried out by X-ray fluorescence spectrometry with calibration according to the method of fundamental parameters on an X-ray fluorescence spectrometer, ARL PFX-101 (Thermo Fisher Scientific (Ecublens) SARL, Ecublens, Switzerland).

2.3.3. Nuclear Magnetic Resonance Spectroscopy

^{31}P NMR spectroscopy was performed on a Bruker AV600 at a working purity of 242 MHz. ^{1}H NMR spectroscopy was carried out on a Bruker AV600 instrument at a working purity of 600 MHz. Deuterochloroform was used as a solvent.

2.3.4. Mass Spectrometric Analysis of MALDI-TOF

Mass spectrometric analysis of MALDI-TOF was carried out on a Bruker Auto Flex II instrument.

2.3.5. Rheology

The temperature profile of viscosity was measured on an MCR 302 rheometer (Anton Paar, Graz, Austria). The measurements were carried out in the continuous shear mode with the measuring plate–plate geometry, shear rate of 10 s^{-1}, and gap of 1 mm. Top plate diameter was 25 mm.

2.4. *Method of Curing*

The calculated amount of hardener, DDS, was added to the epoxy resin, based on the stoichiometric ratio. The resulting composition was stirred and degassed on an oil bath rotor at 125 °C under a vacuum of 100 mm Hg for 20–40 min until the mixture was completely homogenized. The molds were degreased with acetone before pouring and then treated with an anti-adhesive liquid. All samples were cured at 180 °C for 8 h.

2.5. *Methods of Testing*

2.5.1. Tensile Testing

Tensile tests were carried out on a 50ST (Tinius Olsen, Redhill, United Kingdom) universal electromechanical tensile testing machine according to GOST 11262-2017 (ISO 527-2:2012). Samples of the cured binder were cut from a 4 mm-thick plate on a milling machine in accordance with the paragraph 6.1. standard (Figure 2). The stretching speed was 10 mm/min. From the data obtained, the values of tensile strength, tensile modulus, and elongation were calculated.

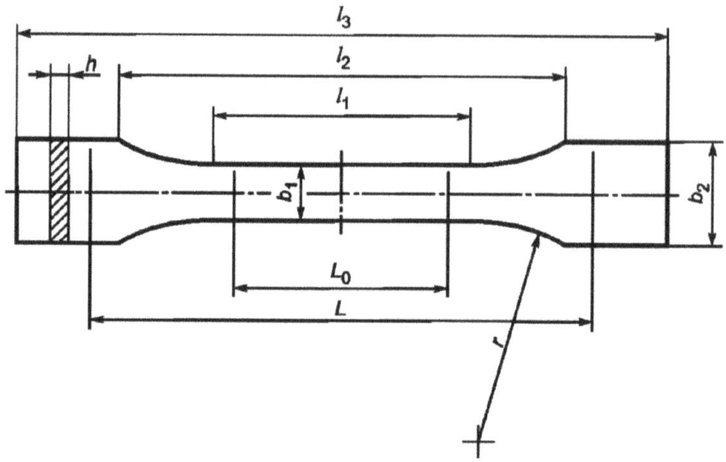

Figure 2. Tensile test sample. Dimensions, in mm: l1 = 80 ± 2, l2 = 109.3 ± 3.2, l3 = 170, L0 = 75 ± 0.5, L = 115 ± 1, b1 = 10 ± 0.2, b2 = 20 ± 0.2, h = 4 ± 0.2, r = 24 ± 1.

2.5.2. Flexural Testing

Flexural tests were carried out on a 50ST electromechanical universal tensile testing machine (Tinius Olsen, Redhill, United Kingdom) according to GOST R 56810-2015 (ASTM D790-10) at a speed of 0.85 mm/min. Samples of the cured binder were cut from a 3.6 mm-thick plate on a milling machine in accordance with the paragraph 6.1. standard. The samples were bars with an aspect ratio of length × width × thickness equal to 80 × 9.8 × 3.6 mm. Span-to-depth ratio was 64:3.6 mm. From the data obtained, the values of flexural strength and flexural modulus were calculated.

2.5.3. Izod Impact Test

Izod impact tests were carried out on an Izod-Charpy K 1053 machine (ATS FAAR, Cassina De' Pecchi, Italy) according to GOST 19109-2017 (ISO 180:2000), "Plastics. Izod impact strength method", which is a modified ISO 180:2000 standard. The samples were bars with an aspect ratio of length × width × thickness equal to 80 × 9.8 × 3.6 mm.

2.5.4. Glass Transition Temperature

The glass transition temperature was determined on a DSC 204 F1 Phoenix differential scanning calorimeter (Netzsch, Selb, Germany) according to GOST R 55135-2012 (ISO 11357-2:1999). Under the dynamic heating mode, the temperature range was 40–270 °C, and the heating rate was 10 °C/min. All tests were carried out under a nitrogen atmosphere with a flow rate of 60–100 mL/min. The weight of the samples was 5–10 mg.

The device was calibrated for temperature and enthalpy for standard metals gallium, indium, tin, bismuth, and zinc.

The analysis of parameters and characteristic points was carried out in the Proteus thermal analysis software.

3. Results and Discussion

3.1. Synthesis and Study of PNA-BPF and PNA-BPF-S

Analysis of the composition and structure of the obtained PhER was carried out using MALDI-TOF spectrometry and ^{31}P NMR spectroscopy.

MALDI-TOF analysis showed the presence in the products of compounds with *m*/*z* values and possible structures, as presented in Table 2.

Table 2. Possible structures and formulas of the main compounds formed during the synthesis of PNA-BPF and PNA-BPF-S.

m/*z*	Structure and Formula [1]	No. of Structure
693	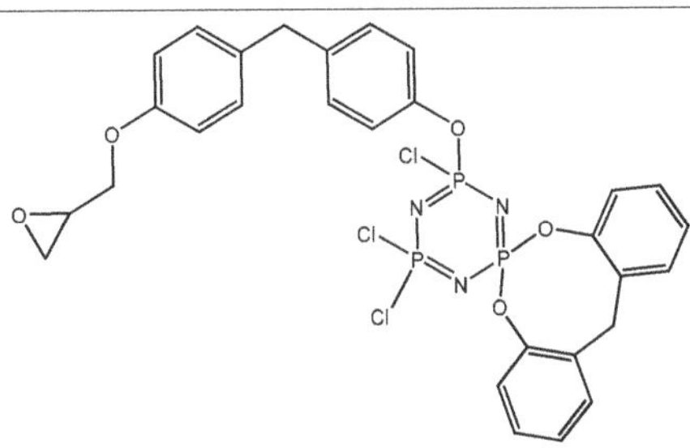 $P_3N_3Cl_3(OArO)(OArOGly)$	I
822	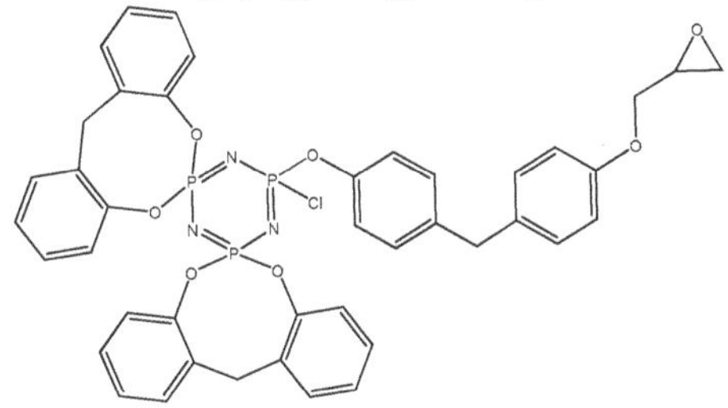 $P_3N_3Cl(OArO)_2(OArOGly)$	II
914	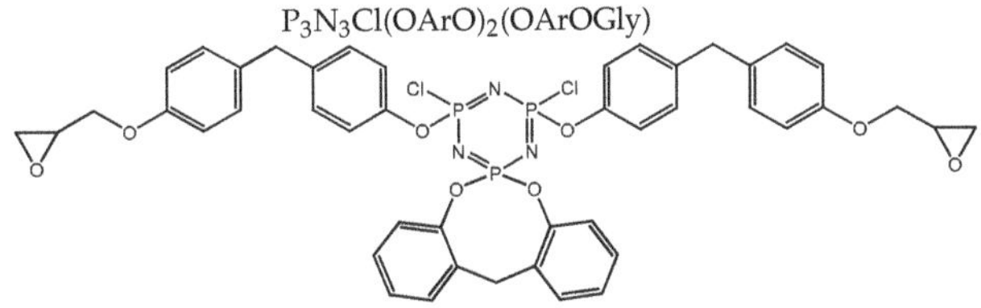 $P_3N_3Cl_2(OArO)(OArOGly)_2$	III
985	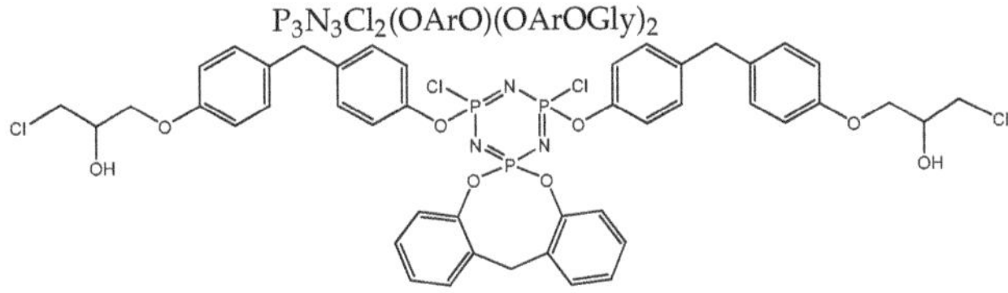 $P_3N_3Cl_2(OArO)(OArOGly')_2$	IV
1007	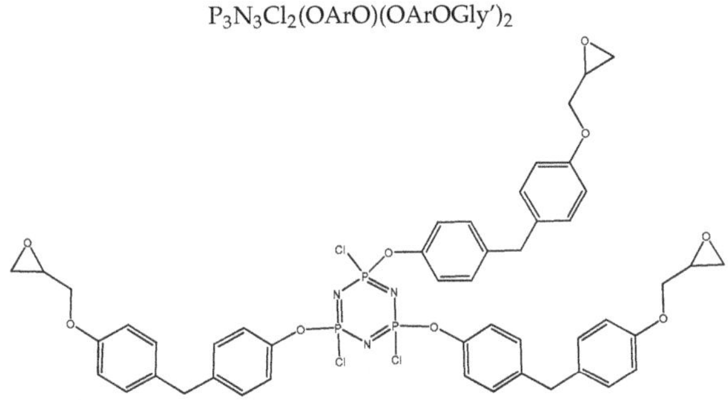 $P_3N_3Cl_3(OArOGly)_3$	V

Table 2. *Cont.*

m/z	Structure and Formula [1]	No. of Structure
1043	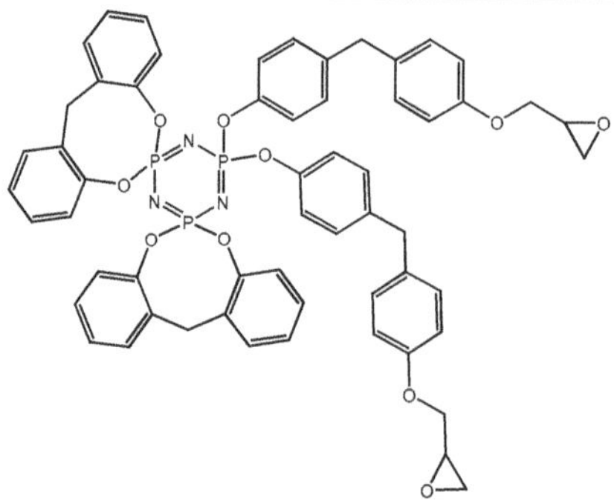 $P_3N_3(OArO)_2(OArOGly)_2$	VI
1135	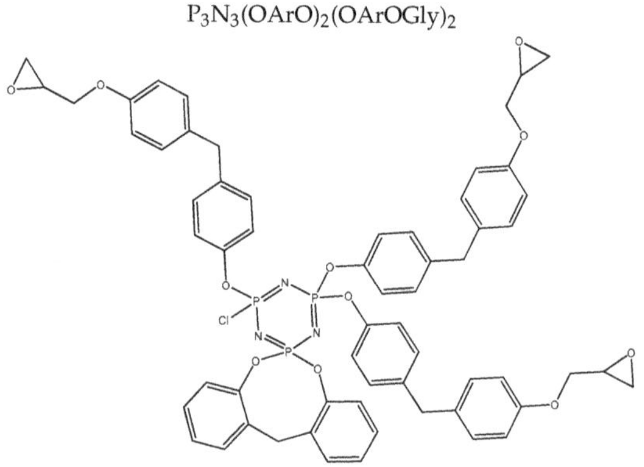 $P_3N_3Cl(OArO)(OArOGly)_3$	VII
1206	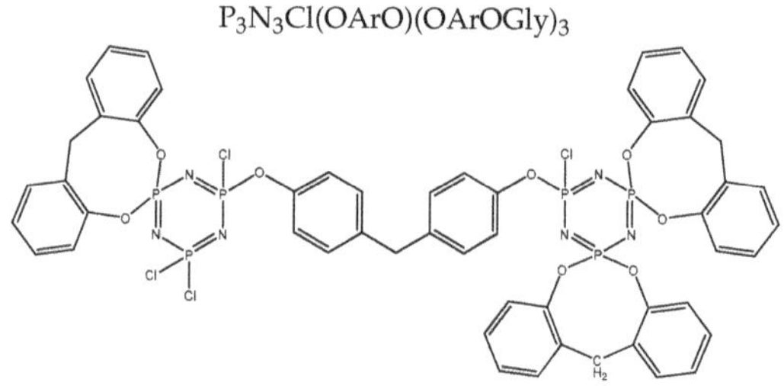 $(P_3N_3)_2Cl_4(OArO)_4$	VIII
1298	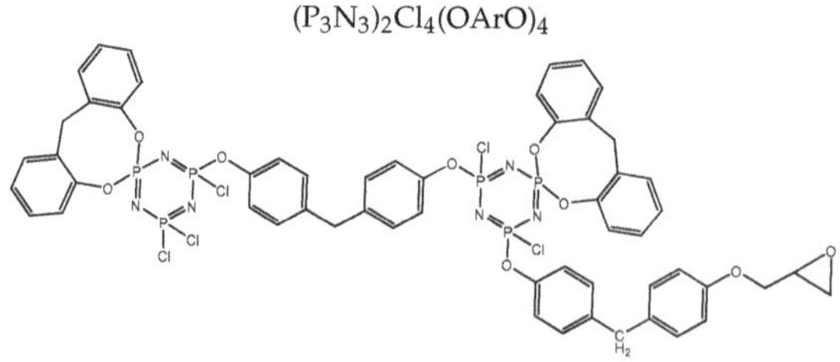 $(P_3N_3)_2Cl_5(OArO)_3(OArOGly)$	IX

Table 2. *Cont.*

m/z	Structure and Formula [1]	No. of Structure
1357	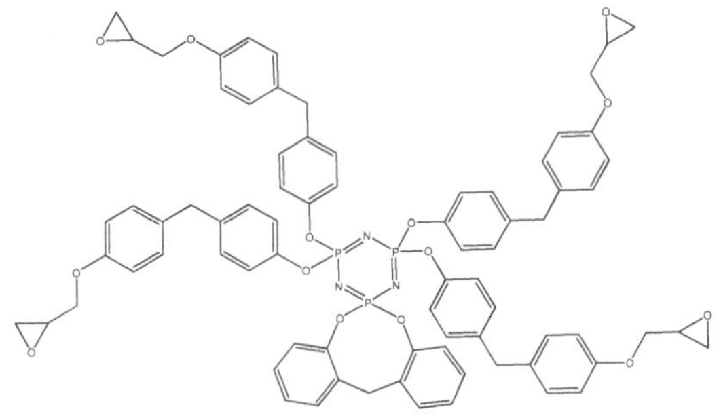 $P_3N_3(OArO)(OArOGly)_4$	X

[1] AR = 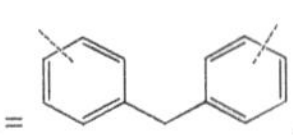, Gly = 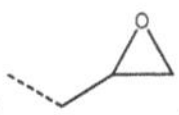, Gly′ = 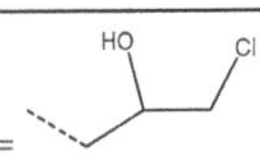.

The formation of VII-membered spirocyclic fragments at the phosphorus atom with the participation of the 2,2-isomer of bisphenol F in almost every component of the product mixture is noteworthy. This leads to a decrease in the average functionality of the mixture in comparison with epoxyphosphazenes based on bisphenol A and resorcinol, which can favorably affect the pot life of the resins.

MALDI-TOF analysis is not an unambiguously quantitative method of analysis without confirmation of its results by other methods, however it can be used to determine the qualitative presence of compounds and trends in their formation depending on the conditions. According to the results of MALDI-TOF, presented in Tables 3 and 4 and Figures 3 and 4, there are pronounced extreme dependencies of the formation of compounds with structures I and V with peaks at a ratio of 1:12. A less pronounced increase in the formation of compounds VI, VIII, and X, and a decline for compounds III, VII, and IX, were observed with peaks at a ratio of 1:18. The amounts of compounds II and IV fluctuate without unambiguous dependencies. Extreme values of the formation of substances are typical for synthesis 4 with a ratio of 1:18.

Table 3. The main compounds formed during the synthesis of PNA-BPF and PNA-BPF-S, the share of which is more than 4% in at least one of the experiments.

m/z	Formula	Content (wt.%)				
		PNA-BPF-1	PNA-BPF-2	PNA-BPF-3	PNA-BPF-4	PNA-BPF-5
693	$P_3N_3Cl_3(OArO)(OArOGly)$ (I)	0	0	8	0	0
822	$P_3N_3Cl(OArO)_2(OArOGly)$ (II)	7	9	8	10	9
914	$P_3N_3Cl_2(OArO)(OArOGly)_2$ (III)	6	5	3	2	4
985	$P_3N_3Cl_2(OArO)(OArOGly')_2$ (IV)	5	6	6	7	6
1007	$P_3N_3Cl_3(OArOGly)_3$ (V)	0	1	5	1	0
1043	$P_3N_3(OArO)_2(OArOGly)_2$ (VI)	3	3	7	10	7
1135	$P_3N_3Cl(OArO)(OArOGly)_3$ (VII)	21	16	14	10	15
1206	$(P_3N_3)_2Cl_4(OArO)_4$ (VIII)	2	3	4	7	5
1298	$(P_3N_3)_2Cl_5(OArO)_3(OArOGly)$ (IX)	8	7	6	5	7
1357	$P_3N_3(OArO)(OArOGly)_4$ (X)	2	2	3	5	4
Total proportion of major compounds		54	52	64	57	57

Table 4. Possible structures and formulas of hexa-substituted phosphazenes formed during the synthesis of PNA-BPF.

Structure and Formula	No. of Structure
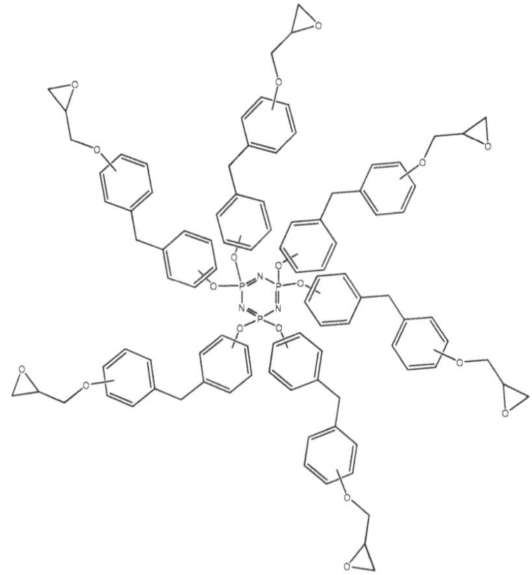	XI
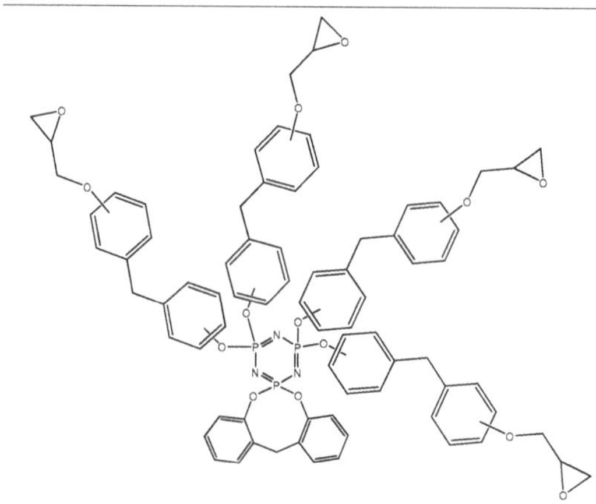	XII

Table 4. *Cont.*

Structure and Formula	No. of Structure
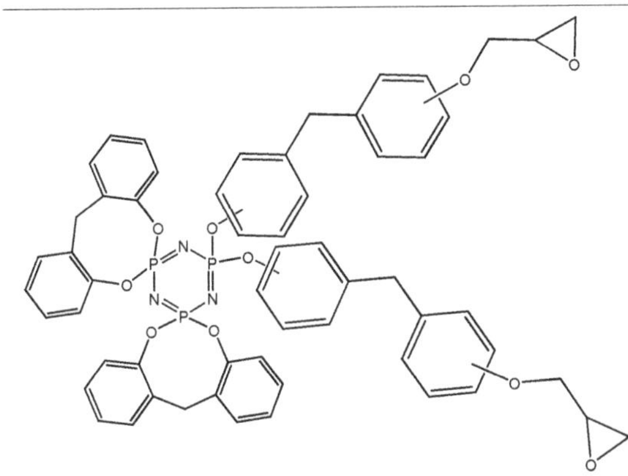	XIII
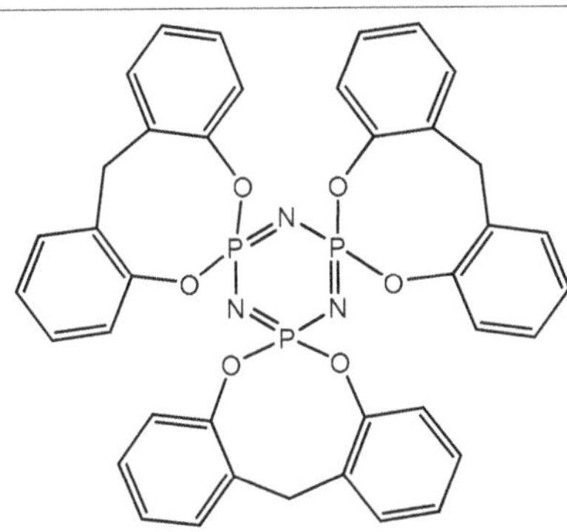	XIV
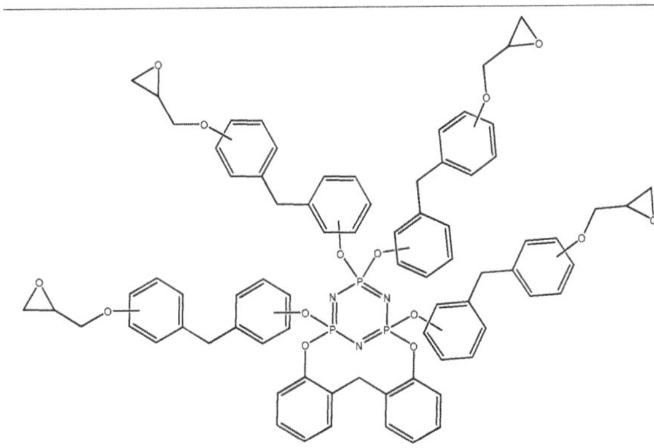	XV
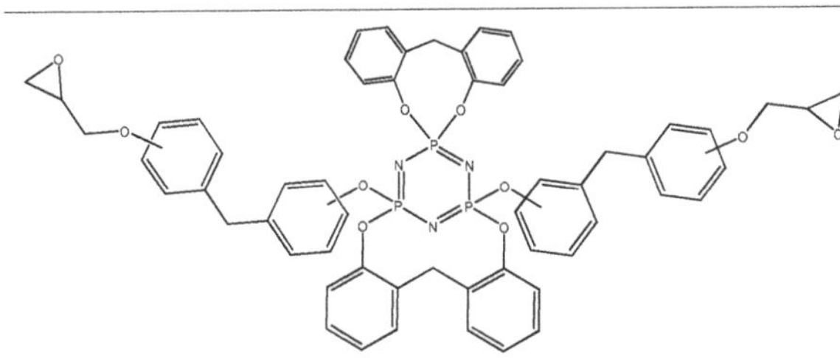	XVI

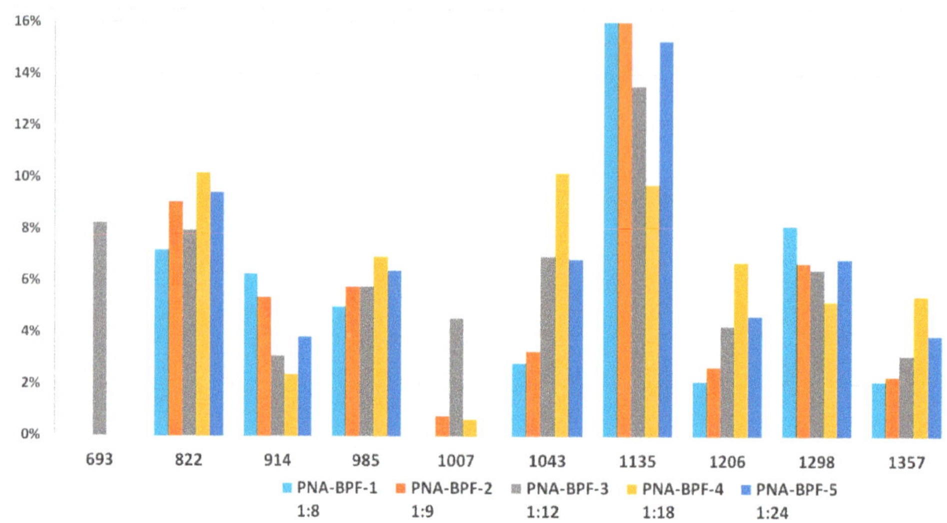

Figure 3. The content of compounds I–X (Table 2) in samples PNA-BPF-1 ÷ 5 ((1)–(5), respectively).

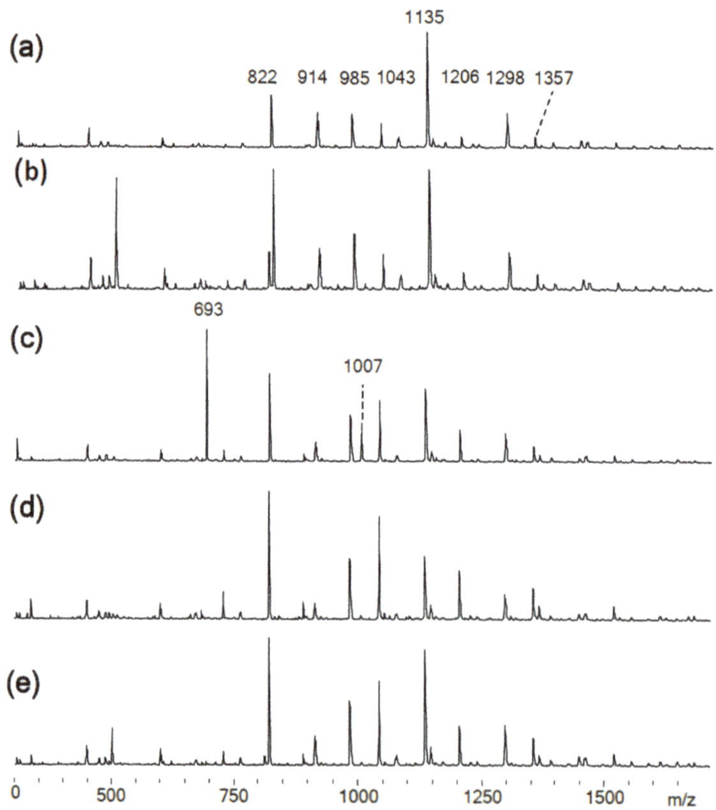

Figure 4. (**a–e**) MALDI-TOF spectra of the obtained PhER syntheses of PNA-BPF-1 ÷ 5, respectively.

^{31}P NMR spectra of the final product (Figure 5) indicate incomplete substitution and the formation of a mixture of tetra-, penta-, and hexa-derivatives of hexachlorocyclotriphosphazene with a predominance of penta-derivatives. Due to the presence of singlet signals in the region of 7.07, 7.51, and 8.08 ppm, it can be assumed that several types of hexa-substituted triphosphazenes are formed (Table 4). However, due to steric difficulties, the formation of compounds XIV, XV, and XVI is unlikely; therefore, the three signals mentioned above probably correspond to compounds XI, XII, and XIII. The presence of compound XIII is also confirmed by MALDI-TOF analysis and corresponds to compound VI. As the BPF excess increases, the yield of hexa-substituted derivatives with a singlet signal of 8.08 ppm decreases, and the yield of hexa-substituted derivatives of the structure increases with signals at 7.51 and 7.07. The penta-derivative penta I corresponds to the doublet (4.88–5.21 ppm) and triplet (20.08–20.75 ppm) systems. The penta-derivative penta II corresponds to the doublet (5.64–5.99 ppm) and triplet (20.75–21.48 ppm) systems. In the penta-derivatives penta I and penta II, the triplet peaks in the region of 20.78 ppm overlap, which is confirmed by the values of the integral areas—the integral area under the peak with overlap is equal to the sum of the areas under the peaks in the region at 21.48 and 20.08 ppm. A tetrasubstituted product corresponds to a triplet (2.96 ppm) and a doublet (19.41–19.74 ppm) signal system. Derivatives of hexa I, II, III, and penta I and II may not be individual substances, but a mixture of derivatives with the same degree of substitution, but different substituents—epoxy, spirocyclic, and bridging—which produce signals in one or close intervals, which can lead to their overlap.

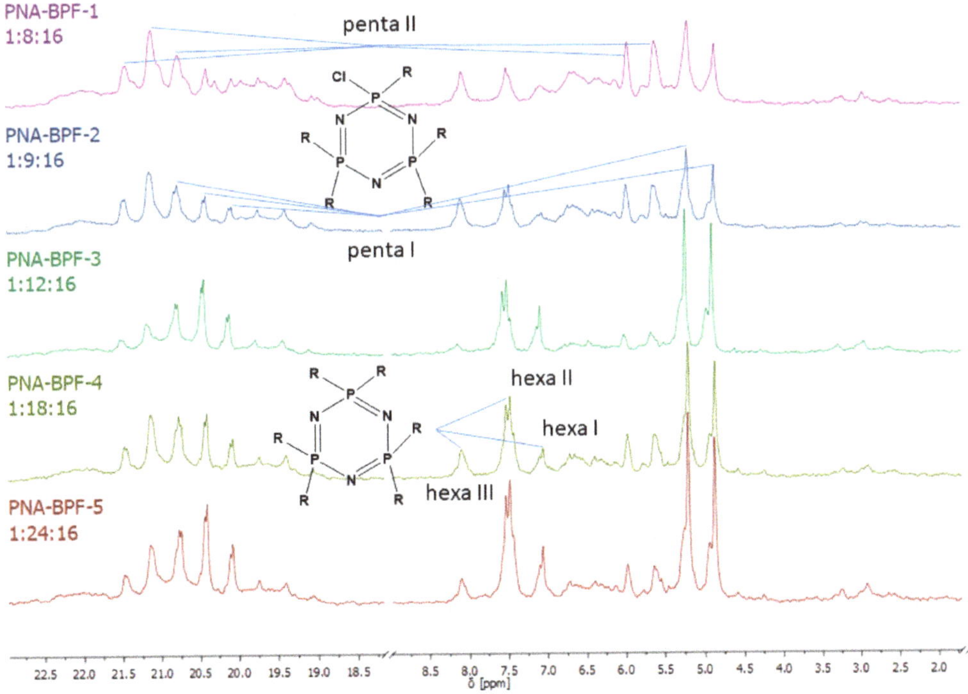

Figure 5. ^{31}P NMR spectra of PNA-BPF.

^{1}H NMR spectra of the obtained epoxy oligomers (Figure 6) are similar to the spectra for dianic epoxy resins (industrial grades ED-20, KER-828, and their analogues); however, the signals in the region of 3.55–3.95 ppm indicate the presence of a small amount of glycol end groups formed as a result of the hydrolysis of epoxy groups, apparently caused by the presence of residual alkali in the reaction medium, which was not completely consumed due to incomplete substitution of chlorine atoms in hexachlorocyclotriphosphazene. The results of the integration of signal systems are consistent with theoretical calculations.

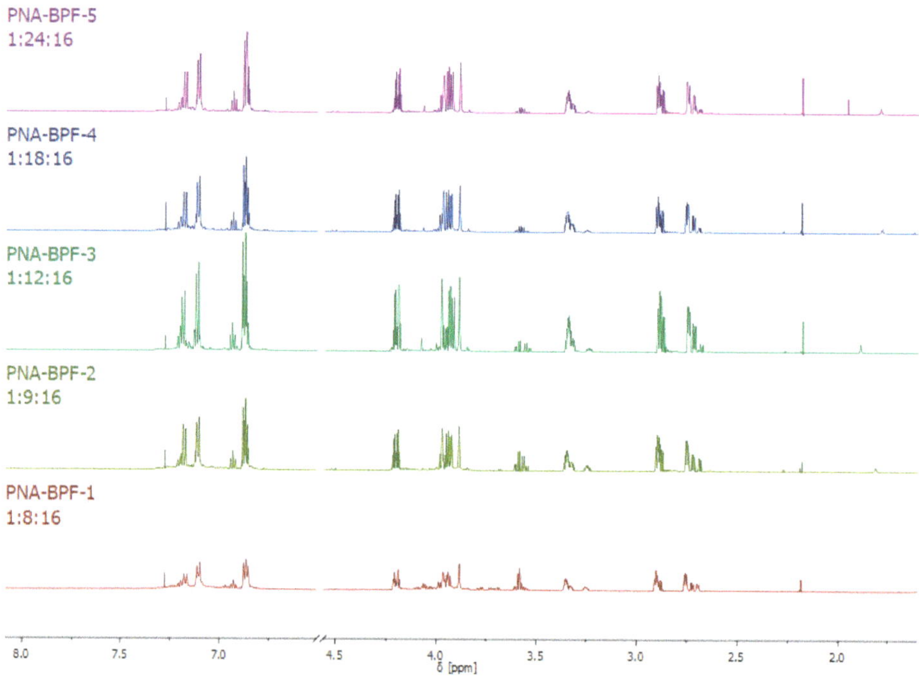

Figure 6. ^{1}H NMR spectra of PNA-BPF.

Table 5 shows the PhER yields depending on the ratios of the initial reagents. The values of the yields of syntheses 1–5 are close and are associated with losses during filtration from salts during the isolation process. The lower yield in the synthesis of PNA-BPF-S-1 is associated with the formation of sparingly soluble oligomers, which were deposited on the filter along with the salt during filtration, which also affected the deviation in the epoxy number.

Table 5. Yield of the obtained PhER.

Sample	Molar Ratio HCP:BPF:ECH	Yield	
		g	%
PNA-BPF-S-1	1:8:16	180	76.0
PNA-BPF-1	1:8:16	216	91.2
PNA-BPF-2	1:9:16	247	92.2
PNA-BPF-3	1:12:16	223	94.0
PNA-BPF-4	1:18:16	220	93.5
PNA-BPF-5	1:24:16	209	94.3

Table 6 presents the technological characteristics of various base and phosphazene-containing epoxy resins. Among all pairs of base and phosphazene-containing epoxy resins based on the same diphenols, there is a pattern of increase in viscosity and decrease in the epoxy number in PhER. An exception can be considered PhER obtained by the interaction of HCP with BPA and phenol. By using phenol, this type of resin reduces the viscosity and increases the phosphorus content, which can improve the fire resistance. However, at the same time, the epoxy number drops significantly for these resins, which will reduce the crosslink density during curing, and may worsen the mechanical properties. The described pattern is observed in the case of PhER based on BPF; first, when epoxyphosphazenes appear in the composition, the mechanical properties increase compared to the base resin, then, with an increase in the proportion of the phosphazene fraction and a decrease in the epoxy number, the mechanical properties either remain at the same level, or fall (Tables 7 and 8).

Table 6. Comparison of technological characteristics of base and phosphazene-containing epoxy resins based on basic diphenols and the content of phosphorus and chlorine between different PhER. The content of P and Cl wt.% in PhER obtained in this work, determined by X-ray fluorescence spectrometry.

Raw Reagents Ratio	Mixture Average Functionality	Content (wt.%)				Viscosity (Pa·s) at the Temperature of (°C)		
		Epoxy Group (%)	P	Cl	Phosphazene Fraction [1]	20	40	70
DGEBA								
	2.0	22.8	-	-	-	41.72	2.17	0.15
DGEBF								
	2.0	24.6–27.0	-	-	-	16.31	0.92	0.09
RDGE								
	2.0	34.4–36.4	-	-	-	1.10	0.11	0.03
HCP:BPA		PhER obtained by interaction of HCP with BPA and ECH [52,53,57]						
1:8	2.5	17.1	3.1	2.7	49	-	220	3
1:9 [2]	2.5	18.1	3.3	2.4	52	5855	85	2
1:12	2.3	20.0	1.8	1.5	30	-	130	2
1:16	2.2	21.4	1.5	1.3	25	440	78	2
HCP:Resorcinol		PhER obtained by interaction of HCP with resorcinol and ECH [54]						
1:12	2.4	21.0	4.0	4.4	43	8.33	6.15	0.36
1:16	2.3	28.6	3.0	2.4	32	2.43	1.94	0.15
1:24	2.2	29.6	2.0	1.9	21	1.71	0.45	0.05
HCP:PhOH:BPA		PhER obtained by interaction of HCP with BPA, phenol, and ECH [55]						
1:2:6	2.2	16.1	4.6	2.2	54	64.6	13.7	0.8
1:3:5	2.0	15.5	5.0	2.3	51	58.6	10.6	0.8
1:4:4	1.9	14.7	5.4	2.7	47	9.4	6.0	0.8
HCP:BPF:ECH		PhER obtained by interaction of HCP with BPF and ECH (this work)						
1:8:16 (NaOH solution)	2.8	15.2	3.7	4.2	56	1942.00	38.68	0.73
1:8:16	2.8	18.1	3.9	2.1	59	2463.11	45.03	1.04
1:9:16	2.7	20.5	3.4	1.9	52	229.24	7.45	0.34
1:12:16	2.4	20.4	2.7	1.6	41	282.99	10.65	0.43
1:18:16	2.2	22.8	1.8	1.4	28	181.12	4.69	0.22
1:24:16	2.2	21.5	1.5	1.3	22	62.65	2.16	0.13

[1] Values for this work are calculated based on the phosphorus content. [2] Synthesis at this ratio was performed within the framework of this work.

Table 7. Tensile and impact properties and glass transition temperature of the cured epoxy resins compared to base epoxy resins.

Sample (HCP:BPF:ECH)	Tensile Strength (MPa)	Tensile Modulus (GPa)	Tensile Elongation at Break (%)	Izod Impact Strength (kJ/m^2)	Tg (°C)
DGEBA	69.8	2.6	4.1	2.1	169.2
DGEBF	50.9	2.7	2.6	2.9	174.8
PNA-BPF-5 (1:24:16)	72.7	2.1	6.1	4.1	156.7
PNA-BPF-4 (1:18:16)	74.5	2.1	5.7	3.7	169.3
PNA-BPF-3 (1:12:16)	70.2	1.7	6.2	2.6	174.7
PNA-BPF-2 (1:9:16)	66.0	2.0	4.5	2.9	153.0
PNA-BPF-1 (1:8:16)	66.7	2.0	5.2	2.2	160.0
PNA-BPF-S (1:8:16)	29.4	2.6	1.2	1.4	141.8

Table 8. Flexural properties of the cured epoxy resins compared to base epoxy resins.

Sample (HCP:BPF:ECH)	Flexural Strength (MPa)	Flexural Modulus (GPa)	Flexural Elongation at Break (%)
DGEBA	125.6	3.02	6.99
DGEBF	121.8	3.13	9.84
PNA-BPF-5 (1:24:16)	131.8	3.28	9.15
PNA-BPF-4 (1:18:16)	133.4	3.27	8.19
PNA-BPF-3 (1:12:16)	130.6	3.17	8.23
PNA-BPF-2 (1:9:16)	146.6	3.63	6.66
PNA-BPF-1 (1:8:16)	128.0	3.35	5.73
PNA-BPF-S (1:8:16)	108.3	4.28	2.68

Figure 7 shows the temperature dependence of the viscosity of PhER compared to base epoxy resins. With a decrease in the higher molecular weight phosphazene component and a change in the substituent from bisphenol A to bisphenol F, when receiving PhER (Table 6), the viscosity also naturally decreases.

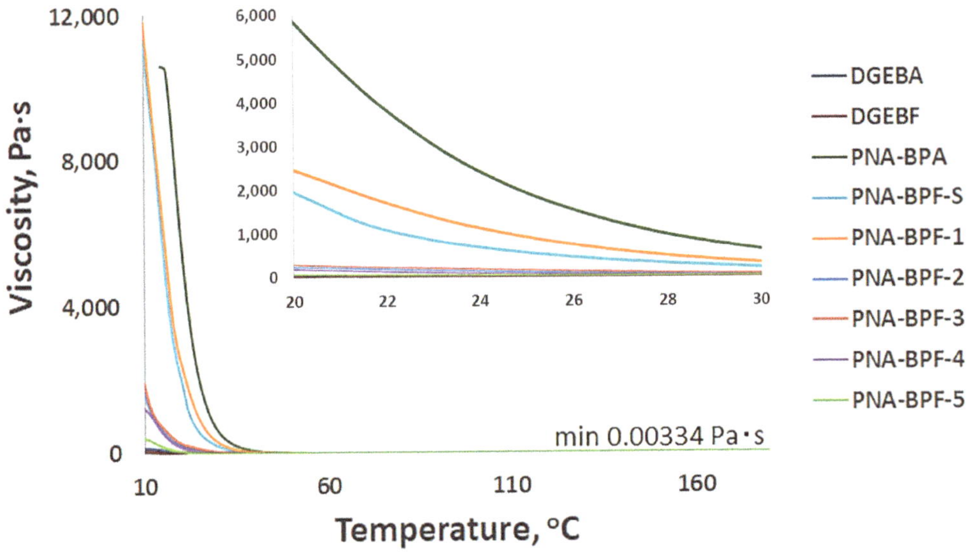

(**a**)

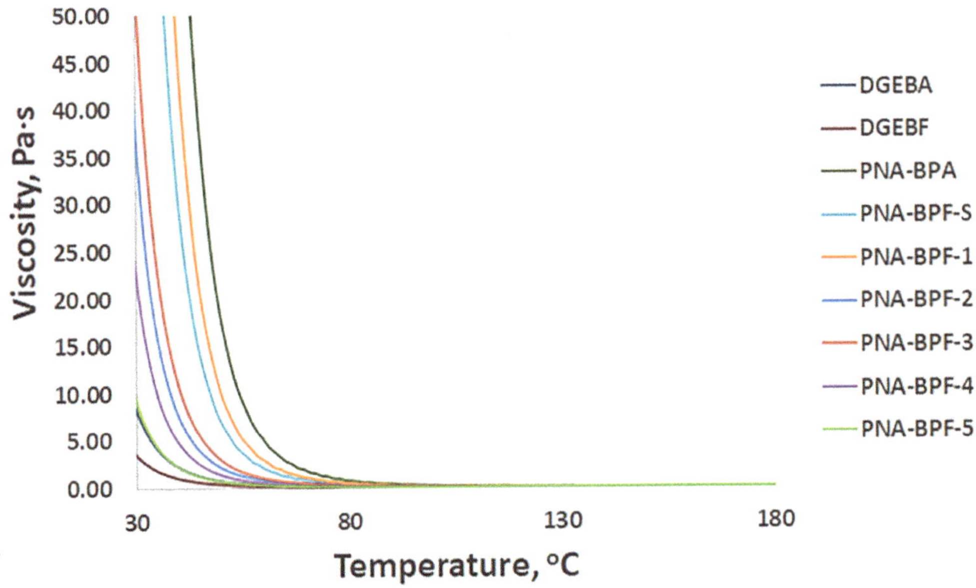

(**b**)

Figure 7. Temperature profile of viscosity versus temperature of obtained PhER, (**a**) from 10 to 180 °C and from 22 to 26 °C, and (**b**) from 30 to 170 °C, respectively. PNA-BPA—epoxy resin obtained by the method in [52] at a ratio of HCP:BPA equal to 1:9.

3.2. *Testing of Cured Compositions PNA-BPF and PNA-BPF-S*

With an increase in the degree of substitution along with an increase in the excess of bisphenol, which is described by the interpretation of ^{31}P NMR spectra (Figure 5), and the overall increase in the content of phosphazene, the dynamics of changes in the physical and mechanical properties of cured epoxy resins with diaminodiphenyl sulfone at 180 °C for 8 h are also complexly correlated. Extreme dependencies of the formation of a number of compounds and fluctuations in their content, which is described in Section 3.1 when interpreting the MALDI-TOF spectra, could affect the fluctuations in the values of mechanical properties (Tables 7 and 8).

With an increase in the phosphazene fraction, first, an increase in the strength and elasticity of the cured PhER during tensile testing was observed, then a tendency to a decrease in tensile modulus (Figure 8). In flexural tests, an increase in strength was also observed, while the trends in elasticity and flexural modulus were different to tensile tests: the flexural elongation at break continuously decreased with increasing content of the phosphazene fraction, while flexural modulus remained unchanged (Figure 9).

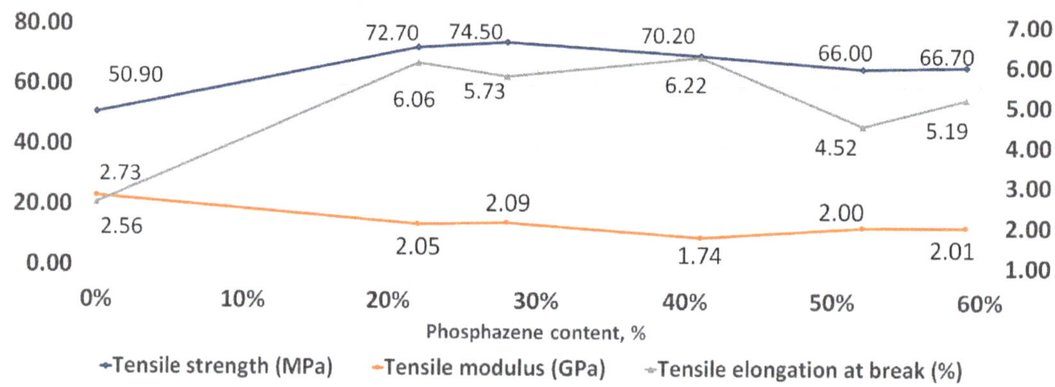

Figure 8. Tensile strength properties of cured resins.

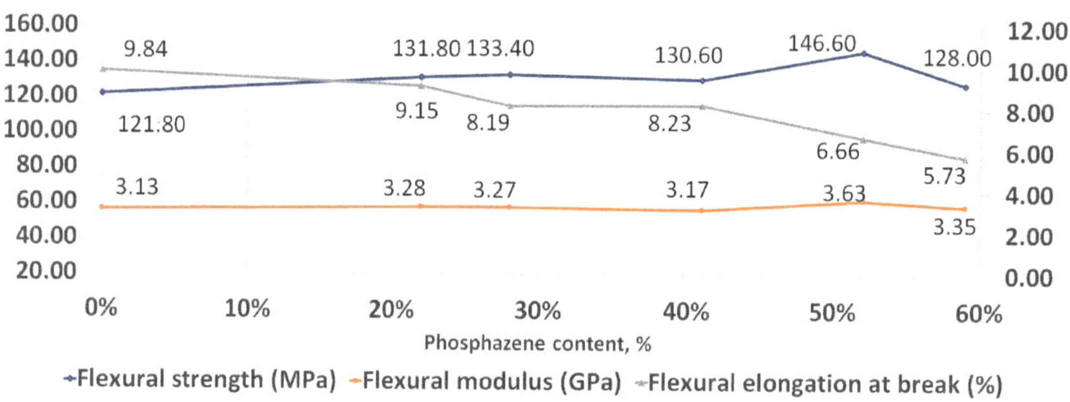

Figure 9. Flexural strength properties of cured resins.

Similar dependencies were also observed when testing for impact strength (Figure 10). With a decrease in the phosphazene fraction, the impact resistance increased to its maximum in the synthesis of PNA-BPF-5, but at the same time it decreased again for the DGEBF epoxy resin, which does not contain epoxyphosphazenes. Based on this, one can talk about the existing modifying property of epoxyphosphazenes based on bisphenol F with respect to the base DGEBF resin. The modifying effect is most likely maximal at the content of the phosphazene fraction in the range of 10–25%. A similar dependence was observed earlier in the study of the properties of PhER based on bisphenol A in [58].

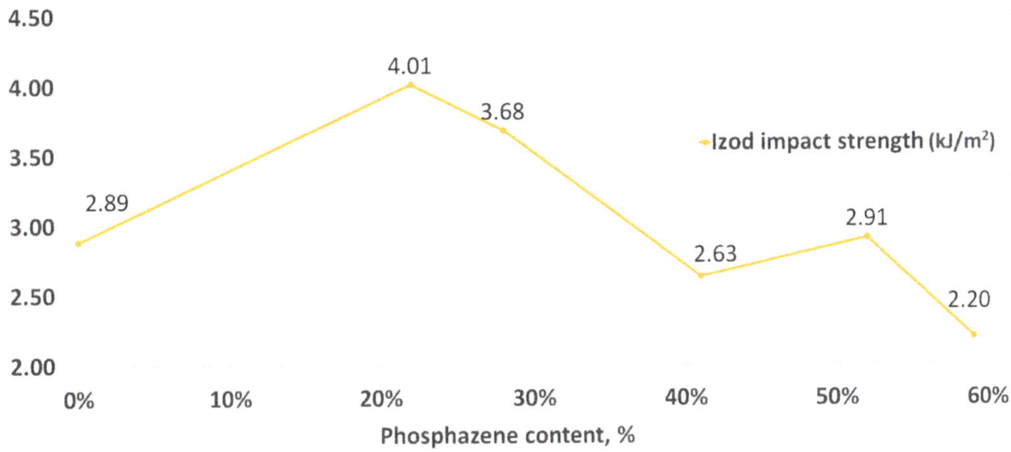

Figure 10. Impact strength according to Izod.

Supplementary Materials show graphs of percentage changes in the mechanical properties of neat and phosphazene-containing resins based on bisphenol F relative to DGEBA.

A different kind of dependence was observed for the glass transition temperature. Its values decreased compared to the DGEBF neat resin (Table 8), with a minimum content of the phosphazene fraction in PNA-BPF-5, after which they passed through a maximum in PNA-BPF-3 with indicators at the level of base epoxy resins, and then with an increased PhER content, continued to drop significantly, which was not observed for PhER of other types [58].

It is likely that with a further decrease in the content of the phosphazene fraction, the glass transition temperature will resume growth, which may coincide with the maxima of the values of impact strength, elasticity in tension and flexion, and viscosity, while maintaining or some increase in the tensile and flexural properties, according to comparison with the base resin DGEBF, which was observed for all experiments during synthesis on the solid alkali.

Despite the sharply different composition of PNA-BPF-4, corresponding to a phosphazene content of 28%, this was not reflected in the physical and mechanical properties. A sharp change in properties was observed immediately at the lowest phosphazene content of 22%, then the properties either remained at the same level or a tendency to decrease was observed. This may be due to a double modifying effect of the phosphazene content: factor 1 is an increase in the total phosphazene content, and factor 2 is a change in the ratio of the various epoxyphosphazenes (compounds I–X) formed depending on the proportions of the initial reagents. Comprehensively, we can speak of an important effect on the improvement of the properties of compounds II and VII, which constitute the main part for all syntheses. For the remaining compounds, more pronounced fluctuations and extreme dependencies in content were observed with an increase in the proportion of phosphazene, which could affect the general trend of a gradual or significant decrease in properties with an increase in the proportion of phosphazene after an initial sharp increase. The difficulty of assessing factor 2 lies in the need to obtain individual substances formed during the synthesis, and for this it is necessary to develop a method for separating the resulting mixture of epoxyphosphazene resins into individual substances, after which it will be possible to evaluate the modifying effect of the content of each epoxyphosphazene. This paper presents a comprehensive assessment of the effect of the phosphazene content in the production of PhER by a one-stage method, which may be of interest from the perspective of commercial production. Based on the results obtained during the tests, to find the best combination of properties, it is worth studying PhER with a content of 10–25% obtained by this method.

4. Conclusions

The phosphazene-containing epoxy resins obtained in this work have the content of epoxy groups of 15.2–22.8%, and phosphorus of 1.5–3.9%. With a decrease in the content of HCP in the initial reagents, an increase in the mechanical properties of the cured resulting epoxy resins was observed. The most successful in terms of their performance were samples with a ratio of initial reagents HCP:BPF:ECH of 1:12:16–1:24:16. They demonstrated an increase in impact strength by 40–80%, relative elongation at fracture in tension and in flexion by 50–100%, while maintaining glass transition temperature and tensile and flexural strength compared to base epoxy resins based on bisphenols A and F. These PhER can be cured with conventional hardeners to obtain materials with mechanical properties and heat resistance above the base commercial grades of epoxy resins and potentially lower combustibility due to the content of phosphorus.

Thus, phosphazene-containing epoxy resins based on bisphenol F, with a low content of the phosphazene fraction, can act as modifiers for traditional epoxy resins, being compatible with them, to increase impact resistance and elasticity while maintaining other basic mechanical and technological characteristics, and can be used as a binder component for composite materials, adhesives, and paints.

Supplementary Materials: The following supporting information can be downloaded at: https://www.mdpi.com/article/10.3390/polym14214547/s1. Figure S1. Flexural properties; Figure S2. Impact strength; Figure S3. Tensile properties.

Author Contributions: Conceptualization, I.S.S.; formal analysis, I.V.T. and I.S.S.; investigation, I.V.T., A.V.O. and R.S.B.; methodology, I.V.T. and I.S.S.; project administration, I.V.T.; writing—original draft, I.V.T.; writing—review and editing, V.V.K. and I.S.S. All authors have read and agreed to the published version of the manuscript.

Funding: This work is supported by the Russian Science Foundation under grant 22-73-10242.

Institutional Review Board Statement: Not applicable.

Informed Consent Statement: Not applicable.

Data Availability Statement: The corresponding author can be approached for the availability of data.

Acknowledgments: We thank Korokhin R.A., Korotkov R.F., Zhidal V.C., Kulemza D.S. and Shutov V.V. for technical assistance in the characterization of materials.

Conflicts of Interest: The authors declare no conflict of interest.

References

1. Jin, F.-L.; Li, X.; Park, S.-J. Synthesis and Application of Epoxy Resins: A Review. *J. Ind. Eng. Chem.* **2015**, *29*, 1–11. [CrossRef]
2. Dodiuk, H.; Goodman, S. *Handbook of Thermoset Plastics*; William Andrew: San Diego, CA, USA, 2014; Volume 254, ISBN 978-1-4557-3107-7.
3. Biron, M. *Thermosets and Composites: Material Selection, Applications, Manufacturing, and Cost Analysis*, 2nd ed.; Elsevier: Amsterdam, The Netherlands, 2013; p. 526.
4. Flick, E.W. *Epoxy Resins, Curing Agents, Compounds, and Modifiers: An Industrial Guide*; William Andrew: San Diego, USA, 2012; ISBN 978-0-8155-1708-5.
5. Petrie, E.M. *Epoxy Adhesive Formulations*; McGraw Hill Professional: New York, NY, USA, 2005; ISBN 978-0-07-158908-6.
6. Visakh, P.M.; Arao, Y. Flame Retardants: Polymer Blends, Composites and Nanocomposites. Springer: Berlin/Heidelberg, Germany, 2015; ISBN 978-3-319-03467-6.
7. Lu, S.-Y.; Hamerton, I. Recent Developments in the Chemistry of Halogen-Free Flame Retardant Polymers. *Prog. Polym. Sci.* **2002**, *27*, 1661–1712. [CrossRef]

8. Green, J. A Review of Phosphorus-Containing Flame Retardants. *J. Fire Sci.* **1992**, *10*, 470–487. [CrossRef]
9. Rodriguez, F.; Cohen, C.; Ober, C.K.; Archer, L.A. *Principles of Polymer Systems*, 6th ed.; CRC Press: Boca Raton, FL, USA, 2014; p. 775. ISBN 978-1-4822-2378-1.
10. Aoa, .. *фоооа аоооа о По аа*; оо а: аа, о, 2004; 372p.
11. Amirova, L.; Andrianova, K. Gradient Polymeric Materials Based on Poorly Compatible Epoxy Oligomers. *J. Appl. Polym. Sci.* **2006**, *102*, 96–103. [CrossRef]
12. Amirova, L.; Stroganov, V.; Sakhabieva, E. Optical Materials the Based on Low Flammable Epoxy Resins. *Int. J. Polym. Mater.* **2000**, *47*, 43–60. [CrossRef]
13. Horrocks, A.R.; Price, D. *Advances in Fire Retardant Materials*; Elsevier: Amsterdam, The Netherlands, 2008; p. 616.
14. Salmeia, K.A.; Gaan, S. An Overview of Some Recent Advances in DOPO-Derivatives: Chemistry and Flame Retardant Applications. *Polym. Degrad. Stab.* **2015**, *113*, 119–134. [CrossRef]
15. Fantin, G.; Medici, A.; Fogagnolo, M.; Pedrini, P.; Gleria, M.; Bertani, R.; Facchin, G. Functionalization of Poly(Organophosphazenes)—III. Synthesis of Phosphazene Materials Containing Carbon-Carbon Double Bonds and Epoxide Groups. *Eur. Polym. J.* **1993**, *29*, 1571–1579. [CrossRef]
16. Allcock, H.; Nelson, C.; Coggio, W. Photoinitiated Graft Poly(Organophosphazenes): Functionalized Immobilization Substrates for the Binding of Amines, Proteins, and Metals. *Chem. Mater.* **1994**, *6*, 516–524. [CrossRef]
17. Chen-yang, Y.W.; Lee, H.-F.; Yuan, C. Flame-Retardant Phosphate and Cyclotriphosphazene-Containing Epoxy Resin: Synthesis and Properties. *J. Polym. Sci. Part A Polym. Chem.* **2000**, *38*, 972–981. [CrossRef]
18. Bertani, R.; Boscolo-Boscoletto, A.; Dintcheva, N.; Ghedini, E.; Gleria, M.; La Mantia, F.; Pace, G.; Pannocchia, P.; Sassi, A.; Scaffaro, R.; et al. New Phosphazene-Based Chain Extenders Containing Allyl and Epoxide Groups. *Des. Monomers Polym.* **2003**, *6*, 245–266. [CrossRef]
19. Scaffaro, R.; Botta, L.; La Mantia, F.P.; Magagnini, P.; Acierno, D.; Gleria, M.; Bertani, R. Effect of Adding New Phosphazene Compounds to Poly(Butylene Terephthalate)/Polyamide Blends. I: Preliminary Study in a Batch Mixer. *Polym. Degrad. Stab.* **2005**, *90*, 234–243. [CrossRef]
20. el gouri, M.; el Bachiri, A.; Hegazi, S.E.; Rafik, M.; Elharfi, A. Thermal Degradation of a Reactive Flame Retardant Based on Cyclotriphosphazene and Its Blend with DGEBA Epoxy Resin. *Polym. Degrad. Stab.* **2009**, *94*, 2101–2106. [CrossRef]
21. Liu, R.; Wang, X. Synthesis, Characterization, Thermal Properties and Flame Retardancy of a Novel Nonflammable Phosphazene-Based Epoxy Resin. *Polym. Degrad. Stab.* **2009**, *94*, 617–624. [CrossRef]
22. el gouri, M.; Cherkaoui, O.; Ziraoui, R.; Elharfi, A. Physico-Chemical Study of DGEBA Epoxy Resin Flame Retarded with an: Ecological Flame Retardant Based on Cyclotriphosphazene. *J. Mater. Environ. Sci.* **2010**, *1*, 157–162.
23. el gouri, M.; el Bachiri, A.; Hegazi, S.E.; Rafik, M.; Elharfi, A. Fireproofing Amelioration of Epoxy Resin Material by Way a Reactive Flame Retardant Based on Cyclophosphazene. *Phys. Chem. News* **2010**, *56*, 128–137.
24. el gouri, M.; Hegazi, S.E.; Rafik, M.; Elharfi, A. Synthesis and Thermal Degradation of Phosphazene Containing the Epoxy Group. *Ann. De Chim. Sci. Des Mater.* **2010**, *35*, 27–39. [CrossRef]
25. Liu, F.; Wei, H.; Huang, X.; Zhang, J.; Zhou, Y.; Tang, X. Preparation and Properties of Novel Inherent Flame-Retardant Cyclotriphosphazene-Containing Epoxy Resins. *J. Macromol. Sci. Part B* **2010**, *49*, 1002–1011. [CrossRef]
26. el gouri, M.; el Bachiri, A.; Hegazi, S.E.; Ziraoui, R.; Rafik, M.; Elharfi, A. A Phosphazene Compound Multipurpose Application -Composite Material Precursor and Reactive Flame Retardant for Epoxy Resin Materials. *J. Mater. Environ. Sci.* **2011**, *2*, 319–334.
27. Gu, X.; Huang, X.; Wei, H.; Tang, X. Synthesis of Novel Epoxy-Group Modified Phosphazene-Containing Nanotube and Its Reinforcing Effect in Epoxy Resin. *Eur. Polym. J.* **2011**, *47*, 903–910. [CrossRef]
28. Bai, Y.; Wang, X.; Wu, D. Novel Cyclolinear Cyclotriphosphazene-Linked Epoxy Resin for Halogen-Free Fire Resistance: Synthesis, Characterization, and Flammability Characteristics. *Ind. Eng. Chem. Res.* **2012**, *51*, 15064–15074. [CrossRef]
29. el gouri, M.; Elharfi, A. Chemical Modification of Hexachlorocyclotriphosphazene—Preparation of Flame Retardants and Ecological Flame Retardant Polymers. *J. Mater. Environ. Sci.* **2012**, *3*, 17–33.
30. Liu, J.; Tang, J.; Wang, X.; Wu, D. Synthesis, Characterization and Curing Properties of a Novel Cyclolinear Phosphazene-Based Epoxy Resin for Halogen-Free Flame Retardancy and High Performance. *RSC Adv.* **2012**, *2*, 5789. [CrossRef]
31. Sun, J.; Wang, X.; Wu, D. Novel Spirocyclic Phosphazene-Based Epoxy Resin for Halogen-Free Fire Resistance: Synthesis, Curing Behaviors, and Flammability Characteristics. *ACS Appl. Mater. Interfaces* **2012**, *4*, 4047–4061. [CrossRef] [PubMed]
32. Feng, H.; Wang, X.; Wu, D. Fabrication of Spirocyclic Phosphazene Epoxy-Based Nanocomposites with Graphene via Exfoliation of Graphite Platelets and Thermal Curing for Enhancement of Mechanical and Conductive Properties. *Ind. Eng. Chem. Res.* **2013**, *52*, 10160–10171. [CrossRef]
33. Huang, X.; Wei, W.; Wei, H.; Li, Y.; Gu, X.; Tang, X. Preparation of Heat-Moisture Resistant Epoxy Resin Based on Phosphazene. *J. Appl. Polym. Sci.* **2013**, *130*, 249–255. [CrossRef]
34. el gouri, M.; Mansouri, A.; Elgouri, R.; Hadik, N.; Cherkaoui, O.; Outzourhit, A.; Elharfi, A. Physical Behaviour of Epoxy Resin Material Flame Retarded with a Reactive Flame Retardant Based on Cyclophosphazene. *J. Mater. Environ. Sci.* **2014**, *5*, 400–407.
35. Lu, L.; Chen, Y.; Wang, S.; Yang, S.; Dong, X. Preparation and Flame Retardancy of MMT Pattern Synergy Intumescent Flame-Retardant Epoxy Resin. *Gaofenzi Cailiao Kexue Yu Gongcheng/Polym. Mater. Sci. Eng.* **2014**, *30*, 139–144.
36. Xu, G.-R.; Xu, M.-J.; Li, B. Synthesis and Characterization of a Novel Epoxy Resin Based on Cyclotriphosphazene and Its Thermal Degradation and Flammability Performance. *Polym. Degrad. Stab.* **2014**, *109*, 240–248. [CrossRef]

37. Huan, L.; Wang, X.; Wu, D. Novel Cyclotriphosphazene-Based Epoxy Compound and Its Application in Halogen-Free Epoxy Thermosetting Systems: Synthesis, Curing Behaviors, and Flame Retardancy. *Polym. Degrad. Stab.* **2014**, *103*, 96–112. [CrossRef]
38. Lakshmikandhan, T.; Sethuraman, K.; Chandramohan, A.; Alagar, M. Development of Phosphazene Imine-Modified Epoxy Composites for Low Dielectric, Antibacterial Activity, and UV Shielding Applications. *Polym. Compos.* **2017**, *38*, E24–E33. [CrossRef]
39. Liu, H.; Wang, X.; Wu, D. Synthesis of a Novel Linear Polyphosphazene-Based Epoxy Resin and Its Application in Halogen-Free Flame-Resistant Thermosetting Systems. *Polym. Degrad. Stab.* **2005**, *118*, 45–58. [CrossRef]
40. Wu, Y.; Zhu, J.; Wang, X.; Shao, Z.; Zhao, D.; Zeng, X. Thermal decomposition kinetics and flame retardance of phosphazene-containing epoxy resin. *Polym. Mater. Sci. Eng.* **2016**, *32*, 54–58. [CrossRef]
41. Takahashi, K.; Yamamoto, T.; Itoh, S.; Harakawa, K.; Kajiwara, M. Mechanical properties of epoxy resins cured by various amines. *Zair. J. Soc. Mater. Sci. Jpn.* **1988**, *37*, 454–459. [CrossRef]
42. Yamamoto, T.; Takahashi, K.; Kon, Y.; Harakawa, K. Curing of Epoxy Resin with Phosphazene Derivatives. *Kobunshi Ronbunshu* **1988**, *45*, 851–856. [CrossRef]
43. Yamamoto, T.; Takahashi, K.; Kon, Y.; Kobayashi, K. Tensile Behavior and Heat Resistance of Epoxy Resin Cured with Phosphazene Derivatives. *Kobunshi Ronbunshu* **1989**, *46*, 177–181. [CrossRef]
44. Takahashi, K.; Ishikawa, N.; Komori, T.; Yoon, H.-S. Curing of an Epoxy Resin with P3N3(NH2)2 (OCH2CF3)4 and Its Mechanical Properties. *Kobunshi Ronbunshu* **1990**, *47*, 727–734. [CrossRef]
45. Takahashi, K.; Ishikawa, N.; Yoon, H.-S. Mechanical Properties of Epoxy Resins Cured with P3N3(NH2)2 (OC6H4Cl)4. *Kobunshi Ronbunshu* **1990**, *47*, 757–762. [CrossRef]
46. Takahashi, K.; Ishikawa, N.; Yoon, H.-S. Resistance of Epoxy Resins Cured with Trichloro-Tridimethylamino-Cyclotriphosphazene against Chemical Substances. *Zair. J. Soc. Mater. Sci. Jpn.* **1990**, *39*, 1001–1006. [CrossRef]
47. Takahashi, K.; Ishikawa, N.; Kohno, T.; Yoon, H.-S. Water Resistance of an Epoxy Resin Cured with P3N3Cl3(N(CH3)2)3 and the Effect of Glass Flake Reinforcement. *Zair. J. Soc. Mater. Sci. Jpn.* **1991**, *40*, 458–463. [CrossRef]
48. Takahashi, K.; Nakashima, J.; Ishiguro, S. Mechanical Properties of Trifunctional Epoxy Resin with Phosphazene Derivatives. *Kobunshi Ronbunshu* **1994**, *51*, 717–723. [CrossRef]
49. Chen, Y.M.; Liao, Y.L.; Lin, J.J. Synergistic Effect of Silicate Clay and Phosphazene-Oxyalkyleneamines on Thermal Stability of Cured Epoxies. *J. Colloid Interface Sci.* **2010**, *343*, 209–216. [CrossRef] [PubMed]
50. Terekhov, I.V.; Filatov, S.N.; Chistyakov, E.M.; Borisov, R.S.; Kireev, V.V. Synthesis of Oligomeric Epoxycyclotriphosphazenes and Their Properties as Reactive Flame-Retardants for Epoxy Resins. *Phosphorus Sulfur Silicon Relat. Elem.* **2017**, *192*, 544–554. [CrossRef]
51. Allcock, H. *Phosphorus-Nitrogen Compounds, Cyclic, Linear and High Polymeric System*; Elsevier: Amsterdam, The Netherlands, 1972; ISBN 978-0-12-050560-9.
52. Sirotin, I.S.; Bilichenko, Y.V.; Brigadnov, K.A.; Kireev, V.V.; Prudskov, B.M.; Borisov, R.S. Single-Stage Synthesis of Phosphazene-Containing Epoxy Oligomers. *Polym. Sci. Ser. B* **2014**, *56*, 471–476. [CrossRef]
53. Brigadnov, K.A.; Bilichenko, Y.V.; Polyakov, V.A.; Borisov, R.S.; Gusev, K.I.; Rudakova, T.A.; Filatov, S.N.; Kireev, V.V. Epoxy Oligomers Modified with Epoxyphosphazenes. *Polym. Sci. Ser. B* **2016**, *58*, 549–555. [CrossRef]
54. Sarychev, I.A.; Sirotin, I.S.; Borisov, R.S.; Mu, J.; Sokolskaya, I.B.; Bilichenko, J.V.; Filatov, S.N.; Kireev, V.V. Synthesis of Resorcinol-Based Phosphazene-Containing Epoxy Oligomers. *Polymers* **2019**, *11*, 614. [CrossRef] [PubMed]
55. Kireev, V.V.; Bilichenko, Y.V.; Borisov, R.S.; Mu, J.; Kuznetsov, D.A.; Eroshenko, A.V.; Filatov, S.N.; Sirotin, I.S. Synthesis of Bisphenol A Based Phosphazene-Containing Epoxy Resin with Reduced Viscosity. *Polymers* **2019**, *11*, 1914. [CrossRef]
56. Sirotin, I.S.; Bilichenko, Y.V.; Suraeva, O.V.; Solodukhin, A.N.; Kireev, V.V. Synthesis of Oligomeric Chlorophosphazenes in the Presence of $ZnCl_2$. *Polym. Sci. Ser. B* **2013**, *55*, 63–68. [CrossRef]
57. Simonov-Emel'yanov, I.D.; Apeksimov, N.V.; Kochergina, L.M.; Bilichenko, Y.V.; Kireev, V.V.; Brigadnov, K.A.; Sirotin, I.S.; Filatov, S.N. Rheological and Rheokinetic Properties of Phosphazene-Containing Epoxy Oligomers. *Polym. Sci. Ser. B* **2016**, *58*, 168–172. [CrossRef]
58. Onuchin, D.V.; Sirotin, I.S.; Sarychev, I.A.; Bornosuz, N.V.; Kireev, V.V.; Gorbunova, I.Y.; Gorbatkina, Y.A. Physicochemical Properties of Epoxy Composites Modified with Epoxyphosphazene. *Polym. Sci. Ser. B* **2019**, *61*, 286–293. [CrossRef]

polymers

Article

Novel Therapeutic Approach in PEGylated Chitosan Nanoparticles of Apigenin for the Treatment of Cancer via Oral Nanomedicine

Md Ali Mujtaba [1,*], Nawaf M. Alotaibi [2], Sultan M. Alshehri [3], Mohammad Yusuf [4], Md Khalid Anwer [5], Mohammad Akhlaquer Rahman [6] and Arshiya Parveen [7]

1 Department of Pharmaceutics, Faculty of Pharmacy, Northern Border University, Rafhaa 73213, Saudi Arabia
2 Department of Clinical Pharmacy, Faculty of Pharmacy, Northern Border University, Rafhaa 73213, Saudi Arabia
3 Department of Pharmaceutics, College of Pharmacy, King Saud University, Riyadh 11451, Saudi Arabia
4 Department of Clinical Pharmacy, College of Pharmacy, Taif University, P.O. Box 11099, Taif 21944, Saudi Arabia
5 Department of Pharmaceutics, College of Pharmacy, Prince Sattam Bin Abdulaziz University, Al-kharj 11942, Saudi Arabia
6 Department of Pharmaceutics and Industrial Pharmacy, College of Pharmacy, Taif University, P.O. Box 11099, Taif 21944, Saudi Arabia
7 Department of Pharmaceutics, School of Pharmaceutical Education and Research, Jamia Hamdard, New Delhi 110062, India
* Correspondence: sajanqa@gmail.com or m.mujtaba@nbu.edu.sa; Tel.: +91-9891611864 or +966-538156614

Citation: Mujtaba, M.A.; Alotaibi, N.M.; Alshehri, S.M.; Yusuf, M.; Anwer, M.K.; Rahman, M.A.; Parveen, A. Novel Therapeutic Approach in PEGylated Chitosan Nanoparticles of Apigenin for the Treatment of Cancer via Oral Nanomedicine. *Polymers* **2022**, *14*, 4344. https://doi.org/10.3390/polym14204344

Academic Editor: Edina Rusen

Received: 11 September 2022
Accepted: 12 October 2022
Published: 15 October 2022

Publisher's Note: MDPI stays neutral with regard to jurisdictional claims in published maps and institutional affiliations.

Copyright: © 2022 by the authors. Licensee MDPI, Basel, Switzerland. This article is an open access article distributed under the terms and conditions of the Creative Commons Attribution (CC BY) license (https://creativecommons.org/licenses/by/4.0/).

Abstract: The goal of this study was to optimize and formulate apigenin (APG)-loaded pegylated chitosan nanoparticles (PEGylated-CNPs) via ionic gelation techniques using the Box–Behnken design (BBD). Three individual variables, X_1(chitosan: TPP concentration), X_2 (PEG-400 concentration), and X_3 (sonication time), were investigated for their influence on response variables (Y_1—particle size (PS); Y_2—drug entrapment efficiency (DEE); and Y_3—zeta potential (ZP). The optimized formula of APG-PEGylated CNPs was picked from the statistical design and was then examined for physical, morphological, release characterization, anti-oxidant, and anti-tumor potential. The average PS, PDI, %DEE, and ZP were found to be 139.63 ± 5.67 nm, 0.296 ± 0.014, 79.55 ± 3.12%, and 24.68 ± 1.84 mV, respectively. The optimized APG formulation was chosen and reformulated based on the desirability function. Results of the observed and predicted values of responses through the BBD process were found to be nearly identical. The resulting APG-PEGylated CNPs were spherical and smooth, according to surface morphology studies. The release study revealed that PEGylated-CNPs exhibited biphasic release patterns distinguished by an initial burst release of APG only at early phases accompanied by a delayed release near 24 h. Furthermore, APG-PEGylated CNPs demonstrated statistically increased antioxidant activities and cytotoxicity against MCF-7 cells compared to pure APG. Based on the findings, it is possible to conclude that BBD was efficient in optimizing the PEGylated CNPs formulation and recognizing the impacts of formulation variables. In conclusion, the developed formulation has a significant potential for anticancer therapy.

Keywords: apigenin; Box–Behnken design; antioxidant; chitosan; nanoparticles

1. Introduction

Natural products persist as incredibly valuable sources of new bioactive molecules with pharmacotherapy utility because of the enormous structural complexity found in them. From a long period of time, plants have been used to treat a wide range of diseases such as cancer with natural sources, accounting for more than 60% of today's anti-cancer medications. Plants, animals, aquatic species, and microorganisms all contribute to drugs, which makes nature an appealing source of new therapeutic entities that could be used as anti-cancer agents [1].

Apigenin (APG), chemically known as 5,7-dihydroxy-2-(4-hydroxyphenyl)-4H-1-benz opyran-4-one, is a flavone found in several vegetables and fruits, which include parsley, celery, onions, oranges, wheat sprouts, corn, rice, tea, and is abundant in chamomile [2]. It has multiple biological effects; in addition, it has potent chemoprotective and chemotherapeutic features for a variety of cancers, including lung, breast, skin, prostate, and colon cancer. The possible mechanism for the anticancer effects of APG could be its capability of inhibiting migration or invasion, endorsement of cell cycle arrest, suppression of proliferation, initiation of apoptosis, or interruption with tumor cellular signal pathways [3,4]. APG has low internal toxicity with no mutagenic properties compared to other plant flavonoids, making it an effective frontrunner for the production of successful cancer therapies [5]. APG is a Biopharmaceutics Classification System (BCS) Class II drug with an aqueous solubility of less than 2.16 g/mL [6]. Unfortunately, its poor water solubility results in poor bioavailability, which significantly limits its clinical use [7]. Therefore, investigating emerging innovations or formulations to improve APG solubility and bioavailability and to enhance antitumor efficacy is, thus, the main issue at present.

Nanoparticles (NPs) are expected to be the key focus in drug-delivery applications; however, NPs are associated with various conjugation properties due to their higher surface area, resulting in in the quickest systemic clearance. To overcome the problem of rapid systemic clearance, NP's surfaces were coated with a hydrophilic polymer-like chitosan (CS) and polyethylene glycol (PEG). CS is a USFDA-approved and popular polymer because of its low cost of production, biocompatibility, biodegradability, and positive charge. The positive charge of CS enhances interaction as well as penetration inside heterogeneous tumors [8]. CS improves drug penetration by opening epithelial tight junctions and facilitating both paracellular and transcellular drug transport [9]. PEG is also a non-toxic and non-immunogenic polymer that changes a molecule's hydrophobicity and enhances its aqueous solubility. PEGylation is a well-known method that enhances stability and decreases proteolysis and renal excretion, thereby minimizing dosing frequency and improving pharmacokinetics. Coating with PEG was intended to improve oral nanoparticle bioavailability because it is known to reduce immune system interactions and can eliminate these interactions, thus further increasing systemic half-life [10,11]. One more feature of PEG coating has been shown to reduce nanoparticle agglomeration and improves the therapeutic agent's safety and efficacy [12,13]. Therefore, the aim of the present study was to create, optimize, and characterize PEGylated chitosan nanoparticles of APG and to evaluate the optimized formulation for its anti-oxidant and anti-tumor potential.

2. Methods and Materials

2.1. Materials

Apigenin (purity 98%), chitosan medium molecular weight (190–310 kDa and degree of acetylation 85%), and polyethylene glycol (PEG) 400 were purchased from Baoji Guokang Bio-Technology Co., Ltd. Baoji, China. Sodium tripolyphosphate (TPP) and all other reagents and chemicals were of analytical grade in this study.

2.2. Preparation of APG-Loaded PEGylated Chitosan Nanoparticles (PEGylated-CNPs)

NPs were prepared by ionotropic gelation and were then homogenized and ultrasonicated [14,15]. For this, CS was dissolved in water that contained 1% *v/v* glacial acetic acid solution. Then, a weighed amount of APG was incorporated to the CS solution, and the added mixture has been homogenized at 12,000 rpm for 10 min. Throughout the process of homogenization, a TPP solution was added drop-wise to the mixture. Based on preliminary research, the CS-to-TPP ratio was determined as 4:1. Instantly, after homogenization, the dispersion was exposed to probe sonication (SONICS Vibra cell VC750, Newton, CT, USA) at 30% amplitude and pulsed at pulse 10 s for 20 min to generate nanoparticles. To keep the temperature increase under control, the dispersion was kept in an ice bath. NPs were coated with PEG 400 immediately after they were formed. Several batches were prepared by using the Box–Behnken design (BBD) to optimize the formulation. The schematic representation for the preparation of PEGylated-CNPs is as follows (Scheme 1).

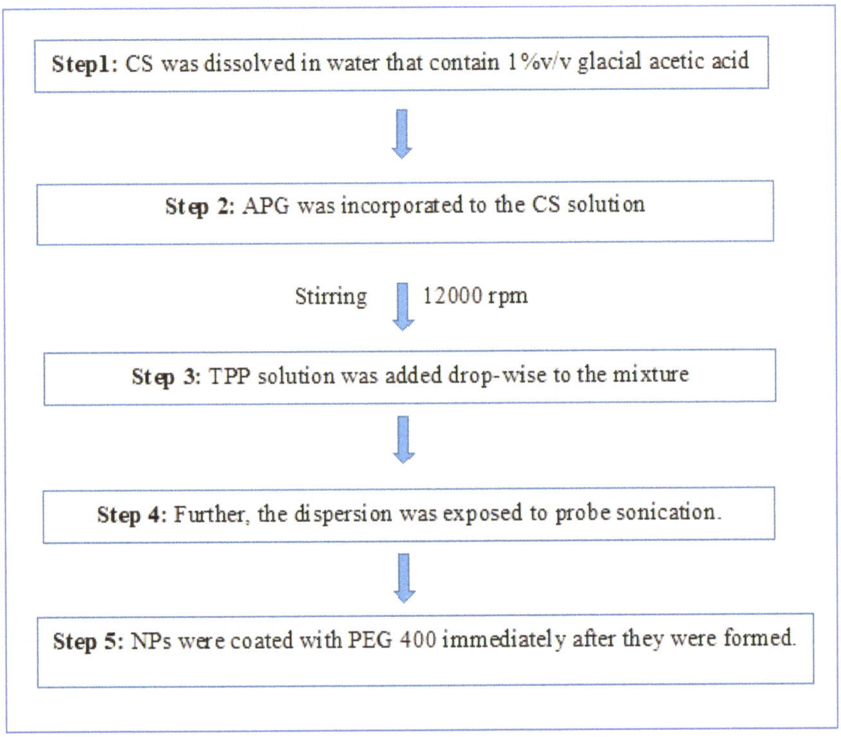

Scheme 1. The schematic representation for the preparation of PEGylated-CNPs.

2.3. *Experimental Design*

The BBD is a 3-factor 3-level design experiment that is recommended over others because it requires minimal experimental runs and provides the desired points under the cuboidal space; therefore, the probability of receiving unsuitable results reduced [16,17]. To examine the formulation variables influencing the studied, a three-factor, three-level BBD was applied, in which three formulation variables (amount of CS: TPP, amount of PEG 400, and sonication time) were differentiated at low (−1), middle (0), and high (+1), explored in Table 1. This design necessitates 15 runs with three replicated center points in order to obtain a more uniform estimation of the prediction variance across the overall experimental design. The responses considered in this studied were particle size (PS) (Y_1), drug entrapment efficiency (DEE) (Y_2), and zeta potential (ZP) (Y_3). The BBD was developed by Design-Expert software (Version 12.0, Stat-Ease Inc., MN, USA), which created and analyzed fifteen experimental runs (Table 2). The ranges for each independent factor were preferred from the preliminary study, which can be seen in Table 1.

In order to identify the most desirable mathematical model by using F-tests, Design-Expert produced linear, two-factor interaction (2FI), and quadratic models. Among these, the quadratic model was considered as the best-fitting model with respect to all responses (Table 3). The predicted and adjusted R^2 should be close with an approximate difference of 0.2 that must be achieved with a "reasonable agreement" [18]. Adequate precision is also one of the significant parameters in predicting the optimum response for a given variable. The chosen model was also subjected to a lack-of-fit test, and the lack of significance observed for this value in comparison to pure errors indicated that the independent variables and their responses had a significant correlation [19]. The polynomial equation obtained by experimental design is expressed mathematically as follows:

$$Y = A_0 + A_1 X_1 + A_2 X_2 + A_3 X_3 + A_{12} X_1 X_3 + A_{13} X_1 X_3 + A_{23} X_2 X_3 + A_{11} X_1^2 + A_{22} X_2^2 + A_{33} X_3^2 \quad (1)$$

where Y is the dependent responses, A_o represents intercept, A_1 to A_{33} represent the regression coefficients of Y, and X_1 to X_3 are presented as independent variables [20]. After producing polynomial equations directly related to dependent and independent variables, the optimization of minimum PS (Y_1), maximum DEE (Y_2), and low ZP (Y_3) was performed using a desirability function to obtain the levels of X_1, X_2, and X_3.

Table 1. Variables in BBD for formulation development of APG-loaded PEGylated-CNPs.

Variable	Levels		
	−1	**0**	**+1**
Independent variables			
X_1 = CS: TPP (% *w/w*)	0.2	0.5	0.8
X_2 = PEG 400 (% *w/w*)	0.5	0.75	1
X_3 = Sonication time (minutes)	10	20	30
Dependent variables			
Y_1 = Particle size (nm)			
Y_2 = Entrapment efficiency (%)			
Y_3 = Zeta potential (mV)			

Table 2. Variables and observed responses in BBD.

Formulation Code	Factors Combinations at Different Levels			Response Variables		
	X_1 = Chitosan: TPP	**X_2 = PEG 400**	**X_3 = Sonication Time**	**Y_1 = Particle Size (nm)**	**Y_2 = Entrapment Efficiency (%)**	**Y_3 = Zeta Potential (mV)**
CNP 1	0.2	0.5	20	140.52	75.11	19.62
CNP 2	0.8	0.5	20	247.5	85.37	24.51
CNP 3	0.2	1	20	158.6	79.72	21.82
CNP 4	0.8	1	20	303.3	90.54	26.56
CNP 5	0.2	0.75	10	248.3	87.61	19.41
CNP 6	0.8	0.75	10	354.2	96.68	24.76
CNP 7	0.2	0.75	30	120.52	65.96	19.62
CNP 8	0.8	0.75	30	214.4	80.76	25.15
CNP 9	0.5	0.5	10	261.5	89.12	23.48
CNP 10	0.5	1	10	314.2	93.41	25.15
CNP 11	0.5	0.5	30	134.1	70.21	23.19
CNP 12	0.5	1	30	153.8	77.52	25.45
CNP 13	0.5	0.75	20	141.2	76.35	23.49
CNP 14	0.5	0.75	20	150.1	80.82	23.92
CNP 15	0.5	0.75	20	153.1	80.91	24.11

Table 3. Summary of results of regression analysis for responses.

Model	R^2	Adjusted R^2	%C.V	S.D	PRESS	Remarks
Response: Particle size (Y_1)						
Linear model	0.7481	0.6899	-	41.57	33,163.03	-
Quadratic model	0.9941	0.9865	4.36	8.66	6423.28	Suggested
Cubic model	0.9985	0.9939		5.85	ND	-
Response: Entrapment efficiency (%) (Y_2)						
Linear model	0.9312	0.9153	-	2.34	105.50	-
Quadratic model	0.9842	0.9638	1.87	1.53	59.49	Suggested
Cubic model	0.9864	0.9457	-	1.87	ND	-
Response: Zeta potential (mV) (Y_3)						
Linear model	0.8494	0.8146	-	0.9122	21.40	-
Quadratic model	0.9933	0.9848	1.12	0.2614	4.38	Suggested
Cubic model	0.9968	0.9874	-	0.2380	ND	-

ND = not defined.

3. Physicochemical Characterization of APG-Loaded PEGylated-CNPs

3.1. PS, and ZP Analysis

The average PS and ZP of 15 formulations (CNP1–CNP15) were determined with zeta sizer (PSS NICOMP Z3000, Port Richey, FL, USA). Initially, 1 mL of samples was diluted ten times with double distilled water and mixed gently by handshaking to obtain an appropriate scattering intensity; it was measured at a fixed scattering angle of 90°. All measurements were taken in triplicate [21].

3.2. Determination of DEE (Drug Entrapment Efficiency) and Loading Capacity (LC)

For determining the DEE, 10 mg of nanoparticles (NPs) was added to 10 mL of acetone to isolate APG into acetone. The samples in acetone were gently ultra-centrifuged at 20,000 rpm for 1 h at 4 °C to collect the supernatant. Then, 10 µL of suspension was diluted to 1 mL with acetone and underwent the process of estimations. The appropriate APG concentrations were obtained with a UV-VIS spectrophotometer (Jenway 6850 double beam spectrophotometer, Cole-Parmer, Saint Neots, UK) [4,22]. The following formula was used to calculate the DEE and LC.

$$\text{DEE }(\%) = \frac{\text{Total amount of APG added } - \text{Total amount of APG in supernatant}}{\text{Total amount of APG added}} \times 100$$

$$\text{LC }(\%) = \frac{\text{Total amount of APG added } - \text{Total amount of APG in supernatant}}{\text{weight of NPs}} \times 100$$

3.3. Transmission Electron Microscope (TEM) Analysis

The sample for TEM was prepared using phosphotungstic acid staining. The TEM sample was developed by placing a drop of the NPs suspension on a carbon-coated copper grid. The sample was dried and examined with TEM (JEOL JEM1010, Tokyo, Japan).

3.4. In Vitro Drug Release

The release study of encapsulated APG from PEGylated-CNPs was studied at two different pH conditions using the dialysis bag diffusion technique, both at 37 °C under sink conditions. The selected dialysis bag (M.W 12 kDa, Sigma–Aldrich, St. Louis, MO, USA) was washed and prepared for study according to the manufacturer's guidelines. A release study was carried out in PBS at pH 6.8 and pH 7.4 to estimate the drug's release in tumor surroundings that became slightly acidic and physiological pH, respectively [8,23]. In both studies, samples containing 10 mg APG were installed in dialysis bags and submerged in a beaker of 100 mL PBS with 1% *w/v* Tween 20 retained at 37 °C and stirring at 200 rpm. The sample's aliquots were withdrawn at predefined time points and examined using a UV-VIS spectrophotometer at 335 nm to estimate the release of APG.

3.5. Storage and pH Stability Studies

Physical and chemical stabilities of the optimized formulation were evaluated at two different temperatures 4 ± 2 °C and 25 ± 5 °C for periods of 4 weeks to analyze any changes with respect to PS, PDI, and ZP. The formulation's stability was also tested at two different pH levels of 6.8 and 7.4.

3.6. Determination of Antioxidant Potential of Optimized APG-Loaded PEGylated-CNPs

The antioxidant potential of plant origin materials is defined as their capacity to catalyze the degradation of free radicals, i.e., 1,1-Diphenyl-2-picrylhydrazyl (DPPH), and the ability to significantly reduce their concentration. According to previously published reports, the study was carried out on optimized APG-loaded PEGylated-CNPs and APG suspension to assess their antioxidant potential [24]. The stock solution (10 mg/mL) was diluted with ethanol up until a concentration of 25–250 g/mL. The collected samples (500 L) were added to DPPH solution (0.02 percent in ethanol). The prepared samples were stirred and kept at 25 °C in a dark place for 1 h to carry the complete reaction process. At the end of the reaction, the violet color of DPPH changed to being colorless. The same experiment was repeated with blank PEGylated-CNPs. The samples were spectrophotometrically examined at 517 nm. The percentage of antioxidant activity was calculated as follows.

$$\text{Antioxidant activity } (\%\ \text{AA}) = \frac{\text{Absorbance of control} - \text{Absorbance of test}}{\text{Absorbance of control}} \times 100$$

3.7. Cytotoxicity Study

The impact of various concentrations of pure APG and APG-loaded PEGylated-CNPs on MCF7 cells was analyzed along with MTT (3-(4,5-dimethylthiazol-2-yl)-2,5-diphenyl tetrazolium bromide). The cytotoxic evaluation via the MTT assay is based on the principle of colorimetric tests, which relies on viable cells' ability to selectively minimize the tetrazolium constituent of MTT to purple-colored formazan crystals [24]. In brief, MCF7 cells were grown into 96-well plates at around 15,000 cells/well density in 100 µL cell media (DMEM, 10% FBS). Seeded cells were kept overnight at 37 °C and 5% CO_2 to enable improved adherence. Various concentrations of pure APG and the formulation were placed in different wells to examine their cytotoxic effects. DMSO was also used to produce stock solutions of the standard and the prepared formulation, which were sequentially diluted in a 96-well plate using serum-free media. The final DMSO concentration was kept below 1% to avoid any unwanted effects on the cells. The same diluted DMSO in the cell culture media was considered as the vehicle control. Following several pilot experiments, a range of concentration was chosen for pure APG and optimized formulations, and the assay was conducted in quadruplicate. After 24 h of exposure to APG, each well, except the blank, was filled with 10 µL MTT solution (5 mg/mL PBS). The treated cells were kept for incubation of 4 h to allow the metabolic process of MTT by viable cells. Cell media was removed from each well, and 100 µL DMSO was incorporated to liquefy the formazan of MTT. After the incubation of 30 min, the plate was read at 570 nm using DMSO as blank. The cellular viability was calculated as the viable cell's percentage compared with the control group [25]. The dose–effect curve was also used to calculate IC_{50} values, which were then expressed as concentrations (µM).

3.8. Statistical Analysis

Data are displayed as Mean ± SD. Raw data were evaluated with Graph Pad InStat demo version (GraphPad Software Inc., La Jolla, CA, USA). Data were examined using one-way ANOVA along with Tukey–Kramer multiple comparison tests to evaluate the statistical significance at different concentration exposures and the control.

4. Results and Discussion

In this study, we optimized and developed APG-loaded PEGylated-CNPs for improving therapeutic potential and for investigating the current use of APG in the treatment of cancer. The use of polymeric NPs as a drug carrier was motivated by the fact that it holds unique potential in cancer treatments. This is due to its biocompatibility and safeness with respect to normal tissues, which are attained by hiding the drug's toxicity. An additional significant benefit of NPs is that they improve drug efficacy at the targeted site because of its permeable properties [8]. Furthermore, CS was chosen as a carrier system for NPs because of its inherent positive charge, as well as other advantages including its biodegradability, biocompatibility, and lower cost. The positively charged surface of NPs targets tumor cells to a greater extent than the negatively charged surface of NPs [26]. They are more readily absorbed by cells, and this may damage cell membranes [27]. The mucoadhesiveness of CS relative to the bio-membrane is due to its adhesive nature. Furthermore, cells preferentially consume positively charged CNPs over negatively or neutrally charged CNPs [28]. The key mechanism is electrostatic interaction between the CS's NH^{3+} groups and negatively charged DNA's phosphoryl groups. The positive surface charge allows binding with negatively charged cell membrane and increases permeation.

To confirm the hypothesis, we used a single-step ionotropic gelation method to make CNPs loaded with APG. CNPs were designed by ionic gelation, which requires two aqueous phases mixture, one of which is the polymer CS and another is a polyanion TPP. The ratio of CS to TPP was optimized to be 1:1. The amino group of CS interacts with the TPP to form a nano-sized complex [29]. Furthermore, CNPs were also coated with PEG400 on the surface to enhance their safety and stability. The surface coatings of CNPs with PEG are based on physical cross-linking using electrostatic interactions. The method used here is a simple and gentle method for the preparation of the formulation. It is preferable to use physical crosslinking rather than chemical crosslinking to prevent excipient toxicity and other undesirable side effects. In particular, this method tries to avoid surfactants and organic solvents, resulting in the preparation of a novel formulation that is both safer and more cost-effective.

Preliminary experiments were carried out to ascertain the composition of PEGylated-CNPs. On the basis of preliminary research, three factors were chosen at three different levels: the amount of CS: TPP (X_1), the amount of PEG 400 (X_2), and the sonication time (X_3). According to selected range of the components, BBD resulted in 15 different APG-loaded PEGylated-CNP formulations, and these were examined with selected response variables, i.e., PS (Y_1), % EE (Y_2), and ZP (Y_3), as seen in Table 2.

4.1. Designed Experiment Statistical Analysis

The surface response methodology along with BBD is an efficient appliance for the examination of independent variables at a different level with limited experimental runs [20]. In this present study, three-factor three-level factorial designs with respect to responses were investigated. The generated BBD was then used to conduct a series of experiments. The result of measured responses for all formulations are reported in Table 2. The contributing factors and its responses were integrated, with statistical evaluation providing the polynomial equation for evaluating the impact of excipient on responses Y_1–Y_3. All dependent variables were analyzed with the quadratic model, and 3D surface plots for responses were created. The quadratic model's statistical study found significant p-values, indicating a better fit for selected responses (Table 4) [30].

Table 4. Statistical ANOVA results of the quadratic model.

			Quadratic Model				
Response	**F-Value**	***p*-Value**	**R^2**	**Adjusted R^2**	**Predicted R^2**	**Lack of Fit**	**Remarks**
PS	131.32	<0.0001	0.9941	0.9865	0.9280	0.1159	Significant

Table 4. *Cont.*

Quadratic Model							
Response	F-Value	*p*-Value	R^2	Adjusted R^2	Predicted R^2	Lack of Fit	Remarks
EE	48.35	<0.0001	0.9842	0.9638	0.9426	0.8763	Significant
ZP	116.01	<0.0001	0.9933	0.9848	0.9390	0.3471	Significant

4.2. Evaluation of Response Surface with Polynomial Equation

The three-dimensional response surface plots shown in Figure 1A–C are helpful for demonstrating the relationship of independent variables with dependent variables. These plots represent the effect of two independent factors on a response at a particular point. Whenever these 3D-plots are critically examined, the characteristic effects of independent variables on each response were observed [16].

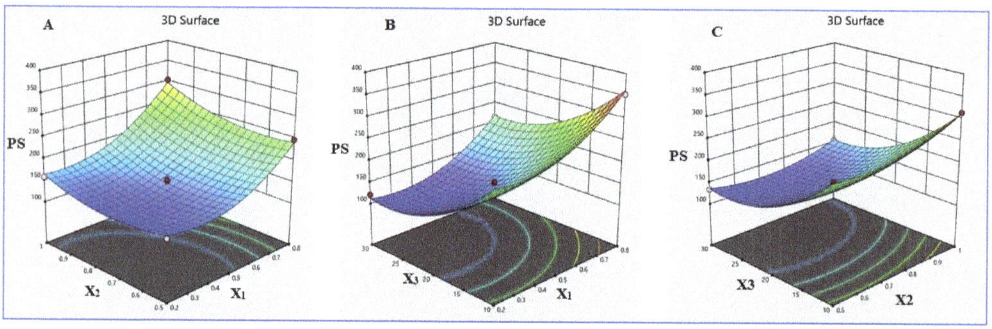

Figure 1. Three-dimensional response surface plot showing the effect of independent variables (**A**) X_1 and X_2; (**B**) X_1 and X_3; and (**C**) X_2 and X_3 on PS.

4.3. Response 1 (Y_1): Effect of Investigated Independent Variables on PS

The PS of APG-PEGylated CNPs was influenced by the selected independent variables; the average PS of multiple batches of formulations ranged from 120.52 nm to 354.2 nm, as shown in the Table 2. The following polynomial equation describes the mathematical relationship between independent variables and PS for APG-PEGylated CNPs. The coded equation assists in identifying the correlation of factors and PS (Equation (2)).

$$Y_1\ (PS) = +145.32 + 56.43\ A + 18.29\ B - 69.42\ C + 9.43\ AB - 3.00\ AC - 8.25\ BC + 42.81\ A^2 + 24.35\ B^2 + 46.23\ C^2 \quad (2)$$

The given equation reflects the quantitative impact of independent variables (A and C) and their interactions in the terms of AB, AC, and BC in the respect of response Y_1. The coefficient's *p*-value (0.05) demonstrated that they had a significant effect on Y_1. The plus sign of the coefficient denotes a synergistic effect, whereas the minus sign denotes the independent variables' antagonistic effect on response. The factor's high coefficient value demonstrates that it has a significant impact on the preferred response. All selected responses admirably corresponded with the quadratic model. ANOVA and multiple correlation tests were used to verify the model's efficiency (R^2). The quadratic model's results, along with ANOVA and the multiple correlation test (R^2), are shown in Table 4.

The F-value 131.32 of this model designated that it is significant. *p*-values less than 0.05 for this model terms are considered as significant. A, B, C, A^2, B^2, and C^2 present as significant model terms, while the model's terms are not significant if the value is greater than 0.10. The F-value of 3.78 for the lack of fit indicates that it is non-significant in comparison to the pure error. There is an 11.59% probability that a large lack-of-fit F-value is caused by noise. A relatively insignificant lack of fit is satisfactory, and we only want the model to fit. The adjusted R^2 of 0.9865 is relatively close to the predicted R^2 of 0.9280; i.e., the difference between them was less than 0.2. The signal-to-noise ratio is calculated by adequate precision. It is better to have a ratio of more than 4. Our signal-to-noise ratio of 37.895 implies an appropriate signal.

Response surface plots were utilized to further clarify the relationship between the dependent and independent variables. Three-dimensional surface response plots were created to better understand the relationship between the variables and PS. The 3D surface plot, illustrated in Figure 1A, showed that increasing the CS:TPP concentration (mass ratio of 1:1) ultimately resulted in the largest PS formation. The relationship between sonication time and PS was found to be negative, with PS decreasing as sonication times increased. The 3D surface plot in Figure 1B demonstrates the influence of CS:TPP concentrations and sonication times on PS, while PEG 400 concentrations and sonication times on the PS of APG-loaded PEGylated CNPs are shown in Figure 1C.

4.4. Response 2 (Y_2): Effect of Independent Variables on % DEE

The outcome of the % DEE study is reported Table 2. It was found that the % DEE ranged between 65.96 and 96.68. The quadratic model was the significant model for the % DEE analysis with a *p*-value of 0.0001, a *p*-value of 0.8763 for the lack of fit, and a % CV of 1.87 after fitting data responses on different models, as shown in Table 3. A polynomial equation (Equation (3)) was used to demonstrate a relationship of different independent factors with % DEE of the APG-loaded PEGylated CNPs.

$$Y_3\ (\%\ DEE) = +79.13 + 5.62\ A + 2.67\ B - 9.05\ C + 0.1400\ AB + 1.43\ AC + 0.7550\ BC + 1.87\ A^2 + +1.68\ B^2 + 1.75\ C^2 \quad (3)$$

It can be concluded from equation 3 that CS:TPP (A) and PEG concentrations (B) both had a positive influence on % DEE, whereas the sonication time (C) had an inverse relationship with % DEE. The final model after ANOVA analysis revealed that A, B, and C factors showed significant effects on the % DEE of APG, as all possessed a *p*-value of <0.05. The model's F-value of 48.35 indicates that it is statistically significant. *p*-values of less than 0.05 are considered significant for model terms. A, B, C, and A^2 are significant model terms in this study. The model terms are not significant if the value is above 0.10. The lack-of-fit F-value of 0.22 indicates that the lack of fit is not significant in comparison to the pure error. The adjusted R^2 of 0.9638 is probably close to the predicted R^2 of 0.9426; i.e., the variation was below 0.2. The signal-to-noise ratio is calculated by adequate precision, and it has a ratio that is more than 4. The ratio of 24.988 indicates that the signal is adequate. The obtained design space is acceptable by using this model.

Interaction patterns are analyzed using a 3D-response surface graph. As shown by the 3D-response surface graph in Figure 2A–C, it is apparent that the percent DEE is significantly affected by CS:TPP concentration, PEG concentration, and sonication time. This could be clarified by the interaction of APG with the hydrophobic group of PEGylated CNPs. Increased CS:TPP and PEG concentrations are assumed to increase the number of hydrophobic groups accessible for interactions with the APG (hydrophobic), resulting in improved drug entrapments in NPs [31].

4.5. Response 3 (Y_3): Effect of Investigated Independent Factors on ZP

The surface charge of the APG-loaded PEGylated CNPs formulations ranged from +19.41 to +26.56 mV (Table 2). The positive charge on the NP's surface indicates the existence of freely ionized amino groups that is necessary for electrostatic repulsion between particles to produce stable nano-dispersions. The measured response, Y_3, has a mathematical concept of a polynomial equation, which is enumerated below.

$$Y_3\ (ZP) = +23.88 + 2.56\ A + 1.02\ B + 0.0762\ C - 0.0375\ AB + 0.0450\ AC + 0.1475\ BC - 1.42\ A^2 + 0.6650\ B^2 - 0.2275\ C^2 \quad (4)$$

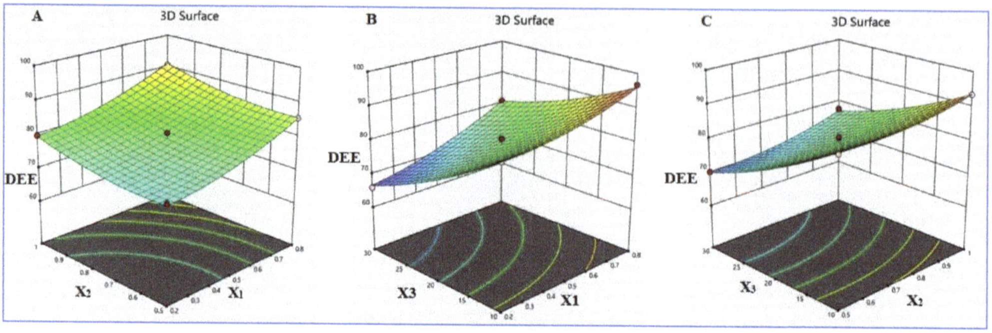

Figure 2. Three-dimensional-response surface plot showing effect of independent variables (**A**) X_1 and X_2; (**B**) X_1 and X_3; and (**C**) X_2 and X_3 on DEE.

The obtained mathematical equation can be used to determine the relative effects of these factors by equating the factor's coefficients. The concentration of CS:TPP and PEG had more significant effects on ZP. The model's F-value of 116.01 suggests that it is really significant. Model terms with p-values below 0.05 are significant. This case show that that A, B, A^2, and B^2 are considered as significant model terms. If the value is higher than 0.10, the model's terms really are not significant. The F-value for the lack of fit is 1.48, which indicates that the lack of fit is insignificant in comparison to the pure error. An insignificant lack of fit is acceptable, and we desired to fit the model. The adjusted R^2 of 0.9848 is probably close to the predicted R^2 of 0.9390; the difference is below 0.2. Adequate precision was discovered at 35.779, indicating that the obtained model is used to processed the design space. 3D surface plots that analyze the impact of independent variables on ZP are reported in Figure 3A–C.

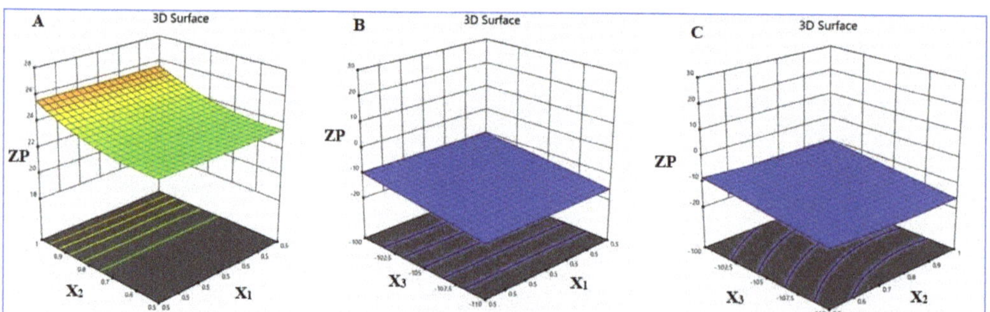

Figure 3. Three-dimensional-response surface plot showing effect of independent variables (**A**) X_1 and X_2; (**B**) X_1 and X_3; and (**C**) X_2 and X_3 on ZP.

4.6. Formulation Optimization for the APG-PEGylated CNPs

The impact of independent variables on responses, as well as the levels of such factors, was decided using the RSM statistical optimization process. Response optimization was used to accomplish an improved formulation with adequate PS, ZP, and high DEE. The selected optimized formulation produce values of independent variables, viz., at X_1 (CS: TPP concentration) of 0.5% *w/v*, X_2 (PEG 400 concentration) of 0.75% *w/v*, and X_3 (sonication time) of 20 min, which granted predicted values for selected responses. Three formulation batches of optimal content were prepared, and their responses for each formulation were examined to confirm the model's appropriateness for prediction. The optimized formulation components, as well as the experimental and predicted values including all dependent responses, are shown in Table 5. The experimental values were found to be very close to the predicted values, denoting the achievement of the BBD coupled with a preferable function for formulation evaluations and optimizations. For predicted versus observed values, the R^2 values (>0.9) showed a linear correlation across all dependent variables, denoting the superior model's reliability (Figure 4).

Table 5. Results of optimization of APG-loaded PEGylated-CNPs prepared under optimum condition.

Independent Variables for Response Optimization		
A- Chitosan: TPP (% *w/w*)	0.50	
B- PEG 400 (% *w/w*)	0.75	
C- Sonication time (minutes)	20	
	Optimized results	
Dependent variables (responses)	Predicted value	Experimental value [a]
Y_1 = Particle size (nm)	145.32	139.63 ± 5.67
Y_2 = Entrapment efficiency (%)	79.12	79.55 ± 3.12
Y_3 = Zeta potential (mV)	23.88	24.68 ± 1.84

[a] Mean ± SD, n = 3.

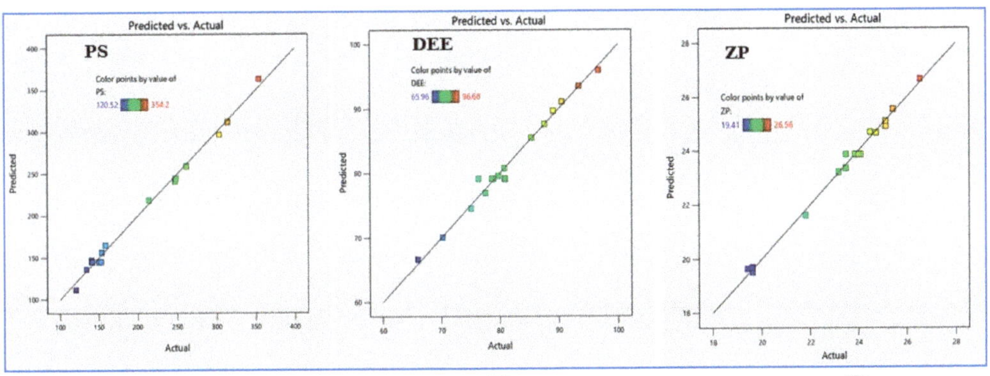

(A) (B) (C)

Figure 4. Predicted vs. actual plot representing the effects of X_1 and X_2 on the measured responses at the mid-level of X_3 on (**A**) PS (Y_1), (**B**) DEE (Y_2), and (**C**) ZP (Y_3).

4.7. Characterization of Optimized APG-PEGylated CNPs

The particle size distribution of NPs play a characteristic features in biodistribution, and as a drug carrier, its required a small particle size (i.e., a low PDI value) [4]. The PS distribution spectrum for the optimized APG-PEGylated CNPs is presented in Figure 5A, expressing the average PS of 139.63 ± 5.67 nm and PDI of 0.296 ± 0.014. The EE% and LC% of the formulation were found at 79.55 ± 3.12% and 37.78 ± 2.46%, respectively. The surface morphology of the optimized nanoparticle was examined by TEM (Figure 5B). The NPs were found to be nearly spheroidal vesicles in shape, with a size <200 nm, which matches the results of Zeta Sizer's PS analysis. The image also revealed that the NPs maintained their shape and that the vesicle's integrity is preserved. The layer seen around the nanoparticle's core could be the PEG molecules tied to the CS polymer.

4.8. In Vitro Dissolution Studies

To estimate the in vivo release profile of APG, in vitro dissolution analyses were conducted in PBS at pH 6.8 and 7.4. In comparison to free drug suspensions, PEGylated CNPs showed controlled drug released, as observed in Figure 6A,B. A biphasic release was also found with an initial burst release of APG in the early phase, followed by a delayed release of APG in the later phase (up to 24 h). The optimized CNPs showed a sustained release, but the rate was a little higher in pH 6.8 than in PBS pH 7.4. In 8 h of study, about 74% of the drug was released at pH 6.8 and about 62% was released at pH 7.4. The burst effect could have been caused by APG elution near the surface during nanoparticle preparation, which then quickly dispersed when nanoparticles interacted with the dissolution medium. Because of the swelling or deterioration of the polymer, APG was slowly released later on. The remaining APG in nanoparticles did not entirely release until particles have been completely eroded or dissolved in the dissolution media because of the interaction between both the residual APG and the few amine groups of chitosan.

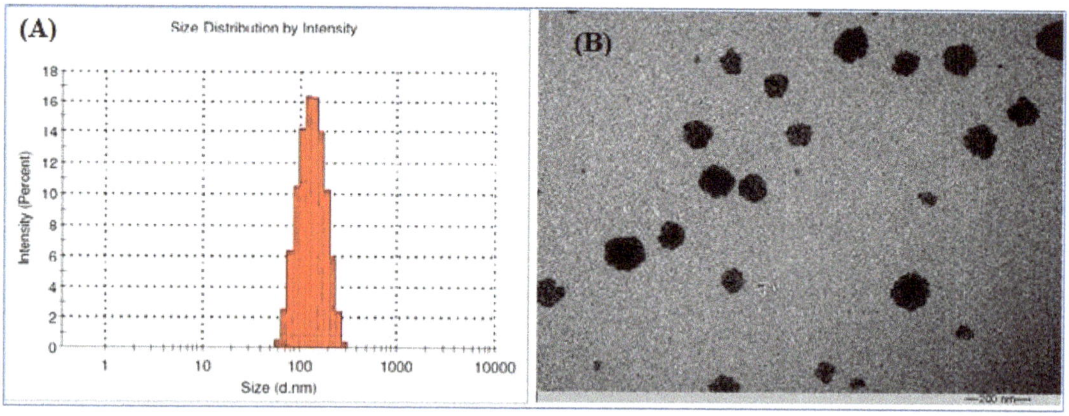

Figure 5. (**A**) Particle size and (**B**) TEM micrograph of optimized APG-PEGylated CNPs formulations.

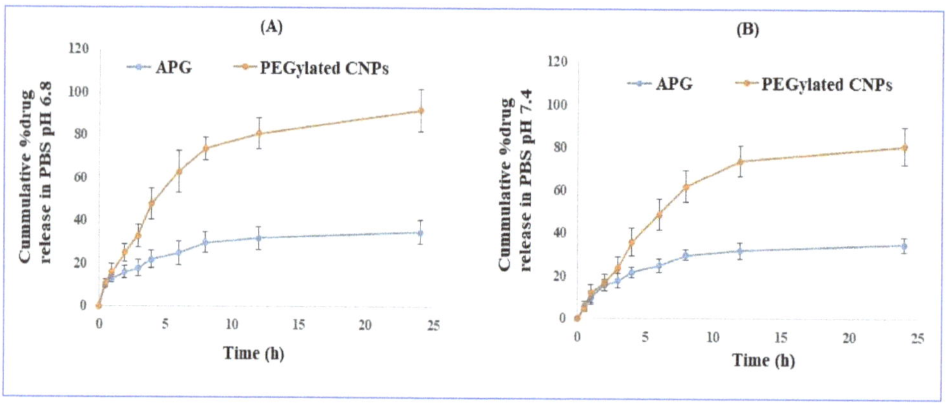

Figure 6. In vitro drug release profile of the optimized APG-PEGylated CNPs in (**A**) pH 6.8 and (**B**) pH 7.4 phosphate buffer.

4.9. Storage and pH-Dependent Stability Studies

Colloidal systems require a high level of stability. Physical instability is frequently a big obstacle to their clinical application. Optimized PEGylated CNPs were stored for three months at temperatures of 5 and 25 °C to assure physical stability. The estimation of stability was carried out based on PS, PDI, and ZP (Table 6). The measured features of prepared formulation were stable and had a statistically insignificant difference, indicating that CNP formulations remained stable during these time periods.

Table 6. Stability data.

CNPs Formulation (5 ± 2 °C)			
Time (Days)	Particle Size ± SD nm	PDI	Zeta Potential (mV)
Initial	139.63 ± 5.74	0.296 ± 0.02	24.68 ± 1.84
30	141.51 ± 7.65	0.294 ± 0.03	24.89 ± 1.96
60	142.19 ± 6.21	0.306 ± 0.03	23.76 ± 2.67
90	142.92 ± 7.58	0.316 ± 0.04	24.13 ± 2.39
CNPs formulation (25 ± 2 °C)			
Initial	139.63 ± 5.74	0.296 ± 0.02	24.68 ± 1.84
30	142.45 ± 6.24	0.301 ± 0.03	23.97 ± 1.88
60	142.98 ± 7.12	0.307 ± 0.04	24.69 ± 2.25
90	143.12 ± 7.28	0.310 ± 0.04	24.13 ± 2.23
Stability at different pH conditions (incubation for 24 h)			
pH condition	Particle size ± SD nm	PDI	Zeta potential (mV)
pH 6.8	138.39 ± 6.14	0.299 ± 0.03	23.98 ± 1.88
pH 7.4	143.23 ± 7.05	0.303 ± 0.03	24.56 ± 2.08

When the formulation comes in direct contact with different pH conditions, it may experience physical changes that cause particle aggregation. As a result, the formulation's physical stability was tested at various pH levels to which it could come into contact. The stability characteristics of CNPs in various experiments is shown in Table 6. The PS, PDI, and zeta potential were all insignificantly different, ensuring the stability of stored formulations under changing pH conditions [32].

4.10. *In Vitro Antioxidant Activity: DPPH Assay*

The biological activity of APG is influenced by antioxidant activities. The DPPH assay was applied to measure the antioxidant activity of the APG-PEGylated CNPs formulations in vitro, which relies on the lowering of the DPPH radical in the existence of an antioxidant molecule. DPPH is an incredibly sensitive and commonly used antioxidant assay that is used to evaluate antioxidant activities of a wide range of drugs and drug-delivery systems [16,33]. The test results are reported in Table 7, and it is undeniable that APG has anti-oxidant properties. Because of the presence of CS, weak antioxidant activities were also identified in the blank formulations (13.8 ± 3.1%). The CS alone showed some scavenging activities in a previous study by Anraku M. et al. [33]. These findings also show that CS has antioxidant properties and point to potential pharmaceutical applications. These findings indicate that the APG-PEGylated-CNPs formulation had higher radical-scavenging actions than the pure APG suspension. Overall, these findings show that APG-loaded PEGyalted-CNPs are more effective at scavenging oxygen- and nitrogen-centered radicals and have higher antioxidant capacities.

Table 7. In vitro antioxidant activity (% AA) of PEG-CNPs formulation in the presence or absence of APG in comparison with a methanolic solution of the APG and calculated as the inhibition percentage of DPPH radical. Data are expressed as mean ± SD (n = 3).

	AA (%)
Pure APG suspension	67.94 ± 4.2
APG loaded PEGylated CNPs	74.11 ± 5.6
PEGylated CNPs (Blank)	13.8 ± 3.1

4.11. *Cytotoxicity Study*

The possible features of the cytotoxic activity study in MCF-7 cells are depicted in Figure 7. When compared to pure APG, PEGylated CNPs composite exhibited significantly greater concentrations and time-dependent cytotoxicity at 24 h of study. Data reveal that the lower concentrations of both APG standard and APG-PEGylated CNPs formulation have MCF7 cell viability, which decreases as the concentration increases (Figure 7). When considering the effects, it seems that the lower concentrations of the APG formulation had improved growth inhibitory effects in MCF7 cells than the standard APG. These influences and comparisons are depicted in Figure 7. After 24 h of treatment against MCF-7 cells, the IC_{50} values of PEGylated CNPs and pure APG were observed to be 162 ± 14.54 µM and 1834.1 ± 55.74 µM, respectively. When compared to standard APG, data from the cell viability assay revealed that the PEGylated-CNPs formulation containing APG significantly enhanced these impacts in terms of lowering the IC_{50}. The APG inhibits the growth of MCF7 cells in a concentration-dependent manner, as reported by cell viability assays or MTT assays. However, that could be time-dependent, and the current study did not require a time-dependent cell viability assay.

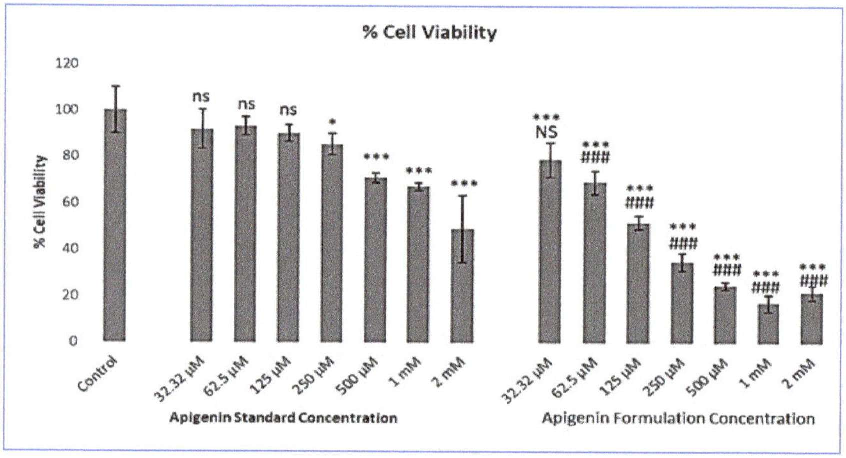

Figure 7. Effect of different concentrations of apigenin standard and apigenin formulation (optimized APG-PEGylated CNPs) on cell viability of A549 cells evaluated by MTT assay. Results are expressed as percentage mean ± SD (n = 3) (ns = not significant when compared with control; * = $p < 0.05$ when compared with control; *** = $p < 0.001$ when compared with control; NS = not significant when compared with same concentration group of apigenin standard; ### = $p < 0.001$ when compared with the same concentration groups of apigenin standard).

This can be explained by the fact that positively charged surface NPs have a higher chance of attacking tumor cells and are more easily gained by cells, potentially causing cell membrane destruction [29,30]. The important mechanism is the electrostatic interaction of the positively charged NH^{3+} groups of CS with the negative charge entity of DNA. They seem to have a total positive surface with respect to NPs, which allows them to bind with plasma membrane and increases the permeation through it [8].

5. Conclusions

In the present inspection, BBD was successfully used to design and formulate PEGylated APG-encapsulated CNPs. BBD had been used to generate design spaces, and three independent factors were used to achieve an optimized formulation that resulted in a smaller size, optimal ZP, and a higher % DEE. The single-step ionotropic gelation technique was used to make CNPs. PEG 400 was also applied to the surface of the NPs to ensure their safety and stability. For APG-PEGylated CNPs, the optimized formula yielded PS, ZP, and % DEE of 139.63 ± 5.67 nm, 24.68 ± 1.84, and 79.55 ± 3.12 respectively. TEM micrographs revealed a smooth surface and a spherical shape. According to the release study, PEGylated CNPs had a sustained release pattern for 24 h of periods. Once compared to pure APG, APG-loaded PEGylated CNPs had significantly higher antioxidant capacity. The formulation showed increased dose-dependent antitumor activities in breast cancer cells in comparison to pure APG. As a result, the study's conclusive result is that PEGylated CNPs improved the therapeutic potential and expanded the current use of novel formulations of APG in cancer treatments.

Author Contributions: M.A.M.: Conceptualization, investigation, methodology, and writing—original draft preparation; N.M.A.: formal analysis and investigation; S.M.A.: methodology and data curation; M.Y.: funding and investigation; M.K.A.: writing—reviewing and editing; M.A.R.: reviewing and editing; A.P.: formal analysis and validation. All authors have read and agreed to the published version of the manuscript.

Funding: The current research is supported by Taif University Researchers Supporting Project number (TURSP—2020/293), Taif University, Taif, Saudi Arabia.

Institutional Review Board Statement: Not applicable.

Informed Consent Statement: Not applicable.

Data Availability Statement: Data are contained within the article.

Acknowledgments: The current research is supported by Taif University Researchers Supporting Project number (TURSP—2020/293), Taif University, Taif, Saudi Arabia. The authors also gratefully acknowledge the Northern Border University for providing necessary facilities for the research study.

Conflicts of Interest: There are no conflicts of interest reported by the authors.

References

1. Mujtaba, M.A.; Akhter, M.H.; Alam, M.S.; Ali, M.D.; Hussain, A. An updated review on therapeutic potential and recent advances in drug delivery of Berberine: Current status and future prospect. *Curr. Pharm. Biotechnol.* **2022**, *23*, 60–71. [CrossRef] [PubMed]
2. Wang, M.; Firrman, J.; Liu, L.; Yam, K. A Review on Flavonoid Apigenin: Dietary Intake, ADME, Antimicrobial Effects, and Interactions with Human Gut Microbiota. *BioMed Res. Int.* **2019**, 7010467. [CrossRef] [PubMed]
3. George, V.C.; Dellaire, G.; Rupasinghe, H.P.V. Plant flavonoids in cancer chemoprevention: Role in genome stability. *J. Nutr. Biochem.* **2017**, *45*, 1–14. [CrossRef] [PubMed]
4. Bhattacharya, S.; Mondal, L.; Mukherjee, B.; Dutta, L.; Ehsan, I.; Debnath, M.C.; Gaonkar, R.H.; Pal, M.M.; Majumdar, S. Apigenin loaded nanoparticle delayed development of hepatocellular carcinoma in rats. *Nanomedicine* **2018**, *14*, 1905–1917. [CrossRef] [PubMed]
5. Salehi, B.; Venditti, A.; Sharifi-Rad, M.; Kręgiel, D.; Sharifi-Rad, J.; Durazzo, A.; Lucarini, M.; Santini, A.; Souto, E.B.; Novellino, E.; et al. The Therapeutic Potential of Apigenin. *Int. J. Mol. Sci.* **2019**, *20*, 1305. [CrossRef] [PubMed]
6. Truzzi, F.; Tibaldi, C.; Zhang, Y.; Dinelli, G.; D'Amen, E. An Overview on Dietary Polyphenols and Their Biopharmaceutical Classification System (BCS). *Int. J. Mol. Sci.* **2021**, *22*, 5514. [CrossRef] [PubMed]
7. Ding, S.M.; Zhang, Z.H.; Song, J.; Cheng, X.D.; Jiang, J.; Jia, X.B. Enhanced bioavailability of apigenin via preparation of a carbon nanopowder solid dispersion. *Int. J. Nanomed.* **2014**, *9*, 2327–2333. [CrossRef] [PubMed]
8. Agrawal, S.; Ahmad, H.; Dwivedi, M.; Shukla, M.; Arya, A.; Sharma, K.; Lal, J.; Dwivedi, A.K. PEGylated chitosan nanoparticles potentiate repurposing of ormeloxifene in breast cancer therapy. *Nanomedicine* **2016**, *11*, 2147–2169. [CrossRef] [PubMed]
9. Mohammed, M.A.; Syeda, J.; Wasan, K.M.; Wasan, E.K. An Overview of chitosan nanoparticles and its application in non-parenteral drug delivery. *Pharmaceutics* **2017**, *9*, 53. [CrossRef]
10. Suk, J.S.; Xu, Q.; Kim, N.; Hanes, J.; Ensign, L.M. PEGylation as a strategy for improving nanoparticle-based drug and gene delivery. *Adv. Drug Deliv. Rev.* **2016**, *99 Pt A*, 28–51. [CrossRef] [PubMed]
11. Alalaiwe, A.; Roberts, G.; Carpinone, P.; Munson, J.; Roberts, S. Influence of PEG coating on the oral bioavailability of gold nanoparticles in rats. *Drug Deliv.* **2017**, *24*, 591–598. [CrossRef]
12. Hinkley, G.K.; Carpinone, P.; Munson, J.W.; Powers, K.W.; Roberts, S.M. Oral absorption of PEG-coated versus uncoated gold nanospheres: Does agglomeration matter? *Part. Fibre. Toxicol.* **2015**, *12*, 9. [CrossRef] [PubMed]
13. Chen, J.; Huang, L.; Lai, H.; Lu, C.; Fang, M.; Zhang, Q.; Luo, X. Methotrexate-loaded PEGylated chitosan nanoparticles: Synthesis, characterization, and in vitro and in vivo antitumoral activity. *Mol. Pharm.* **2014**, *11*, 2213–2223. [CrossRef] [PubMed]
14. Rampino, A.; Borgogna, M.; Blasi, P.; Bellich, B.; Cesàro, A. Chitosan nanoparticles: Preparation, size evolution and stability. *Int. J. Pharm.* **2013**, *455*, 219–228. [CrossRef] [PubMed]
15. Hirpara, M.R.; Manikkath, J.; Sivakumar, K.; Managuli, R.S.; Gourishetti, K.; Krishnadas, N.; Shenoy, R.R.; Jayaprakash, B.; Rao, C.M.; Mutalik, S. Long circulating PEGylated-chitosan nanoparticles of rosuvastatin calcium: Development and in vitro and in vivo evaluations. *Int. J. Biol. Macromol.* **2018**, *107 Pt B*, 2190–2200. [CrossRef] [PubMed]

16. Mujtaba, M.A. Development of Apigenin-Loaded Niosomes Using Ecological Probe Sonication Technique for enhanced oral delivery: Application of Box-Behnken Design. *Curr. Pharm. Biotechnol.* **2022**, *23*, 882–893. [CrossRef] [PubMed]
17. Yadav, P.; Rastogi, V.; Verma, A. Application of Box–Behnken design and desirability function in the development and optimization of self-nanoemulsifying drug delivery system for enhanced dissolution of ezetimibe. *Futur. J. Pharm. Sci.* **2020**, *6*, 7. [CrossRef]
18. Abosabaa, S.A.; ElMeshad, A.N.; Arafa, M.G. Chitosan Nanocarrier Entrapping Hydrophilic Drugs as Advanced Polymeric System for Dual Pharmaceutical and Cosmeceutical Application: A Comprehensive Analysis Using Box-Behnken Design. *Polymers* **2021**, *13*, 677. [CrossRef] [PubMed]
19. Khan, N.; Alruwaili, N.K.; Bukhari, S.N.A.; Alsuwayt, B.; Afzal, M.; Akhter, S.; Yasir, M.; Elmowafy, M.; Shalaby, K.; Ali, A. Improvement of Ocular Efficacy of Levofloxacin by Bioadhesive Chitosan Coated PLGA Nanoparticles: Box-behnken Design, In-vitro Characterization, Antibacterial Evaluation and Scintigraphy Study. *Antibact. Eval. Scintigr. Study Iran. J. Pharm. Res.* **2020**, *19*, 292–311.
20. Soni, K.; Mujtaba, A.; Akhter, M.H.; Zafar, A.; Kohli, K. Optimisation of ethosomal nanogel for topical nano-cur and sulphoraphane delivery in effective skin cancer therapy. *J. Microencapsul.* **2020**, *37*, 91–108. [CrossRef]
21. Kohli, K.; Mujtaba, A.; Malik, R.; Amin, S.; Alam, M.S.; Ali, A.; Barkat, M.A.; Ansari, M.J. Development of natural polysaccharide-based nanoparticles of berberine to enhance oral bioavailability: Formulation, optimization, ex vivo, and in vivo assessment. *Polymers* **2021**, *13*, 3833. [CrossRef]
22. Das, S.; Das, J.; Samadder, A.; Paul, A.; Khuda-Bukhsh, A.R. Efficacy of PLGA-loaded apigenin nanoparticles in Benzo[a]pyrene and ultraviolet-B induced skin cancer of mice: Mitochondria mediated apoptotic signalling cascades. *Food Chem. Toxicol.* **2013**, *62*, 670–680. [CrossRef]
23. Telange, D.R.; Patil, A.T.; Pethe, A.M.; Fegade, H.; Anand, S.; Dave, V.S. Formulation and characterization of an apigenin-phospholipid phytosome (APLC) for improved solubility, in vivo bioavailability, and antioxidant potential. *Eur. J. Pharm. Sci.* **2017**, *108*, 36–49. [CrossRef] [PubMed]
24. Gilani, S.J.; Bin-Jumah, M.; Rizwanullah, M.; Imam, S.S.; Imtiyaz, K.; Alshehri, S.; Rizvi, M.M.A. Chitosan coated luteolin nanostructured lipid carriers: Optimization, in vitro-ex vivo assessments and cytotoxicity study in breast cancer cells. *Coatings* **2021**, *11*, 158. [CrossRef]
25. Alshehri, S.; Imam, S.S.; Rizwanullah, M.; Fakhri, K.U.; Rizvi, M.; Alam, M.; Mahdi, W.; Kazi, M. Effect of chitosan coating on plga nanoparticles for oral delivery of thymoquinone: In vitro, ex vivo, and cancer cell line assessments. *Coatings* **2021**, *11*, e6. [CrossRef]
26. Chung, Y.-I.; Kim, J.C.; Kim, Y.H.; Tae, G.; Lee, S.-Y.; Kim, K.; Kwon, I.C. The effect of surface functionalization of PLGA nanoparticles by heparin-or chitosan-conjugated Pluronic on tumor targeting. *J. Control. Release* **2010**, *143*, 374–382. [CrossRef] [PubMed]
27. Fröhlich, E. The role of surface charge in cellular uptake and cytotoxicity of medical nanoparticles. *Int. J. Nanomed.* **2012**, *7*, 5577–5591. [CrossRef]
28. Yue, Z.-G.; Wei, W.; Lv, P.-P.; Yue, H.; Wang, L.-Y.; Su, Z.-G.; Ma, G.-H. Surface charge affects cellular uptake and intracellular trafficking of chitosan-based nanoparticles. *Biomacromolecules* **2011**, *12*, 2440–2446. [CrossRef]
29. Chandra Hembram, K.; Prabha, S.; Chandra, R.; Ahmed, B.; Nimesh, S. Advances in preparation and characterization of chitosan nanoparticles for therapeutics. *Artif. Cells Nanomed. Biotechnol.* **2016**, *44*, 305–314. [CrossRef] [PubMed]
30. Adena, S.K.R.; Upadhyay, M.; Vardhan, H.; Mishra, B. Development, optimization, and in vitro characterization of dasatinib-loaded PEG functionalized chitosan capped gold nanoparticles using Box-Behnken experimental design. *Drug Dev. Ind. Pharm.* **2018**, *44*, 493–501. [CrossRef] [PubMed]
31. Hassani Najafabadi, A.; Abdouss, M.; Faghihi, S. Synthesis and evaluation of PEG-O-chitosan nanoparticles for delivery of poor water soluble drugs: Ibuprofen. *Mater. Sci. Eng. C Mater. Biol. Appl.* **2014**, *41*, 91–99. [CrossRef] [PubMed]
32. Caddeo, C.; Pucci, L.; Gabriele, M.; Carbone, C.; Fernàndez-Busquets, X.; Valenti, D.; Pons, R.; Vassallo, A.; Fadda, A.M.; Manconi, M. Stability, biocompatibility and antioxidant activity of PEG-modified liposomes containing resveratrol. *Int. J. Pharm.* **2018**, *538*, 40–47. [CrossRef] [PubMed]
33. Anraku, M.; Hiraga, A.; Iohara, D.; Uekama, K.; Tomida, H.; Otagiri, M.; Hirayama, F. Preparation and antioxidant activity of PEGylated chitosans with different particle sizes. *Int. J. Biol. Macromol.* **2014**, *70*, 64–69. [CrossRef] [PubMed]

polymers MDPI

Article

Development of Biodegradable Polyesters: Study of Variations in Their Morphological and Thermal Properties through Changes in Composition of Alkyl-Substituted (ε-DL) and Non-Substituted (ε-CL, EB, L-LA) Monomers

Felipe Robles-González, Teresa Rodríguez-Hernández, Antonio S. Ledezma-Pérez, Ramón Díaz de León, Marco A. De Jesús-Téllez * and Héctor Ricardo López-González *

Centro de Investigación en Química Aplicada, Blvd. Enrique Reyna Hermosillo 140, Saltillo 25294, Coahuila, Mexico
* Correspondence: marco.tellez@ciqa.edu.mx (M.A.D.J.-T.); ricardo.lopez@ciqa.edu.mx (H.R.L.-G.)

Abstract: Three series of polyesters based on monomer combinations of ε-caprolactone (ε-CL), ethylene brassylate (EB), and L-Lactide (LLA) with the alkyl substituted lactone ε-decalactone (ε-DL) were synthesized at different molar ratios. Copolymers were obtained via ring opening polymerization (ROP) employing TBD (1,5,7-triazabicyclo-[4.4.0]-dec-5-ene), an organic catalyst which can be handled under normal conditions, avoiding the use of glove box equipment. The molar monomer composition of resulting copolymers differed from theoretical values due to lower ε-DL reactivity; their M_n and M_w values were up to 14 kDa and 22.8 kDa, respectively, and distributions were ($Đ$) $\leq$ 2.57. The thermal stability of these materials suffered due to variations in their ε-DL molar content. Thermal transitions such as melting (T_m) and crystallization (T_c) showed a decreasing tendency as ε-DL molar content increased, while glass transition (T_g) exhibited minor changes. It is worth mentioning that changes in monomer composition in these polyesters have a strong impact on their thermal performance, as well as in their crystallization degree. Consequently, variations in their chemical structure may have an effect on hydrolyic degradation rates. It should be noted that, in future research, some of these copolymers will be exposed to hydrolytic degradation experiments, including characterizations of their mechanical properties, to determine their adequacy in potential use in the development of soft medical devices.

Keywords: ROP; biodegradable polyesters; TBD; organic catalyst

Citation: Robles-González, F.; Rodríguez-Hernández, T.; Ledezma-Pérez, A.S.; Díaz de León, R.; De Jesús-Téllez, M.A.; López-González, H.R. Development of Biodegradable Polyesters: Study of Variations in Their Morphological and Thermal Properties through Changes in Composition of Alkyl-Substituted (ε-DL) and Non-Substituted (ε-CL, EB, L-LA) Monomers. *Polymers* **2022**, *14*, 4278. https://doi.org/10.3390/polym14204278

Academic Editor: Edina Rusen

Received: 16 September 2022
Accepted: 7 October 2022
Published: 12 October 2022

Publisher's Note: MDPI stays neutral with regard to jurisdictional claims in published maps and institutional affiliations.

Copyright: © 2022 by the authors. Licensee MDPI, Basel, Switzerland. This article is an open access article distributed under the terms and conditions of the Creative Commons Attribution (CC BY) license (https://creativecommons.org/licenses/by/4.0/).

1. Introduction

One of the main research activities in health sciences is the development of biodegradable and biocompatible materials for medical applications. At present, surgical stents and other prosthetics are made of stainless steel, silver and platinum [1]. The use of metals in the construction of implants provides durability and mechanical properties. However, these medical devices suffer from limited control of their degradation rates, resulting in undesirable modifications, compromising functionality and causing adverse systemic reactions in the host [2]. To tackle the limitations of metals in the medical field, the introduction of polymers has brought solutions in terms of biocompatibility and biodegradability issues. Among such materials, aliphatic polyesters have been in the spotlight in recent years [3–5].

Aliphatic polyesters are characterized by their saturated linear structure and the presence of ester groups in their backbone. They are an interesting alternative in biomedical applications due to their renewable nature, physicochemical, thermal and morphological properties, ease of synthesis, processability and customizable degradation rates [6]. Current medical applications of polyesters include, but are not limited, tissue scaffolding, drug delivery systems and bone regeneration. Poly(glycolic acid) (PGA), poly(L-lactide) PLLA and poly(ε-caprolactone) (PCL) are examples of commonly synthesized aliphatic polyesters with adequate chemical, thermal and mechanical properties for the manufacturing of medical devices [5,7]. In tissue engineering, PLLA and PCL are frequently applied in bone, nerve, vascular and skin implants [8], while polyesters with high biodegradability, such as PGA and poly(ε-decalactone) (PDL), have been employed as encapsulations for proteins and antibiotics in drug delivery systems [9].

In ester copolymerization, materials with tailored features can be obtained by combining monomers with specific properties. For instance, inclusion of branched lactones comonomers as ε-decalactone into PLLA which possesses features as brittleness and low thermal stability, result in copolymers with improved mechanical properties, and higher degradation temperatures in comparison with PLLA homopolymer. These properties enhancements are mainly supported in the morphological changes induced by *n*-butyl side groups presented in PDL segments [10]. Regarding copolymers based on L-lactide/ε-caprolactone and ethylene brassylate/δ-valerolactone, variations in molar ratios of monomer were studied in order to establish a correlation between chemical composition, thermal properties and morphology. Copolymers based on L-lactide/ε-caprolactone exhibit enhanced thermal stability using PCL at 50–60 %mol, but their amorphous arrangement disrupts crystal formation [11]. Jin et al. [12] established a T_m eutectic point in ethylene brassylate/δ-valerolactone random copolymers, wherein formulation with δ-valerolactone at 80 %mol induced co-crystallization and isodimorphism. Another example of biodegradable polyesters (M_w up to 285 kDa) includes the use of biobased monomers, such as ethylene brassylate-*b*-δ-hexalactone, wherein a branched comonomer favors an amorphous arrangement, which has an impact on the mechanical properties, giving rise to possible applications in cell culture scaffolds [13]. On the other hand, triblock copolymers consisting of ethylene oxide-*b*-ε-caprolactone-*b*-γ-butyrolactone and ε-caprolactone-*b*-ethylene adipate-*b*-ε-caprolactone have also been reported; these synthesized copolyesters with M_n up to 12.3 and 23.2 kDa, respectively, offer drug delivery system capabilities with specific drug-release profiles [14,15].

There are two main synthesis routes to obtain polyesters: polycondensation (step-growth polymerization) and ring-opening polymerization (ROP). In the first one, the polymerization reaction involves the esterification of diacids and diols, linking both types of molecules through the formation of ester groups. Although this type of polymerization can be performed under moderate conditions, and the required precursors are widely available, one of its most notorious disadvantages is the generation of byproducts, which promote reversible polymer–monomer reactions if they are not eliminated from the reaction media, leading to polymers with low molecular weights and wide dispersity [16]. Taking into consideration the limitations of polycondensation, ROP emerges as an alternative methodology to synthesize aliphatic polyesters which is based in the opening of cyclic monomers (lactones) by usage of catalysts, initiators or specific reaction conditions to obtain linear macromolecules. In contrast to step-growth polymerization, byproducts are not generated through ROP, allowing the synthesis of polymers with high molecular weights and narrow dispersity [17]. For instance, successful ROP of ε-caprolactone, ε-decalactone and pentadecalactone employing a zinc complex with phenoxy-imine ligand as a catalyst has been reported, achieving molecular weights at 21–73 kDa and dispersities of 1.8–2.3 [18]. In addition, some researchers have reported the implementation of stannous octoate and bismuth salts as catalyst in the synthesis of PLLA, PDL and L-lactide/ε-decalactone copolymers with block and random architectures via ROP methodologies, in which molar masses reached 58.7 kDa for PLLA, 86.6 kDa for PDL and 54.1 kDa for L-lactide/ε-decalactone copolymers with dispersity values of 1.2–2.2, although reaction times were higher than 48 h and temperatures were up to 130 °C [10,19].

It is worth mentioning that ROP via organometallic-catalysis represents a feasible technique for aliphatic polyester synthesis of, e.g., PCL and PLA with high values of conversions and molar mass, as well as narrow dispersion [20]; however, some of the major drawbacks are the purity conditions for handling the organometallic inititator and monomers, and the absence of oxygen and environmental humidity in the reaction media to ensure high reaction performance [10,21]. Furthermore, ineffective and expensive purification procedures of the synthesized polymers are required for the removal of metal complex residues coming from catalysts like stannous octoate ($Sn(Oct)_2$), bismuth salts and neodymium isopropoxide. For instance, FDA regulations state that the presence of $Sn(Oct)_2$ cannot exceeded 20 ppm, and so usage of this kind of reagent may be result hazardous in medical, pharmaceutical and food packaging applications [22]. Other drawbacks to consider are depolymerization and oxidation reactions, which increase the toxicity, lower the biocompatibility, and influence the chemical, thermal and mechanical properties [23]. To overcome the consequences of organometallic initiators, polyester synthesis via metal-free ROP is a green and low-cost alternative, featuring milder reaction conditions and undemanding purification techniques of raw materials and products. In this regard, 1,5,7-triazabicyclo[4.4.0]dec-5-ene (TBD) and 1,8-diaza[5.4.0]bicycloundec-7-ene (DBU) have emerged as organo-catalytic options providing versatility in terms of the monomer/catalyst and initiator/catalyst ratios, control over molecular weight and dispersity, ease of storage, handling without requirements of inert atmosphere conditions and consistent biocompatibility due to their easy removal [24]. Therefore, TBD and DBU catalysts are interesting candidates for the synthesis of PLA, PCL, PDL and PEB; previous works have employed these reaction systems for the synthesis of these polyesters, reaching molecular weight values from 2.4 to 85 kDa and dispersities of 1.05–1.90 in periods shorter than 24 h [10,13,25].

In this work, series of polyesters based on combinations of L-lactide/ε-decalactone, ethylene brassylate/ε-decalactone and ε-caprolactone/ε-decalactone at different molar ratios were achieved through the ROP methodology using TBD as a catalyst. It is worth mentioning that there are no previous works reporting the systematic study of the chemical, thermal and morphological properties of this type of copolymer and the influence of ε-decalactone on such properties. Therefore, these aliphatic polyesters may exhibit biodegradable features through variations in their chemical structures, possibly resulting in an interesting alternative for the development of soft medical devices.

2. Materials and Methods

2.1. Reagents

The monomers ε-caprolactone (ε-CL), ε-decalactone (ε-DL) and ethylene brassylate (EB) (Sigma-Aldrich) were purified via a distillation system which included Na^0 as a drier reagent, a vacuum pressure at 15 mm Hg and a temperature of 160 °C to achieve complete removal of humidity and oxygen. L-Lactide (L-LA) (Sigma-Aldrich, St. Louis, MO, USA), was purified through a crystallization process into ethyl acetate (J. T. Baker, Radnor, PA, USA) solution (33.33% weight). Other reagents such as TBD (TCI chemicals, Portland, OR, USA), benzyl alcohol and trioxane (Sigma-Aldrich, St. Louis, MO, USA) were used as received. Toluene and methanol were provided by J. T. Baker. Toluene was washed with H_2SO_4, dried in $CaCl_2$, stirred under reflux conditions with $LiAlH_4$ and distilled using a sodium/benzophenone complex, while methanol was used as received.

2.2. Characterization and Equipment

The chemical composition was determined by proton nuclear magnetic resonance (^1H NMR) at room temperature using a Bruker (Billerica, MA, USA) Advance III 400 MHz spectrometer and deuterated chloroform ($CDCl_3$) as a solvent. Additionally, characteristic functional groups in copolymers were confirmed through Fourier transform infrared spectroscopy (FTIR)–attenuated total reflectance (ATR) mode, recorded from 4000 to 400 cm^{-1} in a FTIR spectrometer model Nicolet iS5 from Thermo-Scientific, (Madison, WI, USA). Gel permeation chromatography (GPC) was used in the determination of molar mass (M_n and M_w) and distribution ($Đ$) in copolymers employing an Agilent (Santa Clara, CA, USA) series 1100 calibrated with polystyrene standards, $CHCl_3$ (HPLC grade) as eluent, a refraction index detector, a set of three columns PL-gel 5 μm, one column 10^6 Å and two column Mixed-C. Thermal stability was recorded in a thermogravimetric analyzer Q500 model from TA Instruments (New Castle, DE, USA); the heating rate was 10 °C min^{-1} from 30 to 600 °C under N_2 atmosphere. Thermal transitions (T_g, T_c and T_m) were obtained with a differential scanning calorimeter (DSC–2500 from TA Instruments) at heating and cooling rates of 10 °C min^{-1} from −70 to 100 °C under a N_2 atmosphere. X-ray diffraction (XRD) provides information about the crystalline arrangements in copolymers; XRD analyses were performed using a Bruker D8 Advance ECO. The radiation frequency selected was the $k_{\alpha1}$ line from Cu (1.5406 Å). A power supply of 40 kV and 25 mA and an angle from 3 to 110° with an increment rate of 0.02° at 0.5 s were used.

2.3. Polyester Copolymers Synthesis

Polyesters were synthesized via ROP using cyclic esters monomers with alkyl side groups (ε-DL) and non-substituted (ε-CL, EB, L-LA). In Series I, the combination of monomers was L-LA and ε-DL. Meanwhile, Series II was done by EB and ε-DL and Series III by ε-CL and ε-DL (see Scheme 1). It should be noted that the chemical arrangement in these copolymers was block architecture, taking into consideration the sequential addition of monomers. In more detail, the first step corresponded to the addition of ε-DL. The non-substituted monomer was added afterwards, following the periods of time mentioned in the Supplementary Information section (Tables S1–S3). Polymerization reactions were done in a 50 mL Schlenk flask, wherein ε-DL, catalyst (TBD), NMR reference (trioxane) and solvent (toluene) were added at the corresponding molar ratios (Tables S1–S3). Each reaction mixture was degassed through three vacuum/N_2 cycles to eliminate oxygen and humidity traces. Thereafter, the reaction system was exposed to heating (100 °C) and stirring (400 rpm) and the initiator (benzyl alcohol, BzOH) was added dropwise. It is worth mentioning that the second monomer (L-LA, EB, ε-CL) was added, taking into consideration the ε-DL percentage loaded in each formulation. The established time for all reactions was fixed at 24 h. Finally, the purification process consisted of the evaporation of toluene, washing the product twice in cold methanol (~0 °C) to eliminate raw materials, filtration and drying in a vacuum oven at 40 °C in a period of 24 h. The reaction conditions implemented in this work come from previous studies, although parameters such as the catalyst, reaction time, and molar ratios were modified [26].

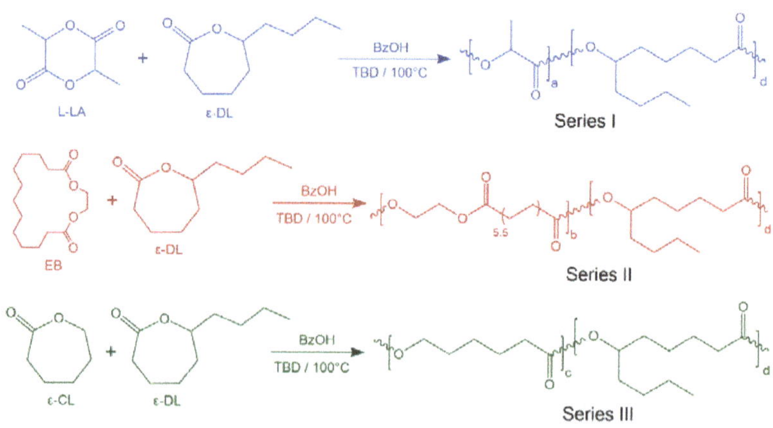

Scheme 1. Synthesis of Series I–III via organocatalytic ROP.

3. Results

3.1. Polyester Copolymers Synthesis

Monomer conversion and the chemical structure in the copolymer series were studied by ^1H NMR in order to corroborate the expected signals and chemical shifts (δ), as well as the percentages of monomers in the synthesized copolymers. The success of the ROP reactions catalyzed by TBD was monitored by changes in δ of representative protons in polymers and monomers, specifically, proton signals in methylene and methine groups, located in the α position of O-C=O (carboxylic groups). Moreover, calculated conversions for the homopolymerization of the four studied monomers were >90% after 24 h, which indicated the high catalytic efficiency of TBD in these polymerizations due to its ability to act as a H-bond donor and acceptor, granted by the secondary and tertiary nitrogen atoms in its structure, thereby favoring the ring-opening mechanism of cyclic esters [27]. Such conversions in the given duration represent an improvement over the reaction time established by Ramos-Durán et al., who employed a neodymium isopropoxide-catalyzed system for the polymerization of ε-CL and ε-DL, among other lactones [28]. Figure 1 shows ^{1}H NMR spectra of copolymers A-3, B-3 and C3. Likewise, the ^{1}H NMR spectra of monomers, as well as those of the rest of the copolymers in Series I–III, are displayed in the Supplementary Materials (Figures S1–S4). In the Series I spectra, the target PLA proton CH α O-CO was fixed at 5.18 ppm. Regarding PDL segments in the copolymers of series I–III, the characteristic proton in the methine group was located at 4.86 ppm, resulting in δ matching the expected values for the proton CH α O-CO in both the PLA and PDL segments. It is worth to mention that ROP in L-LA with TBD as a catalyst and the applied reaction conditions favor the obtention of atactic PLA, which could develop an amorphous arrangement (see Section 3.2). Additionally, the ^{1}H NMR spectra in Series I (Figure S2) reveal multiple signals from 5.27 to 5.14 ppm, suggesting the presence of several stereoregular arrangements. These studies were mentioned by Moins et al., who achieved the synthesis of isotactic PLA via ROP using a TBD catalyst under cryogenic conditions (−72 °C), while non-stereoregular PLA was obtained at 23 °C [29].

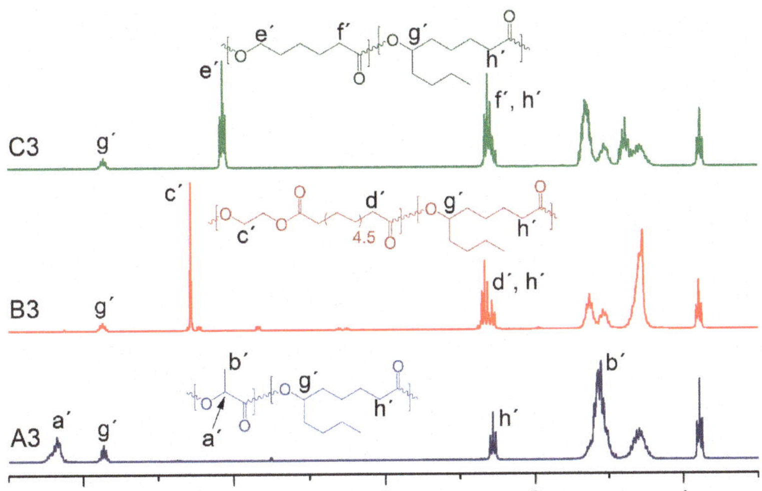

Figure 1. ^{1}H NMR spectra of representative polyesters A3, B3 and C3.

CH_2 α O-CO of PEB segments in Series II and PCL segments in Series III showed the expected δ of 4.28 ppm and 4.06 ppm, respectively, for these polyesters. A composition analysis was performed, taking as reference the integration values of the signals previously mentioned. The calculation methodology employed was previously described by Ramos-Durán and co-workers [28]. The results are displayed in Figure 2 and Tables 1–3. Notably, the molar mass percentage of ε-DL was less than the stoichiometric reference; this was attributed to the minor reactivity ratio in comparison to the non-substituted lactones (ε-CL, L-LA and EB) [30].

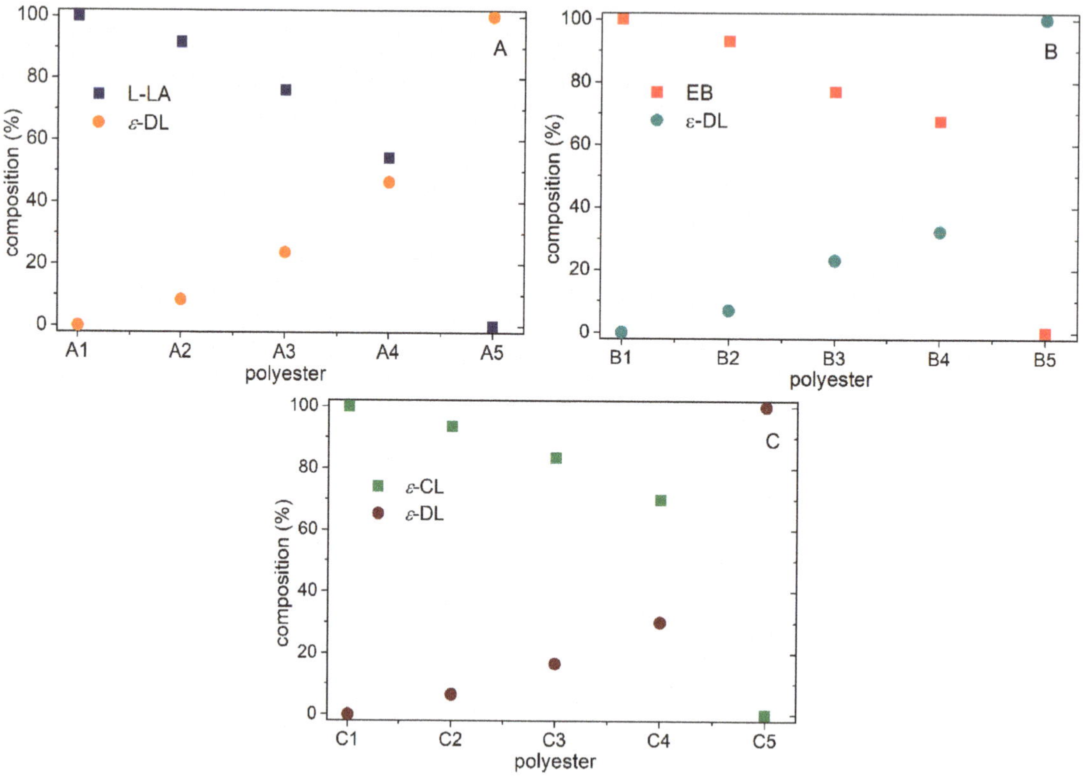

Figure 2. Percentage composition of comonomers determined through ^{1}H NMR in Series I (**A**), Series II (**B**) and Series III (**C**).

The spectra of polyesters A3, B3 and C3 are shown in Figure 3. A complete FTIR characterization of Series I–III is presented in the Supplementary Materials (Figures S5–S7). The characteristic stretching absorption bands for C-H were observed at ~2940 cm^{-1} (CH_3), ~2860 cm^{-1} (CH_2), while the carbonyl group was located at ~1730 cm^{-1} (C=O) and C-O absorption bands exhibited a wide range of wave number, i.e., from 1244 to 1093 cm^{-1}. We draw your attention to the A3 spectrum, in which two absorption bands were observed in the C=O region. Firstly, the signal which corresponded to C=O in PLA was fixed at 1750 cm^{-1}, while the C=O groups related with PDL segments were observed at 1730 cm^{-1}. The overlap in the absorption bands was related to differences in the chemical environment of PLA and PDL, wherein carbonyl groups in PLA underwent a bathochromic effect due to vicinal interactions between polar groups (O-C=O), in contrast with PDL, in which there were longer distances between these groups [31].

Molar mass analysis and the distributions of the obtained polyesters are shown in Tables 1–3. In Series I–III, the theoretical average molecular weight (M_n) was calculated, ranging from 28.5 to 42.5 kDa (Tables S1–S3), while the experimental values of M_n and $Đ$ recorded by GPC gave M_n ranging 4.0 to 14.0 kDa, as shown in Figure 4, denoting that the target molar masses were not achieved through the performed ROP organocatalytic reactions. Taking into consideration the temperature (100 °C) and time established for all reactions (24 h), the low molecular weights were attributed to secondary transesterification reactions which took place during the polymerization of monomers that showed higher reactivity than ε-DL, especially L-LA and ε-CL, leading to the generation of polymeric chains with reduced lengths and high dispersities [32]. Therefore, in Series I and III, a narrowing tendency in the dispersities was observed as the ε-DL content increased. It should be mentioned that these results are approximative, given that the measurements were carried out using polystyrene standard references, which possess a different chemical nature than polyester-based systems. Likewise, chromatograms obtained from GPC allowed us to observe that the molar mass distributions had acceptable monomodal performance ($Đ \geq 1.63$) in all ROP reactions (Figure 5) [33]. In contrast, PLA homopolymer (A-1) showed the formation of an oligomer, observed as a bimodal distribution. This was attributed to the reaction conditions, wherein loss of control in the propagation step was induced, or by the presence of intermolecular transesterification reactions [34] due to long reaction periods (24 h). The sequential addition of the second monomer (L-LA, EB, ε-CL) at specific times, as shown in Tables S1–S3, led us to expect a block arrangement in the copolymers. Through the analysis of the resulting chromatograms, monomodal curves suggested that this sequential addition did not result in the formation of oligomers or low molar mass species.

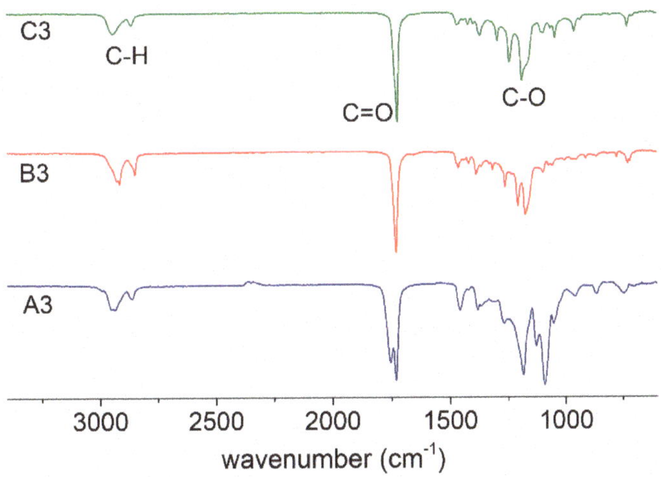

Figure 3. Absorption bands of characteristic functional groups through FTIR in A3, B3 and C3.

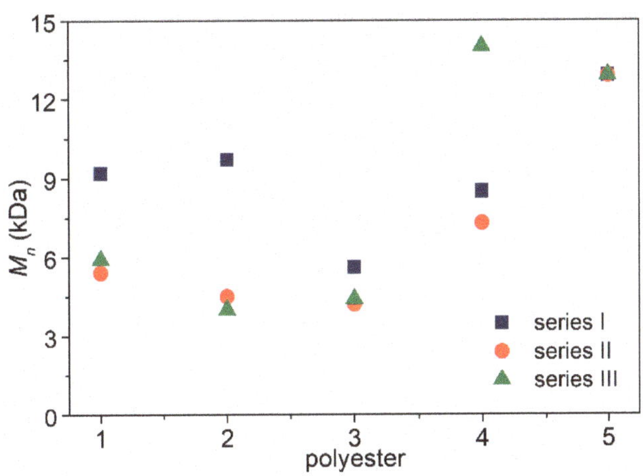

Figure 4. Average number molar mass (M_n) in Series I–III.

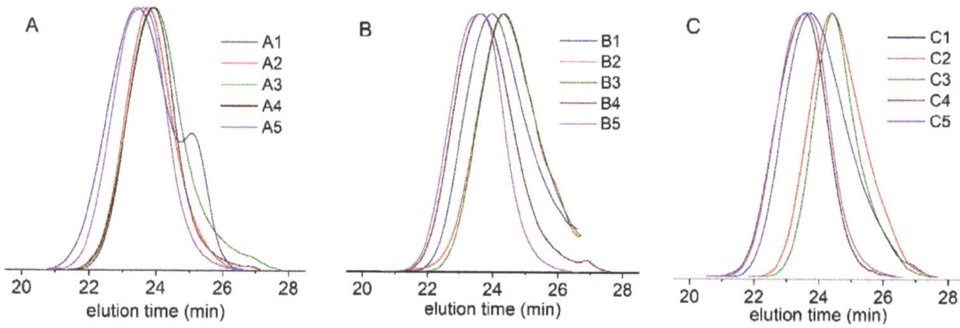

Figure 5. GPC analysis of polyesters of Series I (**A**), Series II (**B**) and Series III (**C**).

3.2. Thermal and Morphological Analysis

Thermal stability in Series I–III was measured to determine the degradation profiles (Supplementary Materials, Figures S8–S10) of these compounds. Data are reported in Tables 1–3, taking a 5% weight loss as a reference. The copolymers in Series I exhibited the lowest stability, in a range from 203 to 237 °C, in comparison to Series II and III (294–324 °C). It should be mentioned that Series I contained PLA blocks, wherein carboxylic groups are closer to each other in comparison to PEB, PCL and PDL blocks. As such, higher quantity of O=C-O groups favored chains scission and the formation of CO_2 as a byproduct at lower temperatures [10]. Likewise, thermal stability is important for setting heating–cooling temperatures in DSC analyses, making it possible to perform tests without compromising the chemical structure of the polyesters.

Regarding DSC, analyses were recorded from −70 to 170 °C in Series I and −70 to 100 °C in Series II and III. Some differences in thermal transitions values were observed during the heating and cooling processes, as shown in Figure 6 and Tables 1–3. For instance, A1-A5 only exhibited second order transitions (T_g), possibly due to the amorphous arrangement in the PLA and PDL blocks (Figure 6A,B). It should be noted that the polyesters in Series I were synthesized at 100 °C, which would have a detrimental effect on the crystallinity of PLA. This fact is supported by the reports mentioned in Section 3.1, where, for example, the PLA obtained using TBD as catalyst at a temperature of 23 °C developed an atactic arrangement [29]. In addition, some works related to block copolymers based on L-LA and ε-DL observed a similar thermal performance, as well as loss of crystalline order caused by microphase separation and perturbation of chain interactions due to side butyl groups in PDL segments [21,30].

In Series II, B1 exhibited endotherms at 57, 62 and 68.8 °C (Figure 6C) that may be related to semicrystalline arrangements. Additionally, the multiplicity in transitions could be associated with differences in the lamellar thickness [35]. This thermal performance could be supported in the structural features of EB, taking in consideration its 11 methylene units between carboxylic groups, and so, the macromolecules can acquire several arrangements that are function of carboxylic interaction and stability in the backbone conformation Likewise, B2–B4 exhibited endotherms in a wide temperature range, while B3 and B4 developed melt recrystallization events, in which the presence of a branched comonomer (ε-DL) had an impact on the crystallization decrease due to decoupled intermolecular interactions. B5 (PDL homopolymer) only displayed T_g (−55.1 °C), which is associated with amorphous domains.

In series III, C1–C3 exhibited endothermic transitions composed of two broad peaks that may have been related to different crystalline arrangements (Figure 6E). These transitions could have a correlation with ordered lamellar domains in the PCL blocks (higher temperature peak) in coexistence with backbone segments that suffer from reduced crystallinity due to their reduced interactions and chain folding [36]. It is worth mentioning that the percentage of ε-DL ≥ 30% mol was sufficient to achieve a transition from semicrystalline to amorphous arrangement, as shown in C4. These morphological changes are of interest for biodegradable applications, where control over degradation rates is required.

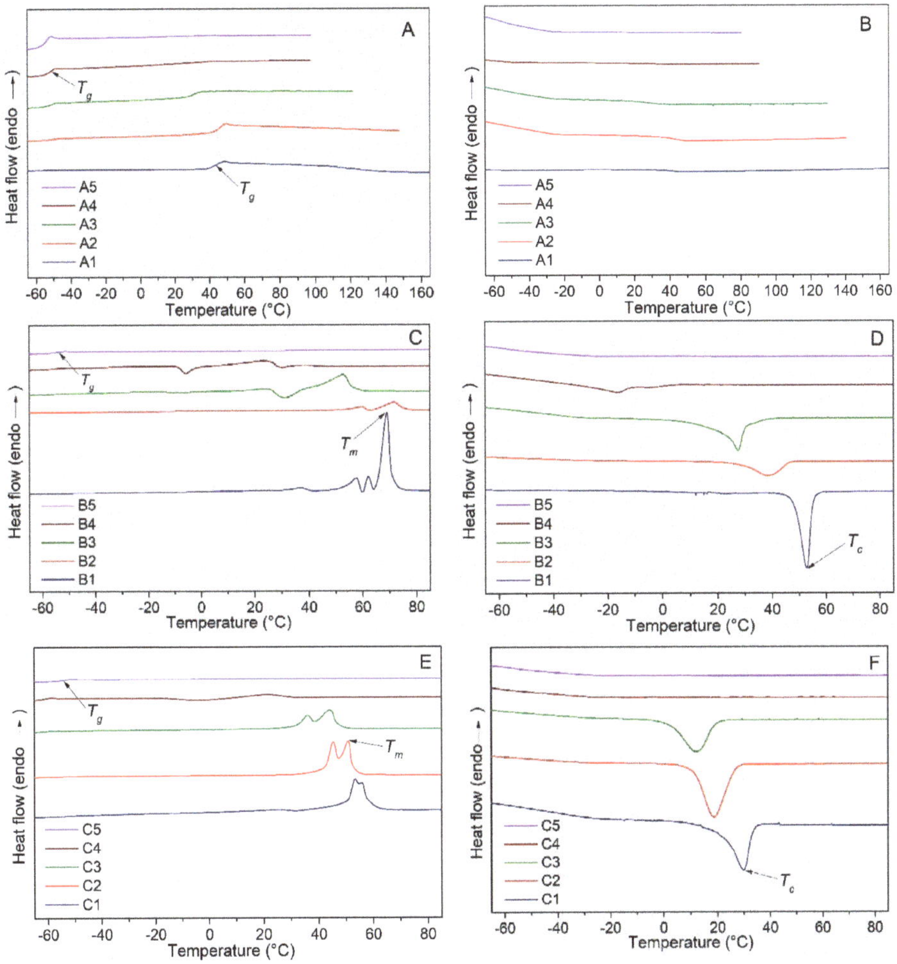

Figure 6. Thermograms of Series I–III. (**A**) heating-Series I; (**B**) cooling-Series I; (**C**) heating-Series II; (**D**) cooling-Series II; (**E**) heating-Series III; (**F**) cooling-Series III.

Regarding the morphological features in Series I–III, XRD patterns were collected to elucidate the structural organization. For instance, diffractograms in Series I exhibited broad undefined peaks in region 2θ = 10–30°, characteristic of non-ordered domains. This amorphous arrangement was predicted in formulations based on L-LA:ε-DL (Figure 7A). In the case of Series II and III, polyesters with low percentages of ε-DL developed a semicrystalline arrangement with similar diffractograms and 2θ values (Figure 7A,B). Herein, they exhibited a main sharp peak at 2θ = 21.45°, corresponding to reflection (110), and a shorter signal at 2θ = 23.8°, corresponding to reflection (200) [37,38]. Additionally, an increase of ε-DL in formulations led to an intensity reduction in the diffraction peaks, denoting a loss of crystalline domains in the polyesters.

Table 1. Comonomer molar percentage, M_n, M_w, Đ, thermal properties and X_c in Series I.

Series I	L-LA (% mol)	ε-DL (% mol)	M_n (kDa)	M_w (kDa)	Đ	T_m (°C)	T_c (°C)	T_g (°C)	TGA (5%w)	[c] X_c
A-1	100	0	9.2	20.8	2.26	-	-	43.1	203	-
A-2	91.6	8.4	9.7	15.9	1.64	-	-	[a] −51.5, [b] 45	247	-
A-3	76.2	23.8	5.6	13.6	2.42	-	-	−51.7, 30	217	-
A-4	54.4	46.6	8.5	14.8	1.74	-	-	−53.2	237	-
A-5	0	100	12.9	21.7	1.68	-	-	−55.1	316	-

[a] T_g in PDL; [b] T_g in PLA; [c] X_c obtained via integration peaks from XRD diffractograms (Figure 7).

Table 2. Comonomer molar percentage, M_n, M_w, Đ, thermal properties and X_c in Series II.

Series II	EB (% mol)	ε-DL (% mol)	M_n (kDa)	M_w (kDa)	Đ	T_m (°C)	T_c (°C)	T_g (°C)	TGA (5%w)	X_c
B-1	100	0	5.4	12.6	2.34	68.8	52.7	-	294	1.0
B-2	93.0	7.0	4.5	9.9	2.2	71.8	38.5	-	321	0.90
B-3	77.0	23.0	4.2	8.9	2.12	52.6	27.5	−56.8	316	0.45
B-4	67.6	32.4	7.3	17.8	2.44	23.5	−16.1	−56.3	318	-
B-5	0	100	12.9	21.7	1.68	-	-	−55.1	3.16	-

Table 3. Comonomer molar percentage, M_n, M_w, Đ, thermal properties and X_c in Series III.

Series III	ε-CL (% mol)	ε-DL (% mol)	M_n (kDa)	M_w (kDa)	Đ	T_m (°C)	T_c (°C)	T_g (°C)	TGA (5%w)	X_c
C-1	100	0	5.9	15.2	2.57	55.8	29.8	-	324	1.0
C-2	93.5	6.5	4.0	8.2	2.04	50.8	18.6	-	317	1.01
C-3	83.4	16.6	4.4	7.8	1.78	44.1	12.1	−61.7	317	0.25
C-4	69.9	30.1	14	22.8	1.63	21.5	-	−62.1	319	-
C-5	0	100	12.9	21.7	1.68	-	-	−55.1	316	-

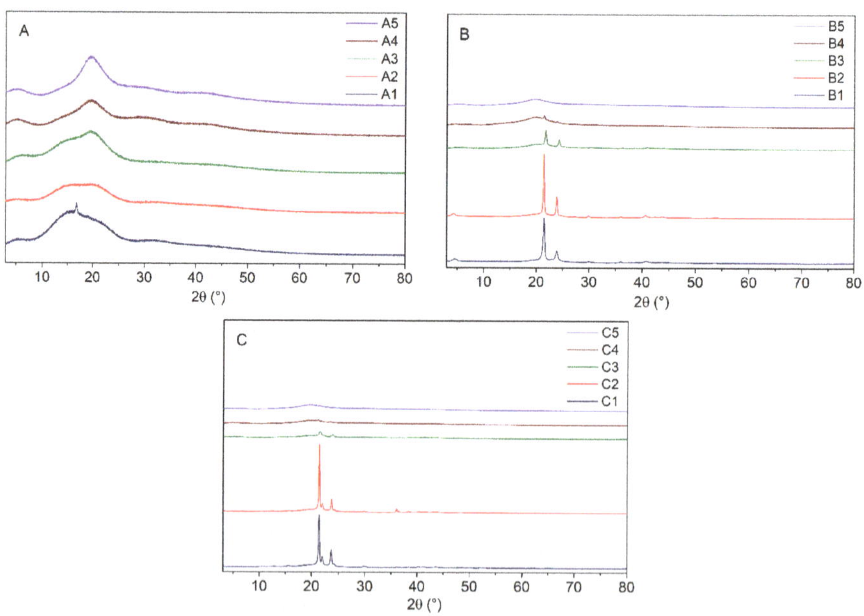

Figure 7. Diffractograms in Series I (**A**), Series II (**B**) and Series III (**C**).

4. Conclusions

In this work, the synthesis of three series of polyesters using a combination of an alkyl substituted monomer (ε-DL) with other non-susbstituted monomers (L-LA, EB, ε-CL) was achieved. The architecture of the copolymers was induced block-type, taking into consideration the sequential addition of monomers. TBD, an organic catalyst, was successfully implemented in ROP reactions without usage of anhydrous handling conditions. Therefore, this methodology may be implemented with other monomers, avoiding the requirement of glove box equipment.

In these copolymers, a reduction of crystallinity was observed as result of an increase in the molar content of ε-DL. In particular, at molar concentrations ≥30%, the remains of crystalline domains were embedded into the amorphous domains. The ROP of L-LA under the established reaction conditions resulted in an amorphous polyester, despite the usage of a reagent with specific stereoregularity.

Generally speaking, the crystallization features of these copolymers can be modified with branched comonomer inclusion in the backbone of the polyesters. Therefore, it is suggested that the degradation rates of this class of materials can be modified by changes in the chemical arrangement of their backbone. In addition, some of the studied polyesters are currently being analyzed under hydrolytic degradation conditions, and encouraging results are being obtained. Likewise, our copolymers are of potential value for the development of soft medical devices with biodegradability and biocompatibility features, as their synthesis does not involve the use of organometallic reagents.

Supplementary Materials: The following supporting information can be downloaded at: https://www.mdpi.com/article/10.3390/polym14204278/s1, Table S1. Nomenclature and molar ratios in synthesis of Series I; Table S2. Nomenclature and molar ratios in synthesis of Series II; Table S3. Nomenclature and molar ratios in synthesis of Series III; Figure S1. ^{1}H NMR spectra of monomers; Figure S2. ^{1}H NMR spectra in polyesters of Series I; Figure S3. ^{1}H NMR spectra in polyesters of Series II; Figure S4. ^{1}H NMR spectra in polyesters of Series III; Figure S5. FTIR-ATR spectra in polyesters of Series I; Figure S6. FTIR-ATR spectra in polyesters of Series II; Figure S7. FTIR-ATR spectra in polyesters of Series III; Figure S8. TGA thermograms in polyesters of Series I; Figure S9. TGA thermograms in polyesters of Series II; Figure S10. TGA thermograms in polyesters of Series III.

Author Contributions: F.R.-G.: Investigation, review and editing; T.R.-H.: Methodology, validation; A.S.L.-P.: Conceptualization, review and editing, supervision; R.D.d.L.: Conceptualization, review and editing, supervision; M.A.D.J.-T.: Investigation, conceptualization, validation, writing original draft; H.R.L.-G.: Conceptualization, funding acquisition, project administration, review and editing. All authors have read and agreed to the published version of the manuscript.

Funding: The authors acknowledge the financial support of the Mexican National Council of Science and Technology (CONACyT) through the Basic Science Project A1-S-34241.

Institutional Review Board Statement: Not applicable.

Informed Consent Statement: Not Applicable.

Data Availability Statement: Not applicable.

Acknowledgments: Authors acknowledge technical support of Judith A. Romero-Cabello, Maricela Garcia-Zamora, Myrna Salinas-Hernández, Guadalupe Mendez-Padilla, Jesus Alfonso Mercado, José Alejandro Díaz, Ricardo Mendoza, and Martha Roa. MAJT acknowledges the financial support from Consejo Nacional de Ciencia y Tecnología (Conacyt) through the Postdoctoral grants program and Conacyt-Project: 76219.

Conflicts of Interest: The authors declare no conflict of interest.

References

1. Williams, D.F. Specifications for innovative, enabling biomaterials based on the principles of biocompatibility mechanisms. *Front. Bioeng. Biotechnol.* **2019**, *7*, 1–10. [CrossRef] [PubMed]
2. Prasad, K.; Bazaka, O.; Chua, M.; Rochford, M.; Fedrick, L.; Spoor, J.; Symes, R.; Tieppo, M.; Collins, C.; Cao, A.; et al. Metallic biomaterials: Current challenges and opportunities. *Materials* **2017**, *10*, 884. [CrossRef] [PubMed]
3. Puppi, D.; Chiellini, F. Biodegradable polymers for biomedical additive manufacturing. *Appl. Mater. Today* **2020**, *20*, 100700. [CrossRef]
4. Ulery, B.D.; Nair, L.S.; Laurencin, C.T. Biomedical applications of biodegradable polymers. *J. Polym. Sci. Part B Polym. Phys.* **2011**, *49*, 832–864. [CrossRef]
5. Song, R.; Murphy, M.; Li, C.; Ting, K.; Soo, C.; Zheng, Z. Current development of biodegradable polymeric materials for biomedical applications. *Drug Des. Dev. Ther.* **2018**, *12*, 3117–3145. [CrossRef]
6. Bu, Y.; Ma, J.; Bei, J.; Wang, S. Surface modification of aliphatic polyester to enhance biocompatibility. *Front. Bioeng. Biotechnol.* **2019**, *7*, 1–10. [CrossRef]
7. Manavitehrani, I.; Fathi, A.; Badr, H.; Daly, S.; Negahi Shirazi, A.; Dehghani, F. Biomedical applications of biodegradable polyesters. *Polymers* **2016**, *8*, 20. [CrossRef]
8. Armentano, I.; Gigli, M.; Morena, F.; Argentati, C.; Torre, L.; Martino, S. Recent advances in nanocomposites based on aliphatic polyesters: Design, synthesis, and applications in regenerative medicine. *Appl. Sci.* **2018**, *8*, 1452. [CrossRef]
9. Dhaliwal, K. Biodegradable polymers and their role in drug delivery systems. *Biomed. J. Sci. Tech. Res.* **2018**, *11*, 8315–8320. [CrossRef]
10. Olsén, P.; Borke, T.; Odelius, K.; Albertsson, A.C. ε-decalactone: A thermoresilient and toughening comonomer to poly(L-lactide). *Biomacromolecules* **2013**, *14*, 2883–2890. [CrossRef]
11. Lin, L.; Xu, Y.; Wang, S.; Xiao, M.; Meng, Y. Ring-opening polymerization of L-lactide and ε-caprolactone catalyzed by versatile tri-zinc complex: Synthesis of biodegradable polyester with gradient sequence structure. *Eur. Polym. J.* **2016**, *74*, 109–119. [CrossRef]
12. Jin, C.; Wei, Z.; Yu, Y.; Sui, M.; Leng, X.; Li, Y. Copolymerization of ethylene brassylate with δ-valerolactone towards isodimorphic random copolyesters with continuously tunable mechanical properties. *Eur. Polym. J.* **2018**, *102*, 90–100. [CrossRef]
13. Fernández, J.; Larrañaga, A.; Etxeberria, A.; Sarasua, J.R. Ethylene brassylate-co-δ-hexalactone biobased polymers for application in the medical field: Synthesis, characterization and cell culture studies. *RSC Adv.* **2016**, *6*, 22121–22136. [CrossRef]
14. Urbánek, T.; Trousil, J.; Rak, D.; Gunár, K.; Konefał, R.; Šlouf, M.; Sedlák, M.; Šebestová Janoušková, O.; Hrubý, M. γ-butyrolactone copolymerization with the well-documented polymer drug carrier poly(ethylene oxide)-block-poly(ε-caprolactone) to fine-tune its biorelevant properties. *Macromol. Biosci.* **2020**, *20*, 1–15. [CrossRef]
15. Atanase, L.I.; Salhi, S.; Cucoveica, O.; Ponjavic, M.; Nikodinovic-Runic, J.; Delaite, C. Biodegradability assessment of polyester copolymers based on poly(ethylene adipate) and poly(ε-caprolactone). *Polymers* **2022**, *14*, 3736. [CrossRef]
16. Hu, Y.; Daoud, W.A.; Cheuk, K.K.L.; Lin, C.S.K. Newly developed techniques on polycondensation, ring-opening polymerization and polymer modification: Focus on poly(lactic acid). *Materials* **2016**, *9*, 133. [CrossRef]
17. Koltzenburg, S.; Maskos, M.; Nuyken, O. Ring opening polymerization. In *Polymer Chemistry*; Koltzenburg, S., Maskos, M., Nuyken, O., Eds.; Springer: Berlin/Heidelberg, Germany, 2017; pp. 321–347. [CrossRef]
18. Jasinska-Walc, L.; Hansen, M.R.; Dudenko, D.; Rozanski, A.; Bouyahyi, M.; Wagner, M.; Graf, R.; Duchateau, R. Topological behavior mimicking ethylene-hexene copolymers using branched lactones and macrolactones. *Polym. Chem.* **2014**, *5*, 3306–3310. [CrossRef]
19. Krukiewicz, K.; Fernandez, J.; Skorupa, M.; Więcławska, D.; Poudel, A.; Sarasua, J.-R.; Quinlan, L.R.; Biggs, M.J.P. Analysis of a poly(ε-decalactone)/silver nanowire composite as an electrically conducting neural interface biomaterial. *BMC Biomed. Eng.* **2019**, *1*, 1–12. [CrossRef]
20. Cichoń, K.; Kost, B.; Basko, M. Synthesis and properties of ABA-triblock copolymers from polyester A-blocks and easily degradable polyacetal B-blocks. *Polym. Chem.* **2022**, *13*, 5243–5255. [CrossRef]
21. Martello, M.T.; Schneiderman, D.K.; Hillmayer, M.A. Synthesis and melt processing of sustainable poly(ε-decalactone)-block-poly(lactide) multiblock thermoplastic elastomers. *ACS Sustain. Chem. Eng.* **2014**, *2*, 2519–2526. [CrossRef]
22. Dzienia, A.; Maksym, P.; Hachuła, B.; Tarnacka, M.; Biela, T.; Golba, S.; Zięba, A.; Chorażewski, M.; Kaminski, K.; Paluch, M. Studying the catalytic activity of DBU and TBD upon water-initiated ROP of ε-caprolactone under different thermodynamic conditions. *Polym. Chem.* **2019**, *10*, 6047–6061. [CrossRef]
23. Bossion, A.; Heifferon, K.V.; Meabe, L.; Zivic, N.; Taton, D.; Hedrick, J.L.; Long, T.E.; Sardon, H. Opportunities for organocatalysis in polymer synthesis via step-growth methods. *Prog. Polym. Sci.* **2019**, *90*, 164–210. [CrossRef]
24. Nifant'ev, I.; Ivchenko, P. DFT modeling of organocatalytic ring-opening polymerization of cyclic esters: A crucial role of proton exchange and hydrogen bonding. *Polymers* **2019**, *11*, 2078. [CrossRef] [PubMed]
25. Lohmeijer, B.G.G.; Pratt, R.C.; Leibfarth, F.; Logan, J.W.; Long, D.A.; Dove, A.P.; Nederberg, F.; Choi, J.; Wade, C.; Waymouth, R.M.; et al. Guanidine and amidine organocatalysts for ring-opening polymerization of cyclic esters. *Macromolecules* **2006**, *39*, 8574–8583. [CrossRef]

26. De Jesús-Téllez, M.A.; Robles-González, F.; Díaz de León, R.; Ledezma-Pérez, A.; López-González, H.R. Síntesis de poliésteres alifáticos vía polimerización por apertura de anillo organocatalítica: Estudio de la influencia de los parámetros de reacción sobre sus propiedades térmicas y estructura molecular. *Cienc. UANL* **2022**, *25*, 39–47. [CrossRef]
27. Thomas, C.; Peruch, F.; Bibal, B. Ring-opening polymerization of lactones using supramolecular organocatalysts under simple conditions. *RSC Adv.* **2012**, *2*, 12851–12856. [CrossRef]
28. Ramos-Durán, G.; del Carmen González-Zarate, A.; Enríquez-Medrano, F.J.; Salinas-Hernández, M.; De Jesús-Téllez, M.A.; Díaz de León, R.; López-González, H.R. Synthesis of copolyesters based on substituted and non-substituted lactones towards the control of their crystallinity and their potential effect on hydrolytic degradation in the design of soft medical devices. *RSC Adv.* **2022**, *12*, 18154–18163. [CrossRef]
29. Moins, S.; Hoyas, S.; Lemaur, V.; Orhan, B.; Delle Chiaie, K.; Lazzaroni, R.; Taton, D.; Dove, A.P.; Coulembier, O. Stereoselective ROP of rac- and meso-lactides using achiral TBD as catalyst. *Catalysts* **2020**, *10*, 620. [CrossRef]
30. Schneiderman, D.K.; Hill, E.M.; Martello, M.T.; Hillmyer, M.A. Poly(lactide)-block-poly(ε-caprolactone-co-ε-decalactone)-block-poly(lactide) copolymer elastomers. *Polym. Chem.* **2015**, *6*, 3641–3651. [CrossRef]
31. Pretsch, E.; Bühlmann, P.; Badertscher, M. IR spectroscopy. In *Structure Determination of Organic Compounds*, 4th ed.; Pretsch, E., Bühlmann, P., Badertscher, M., Eds.; Springer: Berlin/Heidelberg, Germany, 2009; pp. 269–335. [CrossRef]
32. Liu, C.; Lin, S.; Zhou, C.; Yu, W. Influence of catalyst on transesterification between poly(lactic acid) and polycarbonate under flow field. *Polymer* **2013**, *54*, 310–319. [CrossRef]
33. Duda, A.; Kowalski, A. Thermodynamics and kinetics of ring-opening polymerization. In *Handbook of Ring-Opening Polymerization*; Dubois, P., Coulombier, O., Raquez, J.M., Eds.; Wiley: Weinheim, Germany, 2009; pp. 1–53. [CrossRef]
34. Cicogna, F.; Giachi, G.; Rosi, L.; Passaglia, E.; Coiai, S.; Spiniello, R.; Prescimone, F.; Frediani, M. Macromolecular dyes by chromophore-initiated ring opening polymerization of L-lactide. *Polymers* **2020**, *12*, 1979. [CrossRef] [PubMed]
35. Marxsen, S.F.; Song, D.; Zhang, X.; Flores, I.; Fernández, J.; Sarasua, J.R.; Müller, A.J.; Alamo, R.G. Crystallization rate minima of poly(ethylene brassylate) at temperatures transitioning between quantized crystal thicknesses. *Macromolecules* **2022**, *55*, 3958–3973. [CrossRef]
36. Kayser, F.; Fleury, G.; Thongkham, S.; Navarro, C.; Martin-Vaca, B.; Bourissou, D. Reducing the crystallinity of PCL chains by copolymerization with substituted δ/ε-lactones and its impact on the phase separation of PCL-based block copolymers. *Polym. Chem.* **2022**, *13*, 2201–2214. [CrossRef]
37. Xi, X.; Jiang, G.; Wang, X.; Hu, R.; Wang, R. Synthesis, characterization and degradation properties of poly(α-angelica lactone-co-ε-caprolactone) copolymers. *Polym. Renew. Resour.* **2013**, *4*, 49–60. [CrossRef]
38. Wei, Z.; Jin, C.; Xu, Q.; Leng, X.; Wang, Y.; Li, Y. Synthesis, microstructure and mechanical properties of partially biobased biodegradable poly(ethylene brassylate-co-ε-caprolactone) copolyesters. *J. Mech. Behav. Biomed. Mater.* **2019**, *91*, 255–265. [CrossRef] [PubMed]

polymers

MDPI

Review

Facile Synthesis of Novel Disperse Dyes for Dyeing Polyester Fabrics: Demonstrating Their Potential Biological Activities

Alya M. Al-Etaibi [1,*] and Morsy Ahmed El-Apasery [2]

[1] Natural Science Department, College of Health Science, Public Authority for Applied Education and Training, Fayha 72853, Kuwait
[2] Dyeing, Printing and Textile Auxiliaries Department, Textile Research and Technology Institute, National Research Centre, 33 El Buhouth St., Dokki, Cairo 12622, Egypt
* Correspondence: alya.aletaibi@yahoo.com; Tel.: +96-599-807-246

Abstract: Original work showed the composition of the dyes and the antimicrobial/UV protective properties of a series of dyes obtained in our laboratories over the past twelve years in an easy way using microwave technology and their comparisons with conventional methods. The results we obtained clearly indicated that by using the microwave strategy, we were able to synthesize the new disperse dyes in minutes and with a much higher productivity when compared to the traditional methods, which took a much longer time, sometimes up to hours. We also introduced ultrasonic technology in dyeing polyester fabrics at 80 °C for an environmentally friendly approach, which was an alternative to traditional dyeing methods at 100 °C; we obtained a much higher color depth than traditional dyeing methods reaching 102.9%. We presented both the biological activity of the prepared new dyes and the fastness properties and clearly indicated that these dyes possess biological activity and high fastness properties.We presented through the results that when dyeing polyester fabrics with some selected disperse dyes, the color strength of polyester fabrics dyed at high temperatures was greater than the color strength of polyester fabrics dyed at low temperatures by 144%, 186%, 265% and 309%. Finally, we presented that a ZnO or TiO_2 NPs post-dyeing treatment of polyester fabrics is promising strategy for producing polyester fabrics possess multifunction like self-cleaning property, high light fastness, antimicrobial and anti-ultraviolet properties.

Keywords: disperse dyes; polyester fabrics; microwave irradiation; ultraviolet protection factor

Citation: Al-Etaibi, A.M.; El-Apasery, M.A. Facile Synthesis of Novel Disperse Dyes for Dyeing Polyester Fabrics: Demonstrating Their Potential Biological Activities. *Polymers* **2022**, *14*, 3966. https://doi.org/10.3390/polym14193966

Academic Editor: Edina Rusen

Received: 28 August 2022
Accepted: 16 September 2022
Published: 22 September 2022

Publisher's Note: MDPI stays neutral with regard to jurisdictional claims in published maps and institutional affiliations.

Copyright: © 2022 by the authors. Licensee MDPI, Basel, Switzerland. This article is an open access article distributed under the terms and conditions of the Creative Commons Attribution (CC BY) license (https://creativecommons.org/licenses/by/4.0/).

1. Introduction

Since time immemorial, colors have played a major role in human life, as they are deeply linked to their means of living, their traditions, their ideas, their customs and their concepts. Since the dawn of humanity, man has been fascinated by the diversity of colors; therefore, he has been fascinated by beauty of their many colors, which has had far-reaching effects in his religious, emotional and social life. Azo disperse dyes were the most widely used synthetic dyes of the past decade because of their ease of manufacture and wide application in cosmetics, textile dyeing and printing. It is worth noting that heterogeneous, bright-colored. Azo dyes are widely used for dyeing polyester [1–12]. Polyester is one of the polymers containing ester bonds obtained from the polycondensation reaction of diol and dicarboxylic acid (Figure 1) [4].

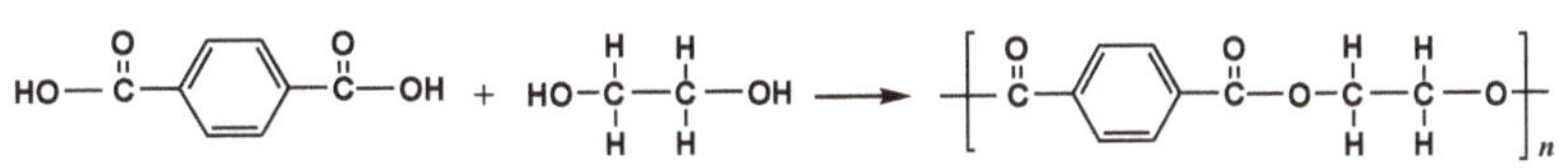

Figure 1. Structure of Polyethylene terephthalate.

Polyester (PET) is the most hydrophobic fiber among ordinary fibers, and this fiber is less prone to wrinkling and has excellent washability and abrasion resistance. It is known that the polyester dyeing process is performed by water dyeing using dispersed dyes in the presence of high temperatures and pressures. It is known that azo dyes are widely used in textile dyeing in view of environmental pollution, so azo dyes harm the environment if their effluents are not treated. Among these treatments are modern techniques using ultrasound, infrared, microwaves, and ultraviolet rays to improve the fabric's absorption capacity for dyes and reduce effluents load from dyeing processes [13–24]. Some recent reports [1,2] have used ultrasound energy to improve the dyeing of polyester fabrics when dyeing with disperse dyes by modifying the surface of the fabric. Additionally, some studies [3] confirmed that UV rays can be used to improve the color fastness properties without compromising the chemical properties of polyester fabric. One of the environmentally friendly advantages of using microwave energy is the rapid heating to high temperatures that allows for greater ease of reactions, because increasing the frequency of molecular vibrations during microwave irradiation speeds up these reactions as well as reduces the use of solvents [25–66]. Skin cancer and sensitive skin are primarily caused by prolonged exposure to ultraviolet radiation from sunlight. Therefore, the best choice is to locate well-designed clothing composed of ultraviolet-blocking materials. One of the qualities cloth should have is the ability to filter ultraviolet rays, either by containing an ultraviolet absorber or by being dyed, as dyed fabrics offer better sun protection than bleached materials. Therefore, it could be claimed that dyed cloths with darker colors may provide better protection from the Sun rays [1,4]. According to a literature survey, pyrazolopyrimidines, pyridones and enaminones are significant key intermediates in the production of new azo dyes, particularly disperse dyes with numerous biological applications. Because of their facile synthesis, low cost, and great efficacy against bacterial infections, they are now widely employed as antimicrobial agents [4]. In general, the kinds of substituted function present in the heterocyclic ring structure, as well as the form of the heterocyclic system, have a significant impact on the antibacterial activity of the entire molecule. The majority of the matching substituted analogues had stronger antibacterial inhibitory activities that were on par with or better than the reference drugs. A lot of researchers are now interested in creating smart fabrics for human wellbeing. The need to create materials that improve defense against microorganisms such as bacteria or fungi is expanding. Due to their propensity to retain moisture, fabrics can act as transporters and ideal environments for the growth of germs such as bacteria, especially when they come into contact with the human body. Therefore, it is essential to create textiles that are microbe-resistant. To produce smart fabrics, we had to improve the functional performance of polyester fabrics by using nanotechnology. To increase their anti-bacterial and UV-protective properties and to achieve this goal, nanoparticles were used, especially metal oxides and metal nanoparticles. The potential mechanism behind titanium dioxide NPs' antibacterial action is their photocatalytic function, which results in the generation of extraordinarily active oxygen species, such as super oxide anions, hydrogen peroxide, singlet oxygen, and hydroxyl radicals, which kill bacteria. Due to the presence of the outer cell wall membrane in the G-ve bacteria, which acts as a barrier to the antibacterial action, the NPs-treated polyester fabric demonstrated greater antibacterial activity against G+ve bacteria in companion with G-ve bacteria. In this review article, we present the contributions of our laboratories over the past twelve years using modern technologies such as ultrasound or microwave technology to synthesize many new disperse dyes while demonstrating their biological activities. We will not only present the use of modern technologies such as ultrasound or microwave technology in dyeing polyester fabrics and compare them with traditional dyeing methods but also discuss the process of treating polyester fabrics dyed with ZnO or TiO_2 NPs and whether it achieved a UV protection factor, as well as the ability of these polyester fabrics to improve the properties of self-cleaning, light fastness and, antimicrobial activities.

2. Materials

It is true that microwave and/or ultrasound technologies have many uses and unique advantages. However, we should also note that with a better understanding of the physical principles underlying the coupling mechanisms between microwave radiation or ultrasound energy and matter, we may be able to expand the applications of these technologies to cutting-edge scientific applications. In 2011, we formed a research team with the intention of synthesizing novel disperse dyes utilizing microwave heating by reacting hydrazine hydrate **4** with hydrazonocyanoacetate 3 [58]. By combining ethyl cyanoacetate **1** and diazoniun salt **2**, hydrazone **3** was formed (Scheme 1) [58,59]. As a result, dye 5 was formed when hydrazone **3** and hydrazine hydrate 4 interacted in ethanol. We found that compound **5** quickly condensed with acetyl acetone **6** to produce the disperse dye 7 under microwave irradiation (c.f. Scheme 1, Figure 2). Compound **5** interacted with enaminones 8a-d through microwave irradiation to create the disperse dyes **9a–d** (c.f. Scheme 1, Figure 3).

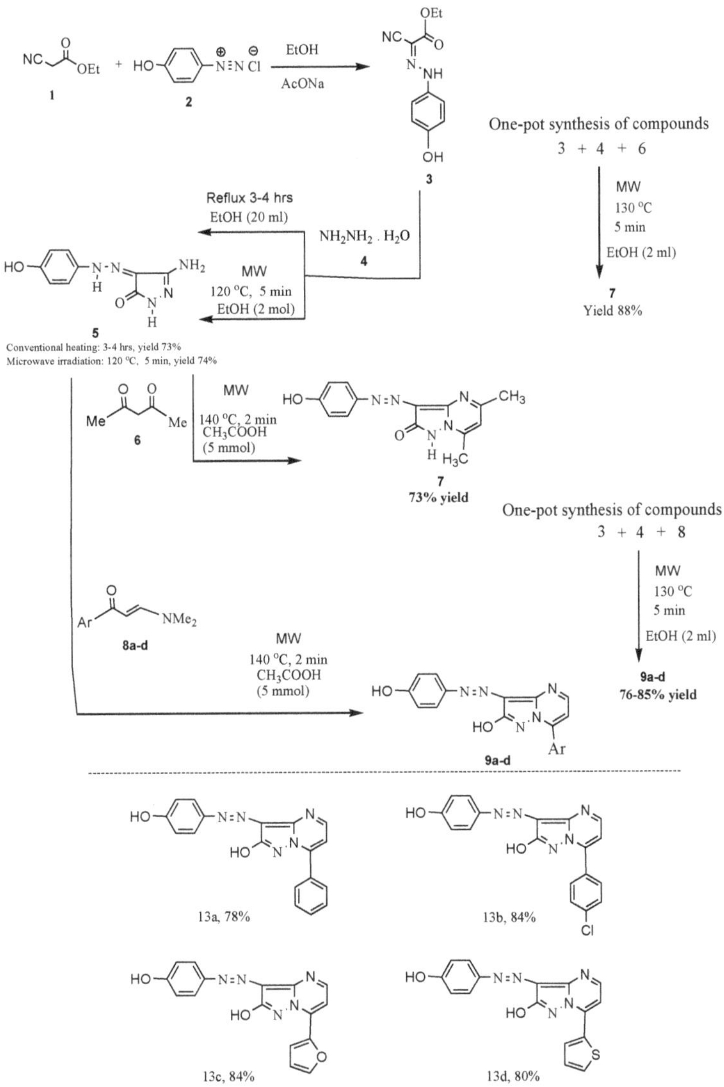

Scheme 1. Preparation of pyrazolopyrimidines disperse dyes 7 and 9.

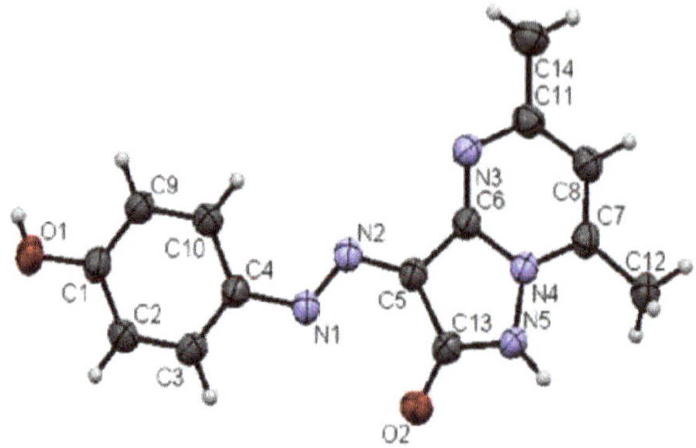

Figure 2. ORTEP of dye 7.

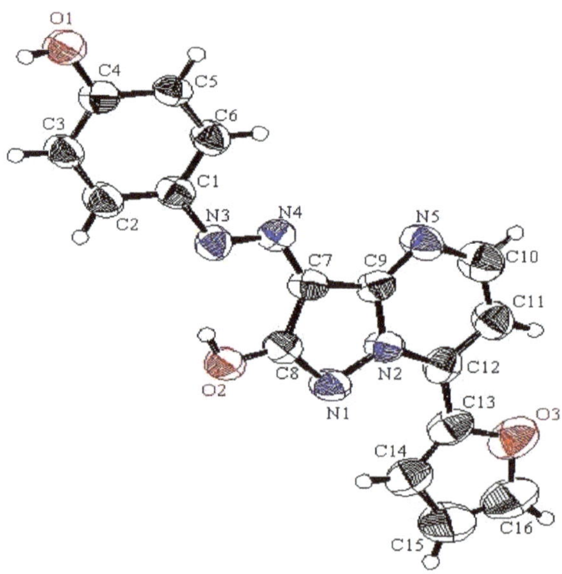

Figure 3. ORTEP of compound **9c**.

In 2012, we were able to explain the one-pot synthesis of compound **7** with a better outcome using microwave irradiation, using hydrazone 3, hydrazine hydrate **4** and acetyl acetone **6**. According to Scheme 1 [58], disperse dye 7 can exist in isomeric forms. We also reported in 2012 [58] that compounds **8a–d**, hydrazone **3**, and hydrazine hydrate **4** could be easily reacted as one-pot synthesis to synthesize compounds **9a–d** (c.f. Scheme 1).

One of the sequences used in the synthesis of the disperse dye **16a–h** was the coupling of malononitrile **10** with diazonium salt **2** to generate compound **12**. (Scheme 2). An increase in the aryl proton signal resulted from irradiating the hydroxyl signal, according to the Nuclear Overhauser Effect (NOE) measurements [59].

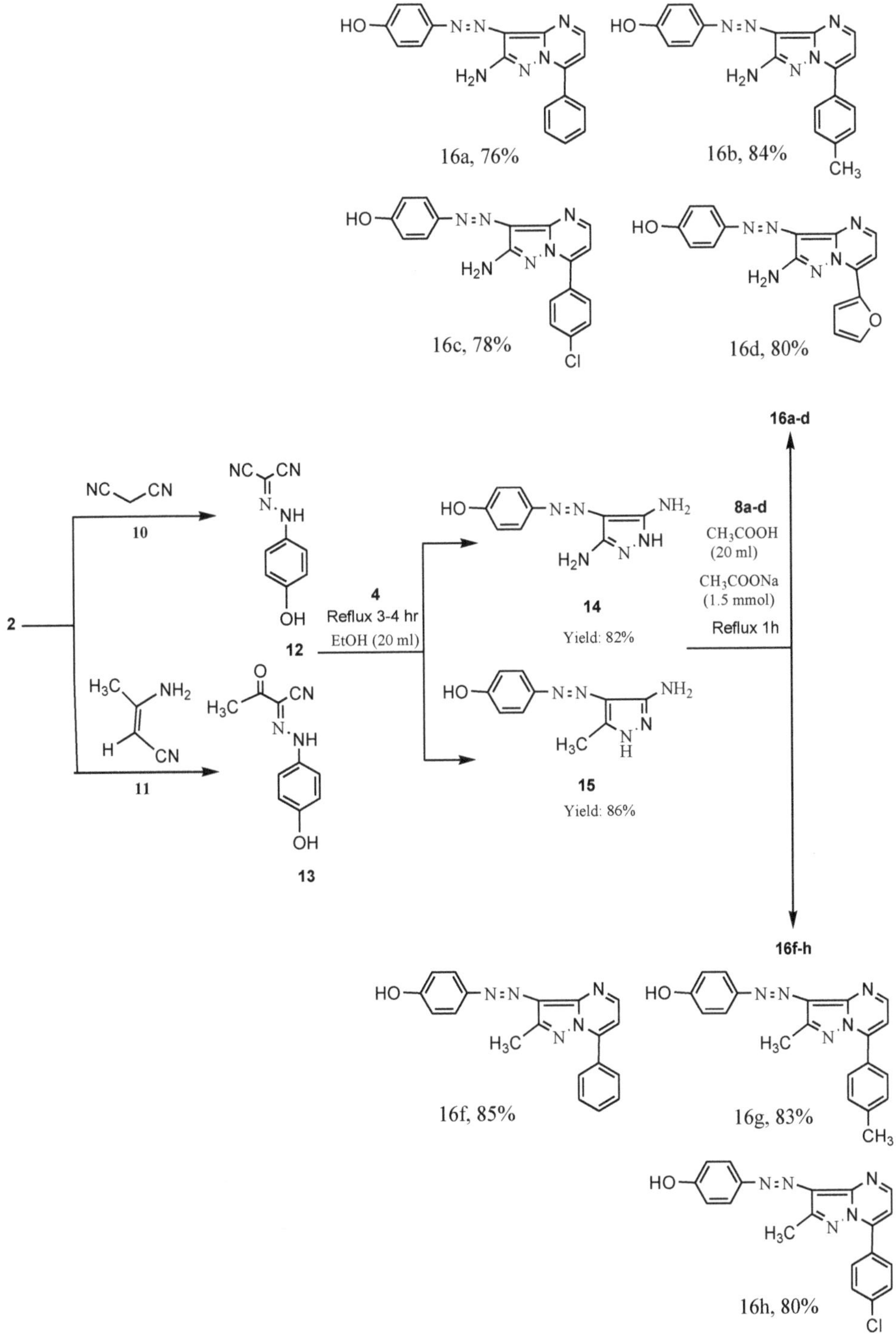

Scheme 2. Synthesis of compound **16**, **19** and **20**.

In order to create the disperse dye **14**, hydrazine hydrate **4** refluxed with compound **12** [60]. When compound **11** was combined with diazonium salt **2** in the presence of ethyl alcohol/sodium acetate, a high yield of hydrazone **13**—85%, was produced. In order to help construct the structure of **13**, NOE experiments were conducted. The results demonstrated that irradiating the NH signal at 12.1 ppm induced an amplification of the aryl proton resonances at 7.39 and 6.80 ppm (Scheme 2) [60].

Hydrazone **13** and hydrazine hydrate were combined to create the disperse dye **15** by refluxing the mixture for four hours while adding ethanol as a solvent. According to NOE difference measurements, irradiation of the NH signal at 11.94 ppm corresponding to compound **15** boosted the methyl proton signal at 2.36 ppm. Through refluxing for an hour in the presence of acetic acid and sodium acetate, dyes **14** or **15** easily condensed with the enaminones **8a–d** to generate disperse dyes **16a–d** or **16f–h** (Scheme 2) [60].

It is well known that dye **22** was created by combining different types of diazonium chloride **21** with diketones **23** to create substituted arylazodiketones **24**, which may then be condensed with cyanoacetamide derivatives **25** using either a traditional heating method or a microwave heating method [67–71]. (Scheme 3).

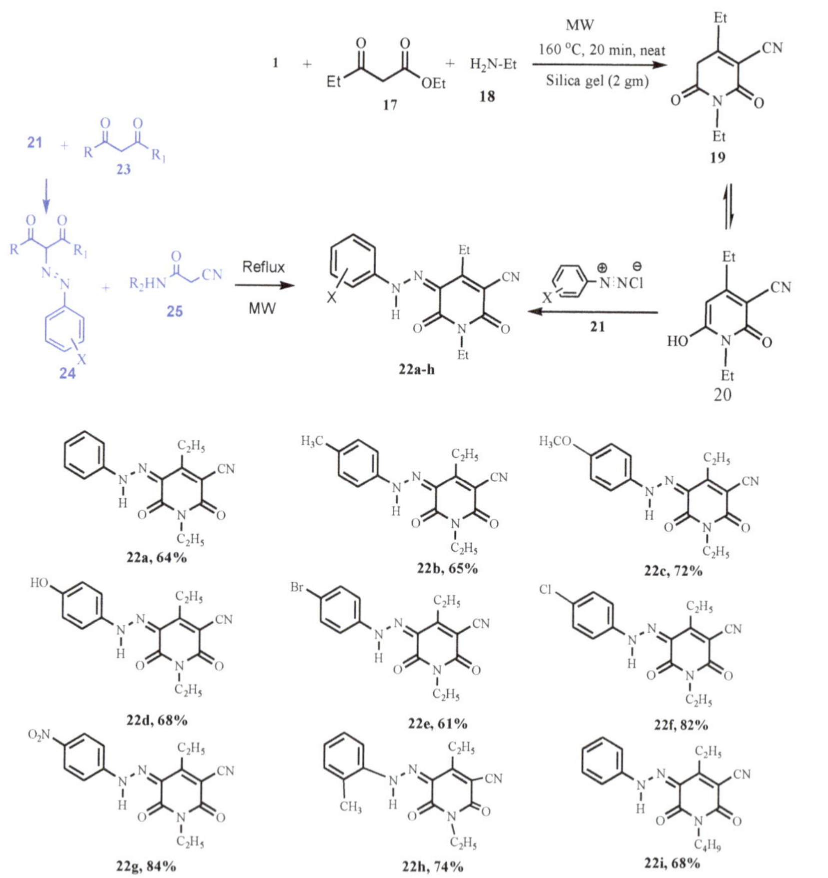

Scheme 3. Synthesis of disperse dyes based on pyridones moieties 22a–i.

On the other hand, the first reaction method for the synthesis of these azo dyes utilized the interaction between diazonium chloride and pyridones. We described a three-component condensation of methyl propionylacetate **17** as -ketoesters, ethyl cyanoacetates **1** and ethyl amines **18** to generate pyridine **19** in 2014 [61] by utilizing microwave irradiation at 160 °C for 20 min. It is important to note that in 2013 [57], we created compound **19** by heating it traditionally for 6 h. Two tautomeric forms, **19** and **20**, of this compound readily equilibrated in solution (Scheme 3).

Based on X-ray crystallographic structure determination, pyridine **19** could be combined with different diazonium salts **21** to create the disperse dyes **22a–i** that exist in the hydrazone tautomeric form, as shown in Scheme 3. (Figures 4 and 5).

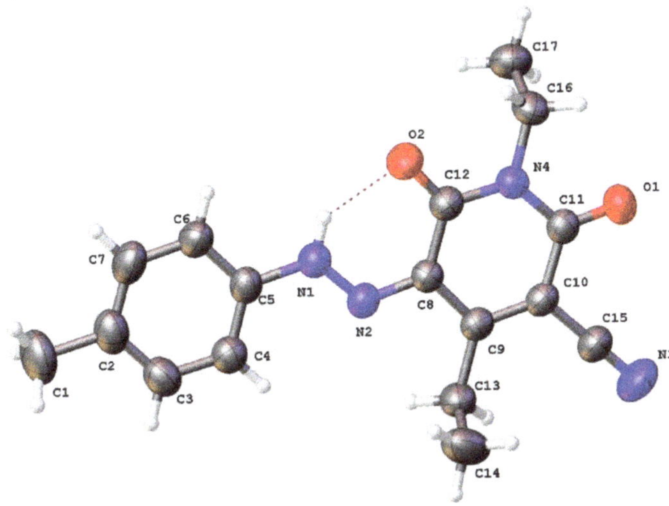

Figure 4. ORTEP of dye **22b**.

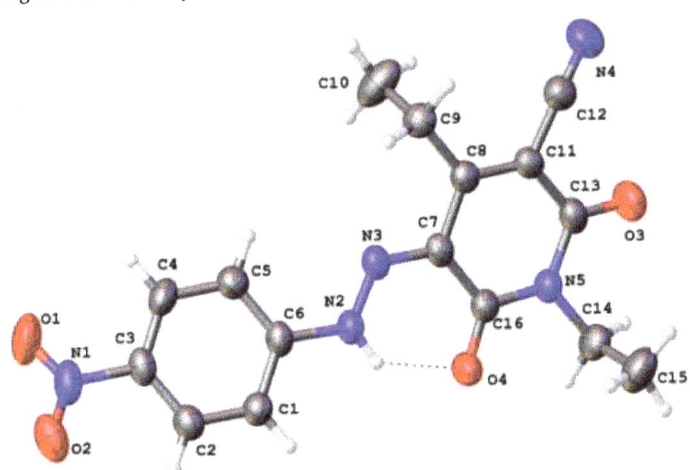

Figure 5. ORTEP of dye 22g.

In 2015 [72–74], we were able to produce novel disperse dyes that were safe for the environment. We accomplished this by reacting enaminones **8e** and **8f** with the phenyl diazonium salt **21a** in an acidic medium to produce the disperse dyes **26a** and **26b** that were 3-oxo-2-(phenylhydrazono)-3-p-arylpropionaldehydes (Scheme 4).

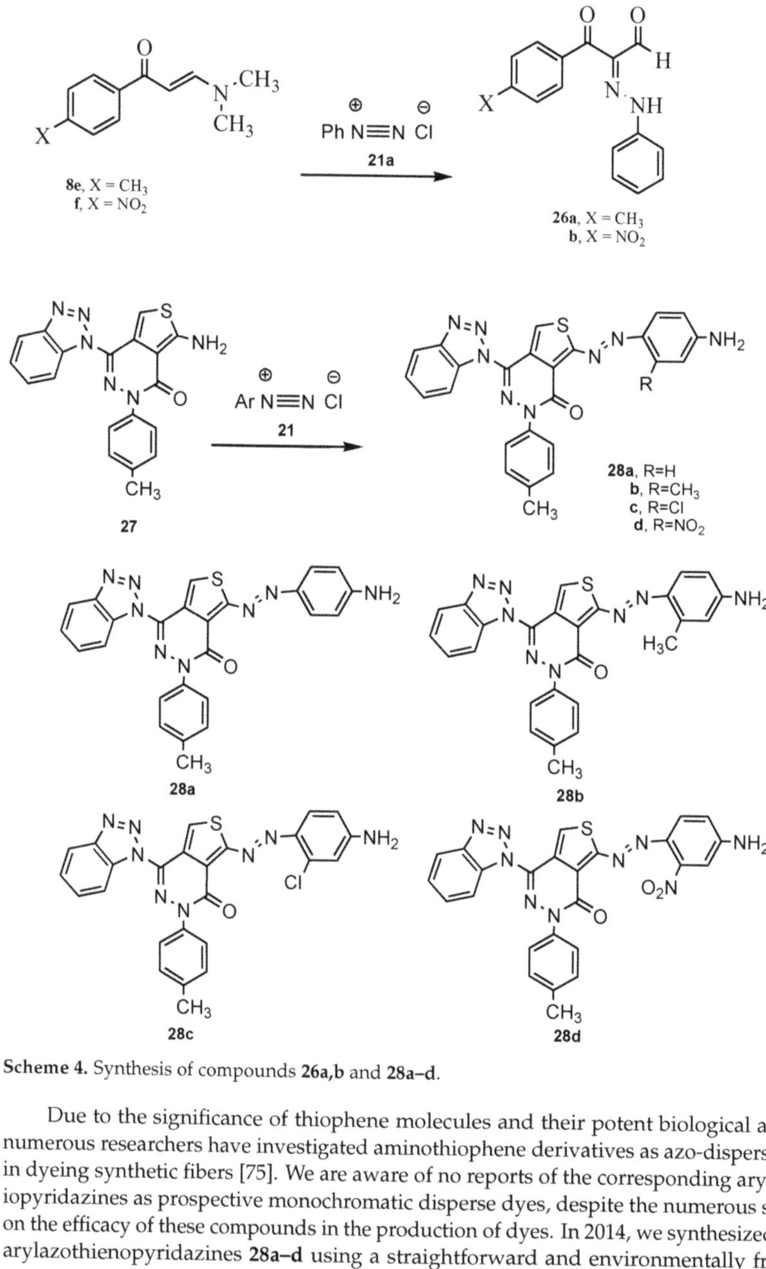

Scheme 4. Synthesis of compounds **26a,b** and **28a–d**.

Due to the significance of thiophene molecules and their potent biological activity, numerous researchers have investigated aminothiophene derivatives as azo-disperse dyes in dyeing synthetic fibers [75]. We are aware of no reports of the corresponding arylazothiopyridazines as prospective monochromatic disperse dyes, despite the numerous studies on the efficacy of these compounds in the production of dyes. In 2014, we synthesized some arylazothienopyridazines **28a–d** using a straightforward and environmentally friendly conventional method by reacting 7-Amino-4-benzotriazol-1-yl-2-p-tolyl-2H-thieno[3,4-d]pyridazin-1-one **27** with various diazonium salts **21**. This work continued the growing interest in the synthesis of arylazothienopyridazinones (Scheme 4) [75].

3. Dyeing

The disperse dyes **5**, **7**, **9a–d**, **14**, **15**, and **16a–h** were used to dye polyester fabrics with hues 2% using microwave heating at 130 °C, producing a range of color shades (Table 1).

Table 1. K/S of polyester fabrics dyed with synthesized dyes 130 °C.

Dye	Color Shade on Polyester	Color Strength (K/S)	Ref.
5	Yellowish-orange	2.12	[58]
7	Pale orange	3.79	
9a	Dark orange	5.95	
9b	Dark orange	5.81	
9c	Orange	4.64	
9d	Orange	4.73	
14	Yellowish brown	0.84	[60]
15	Pale brown	4.35	
16a	Yellow	9.78	
16b	Yellow	15.72	
16c	Yellow	14.92	
16d	Yellowish brown	21.08	
16f	Yellowish orange	23.55	
16g	Orange	20.93	
16h	Orange	16.52	
22a	Yellow	27.39	[61]
22b	Dark orange	30.29	
22c	Orange	30.28	
22d	Orange	8.88	
22e	Yellow	28.91	
22f	Dark Yellow	28.09	
22g	Very dark yellow	27.31	
22h	Deep greenish yellow	19.38	
22i	Orange yellow	12.63	
26a	Greenish-yellow	17.59	[72]
26b	Yellowish-orange	16.69	
28a	Reddish-orange	10.05	[75]
28b	Red	16.03	
28c	Pink	7.84	
28d	Violet	9.79	

Color strength was assessed at the maximum wavelength λ_{max} and given as K/S values. The Kubelka–Munk equation was used to perform K/S [1,21].

$$K/S = \frac{(1-R)^2}{2R} - \frac{(1-R_0)^2}{2R_0}$$

where R is the decimal fraction of the reflectance of the dyed fabric; R_0 is the decimal fraction of the reflectance of the not dyed fabric; K is the absorption coefficient; and S is the scattering coefficient.

The dyes have a good affinity for polyester materials at the stated temperatures of 130 °C, as shown by the color strength values K/S listed in Table 1, giving it brilliant hues. In order to treat the dyeing baths, we decided to reuse the dye bath from the previous study and extend the dyeing time from 60 to 90 min without adding any new dye. It is difficult for the human eye to distinguish the dye's color constancy after the dye bath was reused.

Additionally, it is evident from the results that the color strength K/S values for the initial dyeing process were 2.12, 3.79, 5.95, 5.81, 4.64 and 4.73, and the K/S values for the subsequent dyeing bath process were 1.67, 3.56, 5.47, 3.49, 3.54 and 3.95. After comparing the K/S values of the two methods, we discovered that the K/S for the second dyeing process had reuse rates of 78.77%, 93.93%, 91.93%, 60.06%, 6.29 and 83.5%. Table 1 makes it very evident that dye **16d** yields far stronger colors than dyes **16a–c** and **16h**. Polyester fabrics were dyed using compounds **22a–i**, **26a**, **26b** and **28a–d** at a high pressure, high temperature and 2% shade.

The materials were colored in a variety of hues, from yellow to violet. Following that, the fastness properties of polyester fabrics were used to evaluate the dyeing features of such materials. According to the K/S estimations in Table 1, the **22a–i** hues exhibited a significant attraction for polyester materials, and all color strengths were typically positive.

The International Commission on Illumination (CIE) established the CIELAB (Color space) psychometric coordinates in 1976, where L^* stands for lightness and (C) for chroma. The information in Table 2 shows that almost all colored polyester fabrics conveyed a similar hue when the dye's hue was expressed as (h) values. The positive estimates of b^* showed that the dyed polyester materials'color hues shifted in a reddish direction [61].

Table 2. Color strengths of the untreated polyester fabrics *.

Dye No.	K/S	λ_{max}	L^*	a^*	b^*	c	h	Ref.
			High temperature dyeing at 130 °C					
22h	19.38	445	76.33	4.47	93.78	93.88	87.27	[63]
22i	12.63	450	78.41	−0.56	87.34	87.34	90.37	
26a	17.59	410	77.22	−20.37	53.94	57.66	110.69	[72]
26b	16.69	405	67.58	9.29	61.34	62.04	81.39	
			Low temperature dyeing at 100 °C					
22h	4.74	445	80.06	−4.58	68.35	68.51	93.83	[63]
22i	3.46	450	80.35	−5.20	61.82	62.04	94.80	
26a	12.21	410	84.81	−14.26	52.16	54.08	105.29	[73]
26b	8.97	405	78.08	−0.21	51.20	51.21	90.23	
			US dyeing at 80 °C					
22h	9.07	445	80.58	−3.80	78.86	78.92	92.24	[1]

* Fabrics: bleached 100%polyester fabric (149 g/m^2) were supplied by El-Mahalla El-Kobra Company. The fabrics were scoured in an aqueous solution with a liquor ratioof 1:50 containing 2 g/L nonionic detergent solution (Hostapal, Clariant) and 2 g/L Na_2 CO_3 at 50 °C for 30 min to remove impurities, then rinsed thoroughly in cold tap water and dried at room temperature.Fabric samples (2 g) were introduced into a flask containing a dye bath of 2% (owf) dye shade and Matexil DA-N (supplied by ICI Company, UK) as a dispersing agent (1 g/L) at 130 °C with a 1:20 liquor ratio; during dye bath preparation, the dye was mixed with a few drops of DMF and then mixed with the dispersing agent, and water was added to prepare a homogeneous dispersion of the dye. The pH was adjusted to 4.5 by using acetic acid. The dyeing process was carried out in a microwave oven (Discover [80-mL vessel], CEM) or in a laboratory dyeing machine (Atlas Linitest Plus) for an hour. At the end of the dyeing process, the dyed samples were removed, rinsed in warm water, and subjected to reduction clearing in a solution comprising 2 g l1 of sodium hydrosulphite and 2g/L of sodium hydroxide (caustic soda) for 10 min at 60 °C, with a liquor ratio of 1:40, and the reduction-cleared sample was rinsed thoroughly in water. After rinsing well in cold water and being neutralized with 1 g/L of acetic acid for 5 min at 40 °C, the dyed samples were removed, rinsed in tap water and allowed to dry in the open air. Conventional dyeing: Samples of 100% polyester texture were placed into a container containing 2% disperse dyes 1 or 2 shade, dispersing agent Matexil DA-N, carrier Tanavol EP 2007 at different conventional dyeing temperatures of 80 °C or 100 °C for 1 h. We continued the dyeing method as explained above. Ultrasonic dyeing: Samples of 100% polyester texture were placed into a container containing2% disperse dyes 1 or 2 shade, dispersing agent Matexil DA-N, carrier Tanavol EP 2007 at an 80 °C coloring temperature and a coloring time of 1 h utilizing sonic force level 500 W with wavelength132 kHz (thermostated CREST benchtop 575 HT ultrasonic). The dissolved dye in a few drops of DMF (dimethylformamide) was assorted with Matexil DA-N, then, water was added with liquor ratio 1:50 at pH of 4.5. After the dyeing process, the colored samples were subjected to a reduction clearing process for 15 min at 60 °C.

The total color difference ΔE was measured by using an UltraScan Pro (Hunter Lab, Reston, VA, USA) 10° observer with D65 illuminant, d/2 viewing geometry and measurement area of 2 mm. The total color difference ΔE^* between the sample and the standard was calculated using the following equation:

$$\Delta E* = \sqrt{(\Delta L*)^2 + (\Delta a*)^2 + (\Delta b*)^2}$$

where ΔL^*, Δa^* and Δb^* are the derivatives of corresponding parameters, respectively

The hues **22e–g** were lighter and brighter than the **22b–d** because electron-donating groups were included into the benzene ring, which decreased brightness, while electron withdrawing groups improved lightness and brightness.

Dye Uptake

The colorimetric parameter values obtained for the high temperature and low temperature colored polyesters are listed in Table 2, which demonstrates that the high temperature dyed fabrics were darker than the low temperature dyed fabrics. Color strength (K/S) showed the dye uptake. The K/S values for dyes **22h** and **22i** were 19.38, 12.63, and 4.74 and 3.46, respectively, for materials colored at high and low temperatures. These results show that the dye uptake of fabrics colored at high temperatures was greater than that of materials dyed at low temperatures by 309% and 265%, respectively.

Additionally, the K/S values for dyes **26a** and **26b** were 17.59, 16.69, and 12.21 and 8.97, respectively, for materials colored at high and low temperatures. These results show that the dye uptake of the materials dyed at high temperatures was more than that of low temperature dyeing by 144% and 186%, respectively. According to the results, high temperature dyeing is an environmentally safe technique since it can lessen the pollution load in colored dye effluents, which would otherwise have a negative impact on the environment.

Furthermore, as the rate of dye penetration into the filament increased, the pressure increased along with the dyed fabrics at high temperatures. The kinetic energy of the dye molecules may be increased by the temperature, and the temperature may also cause the polyester fibers to inflate, increasing the dyeing rate relative to low temperature dyeing.

According to Table 2′s findings, the K/S value for dyeing by ultrasound at 80 °C is 9.07, while for dyeing by the traditional method at 100 °C, it is 4.47. This means that the first approach using ultrasound is superior to the traditional method by 102.9% for the disperse dye **22h** [1].

4. Fastness Properties

According to the tests of the American Association of Textile Chemists and Colorists [58], the fastness characteristics of the dyed samples were evaluated against perspiration, washing and light. The information in Table 3 demonstrates that the color fastness characteristics of polyester textiles dyed with dyes **5**, **7**, **9a–d**, **14**, **15**, **16a–h**, **22a–i**, **26a–b** and **28a–d** were measured.

The results for dyed polyester fabrics dyed with dyes **5**, **7**, and **9a–d** are shown in Table 3, where the ratings for color fastness to light and washing were good and very good and the rating for color fastness to perspiration was excellent [60].

Data obtained by testing the color fastness characteristics of polyester fabrics colored with dyes **14**, **15**, and **16a–h** are shown in Table 3. With the exception of dye **16d**, all of the tested dyes had outstanding washing and perspiration fastness, according to the fastness values listed in Table 3.

The polyester colored materials' light fastness showed moderate fastness. The type of substituents in the diazonium component had a considerable impact on the light fastness. As a result, this moiety's fading rate should increase for electron-donating groups, whereas it should decrease for electron-withdrawing groups.

The results (Table 3) show that the inclusion of a methyl group in dyes **16b** and **16g** produced a reduction in light fastness to 3; our idea is consistent with these findings. The chlorine atom in the dyes **16c** and **16h**, on the other hand, was linked to an increase in light fastness to 4 and 6, respectively [61].

Data acquired by assessing the color fastness characteristics of polyester fabrics dyed with dyes **22a–i** are displayed in Table 3 as fastness data. The grades for fastness are listed in Table 3, which demonstrates very good levels of fastness for perspiration and outstanding levels of fastness for washing. The polyester colored fabrics' moderate fastness is shown by their light fastness.

Table 3. Fastness properties of the polyester dyed fabrics with 2% shades of disperse dyes.

Dye No.	Light Fastness	Washing Fastness		Perspiration Fastness				Ref.
				Alkaline		Acidic		
		SC	SW	SC	SW	SC	SW	
5	4	5	5	5	5	4	5	[58]
7	2	5	5	5	5	5	5	
9a	2–3	5	5	5	5	5	5	
9b	3	5	5	5	5	4	5	
9c	2–3	4	4–5	5	5	4–5	4–5	
9d	2–3	4–5	4–5	5	5	5	5	
14	3–4	5	5	5	5	4	5	[60]
15	5–6	5	5	5	5	5	5	
16a	3–4	5	5	5	5	5	5	
16b	3	5	5	5	5	5	5	
16c	3	5	5	5	5	4	5	
16d	3	4	4–5	5	5	5	5	
16f	2–3	5	4–5	5	5	5	5	
16g	3	5	4–5	5	5	5	5	
16h	6	5	5	5	5	5	5	
22a	4	5	5	5	5	5	5	[61]
22b	2	5	5	5	5	5	5	
22c	5	5	5	5	5	5	5	
22d	2	5	5	5	5	5	5	
22e	3–4	5	5	4–5	5	5	5	
22f	3–4	5	5	3	5	3–4	5	
22g	5	5	5	5	5	5	5	
22h	4–5	5	5	5	5	5	5	
22i	4	5	5	5	5	5	5	
26a	3	4–5	4–5	5	5	5	5	[72]
26b	3–4	4–5	4–5	5	5	5	5	
28a	3–4	5	5	5	5	5	5	[75]
28b	4	4–5	4	5	5	5	5	
28c	3–4	5	5	5	5	5	5	
28d	4	5	5	5	5	5	5	

SC = staining on cotton; SW = staining on wool.

The type of substituents in the diazonium component has a considerable impact on the light fastness. The addition of electron-withdrawing (bromine, chorine or nitro) substituents increase the dye **22e**, **22f** and **22g**'s light fastness to (3–4), (3–4) and (5), respectively. In terms of the following factors, the results generally indicated that the dyed fabric might have good fastness: (i) Effective dye molecule diffusion throughout fabric fibers. (ii) The dye molecule's size was regarded as being fairly large. (iii) The solubility and detergency of fabric dyeing were unaffected by any solubilizing groups [61].

5. In Vitro Cytotoxicity Screening

One of the most important markers for biological evaluation in in vitro investigations is cytotoxicity. Different cytotoxicity mechanisms exist for compounds in vitro, such as the suppression of protein synthesis or irreversible binding to receptors [64]. The synthetic disperse dyes **22h** and **22i** were tested for their preliminary anticancer activity against four human cell lines, including HepG-2 cells (for the treatment of hepatocellular carcinoma), MCF-7 cells (for the treatment of breast cancer), HCT-116 cells (for the treatment of colon cancer) and A-549 cells (for the treatment of lung carcinoma). The IC_{50} values, the concentration needed to prevent 50% of the development of the culture when the cells were exposed to the tested disperse dyes for 48 h, were calculated using various concentrations of the two disperse dyes. Disperse dye **22h** displayed significant activity, as shown by Table 4 and Figures 6–9, with IC_{50} values of 23.4, 62.2, 28 and 53.6 g/mL in HePG-2, MCF-7,

HCT-116 and A-549 cells, respectively. In contrast, dispersion dye **22i** showed negligible activity in HePG-2, MCF-7, HCT-116 and A-549 cells with IC_{50} values of 196, 482, 242 and 456 g/mL, respectively.

Table 4. Antitumor and antioxidant activities of disperse dyes 22h and 22i.

Dye No.	Antioxidant Activity (IC_{50})	Cytotoxic Activity (IC_{50})				Ref.
		MCF-7	HepG-2	A-549	HCT-116	
22h	64.5	62.2 ± 4.1	23.4 ± 1.2	53.6 ± 5.8	28 ± 1.9	
22i	191.6	482 ± 8.9	196 ± 3.2	456 ± 7.3	242 ±3.6	
Ascorbic acid	14.2					[64]
Imatinib		24.6			9.7	
Cisplatin			18.4 ± 0.9	19.3 ± 0.8		

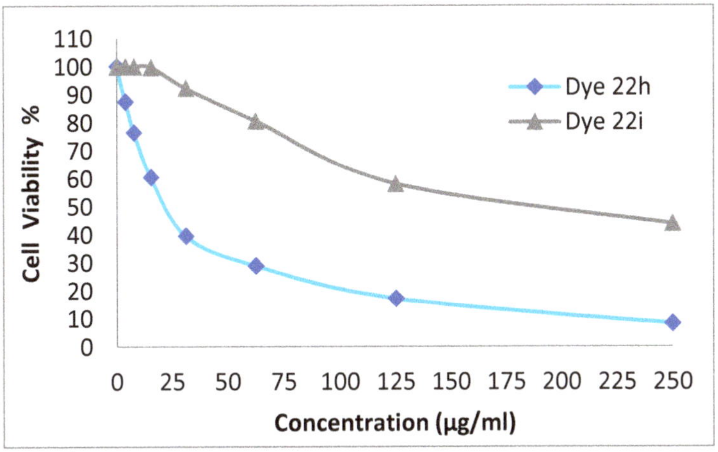

Figure 6. Antitumor (HepG-2) activities of disperse dyes 22h and 22i.

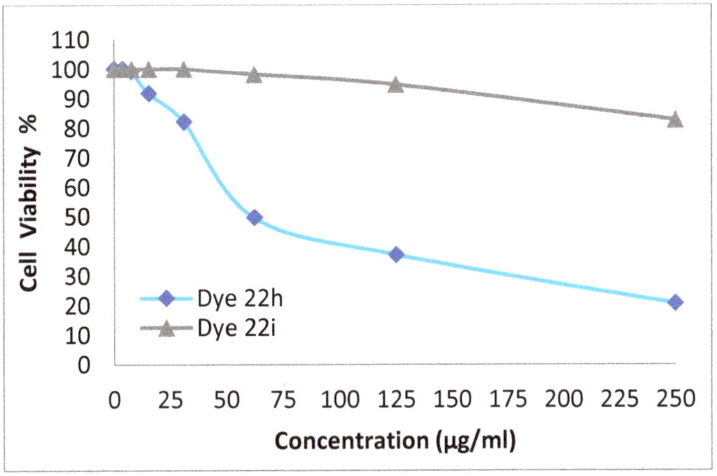

Figure 7. Antitumor (MCF-7) activities of disperse dyes 22h and 22i.

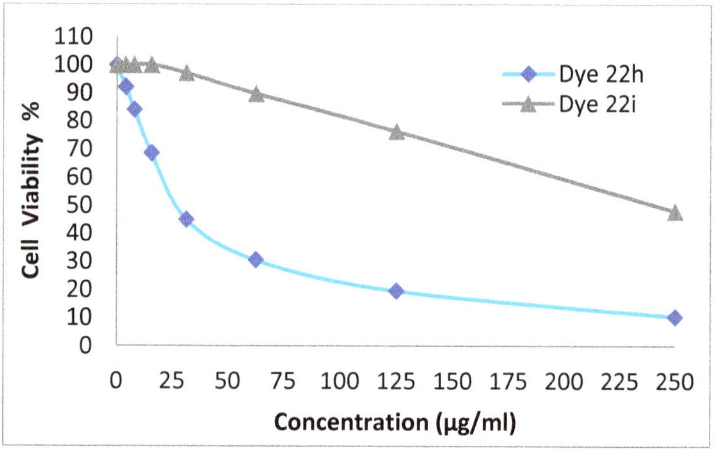

Figure 8. Antitumor (HCT-116) activities of disperse dyes 22h and 22i.

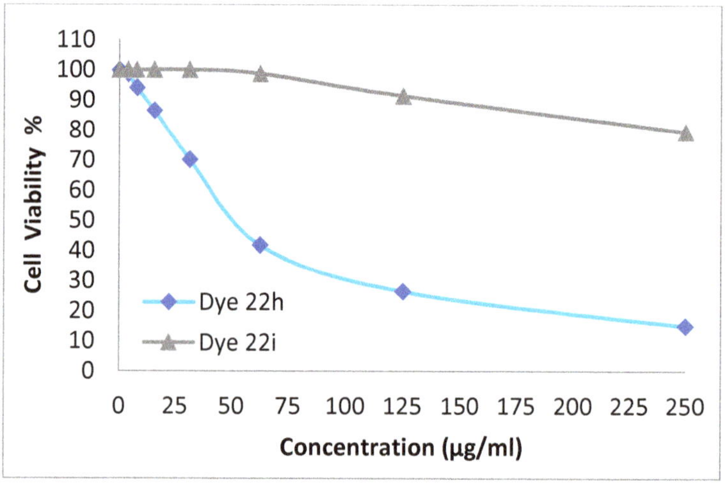

Figure 9. Antitumor (A-549) activities of disperse dyes 22h and 22i.

6. Antioxidant Activity (DPPH Radical Scavenging Activity)

In vitro testing of the two dispersion dyes' antioxidant properties used their capacity to scavenge DPPH free radicals. The IC_{50} value of the dyes—the dose needed to suppress the production of DPPH radicals by 50%, was used to describe their antioxidant activity. Data from Table 4 show that disperse dye **22h** had a moderate antioxidant activity and outperformed ascorbic acid as the standard, which had an IC_{50} of 14.2, while disperse dye **22i** had a poor antioxidant activity with an IC_{50} of 191.6. (Figure 10).

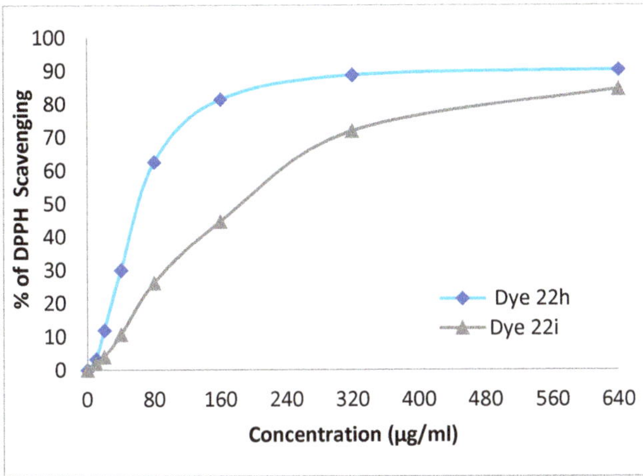

Figure 10. Antioxidant activities of disperse dyes 22h and 22i.

7. Antimicrobial Activities of Dyes 3, 5, 7, 9a–d, 22a–i and 26a,b

The agar well diffusion method was used to investigate the antibacterial effects of the produced dyes **3**, **5**, **7** and **9a–d** against various bacteria and yeast. The information in Table 5 reveals encouragingly effective antibacterial activity. While the other disperse dyes exhibited moderate to poor antibacterial properties, disperse dyes **3** and **5** had strong antibacterial activity against Gram-positive bacteria.

Table 5. Inhibition zones of dyes 3, 5, 7, 9a-d, 22a–i and 26a,b.

Dye No.	Inhibition Zone Diameter (Nearest mm)		
	G− Bacteria	G+ Bacteria	Yeast
	E. coli	*B. subtilus*	*C. albicans*
3	NI	11(0.1)	8(0.3)
5	NI	11(0.2)	1(0.2)
7	NI	4(0.2)	4(0.1)
9a	NI	7(0.2)	1(0.1)
9b	NI	2(0)	2(0.1)
9c	NI	2(0)	3(0.2)
9d	NI	2(0)	3(0.2)
22a	14 (1)	12.2 (0.5)	14 (0.3)
22b	10.8 (1.5)	9 (0.3)	12 (1)
22c	12.1 (0.7)	10.8 (0.5)	19.1 (1.3)
22d	15 (0.5)	14.6 (0.5)	10.1 (0.5)
22e	10 (0.6)	9 (0.2)	12.7 (0.6)
22f	10 (0.4)	NI	20.4 (0.7)
22g	11 (0.6)	9.7 (0.5)	16.2 (1)
22h	8	NI	NI
22i	NI	NI	8
26a	9	NC	NI
26b	10	NC	20

(NI): No inhibition, (NC): Not checked.

All of the substances tested inhibited the development of *Candidia albicans* after six days of incubation. Additionally, the agar well diffusion method was used to test the dispersion dyes **22a–i** and **26a,b** for their antibacterial properties against a range of bacteria and yeast.

In addition, Figure 11 shows that *Candida albicans* re-grew in the formed zone surrounding the wells containing compound **3**. This may reflect the cytostatic effect of the chemicals rather than their cytolytic effects. Note that the plate color changed with the increase in the incubation time indicates a complete diffusion of the chemical used in the agar as the incubation period increased.

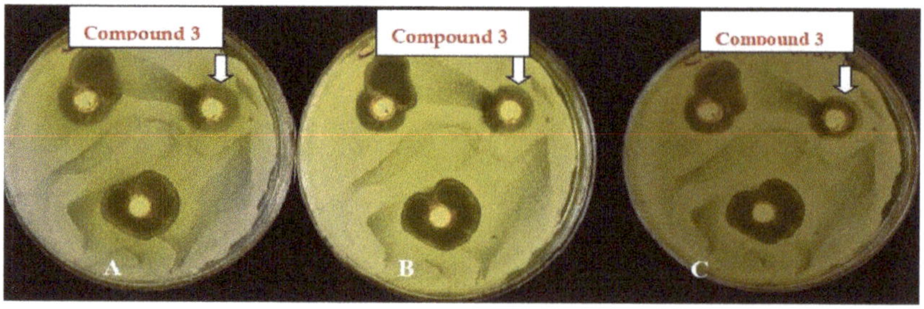

Figure 11. *Candida albicans* treated with 10mg ml^{-1} of compound **3** after one day (**A**), three days (**B**) and six days (**C**) of incubation. It is worth noting that Figure 10 shows the cytolyticeffects of dyes number 5, 7 and 9a on *Candida albican* where after one, three and six days of incubation, the inhibition zone did not change (Figure 12).

While, the cytostatic effect of the same dyes is clear on the plates inoculated with *Bacillus subtilus*. The growth of *B. subtilus* resumed in the inhibited area after six days of incubation, which may indicate that as the concentration/toxicity of dyes number **5**, **7** and **9a** reduce due to possible evaporation of these dyes or diffusion in the media, the effect of these dyes on *B. subtilus* decrease, and the organisms start growing again (Figure 13).

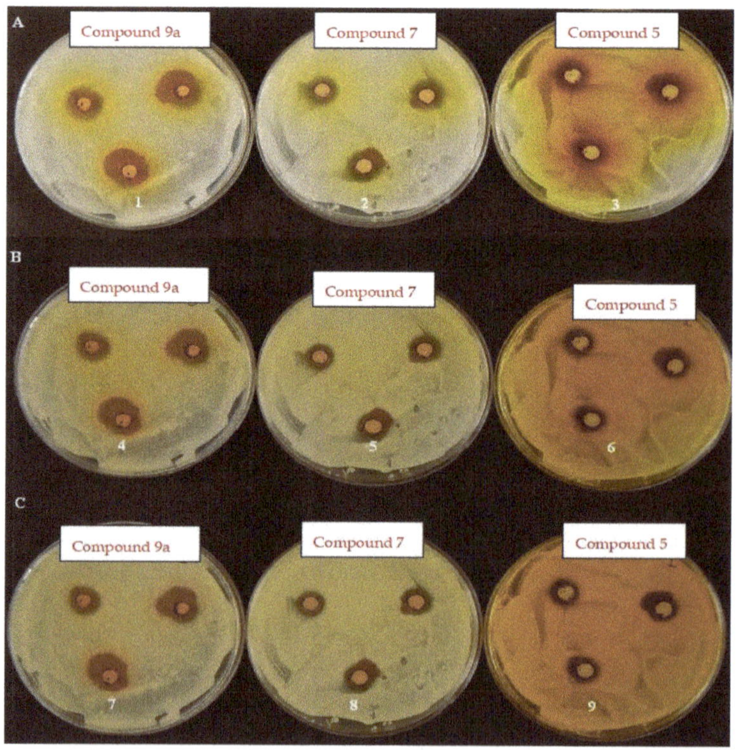

Figure 12. *Bacillus subtilus* treated with 10mg ml^{-1} of dyes 5, 7, and 11a after one day (**A**) (1, 2, 3), three days(**B**) (4, 5, 6) and six days (**C**) (7, 8, 9) of incubation.

Based on data for the inhibition zone diameter for the dispersion dyes **22a–g**, Table 5 shows that all of the tested dyes demonstrated strong positive antibacterial activities against the studied pathogens. For all of the examined bacteria, disperse dye **22a** had a cytolytic effect, with no growth being seen in the inhibition zone.

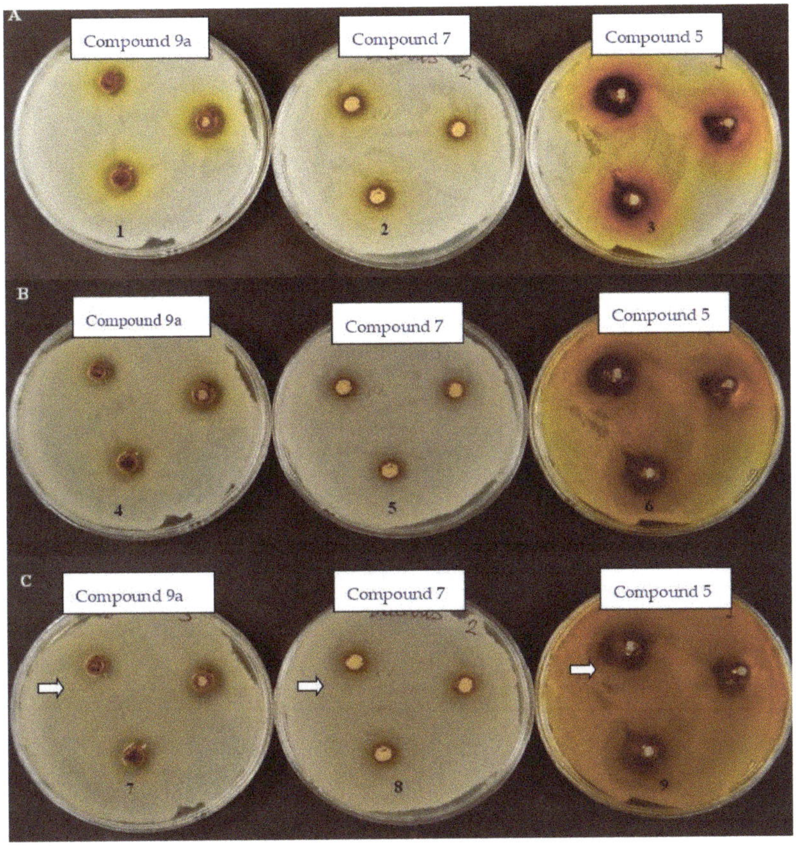

Figure 13. *Candida albican* treated with 10mg ml^{-1} of dyes 5, 7, and 9a after one day (**A**) (1, 2, 3), three days(**B**) (4, 5, 6) and six days (**C**) (7, 8, 9) of incubation.

8. UV Protective Properties of Untreatedand Treated Polyester Fabrics with ZnO or TiO_2 Nanoparticles NPs

The UV protection factor (UPF) was calculated to obtain ultraviolet-protective qualities. UPF is a characteristic that endows materials such as polyester textiles with ultraviolet protection properties. Table 6 shows UV blocking data for polyester fabrics treated with ZnO or TiO_2 nanoparticles. According to the UPF data in Table 6, treated polyester fabrics with ZnO nanoparticles had UPF values of 173.25 for disperse dye **26a** and 190.59 for disperse dye **26b**, respectively. This shows that treated polyester fabrics with dye **26a** have UPF values that are significantly higher than treated polyester fabrics with dye **26b**.Table 6 further demonstrates that treated polyester fabrics have higher UPF values than untreated polyester fabrics, with respective values of 141.88 for dye **26a** and 122.37 for dye **26b**. Additionally, the UPF data in Table 6 shows that treated polyester fabrics with TiO_2 nanoparticles have UPF values of 283.60 for disperse dye **22h** and 34.9 for disperse dye **22i**, respectively. This shows that treated polyester fabrics with dye **22h** have much higher UPF values than treated polyester fabrics with dye **22i**. Table 6 further demonstrates that treated polyester fabrics have higher UPF values than untreated polyester fabrics, with values of 236.2 for dye **22h** and 25.5 for dye **22i**, respectively.

Table 6. UPF of treated polyester fabrics with ZnO or TiO_2 NPs *.

Dye No.	ZnO Treatment	UPF	Dye No.	TiO_2 Treatment	UPF
	Blank	19.42		Blank	8.2
26a	Untreated	141.88	**22h**	Untreated	236.2
	Treated	173.25		Treated	283.60
26b	Untreated	122.37	**22i**	Untreated	25.5
	Treated	190.59		Treated	34.9

* Treatment of polyester fabrics by ZnO or TiO_2 NPs: The dyed polyester fabrics were treated with ZnO or TiO_2 nanoparticles via the exhaustion method. The fabrics were treated with five different percentages of TiO_2 nanoparticles (0.5–2.5% weight of fabric for ZnO or 1–5% weight of fabric for TiO_2) at 80 °C for 20 min, the liquorratio of the exhaustion bath was 1:30.The polyester fabrics were squeezed to remove the excess dispersion and dried in an oven at 70 °C for 15 min. under atmospheric pressure (dry heat).The treated polyester fabrics were cured at 140 °C for 10 min. Finally treated fabrics were washed at 60 °C for 20 min. in an aqueous solution with a liquor ratio of 1:30 containing 3 g/L nonionic detergent solution (Hostapal, Clariant), then left for drying. The drying step was carried out on a horizontal flat surface. Titanium(IV) oxide TiO_2 NPs: Nano powder, 21 nm primary particle size (Tem), ≥99.5% trace metals basis, Lot # MKBV3126V was purchased as 718467-100Gfrom ALDRICH Chemistry.

9. Self-Cleaning of Untreatedand Treated Polyester Fabrics with ZnO or TiO_2 NPs

One of the benefits of polyester fabrics coated with nanoparticles is the conversion of absorbed light into self-cleaning substances to remove stains. To gain the self-cleaning properties of nano ZnO or TiO_2 particles, the photodegradation of methyl red and methylene blue adsorbed on nano ZnO or TiO_2 treated polyester fabrics was explored (Table 7).

Table 7. Effect of Light fastness and Stains * on ZnO or TiO_2 NPs treated polyester fabrics.

Dye No.	ZnO% Treatment	Methyl Red Stain ΔE*	Light Fastness	Dye No.	TiO_2% Treatment	Methylene Blue Stain	Light Fastness
26a	Untreated	67.23	3–4	**22h**	Untreated	No removal	5–6
	0.5	60.20	3		1	80%	5–6
	1.0	59.41	3		2	80%	6
	1.5	60.88	3		3	75%	6
	2.0	55.44	3		4	80%	6
	2.5	60.36	3		5	60%	6
26b	Untreated	50.02	4	**22i**	Untreated	10%	5
	0.5	54.42	5		1	80%	5
	1.0	55.32	5		2	75%	5
	1.5	54.37	4		3	80%	5
	2.0	53.73	4–5		4	70%	5
	2.5	60.18	4–5		5	65%	5–6

* Photo-stimulated color removal on polyester: 60 µL methyl red or methylene blue was marked on both post-treated ZnO or TiO_2 nanoparticles (0.5–2.5% w.o.ffor ZnO and 1–5% w.o.f for TiO_2)-treated polyester and the untreated fabrics. The polyester fabrics were illuminated through exposure to an ultraviolet lamp for 24 h for ZnO and 12 h forTiO_2.

After 24 and 12 h of UV exposure, Table 7 displays the results of methyl red and methylene blue stains on polyester fabrics treated with ZnO or TiO_2 NPs. For polyester fabrics treated with ZnO or TiO_2 NPs, a partial discoloration of methyl red and methylene blue stains brought on by ultraviolet radiation was seen. When polyester fabric is treated with ZnO or TiO_2 NPs, thin layers of ZnO or TiO_2 nanoparticles develop, increasing the fabric's hydrophobic qualities. A hydrophobic surface stops dirt from adhering, keeping the polyester surface clean at all times. The results showed that the highest rates of photodegradation on the surface were 60–70% for methyl red stains treated with ZnO NPs after 24 h and 60–80% after 12 h for methylene blue stains treated with TiO_2 NPs (Table 7).

10. Light Fastness of Untreated and Treated Polyester Fabrics with ZnO or TiO_2 NPs

The light fastness property of all colored polyester fabric samples that were treated with ZnO or TiO_2 NPs of disperse dyes **26a**, **26b**, **22h**, and **22i** were measured, and the results were more significant. Table 7 shows that, with the exception of dye **26a**, using ZnO or TiO_2 NPs more effectively resulted in treated polyester fabrics having stronger light fastness than the untreated samples (Table 7).

11. Antimicrobial Activity of Untreated and Treated Polyester Fabrics with ZnO or TiO_2 NPs

The untreated and treated polyester fabrics are tested against the pathogenic fungi *Aspergillus flavus* and *Penicillium chrysogenum*, as well as the Gram-positive bacteria *Bacillus subtilis* and the Gram-negative bacteria *Klebsiella pneumoniae*. Table 8 revealed that untreated polyester fabrics with ZnO or TiO_2 NPs did not exhibit antibacterial activity against all of the microorganisms [64,73].

Table 8. Inhibition zone diameters of the dyed polyester fabrics of disperse dyes 26a and 26b against Gram-positive, Gram-negative bacteria and yeast.

Dye No.	ZnO% Treatment	Inhibition Zone Diameter (Nearest mm)		Dye No.	TiO_2% Treatment	Inhibition Zone Diameter (Nearest mm) Fungi	
		G+ Bacteria *Bacillus subtilis*	G− Bacteria *Klebsiella pneumoniae*			*Aspergillus flavus*	*Penicillium chrysogenum*
26a	Untreated	NI	NI	**22h**	Untreated	NI	NI
	Treated	11	NI		Treated	21	19
26b	Untreated	NI	NI	**22i**	Untreated	NI	NI
	Treated	8	10		Treated	NI	NI

(NI): No inhibition.

While only *Bacillus subtilis* was the target of the antibacterial activity of nano ZnO treated polyester fabrics of dispersion dye **26a**, nano ZnO treated polyester fabrics of disperse dye **26b** were effective against both *Bacillus subtilis* and *Klebsiella pneumoniae*. ZnO NPs had antibacterial activity against bacteria, and its mechanism involved either ZnO nanoparticles or TiO_2 nanoparticles influencing bacterial membranes to inhibit bacterial growth. Alternatively, nano ZnO may enable the generation of peroxide, which may offer antibacterial properties [65,73].

According to the antifungal screening results given in Table 8, the two harmful fungi *Aspergillus flavus* and *Penicillium chrysogenum* were not resistant to the treated polyester dyed fabrics with TiO_2 NPs of dispersion dye **22i**. *Aspergillus flavus* and *Penicillium chrysogenum* were resistant to *Aspergillus flavus* and the treated polyester colored fabrics with TiO_2 NPs of dispersion dye **22h**.

12. Conclusions

Our original work highlighted that the synthesis of new disperse dyes, with the help of microwaves, could be carried out in the presence of small quantities or sometimes in the absence of any amounts of organic solvents, which are harmful to the environment, in a short time that did not exceed minutes and gave great yields compared to the synthesis of those dyes using traditional methods. It is known that cytotoxicity is one of the most important markers of biological evaluation in laboratory tests, so we presented two examples of new disperse dyes as one of the examples of in vitro cytotoxicity examination.We showed that these dyes possess anticancer activities against some common cancers such as lung, breast, liver and colon cancer. The added value of these new disperse dyes was also discussed and presented by showing the biological activity and clarifying that these dyes have a great biological activity against Gram-positive and Gram-negative bacteria, as well as various fungi and yeasts. It is worth noting here that polyester fabrics dyed with these dyes had biological activity, which makes these fabrics able to be used in many medical activities. We presented and discussed the use of ultrasound energy to dye polyester fabrics, as it gave a high color strength compared to traditional dyeing methods. In this review, we also discussed the methods of treating polyester fabrics with nano-zinc oxide or nano-titanium dioxide and presented the advantages of this strategy in endowing the dyed polyester fabrics with multiple functions, such as self-cleaning property, maximizing light fastness property and maximizing antimicrobial activities.

Author Contributions: Writing—review and editing, A.M.A.-E. and M.A.E.-A. All authors have read and agreed to the published version of the manuscript.

Funding: This research received no external funding.

Institutional Review Board Statement: Not applicable.

Informed Consent Statement: Not applicable.

Data Availability Statement: Not applicable.

Conflicts of Interest: The authors declare no conflict of interest.

References

1. Al-Etaibi, A.M.; El-Apasery, M.A. Ultrasonic Dyeing of Polyester Fabric with Azo Disperse Dyes Clubbed with Pyridonones and Its UV Protection Performance. *Chemistry* **2021**, *3*, 889–895. [CrossRef]
2. Rehman, F.; Adeel, S.; Saif, M.J.; Khosa, M.K.; Anjum, M.N.; Kamran, M.; Zuber, M.; Asif, M. Ultrasonic Assisted Improvement in Dyeing Behaviour of Polyester Fabric Using Disperse Red343. *Pol. J. Environ. Stud.* **2020**, *29*, 261–265. [CrossRef]
3. Adeel, S.; Shahid, S.; Khan, S.G.; Rehman, F.; Muneer, M.; Zube, M.; Akhtar, H. Eco-Friendly Disperse Dyeing of Ultraviolet-Treated Polyester Fabric Using DisperseYellow211. *Pol. J. Environ. Stud.* **2018**, *27*, 1935–1939. [CrossRef]
4. Al-Etaibi, A.M.; El-Apasery, M.A. Microwave-Assisted Synthesis of Azo Disperse Dyes for Dyeing Polyester Fabrics: Our Contributions over the Past Decade. *Polymers* **2022**, *14*, 1703. [CrossRef] [PubMed]
5. Ghaffar, A.; Adeel, S.; Habib, N.; Jalal, F.; Ul-Haq, A.; Munir, B.; Ahmad, A.; Jahangeer, M.; Jamil, Q. Effects of Microwave Radiation on Cotton Dyeing with Reactive Blue 21 Dye. *Pol. J. Environ. Stud.* **2019**, *28*, 1687–1691. [CrossRef]
6. Gedye, R.; Smith, F.; Westaway, K.; Ali, H.; Baldisera, L.; Laberge, L.; Rousell, J. The Use of Microwave Ovens for Rapid Organic Synthesis. *Tetrahedron Lett.* **1986**, *27*, 279–282. [CrossRef]
7. Stefanidis, G.D.; Muñoz, A.N.; Sturm, G.S.J.; Stankiewicz, A. A Helicopter View of Microwave Application to Chemical Processes: Reactions, Separations, and Equipment Concepts. *Rev. Chem. Eng.* **2014**, *30*, 233–259. [CrossRef]
8. Bassyouni, F.A.; Abu-Bakr, S.M.; Abdel Rehim, A. Evolution of microwave irradiation and its application in green chemistry and biosciences. *Res. Chem. Intermed.* **2012**, *38*, 283–322. [CrossRef]
9. Rana, K.K.; Rana, S. Microwave Reactors: A Brief Review on Its Fundamental Aspects and Applications. *Open Access Libr. J.* **2014**, *1*, 686. [CrossRef]
10. Yoon, S.; Kim, H.; Oh, N.; Kim, S.; Kumar, S.; Koh, J. Synthesis and application of high-washability 4-amino-4′-fluorosulfonylazobenzene disperse dyes to cellulose diacetate for high color fastness. *Fibers Polym.* **2021**, *22*, 3075–3081. [CrossRef]
11. Yoon, S.; Choi, B.; Rahman, M.M.; Kumar, S.; Kabir, S.M.M.; Koh, J. Dyeing of polyester with 4-arylazo-5-pyrazolone dyes containing fluoro sulfonyl group and application of environment-friendly after treatment for their high color fastness. *Materials* **2019**, *12*, 4209. [CrossRef]
12. Chandrasekaran, S.; Ramanathan, S.; Basak, T. Microwave food processing—A review. *Food Res. Int.* **2013**, *52*, 243–261. [CrossRef]
13. Crane, C.A.; Pantoya, M.L.; Weeks, B.L. Spatial observation and quantification of microwave heating in materials. *Rev. Sci. Instrum.* **2013**, *84*, 084705. [CrossRef]
14. Rosa, R.; Veronesi, P.; Leonelli, C. A review on combustion synthesis intensification by means of microwave energy. *Chem. Eng. Process. Process Intensif.* **2013**, *71*, 2–18. [CrossRef]
15. Sun, J.; Wang, W.; Yue, Q. Review on Microwave-Matter Interaction Fundamentals and Efficient Microwave-Associated Heating Strategies. *Materials* **2016**, *9*, 231.
16. Orlik, K.; Lorgouilloux, Y.; Marchet, P.; Thuault, A.; Jean, F.; Rguiti, M.; Courtois, C. Influence of microwave sintering on electrical properties of BCTZ lead free piezoelectric ceramics. *J. Eur. Ceram. Soc.* **2020**, *40*, 1212–1216.
17. Ramesh, S.; Zulkifli, N.; Tan, C.; Wong, Y.; Tarlochan, F.; Teng, W.; Sopyan, I.; Bang, L.; Sarhan, A.A.D. Comparison between microwave and conventional sintering on the properties and microstructural evolution of tetragonal zirconia. *Ceram. Int.* **2018**, *44*, 8922–8927. [CrossRef]
18. Chen, Y.; Fan, B.; Yang, B.; Ma, W.; Liu, G.; Li, H. Microwave sintering and fracture behavior of zirconia ceramics. *Ceram. Int.* **2019**, *45*, 17675–17680. [CrossRef]

19. Huang, K.; Zheng, J.; Yuan, W.; Wang, X.; Song, Q.; Li, Y.; Crittenden, J.C.; Wang, L.; Wang, J. Microwave-assisted chemical recovery of glass fiber and epoxy resin from non-metallic components in waste printed circuit boards. *Waste Manag.* **2021**, *124*, 8–16. [CrossRef] [PubMed]
20. Serdar, G.; Demir, E.; Bayrak, S.; Sökmen, M. New Approaches for Effective Microwave Assisted Extraction of Caffeine and Catechins from Green Tea. *Int. J. Second. Metab.* **2016**, *3*, 3–13. [CrossRef]
21. Bacsa, B.; Desai, B.; Dibó, G.; Kappe, C.O. Rapid solid-phase peptide synthesis using thermal and controlled microwave irradiation. *J. Pept. Sci.* **2006**, *12*, 633–638. [CrossRef] [PubMed]
22. Murray, J.K.; Gellman, S.H. Application of microwave irradiation to the synthesis of 14-helical b-peptides. *Org. Lett.* **2005**, *7*, 1517–1520. [CrossRef] [PubMed]
23. Olivos, H.J.; Alluri, P.G.; Reddy, M.M.; Salony, D.; Kodadek, T. Microwave assisted solid-phase synthesis of peptoids. *Org. Lett.* **2002**, *4*, 4057–4059. [CrossRef]
24. Collins, J.M.; Collins, M.J. Novel method for enhanced solid-phase peptide synthesis using microwave energy. *Biopolymers* **2003**, *71*, 361.
25. Mallalcpour, S.E.; Hajipour, A.R.; Zamanlou, M.R. Synthesis of optically active poly(amide-imide)s derived from *N,N′*-(4,4′-carbonyldiphthaloyl)-bis-L-leucine diacid chloride and aromatic diamines by microwave radiation. *J. Polym. Sci. Part A Polym. Chem.* **2001**, *39*, 177–186. [CrossRef]
26. Mallakpour, S.E.; Hajipour, A.R.; Faghihi, K.; Foroughifar, N.; Bagheri, J. Novel Optically Active Poly(amideimide)s with Tetrahydropyrimidinone and Tetrahydro-2-Thioxopynmidine Moieties by Microwave-Assisted Polycondensation. *J. Appl. Polym. Sci.* **2001**, *80*, 2416–2421. [CrossRef]
27. Mallakpour, S.E.; Hajipour, A.R.; Habibi, S. Facile synthesis of new optically active poly(amide-imide)s derived from N,N′-(pyromellitoy1)-bis-l-leucine diacid chloride and aromatic diamines under microwave irradiation. *Eur. Polym. J.* **2001**, *37*, 2435–2442. [CrossRef]
28. Lu, J.; Jiang, Q.; Zhu, X.; Wang, F. Microwave radiation copolymerization of dibutyltin maleate and allyl thiourea. *J. Appl. Polym. Sci.* **2001**, *79*, 312–318.
29. Kappe, C.O.; Dallinger, R.D. The impact of microwave synthesis on drug discovery. *Nat. Rev. Drug Discov.* **2006**, *5*, 51–64. [CrossRef]
30. Kappe, C.O.; Stadler, A.; Dallinger, D.; Strohmeier, G.; Perez, R.; Zbruyev, O.I.; Stiasni, N.; Walla, P.; Gorobets, N.; Yousefi, B.; et al. Adventures in microwave-assisted organic synthesis: Contributions from the Kappe laboratory 2000–2005. *Nato Sci. Ser. Ii-Math.* **2008**, *246*, 225–251.
31. Yin, C. Microwave-assisted pyrolysis of biomass for liquid biofuels production. *Bioresour. Technol.* **2012**, *120*, 273–284. [CrossRef] [PubMed]
32. Bridgwater, A.V. Review of fast pyrolysis of biomass and product upgrading. *Biomass Bioenergy* **2012**, *38*, 68–94. [CrossRef]
33. Wei, R.; Wang, P.; Zhang, G.; Wang, N.; Zheng, T. Microwave-responsive catalysts for wastewater treatment: A review. *Chem. Eng. J.* **2020**, *382*, 122781. [CrossRef]
34. Remya, N.; Lin, J.-G. Current status of microwave application in wastewater treatment—A review. *Chem. Eng. J.* **2011**, *166*, 797–813. [CrossRef]
35. Mudhoo, A.; Sharma, S.K.; Sharma, S.K. Microwave Irradiation Technology in Waste Sludge and Wastewater Treatment Research. *Crit. Rev. Environ. Sci. Technol.* **2011**, *41*, 999–1066. [CrossRef]
36. Vialkova, E.; Obukhova, M.; Belova, L. Microwave Irradiation in Technologies of Wastewater and Wastewater Sludge Treatment: A Review. *Water* **2021**, *13*, 1784. [CrossRef]
37. Dominguez, A.; Menendez, J.A.; Inguanzo, M.; Pis, J.J. Investigations into the characteristics of oils produced from microwave pyrolysis of sewage sludge. *Fuel Process. Technol.* **2005**, *86*, 1007–1020. [CrossRef]
38. Dominguez, A.; Menendez, J.A.; Inguanzo, M.; Pís, J.J. Production of bio-fuels by high temperature pyrolysis of sewage sludge using conventional and microwave heating. *Bioresour. Technol.* **2006**, *97*, 1185–1193. [CrossRef]
39. El-Apasery, M.A. Synthesis of some azo disperse dyes by the use of focused microwave heating. *Pol. J. Appl. Chem.* **2006**, *50*, 75–81.
40. El-Apasery, M.A. Solvent-free one-pot synthesis of some azo disperse dyes under microwave irradiation: Dyeing of polyester fabrics. *J. Appl. Polym. Sci.* **2008**, *109*, 695–699. [CrossRef]
41. Al-Zaydi, K.M.; Borik, R.M.; Elnagdi, M.H. Studies with arylhydrazonopyridinones: Synthesis of new arylhydrazono thieno[3,4-c]pyridinones as novel D2T2 dye class; classical verse green methodologies. *Ultrason. Sonochem.* **2009**, *16*, 660–668. [CrossRef] [PubMed]
42. Devi, A.S.; Kaping, S.; Vishwakarma, J.N. A facile environment-friendly one-pot two-step regioselective synthetic strategy for 3,7-diarylpyrazolo[1,5-a]pyrimidines related to zaleplon and 3,6-diarylpyrazolo[1,5-a]pyrimidine-7-amines assisted by $KHSO_4$ in aqueous media. *Mol. Divers.* **2015**, *19*, 759–771. [CrossRef] [PubMed]
43. El-Bayouki, M.; Abdel Hameed, K.; Asyouni, W.M.; Mohamed, Y.A.; Aly, M.M.; Abbas, S.Y. Novel 4(3H)-quinazolinones containing biologically active thiazole, pyridinone and chromene of expected antitumor and antifungal activities. *Eur. J. Chem.* **2011**, *2*, 455–462. [CrossRef]
44. Okada, Y.; Hihara, T.; Morita, Z. Analysis of the photofading of phenylazo-aniline and phenylazo-pyridone disperse dyes on poly(ethylene terephthalate) substrate using the semiempirical molecular orbital PM5 method. *Dye. Pigment.* **2008**, *79*, 111–125. [CrossRef]

45. Okada, Y.; Hihara, T.; Morita, Z. Photofading of phenylazo-aniline, -pyridone and -quinolone disperse dyes on a nylon 6 substrate. *Color. Technol.* **2009**, *125*, 86–98. [CrossRef]
46. Darwish, E.; Mahmoud, F.F.; Altalbawy, F.M.A. Synthesis and Antimicrobial Evaluation of Some New Pyrazole, Fused Pyrazolo [1,5-a]-pyrimidine and Pyrazolo[1,5-d]pyrimido[4,5-d][1,2,3]triazine Derivatives. *Asian J. Chem.* **2012**, *24*, 2997–3002.
47. Okada, Y.; Hihara, T.; Morita, Z. Analysis of the catalytic fading of pyridone-azo disperse dyes on polyester using the semi-empirical, molecular orbital PM5 method. *Dye. Pigment.* **2008**, *78*, 179–198. [CrossRef]
48. Jang, H.K.; Doh, S.J.; Lee, J.J. Eco-friendly dyeing of poly(trimethylene terephthalate) with temporarily solubilized azo disperse dyes based on pyridone derivatives. *Fibers Polym.* **2009**, *10*, 315–319. [CrossRef]
49. Gadre, J.N.; Periaswamy, R.M.S.; Mulay, M.; Vaze, C.S. Synthesis of pyridone based azo disperse dyes. *Indian J. Heterocycl. Chem.* **2006**, *16*, 43–46.
50. Tsai, P.C.; Wang, I.J. Synthesis and solvatochromic properties of some disazo dyes derived from pyrazolo[1,5-a]pyrimidine derivatives. *Dye. Pigment.* **2005**, *64*, 259–264. [CrossRef]
51. Tsai, P.C.; Wang, I.J. A facile synthesis of some new pyrazolo[1,5-*a*]pyrimidine heterocyclic disazo dyes and an evaluation of their solvatochromic behaviour. *Dye. Pigment.* **2007**, *74*, 578–584. [CrossRef]
52. Balalaie, S.; Kowsari1, E.; Hashtroudi, M.S. An Efficient Method for the Synthesis of 3-Cyano-6-hydroxy-2(1H)-pyridinones under Microwave Irradiation and Solvent-free Conditions. *Mon. Chem.* **2003**, *134*, 453–456. [CrossRef]
53. El-Adasy, A.A.A.M.; Kamel, M.M.; Saleh, M.O.; Hussein, A.M.; El-Apasery, M.A. Disperse Dyes Based on Pyrazolopyrimidinones I: Their Dyeing Applications and Antimicrobial Activities. *Int. J. ChemTech Res.* **2016**, *9*, 31–38.
54. Brittany, H. *Microwave Synthesis: Chemistry at the Speed of Light*; CEM Publishing: Matthews, NC, USA, 2002; pp. 11–27.
55. Díaz-Ortiz, Á.; Prieto, P.; De la Hoz, A. A Critical Overview on the Effect of Microwave Irradiation in Organic Synthesis. *Chem. Rec.* **2019**, *19*, 85.
56. El-Apasery, M.A.; Hussein, A.M.; El-Adasy, A.A.M.; Saleh, M.O.; Kamel, M.M. Microwave Assisted Synthesis of Some Azo Disperse Dyes with Antibacterial Activities. Part 1. *Egypt. J. Chem.* **2019**, *62*, 1253–1259.
57. Al-Etaibi, A.; El-Apasery, M.A.; Al-Awadi, N. The effect of dispersing agent on the dyeing of polyester fabrics with disperse dyes derived from 1,4-diethyl-2,6-dioxo-1,2,5,6-tetrahydropyridine-3-carbonitrile. *Eur. J. Chem.* **2013**, *4*, 240–244. [CrossRef]
58. Al-Etaibi, A.; Al-Awadi, N.A.; El-Apasery, M.A.; Ibrahim, M.R. Synthesis of some novel pyrazolo [1, 5-a] pyrimidine derivatives and their application as disperse dyes. *Molecules* **2011**, *16*, 5182–5193. [CrossRef] [PubMed]
59. Al-Etaibi, A.; El-Apasery, M.A.; Ibrahim, M.R.; Al-Awadi, N.A. A facile synthesis of new monoazo disperse dyes derived from 4-hydroxyphenylazopyrazole-5-amines: Evaluation of microwave assisted dyeing behavior. *Molecules* **2012**, *17*, 13891–13909. [CrossRef]
60. Al-Etaibi, A.; El-Apasery, M.A.; Mahmoud, H.M.; Al-Awadi, N.A. One-pot synthesis of disperse dyes under microwave irradiation: Dyebath reuse in dyeing of polyester fabrics. *Molecules* **2012**, *17*, 4266–4280. [CrossRef]
61. Al-Etaibi, A.; El-Apasery, M.A.; Mahmoud, H.; Al-Awadi, N. Synthesis, characterization and antimicrobial activity, and applications of new azo pyridone disperse dyes on polyester fabric. *Eur. J. Chem.* **2014**, *5*, 321–327.
62. Al-Etaibi, A.M. Synthesis and Antimicrobial Activity of some Disperse Dyes derived from Pyridones. *Int. J. ChemTech Res.* **2019**, *12*, 129–133. [CrossRef]
63. Al-Etaibi, A.M.; El-Apasery, M.A. Dyeing Performance of Disperse Dyes on Polyester Fabrics Using Eco-Friendly Carrier and Their Antioxidant and Anticancer Activities. *Int. J. Environ. Res. Public Health* **2019**, *16*, 4603. [CrossRef] [PubMed]
64. Al-Etaibi, A.M.; El-Apasery, M.A. Nano TiO_2 Imparting Multifunctional Performance on Dyed Polyester Fabrics with some Disperse Dyes Using High Temperature Dyeing as an Environmentally Benign Method. *Int. J. Environ. Res. Public Health* **2020**, *17*, 1377. [CrossRef] [PubMed]
65. Mashaly, H.M.; Abdelghaffar, R.A.; Kamel, M.M.; Youssef, B.M. Dyeing of polyester fabric using nano disperse dyes and improving their light fastness using ZnOnano powder. *Ind. J. Sci. Technol.* **2012**, *7*, 960–967. [CrossRef]
66. Al-Etaibi, A.M.; El-Apasery, M.A. A comprehensive review on the synthesis and versatile applications of biologically active pyridone-based disperse dyes. *Int. J. Environ. Res. Public Health* **2020**, *17*, 4714. [CrossRef]
67. Dostanić, J.; Valentić, N.V.; Uščumlić, G.; Mijin, D. Synthesis of 5-(substituted phenylazo)-6-hydroxy-4-methyl-3–cyano-2-pyridones from ethyl 3-oxo-2-(substituted phenylazo)butanoates. *J. Serb. Chem. Soc.* **2011**, *76*, 499–504. [CrossRef]
68. Mijin, D.Ž.; Baghbanzadeh, M.; Reidlinger, C.; Kappe, C.O. The microwave-assisted synthesis of 5-arylazo-4,6-disubstituted-3-cyano-2-pyridone dyes. *Dyes Pigm.* **2010**, *85*, 73–78. [CrossRef]
69. Gaffer, H.E.; Shkra, S.; Abbas, D.; Allam, E.A. Synthesis and antimicrobial activity of some new sulphonamide disperse dyes and their applications to polyester fibres. *J. Appl. Polym. Res.* **2013**, *9*, 4051–4058.
70. Ahmed, K.A.; Elhennawy, H.M.; Elkashouti, M.A. Microwave Assists the Synthesis of Pyridone azo Dyes and their Application in Polyester Printing. *Res. J. Chem. Sci.* **2012**, *2*, 14–19.
71. Mijin, D.; Uščumlić, G.; Perišić-Janjić, N.; Trkulja, I.; Radetić, M.; Jovančić, P. Synthesis, properties and color assessment of some new 5-(3-and 4-substituted phenylazo)-4, 6-dimethyl-3-cyano-2-pyridones. *J. Serb. Chem. Soc.* **2006**, *71*, 435–444. [CrossRef]
72. Al-Etaibi, A.M.; El-Apasery, M.A.; Kamel, M.M. Dyeing of polyester with disperse dyes: Part 1. Antimicrobial activity and dyeing performance of some disperse dyes. *Int. J. Curr. Microbiol. Appl. Sci.* **2015**, *4*, 923–928.
73. Al-Etaibi, A.M.; El-Apasery, M.A. Dyeing of polyester with disperse dyes: Part 3. Characterization of ZnO nanoparticles treated polyester fabrics for antibacterial, self-cleaning and UV protective. *Int. J. ChemTech Res.* **2016**, *9*, 162–169.

74. Al-Etaibi, A.M.; Alnassar, H.S.; El-Apasery, M.A. Dyeing of polyester with disperse dyes: Part 2. Synthesis and dyeing characteristics of some azo disperse dyes for polyester fabrics. *Molecules* **2016**, *21*, 855. [CrossRef] [PubMed]
75. Al-Etaibi, A.M.; Kamel, M.M.; El-Apasery, M.A. Synthesis and applications of new aminothienopyridazines disperse dyes on polyester fabric. *Int. J. Curr. Microbiol. Appl. Sci.* **2014**, *3*, 826–832.

polymers

MDPI

Article

Improvement of the Polyhydroxyalkanoates Recovery from Mixed Microbial Cultures Using Sodium Hypochlorite Pre-Treatment Coupled with Solvent Extraction

Gabriela Montiel-Jarillo [1,†], Diego A. Morales-Urrea [2,*,†], Edgardo M. Contreras [2], Alex López-Córdoba [3], Edwin Yesid Gómez-Pachón [4], Julián Carrera [1] and María Eugenia Suárez-Ojeda [1,*]

1 GENOCOV Research Group, Department of Chemical, Biological and Environmental Engineering, School of Engineering, Universitat Autònoma de Barcelona, Escola d'Enginyeria. Edifici Q Campus UAB, Bellaterra, 08193 Barcelona, Spain
2 División Catalizadores y Superficies, Instituto de Investigaciones en Ciencia y Tecnología de Materiales, INTEMA (CONICET), Av. Colón 10850, Mar del Plata 7600, Argentina
3 Grupo de Investigación en Bioeconomía y Sostenibilidad Agroalimentaria, Escuela de Administración de Empresas Agropecuarias, Facultad Seccional Duitama, Universidad Pedagógica y Tecnológica de Colombia, Carrera 18 con Calle 22, Duitama 150461, Colombia
4 Grupo de Investigación en Diseño, Innovación y Asistencia Técnica de Materiales Avanzados-DITMAV, Escuela de Diseño Industrial, Universidad Pedagógica y Tecnológica de Colombia-UPTC, Duitama 150461, Colombia
* Correspondence: dmoralesurrea@gmail.com (D.A.M.-U.); mariaeugenia.suarez@uab.cat (M.E.S.-O.)
† These authors contributed equally to this work.

Citation: Montiel-Jarillo, G.; Morales-Urrea, D.A.; Contreras, E.M.; López-Córdoba, A.; Gómez-Pachón, E.Y.; Carrera, J.; Suárez-Ojeda, M.E. Improvement of the Polyhydroxyalkanoates Recovery from Mixed Microbial Cultures Using Sodium Hypochlorite Pre-Treatment Coupled with Solvent Extraction. *Polymers* **2022**, *14*, 3938. https://doi.org/10.3390/polym14193938

Academic Editor: Edina Rusen

Received: 23 July 2022
Accepted: 30 August 2022
Published: 21 September 2022

Publisher's Note: MDPI stays neutral with regard to jurisdictional claims in published maps and institutional affiliations.

Copyright: © 2022 by the authors. Licensee MDPI, Basel, Switzerland. This article is an open access article distributed under the terms and conditions of the Creative Commons Attribution (CC BY) license (https://creativecommons.org/licenses/by/4.0/).

Abstract: The use of mixed microbial cultures (MMC) and organic wastes and wastewaters as feed sources is considered an appealing approach to reduce the current polyhydroxyalkanoates (PHAs) production costs. However, this method entails an additional hurdle to the PHAs downstream processing (recovery and purification). In the current work, the effect of a sodium hypochlorite (NaClO) pre-treatment coupled with dimethyl carbonate (DMC) or chloroform (CF) as extraction solvents on the PHAs recovery efficiency (RE) from MMC was evaluated. MMC were harvested from a sequencing batch reactor (SBR) fed with a synthetic prefermented olive mill wastewaster. Two different carbon-sources (acetic acid and acetic/propionic acids) were employed during the batch accumulation of polyhydroxybutyrate (PHB) and poly(3-hydroxybutyrate-co-3-hydroxyvalerate) (PHBV) from MMC. Obtained PHAs were characterized by ^{1}H and ^{13}C nuclear magnetic resonance, gel-permeation chromatography, differential scanning calorimetry, and thermal gravimetric analysis. The results showed that when a NaClO pre-treatment is not added, the use of DMC allows to obtain higher RE of both biopolymers (PHB and PHBV), in comparison with CF. In contrast, the use of CF as extraction solvent required a pre-treatment step to improve the PHB and PHBV recovery. In all cases, RE values were higher for PHBV than for PHB.

Keywords: polyhydroxyalkanoates; mixed microbial cultures; extraction; sodium hypochlorite; dimethyl carbonate; chloroform

1. Introduction

There is a growing environmental concern as a result of the problems driven from the overuse of petroleum-based plastics. This has caused enormous interest in the research of new bio-based and biodegradable materials that can potentially replace them. According to Kurian et al. [1], as of 2020, the estimated market size for bioplastics and biopolymers in North America and Europe was around 10.5 billion USD, which is projected to rise to 25–30 billion USD by 2025.

Polyhydroxyalkanoates (PHAs) are one of the most promising bioplastics because they are biodegradable and biocompatible and can be obtained from renewable resources. PHAs possess similar mechanical, thermal, and barrier properties to petroleum-based plastics such as polyethylene and polypropylene [2,3]. PHAs have been used in several fields including agriculture, medicine, packaging, and pharmacy [1,4,5].

One of the most important limitations for the commercial application of PHAs is their high production costs compared with petroleum-based plastics [1]. The current price of PHAs is between 2.2 and 5.0 €/Kg. This is around three times higher than the price of petroleum-based plastics such as polystyrene (PS), polypropylene (PP), or polyvinyl chloride (PVC) (<1.0 €/kg) [6]. The main factors affecting the final cost of PHAs are the use of pure cultures (e.g., genetically modified bacterial strains), specific carbon sources and operating conditions, and downstream processing [7].

Currently, a large part of the industrial scale production of PHAs is achieved using pure or genetically modified cultures, which necessitates sterile conditions and the utilization of expensive substrates [8,9]. Therefore, ongoing efforts to lower the cost of PHA manufacturing relies on using MMC. MMC do not require sterile conditions and are flexible enough to adjust to variations in the carbon source, affording the use of a wide variety of waste substrates, such as agricultural and industrial wastes [10,11]. In particular, MMC from waste-activated sludge from wastewater treatment plants have demonstrated a great PHA-accumulating potential under stressing conditions [10,11]. However, it has been reported that different substrates and production methods lead to various types of microbial biomass. Thus, an extraction method that is optimized for one particular MMC may not be effective in all conditions [12]. In addition, the recovery of PHAs from MMC is a more complex process because several microorganisms are coexisting and the different cellular matrices that surround the PHAs' granules hinders their extraction because the cells present different resistance to lysis [13,14]. Therefore, it is often necessary to apply pre-treatment methods that favor cell disruption, such as NaClO digestion [10,15].

Extraction using chlorinated solvents, such as chloroform (CF) and 1, 2-dichloroethane, is one of the most widespread methods to recover PHAs from MCC [7]. In contrast, the use of halogen-free solvents, such as dimethyl carbonate (DMC), in the extraction of PHAs from MMC, has been largely unexplored [7]. DMC has the advantage of being considered a green solvent because it is fully biodegradable, as well as being minimally toxic for the operators and the environment [7]. Samorì et al. (2015) [13] studied the extraction of PHAs from MMC using a NaClO pre-treatment coupled with DMC, finding that polymer recovery increased from 50% to 80%. Recently, Elhami et al. (2022) [16] extracted PHBV from MMC using DMC as an extraction solvent, and ethanol to precipitate the dissolved PHA. They found that the highest PHBV recovery value was 94%. This occurred when biopolymer precipitation was performed following complete evaporation of DMC.

In the current work, the application of an MMC pre-treatment using NaClO as an alternative to increase the extraction of PHAs with DMC or CF was investigated. The extracted biopolymers were characterized in terms of polymer structure, molecular weight, and thermal properties.

2. Materials and Methods

2.1. Production of PHAs

MMC used in the present study were harvested from a sequencing batch reactor (SBR) fed with a synthetic prefermented olive mill wastewaster (Montiel-Jarillo et al., submmited) [17]. The SBR was operated under fully aerobic conditions and under a feast/famine regimen. The biomass was collected at the end of the feast phase. The supernatant was removed by settling, and the biomass was resuspended in a mineral medium devoid of a nitrogen source, as described in Montiel-Jarillo et al. [18]. Before the accumulation experiments, the biomass was aerated overnight in the absence of any substrate. Then, batch PHA-accumulation experiments were performed in 1 L glass stirred reactors according to the protocol described in Mannina et al. (2019) [19]. Briefly, the change of the dissolved oxygen (DO) concentration in the reactor was used as an indicator of the substrate exhaustion. The experiment started with the addition of a pulse of the substrate

(0.2 gCOD L^{-1}). The addition of the substrate caused a noticeable drop in the DO. After a certain time (typically, 5–10 min), the DO increased due to the substrate depletion. Then, a new pulse of the substrate was added. Using this procedure, a total of 30 pulses of the substrate were added to the reactor in 4–5 h of operation. Two different carbon sources were used. In batch A, acetic acid (HAc) was used to obtain PHB, whereas in batch B a mixture of HAc and propionic acid (HPr) in a proportion of 75:25 over the chemical oxygen demand (COD) leads to the accumulation of the copolymer PHBV [18].

During PHA-accumulation experiments, mixed liquor samples were taken to determine the concentration of biomass as total and volatile suspended solids (TSS, VSS), acetate and propionate as volatile fatty acids (VFA), and the specific content of PHA of the biomass. TSS and VSS concentrations were measured according to the APHA methods 2540D and 2540E, respectively [18]. The samples for VFA were filtered through 0.45 μL pore size membranes and transferred to 2 mL septum-capped vials. VFA concentration of the filtrate was measured according to the procedure described by Montiel-Jarillo et al. [18]. To evaluate the specific content of PHA in the biomass, 10 mL of the culture was mixed with 0.4 mL of formaldehyde (37 wt% in H_2O) to inhibit biological activity. Then, at the end of each experiment, 6 mL of formaldehyde (37 wt% in H_2O) was added to the reactor to inhibit biological activity. In all cases, the biomass was harvested by centrifugation at 11× *g* for 45 min. The supernatant was discarded. The PHA-enriched biomass was stored overnight at −80° and then lyophilized during 24 h.

The amount of PHA in the biomass was quantified by gas chromatography (GC) [18]. A known amount of the lyophilized biomass was mixed with benzoic acid as an internal standard. Then, 1.5 mL of butanol and 0.5 mL of hydrochloric acid were added, and the mixture was incubated at 100 °C for 8 h. After that, 2.5 mL of hexane and 4 mL of Milli-Q were added. The tubes were vortexed and left to stand for 15 min to allow the separation of the phases. The organic phase was transferred to clean tubes and 4 mL aliquot of Milli-Q water was added. Then, the tubes were vortexed and centrifuged at 2× *g* for 10 min. The organic phase was filtered through 0.22 μm filters and transferred into GC vials. One microliter of the sample was injected and analyzed in an Agilent Technologies (7820 A) gas chromatograph equipped with an FID detector and an HP-InnoWax column (30 m × 0.53 mm × 1 μm).

2.2. Evaluation PHAs Extraction

PHAs were extracted from the lyophilized biomass using dimethyl carbonate (DMC) and chloroform (CF) as the extraction solvents, according to the method described by Samorì et al. [20]. The PHAs extraction assays were carried-out at boiling temperature of each solvent (DMC at 90 °C and CF at 60 °C) for 1 h, using a lyophilized biomass concentration of 50 g dry-cell/L. These operational conditions were chosen from preliminary experiments in which different biomass concentrations (25 and 100 g dry-cell/L) and extraction time (1 and 3 h) were evaluated (Supplementary Material Table S1).

Figure 1 shows the schematic representation of the proposed PHAs extraction process using DMC and CF. Fifty milligrams of the lyophilized biomass were transferred to Eppendorf vials. Then, the samples were resuspended in 1 mL of Milli-Q and mixed with 2 mL of the tested extraction solvent. The vials were periodically stirred using a vortex. Then, vials were centrifuged at 6.5× *g* for 15 min at room temperature. Three phases were obtained: (1) a supernatant phase composed by water and water-soluble compounds, (2) an intermediate phase that corresponds to the organic fraction containing the solubilized PHAs, and (3) a precipitate composed by biomass and cell debris. The organic fraction (2) was recovered and filtered through 0.45 μm pore size cellulosic membranes (Millipore®, Burlington, MA, USA) to remove any cell debris. Finally, the solvent was removed by evaporation to obtain a dried PHA sample.

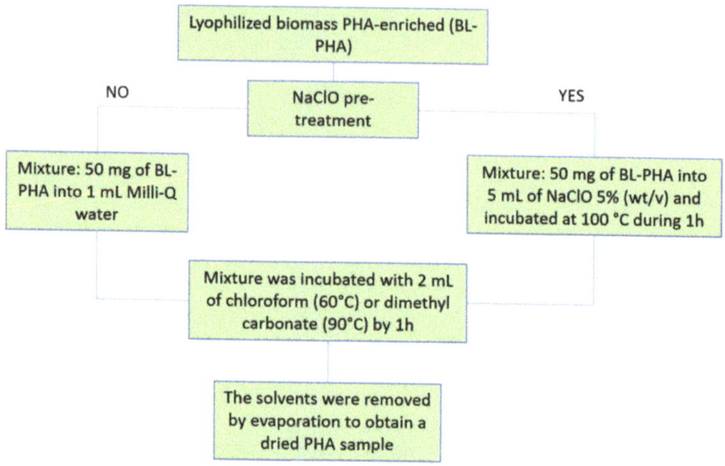

Figure 1. Schematic representation of the proposed PHAs extraction process using DMC and CF.

A second set of experiments to enhance the recovery of PHA from the lyophilized biomass were performed employing a chemical digestion using sodium hypochlorite (NaClO) [13]. In this case, 50 mg of lyophilized biomass was resuspended in 5 mL of NaClO 5% (w/v) and incubated at 100 °C for 1 h. The digested samples were centrifuged at 3× g for 10 min at room temperature. The pellets were washed three times with Milli-Q water and resuspended in 1 mL of Milli-Q water. Then, PHAs were extracted using DMC and CF according to the procedure previously described in this section.

Calculations

To compare the different tested extraction protocols, the overall recovery efficiency (RE, %) was calculated as follows:

$$\text{RE} = 100\frac{Y_E P_E}{PHA_{content}} \quad (1)$$

where Y_E (g extract/g dry-cell) is the extraction yield, $PHA_{content}$ (g PHA/g dry-cell) is the amount of PHA (measured by GC) within the cells, and P_E (g PHA/g extract) is the amount of PHA (measured by GC) per gram of extract. Measured purities were 0.79 g PHA g extract^{-1} for extracts from batch A, and 0.85 g PHA g extract^{-1} for the copolymer obtained from samples of batch B.

2.3. *Characterization of the Obtained Biopolymers*

To obtain enough amounts of PHAs for their further characterization, PHAs were extracted from the lyophilized biomass using a Soxhlet device (Buchi, E-816 SOX). In these cases, 2 g of the lyophilized biomass obtained at the end of each PHA-accumulation experiment was mixed with 220 mL of NaClO 5% (w/v) at 100 °C for 1 h. The digested sample was centrifuged at 3× g for 20 min. The obtained pellet was centrifuged and washed three times with milli-Q water and introduced into the Soxhlet cartridge. PHA was extracted using 90 mL of chloroform (CF) for 1 h. Finally, CF was removed by evaporation.

The chemical structure of the obtained PHAs was determined by quantitative ^{1}H and ^{13}C nuclear magnetic resonance (NMR) spectra using a BRUKER DRX-500 spectrometer. Known amounts (30–40 mg) of the obtained PHA were dissolved in deuterated chloroform ($CDCl_3$). The software Bruker TopSpin3.5pl7 was used to analyze the obtained NMR spectra. Also, gradient-selected ^{1}H/^{13}C heteronuclear single quantum coherence (HSQC) spectra were acquired.

Mass-average (M_w) and number-average (M_n) molecular weights of the biopolymers were measured by Gel Permeation Chromatography (GPC) on a Waters equipment provided with RI and UV detectors. Biopolymer samples were diluted (0.1% w/v) and filtered. Then, 100 μL of this solution were injected and operated using 1,1,1,3,3.3-hexafluoro-2-propanol as a mobile phase with a flow of 0.5 mL min^{-1}. HHR5E and HR2 Waters linear Styragel columns (7.8 mm × 300 mm, pore size 103–104 Å) packed with crosslinked polystyrene and protected with a pre-column were used. Molecular weight was calibrated using poly(methyl methacrylate) as the reference compound.

To evaluate the degradation temperature (T_d), melting temperature (T_m), and enthalpy of fusion (ΔH_m) of the obtained biopolymers, a simultaneous Thermogravimetry (TGA)–Differential Scanning Calorimetry (DSC) analysis (Netzsch STA 449F1 Jupiter®, Gaithersburg, MD, USA) under a nitrogen atmosphere was performed [21]. The crystallinity degree (X, %) of the biopolymer was calculated as follows Dai et al. [22].

$$X\,(\%) = 100\frac{\Delta H}{\Delta H^{o}_{PHB}} \quad (2)$$

where ΔH (J/g) is the melting enthalpy of the analyzed PHA, and ΔH^{o}_{PHB} = 146 J/g is the melting enthalpy of a fully crystalline PHB [23].

3. Results and Discussion

3.1. *Production of PHA Using Mixed Microbial Cultures (MMC)*

Figure 2 shows a typical result corresponding to batch A. When a pulse of HAc was added, pH immediately dropped from a value above 8 to about 6. Also, a quite noticeable decrease of the DO concentration was obtained. While a gradual increase in pH was observed, DO remained below 1 mg L^{-1} for about 6–8 min. Then, a sudden increase in DO indicated the substrate depletion. At this point, the measured soluble COD was negligible, confirming the absence of a substrate. Therefore, a new pulse of HAc was added. Figure 2 also shows that the specific content of PHA in the cells increased as a function of the added pulse.

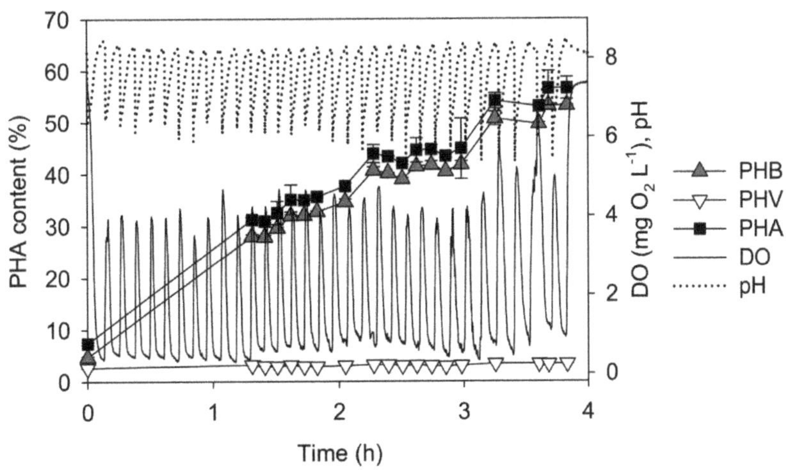

Figure 2. Performance of batch A as a function of time using HAc as the sole carbon source. The black line represents the DO profile, while the black dotted line corresponds to pH. The lines connecting the experimental points were included as a visual aid. Bars indicate the standard deviation of triplicates of PHAs content.

When HAc was tested as the sole carbon source, the specific PHA production rate (q_{PHA}), and the specific substrate consumption rate (q_S) were 0.37 C-mol PHA (C-mol biomass)$^{-1}$ h^{-1} and 0.52 C-mol VFA (C-mol biomass)$^{-1}$ h^{-1}, respectively. Accordingly, the PHA yield (Y_{PHA}) was 0.71 C-mol PHA (C-mol VFA)$^{-1}$. In other words, about 70% of the carbon in form of HAc yielded PHA, while the other 30% was used for biomass synthesis and energy production (e.g., respiration). At the end of the experiment, the specific PHA content of the biomass was about 0.57 gPHA g(dry-cell)$^{-1}$. CG analysis showed that more than 95% (*w*/*w*) of the PHA corresponded to HB units, suggesting that the obtained polymer was mainly PHB.

Figure 3 shows a PHA accumulation assay using a mixture of HAc:HPr as the carbon source (batch B). A similar trend with respect to pH, and DO in comparison with batch A was obtained (Figure 2), however, the PHV increased by up to 35% *w*/*w*. In this case, q_{PHA} and q_S values were 0.47 C-mol PHA (C-mol biomass)$^{-1}$ h^{-1} and 0.55 C-mol VFA (C-mol biomass)$^{-1}$ h^{-1}, respectively. According to these values, Y_{PHA} corresponding to these experiments was 0.85 C-mol PHA (C-mol VFA)$^{-1}$. Thus, the performance of the mixture of HAc:HP, with reference to the production of PHA, was slightly better than that of HAc as the sole carbon source. CG analysis showed that the polymer composition was quite constant during the whole experiment. The average polymer composition of the obtained PHA was 52% (*w*/*w*) of HB and 48% (*w*/*w*) of HV. At the end of the experiment, the PHA content of the biomass reached 0.72 gPHA g dry-cell^{-1}.

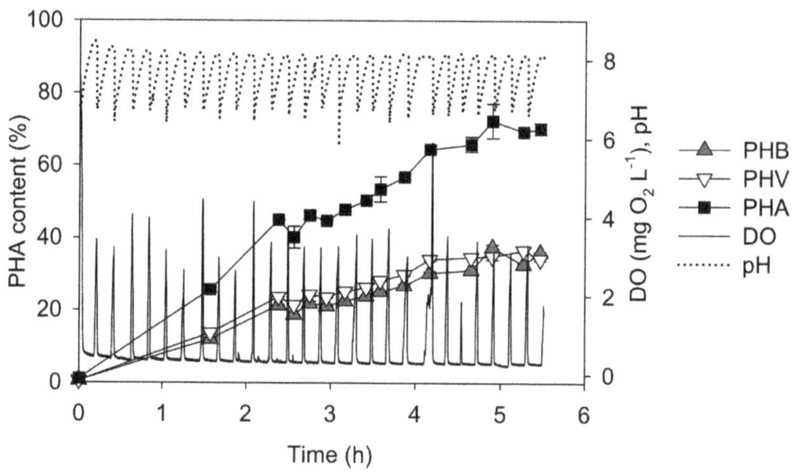

Figure 3. Performance of batch B as a function of time using a mixture of HAc and HPr. The black line represents the DO profile, while the black dotted line corresponds to pH. The lines connecting the experimental points were included as a visual aid. Bars indicate the standard deviation of triplicates of PHAs content.

Kinetic (q_{PHA}, q_S) and stoichiometric (Y_{PHA}) coefficients obtained for both tested carbon sources were within the range reported for the production of PHA using MMC grown on different synthetic substrates such as acetate, propionic, or mixtures of them [24–26]; fermented sugar cane molasses; cheese whey; and waste-activated sludge [27,28].

3.2. Evaluation of PHAs Extraction

Table 1 shows the obtained PHA recovery efficiencies (RE) with DMC or CF as extraction solvent using a biomass concentration of 50 g/L and extraction time of 1 h. In all cases, it was observed that PHBV had higher recovery than PHB, suggesting that PHBV was longer soluble in both solvents. In addition, it was found that when a MMC pre-treatment was not carried out, the use of DMC solvent afforded higher PHB and PHBV recovery than CF. This behaviour was probably because the use of a solvent with a high boiling point promoted cell lysis, leading to better contact of the solvent with the intracellular granules of

PHAs. On the other hand, it was observed that the pre-digestion step successfully enhanced the PHB and PHBV recovery using both solvents. It has been stated that NaClO weakens cell membranes, thus facilitating the subsequent extraction of PHA by solvents [7,8].

Table 1. PHA recovery efficiencies (RE) using dimethyl carbonate (DMC) or chloroform (CF).

Sample	Pre-Treatment with NaClO	RE (%) DMC (at 90 °C)	RE (%) CF (at 60 °C)
Biomass enriched in PHB	No	4 ± 3	2 ± 1
	Yes	8 ± 3	17 ± 1
Biomass enriched in PHBV	No	17 ± 1	5 ± 2
	Yes	25 ± 4	23 ± 2

Depicted values correspond to the average ± one standard deviation of three measurements.

In complementary experiments, the effect of different biomass amounts (25, 50, and 100 g dry-cell/L) and incubation times (1 and 3 h) on PHB and PHBV recovery from MMC was evaluated for comparison purposes. Furthermore, the influence of the ethanol-induced precipitation on the PHAs extraction was studied (Supplementary Material Table S1). It was found that when the lyophilized biomass concentration was 25 g/L, the PHB and PHBV recoveries were negligible. The use of concentrations of 100 g/L led to lower recovery efficiencies than those obtained when using a concentration of 50 g/L. This may be due to the suspension at 100 g/L becoming very viscous, leading to quite low recovery of the biopolymers [29]. Besides, the results showed that an increase in the extraction time from 1 to 3 h and a precipitation step using ethanol did not improve the recovery of the polymers in comparison with the results showed in Table 1. Similarly, de Souza Reis et al. (2020) [29] studied the effect of different extraction times (0.25, 0.5, 1, 1.5, and 2 h) and biomass to DMC solvent proportions (1, 2.5, 5, and 10%) on the PHBV recovery from MMC. They obtained recovery percentages of 32.9 when using a biomass to solvent ratio of 1% with an extraction time of 1.5 h. Abbasi et al. (2022) [30] extracted PHBV from MMC using CF and DMC as extraction solvents, obtaining recovery values of 33.5% and 30.6%, respectively. This variability observed in the recovery percentage of PHBV could be due to the heterogeneity and complex cell structure of the MMCs, meaning that a specific extraction method may not have been efficient for all MMC sources [26,29].

3.3. *Characterization of the Obtained Biopolymers*

3.3.1. Quantitative $^{1}H/^{13}C$ Nuclear Magnetic Resonance (NMR)

Figure 4a shows the ^{1}H-spectrum corresponding to the biopolymer obtained from batch A (PHB), exhibiting three characteristic signals of PHB [31]. The signal corresponding to the methyl group (peak 4 in Figure 4a) is the first doublet at 1.28 ppm (B4 in Figure 4a). Signals at 2.50–2.58 ppm (peak 2 and peak 3 in Figure 4a) correspond to the methylene group (B2 in Figure 4a), while the multiplet at 5.25 ppm (peak 1 in Figure 4a) corresponds to the methine group (B3 in Figure 4a). Other small signals observed in Figure 4a may be due to the presence of small amounts of HV. Figure 4b shows the ^{1}H-spectrum corresponding to the biopolymer from batch B (PHBV). According to the literature, this spectrum has the characteristic signals of a copolymer PHBV [44]. Methyl groups corresponding to HV (V5) and HB (B4) are represented by the resonance peaks at 0.90 (peak 8 in Figure 4b) and 1.28 ppm (peak 7 in Figure 4b), respectively. The peak at 1.63 ppm (peak 6 in Figure 4b) is due to the HV methylene protons (V4). Overlapping resonance peaks within the range of 2.50–2.60 ppm (peaks 3, 4, 5 in Figure 4b) represent the methylene protons of HV (V2) and HB methylene group (B2). Resonance signals at 5.15 and 5.25 ppm (peaks 1,2 in Figure 4b) are characteristic for the methine proton of HV (V3) and HB (B3), respectively.

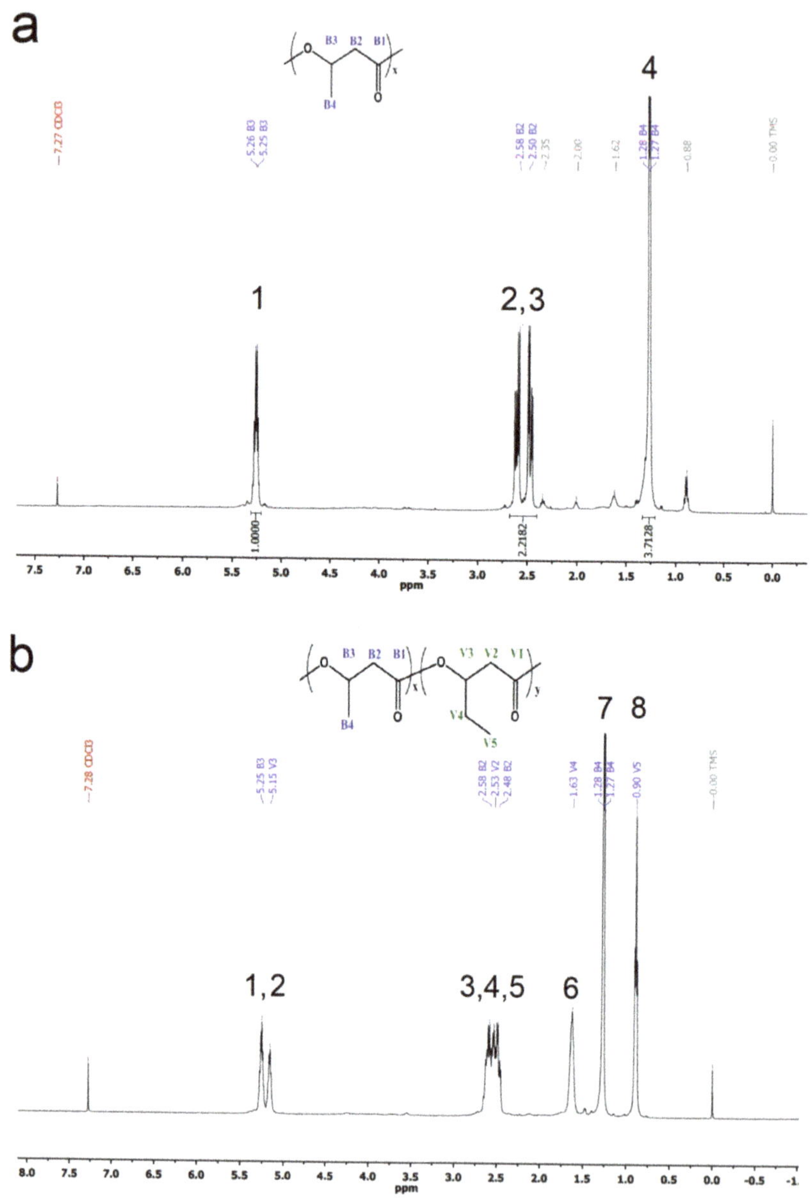

Figure 4. ^{1}H-NMR spectra corresponding to PHB (**a**) and PHBV (**b**), obtained from batch A and B, respectively.

The molar fraction of HB and HV corresponding to the polymer obtained from batch B can be estimated by the intensity of the signals B3 with respect to V3, and V5 to B4 [32] (Table 2). According to the data corresponding to the methine groups, the obtained copolymer was composed of 58% of HB and 42% of HV on a molar basis. A similar composition was obtained from the intensity of the methylene groups (59% and 41% of HB and HV, respectively). It must be noted that these results are similar to those obtained by GC during the accumulation assay (Figure 3), confirming that the biopolymer obtained from batch B was a copolymer of PHBV.

Table 2. Assignment of resonance peaks for ^{13}C-NMR spectra of copolymer PHBV.

Position	Diad/Triad	^{13}C (ppm)
V5	VV VB	9.33 9.36
B4	BB BV	19.73 19.77
V4	VVV BVV VVB BVB	26.75 26.77 26.82 26.85
V2	VVV BVV VVB BVB	38.64 38.65 38.76 38.79
B2	BV BB	40.77 40.78
B3	VB VV	67.61 67.67
V3	BB BV	71.87 71.91
B1	BB BV	169.15 169.32
V1	VB VV	169.35 169.52

B and V represent butyrate and valerate units, respectively.

Figure 5 shows resonance ^{13}C-spectra corresponding to the obtained polymers. The results for the resonance peaks were similar to those documented in the literature [14,32–34]. Only four clearly discernible peaks can be observed in the ^{13}C-spectrum corresponding to PHB (Figure 5a). Those peaks correspond to carbonyl (B1) at 169.12 ppm, methyl carbon (B4) at 19.77, and methylene (B2) and methine (B2) at 40.78 and 67.62 ppm, respectively. All those peaks are characteristic of PHB. Conversely, the ^{13}C-spectrum corresponding to PHBV (Figure 5b) was typical of a copolymer PHBV. Figure 6 shows an expansion of several zones corresponding to the ^{13}C-spectrum depicted in Figure 5b. The methylene moieties of HB and HV are responsible for the peaks between 26 and 41 ppm.

Carbonyl peaks can be observed at around 169 ppm [14,22,34]. These signals contribute to diads of HB and HV units, namely BB, BV, VB, or VV, where B corresponds to butyrate, and V to valerate. The peak at 169.15 ppm is consistent with the carbonyl resonance of PHB, and it was assigned to the sequence BB. According to Doi et al. [35], the sequence VV was assigned for the peak that appears with a difference of 0.38 ppm. Intermediate peaks with low intensities were assigned to the BV and VB sequences. The obtained peaks within the methylene region were consistent with those reported in the literature. First, the peak at 40.78 ppm that corresponds to the methylene of the HB unit (B2) exhibits a shoulder due to the presence of two almost overlapping peaks. The split is due to the diad sequences BV and BB. For the HV unit, the signals for the side-chain methylene group (V4) and main-chain methylene group (V2) may be differentiated, unless both signals are split into four peaks and have almost similar intensities. The main-chain methylene signal is observed between 26.75 and 26.85 ppm and the shift for side-chain methylene is around 38.64–38.79 ppm. In both cases, peaks were assigned for triad sequences of VVV, BVV, VVB, and BVB.

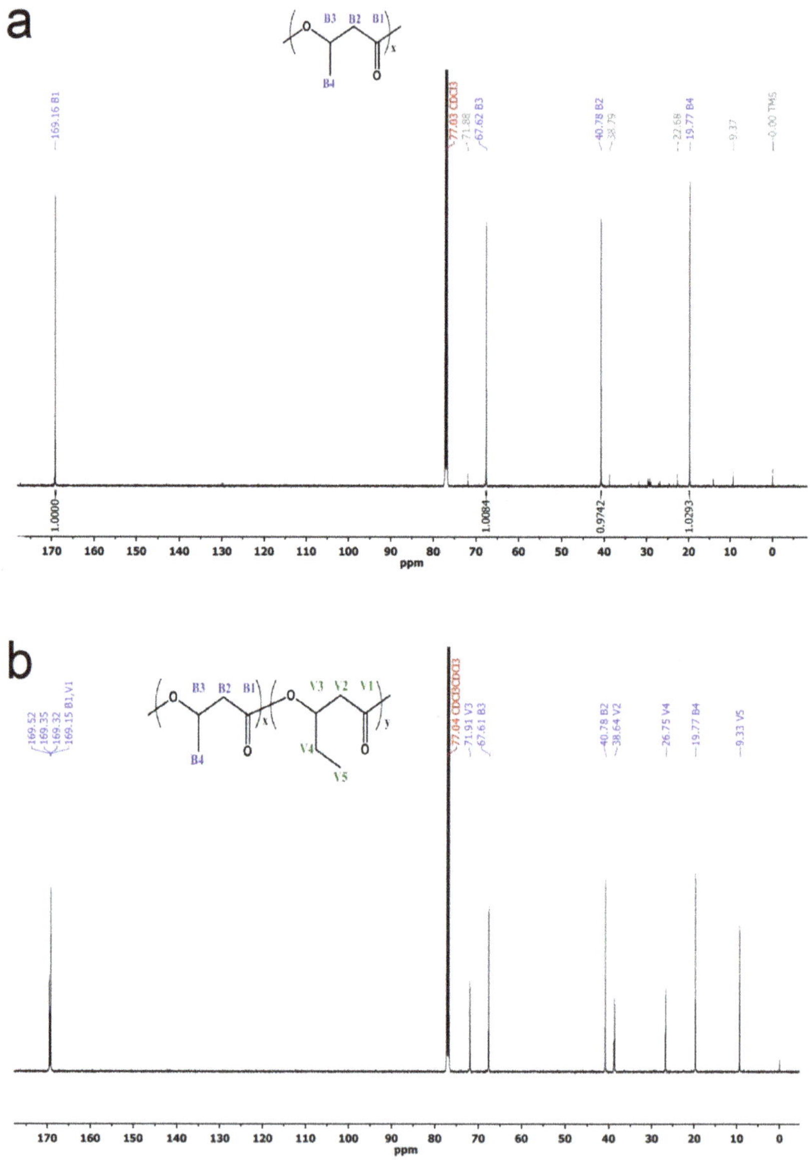

Figure 5. ^{13}C-NMR spectra corresponding to PHB (**a**) and PHBV (**b**) obtained from batch A and B, respectively.

The molar ratio of HV with respect to HB in the copolymer can be obtained from the signals corresponding to the methine carbons at 67.61 ppm for PHB (B3 in Figure 5a), and at 71.91 ppm for PHV (V3 in Figure 5b) [14]. According to these considerations, PHBV was composed of 58.4% of HB and 41.6% of HV. The calculated composition was in agreement with that obtained from the ^{1}H spectrum. Results obtained from the HSQC spectra confirmed the assignment of resonance peaks of the carbons belonging with their corresponding hydrogen signals. The C-H couplings in HSQC are shown in Figure 7 for PHB, and Figure 8 for PHBV.

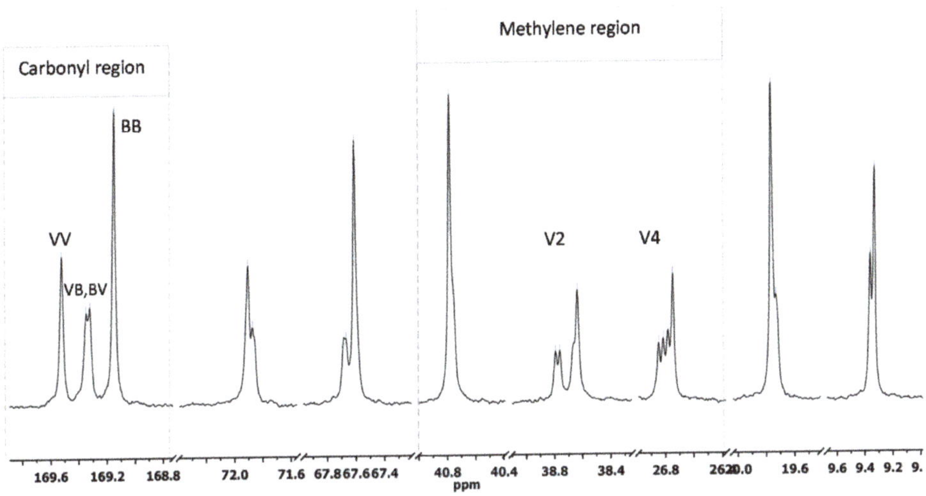

Figure 6. Detail of ^{13}C-NMR spectrum corresponding to PHBV obtained from batch B.

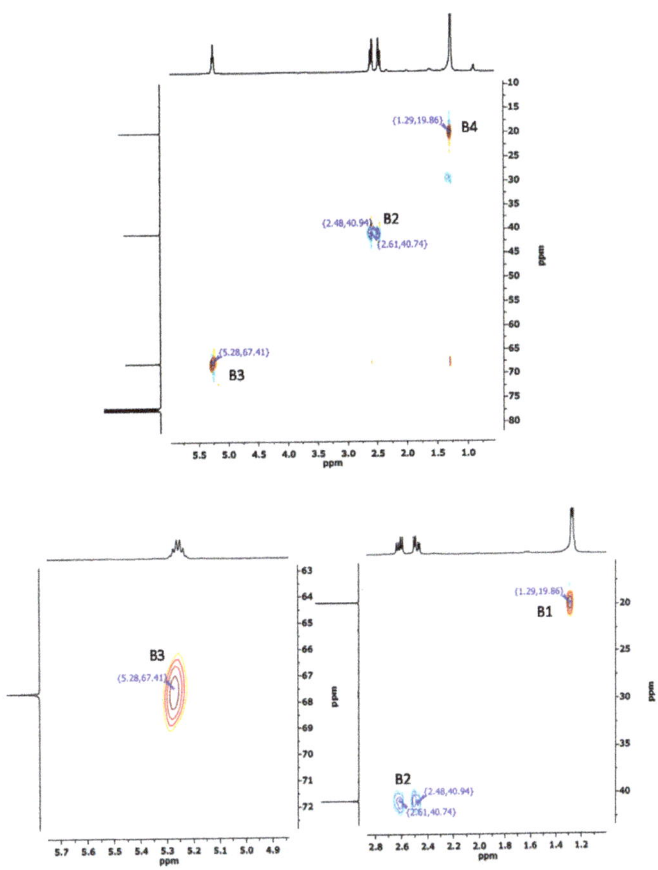

Figure 7. HSQC spectra corresponding to PHB obtained from batch A.

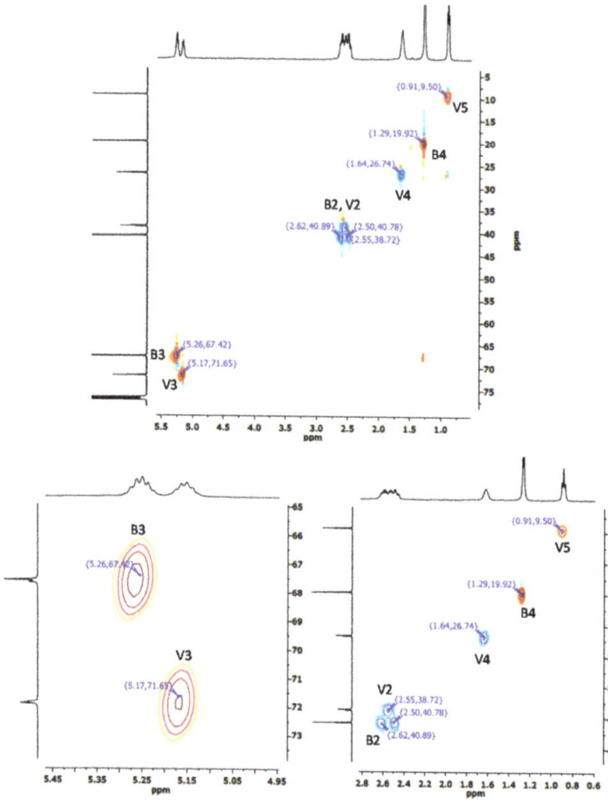

Figure 8. HSQC spectra corresponding to PHBV obtained from batch B.

The comonomer composition distribution (CCD) of a copolymer has been identified as a key factor in determining the physical properties of a biopolymer [34]. To evaluate the degree of randomness of the obtained polymer in batch B, ^{1}H NMR (Figure 4b) and ^{13}C NMR (Figure 5b), spectra were analyzed using the procedure proposed by Kamiya et al. [36] and other authors [34,37]. Two parameters, D and R, can be calculated from the relative peak intensities of ^{1}H and ^{13}C NMR spectra. Statistically random copolymers have D values in the range 0.99–1.5, while $D > 1$ or $D < 1$ define non-random copolymers. Besides, a D value close to 0 is characteristic for alternating nature copolymers, while $D > 1$ indicates "blocky" copolymers [21,36]. However, in cases where D is greater than 1, the copolymer may be a true block copolymer, a mixture of random copolymers, or a mixture of HB and HV homopolymers [36,37]. It must be noted that according to several authors, the parameter R is more sensitive than D for estimating the degree of randomness of a copolymer [34,37]. A value of *R* around 0 indicates a diblock copolymer, while an R value of 1 corresponds to a completely random distribution [21,34,37].

Table 3 shows the relative peak intensities of ^{13}C NMR spectra corresponding to the copolymer PHBV. These were used to calculate the parameter D and R. For the PHBV obtained in this study, the calculated value of D was 4.06. Thus, it can be concluded that the microstructure corresponds to a "blocky" copolymer character. Considering reports found in the literature, PHBV may be considered as a mixture of random copolymers, or a simple block structure [21,34]. Moreover, R value obtained herein was 0.65, supporting the idea that the obtained PHBV is closer to a "blocky" copolymer rather than a random copolymer. As a general rule, "blocky" materials have better elongation properties (e.g., larger Young's module and tensile strength) than random polymers [38,39].

Table 3. Experimental monomer, dyad, and triad sequence mole fractions obtained from $^{1}H^{a}$ and $^{13}C^{b}$ NMR spectra corresponding to the copolymer PHBV.

Monomer	F_B	*0.587* [a] *0.584* [b]
	F_V	*0.413* [a] *0.416* [b]
Dyad	F_{BB}	*0.409*
	F_{BV}	*0.176*
	F_{VB}	*0.151*
	F_{VV}	*0.264*
Triad	F_{VVV}	*0.181*
	F_{VVB}	*0.081*
	F_{BVV}	*0.075*
	F_{BVB}	*0.075*

B: butyrate unit; V: valerate unit. [a] determined by ^{1}H NMR spectra and [b] determined by ^{13}C NMR spectra.

3.3.2. Gel-Permeation Chromatography (GPC)

The results obtained regarding the mass-average molecular weight (M_w), number-average molecular weight (M_n), and polydispersity index (PDI) are summarized in Table 4. These parameters are of great importance as they are responsible for the end-use suitability of a given polymer for specific applications [37,40].

Table 4. Molecular weights corresponding to the obtained biopolymers.

PHA	**M_w ($\times 10^4$ Da)**	**M_n ($\times 10^4$ Da)**	**PDI**
PHB	4.53	1.69	2.7
PHBV	3.49	1.23	2.8

M_w: mass-average molecular weight. M_n: number-average molecular weight. PDI: polydispersity index.

The corresponding M_w and M_n of the polymers obtained in this study were lower than PHA molecular weights frequently reported [40]. The reason for such low molecular weights may be a consequence of the effect of different conditions during the extraction process. On one hand, the first extraction step was the biomass digestion using NaClO for 1 h at 100 °C. Although these conditions are favorable for disrupting cell walls, they also may favor polymer hydrolysis, yielding a decrease of the molecular weight of the obtained polymer [13,41]. In this sense, several authors report that amorphous polymers are weak to alkaline saponification [42]. However, other authors support the use of NaClO in the presence of an extraction solvent, arguing that once PHA is released, it may be immediately dissolved in the extraction solvent, preventing the polymer hydrolysis [42]. Moreover, the temperature and time of the thermal treatment also affect the molecular weight of the obtained polymer. As a general rule, higher temperatures increase the PHA solubility, but also favor its hydrolysis [40]. Additionally, other variables, such as the extraction from dried or wet cells, and fermentation conditions (pH, type and concentration of the carbon source, and nutrients) were reported as factors that affect the molecular weight of the obtained biopolymers [14,37,43].

The polydispersity index (PDI) of a given polymer is a measure of the heterogeneity of the polymer chain lengths [40]. In principle, PDI values range from one to infinity. A polymer composed of molecules with the same chain lengths has PDI values close to one. Conversely, PDI > 1 are characteristic of polymers composed of different chain lengths. While for a typical addition polymerization, PDI can range around 5–20, most probable PDI values for typical step polymerizations are around two. As a general rule, polymers with PDI close to one are preferred, which enables their use in a huge range of applications [41]. Although molecular weights of the obtained polymers were lower than those commonly found in the literature, their PDI values (Table 4) were within the range reported (from 1.84 to 7.12) by other authors during PHAs extraction from MMC [10,21].

3.3.3. Thermogravimetry (TGA) and Differential Scanning Calorimetry (DSC)

Figure 9 shows that TGA profiles corresponding to both obtained polymers were similar. The degradation of both polymers occurred as a single step, starting at around 200 °C. In both cases, the temperature corresponding to the maximum degradation rate was 291 ± 1 °C. The degradation temperature for a 5% weight loss obtained for both polymers was 243 ± 2 °C. This value was within the range reported by other authors [21,23,42].

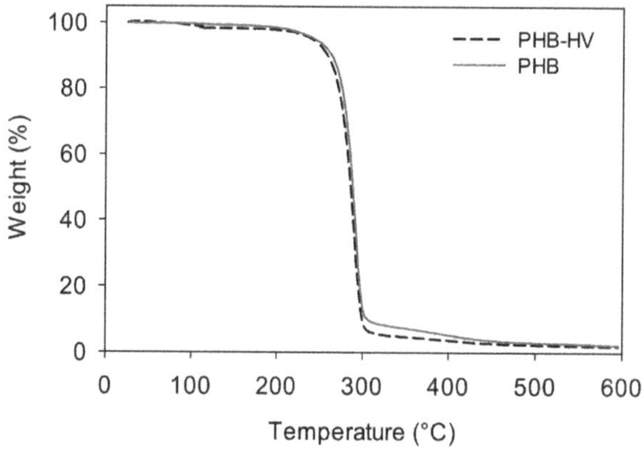

Figure 9. Thermogravimetric curves corresponding to PHB (continuous line) and PHBV (dotted line) obtained from batch A and B, respectively.

Table 5 shows the results obtained with the DSC corresponding to the obtained polymers. PHB exhibited a single melting point at 154.6 °C. This value is within the range of results reported in several studies [42–44]. Conversely, for PHBV, two melting points were observed at 78.3 and 152.9 °C. These melting points were close to those reported by Arcos-Hernández et al. [21] (77.5 and 159.2 °C) corresponding to a PHBV obtained using the same feeding strategy and carbon source as in the experiments reported herein. According to several authors [21,36], blocky copolymers with D > 1.5, such as the PHBV obtained in the present study, can exhibit two or three melting points as a result of the presence of polymers with different structures.

Table 5. Thermal characterization parameters corresponding to the obtained biopolymers.

PHA	$T_{d\text{-}5\%}$	$T_{d\text{-}max}$	T_{m1}	ΔH_{m1}	T_{m2}	ΔH_{m2}	X_c (%)
PHB	244.9		–	–	154.6	47.2	35.8
PHBV	242.3		78.3	15.9	152.9	3.44	14.7

$T_{5\%}$: degradation temperature corresponding to 5% weight loss. T_{max}: temperature corresponding to the maximum degradation rate. X_C: crystallinity (Equation (1)). Temperature values are expressed in °C and ΔH in J g^{-1}.

Based on the melting enthalpy for each biopolymer, the degree of crystallinity (X_c) was obtained using Equation (2). Table 5 shows that the degree of crystallinity corresponding to PHBV was 2.4 times lower than that corresponding to PHB. According to several authors, polymers with higher crystallinity have a greater range of potential applications [23,45].

4. Conclusions

In this study, PHB and PHBV were obtained from MMC using acetic and a mixture of acetic/propionic acids as carbon sources. Overall, the recovery of PHBV was higher than PHB, regardless of which extraction solvent was used. The PHAs recovery without a NaClO pre-treatment step was highest when DMC was employed, while when CF was used, a pre-treatment step was necessary to improve the extraction of PHB and PHBV. PHAs characterisation by ^{1}H and ^{13}C NMR spectra demonstrated that PHBV corresponded to a "blocky" copolymer. In general, thermal properties suggested that the presence of HV units confers desirable thermal characteristics for further PHA processability. The decomposition temperature of both polymers was similar, but the melting point and degree of crystallinity were lower in PHBV than that of PHB. Further studies are necessary to microbiologically characterize the MMC obtained.

Supplementary Materials: The following supporting information can be downloaded at: https://www.mdpi.com/article/10.3390/polym14193938/s1, Table S1: PHA recovery efficiencies (RE) using DMC at 90 °C. Depicted values correspond to the average ± one standard deviation of three measurements.

Author Contributions: Conceptualization and investigation were performed by G.M.-J., D.A.M.-U., E.M.C., J.C. and M.E.S.-O.; methodology, and validation were carried out by G.M.-J., D.A.M.-U., J.C. and M.E.S.-O.; formal analysis, and writing—original draft preparation were performed by G.M.-J., D.A.M.-U., E.M.C., A.L.-C., J.C. and M.E.S.-O.; funding acquisition was carried out by E.M.C., J.C. and M.E.S.-O.; resources, and project administration were carried out by J.C. and M.E.S.-O.; writing—review and editing, visualization, supervision were performed by G.M.-J., D.A.M.-U., E.M.C., A.L.-C., E.Y.G.-P., J.C. and M.E.S.-O. All authors have read and agreed to the published version of the manuscript.

Funding: This research was funded by: TRITON thematic network (316RT0508) from the Programa Iberoamericano de Ciencia y Tecnología para el Desarrollo (CYTED); Minciencias, and the Gobernación de Boyacá through the PATRIMONIO AUTÓNOMO FONDO NACIONAL DE FINANCIAMIENTO PARA LA CIENCIA, LA TECNOLOGÍA Y LA INNOVACIÓN FRANCISCO JOSÉ DE CALDAS (project 110986575000- Conv. 865-2019); Consejo Nacional de Ciencia y Tecnología de México (CONACyT) and Consejo Nacional de Investigaciones Científicas y Técnicas (CONICET) de Argentina.

Institutional Review Board Statement: Not applicable.

Informed Consent Statement: Not applicable.

Data Availability Statement: The data presented in this study are available on request from the corresponding author.

Conflicts of Interest: The authors declare no conflict of interest.

References

1. Kurian, N.S.; Das, B. Comparative Analysis of Various Extraction Processes Based on Economy, Eco-Friendly, Purity and Recovery of Polyhydroxyalkanoate: A Review. *Int. J. Biol. Macromol.* **2021**, *183*, 1881–1890. [CrossRef]
2. Samaniego, K.; Matos, A.; Sánchez-Safont, E.; Candal, M.V.; Lagaron, J.M.; Cabedo, L.; Gamez-Perez, J. Role of Plasticizers on PHB/Bio-TPE Blends Compatibilized by Reactive Extrusion. *Materials* **2022**, *15*, 1226. [CrossRef]
3. Adeleye, A.T.; Odoh, C.K.; Enudi, O.C.; Banjoko, O.O.; Osiboye, O.O.; Toluwalope Odediran, E.; Louis, H. Sustainable Synthesis and Applications of Polyhydroxyalkanoates (PHAs) from Biomass. *Process Biochem.* **2020**, *96*, 174–193. [CrossRef]
4. Kalia, V.C.; Ray, S.; Patel, S.K.S.; Singh, M.; Singh, G.P. The Dawn of Novel Biotechnological Applications of Polyhydroxyalkanoates. In *Biotechnological Applications of Polyhydroxyalkanoates*; Springer: Singapore, 2019; pp. 1–11, ISBN 9789811337598.
5. Chavan, S.; Yadav, B.; Tyagi, R.D.; Drogui, P. A Review on Production of Polyhydroxyalkanoate (PHA) Biopolyesters by Thermophilic Microbes Using Waste Feedstocks. *Bioresour. Technol.* **2021**, *341*, 125900. [CrossRef]
6. Gholami, A.; Mohkam, M.; Rasoul-Amini, S.; Ghasemi, Y. Industrial Production of Polyhydroxyalkanoates by Bacteria: Opportunities and Challenges. *Minerva Biotechnol* **2016**, *28*, 59–74.
7. Pagliano, G.; Galletti, P.; Samorì, C.; Zaghini, A.; Torri, C. Recovery of Polyhydroxyalkanoates From Single and Mixed Microbial Cultures: A Review. *Front. Bioeng. Biotechnol.* **2021**, *9*, 624021. [CrossRef]
8. Koller, M. Established and Advanced Approaches for Recovery of Microbial Polyhydroxyalkanoate (PHA) Biopolyesters from Surrounding Microbial Biomass. *EuroBiotech J.* **2020**, *4*, 113–126. [CrossRef]
9. Pérez-Rivero, C.; López-Gómez, J.P.; Roy, I. A Sustainable Approach for the Downstream Processing of Bacterial Polyhydroxyalkanoates: State-of-the-Art and Latest Developments. *Biochem. Eng. J.* **2019**, *150*, 107283. [CrossRef]
10. Rodrigues, A.M.; Franca, R.D.G.; Dionísio, M.; Sevrin, C.; Grandfils, C.; Reis, M.A.M.; Lourenço, N.D. Polyhydroxyalkanoates from a Mixed Microbial Culture: Extraction Optimization and Polymer Characterization. *Polymers* **2022**, *14*, 2155. [CrossRef] [PubMed]

11. Lorini, L.; Martinelli, A.; Pavan, P.; Majone, M.; Valentino, F. Downstream Processing and Characterization of Polyhydroxyalkanoates (PHAs) Produced by Mixed Microbial Culture (MMC) and Organic Urban Waste as Substrate. *Biomass Convers. Biorefin.* **2021**, *11*, 693–703. [CrossRef]
12. Colombo, B.; Pereira, J.; Martins, M.; Torres-Acosta, M.A.; Dias, A.C.R.V.; Lemos, P.C.; Ventura, S.P.M.; Eisele, G.; Alekseeva, A.; Adani, F.; et al. Recovering PHA from Mixed Microbial Biomass: Using Non-Ionic Surfactants as a Pretreatment Step. *Sep. Purif. Technol.* **2020**, *253*, 117521. [CrossRef]
13. Samorì, C.; Abbondanzi, F.; Galletti, P.; Giorgini, L.; Mazzocchetti, L.; Torri, C.; Tagliavini, E. Extraction of Polyhydroxyalkanoates from Mixed Microbial Cultures: Impact on Polymer Quality and Recovery. *Bioresour. Technol.* **2015**, *189*, 195–202. [CrossRef]
14. Patel, M.; Gapes, D.J.; Newman, R.H.; Dare, P.H. Physico-Chemical Properties of Polyhydroxyalkanoate Produced by Mixed-Culture Nitrogen-Fixing Bacteria. *Appl. Microbiol. Biotechnol.* **2009**, *82*, 545–555. [CrossRef]
15. Madkour, M.H.; Heinrich, D.; Alghamdi, M.A.; Shabbaj, I.I.; Steinbüchel, A. PHA Recovery from Biomass. *Biomacromolecules* **2013**, *14*, 2963–2972. [CrossRef] [PubMed]
16. Elhami, V.; van de Beek, N.; Wang, L.; Picken, S.J.; Tamis, J.; Sousa, J.A.B.; Hempenius, M.A.; Schuur, B. Extraction of Low Molecular Weight Polyhydroxyalkanoates from Mixed Microbial Cultures Using Bio-Based Solvents. *Sep. Purif. Technol.* **2022**, *299*, 121773. [CrossRef]
17. Montiel-Jarillo, G.; Suárez-Ojeda, M.E.; Carrera, J. Production of PHB-Co-PHV from Synthetic Fermented Olive-Mill-Wastewater Containing Polyphenols. *J. Water Process. Eng.* 2022, *submmited*.
18. Montiel-Jarillo, G.; Carrera, J.; Suárez-Ojeda, M.E. Enrichment of a Mixed Microbial Culture for Polyhydroxyalkanoates Production: Effect of PH and N and P Concentrations. *Sci. Total Environ.* **2017**, *583*, 300–307. [CrossRef] [PubMed]
19. Mannina, G.; Presti, D.; Montiel-Jarillo, G.; Suárez-Ojeda, M.E. Bioplastic Recovery from Wastewater: A New Protocol for Polyhydroxyalkanoates (PHA) Extraction from Mixed Microbial Cultures. *Bioresour. Technol.* **2019**, *282*, 361–369. [CrossRef]
20. Samorì, C.; Basaglia, M.; Casella, S.; Favaro, L.; Galletti, P.; Giorgini, L.; Marchi, D.; Mazzocchetti, L.; Torri, C.; Tagliavini, E. Dimethyl Carbonate and Switchable Anionic Surfactants: Two Effective Tools for the Extraction of Polyhydroxyalkanoates from Microbial Biomass. *Green Chem.* **2015**, *17*, 1047–1056. [CrossRef]
21. Arcos-Hernández, M.V.; Laycock, B.; Donose, B.C.; Pratt, S.; Halley, P.; Al-Luaibi, S.; Werker, A.; Lant, P.A. Physicochemical and Mechanical Properties of Mixed Culture Polyhydroxyalkanoate (PHBV). *Eur. Polym. J.* **2013**, *49*, 904–913. [CrossRef]
22. Dai, Y.; Lambert, L.; Yuan, Z.; Keller, J. Characterisation of Polyhydroxyalkanoate Copolymers with Controllable Four-Monomer Composition. *J. Biotechnol.* **2008**, *134*, 137–145. [CrossRef]
23. Rosengart, A.; Cesário, M.T.; de Almeida, M.C.M.D.; Raposo, R.S.; Espert, A.; de Apodaca, E.D.; da Fonseca, M.M.R. Efficient P(3HB) Extraction from Burkholderia Sacchari Cells Using Non-Chlorinated Solvents. *Biochem. Eng. J.* **2015**, *103*, 39–46. [CrossRef]
24. Montano-Herrera, L.; Laycock, B.; Werker, A.; Pratt, S. The Evolution of Polymer Composition during PHA Accumulation: The Significance of Reducing Equivalents. *Bioengineering* **2017**, *4*, 20. [CrossRef]
25. Woraittinun, N.; Suwannasilp, B.B. Polyhydroxyalkanoate Production from Different Carbon Substrates Using Sludge from a Wastewater Treatment Plant: Microbial Communities, Polymer Compositions, and Thermal Characteristics. *Environ. Prog. Sustain. Energy* **2017**, *36*, 1754–1764. [CrossRef]
26. Colombo, B.; Sciarria, T.P.; Reis, M.; Scaglia, B.; Adani, F. Polyhydroxyalkanoates (PHAs) Production from Fermented Cheese Whey by Using a Mixed Microbial Culture. *Bioresour. Technol.* **2016**, *218*, 692–699. [CrossRef] [PubMed]
27. Duque, A.F.; Oliveira, C.S.S.; Carmo, I.T.D.; Gouveia, A.R.; Pardelha, F.; Ramos, A.M.; Reis, M.A.M. Response of a Three-Stage Process for PHA Production by Mixed Microbial Cultures to Feedstock Shift: Impact on Polymer Composition. *New Biotechnol.* **2014**, *31*, 276–288. [CrossRef] [PubMed]
28. Morgan-Sagastume, F.; Karlsson, A.; Johansson, P.; Pratt, S.; Boon, N.; Lant, P.; Werker, A. Production of Polyhydroxyalkanoates in Open, Mixed Cultures from a Waste Sludge Stream Containing High Levels of Soluble Organics, Nitrogen and Phosphorus. *Water Res.* **2010**, *44*, 5196–5211. [CrossRef] [PubMed]
29. de Souza Reis, G.A.; Michels, M.H.A.; Fajardo, G.L.; Lamot, I.; de Best, J.H. Optimization of Green Extraction and Purification of PHA Produced by Mixed Microbial Cultures from Sludge. *Water* **2020**, *12*, 1185. [CrossRef]
30. Abbasi, M.; Coats, E.R.; McDonald, A.G. Green Solvent Extraction and Properties Characterization of Poly(3-Hydroxybutyrate-Co-3-Hydroxyvalerate) Biosynthesized by Mixed Microbial Consortia Fed Fermented Dairy Manure. *Bioresour. Technol. Rep.* **2022**, *18*, 101065. [CrossRef]
31. Sindhu, R.; Binod, P.; Pandey, A. Microbial Poly-3-Hydroxybutyrate and Related Copolymers. In *Industrial Biorefineries and White Biotechnology*; Elsevier: Amsterdam, The Netherlands, 2015; pp. 575–605, ISBN 9780444634535.
32. Ivanova, G.; Serafim, L.S.; Lemos, P.C.; Ramos, A.M.; Reis, M.A.M.; Cabrita, E.J. Influence of Feeding Strategies of Mixed Microbial Cultures on the Chemical Composition and Microstructure of Copolyesters P(3HB-Co-3HV) Analyzed by NMR and Statistical Analysis. *Magn. Reson. Chem.* **2009**, *47*, 497–504. [CrossRef]
33. Irorere, V.U.; Bagheriasl, S.; Blevins, M.; Kwiecień, I.; Stamboulis, A.; Radecka, I. Electrospun Fibres of Polyhydroxybutyrate Synthesized by Ralstonia Eutropha from Different Carbon Sources. *Int. J. Polym. Sci.* **2014**, *2014*, 705359. [CrossRef]
34. Žagar, E.; Kržan, A.; Adamus, G.; Kowalczuk, M. Sequence Distribution in Microbial Poly(3-Hydroxybutyrate-Co-3-Hydroxyvalerate) Co-Polyesters Determined by NMR and MS. *Biomacromolecules* **2006**, *7*, 2210–2216. [CrossRef]

35. Doi, Y.; Kunioka, M.; Nakamura, Y.; Soga, K. Nuclear Magnetic Resonance Studies on Poly (β-hydroxybutyrate) and a Copolyester of P-Hydroxybutyrate and P-Hydroxyvalerate Isolated from Alcaligenes Eutrophus H16. *Macromolecules* **1986**, *19*, 2860–2864. [CrossRef]
36. Kamiya, N.; Yamamoto, Y.; Inoue, Y.; ChiijB, R.; Yoshiharu Doi, J. Microstructure of Bacterially Synthesized Poly (3-Hydroxybutyrate-Co-3-Hydroxyvalerate). *Macromolecules* **1989**, *22*, 1676–1682. [CrossRef]
37. Laycock, B.; Halley, P.; Pratt, S.; Werker, A.; Lant, P. The Chemomechanical Properties of Microbial Polyhydroxyalkanoates. *Prog. Polym. Sci.* **2013**, *38*, 536–583. [CrossRef]
38. Pederson, E.N.; McChalicher, C.W.J.; Srienc, F. Bacterial Synthesis of PHA Block Copolymers. *Biomacromolecules* **2006**, *7*, 1904–1911. [CrossRef] [PubMed]
39. Liu, Z.; Wang, Y.; He, N.; Huang, J.; Zhu, K.; Shao, W.; Wang, H.; Yuan, W.; Li, Q. Optimization of Polyhydroxybutyrate (PHB) Production by Excess Activated Sludge and Microbial Community Analysis. *J. Hazard. Mater.* **2011**, *185*, 8–16. [CrossRef]
40. Fiorese, M.L.; Freitas, F.; Pais, J.; Ramos, A.M.; de Aragão, G.M.F.; Reis, M.A.M. Recovery of Polyhydroxybutyrate (PHB) from Cupriavidus Necator Biomass by Solvent Extraction with 1,2-Propylene Carbonate. *Eng. Life Sci.* **2009**, *9*, 454–461. [CrossRef]
41. Gobi, K.; Vadivelu, V.M. Polyhydroxyalkanoate Recovery and Effect of in Situ Extracellular Polymeric Substances Removal from Aerobic Granules. *Bioresour. Technol.* **2015**, *189*, 169–176. [CrossRef]
42. López-Abelairas, M.; García-Torreiro, M.; Lú-Chau, T.; Lema, J.M.; Steinbüchel, A. Comparison of Several Methods for the Separation of Poly(3-Hydroxybutyrate) from Cupriavidus Necator H16 Cultures. *Biochem. Eng. J.* **2015**, *93*, 250–259. [CrossRef]
43. Yabueng, N.; Napathorn, S.C. Toward Non-Toxic and Simple Recovery Process of Poly(3-Hydroxybutyrate) Using the Green Solvent 1,3-Dioxolane. *Process Biochem.* **2018**, *69*, 197–207. [CrossRef]
44. Bengtsson, S.; Pisco, A.R.; Johansson, P.; Lemos, P.C.; Reis, M.A.M. Molecular Weight and Thermal Properties of Polyhydroxyalkanoates Produced from Fermented Sugar Molasses by Open Mixed Cultures. *J. Biotechnol.* **2010**, *147*, 172–179. [CrossRef] [PubMed]
45. Gahlawat, G.; Soni, S.K. Valorization of Waste Glycerol for the Production of Poly (3-Hydroxybutyrate) and Poly (3-Hydroxybutyrate-Co-3-Hydroxyvalerate) Copolymer by Cupriavidus Necator and Extraction in a Sustainable Manner. *Bioresour. Technol.* **2017**, *243*, 492–501. [CrossRef] [PubMed]

polymers

MDPI

Article

Development and Mechano-Chemical Characterization of Polymer Composite Sheets Filled with Silica Microparticles with Potential in Printing Industry

Sidra Siraj, **Ali H. Al-Marzouqi** * and **Muhammad Z. Iqbal**

Chemical Engineering Department, College of Engineering, United Arab Emirates University, Al Ain 15551, United Arab Emirates
* Correspondence: hassana@uaeu.ac.ae

Abstract: Polymer composite sheets using a low-cost filler (local natural sand) and polymer (high-density polyethylene, HDPE) as a replacement of the traditionally used wood-fiber-based sheets for paper-based applications were developed. The sand/polymer composite sheets were prepared by melt extrusion in a melt blender followed by compression molding. The effects of varying particle size, concentration, and the use of a compatibilizer (polyethylene-grafted maleic anhydride) was studied on the mechano-chemical performance properties of the composite sheets such as morphology, thermal and mechanical properties, and wettability characteristics used in the printing industry. In terms of thermal stability, filler (sand) or compatibilizer addition did not alter the crystallization, melting, or degradation temperatures significantly, thereby promoting good thermal stability of the prepared sheets. Compatibilization improved anti-wetting property with water. Additionally, for the compatibilized sheets prepared from 25 μm sand particles, at 35 wt%, the contact angle with printing ink decreased from 44° to 38.30°, suggesting improved ink-wetting performance. A decrease in the elastic modulus was also observed with the addition of the compatibilizer, with comparable results to commercial stone paper. Results from this study will be considered as a first step towards understanding compatibility of local natural sand and polymers for paper-based application.

Keywords: sustainability; silica; sand; polymer; composite; synthetic; sheets

Citation: Siraj, S.; Al-Marzouqi, A.H.; Iqbal, M.Z. Development and Mechano-Chemical Characterization of Polymer Composite Sheets Filled with Silica Microparticles with Potential in Printing Industry. *Polymers* **2022**, *14*, 3351. https://doi.org/10.3390/polym14163351

Academic Editor: Edina Rusen

Received: 22 July 2022
Accepted: 11 August 2022
Published: 17 August 2022

Publisher's Note: MDPI stays neutral with regard to jurisdictional claims in published maps and institutional affiliations.

Copyright: © 2022 by the authors. Licensee MDPI, Basel, Switzerland. This article is an open access article distributed under the terms and conditions of the Creative Commons Attribution (CC BY) license (https://creativecommons.org/licenses/by/4.0/).

1. Introduction

Traditionally, paper is derived from wood by pressing together moist fibers of cellulose pulp followed by drying. Global use of papers for various applications, such as printing papers, bags, posters, cutlery, etc., is also adding a toll on the environment as it requires a huge amount of water to wash away the initial pulp as well as requires chemicals to bleach such products, thereby also increasing wastewater and sludge disposal problems.

Additionally, increasing population has increased the global production and use of paper and paper products exponentially. An average global growth rate of paper production was almost 3.6% in the 1980s and approximately 3.3% in the 1990s, currently reaching up to a massive 400 million tons per year requirement [1]. On the other hand, increasing use of plastics has become a major cause of concern as it leads to drastic environmental issues due to its non-eco-friendly disposal [2]. An estimate of about 2500 million metric tons (MMT) of plastics are currently in use. It is also reported that over the past 65 years, the cumulative amount of waste generated from primary and recycled plastic waste adds up to a staggering amount of ~6000 MMT. Of this waste, approximately 12% is incinerated and 9% is recycled. Furthermore, over 60% of all plastics produced are directly discarded in nature (marine or in landfills) and continue to accumulate to this date [3]. Figure 1 illustrates a schematic of the worldwide production and disposal of plastics in MMT for the last 65 years. At the current consumption rate of plastics, the Earth will hold an estimate of 3.3 billion tons of plastic by 2050 [4].

Accordingly, both issues require more innovative methods to help tackle current environmental problems as well as benefit the society in a more sustainable way. Hence, the development of synthetic "paper" came into existence. Synthetic paper or sheet produced from thermoplastic polymers (virgin or recycled) and fillers and additives form a thin composite film that can be used in a variety of paper-based applications. Since the synthetic composite sheet is a "tree-free" product, it will contribute towards reducing the environmental impact of the paper industry, and its potential can be even increased when using recycled plastics [5].

The detrimental effects of manufacturing paper and paper-based products from wood on the environment have led to the commercialization of synthetic papers on a global scale. Taiwan Lung Meng industries was the first to commercialize a synthetic paper named "stone paper", which is manufactured using minerals dispersed in polyolefins [6]. The stone paper, which is used to produce a variety of paper-based products such as bags, calendars, labels, etc., has been given several other commercial names, such as Rockstock [7] and mineral paper, and is eco-friendly with the main composition being a varying ratio of a polymer and suitable filler.

Synthetic "paper" or filler-polymer composite sheet has properties similar to the traditional cellulose-based paper [8]. It is developed by extruding polyolefin films which may be multilayered and are filled with inorganic particles, more commonly known as fillers. A large number of inexpensive inorganic particulate materials have been widely used as fillers to enhance the thermal and mechanical properties of polymers such as calcium carbonate ($CaCO_3$), glass, titanium dioxide (TiO_2), and silica (SiO_2) in paper-based applications [9–11]. Out of the mentioned fillers, $CaCO_3$ has been extensively studied and used for paper-based applications as it can provide the properties required such as opaqueness and printability, as well as increase the mechanical properties [2,12]. However, an increase in such a filler leads to an increase in its brittle nature [10,11].

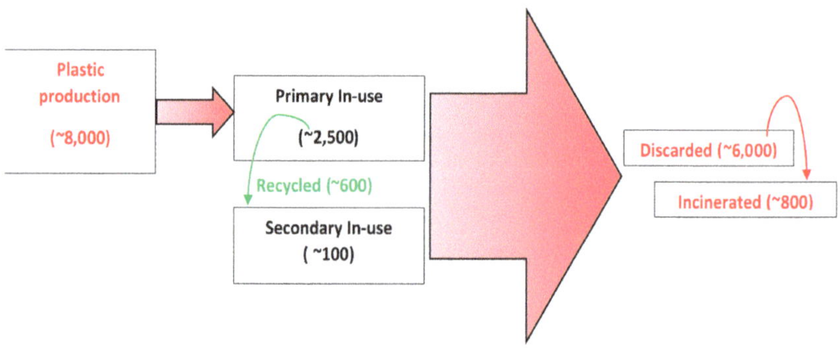

Figure 1. Simplified estimation of worldwide production and disposal of plastics in MMT for the last 65 years [9].

The Middle East, including the UAE, similar to any other onshore country, has naturally abundant sand (silica). The sand available in the UAE varies with geological position in composition, color, size, and surface texture. Due to its high thermal stability, high durability, and unreactive nature, silica is used in various fields such as in 3D printing, construction, textile, electronic applications, etc. [13–20]. Using UAE's sand in writing paper applications can be cost-effective and produce more eco-friendly sheets compared to wood-based paper.

In general, the existing literature indicates the extensive development and commercialization in the production of synthetic composite sheets and its potential replacement for wood-based sheets. The literature represents increased mechanical and durable properties as well as increased thermal stability and degradation, signifying its good potential for such a product [13,14,20–23]. The literature also shows the use of sand in polymers to improve properties in applications such as construction, waste management, and recycling applications [24–28]; however, not much study has been conducted on using natural sand (silica) as a filler to develop sheets for printing applications.

Therefore, the main objective of this research is to utilize the UAE's sand in manufacturing synthetic composite sheets for writing/printing applications using melt extrusion technique followed by compression. The physiochemical properties of developed composite sheets are assessed and compared with commercial stone paper and regular A4 paper for writing applications. The goal is to develop paper-like composite sheets using a natural resource to contribute towards a more sustainable approach of paper production. The presented solution contributes to using sand as a raw material in a useful way and thereby widens its potential for industrial use [29]. Additionally, the novelty of this work lies in attempting to utilize the raw local sand as a potential filler in a composite material that can have a commercial-scale application, such as in the printing industry, which can potentially grow as an alternative to the traditional wood-based fiber sheets which are extensively used.in numerous applications.

2. Materials and Methods

Sand sample was collected from Ras Al-Khaimah, UAE (25.6741° N, 55.9804° E). The local sand was classified into 3 types (carbonates, silicates, and free silica, which is in the form of detrital quartz). The actual relative abundance in terms of the mineral content was analyzed to be about 47 wt% silicates, 26 wt% carbonates, and 14 wt% quartz. The mineral composition of SiO_2 was analyzed to be 36.93 wt% [30]. All chemicals used, unless stated otherwise, were supplied by Sigma Aldrich, Germany and were of appropriate purity for application in this study. High-density polyethylene (HDPE) pellets (ρ = 0.98 g/cm^3) were purchased from Sigma-Aldrich. PE-g-MA (ρ = 1.8 g/cm^3) pellets were also purchased from Sigma-Aldrich. Sand from the UAE was collected and ground to finer particles using heavy-duty grinders. A sieve (200 mm diameter, 25 microns aperture, stainless steel mesh) was further used to refine the fine powder to a particle size of 25 μm. Grinding sand to 5 μm was performed by Retsch, Haan, Germany.

2.1. Experimental Procedure

The sand/polymer composite sheets were prepared via melt-blending, followed by compression molding. For all the experiments, American Standard Test procedures were adopted.

2.2. Synthesis of Sand/Polymer Composite Material via Melt-Blending

Weighed amounts of sand at two different particle sizes (5 and 25 μm) and different fractions (0 wt%, 20 wt%, 35 wt%, and 50 wt%) were mixed with HDPE and melt-blended in a twin-screw extruder (MiniLab HAAKE Rheomex CTW5, Germany) for 15 min at 170 °C and at a screw speed of 100 rpm. These conditions were selected after several trials of varying the time, temperature, and rpm for the process. The total feed amount was kept constant at 4 g as required by the extruder setup [23]. The control sample was prepared by blending pure HDPE using the same conditions as in numerous studies [31–34]. For the samples prepared with the addition of the compatibilizer (C), the composition of the compatibilizer was kept constant at a filler-to-compatibilizer ratio of 2:1. Compatibilizers are used as additives to enhance the properties of the fabricated polymer composites prepared by various techniques. Studies report polyethylene grafted with maleic anhydride (PP-g-MA) as a compatibilizer to increase the dispersion of the clay in the nonpolar polymer matrices and improved properties [14], which is why it was chosen. PE-g-MA and its ratio was also chosen after several trials of varying the compatibilizer type and amount. Table 1 represents chosen weight percentages for the three components for the preparation of the sand/polymer composite sheets.

Table 1. Chosen weight percentages for HDPE, sand, and compatibilizer.

	Sample	HDPE (wt%)	Sand (wt%)	Compatibilizer (wt%)
Sand/polymer composite sheets prepared from 25 μm and 5 μm sand particles	0 wt%	100	0	0
	20 wt%	80	20	0
	35 wt%	65	35	0
	50 wt%	50	50	0
	20 wt% + C	70	20	10
	35 wt% + C	47.5	35	17.5
	50 wt% + C	25	50	25

Synthesis of Sand/Polymer Composite Sheets via Compression Molding

The compounded composites were chopped into small pieces (~1 g) and compressed (hot-pressed) for 10 min using a Carver's press (Carver™ Lab Presses) under 5000 psi pressure at 170 °C to prepare the flat sand/polymer composite sheets. Sand/polymer composite sheets were prepared from two different sizes (25 μm and 5 μm) of sand as the filler. The same sets were prepared with the addition of the compatibilizer as well. Figures 2 and 3 show images for sand/polymer composite sheets prepared from 25 μm sand particles without and with compatibilizer, respectively, whereas Figures 4 and 5 represent the same for sheets prepared with 5 μm, respectively. All the sheets were roughly 1 mm thick and 100 mm in diameter. It could be visually seen that increasing filler addition made the sheets darker in appearance. However, addition of compatibilizer produced sheets that were relatively lighter in appearance and softer. Furthermore, based on a visual inspection, it can be said that the dispersion was random in all sheets, with a more homogeneous dispersion in 20 wt% and 35 wt%, and a non-homogeneous dispersion was observed in both cases of 50 wt% filler before and after compatibilization. The overall process is illustrated in Figure 6. A similar process was followed with the addition of compatibilizer.

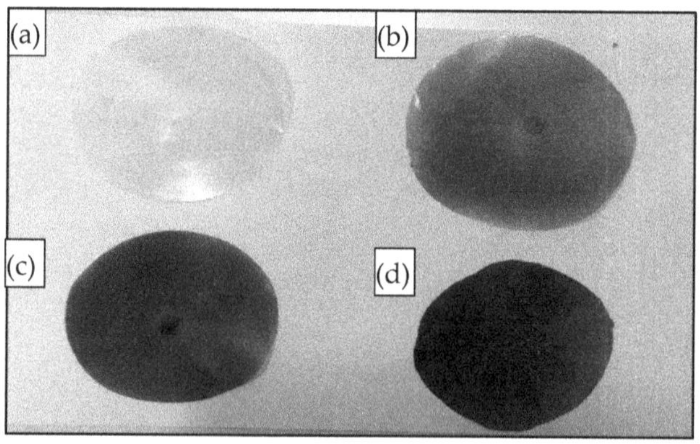

Figure 2. Images of prepared composite sheets for (**a**) pure HDPE, (**b**) 20 wt%, (**c**) 35 wt%, and (**d**) 50 wt% prepared from 25 μm sand particles.

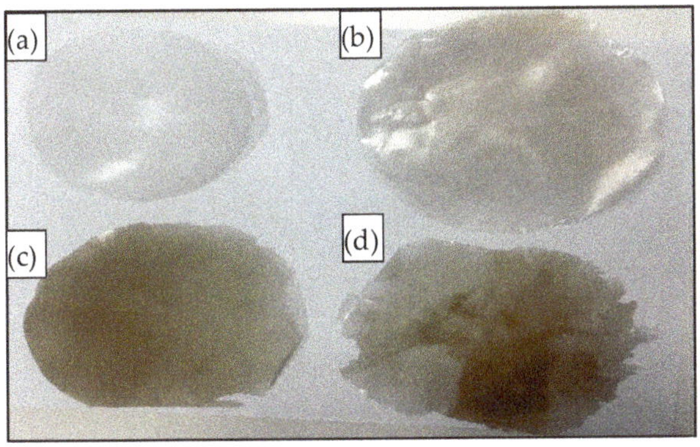

Figure 3. Images of prepared compatibilized composite sheets for (**a**) pure HDPE, (**b**) 20 wt%, (**c**) 35 wt%, and (**d**) 50 wt% prepared from 25 μm sand particles.

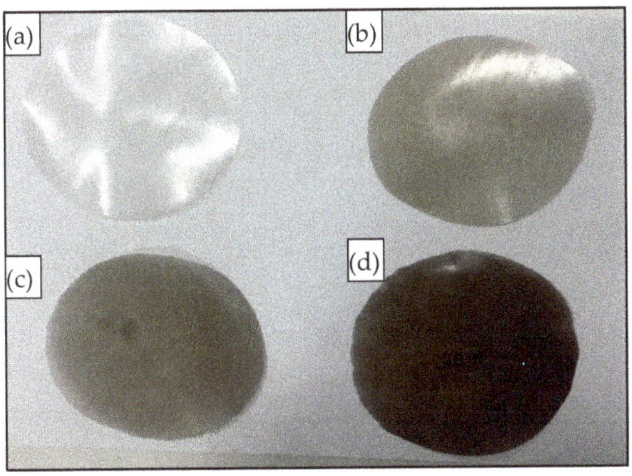

Figure 4. Images of prepared composite sheets: (**a**) pure HDPE, (**b**) 20 wt%, (**c**) 35 wt%, and (**d**) 50 wt% with 5 μm sand particles.

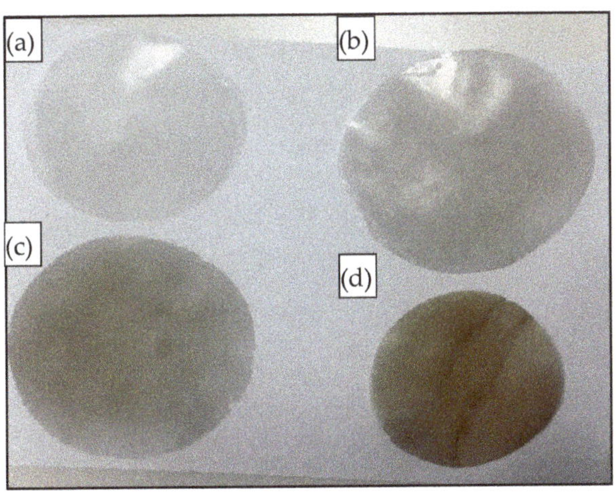

Figure 5. Images of prepared compatibilized composite sheets: (**a**) pure HDPE, (**b**) 20 wt% + C, (**c**) 35 wt% + C, and (**d**) 50 wt% + C, with 5 μm sand particles.

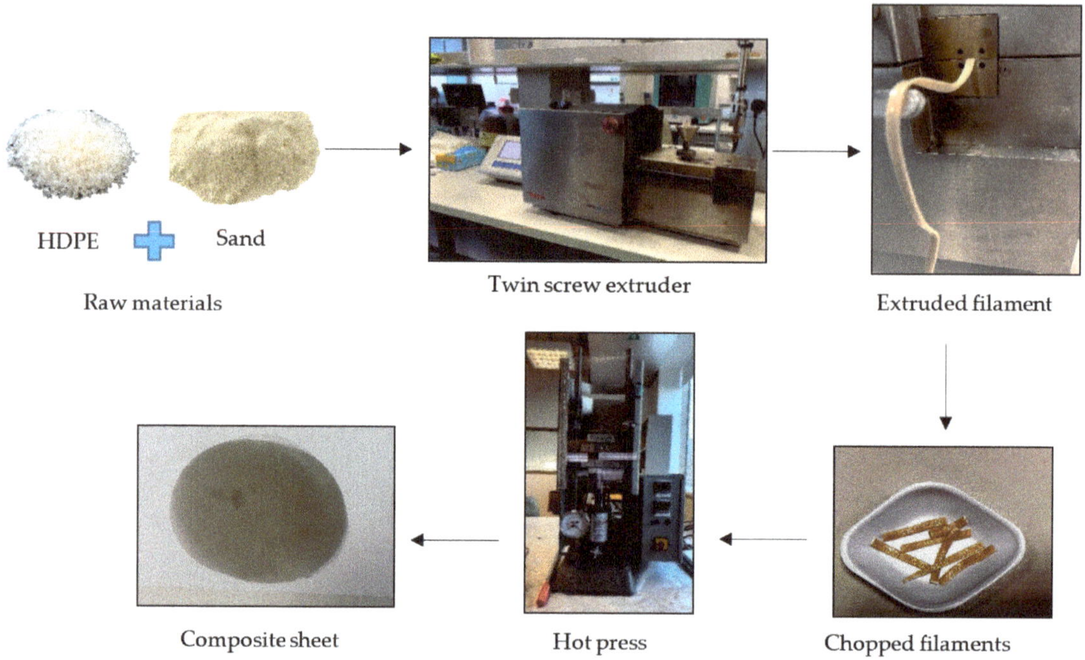

Figure 6. Process of preparing the composite sheets.

2.3. *Characterization*

2.3.1. Scanning Electron Microscope (SEM)

A JEOL/EO scanning electron microscope (SEM), operated at 2 kV, spot size of 40, was used to image the sand of both 25 µm and 5 µm, and neat HDPE at their surface. To improve conductivity and quality of image, samples were coated with Au/C using a vacuum sputter coater. Likewise, SEM analysis was used to observe the surface morphology of the selected composite sheets. Selected composite sheets were placed on an aluminum pin-mount adapter using double-sided carbon tape and then were sputter-coated with gold using a sputter-coater to avoid electrostatic charging during examination. The SEM was operated at high vacuum mode with an acceleration voltage of 15 kV and the images were acquired.

2.3.2. X-ray Diffraction (XRD)

X-ray diffraction (XRD) was performed on sand samples (within 2θ = 5–40°) using a Panalytical X-ray diffraction system (X'Pert3 Powder, Malvern Panalytical, Denver, CO, USA) to confirm the presence of SiO2, which is the major component of sand, as reported in the literature [35]. XRD was also performed on HDPE after the addition of the compatibilizer to determine any alteration in the characteristic peaks of HDPE. Similarly, XRD was performed for selected composite sheets to observe any shifts or alterations due to filler and compatibilizer addition.

2.3.3. Differential Calorimetry Analysis (DSC)

The melting temperatures (Tm) and the crystalline temperatures (Tc) of the control sample, i.e., the neat HDPE sheet and the prepared synthetic composite sheets, as well as the compatibilized composite sheets, were determined using modulated differential scanning calorimetry (Discovery DSC 25, TA Instruments, New Castle, DE, USA). About 5–10 mg of sample was heated from 20–180 °C at a heating rate of 40 °C/min to remove thermal history of the polymer. Once the thermal history was eliminated, all samples were cooled from 180 °C to room temperature (20 °C) at 10 °C/min for recording the crystallization temperature (Tc) followed by subsequent heating scans from 20–180 °C at 10 °C/min to record the melting temperatures (Tm). All experiments were carried out under inert nitrogen atmosphere. The same was also repeated for stone paper and regular A4 paper for comparative analysis.

2.3.4. Thermogravimetric Analysis (TGA)

Thermogravimetric analysis of neat HDPE, the prepared composite sheets, and stone paper, as well as for regular A4 paper, was carried out using TGA (Q500 series, TA Instrument). For each experiment, a sample weight of 6.0 mg (±1.0) was used for thermogravimetric analysis. The heating rate was controlled at 10, 15, 20, and 25 °C/min from 25 °C to 900 °C, using nitrogen as a carrier gas at 20 mL/min. During the thermo-decomposition process, the initial weight was recorded continuously as a function of temperature and time. The derivative (DTG) curve was also plotted for the weight loss of sample per unit time.

2.3.5. Tensile Test

Durability of the prepared composite sheets was tested using mechanical testing. Tensile properties of the composite sheets were determined by the Universal Testing Machine (UTM) using the American Society for Testing and Materials (ASTM)-D 638. Dumbbell-shaped test specimens were prepared from the composite sheets of 35 (l) × 0.5 (w) × 1 (t) mm, and their tensile properties were studied on a Zwick 50 kN. The rate of crosshead motion was set to be 100 mm/min, which was taken from the ASTM D 790 standard. The same was also repeated for stone paper and regular A4 paper for comparative analysis.

2.3.6. Wettability

Contact angle measurement (θ) was used to determine the wettability of composite sheets which can be determined by the standard ASTM D 2578–17 test. This test assumes that the surface energy of a film (γ) is equal to the surface tension of a liquid that wets the film, without merging, when a drop is placed on the surface via contact angle measurements. Each composite sheet was placed in the Teclis tracker tensiometer equipment setup, and a droplet of liquid (water) was ejected out of the micrometer syringe onto the sample using a gauge needle. Images of the droplet on the surface of the sample were taken, and, using the Teclis tracker software, the angle between a tangent drawn on the liquid droplet and the surface of the composite sheet was calculated as reported in the literature [13].

Figure 7 shows samples for the tangent to the surface drawn on a sheet surface for non-polar, extremely non-polar, and polar surfaces, respectively. The same test was also performed using a non-polar solvent, benzene, a mixture of water-non-polar solvent, as well as pure printing ink (Epson ink 664, black, Carrefour, Al Ain, UAE) to test for the wettability characteristics of the prepared sheets for ink-based applications. The wettability tests were also performed on stone paper and regular A4 paper for comparative study.

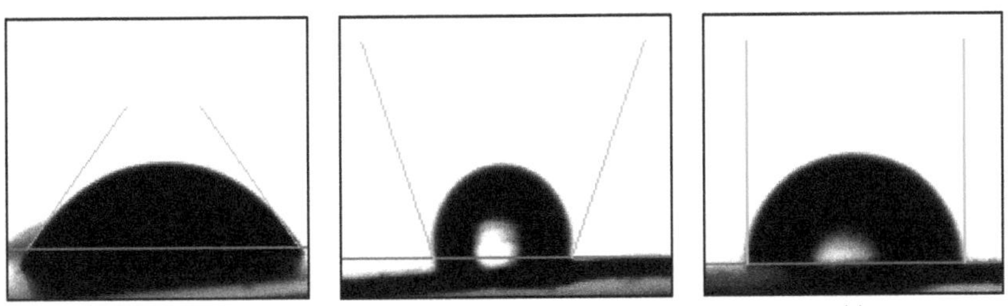

(a) (b) (c)

Figure 7. Contact angle measurements. (**a**) Extremely polar surface: (θ) = 90°, (**b**) Extremely non-polar surface (θ) => 90°, (**c**) Non-polar surface (θ) =< 90° taken via drawing tangent to the liquid droplet and the surface.

2.3.7. Printing Test

Selected composite sheets had no surface treatment and were cleaned with a dry tissue paper. The sheets were then glued on a regular A4 paper, dried for approximately 15 min, and then sent to the printer (RICOH class driver laser printer) as conducted by other studies [2,36]. As a comparison, a printing test was also performed on stone paper and regular A4 paper as well.

2.3.8. Adherence Test

To test the adherence of ink on the prepared sand/polymer sheets, marker pens were used to write on the sheets. After writing on the sheets, the sheets were left to dry for 15 min. A piece of clear adhesive tape (1.5 cm wide) was then applied onto the surface where text was written. Any small bubbles that appeared while applying the clear tape were removed manually, and by using the rolling weight of the hand, the tape was firmly pressed. Consequently, the tape was pulled off each sample, implying that the smaller the amount of ink that was removed by the tape, the better the assessment. The test was repeated twice; once with permanent marker pen and the second time with removable marker pen. The adherence of ink to the selected sand/polymer composite sheet surfaces was assessed qualitatively as excellent, good, regular, or poor, as reported in the literature [13]. The removed tape was placed alongside the sheet to observe the difference. As a comparison, an adherence test was also performed on stone paper and regular A4 paper.

3. Results and Discussion

3.1. Morphology

In order to determine the size of the ground sand samples, SEM analysis was conducted to obtain an idea of the particle size as well as the shape of the particles. Figure 8a,b show the two sand samples, respectively. It can be clearly seen that the 25 μm sand particles have larger, more irregular-shaped distant particles, whereas the 5 μm sand particles have smaller, finer, and closer particles comparatively. The neat HDPE sheet, with 0 wt% filler, which was processed in the same way as the composite sheets was also analyzed using SEM. Figure 8c shows only a fine structure which is expected in a neat polymer film and is in agreement with the literature [37].

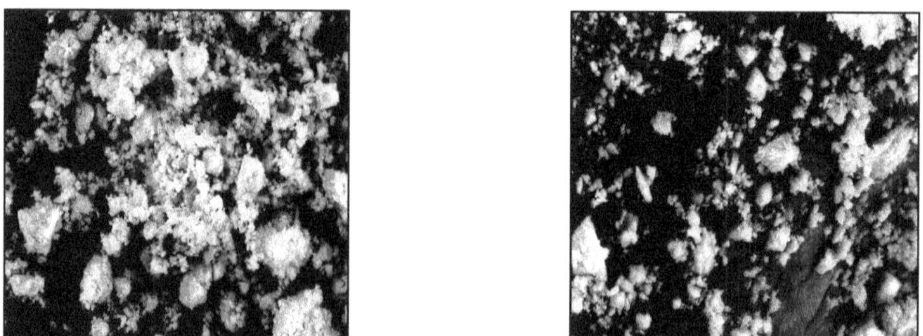

(a) (b)

Figure 8. *Cont.*

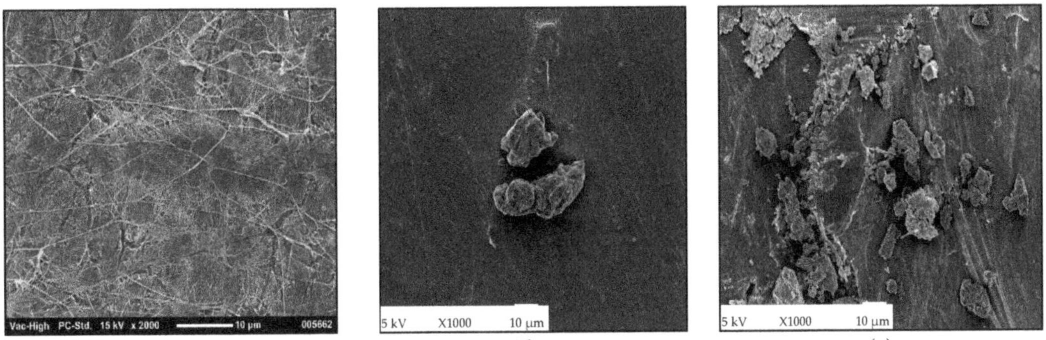

(c) (d) (e)

Figure 8. SEM images of (**a**) 25 µm sand particles and (**b**) 5 µm sand particles; (**c**) neat HDPE; (**d**) 50 wt%, 5 µm sand composite sheet; (**e**) 50 wt%, 5 µm sand compatibilized composite sheet.

Selected composite sheet samples were also chosen to conduct SEM analysis. At 35 wt% prepared from 5 µm, particles were visibly seen. However, after compatibilization, particles had decreased interparticle distance, suggesting improved dispersion and binding ability [21]. The literature reports good appearance at ~30 wt% for composite sheets as well [2]. Similarly, in the case of 50 wt% and 5 µm particle size, compatibilization shows increased presence of particles, thereby indicating decreased interparticle distance. An increase in the filler accumulation can result in agglomeration of particles, which can lead to a brittle material [13]. SEM images of both sets of composite sheets at 50 wt% are shown in Figure 8d,e, respectively.

3.2. XRD Analysis

The successful preparation of the composite sheets with sand as well as with the addition of the compatibilizer (C) was confirmed using XRD analysis. HDPE characteristic peaks at 2θ = ~21.5° and 2θ = ~23.5° were observed, which are supported in the literature as well [38,39]. Addition of the compatibilizer did not alter the characteristic HDPE peaks as shown in Figure 9; however, slight broadening of the peak was observed.

Just as a confirmation, XRD was also performed for selected sand/HDPE composite sheets. Figure 9 also shows that the presence of the 5 µm sand particles and compatibilizer did not alter the characteristic peaks of HDPE, as two distinct diffraction peaks of HDPE were observed for all the prepared sand/polymer composite sheets. As for the sand, two significant characteristic peaks were obtained at 2θ = 21.38° and 23.71°, which are quite comparable to the literature values obtained for precipitated silica as well as sol–gel-produced silicon dioxide, ranging from 21.8° to 23°, respectively [40,41]. The presence of these peaks was noticed for all the prepared samples at values close enough to the 2θ values (with slight shift to the right), showing that the addition of the filler (sand) does not significantly alter the basic structure of the prepared sand/polymer composite sheets at the molecular level as well.

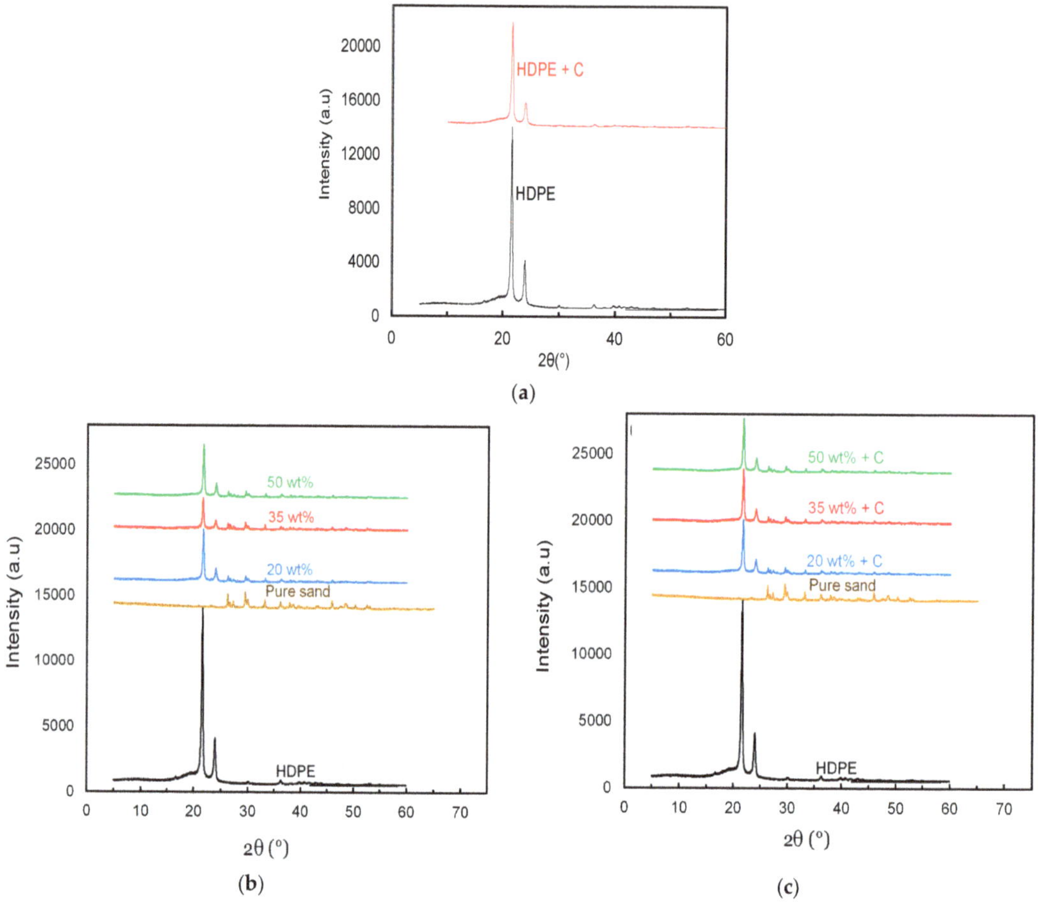

Figure 9. XRD characteristic peak patterns: (**a**) neat HDPE before and after compatibilization; (**b**) sand/polymer composite sheets prepared from 5 μm sand particles; (**c**) sand/polymer compatibilized composite sheets prepared from 5 μm sand particles.

3.3. Thermal Analysis

3.3.1. Melting and Crystallization Behavior

DSC analysis was conducted for all sand/polymer composite sheets prepared from both 25 μm sand particles and 5 μm sand particles at the same heating rate (10 °C/min) to understand the crystallization properties of the prepared sand/polymer composite sheets. Figure 10a shows the cooling (first cycle) and heating (second cycle) profiles for the neat HDPE. The observed peak melting temperature (Tm,peak) and peak crystallization temperature (Tc,peak) values for the neat HDPE were ~133.86 °C and ~117.50 °C, respectively, which is also well reported in the literature [38]. A trend of smooth transition temperatures is seen by the neat HDPE, which also shows the absence of any impurity in the sample.

Figure 10b,c show the cooling and heating profiles for the compatibilized sand/polymer composite sheets prepared from 5 μm sand particles, respectively. The thermal properties for all the sand/polymer composite sheets prepared from 25 μm sand particles and 5 μm sand particles are reported in Table 2.

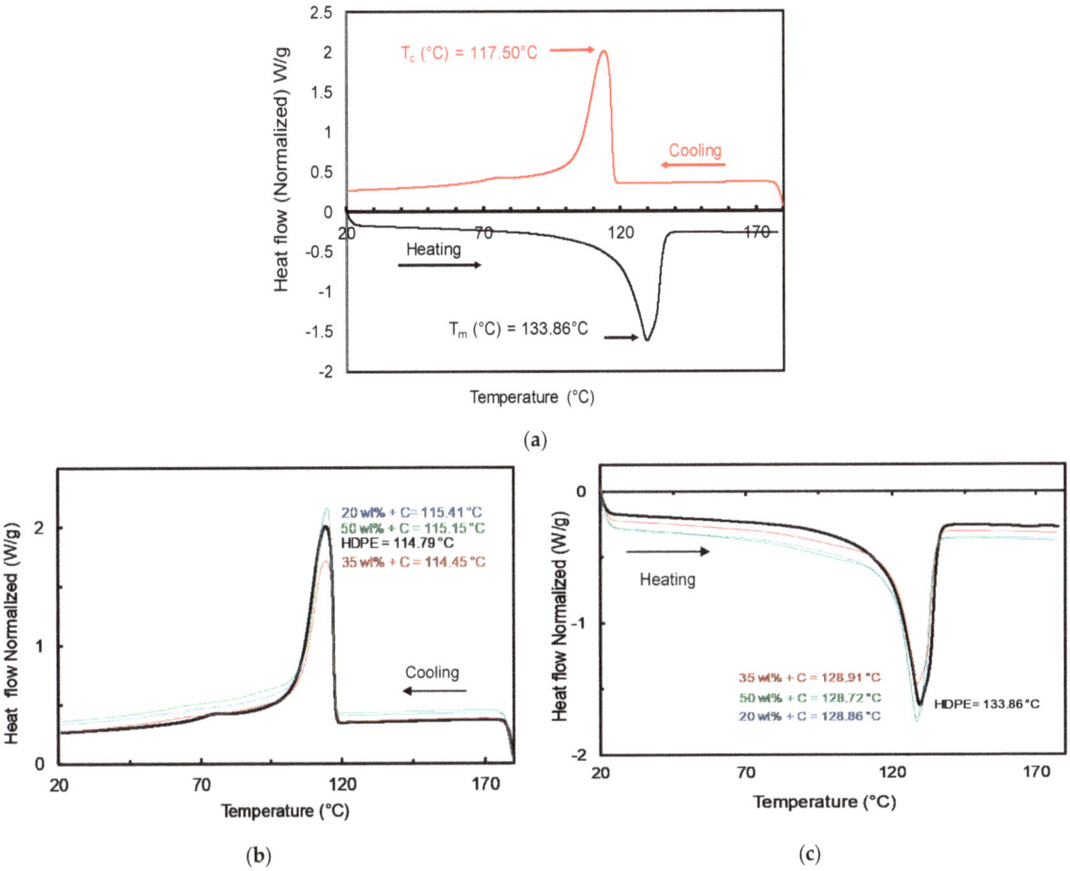

Figure 10. (**a**) DSC thermogram for neat HDPE. (**b**) Cooling profiles for compatibilized sand/polymer composite sheets prepared from 5 μm sand particles. (**c**) Heating profiles for compatibilized sand/polymer composite sheets prepared from 5 μm sand particles.

Table 2. DSC data for sand/polymer composite sheets prepared from 25 μm sand particles and 5 μm sand particles, stone paper, regular A4 paper.

	Sample	1st Cooling Scan			2nd Heating Scan		
		Tc,onset (°C)	Tc,peak (°C)	ΔHc (J/g)	Tm,peak (°C)	ΔHm (J/g)	%Xc
Sand/polymer composite sheets prepared from 25 μm sand particles	0 wt%	117.54 ± 0.01	114.79 ± 0.61	120.62 ± 21	133.86 ± 0.55	148.77 ± 13.9	48.57 ± 8.5
	20 wt%	117.46 ± 0.11	113.91 ± 0.83	121.51 ± 15	135.60 ± 3.5	131.27 ± 9.0	56.00 ± 3.8
	35 wt%	117.82 ± 0.06	114.61 ± 0.88	94.79 ± 14	135.423 ± 3.4	99.791 ± 9.0	52.40 ± 4.7
	50 wt%	118.71 ± 0.09	115.04 ± 0.33	74.38 ± 14	132.11 ± 0.07	86.034 ± 7.5	58.73 ± 5.2
	20 wt% + C	117.50 ± 0.07	115.40 ± 0.39	85.06 ± 9.1	133.58 ± 0.3	110.35 ± 2.5	47.08 ± 1.1
	35 wt% + C	117.82 ± 0.06	115.80 ± 0.40	87.03 ± 1.4	132.31 ± 1.9	96.60 ± 8.2	50.71 ± 4.3
	50 wt% + C	117.68 ± 0.49	114.95 ± 0.13	51.87 ± 16	134.01 ± 1.6	60.11 ± 5.4	41.0 ± 3.7
Sand/polymer composite sheets prepared from 5 μm sand particles	20 wt%	117.59 ± 0.08	115.04 ± 0.58	105.58 ± 10	133.35 ± 3.6	122.877 ± 11	52.42 ± 5.0
	35 wt%	118.23 ± 0.35	114.12 ± 1.32	98.32 ± 5	136.31 ± 4.2	109.07 ± 7.5	57.27 ± 3.4
	50 wt%	118.24 ± 0.05	116.21 ± 0.48	64.30 ± 6.2	133.72 ± 1.5	77.91 ± 5.7	53.18 ± 3.9
	20 wt% + C	117.66 ± 0.15	115.41 ± 0.64	112.23 ± 3.5	134.21 ± 5.1	118.77 ± 14	50.67 ± 6.3
	35 wt% + C	117.95 ± 0.21	115.48 ± 0.91	85.35 ± 3	134.56 ± 2.7	91.32 ± 2.5	47.95 ± 1.3
	50 wt% + C	117.78 ± 50	115.15 ± 0.16	42.49 ± 1.8	132.17 ± 2.7	54.51 ± 3.6	37.42 ± 3.4
	Stone Paper	117.72 ± 0.12	114.62 ± 0.1	27.08 ± 2.3	134.52 ± 0.8	33.26 ± 1.9	41.03 ± 3.7
	Regular A4	-	-	-	-	-	-

The obtained DSC results show that the crystallization temperature of HDPE is not influenced due to the addition of filler (sand) or the compatibilizer. All the samples exhibited a single crystallization exotherm and a corresponding melting endotherm. From the DSC thermograms, the onset crystallization temperatures (Tc,onset), peak temperatures for crystallization exotherms (Tc,peak), and melting endotherms (Tm,peak) for neat HDPE, all the prepared sand/polymer composite sheets, as well as stone paper and regular A4 paper, were evaluated. Determining the Tc,onset, Tc,peak and Tm,peak help in defining the processing temperature range of the polymer. Usually, the processing temperature of polymers is ±40 °C from the melting temperature, as reported by various studies [25,35]. Furthermore, by observing the changes that occur in the Tc,peak and Tm,peak, the values of their respective enthalpies can be calculated, which gives a wider idea of how much heat and energy is needed in the manufacturing process of such polymeric sheets.

Slight changes were observed for all the samples in the Tc,onset and Tc,peak in addition to the peak broadening/stretching during crystallization. For the compatibilized composite sheets prepared from 5 μm sand particles, the peak crystallization temperatures (Tc,peak) increased slightly from ~114 °C at 0 wt% to ~115 °C at 50 wt%. As for the peak melting temperatures (Tm,peak), the temperatures decreased slightly from ~133 °C at 0 wt% to ~132 °C at 50 wt%. Moreover, due to similar melting and crystalline temperatures obtained in all the prepared sand/polymer sheets, it can be inferred that filler or compatibilizer addition does not alter the thermal characteristics of HDPE and promotes their good thermal stability, as reported in literature as well [42,43]. The Tc,onset, Tc,peak, and Tm,peak for neat HDPE and for all the prepared sand/polymer composite sheets, as well as stone paper, are presented in Table 2.

The enthalpies of crystallization (ΔHc) and melting (ΔHm) were calculated by integrating the area under the cooling and heating curves, respectively. The percentage of crystallinity (Xc) in the prepared sand/polymer composite sheets was calculated using the following equation:

$$\mathrm{Xc}(\%) = \frac{\Delta \mathrm{Hm}}{\Delta \mathrm{H}_{100\%}(1-\theta)} \times 100\% \tag{1}$$

where Xc is the percentage of crystallinity of HDPE, ΔHm is the melting enthalpy, $\Delta H_{100\%}$ is the melting enthalpy of a 100% crystalline HDPE, taken as 293 J/g [43], and θ is the mass fraction of the filler (sand). For stone paper, the weight is considered as the manufacturer states, which is 70 wt% [7].

The observed melting enthalpy (ΔHm) in all the prepared sand/polymer composite sheets was always lower than for the pure HDPE (148.77 J/g). Moreover, a general fluctuating value of percent crystallinity (Xc) was observed throughout the prepared sand/polymer composite sheets, ranging from a crystallinity of 40–60%, which is well reported in several filler–polymer composite systems [36,44,45]. This variation is explained by either an insufficient amount of filler (sand) particles at the surface which can cause agglomerates, or an excess accumulation of filler particles which can form a soft layer at the interface present, which tend to decrease the nucleating effect [42].

Moreover, the reduction of the sand/polymer composite melting enthalpies (ΔHm) and crystallinity (Xc) could be further explained by the reduction in the conformational changes available to the macromolecules during crystallization, which is due to the presence of the silica particles in the composites which are not densely packed. According to statistical thermodynamics, particles, in this case, silica, restrict the mobility of macromolecules and reduce the spaces available to be occupied by the macromolecules, thereby restricting the ability to form well-developed crystals. Additionally, the crystalline phase is not densely packed, which results in minimum intermolecular interactions and hence a decrease in the heat of fusion on melting which is also experienced by other HDPE systems such as carbon black/HDPE composites, wood/HDPE composites, and clay/HDPE composites [39,42].

Reduction of ΔHm in industrial terms can be translated to money and power savings during the extrusion/molding process, which encourages a positive attribute for the industrial process. ΔHm, ΔHc, and Xc for all the prepared sand/polymer composite sheets, as well as stone paper, are also presented in Table 2.

3.3.2. Thermal Stability of Composite Sheets

The amount of filler can have a significant impact on the end use properties, for instance, thermal expansion, stiffness, etc., of a final product. The thermal stabilities of the prepared sand/polymer composite sheets were also analyzed by thermal gravimetric analysis (TGA) by observing their onset degradation temperatures ($T_{d,onset}$), peak degradation temperatures ($T_{d,peak}$), and their onset degradation temperatures at 10% weight loss ($T_{d,onset\ @\ 10\%\ weight\ loss}$). By determining the $T_{d,onset}$ and $T_{d,peak}$ values, the upper range of processing temperature and the maximum temperature before degradation of the material can be confirmed, respectively. Hence, the range of processing temperature can be optimized to avoid any degradation of the material to occur which proves to be of substantial value for polymeric industrial scale processes. By noticing the changes occurring to the $T_{d,onset\ @\ 10\%\ weight\ loss}$, the effect of filler percentages and compatibilizer can be evaluated on the initial degradation corresponding to the same weight loss occurring. This can suggest which composition has more impact on the onset of degradation.

The TG curves for 25 μm sand particles and 5 μm sand particles alongside neat HDPE and prepared sand/polymer composite sheets, as well as their compatibilized versions, are shown in Figure 11a–d, respectively. Figure 11 shows that the composites have two degradation steps; the first degradation step is for the polymer, the second degradation is for the sand particles, which is comparable to the neat HDPE, which has only one degradation step, and sand has degradation at higher temperatures [46,47]. HDPE was stable (no significant weight loss) up to 400 °C, confirming that the processing of HDPE and HDPE composites at 170 °C would not degrade the polymer. The complete weight loss was observed near 490 °C, where almost all of the HDPE was burned. The peak degradation temperature ($T_{d,peak}$) for HDPE (obtained from derivative of weight loss curve (DTG)) was observed at ~489 °C.

The degradation steps became more evident at 50 wt% of filler addition. Additionally, it can be seen that increasing the amount of the filler increases the $T_{d,onset}$ and shifts the thermograms to the right, implying improved thermal stability, which is also reported for another study conducted on a SiO_2/polymer composite system [47]. Moreover, in the case of sand/polymer composite sheets prepared from 25 μm sand particles, the $T_{d,peak}$ increased very slightly from ~491 °C at 20 wt% to ~492 °C at 35 wt%, and $T_{d,onset\ @\ 10\%\ weight\ loss}$ increased from 448 °C to 453 °C, respectively. Furthermore, there is also a slight increase in the Tonset for the same set from 20 wt% to 35 wt%, which is also explained in the literature as due to the presence of filler minimizing the permeation of heat [47–49].

In addition to that, it can also be observed that the $T_{d,peak}$ after compatibilization were always lower than their corresponding samples prepared without the addition of the compatibilizer (except for sand/polymer sheet prepared at 20 wt% from 5 μm sand particles). This can be explained as due to the presence of acidic groups of maleic anhydride in the compatibilized sheets possibly interacting with some parts of the filler, resulting in slightly faster degradation [50]. For instance, for the compatibilized sand/polymer composite sheets prepared from 5 μm, the $T_{d,peak}$ decreased very slightly from ~499 °C at 20 wt% to ~488 °C at 50 wt%. However, the overall $T_{d,peak}$ remained comparably close enough to the neat HDPE value (489 °C), hence promoting good thermal stability for the prepared sand/polymer composite sheets. A similar trend was observed for the case of sand/polymer composite sheets prepared from 25 μm sand particles.

The differential rate of weight loss (dW/dt) of all the prepared sand/polymer composite sheets was obtained from differential thermogravimetric analysis (DTG) at a set heating rate of 20 °C/min. Figure 11 also shows the DTG plots (as insets) for neat HDPE (0 wt%) and for sand/polymer composite sheets at varying compositions prepared from 25 μm sand particles and 5 μm sand particles without the addition of the compatibilizer and with the addition of the compatibilizer, respectively.

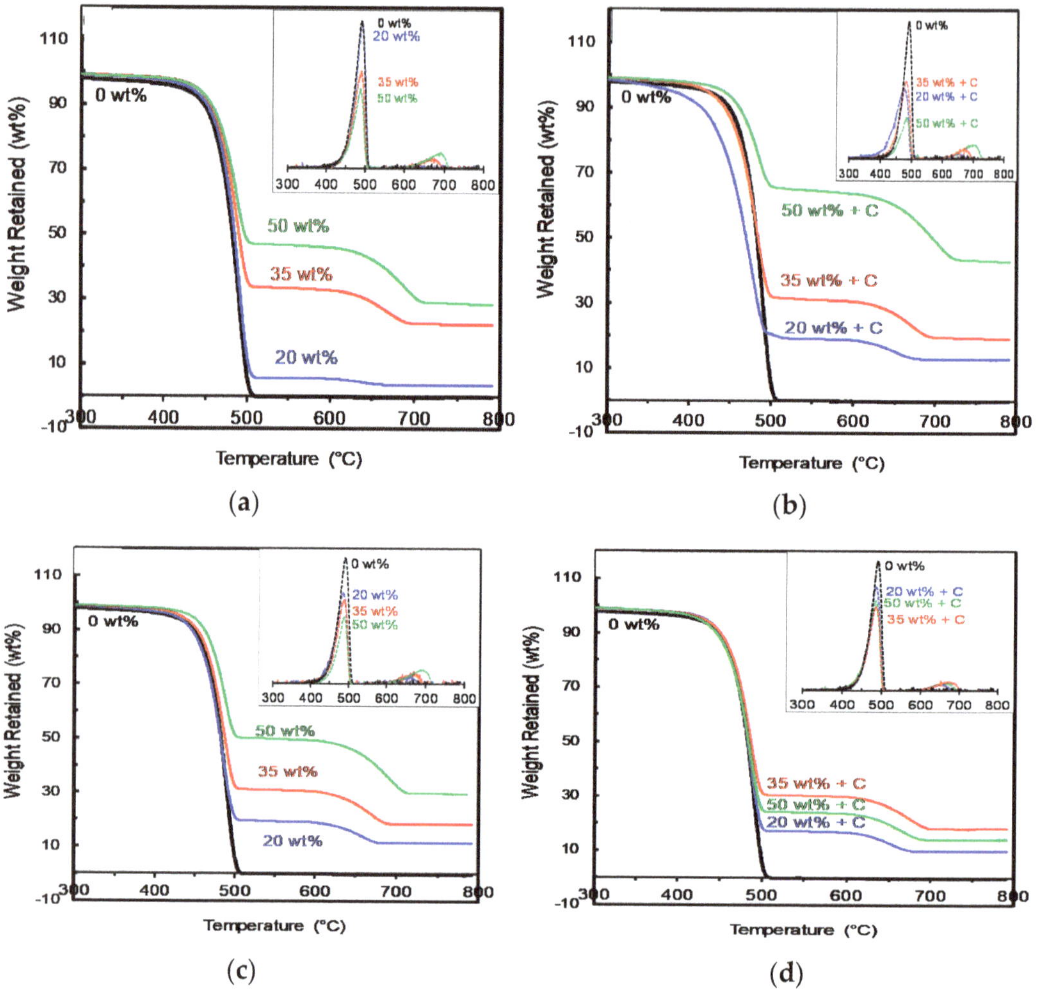

Figure 11. (**a**) TGA thermogram and DTG plot for sand/polymer non-compatibilized composite sheets prepared from 25 µm sand particles. (**b**) TGA thermogram and DTG plot for sand/polymer compatibilized composite sheets prepared from 25 µm sand particles. (**c**) TGA thermogram and DTG plot for sand/polymer non-compatibilized composite sheets prepared from 5 µm sand particles. (**d**) TGA thermogram and DTG plot for sand/polymer compatibilized composite sheets prepared from 5 µm sand particles.

A large fraction of the sand/polymer composite sheets decomposed between 300 °C and 600 °C, and this can be attributed to the decomposition of the HDPE. The thermal decomposition peak between 600 °C and 800 °C was assigned to the decomposition of sand particles which is more visible in samples prepared from 35 wt% composition and 50 wt% composition.

Both the DSC and TG analyses results indicated that the addition of sand particles of both 25 µm particle size and 5 µm particle size, as well as the addition of the compatibilizer, did not affect the thermal properties of the prepared sand/polymer composite sheets significantly, which promotes good thermal stability for the sheets. Comparable results for thermal stability are reported by numerous studies for various filler–polymer composites [36,51]. Table 3 reports the onset degradation temperatures ($T_{d,onset}$), peak degradation temperatures ($T_{d,peak}$), and onset degradation temperatures at 10% weight loss ($T_{d,onset\ @\ 10\%\ weight\ loss}$) for all the prepared sand/polymer composite sheets, stone paper, and regular A4 paper for comparison.

Table 3. TGA data for all the prepared sand/polymer composite sheets, stone paper, and regular A4 paper; 25 μm sand particles, 5 μm sand particles.

	Sample	$T_{d,onset}$ (°C)	$T_{d,peak}$ (°C)	$T_{onset,\ @\ 10\%\ weight\ loss}$ (°C)
Sand/polymer composite sheets prepared from 25 μm sand particles	0 wt%	466.69	489.0	444.86
	20 wt%	458.03	491.0	448.85
	35 wt%	461.74	492.01	453.43
	50 wt%	464.6	490.0	454.66
	20 wt% + C	461.42	479.0	409.42
	35 wt% + C	460.11	481.3	439.66
	50 wt% + C	458.69	488.0	457.61
Sand/polymer composite sheets prepared from 5 μm sand particles	20 wt%	467.78	486.21	443.14
	35 wt%	464.31	490.53	449.48
	50 wt%	462.39	490.10	459.56
	20 wt% + C	437.68	499.01	446.97
	35 wt% + C	454.08	488.41	447.20
	50 wt% + C	456.76	488.06	442.17
	Stone Paper	457.05	489.50	469.72
	Regular A4	347.74	381.60	324.63

3.4. Mechanical Properties

The effect of filler and compatibilizer (C) on the elastic modulus and tensile strength for sand/polymer composite sheets prepared from 25 μm sand particles and 5 μm sand particles was studied with the elastic modulus for both sets illustrated in Figure 12a,b, respectively. As seen in Figure 12, at 0 wt% filler the elastic modulus obtained for the pure HDPE sheet was ~1200 MPa with a corresponding yield stress of 35.15 MPa, respectively, which is comparable to the literature [2,45,52]. Generally, a decrease in the elastic modulus was observed with increasing filler concentration from 1298.33 MPa at 20 wt% to 905. 72 MPa at 50 wt% for the sheets prepared from 25 μm sand particles (in Figure 12a). Similar trends were observed in the case of 5 μm, where the elastic modulus dropped from 950.59 MPa at 20 wt% to 887.47 MPa at 35 wt% (except for an increase in the case of 50 wt%), as seen in Figure 12b.

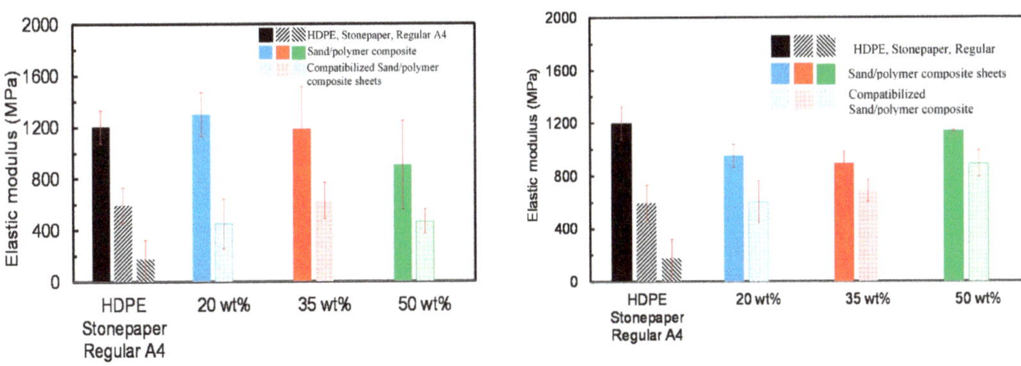

(a) (b)

Figure 12. Elastic modulus for sand/polymer composite sheets and compatibilized sand/polymer composite sheets prepared from (**a**) 25 μm sand particles; (**b**) 5 μm sand particles.

Compatibilization lowered the elastic modulus for all cases compared to their non-compatibilized versions. For instance, compatibilized sand/polymer composite sheets prepared at 35 wt% experienced a sharp reduction in elastic modulus from 1182.23 MPa before compatibilization to 629.95 MPa after compatibilization, and from 887.47 MPa to 687 MPa for sand/polymer composite sheets prepared from 25 μm sand particles and 5 μm sand particles at the same filler composition, respectively (Figure 12a,b). The effect of addition of fillers and compatibilizers on the decreasing value of elastic modulus is observed for many polymer composites and is believed to be caused by the random agglomeration of particles weakening the polymer matrix, causing uneven crystallization leading to their brittle nature, which also contributes to weaker adhesion between filler particles and the polymer matrix [34,47,51,53]. This also contributes to the fact that the compatibilizer (maleic anhydride) is not able to form strong bonds with the composite sheets, leading to a weaker matrix structure, as seen for other blends as well [54,55].

Furthermore, with elastic moduli of 603.54 MPa and 687 MPa, the compatibilized sand/polymer composite sheets prepared with 5 μm sand particles at 20 wt% and 35 wt%, respectively, gave results closest to the stone paper, which yielded an elastic modulus of 596.32 MPa. Moreover, another study on composite sheets also suggested that optimum results were obtained from sheets prepared at 30% by weight [2].

The tensile strength did not change significantly with increasing the filler concentration before and after compatibilization, which is reported in the literature as well [56]. However, the tensile strength decreased significantly with the addition of the compatibilizer for both sets of sand/polymer composite sheets prepared from 25 μm sand particles and 5 μm sand particles. Even though compatibilizers are expected to increase the tensile strength, the decreasing trend could possibly be explained due to the compatibilizer limiting the stress transfer and swamping the surface [57]. Furthermore, as increasing the filler decreases the mechanical properties (such as the elastic modulus), the decreased tensile strength can also be a combined effect of the addition of the compatibilizer and filler, as reported in another study on a filler–polymer composite material [53].

A wide variation of mechanical properties was observed for the case of sheets prepared at 50 wt%, irrelevant of the particle size, which is also reported to occur in thermoplastic-based films subjected to extensive molecular orientation [2]. For filler content of 50 wt%, the lowest tensile strength values were obtained, thereby suggesting increased brittleness for sheets prepared at higher filler percent by weight, which is also supported by the literature [45]. Due to possible formation of hydrogen bonding, crack propagation at weak phase interfaces can be facilitated, resulting in a lower tensile strength of the blend, which is also reported in the literature [54]. Comparatively, one of the highest tensile strength values (15.66 MPa) was measured for the regular A4 paper, which can be corresponded to its fibrous network structure with strong hydrogen bonding [58]. Table 4 reports all the mechanical properties of sand/polymer composite sheets and compatibilized sand/polymer composite sheets prepared from 25 μm sand particles and 5 μm sand particles, as well as stone paper and regular A4 paper.

3.5. Wettability Performance

Surface wettability is one of the most essential properties to determine the use of the material in a specific application, such as in this case for printing-based applications. Wettability is determined based on the contact angle measured on the surface of the material. Typically, any non-polar surface would measure a contact angle of 90° or above. Anything below 90° suggests increased hydrophilicity with the liquid on the surface and results in increased absorption of the liquid in contact. Increasing the filler concentration increased the contact angle slightly from 86.62° at 20 wt% to 94.72° at 50 wt% and 88.88° at 20 wt% to 94.6° at 50 wt% for sand/polymer composite sheets prepared from 25 μm sand particles and 5 μm sand particles, respectively, suggesting improved anti-wetting performance with water for both particle sizes. These results indicate that the nonpolar, hydrophobic blocks of the compatibilizer units may have been arranged on the surface of the composite sheets, as reported for other blends in the literature as well [54]. Contact angles of above 90° were obtained for all the compatibilized sheets. Figure 13 illustrates this data.

Table 4. Mechanical properties of all the prepared sheets, stone paper, and regular A4.

Sample		Elastic Modulus (MPa)	Yield Stress (MPa)	Yield Strain (MPa)	Tensile Strength (MPa)	Tensile Strain (MPa)
Sand/polymer composite sheets prepared from 25 μm sand particles	0 wt%	1200.77 ± 127.3	35.15 ± 2.7	2.1 ± 3.5	33.76 ± 6.2	2.26 ± 0.42
	20 wt%	1298.33 ± 169.8	27.84 ± 4.3	0.0424 ± 0.008	20.23 ± 3.5	0.17 ± 0.15
	35 wt%	1182.33 ± 328.4	23.11 ± 4.8	0.028 ± 0.009	18.96 ± 6.7	0.035 ± 0.01
	50 wt%	905.72 ± 343.1	12.74 ± 5.4	0.022 ± 0.009	9.93 ± 4.7	0.030 ± 0.02
	20 wt% + C	448.78 ± 194.1	11.98 ± 2.4	0.04 ± 0.01	8.01 ± 2.7	0.05 ± 0.01
	35 wt% + C	629.95 ± 138.9	9.56 ± 43.5	0.02 ± 0.004	8.68 ± 3.5	0.02 ± 0.005
	50 wt% + C	465.11 ± 94.9	6.71 ± 1.8	0.02 ± 0.01	4.36 ± 1.9	0.03 ± 0.01
Sand/polymer composite sheets prepared from 5 μm sand particles	20 wt%	950.59 ± 86.6	24.89 ± 1.9	0.04 ± 0.002	21.56 ± 0.6	0.06 ± 0.01
	35 wt%	887.47 ± 96.2	19.11 ± 0.9	0.03 ± 0.002	17.17 ± 2.8	0.05 ± 0.003
	50 wt%	1137.05 ± 8.2	18.64 ± 3.37	0.02 ± 0.002	17.22 ± 1.48	0.02 ± 0.001
	20 wt% + C	603.54 ± 157.9	17.64 ± 2.53	0.05 ± 0.02	15.94 ± 2.76	0.05 ± 0.02
	35 wt% + C	687 ± 80.87	212.48 ± 0.38	0.02 ± 0.00	10.98 ± 1.04	0.03 ± 0.01
	50 wt% + C	890.87 ± 101.6	16.78 ± 3.12	0.03 ± 0.005	15.09 ± 2.18	0.03 ± 0.02
	Stone Paper	596.32 ± 135.14	6.07 ± 0.53	0.11 ± 0.04	6.17 ± 0.73	0.54 ± 0.19
	Regular A4	175.18 ± 146.36	16.17 ± 5.08	0.06 ± 0.03	15.66 ± 5.14	0.06 ± 0.03

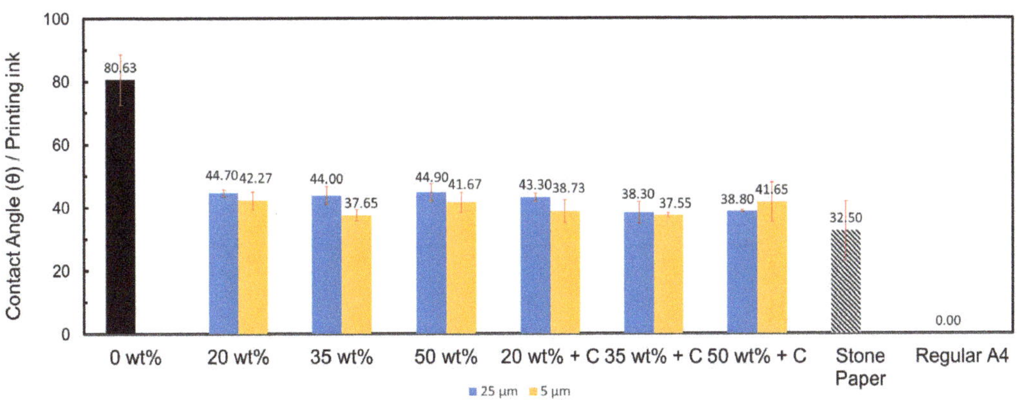

Figure 13. Contact angles of printing ink on sand/polymer composite sheets prepared from 25 μm sand particles and 5 μm sand particles.

In the case of the non-polar solvent, benzene, decreased contact angles were obtained (ranging between 20–35°) for all the sand/polymer composite sheets prepared, suggesting their absorption ability to a certain extent if they were to come into contact with any organic solvents. This can be excepted, as the surface is mainly non-polar and any contact with a non-polar liquid would promote the widely known "like dissolving in like" phenomena. Since maleic anhydride is an organic compound, the oxygen in its functional group binds to the chemical structure of benzene. Values of ~90° were obtained when the sheets were tested with a water–benzene mixture. In addition to that, regular A4 paper also exhibited a relatively higher value of contact angle (82.27°) with water–benzene mixture.

Contact angle values of lower than 45° were obtained with printing ink. Irrespective of the particle size used, a general trend of decreased contact angles was observed with the addition of the compatibilizer compared to their non-compatibilized versions, indicating improved ink-wetting performance. For example, as seen in Figure 13 for sand/polymer composite sheets prepared from 25 μm sand particles, for both at 35 wt% and 50 wt%, the contact angles decreased from ~44° to ~38°. Similar contact angles for ink-wetting have been reported [59]. Moreover, the values obtained were relatively close to the commercial stone paper, which resulted in a contact angle of 32.50° with printing ink. Contact angles below 50° correspond to surface free energies of below 45 dyne/cm, which is preferred for printing [13,60]. A contact angle of 0° was obtained for regular A4 paper with printing ink, which is expected due to its increased hydrophilic surface. The wettability performances with printing ink for all the sand/polymer composite sheets prepared from 25 µm sand particles and 5 µm sand particles are illustrated in Figure 13.

Table 5 summarizes all the measured contact angles for the prepared sand/polymer composite sheets with water, benzene, water–benzene mixture, and printing ink. Contact angles for commercial stone paper and regular A4 paper are reported as well.

Table 5. Contact angle measurements.

Sample		Water	Benzene	Water-Benzene Mixture	Printing Ink
Sand/polymer composite sheets prepared from 25 µm sand particles	0 wt%	97.96 ± 9.2	22.05 ± 1.8	100.93 ± 9.5	80.63 ± 8.2
	20 wt%	86.62 ± 6.3	31.80 ± 2.6	90 ± 0	44.70 ± 1.1
	35 wt%	89.6 ± 0.9	22.83 ± 4.1	90 ± 0	44.0 ± 2.9
	50 wt%	94.72 ± 3.0	37.95 ± 6.9	90 ± 0	44.9 ± 2.8
	20 wt% + C	99.62 ± 6.2	26.03 ± 7.0	90 ± 0	43.40 ± 1.3
	35 wt% + C	92 ± 5.7	21.70 ± 6.9	90 ± 0	38.30 ± 3.4
	50 wt% + C	98.74 ± 8.3	20.17 ± 2.2	91.77 ± 3.1	38.80 ± 0.5
Sand/polymer composite sheets prepared from 5 µm sand particles	20 wt%	86.88 ± 5.5	30.95 ± 6.2	90 ± 0	42.27 ± 2.8
	35 wt%	92.08 ± 4.7	20.37 ± 2.1	90 ± 0	37.65 ± 1.9
	50 wt%	94.6 ± 8.6	27.03 ± 6.7	90 ± 0	41.67 ± 3.2
	20 wt% + C	105.6 ± 9.1	24.23 ± 0.8	90 ± 0	38.73 ± 3.7
	35 wt% + C	99.4 ± 3.9	21.13 ± 2.2	94.83 ± 8.3	37.55 ± 0.78
	50 wt% + C	99.12 ± 3.7	21.87 ± 1.6	102.33 ± 11.4	41.65 ± 6.3
	Stone Paper	105.62 ± 2.1	26.50 ± 4.4	113.30 ± 7.2	32.50 ± 9.5
	Regular A4	83.34 ± 7.8	0 ± 0	83.27 ± 21.8	0 ± 0

3.6. Printing Test

The printing tests showed the capability of the prepared sand/polymer composite sheets of absorbing printer ink. The sheets were printed on and left to dry for over 24 h. It was observed that some parts were still able to wipe off for all the prepared sand/polymer composite sheets, and some parts were able to stay on the surface. Pure HDPE showed the least absorption of the printing ink and was unable to retain the printing ink on the surface. The selected 35 wt% sheet prepared from 25 µm sand particles showed almost similar printability to its compatibilized version. However, for the case of the 35 wt% prepared from 5 µm sand particles, improved printability with the addition of the compatibilizer was noticed. Moreover, by observing the printing tests closely, ink had the ability to form drops yet not spread on the surface of the sand/polymer composite sheets prepared with either 25 µm sand particles or 5 µm sand particles, resulting in a blurred, non-graphic quality, which could be easily smudged. Similar results have been reported in the literature [13]. Figure 14 shows all the sheets after the printing tests (enclosed within the red highlights are the surfaces of the respective sheets).

On the stone paper, slight removal was observed after 24 h in contrast to the regular A4 paper, where the ink was completely dried and absorbed almost immediately. It can be concluded that the quality of the printing improved with the addition of the filler for the case of sand/polymer composite sheets prepared from 25 µm sand particles (compared to pure HDPE), and after compatibilization in the case of sand/polymer composite sheets prepared from 5 µm sand particles.

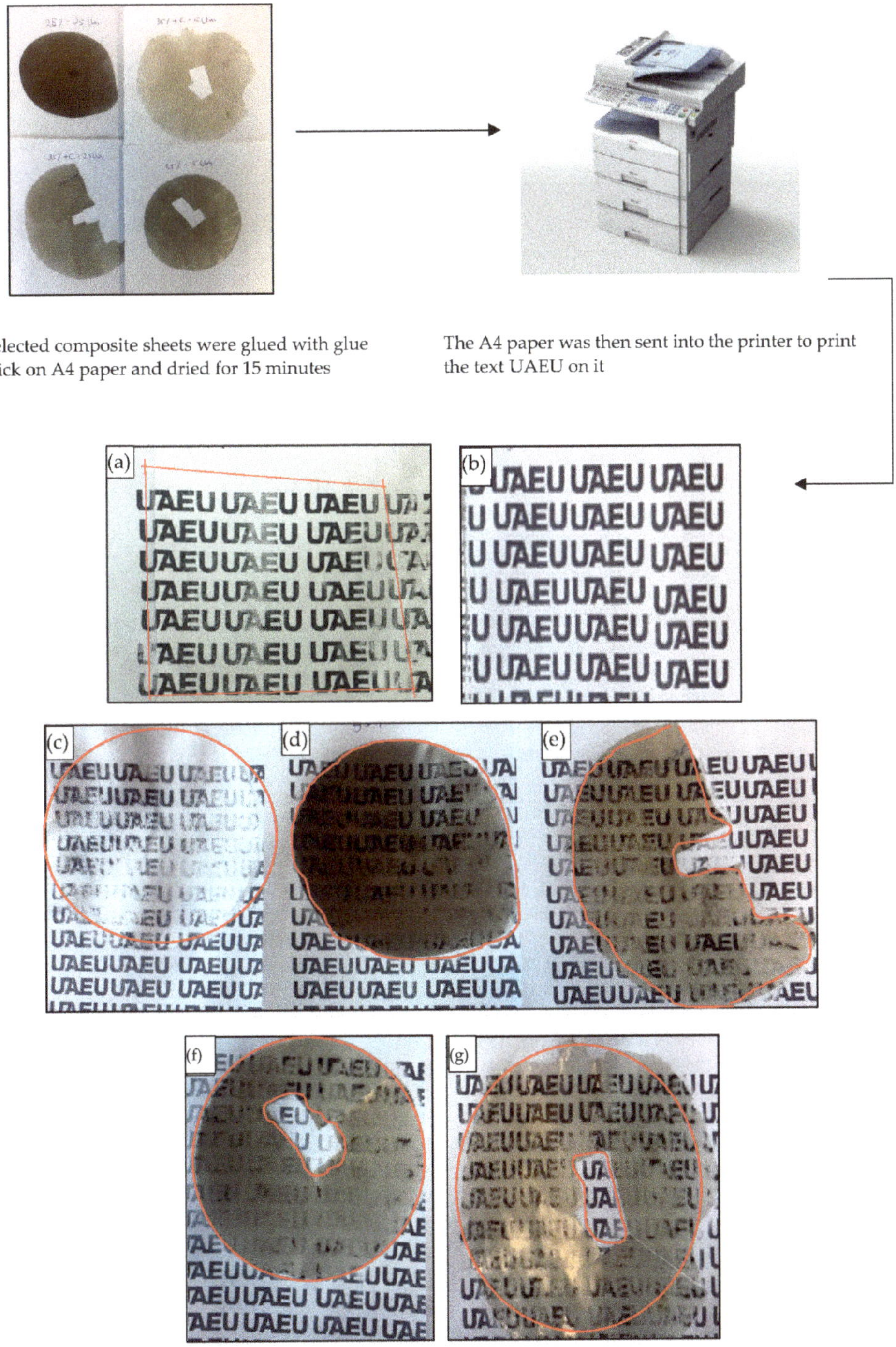

Figure 14. Printing tests after 24 h for (**a**) stone paper, (**b**) regular A4 paper, (**c**) HDPE, (**d**) 35 wt%, 25 μm (**e**) 35 wt% + C, 25 μm, (**f**) 35 wt%, 5 μm, and (**g**) 35 wt% + C, 5 μm.

3.7. Adhesion Test

Selected sand/polymer composite sheet surfaces that were studied for the adherence test using a permanent marker pen are shown in Table 6. Amongst all the sand/polymer composite sheets, including the compatibilized sand/polymer composite sheets, the sand/polymer composite sheet prepared from 5 μm sand particles at 35 wt% showed relatively better adherence. Stone paper and regular A4 paper showed excellent adherence comparatively.

Table 7 shows the sand/polymer composite sheet surfaces before and after the adherence test using removable marker pen. Only stone paper showed excellent adhesion compared to all the other tested sheets, which is reported in the literature as well [13]. For all the sand/polymer composite sheets prepared, the ink was easily removed with the tape irrespective of the particle size used or compatibilizer added. Moreover, the ink could be wiped out manually by swiping of the hand as well, which could be an advantage for applications that require such kind of erasable surfaces.

3.8. Comparative Analysis of Prepared Sand/Polymer Composite Sheets to Stone Paper

In this study, several properties varying from thermal to mechanical and wettability properties with several liquids were analyzed. Printing and adhesion tests were also tried on selected composite sheets as well as on the stone paper. To obtain a complete understanding of which sheets resulted in properties closest to the commercial stone paper, a comparative analysis was conducted.

In terms of percent crystallinity (%Xc), the compatibilized sand/polymer composite sheet prepared from 25 μm sand particles at 50 wt% yielded an almost similar value, of 41%, to stone paper, which resulted in 41.03%. With a peak degradation temperature (T_d) of 490 °C, the sand/polymer composite sheet prepared from 25 μm sand particles at 50 wt% resulted in the closest value to stone paper, which had a peak degradation of 489.5 °C. In terms of mechanical properties, the compatibilized sand/polymer composite sheet prepared from 5 μm at 20 wt% resulted in an elastic modulus (E) of 603.54 MPa, whereas stone paper resulted in an elastic modulus of 596.32 MPa. For the tensile strength, with a value of 4.36 MPa, the compatibilized sand/polymer composite sheet prepared from 25 μm sand particles at 50 wt% gave the closest value to stone paper, which had a tensile strength (TS) of 6.17 MPa.

In terms of the wettability performance, the sand/polymer composite sheet prepared from 5 μm sand particles at 20 wt% gave almost the exact same contact angles (θ) with water to stone paper, each yielding value of 105.61° and 105.62°, respectively. The contact angle (θ) with printing ink that was closest to the stone paper (32.5°) was of the compatibilized sand/polymer composite sheet prepared from 5 μm sand particles at 35 wt% (37.55°).

The printing performance of the sand/polymer composite sheet prepared from 25 μm sand particles at 35 wt% showed the best results amongst all the tested composite sheets. In terms of the adhesion property using permanent marker, the sand/polymer composite sheets prepared from 5 μm sand particles at 35 wt% showed relatively better adherence compared to the other tested composite sheets. As for the adhesion property using a removable marker, all the tested sand/polymer composite sheets showed the same results. The sheets that resulted in values closest to the stone paper with respect to the analyzed properties are tabulated in Table 8.

Table 6. Adherence test using permanent marker pen.

	Stone Paper	Regular A4 Paper	0 wt%	35 wt%, 25 µm	35 wt% + C, 25 µm	35 wt%, 5 µm	35 wt% + C, 5 µm
Before adherence test with permanent marker pen (sheet surface)							
After adherence test with permanent marker pen (sheet surface)							
After adherence test with permanent marker pen (removed adhesive tape)							

Table 7. Adherence test using removable marker pen.

	Stone Paper	Regular A4 Paper	0 wt%	35 wt%, 25 µm	35 wt% + C, 25 µm	35 wt%, 5 µm	35 wt% + C, 5 µm
Before adherence test with removable marker pen (sheet surface)							
After adherence test with removable marker pen (sheet surface)							
After adherence test with removable marker pen (removed adhesive tape)							

Table 8. Comparison of sheets resulting in values closest to stone paper.

Surface / Support		%Xc	T_d	E	TS	θ/Water	θ/Printing Ink	Printing Test	Adhesion Using Permanent Marker
Sand/polymer composite sheets prepared from 25 µm sand particles	20 wt%								
	35 wt%							✓	
	50 wt%		✓						
	20 wt% + C								
	35 wt% + C								
	50 wt% + C	✓			✓				
Sand/polymer composite sheets prepared from 5 µm sand particles	20 wt%					✓			
	35 wt%								✓
	50 wt%								
	20 wt% + C			✓					
	35 wt% + C						✓		
	50 wt% + C								

4. Conclusions

In this work, sand/polymer composite sheets prepared from sand and HDPE were successfully manufactured via melt blending and compression molding at varying filler compositions. Thermal characterization revealed that the crystallization temperatures remained almost constant at ~113–115 °C, and the melting temperatures remained steady at ~132–135 °C for all the prepared sand/polymer composite sheets, promoting their good thermal stability. Moreover, no significant degradation (visual or on the bases of weight loss) at the optimized processing conditions were observed. The maximum degradation temperature was almost constant, ranging from ~489–493 °C for all the prepared sand polymer composite sheets, and similar crystallization, melting, and degradation temperatures were also obtained for the commercial stone paper, giving the prepared sand/polymer composite sheets decent thermal results. The mechanical characterization of all the sand/polymer composite sheets showed a decrease in their strength, as the elastic modulus values decreased significantly with the addition of the compatibilizer. The wettability analysis suggested that increasing the filler composition, as well as addition of the compatibilizer, led to an increase in the contact angles corresponding to the improved anti-wetting performance. Additionally, in the case of contact angles with printing ink, the observed angles were almost half the value obtained for the pure HDPE, varying in the ranges of 30–45°, which was also comparable to the stone paper, suggesting good potential for their ink wettability for printing and paper-based applications. The printing test showed that the prepared sand/polymer composite sheets gave comparatively better results than the pure HDPE sheets, implying some potential for printing with filler and compatibilizer addition. The prepared sheets showed good adherence with the use of permanent marker pen, suggesting potential for possible ink-based applications with such kind of ink. Furthermore, the sheets showed weaker adherence with a removable marker pen, but could possibly be used for any application that requires erasable surfaces (such as erasable sheets/boards for student learning, educational toys, etc.). The results obtained in this research provide information about the potential of the production of local sand/polymer composite sheets and their use in paper-based applications.

Key achievements:

- Local sand was used as a filler to develop composite sheets, which were investigated.

- Addition of a compatibilizer to the composite sheets was also investigated.
- Melt extrusion and compressing molding techniques were used.
- A set of two particle sizes of silica were used for this study.
- Data were compared to regular A4 paper, and stone paper was also analyzed.

Author Contributions: S.S.: formal analysis, investigation, resources, data curation, writing—original draft preparation. A.H.A.-M.: conceptualization, visualization, writing—review and editing, project administration, funding acquisition, supervision. M.Z.I.: methodology, software, validation, supervision, resources. All authors have read and agreed to the published version of the manuscript.

Funding: This research was funded by UAE University, Grant number 21N221.

Institutional Review Board Statement: Not applicable.

Informed Consent Statement: Not applicable.

Data Availability Statement: Not required.

Acknowledgments: The authors would like to thank the University Innovation Program (UIP'18) of Expo Live 2020 for partially funding this research under Grant Number 21N221.

Conflicts of Interest: The authors declare no conflict of interest.

References

1. Julkapli, N.M.; Bagheri, S. Developments in nano-additives for paper industry. *J. Wood Sci.* **2016**, *62*, 117–130. [CrossRef]
2. De Santi, C.R.; Correa, A.C.; Manrich, S. Films of post-consumer polypropylene composites for the support layer in synthetic paper. *Polímeros* **2006**, *16*, 123–128. [CrossRef]
3. Geyer, R.; Jambeck, J.R.; Law, K.L. Production, use, and fate of all plastics ever made. *Sci. Adv.* **2017**, *3*, e1700782. [CrossRef] [PubMed]
4. Rochman, C.M.; Browne, M.A.; Halpern, B.S.; Hentschel, B.T.; Hoh, E.; Karapanagioti, H.K.; Rios-Mendoza, L.M.; Takada, H.; Teh, S.; Thompson, R. Classify plastic waste as hazardous. *Nature* **2013**, *494*, 169–171. [CrossRef]
5. Ahmed, W.; Khenata, R.; Siraj, S.; Al-Douri, Y. Ionic Liquid Potential to Recycle Polymeric Waste: An Experimental Investigation. *Mater. Res.* **2020**, *23*. [CrossRef]
6. Lung, M. Stonepaper, Taiwan Lung Meng Technology Co., Ltd. Available online: http://www.taiwanlm.com/ (accessed on 17 February 2020).
7. Rockstock. "Photo-Degradable Paper Made from Stone". Available online: https://www.stonepaper.co.nz/ (accessed on 17 February 2020).
8. Ahmed, W.; Alnajjar, F.; Zaneldin, E.; Al-Marzouqi, A.H.; Gochoo, M.; Khalid, S. Implementing FDM 3D Printing Strategies Using Natural Fibers to Produce Biomass Composite. *Materials* **2020**, *13*, 4065. [CrossRef]
9. Menendez, H.; White, J.L. A Wide-angle X-ray diffraction method of determining chopped fiber orientation in composites with application to extrusion through dies. *Polym. Eng. Sci.* **1984**, *24*, 1051–1055. [CrossRef]
10. Nguyen, H.X.; Ishida, H. Poly(aryl-ether-ether-ketone) and its advanced composites: A review. *Polym. Compos.* **1987**, *8*, 57–73. [CrossRef]
11. Bhattachrya, S.; Gupta, R.; Kamal, M.R. *Polymeric Nanocomposite: Theory and Practice*; Hanser: Munich, Germany, 2007.
12. Jesionowski, T.; Bula, K.; Janiszewski, J.; Jurga, J. The influence of filler modification on its aggregation and dispersion behaviour in silica/PBT composite. *Compos. Interfaces* **2003**, *10*, 225–242. [CrossRef]
13. Corrêa, A.C.; de Santi, C.R.; Manrich, S. Synthetic Paper from Plastic Waste: The Effect of $CaCO_3$ on Physical, Surface Properties and Printability. In *Macromolecular Symposia*; Wiley Online Library: Hoboken, NJ, USA, 2006; pp. 611–620.
14. Borges, J.R.V.; Santos, A.S.F.E.; Manrich, S. The Influence of Organoclay on Polypropylene Composite Films for Synthetic Paper. *Polym. Technol. Eng.* **2011**, *50*, 1443–1451. [CrossRef]
15. Buddi, T.; Rao, B.N.; Singh, S.K.; Purohit, R.; Rana, R.S. Development and Analysis of High Density Poly Ethylene (HDPE) Nano SiO_2 and Wood Powder Reinforced Polymer Matrix Hybrid Nano Composites. *J. Exp. Nanosci.* **2018**, *13*, S24–S30. [CrossRef]
16. Ahmed, W.; Siraj, S.; Alnajjar, F.; Al Marzouqi, A.H. 3D Printed Implants for Joint Replacement. In *Applications of 3D Printing in Biomedical Engineering*; Sharma, N.R., Subburaj, K., Sandhu, K., Sharma, V., Eds.; Springer: Singapore, 2021; pp. 97–119.
17. Ahmed, W.; Ahmed, S.; Alnajjar, F.; Zaneldin, E. Mechanical Performance of Three-Dimensional Printed Sandwich Composite with a High-Flexible Core. *Proc. Inst. Mech. Eng. Part L J. Mater. Des. Appl.* **2021**, *235*, 1382–1400. [CrossRef]
18. Almazrouei, N.; Siraj, S.; Al-Marzouqi, A.H.; Ahmed, W. Producing Particulate Composite Using 3D Printing Plastics Waste. In Proceedings of the 2022 Advances in Science and Engineering Technology International Conferences (ASET), Dubai, United Arab Emirates, 21–24 February 2022; pp. 1–6.
19. Wang, X.; Wu, L.; Yu, H.; Xiao, T.; Li, H.; Yang, J. Modified Silica-Based Isoprene Rubber Composite by a Multi-Functional Silane: Preparation and Its Mechanical and Dynamic Mechanical Properties. *Polym. Test.* **2020**, *91*, 106840. [CrossRef]

20. Ahmed, W.; Siraj, S.; Al-Marzouqi, A.H. Embracing Additive Manufacturing Technology through Fused Filament Fabrication for Antimicrobial with Enhanced Formulated Materials. *Polymers* **2021**, *13*, 1523. [CrossRef] [PubMed]
21. Siraj, S.; Al Marzouqi, A.; Lqbal, M. Mechanical and Wettability Performance of Sand/HDPE Composite Sheets. *Mater. Sci. Forum* **2020**, *1015*, 9–14. [CrossRef]
22. Siraj, S.; Al-Marzouqi, A.; Ahmed, W. Processing Biodegradable Fused Filament Fabrication Waste with Micro-Silica Particles. *Key Eng. Mater.* **2022**, *907*, 156–162. [CrossRef]
23. Ahmed, W.; Siraj, S.; Al-Marzouqi, A.H. 3D Printing PLA Waste to Produce Ceramic Based Particulate Rein-forced Composite Using Abundant Silica-Sand: Mechanical Properties Characterization. *Polymers* **2020**, *12*, 2579. [CrossRef]
24. Kumar, P.; Das, V.; Gupta, A.; Ameta, N.K. Stabilization of Dune Sand Mixed with Plastic (LDPE) Waste Strips for Design of Flexible Pavement in Construction of Roads. *Am. J. Eng. Res. AJER* **2016**, *5*, 315–320.
25. Bamigboye, G.O.; Ngene, B.U.; Ademola, D.; Jolayemi, J.K. Experimental Study on the Use of Waste Polyethylene Terephthalate (PET) and River Sand in Roof Tile Production. *J. Phys. Conf. Ser.* **2019**, *1378*, 042105. [CrossRef]
26. Ding, T.; Xiao, J.; Zou, S.; Yu, J. Flexural properties of 3D printed fibre-reinforced concrete with recycled sand. *Constr. Build. Mater.* **2021**, *288*, 123077. [CrossRef]
27. Dang, X.; Gao, Y.; Yu, Y.; Li, Q.; Wang, S.; Wu, H.; Wang, H.; Zhao, P. Windproof Efficiency with New Biodegradable PLA Sand Barrier and Traditional Straw Sand Barrier. *J. Beijing For. Univ.* **2015**, *37*, 118–125.
28. Kumi-Larbi, A.; Yunana, D.; Kamsouloum, P.; Webster, M.; Wilson, D.C.; Cheeseman, C. Recycling waste plastics in developing countries: Use of low-density polyethylene water sachets to form plastic bonded sand blocks. *Waste Manag.* **2018**, *80*, 112–118. [CrossRef] [PubMed]
29. Marzouqi, A.H.A.; Ahmed, S.S.; Rabbani, S.F.; Mesmari, H.J.A.Z.A.; Ketbi, S.H.R.A.; Rahman, N.S.H.A. Method for Obtaining Paper from Sand and Products Obtained Therefrom. U.S. Patent No. 10,611,893, 24 October 2019.
30. Alsheriani, T.A.K. Composition and Environmental Assessment of Soils from United Arab Emirates. Master's Thesis, United Arab Emirates University, Abu Dhabi, UAE, 1998.
31. Mittal, V.; Chaudhry, A.U. Effect of Amphiphilic Compatibilizers on the Filler Dispersion and Properties of Polyethylene—Thermally Reduced Graphene Nanocomposites. *J. Appl. Polym. Sci.* **2015**, *132*, 42484–42494. [CrossRef]
32. Soliman, F. Silica-filled,high molecular weight polyolefin. U.S. Patent No. 3,788,923, 29 January 1974.
33. de Carvalho, M.S.; Azevedo, J.B.; Barbosa, J.D.V. Effect of the melt flow index of an HDPE matrix on the properties of composites with wood particles. *Polym. Test.* **2020**, *90*, 106678. [CrossRef]
34. Ahmed, W.; Siraj, S.; Al-Marzouqi, A. Comprehensive Characterization of Polymeric Composites Reinforced with Silica Microparticles Using Leftover Materials of Fused Filament Fabrication 3D Printing. *Polymers* **2021**, *13*, 2423. [CrossRef] [PubMed]
35. Bansal, N.P.; Choi, S.R. Properties of CMAS glass from desert sand. *Ceram. Int.* **2015**, *41*, 3901–3909. [CrossRef]
36. Peşman, E.; Tufan, M. The Effects of $CaCO_3$ Coated Wood Free Paper Usage as Filler on Water Absorption, Me-chanical and Thermal Properties of Cellulose-High Density Polyethylene Composites. *Mater. Sci.* **2016**, *22*, 530–535.
37. Tang, W.; Santare, M.H.; Advani, S.G. Melt Processing and Mechanical Property Characterization of Multi-Walled Carbon Nanotube/High Density Polyethylene (MWNT/HDPE) Composite Films. *Carbon* **2003**, *41*, 2779–2785. [CrossRef]
38. Aggarwal, S.; Sajwan, M.; Singh, R.B. Crystallinity of Hdpe Pipes by Dsc, Xrd and Ftir Spectroscopy-a Forensic Comparison. *Indian J. Criminol. Crim.* **2008**, *2*, 141–148.
39. Lei, Y.; Wu, Q.; Clemons, C.M.; Yao, F.; Xu, Y. Influence of Nanoclay on Properties of HDPE/Wood Composites. *J. Appl. Polym. Sci.* **2007**, *106*, 3958–3966. [CrossRef]
40. Musić, S.; Filipović-Vinceković, N.; Sekovanić, L. Precipitation of Amorphous SiO_2 Particles and Their Properties. *Braz. J. Chem. Eng.* **2011**, *28*, 89–94. [CrossRef]
41. Martínez, J.; Palomares-Sánchez, S.; Ortega-Zarzosa, G.; Ruiz, F.; Chumakov, Y. Rietveld refinement of amorphous SiO_2 prepared via sol–gel method. *Mater. Lett.* **2006**, *60*, 3526–3529. [CrossRef]
42. Jeevananda, T.; Kim, N.H.; Lee, J.H.; Basavarajaiah, S.; Urs, M.D.; Ranganathaiah, C. Investigation of Multi-Walled Carbon Nanotube-Reinforced High-Density Polyethylene/Carbon Black Nanocomposites Using Electrical, DSC and Positron Lifetime Spectroscopy Techniques. *Polym. Int.* **2009**, *58*, 775–780. [CrossRef]
43. Netzch, P. Polymer Manufacturing. Determination of the Degree of Crystallinity If Polymers (PE and PP). Available online: https://www.netzsch.com/en/ (accessed on 26 November 2019).
44. Hitachi High-Tech Science Corporation, 1986, Application Brief-DSC Measurement of Polyethylene: Correlation of Polyethn. Available online: www.hitachihigh-tec-science.com (accessed on 23 February 2020).
45. Brostow, W.; Datashvili, T.; Jiang, P.; Miller, H. Recycled HDPE reinforced with sol–gel silica modified wood sawdust. *Eur. Polym. J.* **2016**, *76*, 28–39. [CrossRef]
46. Liu, Y.-L.; Wei, W.-L.; Hsu, K.-Y.; Ho, W.-H. Thermal Stability of Epoxy-Silica Hybrid Materials by Thermo-gravimetric Analysis. *Thermochim. Acta* **2004**, *412*, 139–147. [CrossRef]
47. Dorigato, A.; D'Amato, M.; Pegoretti, A. Thermo-mechanical properties of high density polyethylene—Fumed silica nanocomposites: Effect of filler surface area and treatment. *J. Polym. Res.* **2012**, *19*, 9889. [CrossRef]
48. Zheng, K.; Yao, X.; Tian, X.; Chen, L.; He, H.; Li, Y. Synthesis and Thermal Behavior of Silica-Graft-Polypropylene Nanocomposites Studied by Step-Scan DSC and TGA. *J. Macromol. Sci. Part B* **2006**, *45*, 493–505. [CrossRef]

49. Elleithy, R.H.; Ali, I.; Ali, M.A.; Al-Zahrani, S.M. High Density Polyethylene/Micro Calcium Carbonate Com-posites: A Study of the Morphological, Thermal, and Viscoelastic Properties. *J. Appl. Polym. Sci.* **2010**, *117*, 2413–2421. [CrossRef]
50. Araujo, J.; Waldman, W.; De Paoli, M.-A. Thermal properties of high density polyethylene composites with natural fibres: Coupling agent effect. *Polym. Degrad. Stab.* **2008**, *93*, 1770–1775. [CrossRef]
51. Cheng, D.; Zan, Y.; Du, J.; Luo, Y. Recycling of Ultrahigh Molecular Weight Polyethylene Waste Used for Pre-paring High Performance Synthetic Paper. *J. Appl. Polym. Sci.* **2016**, *133*, 44159–44165. [CrossRef]
52. Albano, C.; Cataño, L.; Figuera, L.; Perera, R.; Karam, A.; González, G.; Noris, K. Evaluation of a composite based on high-density polyethylene filled with surface-treated hydroxyapatite. *Polym. Bull.* **2009**, *62*, 45–55. [CrossRef]
53. Tasdemir, M.; Ersoy, S. The Mechanical and Morphological Properties of HDPE Composites Filled with SiO_2, ZnO, $Mg(OH)_2$ and $CaCO_3$ Nano Powder. *Rom. J. Mat. RRM* **2013**, *43*, 417–424.
54. Bulatović, V.O.; Mihaljević, A.; Bajsić, E.G. Mechanical and Interfacial Properties of Compatibilized Polyurethane Blends. *Polym. Eng. Sci.* **2018**, *58*, 1911–1922. [CrossRef]
55. Abuoudah, C.K.; Greish, Y.E.; Abu-Jdayil, B.; El-said, E.M.; Iqbal, M.Z. Graphene/Polypropylene Nanocomposites with Improved Thermal and Mechanical Properties. *J. Appl. Polym. Sci.* **2021**, *138*, 50024. [CrossRef]
56. Bula, K.; Jesionowski, T. Effect of Polyethylene Functionalization on Mechanical Properties and Morphology of PE/SiO_2 Composites. *Compos. Interfaces* **2010**, *17*, 603–614. [CrossRef]
57. Rodrigue, D.; Kavianiboroujeni, A.; Cloutier, A. Determination of the optimum coupling agent content for composites based on hemp and high density polyethylene. *AIP Conf. Proc.* **2017**, *1914*, 030003. [CrossRef]
58. Tanpichai, S.; Witayakran, S.; Srimarut, Y.; Woraprayote, W.; Malila, Y. Porosity, Density and Mechanical Properties of the Paper of Steam Exploded Bamboo Microfibers Controlled by Nanofibrillated Cellulose. *J. Mater. Res. Technol.* **2019**, *8*, 3612–3622. [CrossRef]
59. Mendes, L.C.; Cestari, S.P. Printability of HDPE/Natural Fiber Composites with High Content of Cellulosic Industrial Waste. *Mater. Sci. Appl.* **2011**, *2*, 1331. [CrossRef]
60. Testing Machines, Inc. A Practical Means to Measure Surface Treatment Levels of PE Film Using PGX+, a New Portable Contact Angle Instrument. Available online: www.testingmachines.com (accessed on 21 February 2020).

polymers

MDPI

Article

Synthesis and Properties of Thermally Self-Healing PET Based Linear Polyurethane Containing Diels–Alder Bonds

Minghui Xu [1,2,*], Ning Liu [2], Hongchang Mo [2], Xianming Lu [2], Jinkang Dou [2] and Bojun Tan [2]

[1] State Key Laboratory of Fluorine & Nitrogen Chemicals, Xi'an 710065, China
[2] Xi'an Modern Chemistry Research Institute, Xi'an 710065, China
* Correspondence: mhuixu@163.com

Abstract: A Diels–Alder (DA) bond containing poly(tetrahydrofuran)-co-(ethyleneoxide) (PET) based linear polyurethane (PET-DA-PU) was synthesized via a prepolymer process using PET as raw material, DA diol as chain extender agent, and toluene-2,4-diisocyanate (TDI) as coupling agent. The structure of PET-DA-PU was characterized by attenuated total reflectance-Fourier transform-infrared spectroscopy (ATR-FTIR), proton nuclear magnetic resonance spectrometry (^{1}H NMR) and carbon nuclear magnetic resonance spectrometry (^{13}C NMR). The thermal performance and self-healing behavior of PET-DA-PU were investigated by differential scanning calorimetry (DSC), polarized optical microscope, universal testing machine, scanning electron microscopy (SEM) and NMR, respectively. The glass transition temperature of PET-DA-PU was found to be −59 °C. Under the heat treatment at 100 °C, the crack on PET-DA-PU film completely disappeared in 9 min, and the self-healing efficiency that was determined by the recovery of the largest tensile strength after being damaged and healed at 100 °C for 20 min can reach 89.1%. SEM images revealed the micro-cracks along with the blocky aggregated hard segments which were the important reasons for fracture. NMR spectroscopy indicated that the efficiency of retro DA reaction of PET-DA-PU was 70% after 20 min heating treatment at 100 °C. Moreover, the PET-DA-PU/Al/Na_2SO_4 composite was also prepared to simulate propellant formulation and investigated by universal testing machine and SEM; its healing efficiency was up to 87.8% under the same heat treatment process and exhibits good self-healing ability. Therefore, PET-DA-PU may serve as a promising thermally self-healing polymeric binder for future propellant formulations.

Keywords: thermally self-healing; PET; linear polyurethane; Diels–Alder reaction

Citation: Xu, M.; Liu, N.; Mo, H.; Lu, X.; Dou, J.; Tan, B. Synthesis and Properties of Thermally Self-Healing PET Based Linear Polyurethane Containing Diels–Alder Bonds. *Polymers* **2022**, *14*, 3334. https://doi.org/10.3390/polym14163334

Academic Editor: Edina Rusen

Received: 11 July 2022
Accepted: 13 August 2022
Published: 16 August 2022

Publisher's Note: MDPI stays neutral with regard to jurisdictional claims in published maps and institutional affiliations.

Copyright: © 2022 by the authors. Licensee MDPI, Basel, Switzerland. This article is an open access article distributed under the terms and conditions of the Creative Commons Attribution (CC BY) license (https://creativecommons.org/licenses/by/4.0/).

1. Introduction

PET (poly(tetrahydrofuran)-co-(ethyleneoxide)) has attracted considerable attention in recent investigations aimed at developing advanced propellant formulations, owing to its excellent mechanical properties and strong compatibility with energetic materials and nitrate plasticizers [1–3]. Therefore, PET was widely used as a polymeric binder in propellant formulations to achieve higher energetic performance and superior mechanical behavior [4–7]. However, micro-cracks frequently appeared when the propellant suffered from being continuously exposed to the external environment, e.g., mechanical attack, chemical abrasion, and UV radiation [8,9]. These cracks, which are difficult to detect and repair, can adversely affect final properties of the propellant formulations, and even worse, the micro-cracks may develop further into irreversible damage and significantly shorten the service life of propellant formulations [10–12]. Smart polymeric binders with self-healing capability are highly desirable for practical application; however, this demand has not been met yet.

In order to produce polymeric materials with more durable security and reliable properties, various strategies were exploited from renewable resources, including encapsulation, hollow fibers, microvascular networks, supramolecular self-assembly, reversible chemistry, dynamic covalent bonds, etc. [13–18]. Among them, the Diels–Alder (DA) dynamic covalent bond between furan and maleimide (which needs neither additional healing agent nor catalyst) has garnered increasing attention for synthesizing heat activated healable systems [19,20]. The reason is that the maleimide group shows a relatively high reactivity in the DA reaction, due to the more electron-deficient C=C bond, and the formed DA cycloaddition also could be easily cleaved at 90~150 °C. Hence, the self-healing polymeric materials based on DA reaction have high self-healing efficiency and a repeating ability to heal the damage at the same position.

In the last decades, a variety of self-healing polymers based on DA bonds were reported, and the majority of the research was focused on cross-linked networks, polymer gels and thermosetting polymers [21–23]. Typically, the formation of thermally reversible networks based on furan–maleimide DA reaction with maleimide and excess furan were prepared to achieve a satisfying healing efficiency. However, researchers discovered that sufficiently molecular mass polyurethanes were difficult to obtain, because the forward DA reaction gives rise to both endo and exo stereoisomer adducts and does not play a significant role in these macromolecular syntheses, since both participate in the chain growth. Therefore, few teams also concentrated their attention on DA bonds containing linear polyurethane based on furan–maleimide [24–26].

In this paper, PET based linear polyurethane with DA bonds (PET-DA-PU) was synthesized using PET as the raw material, DA diol as chain extender agent, and TDI as coupling agent, via a prepolymer process. The chemical structure and thermal performance of polyurethane were characterized by attenuated total reflectance-Fourier transform-infrared spectroscopy (ATR-FTIR), nuclear magnetic resonance spectrometry (NMR) and differential scanning calorimetry (DSC). The self-healing behaviors of as-synthesized PET-DA-PU were demonstrated by polarized optical microscope (POM), universal testing machine, scanning electron microscopy (SEM) and NMR. Moreover, the PET-DA-PU/Al/Na_2SO_4 composite was also prepared to simulate the propellant formulation and evaluated by universal testing machine and SEM.

2. Experimental

2.1. Materials

PET with M_n of 4000 g mol^{-1} and hydroxyl value of 23.1 mg KOH/g was provided from the Liming Chemical Engineering Research and Design Institute (Luoyang). Dibutyltindilaurate (DBTDL), furfuryl alcohol and 1,1′-(methylenedi-1,4-phenylene) bismaleimide (BMI, 98%) were purchased from Aldrich and used as received. Toluene-2,4-diisocyanate (TDI), N,N-dimethylformamide (DMF), sodium sulfate (Na_2SO_4), chloroform, 1,2-dichloroethane and ethanol were supplied by China National Medicines. Aluminium power with the diameter of 20 μm was self-made. All solvents for the reactions were analytical grade and were dried before use.

2.2. Synthesis of DA Diol

An amount of 2,2′-(methylenebis(4,1-phenylene))bis(4-(hydroxymethyl)-3a,4,7,7a-tetrahydro-1H-4,7-epoxyisoindole-1,3(2H)-dione) (DA diol) was synthesized via cycloaddition reaction using frufuryl alcohol and 1,1′-(methylenedi-1,4-phenylene) bismaleimide as the raw materials, as shown in Scheme 1. Briefly, 1,1′-(methylenedi-1,4-phenylene) bismaleimide (20 g, 55.8 mmol) was dissolved in chloroform (100 mL) in a 250 mL round-bottomed Schlenk flask with a magnetic stirring bar; the reaction solution was heated up to reflux until a uniform system, and then the freshly distilled frufruyl alcohol (11.3 g, 115.2 mmol) was added dropwise into the reaction system. The reaction was allowed to proceed for 30 h. The solution was then poured into 8 fold excess of isopropanol to precipitate the crude product. The precipitation process recrystallization in isopropanol/DMF mixed solvent three times, and the DA diol was obtained after drying under vacuum at 60 °C as a pale yellow solid (22.3 g, yield: 72%).

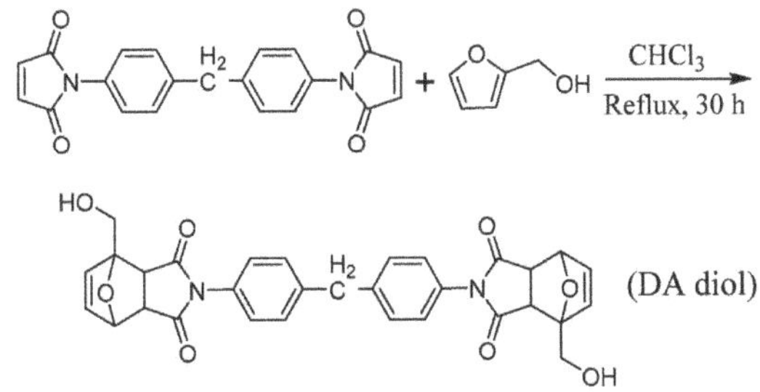

Scheme 1. Synthesis route of DA diol.

2.3. Synthesis of PET-DA-PU

PET based polyurethane with self-healing properties based on Diels–Alder reaction, PET-DA-PU, was synthesized via the successive prepolymerization of PET using TDI as the curing agent, and DA diol as the chain extender, which is illustrated in Scheme 2. In a typical synthesis example, exactly anhydrous PET (20 g, 5.0 mmol) and freshly distilled 1,2-dichloroethane (30 mL) were placed in a 250 mL three-neck round-bottomed Schlenk flask with a mechanical stirrer, condenser and thermometer under N_2 atmosphere, and then heated to 60 °C; TDI (3.48 g, 20 mmol) and DBTDL (20 μL) dissolved in 20 mL 1,2-dichloroethane were added drop by drop into the reaction solution. After stirring for 2 h, DA diol (8.32 g, 15 mmol) was added into the reaction mixture, which was then stirred at 80 °C for an additional 3 h. After the polymerization reaction was finished, the reaction system was precipitated by adding to excess ethanol under vigorous stirring. The product PET-DA-PU was finally obtained after drying under vacuum at 40 °C. (30.97 g, yield: 97.4%).

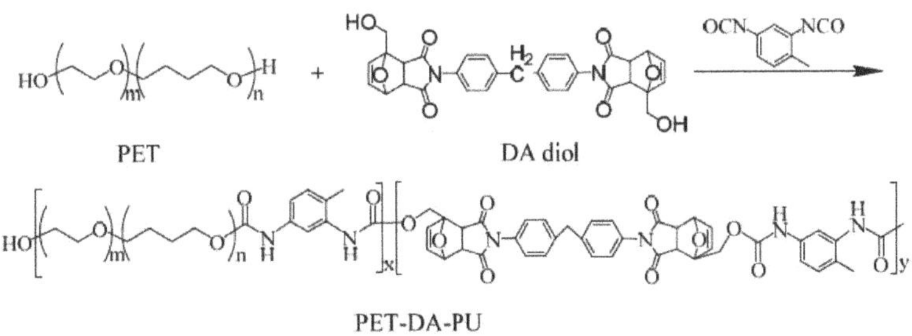

Scheme 2. Synthetic route and chemical structure of PET-DA-PU.

2.4. Synthesis of PET-DA-PU/Al/Na_2SO_4 Composite

The PET-DA-PU/Al/Na_2SO_4 composite was synthesized with a weight ratio of 1/0.5/0.5. Briefly, 20 g of PET-DA-PU was dissolved in ethyl acetate (30 mL) in a 200 mL beaker, Al power (10 g) and Na_2SO_4 (10 g) were added into the solution and the mixture was then left under intensive stirring for 1 h. The mixture was then poured into the Teflon-mold, and the solvent in mixture was removed under vacuum. Finally, the PET-DA-PU/Al/Na_2SO_4 composite was obtained after being dried in vacuum at 40 °C for 24 h.

2.5. Self-Healing Property

The self-healing property of PET-DA-PU was evaluated by three means: (1) The healing process of PET-DA-PU was observed under POM equipped with temperature programmed heating stage. Briefly, 'X' crack with 0.5 mm in depth of PET-DA-PU films was cut with a thin knife, while a sample with a crack was put on the heating stage at 100 °C. The photographs were taken at different time intervals during the healing process.

(2) Evaluating the self-healing efficiency PET-DA-PU films quantitatively by tensile test. In a typical example, the well-prepared PET-DA-PU film was cut off, and heated at 100 °C for 20 min, then cooled down to 60 °C and kept for 48 h to obtain the healed PET-DA-PU film. The tensile properties of pristine and healed films were tested on a universal testing machine, and the self-healing efficiency was calculated by the recovery of the largest tensile strength. (3) NMR spectroscopy was employed to study the thermal reversibility of PET-DA-PU and investigate the degradation rate of DA bonds in PET-DA-PU under heating treatment.

2.6. Characterization

Chemical structure: the chemical structure of PET-DA-PU was examined by ATR-FTIR and NMR. ATR-FTIR was performed on a Bruker Tensor 27 instrument (KBr pellet) in the range 4000–400 cm^{-1}. NMR spectra were performed on a Bruker 500 MHz instrument with $CDCl_3$ as the solvent at 25 °C and tetramethylsilane as the internal standard. Thermal property: the thermal property of PET-DA-PU was investigated by differential scanning calorimetry (DSC). DSC was performed in a TA Instruments Q1000 equipment in the range of −100~0 °C at a heating rate of 10 °C min^{-1} under a nitrogen flow. POM (Zeiss Axioskop 2 plus) with a heating stage was employed to observe the evolution of cracks on PET-DA-PU film. Mechanical property: AG-X Plus universal testing machine (Shimadzu, Kyoto, Japan) was used to measure the tensile property of PET-DA-PU films in accordance with GB/T528-1998 with a loading rate of 500 mm min^{-1}. The specimens were cut into dumbbell-like with the sizes of 20 mm (neck length) × 4 mm (width) × 2 mm (thickness) and kept at 0% humidity for 7 days before measurement. Samples were conducted independently in quintuplicate, and the results were presented with the average values. Fracture morphology of the films which were prepared by freeze-fractured (films were frozen in the liquid nitrogen and ruptured) and tensile-fractured were observed on a Tescan Vega 3 LMU scanning electron microscope (Tescan, Brno, Czech Republic).

3. Result and Discussion

3.1. Preparation and Characterization of PET-DA-PU

PET-DA-PU was synthesized via a prepolymer process of PET, using DA diol which prepared though cycloaddition reaction as chain extender, and TDI as a curing agent, as demonstrated in Schemes 1 and 2. The chemical structure of as-synthesized PET-DA-PU was first investigated by ATR-FTIR (as shown in Figure 1). In the ATR-FTIR of PET-DA-PU, the shoulder peaks at 1775 and 1513 cm^{-1} specific were due to the characteristic absorption of DA ring on PET-DA-PU [27]. The peak at 1107 cm^{-1} is ascribed to the characteristic absorption of C–O from PET. As compared to PET, the intensity decreases in –OH peak (3490 cm^{-1}) of PET and the absence of adsorption peak at around 2270 cm^{-1} indicated that there were no residual hydroxyl and isocyanate groups in PET-DA-PU. It is worth noting the additional peaks at 3295, 1714 and 1540 cm^{-1} which were due to the –NH stretching vibration, –C=O band and –N–H stretching bands of the urethane group on PET-DA-PU, respectively [28]. Therefore, it could be confirmed that PET based polyurethane containing DA bonds had been synthesized successfully.

The chemical structure of as-synthesized PET-DA-PU was also characterized by ^1H NMR and ^{13}C NMR (as shown in Figure 2). The peaks at 1.61, 3.46 and 3.62 ppm were assigned to the methylene protons of PET as shown in Figure 2a, and the corresponding carbon atom signals appearing at 26.6 and 70.3 ppm in Figure 2b. Meanwhile, the peaks at 3.04, 3.15, 5.15 and 6.54 ppm due to DA ring on the PET-DA-PU. The signals at 1.25 and 7–8 ppm were attributed to the methyl protons and hydrogen protons on the benzene rings of TDI, respectively, and the corresponding carbon signals appeared at 16.9 ppm and 127–135 ppm [29]. All of the results above confirmed that PET-DA-PU has been synthesized successfully via the prepolymerization of PET.

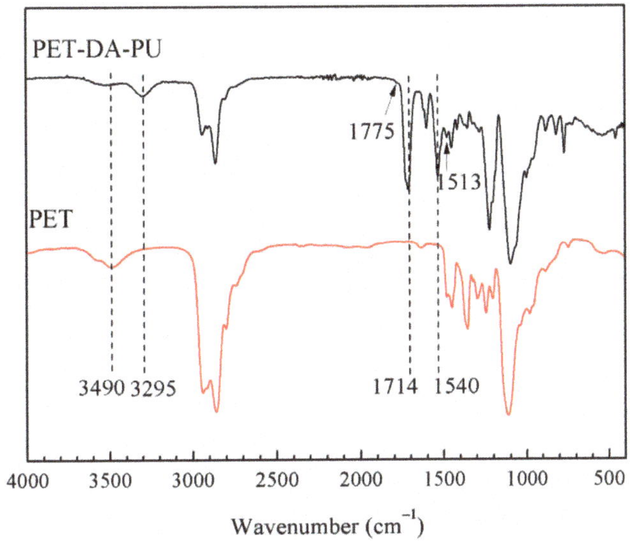

Figure 1. ATR-FTIR spectra of PET-DA-PU and PET.

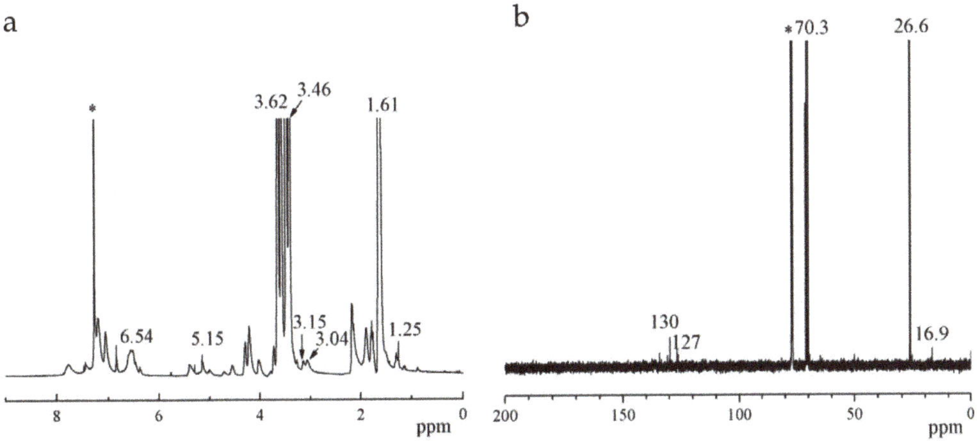

Figure 2. ^{1}H NMR (**a**) and ^{13}C NMR (**b**) spectra of PET-DA-PU. ('*' for $CDCl_3$).

3.2. Glass Transition Temperature of PET-DA-PU

Glass transition temperature (T_g) is an important property of polymeric binders as it determines the application temperature range. In this study, DSC was used to investigate the T_g of PET-DA-PU, and the T_g was measured by the cooling DSC curves of the PET-DA-PU. As shown in Figure 3, in comparison of PET prepolymer (T_g = −84 °C), the PET-DA-PU showed a slightly higher T_g of −59 °C, which allows it to work in a low-temperature environment. It is due to the decrease in chain flexibility of PET as to the introduction of DA diol and TDI, and the formation of urethane group [30]. Moreover, PET-DA-PU exhibits only one T_g, which may be attribute to the fact that PET and TDI-extended DA diol segments are well-dispersed and blended within the films, thus the polyurethane consists of one phase. The same phenomenon was also observed in other literature [31].

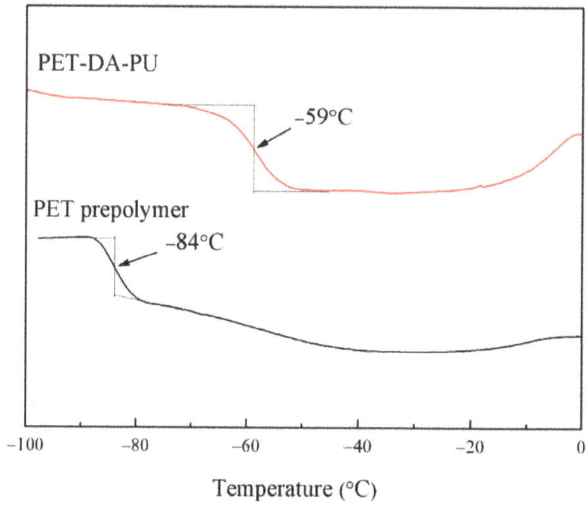

Figure 3. DSC curves of PET-DA-PU and PET prepolymer.

3.3. Self-Healing Property of PET-DA-PU

The self-healing property of PET-DA-PU under heat treatment was firstly investigated by observing the healing process with POM, as shown in Figure 4 [32]. The PET-DA-PU film sample with 1 mm thickness was prepared, and an 'X' crack with 0.5 mm in depth was produced by a thin knife on the film. The PET-DA-PU film sample was put on the heating stage and heated for 100 °C and observed by POM.

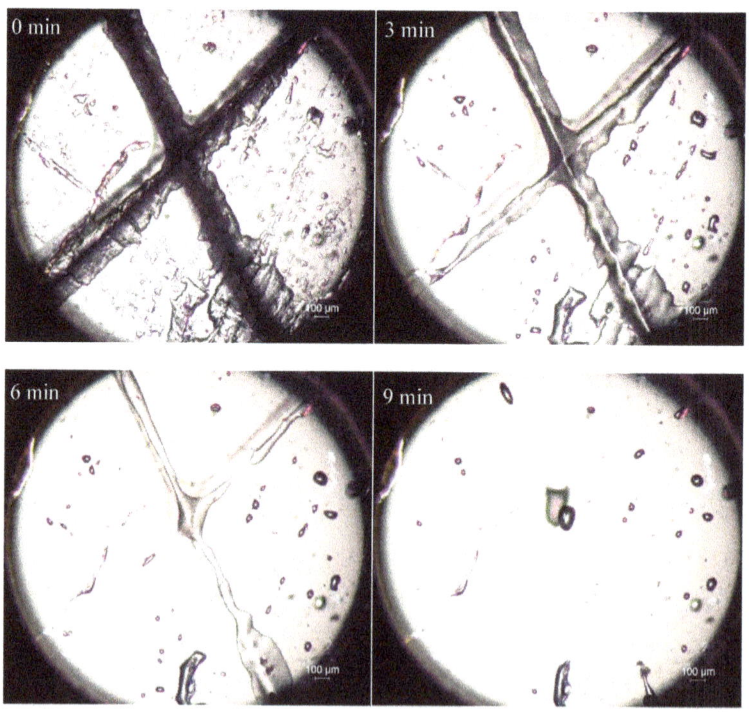

Figure 4. The healing process of PET-DA-PU film upon heat treatment at 100 °C.

As shown in Figure 4, the width of the crack in PET-DA-PU film decreased clearly with the heat treatment time increasing, and the crack on the film had completely disappeared in 9 min [33]. The results indicated good self-healing properties of PET-DA-PU.

The healing efficiency of the PET-DA-PU was measured quantitatively by the universal testing machine [34]. The PET-DA-PU film samples were prepared and tested according to GB/T528-1998. The PET-DA-PU film was cut off by a thin knife, and the film was heated at 100 °C for 20 min, then cooled down to 60 °C and kept for 48 h, and the healed PET-DA-PU film was obtained. As shown in Figure 5, compared with the PET-DA-PU (the tensile strength was 0.92 MPa, with an elongation at 151.3%), the tensile strength of repaired PET-DA-PU was 0.82 MPa with an elongation at 137.5%. Thus, the healing efficiency of PET-DA-PU was 89.1%. The results reveal that the PET-DA-PU has a satisfactory healing efficiency.

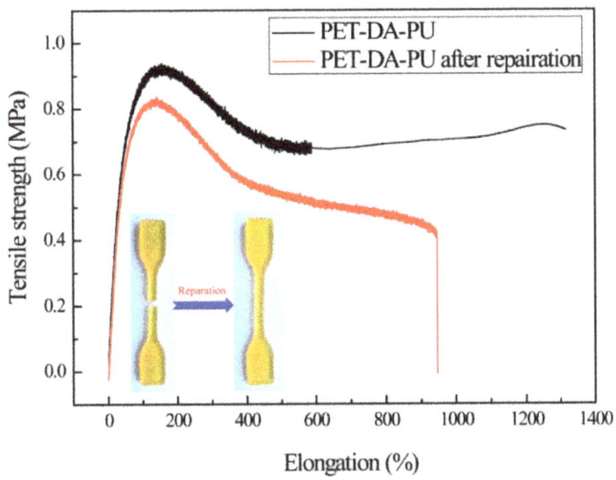

Figure 5. Mechanical property of PET-DA-PU and its damaged and healed samples.

To investigate the rupture mechanism of PET-DA-PU in the tension process, the fracture morphologies of the PET-DA-PU films which were prepared by freeze-fractured and tensile-fractured were studied by SEM. As shown in Figure 6a,b, the wrinkles are marked by red restangular frames representative of aggregated hard segments, and the ravines are marked by blue circle frames representative of non-homogeneous soft segments. Clearly, the phase separation exists between the soft and hard segments and is well dispersed within the whole vision [35–38]. However, the hard segments (marked as red rectangular frames) in tension-fractured PET-DA-PU film display aggregated as blocky shapes in Figure 6c,d, and the micro-cracks along with the blocky aggregated hard segments which were the important reasons for fracture were also observed [39]. These results revealed the fracture of hard segments and the formation of micro-cracks resulting in the fatigue fracture of PET-DA-PU film.

3.4. Thermal Reversibility of PET-DA-PU

To study the thermal reversibility of PET-DA-PU, NMR was employed to investigate the changes in ^{1}H NMR spectroscopy after the heat treatment in Figure 7 [40]. It is well known that BMI and furan prepolymer formed in the retro-DA reaction of PET-DA-PU under heat treatment. As marked in Figure 7, the chemical shift at 3.17 ppm (labeled as a), 4.93 and 4.96 ppm (labeled as b), 6.54 ppm (labeled as c), 6.63ppm (labeled as d), 5.24 ppm (labeled as e), 7.14 ppm (labeled as g), 7.35 ppm (labeled as f) decreased visibly, and the new chemical shifts at 7.25 ppm (labeled as a'), 7.28 ppm (labeled as f'), 7.15 ppm (labeled as g') owing to BMI, and chemical shifts at 7.69 ppm (labeled as b'), 6.46 ppm (labeled as c'),

6.54 ppm (labeled as d′), 5.09 ppm (labeled as e′) owing to furan prepolymer appeared after heat treatment [41–44]. It is indicated that DA bonds in PET-DA-PU de-bonded under heat treatment.

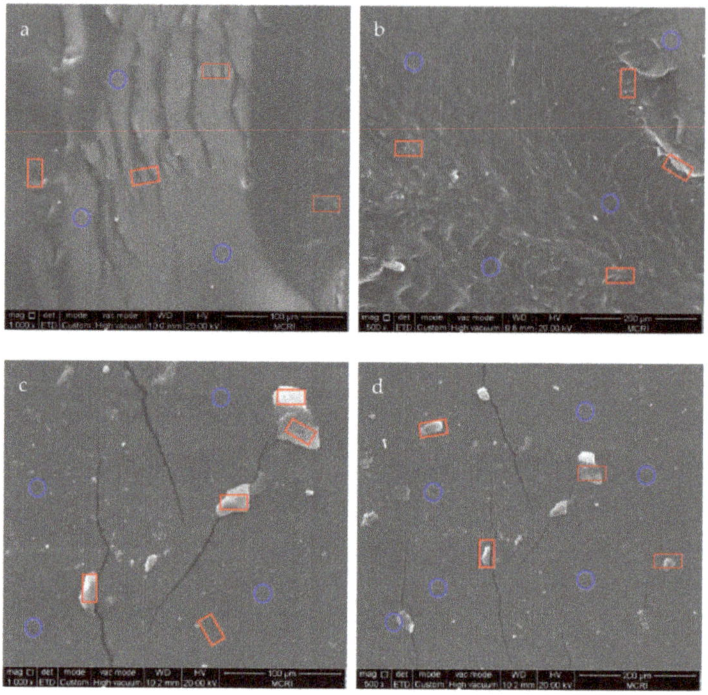

Figure 6. SEM images of freeze-fractured (**a**,**b**) and tension-fracture (**c**,**d**) surface of the PET-DA-PU gels. (wrinkles were marked as red square frame, and ravines were marked as blue circle frame).

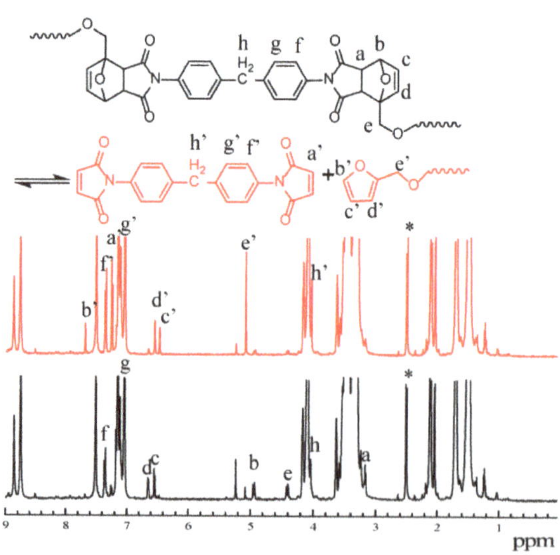

Figure 7. ^{1}H NMR spectra of PET-DA-PU during the heating procedure at 100 °C for 20 min. (PET-DA-PU was marked as black, and PET-DA-PU after heat treatment was marked as red, '*' for DMSO-d_6).

To investigate the efficiency of retro DA reaction under heat treatment quantitatively, the changes in ^{1}H NMR spectroscopy of DA diol were also studied [45]. As shown in Figure 8, the chemical shifts at 5.18 ppm (labeled as a), 4.08 and 3.77 ppm (labeled as b), 6.57 ppm (labeled as c), 5.0 ppm (labeled as d), 3.02 and 3.18 ppm (labeled as e), 7.35 ppm (labeled as f), 7.14 ppm (labeled as g) owing to DA diol were significantly decreased after heat treatment, and the new chemical shifts at 5.19 ppm (labeled as a′), 4.38 ppm (labeled as b′), 6.27 and 6.38 ppm (labeled as c′), 7.57 ppm (labeled as d′) owing to frufuryl alcohol, and 7.36 ppm (labeled as e′), 7.26 ppm (labeled as f′), 7.15 ppm (labeled as g′), 4.03 ppm (labeled as h′) owing to BMI were observed. Based on the changes in the ^{1}H NMR spectroscopy peak areas of DA diol after heat treatment, conversion x can be calculated from the following formula:

$$x = \frac{A_{c'} + A_{c''}}{A_c + A_{c'} + A_{c''}} \tag{1}$$

A_c, $A_{c'}$, $A_{c''}$ integral areas of peaks c, c′, c″, respectively.

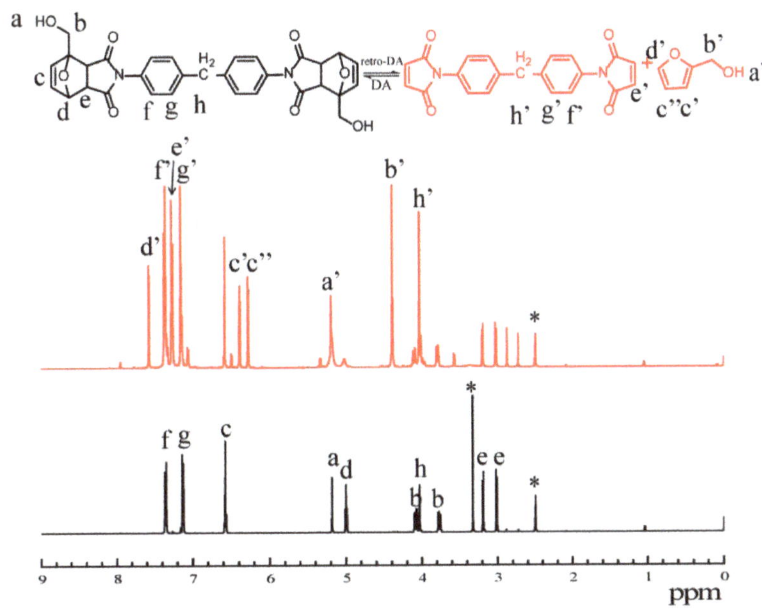

Figure 8. ^{1}H NMR spectra of DA diol during the heating procedure at 100 °C for 20 min. (DA diol was marked as black, and DA diol after heat treatment was marked as red, '*' for DMSO-d_6).

According to the formula, the conversion of DA diol was 70% after 20 min heating treatment at 100 °C. It demonstrated that DA bonds in PET-DA-PU can degrade into BMI and furan prepolymer in a short time via retro DA reaction, and it is good benefit for repairing of PET-DA-PU. All of these results indicate that PET-DA-PU has a good thermal reversibility.

3.5. Healing Behavior of PET-DA-PU Based Composites

To evaluate the healing efficiency of PET-DA-PU in propellant, the PET-DA-PU/Al/Na_2SO_4 composite was prepared and measured by the universal testing machine. The PET-DA-PU/Al/Na_2SO_4 composite was prepared in a ratio of 1/0.5/0.5, and the obtained composite film was cut into dumbbell-shaped specimens and tested according to GB/T528-1998. The PET-DA-PU/Al/Na_2SO_4 composite film was cut off and heated at 100 °C for 20 min, then cooled down to 60 °C and kept for 48 h and the healed PET-DA-PU/Al/Na_2SO_4 composite films were obtained. As shown in Figure 9 and Table 1, in comparison with the PET-DA-PU/Al/Na_2SO_4 composite (the tensile strength was 0.82 MPa, with an elongation at 138.8%), the tensile strength of repaired PET-DA-PU/Al/Na_2SO_4 composite was 0.72 MPa with an elongation at 113%. Hence, the healing efficiency of PET-DA-PU/Al/Na_2SO_4 composite was 87.8%, which was slightly lower than PET-DA-PU. It is due to the reduction in DA bond density in composite as to the introduction of Al and Na_2SO_4 and decrease the healing efficiency of the PET-DA-PU/Al/Na_2SO_4 composite. However, the PET-DA-PU/Al/Na_2SO_4 composite still has a high healing efficiency exceeding 85% and exhibits outstanding self-healing performance.

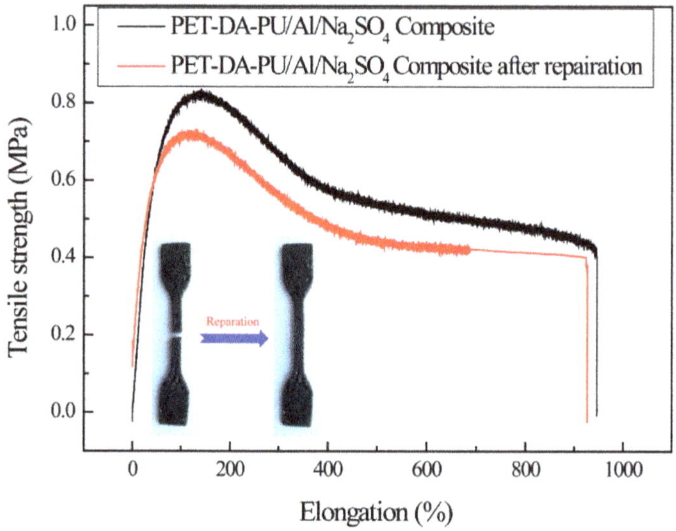

Figure 9. Mechanical property of PET-DA-PU/Al/Na_2SO_4 composite and its damaged and healed samples.

Table 1. Tensile testing of PET-DA-PU/Al/Na_2SO_4 composites before and after reparation.

Sample	E Modulus/MPa	Max Tensile Strength/MPa	Elongation at Max Stress/%	Elongation at Break/%
PET-DA-PU/Al/Na_2SO_4 composite	1.72	0.82	138.8	945.8
PET-DA-PU/Al/Na_2SO_4 composite after repairation	1.56	0.72	113	926.2

The fracture morphologies of the PET-DA-PU/Al/Na_2SO_4 composite which were prepared by freeze-fracturing and tensile-fracturing were also observed by SEM. As shown in Figure 10a,b, the crack surface of the composite was smooth and the interface between the exposed particles and polymeric matrix was fuzzy, meaning that the interface adhesion property between solid filler and matrix is strong. However, a number of exposed particles and cracks appeared in Figure 10c,d, and voids were found between the exposed particles and polymeric matrix, meaning that the interface adhesion property between solid fillers and PET-DA-PU matrix became poor during the tensile process [46]. Generally, the cracks and voids also occurred in aging stages of polymeric binder. Therefore, the PET-DA-PU polymeric binder which can repair the micro-cracks within the matrix under heat treatment could prolong its service life.

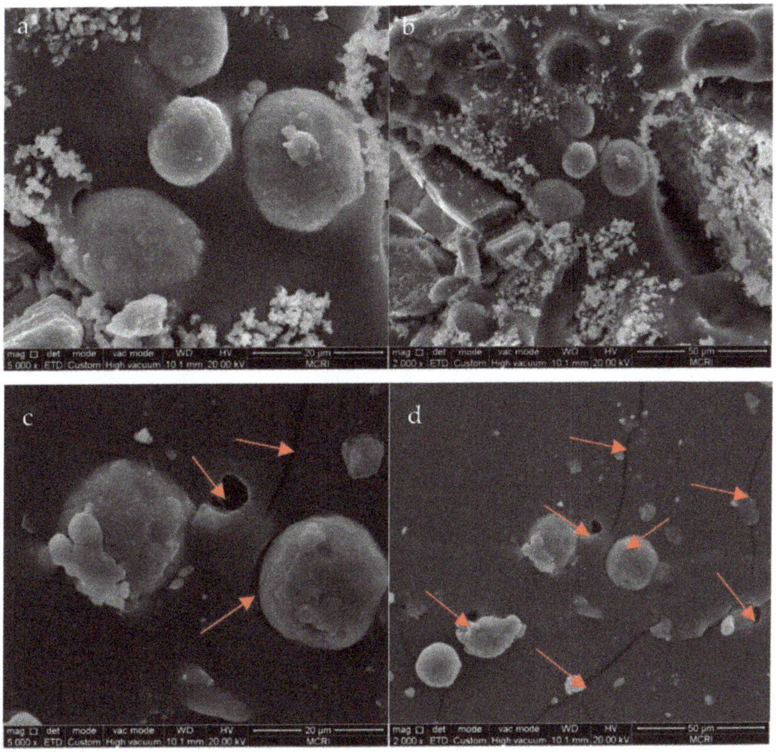

Figure 10. SEM images of freeze-fractured (**a**,**b**) and tension-fracture (**c**,**d**) surface of the PET-DA-PU/Al/Na_2SO_4 composite. (the cracks and voids were marked as red arrows).

4. Conclusions

A Diels–Alder bond containing PET based linear polyurethane, PET-DA-PU, was synthesized using PET as raw material, DA diol as chain extender agent, and TDI as coupling agent. From ATR-FTIR, ^{1}H NMR and ^{13}C NMR results, the PET-DA-PU was synthesized successfully via prepolymer process. The DSC curves indicated that PET-DA-PU had a low glass transition temperature of −59 °C, which allows it to work in a low temperature environment. The evolution of cracks on PET-DA-PU was observed by POM, and the results indicated that the cracks on PET-DA-PU film completely disappeared in 9 min under the heat treatment at 100 °C. A tensile test was used to determine the self-healing efficiency by the recovery of the largest tensile strength after being damaged and healed; the self-healing efficiency of PET-DA-PU can reach 89.1% after 20 min heating treatment at 100 °C. NMR spectroscopy indicated that the efficiency of retro DA reaction of PET-DA-PU can reach up to 70% after 20 min heating treatment at 100 °C. SEM images were used to investigate the fracture morphologies of the PET-DA-PU film, and the results revealed the micro-cracks along with the blocky aggregated hard segments which were the important reasons for fracture. Moreover, the PET-DA-PU/Al/Na_2SO_4 composite was also prepared to simulate the propellant formulation, and its healing efficiency was 87.8% under the same heat treatment. In the SEM images of tensile-fractured PET-DA-PU/Al/Na_2SO_4 composite, exposed particles, cracks and voids were observed, meaning poor interface adhesion property between solid fillers and PET-DA-PU matrix. Consequently, PET may find its application as a novel self-healing binder in propellant formulations.

Author Contributions: Conceptualization, M.X. and N.L.; methodology, H.M.; software, X.L.; validation, N.L., X.L., J.D. and B.T.; formal analysis, J.D.; investigation, M.X.; resources, B.T.; data curation, N.L.; writing—original draft preparation, M.X.; writing—review and editing, N.L. and X.L.; visualization, J.D. and B.T.; supervision, H.M.; project administration, N.L. All authors have read and agreed to the published version of the manuscript.

Funding: This research received no external funding.

Institutional Review Board Statement: Not applicable.

Informed Consent Statement: Not applicable.

Data Availability Statement: The data presented in this study are available on request from the corresponding author.

Conflicts of Interest: The authors declare no conflict of interest.

References

1. Zhang, Q.; Liu, N.; Mo, H.; Lu, X.; Wang, Y.; Xu, M.; Shu, Y. Facile preparation and properties of crosslinked copolyether elastomers with 1,2,3-triazole and urethane subunit via click polymerization. *Chem. Open* **2019**, *8*, 571–579. [CrossRef] [PubMed]
2. Gong, L.; Li, Y.; Guo, Y.; Li, J.; Yang, R. Effect of morphology for ammonium dinitramide on the mechanical and combustion properties of composite propargyl-terminated copolyether propellant. *Propellants Exp. Pyro.* **2020**, *45*, 864–870. [CrossRef]
3. Pal, Y.; Mahottamananda, S.N.; Palateerdham, S.K.; Subha, S.; Ingenito, A. Review on the regression rate-improvement techniques and mechanical performance of hybrid rocket fuels. *Fire Phys. Chem.* **2021**, *1*, 272–282. [CrossRef]
4. Awad, S.A.; Khalaf, E.M. Improvement of the chemical, thermal, mechanical and morphological properties of polyethylene terephthalate–graphene particle composites. *Bull. Mater. Sci.* **2018**, *41*, 67–72. [CrossRef]
5. Monti, M.; Scrivani, M.T.; Kociolek, I.; Larsen, Å.G.; Olafsen, K.; Lambertini, V. Enhanced impact strength of recycled PET/glass fiber composites. *Polymers* **2021**, *13*, 1471. [CrossRef]
6. Mondadori, N.; Nunes, R.; Canto, L.; Zattera, A. Composites of recycled PET reinforced with short glass fiber. *J. Thermoplast. Compos. Mater.* **2012**, *25*, 747–764. [CrossRef]
7. Cunha, B.; Mattos, E.; Rocco, J. Influence of storage time before casting on viscosities and mechanical properties of AP/HTPB/Al solid composite propellants. *Fire Phys. Chem.* **2022**, *2*, 50–55. [CrossRef]
8. Junling, L.; Hua, F.; Duowang, T.; Fangyun, L.; Rong, C. Fracture behaviour investigation into a polymer-bonded explosive. *Strain* **2012**, *48*, 463–473. [CrossRef]
9. Liu, C.; Thompson, D.G. Crack Initiation and Growth in PBX 9502 High Explosive Subject to Compression. *J. Appl. Mech. Trans. ASME* **2014**, *81*, 101004. [CrossRef]
10. Murphy, E.B.; Wudl, F. The world of smart healable materials. *Prog. Polym. Sci.* **2010**, *35*, 223–251. [CrossRef]
11. Xia, D.; Wang, P.; Ji, X.; Khashab, N.M.; Sessler, J.L.; Huang, F. Functional supramolecular polymeric networks: The marriage of covalent polymers and macrocycle-based host-guest interactions. *Chem. Rev.* **2020**, *120*, 6070–6123. [CrossRef] [PubMed]
12. Sengezer, E.C.; Seidel, G.D. Real time In-situ sensing of damage evolution in nanocomposite bonded surrogate energetic materials. In *Behavior and Mechanics of Multifunctional Materials and Composites, Proceedings of the SPIE Smart Structures and Materials + Nondestructive Evaluation and Health Monitoring, Las Vegas, NV, USA, 20–24 March 2016*; SPIE: Bellingham, WA, USA, 2016; Volume 9800, pp. 1–10.
13. Scheiner, M.; Dickens, T.J.; Okoli, O. Progress towards self-healing polymers for composite structural applications. *Polymer* **2016**, *83*, 260–282. [CrossRef]
14. Zhu, D.; Rong, M.; Zhang, M. Self-healing polymeric materials based on microencapsulated healing agents: From design to preparation. *Prog. Polym. Sci.* **2015**, *49–50*, 175–220. [CrossRef]
15. Aida, T.; Meijer, E.W.; Stupp, S.I. Functional Supramolecular Polymers. *Science* **2012**, *335*, 813–817. [CrossRef] [PubMed]
16. Huang, Z.; Chen, X.; O'Neill, S.J.K.; Wu, G.; Whitaker, D.J.; Li, J.; McCune, J.A.; Scherman, O.A. Highly compressible glass-like supramolecular polymer networks. *Nat. Mater.* **2022**, *21*, 103. [CrossRef] [PubMed]
17. Wang, S.; Urban, M.W. Self-healing polymers. *Nat. Rev. Mater.* **2020**, *5*, 562–583. [CrossRef]
18. Chakma, P.; Konkolewicz, D. Dynamic covalent bonds in polymeric materials. *Angew. Chem.-Int. Ed.* **2019**, *58*, 9682–9695. [CrossRef]
19. Liu, Y.; Chuo, T. Self-healing polymers based on thermally reversible Diels-Alder chemistry. *Polym. Chem.* **2013**, *4*, 2194–2205. [CrossRef]
20. Mangialetto, J.; Gorissen, K.; Vermeersch, L.; Van Mele, B.; Van den Brande, N.; De Vleeschouwer, F. Hydrogen-bond-assisted Diels-Alder kinetics or self-healing in reversible polymer networks? A combined experimental and theoretical study. *Molecules* **2022**, *27*, 1961. [CrossRef]
21. Krishnakumar, B.; Sanka, R.V.S.P.; Binder, W.H.; Parthasarthy, V.; Rana, S.; Karak, N. Vitrimers: Associative dynamic covalent adaptive networks in thermoset polymers. *Chem. Eng. J.* **2020**, *385*, 123820. [CrossRef]

22. Lu, C.; Guo, X.; Wang, C.; Wang, J.; Chu, F. Integration of metal-free ATRP and Diels-Alder reaction toward sustainable and recyclable cellulose-based thermoset elastomers. *Carbohydr. Polym.* **2020**, *242*, 116404. [CrossRef] [PubMed]
23. Kim, S.Y.; Lee, T.H.; Park, Y.I.; Nam, J.H.; Noh, S.M.; Cheong, I.W.; Kim, J.C. Influence of material properties on scratch-healing performance of polyacrylate-graft-polyurethane network that undergo thermally reversible crosslinking. *Polymer* **2017**, *128*, 135–146. [CrossRef]
24. Liu, X.; Du, P.; Liu, L.; Zheng, Z.; Wang, X.; Joncheray, T.; Zhang, Y. Kinetic study of Diels-Alder reaction involving in maleimide-furan compounds and linear polyurethane. *Polym. Bull.* **2013**, *70*, 2319–2335. [CrossRef]
25. Li, Y.; Li, G.; Li, J.; Luo, Y. Preparation and properties of semi-interpenetrating networks combined by thermoplastic polyurethane and a thermosetting elastomer. *New J. Chem.* **2018**, *42*, 3087–3096. [CrossRef]
26. Du, P.; Liu, X.; Zheng, Z.; Wang, X.; Joncheray, T.; Zhang, Y. Synthesis and characterization of linear self-healing polyurethane based on thermally reversible Diels-Alder reaction. *Rsc. Adv.* **2013**, *3*, 15475–15482. [CrossRef]
27. Buonerba, A.; Speranza, V.; Capacchione, C.; Milione, S.; Grassi, A. Improvement of tensile properties, self-healing and recycle of thermoset styrene/2-vinylfuran copolymers via thermal triggered rearrangement of covalent crosslink. *Eur. Polym. J.* **2018**, *99*, 368–377. [CrossRef]
28. Sugane, K.; Yoshioka, Y.; Shimasaki, T.; Teramoto, N.; Shibata, M. Self-healing 8-armed star-shaped epsilon-caprolactone oligomers dually crosslinked by the Diels-Alder and urethanization reactions. *Polymer* **2018**, *144*, 92–102. [CrossRef]
29. Xu, M.; Lu, X.; Liu, N.; Zhang, Q.; Mo, H.; Ge, Z. Fluoropolymer/Glycidyl Azide Polymer (GAP) Block Copolyurethane as New Energetic Binders: Synthesis, Mechanical Properties, and Thermal Performance. *Polymers* **2021**, *13*, 2706. [CrossRef]
30. Xu, M.; Yang, W.; Lu, X.; Mo, H.; Ge, Z. Cryogenic Mechanical Properties of GAP/PET Block Thermoplastic Elastomers. *Chin. J. Explos. Propellants* **2022**, *45*, 67–72.
31. Zhang, W.; Li, J.; Luo, Y. Morphology and properties of novel thermoplastic polyurethane elastomer. *Chin. J. Explos. Propellants* **2010**, *6*, 39–42.
32. Lee, H.-Y.; Cha, S.-H. Enhancement of self-healing property by introducing ethylene glycol group into thermally reversible Diels-Alder reaction based self-healable materials. *Macromol. Res.* **2017**, *25*, 640–647. [CrossRef]
33. Jung, S.; Liu, J.T.; Hong, S.H.; Arunbabu, D.; Noh, S.M.; Oh, J.K. A new reactive polymethacrylate bearing pendant furfuryl groups: Synthesis, thermoreversible reactions, and self-healing. *Polymer* **2017**, *109*, 58–65. [CrossRef]
34. Appuhamillage, G.A.; Reagan, J.C.; Khorsandi, S.; Davidson, J.R.; Voit, W.; Smaldone, R.A. 3D printed remendable polylactic acid blends with uniform mechanical strength enabled by a dynamic Diels-Alder reaction. *Polym. Chem.* **2017**, *8*, 2087–2092. [CrossRef]
35. Lin, C.; Sheng, D.; Liu, X.; Xu, S.; Ji, F.; Dong, L.; Zhou, Y.; Yang, Y. NIR induced self-healing electrical conductivity polyurethane/graphene nanocomposites based on Diels-Alder reaction. *Polymer* **2018**, *140*, 150–157. [CrossRef]
36. Zhao, X.; Guo, S.; Li, H.; Liu, J.; Su, C.; Song, H. One-pot synthesis of self-healable and recyclable ionogels based on polyamidoamine (PAMAM) dendrimers via Schiff base reaction. *RSC Adv.* **2017**, *7*, 38765–38772. [CrossRef]
37. Lee, W.J.; Cha, S.H. Improvement of mechanical and self-healing properties for polymethacrylate derivatives containing maleimide modified graphene oxide. *Polymers* **2020**, *12*, 603. [CrossRef]
38. Zirnstein, B.; Schulze, D.; Schartel, B. Mechanical and fire properties of multicomponent flame retardant EPDM rubbers using aluminum trihydroxide, ammonium polyphosphate, and polyaniline. *Materials* **2019**, *12*, 1932. [CrossRef]
39. Xiao, C.; Zhu, Y.; Chen, J.; Zhang, S. Synthesis of emulsion-templated macroporous materials via Diels-Alder polymerization. *Polymer* **2017**, *110*, 74–79. [CrossRef]
40. Stirn, Z.; Rucigaj, A.; Krajnc, M. Innovative approach using aminomaleimide for unlocking phenolic diversity in high-performance maleimidobenzoxazine resins. *Polymer* **2017**, *120*, 129–140. [CrossRef]
41. Zhong, Y.; Wang, X.; Zheng, Z.; Du, P. Polyether-maleimide-based crosslinked self- healing polyurethane with Diels-Alder bonds. *J. Appl. Polym. Sci.* **2015**, *132*, 41944–41952. [CrossRef]
42. Luo, K.; Li, J.; Duan, G.; Wang, Y.; Yu, J.; Zhu, J.; Hu, Z. Comb-shaped aromatic polyamide cross-linked by Diels-Alder chemistry: Towards recyclable and high-performance thermosets. *Polymer* **2018**, *142*, 33–42. [CrossRef]
43. Sannikov, O.; Hanson, S.; Saunders, P.; Branch, K.L.; Merbouh, N. Introducing complex NMR mixtures at the undergraduate level: Analysis of the diels-alder reaction between methylcyclopentadiene and maleic anhydride (Part I). *J. Lab. Chem. Educ.* **2019**, *7*, 8–18.
44. Moura, O.; Castro, T.; Franca, D.; Tavares, O.; Silva, C.; Pereira, M.; Grarcia, A. Spectroscopic techniques combined with chemometrics to study organic matter in tropical soils with different degrees of pedogenetic evolution. *Química. Nova.* **2021**, *45*, 40–47. [CrossRef]
45. Nasresfahani, A.; Zelisko, P.M. Synthesis of a self-healing siloxane-based elastomer cross-linked via a furan-modified polyhedral oligomeric silsesquioxane: Investigation of a thermally reversible silicon-based cross-link. *Polym. Chem.* **2017**, *8*, 2942–2952. [CrossRef]
46. Bode, S.; Enke, M.; Hernandez, M.; Bose, R.K.; Grande, A.M.; van der Zwaag, S.; Schubert, U.S.; Garcia, S.J.; Hager, M.D. Characterization of self-healing polymers: From macroscopic healing tests to the molecular mechanism. *Self Health Mater.* **2016**, *273*, 113–142.

polymers
MDPI

Article

Spatiotemporal Temperature Distribution of NIR Irradiated Polypyrrole Nanoparticles and Effects of pH

Omar Peñuñuri-Miranda [1], Miguel Olivas-Martinez [1], José Alberto Ibarra-Espinoza [1], Rosalva Josefina Rodríguez-Córdova [1,2], Karol Yesenia Hernández-Giottonini [1,2], Daniel Fernández-Quiroz [1], Paul Zavala-Rivera [1,2] and Armando Lucero-Acuña [1,2,*]

[1] Department of Chemical and Metallurgical Engineering, University of Sonora, Hermosillo 83000, Mexico; a212205033@unison.mx (O.P.-M.); miguel.olivas@unison.mx (M.O.-M.); a217201103@unison.mx (J.A.I.-E.); rosalva.rodriguez@unison.mx (R.J.R.-C.); yesenia.hgio@hotmail.com (K.Y.H.-G.); daniel.fernandez@unison.mx (D.F.-Q.); paul.zavala@unison.mx (P.Z.-R.)
[2] Nanotechnology Graduate Program, Department of Physics, University of Sonora, Hermosillo 83000, Mexico
* Correspondence: armando.lucero@unison.mx; Tel.: +52-662-259-2105

Abstract: The spatiotemporal temperature distributions of NIR irradiated polypyrrole nanoparticles (PPN) were evaluated by varying PPN concentrations and the pH of suspensions. The PPN were synthesized by oxidative chemical polymerization, resulting in a hydrodynamic diameter of 98 ± 2 nm, which is maintained in the pH range of 4.2–10; while the zeta potential is significantly affected, decreasing from 20 ± 2 mV to -5 ± 1 mV at the same pH range. The temperature profiles of PPN suspensions were obtained using a NIR laser beam (1.5 W centered at 808 nm). These results were analyzed with a three-dimensional predictive unsteady-state heat transfer model that considers heat conduction, photothermal heating from laser irradiation, and heat generation due to the water absorption. The temperature profiles of PPN under laser irradiation are concentration-dependent, while the pH increase only induces a slight reduction in the temperature profiles. The model predicts a value of photothermal transduction efficiency (η) of 0.68 for the PPN. Furthermore, a linear dependency was found for the overall heat transfer coefficient (U) and η with the suspension temperature and pH, respectively. Finally, the model developed in this work could help identify the exposure time and concentration doses for different tissues and cells (pH-dependent) in photothermal applications.

Keywords: polypyrrole nanoparticles; NIR laser irradiation; photothermal modeling; photothermal transduction efficiency; overall heat transfer coefficient

Citation: Peñuñuri-Miranda, O.; Olivas-Martinez, M.; Ibarra-Espinoza, J.A.; Rodríguez-Córdova, R.J.; Hernández-Giottonini, K.Y.; Fernández-Quiroz, D.; Zavala-Rivera, P.; Lucero-Acuña, A. Spatiotemporal Temperature Distribution of NIR Irradiated Polypyrrole Nanoparticles and Effects of pH. *Polymers* **2022**, *14*, 3151. https://doi.org/10.3390/polym14153151

Academic Editor: Edina Rusen

Received: 19 July 2022
Accepted: 29 July 2022
Published: 2 August 2022

Publisher's Note: MDPI stays neutral with regard to jurisdictional claims in published maps and institutional affiliations.

Copyright: © 2022 by the authors. Licensee MDPI, Basel, Switzerland. This article is an open access article distributed under the terms and conditions of the Creative Commons Attribution (CC BY) license (https://creativecommons.org/licenses/by/4.0/).

1. Introduction

Nanoparticles based on conductive polymers have emerged as a promising material for applications in targeted photothermal medical therapies [1,2]. In this therapy, nanoparticles could be placed at different tissue-penetration depths and convert irradiated energy from the near-infrared (NIR) region into thermal energy [3,4]. Several conductive polymers, such as polyaniline, polypyrrole, etc., could be used for photothermal therapies due to their biocompatibility and excellent photothermal conversion performance [5–8]. Within these conductive polymers, polypyrrole is one of the most widely used for developing nanomaterials for biomedical applications [9]. Furthermore, polypyrrole-based nanostructures are being studied because of their strong absorption in the NIR region (λ = 700–1200 nm) [2]. For instance, polypyrrole nanoparticles (PPN) could generate heat suitable to ablate cancer cells through their effective photon-to-thermal energy transfer [10–12]. Moreover, PPN could be used singly or combined with chemotherapy agents [11].

Furthermore, PPN shows high specificity and excellent anti-cancer effects in cellular and animal experiments [13]. The elevated temperature could thus target cancer cells while avoiding significant side effects on normal cells. The latter is based on the higher heat tolerance of normal cells compared to cancer cells [14]. However, the photothermal application of polypyrrole-based materials is still in its early stages. Further research is required to understand and improve its potential performance under diverse in vitro or in vivo scenarios. In this context, one of the most common PPN preparation methods is oxidative chemical polymerization, consisting of a facile one-step aqueous dispersion polymerization that could be used with poly(vinyl alcohol) as stabilizing agent [15]. This method allows, up to some degree, the tuning of PPN properties by changing the stabilizing agent, varying the stabilizing agent concentration, and adjusting the molar ratios of the oxidative agent and the pyrrole monomer [16].

The modeling of heat dissipation from suspended nanoparticles irradiated at different wavelengths has been reported previously [17–19]. Roper et al. determined the efficiency of transducing resonant continuous-wave irradiation to bulk heat by a gold NP suspension by analyzing thermal energy transfer in a thermally isolated system with a macroscopic linearized heat transfer model that incorporates radiation and conduction [20]. The calculation of the photothermal transduction efficiency requires theoretical analysis of the experimental data [20–23]. Huang et al. analyzed the temperature distribution with a two-dimensional model for either fluid or tissue containing gold nanorods, based on the Pennes' bioheat equation, with an additional term to account for energy released by the gold nanorods [24].

In this work, the chemical oxidative polymerization technique was used to prepare PPN, and their properties at different pH values were studied. Furthermore, the spatiotemporal temperature distributions of NIR-irradiated PPN were experimentally evaluated under different particle concentrations and pH values. An unsteady-state three-dimensional heat transfer model that describes the temperature behavior of the suspensions by considering conduction, photothermal heating from laser irradiation, and heat generation due to the water absorption was developed. The model was validated with experimentally obtained temperature measurements of PPN suspensions irradiated in the NIR region. This mathematical approach allows the prediction of the nanoparticle spatiotemporal temperature distributions and could help improve the doses of bioactive NPs and times of light irradiation at different pH values to reach the desired temperatures in photothermal therapies.

2. Materials and Methods

2.1. Materials

Pyrrole (reagent grade, 98%), iron (III) chloride hexahydrate (reagent grade, 97%, $FeCl_3 \cdot 6H_2O$), and polyvinyl alcohol (86.7–88.7% hydrolysis, Mw~31,000 a.m.u, PVA) were purchased from Sigma Aldrich, Inc., St. Louis, MO, USA. The pyrrole monomer was purified by distillation under vacuum and inert atmosphere conditions before use.

2.2. Synthesis of PPN

PPN were synthesized by the oxidative chemical polymerization of pyrrole using ferric chloride as oxidant and PVA as stabilizer [15,25,26]. A schematic of the synthesis of PPN is presented in Figure 1. A total of 20 mL of an aqueous solution of 7.5% of PVA (% *w/v*) was kept under magnetic stirring at 500 pm, at 5 °C (Figure 1A). Then, 4.6 mmol of ferric chloride ($[FeCl_3]/[\text{monomer}] = 2.3$) was added to a PVA solution. The mixture reacts for 1 h to reach equilibrium, forming the PVA/iron cation complex (orange solution) (Figure 1B). Then, 2 mmol of pyrrole monomer was added to the PVA/iron cation complex solution. The solution turned black-green, indicating that polymerization started (Figure 1C), and the polymerization process was allowed to react for 4 h with continuous stirring and at 5 °C. Then, PPN was washed by three centrifugation cycles at 22,096× *g* (14,500 rpm) for 35 min, where supernatant was discarded, and the precipitate was resuspended in deionized water. Finally, PPN was collected and stored at 4 °C for further use. All experiments were performed in triplicate.

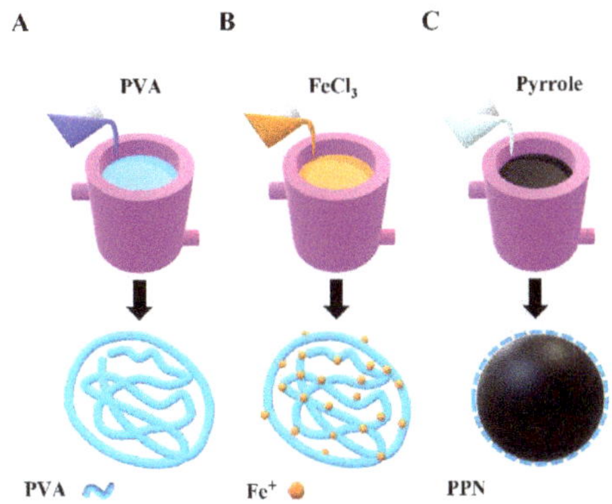

Figure 1. PPN synthesis scheme by chemical oxidation of pyrrole. (**A**) PVA solution at 5 °C; (**B**) formation of PVA/iron complex, and (**C**) polymerization of pyrrole and formation of PPN.

2.3. *pH Effects over PPN*

The polymerization of pyrrole with $FeCl_3$ and PVA results in acidic PPN suspensions. Accordingly, the pH values of the PPN suspensions must be tuned to be used in biological applications. Therefore, it is necessary to evaluate the influence of pH on the properties of the PPN. The pH values studied were 4.2 (pH obtained from synthesis), 6, 7.4, 8, and 10; adjusted using solutions of 0.1 M NaOH.

2.4. *Characterization of PPN*

Hydrodynamic size, polydispersity index (PDI), and zeta potential (ζ) were measured using a Zetasizer (Zetasizer Nano ZS Malvern Instruments Ltd., Worcestershire, UK). Hydrodynamic size measurements of PPN were performed by the Dynamic Light Scattering technique (DLS). A refractive index of 1.33 and water as a dispersant were used in the analysis. Each sample was measured three times with at least 10 runs at 25 °C for the size and ζ analysis. The ζ was obtained by using the laser Doppler electrophoresis technique. The morphology of PPN was observed with a field transmission electron microscopy (TEM) JEOLTEM-2010F with an operating voltage of 200 kV (JEOL, Ltd., Tokyo, Japan). For TEM observations, a drop of the sample solution was spread on a smooth film of a carbon-coated copper TEM grid and dried. After that, 1% phosphotungstic acid was applied on the grid for 1 min; the excess was absorbed using filter paper. All samples were subsequently dried in a vacuum before observation. The optical absorption spectra of PPN were acquired in a UV6300PC spectrophotometer (VWR, Radnor, PA, USA). The characteristic absorption band of PPN in the NIR region was verified. A standard curve of PPN suspensions was built at 808 nm to determine the PPN concentration. All experiments were performed in triplicate.

2.5. *Photothermal Evaluation of PPN*

The temperature variations of PPN suspensions irradiated with an NIR laser beam with an output power of 1.5 W centered at 808 nm (Changchun New Industries Optoelectronics Tech Co., Ltd., Model No. PSU-III-LED, Changchun, China) were measured. PPN suspensions were placed into a rectangular quartz cuvette cell and irradiated with the NIR laser. The experimental setup for the analysis and data acquisition is shown in Figure 2. The temperature variations were measured using a K-Type thermocouple connected to a multimeter that acted as a data acquisitor. Suspensions of PPN from 0 to 30 µg/mL and the pH of nanoparticle suspensions from 4.2 to 10 were analyzed. As a control, the temperature of a nonirradiated cuvette cell with water was measured simultaneously to calculate the temperature increments of the suspensions; that is, to adjust to the ambient temperature variations.

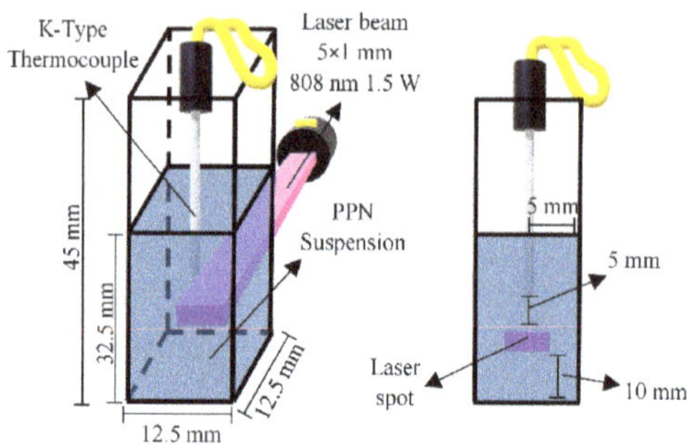

Figure 2. Scheme of heating experiments with the laser irradiation, indicating the dimensions of the quartz cuvette cell and the suspension volume (**left**), and a frontal view with the approximate position of the laser spot and the thermocouple (**right**).

2.6. Spatiotemporal Temperature Distribution Analysis

The spatiotemporal temperature distribution of irradiated PNP suspensions was mathematically analyzed by modeling PPN suspensions inside a quartz cuvette. The modeled PNP suspension volume is comprised of two regions: the laser region, where the laser irradiates and the heat is produced, then the suspension region, where the heat is transported only by heat conduction. The energy equation in the laser region is:

$$\rho C_P \frac{\partial T}{\partial t} = k\nabla^2 T + \eta \frac{I}{V}\left(1 - 10^{-OD}\right) + S_W \tag{1}$$

where the term on the left side is the thermal energy accumulation. The first term on the right-hand side is the heat conduction, and the second term represents the photothermal heating from laser irradiation. This second term $\eta \frac{I}{V}\left(1 - 10^{-OD}\right)$ consists of heat generation due to the PPN. Furthermore, the heat generation due to the water absorption is considered in the third term of the right side (S_w). The energy equation for the suspension region is

$$\rho C_P \frac{\partial T}{\partial t} = k\nabla^2 T \tag{2}$$

The equation shows the accumulation term on the left-hand side, and on the right-hand side, only the conduction term appears.

In these equations ρ (kg/m^3) is the density and C_P (J/kg·K) is the heat capacity of the suspension, respectively, k is the thermal conductivity (W/m·K), η counts for the photothermal transduction efficiency, I (W) is the laser power, V (m^3) stands for the irradiated volume, OD is the optical density of the suspension and S_W (W/m^3) is the water heat generation in the irradiated volume.

To solve Equations (1) and (2), the following boundary condition was applied to the cell walls and the air/suspension interface:

$$q = U\left(T_f - T\right) + \varepsilon\sigma\left(T_\infty^4 - T^4\right) \tag{3}$$

where q is the heat flux at the system boundaries (W/m^2), U is the overall heat transfer coefficient (W/m^2·K), T_f is the temperature of media surrounding the cell (air) (K), T is the suspension temperature (K), ε is the emissivity of either the cell wall or the air/solution interface, σ is the Stefan–Boltzmann constant, and T_∞ is the temperature of the radiation sink on the exterior of the domain (K). The overall heat transfer coefficient U incorporates the thermal resistance of the cell wall and the external free convection thermal resistance. Furthermore, the bottom of the quartz cuvette was considered adiabatic, resulting in the following boundary condition:

$$q = 0 \; at \; z = 0 \tag{4}$$

The main physical parameters used in the mathematical model were obtained from the experimental setup and the literature [27], as presented in Table 1. The model was numerically solved using Ansys Fluent R3 [20,21] using a hexahedral mesh of 85,536 elements. A semi-implicit method for pressure-linked equations was used with a time step of 0.4 s and 3750-time elements.

Table 1. Main physical parameters used in the mathematical model.

Parameter	Description	Unit	Value
ρ	Dispersion density	kg/m^3	998.2
C_P	Heat capacity at constant pressure	J/kg·K	4182
k	Thermal conductivity	W/m·K	0.6
I_{laser}	Laser power	W	1.5
V_{laser}	Irradiated volume	m^3	6.25×10^{-8}
T_f	Temperature of surrounding media (air)	K	297.15
T_∞	Temperature of the domain exterior	K	297.15
ε_{cell}	Emissivity of the cell walls	–	0.925 *
ε_{water}	Emissivity of the air/solution interface	–	0.96 *

* The emissivity of glass at 300 K is between 0.90 and 0.95; the emissivity of water at 300 K is 0.96 [27].

Determination of Model Parameters: η, U, and S_W

The values of parameters η, U and S_W were obtained by fitting the model temperature predictions to the experimental measurements during laser heating. A two-step approach was followed: first, the order of magnitude η and U was obtained by analyzing a hypothetical case of the adiabatic cell and using a heat transfer correlation, respectively. The second step consisted of minimizing the sum of squares of differences between the experimental and predicted temperature increments using the heat model.

The order of magnitude of η was determined from the maximum temperature in the irradiated system (analyzed by considering a hypothetical case of an adiabatic cell), where all the energy absorbed by the system is accumulated, resulting in the maximum possible temperature at the given laser power and time of irradiation. The value of η defined as baseline must satisfy two conditions at a specific PPN concentration. The first condition is that the temperature increment in the adiabatic case be higher than the experimental one. The second condition is that the initial heating rate be faster than the experimental heating rate (Figure S1).

The order of magnitude of U was obtained using a correlation for free convection on a vertical surface. The properties of air used were estimated from an average temperature between the temperature of the surrounding media and the temperature measured by the thermocouple in the heating plateau (Figure S2).

The second step for determining η and U was implemented as follows: a set of values of U for each concentration and a single value of η equal to or greater than the baseline values were used to solve the model. Then the model was solved using the parameters and conditions described above. Next, the sum of squares was calculated for the values following Equation (5). The process was repeated at least three times to find the values of η and U and the minimum sum of squares. The term S_W was obtained by fitting the heat model to the experimental data at a concentration of 0 µg/mL nanoparticles.

$$\sigma^2 = \sum_{i=1}^{n} \frac{\left(\Delta T_{exp} - \Delta T_{sim}\right)^2}{n-k} \quad (5)$$

2.7. Statistical Data Analysis

The statistical analysis for PPN hydrodynamic size, PDI, and ζ measurements was performed by the software OriginPro (ver. 9). Results were analyzed by ANOVA followed by Tukey's HSD test ($\alpha = 0.05$). The differences were considered statistically significant when the p-values were minor or equal to 0.05.

3. Results and Discussion

3.1. Characterization of PPN Synthesis

PPN was synthesized by chemical oxidizing polymerization with hydrodynamic average sizes of 98 ± 2 nm from three independent preparations, verifying the method's reproducibility. Furthermore, the DLS results show a narrow size distribution with a PDI of 0.04 ± 0.02 (as shown in Figure 3A). Different sizes can be obtained by varying the concentrations and molecular weight of PVA and the molar ratio between $FeCl_3$ and pyrrole monomer [16]. Furthermore, using a different stabilizer in the chemical oxidizing polymerization could influence the PPN sizes. The literature has reported polyvinylpyrrolidone (PVP) as a stabilizer, resulting in PPN sizes ranging from 151.5 to 93.5 nm in the function of the PVP concentration increment [28]. The ζ distribution of PPN that resulted from the synthesis shows a single peak with a narrow width and a positive value, as seen in Figure 3B. This value, obtained from three independent preparations of PPN, was 20.0 ± 2.1 mV, indicating good nanoparticle stability due to electrostatic particle repulsion and preventing further particle aggregation [29]. The positive charge of these PPN is attributed to the presence of amino groups in the polypyrrole polymeric chain [30]. Other authors using a similar synthesis method of PPN report ζ values of 14.0 ± 0.4 mV for particles of 106 ± 1 nm [31]. The slight differences in the ζ values may be due to the concentrations of PVA used in the synthesis since the PPN surface covered by this stabilizer changes with the concentration used in the synthesis. Then, as the PVA concentration increases in the synthesis, the PPN surface-exposed hydroxyl groups increase, producing variations in the PPN surface charge.

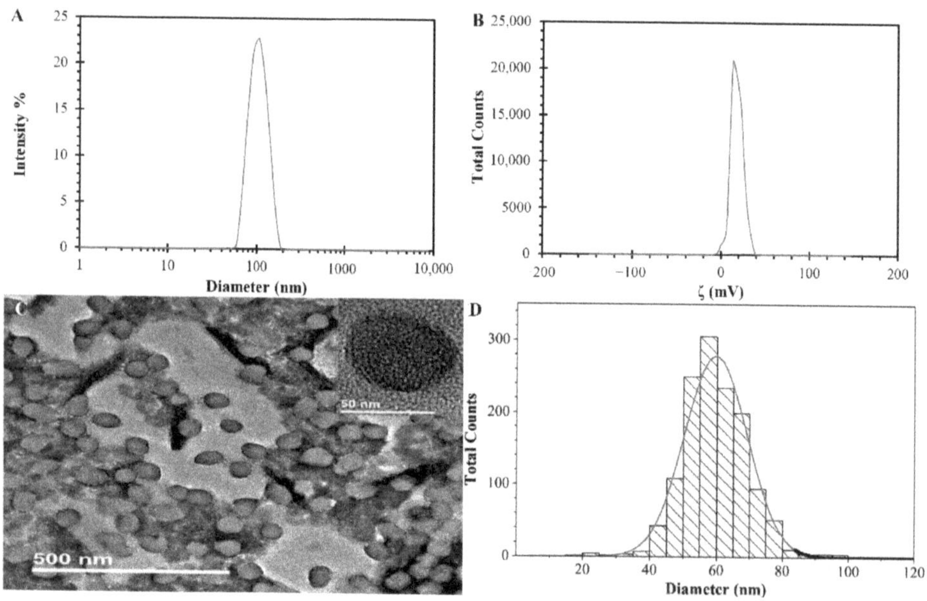

Figure 3. PPN properties were obtained from the synthesis (pH value from the synthesis of 4.2). (**A**) Diameter size distribution from DLS measurements. Means ± SD, n = 3. (**B**) ζ distribution. Means ± SD, n = 3. (**C**) TEM image of PPN with an insert of a single particle. (**D**) Histogram of the size distribution of PPN from TEM images (analysis with a number of particles >700).

The PPN morphologies and sizes can be observed in Figure 3C,D, respectively. The PPN has a uniform quasi-spherical morphology, with diameters of ~60 nm, as shown in Figure 3C and the insert. These morphologies and sizes agree with previous literature reports [32]. Furthermore, the excellent monodispersity can be confirmed by the narrow histogram distribution presented in Figure 3D, obtained from TEM pictures using the free software ImageJ version 1.53e. The difference in the sizes obtained from the DLS technique and the TEM analysis, as known, is related to the hydrodynamic size of PPN in the suspension (DLS) and the actual size of dried particles (TEM). Furthermore, in the literature, it was found that PPN synthesized with different methods showed similar amorphous structures [33,34].

The optical properties of the aqueous PPN suspensions were analyzed by UV-vis spectroscopy. The UV-vis absorption spectra were recorded at different PPN concentrations in the wavelength range of 300–1000 nm, as shown in Figure 4. PPN exhibits the typical π–π^* transition band of the polypyrrole around 420 nm, in agreement with the literature [35,36]. At the same time, the observed band above ~650 nm is attributable to the sizeable π-conjugated structure of polypyrrole chains [37]. As expected, the absorbance increased proportionally to the PPN concentration. Similarly, Zha et al. reported that the absorbance increases linearly as the concentration of PPN in water is elevated, indicating the excellent dispersity of the aqueous PPN solution [15]. Furthermore, a calibration curve was obtained from this linear behavior, with a resulting extinction coefficient (ε) of 0.0559 a.u./(μg/mL·cm) and a determination coefficient (R^2) of 0.994 (Figure S3). It should be mentioned that the absorption spectra of PPN also depend on the nature of the solvent [38].

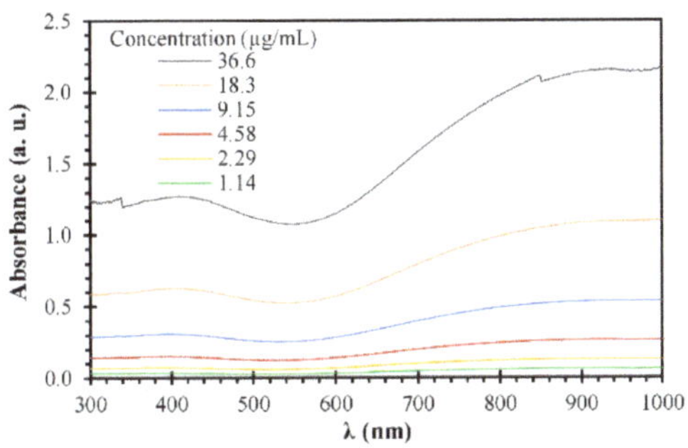

Figure 4. Optical absorption spectra of PPN suspensions at pH values of 4.2 and different concentrations of particles.

3.2. *pH Effects over PPN*

The pH effect over the hydrodynamic particle size, ζ, and optical spectra of PPN suspensions are presented in Figure 5. The pH values of PPN suspensions were: 4.2, 6.0, 7.4, 8.0, and 10.0. In this range of pH, the PPN size presents a quadratic behavior with respect to the pH of the suspension (Figure 5A). However, the size differences are not significant between the pH values except for a pH of 10, where a significant difference resulted, compared with all the other pH values. This behavior is similar to the PDI values obtained, the results of which indicate monodisperse particles (PDI < 0.1) at pH values below 10 (Figure 5A). As observed for the size results, the only significant difference in PDI was at a pH of 10, compared with all the other PDI values. The ζ results for the different pH suspensions could be described with a quadratic relationship, as presented in Figure 5B. The ζ values for suspensions with pH of 4.2 and 6.0 do not present significant variations, with values around 20 mV. However, when the pH value of the suspension increases to 7.4 and 8.0, the ζ decreases to values of 9.8 ± 0.4 and 9.3 ± 1.1 mV, correspondingly, with significant differences for the pH values of 4.2 and 6.0.

Nevertheless, the ζ differences between pH 7.4 and 8.0 are not significant. The ζ values of suspensions of PPN at pH 10 resulted in −4.7 ± 1.3 mV, with substantial differences from all the other ζ values. The changes in the ζ of PPN are related to the deprotonation process that suffers the polymeric chain of PPN as the pH rises [39,40]. This deprotonation affects the surface charge due to changes in the PPN surface composition. Other authors have reported that iron nanoparticles coated with polypyrrole change from a positive ζ

value to a negative one at a pH higher than 9 [41,42]. The significant differences observed in size and PDI measurement results could be attributed to the change in the surface charge character of the PPN, from positive to negative, indicating that some disturbance was obtained by passing throughout the isoelectric point of PPN (~pH of 9.4).

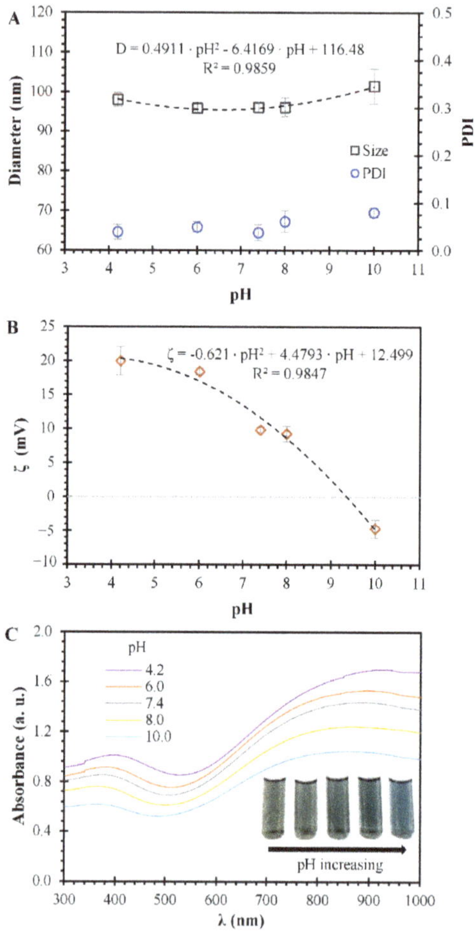

Figure 5. PPN properties at different pH values. (**A**) Diameter size distributions and PDI of PPN; Squares represent nanoparticle diameter sizes, the continuous line represents the quadrating fitting to the diameter size, and circles represent the polydispersity index. Means ± SD, $n = 3$. (**B**) ζ of PPN; squares represent nanoparticle ζ, and the continuous line represents the quadrating fitting. Means ± SD, $n = 3$. (**C**) Optical absorption spectra of PPN suspensions; insert shows the color of suspensions at different pH values, ranging from 4.2 to 10.

The optical spectra of PPN at different pH values are presented in Figure 5C, with wavelength scans ranging from 300 to 1000 nm. The characteristic optical spectra bands of PPN suspensions are similar for all the pH values. However, when the pH of the PPN suspension increases, a shift to a shorter wavelength is produced in the spectra of PPN. Furthermore, a change in the color of the PPN solutions is observed, from a black-green color in acidic conditions to a light blue in alkaline conditions (insert of Figure 5C). These wavelength shifts are related to the surface charge of the PPN, as described in Figure 5B.

As the PPN is deprotonated (pH rises), the hydroxyl groups cover the surface, and a shift to shorter wavelengths is observed.

3.3. Determination of Model Parameters: η, U, and S_W

The order of magnitude values of η (adiabatic cells) and U were 0.65 and 7 W/m^2·K, respectively. These values were used to initiate the minimization of the sum of squares of differences between the experimental and predicted temperature increments. The resulting value of η was considered constant for all the nanoparticle concentrations. This consideration is justified by the low PPN polydispersity (monodisperse particles), the homogeneous nanoparticle morphology, and the range of particle concentrations used [21,43]. The parameter U changed proportionally to the nanoparticle concentration. This effect could be explained by the coefficient U's dependence on the media's temperature after irradiation [27].

The resulting value η, from the minimization of the sum of squares, was 0.68. This value is greater than other similar systems of PPN prepared using PVP and the same oxidant [2,5] and similar to gold nanoparticles of similar sizes [44]. Other works using $FeCl_3$ as oxidant have shown remaining Fe at the PPN surface in the form of iron oxides, improving the electrical conductivity and optical properties of the particles by the formation of a semiconductor–semiconductor heterojunction [45]. For instance, these remaining oxides could also improve the η value. The values of U for different PPN concentrations in the suspension are shown in Table 2. Furthermore, a value of 4.7×10^5 (W/m^3) was fitted to S_W by simulation of the particle-free suspension. These values agree with other values reported in the literature [46,47].

Table 2. Maximum temperature increments and the corresponding values of U as a function of the PPN concentration.

PPN Concentration (μg/mL)	ΔT_{max} (K)	U (W/m^2·K)
30	27.7	24
15	24.9	22
7.5	17.8	19
3.75	11.7	17
1.875	6.1	15

3.3.1. Photothermal Effect of PPN at Different Concentrations

The photothermal effect of PPN at different concentrations under laser irradiation and the respective heat transfer model fittings are presented in Figure 6. The heating analysis as a function of PPN concentration was evaluated at a fixed pH value of 7.4 to simulate biological conditions. The concentrations of PPN evaluated were 30, 15, 7.5, 3.75, 1.875, 0 μg/mL, equivalent to the absorbances of 1.677, 0.839, 0.419, 0.210, and 0.105, respectively (Figure 6A–E). The spatiotemporal temperature distributions of irradiated PNP suspensions show the same heating profile, but different temperature increments are reached proportionally to the nanoparticle concentration increase. The profiles started with a period of rapid heating until a plateau was reached, indicating a steady state. Results show a direct dependency between the PPN concentration and the temperature increment obtained, with maximum temperature increments of 27.6, 24.8, 17.7, 11.7, and 6 °C for 30, 15, and 7.5, 3.75, and 1.875 μg/mL, respectively. Furthermore, a slight temperature increment for the 0 μg/mL solution is presented after irradiation, confirming that most of the temperature increments are due to the laser interactions with the PPN. Comparable results have been reported in the literature for the spatiotemporal temperature distributions of other irradiated materials, such as gold nanoparticles [48], gold nanorods [24], carbon nanotubes [49], iron oxide nanoparticles [50], etc.

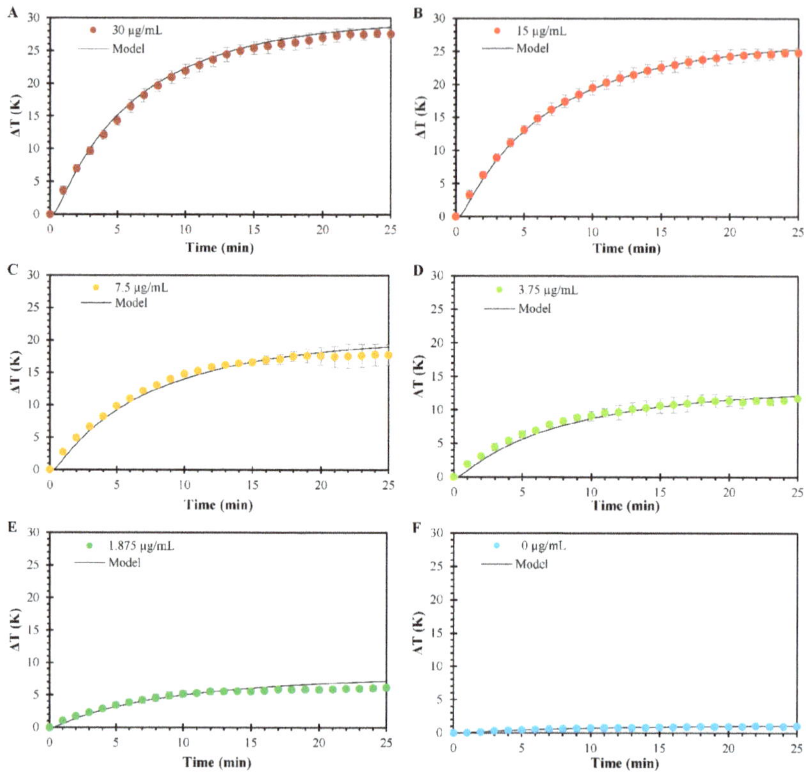

Figure 6. Temperature profiles of PPN suspensions at pH values of 7.4 and model fitting at concentrations of (**A**) 30, (**B**) 15, (**C**) 7.5, (**D**) 3.75, (**E**) 1.875, and (**F**) 0 µg/mL. Circles represent experimental data, and the continuous solid lines represent the model fitting. Means ± SD, n = 3.

The mathematical analysis for the spatiotemporal temperature distributions of irradiated PNP at different concentrations shows good agreement with the experimental data. The model predicted the temperatures with an average absolute error below 1 °C, compared to the experimental data, as shown in Figure S4. The best agreement is found at higher PPN suspension concentrations. However, with lower nanoparticle concentrations, the model underestimates at the early stages of heating, overestimating the temperature change when reaching the thermal plateau. Furthermore, a linear correlation between the overall heat transfer coefficient and temperature increment is shown in Figure 7, contrasting with the non-linear dependency predicted by the correlation used as a starting point (Figure S2).

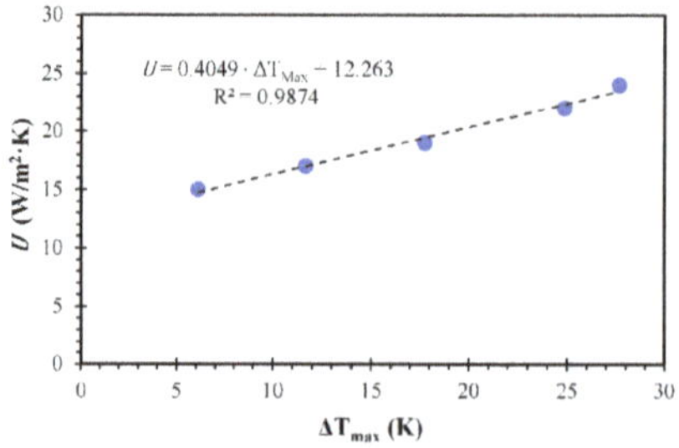

Figure 7. Estimated overall heat transfer coefficient. Circles represent the overall heat transfer coefficient, and the continuous dotted line represents a linear fitting.

3.3.2. Photothermal Effect of PPN at Different pH

The photothermal effect of PPN at different pH was performed at an average nanoparticle concentration of 28.5 μg/mL. Figure 8A–E present the measured temperature profiles and their model predictions at solution pH values of 4.2, 6.0, 7.4, 8.0, and 10, respectively. The mathematical analysis at different pH values agrees well with the experimental data. The model predicted the temperatures with an average absolute error below 1 °C compared to the experimental data (Figure S5). The heating profiles in Figure 8 present similar behavior within them, observing a gradual temperature increase until a steady temperature is reached.

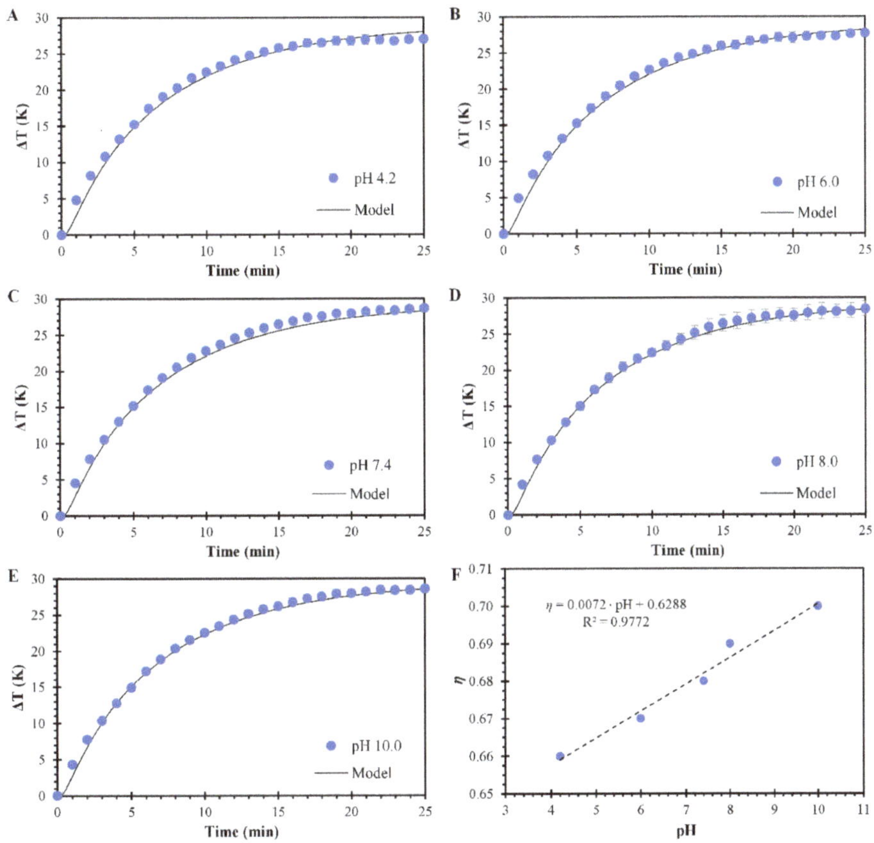

Figure 8. Laser irradiation effects of PPN suspensions as a function of the pH. Temperature profiles of PPN suspensions and model fitting at pH values of (**A**) 7.4, (**B**) 6.0, (**C**) 7.4, (**D**) 8.0, and (**E**) 10. Circles represent the experimental data of PPN suspensions, and the continuous solid lines represent model fitting. Means ± SD, n = 3. (**F**) The efficiency of transducing resonant irradiation of light to heat (η) for a PPN suspension of 28 μg/mL as a function of the pH. Circles represent the η value obtained, and the dotted line the linear adjustment.

Furthermore, a small increment in the suspensions' maximum temperature occurs when pH increases. This increment in the maximum temperature is not related to the absorbance of the suspensions but to variations in the efficiency of transducing resonant irradiation of light to heat. As a result, the model predicts that this efficiency increases linearly with the pH values of the suspension (Figure 8F). This finding is supported by the ζ value measurements presented in Figure 5B, in which a decrease in the ζ value occurs as the pH increases, indicating that the nanoparticle surface charge affects the transduction of light.

The predicted temperature contours for a concentration of PPN of 3.75 μg/mL and pH 7.4 are presented in Figure 9. At a glance, no matter which contour is selected, the laser region is the hottest spot in the contour, unlike the suspension/air interface in the suspension region, which remains almost at room temperature during the experiment. The temperature of ~51 °C is reached in the laser region, with an exposure time of 25 min. The contour analysis helps to understand and control the maximum temperature reached in the irradiated volume as a function of time, which is necessary for all biological applications to avoid undesirable effects, for example, necrosis, damage surrounding healthy tissue, and the possible effects of prolonged laser exposure.

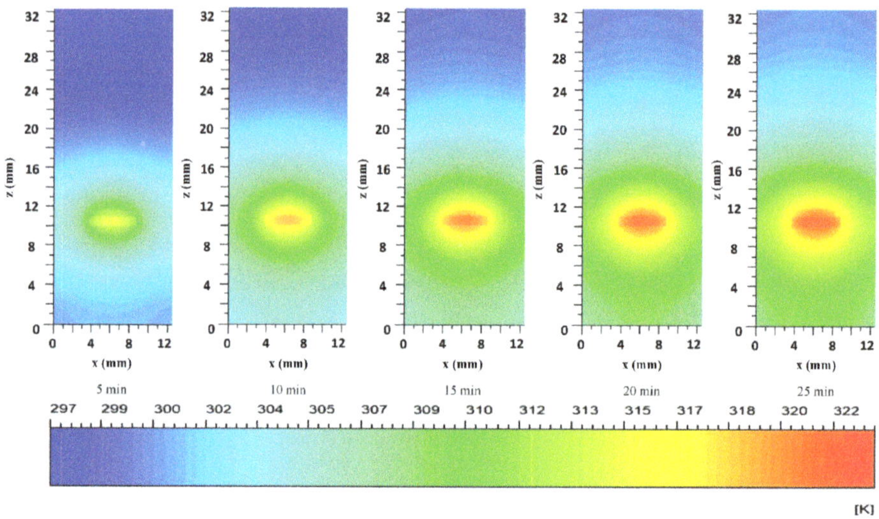

Figure 9. Computed temperature contours at different times during the suspensions heating at PPN of 3.75 μg/mL and pH of 7.4. Plane xz at y = 6.3 mm located at the midpoint of the cuvette depth. The color scale represents temperatures in K.

4. Conclusions

Polypyrrole nanoparticles (PPN) were synthesized by oxidative chemical polymerization, resulting in an appropriate size, good stability, and strong absorption in the NIR region, suitable for applications in photothermal therapies. The zeta potential (ζ) of PPN could be tuned from positive to negative values by controlling the pH of the suspensions while maintaining similar sizes. A three-dimensional mathematical model that considers the extinct radiation of suspensions of PPN and the photothermal transduction efficiency was developed. A direct approach was proposed to determine the order of magnitude of the photothermal transduction efficiency and the overall heat transfer coefficient. The heat transfer model shows good agreement between the experimental and the predicted temperature changes. Furthermore, a linear dependency of the overall heat transfer coefficient with the temperature was found. For PPN of 60 nm dispersed at a pH of 7.4, the photothermal transduction efficiency had a value of 0.68.

Additionally, a linear dependency was found between the photothermal transduction efficiency and the pH of the suspensions. The model could predict temperature zones with potential photothermal therapy use around the laser region regarding nanoparticle concentration and power. The current approach for modeling the conversion of NIR light into heat by using nanoparticle suspensions could contribute to the analysis and design of systems/devices for photothermal ablation of cells and tissues.

Supplementary Materials: The following supporting information can be downloaded at: https://www.mdpi.com/article/10.3390/polym14153151/s1, Figure S1: Temperature gradients at the thermocouple position (6.25, 6.25, 16 mm) for the adiabatic cell for different photothermal transduction efficiency (η). The experimental data correspond to a concentration of PPN of 30 μg/mL and a pH value of 7.4; Figure S2: Calculated temperature gradients for different values of overall heat transfer coefficient (U). The solid dots correspond to the experimental heating data at a concentration of PPN of 30 μg/mL and a pH value of 7.4; Figure S3: PPN calibration curve at a wavelength of 808 nm. The mass extinction coefficient (ε) of PPN is 0.0559 $a.u./\left(\frac{\mu g}{mL}\cdot cm\right)$. Mean ±SD, $n = 3$; Figure S4: Comparison of simulated temperature gradients and experimental temperature gradients for the different concentrations of PPN at a pH value of 7.4; Figure S5: Comparison of simulated temperature gradients and experimental temperature gradients for the different pH values at a concentration of PPN of 30 μg/mL.

Author Contributions: Conceptualization, A.L.-A.; methodology, A.L.-A., O.P.-M. and M.O.-M.; formal analysis, M.O.-M., J.A.I.-E. and K.Y.H.-G.; investigation, O.P.-M. and R.J.R.-C.; resources, A.L-A., M.O.-M. and P.Z.-R.; data curation R.J.R.-C. and D.F.-Q.; writing—original draft preparation, A.L.-A., O.P.-M., M.O.-M. and K.Y.H.-G.; writing—review and editing, A.L.-A., D.F.-Q. and P.Z.-R.; supervision, A.L.-A.; project administration, A.L.-A.; funding acquisition, A.L.-A., P.Z.-R., M.O.-M., R.J.R.-C. and D.F.-Q. All authors have read and agreed to the published version of the manuscript.

Funding: This research received no external funding.

Institutional Review Board Statement: Not applicable.

Informed Consent Statement: Not applicable.

Data Availability Statement: Not applicable.

Acknowledgments: Omar Peñuñuri-Miranda, Karol Yesenia Hernández-Giottonini, and Rosalva Josefina Rodríguez-Córdova thank the National Council of Science and Technology of Mexico (CONACYT) for the graduate scholarships 638,788, 302,373, and 281,669, respectively. We are grateful for the use of the transmission electron microscopy (TEM) facilities of the University of Sonora.

Conflicts of Interest: The authors declare no conflict of interest.

References

1. Zhou, J.; Lu, Z.; Zhu, X.; Wang, X.; Liao, Y.; Ma, Z.; Li, F. NIR Photothermal Therapy Using Polyaniline Nanoparticles. *Biomaterials* **2013**, *34*, 9584–9592. [CrossRef]
2. Chen, M.; Fang, X.; Tang, S.; Zheng, N. Polypyrrole Nanoparticles for High-Performance in vivo near-Infrared Photothermal Cancer Therapy. *Chem. Commun.* **2012**, *48*, 8934–8936. [CrossRef]
3. Baez-Castillo, L.; Ortiz-Rascón, E.; Carrillo-Torres, R.C.; Bruce, N.C.; Garduño-Mejia, J.; Lucero-Acuña, A.; Álvarez-Ramos, M.E. Deep Photothermal Effect Induced by Stereotactic Laser Beams in Highly Scattering Media. *Opt. Lett.* **2021**, *46*, 2–5. [CrossRef]
4. Yang, J.; Zhai, S.; Qin, H.; Yan, H.; Xing, D.; Hu, X. NIR-Controlled Morphology Transformation and Pulsatile Drug Delivery Based on Multifunctional Phototheranostic Nanoparticles for Photoacoustic Imaging-Guided Photothermal-Chemotherapy. *Biomaterials* **2018**, *176*, 1–12. [CrossRef]
5. Guo, B.; Zhao, J.; Wu, C.; Zheng, Y.; Ye, C.; Huang, M.; Wang, S. One-Pot Synthesis of Polypyrrole Nanoparticles with Tunable Photothermal Conversion and Drug Loading Capacity. *Colloids Surfaces B Biointerfaces* **2019**, *177*, 346–355. [CrossRef]
6. Balint, R.; Cassidy, N.J.; Cartmell, S.H. Conductive Polymers: Towards a Smart Biomaterial for Tissue Engineering. *Acta Biomater.* **2014**, *10*, 2341–2353. [CrossRef]
7. Liu, Y.; Yu, Q.; Chang, J.; Wu, C. Nanobiomaterials: From 0D to 3D for Tumor Therapy and Tissue Regeneration. *Nanoscale* **2019**, *11*, 13678–13708. [CrossRef]
8. Vines, J.B.; Lim, D.J.; Park, H. Contemporary Polymer-Based Nanoparticle Systems for Photothermal Therapy. *Polymers* **2018**, *10*, 1357. [CrossRef]
9. Vaitkuviene, A.; Kaseta, V.; Voronovic, J.; Ramanauskaite, G.; Biziuleviciene, G.; Ramanaviciene, A.; Ramanavicius, A. Evaluation of Cytotoxicity of Polypyrrole Nanoparticles Synthesized by Oxidative Polymerization. *J. Hazard. Mater.* **2013**, *250–251*, 167–174. [CrossRef]
10. Xiao, Z.; Xu, C.; Jiang, X.; Zhang, W.; Peng, Y.; Zou, R.; Huang, X.; Liu, Q.; Qin, Z.; Hu, J. Hydrophilic Bismuth Sulfur Nanoflower Superstructures with an Improved Photothermal Efficiency for Ablation of Cancer Cells. *Nano Res.* **2016**, *9*, 1934–1947. [CrossRef]
11. Wang, J.; Lin, F.; Chen, J.; Wang, M.; Ge, X. The Preparation, Drug Loading and in vitro NIR Photothermal-Controlled Release Behavior of Raspberry-like Hollow Polypyrrole Microspheres. *J. Mater. Chem. B* **2015**, *3*, 9186–9193. [CrossRef]
12. Song, X.; Chen, Q.; Liu, Z. Recent Advances in the Development of Organic Photothermal Nano-Agents. *Nano Res.* **2015**, *8*, 340–354. [CrossRef]

13. Yang, K.; Xu, H.; Cheng, L.; Sun, C.; Wang, J.; Liu, Z. In vitro and in vivo Near-Infrared Photothermal Therapy of Cancer Using Polypyrrole Organic Nanoparticles. *Adv. Mater.* **2012**, *24*, 5586–5592. [CrossRef]
14. Chen, J.; Ning, C.; Zhou, Z.; Yu, P.; Zhu, Y.; Tan, G.; Mao, C. Nanomaterials as Photothermal Therapeutic Agents. *Prog. Mater. Sci.* **2019**, *99*, 1–26. [CrossRef]
15. Zha, Z.; Yue, X.; Ren, Q.; Dai, Z. Uniform Polypyrrole Nanoparticles with High Photothermal Conversion Efficiency for Photothermal Ablation of Cancer Cells. *Adv. Mater.* **2013**, *25*, 777–782. [CrossRef]
16. Hong, J.Y.; Yoon, H.; Jang, J. Kinetic Study of the Formation of Polypyrrole Nanoparticles in Water-Soluble Polymer/Metal Cation Systems: A Light-Scattering Analysis. *Small* **2010**, *6*, 679–686. [CrossRef]
17. Andreozzi, A.; Brunese, L.; Iasiello, M.; Tucci, C.; Vanoli, G.P. Modeling Heat Transfer in Tumors: A Review of Thermal Therapies. *Ann. Biomed. Eng.* **2019**, *47*, 676–693. [CrossRef]
18. Singh, S.; Melnik, R. Thermal Ablation of Biological Tissues in Disease Treatment: A Review of Computational Models and Future Directions. *Electromagn. Biol. Med.* **2020**, *39*, 49–88. [CrossRef]
19. Qin, Z.; Bischof, J.C. Thermophysical and Biological Responses of Gold Nanoparticle Laser Heating. *Chem. Soc. Rev.* **2012**, *41*, 1191–1217. [CrossRef]
20. Roper, D.K.; Ahn, W.; Hoepfner, M. Microscale Heat Transfer Transduced by Surface Plasmon Resonant Gold Nanoparticles. *J. Phys. Chem. C* **2007**, *111*, 3636–3641. [CrossRef]
21. Tian, Q.; Jiang, F.; Zou, R.; Liu, Q.; Chen, Z.; Zhu, M.; Yang, S.; Wang, J.; Wang, J.; Hu, J. Hydrophilic Cu_9S_5 Nanocrystals: A Photothermal Agent with a 25.7% Heat Conversion Efficiency for Photothermal Ablation of Cancer Cells in Vivo. *ACS Nano* **2011**, *5*, 9761–9771. [CrossRef]
22. Baffou, G.; Cichos, F.; Quidant, R. Applications and Challenges of Thermoplasmonics. *Nat. Mater.* **2020**, *19*, 946–958. [CrossRef]
23. Liu, Y.; Kangas, J.; Wang, Y.; Khosla, K.; Pasek-Allen, J.; Saunders, A.; Oldenburg, S.; Bischof, J. Photothermal Conversion of Gold Nanoparticles for Uniform Pulsed Laser Warming of Vitrified Biomaterials. *Nanoscale* **2020**, *12*, 12346–12356. [CrossRef]
24. Huang, H.C.; Rege, K.; Heys, J.J. Spatiotemporal Temperature Distribution and Cancer Cell Death in Response to Extracellular Hyperthermia Induced by Gold Nanorods. *ACS Nano* **2010**, *4*, 2892–2900. [CrossRef]
25. Zayan, S.E.; El-Shazly, A.H.; El-Kady, M.F. Assessment of Polypyrrole Nanoparticles Synthesized in Presence and Absence of Surfactant for Heavy Metals Decontamination. In Proceedings of the AIP Conference Proceedings, Technologies and Materials for renewable Energy, Environment and Sustainability: TMREES19Gr, Athens, Greece, 4–6 September 2019; Haider, A.J., Jabur, A.R., Salame, C., Vokas, G., Eds.; American Institute of Physics: College Park, MD, USA; Volume 2190, p. 020025-1–020025–020026.
26. Wang, M. Emerging Multifunctional NIR Photothermal Therapy Systems Based on Polypyrrole Nanoparticles. *Polymers* **2016**, *8*, 373. [CrossRef]
27. Incropera, F.P.; DeWitt, D.P. *Fundamentos de Transferencia de Calor*, 4th ed.; Prentice Hall: Mexico, 1999; ISBN 970-17-0170-4.
28. Wen, J.; Tian, Y.; Mei, Z.; Wu, W.; Tian, Y. Synthesis of Polypyrrole Nanoparticles and Their Applications in Electrically Conductive Adhesives for Improving Conductivity. *RSC Adv.* **2017**, *7*, 53219–53225. [CrossRef]
29. Pan, H.; Marsh, J.N.; Christenson, E.T.; Soman, N.R.; Ivashyna, O.; Lanza, G.M.; Schlesinger, P.H.; Wickline, S.A. Postformulation Peptide Drug Loading of Nanostructures. In *Methods in Enzymology*, 1st ed.; Abelson, J.N., Simon, M.I., Eds.; Elsevier Inc.: San Diego, CA, USA, 2012; Volume 508, pp. 17–39. ISBN 9780123918604.
30. Samanta, D.; Meiser, J.L.; Zare, R.N. Polypyrrole Nanoparticles for Tunable, pH-Sensitive and Sustained Drug Release. *Nanoscale* **2015**, *7*, 9497–9504. [CrossRef]
31. Liu, H.; Li, W.; Cao, Y.; Guo, Y.; Kang, Y. Theranostic Nanoplatform Based on Polypyrrole Nanoparticles for Photoacoustic Imaging and Photothermal Therapy. *J. Nanopart. Res.* **2018**, *20*, 57. [CrossRef]
32. Kisiel, A.; Korol, D.; Michalska, A.; Maksymiuk, K. Polypyrrole Nanoparticles of High Electroactivity. Simple Synthesis Methods and Studies on Electrochemical Properties. *Electrochim. Acta* **2021**, *390*, 138787. [CrossRef]
33. Zhang, H.; Zhong, X.; Xu, J.; Chen, H. Fe_3O_4/Polypyrrole/Au Nanocomposites with Core/Shell/Shell Structure: Synthesis, Characterization, and Their Electrochemical Properties. *Langmuir* **2008**, *24*, 13748–13752. [CrossRef]
34. Tiwari, A.P.; Hwang, T.I.; Oh, J.M.; Maharjan, B.; Chun, S.; Kim, B.S.; Joshi, M.K.; Park, C.H.; Kim, C.S. PH/NIR-Responsive Polypyrrole-Functionalized Fibrous Localized Drug-Delivery Platform for Synergistic Cancer Therapy. *ACS Appl. Mater. Interfaces* **2018**, *10*, 20256–20270. [CrossRef]
35. Liu, Y.; Chu, Y.; Yang, L. Adjusting the Inner-Structure of Polypyrrole Nanoparticles through Microemulsion Polymerization. *Mater. Chem. Phys.* **2006**, *98*, 304–308. [CrossRef]
36. Zeng, G.; An, Y.; Fei, H.; Yuan, T.; Qing, S.; Ci, L.; Xiong, S.; Feng, J. Green and Facile Synthesis of Nanosized Polythiophene as an Organic Anode for High-Performance Potassium-Ion Battery. *Funct. Mater. Lett.* **2018**, *11*, 1–3. [CrossRef]
37. Li, X.-G.; Li, A.; Huang, M.-R.; Liao, Y.; Lu, Y.-G. Efficient and Scalable Synthesis of Pure Polypyrrole Nanoparticles Applicable for Advanced Nanocomposites and Carbon Nanoparticles. *J. Phys. Chem. C* **2010**, *114*, 19244–19255. [CrossRef]
38. Hazarika, J.; Kumar, A. Controllable Synthesis and Characterization of Polypyrrole Nanoparticles in Sodium Dodecylsulphate (SDS) Micellar Solutions. *Synth. Met.* **2013**, *175*, 155–162. [CrossRef]
39. Zhang, X.; Bai, R. Surface Electric Properties of Polypyrrole in Aqueous Solutions. *Langmuir* **2003**, *19*, 10703–10709. [CrossRef]
40. Pei, Q.; Qian, R. Protonation and Deprotonation of Polypyrrole Chain in Aqueous Solutions. *Synth. Met.* **1991**, *45*, 35–48. [CrossRef]

41. Shanehsaz, M.; Seidi, S.; Ghorbani, Y.; Shoja, S.M.R.; Rouhani, S. Polypyrrole-Coated Magnetic Nanoparticles as an Efficient Adsorbent for RB19 Synthetic Textile Dye: Removal and Kinetic Study. *Spectrochim. Acta-Part A Mol. Biomol. Spectrosc.* **2015**, *149*, 481–486. [CrossRef]
42. Bai, L.; Li, Z.; Zhang, Y.; Wang, T.; Lu, R.; Zhou, W.; Gao, H.; Zhang, S. Synthesis of Water-Dispersible Graphene-Modified Magnetic Polypyrrole Nanocomposite and Its Ability to Efficiently Adsorb Methylene Blue from Aqueous Solution. *Chem. Eng. J.* **2015**, *279*, 757–766. [CrossRef]
43. Alrahili, M.; Peroor, R.; Savchuk, V.; McNear, K.; Pinchuk, A. Morphology Dependence in Photothermal Heating of Gold Nanomaterials with Near-Infrared Laser. *J. Phys. Chem. C* **2020**, *124*, 4755–4763. [CrossRef]
44. Jiang, K.; Smith, D.A.; Pinchuk, A. Size-Dependent Photothermal Conversion Efficiencies of Plasmonically Heated Gold Nanoparticles. *J. Phys. Chem. C* **2013**, *117*, 27073–27080. [CrossRef]
45. Gogoi, R.; Singh, A.; Moutam, V.; Sharma, L.; Sharma, K.; Halder, A.; Siril, P.F. Revealing the Unexplored Effect of Residual Iron Oxide on the Photoreforming Activities of Polypyrrole Nanostructures on Plastic Waste and Photocatalytic Pollutant Degradation. *J. Environ. Chem. Eng.* **2022**, *10*, 106649. [CrossRef]
46. Mirrahimi, M.; Hosseini, V.; Kamrava, S.K.; Attaran, N.; Beik, J.; Kooranifar, S.; Ghaznavi, H.; Shakeri-Zadeh, A. Selective Heat Generation in Cancer Cells Using a Combination of 808 nm Laser Irradiation and the Folate-Conjugated Fe_2O_3 @Au Nanocomplex. *Artif. Cells Nanomed. Biotechnol.* **2018**, *46*, 241–253. [CrossRef]
47. Sun, C.; Ji, C.; Li, Y.; Kuang, J.; Wu, J. A Comparison Study of Photothermal Effect between Moxibustion Therapy and Laser Irradiation on Biological Tissue. *Int. J. Therm. Sci.* **2021**, *164*, 106924. [CrossRef]
48. Lu, J.; Cai, L.; Dai, Y.; Liu, Y.; Zuo, F.; Ni, C.; Shi, M.; Li, J. Polydopamine-Based Nanoparticles for Photothermal Therapy/Chemotherapy and Their Synergistic Therapy with Autophagy Inhibitor to Promote Antitumor Treatment. *Chem. Rec.* **2021**, *21*, 781–796. [CrossRef]
49. Zhao, Y.; Zhao, T.; Cao, Y.; Sun, J.; Zhou, Q.; Chen, H.; Guo, S.; Wang, Y.; Zhen, Y.; Liang, X.J.; et al. Temperature-Sensitive Lipid-Coated Carbon Nanotubes for Synergistic Photothermal Therapy and Gene Therapy. *ACS Nano* **2021**, *15*, 6517–6529. [CrossRef]
50. Estelrich, J.; Antònia Busquets, M. Iron Oxide Nanoparticles in Photothermal Therapy. *Molecules* **2018**, *23*, 1567. [CrossRef]

polymers

MDPI

Article

A New Approach Utilizing Aza-Michael Addition for Hydrolysis-Resistance Non-Ionic Waterborne Polyester

Hao Fu, Linbo Gong * and Shuling Gong *

College of Chemistry and Molecular Sciences, Wuhan University, Wuhan 430072, China; haofu@whu.edu.cn
* Correspondence: gonglb@whu.edu.cn (L.G.); gongsl@whu.edu.cn (S.G.)

Abstract: This work first synthesized a series of linear polyesters by step-growth polycondensation, then an amino-terminated hydrophilic polyether was grafted to the polyester as side-chains through aza-Michael addition to prepare a self-dispersible, non-ionic waterborne comb-like polyester (NWCPE). In contrast to traditional functionalization methods that usually require harsh reaction conditions and complex catalysts, the aza-Michael addition proceeds efficiently at room temperature without a catalyst. In this facile and mild way, the NWCPE samples with number-average molecular weight (M_n) of about 8000 g mol^{-1} were obtained. All dispersions showed excellent storage stability, reflected by no delamination observed after 6 months of storage. The NWCPE dispersion displayed better hydrolysis resistance than an ionic waterborne polyester, as was indicated by a more slight change in pH value and M_n after a period of storage. In addition, the film obtained after the NWCPE dispersion was cross-linked with the curing agent, it exhibited good water resistance, adhesion, and mechanical properties.

Keywords: aza-Michael addition; comb-like polymer; non-ionic waterborne polyester; hydrolysis resistance

Citation: Fu, H.; Gong, L.; Gong, S. A New Approach Utilizing Aza-Michael Addition for Hydrolysis-Resistance Non-Ionic Waterborne Polyester. *Polymers* **2022**, *14*, 2655. https://doi.org/10.3390/polym14132655

Academic Editor: Edina Rusen

Received: 28 May 2022
Accepted: 26 June 2022
Published: 29 June 2022

Publisher's Note: MDPI stays neutral with regard to jurisdictional claims in published maps and institutional affiliations.

Copyright: © 2022 by the authors. Licensee MDPI, Basel, Switzerland. This article is an open access article distributed under the terms and conditions of the Creative Commons Attribution (CC BY) license (https://creativecommons.org/licenses/by/4.0/).

1. Introduction

The emission of volatile organic compounds (VOC) is harmful to the environment and human health, causing many countries and regions have formulated strict restrictions on it. This has driven the development of waterborne polymer materials [1]. In addition to the characteristics of traditional polyester (including high gloss, good weather resistance, and biocompatibility), waterborne polyester (WPE) also has the advantages of low VOC, low toxicity, lack of odor, nonflammability, ease of disposal, and so on [2,3]. These characteristics make waterborne polyester attract lots of attention in coatings, adhesives, biomedical applications, and other fields [4–6].

According to whether an emulsifier is used in the preparation process, WPEs can be divided into two groups: external-emulsifying polyester and self-emulsifying polyester. The emulsifiers are required in the preparation of external-emulsifying WPE. There are often residual emulsifiers in the prepared emulsion which would migrate to the surface of the emulsion in the film-forming process. This event would affect the performance of emulsions, as well as reduce the surface evenness and the water resistance of cross-linked films [7]. Self-emulsifying WPE can be roughly divided into ionic and nonionic types according to their hydrophilic components. The ionic WPE usually contains carboxyl groups or sulfonic acid groups. When Ma et al. [8] prepared WPE with trimellitic anhydride as terminal groups, the polyester dispersed well in water after neutralization, but the molecular weight was below 2000 g mol^{-1}, which would limit its application scope. Zhang et al. [9] introduced 5-sulfoisophthalic acid monosodium salt into the molecular backbone and obtained a WPE with a solid content of 10%. Although the ionic polyesters are facile in synthesis and dispersing procedures, there are still defects to be resolved. The anionic polyester is usually difficult to apply in acidic environments and has poor compatibility with certain additives and inert pigments, [10] and the sulfonate-based polyesters are corrosive to metal substrates [11]. Waterborne polyester prepolymer with a molecular weight of around 6000 g mol^{-1} usually exhibits better comprehensive properties and is suitable for the classical hardening formulations, [12] but it was difficult for ionic WPE to achieve such a molecular weight. For the carboxyl-terminated polyester, decarboxylation is unavoidable at high temperatures; moreover, only terminal carboxyl groups cannot provide sufficient hydrophilic capacity for high molecular weight polyester [13]. For the multi-carboxyl containing polyester, the uncontrolled branching reaction between carboxyl groups and polyols would limit the growth of molecular weight [14]. For the sulfonate-based polyester, the molecular weight of polymer decreases when a sulfonic acid monomer is involved in copolymerization [15]. Moreover, ionic WPEs are usually poor in hydrolysis resistance. Non-ionic WPEs, which rely on polyethylene glycol (PEG) segments on the main chain to achieve water-dispersibility, would sacrifice the hydrolysis resistance and mechanical properties, as well as the hydrophilicity of the obtained polymer is usually weak. [16,17].

Therefore, researchers have made some attempts to explore other methods for the preparation of water solubility polymers [18]. It was found that by grafting the hydrophilic segments as side chains to the molecular backbone, the molecular weight of the WPE was improved, as was the mechanical property, etc. [19,20] Li et al. [20] used dimethylolpropionic acid modified by PEG to synthesize a polyester prepolymer. After conducting a reaction with isocyanate, a series of non-ionic waterborne comb-like polyurethane were obtained. With the increase in the comb-like polyester segment, the elongation at break (ε) of the product increased by 30%, and the solid content increased from 20% to 30%. However, the modification process has to undergo tedious protection and deprotection procedures. Taniguchi et al. [21] prepared a water-soluble, comb-like polyester through a chemoselective reaction between PEO side chains and ketone-bearing polyester; however, the water-solubility and mechanical properties of these polymers were not discussed in detail, and the synthetic pathways were complex, involving multiple steps with multiple catalysts. Furthermore, when side chains with a certain steric hindrance were grafted to the molecular skeleton to form a comb-like structure, the ester bonds of the polyester were protected according to the "rule of six" theory, and the hydrolysis resistance of the WPE would be improved [22]. However, the synthesis of polyesters with side chains or side chain functionalized groups usually require complex multistep reactions under harsh conditions including multiple expensive catalysts and long reaction times, which usually leads to the degradation of polyester chains in deprotection reactions [21,23]. In recent years, the aza-Michael addition, which is considered as a very efficient method to create new C-N bond and β-amino carbonyl derivatives, has attracted the attention of researchers in the post-polymerization modification of polymers [24–26]. The aza-Michael addition proceeds between a nucleophilic amine and an electron-deficient alkene. Interestingly, dimethyl maleate, a common structural unit in polyester, has been confirmed to be a class of Michael receptors with high reactivity and selectivity [27]. Bosica et al. [28] demonstrated that dimethyl maleate could undergo an efficient Michael addition reaction with various aliphatic amines at room temperature without any catalyst. Yu et al. [29] also confirmed that the primary amines with different chain lengths react efficiently with *cis*-methyl maleate and that the obtained sample was highly functionalized, with the addition reaction being almost irreversible. These works inspire us for preparing NWCPE via green aza-Michael addition.

This work aims to develop a facile and green methodology for preparing self-dispersible NWCPE. By step-growth polycondensation and aza-Michael addition, the NWCPE with hydrophilic polyether-amines Jeffamine M-1000 as side chains was prepared. In this way, the properties of waterborne polyester, such as molecular weight, storage stability, hydrolysis resistance, and mechanical properties have been improved. The results exhibited that the NWCPE has excellent application prospects in eco-friendly waterborne coating.

2. Materials and Methods

2.1. Materials

Maleic acid (MA), hexanedioicacid (HA), n-butyl titanate ($Ti(OBu)_4$), methanol, chloroform, and p-toluene sulfonic acid (TsOH) were obtained from Sinopharm Chemical Reagent Co., Ltd. (Shanghai, China). In addition, 1,4-cyclohexanedimethanol (CHDM), 1,6-hexanediol (HG), eopentyl glycol (NPG), and dibutyltin dilaurate (DBTDL) were purchased from Aladdin Reagents Co., Ltd. (Shanghai, China). Jeffamine® M-1000 was commercialized by Huntsman Corp. (Texas, USA). Waterborne non-ionic hexamethylene diisocyanate trimer (WHDIT, with a -NCO content of 19.5 ± 0.5%) was purchased from Habo Chemical Co., Ltd. (Wuhan, China).

2.2. Methods

2.2.1. Synthesis of Linear Polyesters

The mixture of monomers (MA, HA, HG, CHDM, and NPG), catalyst (0.6 mol% with respect to the diacids), and antioxidant 4-methoxy phenol (0.5 wt% related to all reactants) were added to a 100 mL three-neck round-bottom flask, which was equipped with a mechanical agitator, a nitrogen inlet, and a condenser. Firstly, the mixture was stirred at 180 °C (760 mmHg) under N_2 atmosphere for 5h. Secondly, the pressure of the system was gradually decreased to 0.3–2 mmHg, and the mixture was stirred under vacuum until an apparent Weissenberg effect (the viscoelastic liquids climb up to a rotating rod) was observed. The product was dissolved in chloroform, followed by precipitation with a large amount of cold methanol. Following this, a pale yellow, transparent, viscous, linear polyester was obtained.

2.2.2. Synthesis of Comb-like Polyester

Once the linear polyester was dissolved in chloroform, the Jeffamine M-1000 was added to the solution at a ratio of 10:3 ([maleate units]/[Jeffamine], mol/mol), and the mixture reacted at room temperature for 10 h. The mixture was precipitated by being poured into diethyl ether and was then dried in a vacuum oven. The obtained comb-like polyester was a light yellow fluid and its viscosity was lower than the linear polymer prepared in previous step. The synthesis route of comb-like polyester was displayed in Scheme 1.

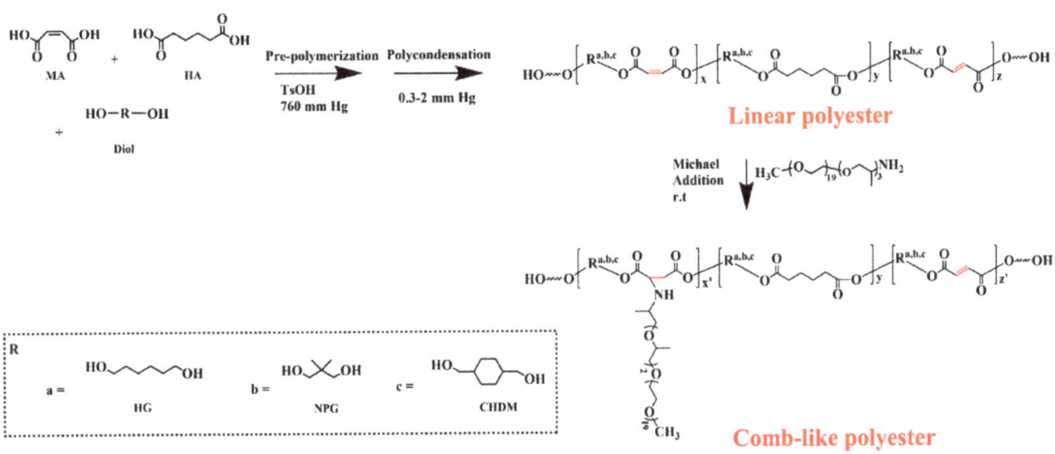

Scheme 1. The synthesis route of comb-like polyester.

2.2.3. Preparation of Non-Ionic Waterborne Comb-like Polyester (NWCPE) Dispersion

The dried comb-like polyester was dispersed by adding deionized water dropwise at 60 °C under a mechanical dispersion process at 1500 rpm, after which a light yellow and transparent NWCPE dispersion with 40% solid content was obtained. The formulations of different linear polyester and NWCPEs were listed at Table 1.

Table 1. The formulation and synthesis condition of NWCPEs.

Sample [(a)]	Catalyst	Reaction Temperature (°C)	Diacid Dosage (mol)		Diol Dosage (mol)			Functionality [(b)]
			MA	HA	CHDM	HG	NPG	
Entry-1L	DBTDL	180	0.05	0.05	0	0.084	0.021	/
Entry-2L	$Ti(OBu)_4$	180	0.05	0.05	0	0.084	0.021	/
Entry-3L Entry-3C	TsOH	180	0.05	0.05	0	0.084	0.021	/ 4.94
Entry-4L	TsOH	140	0.05	0.05	0	0.084	0.021	/
Entry-5L	TsOH	170	0.05	0.05	0	0.084	0.021	/
Entry-6L	TsOH	190	0.05	0.05	0	0.084	0.021	/
Entry-7L Entry-7C	TsOH	180	0.03	0.07	0	0.084	0.021	/ 3.52
Entry-8L Entry-8C	TsOH	180	0.04	0.06	0	0.084	0.021	/ 4.16
Entry-9L Entry-9C	TsOH	180	0.06	0.04	0	0.084	0.021	/ 5.63
Entry-10L Entry-10C	TsOH	180	0.07	0.03	0	0.084	0.021	/ 7.48
Entry-11L Entry-11C	TsOH	180	0.05	0.05	0.021	0.063	0.021	/ 4.48
Entry-12L Entry-12C	TsOH	180	0.05	0.05	0.042	0.042	0.021	/ 4.82
Entry-13L Entry-13C	TsOH	180	0.05	0.05	0.063	0.021	0.021	/ 5.15
Entry-14L Entry-14C	TsOH	180	0.05	0.05	0.084	0	0.021	/ 5.21

(a) The linear and comb-like polyesters in same formula were distinguished by suffix letters: Entry-L represented the linear polyester and Entry-C represented the comb-like polyester; (b) the functionality (i.e., the number of side chains on a molecular chain) was calculated with 1H NMR spectroscopy and GPC data. (The calculation method was based on the protons of HA units ($-OOCC\mathit{H}_2(CH_2)_2C\mathit{H}_2COO-$) and the protons of the residual *trans* MA units.)

2.2.4. Cross-Linking of the NWCPE Dispersions

The synthesized NWCPE dispersion and curing agent WHDIT were fed in a ratio of -NCO/-OH = 1.5:1.The mixture was diluted to solid content of 35% with deionized water under stirring. The mixture was stirred for 15 min and then coated on a glass plate or clean tinplate After drying to constant weight at room temperature, the film was taken out for further testing.

2.3. Characterizations

Fourier transform infrared spectroscopy. FTIR of samples was performed by Thermo Nicolet Fourier transform infrared spectroscopy (Thermo Fisher, Waltham, MA, USA) with the attenuated total reflectance (ATR) method. The scanning range was 4000–500 cm^{-1}.

Nuclear magnetic resonance. ^1H NMR spectra of samples were taken on a Bruker Avance-400 MHz spectrometer in deuterated chloroform.

Particle size. Malvern Zetasizer Nano Model ZS90 was used to measure the particle size and distribution (PDI). The sample was diluted before testing (<0.01 wt %).

Solid content. The solid content of the sample was determined according to GB/T 1725-1979:

$$\text{Solid\%} = \frac{\text{Dried sample weight}}{\text{Initial sample weight}} \times 100\% \quad (1)$$

Gel permeation chromatography (GPC). The Waters-515 gel permeation chromatograph (US WATERS Corporation) equipped with a Waters-2414 refractive index detector was used to measure the number-average molecular weight (M_n) and weight-average molecular weight (M_w) of the samples. THF was selected as eluent with polystyrene as calibration.

Viscosity. The viscosity of NWCPE dispersions was measured by DV-79 digital viscometer. The E-type rotor was selected and tested at 75 r min^{-1}.

Water resistance. According to the national standard GB/T 1733-1993, the film was boiled in boiling deionized water for 2 h. The state of the film was observed.

Scratch experiment. The adhesion of the film coated on the tinplate was evaluated according to GB/T 9286-1998.

Pencil hardness. The pencil hardness of the film coated on the tinplate was tested according to GB/T 6739-2006.

Mechanical properties measurement. The tensile property of the film was tested according to the ASTM D 638 by using the Electronic Universal Testing Machine from MTS SYSTEMS (China) Co., Ltd. (Shanghai, China), at a crosshead speed of 10 mm min^{-1}.

Light transmittance. The transmittance of the cross-linked NWCPE sample was tested by an ultraviolet−visible spectrometer (UV-3600, Shimadzu, Japan).

Scanning electron microscopy (SEM). The film was coated with gold at a voltage of 5 kV, and the sample was observed on a field emission scanning electron microscope (SEM, Zeiss Sigma, Germany).

Differential scanning calorimetry (DSC). TA Q2000 DSC Instrument (TA Instruments, New Castle, DE, USA) was used to test the glass-transition temperature (T_g) and the measurement was carried out in a nitrogen atmosphere at a temperature range of −80 °C to 100 °C at a rate of 10 °C min^{-1}.

Thermogravimetric analysis (TGA). The thermal stability of the sample was measured with a TG 209 F1 thermogravimetric analyzer from NETZSCH (Selb, Germany). The measurement was carried out in a nitrogen atmosphere at a temperature range of 30 °C to 650 °C and at the heating rate of 10 °C min^{-1}.

3. Results and Discussion

In this work, a MA units containing linear polyester was synthesized by step-growth polycondensation. Following this, the hydrophilic polyether was grafted to the molecular skeleton as side chains through aza-Michael addition to prepare NWCPE.

3.1. Structure Characterizations

The ^{1}H NMR data of Entry 3L and Entry 3C are given as a representative in Figure 1. The characteristic signal at 6.24 ppm originates from *cis* double bonds and the signal at 6.85 ppm belongs to *trans* double bonds. After the completion of the step-growth polycondensation, the integral area ratio of *cis* double bonds to *trans* double bonds was about 3:1, indicating that 25% of the *cis* units isomerized to *trans* units. The signal of the *cis* double bonds (δ = 6.24 ppm) disappeared and the signal of the *trans* double bonds (δ = 6.85 ppm) increased slightly after aza-Michael addition. The reason for this was that part of the *cis* units participated in the addition reaction and partially isomerized into the *trans* units [29]. The newly emerging signals at 4.02–3.98 ppm ($-OOCCH_2\boldsymbol{CH}NHCH_2-$) and 2.69 ppm ($-OCOC\boldsymbol{H_2}CHNHCH_2-$) belong to the reacted *cis* units' proton. According to the integration analyses of the ratios between the two new signals and the trans signal peaks, the functionalization extent of the polyester was about 31%.

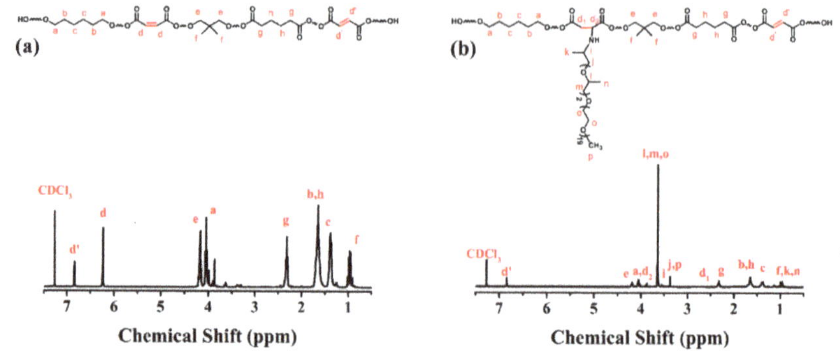

Figure 1. The ^{1}H NMR of linear polyester (**a**), and comb-like polyester (**b**).

The FTIR spectrum and the second derivative spectrum of both Entry 3L and Entry 3C are shown in Figure 2. The broad absorption peak at 3545 cm^{-1} is the stretching vibration peak of O−H, and the absorption peak at 3420 cm^{-1} in the spectrum of comb-like polyester is the stretching vibration peak of N−H. The stretching vibration peaks of $-CH_3$ and $-CH_2$ appear at 2936 cm^{-1} and 2862 cm^{-1}, and the peak at 1459 cm^{-1} belongs to $-CH_3$ and $-CH_2$ bending vibration. The difference of the peak at 1723 cm^{-1} was amplified by the second derivative spectrum. The peak at 1740 cm^{-1} belongs to the carbonyl ester C=O, which is attached to C−C bonds. The peak that appears at 1700 cm^{-1} is ascribed to the carbonyl ester C=O attached to the C=C bonds. The two peaks at 1253 cm^{-1} and 1102 cm^{-1} belong to the asymmetric stretching vibration and symmetric stretching vibration of the carboxyl group C−O−C, respectively [30]. After the aza-Michael addition was accomplished, the ratio of the C−O−C peak intensity to the C=O peak absorption intensity increased, which indicated that the addition reaction was successful. The two peaks observed at 1640 cm^{-1} and 976 cm^{-1} are ascribed to the C=C tensile vibration and the *trans* C=C−H out-of-plane bending vibration, respectively. The peak intensity at 976 cm^{-1} was enhanced after aza-Michael addition, which indicates that cis units reduced in the addition process.

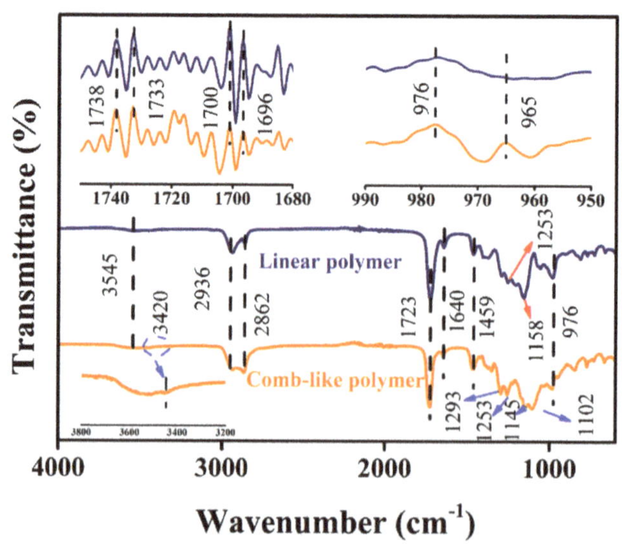

Figure 2. The FTIR spectrum and the second derivative spectrum of linear polyester and comb-like polyester.

3.2. The Influence of Various Factors on Polyester and Polyester Dispersion

The aza-Michael addition proceeds between a nucleophilic amine and an electron deficient alkene. In this work, the electron deficient alkene was the *cis* MA units, and the Jeffamine M-1000 acted as the nucleophilic amine. The *cis* double bonds of MA would isomerize into *trans* units in both polymerization and aza-Michael addition process, [31] but the amino-terminated hydrophilic polyether is more likely to react with *cis* double bonds in addition reaction [29,32]. The amount of the side chains was closely related to the *cis* MA units' content on the molecular skeleton, which determined the hydrophilic nature and stability of the NWCPE dispersions. Moreover, in the polycondensation process a side reaction called Ordelt saturation took place between the hydroxyl group of the diol monomers and the double bonds of the unsaturated diacid. This interaction should be avoided, because it would lead to uncontrollable branching [33,34]. The routes of Ordelt saturation were described in Scheme 2. In the polycondensation process, the isomerization extent is related to various factors including reaction temperature and the type of catalyst. At the same time, the Ordelt saturation reaction is more likely to proceed at high temperatures. In addition, the properties of the NWCPE and the cross-linked film are related to the type and dosage of the monomers. Hence, it was worth studying the influence of the synthesis conditions such as catalyst type, reaction temperatures, and the dosage of different monomers on the polymer.

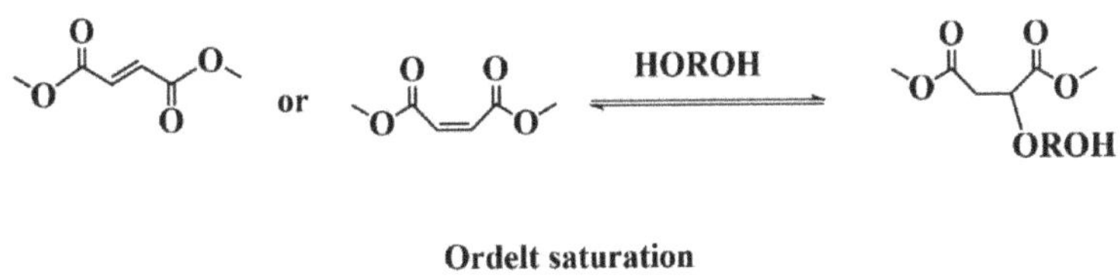

Scheme 2. The Ordelt reaction, *cis* and *trans* C=C bonds saturate with diol.

3.2.1. Effect of the Catalyst Type

In previous works, various catalysts have been explored for the preparation of polyesters, including organometallic compounds $Ti(OBu)_4$ [35,36], DBTDL [37], and acid TsOH [38,39]. In this paper, we investigated the influence of the catalyst type on the isomerization extent and molecular weight of the polyester. As shown in Table 2, when the $Ti(OBu)_4$ was selected as the catalyst, higher molecular weight polyester was obtained, but the severe isomerization made it unsuitable for this system. Although the isomerization extent was well controlled with DBTDL as the catalyst, the molecular weight (2820 g mol^{-1}) could be further improved. When TsOH was selected, the M_n of the synthesized polyester was about 5870 g mol^{-1}, the isomerization extent was below 25%. Because the catalytic activity of $Ti(OBu)_4$ was the higher than both TsOH and DBTDL, the Entry 2L was the largest in molecular weight, but the excessive degradation reactions of $Ti(OBu)_4$ during polymerization process resulted the darkest color of the sample [37]. Moreover, the catalysts play an important in the isomerization process, which was usually used to manufacture fumarate from maleate [40], and the different isomerization extent using different catalysts may be related to the nucleophilicity [41] and rigidity [42] of the catalyst. In general, TsOH was more suitable as a polycondensation reaction catalyst in this work.

Table 2. Effect of catalyst type on the polymerization and properties of polyester.

Sample	Catalyst	*cis:trans*	M_n (g mol^{-1})	M_w/M_n	Appearance
Entry-1L	DBTDL	83:17	2820	2.12	Light yellow transparent
Entry-2L	$Ti(OBu)_4$	62:38	14330	2.25	Dark brown transparent
Entry-3L	TsOH	75:25	5870	2.05	Yellow transparent

3.2.2. Effect of the Reaction Temperature

The reaction temperature is a significant yet complicated factor in polycondensation. High reaction temperatures were usually accompanied by some negative influence, including unwanted Ordelt saturation reaction and severe isomerization, whereas the polyester obtained at low temperature was usually lower in molecular weight [29,43]. As depicted in Figure 3, four samples were prepared under different reaction temperatures ranging from 140 °C to 190 °C. During the polymerization, the temperature must exceed 140 °C to maintain the molten state of the monomers. As shown in Figure 3a, the isomerization extent increased from 7% to 42% when the reaction temperature gradually increased from 140 °C to 190 °C. The driving force of the isomerization behavior is regarded to be that the *trans* isomer possesses higher energetic stability. The sufficiently thermodynamic driving force was obtained at a high temperature, which might lead to uncontrolled isomerization [44]. The *cis* units served as grafting sites for hydrophilic side chains, and its content was closely related to the water dispersibility of the synthesized polyester. An excessive degree of isomerization would result in insufficient grafting sites, leading to polyester that is difficult to disperse in water. In addition, even in the presence of antioxidants, when the temperature reached 190 °C, obvious gelation caused by Ordelt saturation reaction was observed, leading to the low yield. As shown in Figure 3b, the M_n of polyester ranged from 3410 g mol^{-1} to 6410 g mol^{-1} with the reaction temperature rising. The cause might be that the viscosity decreases with the increasing temperature, which accelerated the discharge of water. This drove the polymerization reaction forward further. In addition, 4-methoxyphenol was selected as a radical quencher to suppress the Ordelt saturation reaction in this manuscript. Satisfactorily, this was successful, as was confirmed by the relatively low molecular weight distribution value as shown in Figure 3b, which indicated the Ordelt saturation reaction was effectively suppressed [29]. On the whole, it was more reasonable to set the reaction temperature to 180 °C.

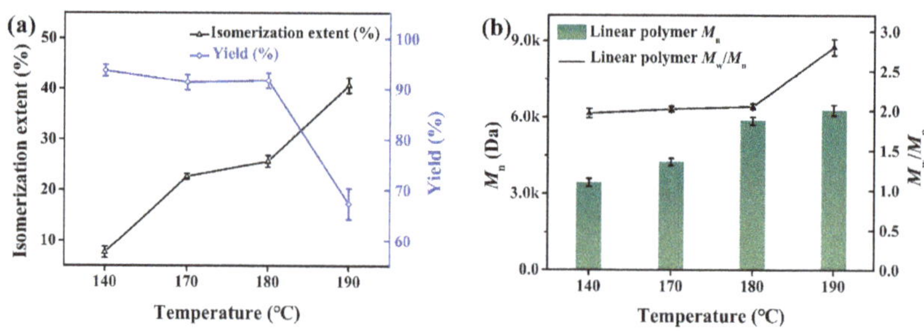

Figure 3. Effect of reaction temperatures on the isomerization extent and yield (**a**), M_n and M_w/M_n (**b**) of polyester.

3.2.3. The Influence of MA/HA Monomer Ratio

The water dispersibility and storage stability of the NWCPE dispersions were affected by the density of side chains, which was related to the dosage of MA. The influence of the MA content was investigated with all other reaction parameters (the formulation of monomers, catalyst, inhibitor, and the reaction time), which were kept invariable. As we can see in Figure 4a, at the same reaction temperature the isomerization extent of the polyester was not significantly affected by the MA content, which stayed at 25−30%. As shown in Figure 4b, the M_n of the linear polyester increased from 4950 g mol^{-1} to 7230 g mol^{-1}, with the content of MA increasing from 30% to 70%. This behavior might be related to the higher reactivity of MA as compared to HA during the polycondensation. In the same reaction time, it was found that the more reactive the monomer, the greater the polymerization extent, resulting in a larger molecular weight. All samples exhibited a reasonable polydispersity value, ranging between 2.05–2.79 for linear polyester. This indicates that no branched polymers were brought during the polymerization [33]. After functionalization was complete, the corresponding comb-like polyesters were obtained. As shown in Figure 4c, the M_n of the comb-like polyester changed from 6530 g mol^{-1} to 8950 g mol^{-1} with polydispersity ranging from 2.34 to 2.94. Because of the extra volume gained from the side chains, the M_n of comb-like polyesters increased slightly after the addition, and the polydispersity also became wider. In addition, the longer polyester possessed more side chains, which was related to the hydrophilic of NWCPE dispersion.

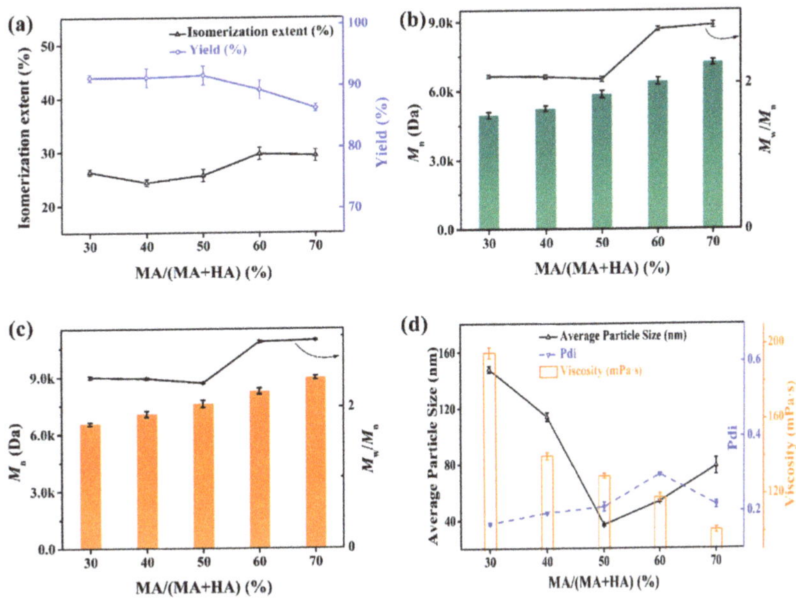

Figure 4. Isomerization extent and yield (**a**), M_n and M_w/M_n of linear polyester (**b**) and comb-like polymer (**c**), particle size and viscosity (**d**) of polyester dispersions in different MA/(MA + HA).

The self-dispersing ability in water of NWCPE dispersions with different MA content is shown in Figure 4d and Table 3. Due to the free hydrophilic side chains, Jeffamine M-1000 swelled well in water and provided good dispersion capability, and all of the polyesters were able to be self-dispersed in the water. With the content of MA improved from 30% to 50%, the amount of side chains increased from 3.52 to 4.94, the hydrophilicity of the polyester was significantly enhanced, and the particle size gradually decreased. However, with the further increase in the number of side chains, the particle size showed an increasing trend. The reason might be due to the following factors. As the number of side chains increased from 3.52 to 4.94, the hydrophilicity of the polymer increased. At the same time, owing to the increase in side chains, the intermolecular repulsion increased and the intermolecular aggregation decreased. All of these led to a decrease in particle size. On the other hand, as the MA dosage continued to increase, the molecular weight of the polymer increased slightly. Meanwhile, the polyester particles were greatly softened with an increase in flexible polyether-amine segments. The soft particles tended to coalesce under the external shear force during the emulsification process [45,46]. Hence, the particle size of dispersions decreased and then increased with the increase in the number of side chains. The storage stability of dispersions is related to the average particle size and viscosity [47]. The prepared dispersions were metastable systems because of the small average particle size [48]. In addition, with the increase in the density of side chains, the viscosity of the dispersion decreased. The reason might be that, with the increase in the number of side chains, the hydrophilic components of the polyester increased, and the hydrodynamic volume of particles decreased [49–51]. When the number of side chains exceeded 4.94, the viscosity was lower than 120 mPa·s and the particle size of the dispersion was below

100 nm. According to Stokes' law, dispersions with small particle size and low viscosity are more prone to precipitate [47]. As a result, the storage stability of the dispersions deteriorated seriously and obvious flocculation was observed in less than 30 days. As for the coating properties, with the MA content increased from 30% to 70%, the pencil hardness of cured films was maintained at B, the adhesive property was kept at level 1, and all cured films exhibited good water resistance. In summation, when the content of MA was 50%, the prepared NWCPE dispersion and cross-linked film possessed good performance.

Table 3. The properties of NWCPE dispersions and cured films with different MA dosage.

Sample	MA Content	Dispersion Appearance	Pencil Hardness	Adhesion	Water Resistance	Storage Stability
Entry-7C	30%	Milky white, untransparent	B	1°	Transparent	>180 d
Entry-8C	40%	Light yellow, transparent	B	1°	Transparent	>180 d
Entry-3C	50%	Light yellow, transparent	B	1°	Transparent	>180 d
Entry-9C	60%	Flocculation	/	/	/	<30 d
Entry-10C	70%	Flocculation	/	/	/	<30 d

3.2.4. The Influence of the Rigid Monomer Content

Since our objective is to synthesize a co-polyester with good comprehensive properties, mechanical properties are of great concern to us. As such, the toughness monomer CHDM was introduced into the molecular skeleton [52–54]. Although the transformation of isomerization is mainly affected by the steric hindrance effect of the polymer, the synthesized polyesters exhibited almost no difference in isomerization extent (all were around 25%) with the increase in CHDM dosage. The reason for this might be that the two primary hydroxyl groups of CHDM are widely separated [31]. The result indicated that this method is universal and applicable to a wide variety of diols. As can be gathered from Figure 5a,b, the M_n of all linear polymer ranged from 5850 g mol^{-1} to 6810 g mol^{-1} with the polydispersity ranging between 2.03–2.27, the M_n of comb-like polymer ranged from 7590 g mol^{-1} to 8810 g mol^{-1}, with the polydispersity ranging from 2.32 to 2.48. All synthesized polyesters were similar in molecular weight and polydispersity, indicating that the step-growth polycondensation of the system was not significantly affected when the HG was replaced with CHDM.

As shown in Figure 5c, all synthesized comb-like polyesters only showed one glass transition temperature, indicating that the polyesters were random co-polyesters. As the content of rigid monomer CHDM increased, the T_g of the polyester rose slightly. Furthermore, the crystal behavior existed in some samples, and it was found that the higher the CHDM content, the higher the crystallization temperature. The mobility of the NWCPE molecule was limited when the soft monomer was replaced by more than 60%. Therefore, the crystal behavior was not observed in samples CHDM 60% and CHDM 80% [55]. After the NWCPE was cross-linked with the WHDIT, the T_g increased significantly, and, due to further restricted in the movement of polymer chains, no crystallization melting behavior was observed.

As we can see from Figure 5d, after dispersing into water the average particle size of NWCPE dispersions ranged from 30.3 nm to 84.4 nm with the pdi value below 0.25. When the CHDM was introduced to the polyester sparingly, there was nearly no difference in average particle size, but the polyester particle size increased significantly when the CHDM content exceeded 40%. It might be that when the soft monomer HA was largely replaced by the more rigid monomer CHDM, the movement of the polymer chains was limited, resulting in poorer swelling of the NWCPE. Owning to the similar molecular weight and the density of side chains, the hydrodynamic volume of the NWCPEs in water with different CHDM content was relatively close, and the viscosity of all the dispersions was at about 130 mPa·s [56]. The storage stability would be similar naturally when the dispersions were closely in the particle size and viscosity.

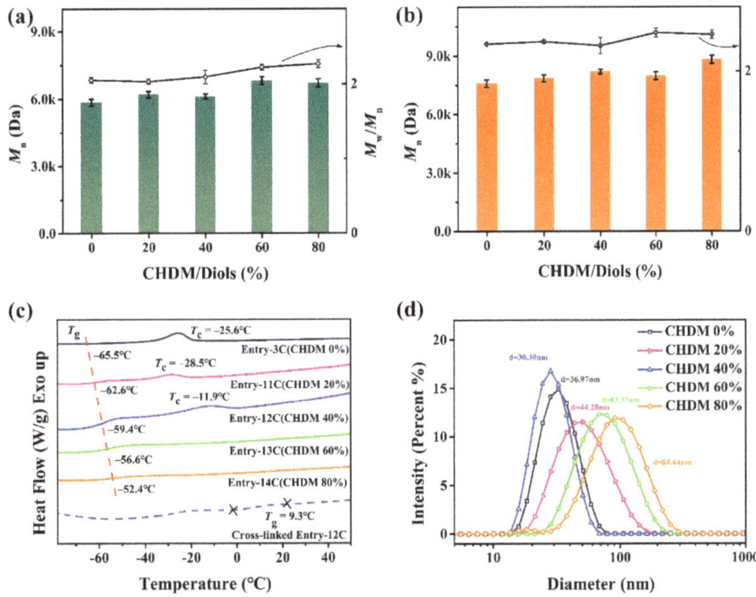

Figure 5. Molecular weight and polydispersity for linear polyester (**a**) and comb-like polyester (**b**), DSC curves of comb-like polyester and part of cross-linked polyester (**c**), and average particle size of dispersions (**d**) in different CHDM content.

As shown in Figure 6a, all polyester dispersions were alkaline because of the polyetheramine side chains, and the pH value of dispersions decreased during storage. In this work, the ΔpH of prepared dispersions was below 0.8 after 2 months of storage at room temperature. As mentioned in previous work, the pH of polyester dispersions changed above 1.5 after just 20 days of storage [57]. As we can see from Figure 6b, after 6 months of storage, the M_n of the polymer just decreased by a few hundred Da, and the change was less than 10%, while the M_n of polyester which, in other work, dropped by 40% after only being stored for one week [58]. This result indicates that the prepared NWCPE dispersions possess stronger hydrolysis resistance than waterborne ionic polyester. Furthermore, after the CHDM was utilized as a monomer, the decline in pH value and M_n were arrested, which was also related to the hydrolysis resistance of CHDM.

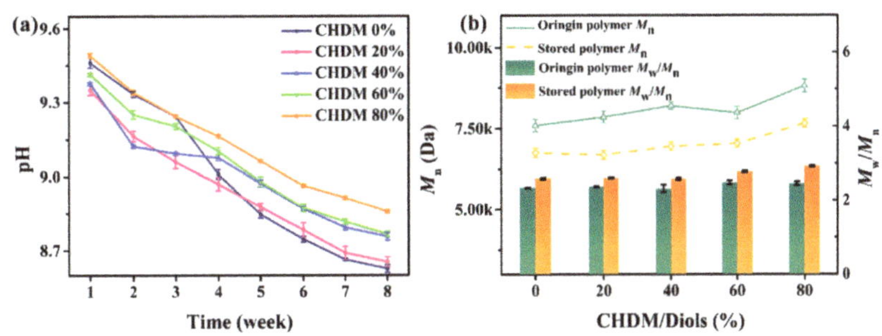

Figure 6. The ΔpH of dispersions (**a**), and ΔM_n of comb-like polyester after storage for 6 months (**b**) in different CHDM content.

The hardness of the coatings is related to the cross-linking density of networks and the chemical structure of the prepolymers [59]. As displayed in Table 4, the pencil hardness of the cross-linked coating increased with the increase in the CHDM content. The reason might be that the rigidity and the intermolecular force of the cured polyester increased with the increasing rigidity of diol. However, when the CHDM content reached 40%, the improvement of the rigid monomer dosage showed little effect on the hardness of the film. After the cross-linking process was accomplished, many urethane bonds were formed on the molecule, which grants that all cross-linked coatings exhibited good adhesion on the tinplate substrate. With the increasing dosage of CHDM, the molecular rigidity increased as well. The aliphatic chain length decreased compared to HA, but due to the steric hindrance of the side chains, the internal stress displayed almost no change [60]. As a result, the adhesion did not decrease with the utilization of CHDM. Because of the hydrophilic polar groups, the waterborne coatings are more sensitive to moisture, hence demonstrating that water resistance is an important factor in determining their application. Because the hydrophilic segments of the waterborne non-ionic polymer acted as side chains, all cross-linked films displayed good water resistance.

Table 4. The properties of NWCPE dispersions and cured films with different CHDM dosage.

Sample	CHDM Content	Dispersion Appearance	Pencil Hardness	Adhesion	Water Resistance	Storage Stability
Entry-3C	0%	Light yellow, transparent	B	1°	Transparent	>180 d
Entry-11C	20%	Light yellow, transparent	HB	1°	Transparent	>180 d
Entry-12C	40%	Light yellow, transparent	H	0°	Transparent	>180 d
Entry-13C	60%	Light yellow, transparent	H	0°	Transparent	>180 d
Entry-14C	80%	Dark yellow, transparent	H	0°	Transparent	>180 d

3.3. Thermogravimetric Analysis

The thermal degradation behaviors and DTG curves of Entry 12C are shown in Figure 7, and some crucial data are also displayed. The degradation curve of the cross-linked polyester in the first step was almost the same as the pure polyester, which was reflected by the same 5% weight loss temperature (327 °C) and the maximum weight loss rate temperature (403 °C). This step degradation was attributed to the ester and aliphatic moieties. The Tmax2 (455 °C) was observed in the cross-linked sample, which indicated the cross-linked polyester degraded by a multistep process. The reason might be that the hard segment urethane bonds formed after cross-linking, which reduced the migration of the chain segments. The thermal stability of the material was enhanced after cross-linking [61].

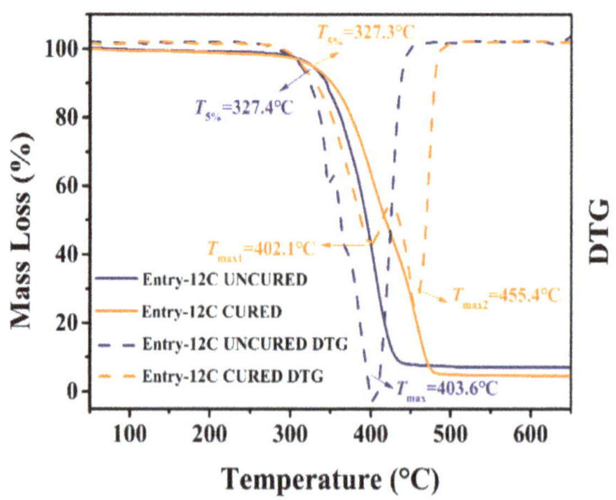

Figure 7. TG and DTG of NWCPE and cured polyester film.

3.4. Tensile Properties of the Cross-Linked NWCPE Films

The tensile properties of cross-linked NWCPE films were summarized in Table 5, and the stress–strain cures were listed in Figure 8. Mechanical properties are crucial for polymer materials which directly affect their industrial applications. The tensile strength increased from 3.1 MPa to 5.9 MPa after CHDM being introduced into the molecule. But when the content of CHDM reached 40%, the strength of the polymer no longer increased with the increase in the CHDM content. In addition, the toughness even decreased slightly due to the excessively high content of this rigid monomer. The result may be related to the following factors. First, the utilization of CHDM enhanced the rigidity of the polymer which could restrict the movement of the molecular chains, and the intermolecular force was increased, resulting in improved mechanical properties. Secondly, the molecular weight of the polymers were slightly different. Polyester with minor M_n contained more terminal hydroxyl groups in the same mass, resulting in a slightly larger cross-linking density during the curing process. The stress–strain curves of all of the films showed typical plastic deformation, and the elongation at the break (ε) of the films were all basically above 70%. The cause might be the following two factors. First, the synthesized polyester has low crystallinity, leading to high mobility between molecular chains. Secondly, the polyether-amine side chains are similar to a spring and increased the ductility of the molecular. The tensile properties of cured NWCPE were better than some cured ionic waterborne polyesters, some of which were about 2–3 MPa [3,62,63].

Table 5. Tensile properties of cross-linked films.

CHDM Content	0%	20%	40%	60%	80%
σ (MPa)	3.1 ± 0.2	5.2 ± 0.2	5.9 ± 0.2	5.8 ± 0.3	5.2 ± 0.3
E (MPa)	4.0 ± 0.4	10.1 ± 0.3	5.8 ± 0.5	6.7 ± 0.8	4.8 ± 0.7
ε (%)	70.3 ± 0.8	51.6 ± 0.6	88.1 ± 0.4	73.2 ± 0.9	90.5 ± 1.2

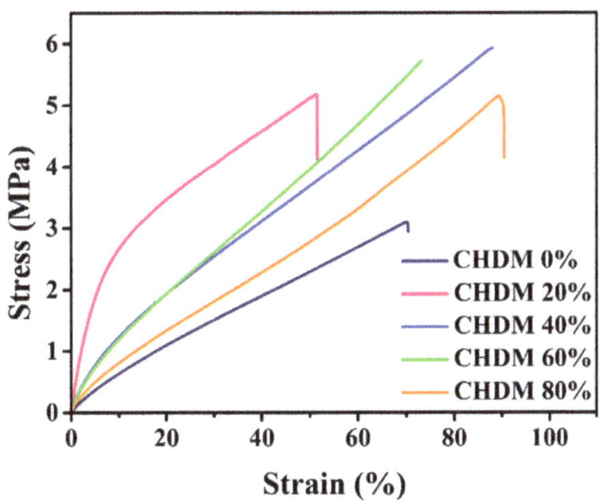

Figure 8. Stress-strain profiles of cross-linked films.

3.5. The Morphology and Light Transmittance of the Cured Film

As shown in Figure 9a,c, the neat cross-linked film was transparent, and, at the entire visible wavelength range, the light transmittance of the film was high. In addition, owing to the absence of emulsifier, no surfactant molecules migration during the film-forming process was observed, and the surface of the film was smooth and flat according to Figure 9b [64]. At the same cross-linking density, the CHDM content exerted almost no influence on the transmittance of the film. The SEM and photographic images of cured

Entry 12C (CHDM 40%) were taken as examples. The high transparency was attributed to the following reasons. First, the synthesized NWCPEs have good compatibility with the curing agent WHDIT and there was not significant macrophase separation during the cross-linking procedure. Second, the surface of the film was smooth, which was hard for scattering light [65]. As such, the low crystallinity of the WCPE may also contribute to the high transmittance of the film.

Figure 9. The light transmittance (**a**), SEM image (**b**), and photograph (**c**) of cross-linked films.

4. Conclusions

In this paper, MA, HA, CHDM, HG, and NPG were first used as monomers to synthesize linear polyesters by step-growth polycondensation. After this, the hydrophilic segment Jeffamine M-1000 was grafted to the polymer backbone by aza-Michael addition to prepare the NWCPE. The optimal reaction conditions were verified as TsOH as the selected catalyst, a reaction temperature of 180 °C, a MA/HA ratio of 5:5, and the content of the rigid monomer CHDM at 40%. In this condition, NWCPE with a molecular weight of about 8000 g mol^{-1} was prepared. After dispersing into water, the dispersion with a solid content of 40% and the particle size was appropriate with narrow distribution. The dispersion was good in storage stability and remained stable after 6 months. The good hydrolysis resistance was identified by the slight drop in pH after two months and little change in M_n of the dispersion. After cross-linking with the curing agent, a transparent film with good mechanical properties and water resistance was obtained. The hardness of the film was H, and the adhesion level on the tinplate reaches level 0. The tensile strength of the film was 5.9 MPa, and the ε was about 88%. This work provides ideas for the synthesis and functionalization of waterborne polyesters. These synthesized polymers have wide application prospects in environmentally friendly polymer materials.

Author Contributions: Conceptualization, methodology, and validation, H.F., L.G. and S.G.; writing—original draft preparation, H.F.; writing—review and editing, H.F., L.G., and S.G.; supervision, and funding acquisition, L.G. and S.G. All authors have read and agreed to the published version of the manuscript.

Funding: This research received no external funding.

Institutional Review Board Statement: Not applicable.

Informed Consent Statement: Not applicable.

Data Availability Statement: Not applicable.

Conflicts of Interest: The authors declare no conflict of interest.

References

1. Islam, S.; Inglefield, D.L.; Velev, O.D. Revisiting the colloidal fundamentals of water-dispersible polyesters: Interactions and self-assembly of polymer nanoaggregates in water. *Soft Matter* **2018**, *14*, 2118–2130. [CrossRef] [PubMed]
2. Ma, G.; Guan, T.; Hou, C.; Wu, J.; Wang, G.; Ji, X.; Wang, B. Preparation, properties and application of waterborne hydroxyl-functional polyurethane/acrylic emulsions in two-component coatings. *J. Coat. Technol. Res.* **2015**, *12*, 505–512. [CrossRef]

3. Geeti, D.K.; Niranjan, K. Environmentally benign bio-based waterborne polyesters: Synthesis, thermal- and bio-degradation studies. *Prog. Org. Coat.* **2019**, *127*, 419–428. [CrossRef]
4. Muto, A.; Toshin, K.; Shibao, F.; Ueda, K. Effect of melamine enrichment on the structure of water-based polyester/melamine films and properties of pre-painted steel sheets. *Prog. Org. Coat.* **2021**, *150*, 105963. [CrossRef]
5. Wei, L.; Yue, G.; Hongjian, P.; Hongying, C. Dispersion stability of titanium dioxide in aqueous isopropanol with polymer dispersant. *J. Coat. Technol. Res.* **2020**, *17*, 1083–1090. [CrossRef]
6. Washington, K.E.; Kularatne, R.N.; Karmegam, V.; Biewer, M.C.; Stefan, M.C. Recent advances in aliphatic polyesters for drug delivery applications. *WIREs. Nanomed. Nanobiotechnol.* **2017**, *9*, e1446. [CrossRef]
7. Pierlot, C.; Ontiveros, J.F.; Royer, M.; Catté, M.; Salager, J.-L. Emulsification of viscous alkyd resin by catastrophic phase inversion with nonionic surfactant. *Colloid Surf. A* **2018**, *536*, 113–124. [CrossRef]
8. Ma, S.; Qian, J.; Zhuang, Q.; Li, X.; Kou, W.; Peng, S. Synthesis and application of water-soluble hyperbranched polyester modified by trimellitic anhydride. *J. Macromol. Sci. A.* **2018**, *55*, 414–421. [CrossRef]
9. Zhang, S.X.; Wu, X.L.; Hao, T.H.; Hu, G.H.; Jiang, T.; Zhang, Q.C.; Zhao, H. Structure design, fabrication and property investigation of water-based polyesters with notable surface hydrophilicity. *New J. Chem.* **2018**, *42*, 20015–20023. [CrossRef]
10. Torres G, M.Y.; Bravo L, M.K.; Flórez M, S.A.; Simon, P.; Aguirre P, J.J.; Macías L, M.A.; Gauthier, G.H. Study of applicability in an aqueous paint of the blue pigment YIn0.95Mn0.05O3. *Dye. Pigment.* **2018**, *156*, 17–25. [CrossRef]
11. Zhou, Y.X.; Yuan, Q.X.; Gong, S.L. Preparation and characterization of self-emulsifying poly(ethylene glycol) methyl ether methacrylate grafted polyacrylate copolymers modified by waterborne polyester. *J. Appl. Polym. Sci.* **2021**, *139*, e51988. [CrossRef]
12. Alemdar, N.; Erciyes, A.T.; Bicak, N. Preparation of unsaturated polyesters using boric acid as mild catalyst and their sulfonated derivatives as new family of degradable polymer surfactants. *Polymer* **2010**, *51*, 5044–5050. [CrossRef]
13. Rokicki, G.; Wodzicki, H. Waterborne unsaturated polyester resins. *Macromol. Mater. Eng.* **2000**, *278*, 17–22. [CrossRef]
14. Hazarika, D.; Karak, N. Waterborne Sustainable Tough Hyperbranched Aliphatic Polyester Thermosets. *ACS Sustain. Chem. Eng.* **2015**, *3*, 2458–2468. [CrossRef]
15. Zheng, Y.; Zhu, P.; Cheng, F.; Zhou, M.; Zhang, K.; Wang, T. Preparation of waterborne elastic polyesters by chain extension with isophorone diisocyanate as a chain extender. *J. Appl. Polym. Sci.* **2019**, *137*, 48453. [CrossRef]
16. Hsiao, K.J.; Kuo, J.L.; Tang, J.W.; Chen, L.T. Physics and kinetics of alkaline hydrolysis of cationic dyeable poly(ethylene terephthalate) (CDPET) and polyethylene glycol (PEG)–modified CDPET polymers: Effects of dimethyl 5-sulfoisophthalate sodium salt/PEG content and the number-average molecular weight of the PEG. *J. Appl. Polym. Sci.* **2005**, *98*, 550–556. [CrossRef]
17. Tsai, H.C.; Hong, P.D.; Yen, M.S. Preparation and physical properties of nonionic aqueous polyurethane coatings containing different side chain PEGME length. *J. Appl. Polym. Sci.* **2008**, *108*, 2266–2273. [CrossRef]
18. Nabiyan, A.; Max, J.B.; Schacher, F.H. Double hydrophilic copolymers-synthetic approaches, architectural variety, and current application fields. *Chem. Soc. Rev.* **2022**, *51*, 995–1044. [CrossRef]
19. Mehravar, E.; Leiza, J.R.; Asua, J.M. Performance of latexes containing nano-sized crystalline domains formed by comb-like polymers. *Polymer* **2016**, *96*, 121–129. [CrossRef]
20. Li, B.; Peng, D.; Zhao, N.; Mu, Q.; Li, J. The physical properties of nonionic waterborne polyurethane with a polyether as side chain. *J. Appl. Polym. Sci.* **2013**, *127*, 1848–1852. [CrossRef]
21. Taniguchi, I.; Mayes, A.M.; Chan, E.W.L.; Griffith, L.G. A Chemoselective Approach to Grafting Biodegradable Polyesters. *Macromolecules* **2005**, *38*, 216–219. [CrossRef]
22. Newman, M.S. Some Observations Concerning Steric Factors. *J. Am. Chem. Soc.* **1950**, *72*, 4783–4786. [CrossRef]
23. Riva, R.; Schmeits, S.; Jérôme, C.; Jérôme, R.; Lecomte, P. Combination of Ring-Opening Polymerization and "Click Chemistry": Toward Functionalization and Grafting of Poly(ε-caprolactone). *Macromolecules* **2007**, *40*, 796–803. [CrossRef]
24. Genest, A.; Portinha, D.; Fleury, E.; Ganachaud, F. The aza-Michael reaction as an alternative strategy to generate advanced silicon-based (macro)molecules and materials. *Prog. Polym. Sci.* **2017**, *72*, 61–110. [CrossRef]
25. Robert, T.; Friebel, S. Itaconic acid—A versatile building block for renewable polyesters with enhanced functionality. *Green Chem.* **2016**, *18*, 2922–2934. [CrossRef]
26. Rulev, A.Y. Aza-Michael reaction: Achievements and prospects. *Russ. Chem. Rev.* **2011**, *80*, 197–218. [CrossRef]
27. Noordzij, G.J.; Wilsens, C. Cascade aza-Michael Addition-Cyclizations; Toward Renewable and Multifunctional Carboxylic Acids for Melt-Polycondensation. *Front. Chem.* **2019**, *7*, 729. [CrossRef]
28. Bosica, G.; Debono, A.J. Uncatalyzed, green aza-Michael addition of amines to dimethyl maleate. *Tetrahedron* **2014**, *70*, 6607–6612. [CrossRef]
29. Yu, Y.; Wei, Z.; Leng, X.; Li, Y. Facile preparation of stereochemistry-controllable biobased poly(butylene maleate-co-butylene fumarate) unsaturated copolyesters: A chemoselective polymer platform for versatile functionalization via aza-Michael addition. *Polym. Chem-UK* **2018**, *9*, 5426–5441. [CrossRef]
30. Li, R.Q.; Huang, T.; Gong, S.L. Preparation of high hydroxyl self-emulsifying polyester and compounding with acrylate. *J. Appl. Polym. Sci.* **2019**, *137*, 48278. [CrossRef]
31. Grobelny, J.N.m.r. study of maleate (cis)—fumarate (trans) isomerism in unsaturated polyesters and related compounds. *Polymer* **1995**, *36*, 4215–4222. [CrossRef]
32. Tang, T.; Takasu, A. Facile synthesis of unsaturated polyester-based double-network gels via chemoselective cross-linking using Michael addition and subsequent UV-initiated radical polymerization. *RSC Adv.* **2015**, *5*, 819–829. [CrossRef]

33. Farmer, T.J.; Clark, J.H.; Macquarrie, D.J.; Ogunjobi, J.K.; Castle, R.L. Post-polymerisation modification of bio-derived unsaturated polyester resins via Michael additions of 1,3-dicarbonyls. *Polym. Chem.* **2016**, *7*, 1650–1658. [CrossRef]
34. Lehtonen, J.; Salmi, T.; Immonen, K.; Paatero, E.; Nyholm, P. Kinetic Model for the Homogeneously Catalyzed Polyesterification of Dicarboxylic Acids with Diols. *Ind. Eng. Chem. Res.* **1996**, *35*, 3951–3963. [CrossRef]
35. Kang, H.; Li, M.; Tang, Z.; Xue, J.; Hu, X.; Zhang, L.; Guo, B. Synthesis and characterization of biobased isosorbide-containing copolyesters as shape memory polymers for biomedical applications. *J. Mater. Chem. B* **2014**, *2*, 7877–7886. [CrossRef]
36. DeRosa, C.A.; Kua, X.Q.; Bates, F.S.; Hillmyer, M.A. Step-Growth Polyesters with Biobased (R)-1,3-Butanediol. *Ind. Eng. Chem. Res.* **2020**, *59*, 15598–15613. [CrossRef]
37. Terzopoulou, Z.; Karakatsianopoulou, E.; Kasmi, N.; Tsanaktsis, V.; Nikolaidis, N.; Kostoglou, M.; Papageorgiou, G.Z.; Lambropoulou, D.A.; Bikiaris, D.N. Effect of catalyst type on molecular weight increase and coloration of poly(ethylene furanoate) biobased polyester during melt polycondensation. *Polym. Chem.* **2017**, *8*, 6895–6908. [CrossRef]
38. Dai, J.; Ma, S.; Wu, Y.; Han, L.; Zhang, L.; Zhu, J.; Liu, X. Polyesters derived from itaconic acid for the properties and bio-based content enhancement of soybean oil-based thermosets. *Green Chem.* **2015**, *17*, 2383–2392. [CrossRef]
39. Chen, T.; Tian, S.; Xie, Z.; Guo, Z.-X.; Xu, J.; Guo, B.-H. Two new approaches based on dynamic carboxyl–hydroxyl or hydroxyl–carboxyl transformation for high molecular weight poly(butylene maleate). *Polym. Chem.* **2020**, *11*, 5884–5892. [CrossRef]
40. Verissimo Lobo, V.T.; Pacheco Ortiz, R.W.; Gonçalves, V.O.O.; Cajaiba, J.; Kartnaller, V. Kinetic Modeling of Maleic Acid Isomerization to Fumaric Acid Catalyzed by Thiourea Determined by Attenuated Total Reflectance Fourier-Transform Infrared Spectroscopy. *Org. Process. Res. Dev.* **2020**, *24*, 988–996. [CrossRef]
41. Li, Q.; Tao, W.; Li, A.; Zhou, Q.; Shuang, C. Poly (4-vinylpyridine) catalyzed isomerization of maleic acid to fumaric acid. *Appl. Catal. A-Gen.* **2014**, *484*, 148–153. [CrossRef]
42. Lam, Y.P.; Lam, Z.; Yeung, Y.Y. Zwitterion-Catalyzed Isomerization of Maleic to Fumaric Acid Diesters. *J. Org. Chem.* **2021**, *86*, 1183–1190. [CrossRef]
43. Zheng, L.; Wang, Z.; Li, C.; Xiao, Y.; Zhang, D.; Guan, G.; Zhu, W. Synthesis, characterization and properties of novel linear poly(butylene fumarate) bearing reactive double bonds. *Polymer* **2013**, *54*, 631–638. [CrossRef]
44. Dunjic, B.; Sepulchre, M.O.; Sepulchre, M.; Spassky, N.; Djonlagic, J. Synthesis and rheological study of some maleic acid and fumaric acid stereoregular polyesters, 10-Synthesis and characterization of alpha, omega-dihydroxyoligo(alkylene maleate)s. *Macromol. Chem. Phys.* **1998**, *199*, 1051–1055. [CrossRef]
45. Wu, C.H.; Huang, Y.C.; Lai, T.H.; Chiu, S.H.; Uchibe, N.; Lin, H.W.; Chiu, W.Y.; Tung, S.H.; Jeng, R.J. Facile synthesis toward self-dispersible waterborne comb-like Poly(hydroxyaminoethers). *Polymer* **2020**, *196*, 122464. [CrossRef]
46. Martinet, F.; Guillot, J. Copolymerization of α-methylstyrene with methyl methacrylate. III. Emulsion process: Experimental data on kinetics, particle size, composition, molecular weight, and glass transition temperature. *J. Appl. Polym. Sci.* **1999**, *72*, 1627–1643. [CrossRef]
47. Querol, N.; Barreneche, C.; Cabeza, L. Storage Stability of Bimodal Emulsions vs. Monomodal Emulsions. *Appl. Sci.* **2017**, *7*, 1267. [CrossRef]
48. De Oca-Ávalos, J.M.M.; Candal, R.J.; Herrera, M.L. Nanoemulsions: Stability and physical properties. *Curr. Opin. Food Sci.* **2017**, *16*, 1–6. [CrossRef]
49. Wajnryb, E.; Dahler, J.S. The Viscosity of Polymerically Stabilized Dispersions of Spherical Colloid Particles. *J. Colloid Interface Sci.* **1999**, *217*, 259–268. [CrossRef]
50. Wever, D.A.Z.; Picchioni, F.; Broekhuis, A.A. Branched polyacrylamides: Synthesis and effect of molecular architecture on solution rheology. *Eur. Polym. J.* **2013**, *49*, 3289–3301. [CrossRef]
51. Lei, L.; Xia, Z.; Ou, C.; Zhang, L.; Zhong, L. Effects of crosslinking on adhesion behavior of waterborne polyurethane ink binder. *Prog. Org. Coat.* **2015**, *88*, 155–163. [CrossRef]
52. Koo, J.M.; Hwang, S.Y.; Yoon, W.J.; Lee, Y.G.; Kim, S.H.; Im, S.S. Structural and thermal properties of poly(1,4-cyclohexane dimethylene terephthalate) containing isosorbide. *Polym. Chem.* **2015**, *6*, 6973–6986. [CrossRef]
53. Cai, X.; Yang, X.; Zhang, H.; Wang, G. Modification of biodegradable poly(butylene carbonate) with 1,4-cyclohexanedimethylene to enhance the thermal and mechanical properties. *Polym. Degrad. Stab.* **2017**, *143*, 35–41. [CrossRef]
54. Hahm, S.; Kim, J.-S.; Yun, H.; Park, J.H.; Letteri, R.A.; Kim, B.J. Bench-Scale Synthesis and Characterization of Biodegradable Aliphatic–Aromatic Random Copolymers with 1,4-Cyclohexanedimethanol Units Toward Sustainable Packaging Applications. *ACS Sustain. Chem. Eng.* **2019**, *7*, 4734–4743. [CrossRef]
55. Bikiaris, D.N.; Papageorgiou, G.Z.; Giliopoulos, D.J.; Stergiou, C.A. Correlation between chemical and solid-state structures and enzymatic hydrolysis in novel biodegradable polyesters. The case of poly(propylene alkanedicarboxylate)s. *Macromol. Biosci.* **2008**, *8*, 728–740. [CrossRef]
56. Yang, Z.H.; Wu, G.F.; Zhang, H.X. Effects of the reagent molar ratio on the phase separation and properties of waterborne polyurethane for application in a water-based ink binder. *J. Appl. Polym. Sci.* **2017**, *134*, 45406. [CrossRef]
57. Wang, L.; Yan, F.A. Research on Synthesis of Hydroxy Waterborne Polyester-acrylic Resin Hybrid. *China Coatings* **2019**, *34*, 41–45. [CrossRef]
58. Rowe, M.D.; Eyiler, E.; Walters, K.B. Hydrolytic degradation of bio-based polyesters: Effect of pH and time. *Polym. Test.* **2016**, *52*, 192–199. [CrossRef]

59. Bayramoğlu, G.; Kahraman, M.V.; Kayaman-Apohan, N.; Güngör, A. The coating performance of adipic acid modified and methacrylated bisphenol- A based epoxy oligomers. *Polym. Adv. Technol.* **2007**, *18*, 173–179. [CrossRef]
60. Dai, J.; Ma, S.; Wu, Y.; Zhu, J.; Liu, X. High bio-based content waterborne UV-curable coatings with excellent adhesion and flexibility. *Prog. Org. Coat.* **2015**, *87*, 197–203. [CrossRef]
61. Dong, X.; Ren, J.; Duan, Y.; Wu, D.; Lin, L.; Shi, J.; Jia, R.; Xu, X.; He, X. Preparation and properties of green UV-curable itaconic acid cross-linked modified waterborne polyurethane coating. *J. Appl. Polym. Sci.* **2021**, *139*, e52042. [CrossRef]
62. Gogoi, G.; Gogoi, S.; Karak, N. Dimer acid based waterborne hyperbranched poly(ester amide) thermoset as a sustainable coating material. *Prog. Org. Coat.* **2017**, *112*, 57–65. [CrossRef]
63. Gao, C.; Han, S.; Zhang, D.; Wang, B.; Wang, C.; Wu, Y.; Liu, Y. A facile preparation of UV-cured films from waterborne unsaturated polyester via click reaction. *Prog. Org. Coat.* **2018**, *124*, 232–239. [CrossRef]
64. Scalarone, D.; Lazzari, M.; Castelvetro, V.; Chiantore, O. Surface Monitoring of Surfactant Phase Separation and Stability in Waterborne Acrylic Coatings. *Chem. Mater.* **2007**, *19*, 6107–6113. [CrossRef]
65. Zhong, X.; Hu, H.; Yang, L.; Sheng, J.; Fu, H. Robust Hyperbranched Polyester-Based Anti-Smudge Coatings for Self-Cleaning, Anti-Graffiti, and Chemical Shielding. *ACS Appl. Mater. Inter.* **2019**, *11*, 14305–14312. [CrossRef] [PubMed]

polymers

Review

Development of Synthesis and Application of High Molecular Weight Poly(Methyl Methacrylate)

Ming Yuan *, Dayun Huang and Yixuan Zhao

Department of Chemistry and Chemical Engineering, College of Ecology, Lishui University, Lishui 323000, China; dayunhuang@lsu.edu.cn (D.H.); zyx13566196913@163.com (Y.Z.)
* Correspondence: ym@lsu.edu.cn; Tel.: +86-0578-2271-458

Abstract: Poly(methyl methacrylate) (PMMA) is widely used in aviation, architecture, medical treatment, optical instruments and other fields because of its good transparency, chemical stability and electrical insulation. However, the application of PMMA largely depends on its physical properties. Mechanical properties such as tensile strength, fracture surface energy, shear modulus and Young's modulus are increased with the increase in molecular weight. Consequently, it is of great significance to synthesize high molecular weight PMMA. In this article, we review the application of conventional free radical polymerization, atom transfer radical polymerization (ATRP) and coordination polymerization for preparing high molecular weight PMMA. The mechanisms of these polymerizations are discussed. In addition, applications of PMMA are also summarized.

Keywords: poly(methyl methacrylate); high molecular weight; polymerization; application

Citation: Yuan, M.; Huang, D.; Zhao, Y. Development of Synthesis and Application of High Molecular Weight Poly(Methyl Methacrylate). *Polymers* **2022**, *14*, 2632. https://doi.org/10.3390/polym14132632

Academic Editor: Edina Rusen

Received: 10 June 2022
Accepted: 24 June 2022
Published: 28 June 2022

Publisher's Note: MDPI stays neutral with regard to jurisdictional claims in published maps and institutional affiliations.

Copyright: © 2022 by the authors. Licensee MDPI, Basel, Switzerland. This article is an open access article distributed under the terms and conditions of the Creative Commons Attribution (CC BY) license (https://creativecommons.org/licenses/by/4.0/).

1. Introduction

Poly(methyl methacrylate) (PMMA), commonly known as plexiglass, is a kind of polymer synthesized by free radical polymerization, ionic polymerization and coordination polymerization [1–3]. Compared with ordinary glass (silica glass), PMMA not only has a relatively low density, but also has a better crushing resistance. PMMA can be used as an excellent functional polymer material due to its good biocompatibility [4]. Moreover, PMMA is also widely used in medical technologies and photoelectronic devices as well as other areas [5].

In general, the PMMA with a number-average molecular weight exceeding 10^6 Da can be defined as ultrahigh molecular weight PMMA and the PMMA with a number-average molecular weight exceeding 10^5 Da can be defined as high molecular weight PMMA [6,7]. Low molecular weight PMMA has limited applications because of its poor heat resistance, poor wear resistance, poor organic solvent resistance, low hardness and easy combustion features. It has long been known that the mechanical properties of PMMA largely depend on its molecular weight [8–12]. Consequently, it is of great importance to develop feasible polymerization routes for the synthesis of high molecular weight PMMA. Conventional free radical polymerizations including emulsion polymerization and suspension polymerization can be used to synthesize high molecular weight PMMA [13]. Since it was discovered by Matyjaszewski in 1995 [14], atom transfer radical polymerization (ATRP) has become a successful approach for the preparation of well-defined polymers. The synthesis of high molecular PMMA by ATRP can also be realized. Other synthesis routes such as coordination polymerization, reversible addition-fragmentation chain transfer radical polymerization (RAFT) and plasma-initiated polymerization are also employed to prepare high molecular weight PMMA [15].

In this review, synthesizing high molecular weight PMMA by conventional radical polymerization, atom transfer radical polymerization and coordination polymerization from the aspects of catalytic activity, initiation activity, molecular weight and mechanism are discussed. Other polymerization methods, including reversible addition-fragmentation chain transfer radical polymerization (RAFT) and plasma-initiated polymerization, are also described. Applications of high molecular weight PMMA and its copolymers on medical, electricity and other areas are summarized. In addition, we briefly introduce some achievements of our group in the synthesis of ultra-high molecular weight vinyl polymers.

2. Synthesis of High Molecular Weight PMMA

2.1. Free Radical Polymerization

Free radical polymerization is one of the most important technologies for industrial production of polymers, owing to its simple operation and good reproducibility property. Free radical polymerization including conventional radical polymerization, ATRP and RAFT are effective methods for preparing high molecular weight PMMA. In this part, applications of conventional radical polymerization and ATRP for synthesizing high molecular weight PMMA are mainly discussed.

2.1.1. Conventional Radical Polymerization

It is difficult to obtain the polymer with controlled molecular weight and narrow molecular weight distribution by conventional radical polymerization because of the slow initiation, fast growth and fast termination characteristics of conventional free radical polymerization. However, conventional free radical polymerization has an excellent advantage for preparing high molecular weight polymers.

In general, the rate of free radical polymerization of vinyl monomer and the molecular weight of its vinyl polymer mainly depend on initiation, propagation, and termination steps. In a free radical polymerization, the rate of polymerization (R_p) should follow the equation of $R_p = k_p[M][I]^{0.5}(fk_d/k_t)^{0.5}$ (k_p is the propagation rate constant; k_d is the decomposition rate constant; k_t represents the termination rate constant; f represents the initiator efficiency) and the kinetic chain length (υ) should follow the equation of $\upsilon = k_p[M]/2(fk_dk_t[I])^{1/2}$ [16]. It indicates that the molecular weight of the polymer is increased with the decrease in initiator concentration. Therefore, a low initiator concentration is needed for preparing high molecular weight polymers. However, a low initiator concentration will cause the decrease in polymerization rate (R_p) and it needs much longer reaction time to realize the preparation of high molecular weight polymer with a high yield. Moreover, the polymerization will not occur under an extremely low initiator concentration [16]. Therefore, a suitable reaction system is essential for the synthesis of high molecular weight polymers.

Azodiisobutyronitrile (AIBN) and benzoyl peroxide (BPO) are the most commonly used initiators in conventional radical polymerization. Figure 1 shows the mechanism of conventional free radical polymerization of MMA initiated by BPO.

In a conventional free radical polymerization of MMA, PMMA can be synthesized by bulk polymerization, emulsion polymerization and suspension polymerization. High molecular weight PMMA can be prepared by bulk polymerization. However, it is difficult to remove the generated heat in the polymerization, resulting in a generation of gel effect and side reactions. Therefore, there is still obstacles to preparing high molecular weight PMMA with high yield through bulk polymerization. Compared with ionic polymerization, conventional free radical polymerization is more resistant to water. Consequently, dispersion polymerization, suspension polymerization and emulsion polymerization using water or other solvents as medium are proposed and used to synthesize high molecular weight PMMA.

Commonly, in a free radical polymerization, the initiating and chain growing free radicals are easily quenched by oxygen (O_2), generating stable and inactive peroxyl radicals. Therefore, it is important to solve the problem of oxygen inhibition in a radical polymerization. Degassing of the polymerization environment with an inert inactive gas, such as argon or nitrogen, is an effective method to prevent oxygen inhibition.

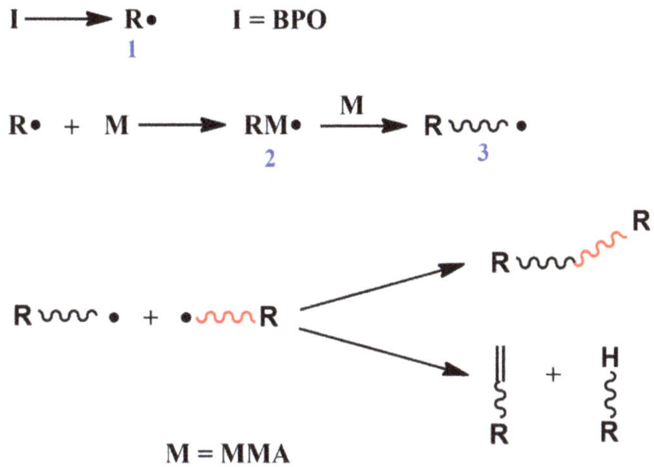

Figure 1. Conventional radical polymerization of MMA initated by BPO.

Hsiao et al. [17] report a polymerization of MMA by using poly(1,1-dihydroperfluorooctyl acrylate) [poly(FOA)] as a stabilizer, AIBN as an initiator, and supercritical CO_2 as a reaction medium. Due to the special properties of supercritical CO_2, the dissolution capacity of the solution is easily controlled by changing temperature and pressure. They find that the molecular weight of PMMA is influenced by stabilizer concentration and monomer concentration. A relatively higher monomer concentration is beneficial for preparing high molecular weight PMMA, and the PMMA with a number-average molecular weight (M_n) of 3.16×10^5 Da has been synthesized. Subsequently, Hsiao et al. [18] also report a dispersion polymerization of MMA with poly(FOA) as a stabilizer, AIBN as an initiator, and supercritical CO_2 as a reaction medium in the presence of helium. The high molecular weight PMMA with a M_n of 3.65×10^5 Da is synthesized. The solvency of supercritical CO_2 has an important influence on the particle size and particle size distribution of the synthesized PMMA. The solvency of supercritical CO_2 can be adjusted by the addition of helium. Consequently, the use of helium offers an attractive method for controlling the particle size and size distribution.

Lepilleur et al. [19] report a dispersion polymerization of MMA with poly(methyl methacrylate-*co*-hydroxyethyl methacrylate)-*g*-poly(perfluoropropylene oxide) [P(MMA-*co*-HEMA)-*g*-PFPO] as a stabilizer, AIBN as an initiator and supercritical CO_2 as a reaction medium. The effect of P(MMA-*co*-HEMA)-*g*-PFPO hyperbranched polymers with different molecular weights on molecular weight of the synthesized PMMA is investigated, and the PMMA with a high molecular weight of 3.55×10^5 Da has been synthesized. Actually, the solubility of P(MMA-*co*-HEMA)-*g*-PFPO hyperbranched polymers has a significant effect on the molecular weight distribution of PMMA, and P(MMA-*co*-HEMA)-*g*-PFPO with good solubility is beneficial for synthesizing narrow molecular weight distribution PMMA.

Although the addition of stabilizers improves the dispersing ability of the monomer in the polymerization system, it brings increased costs of production and post-processing. Wang and coworkers find that the dispersion polymerization of MMA using AIBN as an initiator and supercritical CO_2 as a reaction medium in the absence of stabilizers can realize the synthesis of high molecular weight PMMA [7]. It seems that the vapor-liquid equilibrium (VLE) is one of the key factors of the molecular weight. For instance, the polymerization conducted under 28.5–11.8 MPa, the PMMA with a low molecular weight is obtained and MMA-CO_2 mixture is in a homogeneous phase. However, when the polymerization conducted below 9.2 MPa, the MMA-CO_2 mixture is at VLE and the high molecular weight PMMA ($M_n = 1.31 \times 10^5$ Da) is synthesized.

Similarly to supercritical CO_2, ionic liquids (ILs) are also an environmentally friendly reaction medium. IL is a kind of low melting point (<100 °C) salt composed of a large volume of organic cation and an organic or inorganic anion [20,21]. ILs are widely investigated because of their advantages such as being non-volatile and non-flammable, and having high solubility and recyclability. ILs can be used as plasticizers for PMMA and poly(vinylidene fluoride) [22–24], and can also affect the formation of functional polymer materials such as coordination polymers [25], silicon-containing gels [26], and macroporous membranes [27].

Vygodskii et al. [28] report a free radical polymerization of MMA with AIBN as an initiator and ILs as a solvent, and ultrahigh molecular weight PMMA with a weight-average molecular weight (M_w) of 5.77×10^6 Da is synthesized. The used ILs are mainly composed of 1,3-dialkylimidazole cations and BF_4^- PF_6^-, SbF_6^-, $CF_3SO_3^-$, $(CF_3SO_3)_2N^-$, CH_3COO^-, $(CF_3CF_2)_3PF_3^-$ and other anions. They find that the molecular weight is affected by ILs, and ILs with a long carbon chain length are not beneficial forpreparing high molecular weight PMMA.

Suspension polymerization has a similar polymerization mechanism for bulk polymerization. However, it is easier to control the heat and the viscosity in suspension polymerization. Therefore, suspension polymerization is more promising for synthesizing high molecular weight PMMA with high monomer conversion. Moreover, suspension polymerization is also applied to prepare polymer-inorganic nanocomposites, which is widely used in optics [29], photoconductors [30,31], and electronic equipment areas [32]. Regrettably, the polymer produced by suspension polymerization inevitably contains a small amount of dispersant residue, which will affect the electrical properties of the product.

In fact, alongside a high molar ratio of monomer to initiator, a relatively low polymerization temperature is also needed for preparing high molecular weight polymers using suspension polymerization. This is because of the suppression of chain termination reaction and chain transfer reaction under low polymerization temperature. Therefore, a low-temperature initiator such as 2,2′-azobis (2,4-dimethylvaleronitrile) (ADMVN) can use to initiate the suspension polymerization of MMA at a low temperature.

Lyoo et al. [16] report a suspension polymerization of MMA conducted at 25 °C with poly (vinyl alcohol) (PVA) as a suspending agent and 2,2′-azobis (2,4-dimethylvaleronitrile) (ADMVN) as an initiator. The ultrahigh molecular weight PMMA with a M_n of 3.61×10^6 Da and monomer conversion of 83% is synthesized at 25 °C with a polymerization time of 96 h. The feed ratio has an observable effect on polymerization and the molecular weight of PMMA is increased with the decrease in the molar ratio of initiator to monomer. However, the polymerization fails to be realized when the molar ratio of initiator to monomer decreased to 5×10^{-5}. In the presence of silver nanoparticles, ultrahigh molecular weight ($M_\upsilon = 3.70 \times 10^6$ Da; M_υ is viscosity-average molecular weight) PMMA is also synthesized by suspension polymerization with PVA as a suspending agent and ADMVN as an initiator [33]. Actually, the advantage of addition of silver microspheres not only promotes the formation of PMMA/silver microspheres but is also beneficial in improving the polymerization rate.

Emulsion polymerization is one of the most commonly used methods to synthesize high molecular weight polymer owing to its high reactivity and property of simultaneous increase in the polymerization rate and molecular weight. However, it is unavoidable to add a large amount of emulsifier in emulsion polymerization, resulting in a decrement in the properties of synthesized polymers. Consequently, both differential microemulsion polymerization and soap free emulsion polymerization are developed and applied to the synthesis of high molecular weight polymers.

Jiang et al. [34]. propose a microemulsion polymerization of MMA using sodium dodecylsulfate (SDS) as a surfactant, 1-pentanol (n-Pt) as a co-surfactant, and ammonium persulfate (APS) as an initiator. The dosage of surfactant is obviously decreased, which provides beneficial conditions for the interaction between the free radicals and polymer particles. Therefore, it is easier to synthesize PMMA with narrow molecular weight distribution and high molecular weight. By optimizing the conditions, the high molecular weight PMMA with a weight-average molecular weight of 7.60×10^5 Da and a molecular weight distribution (M_w/M_n, PDI) of 1.40 has been successfully synthesized in their group. Moreover, this polymerization method provides a new route for preparing multi-chain PMMA particles. Norakankorn et al. [35] report a differential microemulsion polymerization of MMA with SDS as surfactant and AIBN as initiator, and PMMA with a high number-average molecular weight of 1.00×10^6 Da has been successfully prepared under a condition of [MMA]:[SDS] = 130:1 (weight ratio). As an oil-soluble initiator is used in this reaction, the particle nucleation is believed to take place in micelles and follow the heterogeneous nucleation mechanism.

In recent years, the semiconductor nanoparticles composed of transition metal ions and group VI and group VII ions have attracted much attention of chemists due to their wide applications in solar cells, photovoltaics, biosensors, and environmental protection areas [36,37]. Based on this, the semiconductor nanoparticles mediated photopolymerization of vinyl monomers has attracted the great attention of chemists [38,39]. In addition, the use of O_2 in the initiation mechanism has also been investigated in photopolymerization [40,41].

In the presence of O_2, Dadashi-Silab et al. [37] find that the ZnO or Fe/ZnO semiconductor nanoparticles can be used as photoinitiators for free radical polymerization of MMA under the irradiation of UV light. The addition of triethylamine (TEA) and diphenyliodonium hexafluorophosphate ($Ph_2I^+PF_6^-$) can promote the polymerization, and the PMMA with a M_n of 2.10×10^5 Da is synthesized via Fe/ZnO/TEA system. They believe that the addition of TEA and $Ph_2I^+PF_6^-$ will promote the formation of radical capable of initiating the polymerization. Mandal et al. [42] report a photopolymerization of MMA with ZnO or ZnO/Ag nanoparticles as a catalyst under the irradiation of UV light. A relatively high polymerization rate is achieved in the ZnO/Ag catalyst system, and the monomer conversion is reached to 82% after 9 h. In the ZnO catalyst system, the polymerization rate is comparatively slower, and the monomer conversion is only upped to 50% after 10 h. The molecular weight of PMMA produced in the ZnO/Ag system is higher than in the ZnO system, and the PMMA with a M_n of 1.94×10^5 Da is prepared. The mechanism has also been investigated, finding that the hydroxyl radical generated due to the photolysis of water by ZnO/Ag nanoparticle is the active species for the polymerization.

2.1.2. Atom Transfer Radical Polymerization (ATRP)

The high molecular weight polymer with a controlled structure is widely used in crystallization characteristics control, rheology modification, mechanical performance and morphology development [43]. Conventional free radical polymerization has the advantage of preparing high molecular weight polymers. However, it is difficult to prepare polymers with a controlled structure and a narrow molecular weight distribution through conventional free radical polymerization. In 1995, Matyjaszewski and coworkers firstly report a "living"/controlled polymerization of styrene via atom transfer radical polymerization (ATRP) [14]. In 2001, Matyjaszewski and coworkers systematically summarized the development of ATRP [44]. In this period, a series of catalytic systems were developed and then used for the preparation of well-defined polymers. Nowadays, ATRP also becomes an effective route for preparing high molecular weight polymers [45,46]. The mechanism of copper mediated normal ATRP of MMA is shown in Figure 2.

As shown in Figure 2, under the catalysis of a transition metal complex (Cu^IX/L), the radical (R•) is formed via a reversible redox process which undergoes an abstraction of a halogen atom (X) from a dormant species (R-X). The chain growth reaction is realized via the addition of monomer to the active species (R• or RM_n•). There exists reversible dynamic equilibrium (k_a/k_d) between active specie (RM_n•) and dormant specie (RM_nX). Owing to this, the molecular weight of the polymer is increased with the increase in monomer conversion. Actually, similarly to conventional free radical polymerization, termination reactions such as radical coupling and disproportionation also occur in ATRP. Differently, there is no more than a few percent of the polymer chains which undergo termination in a well-controlled ATRP [44]. Oxidized metal complexes ($Cu^{II}X_2/L$) generated in the system have a significant effect on the chain termination during the initial stage of the polymerization. The formation of $Cu^{II}X_2/L$ will reduce the chain growth radical concentration and thereby decrease the occurrence of termination reaction.

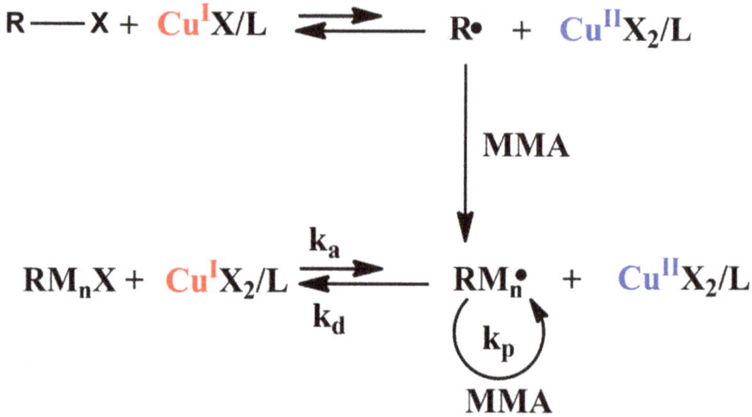

Figure 2. Mechanism of copper mediated normal ATRP of MMA [44].

Typically, the normal ATRP system is composed of monomer, organic halide initiator and catalyst. A variety of monomers such as styrenes and (meth)acrylates are successfully polymerized via ATRP. Each monomer has its own suitable atom transfer equilibrium constant ($K_{eq} = k_a/k_d$) for its active and dormant species. ATRP will occur very slowly or even fail if the K_{eq} is too small. Conversely, a large K_{eq} will cause a high radical concentration and thereby cause a large amount of termination [45]. Therefore, it is important to control the active species concentration and radical deactivation rate in the polymerization system.

The initiator is also a key factor in ATRP owing to its control of the number of macromolecular growth chain. Using a good ATRP initiator, the initiation should be fast and quantitative. Therefore, the number of macromolecular growth chain can be controlled. In addition, the selection of initiator will affect the formation of polymer. One of the reasons is that the migration speed of the halide group (X) will affect the PDI of synthesized polymer and a fast migration speed of the halide group between the chain growth radical and the transition-metal complex is beneficial to obtain a polymer with small PDI. It means that a successful ATRP should have features such as a small amount of terminated chains and a uniform growth of all the chains.

In order to prepare high molecular weight PMMA, we should firstly know the kinetics of ATRP. Typically, in a copper mediated normal ATRP, the polymerization rate (R_p) follows the equation of $R_p = k_p[M][P^*] = k_p K_{eq}[M][I_0] \times [Cu^I]/[Cu^{II}]$ and the degree of polymerization (DP) follows the equation of $DP = \Delta M/[I_0]$ [44,47]. Therefore, we can know that a high molar ratio of monomer to initiator is necessary for preparing high molecular weight polymers. In addition, the decrease in initiator concentration can also decrease the occurrence of termination reaction [43]. However, this will inevitably cause a decrease in polymerization rate.

Agarwal et al. [43] report a ATRP of MMA with phenyl 2-bromo-2-methylpropionate (BMPE) as an initiator, CuCl as a catalyst and *N,N,N′,N″,N″*-pentamethyldiethylene triamine (PMDETA) or 4,4′-dinonyl-2,2′-bipyridine (dnNbpy) as a ligand. The molecular weight synthesized in PMDETA system is deviated from the theoretical values, which may be caused by coupling termination of free radicals. High molecular weight PMMA with a M_n of 3.67×10^5 Da and molecular weight distribution of 1.20 is prepared in dnNbpy system with a molar ratio of [MMA]:[BMPE]:[CuCl] = 6400:1:6.

The star polymer attracts a lot of interests because of its branched structure. Similarly to the preparation of linear polymers, the preparation of high molecular weight star polymers also requires a low extent of chain transfer and chain termination, high initiation efficiency, and moderate polymerization rate. Agarwal and coworkers [48] report a ATRP of MMA using 2,3,6,7,10,11-triphenylhexyl-2-bromo-2-methylpropionate (TP6Br) or ethylene glycol bis[3,4,5-tris(2-Bromo-2-methylpropionate)]benzoate ($EG(Bz3Br)_2$) as an initiator, CuCl as a catalyst, and dnNbpy as a ligand. The high molecular weight six-arm star PMMA with a M_w of 2.35×10^6 Da is synthesized using TP6Br as initiator. In addition, a six-arm star PMMA with a M_w of 2.10×10^6 Da is successfully synthesized using $EG(Bz3Br)_2$ as initiator.

A high concentration of catalyst is often required to ensure the controllability of normal ATRP, which will cause an increment in side reaction and the decrement in molecular weight. The concentration of the catalyst can be significantly decreased in an electron transfer activation regeneration catalyst atomic transfer radical polymerization (ARGET ATRP), resulting in a decrease in side reactions. Therefore, the ARGET ATRP may be applied to prepare higher molecular weight polymers. The mechanism of copper mediated ARGET ATRP of MMA is shown in Figure 3.

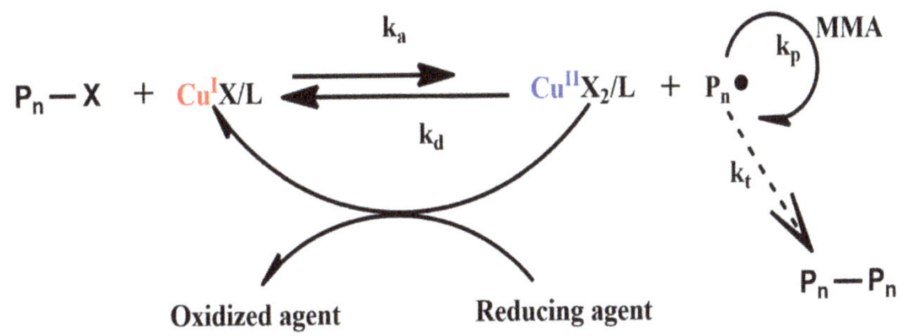

Figure 3. Mechanism of copper mediated ARGET ATRP of MMA (Reprinted with permission from Ref. [49]. Copyright 2007 American Chemical Society).

Different from normal ATRP, in a copper mediated ARGET ATRP, the transition metal complex in its higher oxidation state is added to the system. The activator species (Cu^I) are constantly regenerated from the transition metal complex in its higher oxidation state (Cu^{II}) by the reduction in reducing agent. Therefore, the amount of Cu(II) species is much lower than the amount of reducing agent. Actually, the use of Cu(II) species can be reduced to around 10 ppm. The extent of the side reaction is reduced owing to the drastically smaller amounts of required Cu(II) species and thereby high molecular weight polymer can be prepared via ARGET ATRP [50].

Alkyl dithioester is a commonly used RAFT chain transfer agent (CTA), which can be activated by free radicals or ATRP catalysts to initiate the polymerization of MMA. Therefore, CTA acted as an alkyl pseudohalide is also an ATRP initiator. Matyjaszewski et al. [50] report an ARGET ATRP of MMA by using cumyl dithiobenzoate (CDB) as an initiator, CuBr as a catalyst, copper powder or copper wire as a reducing agent. The effects of catalyst concentration on the polymerization rate and the molecular weight of PMMA are investigated. Within a certain catalyst concentration range (0.33–50 ppm), increasing the catalyst concentration will cause an increment in the polymerization rate and a decrement in the molecular weight. The high molecular weight PMMA ($M_n = 1.25 \times 10^6$ Da) with a narrow molecular weight distribution (PDI = 1.21) is synthesized. Moreover, the high molecular weight ($M_n = 2.22 \times 10^5$ Da) poly(styrene) (PS) and high molecular weight ($M_n = 1.80 \times 10^6$ Da) poly(methyl methacrylate)-*b*-poly(butyl methacrylate) (PMMA-*b*-PBMA) are also synthesized. Matyjaszewski et al. [51] also report a surface-initiated ATRP (SI-ATRP) of MMA using SiO_2-Br as an initiator. The polymerization of MMA can be successfully realized with a low concentration of catalysts (12.5 ppm) and producing a high molecular weight ($M_n = 1.70 \times 10^6$ Da) SiO_2-*b*-PMMA.

Youk et al. [52] report that the high molecular weight 3-arm PMMA can be synthesized via ARGET ATRP by using 1,3,5-tris(2-bromoisobutoxy)benzene (TBIB) as an initiator,

$CuBr_2$ as a catalyst, PMDETA as a ligand and tin(II) 2-ethylhexanoate [$Sn(EH)_2$] as a reducing agent. The effects of initiator concentration, catalyst concentration and reducing agent concentration on the molecular weight of PMMA have been investigated, and the high molecular weight PMMA ($M_n = 5.70 \times 10^5$ Da, PDI = 1.36) is successfully prepared under a molar ratio of [$CuBr_2$]:[PMDETA]:[$Sn(EH)_2$] = 0.1:1.5:1.5.

In a free radical polymerization system, the molecular weight of synthesized polymers is proportional to the ratio of $k_p{}^2/k_t$, and the propagation rate constant (k_p) can be improved by increasing the pressure [53,54]. Therefore, increasing polymerization pressure is beneficial to prepare high molecular weight polymer.

Matyjaszewski et al. [55] report a ATRP of MMA by using $CuBr_2$ as a catalyst, EBiB as an initiator, tris(2-pyridylmethyl)amine (TPMA) as a ligand and ascorbic acid (AsAc) as a reducing agent. The effect of pressure (10^{-3}, 3, 6, and 10 kbar) on the polymerization is investigated. The polymerization proceeded at 3 kbar has a relatively higher polymerization rate than the polymerization conducted at atmospheric pressure. The high molecular weight ($M_n = 5.93 \times 10^5$ Da) PMMA can be synthesized at 10 kbar, but a relatively low monomer conversion (46%) and high molecular weight distribution (PDI = 1.96) are obtained. The polymerization rate and molecular weight can both be increased at 6 kbar, giving a well-defined PMMA with a M_n of 1.88×10^6 Da.

In the presence of a low catalyst concentration ([CuCl] = 14 mM), Fukuda et al. [53] systematically investigated the effect of pressure (0.1-500 MPa) on the polymerization of MMA. The controllability of the polymerization is increased by increasing the pressure, producing PMMA with a narrow molecular weight distribution (PDI = 1.20–1.60). Under a 500 MPa pressure condition, the polymerization rate is 20 times faster than that at ambient pressure and produces the PMMA with a M_n of 5.00×10^5 Da and PDI of 1.25. However, the controllability of the system will decrease at higher monomer conversion. In order to improve the controllability, 0.03 mM $CuCl_2$ used as a deactivator is added into the original feed. The polymerization is also conducted at 500 MPa, finding that the polymerization rate and PDI are both decreased, and giving a well-defined high molecular weight PMMA ($M_n = 1.50 \times 10^6$ Da, PDI = 1.25). The effect of initiator concentration on the molecular weight is also investigated, and the PMMA with a high molecular weight ($M_n = 3.60 \times 10^6$ Da) is synthesized with a low initiator concentration of 0.047 mM.

2.2. *Coordination Polymerization*

Coordination polymerization is also an effective method for preparing high molecular weight PMMA [56]. The mechanism of coordination polymerization of MMA is shown in Figure 4.

In a coordination polymerization, the catalytic activity and the polymer molecular weight are significantly affected by metal, ligand, temperature, cocatalyst and molar ratio of monomer to catalyst [57–65]. Generally, increasing the molar ratio of monomer/catalyst will cause an increment in the molecular weight. The steric encumbrance in the ligand around the metal center has an obvious influence on the catalytic activity. For instance, for a pyrazole-based ligand, if the pyrazole moiety has a substituent, the catalytic activity will be enhanced in the polymerization of MMA. The bulkier the substituents on pyrazole moiety, the higher the catalytic activity for MMA polymerization. The solubility of the catalyst will also affect the catalytic activity. For a typical Ziegler-Natta catalyst such as $TiCl_4$-AlR_3, if the halogen has been substituted by cyclopentadienyl (Cp), the obtained catalyst (Cp_2TiCl_2-AlR_3) will be soluble in the organic solvent. The polymerization rate can be increased by increasing the temperature. However, this will cause a decrement in the molecular weight. According to the literature [57–69], temperature range from 30 °C to 80 °C might be conducive to the synthesis of high molecular weight PMMA in a coordination polymerization. The cocatalyst (AlR_3) is also a key factor for the polymerization of MMA. Within a certain range, both polymerization rate and polymer molecular weight can be increased by increasing the concentration of cocatalyst. However, a high concentration of cocatalyst will cause a decrement in molecular weight, owing to the chain transfer reaction.

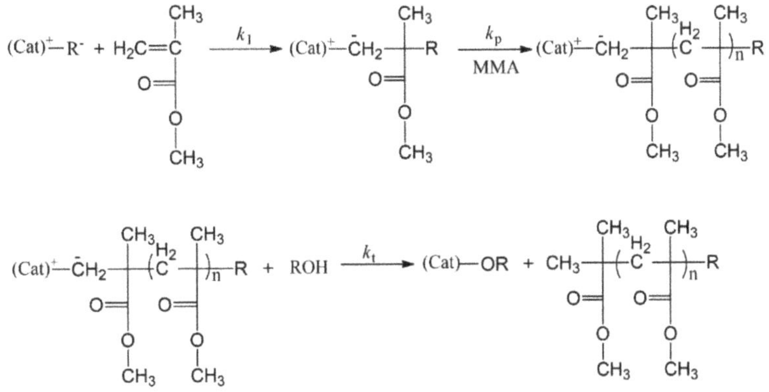

Figure 4. Mechanism of coordination polymerization of MMA.

The electron-donating atoms such as O and N in polar monomers are easy to coordinate with the catalyst, forming a stable complex and destroying the catalytic ability. Consequently, the typical Ziegler-Natta catalyst is not quite suitable for the polymerization of polar monomers such as AN and MMA. Homogeneous metallocene activated by aluminum alkyl (AlR_3) can induce a coordination polymerization of MMA but showing a relatively poorer efficiency [57]. Lanthanocene catalysts have high activity in the coordination polymerization of MMA, and easily produce high molecular weight PMMA [56].

Sun et al. [57] report a coordination polymerization of MMA by using ethylene bridged of titanocene chloride and samarocene (Sm-Ti) as a single component catalyst or using the triisobutylaluminum (TIBA) activated Sm-Ti as a catalyst. Both the binary catalyst and the single component catalyst are soluble in the polymerization system. The binary catalyst shows a higher activity than single component catalyst but provides PMMA with a relatively lower molecular weight. In the single component catalytic system, the molecular weight of PMMA is increased with the increase in temperature. However, the molecular weight of PMMA synthesized in binary catalytic system is decreased with the increase in temperature. By optimizing the reaction conditions, the PMMA with a high M_υ of 1.50×10^6 Da and monomer conversion of 87% has been synthesized with a polymerization time of 20 h and temperature of 80 °C.

Schiff base is an important nitrogen donor ligand, and a large number of schiff bases are investigated due to their activity to reversibly bind oxygen [58]. Yousaf et al. [59] propose that lanthanocene complexes with schiff base derivatives or methoxyethylindenyl derivatives as ligands can be used as an effective catalyst for the polymerization of MMA. The catalyst is easily dissolved in toluene and thus the homogeneous catalysis can be realized. The effects of the catalyst such as [(CP)(Cl)Ln Schiff base (THF)] (Ln = Sm, Dy, Er, Y) and (COT)Ln (methoxyethylindenyl) (THF) (Ln = La, Nd, Sm, Dy, Er) on the polymerization are investigated. Compared to (COT)Ln (methoxyethylindenyl) (THF), the [(CP)(Cl)Ln Schiff base (THF)] shows a better activity in the polymerization. The effects of temperature and feed ratio on the polymerization have also been investigated, and the high molecular weight PMMA with a M_υ of 4.16×10^5 Da has been synthesized with a polymerization time of 20 h and temperature of 70 °C.

Compared to the early transition metal, the later transition metal has a lower oxygen affinity and higher functional group tolerance. Therefore, the later transition metal can also be applied to the polymerization of polar monomers. For instance, late transition metals such as Cu(II), Pd(II), Zn(II), Cd(II) and Co(II) complexed with *N*-substituted pyridylamine ligands are used as the catalysts in the polymerization of MMA [60,61]. Lee and coworkers report that Zn(II) complexes bearing 2-iminomethylquinoline and 2-iminomethylpyridine based ligands can be applied to the polymerization of MMA [62]. The bidentate $[(NN')ZnCl_2]$ complexed with modified methylaluminoxane (MMAO) is used as a catalyst and toluene is used as a solvent in the polymerization of MMA. The PMMA with a M_w of 9.62 × 10^5 Da is prepared at 60 °C with a polymerization time of 2 h.

Lavastre et al. [63] report that $CuCl_2$ or $Cu(OAc)_2$ combined with *N*-tripod ligand is an effective catalyst for the polymerization of MMA in the presence of methylaluminoxane (MAO). The commonly used *N*-tripod ligands are shown in Figure 5.

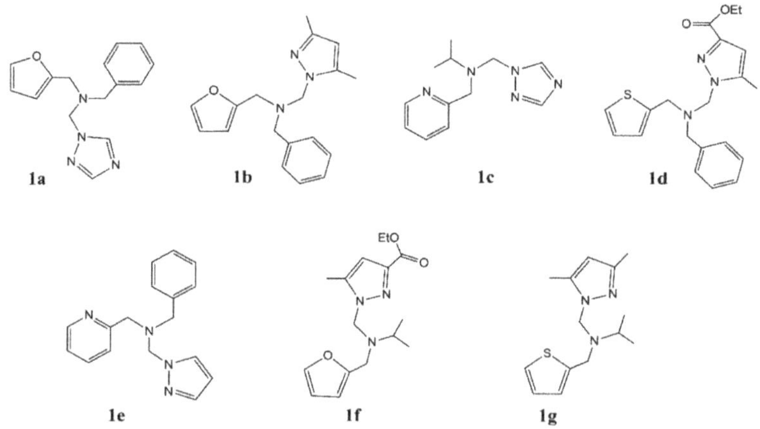

Figure 5. The structure of *N*-tripod ligands (Reprinted with permission from Ref. [63]. Copyright 2007 Elsevier B.V.).

$CuCl_2$ or $Cu(OAc)_2$ alone does not induce the polymerization, indicating that the *N*-tripod ligand is also a key factor for the polymerization. In $Cu(OAc)_2$ catalyst systems, the PMMA is prepared with a yield of 23% to 40%. $CuCl_2$ combined with ligands **1b** or **1g** gave a PMMA with a yield of 60%. The molecular weight of PMMA prepared in $Cu(OAc)_2$ system is higher than that produced in $CuCl_2$ system. In the presence of toluene, the PMMA with a M_n of 3.39 × 10^5 Da is synthesized in $Cu(OAc)_2$/**1b** system with a polymerization time of 4 h and temperature of 30 °C.

Lee and coworkers [64] report a coordination polymerization of MMA using the cobalt (II) complex combined with *N*,*N*-bis{(1-pyrazolyl)methyl}aniline (bpmaL1) ligand as a catalyst in the presence of MMAO. A series of bidentate [*N*,*N*]-cobalt (II) dichloride complexes are synthesized and presented in Figure 6.

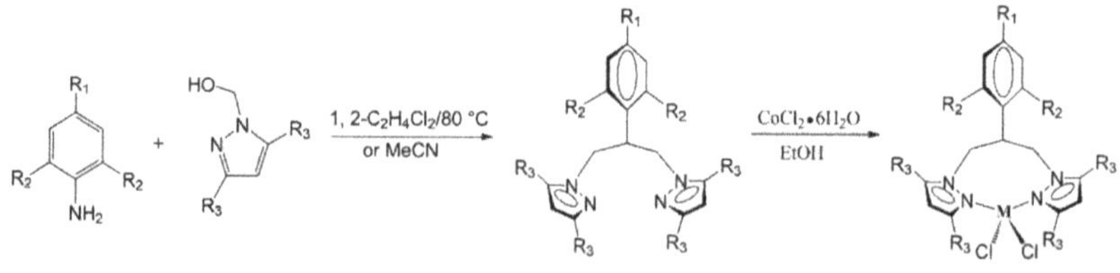

Figure 6. Synthesis of cobalt (II) complex catalyst (Reprinted with permission from Ref. [64]. Copyright 2010 Elsevier B.V.).

Effects of catalysts on the polymerization rate and molecular weight have been studied. The 4w/MMAO (1.14 × 10^6 g/mol-Co·h) shows the highest activity, 2w/MMAO (8.00 × 10^5 g/mol-Co·h) and 3w/MMAO (8.73 × 10^5 g/mol-Co·h) have a moderate activity, and 1w/MMAO (6.17 × 10^4 g/mol-Co·h) presents the lowest activity. However, the

1w/MMAO system is more suitable for preparing high molecular weight PMMA. With a polymerization temperature of 60 °C, the PMMA with a M_n of 1.13 × 10^6 Da and PDI of 1.75 has been synthesized in 1w/MMAO system. Later, Lee et al. [65] also report a polymerization of MMA by using 4-coordinate cobalt(II) complexes bearing *N′*-substituted *N,N′,N*-bis((1H-pyrazol-1-yl)methyl)amine derivatives ligands as catalysts in the presence of MMAO. In the presence of toluene, the high molecular weight PMMA with a M_w of 1.05 × 10^6 Da is prepared at 60 °C with a polymerization time of 2 h.

Half-sandwich nickel(II) complexes [Ni(■^{5}-C_5H_4R)(X)(NHC) (R = H or alkyl, X = Cl, Br, I)] (NHC = N-heterocyclic carbenes) are an efficient catalyst for the polymerization of styrene and phenylacetylene [66,67]. The Ni(II) complexes such as Ni(acac)$_2$ can be used as a catalyst for the polymerization of MMA in the presence of MAO [68]. In the presence of MAO, Buchowicz and coworkers find that the [Ni(■^{5}-C_5H_4R)(X)(NHC) (R = H or alkyl, X = Cl, Br, I)] complex can also be used as a catalyst for the polymerization of MMA [69]. The Ni(II) complex or MAO alone does not induce the polymerization, indicating the polymerization must be conducted by both the participation of Ni(II) complexes and MAO. The effect of catalyst on the molecular weight of PMMA is investigated. After a polymerization time of 3 h at 50 °C, the ultrahigh molecular weight PMMA with a M_n of 1.50 × 10^6 Da has been successfully prepared in the presence of toluene.

2.3. Other Polymerizations

Except for conventional free radical polymerization, atom transfer radical polymerization (ATRP) and coordination polymerization, other polymerizations such as reversible addition-fragmentation chain transfer polymerization (RAFT), anionic polymerization and plasma-initiated polymerization are also used to prepare high molecular weight PMMA. However, reports of RAFT, anionic polymerization and plasma-initiated polymerization for preparing high molecular weight PMMA are much lower than radical polymerization.

Rzayev et al. report a RAFT of MMA by using AIBN as an initiator and cyanoisopropyl dithiobenzoate as a RAFT [6]. The effect of pressure (5 or 9 kbar) on the molecular weight of PMMA is investigated. The PMMA with M_n of 1.50 × 10^5 Da and 2.02 × 10^5 Da are synthesized under 9 kbar and 5 kbar, respectively. The effect of feeding ratio on the molecular weight is also investigated. Moreover, the ultrahigh molecular weight PMMA with a M_n of 1.25 × 10^6 Da is synthesized with a molar ratio of [MMA]:[RAFT agent]:[AIBN] = 1.20 × 10^4:1:0.2.

Kitayama and coworkers report an anionic polymerization of MMA using α-lithioisobutyrate (Li-iPrIB) as initiator in the presence of lithium trimethylsilanolate (Me_3SiOLi) [70]. The molecular weight of PMMA is significantly influenced by the feeding ratio. Moreover, the PMMA with a high molecular weight (M_n = 8.25 × 10^5 Da) and low polydispersity index (PDI = 1.16) has been prepared under a molar ratio of [MMA]:[initiator] = 3000:1.

Since it was discovered by Osada and coworkers [71], the plasma-initiated polymerization has attracted a lot of attention of the chemists. It is easy to synthesize ultrahigh molecular weight PMMA with a M_n upped to 10^7 Da through plasma-initiated polymerization [71,72]. The plasma-initiated polymerization has a complicated mechanism, and the most accepted mechanism at present is the radical polymerization mechanism. Plasma-initiated polymerization possesses an outstanding advantage for preparing high molecular weight polymers due to its little number of macromolecular chains. Cheng et al. report that the combination of RAFT and plasma-initiated polymerization can realize the polymerization of MMA [73]. The 2-cyanoprop-2-yl-1-dithionaphthalate (CPDN) is synthesized in their group, and then using as a RAFT reducing agent in the plasma-initiated RAFT polymerization of MMA without the use of any other thermal initiators or photoinitiators. The effect of CPDN concentration on the polymerization of MMA is investigated, finding that the controllability of the polymerization is increased with the increase in CPDN concentration. The polymerization of MMA is also realized in the absence of CPDN, and produces PMMA with an ultrahigh molecular weight of 1.41 × 10^6 Da.

Nanoparticles have high catalytic activity due to their large surface energy. In our group, the polymerization of MMA with organic halide as initiator in the presence of palladium nanoparticles has been investigated. The ultrahigh molecular weight PMMA with a M_n of 4.65 × 10^6 Da and a M_w of 8.08 × 10^6 Da is synthesized at 70 °C using 2-bromoisobutyric acid ethyl ester (EBiB) as an initiator in the presence of catalytical amount (10.1 ppm) of Pd NPs [74].

As the development of research, much more polymerization methods are reported and used for the preparation of high molecular weight PMMA [75,76]. Some typical experimental results for the preparation of high molecular weight PMMA are summarized in Table 1.

Table 1. The results of the polymerization of high molecular weight PMMA.

Entry	Catalyst	Ligand/Additive	Initiator	Temp. (°C)	Time (h)	Conv. (%)	M_n (g/mol)	M_w/M_n	Refs.
1	/	cyanoisopropyl dithiobenzoate	AIBN	65	7.0	99.0	1.25 × 10^6	1.03	[6]
2	/	Supercritical CO_2	AIBN	65	10.0	100.0	1.31 × 10^5	2.54	[7]
3	/	PVA	ADMVN	25	96.0	83.0	3.61 × 10^6	2.40	[16]
4	/	Poly(FOA)	AIBN	N/A	4.0	92.0	3.16 × 10^5	2.09	[17]
5	/	Poly(FOA)	AIBN	65	4.0	88.0	3.65 × 10^5	2.48	[18]
6	/	P(MMA-co-HEMA)-g-PFPO	AIBN	65	10.0	90.0	3.55 × 10^5	1.70	[19]
7	/	SDS	APS	70.0	N/A	N/A	5.30 × 10^5	1.43	[34]
8	/	SDS	AIBN	70	N/A	50.0	1.00 × 10^6	1.80	[35]
9	/	TEA	Fe/ZnO	N/A	N/A	36.0	2.10 × 10^5	2.85	[37]
10	/	poly(1-vinyl-3-butylimidazolium ascorbate)	ZnO/Ag	30	9.0	82.0	1.94 × 10^5	1.38	[42]
11	CuCl	dnNbpy	BMPE	90	2.0	N/A	3.67 × 10^5	1.20	[43]
12	CuBr	Copper powder	CDB	80	125.0	43.9	1.25 × 10^6	1.21	[50]
13	$CuBr_2$	Tris(2-dimethylaminoethyl)amine	SiO_2-Br	60	24.0	40.5	1.70 × 10^6	1.31	[51]
14	$CuBr_2$	PMDETA	TBIB	90	4.5	53.0	5.70 × 10^5	1.36	[52]
15	CuCl	dnNbpy	EBiB	60	24.0	N/A	3.60 × 10^6	1.24	[53]
16	$CuBr_2$	TPMA	EBiB	20	15.0	57.0	1.88 × 10^6	1.25	[55]
17	$Cu(OAc)_2$	MAO	/	30	2.0	31.0	3.39 × 10^5	N/A	[63]
18	$CoCl_2$	*N,N*-bis{(1-pyrazolyl)methyl}aniline	/	60	N/A	N/A	1.13 × 10^6	1.75	[64]
19	Ni(II) complexes	MAO	/	50	3.5	20.0	1.50 × 10^6	N/A	[69]
20	/	Me_3SiOLi	Li-iPrIB	−78	1.0	100.0	8.25 × 10^5	1.16	[70]
21	/	/	/	25	16	5.0	1.41 × 10^6	2.03	[73]
22	Palladium nanoparticles	/	EBiB	70	24	82.8	4.65 × 10^6	1.73	[74]

3. Applications of High Molecular Weight PMMA

High molecular weight PMMA and its copolymers are important precursors for the preparation of polymer materials with high light transmittance, excellent electrochemical performance and mechanical properties. PMMA and its copolymers or composites are widely used in electronic equipment, medical technology, and polymer membranes areas. The relationship between molecular weight and physical properties is deeply investigated [77,78]. For instance, the mechanical properties of PMMA are significantly affected by the molecular weight (Figure 7). It shows that the tensile strength, fracture surface energy, shear modulus and Young's modulus increase with the increase in molecular weight (M_υ) up to 10^6 Da. Therefore, the synthesized high molecular weight PMMA will further expand the application value of PMMA.

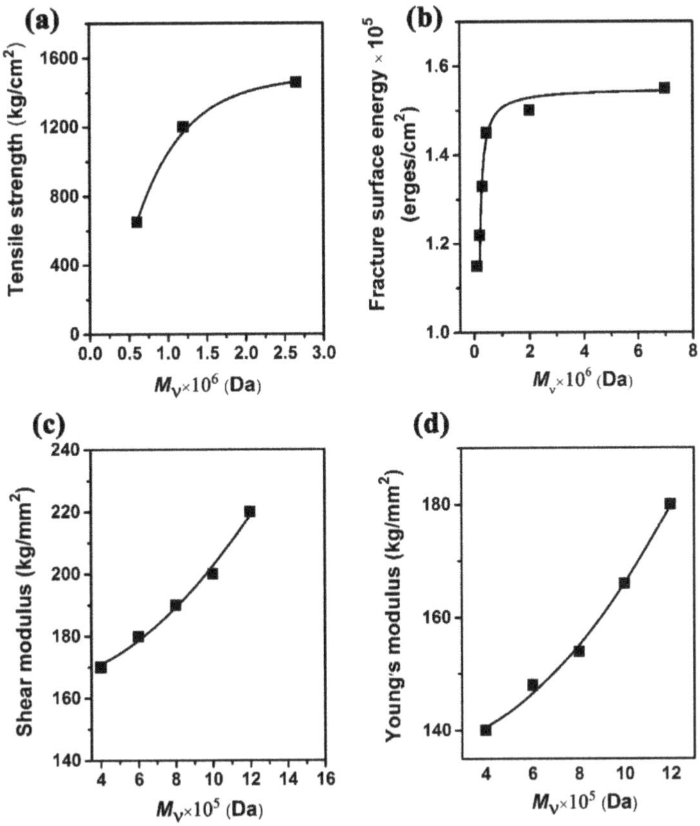

Figure 7. Relationships between mechanical strength ((**a**): Tensile strength; (**b**): Fracture surface energy; (**c**): Shear modulus; (**d**): Young's modulus) of PMMA and its molecular weight [77,78].

3.1. Application of High Molecular Weight PMMA in Medical Field

3.1.1. PMMA Bone Cements

Osteoporotic vertebral compression fracture (OVCF) is one of the most common complications of osteoporosis. At present, percutaneous vertebroplasty (PVP) is widely used for the treatment of OVCF. PMMA bone cement shows an important function in PVP. In a PVP surgery, PMMA bone cement is injected into the injured vertebra of the patient, which can quickly relieve pain, stabilize and strengthen the injured vertebra, and restore the height and angle of the injured vertebra [79–81]. However, the strength of ordinary PMMA bone cement is weaker than the bone [82]. Therefore, high molecular weight PMMA can use to prepare higher strength PMMA bone cement. PMMA with viscosity-average molecular weights (M_υ) range from 1.7×10^5 Da to 7.5×10^5 Da are successfully used for the preparation of PMMA bone cements [83], indicating that high molecular weight PMMA might be an excellent raw material for PMMA bone cements.

Inadequate strength at the cement/bone interface is one of the main drawbacks of PMMA bone cement in the current orthopedic surgeries. PMMA cement strength, surface roughness properties and osteo-blast cell growth can be improved by incorporating additives such as MgO, chitosan and hydroxyapatite to PMMA [84–86]. Khandaker and coworkers have investigated the fracture toughness (K_{IC}) of bone-PMMA with nano MgO particles or micro MgO particles, finding that the K_{IC} of bone-PMMA with nano MgO particles and bone-PMMA with micro MgO particles are much higher than the K_{IC} of bone-PMMA [87].

Lin and coworkers report that the bone/cement interfacial strength can be enhanced by partially degradable PMMA/Mg composite bone cement (PMC) [88]. This reinforcement is accomplished via the increase in the osteo-conductivity of PMMA and the enhancement of the mechanical interlocking between bone tissue and the porous PMMA surface. The effects of Mg composition, particle size and content on the injectability, biocompatibility, mechanical, and degradation properties of PMCs are investigated. They find that the biocompatibility, mechanical and degradation properties of PMCs are influenced by the particle size (75–550 μm), concentration (9–17 wt%) and alloy composition of Mg particles. Moreover, antibacterial capabilities are increased owing to the degradation of Mg, resulting in a decrement in the infection rate.

Roldo and coworkers report a preparation and characterization of antibacterial PMMA composite cement [89]. The chitosan (CS) and methacryloyl chitosan (CSMCC) with concentrations ranging from 10 to 30% *w/w* are added to PMMA cement, finding that the mechanical behavior of PMMA cement can be modified by the addition of CS and CSMCC. Bioactive PMMA surfaces at the site of implantation are obtained via the addition of amphiphilic molecule phosphorylated 2-hydroxyethylmethacrylate (HEMA-P) in PMMA bone cement [90]. The addition of HEMA-P shows a positive effect with respect to differentiation and proliferation of the osteoblast-like cell (SaOs-2) without the detrimental changes in other properties. The effects of the addition of soluble calcium and carbonate salts on the properties of PMMA bone cement are investigated. A small amount (1–5%) of soluble salts can enhance the clinical performance of bone cement.

3.1.2. PMMA Denture

PMMA resin is commonly used dental material because of the low cost and lightweight performances. However, the properties such as tenacity, fracture strength of PMMA used for dental material should meet a suitable value. Kusy and coworkers find that the fractions of viscosity-average molecular weight less than 10^5 make no contribution to the plastic toughening of the material [91]. Huggett and coworkers have investigated the relationship between molecular weight and properties of PMMA denture base, finding that denture base systems with weight-average molecular weights >10^5 display optimum fracture strength properties [92]. Therefore, the high molecular weight PMMA ($M_n > 10^5$ Da) should be a suitable candidate for dental material.

Both heat- and cold-cured PMMA materials are used for the relining of dentures [93]. Heat-cured PMMA has good bonding strength and wear resistance. However, the roughening of the surface results in a difficulty in cleaning of dental material. Similarly, cold-cured PMMA has poor mechanical properties, leaching of monomers, and associated biocompatibility issues [93,94].

Acrylic (PMMA) teeth are a novel dental material which is manufactured by compression or injection molding techniques. Compared to heat-cured PMMA, acrylic teeth are less brittle owing to the high resilience and flexibility. Distinct from porcelain teeth, acrylic teeth are lightweight and do not cause clicking sounds [95]. However, the strength and adhesion of acrylic teeth are still to be considered. The strength can be improved by the use of additive (nanofillers) or using silanized, feldspar-reinforced PMMA.

Alaa Mohammed and coworkers prepare a new composite material via the mixing of eggshell powder and PMMA resin [96]. The effects of different eggshell powder concentration (1%, 3%, 5% and 7% *w/w*) on the property of the material have been studied. The tensile properties and fracture toughness are enhanced by the addition of 7% *w/w* of eggshell powder, while the elongation percentage at break and impact strength are decreased compared with other specimens. Eggshell has a poor dispersion ability in PMMA and this may cause a formation of agglomerates in the PMMA matrix. This might be a main reason for the decrement in percentage at break and impact strength. A similar result is obtained when using some other nanofillers as additives [97].

The addition of silica to PMMA has a negative impact on flexural strength. However, the flexural strength can be slightly improved by using silanized feldspar as an additive. Raszewski and coworkers find that properties such as Brinell hardness, elastic modulus, maximal displacement, and flexural strength of PMMA modified with silanized feldspar are obviously improved [97]. In addition, the PMMA modified with silanized feldspar has no adverse effect on Isolde impact resistance compared with the conventional acrylic resin. When using silica filler as the additive, the Brinell hardness and elastic modulus of PMMA are increased. However, this causes a significant decrease in the flexural strength and Isolde impact resistance.

Various types of PMMA materials such as heat-cured PMMA, cold-cured PMMA and light-cured PMMA are used for denture repair [98]. The heat-cured PMMA has a better mechanical property than cold-cured PMMA. However, the heat-cured PMMA has some disadvantages such as time consuming and denture warpage. Compared with heat-cured PMMA and cold-cured PMMA, the light-cured PMMA has some advantages of ease of manipulation, controlled polymerization time, no monomer issues, and better mechanical properties [99,100]. In addition, light-cured PMMA has a better repair strength (40–44 MPa) than heat-cured PMMA (21–34 MPa) and cold-cured PMMA (~13 MPa) [101].

3.2. *Applications of High Molecular Weight PMMA in Optical and Electricity Area*

The PMMA-based polymer nanocomposites have attracted the considerable interest of chemists due to their excellent optical, mechanical and electrical properties [102–104].

Polymer field-effect transistors (PFETs) have attracted the significant attention of scientists owing to their potential applications in smart card, displays and sensor [105,106]. The property of a polymer thin film transistor is not only affected by semiconductors, but also by gate insulating film. Therefore, suitable insulating gate dielectric film is very important for the investigation of high-performance polymer field-effect transistors. As the development of donor-acceptor (D-A) copolymers, it has long been known that the molecular weights of polymeric semiconductors play significant roles in enhancing performances of polymer field-effect transistors [107]. PMMA is a commonly used dielectric material owing to its excellent electrical properties.

Mao and coworkers reported that the electron and hole mobilities in polymer field-effect transistors can be enhanced by tailoring the molecular weight of polymeric dielectric [107]. PMMA with different molecular weights are used to investigate the electrical properties of polymer field-effect transistors (Table 2).

Table 2. Performance parameters of PFET devices using the PMMA dielectric with different molecular weights [107].

Entry	M_w (Da)	*n*-Channel $\mu_{e,max}$ (cm^2 V^{-1} S^{-1})	*p*-Channel $\mu_{h,max}$ (cm^2 V^{-1} S^{-1})	*n*-Channel N_{trap} ($\times10^{11}$ cm^{-2})	*p*-Channel N_{trap} ($\times10^{11}$ cm^{-2})
1	1.2×10^5	0.30	0.01	5.46	3.64
2	5.5×10^5	0.55	0.18	4.67	2.33
3	1.0×10^6	0.85	0.35	1.38	1.10

It shows that the PFETs based on PMMA ($M_w = 1.2 \times 10^5$ Da) exhibit a large electron mobility of 0.30 cm^2 V^{-1} s^{-1} but show a low hole mobility of 0.01 cm^2 V^{-1} s^{-1} (Table 2, entry 1). When the molecular weight of PMMA increases to 5.5×10^5 Da, hole mobility of the PFETs is greatly increased to 0.18 cm^2 V^{-1} s^{-1} and electron mobility is also improved to 0.55 cm^2 V^{-1} s^{-1} (Table 2, entry 2). Moreover, electron mobility and hole mobility are increased to 0.85, 0.35 cm^2 V^{-1} s^{-1} respectively when using PMMA with a molecular weight of 1.0×10^6 Da as the dielectric (Table 2, entry 3). In addition, when the molecular weight of PMMA increases from 1.2×10^5 to 1.0×10^6 Da, the trap density (N_{trap}) for electron traps decrease from 5.46 to 1.38×10^{11} cm^{-2} and the trap density (N_{trap}) for hole traps decrease from 3.64 to 1.10×10^{11} cm^{-2}. Therefore, using high molecular weight PMMA as dielectric is beneficial for simultaneously enhancing electron and hole mobilities. Consequently, high molecular weight PMMA is an excellent candidate for electrical applications.

Organic field effect transistors (OFETs) have also attracted significant attention owing to their potential applications in electronics [108]. Dinaphtho[2,3-b:2′,3′-f]thieno [3,2-b]thiophene (DNTT) based OFET devices with a bilayer dielectric system comprising of poly (vinyl alcohol) (PVA) and poly (methyl methacrylate) (PMMA) are fabricated by Dhar and coworkers [109]. The influence of molecular weight of PMMA on the property of DNTT based OFET is investigated. They find that high molecular weight PMMA devices are more effective for achieving high photosensitivity and responsivity from the transistors.

Recently, the rare earth luminescent materials have attracted considerable attention owing to their potential applications in optoelectronic devices [110]. However, properties such as processing features, thermal stability and mechanical strength of lanthanide complexes are relatively poor. The addition of polymer is an effective method for the improvement in thermal stability and mechanical strength of lanthanide complexes. Kara and coworkers find that the high molecular weight PMMA (M_w = 3.50 × 10^5 Da) is an excellent material for the preparation of Sm/PMMA luminescent composite fiber [111]. Compared to the pure Sm(III) complex, the photostability and temperature stability of Sm/PMMA composite fibers are enhanced due to the modification via PMMA matrix. PMMA provides a rigid environment to prevent the decomposition of the pure Sm(III) complex under high temperature and UV irradiation. The luminescent spectra of Sm/PMMA composite fibers display intense characteristic emissions of the Sm^{3+} ion.

The Co doped ZnO nanoparticle (NP) is used to prepare PMMA (M_w = 3.50 × 10^5 Da) and poly (ethyl methacrylate) (PEMA) nanocomposite via casting method [112]. The conductivity of the nanocomposite is increased with the increase in nanofiller contents due to the formation of charge transfer complexes. Furtheemore, the dielectric constancy of the nanocomposite is increased with the increase in temperature. Moreover, the nanocomposite shows excellent thermal and electric properties. Therefore, PEMA/PMMA-Co/ZnO polymer electrolyte is a promising candidate for applications in electrochemical devices.

3.3. Applications of High Molecular Weight PMMA in Other Areas

Metal injection molding (MIM) is an important method for the production of small parts with complex shape. The key factor in MIM is the selection of the binder, which should permit the mixing and injection molding of feedstocks with high powder loading. Ideally, the debinding should occur in a short time without causing defects. Bakan et al. report that the complexes consisted of PMMA (M_n = 1.00 × 10^6 Da), PEG and stearic acid can be used as a binder for the MIM of 316L stainless steel powder [113]. The viscosity of the system can be reduced by the decrease in the molar ratio of PMMA/PEG, which is beneficial to the molding of feedstocks with high powder loading. However, the strength, stiffness and toughness of the molding are decreased after the binder solidifies. Therefore, the PMMA can harden the solidified molding product.

Membrane gas separation technology is widely used in nitrogen purification, oxygen enrichment, hydrogen recovery from reactor purge gas, and stripping of carbon dioxide from natural gas [114]. Compared to traditional gas separation technologies, the membrane gas separation technology is more competitive. The main subject of membrane gas separation technology is the improvement in the permeability and selectivity of polymer membranes [114,115]. PMMA membrane has a good O_2/N_2 selectivity, but it is difficult to operate under a high-pressure condition due to its brittleness [116]. Fu and coworkers report that the high molecular weight PMMA films (M_w = 9.96 × 10^5 Da) with different free volumes (FFV) are synthesized in different solvents (dichloromethane, ethyl acetate, tetrahydrofuran, butyl acetate, methyl isobutyl ketone) [114]. This membrane can be operated under a high pressure (30 atm) condition. The permeability coefficients of different gases (He, O_2, N_2, CO_2) in the PMMA membrane are studied, finding that the gas permeability is increased with the increase in FFV. Moreover, they have found that the high molecular weight PMMA membranes possess higher gas permeability than the low molecular weight membranes.

4. Summary and Outlook

Compared to low molecular weight PMMA, the high molecular weight PMMA has a better mechanical strength. The application value of PMMA is further expanded due to the synthesis of high molecular weight PMMA. This review summarized the application of conventional free radical polymerization, ATRP, RAFT, plasma-initiated polymerization and ionic polymerization on the preparation of high molecular weight PMMA. A series of linear PMMA, star-shaped PMMA and block polymers with molecular weights between 10^5–10^7 Da have been prepared and reported by chemists. The mechanisms of some polymerization systems have also been summarized. Finally, this work also proides a brief description of the application of high molecular weight PMMA in medical, optical and electricity, and other areas. We hope that this work can bring more research insights into this field.

Author Contributions: The preparation of the original manuscript, M.Y. and Y.Z.; Review and editing, M.Y. and D.H.; Funding acquisition, D.H. All authors have read and agreed to the published version of the manuscript.

Funding: The work is funded by the Natural Science Foundation of Zhejiang Province (LQ18B020001).

Institutional Review Board Statement: Not applicable.

Informed Consent Statement: Not applicable.

Data Availability Statement: Not applicable.

Conflicts of Interest: The authors declare no conflict of interest.

References

1. Khezri, K.; Fazli, Y. ATRP of Methyl methacrylate in the presence of HMDS-modified silica aerogel: ARGET approach. *J. Inorg. Organomet. Polym. Mater.* **2019**, *29*, 608–616. [CrossRef]
2. Otozawa, N.; Hamajima, R.; Yoshioka, M.; Kato, R.; Tanaka, A.; Fukuma, H.; Terao, T.; Manabe, K.; Fujii, S.; Nakamura, Y.; et al. Preparation of polymethyl methacrylate with well-controlled stereoregularity by anionic polymerization in an ionic liquid Solvent. *J. Polym. Sci.* **2020**, *58*, 1960–1964. [CrossRef]
3. Sheng, Y.; Huang, H.; Yu, M.; Zhang, X.; Cheng, L.; Liu, Z.; Liu, W.; Huang, Q.; Yi, J.; Yang, W. Synthesis of 4-arms hydroxy-functionalized PMMA-b-PE through combining free radical polymerization with coordination polymerization. *J. Ind. Eng. Chem.* **2014**, *20*, 419–425. [CrossRef]
4. Archana, R.; Baldia, M.; Jeeva, J.B.; Joseph, M. Strength analysis of cranioplasty PMMA flap material. *Mater. Today Proc.* **2019**, *15*, 167–172. [CrossRef]
5. Kaufmann, T.J.; Jensen, M.E.; Ford, G.; Gill, L.L.; Marx, W.F.; Kallmes, D.F. Cardiovascular effects of polymethylmethacrylate use in percutaneous vertebroplasty. *Am. J. Neuroradiol.* **2002**, *23*, 601–604.
6. Rzayev, J.; Penelle, J. HP-RAFT: A free-radical polymerization technique for obtaining living polymers of ultrahigh molecular weights. *Angew. Chem. Int. Ed.* **2004**, *43*, 1691–1694. [CrossRef]
7. Wang, Z.; Yang, Y.J.; Dong, Q.Z.; Hu, C.P. Free radical polymerizations of some vinyl monomers in carbon dioxide fluid. *J. Appl. Polym. Sci.* **2008**, *110*, 468–474. [CrossRef]
8. Genix, A.C.; Bocharova, V.; Kisliuk, A.; Carroll, B.; Zhao, S.; Oberdisse, J.; Sokolov, A.P. Enhancing the mechanical properties of glassy nanocomposites by tuning polymer molecular weight. *Appl. Mater. Interfaces* **2018**, *10*, 33601–33610. [CrossRef]
9. Liu, D.; Chen, H.; Yin, P.; Hao, Z.; Fan, L. Use of Gd powder as catalyst for single electron transfer-living radical polymerization: Applications for synthesis of high molecular weight polymethyl methacrylate. *J. Polym. Sci. Part A Polym. Chem.* **2012**, *50*, 4809–4813. [CrossRef]
10. Jayabalan, M. Thermal modification of poly(vinyl chloride) and formation of dehydrochlorinated poly (vinyl chloride)-poly(methyl Methacrylate) polymer blend in the process of foaming: Identification and application. *J. Appl. Polym. Sci.* **1982**, *27*, 43–52. [CrossRef]
11. Muthukrishnan, S.; Mitra, S.; Ranjith, P.; Gopalakrishnan, S.; Parthipan, P.; Al-Farhood, B. High heat resistant blends of poly(methyl methacrylate) and styrenic copolymers via post reactor modification. *J. Appl. Polym. Sci.* **2018**, *135*, 46220. [CrossRef]
12. Gu, D.; Zhang, L.; Chen, S.; Song, K.; Pan, D.; Yang, B.; Liu, S. Heat treatment to improve the wear resistance of PTFE/PMMA composites. *RSC Adv.* **2019**, *9*, 22289–22294. [CrossRef] [PubMed]

13. Bhattacharjee, M.; Patra, B.N. [Cp_2TiCl_2] catalyzed polymerization in water: Polymerization of methylmethacrylate to a high molecular weight polymer. *Polymer* **2004**, *45*, 3111–3114. [CrossRef]
14. Wang, J.S.; Matyjaszewski, K. Controlled/"living" radical polymerization. Atom transfer radical polymerization in the presence of transition-metal complexes. *J. Am. Chem. Soc.* **1995**, *117*, 5614–5615. [CrossRef]
15. Johnson, D.R.; Osada, Y.; Bell, A.T.; Shen, M. Studies of the mechanism and kinetics of plasma-initiated polymerization of methyl methacrylate. *Macromolecules* **1981**, *14*, 118–124. [CrossRef]
16. Lyoo, W.S.; Noh, S.K.; Yeum, J.H.; Kang, G.C.; Ghim, H.D.; Lee, J.; Ji, B.C. Preparation of high molecular weight poly(methyl methacrylate) with high yield by room temperature suspension polymerization of methyl methacrylate. *Fibers Polym.* **2004**, *5*, 75–81. [CrossRef]
17. Hsiao, Y.L.; Maury, E.E.; DeSimone, J.M. Dispersion polymerization of methyl methacrylate stabilized with poly(1,1-dihydroperfluorooctyl acrylate) in supercritical carbon dioxide. *Macromolecules* **1995**, *28*, 8159–8166. [CrossRef]
18. Hsiao, Y.L.; DeSimone, J.M. Dispersion polymerization of methyl methacrylate in supercritical carbon dioxide: Influence of helium concentration on particle size and particle size distribution. *J. Polym. Sci. A Polym. Chem.* **1997**, *35*, 2009–2013. [CrossRef]
19. Lepilleur, C.; Beckman, E.J. Dispersion polymerization of methyl methacrylate in supercritical CO_2. *Macromolecules* **1997**, *30*, 745–756. [CrossRef]
20. Huddleston, J.G.; Visser, A.E.; Reichert, W.M.; Willauer, H.D.; Broker, G.A.; Rogers, R.D. Characterization and comparison of hydrophilic and hydrophobic room temperature ionic liquids incorporating the imidazolium cation. *Green Chem.* **2001**, *3*, 156164. [CrossRef]
21. Seddon, K.R. Ionic liquids for clean technology. *J. Chem. Tech. Biotechnol.* **1997**, *68*, 351–356. [CrossRef]
22. Scott, M.P.; Rahman, M.; Brazel, C.S. Application of ionic liquids as low-volatility plasticizers for PMMA. *Eur. Polym. J.* **2003**, *39*, 1947–1953. [CrossRef]
23. Scott, M.P.; Brazel, C.S.; Benton, M.G.; Mays, J.W.; Holbreyc, J.D.; Rogers, R.D. Application of ionic liquids as plasticizers for poly(methyl methacrylate). *Chem. Commun.* **2002**, *38*, 1370–1371. [CrossRef] [PubMed]
24. Yeon, S.H.; Kim, K.S.; Choi, S.; Cha, J.H.; Lee, H.; Oh, J.; Lee, B.B. Poly(vinylidenefluoride)-hexafluoropropylene gel electrolytes based on *N*-(2-hydroxyethyl)-*N*-methyl morpholinium ionic liquids. *Korean J. Chem. Eng.* **2006**, *23*, 940–947. [CrossRef]
25. Tomás-Alonso, F.; Rubio, A.M.; Giménez, A.; De los Ríos, A.P. Influence of ionic liquid composition on the stability of polyvinyl chloride-based ionic liquid inclusion membranes in aqueous solution. *AIChE J.* **2017**, *63*, 770–780. [CrossRef]
26. Dai, S.; Ju, Y.H.; Gao, H.J.; Lin, J.S.; Pennycook, S.J.; Barnes, C.E. Preparation of silica aerogel using ionic liquids as solvents. *Chem. Commun.* **2000**, *36*, 243–244. [CrossRef]
27. Carlin, R.T.; Fuller, J. Ionic liquid-polymer gel catalytic membrane. *Chem. Commun.* **1997**, *33*, 1345–1346. [CrossRef]
28. Vygodskii, Y.S.; Mel'nik, O.A.; Lozinskaya, E.I.; Shaplov, A.S.; Malyshkina, I.A.; Gavrilova, N.D.; Lyssenko, K.A.; Antipin, M.Y.; Golovanov, D.G.; Korlyukov, A.A.; et al. The influence of ionic liquids nature on free radical polymerization of vinyl monomers and ionic conductivity of the obtained polymeric materials. *Polym. Adv. Technol.* **2007**, *18*, 50–63. [CrossRef]
29. Carotenuto, G.; Her, Y.S.; Matijević, E. Preparation and characterization of nanocomposite thin films for optical devices. *Ind. Eng. Chem. Res.* **1996**, *35*, 2929–2932. [CrossRef]
30. Wang, Y.; Herron, N. Photoconductivity of CdS nanocluster-doped polymers. *Chem. Phys. Lett.* **1992**, *27*, 71–75. [CrossRef]
31. Hailstone, R.K. Computer simulation studies of silver cluster formation on AgBr microcrystals. *J. Phys. Chem.* **1995**, *99*, 4414–4428. [CrossRef]
32. Lira-Cantú, M.; Gómez-Romero, P. Electrochemical and chemical syntheses of the hybrid organic-inorganic electroactive material formed by phosphomolybdate and polyaniline. Application as cation-insertion electrodes. *Chem. Mater.* **1998**, *10*, 698–704. [CrossRef]
33. Yeum, J.H.; Deng, Y. Synthesis of high molecular weight poly(methyl methacrylate) microspheres by suspension polymerization in the presence of silver nanoparticles. *Colloid Polym. Sci.* **2005**, *283*, 1172–1179. [CrossRef]
34. Jiang, W.; Yang, W.; Zeng, X.; Fu, S. Structure and properties of poly(methyl methacrylate) particles prepared by a modified microemulsion polymerization. *J. Polym. Sci. Part A Polym. Chem.* **2004**, *42*, 733–741. [CrossRef]
35. Norakankorn, C.; Pan, Q.; Rempel, G.L.; Kiatkamjornwong, S. Synthesis of poly(methyl methacrylate) nanoparticles initiated by 2,2′-azoisobutyronitrile via differential microemulsion polymerization. *Macromol. Rapid Commun.* **2007**, *28*, 1029–1033. [CrossRef]
36. Schmidt-Mende, L.; MacManus-Driscoll, J.L. ZnO—nanostructures, defects, and devices. *Mater. Today* **2007**, *10*, 40–48. [CrossRef]
37. Dadashi-Silab, S.; Asiri, A.M.; Khan, S.B.; Alamry, K.A.; Yagci, Y. Semiconductor nanoparticles for photoinitiation of free radical polymerization in aqueous and organic media. *J. Polym. Sci. Part A Polym. Chem.* **2014**, *52*, 1500–1507. [CrossRef]
38. Strandwitz, N.C.; Khan, A.; Boettcher, S.W.; Mikhailovsky, A.A.; Hawker, C.J.; Nguyen, T.Q.; Stucky, G.D. One- and two-photon induced polymerization of methylmethacrylate using colloidal CdS semiconductor quantum dots. *J. Am. Chem. Soc.* **2008**, *130*, 8280–8288. [CrossRef]
39. Stroyuk, A.L.; Granchak, V.M.; Korzhak, A.V.; Kuchmii, S.Y. Photoinitiation of buthylmethacrylate polymerization by colloidal semiconductor nanoparticles. *J. Photochem. Photobiol. A Chem.* **2004**, *162*, 339–351. [CrossRef]
40. Balta, D.K.; Arsu, N.; Yagci, Y.; Sundaresan, A.K.; Jockusch, S.; Turro, N.J. Mechanism of photoinitiated free radical polymerization by thioxanthone—anthracene in the presence of air. *Macromolecules* **2011**, *44*, 2531–2535. [CrossRef]
41. Pynaert, R.; Buguet, J.; Croutxe-Barghorn, C.; Moireau, P.; Allonas, X. Effect of reactive oxygen species on the kinetics of free radical photopolymerization. *Polym. Chem.* **2013**, *4*, 2475–2479. [CrossRef]

42. Dule, M.; Biswas, M.; Biswas, Y.; Mandal, T.K. Redox-active poly(ionic liquid)-engineered Ag nanoparticle-decorated ZnO nanoflower heterostructure: A reusable composite catalyst for photopolymerization into high-molecular-weight polymers. *Polymer* **2017**, *133*, 223–231. [CrossRef]
43. Xue, L.; Agarwal, U.S.; Lemstra, P.J. High molecular weight PMMA by ATRP. *Macromolecules* **2002**, *35*, 8650–8652. [CrossRef]
44. Matyjaszewski, K.; Xia, J. Atom transfer radical polymerization. *Chem. Rev.* **2001**, *101*, 2921–2990. [CrossRef] [PubMed]
45. Queffelec, J.; Gaynor, S.G.; Matyjaszewski, K. Optimization of atom transfer radical polymerization using Cu(I)/tris(2-(dimethylamino)ethyl)amine as a catalyst. *Macromolecules* **2000**, *33*, 8629–8639. [CrossRef]
46. Mao, B.W.; Gan, L.H.; Gan, Y.Y. Ultra high molar mass poly[2-(dimethylamino)ethyl methacrylate] via atom transfer radical polymerization. *Polymer* **2006**, *47*, 3017–3020. [CrossRef]
47. Matyjaszewski, K. The importance of exchange reactions in controlled/living radical polymerization in the presence of alkoxyamines and transition metals. *Macromol. Symp.* **1996**, *111*, 47–61. [CrossRef]
48. Xue, L.; Agarwal, U.S.; Zhang, M.; Staal, B.B.P.; Mu1ller, A.H.E.; Bailly, C.M.E.; Lemstra, P.J. Synthesis and direct topology visualization of high-molecular-weight star PMMA. *Macromolecules* **2005**, *38*, 2093–2100. [CrossRef]
49. Min, K.; Gao, H.; Matyjaszewski, K. Use of ascorbic acid as reducing agent for synthesis of well-defined polymers by ARGET ATRP. *Macromolecules* **2007**, *40*, 1789–1791. [CrossRef]
50. Nicolaÿ, R.; Kwak, Y.; Matyjaszewski, K. A green route to well-defined high-molecular-weight (Co)polymers using ARGET ATRP with alkyl pseudohalides and copper catalysis. *Angew. Chem.* **2010**, *122*, 551–554. [CrossRef]
51. Wang, Z.; Liu, T.; Lin, K.C.; Li, S.; Yan, J.; Olszewski, M.; Sobieski, J.; Pietrasik, J.; Bockstaller, M.R.; Matyjaszewski, K. Synthesis of ultra-high molecular weight SiO_2-g-PMMA particle brushes. *J. Inorg. Organomet. Polym. Mater.* **2000**, *30*, 174–181. [CrossRef]
52. Jeon, H.J.; Youk, J.H.; Ahn, S.H.; Choi, J.H.; Cho, K.S. Synthesis of high molecular weight 3-arm star PMMA by ARGET ATRP. *Macromol. Res.* **2009**, *17*, 240–244. [CrossRef]
53. Arita, T.; Kayama, Y.; Ohno, K.; Tsujii, Y.; Fukuda, T. High-pressure atom transfer radical polymerization of methyl methacrylate for well-defined ultrahigh molecular-weight polymers. *Polymer* **2008**, *49*, 2426–2429. [CrossRef]
54. Olaj, O.F.; Kornherr, A.; Zifferer, G. On the (im)possibility of evaluating correct individual rate constants of chain propagation k_p and chain termination k_t by combining k_p^2/k_t and k_p/k_t data for chain-length dependent termination. *Macromol. Theory Simul.* **2000**, *9*, 131–140. [CrossRef]
55. Kwiatkowski, P.; Jurczak, J.; Pietrasik, J.; Jakubowski, W.; Mueller, L.; Matyjaszewski, K. High molecular weight polymethacrylates by AGET ATRP under high pressure. *Macromolecules* **2008**, *41*, 1067–1069. [CrossRef]
56. Sun, J.; Pan, Z.; Zhong, Y.; Hu, W.; Yang, S. Polymerization of methyl methacrylate by single component catalyst of lanthanocene chloride $O(C_2H_4C_5H_3CH_3)_2LnCl$ (Ln = Y, Nd, Sm). *Eur. Polym. J.* **2000**, *36*, 2375–2380.
57. Sun, J.; Pan, Z.; Hu, W.; Yang, S. Polymerization of methyl methacrylate with ethylene bridged heterodinuclear metallocene of samarium and titanium. *Eur. Polym. J.* **2002**, *38*, 545–549.
58. Jones, R.D.; Summerville, D.A.; Basolo, F. Synthetic oxygen carriers related to biological systems. *Chem. Rev.* **1979**, *79*, 139–179. [CrossRef]
59. Yousaf, M.; Chatha, A.A.; Jabbar, A.; Ahmad, H.B.; Zahoor, A.F.; Khosa, M.K.; Alam, S.S. Role of tridentate schiff base and methoxyethylindenyl derivatives of lanthanocene complexes for the synthesis of high molecular weight polymethyl methacrylate (PMMA). *Turk. J. Chem.* **2007**, *31*, 471–477.
60. Kim, D.; Song, Y.; Kim, S.; Lee, H.J.; Lee, H. *N,N′*, X-bidentate versus *N,N′*, X-tridentate *N*-substituted 2-iminomethylpyridine- and 2-iminomethylquinoline-coordinated palladium(II)complexes. *J. Coord. Chem.* **2014**, *67*, 2312–2329. [CrossRef]
61. Song, Y.; Kim, D.; Lee, H.J.; Lee, H. Cd(II) and Zn(II) complexes containing *N,N′*-bidentate *N*-(pyridin-2-ylmethylene)cyclopentanamine: Synthesis, characterisation and methyl methacrylate polymerisation. *Bull. Korean Chem. Soc.* **2014**, *35*, 2929–2934. [CrossRef]
62. Heo, J.; Lee, H.; Nayab, S. Polymerizations of methyl methacrylate and rac-lactide by Zinc(II) precatalyst containing *N*-substituted 2-iminomethylpyridine and 2-iminomethylquinoline. *J. Coord. Chem.* **2017**, *70*, 3837–3858. [CrossRef]
63. Lansalot-Matras, C.; Bonnette, F.; Mignard, E.; Lavastre, O. *N*-tripodal ligands to generate copper catalysts for the syndiotactic polymerization of methyl-methacrylate: Crystal structures of copper complexes. *J. Organomet. Chem.* **2008**, *693*, 393–398. [CrossRef]
64. Yang, M.; Park, W.J.; Yoon, K.B.; Jeong, J.H.; Lee, H. Synthesis, Characterization, and MMA polymerization activity of tetrahedral Co (II)complex bearing *N, N*-bis(1-pyrazolyl)methyl ligand based on aniline moiety. *Inorg. Chem. Commun.* **2011**, *14*, 189–193. [CrossRef]
65. Shin, S.; Nayab, S.; Lee, H. Polymerizations of methyl methacrylate and rac-lactide by 4-coordinate cobalt(II) complexes supported by *N′*-substituted *N,N′,N*-bis((1H-pyrazol-1-yl)methyl)amine derivatives. *Polyhedron* **2018**, *141*, 309–321. [CrossRef]
66. Buchowicz, W.; Kozioł, A.; Jerzykiewicz, L.B.; Lis, T.; Pasynkiewicz, S.; Pecherzewska, A.; Pietrzykowski, A. *N*-heterocyclic carbene complexes of cyclopentadienylnickel(II): Synthesis, structure and catalytic activity in styrene polymerization. *J. Mol. Catal. A Chem.* **2006**, *257*, 118–123. [CrossRef]
67. Buchowicz, W.; Wojtczak, W.; Pietrzykowski, A.; Lupa, A.; Jerzykiewicz, L.B.; Makal, A.; Woźniak, K. Synthesis, structure, and polymerization activity of cyclopentadienylnickel(II) *N*-heterocyclic carbene complexes: Selective cross-metathesis in metal coordination spheres. *Eur. J. Inorg. Chem.* **2010**, *2010*, 648–656. [CrossRef]
68. Ihara, E.; Fujimura, T.; Yasuda, H.; Maruo, T.; Kanehisa, N.; Kai, Y. High polymerization of methyl methacrylate with organonickel/MMAO systems. *J. Polym. Sci. Part A Polym. Chem.* **2000**, *38*, 4764–4775. [CrossRef]

69. Buchowicz, W.; Conder, J.; Hryciuk, D.; Zachara, J. Nickel-mediated polymerization of methyl methacrylate. *J. Mol. Catal. A Chem.* **2014**, *381*, 16–20. [CrossRef]
70. Kitaura, T.; Kitayama, T. Anionic polymerization of methyl methacrylate with the aid of lithium trimethylsilanolate (Me_3SiOLi)—superior control of isotacticity and molecular weight. *Macromol. Rapid Commun.* **2007**, *28*, 1889–1893. [CrossRef]
71. Osada, Y.; Bell, A.T.; Shen, M. Plasma-initiated polymerization of methyl methacrylate. *J. Polym. Sci. Polym. Lett. Ed.* **1978**, *16*, 309–311. [CrossRef]
72. Gavens, L.M.; Wu, B.J.; Hess, D.W.; Bell, A.T.; Soong, D.S. Ultrahigh molecular weight poly(methyl methacrylate) as an electron-beam resist. *J. Vac. Sci. Technol. B* **1983**, *1*, 481–486. [CrossRef]
73. Chen, G.; Zhu, X.; Zhu, J.; Cheng, Z. Plasma-initiated controlled/living radical polymerization of methyl methacrylate in the presence of 2-cyanoprop-2-yl 1-dithionaphthalate (CPDN). *Macromol. Rapid Commun.* **2004**, *25*, 818–824. [CrossRef]
74. Yuan, M.; Xu, L.; Cui, X.; Lv, J.; Zhang, P.; Tang, H. Facile synthesis of ultrahigh molecular weight poly(methyl methacrylate) by organic halides in the presence of palladium nanoparticles. *Polymers* **2020**, *12*, 2747. [CrossRef]
75. Zhang, Y.; Miyake, G.M.; Chen, E.Y.X. Alane-based classical and frustrated lewis pairs in polymer synthesis: Rapid polymerization of MMA and naturally renewable methylene butyrolactones into high-molecular-weight polymers. *Angew. Chem. Int. Ed.* **2010**, *49*, 10158–10162. [CrossRef]
76. Demirel, G.; Akovali, G. Some studies on plasma-initiated polymerization of (MMA) and on plasma polymerized (PMMA). *J. Polym. Sci. Polym. Chem. Ed.* **1985**, *23*, 2377–2383. [CrossRef]
77. Martin, J.R.; Johnson, J.F.; Cooper, A.R. Mechanical properties of polymers: The influence of molecular weight and molecular weight distribution. *J. Macromol. Sci. Rev. Macromol. Chem.* **1972**, *8*, 57–199. [CrossRef]
78. Kusy, R.P.; Greenberg, A.R. Influence of molecualr weight on the dynamic mechanical properties of poly(methyl methacrylate). *J. Therm. Anal.* **1980**, *18*, 117–126. [CrossRef]
79. Ahn, Y.; Lee, J.H.; Lee, H.Y.; Lee, S.H.; Keem, S.H. Predictive factors for subsequent vertebral fracture after percutaneous vertebroplasty. *J. Neurosurg. Spine* **2008**, *9*, 129–136. [CrossRef]
80. Zhang, W.; Liao, S.S.; Cui, F.Z. Hierarchical self-assembly of nano-fibrils in mineralized collagen. *Chem. Mater.* **2003**, *15*, 3221–3226. [CrossRef]
81. Yu, X.; Xu, L.; Cui, F.Z.; Qu, Y.; Lian, X.J.; Wang, X.M.; Wang, Y. Clinical evaluation of mineralized collagen as a bone graft substitute for anterior cervical intersomatic fusion. *J. Biomater. Tissue Eng.* **2012**, *2*, 170–176. [CrossRef]
82. Soha, S.; Pal, S. Mechanical properties of bone cement: A review. *J. Biomed. Mater. Res.* **1984**, *18*, 435–462. [CrossRef] [PubMed]
83. Kuhn, K.D. *PMMA Cements*; Springer: Berlin/Heidelberg, Germany, 2014.
84. Heo, S.J.; Shin, H.J.; Park, S.A.; Lee, Y.J.; Yoon, T.R.; Seo, H.Y.; Ahn, K.C.; Kim, S.E.; Shin, J.W. Evaluation of bonding stress for the newly suggested bone cement: Comparison with currently used PMMA through animal studies. *Key Eng. Mater.* **2007**, *342*, 373–376. [CrossRef]
85. Lewis, G. Alternative acrylic bone cement formulations for cemented arthroplasties: Present status, key issues, and future prospects. *J. Biomed. Mater. Res. Part B Appl. Biomater.* **2007**, *84*, 301–319. [CrossRef]
86. Liu, H.; Webster, T.J. Nanomedicine for implants: A review of studies and necessary experimental tools. *Biomaterials* **2007**, *28*, 354–369. [CrossRef]
87. Khandaker, M.; Li, Y.; Morris, T. Micro and nano MgO particles for the improvement of fracture toughness of bone-cement interfaces. *J. Biomech.* **2013**, *46*, 1035–1039. [CrossRef]
88. Lin, X.; Chan, A.; Tan, X.; Yang, H.; Yang, L. Fabrication and characterizations of metallic Mg containing PMMA-based partially degradable composite bone cements. *Acta Metall. Sin.* **2019**, *32*, 808–816. [CrossRef]
89. Mori, A.D.; Gregorio, E.D.; Kao, A.P.; Tozzi, G.; Barbu, E.; Sanghani-Kerai, A.; Draheim, R.R.; Roldo, M. Antibacterial PMMA composite cements with tunable thermal and mechanical properties. *ACS Omega* **2019**, *4*, 19664–19675. [CrossRef]
90. Wolf-Brandstetter, C.; Roessler, S.; Storch, S.; Hempel, U.; Gbureck, U.; Nies, B.; Bierbaum, S.; Scharnweber, D. Physicochemical and cell biological characterization of PMMA bone cements modified with additives to increase bioactivity. *J. Biomed. Mater. Res. Part B Appl. Biomater.* **2013**, *101*, 599–609. [CrossRef]
91. Kusy, R.P.; Turner, D.T. Influence of the molecular weight of poly(methyl methacrylate) on fracture surface energy in notched tension. *Polymer* **1976**, *17*, 161–166. [CrossRef]
92. Huggett, R.; Bates, J.F.; Packham, D.E. The effect of the curing cycle upon the molecular weight and properties of denture base materials. *Dent. Mater.* **1987**, *3*, 107–112. [CrossRef]
93. Nejatian, T.; Pezeshki, S.; Yaqin Syed, A.U. Acrylic denture base materials. In *Advanced Dental Biomaterials*; Khurshid, Z., Najeeb, S., Zafar, M.S., Sefat, F., Eds.; Woodhead Publishing: Cambridge, UK, 2019; pp. 79–104.
94. Habib, S.R.; Vohra, F.A. Replacing existing dentures by copy-denture technique for geriatric patients: A case report. *JPDA* **2013**, *22*, 265–270.
95. Zafar, M.S. Prosthodontic applications of polymethyl methacrylate (PMMA): An update. *Polymers* **2020**, *12*, 2299. [CrossRef] [PubMed]
96. Mohammed, R.A. PMMA-eggshell composite preparation and studying mechanical properties as a dental material. *J. Eng. Sustain. Dev.* **2020**, *24*, 30–47. [CrossRef]

97. Raszewski, Z.; Nowakowska-Toporowska, A.; Wezgowiec, J.; Nowakowska, D.; Wieckiewicz, W. Influence of silanized silica and silanized feldspar addition on the mechanical behavior of polymethyl methacrylate resin denture teeth. *J. Prosthet. Dent.* **2020**, *123*, 647–653. [CrossRef]
98. Gad, M.M.; Rahoma, A.; Al-Thobity, A.M.; ArRejaie, A.S. Influence of incorporation of ZrO_2 nanoparticles on the repair strength of polymethyl methacrylate denture bases. *Int. J. Nanomed.* **2016**, *11*, 5633–5643. [CrossRef]
99. Shifman, A. Clinical applications of visible light-cured resin in maxillofacial prosthetics. part I: Denture base and reline material. *J. Prosthet. Dent.* **1990**, *64*, 578–582. [CrossRef]
100. Fellman, S. Visible light-cured denture resin used in making dentures with conventional teeth. *J. Prosthet. Dent.* **1989**, *62*, 356–359. [CrossRef]
101. Polyzois, G.L.; Handley, R.W.; Stafford, G.D. Repair strength of denture base resins using various methods. *Eur. J. Prosthodont. Restor. Dent.* **1995**, *3*, 183–186.
102. Khodair, Z.T.; Saeed, M.H.; Abdul-Allah, M.H. Study of optical properties of (PMMA) doped by methyl red and methyl blue films. *Iraqi J. Phys.* **2014**, *12*, 47–51.
103. Al-Ammar, K.; Hashim, A.; Husaien, M. Synthesis and study of optical properties of (PMMA-$CrCl_2$) composites. *Chem. Mater. Eng.* **2013**, *1*, 85–87. [CrossRef]
104. Gupta, A.K.; Bafna, M.; Vijay, Y.K. Study of optical properties of potassium permanganate ($KMnO_4$) doped poly(methylmethacrylate) (PMMA) composite films. *Bull. Mater. Sci.* **2018**, *41*, 160. [CrossRef]
105. Yan, H.; Chen, Z.; Zheng, Y.; Newman, C.; Quinn, J.R.; Dötz, F.; Kastler, M.; Facchetti, A. A high-mobility electron-transporting polymer for printed transistors. *Nature* **2009**, *457*, 679–686. [CrossRef]
106. Wu, W.; Liu, Y.; Zhu, D. π-Conjugated molecules with fused rings for organic field-effect transistors: Design, synthesis and applications. *Chem. Soc. Rev.* **2010**, *39*, 1489–1502. [CrossRef]
107. Mao, Z.; Guo, Y.; Chen, H.; Zhang, W.; Yu, G. Tailoring molecular weight of polymeric dielectric to enhance electron and hole mobilities in polymer field-effect transistors. *Polymer* **2016**, *99*, 496–502. [CrossRef]
108. Ren, H.; Cui, N.; Tang, Q.; Tong, Y.; Liu, Y. High-performance, ultrathin, ultraflexible organic thin-film transistor array via solution process. *Small* **2018**, *14*, 1801020. [CrossRef]
109. Panigrahi, D.; Dhar, A. Tuning dielectric-semiconductor interfacial properties in organic phototransistors for ultralow light detection. *Org. Electron.* **2019**, *70*, 107–112. [CrossRef]
110. Bibi, S.; Mohamad, S.; Manan, N.S.A.; Ahmad, J.; Kamboh, M.A.; Khor, S.M.; Yamin, B.M.; Abdul Halim, S.N. Synthesis, characterization, photoluminescence, and electrochemical studies of novel mononuclear Cu (II) and Zn (II) complexes with the 1-benzylimidazolium ligand. *J. Mol. Struct.* **2017**, *1141*, 31–38. [CrossRef]
111. Kara, H.; Oylumluoglu, G.; Coban, M.B. Photoluminescence properties of a new Sm(III) complex/PMMA electrospun composite fibers. *J. Clust. Sci.* **2020**, *31*, 701–708. [CrossRef]
112. Rajeh, A.; Ragab, H.M.; Abutalib, M.M. Co doped ZnO reinforced PEMA/PMMA composite: Structural, thermal, dielectric and electrical properties for electrochemical applications. *J. Mol. Struct.* **2020**, *1217*, 128447. [CrossRef]
113. Bakan, H.I.; Jumadi, Y.; Messer, P.F.; Davies, H.A.; Ellis, B. Study of processing parameters for MIM feedstock based on composite PEG-PMMA binder. *Powder Metall.* **1998**, *41*, 289–291. [CrossRef]
114. Fu, Y.J.; Hu, C.C.; Lee, K.R.; Tsai, H.A.; Ruaan, R.C.; Lai, J.Y. The correlation between free volume and gas separation properties in high molecular weight poly(methyl methacrylate) membranes. *Eur. Polym. J.* **2007**, *43*, 959–967. [CrossRef]
115. Ghosal, K.; Chern, R.T.; Freeman, B.D. Effect of basic substituents on gas sorption and permeation in polysulfone. *Macromolecules* **1996**, *29*, 4360–4369. [CrossRef]
116. Wright, C.T.; Paul, D.R. Gas sorption and transport in poly(tertiary-butyl methacrylate). *Polymer* **1997**, *38*, 1871–1878. [CrossRef]

polymers

Article

Synthesis and Study of Physical and Mechanical Properties of Urethane-Containing Elastomers Based on Epoxyurethane Oligomers with Controlled Crystallinity

Alexey Slobodinyuk *, Vladimir Strelnikov, Nadezhda Elchisheva, Dmitriy Kiselkov and Daria Slobodinyuk

Institute of Technical Chemistry Ural Branch of the Russian Academy of Sciences, Ac. Korolev 3, 614130 Perm, Russia; svn@itcras.ru (V.S.); env-1981@mail.ru (N.E.); dkiselkov@yandex.ru (D.K.); selivanovadg@gmail.com (D.S.)

* Correspondence: slobodinyuk.aleksey.ktn@mail.ru; Tel.: +7-(342)-237-8256

Abstract: The influence of the molecular weight of oligoamine, oligoether, and the type of diisocyanate on the physical and mechanical properties of elastomers with urethane hydroxyl hard segments was studied. For this purpose, oligoetherdiamines with molecular weights ~1008 and ~1400 g mol^{-1} were synthesized by a three-stage method. Epoxyurethane oligomers were synthesized according to a two-step route with an oligodiisocyanate as an intermediate product. A series of 12 elastomers with controlled crystallinity were synthesized from these elastomers and amines. The deformation and strength properties of the elastomers were studied.

Keywords: curing agent; epoxyurethane oligomer; oligotetramethylene oxide diol; synthesis; bromination; Gabriel reaction

Citation: Slobodinyuk, A.; Strelnikov, V.; Elchisheva, N.; Kiselkov, D.; Slobodinyuk, D. Synthesis and Study of Physical and Mechanical Properties of Urethane-Containing Elastomers Based on Epoxyurethane Oligomers with Controlled Crystallinity. *Polymers* **2022**, *14*, 2136. https://doi.org/10.3390/polym14112136

Academic Editor: Edina Rusen

Received: 30 April 2022
Accepted: 23 May 2022
Published: 24 May 2022

Publisher's Note: MDPI stays neutral with regard to jurisdictional claims in published maps and institutional affiliations.

Copyright: © 2022 by the authors. Licensee MDPI, Basel, Switzerland. This article is an open access article distributed under the terms and conditions of the Creative Commons Attribution (CC BY) license (https://creativecommons.org/licenses/by/4.0/).

1. Introduction

Smart materials are the next generation of functional materials, after natural, man-made, and synthetic materials [1,2]. These high-tech materials combine the advantages of structural and functional materials [3–5].

Shape memory polymers (SMPs) are a type of smart materials that can recover from a deformed temporary shape to a "memorized" permanent shape in response to an external environmental stimulus, for example, light, temperature, solvent, pH, electrical or magnetic field [6–10]. Shape-memory polymers can be both thermoplastic and thermosetting ones. The mechanical properties, thermal stability, fixation ratio, and recovery rate of the thermosets are higher than those of the thermoplastics [11]. At present, these polymers are used in aerospace structures and biomedical devices [12–16].

Recently, segmented polyurethanes have been increasingly considered as shape memory materials. This class of polymers is characterized by a high shape recovery ability, controlled shape recovery temperature, high stress–strain properties, wear resistance, and biocompatibility.

In general, segmented urethane-containing elastomers (SUE), such as polyurethanes and polyurethane ureas, can be considered to be a unique class of polymers. The structure and properties of these polymers can be regulated over a wide range [17]. Segmented polyurethanes (SPU) and polyurethane ureas (SPUM) are block copolymers with macromolecular chains consisting of the alternating soft and hard segments, SS and HS, respectively. Hard segments are formed as a result of the reaction of a diisocyanate with low molecular weight chain extenders, such as diamines and diols. The soft segment structure is determined by the oligomers used in the synthesis of an SUE [18].

The difference in the polarity of soft and hard segments results in the microphase separation followed by the formation of hard domains [19–21]. In the microdomains, the hydrogen bonds play an important role in the stabilization of the hard phase structure [22].

Solid domains act as a reinforcing filler. These domains are in the nodes of a specific physical network stable in a wide range of positive temperatures [23].

It should be noted that the shape memory effect of polyurethane elastomers is largely dependent on the degree of crystallinity and hydrogen bonds of the polymer, as well as its microphase separation [24]. These parameters can be varied by using polyesters with different molecular weights, different types of diisocyanates, etc.

Despite all the advantages of polyurethanes and polyurethane ureas, the deformation and strength properties of these elastomers depend on humidity. This is due to the ability of the terminal isocyanate groups, present in oligodiisocyanates, to react with water. To solve this problem, isocyanate groups should be "blocked" by some reactive substance. From this point of view, the most promising reactants are epoxy alcohols, for example, 2,3-epoxy-1-propanol [25]. In this case, the isocyanate group of the oligodiisocyanate reacts with the hydroxyl group of 2,3-epoxy-1-propanol to provide an epoxyurethane oligomer. In addition, the toxicity of isocyanate terminal compounds is substantially reduced. For the elastomers based on these oligomers, no deterioration of the stress–strain properties is observed when subjected to humidity. Amines, anhydrides of dicarboxylic acids, carboxyl-containing compounds, and imidazoles are used as curing agents for epoxyurethane oligomers [26–31].

The epoxyurethane elastomers are characterized by high dielectric properties and mechanical characteristics, and, not least, their properties are independent of humidity.

The elastomers, synthesized using polyester-based epoxyurethane elastomers, are crystallizable [25] and can be used as shape memory materials. However, the use of these materials at low temperatures is limited by a high glass transition temperature of −30–40 °C.

The glass transition temperature of polyether-based epoxyurethane elastomers is up to −72–74 °C. Thus, at a low temperature, it is advisable to use this class of elastomers. However, these elastomers are non-crystallizable, and this property limits the use of these polymers as shape memory materials.

In our opinion, for the preparation of crystallizable polyether-based elastomers, linear oligomer curing agents should be used.

In literature, a wide variety of synthetic methods for oligoethylene glycol diamine have been described, and the synthesis of oligopropylene glycol diamine (trademark Jeffamine) is also well-developed. Such oligomers can be prepared from oligoglycols by a variety of synthetic routes. The most common synthetic route is a three-step procedure that includes mesylation, azidation, and reduction [32]. In addition, in [33], another preparation method for oligoether diamines is described. According to this procedure, amino terminated oligomers can be prepared via the two-step route: substitution of oligoglycol hydroxyls with phthalimide moieties, followed by the hydrazinolysis reaction. Synthetic methods for oligoetherdiamines, including the step of catalytic amination under elevated pressure and temperature conditions, have been also studied [34,35]. As reported by J. M. Harris and colleagues, oligoethylene glycol diamine was synthesized via oxidation of oligoethylene glycol followed by reduction of carbonyl compound [36]. In our opinion, the elastomers cured with these amines are likely to be non-crystallizable. In the first case, when using methyl branched Jeffamine, it is due to the steric hindrance, and in the second—to high oxygen content.

In the present paper, the elastomers cured with oligoamine, synthesized from oligotetramethylene oxide diol, are studied. The authors consider this oligoamine to be the most promising curing agent for the preparation of crystallizable elastomers based on epoxyurethane oligomers.

The aim of the present study was the preparation and characterization of cold-resistant crystallizable elastomers. Four epoxy-terminated oligomers based on oligotetramethylene oxide diols of different molecular weights were synthesized. A novel synthetic method for oligoamines was developed. Two amino terminated oligomers with molecular weights of 1000 and 1400 were used as curing agents. The elastomers based on oligoethers and oligoamines were synthesized and characterized.

2. Materials and Methods

2.1. Materials and Synthesis

2.1.1. Materials

Phosphorus tribromide 99% (Sigma-Aldrich Co., St. Louis, MO, USA), Phthalimide potassium salt 98% (Sigma-Aldrich Co., St. Louis, MO, USA), Hydrazine monohydrate 98% (Sigma-Aldrich Co., St. Louis, MO, USA), 2,4-toluene diisocyanate (TDI) (BASF, Ludwigshafen, Germany), isophorone diisocyanate (IPDI) (Evonik Chemistry Ltd., Essen, Germany), oligotetramethylene oxide diol (OTMO; BASF, Ludwigshafen, Germany) with $M_n \sim 1008$ g·mol^{-1}, $M_n \sim 1400$ g·mol^{-1}, glycidol (grade pure, 99.0%, Research Institute of Polymer Materials, Perm, Russia), Dibutyltin dilaurate (grade pure, 99.8%) were used without purification.

2.1.2. Synthesis of OTMO-Diamines

The three-step synthetic route for preparation of OTMO-diamines is presented in Figure 1.

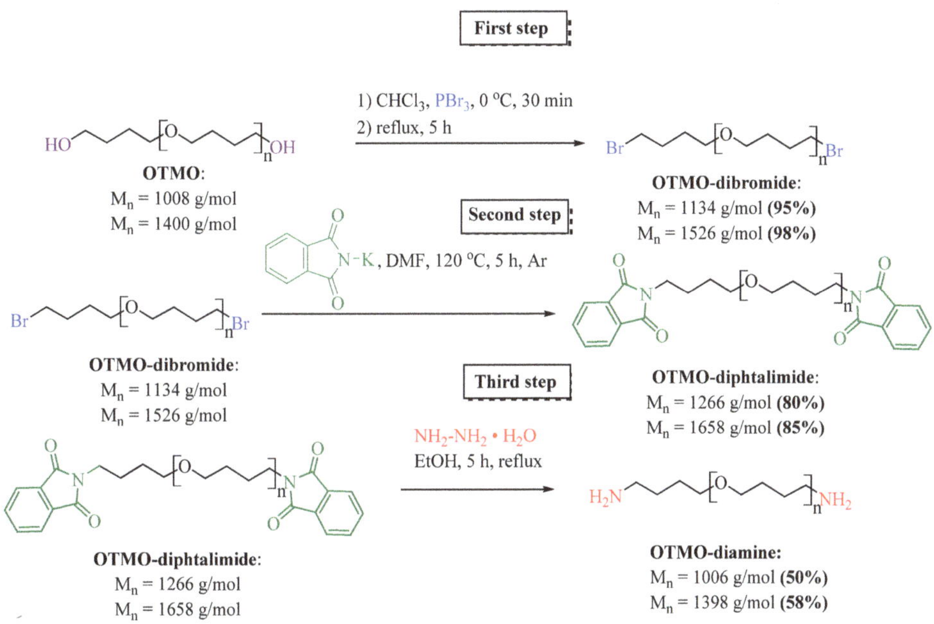

Figure 1. Synthetic route for amino terminated oligo tetramethylene oxide (OTMO-diamine).

Nucleophilic substitution of polyfurite hydroxyls with bromine was carried out using phosphorus tribromide. This approach excludes the use of toxic thionyl chloride, as in the methods of polyethylene glycol halogenation described in [37]. Bromination of polyfurite was carried out in a round-bottomed three-necked flask equipped with a mechanical stirrer and a thermometer. First, a weight of polyfurite (23 mmol, $M_n \sim 1008$ g·mol^{-1}, $M_n \sim 1400$ g·mol^{-1}) was dissolved in chloroform (50 mL), and the solution was introduced into the flask and cooled to 0 °C. Then phosphorus tribromide (1.82 mL, 19 mmol) was added to the solution dropwise, and the reaction mixture was heated to room temperature and then boiled for 5 h. After the reaction was completed, reaction mixture was cooled and added to a saturated aqueous sodium hydrocarbonate solution (50 mL). The organic phase was washed with water (3 × 50 mL) and dried over anhydrous magnesium sulfate. The excessive solvent was removed at a rotary evaporator. As a result, OTMO-dibromides were obtained in high yields (95–98%).

In the second step, for preparation of OTMO-diamines, potassium phthalimide (24.5 g, 13.2 mmol) was added to a solution of an OTMO-dibromide (22 mmol, M_n ~ 1134 g·mol^{-1}, M_n ~ 1526 g·mol^{-1}) in DMF (180 mL) and the reaction mixture was stirred under argon at 120 °C for 5 h. Then the solution was cooled and filtered, and DMF was evaporated in vacuum. For purification, the product was dissolved in DCM, filtered, and the excess solvent was removed at a rotary evaporator. OTMO-diphtalimides were obtained in the form of a white viscous mass in high yields (80–85%).

At the final step, hydrazine hydrate (180 mmol, 8.75 mL) was added to OTMO-diphtalimide (18 mmol) dissolved in absolute ethanol (100 mL). The reaction mixture was stirred at reflux under argon atmosphere for 5 h. Then the solution was cooled and filtered, and ethanol was evaporated under vacuum. For purification, the product was dissolved in DCM and filtered, and the excess solvent was removed at a rotary evaporator. The yield of OTMO-diamines was 50–58%.

2.1.3. Synthesis of Epoxyurethane Oligomers

A two-step synthetic route for epoxyurethane oligomers via oligodiisocyanate formation is shown in Figure 2.

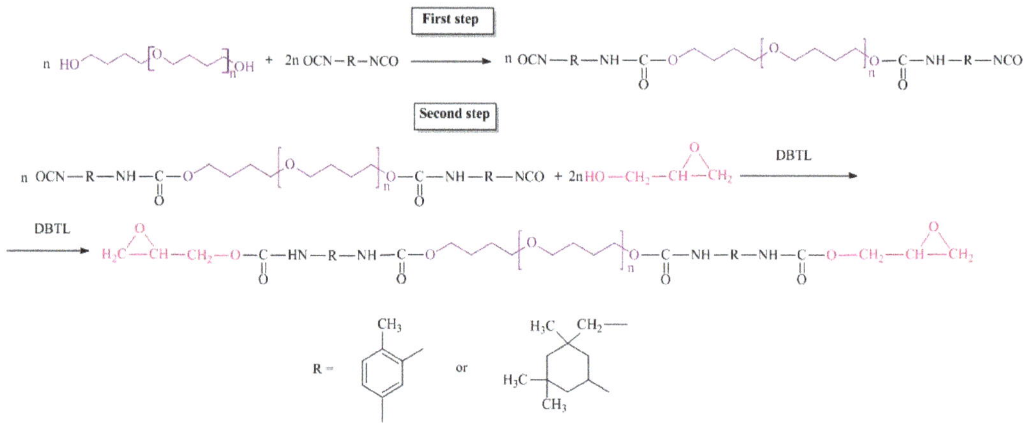

Figure 2. Synthetic route for epoxyurethane oligomers.

The pre-synthesized polyesters were dried at 90 °C for 7 h. Oligodiisocyanates were obtained via interaction of oligodiols and diisocyanate (NCO/OH = 2.03) at constant temperature of the reaction mixture of 1 h at 60 °C and 80 °C at stirring for 6 h. The content of NCO groups in the prepolymers was determined by titration with n-butylamine (standard method ASTM D 2572-97). Then the reaction mixture was cooled to 40 °C, and the catalyst, di-n-butyl tin dilaurate, and the calculated amount of glycidol were added. The catalyst amount was 0.03 wt.% of the reaction mixture. Then the reaction mixture was heated to 70 °C and stirred for 8 h. The completeness of the reaction was confirmed by IR spectroscopy. No band at 2270 cm^{-1}, characteristic of isocyanate group [38], was observed in the IR spectra of the reaction products. The content of free epoxy groups was determined according to the technique described in [39]. The composition and properties of the synthesized oligomers are summarized in Table 1.

Table 1. Composition and properties of the synthesized oligomers.

Product Code	Molecular Weight of Initial EUO OTMO	Diisocyanate Type	Content of Free Isocyanate Groups, wt %		Content of Free Epoxy Groups, wt %	
			Calculated	Determined	Calculated	Determined
FP-1	1008	2,4-toluene diisocyanate	6.23	6.31 ± 0.03	5.82	5.71 ± 0.03
FP-2	1008	isophorone diisocyanate	5.82	5.85 ± 0.03	5.43	5.31 ± 0.03
FP-3	1400	2,4-toluene diisocyanate	4.81	4.92 ± 0.03	4.64	4.72 ± 0.03
FP-4	1400	isophorone diisocyanate	4.56	4.62 ± 0.03	4.37	4.45 ± 0.03
FP-5	2000	2,4-toluene diisocyanate	3.57	3.65 ± 0.03	3.51	3.47 ± 0.03
FP-6	2000	isophorone diisocyanate	3.43	3.53 ± 0.03	3.54	3.45 ± 0.03

2.1.4. Polymer Synthesis

At the final synthetic step, epoxyurethane oligomers were mixed with a synthesized curing agent for 10 min under vacuum (1–2 kPa) at 40 °C. The resulting reaction mixture was cured for 48 h at 30 °C. Cure monitoring by FTIR was used to determine the required cure time. Disappearance of the absorption band at 910 cm^{-1} indicated the completeness of the epoxy group conversion [40]. The synthetic route is demonstrated in Figure 3.

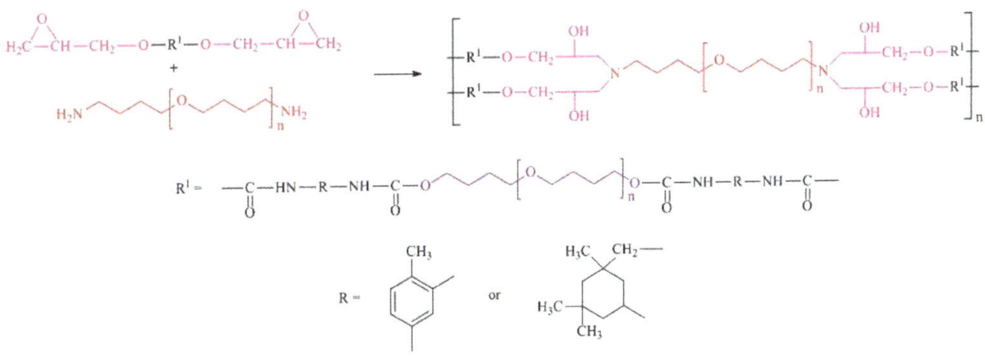

Figure 3. Synthetic route of polymers.

The sample compositions are provided in Table 2.

Table 2. Compositions for preparation of the polymers.

Composition Code	Oligomer Code	Curing Agent
C-1	FP-1	OTMODA 1000
C-2	FP-3	OTMODA 1000
C-3	FP-5	OTMODA 1000
C-4	FP-1	OTMODA 1400
C-5	FP-3	OTMODA 1400
C-6	FP-5	OTMODA 1400
C-7	FP-2	OTMODA 1000
C-8	FP-4	OTMODA 1000
C-9	FP-6	OTMODA 1000
C-10	FP-2	OTMODA 1400
C-11	FP-4	OTMODA 1400
C-12	FP-6	OTMODA 1400

2.2. *Methods*

2.2.1. ^{1}H- and ^{13}C-NMR Spectroscopy

^{1}H and ^{13}C NMR spectra were recorded using a Bruker, Moccow, Russia) Avance Neo III spectrometer (^{1}H: 400 MHz, ^{13}C: 75 MHz); tetramethyl silane was used as an internal standard. NMR chemical shifts were calibrated using the deuterium signal of $CDCl_3$ (7.26 ppm for ^{1}H and 77.16 ppm for ^{13}C).

2.2.2. Elemental Analysis

Elemental analysis was carried out using a LECO, Moscow, Russia CHNS-932 analyzer.

2.2.3. Gel Permeation Chromatography

The molecular mass of the oligomers obtained was determined by gel permeation chromatography using an ULTIMATE 3000 chromatograph (Dionix Thermo Scientific, Moscow, Russia) equipped with a RefractoMax 521 refractometric detector according to [41].

2.2.4. FTIR Spectroscopy

FTIR spectra in the area of carbonyl valence vibrations (between wave numbers ν = 1600 and 1760 cm^{-1}) of the investigated samples were recorded using an IFS-66/S spectrometer (Bruker, Moscow, Russia) with spectral resolution of 1 cm^{-1}. The spectra were normalized with respect to the band at 2860 cm^{-1}, corresponding to symmetric vibrations of aliphatic –CH_2 groups [42].

2.2.5. Differential Scanning Calorimetry (DSC)

Endothermic and exothermic effects in the samples within the temperature range from −100 °C to +100 °C were recorded using a Mettler Toledo MDSC Q100 calorimeter (Mettler Toledo, Moscow, Russia). Heating and cooling rates were 5 K min^{-1}.

2.2.6. Mechanical Tests

Mechanical tests of specimens of the materials obtained were performed with an Instron 3365 (Moscow, Russia) testing machine at the extension velocity υ = 0.417 s^{-1} and a temperature of 25 ± 1 °C by the standard procedure. The following characteristics were measured: the nominal strength σk (MPa), i.e., the maximal stress per initial specimen cross section; the relative critical strain ε_k (%); the nominal elastic modulus E_{100} (stress at the relative strain ε = 100%). The synthesized polymer was subjected to 5 tests.

3. Results

This section is divided into subheadings. It should provide a concise and precise description of the experimental results, their interpretation, as well as the experimental conclusions that can be drawn.

3.1. *NMR Spectra of Functionalized Oligotetramethylene Oxides*

The transformations of the terminal fragments of the initial polyfurites were confirmed by NMR data (Figure 4a–d). In the ^{1}H NMR spectra, the protons of the methylene groups of the initial polyfurites (M_n ~ 1008 g·mol^{-1}, M_n ~ 1400 g·mol^{-1}) are observed at 1.51–1.62 (O-CH_2-CH_2-CH_2-CH_2-O (a)) and 3.31–3.39 (O-CH_2-CH_2-CH_2-CH_2-O (b)) ppm, respectively. The triplet at 3.56 (c) ppm can be attributed to the terminal hydroxy methylene groups. The signals of the two hydroxy group protons appear at 2.31 ppm (M_n ~ 1008 g·mol^{-1}) or at 2.60 ppm (M_n ~ 1400 g·mol^{-1}). Upon the substitution of hydroxy groups with bromine, these signals disappeared. In addition, the signals attributed to the protons of the methylene groups in the vicinity of halogen atoms were found to be shifted. Further, the substitution of bromine with phthalimide groups resulted in the appearance of phthalimide proton signals in a weak field region (δ = 7.63 ppm and 7.76 ppm), and again a shift of the methylene proton signals in the vicinity of the substituent, in this case, phthalimide, is observed. In the next step, in OTMO diamine spectra, there are the peaks attributed to the protons of two amino groups (δ = 2.87 ppm for M_n ~ 1008 g·mol^{-1} and δ = 2.67 ppm for M_n ~ 1400 g·mol^{-1}). The strong field shift of the signals of the functionalized methylene groups is observed.

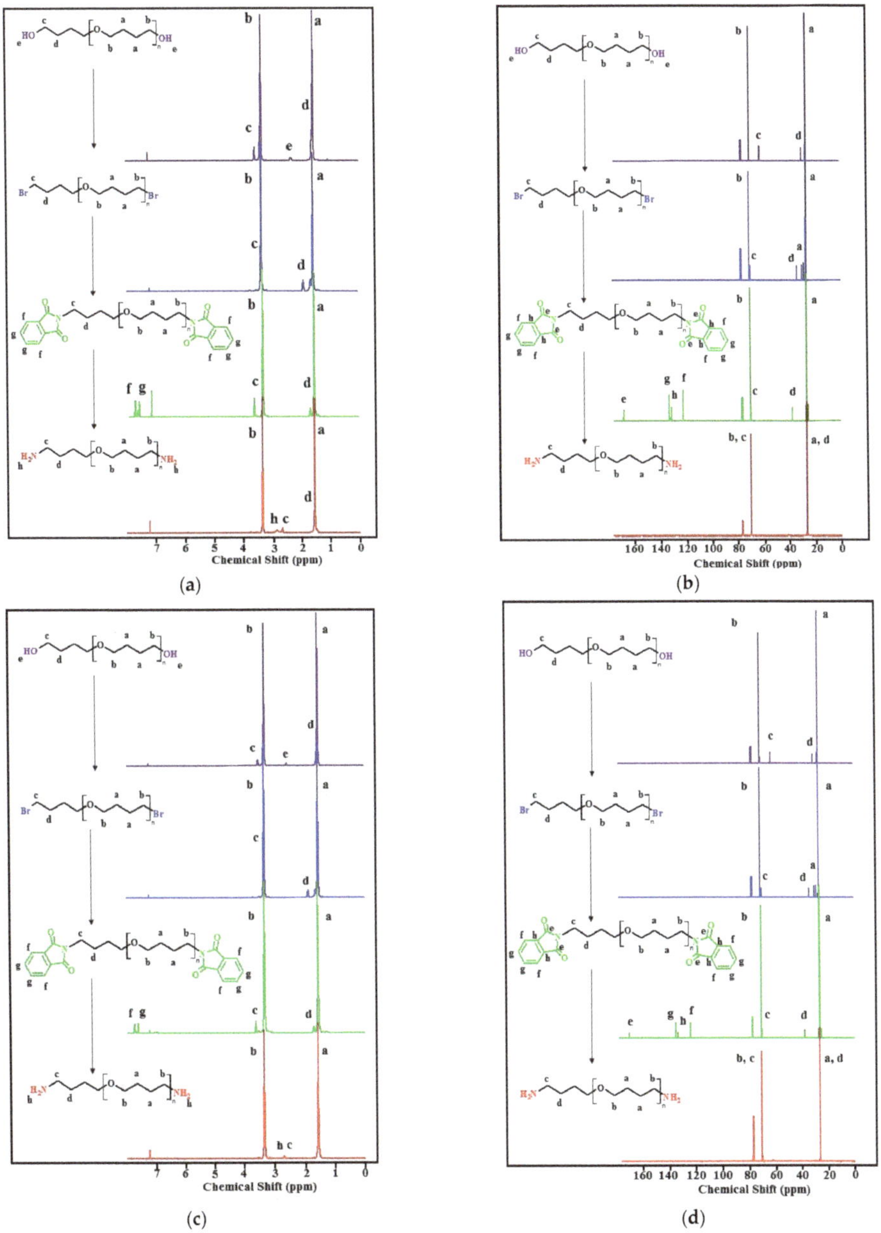

Figure 4. ^{1}H NMR spectra: (**a**)—with M_n ~ 1008 g·mol^{-1} and its derivatives, (**c**) OTMO with M_n ~ 1400 g·mol^{-1} and its derivatives; ^{13}C NMR spectra: (**b**)—with M_n ~ 1008 g·mol^{-1} and its derivatives, (**d**) OTMO with M_n ~ 1400 g·mol^{-1} and its derivatives.

In ^{13}C NMR spectra, at each step of the synthetic route, a shift of the signals of the terminal methylene carbon (c) and (d) can be distinguished (Figure 4b,d). In addition, weak field signals (δ = 123, 132, 134, and 168 ppm) appear in the spectra of OTMO-diphtalimides. These signals disappear after the aminolysis is completed.

Number-average molecular weight (M_n) was determined from ^1H NMR spectroscopy by comparing the integration of the end-group proton resonances to the repeating unit proton resonances. The results are presented below.

OTMO-dibromide (M_n = 1134 g/mol): ^1H NMR (400 MHz, $CDCl_3$, δ, ppm): 3.35 (br m, 55H, $OCH_2CH_2CH_2CH_2Br$, $OCH_2CH_2CH_2CH_2O$ main chain), 1.88 (t, 4H, $OCH_2CH_2CH_2CH_2Br$), 1.64 (t, 4H, $OCH_2CH_2CH_2CH_2O$ main chain), 1.55 (br m, 47H, $OCH_2CH_2CH_2CH_2O$ main chain). ^{13}C NMR (75 MHz, $CDCl_3$, δ, ppm): 26.5 ($OCH_2CH_2CH_2CH_2O$), 28.3 ($OCH_2CH_2CH_2CH_2O$), 29.7 ($OCH_2CH_2CH_2CH_2O$), 33.6 ($OCH_2CH_2CH_2CH_2Br$), 69.6 (O $CH_2CH_2CH_2CH_2Br$), 70.5 ($OCH_2CH_2CH_2CH_2O$).

OTMO-dibromide (M_n = 1526 g/mol): ^1H NMR (400 MHz, $CDCl_3$, δ, ppm): 3.35 (br m, 80H, $OCH_2CH_2CH_2CH_2Br$, $OCH_2CH_2CH_2CH_2O$ main chain), 1.88 (t, 4H, $OCH_2CH_2CH_2CH_2Br$), 1.64 (t, 4H, $OCH_2CH_2CH_2CH_2O$ main chain), 1.55 (br m, 72H, $OCH_2CH_2CH_2CH_2O$ main chain). ^{13}C NMR (75 MHz, $CDCl_3$, δ, ppm): 26.4 ($OCH_2CH_2CH_2CH_2O$), 28.3 ($OCH_2CH_2CH_2CH_2O$), 29.7 ($OCH_2CH_2CH_2CH_2O$), 33.6 ($OCH_2CH_2CH_2CH_2Br$), 69.6 (O $CH_2CH_2CH_2CH_2Br$), 70.5 ($OCH_2CH_2CH_2CH_2O$).

OTMO-diphtalimide (M_n = 1266 g/mol): ^1H NMR (400 MHz, $CDCl_3$, δ, ppm): 7.76 (m, 4H, Pthalimide), 7.63 (m, 4H, Pthalimide), 3.65 (t, 4H, -CH_2CH_2-Phtalimide), 3.35 (br m, 51H, $OCH_2CH_2CH_2CH_2O$ main chain), 1.70 (t, 4H, -CH_2CH_2-Phtalimide), 1.55 (br m, 51H, $OCH_2CH_2CH_2CH_2O$ main chain). ^{13}C NMR (75 MHz, $CDCl_3$, δ, ppm): 25.3 ($OCH_2CH_2CH_2CH_2O$), 26.4 ($OCH_2CH_2CH_2CH_2O$), 27.0 ($OCH_2CH_2CH_2CH_2O$), 37.7 ($OCH_2CH_2CH_2CH_2$Phtalimide), 69.9 ($OCH_2CH_2CH_2CH_2$Phtalimide), 70.4 ($OCH_2CH_2CH_2$ CH_2O), 123.0 (Pthalimide), 132.0 (Pthalimide), 133.7 (Pthalimide), 168.2 (Pthalimide).

OTMO-diphtalimide (M_n = 1658 g/mol): ^1H NMR (400 MHz, $CDCl_3$, δ, ppm): 7.76 (m, 4H, Pthalimide), 7.63 (m, 4H, Pthalimide), 3.64 (t, 4H, -CH_2CH_2-Phtalimide), 3.35 (br m, 76H, $OCH_2CH_2CH_2CH_2O$ main chain), 1.70 (t, 4H, -CH_2CH_2-Phtalimide), 1.55 (br m, 76H, $OCH_2CH_2CH_2CH_2O$ main chain). ^{13}C NMR (75 MHz, $CDCl_3$, δ, ppm): 25.3 ($OCH_2CH_2CH_2CH_2O$), 26.4 ($OCH_2CH_2CH_2CH_2O$), 27.0 ($OCH_2CH_2CH_2CH_2O$), 37.7 ($OCH_2CH_2CH_2CH_2$Phtalimide), 69.9 ($OCH_2CH_2CH_2CH_2$Phtalimide), 70.5 ($OCH_2CH_2CH_2$ CH_2O), 123.0 (Pthalimide), 132.1 (Pthalimide), 133.7 (Pthalimide), 168.2 (Pthalimide).

OTMO-diamines (M_n = 1006 g/mol): ^1H NMR (400 MHz, $CDCl_3$, δ, ppm): 3.35 (br m, 51H, $OCH_2CH_2CH_2CH_2O$ main chain), 2.87 (br s, 4H, NH_2), 2.66 (t, 4H, $OCH_2CH_2CH_2CH_2NH_2$), 1.55 (br m, 55H, $OCH_2CH_2CH_2CH_2O$ main chain). ^{13}C NMR (75 MHz, $CDCl_3$, δ, ppm): 26.4 ($OCH_2CH_2CH_2CH_2O$, $OCH_2CH_2CH_2CH_2NH_2$), 70.4 ($OCH_2CH_2CH_2CH_2NH_2$, OCH_2CH_2 CH_2CH_2O).

OTMO-diamines (M_n = 1398 g/mol): ^1H NMR (400 MHz, $CDCl_3$, δ, ppm): 3.35 (br m, 80H, $OCH_2CH_2CH_2CH_2O$ main chain, $OCH_2CH_2CH_2CH_2NH_2$), 2.67 (br s, 4H, NH_2), 1.55 (br m, 80H, $OCH_2CH_2CH_2CH_2O$ main chain). ^{13}C NMR (75 MHz, $CDCl_3$, δ, ppm): 26.5 ($OCH_2CH_2CH_2CH_2O$, $OCH_2CH_2CH_2CH_2NH_2$), 70.6 ($OCH_2CH_2CH_2CH_2NH_2$, OCH_2CH_2 CH_2CH_2O).

3.2. Elemental Analysis

The data obtained in the course of elemental analysis are provided in the Table 3. The closeness of the indices to the theoretical values confirms the structure of the synthesized compounds.

Table 3. Elemental analysis of the synthesized compounds.

	C, %		H, %		N, %	
	Founded	Calculated	Founded	Calculated	Founded	Calculated
OTMO-dibromide (M_n = 1134 g/mol)	58.41	58.20	9.87	9.70	-	-
OTMO-dibromide (M_n = 1526 g/mol)	60.58	60.38	10.19	10.06	-	-
OTMO-diphtalimide (M_n = 1266 g/mol)	67.48	67.30	9.49	9.32	2.37	2.21
OTMO-diphtalimide (M_n = 1658 g/mol)	67.32	67.15	9.90	9.74	1.84	1.69
OTMO-diamines (M_n = 1006 g/mol)	65.75	65.61	11.46	11.33	2.90	2.78
OTMO-diamines (M_n = 1398 g/mol)	66.05	65.90	11.40	11.27	2.17	2.00

3.3. Gel Permeation Chromatography of Functionalized Oligotetramethylene Oxides

In the determination of the molecular mass of compounds, the retention time was from 4.98 to 5.75 min (Figure 5a,b). The small width of the peaks corresponds to the narrow molecular-mass distribution of the oligomer. The obtained values of the average compound molecular mass agree with the theoretical values (Table 4). Furthermore, NMR spectroscopy and GPC revealed that the M_n values of compounds in both series remained constant throughout the end-group transformations (Table 4).

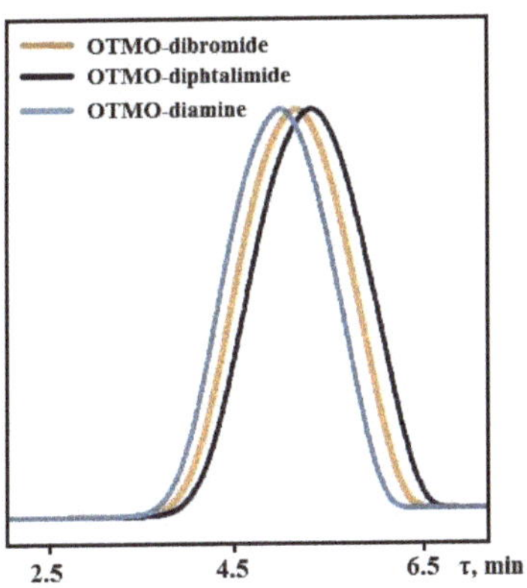

(a)

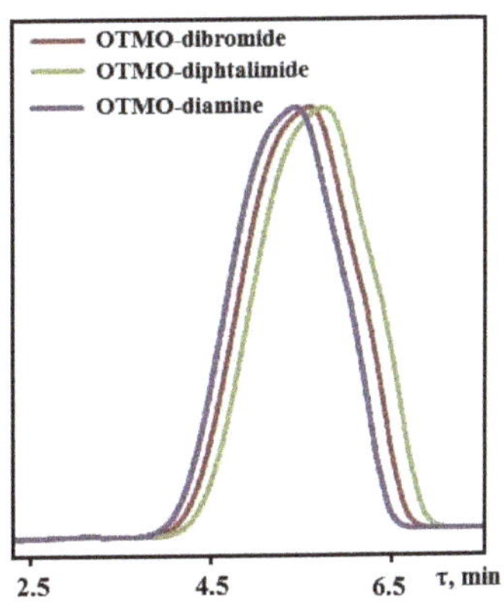

(b)

Figure 5. GPC chromatograms: (**a**) OTMO with M_n ~ 1008 g·mol^{-1} and its derivatives; (**b**) OTMO with M_n ~ 1400 g·mol^{-1} and its derivatives recorded using THF as the mobile phase.

Table 4. Molecular weight characteristics of the compounds determined via ^{1}H NMR spectroscopy, and GPC.

	M_n ^{1}		M_n ^{2}	
	GPC	^{1}H	GPC	^{1}H
OTMO-dibromide	1121 (5.105 *)	1134	1480 (5.550 *)	1526
OTMO-diphtalimide	1250 (5.265 *)	1266	1640 (5.748 *)	1658
OTMO-diamines	1020 (4.980 *)	1006	1380 (5.426 *)	1398

M_n ^{1}—Number-average molecular weight of compounds synthesized on the basis of OTMO with M_n ~ 1008 g·mol^{-1}, M_n ^{2}—Number-average molecular weight of compounds synthesized on the basis of OTMO with M_n ~ 1400 g·mol^{-1}, *—Retention time, min.

3.4. FTIR Spectroscopy

3.4.1. FTIR Spectra of Functionalized Oligotetramethylene Oxides

The transformation of the terminal hydroxyls in polyfurites into amino groups was also demonstrated by FTIR spectra (Figure 6).

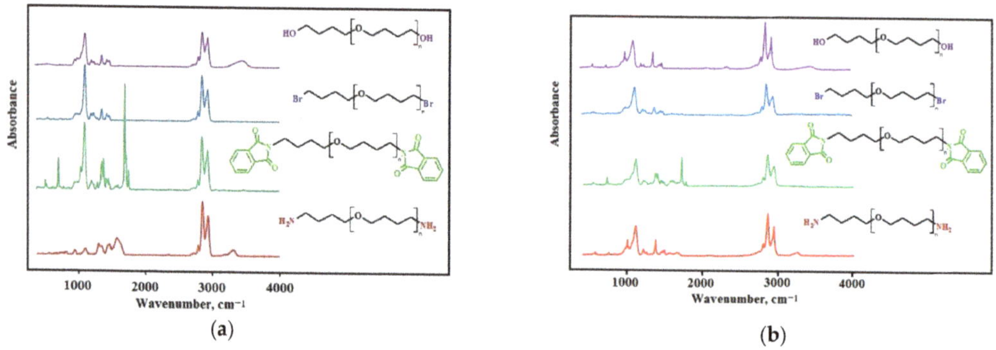

Figure 6. The FTIR spectra for the samples: (**a**) OTMO with M_n ~ 1008 g·mol^{-1} and its derivatives; (**b**) OTMO with M_n ~ 1400 g·mol^{-1} and its derivatives.

The absorption band at 665 cm^{-1} (C-Br) is observed in polyfurite spectra after nucleophilic substitution of hydroxy groups with bromine. At the same time, there are no hydroxyl group bands at 3300–3500 cm^{-1}, usually present in polyfurite spectra. In the OTMO-diphtalimide spectrum, a characteristic imide peak appears at 1720 cm^{-1}. An absorption band of amino groups at 3300–3600 cm^{-1} is observed in the FTIR spectra of OTMO diamines. The rest of the bands of the intermediates and the end product, OTMO-diamine, are identical to those of the initial polyfurite. Thus, it can be concluded that the main chain structure of the oligomer remained unchanged, and only the terminal groups were involved in the reactions.

3.4.2. FTIR Spectra of the Synthesized Elastomers

The FTIR spectra of the synthesized elastomers are shown in Figure 7a,b. The NH band of urethane can be found at 3350 cm^{-1} as a broad absorption. A broad band with the center at 2950 cm^{-1} and the one at 2860 cm^{-1} were assigned to the CH asymmetric stretching and the symmetric one in the CH_2 groups. The absorption bands at 1542, 1454, and 1412 cm^{-1} were assigned to the amide–NH stretching. The elastomers synthesized from 2,4-toluene diisocyanate (C-1–C-6) have absorption bands at 1600 cm (aryl ring) and also at 1612 cm^{-1}. This band is typical for urethane-containing elastomers synthesized on the basis of this diisocyanate. For elastomers synthesized from isophrondiisocyanate, these bands do not appear.

The analysis of the FTIR spectra in the range of carbonyl stretching vibrations (1600–1760 cm^{-1}) reveals the important features of the structural organization of the synthesized elastomers. It is known that the microphase separation of soft and hard segments of elastomers with urethane hydroxyl hard blocks is characterized by an absorption band at 1695 cm^{-1} when using isophorone diisocyanate and 1705 cm^{-1} when using 2,4-toluene diisocyanate [43].

The samples synthesized from 2,4-toluylene diisocyanate (Figure 8a,b) in the vibration range of 1600–1760 cm^{-1} have a strong absorption band at 1705 cm^{-1} that corresponds to the hydrogen bond between the hard segments of the elastomer showing the high degree of microphase separation. With an increase in the molecular weight of the oligodiol used in the synthesis of the epoxyurethane oligomer, the intensity of this band decreases. This fact indicates that the degree of microphase separation decreases with an increase in the molecular weight of oligodiol. In the case of using isophorone diisocyanate, the same regularity appears (Figure 9). It should be noted that the degree of microphase separation is higher for samples synthesized from isophorone diisocyanate.

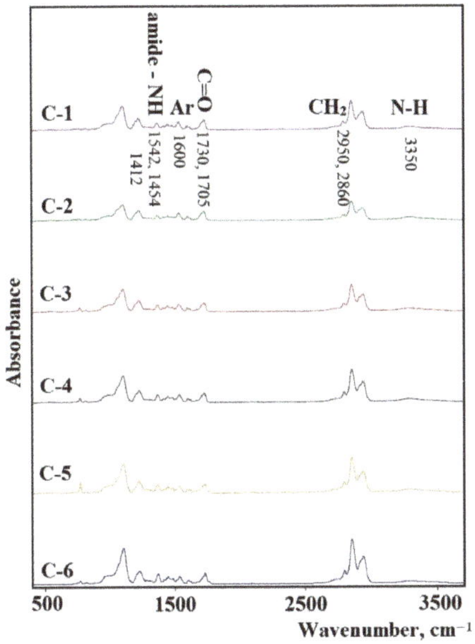

(**a**)

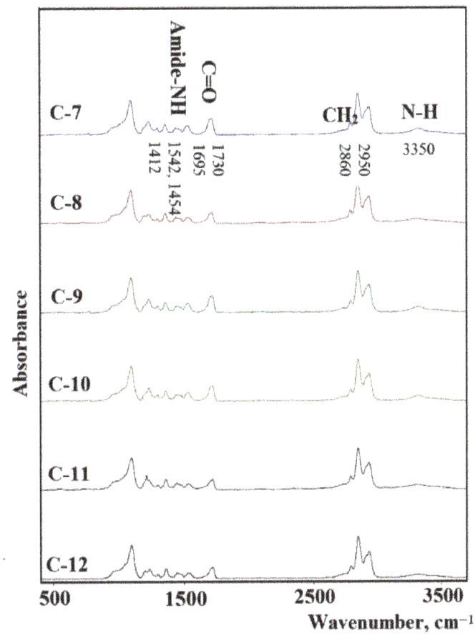

(**b**)

Figure 7. The FTIR spectra for the samples: (**a**) C-1, C-2, C-3, C-4, C-5, C-6; (**b**) C-7, C-8, C-9, C-10, C-11, C-12.

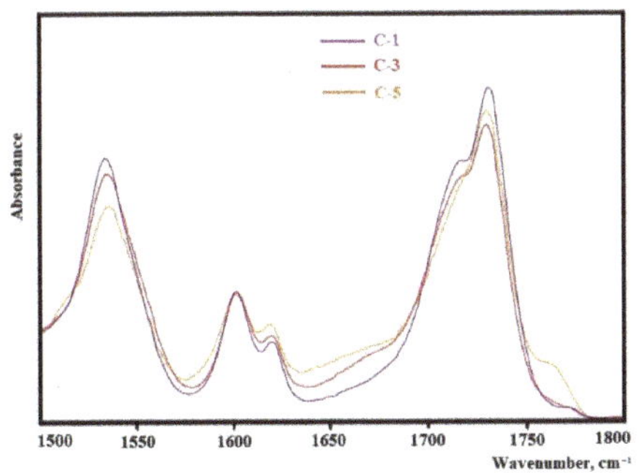

(**a**)

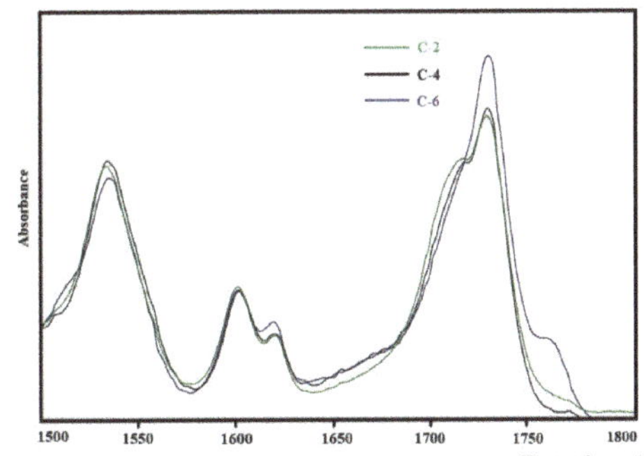

(**b**)

Figure 8. Sections of the FTIR spectra for the samples: (**a**) C-1, C-3, C-5; (**b**) C-2, C-4, C-6.

3.5. Differential Scanning Calorimetry Data

The thermal properties of the synthesized elastomers were studied by differential scanning calorimetry. First, the samples were heated to 150 °C, then cooled to 100 °C below zero, kept for 30 min, and heated at a heating rate of 5 °C/min. In Figure 10a,b, the reheating thermograms of the samples C1–C12 are shown. The thermophysical properties of the synthesized elastomers are shown in Table 5.

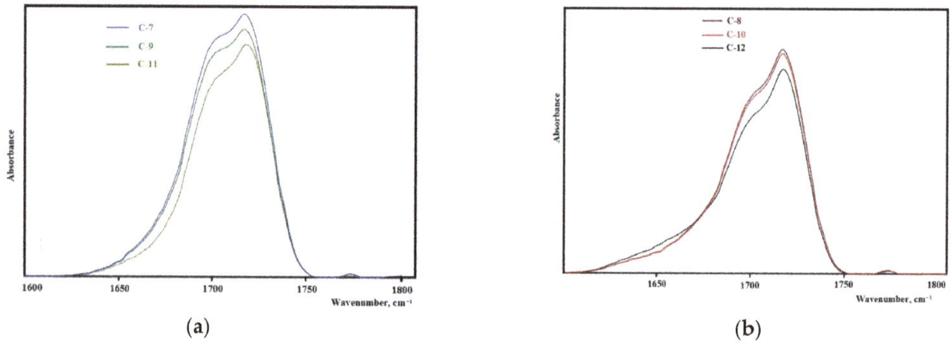

Figure 9. Sections of the FTIR spectra for the samples: (**a**) C-7, C-9, C-11; (**b**) C-8, C-10, C-12.

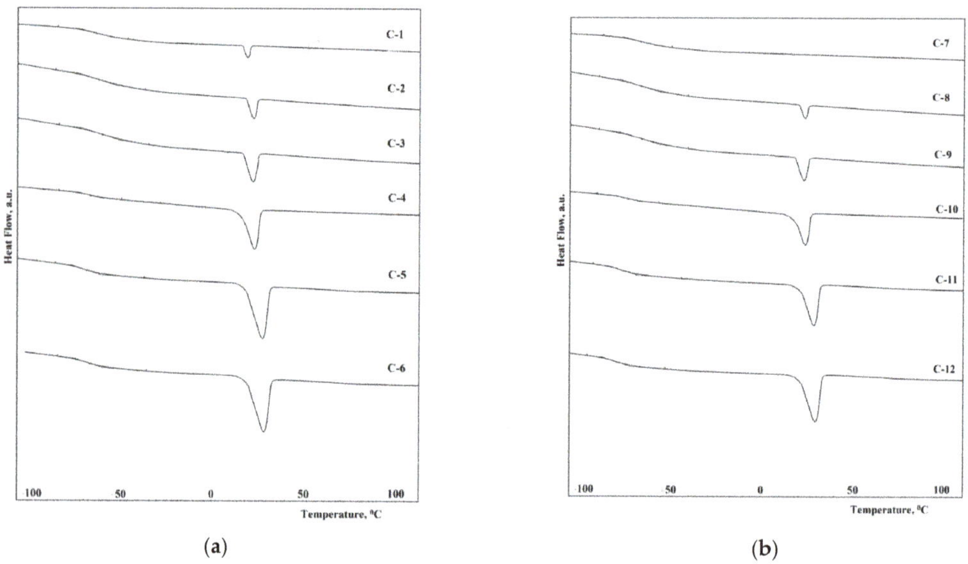

Figure 10. DSC-curves of the epoxy urethane samples from: (**a**) C-1, C-2, C-3, C-4, C-5, C-6; (**b**) C-7, C-8, C-9, C-10, C-11, C-12.

Table 5. Thermophysical properties of synthesized elastomers.

Composition Code	Glass Transition Temperature of the Soft Phase, °C	Melting Temperature of the Soft Phase, °C
C-1	−63	29
C-2	−61	29
C-3	−59	29
C-4	−63	29
C-5	−65	29
C-6	−64	29
C-7	−67	-
C-8	−66	30
C-9	−65	30
C-10	−73	30
C-11	−71	30
C-12	−70	30

From the presented thermograms, it can be seen that when using the hardener with a molecular weight of 1000, the glass transition temperature is 5–8 °C higher than when using a hardener with a molecular weight of 1400. This is due to an increase in the segmental mobility of polymer chains. On the other hand, the use of a hardener with a higher molecular weight provides the prerequisites for the crystallization of the polymer, which should reduce the segmental activity of the chains.

The elastomers synthesized from isophorone diisocyanate have a glass transition temperature lower by 5–10 °C than elastomers from 2,4-toluene diisocyanate. This is due to the lower degree of microphase separation of soft and hard segments in elastomers (Figures 8a,b and 9a,b). In this case, an increase in the degree of microphase separation leads to a decrease in the degree of crystallinity of elastomers. It should be noted that the melting temperature of the flexible phase of all the elastomers except C-7 (there is no melting in sample C-7) is 29–30 °C.

3.6. Deformation and Strength Characteristics

According to the data obtained during mechanical tests (Table 6), the deformation–strength characteristics depend on both the molecular and supramolecular structure of the elastomers.

Table 6. Physical–mechanical characteristics of the synthesized elastomers.

Composition Code	σ_k, MPa	ε_k, %	E_{100}, MPa
C-1	6.2 ± 0.3	175 ± 7	5.3 ± 0.3
C-2	6.1 ± 0.3	196 ± 7	5.7 ± 0.3
C-3	6.3 ± 0.3	224 ± 7	6.0 ± 0.3
C-4	6.1 ± 0.3	115 ± 5	5.8 ± 0.3
C-5	7.2 ± 0.4	121 ± 5	6.0 ± 0.3
C-6	6.9 ± 0.3	141 ± 5	6.2 ± 0.3
C-7	6.5 ± 0.3	205 ± 7	4.1 ± 0.2
C-8	10.2 ± 0.5	221 ± 7	4.2 ± 0.2
C-9	9.3 ± 0.4	256 ± 10	4.6 ± 0.2
C-10	6.2 ± 0.3	196 ± 7	4.9 ± 0.2
C-11	6.4 ± 0.3	206 ± 7	5.0 ± 0.3
C-12	7.0 ± 0.3	261 ± 10	5.5 ± 0.3

The elastomers synthesized from 2,4-toluene diisocyanate show a higher Young's modulus (E_{100}), which is explained by a higher degree of crystallinity of the elastomers. However, elastomers synthesized from isophrondiisocyanate, due to a higher degree of microphase separation of soft and hard segments, are characterized by higher strength characteristics. With an increase in the molecular weight of the used hardener, Young's modulus also increases. Increasing the molecular weight of EUO when using a higher molecular weight OTMO in its synthesis leads to an increase in the deformation characteristics of the cured elastomer.

4. Conclusions

For the first time, a method has been developed for the synthesis of oligotetramethylene oxides with terminal amino groups, including the initial bromination of oligotetramethylene oxide diols, followed by the Gabriel reaction.

Six epoxyurethane oligomers were prepared using the oligotetramethylene oxide diol with $M_n \sim 1008$ g·mol^{-1}, $M_n \sim 1400$ g·mol^{-1}, isophorone diisocyanate, 2,4-toluene diisocyanate, and epoxy alcohol-glycidol.

Twelve elastomers from oligomers with urethane hydroxyl hard segments were prepared using synthesized amines.

For the first time, on the basis of epoxyurethane oligomers, synthesized on the basis of polyethers, crystallizable elastomers have been obtained.

The degree of microphase separation is higher for samples synthesized from isophorone diisocyanate.

It has been shown that the use of new oligoamines makes it possible to obtain elastomers with a controlled degree of crystallinity, which allows them to be used as shape memory materials. At the same time, the glass transition temperature of elastomers −60–70 °C allows them to be used in extreme conditions of the far North.

The elastomers synthesized from 2,4-toluene diisocyanate exhibit a higher Young's modulus (E_{100}) due to a higher degree of crystallinity of the elastomers. However, elastomers synthesized from isophrondiisocyanate, due to a higher degree of microphase separation of soft and hard segments, are characterized by higher strength characteristics.

Author Contributions: Conceptualization, A.S.; methodology, D.K.; investigation, A.S. and D.S.; resources, N.E.; writing—original draft preparation, A.S.; writing—review and editing, A.S. and D.S.; project administration, V.S. All authors have read and agreed to the published version of the manuscript.

Funding: The reported study was funded by RFBR and Perm Territory, project number No. 20-43-596010. The work was carried out within the framework of the State Assignment (theme state registration number 122011900165-2) using the equipment of the Center for Collective Use "Investigations of materials and substances" of Perm Federal Research Center of the Ural Branch of the Russian Academy of Sciences.

Institutional Review Board Statement: Not applicable.

Informed Consent Statement: Not applicable.

Data Availability Statement: The most significant data generated or analyzed during this study are included in this published article. Further results obtained during the current study are available from the corresponding author on reasonable request.

Acknowledgments: The study was performed using the equipment of the Center for Shared Use Studies of Materials and Substances at the Perm Federal Research Center Ural Branch Russian Academy of Sciences.

Conflicts of Interest: The authors declare no conflict of interest.

References

1. Bahl, S.; Nagar, H.; Singh, I.; Sehgal, S. Smart materials types, properties and applications: A review. *Mater. Today Proc.* **2020**, *28*, 1302–1306. [CrossRef]
2. Shahinpoor, M. *Fundamentals of Smart Materials*; Royal Society of Chemistry: London, UK, 2020.
3. Cheng, F.; Liang, J.; Tao, Z.; Chen, J. Functional Materials for Rechargeable Batteries. *Adv. Mater.* **2011**, *23*, 1695–1715. [CrossRef] [PubMed]
4. Cornwell, D.J.; Smith, D.K. Expanding the scope of gels—Combining polymers with low-molecular-weight gelators to yield modified self-assembling smart materials with high-tech applications. *Mater. Horiz.* **2015**, *2*, 279–293. [CrossRef]
5. Singh, A.V.; Rahman, A.; Kumar, N.S.; Aditi, A.; Galluzzi, M.; Bovio, S.; Barozzi, S.; Montani, E.; Parazzoli, D. Bio-inspired approaches to design smart fabrics. *Mater. Des.* **2012**, *36*, 829–839. [CrossRef]
6. Leonardi, A.B.; Fasce, L.A.; Zucchi, I.; Hoppe, C.E.; Soulé, E.R.; Perez, C.J.; Williams, R.J. Shape memory epoxies based on networks with chemical and physical crosslinks. *Eur. Polym. J.* **2011**, *47*, 362–369. [CrossRef]
7. Xiao, X.; Kong, D.; Qiu, X.; Zhang, W.; Liu, Y.; Zhang, S.; Zhang, F.; Hu, Y.; Leng, J. Shape memory polymers with high and low temperature resistant properties. *Sci. Rep.* **2015**, *5*, 14137. [CrossRef]
8. Zhang, F.H.; Zhang, Z.C.; Luo, C.J.; Lin, I.-T.; Liu, Y.; Leng, J.; Smoukov, S.K. Remote, fast actuation of programmable multiple shape memory composites by magnetic fields. *J. Mater. Chem. C* **2015**, *3*, 11290–11293. [CrossRef]
9. Xiao, R.; Guo, J.; Safranski, D.L.; Nguyen, T.D. Solvent-driven temperature memory and multiple shape memory effects. *Soft Matter* **2015**, *11*, 3977–3985. [CrossRef]
10. Dong, Y.; Xia, H.; Zhu, Y.; Ni, Q.-Q.; Fu, Y. Effect of epoxy-graft-polyoxyethylene octyl phenyl ether on preparation, mechanical properties and triple-shape memory effect of carbon nanotube/water-borne epoxy nanocomposites. *Compos. Sci. Technol.* **2015**, *120*, 17–25. [CrossRef]
11. Dong, Y.; Ni, Q.-Q.; Fu, Y. Preparation and characterization of water-borne epoxy shape memory composites containing silica. *Compos. Part A Appl. Sci. Manuf.* **2015**, *72*, 1–10. [CrossRef]
12. Lendlein, A.; Langer, R. Biodegradable, Elastic Shape-Memory Polymers for Potential Biomedical Applications. *Science* **2002**, *296*, 1673–1676. [CrossRef]

13. Zarek, M.; Layani, M.; Cooperstein, I.; Sachyani, E.; Cohn, D.; Magdassi, S. 3D Printing of Shape Memory Polymers for Flexible Electronic Devices. *Adv. Mater.* **2015**, *28*, 4449–4454. [CrossRef] [PubMed]
14. Hu, J.; Zhu, Y.; Huang, H.; Lu, J. Recent advances in shape–memory polymers: Structure, mechanism, functionality, modeling and applications. *Prog. Polym. Sci.* **2012**, *37*, 1720–1763. [CrossRef]
15. Wang, Y.; Tian, W.; Xie, J.; Liu, Y. Thermoelectric Responsive Shape Memory Graphene/Hydro-Epoxy Composites for Actuators. *Micromachines* **2016**, *7*, 145. [CrossRef]
16. Hu, J.; Meng, H.; Li, G.; Ibekwe, S.I. A review of stimuli-responsive polymers for smart textile applications. *Smart Mater. Struct.* **2012**, *21*, 053001. [CrossRef]
17. Akindoyo, J.O.; Beg, M.D.H.; Ghazali, S.; Islam, M.R.; Jeyaratnam, N.; Yuvaraj, A.R. Polyurethane types, synthesis and applications—A review. *RSC Adv.* **2016**, *6*, 114453–114482. [CrossRef]
18. Kojio, K.; Nozaki, S.; Takahara, A.; Yamasaki, S. Influence of chemical structure of hard segments on physical properties of polyurethane elastomers: A review. *J. Polym. Res.* **2020**, *27*, 140. [CrossRef]
19. Shokrolahi, F.; Yeganeh, H. Soft segment composition and its influence on phase-separated morphology of PCL/PEG-based poly(urethane urea)s. *Iran. Polym. J.* **2014**, *23*, 505–512. [CrossRef]
20. Zhang, L.; Zhang, C.; Zhang, W.; Zhang, H.; Hou, Z. Synthesis and properties of biodegradable poly(ester-urethane)s based on poly(ε-caprolactone) and aliphatic diurethane diisocyanate for long-term implant application: Effect of uniform-size hard segment content. *J. Biomater. Sci. Polym. Ed.* **2019**, *30*, 1212–1226. [CrossRef]
21. Hou, Z.; Xu, J.; Teng, J.; Jia, Q.; Wang, X. Facile preparation of medical segmented poly(ester-urethane) containing uniformly sized hard segments and phosphorylcholine groups for improved hemocompatibility. *Mater. Sci. Eng. C* **2019**, *109*, 110571. [CrossRef]
22. Kasprzyk, P.; Benes, H.; Donato, R.K.; Datta, J. The role of hydrogen bonding on tuning hard-soft segments in bio-based thermoplastic poly(ether-urethane)s. *J. Clean. Prod.* **2020**, *274*, 122678. [CrossRef]
23. Tereshatov, V.V.; Makarova, M.; Senichev, V.Y.; Volkova, E.; Vnutskikh, Z.A.; Slobodinyuk, A.I. The role of the soft phase in the hardening effect and the rate dependence of the ultimate physico-mechanical properties of urethane-containing segmented elastomers. *Colloid Polym. Sci.* **2014**, *293*, 153–164. [CrossRef]
24. Xiao, M.; Zhang, N.; Zhuang, J.; Sun, Y.; Ren, F.; Zhang, W.; Hou, Z. Degradable Poly(ether-ester-urethane)s Based on Well-Defined Aliphatic Diurethane Diisocyanate with Excellent Shape Recovery Properties at Body Temperature for Biomedical Application. *Polymers* **2019**, *11*, 1002. [CrossRef]
25. Slobodinyuk, A.; Strelnikov, V.; Senichev, V.Y.; Slobodinyuk, D. Preparation, Structure and Properties of Urethane-Containing Elastomers Based on Epoxy Terminal Oligomers. *Polymers* **2022**, *14*, 524. [CrossRef] [PubMed]
26. Doğan, M.; Unlu, S.M. Flame retardant effect of boron compounds on red phosphorus containing epoxy resins. *Polym. Degrad. Stab.* **2014**, *99*, 12–17. [CrossRef]
27. Levchik, S.; Piotrowski, A.; Weil, E.; Yao, Q. New developments in flame retardancy of epoxy resins. *Polym. Degrad. Stab.* **2005**, *88*, 57–62. [CrossRef]
28. Izadi, M.; Mardani, H.; Roghani-Mamaqani, H.; Salami-Kalajahi, M.; Khezri, K. Hyperbranched Poly(amidoamine)-Grafted Graphene Oxide as a Multifunctional Curing Agent for Epoxy-Terminated Polyurethane Composites. *ChemistrySelect* **2021**, *6*, 2692–2699. [CrossRef]
29. Slobodinyuk, A.; Strelnikov, V.; Kiselkov, D.; Slobodinyuk, D. Synthesis of oligotetramethylene oxides with terminal amino groups as curing agents for an epoxyurethane oligomer. *Z. Nat. B* **2021**, *76*, 511–515. [CrossRef]
30. Byczyński, L.; Dutkiewicz, M.; Maciejewski, H. Synthesis and properties of high-solids hybrid materials obtained from epoxy functional urethanes and siloxanes. *Prog. Org. Coat.* **2015**, *84*, 59–69. [CrossRef]
31. Wang, J.-S.; Liu, Y.; Zhao, H.-B.; Liu, J.; Wang, D.-Y.; Song, Y.-P.; Wang, Y.-Z. Metal compound-enhanced flame retardancy of intumescent epoxy resins containing ammonium polyphosphate. *Polym. Degrad. Stab.* **2009**, *94*, 625–631. [CrossRef]
32. Yan, J.; Marina, P.; Blencowe, A. A Facile Strategy for the High Yielding, Quantitative Conversion of Polyglycol End-Groups to Amines. *Polymers* **2021**, *13*, 1403. [CrossRef] [PubMed]
33. Brandl, F.; Henke, M.; Rothschenk, S.; Gschwind, R.; Breunig, M.; Blunk, T.; Tessmar, J.; Göpferich, A. Poly(Ethylene Glycol) Based Hydrogels for Intraocular Applications. *Adv. Eng. Mater.* **2007**, *9*, 1141–1149. [CrossRef]
34. Chenault, H.K. Method for Preparing Multi-Arm Poly(Ethylene Glycol) Amines. U.S. Patent 7 868 132, 11 January 2011.
35. Marcduval, J.; Delestre, C.; Carré, M.; Hubert, P.; Dellacherie, E. Synthesis and characterization of some covalent dextran-polyoxyethyleneglycol derivatives. *Carbohydr. Polym.* **1991**, *15*, 233–242. [CrossRef]
36. Harris, J.M.; Struck, E.C.; Case, M.G.; Paley, M.S.; Yalpani, M.; Van Alstine, J.M.; Brooks, D.E. Synthesis and characterization of poly(ethylene glycol) derivatives. *J. Polym. Sci. Polym. Chem. Ed.* **1984**, *22*, 341–352. [CrossRef]
37. Wang, L.; Wang, S.; Bei, J.Z. Synthesis and characterization of macroinitiator-amino terminated PEG and poly (γ-benzyl-l-glutamate)-PEO-poly (γ-benzyl-l-glutamate) triblock copolymer. *Polym. Adv. Technol.* **2004**, *15*, 617–621. [CrossRef]
38. Yeganeh, H.; Lakouraj, M.M.; Jamshidi, S. Synthesis and characterization of novel biodegradable epoxy-modified polyurethane elastomers. *J. Polym. Sci. Part A Polym. Chem.* **2005**, *43*, 2985–2996. [CrossRef]
39. Lee, H.; Neville, K. *Handbook of Epoxy Resins*; McGraw-Hill: New York, NY, USA, 1967.
40. Guadagno, L.; Vertuccio, L.; Sorrentino, A.; Raimondo, M.; Naddeo, C.; Vittoria, V.; Iannuzzo, G.; Calvi, E.; Russo, S. Mechanical and barrier properties of epoxy resin filled with multi-walled carbon nanotubes. *Carbon* **2009**, *47*, 2419–2430. [CrossRef]

41. Stefani, P.M.; Moschiar, S.M.; Aranguren, M.I. Epoxy-urethane copolymers: Relation between morphology and properties. *J. Appl. Polym. Sci.* **2001**, *82*, 2544–2552. [CrossRef]
42. Socrates, G. *Infrared and Raman Characteristic Group Frequencies: Tables and Charts*, 3rd ed.; Wiley: New York, NY, USA, 2004.
43. Strel'Nikov, V.N.; Senichev, V.Y.; Slobodinyuk, A.I.; Savchuk, A.V.; Volkova, E.R.; Makarova, M.A.; Nechaev, A.I.; Krasnosel'Skikh, S.F.; Ukhin, K.O. Preparation and Properties of Frost-Resistant Room-Temperature-Curable Compounds Based on Oligoethertetraurethane Diepoxides of Various Chemical Structures. *Russ. J. Appl. Chem.* **2018**, *91*, 463–468. [CrossRef]

polymers

MDPI

Article

Mathematical Description of the RAFT Copolymerization of Styrene and Glycidyl Methacrylate Using the Terminal Model

José Alfredo Tenorio-López [1], Juan José Benvenuta-Tapia [1,*], Norma García-Navarro [1], Eduardo Vivaldo-Lima [2,*], Pascale Champagne [3] and Enrique Saldívar-Guerra [4,*]

1 Facultad de Ciencias Químicas, Universidad Veracruzana (UV), Coatzacoalcos 96535, Mexico; altenorio@uv.mx (J.A.T.-L.); nogarcia@uv.mx (N.G.-N.)
2 Facultad de Química, Departamento de Ingeniería Química, Universidad Nacional Autónoma de México, Ciudad de México 04510, Mexico
3 Faculty of Engineering and Applied Science, Queen's University, Kingston, ON K7L3N6, Canada; pascale.champagne@queensu.ca
4 Polymer Synthesis Department, Centro de Investigación en Química Aplicada (CIQA), Saltillo 25294, Mexico
* Correspondence: jbenvenuta@uv.mx (J.J.B.-T.); vivaldo@unam.mx (E.V.-L.); enrique.saldivar@ciqa.edu.mx (E.S.-G.)

Abstract: A mathematical model for the kinetics, composition and molar mass development of the bulk reversible addition-fragmentation chain transfer (RAFT) copolymerization of glycidyl methacrylate (GMA) and styrene (St), at several GMA molar feed fractions at 103 °C, in the presence of 2-cyano isopropyl dodecyl trithiocarbonate as the RAFT agent and 1,1′-azobis(cyclohexane carbonitrile), as the initiator, is presented. The copolymerization proceeded in a controlled manner and dispersities of the copolymers remained narrow even at high conversions. Experimental data and calculated profiles of conversion versus time, composition versus conversion and molar mass development for the RAFT copolymerization of St and GMA agreed well for all conditions tested, including high-conversion regions. The kinetic rate constants associated with the RAFT- related reactions and diffusion-controlled parameters were properly estimated using a weighted nonlinear multivariable regression procedure. The mathematical model developed in this study may be used as an aid in the design and upscaling of industrial RAFT polymerization processes.

Keywords: glycidyl methacrylate; styrene; kinetic modeling; RAFT copolymerization

Citation: Tenorio-López, J.A.; Benvenuta-Tapia, J.J.; García-Navarro, N.; Vivaldo-Lima, E.; Champagne, P.; Saldívar-Guerra, E. Mathematical Description of the RAFT Copolymerization of Styrene and Glycidyl Methacrylate Using the Terminal Model. *Polymers* **2022**, *14*, 1448. https://doi.org/10.3390/polym14071448

Academic Editor: Nicolas Sbirrazzuoli

Received: 31 January 2022
Accepted: 27 March 2022
Published: 1 April 2022

Publisher's Note: MDPI stays neutral with regard to jurisdictional claims in published maps and institutional affiliations.

Copyright: © 2022 by the authors. Licensee MDPI, Basel, Switzerland. This article is an open access article distributed under the terms and conditions of the Creative Commons Attribution (CC BY) license (https://creativecommons.org/licenses/by/4.0/).

1. Introduction

Functionalized copolymers are relevant at the industrial level due to their participation in chain extension, crosslinking, and polymer grafting reactions. The applications of these copolymers are diverse. They can be used as dispersants, surfactants, surface modifiers, compatibilizers, and drug delivery matrices [1]. Poly(styrene-co-glycidyl methacrylate) (St-GMA copolymers) are very interesting functional polymers. The epoxide group is useful for the chemical modification of copolymers, leading to a wide variety of applications. The glycidyl functional group has potential applications in functionalization with amines [1], compatibilized polymer blends [2], acids [3], cation-exchange adsorbents [4], and the chain extension of polyesters [5,6], leading to the production of materials with increased melt viscosities and strengths.

Industrial production of St-GMA copolymers is usually carried out by conventional free-radical polymerization (FRP) processes [7,8]. FRP has the advantages of undemanding operation conditions and versatility of monomers that can be used. The negative aspects of FRP include poor control of the molar mass and polymer microstructure of the product. St-GMA copolymers synthesized by FRP typically possess broad molar mass distributions [9].

Another disadvantage of FRP processes is temperature control. For instance, the free radical copolymerization of St and GMA proceeds very rapidly and exothermically.

Therefore, efficient heat removal is required to avoid reaching reactor-runaway conditions, which may result in the production of out-of-specification materials.

Reversible-deactivation radical polymerization (RDRP), also known as controlled radical polymerization (CRP), is important because it allows the synthesis of copolymers with not only narrow molar mass distributions and well-defined microstructures, but also more homogeneous composition distributions, compared to FRP [10].

The main RDRP techniques, namely, nitroxide-mediated polymerization (NMP) [11,12], atom transfer radical polymerization (ATRP) [13], and reversible addition-fragmentation chain transfer polymerization [14], have allowed the synthesis of functional polymers with predefined molar masses, low dispersity values, and defined microstructures.

Recently, the incorporation to RAFT polymerization of photoinduced electron/energy transfer (PET) and polymerization-induced self-assembly (PISA) methodologies into RAFT polymerization has been applied to the polymerization of various functional monomers, resulting in good control of molecular weight and molecular weight distributions, similar to thermally initiated systems, with supplementary advantages, such as mild reaction conditions and low energy consumption [15,16]. PET-RAFT polymerization was successfully conducted, obtaining polymer products with controlled molecular weights and narrow dispersities (Đ = 1.02–1.13) [15], while RAFT-PISA was successfully applied to the synthesis of block copolymer nano-objects with different morphologies [16].

Few experimental reports on the RAFT copolymerization of St and GMA are available, and to the best of our knowledge there are no reported studies on the production of functional St-GMA copolymers synthesized by RAFT polymerization above 100 °C [17], which is the range of interest in the synthesis of acrylate-containing copolymers.

As far as we are aware, only the syntheses of St-GMA copolymers by atom transfer radical polymerization using copper-based catalyst systems in bulk and in toluene, at 60 °C [18], and by nitroxide mediated polymerization in 50 wt.% 1,4-dioxane solution, at 90 °C, have been reported [19].

The control of polymer microstructural parameters, such as copolymer composition, copolymer sequence distributions and molar mass dispersities, is important in copolymerization processes. Parameter estimation of kinetic rate constants associated with copolymerization processes is important for accurate calculation of polymerization rates and copolymer sequence distributions, which are fundamental to the production of copolymers with predefined properties.

The application of fundamental polymerization models, including kinetic and reactor models, is necessary to understand the mechanisms and phenomena behind these processes. Kinetic studies are particularly important because they provide better control strategies during the production of polymers at an industrial scale [20]. These models can be used to calculate monomer conversion, polymer microstructure, molar mass averages, and full molar mass distributions (MMD) under various operating conditions. Furthermore, mathematical modeling may be used to better understand and operate polymerization processes, allowing the prediction of the effect of operating conditions on polymerization rate and polymer properties; a good model may simplify experimental programs. The industrial production of polymer materials usually involves high-temperature processes. Therefore, it is important to carry out kinetic studies at similarly high temperatures. To the best of our knowledge, the kinetic modeling of the RAFT copolymerization of St and GMA has not been reported so far. It is known that RAFT agents alter the concentration profiles of active species, compared to FRP [20–22]. This situation may favor the preferential incorporation of one of the comonomers, affecting both, copolymer composition and polymerization kinetics [21].

In this contribution, the kinetic modeling of the RAFT bulk copolymerization of St and GMA using 2-cyano-2-propyl dodecyl trithiocarbonate (CPDT) as a RAFT agent, and 1,1′-azobis (cyclohexane carbonitrile) as an initiator, at different levels of GMA content, is reported. Our objectives were to estimate the rate coefficients for RAFT reactions involved in the polymerization scheme and evaluate the performance of the model by comparing model predictions against experimental data generated in our laboratory. In developing the model, it was assumed that the terminal model is valid and that the RAFT activation and transfer cycles proceed with the same kinetic parameters. Diffusion-controlled termination was considered using a model based on free-volume theory.

2. Experimental

2.1. Reagents

GMA (97%) and styrene (99%) were purchased from Aldrich (Saint Louis, MO, USA) and purified as explained in one of our earlier studies. [9] 1,1′-azobis(cyclohexane carbonitrile) (ACHN, 98%) and 2-cyano-2-propyl dodecyl trithiocarbonate (97%) were both purchased from Aldrich (Saint Louis, MO, USA) and used as received.

2.2. Size Exclusion Chromatography (SEC) Characterization Method

Number- and weight-average molar masses (M_n and M_w, respectively), and molar mass dispersity (Đ = M_w/M_n) of the synthesized polymers were measured using a Waters 1515 gel permeation chromatograph (GPC, Waters, Milford, MA, USA) equipped with a refractive-index detector as well as HR 1, HR 3, and HR 4 columns. Calibration procedure and operating conditions were the same as described in an earlier report from our group [19].

2.3. ^{1}H-NMR Characterization Method

^{1}H-nuclear magnetic resonance (^{1}H-NMR) spectra were obtained using a Varian 300 MHz spectrometer (Varian, Santa Clara, CA, USA) using deuterated chloroform ($CDCl_3$) as a solvent and tetramethylsilane (TMS) as an internal standard, at room temperature. Sample preparation and measurement proceeded as reported in our previous studies. [14] Data analysis for determination of copolymer composition and sequence distribution of monomer units from ^{1}H-NMR characteristic signals was carried out using the appropriate equations [23].

2.4. Copolymerization Reactions

Bulk copolymerizations of St and GMA, at 103 °C, proceeded in a 1-L high-pressure stainless-steel jacketed reactor (Parr Model 4523, Moline, IL, USA) with temperature control, as well as pressure and stirring sensors. Ultra-high-purity nitrogen was used to provide an inert environment. Appropriate amounts of GMA and St were then added to the reactor, followed by the initiator and RAFT agent. A mixing rate of 150 rpm was used. The polymerizations proceeded under nitrogen atmosphere, at 4.14 bar. Then, temperature was increased to 103 °C using a Huber Unistat 815w thermoregulator, in cascade mode. The thermal oil flew directly to the reactor jacket and to the coil, which ensured temperature control with a precision of ±1 °C. Sample withdrawal, preparation and analyses proceeded as detailed in one of our earlier studies [19].

Monomer conversion was not measured directly. We measured polymer yield, which was calculated gravimetrically, as the ratio of mass of produced polymer to mass of initial total monomer. Therefore, although we refer to monomer conversion in the figures of this contribution, when referring to experimental data, it is strictly polymer yield.

3. Model Development

A kinetic mathematical model was developed to calculate polymerization rate, evolution of molar mass averages, and copolymer composition. The model contains the following assumptions: (a) penultimate effects were neglected; (b) while in some reactions the intermediate macroradical species formed during the additional step of the RAFT process may be stable enough to delay polymerization (maximum lifetime of 1 s), it is not considered here to initiate new species and terminate [24–27]; and (c) branching has also been neglected.

The starting polymerization scheme is shown in Table 1. Three polymer populations are involved: propagating radical or active (living) polymer molecules with terminal units A or B (P_n and Q_n), dormant polymer molecules with terminal units A or B (TP_n and TQ_n), and dead polymer molecules (M_n), where subscript n is the number of monomeric units in the macromolecule. A and B represent St and GMA terminal units, respectively. T is the RAFT agent.

Table 1. Detailed polymerization scheme used in this study.

Initiation	Propagation	
$I \overset{k_{d}f}{\rightarrow} 2R^{\bullet}$		
$R^{\bullet} + A \overset{k_{i1}}{\rightarrow} P_1$	$P_n + A \overset{k_{p11}}{\rightarrow} P_{n+1}$	$Q_n + A \overset{k_{p21}}{\rightarrow} P_{n+1}$
$R^{\bullet} + B \overset{k_{i2}}{\rightarrow} Q_1$	$P_n + B \overset{k_{p12}}{\rightarrow} Q_{n+1}$	$Q_n + B \overset{k_{p22}}{\rightarrow} Q_{n+1}$
		RAFT transfer
		$P_n + TP_r \underset{k_{ft1}}{\overset{k_{at1}}{\rightleftarrows}} TP_n + P_r$
RAFT activation		
$T + P_n \underset{k_{fa1}}{\overset{k_{aa1}}{\rightleftarrows}} R^{\bullet} + TP_n$		$P_n + TQ_r \underset{k_{ft3}}{\overset{k_{at3}}{\rightleftarrows}} TP_n + Q_r$
$T + Q_n \underset{k_{fa2}}{\overset{k_{aa2}}{\rightleftarrows}} R^{\bullet} + TQ_n$		$Q_n + TQ_r \underset{k_{ft2}}{\overset{k_{at2}}{\rightleftarrows}} TQ_n + Q_r$
		$Q_n + TP_r \underset{k_{ft4}}{\overset{k_{at4}}{\rightleftarrows}} TQ_n + P_r$
	Termination	
$P_n + P_r \overset{k_{tc1}}{\rightarrow} M_{n+r}$		$P_n + P_r \overset{k_{td1}}{\rightarrow} M_n + M_r$
$Q_n + Q_r \overset{k_{tc2}}{\rightarrow} M_{n+r}$		$Q_n + Q_r \overset{k_{td2}}{\rightarrow} M_n + M_r$
$P_n + Q_r \overset{k_{tc3}}{\rightarrow} M_{n+r}$		$P_n + Q_r \overset{k_{td3}}{\rightarrow} M_n + M_r$

The mathematical model developed in this contribution is based on the method of moments. The definitions of moments of several polymer species are shown in Table 2.

Table 2. Definition of moments of the several polymer species.

$\mathbf{Y_i^a} = \Sigma_{n=0}^{\infty} n^i\,[P_n]$	$\mathbf{Z_i^a} = \Sigma_{n=0}^{\infty} n^i\,[TP_n]$	$\mathbf{D_i} = \Sigma_{n=0}^{\infty} n^i\,[M_n]$
$\mathbf{Y_i^b} = \Sigma_{n=0}^{\infty} n^i[Q_n]$	$\mathbf{Z_i^b} = \Sigma_{n=0}^{\infty} n^i[TQ_n]$	
$i = 0, 1, 2$		

The detailed kinetic equations for low molar mass and polymer species are summarized in Table 3.

Table 3. Kinetic equation for species.

$\frac{dA}{dt} = -k_{i1}[\mathbf{R^{\bullet}}][A] - k_{p11}[A][\mathbf{Y_o^a}] - k_{p21}[A][\mathbf{Y_o^b}]$	St monomer	(1)
$\frac{dB}{dt} = -k_{i2}[\mathbf{R^{\bullet}}][B] - k_{p12}[B][\mathbf{Y_o^a}] - k_{p22}[B][\mathbf{Y_o^b}]$	GMA monomer	(2)
$\frac{dI}{dt} = -k_d[I]$	initiator	(3)
$\frac{dR^{\bullet}}{dt} = 2k_d f[I] - k_{i1}[A][\mathbf{R\bullet}] - k_{i2}[B][\mathbf{R^{\bullet}}] - k_{fa1}[\mathbf{Z_o^a}][\mathbf{R^{\bullet}}] + k_{aa1}[\mathbf{Y_o^a}][T] - k_{fa2}[\mathbf{Z_o^b}][R^*] + k_{aa2}[\mathbf{Y_o^b}][T]$		(4)
$\frac{dt}{dt} = k_{fa2}[\mathbf{R^{\bullet}}][\mathbf{Z_o^b}] - k_{aa2}[T][\mathbf{Y_o^b}] + k_{fa1}[\mathbf{R^{\bullet}}][\mathbf{Z_o^a}] - k_{aa1}[T][\mathbf{Y_o^a}]$	RAFT agent	(5)
$\frac{dP_n}{dt} = k_{i1}[\mathbf{R^{\bullet}}][A] + k_{fa1}[\mathbf{R^{\bullet}}][TP_n] - k_{aa1}[T][P_n] - k_{at3}[\mathbf{Z_o^b}][P_n] + k_{ft3}[\mathbf{Y_o^b}][TP_n] + k_{ft1}[\mathbf{Y_o^a}][TP_n] - k_{at1}[\mathbf{Z_o^a}][P_n] - k_{p11}[A][P_n] - k_{p12}[B][P_n] + k_{p11}[A][P_{n-1}] + k_{p21}[A][Q_{n-1}] - k_{tc3}[\mathbf{Y_o^b}][P_n] - k_{td3}[\mathbf{Y_o^b}][P_n] - k_{tc1}[\mathbf{Y_o^a}][P_n] - k_{td1}[\mathbf{Y_o^a}][P_n]$		(6)
$\frac{dQ_n}{dt} = k_{i2}[\mathbf{R^{\bullet}}][B] + k_{fa2}[R^*][TQ_n] - k_{aa2}[T][Q_n] - k_{at4}[\mathbf{Z_o^a}][Q_n] + k_{ft4}[\mathbf{Y_o^a}][TQ_n] + k_{ft2}[\mathbf{Y_o^b}][TQ_n] - k_{at2}[\mathbf{Z_o^b}][Q_n] - k_{p21}[A][Q_n] - k_{p22}[B][Q_n] + k_{p12}[B][P_{n-1}] + k_{p22}[B][Q_{n-1}] - k_{tc3}[\mathbf{Y_o^a}][Q_n] - k_{td3}[\mathbf{Y_o^a}][Q_n] - k_{tc2}[\mathbf{Y_o^b}][Q_n] - k_{td2}[\mathbf{Y_o^b}][Q_n]$		(7)
$\frac{dTP_n}{dt} = -k_{fa1}[\mathbf{R^{\bullet}}][TP_n] + k_{aa1}[T][P_n] + k_{at3}[\mathbf{Z_o^b}][P_n] - k_{ft3}[\mathbf{Y_o^b}][TP_n] + k_{at1}[\mathbf{Z_o^a}][P_n] - k_{ft1}[\mathbf{Y_o^a}][TP_n]$		(8)
$\frac{dTQ_n}{dt} = -k_{fa2}[\mathbf{R^{\bullet}}][TQ_n] + k_{aa2}[T][Q_n] - k_{ft2}[\mathbf{Y_o^b}][TQ_n] + k_{at2}[\mathbf{Z_o^b}][Q_n] + k_{at4}[\mathbf{Z_o^a}][Q_n] - k_{ft4}[\mathbf{Y_o^a}][TQ_n]$		(9)
$\frac{dM_n}{dt} = k_{tc3}(\Sigma_{a=0}^{n}\mathbf{P_a Q_{n-a}}) + \frac{1}{2} k_{tc1}(\Sigma_{a=0}^{n}\mathbf{P_a P_{n-a}}) + \frac{1}{2} k_{tc2}(\Sigma_{a=0}^{n}\mathbf{Q_a Q_{n-a}}) + k_{td3}[\mathbf{Y_o^b}][P_n] + k_{td1}[\mathbf{Y_o^a}][P_n] + k_{td2}[\mathbf{Y_o^b}][Q_n] + k_{td3}[\mathbf{Y_o^a}][Q_n]$		(10)

Table 4 shows the obtained moment equations.

Table 4. Moment equations for the species present in the St-GMA copolymerization.

Zeroth-order moments	
$\frac{d\mathbf{Y_o^a}}{dt} = k_{i1}[\mathbf{R^\bullet}][A] + k_{fa1}[\mathbf{R^\bullet}][\mathbf{Z_o^a}] - k_{aa1}[T][\mathbf{Y_o^a}] - k_{at3}[\mathbf{Z_o^b}][\mathbf{Y_o^a}] + k_{ft3}[\mathbf{Y_o^b}][\mathbf{Z_o^a}] + k_{ft1}[\mathbf{Y_o^a}][\mathbf{Z_o^a}] - k_{at1}[\mathbf{Z_o^a}][\mathbf{Y_o^a}] - k_{p11}[A][\mathbf{Y_o^a}] - k_{p12}[B][\mathbf{Y_o^a}] + k_{p11}[A][\mathbf{Y_o^a}] + k_{p21}[A][\mathbf{Y_o^b}] - k_{tc3}[\mathbf{Y_o^b}][\mathbf{Y_o^a}] - k_{td3}[\mathbf{Y_o^b}][\mathbf{Y_o^a}] - k_{tc1}[\mathbf{Y_o^a}][\mathbf{Y_o^a}] - k_{td1}[\mathbf{Y_o^a}][\mathbf{Y_o^a}]$	(11)
$\frac{d\mathbf{Y_o^b}}{dt} = k_{i2}[\mathbf{R^\bullet}][B] + k_{fa2}[\mathbf{R^\bullet}][\mathbf{Z_o^b}] - k_{aa2}[T][\mathbf{Y_o^b}] - k_{at4}[\mathbf{Z_o^a}][\mathbf{Y_o^b}] + k_{ft4}[\mathbf{Y_o^a}][\mathbf{Z_o^b}] + k_{ft2}[\mathbf{Y_o^b}][\mathbf{Z_o^b}] - k_{at2}[\mathbf{Z_o^b}][\mathbf{Y_o^b}] - k_{p21}[A][\mathbf{Y_o^b}] - k_{p22}[B][\mathbf{Y_o^b}] + k_{p12}[B][\mathbf{Y_o^a}] + k_{p22}[B][\mathbf{Y_o^b}] - k_{tc3}[\mathbf{Y_o^a}][\mathbf{Y_o^b}] - k_{td3}[\mathbf{Y_o^a}][\mathbf{Y_o^b}] - k_{tc2}[\mathbf{Y_o^b}][\mathbf{Y_o^b}] - k_{td2}[\mathbf{Y_o^b}][\mathbf{Y_o^b}]$	(12)
$\frac{d\mathbf{Z_o^a}}{dt} = -k_{fa1}[\mathbf{R^\bullet}][\mathbf{Z_o^a}] + k_{aa1}[T][\mathbf{Y_o^a}] + k_{at3}[\mathbf{Z_o^b}][\mathbf{Y_o^a}] - k_{ft3}[\mathbf{Y_o^b}][\mathbf{Z_o^a}] + k_{at1}[\mathbf{Z_o^a}][\mathbf{Y_o^a}] - k_{ft1}[\mathbf{Y_o^a}][\mathbf{Z_o^a}]$	(13)
$\frac{d\mathbf{Z_o^b}}{dt} = -k_{fa2}[\mathbf{R^\bullet}][\mathbf{Z_o^b}] + k_{aa2}[T][\mathbf{Y_o^b}] - k_{ft2}[\mathbf{Y_o^b}][\mathbf{Z_o^b}] + k_{at2}[\mathbf{Z_o^b}][\mathbf{Y_o^b}] + k_{at4}[\mathbf{Z_o^a}][\mathbf{Y_o^b}] - k_{ft4}[\mathbf{Y_o^a}][\mathbf{Z_o^b}]$	(14)
$\frac{d\mathbf{D_o}}{dt} = ktc_3[\mathbf{Y_o^a}][\mathbf{Y_o^b}] + \frac{1}{2}ktc_1[\mathbf{Y_o^a}][\mathbf{Y_o^a}] + \frac{1}{2}ktc_2[\mathbf{Y_o^b}][\mathbf{Y_o^b}] + ktd_3[\mathbf{Y_o^b}][\mathbf{Y_o^a}] + ktd_1[\mathbf{Y_o^a}][\mathbf{Y_o^a}] + ktd_2[\mathbf{Y_o^b}][\mathbf{Y_o^b}] + ktd_3[\mathbf{Y_o^b}][\mathbf{Y_o^a}]$	(15)
First-order moments	
$\frac{d\mathbf{Y_1^a}}{dt} = k_{i1}[\mathbf{R^\bullet}][A] + k_{fa1}[\mathbf{R^\bullet}][\mathbf{Z_1^a}] - k_{aa1}[T][\mathbf{Y_1^a}] - k_{at3}[\mathbf{Z_o^b}][\mathbf{Y_1^a}] + k_{ft3}[\mathbf{Y_o^b}][\mathbf{Z_1^b}] + k_{ft1}[\mathbf{Y_o^a}][\mathbf{Z_1^a}] - k_{at1}[\mathbf{Z_o^a}][\mathbf{Y_1^a}] - k_{p11}[A][\mathbf{Y_1^a}] - k_{p12}[B][\mathbf{Y_1^a}] + k_{p11}[A][\mathbf{Y_1^a} + \mathbf{Y_o^a}] + k_{p21}[A][\mathbf{Y_1^b} + \mathbf{Y_o^b}] - k_{tc3}[\mathbf{Y_o^b}][\mathbf{Y_1^a}] - k_{td3}[\mathbf{Y_o^b}][\mathbf{Y_1^a}] - k_{tc1}[\mathbf{Y_o^a}][\mathbf{Y_1^a}] - k_{td1}[\mathbf{Y_o^a}][\mathbf{Y_1^a}]$	(16)
$\frac{d\mathbf{Y_1^b}}{dt} = k_{i2}[\mathbf{R^\bullet}][B] + k_{fa2}[\mathbf{R^\bullet}][\mathbf{Z_1^b}] - k_{aa2}[T][\mathbf{Y_1^b}] - k_{at4}[\mathbf{Z_o^a}][\mathbf{Y_1^b}] + k_{ft4}[\mathbf{Y_o^a}][\mathbf{Z_1^b}] + k_{ft2}[\mathbf{Y_o^b}][\mathbf{Z_1^b}] - k_{at2}[\mathbf{Z_o^b}][\mathbf{Y_1^b}] - k_{p21}[A][\mathbf{Y_1^b}] - k_{p22}[B][\mathbf{Y_1^b}] + k_{p12}[B][\mathbf{Y_1^a} + \mathbf{Y_o^a}] + k_{p22}[B][\mathbf{Y_1^b} + \mathbf{Y_o^b}] - k_{tc3}[\mathbf{Y_o^a}][\mathbf{Y_1^b}] - k_{td3}[\mathbf{Y_o^a}][\mathbf{Y_1^b}] - k_{tc2}[\mathbf{Y_o^b}][\mathbf{Y_1^b}] - k_{td2}[\mathbf{Y_o^b}][\mathbf{Y_1^b}]$	(17)
$\frac{d\mathbf{Z_1^a}}{dt} = -k_{fa1}[\mathbf{R^\bullet}][\mathbf{Z_1^a}] + k_{aa1}[T][\mathbf{Y_1^a}] + k_{at3}[\mathbf{Z_o^b}][\mathbf{Y_1^a}] - k_{ft3}[\mathbf{Y_o^b}][\mathbf{Z_1^a}] + k_{at1}[\mathbf{Z_o^a}][\mathbf{Y_1^a}] - k_{ft1}[\mathbf{Y_o^a}][\mathbf{Z_1^a}]$	(18)
$\frac{d\mathbf{Z_1^b}}{dt} = -k_{fa2}[\mathbf{R^\bullet}][\mathbf{Z_1^b}] + k_{aa2}[T][\mathbf{Y_1^b}] - k_{ft2}[\mathbf{Y_o^b}][\mathbf{Z_1^b}] + k_{at2}[\mathbf{Z_o^b}][\mathbf{Y_1^b}] + k_{at4}[\mathbf{Z_o^a}][\mathbf{Y_1^b}] - k_{ft4}[\mathbf{Y_o^a}][\mathbf{Z_1^b}]$	(19)
$\frac{d\mathbf{D_1}}{dt} = ktc_3[\mathbf{Y_o^a}\mathbf{Y_1^b} + \mathbf{Y_1^a}\mathbf{Y_o^b}] + \frac{1}{2}ktc_1[\mathbf{Y_o^a}\mathbf{Y_1^a} + \mathbf{Y_1^a}\mathbf{Y_o^a}] + \frac{1}{2}ktc_2[\mathbf{Y_o^b}\mathbf{Y_1^b} + \mathbf{Y_1^b}\mathbf{Y_o^b}] + ktd_3[\mathbf{Y_o^b}][\mathbf{Y_1^a}] + ktd_1[\mathbf{Y_o^a}][\mathbf{Y_1^a}] + ktd_2[\mathbf{Y_o^b}][\mathbf{Y_1^b}] + ktd_3[\mathbf{Y_o^a}][\mathbf{Y_1^b}]$	(20)
Second-order moments	
$\frac{d\mathbf{Y_2^a}}{dt} = k_{i1}[\mathbf{R^\bullet}][A] + k_{fa1}[\mathbf{R^\bullet}][\mathbf{Z_2^a}] - k_{aa1}[T][\mathbf{Y_2^a}] - k_{at3}[\mathbf{Z_o^b}][\mathbf{Y_2^a}] + k_{ft3}[\mathbf{Y_o^b}][\mathbf{Z_2^a}] + k_{ft1}[\mathbf{Y_o^a}][\mathbf{Z_2^a}] - k_{at1}[\mathbf{Z_o^a}][\mathbf{Y_2^a}] - k_{p11}[A][\mathbf{Y_2^a}] - k_{p12}[B][\mathbf{Y_2^a}] + k_{p11}[A][\mathbf{Y_2^a} + 2\mathbf{Y_1^a} + \mathbf{Y_o^a}] + k_{p21}[A][\mathbf{Y_2^b} + 2\mathbf{Y_1^b} + \mathbf{Y_o^b}] - k_{tc3}[\mathbf{Y_o^b}][\mathbf{Y_2^a}] - k_{td3}[\mathbf{Y_o^b}][\mathbf{Y_2^a}] - k_{tc1}[\mathbf{Y_o^a}][\mathbf{Y_2^a}] - k_{td1}[\mathbf{Y_o^a}][\mathbf{Y_2^a}]$	(21)
$\frac{d\mathbf{Y_2^a}}{dt} = k_{i2}[\mathbf{R^\bullet}][B] + k_{fa2}[\mathbf{R^\bullet}][\mathbf{Z_2^b}] - k_{aa2}[T][\mathbf{Y_2^b}] - k_{at4}[\mathbf{Z_o^a}][\mathbf{Y_2^b}] + k_{ft4}[\mathbf{Y_o^a}][\mathbf{Z_2^b}] + k_{ft2}[\mathbf{Y_o^b}][\mathbf{Z_2^b}] - k_{at2}[\mathbf{Z_o^b}][\mathbf{Y_2^b}] - k_{p21}[A][\mathbf{Y_2^b}] - k_{p22}[B][\mathbf{Y_2^b}] + k_{p12}[B][\mathbf{Y_2^a} + 2\mathbf{Y_1^a} + \mathbf{Y_o^a}] + k_{p22}[B][\mathbf{Y_2^b} + 2\mathbf{Y_1^b} + \mathbf{Y_o^b}] - k_{tc3}[\mathbf{Y_o^a}][\mathbf{Y_2^b}] - k_{td3}[\mathbf{Y_o^a}][\mathbf{Y_2^b}] - k_{tc2}[\mathbf{Y_o^b}][\mathbf{Y_2^b}] - k_{td2}[\mathbf{Y_o^b}][\mathbf{Y_2^b}]$	(22)
$\frac{d\mathbf{Z_2^a}}{dt} = -k_{fa1}[\mathbf{R^\bullet}][\mathbf{Z_2^a}] + k_{aa1}[T][\mathbf{Y_2^a}] + k_{at3}[\mathbf{Z_o^b}][\mathbf{Y_2^a}] - k_{ft3}[\mathbf{Y_o^b}][\mathbf{Z_2^a}] + k_{at1}[\mathbf{Z_o^a}][\mathbf{Y_2^a}] - k_{ft1}[\mathbf{Y_o^a}][\mathbf{Z_2^a}]$	(23)
$\frac{d\mathbf{Z_2^b}}{dt} = -k_{fa2}[\mathbf{R^\bullet}][\mathbf{Z_2^b}] + k_{aa2}[T][\mathbf{Y_2^b}] - k_{ft2}[\mathbf{Y_o^b}][\mathbf{Z_2^b}] + k_{at2}[\mathbf{Z_o^b}][\mathbf{Y_2^b}] + k_{at4}[\mathbf{Z_o^a}][\mathbf{Y_2^b}] - k_{ft4}[\mathbf{Y_o^a}][\mathbf{Z_2^b}]$	(24)
$\frac{d\mathbf{D_2}}{dt} = ktc_3[\mathbf{Y_o^a}\mathbf{Y_2^b} + 2\mathbf{Y_1^a}\mathbf{Y_1^b} + \mathbf{Y_2^a}\mathbf{Y_o^b}] + \frac{1}{2}ktc_1[\mathbf{Y_o^a}\mathbf{Y_2^a} + 2\mathbf{Y_1^a}\mathbf{Y_1^a} + \mathbf{Y_2^a}\mathbf{Y_o^a}] + \frac{1}{2}ktc_2[\mathbf{Y_o^b}\mathbf{Y_2^b} + 2\mathbf{Y_1^b}\mathbf{Y_1^b} + \mathbf{Y_2^b}\mathbf{Y_o^b}] + ktd_3[\mathbf{Y_o^b}][\mathbf{Y_2^a}] + ktd_1[\mathbf{Y_o^a}][\mathbf{Y_2^a}] + ktd_2[\mathbf{Y_o^b}][\mathbf{Y_2^b}] + ktd_3[\mathbf{Y_o^a}][\mathbf{Y_2^b}]$	(25)

Overall monomer conversion, copolymer composition and average molar masses, M_n and M_w are calculated using Equations (26)–(32). M in Equations (26) and (27) stands for monomer content; subscripts 1, 2 and 0 stand for monomer 1, monomer 2, and initial conditions, respectively.

$$\text{Overall conversion} = (M_{1o} + M_{2o} - (M_1 + M_2))/(M_{1o} + M_{2o}) \quad (26)$$

$$\text{Copolymer composition: } F_1 = (M_{1o} - M_1)/((M_{1o} - M_1) + (M_{2o} - M_2)) \quad (27)$$

$$\text{Number-average chain length}: \; r_N = \frac{Y_1^a + Y_1^b + Z_1^a + Z_1^b + D_1}{Y_0^a + Y_0^b + Z_0^a + Z_0^b + D_0} \tag{28}$$

$$\text{Weight-average chain length}: \; r_w = \frac{Y_2^a + Y_2^b + Z_2^a + Z_2^b + D_2}{Y_1^a + Y_1^b + Z_1^a + Z_1^b + D_1} \tag{29}$$

$$\text{Dispersity}: \; Đ = \frac{r_w}{r_N} \tag{30}$$

$$M_n = r_N \left(F_1\, PM_1 + F_2\, PM_2\right) \tag{31}$$

$$M_w = Đ\, M_n \tag{32}$$

The kinetic rate constants and parameters required by the model are summarized in Table 5. The values of the reactivity ratios for the RAFT copolymerization of St and GMA were obtained using a weighted non-linear multivariable regression approach, using software RREVM [19]. These values are also provided in Table 5.

Table 5. Kinetic constants for RAFT copolymerization of St (1)-GMA (2).

PM_1 = 104.15 $gmol^{-1}$		
PM_2 = 142.20 $gmol^{-1}$		
k_d (s^{-1}) = 2.2798 × 10^{19} exp(−166.9 $kJmol^{-1}$/RT)	initiator	[28]
f = 0.60		
k_{p11} (L mol^{-1} s^{-1}) = 4.266 × 10^7 exp(−3909.61/T)	St	[29]
k_{p22} (L mol^{-1} s^{-1}) = 5.100 × 10^6 exp(−2754/T)	GMA	[30]
k_{i1} (L mol^{-1} s^{-1}) = k_{p11}		
k_{i2} (L mol^{-1} s^{-1}) = k_{p22}		
r_1 (k_{p11}/k_{p12}) = 0.3724		[19]
r_2 (k_{p22}/k_{p21}) = 0.6834		[19]
k_{tc1} (L mol^{-1} s^{-1}) = 2.0 × 10^{10} exp(−1553.01/T)	St	[31]
k_{tc2} (L mol^{-1} s^{-1}) = 2.57 × 10^8 exp(−292.0/T) (a)	GMA	[32]
ktc_3 (L mol^{-1} s^{-1}) = $\sqrt{k_{tc1}k_{tc2}}$		
k_{td} = k_{tc}		

(a) Assumed equal to the corresponding value for butyl acrylate (see [32]).

The mobility of high-molar-mass macromolecules is reduced at high conversions in FRP. Consequently, the rates of termination, propagation and RAFT reactions involving polymer molecules change throughout the reaction. In this study, diffusion-controlled (DC) effects were considered only for the termination reactions (auto-acceleration (AA) effect), using Equations (33) and (34), where k_t is an effective kinetic rate constant, k_t^o is the corresponding intrinsic kinetic rate constant, V_{fo} and V_f are the initial and final free-volume fractions, respectively, and βk_{kt} is a free-volume parameter. It was assumed that $\beta k_{tc} = \beta k_{td}$, parameters to be evaluated as AA effect.

$$k_{tc} = k_{tc}^o \exp\left[-\beta k_{tc}\left(\frac{1}{V_f} - \frac{1}{V_{fo}}\right)\right] \tag{33}$$

$$k_{td} = k_{td}^o \exp\left[-\beta k_{td}\left(\frac{1}{V_f} - \frac{1}{V_{fo}}\right)\right] \tag{34}$$

The free-volume fraction, V_f, is calculated using Equation (35) [33].

$$V_f = \left[0.025 + \alpha_p\left(T - T_{g,p}\right)\right]\varphi_p + \left[0.025 + \alpha_{M_1}\left(T - T_{g,M_1}\right)\right]\varphi M_1 + \left[0.025 + \alpha_{M_2}\left(T - T_{g,M_2}\right)\right]\varphi M_2 \tag{35}$$

α in Equation (35) is the thermal expansion coefficient, φ is the volume fraction, and T_g is the glass transition temperature. Subscripts p and M_i denote polymer and monomer i, respectively. T_{gp} is estimated using the Fox expression, given by Equation (36) [34].

$$T_{g,p} = 1/\left[\frac{f_{P1}}{T_{g,P_1}} + \frac{f_{P2}}{T_{g,P_2}}\right] \quad (36)$$

f_p in Equation (36) is the weight fraction of the polymer. Table 6 shows the physical properties of monomers and polymers used.

Table 6. Physical properties of monomers and polymer for calculation of fractional free volume.

ρ_{M_1} (g mL^{-1}) = 0.9193 − 0.000665 T/°C	[35]
ρ_{M_2} (g mL^{-1}) = 1.09428 − 0.001041 T/°C	[36]
ρ_{P_1} (g mL^{-1}) = 1.09926 − 2.65 × 10^{-4} T/°C	[35]
ρ_{P_2} (g mL^{-1}) = 1.08	This work
T_{g,M_1} (K) = 156.15	[37]
T_{g,M_2} (K) = 348.15	[38]
T_{g,P_1} (K) = 373.15	[39]
T_{g,P_2} (K) = 347.0	[40]
α_{M_1} (K^{-1}) = 0.00062	[41]
α_{M_2} (K^{-1}) = 0.0004	This work
α_{P_1} (K^{-1}) = 0.00045	[39]
α_{P_2} (K^{-1}) = 0.00048 [(a)]	[41]
α_p (K^{-1}) = $F_1\ \alpha_{P_1} + F_2\ \alpha_{P_2}$	[42]

[(a)] Assumed equal to the corresponding value for butyl acrylate (see [36]).

Although there are many mathematical models for DC effects in FRP and step-growth polymerization processes available in the literature, it is difficult to adequately describe the performance of different monomers under wide ranges of operating conditions using a single model with a single set of parameters. One such model is the Marten-Hamielec (MH) model [43,44], but it has the disadvantage of being discontinuous and requires an onset trigger criterion. Attempts to remove the trigger criterion resulted in a simpler, but less accurate model [45]. Therefore, in this study, we used the simplified version of the MH model [45], with a simpler onset trigger criterion, which causes it to be closer to the original model.

The assumptions summarized in Table 7 allow the determination the kinetic rate constants of RAFT activation and transfer for homopolymerizations of St and GMA. The assumptions indicated in the columns of Table 7 (e.g., $k_{aa1} = k_{at1}$) are necessary, due to the absence of experimental data to isolate the contributions of the two RAFT cycles to the properties of the produced polymer. Some authors have argued that the kinetic constants of the RAFT activation and RAFT transfer cycles may be different [46–48]; in RAFT polymerization modeling work, the equality of the kinetic constants of the RAFT cycles is supported [49–52].

Table 7. Simplifications made about the RAFT activation and transfer kinetic rate constants.

		St homopolymerization		
RAFT activation	$k_{aa1} = k_{fa1}$	RAFT transfer	$k_{at1} = k_{ft1}$	$k_{aa1} = k_{at1}$
		GMA homopolymerization		
RAFT activation	$k_{aa2} = k_{fa2}$	RAFT transfer	$k_{at2} = k_{ft2}$	$k_{aa2} = k_{at2}$

Another assumption is that the kinetic rate constants associated with the dormant species $[TP_n]$ and $[TQ_n]$ (k_{at3}, k_{ft3}, k_{at4}, k_{ft4}) can be approximated from the Mayo-Lewis terminal model [53]. This is achieved by considering the four reactions present in the RAFT transfer cycle and performing only consumption balances for the $[TP_n]$ and $[TQ_n]$ species, which are complemented with consumption balances for the $[TP_r]$ and $[TQ_r]$ species to complete the cycle. By calculating this, Equations (37)–(42) were obtained.

$$\frac{d\,[TP_n]}{d\,[TQ_n]} = \frac{k_{ft3}[TP_n](r_3[P_r] + [Q_r])}{k_{ft4}[TQ_n](r_4[Q_r] + [P_r])} \quad (37)$$

where:

$$r_3 = \frac{k_{ft1}}{k_{ft3}} \quad (38)$$

$$r_4 = \frac{k_{ft2}}{k_{ft4}} \quad (39)$$

$$\frac{d\,[TP_r]}{d\,[TQ_r]} = \frac{k_{at4}[TP_r](r_5[P_n] + [Q_n])}{k_{at3}[TQ_r](r_6[Q_n] + [P_n])} \quad (40)$$

where:

$$r_5 = \frac{k_{at1}}{k_{at4}} \quad (41)$$

$$r_6 = \frac{k_{at2}}{k_{at3}} \quad (42)$$

Additionally, $r_1 = r_3 = r_5$ and $r_2 = r_4 = r_6$. This is due to the application of the terminal model to the RAFT-transfer cycle. The RAFT-activation and -transfer kinetic parameters corresponding to homopolymerization of styrene and GMA were estimated using homopolymerization data for each monomer.

RAFT-related kinetic rate constants (optimization A) and AA effect parameters (optimization B) were estimated from overall conversion (X)-time and M_n-time experimental results, using a weighted non-linear multivariable regression procedure where the residual variance was minimized. The objective function is defined in Equation (43).

$$\text{Objective function} = \min\left[\Sigma_i^n \frac{1}{\sigma_X^2}(X_i^e - X_i^c)^2 + \Sigma_i^n \frac{1}{\sigma_M^2}(M_i^e - M_i^c)^2\right] \quad (43)$$

Superscripts e and c in Equation (43) stand for experimental and calculated values, respectively; ${\sigma_X}^2$ and ${\sigma_M}^2$ are variances of conversion and molar mass data, respectively; and n is the number of data points in each experimental data set. However, for simplicity, both variances were assumed equal to one. Optimization A was carried out using St and GMA homopolymerization data only, using the model without the AA terms. Optimization B was conducted for each copolymerization data set. The parameters obtained from each data set were regressed to obtain the final estimates. The flow chart that describes the modeling and parameter estimation strategies used in this contribution is shown in Figure 1. The model equations were solved using an in-house Fortran code. The optimization procedure for parameter estimation was carried out with the subroutine UWHAUS [54]. The system of ordinary differential equations was solved using subroutine DDASSL [55].

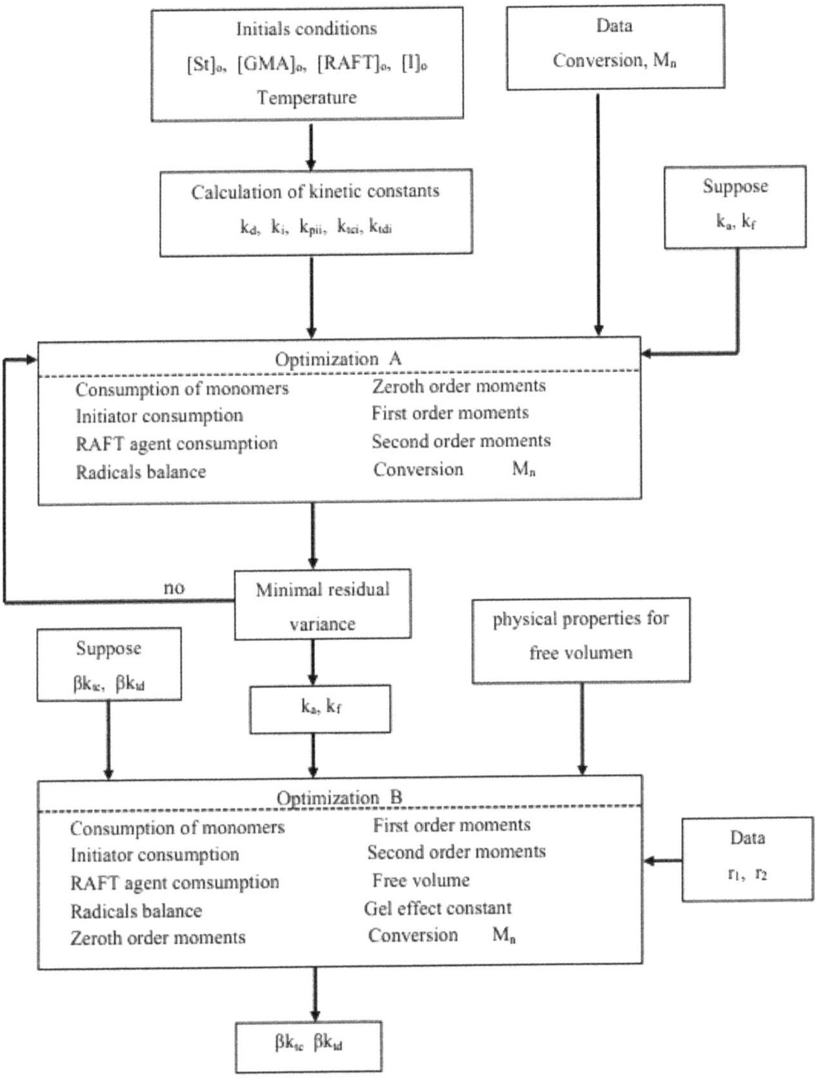

Figure 1. Flow chart for the estimation of kinetic and AA effect parameters.

4. Results

RAFT Synthesis and Characterization of Reactive Copolymers

St-GMA copolymers of different compositions (f_{GMA} = 0.10, 0.15, 0.30, and 0.40) were synthesized by RAFT bulk copolymerization of the monomers, at 103 °C, according to Figure 2. A M_n of ~30,000 g mol^{-1} was sought for all polymers. Final overall monomer conversions in a range of 85–90% were obtained. The experimental conditions used in this study are reported in Table 8.

Molar compositions of the St-GMA copolymers synthesized in this study were determined from the relative areas of the ^1H NMR characteristic signals [23,30]. ^1H NMR spectra for some of the obtained St-GMA copolymers are shown in Figure 2. Chemical shifts from phenyl protons in the region of 6.6–7.3 ppm, and methylene oxy (–OCH_2–) protons and methyl protons of GMA units at 3.5–4.5 and 0.5–1.2 ppm, respectively, are observed in Figure 3. The mole fraction of GMA in the copolymer was calculated as: $F_2 = 5 A_3/(5 A_3 + 3 A_2)$, where A_2 and A_3 are peak areas of phenyl and methyl protons, respectively. This method was used in this work due to the distinct NMR resonance of the GMA methyl group even at low GMA mole fractions in the copolymer [30].

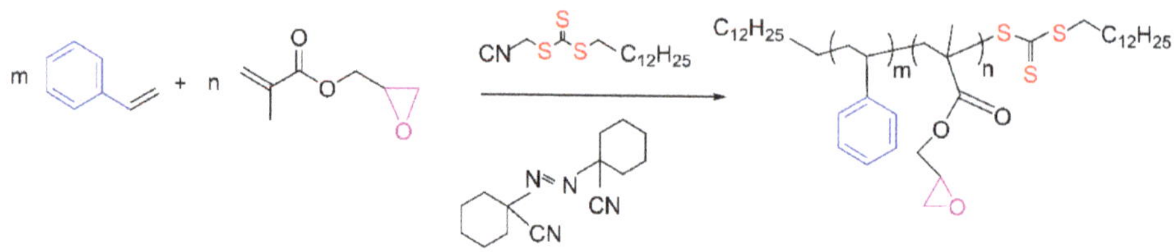

Figure 2. Simplified polymerization scheme.

Table 8. Summary of experimental conditions used in this study; T = 103 °C; $[St + GMA]_o$:$[CPDT]_o$:$[ACHN]_o$ = R_3:R_2:R_1.

Copolymer Identifier	Styrene $[St]_o$ (mol L^{-1})	GMA $[GMA]_o$ (mol L^{-1})	RAFT $[CPDT]_o$ (mol L^{-1})	Initiator $[ACHN]_o$ (mol L^{-1})	R_3:R_2:R_1
St-GMA 100-00	8.6445	–	0.0297	0.0074	291:4:1
St-GMA 90-10	7.6091	0.8455	0.0304	0.0076	278:4:1
St-GMA 85-15	7.1370	1.2595	0.0307	0.0077	273:4:1
St-GMA 70-30	5.7546	2.4662	0.0316	0.0079	260:4:1
St-GMA 60-40	4.8669	3.2446	0.0322	0.0080	251:4:1
St-GMA 00-100	–	7.5949	0.0360	0.0090	211:4:1

The St-GMA copolymers were characterized by SEC. They had M_n ~22,200–26,300 g mol^{-1} and Đ~1.21–1.28, which suggests that no side reactions took place and that most of the active polymer molecules remained living until the end of the polymerization.

Several kinetic models have been developed for RAFT homo- [56–58] and copolymerization of a few monomers [59–62]. As stated earlier, in our polymerization scheme we assumed that no branches to the adduct were produced, making it easier to model our RAFT copolymerization system using the terminal model [53], which is given by Equations (37)–(42). Therefore, the RAFT homo- and cross-propagation kinetic rate constants for the copolymerization system were determined by the corresponding values of RAFT homopolymerizations of St and GMA, and from reported values of r_1 y r_2 for the same copolymerization system [19].

Even though the RAFT polymerization mechanism is well-established and accepted [63–67], the parameters involved, such as addition, fragmentation, and termination kinetic rate constants, are not always reliable even in well-known systems, such as the RAFT homopolymerizations of methyl methacrylate and St [68]. The activation and transfer kinetic rate constants evaluated in this study for RAFT copolymerization of St and GMA are summarized in Table 9.

The profiles obtained with the parameters reported in Table 9 are not included due to space restrictions, but very good agreement is obtained in the low-conversion region, where DC effects are not observed. Although it has been reported that DC effects are important in all the reactions where polymer molecules are involved, in RAFT polymerizations [64], we considered DC termination only [69], to capture the phenomenon without adding too many additional parameters that required estimation. The AA termination parameters evaluated are provided in Table 10. As observed in Table 10 the higher the content of St in the copolymer, the higher the value of the AA termination parameter.

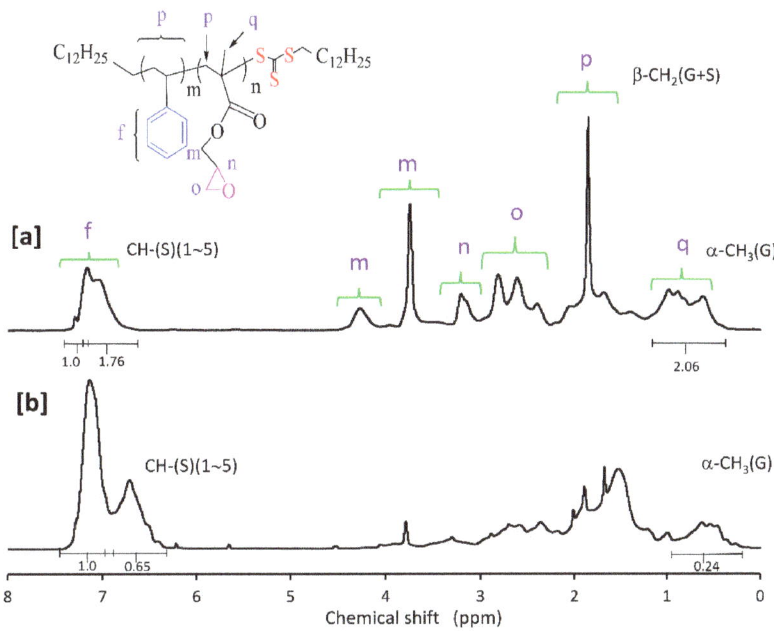

Figure 3. ^{1}H NMR spectrum of a RAFT synthesized St-GMA copolymer with (**a**) 55% and (**b**) 20% mole fractions of GMA in the feed mixture, using CPDT and ACHN. (f) phenyl protons of styrene; (m, n, o) methylene oxy (–OCH$_2$–) protons of GMA; (p, q) methyl protons of GMA.

Figure 4 shows a first order behavior plot. A comparison of experimental data and calculated profiles of conversion, M_n, M_w, and dispersity versus time (or conversion, in one case) is shown in Figures 5–10.

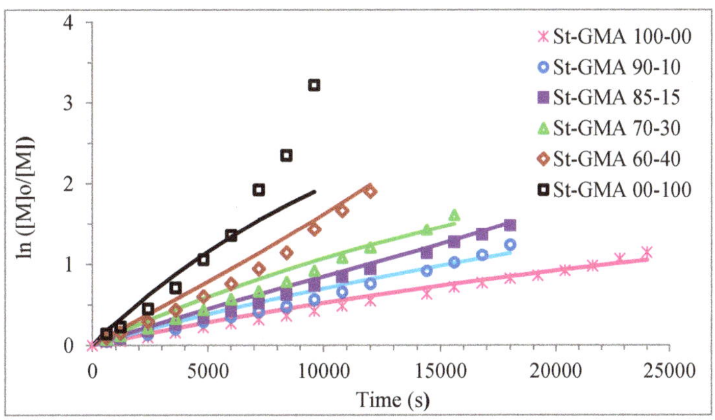

Figure 4. First-order of different St-GMA samples. Symbols represent experimental data, whereas solid lines correspond to model predictions.

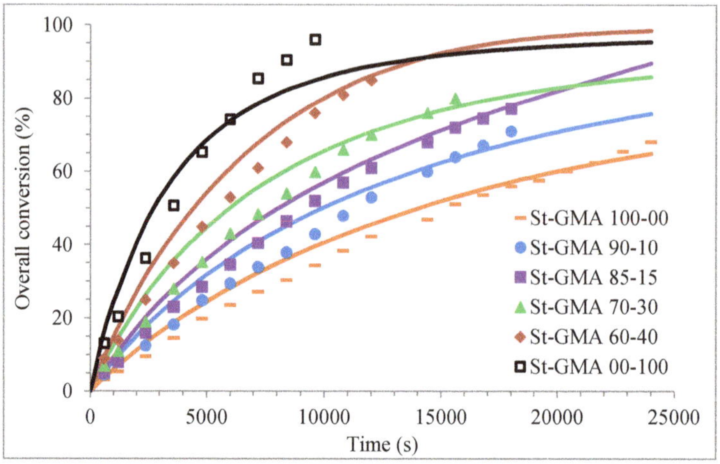

Figure 5. Profiles of overall conversion versus time for the different St-GMA samples. Symbols represent experimental data, whereas solid lines correspond to model predictions.

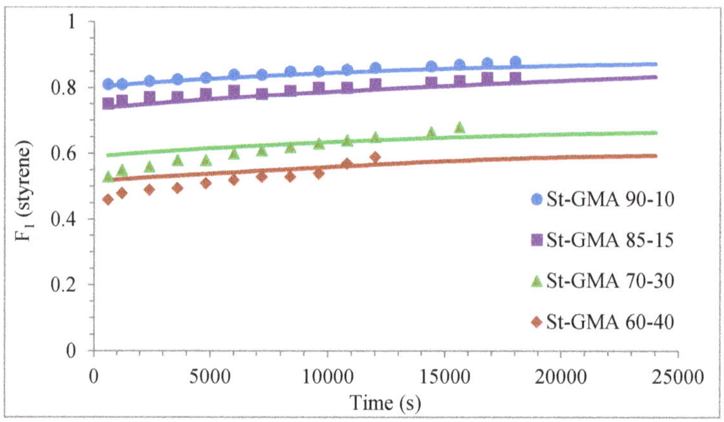

Figure 6. Copolymer composition versus time profiles for the different St-GMA samples. Symbols represent experimental data, whereas solid lines correspond to model predictions.

Table 9. Activation and transfer kinetic rate constants evaluated in this study.

RAFT activation, L mol^{-1} s^{-1}	
k_{fa1} = 101,113 ± 2690	k_{aa1} = 101,113 ± 2690
k_{fa2} = 484,000 ± 1200	k_{aa2} = 484,000 ± 1200
RAFT transfer, L mol^{-1} s^{-1}	
k_{ft1} = 101,113 ± 2690	k_{at1} = 101,113 ± 2690
k_{ft2} = 484,000 ± 1200	k_{at2} = 484,000 ± 1200
k_{ft3} = 271,517 ± 7223	k_{at3} = 708,224 ± 1756
k_{ft4} = 708,224 ± 1756	k_{at4} = 271,517 ± 7223

Table 10. AA termination parameters determined in this study.

Sample	St-GMA					
	100-00	90-10	85-15	70-30	60-40	00-100
$\beta k_{tc} = \beta k_{td}$	0.3836 ± 0.0023	0.3581 ± 0.0214	0.3979 ± 0.0001	$0.23045 \pm 5 \times 10^{-5}$	$0.2925 \pm 5 \times 10^{-5}$	$0.00081 \pm 7 \times 10^{-7}$

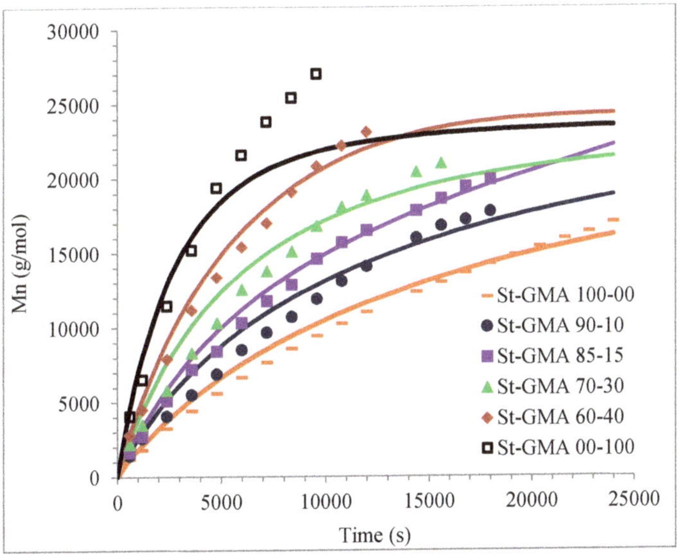

Figure 7. Profiles of M_n versus time for St-GMA samples. Symbols and solid lines correspond to experimental and calculated profiles, respectively.

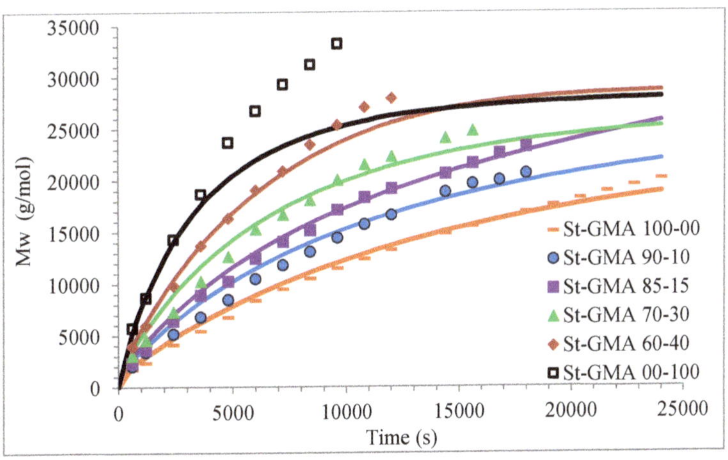

Figure 8. Profiles of M_w versus time for the different St-GMA samples. Symbols and solid lines correspond to experimental and calculated profiles, respectively.

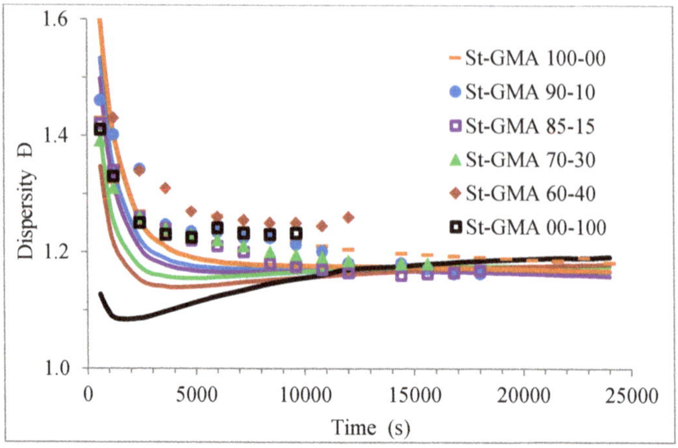

Figure 9. Profiles of dispersity versus time for the different St-GMA samples. Symbols and solid lines correspond to experimental and calculated profiles, respectively.

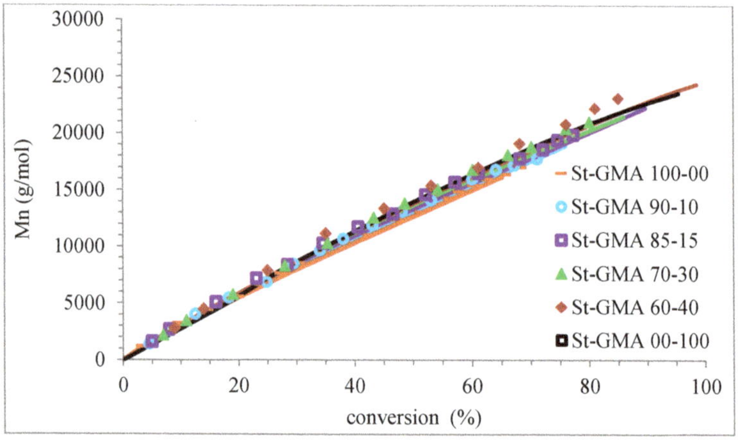

Figure 10. Profiles of M_n versus conversion for the different St-GMA samples. Symbols and solid lines correspond to experimental and calculated profiles, respectively.

The AA effect occurs at high conversions for the copolymerization reactions of St-GMA. Considering this, the determination of the kinetic rate constants associated with the RAFT cycles and the parameters associated with the termination reactions were carried out independently, which helped minimize possible correlations.

Except for sample St-GMA 00-100 where some discrepancies between experimental data and calculated profiles of conversion versus time were obtained (see Figure 5), the agreement is good in all other cases. Regarding copolymer composition, the agreement between calculated and experimental profiles of F_1 versus conversion is good when St content is high, but some deviations are observed in low conversions when its content decreases (see Figure 6).

Calculated and experimental profiles of M_n and M_w versus time are compared in Figures 7 and 8, respectively. Once again, the agreement is good except for sample S-GMA 00-100, where some discrepancies are observed. Considering the results obtained with the adjustment in high conversions, it is not enough to consider only diffusion-controlled effects on the termination kinetic rate constants for GMA homopolymerization (St-GMA 00-100).

Figure 9 shows how the dispersity of the produced copolymers evolves over time. Figures 9 and 10 show that the model can describe the controlled/living behavior of this RDRP system. Figure 10 shows the typical linear behavior of an M_n versus conversion profile for an RDRP system.

The effect of GMA content on the AA-termination parameter (βk_t) is shown in Figure 11. A linear behavior of the AA termination parameter as a function of GMA content is observed, with an R^2 correlation of 0.9237.

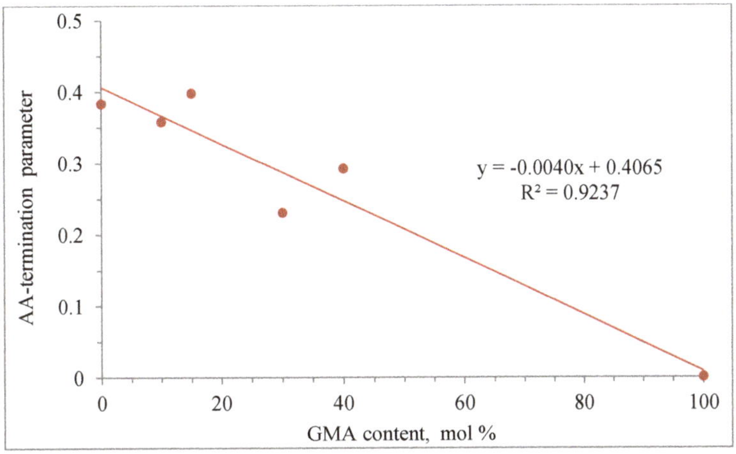

Figure 11. Linear regression of the AA-termination parameters vs. GMA content.

As observed from this modeling study, AA-termination parameters for each copolymerization case were needed to obtain good results, but as observed in Figure 11, a linear trend was obtained for the GMA homopolymerization case.

5. Conclusions

Our model for the RAFT copolymerization of St and GMA agrees very well with the experimental data generated in our laboratory and also reported in this contribution. The reactivity ratios determined for this copolymerization system were used considering that the terminal model was also fulfilled in the RAFT activation and transfer reactions. DC-termination using a simple free-volume model was sufficient to capture the effect of DC reactions in this system.

Unlike other modeling studies where neglection of the intermediate adduct results in qualitatively correct, but quantitatively inaccurate, predictions of the behavior of a RAFT polymerization system (Model 3 of [62]), the use of activation and transfer RAFT reactions resulted in our case in both qualitatively and quantitatively correct representations of the RAFT copolymerization of St and GMA. This model can be applied to other homo- and copolymerizations, and also extended to other systems, such as photo-RAFT polymerizations [15,70–72].

Author Contributions: J.J.B.-T. conceived and designed experiments; J.A.T.-L. and J.J.B.-T. wrote the paper, carried out the modeling, analyzed and interpreted data; J.J.B.-T. and N.G.-N. performed the experiments; E.V.-L. wrote, reviewed the paper and also helped in data analysis and interpretation; P.C. reviewed and edited the paper; E.S.-G. reviewed and edited the paper, providing insightful technical comments. We are indebted to Alex Penlidis, from the University of Waterloo, for technical discussions and for allowing E.V.-L. to check the manuscript for similarity through his UW affiliation. All authors have read and agreed to the published version of the manuscript.

Funding: This research was funded by: (a) Dirección General de Asuntos del Personal Académico (DGAPA), Universidad Nacional Autónoma de México (UNAM), México, Projects PAPIIT IV100119 and IG100122; (b) UNAM-UV collaboration agreement; and (c) CIQA-UV collaboration agreement.

Institutional Review Board Statement: Not applicable.

Informed Consent Statement: Not applicable.

Data Availability Statement: Data is contained within the article.

Acknowledgments: J.A.T.-L. acknowledges the Collaboration Agreement CIQA-UV; the financial support from UNAM within collaboration agreement UNAM-UV; and L324B-FQ-UNAM, CIQA and Queen's University for hosting a research visit. E.V.-L. acknowledges financial support from DGAPA, UNAM, Projects PAPIIT IV100119 and IG100122.

Conflicts of Interest: The authors declare no conflict of interest.

References

1. Hainey, P.; Sherrington, D.C. Oligoamine-functionalised poly(glycidyl methacrylate ethyleneglycol dimethacrylate) resins as moderate base extractants for gold from cyanide solutions. *React. Funct. Polym.* **2000**, *43*, 195–210. [CrossRef]
2. Zhang, H.; Ruckenstein, E. One-pot three-step synthesis of amphiphilic comb-like copolymers with hydrophilic backbone and hydrophobic side chains. *Macromolecules* **2000**, *33*, 814–819. [CrossRef]
3. Martin, P.; Devaux, J.; Legras, R.; Gurp, M.V.; Duin, M.V. Competitive reactions during compatibilization of blends of polybutyleneterephthalate with epoxide-containing rubber. *Polymer* **2000**, *42*, 2463–2478. [CrossRef]
4. Bondar, Y.; Kim, H.J.; Yoon, S.H.; Lim, Y.J. Synthesis of cation-exchange adsorbent for anchoring metal ions by modification of poly(glycidyl methacrylate) chains grafted onto polypropylene fabric. *React. Funct. Polym.* **2004**, *58*, 43–51. [CrossRef]
5. Benvenuta-Tapia, J.J.; Tenorio-López, J.A.; Martínez-Estrada, A.; Guerrero-Sánchez, C. Application of RAFT-synthesized reactive tri-block copolymers for the recycling of post-consumer R-PET by melt processing. *Mater. Chem. Phys.* **2019**, *229*, 474–481. [CrossRef]
6. Benvenuta-Tapia, J.J.; Champagne, P.; Tenorio-López, J.A.; Vivaldo-Lima, E.; Guerrero-Santos, R. Improving recycled poly(lactic acid) biopolymer properties by chain Extension using block copolymers synthesized by nitroxide-mediated polymerization (NMP). *Polymers* **2021**, *13*, 2791. [CrossRef]
7. Maa, C.-T.; Chang, F.-C. In situ compatibilization of PET/PS blends through reactive copolymers. *J. Appl. Polym. Sci.* **1993**, *49*, 913–924. [CrossRef]
8. Fakirov, S. (Ed.) *Handbook of Thermoplastic Polyesters*; Wiley-VCH: Weinheim, Germany, 2002.
9. Lee, H.; An, S.; Kim, S.; Jeon, B.; Kim, M.; Kim, I.S. Readily Functionalizable and Stabilizable Polymeric Particles with Controlled Size and Morphology by Electrospray. *Sci. Rep.* **2018**, *8*, 15725. [CrossRef]
10. Barner-Kowollik, C. *Handbook of RAFT Polymerization*; Wiley-VCH: Weinheim, Germany, 2008; pp. 110–122.
11. Nicolas, J.; Guillaneuf, Y.; Lefay, C.; Bertin, D.; Gigmes, D.; Charleux, B. Charleux; Nitroxide-mediated polymerization. *Prog. Polym. Sci.* **2013**, *38*, 63–235. [CrossRef]
12. Vivaldo-Lima, E.; Jaramillo-Soto, G.; Penlidis, A. Nitroxide-mediated polymerization (NMP). In *Encyclopedia of Polymer Science and Technology, Herman*; Mark, F., Ed.; John Wiley & Sons: Hoboken, NJ, USA, 2016; pp. 1–48. [CrossRef]
13. Matyjaszewski, K.; Xia, J. Atom transfer radical polymerization. *Chem. Rev.* **2001**, *101*, 2921–2990. [CrossRef]
14. Barner-Kowollik, C.; Davis, T.P.; Heuts, J.P.A.; Stenzel, M.H.; Vana, P.; Whittaker, M. RAFTing Down under: Tales of missing radicals, fancy architectures, and mysterious holes. *J. Polym. Sci. Part A Polym. Chem.* **2003**, *41*, 365–375. [CrossRef]
15. Zhu, Y.; Liu, Y.; Miller, K.A.; Zhu, H.; Egap, E. Lead Halide Perovskite Nanocrystals as Photocatalysts for PET-RAFT Polymerization under Visible and Near-Infrared Irradiation. *ACS Macro Lett.* **2020**, *9*, 725–730. [CrossRef]
16. Wan, J.; Fan, B.; Liu, Y.; Hsia, T.; Qin, K.; Junkers, T.; Teo, B.M.; Thang, S.H. Room temperature synthesis of block copolymer nano-objects with different morphologies via ultrasound initiated RAFT polymerization-induced self-assembly (sono-RAFT-PISA). *Polym. Chem.* **2020**, *11*, 3564–3572. [CrossRef]
17. Benvenuta-Tapia, J.J.; Tenorio-López, J.A.; Vivaldo-Lima, E. Estimation of reactivity ratios in the RAFT copolymerization of styrene and glycidyl methacrylate. *Macromol. React. Eng.* **2018**, *12*, 5. [CrossRef]
18. Brar, A.S.; Kumar, A.A.; Goyal, A.K. Characterization and optimization of poly (glycidyl methacrylate-co-styrene) synthesized by atom transfer radical polymerization. *Eur. Polym. J.* **2008**, *44*, 4082–4091. [CrossRef]
19. Moayeri, A.; Lessard, B.; Maric, M. Nitroxide mediated controlled synthesis of glycidyl methacrylate-rich copolymers enabled by SG1-based alkoxyamines bearing succinimidyl ester groups. *Polym. Chem.* **2011**, *2*, 2084–2092. [CrossRef]
20. Stenzel-Rosenbaum, M.H.; Davis, T.P.; Fane, A.G.; Chen, V. Porous polymer films and honeycomb structures made by the self-organization of well-defined macromolecular structures created by living radical polymerization techniques. *Angew. Chem.* **2001**, *40*, 3428–3432. [CrossRef]
21. Favier, A.; D'Agosto, F.; Charreyre, M.-T.; Pichot, C. Synthesis of N-acryloxysuccinimide copolymers by RAFT polymerization, as reactive building blocks with full control of composition and molecular weights. *Polymer* **2004**, *45*, 7821–7830. [CrossRef]

22. Feldermann, A.; Toy, A.A.; Phan, H.; Stenzel, M.; Davis, T.; Barner-Kowollik, C. Reversible addition fragmentation chain transfer copolymerization: Influence of the RAFT process on the copolymer composition. *Polymer* **2004**, *45*, 3997–4007. [CrossRef]
23. Benvenuta-Tapia, J.J.; Vivaldo-Lima, E.; Guerrero-Santos, R. Effect of copolymers synthesized by nitroxide-mediated polymerization as chain extenders of postconsumer poly (ethylene terephthalate) waste. *Polym. Eng. Sci.* **2019**, *59*, 2255–2264. [CrossRef]
24. Braunecker, W.A.; Matyjaszewski, K. Controlled/living radical polymerization: Features, developments, and perspectives. *Prog. Polym. Sci.* **2007**, *32*, 93–146. [CrossRef]
25. Chernikova, E.; Morozov, A.; Leonova, E.; Garina, E.; Golubev, V.; Bui, C.; Charleux, B. Controlled free-radical polymerization of n-butyl acrylate by reversible addition−fragmentation chain transfer in the presence of tert-butyl dithiobenzoate. A kinetic study. *Macromolecules* **2004**, *37*, 6329–6339. [CrossRef]
26. Barner-Kowollik, C.; Coote, M.L.; Davis, T.P.; Radom, L.; Vana, P. The reversible addition-fragmentation chain transfer process and the strength and limitations of modeling: Comment on "the magnitude of the fragmentation rate coefficient. *J. Polym. Sci. Part A Polym. Chem.* **2003**, *41*, 2828–2832. [CrossRef]
27. Zargar, A.; Schork, F.J. Copolymer sequence distributions in controlled radical polymerization. *Macromol. React. Eng.* **2009**, *3*, 118–130. [CrossRef]
28. Dixon, K.W. Decomposition rates of organic free radical initiatiors. In *Polymer Handbook*, 4th ed.; Brandrup, J., Immergut, E.H., Grulke, E.A., Abe, A., Bloch, D.R., Eds.; John Wiley & Sons: Hoboken, NJ, USA, 1998; pp. II/1–II/76, ISBN 9780471166283.
29. Buback, M.; Gilbert, R.G.; Hutchinson, R.A.; Klumperman, B.; Kuchta, F.-D.; Manders, B.G.; O'Driscoll, K.F.; Russell, G.T.; Schweer, J. Critically evaluated rate coefficients for free-radical polymerization, 1. Propagation rate coefficient for styrene. *J. Macromol. Chem. Phys.* **1995**, *196*, 3267–3280. [CrossRef]
30. Wang, W.; Hutchinson, R.A. PLP/SEC/NMR study of free copolymerization of styrene and glycidyl methacrylate. *Macromolecules* **2008**, *41*, 9011–9018. [CrossRef]
31. Hui, A.W.; Hamielec, A.E. Thermal polymerization of styrene at high conversions and temperatures. An experimental study. *J Appl. Polym. Sci.* **1972**, *16*, 749–769. [CrossRef]
32. Beuermann, S.; Buback, M.; Schmaltz, C. Termination rate coefficients of butyl acrylate free-radical homopolymerization in supercritical CO_2 and in bulk. *Ind. Eng. Chem. Res.* **1999**, *38*, 3338–3344. [CrossRef]
33. Achilias, D.; Kiparissides, C. Modeling of diffusion-controlled free-radical polymerization reactions. *J. Appl. Polym. Sci.* **1988**, *35*, 1303–1323. [CrossRef]
34. Fox, T.G.; Flory, P.J. Second-order transition temperatures and related properties of polystyrene. *J. Appl. Phys.* **1950**, *21*, 581–591. [CrossRef]
35. Li, D.; Li, N.; Hutchinson, R.A. High-temperature free Radical copolymerization of styrene and butyl methacrylate with depropagation and penultimate kinetic effects. *Macromolecules* **2006**, *39*, 4366–4373. [CrossRef]
36. Hutchinson, R.A.; Beuermann, S.; Paquet, D.A.; McMinn, J.H.; Jackson, C. Determination of free-Radical propagation rate coefficients for cycloalkyl and functional methacrylates by pulsed-laser polymerization. *Macromolecules* **1998**, *31*, 1542–1547. [CrossRef]
37. Bhattacharya, D.; Hamielec, A. Bulk thermal copolymerization of styrene p-methylstyrene: Modelling diffusion-controlled termination and propagation using free-volume theory. *Polymer* **1986**, *27*, 611–618. [CrossRef]
38. Dow The Materials Science Company Explore Products. Available online: https://www.dow.com (accessed on 12 October 2015).
39. Soh, S.K.; Sundberg, D.C. Diffusion-controlled vinyl polymerization. I. The gel effect. *J. Polym. Sci. Polym. Chem. Ed.* **1982**, *20*, 1299–1313. [CrossRef]
40. Narasimhaswamy, T.; Sumathi, S.C.; Reddy, B.S.R. 2,4,6-Tribromophenyl acrylate-co-glycidyl methacrylate polymers: Synthesis, characterization, and reactivity ratios. *J. Polym. Sci. Part A Polym. Chem.* **1992**, *30*, 2165–2172. [CrossRef]
41. Dube, M.A.; Rilling, K.; Penlidis, A. A kinetic investigation of butyl acrylate polymerization. *J. Appl. Polym. Sci.* **1991**, *43*, 2137–2145. [CrossRef]
42. Keramopoulos, A.; Kiparissides, C. Mathematical modeling of diffusion-controlled free-radical terpolymerization reactions. *J. Appl. Polym. Sci.* **2003**, *88*, 161–176. [CrossRef]
43. Marten, F.L.; Hamielec, A.E. High conversion diffusion Controlled polymerization. *ACS Symp. Ser.* **1979**, *104*, 43–70. [CrossRef]
44. Marten, F.L.; Hamielec, A.E. High conversion diffusion controlled polymerization of styrene I. *J. Appl. Polym. Sci.* **1982**, *27*, 489–505. [CrossRef]
45. Vivaldo-Lima, E.; Hamielec, A.; Wood, P. Auto-acceleration effect in free radical polymerization. A Comparison of the CCS and MH models. *Polym. React. Eng.* **1994**, *2*, 17–85. [CrossRef]
46. Moad, G.; Rizzardo, E.; Thang, S.H. Living radical polymerization by the RAFT process–A second update. *Aust. J. Chem.* **2009**, *82*, 1402–1472. [CrossRef]
47. Chernikova, E.; Golubev, V.; Filippov, A.; Lin, C.Y.; Coote, M.L. Use of spin traps to measure the addition and fragmentation rate coefficients of small molecule RAFT-aducts radicals. *Polym. Chem.* **2010**, *1*, 1437–1440. [CrossRef]
48. Perrier, S.; Takolpuckdee, P. Macromolecular design via reversible addition-fragmentation chain transfer (RAFT)/xanthates (MACIX) polymerization. *J. Polym. Chem.* **2005**, *43*, 5347–5393. [CrossRef]

49. De Rybel, N.; Van Steenberge, P.H.M.; Reyniers, M.-F.; Barner-Kowollik, C.; D'Hooge, D.R.; Marin, G.B. An update on the pivotal role of kinetic modeling for the mechanistic understanding and design of bulk and RAFT polymerization. *Macromol. Theory Simul.* **2017**, *26*, 1. [CrossRef]
50. Zapata-González, I.; Saldívar-Guerra, E.; Ortiz-Cisneros, J. Full molecular in RAFT polymerization. New mechanistic insight by direct integration of the equations. *Macromol. Theory Simul.* **2011**, *20*, 370–388. [CrossRef]
51. Zapata-González, I.; Saldívar-Guerra, E.; Licea-Claverie, A. A kinetic modeling of RAFT polymerization via dithiobenzoate agents considering the missing step. *Theory. Chem. Eng. J.* **2017**, *326*, 1242–1254. [CrossRef]
52. Fortunatti, C.; Sarmonia, C.; Brandolin, A.; Asteasuain, M. Modeling of RAFT polymerization using probability generating functions. Detailed prediction of full molecular weigth distributions and sensitivity analysis. *Macromol. React. Eng.* **2014**, *8*, 781–795. [CrossRef]
53. Mayo, F.R.; Lewis, F.M. Copolymerization. I. A Basis for comparing the behavior of monomers in copolymerization; the copolymerization of styrene and methyl Methacrylate. *J. Am. Chem. Soc.* **1944**, *66*, 1594–1601. [CrossRef]
54. Meeter, D.A.; Wolfe, P.J. *UWHAUS Nonlinear Least Squares Fitting and Function Minimization*; University of Wisconsin Computing Center: Madison, WI, USA, 1965.
55. Breman, K.E.; Campbell, S.L.; Petzold, L.R. *Numerical Solution of Initial-Value Problems in Differential-Algebraic Equations*; Elsevier Science Publishing Company: New York, NY, USA, 1989.
56. Bitsch, B.; Barner-Kowollik, C.; Zhu, S. Modeling the effects of reactor backmixing on RAFT polymerization. *Macromol. React. Eng.* **2011**, *5*, 55–68. [CrossRef]
57. Wang, A.R.; Zhu, S. Modeling the reversible addition-fragmentation transfer polymerization process. *J. Polym. Sci. Part A Polym. Chem.* **2003**, *41*, 1553–1566. [CrossRef]
58. Gao, X.; Zhu, S. Modeling analysis of chain transfer in reversible addition-fragmentation chain transfer polymerization. *J. Appl. Polym. Sci.* **2011**, *12*, 497–508. [CrossRef]
59. Sun, X.; Luo, Y.; Wang, R.; Li, B.-G.; Liu, A.B.; Zhu, S. Programmed synthesis of composition with controlled chain Composition distribution via semibatch RAFT copolymerization. *Macromolecules* **2007**, *40*, 849–859. [CrossRef]
60. Jiang, J.; Wang, W.-J.; Li, B.-G.; Zhu, S. Modeling and experimentation of RAFT solution copolymerization of styrene and butyl acrylate effect of chain transfer reaction on polymer molecular weight distribution. *Macromol. React. Eng.* **2017**, *11*, 6. [CrossRef]
61. Wang, R.; Luo, Y.; Li, B.; Sun, X.; Zhu, S. Design and control of copolymer composition distribution in living radical polymerization using semi-batch feeding police: A model simulation. *Macromol. Theory Simul.* **2006**, *15*, 356–368. [CrossRef]
62. Pallares, J.; Jaramillo-Soto, G.; Flores-Cataño, C.; Vivaldo-Lima, E.; Lona, L.M.F.; Penlidis, A. A comparison of reaction mechanisms for reversible addition-fragmentation chain transfer polymerization using modeling tools. *J. Macromol. Sci. Part A Pure Appl. Chem.* **2006**, *43*, 1293–1322. [CrossRef]
63. Moad, G.; Chiefari, J.; Chong, Y.K.; Krstina, J.; Mayadunne, R.T.A.; Postma, A.; Thang, S.H. Living free radical polymerization with reversible addition–Fragmentation chain transfer (the life of RAFT). *Polym. Int.* **2000**, *49*, 993–1001. [CrossRef]
64. Kwak, Y.; Goto, A.; Tsujii, Y.; Murata, Y.; Komatsu, K.; Fukuda, T. A kinetic study on the rate retardation in radical polymerization of styrene with addition−fragmentation chain transfer. *Macromolecules* **2002**, *35*, 3026–3029. [CrossRef]
65. Kwak, Y.; Goto, A.; Fukuda, T. Rate retardation in reversible addition−Fragmentation chain transfer (RAFT) polymerization: Further evidence for cross-termination producing 3-arm star chain. *Macromolecules* **2004**, *37*, 1219–1225. [CrossRef]
66. Barner-Kowollik, C.; Quinn, J.; Morsley, D.R.; Davis, T.P. Modeling the reversible addition–Fragmentation chain transfer process in cumyl dithiobenzoate-mediated styrene homopolymerizations: Assessing rate coefficients for the addition–fragmentation equilibrium. *J. Polym. Sci. Polym. Chem.* **2001**, *39*, 1353–1365. [CrossRef]
67. Feldermann, A.; Coote, M.L.; Stenzel, M.H.; Davis, T.P.; Barner-Kowollik, C. Consistent experimental and theoretical evidence for long-lived intermediate radicals in living free radical polymerization. *J. Am. Chem. Soc.* **2004**, *126*, 15915–15923. [CrossRef]
68. Wang, A.R.; Zhu, S.; Kwak, Y.; Goto, A.; Fukuda, T.; Monteiro, M.S. A difference of six orders of magnitude: A reply to "the magnitude of the fragmentation rate coefficient". *J. Polym. Sci. Part A Polym. Chem.* **2003**, *41*, 2833–2839. [CrossRef]
69. Wang, A.R.; Zhu, S. Effects of diffusion-controlled radical reactions on RAFT polymerization. *Macromol. Theory Simul.* **2003**, *12*, 196–208. [CrossRef]
70. Bellotti, V.; Simonutti, R. New Light in Polymer Science: Photoinduced Reversible Addition-Fragmentation Chain Transfer Polymerization (PET-RAFT) as Innovative Strategy for the Synthesis of Advanced Materials. *Polymers* **2021**, *13*, 1119. [CrossRef] [PubMed]
71. Doerr, A.M.; Burroughs, J.M.; Gitter, S.R.; Yang, X.; Boydston, A.J.; Long, B.K. Advances in Polymerizations Modulated by External Stimuli. *ACS Catal.* **2020**, *10*, 14457–14515. [CrossRef]
72. Zhu, Y.; Egap, E. Light-Mediated Polymerization Induced by Semiconducting Nanomaterials: State-of-the-Art and Future Perspectives. *ACS Polym. Au* **2021**, *1*, 76–99. [CrossRef]

polymers

Article

Characterization of a Radiofluorogenic Polymer for Low-Energy Electron Beam Penetration Depth Visualization

Magdalena Maria Skowyra [1,*], Christina Ankjærgaard [1], Liyun Yu [2], Lars René Lindvold [1], Anne Ladegaard Skov [2] and Arne Miller [1]

[1] Department of Health Technology, Technical University of Denmark, Frederiksborgvej 399, 4000 Roskilde, Denmark; cank@dtu.dk (C.A.); lali@dtu.dk (L.R.L.); armi@dtu.dk (A.M.)
[2] Department of Chemical and Biomedical Engineering, Technical University of Denmark, Søltofts Plads 228A, 2800 Kgs. Lyngby, Denmark; lyyu@kt.dtu.dk (L.Y.); al@kt.dtu.dk (A.L.S.)
* Correspondence: magsk@dtu.dk

Abstract: Low-energy (80–300 keV) electron beam accelerators are gaining in importance in the radiation processing industry due to their ease of use and wide range of applications (e.g. product surface sterilizations or polymer curing and cross-linking). Due to their very low penetration depth (tens to hundreds of microns), currently used film dosimeters exhibit dose gradients over their thickness and do not resolve the dose response in the first microns of the irradiated material. Hence, the surface dose, defined as the dose in the first micron D_μ, cannot be measured directly. This study presents a polymer material as a dosimeter candidate for high-dose low-energy electron beam irradiations. The readout of the dose-dependent fluorescence intensity, originating from a pararosaniline dye reaction when irradiated, is measured using fluorescence microscopy. So far, no in-depth characterization of the material has been performed, leaving the stability and fluorescence properties of the material not fully optimized. We describe the improvements in polymer composition and the fabrication method, and characterize the material properties in terms of the thermal stability, glass transition temperature, refractive index, hardness, rheological behavior, and water affinity. All of these create a complex set of requirements a polymer needs to fulfill to become an effective dosimeter when measuring using confocal microscopy. The fluorescence readout procedure will be addressed in further studies.

Keywords: low-energy electron beam; fluorescence; polymer dosimeter; radiofluorogenic; 3D dosimetry

Citation: Skowyra, M.M.; Ankjærgaard, C.; Yu, L.; Lindvold, L.R.; Skov, A.L.; Miller, A. Characterization of a Radiofluorogenic Polymer for Low-Energy Electron Beam Penetration Depth Visualization. *Polymers* **2022**, *14*, 1015. https://doi.org/10.3390/polym14051015

Academic Editors: Bożena Jarząbek and Mohammad Afsar Uddin

Received: 28 November 2021
Accepted: 28 February 2022
Published: 3 March 2022

Publisher's Note: MDPI stays neutral with regard to jurisdictional claims in published maps and institutional affiliations.

Copyright: © 2022 by the authors. Licensee MDPI, Basel, Switzerland. This article is an open access article distributed under the terms and conditions of the Creative Commons Attribution (CC BY) license (https://creativecommons.org/licenses/by/4.0/).

1. Introduction

Having its beginnings in the 1950s, electron beam (e-beam) industrial processing has been developing rapidly and used extensively in many areas of industry, where there is a need for enhancement of the physical and chemical properties of materials or a reduction of undesirable contaminants [1]. Low-energy electron beams, which are typically of energies ranging between 80 and 300 keV [2], have found several applications in the radiation processing industry, starting from polymer curing [3], cross-linking of coatings and inks [4] and polymer functionalization by grafting [5], through decontamination of packaging materials [1], up to sterilization of surfaces [6,7] and pharmaceutical components [8]. Low-energy e-beam accelerators offer a number of advantages, such as (1) electrical energy conversion of the order of 70% [9], (2) high dose rates (~100 kGy/s), allowing rapid processing [1], (3) radiation protection in a form of self-shielded equipment [1], (4) environmental benefits, as they reduce the emission of volatile organic compounds and air pollutants [10], and (5) low costs [1].

However, low-energy electrons are able to penetrate matter only up to a few hundred micrometers [2], which in fact becomes challenging in terms of absorbed dose determination. The dosimetry requirements are especially rigorous when performing the surface sterilization process of, for example, pharmaceutical or medical products [11]. There, only the external surfaces of, for example, pre-sterilized tubs (of vials or syringes) should be irradiated before they can enter an aseptic filling area. In this way, the production of radiolysis products within the volume of a tub is avoided [12].

Several dosimetry systems are used in the radiation processing industry, such as alanine-EPR [13], calorimeters [14] or radiochromic dye films [13,15]. They are all excellent tools for high-dose measurements (~kGy). However, at low electron energies, dose gradients are introduced over the thickness of a dosimeter, and the measured average doses will depend on the chosen system [16]. In addition, they are not able to measure the dose in the first microns of irradiated material, the so-called surface dose. Thus far, the surface dose has been extrapolated from obtained depth–dose curves and corrected to the average dose in the first micrometer—D_{μ} [16]. Schuster et al. [17] proposed a thin dosimetric spray-coating material for surface dose measurements of complex surface geometries (exemplified by PET bottles). The concept relies on a readout of the dose-dependent luminescence decay time of an inorganic phosphor under near-infrared excitation. However, fading of the phosphor was a limiting factor that this dosimetry system so far has not been able to overcome, and hence, the challenge of directly measuring the dose in the first micrometer of a material is still to be addressed.

Here, we investigate a new type of material that could be applied as a low-energy electron beam dosimeter, focusing on surface dose verification. As a starting point, a radiochromic and radiofluorogenic polymer dosimeter developed by Bernal-Zamorano et al. [18–20], which was initially intended to work as a dosimeter for the medical dosimetry dose range (~Gy), was selected. However, industrial radiation processing applies doses of the order of tens or even hundreds of kGy, so the material had to be appropriately modified. Furthermore, the material's composition, its mechanical, optical and thermal properties, hydrophilicity as well as the photocuring process had not yet been fully optimized or characterized. The system previously developed by Skowyra et al. [21] was a proof of concept where the solid state polymer material responded to kilogray radiation doses of low-energy electron beam irradiations by its change in fluorescence intensity, measured using an epi-fluorescence widefield microscope. However, issues related to the background fluorescence signal, shelf life and environmental stability of the material prompted further investigation into establishing and demonstrating suitable material requirements before optimizing the readout technique into confocal laser scanning microscopy.

In this paper, we investigate the influence of the polymer material's composition on the properties of the material. This includes the polymer matrix and the optimal solvent for the radiation-sensitive dye. We present the material's properties in terms of its thermomechanical behavior, particularly the decomposition process and glass transition temperature, as well as hardness and rheological studies. Additionally, the optical quality of the polymer, with a focus on refractive index measurement, was determined, and for fluorescence imaging purposes, the water affinity of the material was evaluated. Finally, we show the absorbance and fluorescence response spectra of pararosaniline leuco-dye in both a solution and solid material, and demonstrate the spectral response of the material irradiated with a low-energy electron beam.

2. Materials and Methods

2.1. Composition

The dosimeter material consisted of (1) host polymer–PPGDA (poly(propylene glycol) diacrylate, 1.12 M, Sigma-Aldrich, Darmstadt, Germany, CAS: 52496-08-9, inhibitors: 100 ppm BHT and 100 ppm MEHQ); (2) photoinitiator–TPO (diphenyl (2,4,6-trimethylbenzoyl) phosphine oxide, 28 mM, Sigma-Aldrich, Germany, CAS: 75980-60-8); (3) leuco-dye (4′,4′,4″-triamino-triphenylaceto-nitrile, 6.89 mM, NCK A/S, Farum, Denmark); and (4) acetonitrile (0.99 M, Sigma-Aldrich, Germany, CAS: 75-05-8, max. 0.005% H_2O).

The solvents tested for solubility were acetone (CAS: 67-64-1, ≥99.5%), acetonitrile (CAS: 75-05-8) and ethanol (CAS number: 64-17-5, ≥99.9%) supplied by Sigma-Aldrich

(Merck, Germany) and 2-ethoxy ethanol (CAS: 110-80-5, ≥98.5%) provided by Fluka Chemie (Buchs, Switzerland). The concentration of the dye solution was 20 mg/mL.

2.2. Fabrication

The leuco-dye, the selected solvent (acetonitrile) and the host polymer were mixed at room temperature using a magnetic stirrer in a closed, UV-blocking glass vial for 30 min. After that, the photoinitiator was added, and the solution was mixed for another half an hour. The mixed composition was then placed in a refrigerator at 4 °C until the photocuring was carried out to lower the oxygen diffusion and avoid a viscosity increase [22]. The samples were prepared by means of a photopolymerization reaction using a LED lamp of a 385 nm wavelength as described in [21]. Three different sample types were manufactured: (1) cuboids of dimensions 3 × 3 × 45 mm, (2) cuboids of dimensions 1 × 3 × 45 mm and (3) cylinders 10 mm in diameter and 3 mm in thickness. The samples were cured in silicone molds [21] of corresponding shapes with a plastic foil placed on the top surface to block oxygen inhibition during the reaction [23]. The illumination time was 5 min, with the intensity of the UV light measured to be 15.0 ± 0.3 mW/cm^2. Unless otherwise stated, the uncertainties throughout the performed measurements were given at one standard deviation (k = 1).

Two resins for 3D-printed molds were tested as alternatives for the silicone mold: (1) traditional resin (black, ELEGOO standard [24]) and (2) acrylonitrile butadiene styrene (ABS)-like resin (transparent, ELEGOO ABS-like [25]). The compositions of the resins were: epoxy resin (50%), hexamethylene diacrylate (40%) monomer, N-(dimethylcarbamoyl) glycine (5%) and hydroxycyclohexyl phenyl ketone (5%) photoinitiator. The molds were 3D printed using stereolithography (SLA) technology.

2.3. Irradiation Process

Irradiations were carried out using a low-energy electron beam accelerator at the Risø High Dose Reference Laboratory (e-beam, EBLab-200, COMET AG, Wünnewil-Flamatt, Switzerland), maintaining a 15-m/min conveyor speed and 15 mm air gap between the sample surface and electron beam window, at laboratory conditions with an average temperature of 21 °C and a humidity level between 30 and 50%. The samples were placed on a polystyrene backplate and fastened with tape to ensure surface evenness. The irradiation process parameters used throughout for these measurements are presented in Table 1.

Table 1. Specification of the irradiation process parameters.

Nominal Irradiation Dose (kGy)	Voltage (kV)		
	80	120	200
	Beam Current (mA)		
10	1.746	-	-
15	2.619	-	-
20	3.492	-	-
30	5.238	3.065	4.639
40	6.894	-	-
50	8.730	5.108	7.733

2.4. Thermogravimetric Analysis (TGA)

In order to establish the thermal stability of the polymer material, thermogravimetric analysis was performed. The measurements of weight loss versus temperature were carried out on a Discovery TGA (TA Instruments, New Castle, DE, USA). The TGA data also helped in selecting the experimental conditions for the differential scanning calorimetry measurements. 10 mg of the polymer (top surface part) was placed in a platinum pan and heated in a nitrogen atmosphere from room temperature to 900 °C, with a heating rate of 10 °C/min. The readout of the sample weight versus temperature was performed every 0.01 s. The analysis was performed on both non-irradiated and irradiated samples (120 keV e-beam, 30 kGy), and the results were analyzed by TRIOS software (TA Instruments, New Castle, DE, USA). The accuracy of the weight determination was <±0.1% of a measured value.

2.5. Differential Scanning Calorimetry (DSC)

Differential scanning calorimetry (DSC) was conducted to determine the phase transitions of the polymer by measuring the heat flow produced in a sample as a function of the temperature. Particularly important was the determination of the glass transition temperature (T_g), as this indicates what physical properties of the polymer can be expected in a specified range of temperatures and how they will later influence the dosimetric properties of the material. A total of 2–5 mg of the polymer was cut into small pieces and placed in the T_{zero} aluminum hermetic pan. The masses of the pieces were determined with the uncertainty of ±0.01 mg. Both the non-irradiated and irradiated material was tested. DSC measurements were performed in a nitrogen atmosphere on the Discovery TGA with a heating and cooling rate of 10 °C/min. Each measurement consisted of a first heating step from −90 °C to 200 °C (isothermal at −90 °C for 5 min), a cooling step from 200 °C to −90 °C (isothermal at −90 °C for 5 min) and finally a second heating step from −90 °C to 300 °C.

The glass transition temperature was measured from DSC curves as the endothermic transition from the amorphous glassy state into the viscous rubbery state [26] by means of the second heating scan. T_g was determined using TRIOS software (TA Instruments, New Castle, DE, USA) as the midpoint (i.e. halfway point) temperature of the jump in heat capacity in the chosen temperature range [27]. The temperature accuracy of the equipment was ±0.1 °C, and the temperature precision was ±0.05 °C.

2.6. Refractive Index

For the purpose of verifying the homogeneity of the material and finding a refractive index matching medium for fluorescence microscopy readouts, optical coherence tomography (OCT) was used. The refractive index (RI) measurements of a cured polymer sample were measured using Spectral Domain Optical Coherence Tomography (Telesto-II, Thorlabs Inc., Newton, NJ, USA) by applying the optical path shifting method [28,29] at a wavelength of 1310 nm. At this wavelength, no absorption of light was observed by the polymer material. Only when the optical path length of the sample and reference arm match, the observed interference signal reach a maximum. Hence, it can be used to measure the distance precisely. First, an image of the reflection coming from the scanning stage (glass) was acquired. While keeping the distance between the probe and the scanning stage fixed, the cuboid sample (Type 2) was placed on the scanning stage, and another image was acquired. In OCT measurements, the measured value of the thickness does not present its real thickness but its optical thickness OT [29]:

$$OT = n \times T \tag{1}$$

where n and T are the refractive index and the physical thickness of the sample, respectively. After the sample placement, the position of the scanning stage shifts down due to imaging of material with a refractive index larger than 1, and that optical shift OS can be expressed as follows [29]:

$$OS = OT - T \tag{2}$$

Consequently, the refractive index n of the sample was calculated using the following equation:

$$n = \frac{OS}{T} + 1 \tag{3}$$

The RI of three samples was measured, and the uncertainty of one standard deviation (k = 1) of the result was determined. The physical thickness of the samples was measured using a Millitron Meter (Mahr Feinprüf, Germany, Type 5312340 230 V) with an accuracy of ±1 µm.

2.7. Hardness

Measurements of the hardness of the polymer material were conducted in order to test if additional post-processing high-temperature heating was needed. The hardness of the polymer material was measured using a Shore A durometer (AD-300, Checkline Europe, Enschede, The Netherlands). Cured, non-irradiated samples (Type 2), as well as samples heated in an oven at 180 °C for 10 min, were tested. Six samples, each approximately 1 mm thick, were stacked on top of each other to fulfill the minimal thickness requirement of 6 mm for the measurement [30]. Five measurements of both sample types were carried out, each with a time duration of 15 s.

2.8. Rheology

Rheological studies of the polymer material were performed in order to determine its viscoelastic properties that contribute to the optimal dosimetric properties. The viscoelastic properties of the polymer material were tested using a rheometer (DHR-1, TA Instruments, New Castle, DE, USA) with a 20-mm parallel plate (Peltier plate steel 106669), a gap of the sample's thickness (1 mm) and temperature of 25 °C. Some 1-mm thick cylinder samples were cut out of cuboid samples (Type 2). Amplitude sweep tests were performed for three different samples by applying an angular frequency of 10 rad/s in the strain region from 0.1 to 10% in order to determine the linear viscoelastic region of the polymer material. For the chosen strain of 0.25%, the frequency sweep test of those samples was run in the angular frequency region from 100 to 0.1 rad/s. In addition, a temperature sweep measurement was performed by applying 0.25% strain at an angular frequency of 10 rad/s in the temperature range of 10–100 °C. A frequency sweep test was also run for samples irradiated with a low-energy e-beam of energies 80, 120 and 200 keV and a dose of 50 kGy, applying 0.25% strain in the frequency region from 100 to 0.1 rad/s.

2.9. Water Affinity

The water affinity of the polymer material was tested in order to determine the material swelling properties and the possible use of a water immersion objective for fluorescence imaging. Water contact angle (WCA) measurements were conducted on an OCA20 Contact Angle System (DataPhysics Instruments GmbH, Filderstadt, Germany) using a sessile drop technique at room temperature. A 6 µl droplet of MilliPore water was dispersed through a needle and placed on the surface of the sample. The non-irradiated and irradiated (200 keV, 30 kGy) samples' top, bottom and side surfaces were tested. The WCA was determined using SCA20 software (DataPhysics Instruments GmbH) as an average value of three measurement points on each of the sides.

2.10. Absorbance and Fluorescence Spectra

The absorbance and fluorescence spectra were measured using a Tecan Spark M10 multimode plate reader (Tecan Trading AG, Männedorf, Switzerland) with a 12-well plate (VWR Tissue Culture Plates). The absorbance spectra were measured as the optical density (OD) in the wavelength range of 300–800 nm with a wavelength step size of 2 nm. The fluorescence (emission) spectra were recorded in the wavelength range of 560–800 nm with a step size of 1 nm, after excitation with 555 nm light of a 5 nm bandwidth. The fluorescence signals were quantified as relative fluorescence units (RFUs) while keeping a constant gain of 100 (the amplification factor for the photomultiplier).

For this, 0.5% w/v of the leuco-dye was dissolved in the selected solvent (acetonitrile) and divided into two parts: (1) a non-irradiated liquid solution (kept in laboratory conditions up to the measurement time) and (2) an irradiated liquid solution (200 keV e-beam, 30 kGy). The liquid solution was irradiated in a thin (0.5 mm) metal petri dish. The solid samples were prepared as described in Section 2.2 and divided into: (1) a non-irradiated solid and (2) an irradiated solid (200 keV e-beam, 30 kGy). Additionally, the absorbance and fluorescence spectra of polymer material, irradiated with a low-energy e-beam of 80 keV and 5 doses (15, 20, 30, 40 and 50 kGy), were measured. The spectra were normalized to the absorbance and fluorescence peak value of the 50 kGy sample.

3. Results

3.1. Dosimeter Composition

3.1.1. Solubility of the Dye

Pararosaniline leuco-dye, used as a radiation-sensitive medium, needs to be fully dissolved before it is mixed with other chemicals. This step is particularly important in order to ensure the uniformity of the samples and their later response to the radiation dose, measured as fluorescence intensity values. Accumulation of undissolved dye might also impair the fluorescence readouts by introducing shadowing or scattering effects. The composition developed by Bernal-Zamorano et al. [19] used ethanol as a solvent. However, our observations indicated that the dye was sparingly soluble, which is also in agreement with the literature values of triphenylmethane dye solubility in ethanol [31]. The dye formed a precipitate and rapidly changed its color once dissolved in ethanol. The dye is reported to be very soluble in solvents like diethyl ether, pyridine, chloroform and benzene [31], but none of these were considered here due to their toxicity. The commercially used Risø B3 radiochromic film [32], which contains the same radiation-sensitive dye as our dosimeter material, uses 2-ethoxy ethanol as a solvent in the fabrication process, and hence it was also tested. Additionally, acetone and acetonitrile were tested due to their lower toxicity and versatility.

All the tested solvents—2-ethoxy ethanol, acetone and acetonitrile—successfully dissolved the required amount of dye. However, the 2-ethoxy ethanol solution very quickly changed its color to magenta and was not further tested, which was also due to its toxicity. Acetone and acetonitrile solutions, after mixing with the rest of the composition and being stored in the refrigerator for 5 days, showed much less coloration. However, the acetone solution started to precipitate after a few days. Due to acetone being more volatile [33] and less polar than acetonitrile [34], acetonitrile was chosen as a solvent for the composition.

3.1.2. Host Polymer: PEGDA vs. PPGDA

The polymer matrix, which is formed by cross-linking of a diacrylate host polymer, must provide stiffness for the embedded leuco-dye so that it can undergo a radiofluorogenic reaction and efficiently emit the fluorescence signal. The environment of the fluorescent molecule should hinder the mobility of the benzene groups of the leuco-dye. The fluorescence is known to be favored in a more viscous medium [35–37], as the energy released by other means, such as vibrations and rotation, becomes suppressed. Previous systems [19,21] employed poly(ethylene glycol) diacrylate (PEGDA) as the host polymer with the addition of 2-hydroxyethyl methacrylate (HEMA) as an additional polymer for increased mechanical stability. However, we observed that the polymer prepared using that composition was very sensitive to UV light, and the coloration process of the dye would start even in daylight conditions after a few minutes. Moreover, HEMA seemed to introduce an increase in the autofluorescence signal seen in the fluorescence spectra, which was due to the hydroxyl group present in the backbone that may undergo a deprotonation process in the presence of amino groups in the leuco-dye structure. Additionally, in order to limit the rotational freedom of the phenyl groups of the dye and improve the rigidity, bulkier substituents should be present in the polymer chain [38], as in the case of poly(propylene glycol) diacrylate (PPGDA), where a single methylene group is introduced. However, polymers with longer side chains are not recommended, since they might favor flexibility and cause a plasticizing effect. The same applies to branched substituent groups—even though they might support the stiffness requirement, the prepolymers were usually extremely viscous, and the cured samples showed restricted diffusion of oxygen, preventing the reaction of the leuco-dye. Hence, the new composition contained PPGDA as a host polymer, which is UV transparent and has good polarity [39], stability and solubility properties [40].

3.2. Sample Preparation: Curing Mold

The photopolymerization process from a liquid solution into a solid sample took place in a specially designed silicone mold, like the one shown in Figure 1a. We observed that the lifetime of the silicone mold was limited to approximately 2 months of daily use. After that, the high-intensity UV light used for curing caused damage to the silicone [41], which started interacting with the polymer solution and drastically worsened the optical quality of the sample surfaces. Therefore, we tested two different 3D-printed molds, presented in Figure 1b. Both of them, however, exhibited sticking of the cured solution inside the 3D-printed mold, with no possibility of removal because of the lack of flexibility of the mold. This was probably due to the polymer solution reacting with some part of the residue resin that was not fully cured during 3D printing. The photopolymerization process of polymer samples used UV light of a wavelength close to the excitation wavelength of the printed resin's photoinitiator. Therefore, it was decided to continue using the silicone mold for the specified lifetime of 2 months, being approximately 100 curing procedures. Nonetheless, other 3D printing resins with higher flexibility and lower hardness could be tested, but this was beyond the scope of this study.

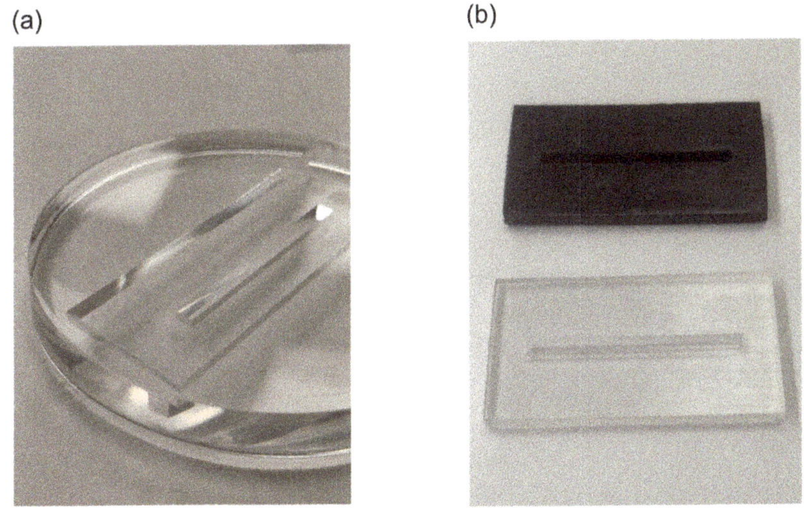

Figure 1. Manufacturing molds: (**a**) silicone mold, (**b**) 3D-printed molds: transparent (bottom) = ABS-like resin; black (top) = traditional resin.

3.3. Material Characteristics

3.3.1. Thermal Stability

TGA measurement of the non-irradiated and irradiated (120 keV e-beam, 30 kGy) samples was carried out to determine the decomposition temperature of the polymer, which was further needed for the selection of experimental conditions (temperature range) for the DSC measurements. The weight loss versus temperature curves for the non-irradiated and irradiated material can be seen in Figure 2a,b, respectively. The thermal treatment caused the mass of the polymer to change continuously, and rapid weight loss of the material started to occur at around 350 °C, followed by a complete, one-step decomposition at around 380 °C for both the non-irradiated and irradiated samples. In the end, after heating to 900 °C, there were no residues left in the crucible, leaving the material fully degraded.

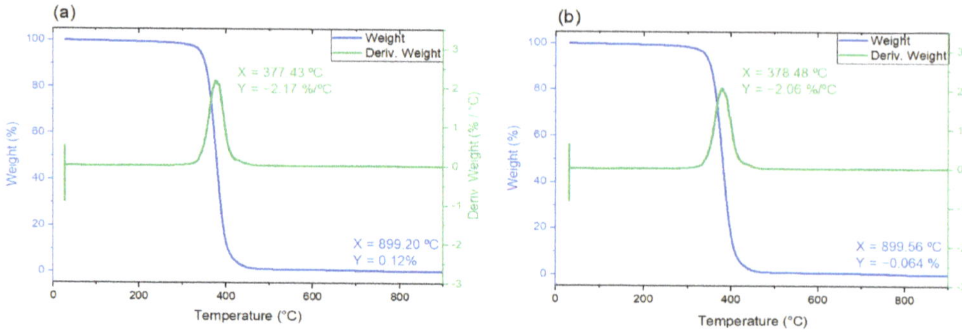

Figure 2. TGA curves of: (**a**) non-irradiated and (**b**) irradiated (120 keV, 30 kGy) polymer material. Blue = weight (%) vs. temperature (°C); green = the derivative of weight with respect to temperature (%/°C). Green coordinates illustrate the maximum of the derivative of weight, and blue coordinates represent the end weight of the tested sample.

3.3.2. Glass Transition Temperature

The measurement of the heat flow versus temperature of the non-irradiated and irradiated (80 keV e-beam, 30 kGy) polymer material can be seen in Figure 3a,b, respectively. The first heating cycle erased the thermal history of the material and left the determination of the glass transition temperature in the second heating cycle more distinct. The calculated T_g of both the non-irradiated and irradiated material was around −45 °C, with no changes introduced during the irradiation process. The glass transition temperature of the material is below room temperature, which means that the polymer will be in a rubbery state while working at the intended temperature. Above T_g, it will remain soft and flexible [38], and because large elastic deformations are possible, the polymer is actually both tougher and more pliable [26]. This allows for a more profound oxygen diffusion [42], which is needed for the reaction of the leuco-dye to proceed. Furthermore, it provides a rigid polymer matrix, which facilitates the fluorescence emission of the dye [35].

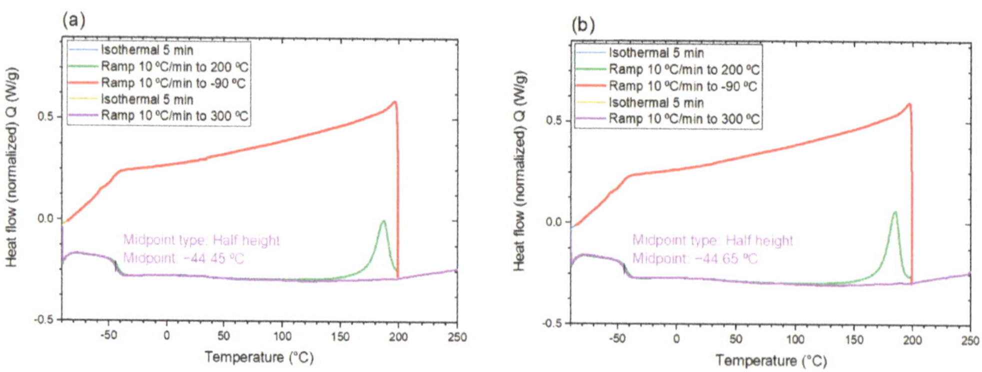

Figure 3. DSC curves of: (**a**) non-irradiated and (**b**) irradiated (80 keV, 30 kGy) polymer material. Green = 1st heating cycle; red = cooling step; violet = 2nd heating cycle. Tg value (midpoint) is presented on the graph in purple.

An exothermic peak appeared during the first heating cycle for both the non-irradiated and irradiated material, with its maximum at around 180 °C. It could have its origin in some additional high-temperature curing reaction of a residual monomer or a reaction commencing due to free radicals trapped in the polymer matrix [43], since the obtained double-bond conversion was of the order of 80%. In Section 3.3.4, it is tested how a post-manufacturing heating at 180 °C may influence the hardness, and consequently the material stiffness.

Additionally, a measurement of the glass transition temperature of the polymer samples irradiated with an 80 keV e-beam and different nominal doses of 10, 30 and 50 kGy was performed. Their second heating cycles are presented in Figure 4. No influence of the irradiation dose on the glass transition temperature was observed, meaning that the polymer material did not undergo any significant chain scission, chain reorganization or cross-linking while exposed to high-dose e-beam radiation. The glass transition temperature is a function of the polymer's molecular weight [44], and it may be altered after exposure to radiation, which is known to cause chain scission and increase the polydispersity of the molecular weight [45].

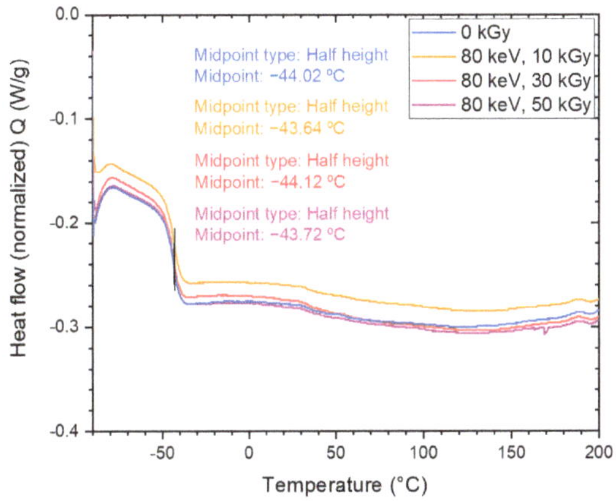

Figure 4. DSC curves: second heating cycle for polymer material irradiated with 80 keV electron beam for 3 doses: 10, 30 and 50 kGy. A curve for non-irradiated polymer (0 kGy) is also presented for comparison.

3.3.3. Refractive Index

The refractive index of a cured polymer sample gives information about what kind of immersion medium has to be used during fluorescence imaging in order to minimize the refractive index mismatch [46]. A mismatch might lead to poor imaging quality, and therefore a correction factor has to be introduced to the axial intensity distribution. A schematic of the readout geometry is presented in Figure 5a, where a sample (magenta) is placed on a scanning stage. The OCT axial scan (B-scan) of the single field of view (FOV) is shown in Figure 5b. The calculated refractive index of the polymer material was $n = 1.47 \pm 0.01$. This finding suggests that fluorescence imaging should be performed in an immersion medium of a refractive index between 1.46 and 1.48, (e.g. glycerol) [47] or a fluorescence immersion oil (Type FF, Cargille-Sacher Laboratories). Another possibility would be to correct the introduced axial distortion, caused by the difference between the observed optical thickness and actual thickness, by multiplying the axial sampling by 1.47 (assuming air of $n = 1$ as the surrounding medium). It was also observed that the material was optically clear, and its properties were not dominated by scattering. The thin line visible in the middle of the sample (Figure 5b) might be a reflection of the top of the sample from the glass scanning stage.

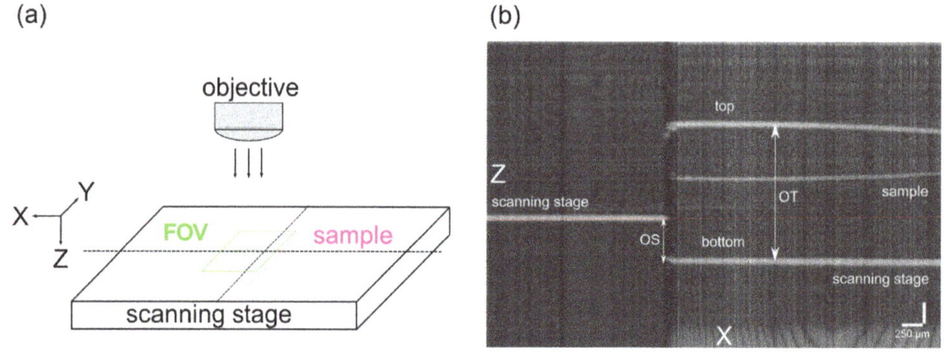

Figure 5. (**a**) The scheme of the OCT readout. Sample (magenta) placed on a scanning stage, and XY field of view (FOV, green) is chosen for the axial profile measurement. (**b**) OCT B-scan of the polymer sample for RI measurement using optical path shifting method. OT = optical thickness of the sample; OS = optical shift of the scanning stage.

However, the refractive index value is dependent on the wavelength [48], a phenomenon known as dispersion. Hence, the value measured at 1300 nm might be different, in this case lower from the value at the excitation wavelength of 555 nm. This discrepancy may be a cause of additional uncertainty introduced due to imaging in a medium with a different refractive index or when correcting for the axial distortion.

3.3.4. Hardness

The hardness of the polymer material was tested to evaluate if additional post-fabrication heating at a high temperature (180 °C) would be needed for full curing (monomer conversion) and stabilization of the remaining free radicals. The expectation was that the additional heating step would increase the stiffness of the polymer by increasing the crosslink density [49] and ultimately increase its hardness. As can be seen in Figure 6a, the material's Shore A value was measured to be 57 ± 3, and after heating at 180 °C for 10 min, the value was 60 ± 6. No significant difference in hardness was observed. Moreover, an undesirable color formation was obtained after heating, which could worsen the optical properties of the material and hinder the leuco-dye reaction when irradiated. Hence, high-temperature heating after manufacturing was contraindicated. It is possible that yellowing of the sample was caused by the formation of benzil from benzoyl radicals, which were produced during the photopolymerization process of TPO [50].

3.3.5. Rheology

In order to evaluate the structural and viscoelastic characteristics of the polymer material, amplitude and frequency sweeps were performed, as seen in Figure 7a,b, respectively. It can be noted that the storage modulus G' was larger than the loss modulus G'' for all applied oscillation strains (Figure 7a), indicating that the elastic behavior of the material dominated over the viscous behavior. The permissible maximum strain frequency sweep measurement was chosen to be $\gamma = 0.25\%$, and the value agreed with the estimation for cross-linked polymers of $\gamma \leq 1\%$ [51,52]. Higher strains could cause an irreversible deformation, with a tendency toward viscous-like behavior. The variation in loss modulus values might be due to axial force changes or thickness variation of the sample introduced during the photocuring step. The differences in thickness could be a result of resin shrinkage during the photopolymerization reaction, which was caused by the reduction of distance between the monomers while shifting from longer double-bonds (van der Waals) to shorter covalent bonds [53].

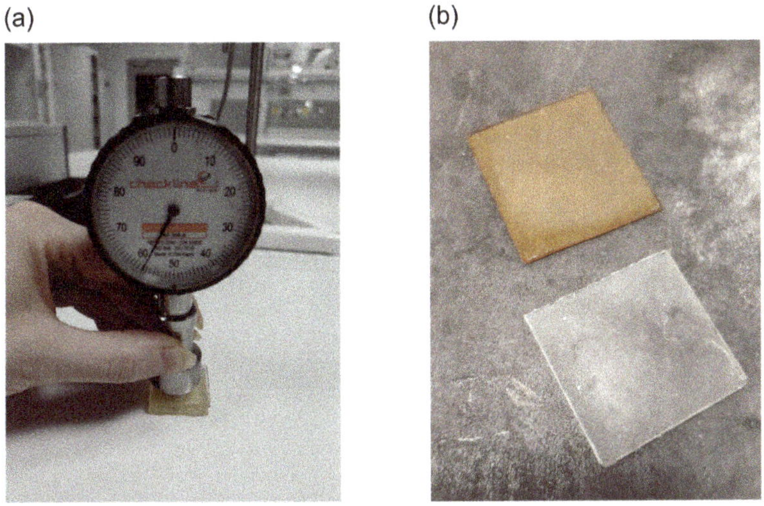

Figure 6. (**a**) Shore A measurement of cured, non-irradiated polymer material. (**b**) Polymer before (bottom) and after (top) post-manufacturing heating at 180 °C for 10 min.

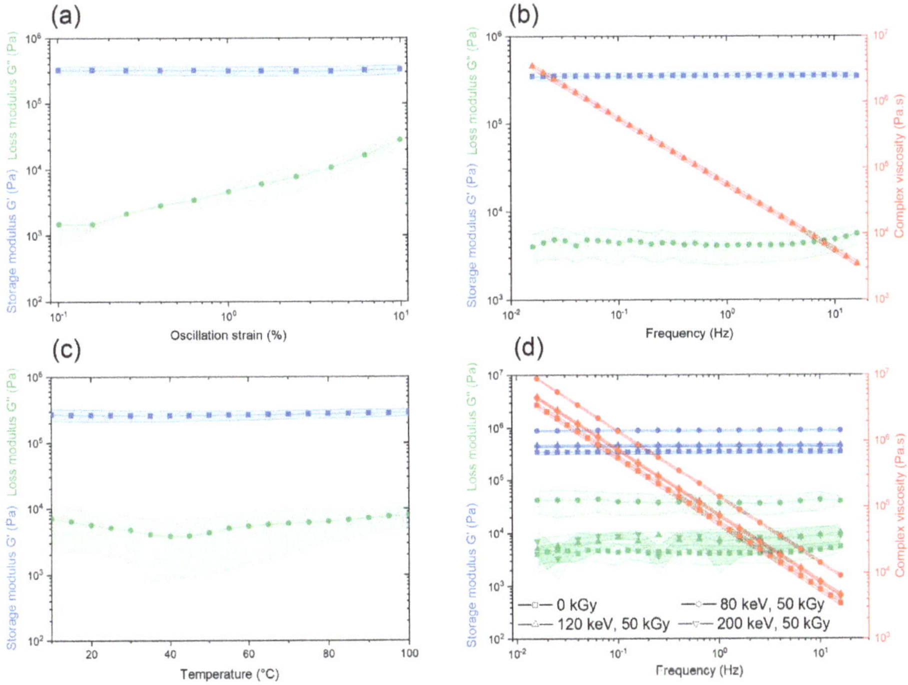

Figure 7. Rheology results: (**a**) amplitude sweep of non-irradiated polymer material with measurement performed in triplicate, (**b**) frequency sweep of non-irradiated polymer material with measurement performed in triplicate, (**c**) temperature sweep of non-irradiated polymer material with measurement performed in triplicate and (**d**) frequency sweep of irradiated polymer material with measurement performed in triplicate. Blue = storage modulus G′ (Pa); green = loss modulus G″ (Pa); red = complex viscosity (Pa·s). Error bar areas are filled under the curve.

Figure 7b shows the dependency of the storage and loss moduli on the frequency at a strain of 0.25%. Both G′ and G″ displayed almost parallel lines in the whole frequency range, showing the behavior of closely cross-linked polymers [51]. Moreover, G′ was high in relation to G″, and the loss tangent was tan $\varphi = \frac{G''}{G'} \approx 0.01$. Even at lower frequencies, the values of G′ and G″ remained constant, showing a high structural strength (rigidity). From the polymer point of view, rigidity of the polymer matrix is required for efficient fluorescence emission of the immobilized dye [19,35].

The variations of the elastic and viscous moduli versus temperature for the non-irradiated polymer were also tested, and the curves are presented in Figure 7c. Here, G′ > G″ also remained valid over the whole temperature range. Above the glass transition temperature of the material, only a limited deformation of the polymer network is possible, depending on the length and density of the bonds. The material did not reach a liquid state even at high temperatures due to the strong chemical bonds that form the polymer network.

The frequency sweep of an irradiated polymer material was also tested, as presented in Figure 7d, for the samples irradiated with 50 kGy and electron beam energies of 80, 120 and 200 keV. Here as well, both the storage and loss moduli took the form of parallel lines in the chosen frequency range, with G′ always being much greater than G″. The irradiation process did not cause any significant structural changes to the material. A slight increase in G′ could be observed for the irradiated samples, which could be an indication of probable cross-linking reactions taking place at these very high doses [54] and a lack of chain scissions caused by the irradiation process [55,56]. All of this demonstrates the suitability of the chosen polymer matrix as a radiation dosimeter for low-energy electron beams.

3.3.6. Water Affinity

Determination of the wettability of the polymer samples is extremely important for fluorescence microscopy measurements when using an immersion medium, because an inadequate choice of a surrounding medium might result in sample swelling or altering of its properties. As a result, additional uncertainties in fluorescence intensity measurements would be introduced. Additionally, when air is utilized as an imaging medium, the ability of the microscope objective to gather light and resolve fine details is limited by its numerical aperture, which in practice cannot be more than 1. However, by increasing the imaging medium's refractive index to, for example, 1.33 for water, it is possible to obtain a higher numerical aperture, and ultimately, more details can be distinguished in a sample. Hence, we determined the water affinity of the polymer surfaces by measuring the water contact angle (WCA) as shown in Figure 8a. The WCA was measured for both the non-irradiated and irradiated (200 keV, 30 kGy) material, and the average values are presented in Figure 8b. It can be seen that none of the values exceeded 90°, indicating that the surfaces were hydrophilic [57] and had a high affinity with water. There was no significant difference between the top, bottom and side of the sample, indicating uniformity of the manufactured samples. Such a hydrophilic polymer network is similar to a hydrogel, which exhibits a swelling ability in water but will not dissolve when immersed in it [58]. Moreover, the affinity to water did not change after irradiation, which is sometimes introduced deliberately for other polymers by using electron beam irradiation [59]. Based on the obtained results, it is clear that the use of water immersion objectives should be avoided. It also suggests that the polymer material should be stored in a low-humidity environment.

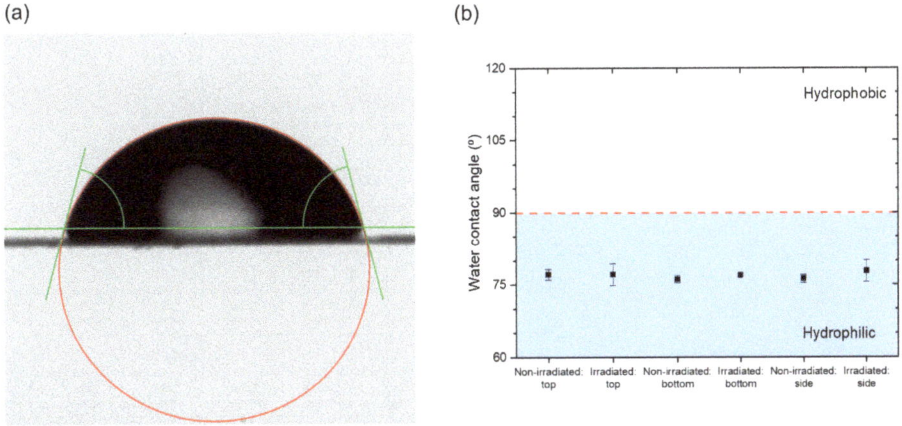

Figure 8. Water contact angle (WCA) measurements. (**a**) A single water drop placed on a polymer surface. The green horizontal line indicates a baseline, and the red circle is the circle of interest. The WCAs (left and right) are marked. (**b**) WCAs of non-irradiated and irradiated (200 keV, 30 kGy) polymer material (top, bottom and side). The red dashed line marks a 90° angle where the transition from hydrophilic to hydrophobic properties takes place [57].

3.4. Fluorescence Signal

3.4.1. Fluorescence from the Dye Dissolved in a Liquid and Solid

As a proof of concept, the absorbance and fluorescence spectra of the dye dissolved in a liquid and a solid sample were obtained, as can be seen in Figure 9a,b, respectively. The non-irradiated acetonitrile solution did not absorb or fluoresce, as expected, as is shown in the dashed black and red curves in Figure 9a. Then, once irradiated with a 200 keV e-beam to 30 kGy, an absorption peak with a maximum at around 555 nm was present, but there was no fluorescence emission (solid lines of the same figure). This observation agrees with the theory that triphenylmethane dyes need to be held in some fixed configuration when excited [35], and this was supported by the observation that after placing the dye in a stiff polymer matrix, a fluorescence signal with a peak at around 600 nm was detected, as presented in Figure 9b (solid red line). Additionally, a slight increase in the fluorescence signal for the non-irradiated sample was observed (dashed red line), which might have been due to the background absorbance of the sample at the excitation wavelength [60]. At the peak fluorescence value, the fluorescence intensity of the non-irradiated sample was 5% of the value of the sample irradiated with 30 kGy. This observation may have an influence on the dose-dependent fluorescence signal and the assessment of the dose, and it should be investigated in a separate study.

3.4.2. Absorbance and Fluorescence Spectra from an 80 keV E-Beam-Irradiated Polymer

The excitation and emission spectra of the dosimeter material were measured for the samples (Type 3) irradiated with an 80 keV e-beam to doses of 15, 20, 30, 40 and 50 kGy, as shown in Figure 10a. The absorbance and fluorescence peak values increased with the absorbed dose. It should be noted that the penetration depth of the 80 keV e-beam is very small, of the order of 60 μm in water [2], and so the material needs to have good sensitivity. The noise present in the spectra may have its origin in a very high resolution (1 nm) set for the measured signal while keeping the excitation light bandwidth comparatively broad (5 nm). The background subtraction of the well plate and normalization process also introduced unwanted signal fluctuations. A possible way to reduce these would be to perform additional measurements, which could average out the noise.

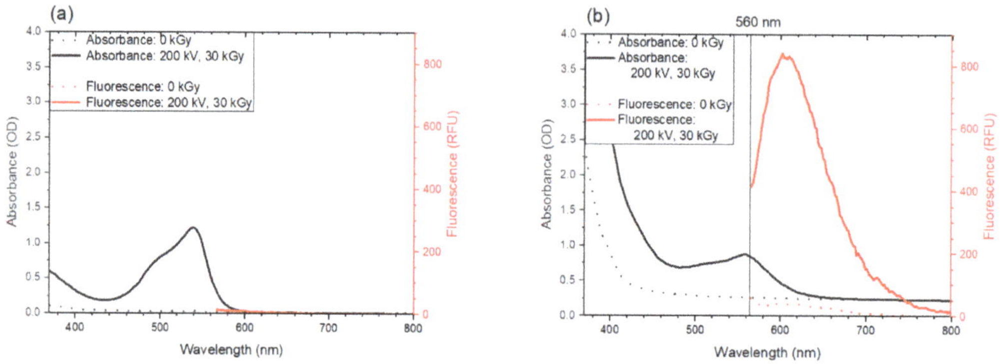

Figure 9. Absorbance and fluorescence spectra of the dye placed in: (**a**) a liquid solution of acetonitrile and (**b**) a polymer (solid) sample. Absorbance spectra are depicted as black solid and dashed lines for irradiated and non-irradiated samples, respectively. Fluorescence spectra are presented as red solid and dashed lines for irradiated and non-irradiated samples, respectively.

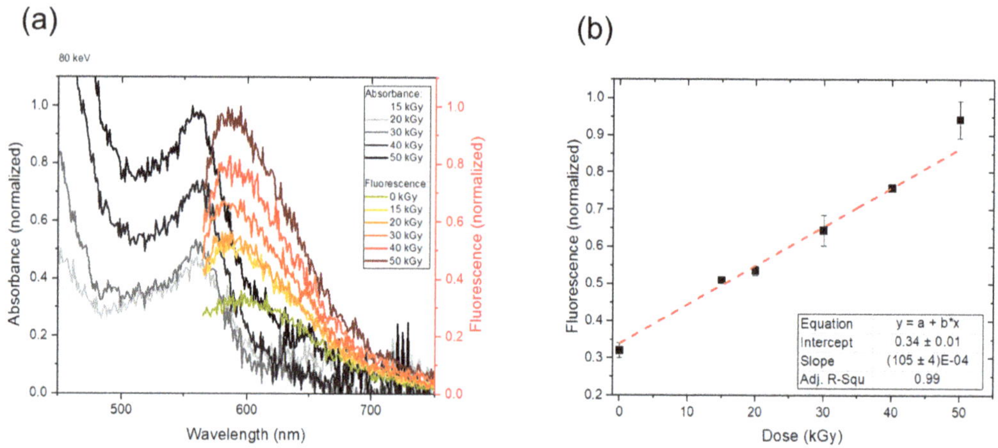

Figure 10. (**a**) Absorbance (grayscale) and fluorescence (red scale) spectra of polymer material irradiated with 80 keV electron beam of different doses. The spectra were normalized to the maximum values of the 50 kGy sample. (**b**) Peak fluorescence intensity versus dose for the polymer irradiated with 80 keV electron beam. The data points were fitted with a linear function of the parameters presented in the Table.

The peak fluorescence intensity at 600 nm increased linearly with the dose as shown in Figure 10b, demonstrating the proof of concept of the dosimeter material. The trend is in agreement with the theory that at low concentrations of analyte, the emission intensity is proportional to its concentration [61]. However, a significant part of the fluorescence signal could also be observed for a non-irradiated material, introducing a non-zero background fluorescence signal. This would need to be subtracted from the fluorescence intensity measurements for correct dose estimation. However, this will cause the signal-to-noise ratio to decrease, and a part of the meaningful information might be lost. Another possibility, therefore, would be to include the background signal in the calibration of the material and thereby correct for the introduced offset. A following paper will describe the fluorescence imaging results, including the background information concerns.

4. Discussion

The development of a polymer material for its use as a radiofluorogenic dosimeter consisted of several interrelated steps, and only careful evaluation of each of them allowed for the choice of a final material with the necessary characteristics needed for fluorescence imaging using confocal laser scanning microscopy.

In terms of composition, the choice of a proper solvent for the radiation-sensitive dye was important, and acetonitrile showed the best performance. The dose-dependent fluorescence intensity is a function of the dye concentration, so any errors introduced already at this stage would propagate further and cause artifacts such as shadowing and saturation problems in fluorescence microscopy imaging.

During the selection of a host polymer, a compromise between rigidity and diffusion had to be met for an efficient reaction of the leuco-dye. Both acrylates and methacrylates were initially considered as candidates. However, the photopolymerization rates for methacrylates were much lower due to the stabilizing effect of the α-methyl group [62]. The functionality of a monomer was also a deciding factor, as higher functionalities led to higher reaction rates and faster vitrification. However, a higher cross-link density lowered the conversion, so more of the uncured monomer remained trapped between the cross-links [63]. Higher functionality of a monomer also means its viscosity is greater, which limits the easy handling of a solution. Therefore, lower functionalities were chosen, which due to slower vitrification enabled in-depth curing with less shrinkage and enhanced mobility of the system [64]. The length and nature of a spacer group were also considered, with a conclusion of a not-too-bulky and long spacer group between unsaturations. Poly(propylene glycol) diacrylate turned out to be an excellent candidate for the host polymer, but similar monomers could be also considered.

The photopolymerization process is responsible for the quality and uniformity of the produced samples. Hence, the lifetime of used silicone molds was limited to 2 months of use, corresponding to approximately 100 curing procedures. Even though the attempted use of 3D-printed molds was not successful here, they presented a great potential for a new method of sample production by using, for example, highly flexible 3D printing resins.

The polymer's glass transition temperature, measured to be −44.45 °C, indicated that the material remained in a rubbery state at room temperature. Low T_g is also a sign of a higher cure extent [65]. Such an elastic material could be compared by its hardness of 57 ± 3 Shore A to a pencil eraser, indicating both toughness and pliability. The domination of elasticity was confirmed in the rheology measurements, where the elastic portion (G') was around 100 times greater than the viscous one (G'') in the chosen frequency range, a feature similar to thermosets and closely cross-linked materials. Additionally, high thermal stability (up to 350 °C) was achieved in an inert atmosphere. This would be lowered once oxygen is introduced into the system. It is a great advantage that none of these features were influenced by the absorbed dose, meaning that the radiation-induced defects were not the most significant obstacles.

For fluorescence imaging, the assessment of the refractive index and the water affinity is crucial. Thus, a refractive index matching medium cannot be water due to the hydrophilic properties of the polymer, and it should be chosen so that $n = 1.47 \pm 0.01$. The refractive index was measured at 1300 nm, and it might be somewhat greater at the excitation wavelength of 555 nm due to chromatic dispersion [48].

All of the above had an impact on the observed fluorescence signal; a molecule non-fluorescent in liquid became fluorescent once embedded in a rigid polymer matrix. The solid state polymer restricted the motion of the benzene groups of the leuco-dye. Irradiation with the low-energy electron beam at high doses caused the fluorescence signal to increase linearly with the dose, as expected, for increasing concentration of excited fluorophores. The background signal will have to be taken into account during fluorescence imaging, which can be solved by either its subtraction or inclusion in calibration. The first one may limit the dynamic range and lower the signal-to-noise ratio, whereas the second might complicate the readout procedure.

There exists a demand for the development of new radiofluorogenic dosimeters, especially in the medical dosimetry field [66–70]. Typically, radiofluorogenic gels are investigated, which show a linear increase in fluorescence intensity with the dose (Gy) but often lack system stability due to diffusion of the radiation products, oxygen interference and dose rate dependency. Our polymer material, even though intended to work with much higher doses, brings advantages in terms of mechanical and thermal stability, as well as the possibility of 3D signal detection when imaged using confocal laser scanning microscopy. The limitations of our study, such as low measurement reproducibility and a limited number of tested compositions and characterization techniques, should be accounted for in future studies of the radiofluorogenic polymer dosimeter.

5. Conclusions

The present study characterized a novel polymer material as a dosimeter candidate for high-dose, low-energy electron beam irradiations widely used for product surface sterilizations. The current lack of a dosimetry system being able to resolve the information about the surface dose might be overcome with the proposed solid state radiofluorogenic polymer material, which responds linearly to the absorbed dose and shows properties not harmed by it. This material is the first of its kind to deliver information about the absorbed dose distribution in 3D, which is to be measured using confocal microscopy.

The properties of the proposed polymer material were analyzed in order to establish the required characteristics for a reliable solid state dosimeter. Acetonitrile showed the best performance in dissolving the leuco-dye. Improved rigidity was achieved by using poly(propylene) glycol diacrylate as a host polymer. The thermal decomposition of the polymer was measured by TGA to occur at 350 °C and was not influenced by the irradiation process. The glass transition temperature determined from DSC measurement was -44.45 ± 1.35 °C with no changes after irradiation. A 57 ± 3 Shore A hardness was determined, and high-temperature heating was not needed for improved conversion. The refractive index of the material, evaluated using OCT, was equal to $n = 1.47 \pm 0.01$. A closely cross-linked polymer, with a dominating storage modulus ($G' > G''$) in the analyzed frequency and temperature range, was found. A water contact angle of less than 90° was measured, indicating a hydrophilic material. The fluorescence intensity of the dye, which was not seen for the liquid system, increased linearly with the absorbed dose, although it exhibited a non-zero background signal. To conclude, the polymer demonstrated promising characteristics for use as a dosimeter in low-energy electron beams. The readout of the fluorescence signals will be further investigated in a following paper.

Author Contributions: Conceptualization, M.M.S., L.R.L., A.L.S. and A.M.; methodology, L.Y.; validation, L.Y.; formal analysis, M.M.S.; investigation, M.M.S., L.Y. and C.A.; resources, L.R.L.; writing—original draft preparation, M.M.S.; writing—review and editing, C.A., L.R.L., A.L.S., A.M. and L.Y.; visualization, M.M.S.; supervision, L.R.L., C.A., A.L.S. and A.M.; project administration, L.R.L. All authors have read and agreed to the published version of the manuscript.

Funding: This research received no external funding.

Institutional Review Board Statement: Not applicable.

Informed Consent Statement: Not applicable.

Data Availability Statement: The data presented in this study are available on request from the corresponding author.

Acknowledgments: We acknowledge 3DWayTech for their help with the 3D printing of the sample molds, DTU Health Tech for access to the Tecan Spark M10 plate reader, DTU Chemical Engineering for permission to use the equipment needed for material characterization—TGA, DSC, Rheometry and WCA, DTU Health Tech Biophotonic Imaging Group for access to the OCT, and R&D Engineer M. Bailey for proofreading the article.

Conflicts of Interest: The authors declare no conflict of interest.

References

1. Auslender, V.L.; Berejka, A.J.; Bol, J.L.; Brinston, R.; Bryazgin, A.; Calvo, W.A.P.; Chmielewski, A.; Cleland, M.R.; Cokragan, A.; Ehlermann, D.A.E.; et al. *Industrial Radiation with Electron Beams and X-rays*; IIA: Shropshire, UK, 2011; Available online: http://iiaglobal.com/uploads/documents/Industrial_Radiation_eBeam_Xray.pdf (accessed on 19 December 2021).
2. *ISO/ASTM 51818*; Standard Practice for Dosimetry in an Electron Beam Facility for Radiation Processing at Energies between 80 and 300 keV. ISO: Geneva, Switzerland, 2020.
3. Ando, M.; Uryu, T. Synthesis of polymer materials by low energy electron beam. IV. EB-polymerized urethane-acrylate, -methacrylate and -acrylamide. *Int. J. Radiat. Appl. Instrum. Part C. Radiat. Phys. Chem.* **1988**, *31*, 607–614. [CrossRef]
4. Läuppi, U.V. Electron-Beam Processing for Industrial Inkjet Printing: Cross-Linking and Curing. In *Handbook of Industrial Inkjet Printing*; Wiley-VCH Verlag GmbH & Co. KGaA: Weinheim, Germany, 2017; pp. 543–556. Available online: https://onlinelibrary.wiley.com/doi/10.1002/9783527687169.ch31 (accessed on 1 March 2022).
5. Scharf, W.; Wieszczycka, W. Electron accelerators for industrial processing—A review. *AIP Conf. Proc.* **1999**, *475*, 949–952. [CrossRef]
6. Wetzel, C.; Schönfelder, J.; Schwarz, W.; Funk, R.H.W. Surface modification of polyurethane and silicone for therapeutic medical technics by means of electron beam. *Surf. Coat. Technol.* **2010**, *205*, 1618–1623. [CrossRef]
7. Gotzmann, G.; Portillo, J.; Wronski, S.; Kohl, Y.; Gorjup, E.; Schuck, H.; Rögner, F.H.; Müller, M.; Chaberny, I.F.; Schönfelder, J.; et al. Low-energy electron-beam treatment as alternative for on-site sterilization of highly functionalized medical products—A feasibility study. *Radiat. Phys. Chem.* **2018**, *150*, 9–19. [CrossRef]
8. Berejka, A.J.; Cleland, M.R.; Walo, M. The evolution of and challenges for industrial radiation processing—2012. *Radiat. Phys. Chem.* **2014**, *94*, 141–146. [CrossRef]
9. Berejka, A.J. Irradiation processing in the '90's: Energy savings and environmental benefits. *Radiat. Phys. Chem.* **1995**, *46*, 429–437. [CrossRef]
10. Felis, K.P.; Avnery, T.; Berejka, A.J. Innovative energy efficient low-voltage electron beam emitters. *Radiat. Phys. Chem.* **2002**, *63*, 605–608. [CrossRef]
11. *ISO 11137-3*; Sterilization of Health Care Products—Radiation—Part 3: Guidance on Dosimetric Aspects of Development, Validation and Routine Control. ISO: Geneva, Switzerland, 2017.
12. Miller, A.; Helt-Hansen, J.; Gondim, O.; Tallentire, A. *Guide on the Use of Low Energy Electron Beams for Microbiological Decontamination of Surfaces*; Panel on Gamma and Electron Irradiation: London, UK, 2013; pp. 1–34.
13. ICRU. *ICRU Report 80: Dosimetry Systems for Use in Radiation Processing*; ICRU: Washington, DC, USA, 2008; Volume 8. [CrossRef]
14. Helt-Hansen, J.; Miller, A.; Duane, S.; Sharpe, P.; McEwen, M.; Clausen, S. Calorimetry for dose measurement at electron accelerators in the 80-120 keV energy range. *Radiat. Phys. Chem.* **2005**, *74*, 354–371. [CrossRef]
15. Helt-Hansen, J.; Miller, A.; Sharpe, P. Dose response of thin-film dosimeters irradiated with 80-120 keV electrons. *Radiat. Phys. Chem.* **2005**, *74*, 341–353. [CrossRef]
16. Helt-Hansen, J.; Miller, A.; Sharpe, P.; Laurell, B.; Weiss, D.; Pageau, G. Dμ-A new concept in industrial low-energy electron dosimetry. *Radiat. Phys. Chem.* **2010**, *79*, 66–74. [CrossRef]
17. Schuster, C.; Kuntz, F.; Strasser, A.; Härtling, T.; Dornich, K.; Richter, D. 3D relative dose measurement with a μm thin dosimetric layer. *Radiat. Phys. Chem.* **2021**, *180*, 109238. [CrossRef]
18. Bernal-Zamorano, M.R.; Sanders, N.H.; Lindvold, L.R.; Andersen, C.E. Radiochromic and radiofluorogenic 3D solid polymer dosimeter: Initial results for high doses. *J. Phys. Conf. Ser.* **2017**, *847*, 012016. [CrossRef]
19. Bernal-Zamorano, M.R.; Sanders, N.H.; Lindvold, L.; Andersen, C.E. Radiochromic and radiofluorogenic 3D solid polymer dosimeter; effect of the photoinitiator. *Radiat. Meas.* **2017**, *106*, 192–195. [CrossRef]
20. Bernal-Zamorano, M.R.; Sanders, N.H.; Lindvold, L.; Andersen, C.E. Radiochromic and radiofluorogenic solid state polymer dosimeter; a third signal: Electron Paramagnetic Resonance (EPR). *Radiat. Phys. Chem.* **2019**, *161*, 72–76. [CrossRef]
21. Skowyra, M.M.; Lindvold, L.R. Radio-fluorogenic, solid-state polymer dosimeter for recording depth-dose profiles related to radiation processing of surfaces. *Radiat. Meas.* **2020**, *136*, 106383. [CrossRef]
22. Ligon, S.C.; Husár, B.; Wutzel, H.; Holman, R.; Liska, R. Strategies to reduce oxygen inhibition in photoinduced polymerization. *Chem. Rev.* **2014**, *114*, 577–589. [CrossRef]
23. Decker, C.; Moussa, K. Real-Time Kinetic Study of Laser-Induced Polymerization. *Macromolecules* **1989**, *22*, 4455–4462. [CrossRef]
24. ELEGOO Standard Resin. Available online: https://www.elegoo.com/products/elegoo-standard-resin (accessed on 8 October 2021).
25. ELEGOO ABS-like Resin. Available online: https://www.elegoo.com/products/elegoo-abs-like-resin (accessed on 8 October 2021).
26. Ravve, A. *Principles of Polymer Chemistry*, 3rd ed.; Springer: Niles, IL, USA, 2012; ISBN 9781461422112.
27. *ASTM E1356-08*; Standard Test Method for Assignment of the Glass Transition Temperatures by Differential Scanning Calorimetry. ASTM: West Conshohocken, PA, USA, 2008.
28. Wang, X.; Zhang, C.; Zhang, L.; Xue, L.; Tian, J. Simultaneous refractive index and thickness measurements of bio tissue by optical coherence tomography. *J. Biomed. Opt.* **2002**, *7*, 628–632. [CrossRef]
29. Kim, Y.L.; Walsh, J.T.; Goldstick, T.K.; Glucksberg, M.R. Variation of corneal refractive index with hydration. *Phys. Med. Biol.* **2004**, *49*, 859–868. [CrossRef]

30. *ASTM D2240-15*; Standard Test Method for Rubber Property—Durometer Hardness. ASTM: West Conshohocken, PA, USA, 2021.
31. Haynes, W.M.; Lide, D.R.; Bruno, T.J. *CRC Handbook of Chemistry and Physics*, 95th ed.; CRC Press: Boca Raton, FL, USA, 2014.
32. Miller, A.; McLaughlin, W.L. *On a Radiochromic Dye Dosemeter*; Risø National Laboratory: Roskilde, Denmark, 1980; pp. 1–34. ISBN 8755007376.
33. National Center for Biotechnology Information. PubChem Compound Summary for CID 180, Acetone. Available online: https://pubchem.ncbi.nlm.nih.gov/compound/Acetone (accessed on 21 June 2021).
34. Reichardt, C. Solvent Effects on the Absorption Spectra of Organic Compounds. In *Solvents and Solvent Effects in Organic Chemistry*; Wiley-VCH Verlag GmbH & Co. KGaA: Weinheim, Germany, 2003; pp. 329–388. ISBN 3527306188. Available online: https://onlinelibrary.wiley.com/doi/10.1002/3527601791.ch6 (accessed on 1 March 2022).
35. Oster, G.; Joussot-Dubien, J.; Broyde, B. Photoreduction of Dyes in Rigid Media. I. Triphenylmethane Dyes. *J. Am. Chem. Soc.* **1959**, *81*, 1869–1872. [CrossRef]
36. Valeur, B.; Berberan-Santos, M.-N. *Molecular Fluorescence. Principles and Applications*, 2nd ed.; Wiley-VCH Verlag GmbH & Co.: Weinheim, Germany, 2012.
37. Vogel, M.; Rettig, W. Excited State Dynamics of Triphenylmethane-Dyes used for Investigation of Microviscosity Effects. *Ber. Bunsenges. Phys. Chem.* **1987**, *91*, 1241–1247. [CrossRef]
38. Stevens, M.P. *Polymer Chemistry: An Introduction*, 3rd ed.; Oxford University Press: New York, NY, USA, 1999; ISBN 978-0-19-512444-6.
39. Holmes, E.O. The effect of the properties of solvents of various dielectric constants and structures on the photoionization of the leucocarbinols and leucocyanides of malachite green, crystal violet, and sunset orange and related phenomena. *J. Phys. Chem.* **1966**, *70*, 1037–1046. [CrossRef]
40. Malucelli, G.; Gozzelino, G.; Ferrero, F.; Bongiovanni, R.; Priola, A. Synthesis of poly(propylene-glycol-diacrylates) and properties of the photocured networks. *J. Appl. Polym. Sci.* **1997**, *65*, 491–497. [CrossRef]
41. Efimenko, K.; Wallace, W.E.; Genzer, J. Surface modification of Sylgard-184 poly(dimethyl siloxane) networks by ultraviolet and ultraviolet/ozone treatment. *J. Colloid Interface Sci.* **2002**, *254*, 306–315. [CrossRef] [PubMed]
42. Compañ, V.; Lopez, M.L.; Monferrer, L.; Garrido, J.; Riande, E.; San Roman, J. Determination of the glass transition temperature of poly(cyclohexyl acrylate) from oxygen permeability measurements. *Polymer* **1993**, *34*, 2971–2974. [CrossRef]
43. Chartoff, R. Thermal characteristics of thermosets formed by free radical photocuring. *J. Therm. Anal. Calorim.* **2006**, *85*, 213–217. [CrossRef]
44. Fox, T.G.; Flory, P.J. Second-order transition temperatures and related properties of polystyrene. I. Influence of molecular weight. *J. Appl. Phys.* **1950**, *21*, 581–591. [CrossRef]
45. Dalnoki-Veress, K.; Forrest, J.A.; Murray, C.; Gigault, C.; Dutcher, J.R. Molecular weight dependence of reductions in the glass transition temperature of thin, freely standing polymer films. *Phys. Rev. E* **2001**, *63*, 031801. [CrossRef]
46. Jacobsen, H.; Hell, S.W. Effect of the specimen refractive index on the imaging of a confocal fluorescence microscope employing high aperture oil immersion lenses. *Bioimaging* **1995**, *3*, 39–47. [CrossRef]
47. Birkhoff, R.D.; Painter, L.R.; Heller, J.M. Optical and dielectric functions of liquid glycerol from gas photoionization measurements. *J. Chem. Phys.* **1978**, *69*, 4185–4188. [CrossRef]
48. Hartmann, P. Refractive Index and Dispersion. In *Optical Glass*; Society of Photo-Optical Instrumentation Engineers: Bellingham, WA, USA, 2014; pp. 33–76. Available online: http://ebooks.spiedigitallibrary.org/content.aspx?bookid=968&sectionid=65642104 (accessed on 1 March 2022).
49. Gilman, J.J. *Chemistry and Physics of Mechanical Hardness*; John Wiley & Sons, Inc.: Hoboken, NJ, USA, 2009; ISBN 9780470446836.
50. Arsu, N.; Stephen Davidson, R.; Holman, R. Factors affecting the photoyellowing which occurs during the photoinitiated polymerization of acrylates. *J. Photochem. Photobiol. A Chem.* **1995**, *87*, 169–175. [CrossRef]
51. Mezger, T.G. *The Rheology Handbook*, 2nd ed.; Vincentz Network: Hannover, Germany, 2006; ISBN 3-87870-174-8.
52. *ASTM D4473-03*; Standard Test Method for Plastics: Dynamic Mechanical Properties: Cure Behavior. ASTM: West Conshohocken, PA, USA, 2003.
53. Koseki, K.; Sakamaki, H.; Jeong, K.M. In situ measurement of shrinkage behavior of photopolymers. *J. Photopolym. Sci. Technol.* **2013**, *26*, 567–572. [CrossRef]
54. Ferry, M.; Ngono-Ravache, Y.; Aymes-Chodur, C.; Clochard, M.C.; Coqueret, X.; Cortella, L.; Pellizzi, E.; Rouif, S.; Esnouf, S. Ionizing Radiation Effects in Polymers. In *Reference Module in Materials Science and Materials Engineering*; Elsevier: Amsterdam, The Netherlands, 2016; Available online: https://linkinghub.elsevier.com/retrieve/pii/B9780128035818020956 (accessed on 9 September 2020).
55. Miller, A.A.; Lawton, E.J.; Balwit, J.S. Eflect of Chemical Structure of Vinyl Polymers on Crosslinking and Degradation by Ionizing Radiation. *J. Polym. Sci.* **1954**, *XIV*, 503–504. [CrossRef]
56. Lyons, B.J.; Weir, F.E. The Effect of Radiation on the Mechanical Properties of Polymers. In *The Radiation Chemistry of Macromolecules*; Elsevier: Amsterdam, The Netherlands, 1973; pp. 281–306. Available online: https://linkinghub.elsevier.com/retrieve/pii/B9780122198021500216 (accessed on 1 March 2022).
57. Law, K.-Y. Definitions for Hydrophilicity, Hydrophobicity, and Superhydrophobicity: Getting the Basics Right. *J. Phys. Chem. Lett.* **2014**, *5*, 686–688. [CrossRef]
58. Ahmed, E.M. Hydrogel: Preparation, characterization, and applications: A review. *J. Adv. Res.* **2015**, *6*, 105–121. [CrossRef]

59. Clough, R.L. High-energy radiation and polymers: A review of commercial processes and emerging applications. *Nucl. Instrum. Methods Phys. Res. Sect. B* **2001**, *185*, 8–33. [CrossRef]
60. Kasha, M. Characterization of electronic transitions in complex molecules. *Discuss. Faraday Soc.* **1950**, *9*, 14–19. [CrossRef]
61. Harris, D.C. *Quantitative Chemical Analysis*, 8th ed.; W.H. Freeman and Co.: New York, NY, USA, 2007; ISBN 9781429218153.
62. Odian, G. *Principles of Polymerization*, 4th ed.; John Wiley & Sons, Inc: New York, NY, USA, 2004; ISBN 3175723993.
63. Andrzejewska, E. Photopolymerization kinetics of multifunctional monomers. *Prog. Polym. Sci.* **2001**, *26*, 605–665. [CrossRef]
64. Kurdikar, D.L.; Peppas, N.A. A kinetic study of diacrylate photopolymerizations. *Polymer* **1994**, *35*, 1004–1011. [CrossRef]
65. Andrzejewska, E. Free-radical photopolymerization of multifunctional monomers. In *Three-Dimensional Microfabrication Using Two-Photon Polymerization*; Elsevier: Amsterdam, The Netherlands, 2020; pp. 77–99. ISBN 9780128178270.
66. Sandwall, P.A.; Bastow, B.P.; Spitz, H.B.; Elson, H.R.; Lamba, M.; Connick, W.B.; Fenichel, H. Radio-fluorogenic gel dosimetry with coumarin. *Bioengineering* **2018**, *5*, 53. [CrossRef]
67. Maeyama, T.; Kato, A.; Mochizuki, A.; Sato, N.; Watanabe, Y. Dose-rate-independent and diffusion-free nanoclay-based radio-fluorogenic gel dosimeter. *Sens. Actuators A* **2019**, *298*, 111435. [CrossRef]
68. Luthjens, L.H.; Yao, T.; Warman, J.M. A Polymer-Gel Eye-Phantom for 3D Fluorescent Imaging of Millimetre Radiation Beams. *Polymers* **2018**, *10*, 1195. [CrossRef] [PubMed]
69. Warman, J.M.; de Haas, M.P.; Luthjens, L.H. High-energy radiation monitoring based on radio- fluorogenic co-polymerization. I: Small volume in situ probe. *Phys. Med. Biol.* **2009**, *54*, 3185–3200. [CrossRef] [PubMed]
70. Warman, J.M.; Luthjens, L.H.; de Haas, M.P. High-energy radiation monitoring based on radio-fluorogenic co-polymerization II: Fixed fluorescent images of collimated X-ray beams using an RFCP gel. *Phys. Med. Biol.* **2011**, *56*, 1487–1508. [CrossRef] [PubMed]

polymers

MDPI

Article

Tuning the Morphology in the Nanoscale of NH_4CN Polymers Synthesized by Microwave Radiation: A Comparative Study

Cristina Pérez-Fernández [1], Pilar Valles [2], Elena González-Toril [1], Eva Mateo-Martí [1], José Luis de la Fuente [2] and Marta Ruiz-Bermejo [1,*]

[1] Centro de Astrobiología (INTA-CSIC), Departamento de Evolución Molecular, Ctra. Torrejón-Ajalvir, km 4, Torrejón de Ardoz, 28850 Madrid, Spain; crisperez@cab.inta-csic.es (C.P.-F.); gonzalezte@cab.inta-csic.es (E.G.-T.); mateome@cab.inta-csic.es (E.M.-M.)

[2] Instituto Nacional de Técnica Aeroespacial "Esteban Terradas" (INTA), Ctra. Torrejón-Ajalvir, km 4, Torrejón de Ardoz, 28850 Madrid, Spain; vallesgp@inta.es (P.V.); fuentegj@inta.es (J.L.d.l.F.)

* Correspondence: ruizbm@cab.inta-csic.es; Tel.:+34-915206458

Abstract: A systematic study is presented to explore the NH_4CN polymerization induced by microwave (MW) radiation, keeping in mind the recent growing interest in these polymers in material science. Thus, a first approach through two series, varying the reaction times and the temperatures between 130 and 205 °C, was conducted. As a relevant outcome, using particular reaction conditions, polymer conversions similar to those obtained by means of conventional thermal methods were achieved, with the advantage of a very significant reduction of the reaction times. The structural properties of the end products were evaluated using compositional data, spectroscopic measurements, simultaneous thermal analysis (STA), X-ray diffraction (XRD), and scanning electron microscopy (SEM). As a result, based on the principal component analysis (PCA) from the main experimental results collected, practically only the crystallographic features and the morphologies in the nanoscale were affected by the MW-driven polymerization conditions with respect to those obtained by classical syntheses. Therefore, MW radiation allows us to tune the morphology, size and shape of the particles from the bidimensional C=N networks which are characteristic of the NH_4CN polymers by an easy, fast, low-cost and green-solvent production. These new insights make these macromolecular systems attractive for exploration in current soft-matter science.

Keywords: HCN polymers; cyanide polymerization; microwave-driven polymerization; nanoparticles; nanofibers; multifunctional materials

Citation: Pérez-Fernández, C.; Valles, P.; González-Toril, E.; Mateo-Martí, E.; de la Fuente, J.L.; Ruiz-Bermejo, M. Tuning the Morphology in the Nanoscale of NH_4CN Polymers Synthesized by Microwave Radiation: A Comparative Study. *Polymers* **2022**, *14*, 57. https://doi.org/10.3390/polym14010057

Academic Editor: Edina Rusen

Received: 2 November 2021
Accepted: 20 December 2021
Published: 24 December 2021

Publisher's Note: MDPI stays neutral with regard to jurisdictional claims in published maps and institutional affiliations.

Copyright: © 2021 by the authors. Licensee MDPI, Basel, Switzerland. This article is an open access article distributed under the terms and conditions of the Creative Commons Attribution (CC BY) license (https://creativecommons.org/licenses/by/4.0/).

1. Introduction

The development of new smart and multifunctional materials is currently encouraged to be constrained toward the design of low cost and free-solvent synthetic processes, or at least green-solvents, and easy, fast, effective and robust productions. The MW-driven polymerization of NH_4CN may be considered as a good example of this kind of processes for the generation of polymeric macrostructures with potentially versatile properties. One of the advantages of this polymerization technique is to reduce the reaction times with respect to the conventional heating processes. However, the kinetic variations during the syntheses due to the effect of the MW radiation must be taken in consideration, as they may change the final properties of the products. In the case of the HCN-derived polymers, it is well established that their properties are strongly conditioned by the synthetic conditions and for extension, their potential applications are directly dependent on their chemical composition and structure, thermal stability, particle shape and size, among others.

Recently, it has been shown that the particle morphology of the known HCN-derived polymers can be tuned by the choice of the synthetic conditions [1–3]. Some polymers of this heterogeneous family [4] have been proposed as emergent materials with different applications, such as photocatalysts [5], semiconductors, nanowires, ferroelectric materials [6], capacitors [1], coatings with potential biomedical applications [2,7–9], protective films against corrosion [10,11], or for the development of antimicrobial media for passive filtration [12].

Different HCN-derived polymers present some structural characteristics resembling those of the extensively researched carbon nitrides [1,13,14], which are used as carbon materials in multiple fields [15–18]. Moreover, it has been proposed that the incorporation of HCN-derived polymeric nanoparticles as novel fillers in different polymeric matrixes could lead to the creation of new composite materials [1]. However, only one work has described the production of nanoparticles derived from HCN [13]. In that case, the aqueous polymerization assisted by the MW radiation of cyanide lead to the simultaneous generation of nanoparticles with different sizes and shapes, together with isolated long nanofibers. In this context, it is important to note that HCN polymer nanofibers have attracted considerable attention due to their attractive features, e.g., their high visible-light photocatalytic activity. These materials were prepared from formamide, the hydrolysis product of HCN, through conventional thermal heating at 200 °C and a prolonged reaction time of 48 h [5]. Therefore, these polymer materials, at the nanoscale, present an extraordinary potential in the different areas mentioned previously.

Taking into account the high potential of HCN-derived polymers for the development of multifunctional materials, as well as the key role of their polymeric morphology for these potential applications, several aspects related with the MW-driven synthesis of NH_4CN polymers are explored here in detail, paying special attention to the relationships between the experimental conditions and the features and properties of the final polymeric products. In this way, two series of NH_4CN polymers synthetized using MW radiation were compared between themselves and also against a control polymer produced under conventional heating using equivalent reaction times or by fixing this time, in order to determinate the influence of the MW radiation on the features of the insoluble polymers obtained.

Moreover, statistical methods were used for a better interpretation of the results reported in the present work. The obtained data were comparatively discussed in order to obtain useful synthetic conditions for the production of cyanide polymers using MW radiation, taking into special consideration the structural characteristic of the polymers and their morphological and textural properties for the further development of multifunctional materials.

2. Materials and Methods

2.1. Synthesis of the NH_4CN Polymers

All of the NH_4CN polymers synthetized in this work were produced as was previously described in [13] using the reaction times and temperatures indicated in Table 1. The initial equimolar concentrations of NaCN and NH_4Cl always were 1 M. Figure 1 shows some experimental details of the synthesis process.

2.2. Characterization of the NH_4CN Polymers

The Fourier transform infrared (FTIR) spectra were recorded as was reported in [1], but using a Nicolet Is50-FTIR spectrometer (Thermo Scientific, Waltham, MA, USA). The powder X-ray difracction (XRD) was performed as in [1], but by scanning the samples from 10° to 50° (2θ) and with a count time of 3 s. All of the parameters and the equipment for the measurements of the elemental analysis, thermal analysis and ^{13}C solid-state cross polarization/magic angle spinning nuclear magnetic resonance (CP/MAS) NMR were described in [1]. The X-ray Photoelectron Spectroscopy (XPS) analysis of the samples (pure pellets of the polymers) was carried out in an ultra-high vacuum chamber equipped with a hemispherical electron analyser, using an Al Kα X-ray source (1486.6 eV) with an aperture of 4 mm × 7 mm (Specs company, Berlin, Germany). The base pressure in the chamber was 5×10^{-9} mbar, and the experiments were performed at room temperature. The following core level peaks were recorded under the same experimental conditions: C (1 s), N (1 s) and O (1 s). The pass energy applied to take the overview sample was 30 eV, while 20 eV pass energy was applied for the fine analysis of the core level spectra. The surface morphologies of the polymers were determined by a ThermoScientific Apreo C-LV field emission electron microscope (FE-SEM) equipped with an Aztec Oxford energy-dispersive X-ray microanalysis system (EDX). The samples were coated with 4 nm of chromium using a sputtering Leica EM ACE 600. The images were obtained at 10 kV. This SEM study was analogous to those detailed in [13]. The values of the Z-average and polydispersity index (PdI) were registered using a Zetasizer Nano instrument (Malvern Instruments Ltd., Almelo, Netherlands), using ethanol as a solvent to scatter the samples.

Table 1. Experimental conditions used for the synthesis of NH_4CN polymers assisted by MW radiation, as well as some of their properties. Note that polymer **2** is present in both series. [1] Maximum pressure reached along the syntheses. * No reproducible measurements.

Polymer	T (°C)	t (min)	P [1] (bar)	Z-Average (nm)	PdI	Crystallinity (%)	Moisture (%)	Char (%)
				Series 1				
1	130	309	3	506	0.23	71	10	20
2	150	67	5	342	0.25	66	10	20
3	170	18	10	318	0.22	64	9	19
4	190	4	14	276	0.22	65	10	19
5	205	2.5	19	279	0.23	64	9	22
				Series 2				
6	130	67	3	724	0.75	61	10	13
7	170	67	10	383	0.25	66	10	22
8	190	67	14	321	0.34	63	10	24
9	205	67	19	601 *	0.73 *	67	10	25

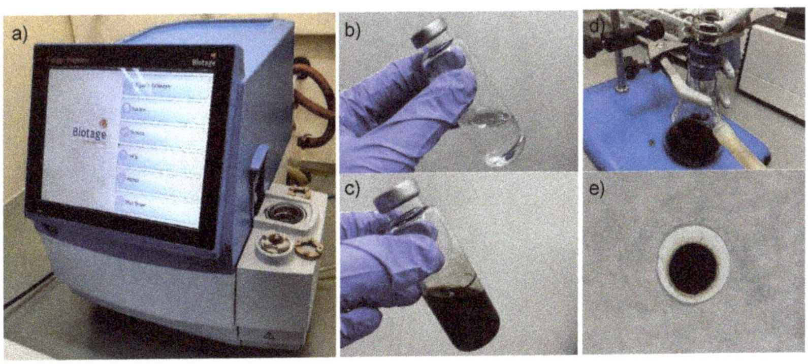

Figure 1. (**a**) Microwave reactor using in the present work; (**b**) equimolar water solution of NaCN and NH_4Cl; (**c**) raw reaction. A representative suspension of the cyanide polymers; (**d**) filtration sys-tem. In the bottom part, dark solution or sol fraction which will be freeze-dried. In the upper part, black insoluble solid or gel fraction; (**e**) characteristic cyanide polymer over a glass fiber filter after drying by reduced pressure.

2.3. Statistical Analysis

Three principal component analyses (PCAs) using the polymers' physicochemical characteristics as variables were carried out. The tests were performed using the multivariate data analysis software CANOCO 4.5 (Microcomputer Power, Ithaca, NY, USA) [19]. The program CANODRAW 4.0 (in the Canoco package) was used for the graphical presentation.

3. Results and Discussion

3.1. MW Radiation Effect in the Production of NH_4CN Polymers

In order to explore the effect of the MW radiation in the aqueous polymerization of NH_4CN, two set of experiments were carried out. The first one used equivalent reaction times (polymers **1–5**, Table 1); the second one fixed the reaction time at 67 min (polymers **2**, **6–9**, Table 1). For the first series of syntheses (series 1), the reaction times were chosen with a comparative proposal in a relationship with the NH_4CN polymer obtained at 80 °C using conventional heating and a reaction time of 144 h (control polymer) [1]. Based on the manufacturer specification of the MW reactor, 6 days under a classical thermal treatment is equivalent to the short reaction times using MW radiation indicated in Table 1 (series 1). On the other hand, the fixed time for the second series (series 2) was based on the fact that, in order to probe a polymerization time shorter than 309 min (polymer **1**) but longer than those explored previously in the NH_4CN polymerization assisted by MW at 180 °C, with a reaction time of up to 30 min [13]. Thus, the reaction time for polymer **2** was chosen for polymeric series 2.

The conversion degrees, α (%), reached for each polymer are displayed in Figure 2a, and they were calculated as is described in [1] by means of gravimetric measurements of the gel fraction and the insoluble product. The use of conventional heating demonstrated that the increase in the temperature leads to a decrease in the yield of the NH_4CN polymers [14]. For the control polymer, the conversion was about ≈38%, while this value decreased to ≈17% when the polymerizations are conducted at 180 °C using MW radiation [13]. Note that the higher conversion was found in series 1, around 25%, at the lowest temperature under study 130 °C, and that the data of series 1 is in agreement with these previous results. However, a slight increase in the conversion for the polymerizations was observed at temperatures higher than 180 °C. For series 2, at 130 °C, a decrease of practically five times in the polymerization time produced a relevant and unexpected increase in conversion. On the contrary, an increase in the reaction time at temperatures of 170 and 205 °C causes an increase in the yield of insoluble polymer. The conversion values for polymers **6**, **7** and **9** was really unexpected, reaching yields similar to those obtained by means of a traditional heating process [1,14]. This behavior of the polymerization in the presence of MW radiation seems to indicate significant differences with respect to the kinetic approach reported using conventional heating, in which a clear limit conversion was observed after a concrete reaction time for a fixed temperature [4]. Further works are in progress to study in detail the kinetic behavior of the MW-driven polymerization of cyanide.

In addition, the pathways proposed to explain the formation of NH_4CN polymers in aqueous media suggest collateral reactions during the course of this precipitation polymerization, such as deaminations, denitrogenations, dehydrocyanations, hydrolysis and oxidations. These last processes are responsible for the incorporation of oxygen in the macrostructures [4], which would lead to a gain of weight considering the total initial mass input in the system (the initial weight of NaCN plus NH_4Cl). Figure 2b shows the weight balance found in both series, considering the reaction conditions shown in Table 1, based on the following formula: Weight Balance (%) = [(weight NaCN + weight NH_4Cl) − (gel fraction weight + freeze-dry sol content)]/(weight NaCN + weight NH_4Cl)]·100. In any case, no direct relationship could be found between the conversion values and the weight balance, as can be observed when comparing Figure 2a,b. In series 1, an increasing weight loss was found as the temperature increased; however, in series 2, when the polymerization time was kept constant, a dissimilar behavior was observed. In this case, at low temperatures, the weight loss, around 5%, was maintained throughout these polymerizations. However, the MW reactions at 170, 190 and 205 °C occurred with substantial weight gain, close to 15% or higher. The increase of weight might be related with a minor loss of nitrogen and/or to a higher increase of mass by oxygenation, but the results of the elemental analysis were inconsistent with this proposal (see below), which suggests that complex secondary processes have no reflex directly in the generation of the gel fraction macrostructures.

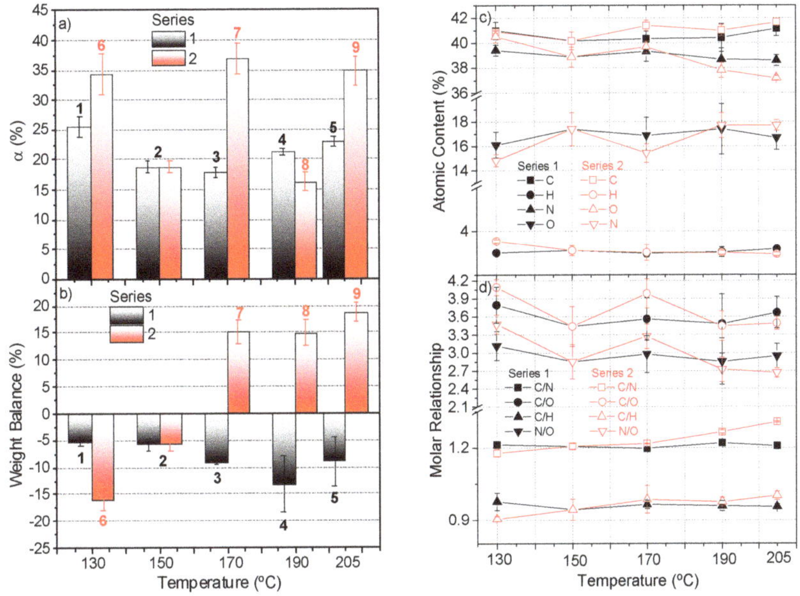

Figure 2. (**a**) Conversion degree, α (%), for the insoluble polymers synthesized according to the reaction conditions shown in Table 1. Above each column is indicated the number of the corresponding polymer. Note that the polymer **2** is included in both series. The conversions for polymers **7** and **9** were calculated using the values obtained in seven independent experiments; for the rest of the polymers, at least three independent syntheses were carried out. (**b**) Weight balance for the global NH_4CN polymerization reactions indicated in Table 1, according with the following expression: Weigh balance (%) = [(mg gel fraction + mg freeze-dried solution) − (mg NaCN + mg NH_4Cl)]/(mg NaCN + mg NH_4Cl)]·100. (**c**) Elemental analysis data for the polymers synthetized according to the conditions indicated in Table 1. At least three representative independent samples were measured for each point in the graph; (**d**) corresponding molar ratios. Note that the %O was calculated by the difference. For the C/O and N/O molar ratios, we considered the %O present in the macrostructure after subtracting this percentage from the water absorbed for the polymers as moisture, which was about ~10% for all of the samples. This amount of water was estimated by the TG data discussed below.

The elemental analysis data displayed in Figure 2c indicates that the C percentage did not present significant differences among the two series under study, and it was about 41%. The same behavior was observed for the %O, at around 17%. Only a slight decrease in the nitrogen content was observed in series 2 with the increase of the temperature, from 41 to 37%. This result was also reflexed in the C/H and C/N molar relationship for this series, when the reaction time was fixed, going from 1.18 to 1.31 for this last ratio (Figure 2d). Therefore, it seems that the increase in the temperature favors the deamination processes during the polymerization based on the lower %N and %H. This fact was in agreement with previous results [13]. In contrast, for series 1, no significant variations were found in the molar relationships with the increasing of the polymerization temperature.

As a result, it can be concluded that a temperature of 170 °C seems to be a key parameter in these polymerizations under MW radiation, and also that low polymerization temperatures, such as 130 °C, present a singular behavior. However, the compositional data do not give a clear response to the high weight gain observed in series 2.

At this point, it was necessary to point out that the lower yields for the insoluble polymers may be due to the hydrothermal conditions achieved by MW radiation, which favors the oxidation, hydrolysis and decomposition processes of the NaCN and/or other intermediate reaction products, as in Scheme 1, decreasing the availability of them, and therefore reducing the final amount of insoluble polymers collected. Note that all of the reactions were prepared under ambient conditions of pressure and moisture, and neither oxygen or CO_2 were removed from the reaction vessel or the solvent.

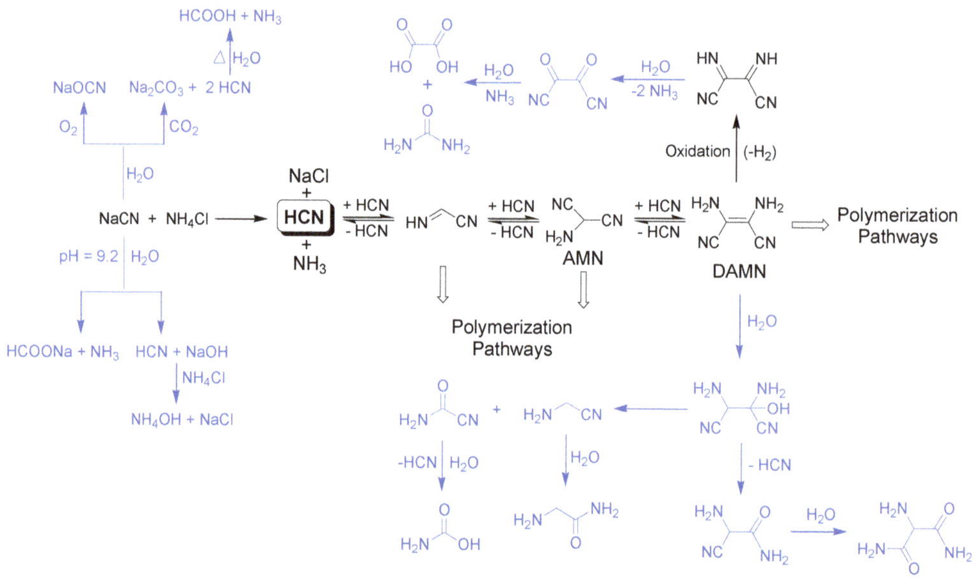

Scheme 1. Possible oxidation and hydrolysis processes likely favored by the MW radiation, which can be undergone by cyanide and by DAMN. These processes lead to lower conversion values for the NH_4CN polymerization assisted by MW radiation than for polymerization carried out under conventional heating due to the decrease of the available cyanide and DAMN [20,21].

The left part of Scheme 1 shows the possible processes which can decrease the available amount of cyanide to polymerize. Atmospheric oxygen can lead to the partial oxidation of cyanide at elevated temperatures, generating cyanate. CO_2 can react with cyanide to produce carbonate and HCN. In addition, the hydrolysis of the cyanide solutions under heating can produce formiate and ammonium. Under the conditions considered herein, the initial cyanide solutions have a pH of 9.2, and a ~50% of HCN was present in the solutions, which can be hydrolyzed to give formic acid and ammonia. The generation of volatile compounds such as HCOOH, NH_3 and even HCN may explain the negative weight balances observed in the NH_4CN polymerization above studied. However, the increase observed for the production of insoluble polymers **6**, **7** and **9** might be due to the recycling of the delivered HCN and NH_3, generated first as by-products, that finally were implicated in new processes of oligomerization/polymerization, as a closed system was used in the described reactions. On the other hand, the positive weight balances can be due to the formation of no volatile oxygenated molecules such as carbonate and cyanate, as this takes place at high temperatures within series 2. Besides this, in order to explore whether the ambient conditions have a significant influence in these plausible side reactions, where the atmospheric oxygen and the CO_2 were implicated, two additional reactions were carried out. Polymers **3** and **7** were again synthetized, but using an inert atmosphere of nitrogen and carefully removing the CO_2 and O_2 of the water by an N_2 stream. The following conversion values were obtained (syntheses were made twice) $\alpha(\%) = 7.7 \pm 0.1$ and $\alpha(\%) = 3.8 \pm 0.3$ for the analogous polymers **3** and **7**, respectively. Furthermore, in both cases a loss of mass was observed in the total weight balance. Note that the effect of the absence of air in the conversion obtained was outstandingly significant for the analogous polymer **7**. Therefore, the lack of air in the reaction environment does not seem to prevent the side reactions in the cyanide polymerization assisted by MW radiation; on the contrary, the yield of the reaction was remarkably lower. Due to the great impact of the air in the microwave-cyanide polymerizations, new works will be developed to obtain a full understanding of this factor in these highly complex polymerization reactions.

In addition, we might take into consideration that other hydrolysis and/or oxidations of some the intermediate products during the polymerization process finally lead to the lack of formation of the insoluble polymers, as is the case for diaminomaleonitrile (DAMN). DAMN, the tetramer of the HCN, was considered to be a main intermediate in the generation of the HCN polymers (Scheme 1) (see, e.g., [4]). However, this compound can be oxidized to diiminosuccinonitrile in water, and this is finally decomposed into oxalic acid and urea (upper part of the Scheme 1). Furthermore, DAMN in an aqueous medium can be transformed into formamide, glycine and aminomanolic acid (the bottom part the Scheme 1), as was revealed by Ferris and Edelson [20].

The higher yields (~75%) in the hydrothermal polymerizations of DAMN than were the obtained in analogous NH_4CN polymerizations (35–40%) using conventional heating for the production of black insoluble HCN-derived polymers [1], and the significant diminution of these yields in the MW-driven polymerization of DAMN (~30%) (a further work is in progress; no data are shown here) allows us to reaffirm the following points: (i) the DAMN was a main intermediate product cyanide polymerization, as similar—although not identical—final products can be obtained in both ways, i.e., the cyanide and DAMN polymerizations produced resemble—but are not identical—to HCN-derived polymers [1]. The possible polymerization of cyanide by other different DAMN pathways through its dimer and/or its trimer, aminomalononitrile (AMN), for the generation of the suggested C=N networks of the HCN-derived polymers must also be taken in consideration (Scheme 2); (ii) the higher yields obtained for the DAMN polymers seem to indicate that the cyanide undergoes several processes in water, which prevent the generation of DAMN and the subsequent polymerization of this oligomer. Therefore, it seems that the cyanide polymerization via the DAMN pathway was the main process.

As a summary of the above, one can say that: (i) the lack of air in the cyanide polymerization assisted by MW radiation does not seem to inhibit the side reactions which prevent the generation of DAMN, and (ii) the decomposition processes of the DAMN are related to the presence of water. The increase of the temperate increases the ratio of DAMN decomposition in water environments. Both aspects are the main focus of other works in progress.

The FTIR spectra of both series closely resembled those reported from the NH_4CN polymerizations under a classical heating and also assisted by MW at 180 °C (Figure 3a,b) [1,4,13,14]. In these previous works, analyses of the features for these spectra were discussed in detail. The greatest difference between the spectral features of series 1 and 2 was the relative intensity of the bands related with the nitrile groups (-C≡N), around 2200 cm^{-1}. This difference was clearly observed when the EOR values (the extension of the reaction, EOR = $[I_{1645}/(I_{1645} + I_{2200})]$ 100) [22] were represented (Figure 3c). For series 1, this value decreased with the temperature, and by the contrary, for series 2 the behavior was the opposite. This quantitative spectroscopic parameter was directly related to the progress of the polymerization reaction under traditional heating conditions, i.e., the increase of the monomer conversion along the reaction time gives noticeably higher EOR values [22]. However, there was no observed clear relationship between the EOR and the conversion degree, especially in series 2 (Figure 3d, upper part). This behavior was also observed from the NH_4CN polymers synthetized at 180 °C by MW [13].

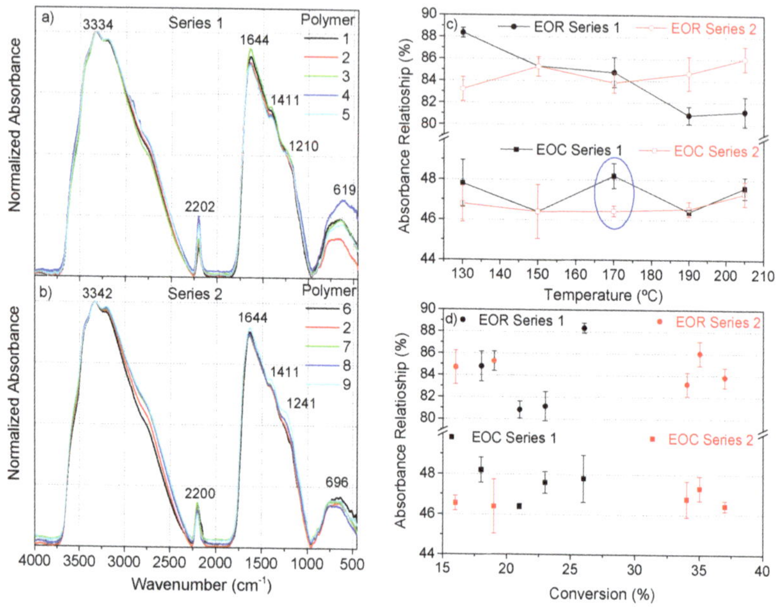

Figure 3. (**a**) Representative FTIR spectra of polymers **1–5** (series 1), and (**b**) of polymers **2**, **6–9** (series 2); (**c**) relationship of the EOR (extension of the reaction) and the EOC (extension of the conjugation) with the temperature; (**d**) dependence of both the EOR and EOC relationships on the conversion degree. (**c**,**d**) Each point was calculated from at least three FTIR spectra from independent samples.

In addition, the calculation of the EOC (extension of the conjugation, EOC = $[I_{1645}/(I_{1645} + I_{3330})]\cdot 100$) [1,13] indicated no significant differences between series 1 and 2, with the exception of the values determined at 170 °C (Figure 3c). Likewise, there was no clear trend found when these values were confronted with the conversion. Again, these spectroscopic data revealed that 170 °C shows a remarkable singularity with respect to the rest of the temperatures under study.

Simultaneous thermal analysis (STA) for representative samples of polymers **1–9** was carried out (Figure 4), and some parameters obtained from these measurements are listed in Table 1. All of the samples showed the same amount of moisture, at about 10%; however, while the NH_4CN polymers from series 1 presented a similar content of char residue, 19–22%, the samples from series 2 exhibited an increase of the char from 13 to 25% with the increase of the temperature. Polymer **9** was the sample that presented the highest percentage of char, and by the contrary, polymer **6** registered the lowest value. In general, a greater amount of char was related with highly cross-linking macrostructures and rigid chains; in particular, NH_4CN polymers were associated with the presence of cross-linking oxygenated groups, such as inter- and intramolecular amide bonds [4]. The C/H, C/N, C/O and N/O molar ratios were 0.9, 1.2, 4.1 and 3.5, respectively, for polymer **6**; and 1.0, 1.3, 3.5 and 2.7, respectively, for polymer **9**. Thus, it seems that the higher thermal stability of sample **9** was related with greater oxygenated macrostructures, and with a minor amount of nitrogenized groups, but not with a more conjugated structure, considering the C/H values and the EOC data. Therefore, it might be proposed that this stability may be due to the presence of oxygenated cross-linking groups.

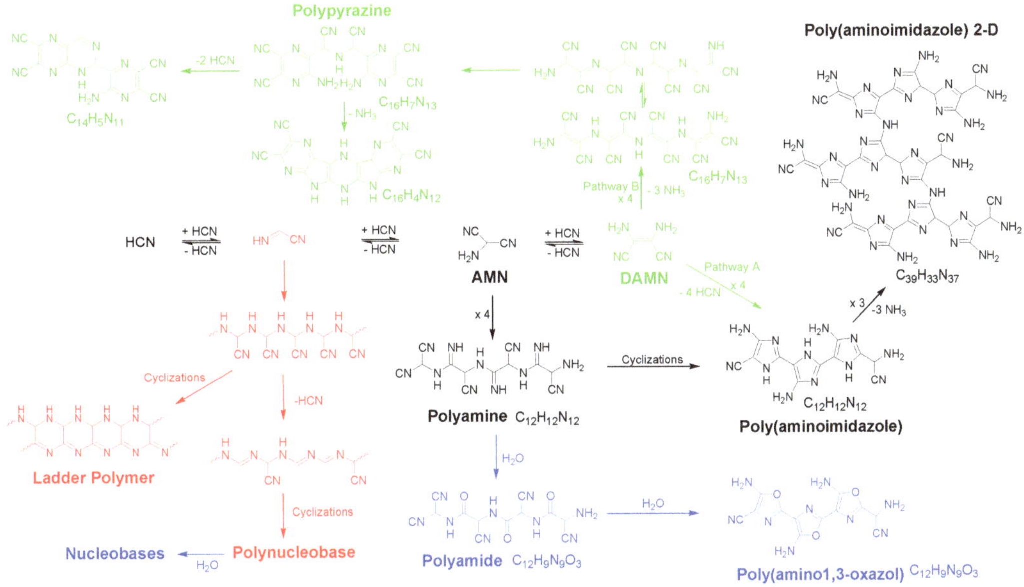

Scheme 2. General pathways for the generation of HCN-derived polymers in water. In this scheme is shown the oligomer formation as far as the tetramer DAMN, as well as proposed macromolecular structures from the dimer (red color), the trimer (AMN) (black color) and DAMN (green color) according to recent works [1,3,4,23]. Note that the incorporation of oxygen in some of the structures for the suggested macromolecular systems was due to hydrolytic processes (blue color).

On the other hand, the DTG curves showed similar profiles to other NH_4CN polymers synthetized at 80 °C [1]. Five characteristic peaks were observed: (i) at ~70 °C (water desorption); (ii) at 220–260 °C and at ~400 °C (thermal break of the weakest bonds); and (iii) at ~670–680 °C and 820–840 °C (high thermal decomposition, carbonization stage). Only a slight difference was found for polymer **5** with an additional degradation peak at 477 °C, and the decrease of the relative peak intensity at ~260 °C in series 2 with the increase of the temperature. In addition, the DSC curves were really dissimilar to those registered for NH_4CN polymers synthetized at lower temperatures [1,4], proving that the thermal analysis techniques provide excellent fingerprints to distinguish NH_4CN polymers with very similar FTIR spectra [4]. For both series of experiments, the first endothermic peak, at 75–85 °C, can be related to the loss of the absorbed water in the polymeric matrix, and the rest of peaks can be related to degradative processes.

Some morphological aspects, such as the hydrodynamic diameter, were measured for representative polymers **1–9** (Z-average, Table 1). Considering this parameter, for series 1, the particle size was decreased with the increase of the temperature, reaching nanoparticles with a diameter of ~280 nm and with relatively low PdI = 0.22–0.25. On the contrary, in series 2, although an apparent decrease in the molecular size may be related with the increase of the temperature, the PdI values were higher than those in series 1, indicating a higher degree of heterogeneity. Indeed, the data from sample **9** were not properly registered due to a presumable aggregation of the particles.

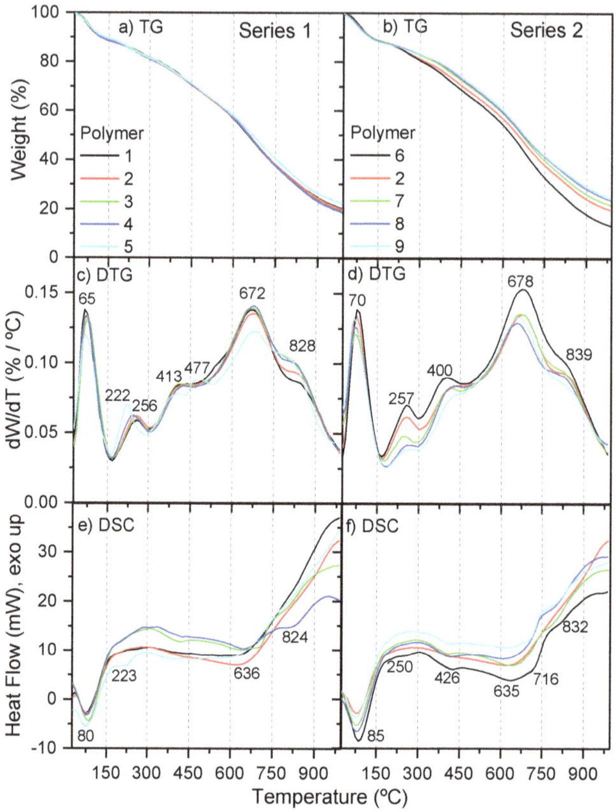

Figure 4. Thermal analyses for representative samples of polymers **1–9**. (**a**,**b**) TG curves; (**c**,**d**) DTG curves; (**e**,**f**) DSC curves.

The XRD pattern profiles for NH_4CN polymers **1–9** have the same look as those previously reported for analogous polymers synthetized using a conventional hydrothermal procedure or MW radiation [13,14] (Figure 5a,b). It was expected that an increase in the temperature would lead to a more ordered macrostructure [13,14]. This is the unique observed peak (2θ ~27°) related to graphitic-like two-dimensional (2-D) structures, such as layered g-C_3N_4 [16,24], would be higher and narrower at the higher polymerization temperatures. However, this prediction was not valid for the range of temperatures under study when analyzing the data of the crystallinity of the different samples collected in Table 1. There was no clear relationship between the order level of the macrostructures with the increase of the temperature or the reaction time. All the samples exhibited an average crystallinity values of 65%, with the exception of polymers prepared at 130 °C, with values higher and lower values than this; showing a particular behavior of the cyanide polymerization at this temperature at least related with the internal order of the macrostructure.

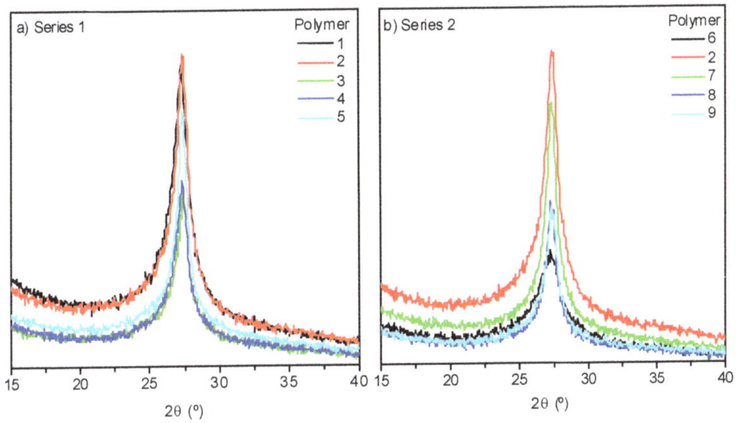

Figure 5. XRD patterns of representative samples from both series, (**a**) and (**b**), synthesized by a MW method.

It was clearly shown above that the heating by MW radiation has an unexpected and significant influence in the behavior of the cyanide polymerization, providing results which were not previously observed with conventional heating, i.e., there was not a proportional direct relationship between the reaction time and the conversion degree (%) and EOC [14,22], or temperature and crystallinity [14], for example. However, the final reaction products present resemble the spectroscopic and thermal characteristics of analogous HCN-derived polymers synthetized using conventional heating; for the size of the particles and the higher order of the macrostructures, this was the main difference found between the conventional heating against MW radiation. No nanoparticles were identified in NH_4CN polymers synthetized using conventional heating (the Z-average and PdI could not be measured; the data are not shown; however, see [1] for SEM images); at the lower temperature herein considered, the greater size of particles were observed. In addition, at 130 °C and 170 °C, particular features were observed with respect to the syntheses carried out at the other temperatures. The global and proper interpretation of all of these results is not simple and trivial. Thus, multivariate analysis could be an excellent analytical tool to reach a better and more objective interpretation of the data presented above, and could help us to choose preferential conditions to tune the nanoscale of the NH_4CN polymeric particles. In this way, a PCA was carried out to reach a first full overview of the MW radiation effect in the NH_4CN polymerization behavior (Figure 6). For this analysis, the total data for the nine polymers (Table 1) were considered. The conversion degree, the balance weight, and all of the molar ratios were calculated; the EOR, the EOC, the crystallinity, the char content, the moisture percentage, the Z-average and the PdI were taken in consideration, explaining 99% of the variance. The first axis shows the highest positive correlation among a wide number of variables, with the exception of EOC and the balance weight. The right-hand portion of the first axis was predominantly occupied by EOR, moisture, monomer conversion, C/O molar ratio, Pdl, Z-average and N/O molar ratio; the left-hand portion was occupied by crystallinity (%) and the C/N molar ratio. The second axis indicates the highest positive correlation among char (%) and the C/H molar ratio. By contrast, the EOC and balance weight exhibited the highest negative correlation with this axis. From these statistical results, one can say that the there is a direct relationship between the conversion degree and the EOR values, as was expected, although a clear relationship was not observed in Figure 3d. In addition, a higher conversion degree was directly related to less-oxidized structures; interestingly, this minor oxidation degree was directly related with a higher size of the particle and a higher polydispersity degree. On the other hand, the percentage of char (%) was strongly correlated with the molar C/H ratio, indicating that the more conjugated structures are the more thermally stable, but apparently independent of the %O in the structure. A higher internal order based on the percentage of crystallinity is directly related with the C/N ratio and with the char (%), i.e., the more carbonaceous and more conjugated structures are the more crystalline and also the more thermally stable, but the Z-average and crystallinity (%) are independent variables. Therefore, if the MW radiation is considered to obtain nanoparticles of NH_4CN polymers, as under conventional heating these are not formed and MW notably reduces the reaction times, it seems that reaction conditions which lead preferentially to the oxidization of structures but are highly conjugated should be considered.

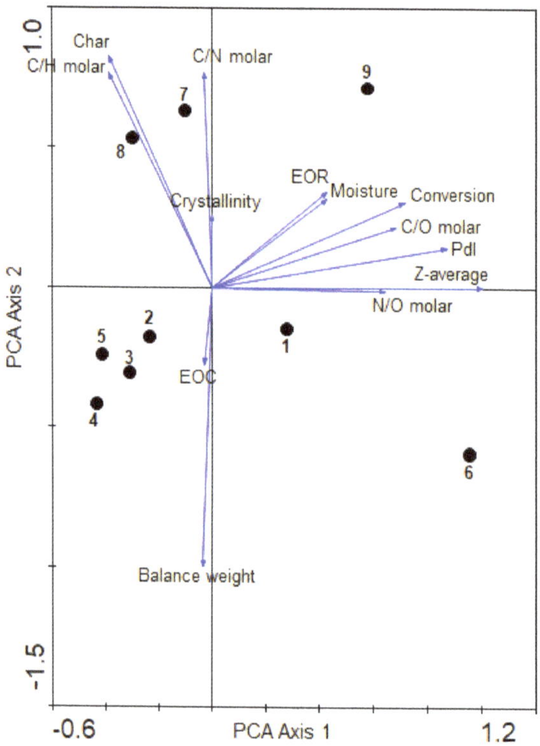

Figure 6. PCA biplot based on the physicochemical variables: C/N, C/O and N/O molar ratios; char (%); crystallinity (%); α (%) conversion; EOC and EOR; balance weight; PdI; moisture (%); and Z-average (nm) for the polymers obtained. The relative proportions of the different variables are indicated by arrows, and polymers are indicated by dots.

With respect to the polymeric samples, PCA analysis indicated groupings of the NH_4CN polymers into well-differentiated sets: (i) one corresponding to series 1 (polymers **2**, **3**, **4** and **5**), with the exception of polymer **1**; (ii) another one with polymer **7** and **8**; (iii) and polymers **1**, **6** and **9** isolated from the rest, indicating a particular behavior, as was expected especially for **1** and **6**, as they were the polymers synthetized at 130 °C, although polymer **1** was next to the group of series 1, in agreement with a relative resemblance with the samples of this group. In relation to the first group, the clear grouping of polymers **2**, **3**, **4** and **5** indicated that, statistically, the choice of the equivalent reaction times for different temperatures according to the instruction of the manufacturer of the MW reactor, and considering the experimental results indicated above, gives NH_4CN polymers with similar characteristics. The second group, formed by polymers **7** and **8**, showed higher char (%), C/H and C/N molar ratios and crystallinity (%), indicating that these polymers are thermally stables carbonaceous structures. For the first, the polymers **6** and **9**, and to a lesser extent **1**, showed a higher EOR, moisture (%), α (%), C/O molar ratio, PdI, Z-average and N/O molar, i.e., they presented the polymeric particles with the highest size and lowest oxygen content. Therefore, the lowest and the highest temperature here studied do not seem adequate for the development of nanoparticles.

These well-defined groupings by similar and resemblance features revealed by the PCA analysis, in a certain way, could also be observed for the shape differences found by SEM (Figure 7). Thus, polymers **2**, **3**, **4** and **5** presented rice-shaped nanoparticles and other stacking oval particles together with isolated nanofibers. Polymers **7** and **8** showed clear groupings of nanofibers together with other particles with other shapes (please see Figures S1–S9 from the supplementary information for details), whereas polymer **9** was estranged from this set in the PCA analysis, showing isolated nanofibers and undefined-shape particles. Polymers **1** and **6**, synthetized at the lowest temperature, presented a particular behavior. Polymer **1** displayed similar rice-shaped nanoparticles and staking oval particles to the rest of the polymers from series 1, but also spherical particles similar to those identified in the control polymer [1]. The particles observed in polymer **6** were similar to the particles found for series 1, although the staking oval particles were not identified. These SEM images also seemed to indicate that 130 °C and 205 °C are not optimal temperatures to produce nanoparticles from cyanide, which is in agreement with the PCA results.

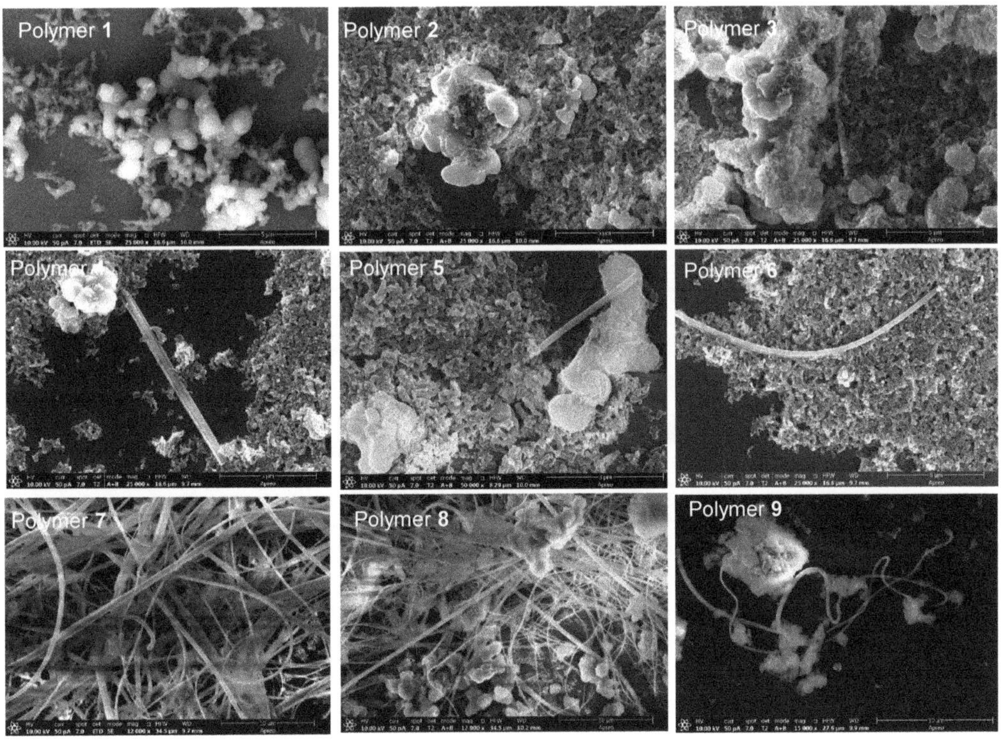

Figure 7. SEM images for NH_4CN polymers representative samples from series 1 and 2, Table 1.

3.2. Structural Comparative Study between NH_4CN Polymers

In order to complete this study about the effect of MW radiation on cyanide polymerization, a detailed structural comparison between the control polymer and polymer **3**, as a representative sample of series 1, was carried out. The previous results for this series indicated that polymers **2**, **3**, **4** and **5** are very similar, as was expected considering the equivalence of the reaction times; they resembled polymer **1**, but were at least morphologically very different to the control polymer. Thus, this section can help us to obtain a comprehensive knowledge about the MW heating role in the polymerization of the NH_4CN when equivalent polymerization times are considered on the spectroscopic and thermal properties, as the morphological differences clearly showed above.

The comparison of the data from the control polymer and from polymer **3** indicated that the conversion degree decreases notably when the NH_4CN polymerization is assisted by MW radiation, as it was explained above, but no significant elemental composition variations were observed. For the control polymer, the elemental compositional data were %C 41.4 $\pm$ 0.3, %H 3.8 $\pm$ 0.2, %N 40.1 $\pm$ 0.4 and %O 14.7 $\pm$ 0.7, and for polymer **3** they were %C 40.3 $\pm$ 0.6, %H 3.5 $\pm$ 0.1, %N 39.3 $\pm$ 0.8, %O 16.7 $\pm$ 1.5 (taking into consideration at least three samples synthetized independently). The subtraction of the normalized FTIR spectra of both samples does not indicate significant differences among them, except for a few low-intensity features (Figure 8a). Some of these bands can be related with the resonances found when the corresponding ^{13}C NMR spectra were subtracted (Figure 8b). The FTIR band centered at 2163 cm^{-1} can be related with the resonance at 115 ppm assigned to nitrile groups, the band at 1720 cm^{-1} with the resonance at 151 ppm associated with carbonyl groups, and the bands at 3615 and 3495 cm^{-1} with the signal at 51 ppm related to hydroxyl groups. However, the relative intensity of these FTIR bands and resonances seems to point to there being no great differences between the two polymers.

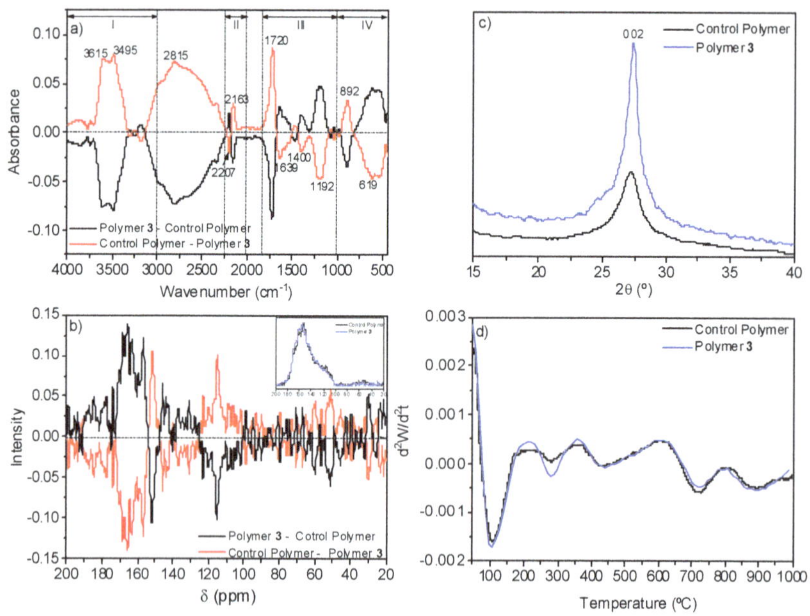

Figure 8. Comparative spectroscopic and thermal studies between polymer **3** and a control polymer synthetized using conventional heating. (**a**) Substation of the FTIR spectra; (**b**) substation of the ^{13}C NMR spectra; (**c**) XRD patters; (**d**) second derivative of the TG curves.

The XRD analyses showed the same diffraction, but in the case of the polymer **3**, this peak was higher and narrower, indicating a more ordered structure (Figure 8c). In addition, the analysis of the second derivative of their corresponding TG curves showed a very resemble thermal behavior (Figure 8d) which would indicate similar macrostructures, as both polymers present the same thermal behavior. Only the decomposition step at 278 °C may be more noticeable for the polymer **3**. This thermal decomposition step would be related, based on the TG-MS curves (data no shown), with the fragment *m*/*z* 44 which can be assigned to the loss of CO_2 and/or HC(=NH)NH_2 or HCONH- (a detailed discussion of the TG-MS results is out of the scope of the present work, and it will be given in a further paper). The slightly higher delivery of CO_2 or HCONH- for polymer **3** is in good agreement with the spectroscopic data and elemental compositional data indicated above, indicating a higher content of oxygen in polymer **3**.

In the light of these results, the control polymer and the polymer **3** seem to resemble one another, except for very little differences related with the amount of oxygenated functional groups. Thus, detailed XPS analyses of these two samples were made in order to provide further information about them. Figure 9 shows the core-level spectra of the C (1s), N (1s) and O (1s) peaks, as well as their deconvolutions on different components of the control polymer and polymer **3** samples.

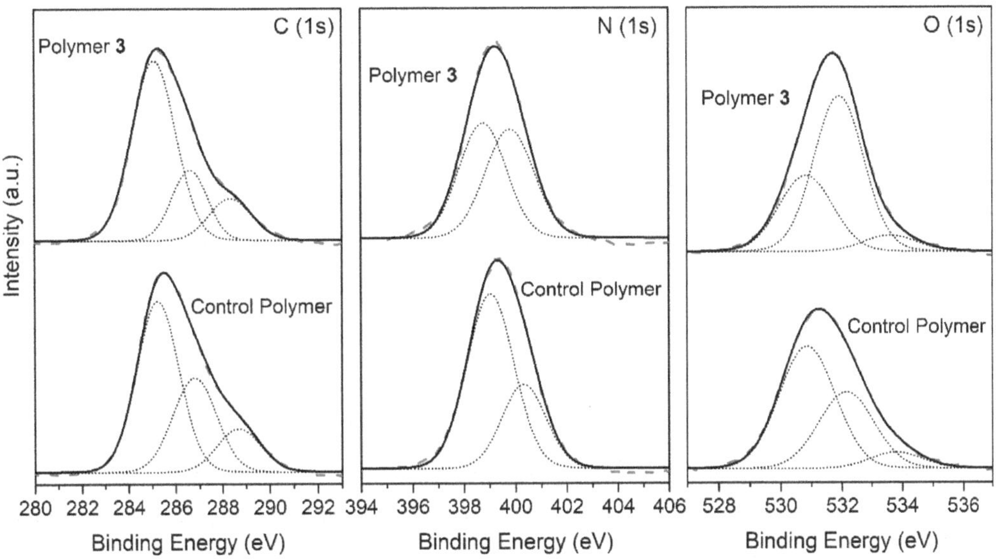

Figure 9. XPS photoemission spectra of the C (1 s), N (1 s) and O (1 s) core level peaks, and the deconvolution of a control polymer and polymer **3**.

A deconvolution study of the C (1 s) peak showed three components for both cases: the first component at 285.1 eV (binding energy) was attributed to the C adventitious, C–H and C–C group; the second component at 286.7 eV corresponded to C-N, C-O, C=N and amide groups; whereas the third component was observed at 288.5 eV, and was assigned to the C=O and nitriles groups. Both samples showed similar carbon components, and the ratio between the components was also comparable; polymer **3** showed a slight increase of 10% for the first component and a decrease of 10% for the second. Thus, the resemblance between spectra C (1 s) seems to indicate very similar macromolecular structures for polymer **3** and the control polymer. The N(1 s) peaks of both samples were resolved into two components, the first one at 398.9 eV being assigned to -$CONH_2$ and imines (-N=C<), which were predominant in the control polymer sample, and the second one at 400.0 eV corresponding to -CONH- groups, amides and nitriles; both nitrogen components showed a similar percentage for polymer **3's** case. Regarding the O (1 s) peak, we fitted the experimental data points using three components. The first component occurred at 530.9 eV, which was possibly assigned to the carboxylate group and to the amide group (-CONH-), which was predominant in the control polymer's case; the second one appeared at 532.0 eV, but it was mainly a contribution from contamination during the sample preparation in air instead of under UHV conditions; the main component for the polymer **3** sample, and finally a third component at 533.7 eV assigned to C=O and COOH groups, were similar in both cases. Therefore, this comparison study did not show large differences between both samples. The overlapping of several functional groups at similar binding energies did not help us to make an unambiguous assignment for the complex functional group mixture present in the NH_4CN synthetized polymer's structure. Nevertheless, the carbon and oxygen components related to the adventitious are more intense for polymer **3**, whereas for the control polymer, the carboxylate and C=N functional groups seems to be the principal component of the analysis.

As a result, taking into account the comparative results between the control polymer and polymer **3**, the more significant effects of the NH_4CN polymerization assisted by MW radiation are the decrease on the conversion degree and variations in the textural and morphological properties of the final products (as was indicated above, please see the SI of [1] for the look of the control polymer particles; these ones were not nanoparticles). The MW radiation leads to the generation of nanoparticles and/or nanofibers of cyanide polymers in minor yields, but with similar compositional/structural characteristics and the same thermal properties with respect to those microparticles produced under conventional heating conditions and with a more ordered macrostructure. Thus, nanoparticles/nanofibers can be obtained using MW radiation by aqueous cyanide polymerization. The size, shape and polydispersity of these particles seems to be directly related with the reaction time and with the temperature. In order to deepen this result, in the next section, analyses of the morphology of a series of NH_4CN polymers synthetized at 170 °C and distinct reaction times are given.

3.3. Polymeric Particles' Morphology Variations along the Reaction Time

A relative study of samples synthetized at 170 °C was completed by SEM using different reaction times, from 5 to 120 min. Representative images of these new polymeric series are shown in Figure 10 (for more details, please see Figures S11–S17 in the supplementary information section). We focused on this temperature based on the data reported in the first part of this work, as greater yields were obtained at this temperature; 130 °C and 205 °C were ruled out to explore the production of nanoparticles/nanofibers based on the PCA results, and also due to the easier dispersion of the polymer synthetized at 170 °C in EtOH to prepare the samples for the SEM measurements compared to those synthetized at higher temperatures, i.e., 190 °C and 205 °C. Note that the values of the PdI and Z-average increase with the increase of the temperature and with the reaction time (Table 1), leading finally to molecular aggregates.

All of the samples from the 170 °C series present isolated long nanofibers, with the exception of polymer **7** (reaction time ~ 67 min), which showed a clear grouping of long nanofibers (Figure 7 and Figure S7). Other shapes observed were spherical/oval particles, irregular and planar stacking, rice-shaped nanoparticles and short nanofiber networks. In all of the samples studied, on general lines, there was a dominant morphology against others depending on the reaction time, as is qualitatively summarized in Table 2. Interestingly, it is the generation of short nanofibers at 36 and 52 min which was not observed previously for this type of polymer. On the other hand, the production of rice-shaped nanoparticles was specially improved at 105 min, and was practically the only shape observed. Thus, to highlight, short nanofibers, long nanofibers and nanoparticles were obtained at 170 °C with reaction times of 36, 67 and 105 min, respectively. Therefore, it seems that there is a clear effect of the reaction time on the shape and size of the NH_4CN polymers synthetized in the presence of MW radiation.

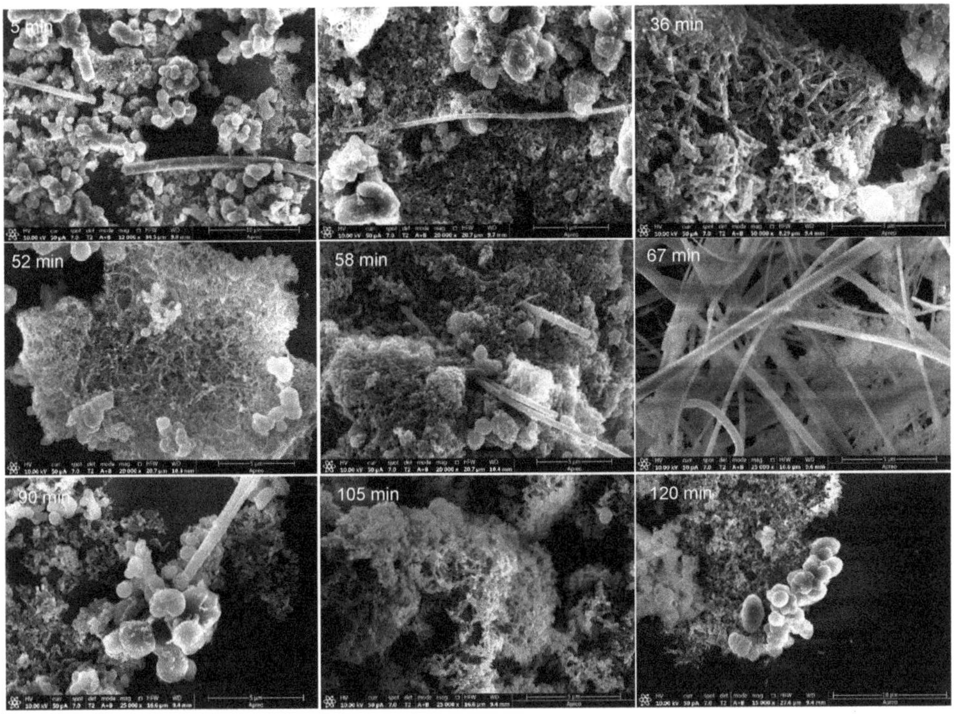

Figure 10. SEM images for representative samples from NH_4CN polymers synthetized at 170 °C. Note the variation of the morphology with the reaction time indicated in each picture.

Table 2. Qualitative summary for the observed shapes based on the SEM images from the NH_4CN polymers synthetized at 170 °C (please see the images collected in the supplementary information). – = not detected; + = detected but very minority and poorly representative; ++ and +++ = observed; ++++ = main shape observed.

	Reaction Time (min)									
Shape of Particles	**5**	**18**	**36**	**52**	**58**	**67**	**72**	**90**	**105**	**120**
Long isolated nanofibers	+	+	+	+	+	+++	+	+	+	+
Short nanofibers networks	–	–	++++	++++	–	–	–	–	–	–
Spherical/oval nanoparticles	+++	++	–	++	+	–	++	+++	+	++
Irregular and planar stacking	++	–	+	–	–	+	–	–	–	
Rice-shape like nanoparticles	+++	+++	+	–	++++	++++	+++	+++	++++	++++

Based on the PCA results discussed above, the more oxidized and conjugated macrostructures derived from the MW-driven cyanide polymerization would be nano-sized. This statistical result could be experimentally proven. Thus, Figure 11 shows the evolution in the composition along the time of the cyanide polymers synthetized at 170 °C. It can clearly be seen that the two samples with a greater content in oxygen, in this case those obtained at 36 and 52 min, present short nanofiber networks, with these being apparently the smallest nanoparticles observed. It is also interesting that the sample obtained at 105 min, with the highest content in oxygen, showed the tiniest particles observed in all of the cases here studied. On the contrary, the sample prepared using a reaction time of 67 min, polymer **7**, presented the lower content in oxygen, and in this case long nanofiber networks were found. Therefore, taking into account these results, it seems possible to tune the morphologies of the NH_4CN polymers to obtain mainly short or long nanofibers or nanoparticles for the development of new families of polymeric materials. However, the lack of a direct relationship between the conversion degree, elemental composition and reaction time, as was observed in Figure 11, encourages us to carry out comprehensive studies about the cyanide polymerization promoted by MW radiation.

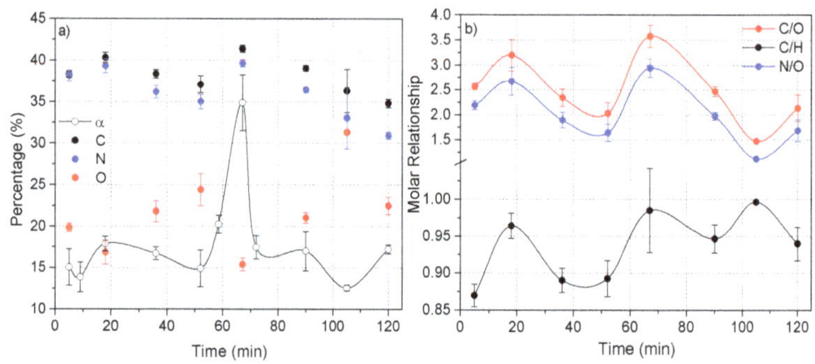

Figure 11. Evolution of the polymerization of NH_4CN with the time at 170 °C. (**a**) Variation of the conversion degree, α (%), and the elemental composition of the cyanide polymers; (**b**) variation of the molar relationships.

4. Conclusions and Outlooks

This is the first systematic study regarding the MW-assisted polymerization of cyanide, addressing a wide range of temperatures and reaction times. As the main results, the MW radiation has a notable influence on the yields of the insoluble polymers obtained. This fact could be due to the likely increase of the decomposition processes for the cyanide, and of its main oligomer, the DAMN, hindering the polymerization pathways proposed for the production of the extended C=N polymeric networks, as postulated for the HCN-derived polymers. However, using equivalent reaction times, polymers with very similar compositional, spectroscopic and thermal properties were obtained. In addition, note that for particular reaction conditions at higher temperatures, the conversions achieved were similar to those using traditional heating systems. Moreover, it is highly informative that the MW radiation allows the generation of HCN-derived polymeric particles at the nanoscale which were not observed previously under any experimental condition using conventional methods; this technique could be successful for the tuning of their morphological properties, and for extension to obtain a new promising family of nanomaterials, taking into account the recent potential revealed by these polymeric systems. Due to the unexpected behavior of the cyanide MW polymerization, the multivariate analysis has turned out to be a successful tool to obtain a global interpretation of the results obtained herein and beyond to help us to find appropriate reaction conditions which lead to the generation of materials with concrete structural and morphological properties.

Being a very fast, robust, easy, low cost and green-solvent process, and the possibility to tune properly the properties of the final products makes the aqueous microwave-driven cyanide polymerization a highly attractive and promising methodology for the generation of new multifunctional materials. However, due to the apparently random behavior of the system along the reaction time, mainly due to the experimental variables to tune the properties of the polymers, it is necessary to carry out exhaustive synthetic and structural studies, and to examine the results under the light of statistical approaches to develop properly cyanide-based materials.

Supplementary Materials: The following are available online at https://www.mdpi.com/article/10.3390/polym14010057/s1. Figures S1–S17: SEM images of all of the cyanide polymers synthetized herein.

Author Contributions: C.P.-F., validation, formal analysis and investigation; P.V., formal analysis and resources; E.G.-T., formal analysis and writing—original draft preparation; E.M.-M., formal analysis, resources, writing—original draft preparation and funding acquisition; J.L.d.l.F., writing—review and editing, and supervision; M.R.-B., conceptualization, methodology, formal analysis, writing—original draft preparation, writing—review and editing. All authors have read and agreed to the published version of the manuscript.

Funding: This research was funded by the projects PID2019-104205GB-C21, PID2019-107442RBC32 and PID2019-104205GB-C22 from the Spanish Ministerio de Ciencia e Innovación, and by the Spanish State Research Agency, project MDM-2017-0737 Centro de Astrobiología (CSIC-INTA), Unidad de Excelencia María de Maeztu. C.P.-F. was supported by a research training grant from the Spanish Ministerio de Ciencia e Innovación, PRE2018-085781.

Institutional Review Board Statement: Not applicable.

Informed Consent Statement: Not applicable.

Data Availability Statement: Not applicable.

Acknowledgments: C.P.-F., E.G.-T., M.R.-B. and E.M.-M. used the research facilities of the Centro de Astrobiología (CAB), and were supported by the Instituto Nacional de Técnica Aeroespacial "Esteban Terradas" (INTA). Additionally, the authors are grateful to Mª Teresa Fernández for performing the FTIR spectra and the XRD measurements, the "Servicio de Análisis Térmico" of ICMM (CSIC, Spain), and also to the "Servicio de Resonancia Magnética Nuclear" of ICMM (CSIC, Spain).

Conflicts of Interest: The authors declare no conflict of interest. The funders had no role in the design of the study; in the collection, analyses, or interpretation of data; in the writing of the manuscript, or in the decision to publish the results.

References

1. Ruiz-Bermejo, M.; de la Fuente, J.L.; Carretero-González, J.; García-Fernández, L.; Aguilar, M.R. A Comparative Study on HCN Polymers Synthesized by Polymerization of NH_4CN or Diaminomaleonitrile in Aqueous Media: New Perspectives for Prebiotic Chemistry and Materials Science. *Chem.—Eur. J.* **2019**, *25*, 11437–11455. [CrossRef]
2. Thissen, H.; Evans, R.A.; Ball, V. Films and Materials Derived from Aminomalononitrile. *Processes* **2021**, *9*, 82. [CrossRef]
3. Toh, R.J.; Evans, R.; Thissen, H.; Voelcker, N.H.; d'Ischia, M.; Ball, V. Deposition of Aminomalononitrile-Based Films: Kinetics, Chemistry, and Morphology. *Langmuir* **2019**, *35*, 9896–9903. [CrossRef]
4. Ruiz-Bermejo, M.; de la Fuente, J.L.; Pérez-Fernández, C.; Mateo-Martí, E. A Comprehensive Review of HCN-Derived Polymers. *Processes* **2021**, *9*, 597. [CrossRef]
5. Zhou, X.; Fang, Y.; Su, Y.; Ge, C.; Jin, B.; Li, Z.; Wu, S. Preparation and Characterization of Poly-Hydrogen Cyanide Nanofibers with High Visible Light Photocatalytic Activity. *Catal. Commun.* **2014**, *46*, 197–200. [CrossRef]
6. Rahm, M.; Lunine, J.I.; Usher, D.A.; Shalloway, D. Polymorphism and Electronic Structure of Polyimine and Its Potential Significance for Prebiotic Chemistry on Titan. *Proc. Natl. Acad. Sci. USA* **2016**, *113*, 8121–8126. [CrossRef]
7. Cheng, S.-Y.; Chiang, Y.-L.; Chang, Y.-H.; Thissen, H.; Tsai, S.-W. An Aqueous-Based Process to Bioactivate Poly (ε-Caprolactone)/Mesoporous Bioglass Composite Surfaces by Prebiotic Chemistry-Inspired Polymer Coatings for Biomedical Applications. *Colloids Surf. B Biointerfaces* **2021**, *205*, 111913. [CrossRef]
8. Thissen, H.; Koegler, A.; Salwiczek, M.; Easton, C.D.; Qu, Y.; Lithgow, T.; Evans, R.A. Prebiotic-Chemistry Inspired Polymer Coatings for Biomedical and Material Science Applications. *NPG Asia Mater.* **2015**, *7*, e225. [CrossRef]
9. d'Ischia, M.; Manini, P.; Moracci, M.; Saladino, R.; Ball, V.; Thissen, H.; Evans, R.A.; Puzzarini, C.; Barone, V. Astrochemistry and Astrobiology: Materials Science in Wonderland? *Int. J. Mol. Sci.* **2019**, *20*, 4079. [CrossRef] [PubMed]
10. Ball, V. Antioxidant Activity of Films Inspired by Prebiotic Chemistry. *Mater. Lett.* **2021**, *285*, 129050. [CrossRef]
11. Pérez-Fernández, C.; Ruiz-Bermejo, M.; Gálvez-Martínez, S.; Mateo-Martí, E. An XPS Study of HCN-Derived Films on Pyrite Surfaces: A Prebiotic Chemistry Standpoint towards the Development of Protective Coatings. *RSC Adv.* **2021**, *11*, 20109–20117. [CrossRef]
12. Jung, J.; Menzies, D.J.; Thissen, H.; Easton, C.D.; Evans, R.A.; Henry, R.; Deletic, A.; McCarthy, D.T. New Prebiotic Chemistry Inspired Filter Media for Stormwater/Greywater Disinfection. *J. Hazard. Mater.* **2019**, *378*, 120749. [CrossRef]
13. Hortal, L.; Pérez-Fernández, C.; de la Fuente, J.L.; Valles, P.; Mateo-Martí, E.; Ruiz-Bermejo, M. A Dual Perspective on the Microwave-Assisted Synthesis of HCN Polymers towards the Chemical Evolution and Design of Functional Materials. *Sci. Rep.* **2020**, *10*, 22350. [CrossRef]
14. Mas, I.; de la Fuente, J.L.; Ruiz-Bermejo, M. Temperature Effect on Aqueous NH4CN Polymerization: Relationship between Kinetic Behaviour and Structural Properties. *Eur. Polym. J.* **2020**, *132*, 109719. [CrossRef]

15. Audebert, P.; Kroke, E.; Posern, C.; Lee, S.-H. State of the Art in the Preparation and Properties of Molecular Monomeric S-Heptazines: Syntheses, Characteristics, and Functional Applications. *Chem. Rev.* **2021**, *121*, 2515–2544. [CrossRef] [PubMed]
16. Inagaki, M.; Tsumura, T.; Kinumoto, T.; Toyoda, M. Graphitic Carbon Nitrides (g-C_3N_4) with Comparative Discussion to Carbon Materials. *Carbon* **2019**, *141*, 580–607. [CrossRef]
17. Cao, Q.; Kumru, B.; Antonietti, M.; Schmidt, B.V.K.J. Grafting Polymers onto Carbon Nitride via Visible-Light-Induced Photofunctionalization. *Macromolecules* **2019**, *52*, 4989–4996. [CrossRef]
18. Kumru, B.; Shalom, M.; Antonietti, M.; Schmidt, B.V.K.J. Reinforced Hydrogels via Carbon Nitride Initiated Polymerization. *Macromolecules* **2017**, *50*, 1862–1869. [CrossRef]
19. Ter Braak, C.J.F.; Smilauer, P. *CANOCO Reference Manual and Canodraw for Windows User's Guide: Software for Canonical Community Ordination*; Microcomputer Power: New York, NY, USA, 2002.
20. Ferris, J.P.; Edelson, E.H. Chemical Evolution. 31. Mechanism of the Condensation of Cyanide to Hydrogen Cyanide Oligomers. *J. Org. Chem.* **1978**, *43*, 3989–3995. [CrossRef]
21. Gail, E.; Gos, S.; Kulzer, R.; Lorösch, J.; Rubo, A.; Sauer, M.; Kellens, R.; Reddy, J.; Steier, N.; Hasenpusch, W. Cyano Compounds, Inorganic. In *Ullmann's Encyclopedia of Industrial Chemistry*; Wiley-VCH Verlag GmbH & Co. KGaA: Weinheim, Germany, 2011; pp. 673–710, ISBN 978-3-527-30673-2.
22. Dalton, S.; Heatley, F.; Budd, P.M. Thermal Stabilization of Polyacrylonitrile Fibres. *Polymer* **1999**, *40*, 5531–5543. [CrossRef]
23. Hortelano, C.; Ruiz-Bermejo, M.; de la Fuente, J.L. Solid-State Polymerization of Diaminomaleonitrile: Toward a New Generation of Conjugated Functional Materials. *Polymer* **2021**, *223*, 123696. [CrossRef]
24. Yang, C.; Wang, B.; Zhang, L.; Yin, L.; Wang, X. Synthesis of Layered Carbonitrides from Biotic Molecules for Photoredox Transformations. *Angew. Chem. Int. Ed.* **2017**, *56*, 6627–6631. [CrossRef] [PubMed]

polymers

Review

Copolymerization of Ethylene with Selected Vinyl Monomers Catalyzed by Group 4 Metal and Vanadium Complexes with Multidentate Ligands: A Short Review

Marzena Białek * and Julia Fryga

Institute of Chemistry, University of Opole, Oleska 48, 45-052 Opole, Poland; julia.herud@gmail.com
* Correspondence: marzena.bialek@uni.opole.pl

Abstract: This paper gives a short overview of homogeneous post-metallocene catalysts based on group 4 metal and vanadium complexes bearing multidentate ligands. It summarizes the catalytic behavior of those catalysts in copolymerization of ethylene with 1-olefins, with styrenic monomers and with α,ω-alkenols. The review is focused on finding correlations between the structure of a complex, its catalyst activity and comonomer incorporation ability, as well as the microstructure of the copolymer chains.

Keywords: post-metallocene catalyst; copolymerization; ethylene; 1-olefin; styrenic monomer; α,ω-alkenol

Citation: Białek, M.; Fryga, J. Copolymerization of Ethylene with Selected Vinyl Monomers Catalyzed by Group 4 Metal and Vanadium Complexes with Multidentate Ligands: A Short Review. *Polymers* **2021**, *13*, 4456. https://doi.org/10.3390/polym13244456

Academic Editor: Edina Rusen

Received: 21 November 2021
Accepted: 16 December 2021
Published: 19 December 2021

Publisher's Note: MDPI stays neutral with regard to jurisdictional claims in published maps and institutional affiliations.

Copyright: © 2021 by the authors. Licensee MDPI, Basel, Switzerland. This article is an open access article distributed under the terms and conditions of the Creative Commons Attribution (CC BY) license (https://creativecommons.org/licenses/by/4.0/).

1. Introduction

According to statistics, the global production of plastics in 2019 reached 368 million tons [1]. The production of polyethylene is estimated to account for about one third of all plastics manufactured, of which nearly 80% is formed in the catalytic processes [2]. The catalysts based on transition metal compounds were, for the first time, employed in the ethylene polymerization in the first half of 1950s. These were the Philips supported chromium catalyst and Ziegler–Natta catalysts comprising titanium chlorides and organoaluminium compounds which were converted into supported systems in time [3]. These multi-site catalysts produce polymers which are mixtures of macromolecules with different lengths and, in the case of copolymers, also with different compositions. Then, single-site metallocene and constrained geometry catalysts (CGC) were developed, which made it possible to synthesize polyolefins with more uniform macromolecules with narrow molecular weight distribution and more homogeneous comonomer distribution [4–6]. Moreover, they afford the incorporation of bulky olefins into polyethylene chains, as well as allowing production of copolymers with higher comonomer content than Ziegler–Natta catalysts and enable the synthesis of a variety of other materials, such as homopolymers and copolymers of cycloolefins, syndiotactic polystyrene, ethylene/styrene copolymers, etc. Since the 1990s, significant interest has arisen in catalysts based on well-defined transition metal complexes with non-cyclopentadienyl type ligands. Those catalysts with ligands containing donor heteroatoms, such as oxygen, nitrogen, sulfur, etc., termed "post-metallocene catalysts", enable fine-tuning of chain microstructure of polymers by ligand design [7,8]. The presence of many donor atoms in a ligand usually has a good stabilizing effect on active sites. Moreover, the ligand structure is decisive for the geometry of the complex to be formed, for metal coordination number, and for the formal oxidation state of the transition metal [9].

The catalytic polymerization of ethylene leads to high density polyethylene (HDPE), which is a thermoplastic polymer with high softening point and high crystallinity, and has very good chemical resistance [10]. Its properties are strongly affected by molecular weight, and higher molecular weights are advantageous inter alia for mechanical properties of polyethylene. Classifications of PE materials depending on their molecular weights are presented in Table 1 [11,12]. Molecular weight distribution (MWD) of polyethylene is another factor which influences the properties of polyethylene. For example, broad MWD polyethylene offers good impact resistance and good processability, while narrow MWD polyethylene will have greater toughness at low temperatures and higher resistance to environmental stress-cracking [13,14].

Table 1. Classification of linear polyethylene by its weight [11].

PE Type	M_w, kg/mol	T_m, °C
HDPE	100–500	125–135
HMWPE	200–1000	130–135
UHMWPE [2]	>3000 [1]	135–148

HMWPE—high molecular weight polyethylene, UHMWPE—ultra high molecular weight polyethylene, [1]—viscosity average molecular weight (M_v), and [2]—there is no single definition for that polymer. According to ISO 11542-1, this is PE with molecular weight of at least one million g/mol, while ASTM D 4020-00a specifies that the molecular weight value for UHMWPE is over 3.1 million g/mol [15].

Among the commercially available various grades of polyethylene (ref. to Tables 2 and 3), of great importance are polyethylenes which, from the viewpoint of their chemical structures, are copolymers of ethylene with other monomers, mostly with 1-olefins. They are usually statistical copolymers differing in the amounts of comonomer incorporated and chemical compositions distributions. That group covers: linear low density polyethylene (LLDPE and produced by metallocene catalysts mLLDPE), very low density polyethylene (VLDPE) and ultra-low density polyethylene (ULDPE). Ethylene and 1-olefin copolymers are commercially produced with the use of Ziegler–Natta catalysts and metallocene catalysts including constrained geometry catalysts. Multiblock copolymers of ethylene and 1-olefins, called olefin block copolymers (OBC), are also commercially available, and are synthesized with the use of post-metallocene catalysts and a new coordination polymerization method, i.e., Chain Schuttling Polymerization (CSP) [7,16]. The amount of 1-olefin incorporated into the copolymer, as well as the microstructure of the copolymer, influences the product density and the physicochemical properties which are determining factors for its final application.

Table 2. Structures of main grades of polyethylene and ethylene copolymers.

Ethylene Homopolymers		Ethylene Copolymers		
HDPE	UHMWPE	LLDPE	mLLDPE	VLDPE
			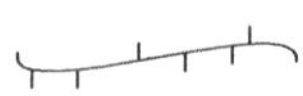	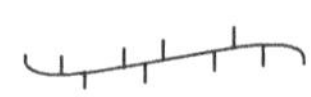
LDPE		ULDPE	OBC	COC
			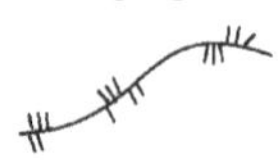	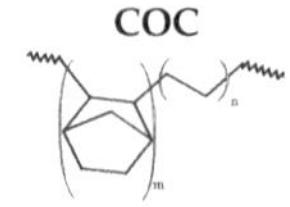

The properties of polyethylene can be modified, not only by introducing 1-olefin units, but also by copolymerization of ethylene with other compounds like norbornene, styrene, or polar monomers. The ethylene/norbornene copolymers, called cyclic olefin copolymers (COC), are successfully produced commercially and show high transparency, high humidity and thermal-resistance, and high water-vapor barrier properties. They are resistant to polar solvents and chemicals, too [17,18]. The introduction of styrene units into the polyethylene macromolecules causes changes in the viscoelastic behavior and thermo-mechanical properties of HDPE [19]. Functionalization of polyethylene through copolymerization of ethylene with polar comonomers is advantageous for the produced material from the viewpoint of its adhesion, toughness, printability/paintability, and rheological properties [20]. The copolymers of ethylene with polar comonomers or with styrene offer improved miscibility with other polymers [19,20]. It should be noted, however, that catalytic copolymerization of ethylene with polar monomers, for example, α,ω-alkenols, or styrene and its derivatives, faces some problems which result from the nature of those comonomers. Early transition metal catalysts have a tendency to form stable σ-complexes with the polar group of comonomer due to high oxophilicity of metal centers and, as a consequence, they can be easily poisoned [21]. Copolymerization of ethylene and styrenic monomers, on the other hand, often in addition to copolymer, yields homopolymer as contaminants. However, in the late 90s, production of ethylene/styrene copolymers, called pseudorandom interpolymers, was commercialized with the use of type CGC catalysts [22].

Table 3. Properties of LDPE and main polyethylene and ethylene copolymers grades produced using transition metal catalysts.

Type of (Co)Polymer	d [1], g/cm^3	T_m [2], °C	χ [3], %	MW [4], kg/mol	CC [5], wt%	Ref.
UHMWPE	0.93–0.94	130–140	60–75	2000–6000	-	[23]
HDPE	0.94–0.97	130–135	60–80	200–500	0–2.5	[23,24]
LDPE	0.91–0.93	100–110	40–50	≤600	0	[23,24]
LLDPE	0.91–0.94	120–125	40–60	≤600	5.0–12.0	[23–25]
VLDPE	0.89–0.91	118–122	25–40	-	10.0–35.0	[24,25]
ULDPE	0.86–0.89	40–84	-	-	-	[26,27]
OBC	~0.865–0.935	~120–130	-	-	-	[25]
COC	1.01–1.02	100–180 [6]	-	-	-	[23]

[1] d—density, [2] T_m—melting temperature, [3] χ—crystallinity, [4] molecular weight, [5] CC—comonomer content in copolymer, and [6] glass temperature (T_g, °C).

2. Copolymerization of Ethylene with Vinyl Monomers

2.1. Copolymerization of Ethylene with 1-Olefins

Copolymerization of ethylene with 1-olefins leads to structural changes in macromolecules of linear polyethylene, i.e., incorporated 1-olefins produce short branches the lengths of which are dependent on the type of the higher 1-olefin used. Such modifications cause changes in the crystallinity and density of polyethylene, and other physical properties, e.g., melting point, optical properties, as well as mechanical properties [25]. The ethylene/1-olefin copolymers show good strength, toughness, and sealing properties. Moreover, they offer advantageous optical properties which make them widely applicable in the packaging industry [24]. In addition to type of 1-olefin used, the properties of those materials depend on the amount of comonomer incorporated, as well as its distribution along the copolymer chains (inter- and intramolecular). The microstructure of copolymers, and other characteristics, like molecular weight and its distribution, are strictly related to the type of catalyst used for copolymerization. The metallocene catalysts make it possible to produce copolymers with higher degrees of comonomer incorporation, with more uniform comonomer distribution, and with narrower molecular weight distribution compared to the copolymers obtained with the use of Ziegler–Natta heterogeneous catalysts [26,28]. That results from existence of single site in metallocene catalysts and multiple active sites in the Ziegler–Natta catalysts with different reactivities toward the monomers, and from different accessibility of catalytic active sites for the comonomer molecules [26,28]. In addition to statistical ethylene/1-olefin copolymers, post-metallocene catalysts offer copolymers with new architecture, block copolymers, and multiblock copolymers. It became possible due to innovation in living polymerization of olefins and development of chain shuttling polymerization.

2.1.1. Complexes of Group 4 Transition Metals

The post-metallocene catalysts, which are based on the complexes of group 4 transition metals, have been extensively examined in copolymerization of ethylene with 1-olefins. It is not possible to present the catalytic performance of all the groups of the investigated complexes within this short review. Hence, the survey is limited only to the most representative catalysts producing copolymers with different macromolecular structures.

The titanium and zirconium complexes with tetradentate ligands make one of the groups most often used in olefin copolymerization. S.E. Reybuck et al. [29] compared the effects of the symmetry of complexes, steric bulk of the substituent at the aryl ring, and the type of activator on the ethylene/1-hexene copolymerization behavior of zirconium diamino-bis(phenolate) complexes **Zr-1–Zr-5** (Figure 1). The complexes with C_s symmetry (**Zr-2–Zr-5**), i.e., those with the amino-bis(phenolate) ligands with the pendant donor group, were demonstrated to be more active and to offer a better ability to incorporate comonomer than those with C_2 symmetry (**Zr-1**). The copolymer produced by **Zr-2**/MMAO (MMAO- modified methylaluminoxane) at $X_E/X_H = 0.11$ (mol/mol) contained 44.9 mol% 1-hexene, while the copolymer synthesized with **Zr-1**/MMAO at the same feed ratio of comonomers contained only 8.7 mol% of 1-hexene. The symmetry of the complex also affected the distribution of comonomer. The product of the reactivity ratios for copolymerization with **Zr-1** was higher than 1 ($r_e r_h = 2.6$) and for copolymerization with **Zr-2** below 1 ($r_e r_h = 0.18$), which suggested the presence of blocky and slightly alternating sequences of the comonomer units, respectively. It was also shown that the 1-hexene incorporation ability of the C_s symmetric complexes increases in the order of decreasing steric bulk at the 2,4-aryl positions. However, bulky substituents like *tert*-butyl groups (**Zr-2**) or phenyl groups (**Zr-3**) were found to improve the complex activity at the same time. The copolymerization behavior of the complexes was also sensitive to the nature of the activator. Copolymers produced by C_s-symmetric complexes activated with MMAO had higher 1-hexene incorporation by about 10 mol% than those synthesized with borates as activators. All the copolymers synthesized with the use of **Zr-1–Zr-5** complexes showed monomodal distribution and narrow molecular weight distribution ($M_w/M_n = 1.3$–1.9), which is indicative for homogeneity of active sites in the catalysts [29].

The dimeric diamine-bisphenolate complex **Zr-6** activated with $Al(iBu)_3/Ph_3CB(C_6F_5)_4$ produced copolymers with very high 1-octene incorporation, up to ≈20 mol% at 0.70 mol/L of comonomer in the feed, with moderate molecular weight and dispersity (M_w = 14,000–42,000 g/mol, $M_w/M_n = 2.1$–3.2). In addition, the activating effect of the comonomer on the copolymerization yield was observed independently on the comonomer concentration. Its activity was over four times higher in ethylene/1-octene copolymerization than in ethylene homopolymerization [30]. In contrast, the titanium complexes **Ti-1–Ti-4** bearing the same type of tetradentate ligand in conjunction with $Al(iBu)_3/Ph_3CB(C_6F_5)_4$ exhibited the negative comonomer effect and lower activity than the zirconium catalyst. The comonomer incorporation ability for these catalysts was also lower than for **Zr-6** and it was strictly dependent on the ligand structure. The complexes **Ti-1** and **Ti-4** with the NMe_2 donor group gave copolymers with 9.0 and 12.0 mol% of 1-octene at the comonomer concentration equal to 0.70 mol/L. Interestingly, **Ti-4** produced the ethylene/1-octene copolymer, not only with a high incorporation level, but at the same time with the very high molecular weight and very narrow molecular weight distribution (M_w = 755,000 g/mol, $M_w/M_n = 1.7$). The introduction of sterically bulkier $N(iPr)_2$ pendant donor group in the complexes **Ti-2** and **Ti-3** resulted in lower 1-octene contents: 5.1 mol% and 2.8 mol%, respectively [30].

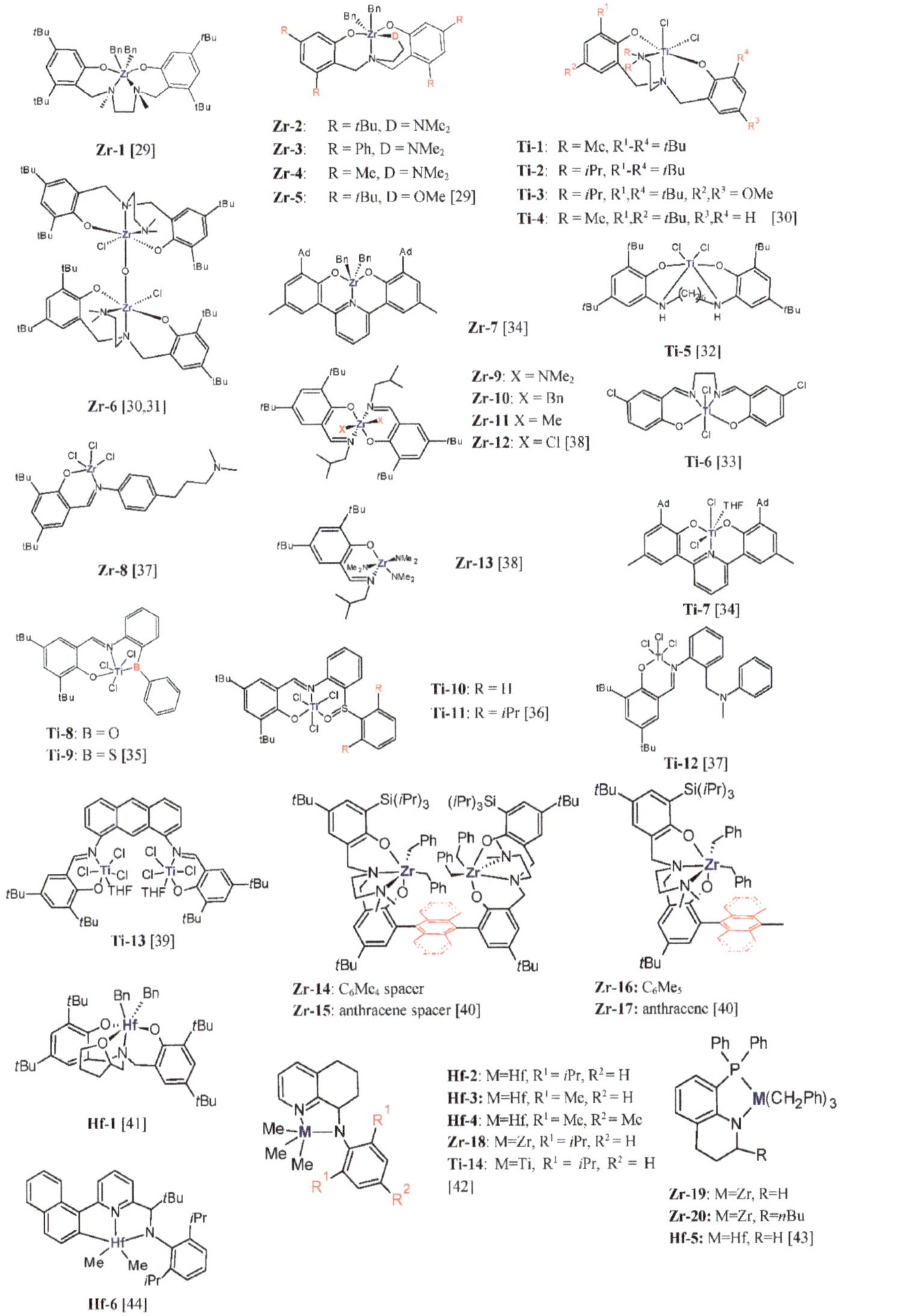

Figure 1. Titanium, zirconium and hafnium complexes investigated as precatalysts for ethylene/1-olefin copolymerization [29–44].

Further studies [31] showed that the presence of $AlMe_3$ in the feed during the copolymerization catalyzed by **Zr-6**/Al(*i*Bu)$_3$/$Ph_3CB(C_6F_5)_4$ lowered the catalyst activity and greatly increased the amount of comonomer content in the copolymer from about 10 to approximately 40 mol% for 1-octene and 31 mol% for 1-hexene when the $AlMe_3$/Zr molar ratio was equal to 100/1. The increase in the $AlMe_3$/Zr molar ratio to 150:1 resulted in the production of ethylene/1-octene copolymer with the comonomer content of 65 mol%. The presence of $AlMe_3$ significantly affected the microstructure of the produced copolymers which possessed both blocky and alternating sequences, and decreased the molecular weight of the produced copolymers. However, other alkylaluminium compounds, $AlEt_3$ and Al(*i*Bu)$_3$, changed molecular weight of copolymers but did not cause any important changes in the catalytic activity, composition, or microstructure of the copolymerization products. Hence, only $AlMe_3$ interacts with the active sites to modify their catalytic properties [31].

Titanium catalyst **Ti-5**/MAO (MAO-methylaluminoxane), based on a complex with tetradentate salan type ligand, was investigated in ethylene/1-octene copolymerization as well. At the comonomer concentration of 0.73 mol/L, it gave the copolymer with moderate incorporation (3.7 mol%) and its activity was low (3.7 kg/(mol$_{Ti}$·h)) [32]. Similarly, low activity and moderate incorporation ability were exhibited in ethylene/1-octene copolymerization by the titanium catalyst **Ti-6**/MAO, which was based on the tetradentate salen type ligand [33]. At the highest investigated comonomer concentration, 1.340 mol/L, the copolymer with 5.5 mol% of 1-octene was obtained and the catalytic activity was 14.7 kg/(mol$_{Ti}$·0.5 h). The copolymers produced by **Ti-6** were characterized by very high molecular weights (M_w = 1023 × 10^3–1545 × 10^3 g/mol) and very broad molecular weight distributions.

Among the large number of group 4 post-metallocene catalysts reported for ethylene copolymerization, are those based on complexes with tridentate ligands. That group covers **Ti-7** and **Zr-7** complexes bearing the bisphenolate ligand with the pyridinediyl linker which were tested in ethylene/1-octene copolymerization [34]. Both complexes in conjunction with MAO turned out ineffective in that process. The titanium catalyst exhibited very low, 3 kg/(mol$_{Ti}$·h), activity and **Zr-7**/MAO gave the copolymer which contained only 0.5 mol% of comonomer, despite low ethylene pressure (1 atm) and high concentration of 1-olefin (~50% vol.) in the feed. In contrast, the titanium complexes **Ti-8–Ti-10** bearing the tridentate monoanionic [ONX] (X = O, S) ligands, after activation with MMAO, showed much better capability to copolymerize ethylene and 1-hexene [35,36]. The complex **Ti-9** with S in the sidearm was both highly active (387 kg/[mol$_{Ti}$·h·atm]) and gave a copolymer with the very high 1-hexene content (30 mol%). The corresponding complex with an O-donor as the sidearm (**Ti-8**) under the same reaction conditions produced copolymers with lower activity and comonomer incorporation equal to 23 kg/(mol$_{Ti}$·h·atm) and 14.1 mol%, respectively. The excellent copolymerization behavior showed the complex **Ti-10** with sulfoxide group [36] which produced copolymer with incorporation levels as high as 38.8 mol% and the activity of 320 kg/(mol$_{Ti}$·h·atm). These results show that pendant donor group in phenoxy-imine ligand can easily tune the activity of catalyst and its capability for copolymerization. Further investigations showed that substituent on the sidearm also influenced strongly the copolymerization behavior of the complexes. In contrast to **Ti-10**, complex **Ti-11,** due to the steric effects of the isopropyl group, produced copolymer with very low comonomer content (0.6 mol%) [36]. Those results demonstrate that the catalytic properties of **Ti-8–Ti-11** can be well correlated with the steric and electronic properties of the side arm donor group [35].

When the zirconium **Zr-8** and titanium **Ti-12** complexes bearing potentially tridentate phenoxy-imine ligands were employed in ethylene/1-octene copolymerization in the presence of MMAO, their activities were very high (≈12,300 kg/[mol$_{Zr}$·h]) and moderate (≈450 kg/[mol$_{Ti}$·h]). However, their comonomer incorporation abilities were very low (up to 0.1 mol%) and low (up to 1.5 mol%), respectively [37]. When activated with Al(*i*Bu)$_3$/$Ph_3CB(C_6F_5)_4$, these complexes were able to produce copolymers with higher comonomer contents (1.6 and 3.8 mol%, respectively), yet their activities were much lower. Interestingly, the produced copolymers showed broad chemical composition distributions and very broad molecular weight distributions, as well as high molecular weights irrespective of the comonomer concentration in the feed. At the 1-octene concentration of 0.58 mol/L, **Zr-8** produced copolymers with M_w = 440,000 g/mol and M_w/M_n = 116 (trimodal distribution), whereas **Ti-12** gave the product with M_w = 690,000 g/mol and M_w/M_n = 26.5 (bimodal distribution) [37].

Y. Gao et al. [38] studied ethylene/1-octene copolymerization with the use of phenoxy-imine complexes of zirconium (**Zr-9–Zr-12**). In contrast to other studies, they did not focus on modification of ancillary ligand structure, but on changing the type of monodentate σ-ligands. The synthesized complexes had chloride, methyl, benzyl, and dimethylamine as labile ligands. Such modification of precatalyst **Zr-9–Zr-12** was found to have considerable effect on their copolymerization behavior. Poly(ethylene-*co*-1-octene) produced by **Zr-9**/$AlMe_3$/$Ph_3CB(C_6F_5)_4$ contained 7.2 mol% of comonomer. Under the same conditions (5 mL of 1-octene, 4 atm of ethylene, 40 °C, Zr/$AlMe_3$ = 40), complexes with benzyl and methyl σ-ligands, **Zr-10** and **Zr-11**, gave copolymers with much lower 1-olefin incorporation, ≤1 mol%. Then, the complex **Zr-12** with the chloride labile ligand turned out inactive in copolymerization of ethylene and 1-octene. The comparison of the copolymerization results for the bis-FI and mono-FI complexes **Zr-9** and **Zr-13** revealed that both complexes are capable of producing copolymers with high comonomer incorporation, with 6.3 and 7.2 mol% of 1-octene, respectively, albeit with different activities. The activity of **Zr-9** with two phenoxy-imine ligands was 2846 kg/(mol·h·atm) while that of **Zr-13** was equal to 1812 kg. These two complexes behaved even more similarly at Zr/$AlMe_3$ = 120 which could indicate that both those complexes form similar or identical cationic mono-FI-Zr catalytic species [38].

Recently, dinuclear titanium complex bearing anthracene-briged bisphenoxyimine ligand (**Ti-13**) activated with MAO was investigated as a precatalyst for ethylene/1-hexene copolymerization [39]. However, it was found to exhibit rather low activity and comonomer incorporation ability (303 kg/(mol_{Ti}·h), 1.8 mol%). In contrast, C_2 symetric dinuclear zirconium catalysts supported by bisphenolate ligands with bulky Si(*i*Pr)$_3$ substituent, **Zr-14** and **Zr-15**, in conjunction with MAO showed high activity and produced copolymers with good 1-hexene incorporation. At 3 bar of ethylene and 20 mL of comonomer in the feed, their activity and 1-hexene incorporation were equal to 2000 kg/(mol_{Zr}·h), 22 mol% and 1900 kg/(mol_{Zr}·h), 30 mol%, respectively. Their monometallic analogues, **Zr-16** and **Zr-17**, showed slightly lower incorporation ability (19 mol% and 28 mol%) [40]. It was also revealed that **Zr-14–Zr-17** incorporate the longer chain 1-olefin, 1-tetradecene, at lower level (3 mol%–10 mol%).

Ethylene and 1-hexene were also copolymerized by hafnium-amine bis(phenolate) complex with tetrahydrofuran pendant arm (**Hf-1**), activated by $B(C_6F_5)$ [41]. The catalyst was able to produce copolymers over a wide range of 1-hexene incorporation, from 25 mol% to 85 mol%, narrow dispersity, and M_n = 5600–47500 g/mol by changing comonomer concentration from 0.1 to 2.5 mol/L and reaction time from 10 to 90 min. However, its activity was modest (5.4–41.6 g/($mmol_{Hf}$·h)). Amido-trihydroquinoline hafnium complexes **Hf-2, Hf-3** and **Hf-4** with $Ph_3CB(C_6F_5)_4$ as the activator showed both high activity and ability to produce copolymers with high 1-octene content and high molecular weights (224,000–1220,000 g/mol) at high temperature (100 °C). At 1-octene concentration equal to 1 mol/L their activity decreased in the order **Hf-2**, 16,200 kg/(mol_{Hf}·h) >> **Hf-4**, 2520 kg/(mol_{Hf}·h) **>** **Hf-3**, 2340 kg/(mol_{Hf}·h) and incorporation ability was changed as follows: **Hf-2** (15.6 mol%) > **Hf-3** (7.8 mol%) > **Hf-4** (4.8 mol%), indicating significant substituent steric effect on catalytic properties of hafnium catalysts [42]. Zirconium complex **Zr-18** bearing the same ligand as complex **Hf-2** was also highly efficient in ethylene/1-octene copolymerization (13,500 kg/(mol_{Zr}·h), 12.4 mol%) but the exchange of Hf by Ti (**Ti-14**) led to catalyst with lower activity and low incorporation ability (2160 kg/(mol_{Ti}·h), 1.7 mol%) [42].

Phosphine–amido hafnium and zirconium complexes bearing various substituents at the 2-position of the tetrahydroquinoline framework (R = H, Me, *i*Pr, *n*Bu) were evaluated in ethylene/1-octene copolymerization with $[HNMe(C_{18}H_{37})_2][B(C_6F_5)_4]$ as activator and under the following conditions: 1 mol/L of 1-octene, 30 bar of ethylene, 100 °C (initial reaction temperature), and 3 min [43]. Moderate 1-octene incorporation (7.7 mol%) showed zirconium complex **Zr-19**, however its Hf counterpart (**Hf-5**) and the other Zr and Hf complexes exhibited low comonomer incorporation (2.1 mol%–3.6 mol%). Activities of complexes fell in the range of 7000–29,000 kg/(mol_M·h), with the highest and lowest activity shown by **Zr-20** and **Hf-5**, respectively.

One of the interesting classes of postmetallocene catalysts are catalysts based on pyridylamido hafnium complexes which were shown to have high incorporation ability in ethylene/1-olefin copolymerization at high temperature [44–46]. C_1-symmetric *tert*-butyl substituted pyridylamido hafnium complex **Hf-6** activated with $Ph_3CB(C_6F_5)_4/Al(i\text{Bu})_3$ was able to incorporate 6.0 mol% of 1-octene and 8.7 mol% of 1-hexene in ethylene/1-olefin copolymerization carried out with the addition of 0.1 mol of comonomer, at 80 °C for 10 min under 10 atm of ethylene. The produced copolymers had broad MWD with bimodal distribution and high molecular weight equal to 262,000 g/mol and 931,000 g/mol (copolymer with 1-octene and 1-hexene, respectively) [44]. Selected results of ethylene/1-olefin copolymerization with group 4 metal complexes are displayed in Table 4.

Fluorinated bis(phenoxy-imine)titanium complexes **Ti-15**, **Ti-16**, **Ti-17**, **Ti-18**, and **Ti-19** (Figure 2) activated by MAO were also applied in ethylene/1-olefin copolymerization (1-olefin = 1-hexene, 1-octene, 1-decene) [47]. All the catalysts were shown to offer high productivity. However, they had significantly different abilities to incorporate 1-olefins; this was dependent on the steric hindrance of the *ortho*-substituent. Reduction in the bulkiness of the substituent R enhanced reactivity of the catalysts towards higher 1-olefins. **Ti-18** (R^1 = H) displayed the highest 1-hexene incorporation ability (22.6 mol%), whereas **Ti-16** (R^1 = *t*Bu) showed the lowest efficiency (3.2 mol%). Moreover, the produced copolymers had very narrow molecular weight distributions (M_w/M_n = 1.07–1.22) which indicated that the living ethylene/1-olefin copolymerization took place in the presence of those catalysts.

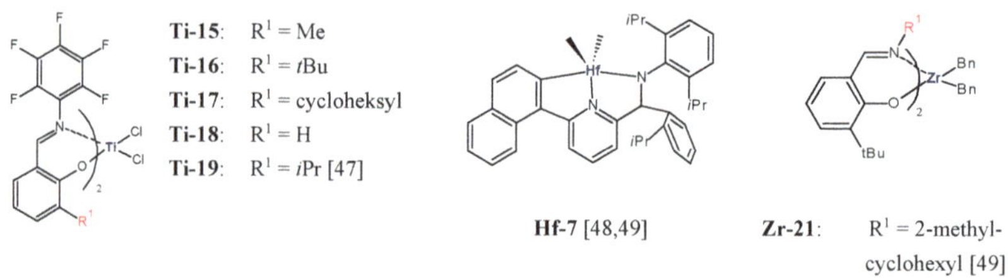

Figure 2. Fluorinated bis(phenoxy-imine) titanium complexes as well as bis(phenoxy-imine) zirconium and pyridylamide hafnium complexes used for synthesis of multiblock copolymers [47–49].

Table 4. Results of ethylene/1-olefin copolymerization with group 4 metal complexes.

Complex	[M], μmol	Activator	E, bar	Time, min	T, °C	Comonomer		Activity, kg/(mol$_M$·h)	CI, mol%	M_w, kg/mol	M_w/M_n	Ref.
						Type	mol/L					
Zr-2	5	MMAO	2.8	[1]	20	1-Hex	0.0666 [2]	798	49.9	209 [3]	1.7	[29]
Zr-4	100	MMAO	3.6	[1]	20	1-Hex	0.0809 [2]	460	64.1	114 [3]	1.6	[29]
Zr-6	80	Al(*i*Bu)$_3$/B	5	30	60	1-Oct	0.19	405.0	6.5	42	2.1	[30]
Zr-6 [4]	50	Al(*i*Bu)$_3$/B	5	30	60	1-Oct	0.27	28.3	39.5	15.8	5.5	[31]
Ti-1	80	Al(*i*Bu)$_3$/B	5	30	60	1-Oct	0.19	63.8	2.7	235	2.4	[30]
Ti-4	80	Al(*i*Bu)$_3$/B	5	30	60	1-Oct	0.19	44.8	3.2	292	3.9	[30]
Ti-4	80	Al(*i*Bu)$_3$/B	5	30	60	1-Oct	0.70	42.0	12.0	755	1.7	[30]
T-5	70	MAO	5	30	30	1-Oct	0.73	78.2	3.7	229	17.6	[32]
Ti-8	7	MMAO	1	15	25	1-Hex	0.7	23.0	14.1	229	2.5	[35]
Ti-9	10	MMAO	1	15	25	1-Hex	0.7	387.0	30.0	146	2.36	[35]
Ti-10	3	MMAO	1	60	50	1-Hex	0.34	100.0	11.0	79	2.0	[36]
Ti-11	3	MMAO	1	10	50	1-Hex	0.34	220.0	0.6	143	3.8	[36]
Ti-12	25	Al(*i*Bu)$_3$/B	5	20	60	1-Oct	0.58	87	3.8	690	26.5	[37]
Zr-9	10	$AlMe_3$/B	4	2	40	1-Oct	0.64	2846 [5]	7.2	6.0	1.5	[38]
Zr-10	10	$AlMe_3$/B	4	0.5	40	1-Oct	0.64	5649 [5]	1.0	5.0	1.3	[38]
Hf-1	3 [6]	$B(C_6F_5)_3$	1	30	25	1-Hex	0.1	5.4	25	5.6 [3]	1.31	[41]
Hf-1	3 [6]	$B(C_6F_5)_3$	1	30	25	1-Hex	0.5	20.3	72	14 [3]	1.18	[41]
Hf-6	1	Al(*i*Bu)$_3$/B	10	10	80	1-Hex	1.0	5280	8.7	931	4.3	[44]

1-Hex = 1-hexene, 1-Oct = 1-octene, E = ethylene, CI = comonomer incorporation, B = $Ph_3CB(C_6F_5)_4$, [1] 20–30 min, [2] feed ratio of comonomers $X_E/X_{1\text{-}Hex}$, [3] M_n, kg/mol, [4] copolymerization in the presence of $AlMe_3$ ($AlMe_3$/Zr = 100), [5] kg/(mol$_M$·h·atm), and [6] mmol/L.

The living polymerization which occurs with rapid initiation and negligible chain termination or chain transfer makes it possible to synthesize copolymers with well-defined blocks [50]. Most of the catalysts displayed the living olefin polymerization characteristics at low temperatures which resulted in low activity and insufficient molecular weights of the polymers [51,52]. Bis(phenoxy-imine) titanium catalysts which incorporate fluorine atom(s) ortho to the imine-nitrogen can catalyze the living (co)polymerization of olefins at ambient and elevated temperatures [51]. In the presence of **Ti-15**/MAO, a number of block copolymers of type PE-*block*-poly(ethylene-*co*-1-hexene) were synthesized using the sequential addition polymerization procedure. Linear polyethylene macromolecules (M_n = 38,300 g/mol, M_w/M_n = 1.11) were obtained initially and then a mixture of ethylene and 1-hexene monomers was fed to obtain another block. The produced block copolymers contained from 3.3 mol% to 15.0 mol% of 1-hexene and were characterized by high molecular weights (M_n = 61,500–121,000 g/mol), narrow MWD (M_w/M_n = 1.21–1.31), and melting points from 117 °C to 130 °C [47]. The block copolymers of that type were shown to possess a good combination of extensibility and toughness compared to the corresponding random copolymers.

Further, the coordinative chain transfer copolymerization, CCTcoP, of ethylene with 1-olefins was developed. It was aimed at reducing the amounts of transition metal catalysts needed and controlling the molecular weight of polymers. In this approach, the growing polymer chain is capable of transferring from the catalyst to the chain transfer agent (CTA), which is a dormant species in the course of copolymerization [53]. P.D. Hustad et al. reported the method for the synthesis of block copolymers in the continuous process using CCT*co*P, with **Hf-7**/MAO as a catalyst (Figure 2) and $ZnEt_2$ as CTA [48]. Two reactors were employed in the process (Figure 3). Ethylene homopolymerization took place in reactor I, while copolymerization of ethylene and 1-olefin was carried out in reactor II. When the HDPE-*block*-VLDPE copolymers obtained with CTA were compared with the products synthesized with no CTA (physical blend obtained in a dual-reactor), it was found that the copolymer produced with the use of CTA showed higher molecular weight, narrower MWD (M_n = 44.5 kg/mol, M_w/M_n = 1.67), and somewhat lower melting point (T_m = 122 °C) than the product obtained with no use of CTA (M_n = 25.9 kg/mol, M_w/M_n = 4.42, T_m = 126 °C). In that type of copolymerization, both the block composition and the comonomer content in each block can be easily tailored by changing the production rate in reactor I and reactor II, and by feed composition, respectively [48].

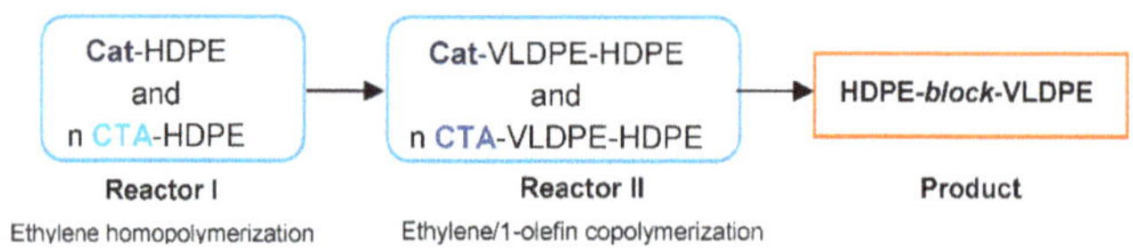

Figure 3. Synthesis of HDPE-*block*-VLDPE copolymer using coordinative chain transfer copolymerization and two reactors. Adopted from [48].

Block copolymers with novel architecture, showing excellent elastomeric properties, were introduced by Arriola et al. [49]. Linear multiblock ethylene/1-octene copolymers, with sequential crystallizable segments (low 1-octene content) and amorphous segments (high 1-octene content) were synthesized via chain shuttling polymerization. That type of polymerization involves two catalysts with different comonomer incorporation abilities and a chain shuttling agent (CSA) which is able to pass the growing polymer chain between catalytic sites so that different parts of a single polymer molecule grow on different catalysts [49,53]. The pyridylamide complex **Hf-7**, with the high ability to incorporate 1-octene, and the bis(phenoxy-imine) complex **Zr-21** (Figure 2), with low incorporation ability, combined with $ZnEt_2$ as a CSA, made it possible to obtained multiblock copolymers which were characterized by very low densities (0.879–0.883 g/cm^3) and high melting points (T_m = 120–124 °C) at the same time. Moreover, unlike the copolymer prepared without $ZnEt_2$, which was bimodal with M_w/M_n = 13.8, they show monomodal and narrow distributions of molecular weights (M_w/M_n = 1.98–3.22) [49].

2.1.2. Complexes of Vanadium

The vanadium catalysts for the olefin polymerization have been known since the 1950s and such classical Ziegler–Natta ones possess some interesting properties, which made them applicable in the commercial production of ethylene/propylene/diene and ethylene/cyclic olefins copolymers [54]. Although metallocene-type vanadium catalysts did not attract much attention of the researchers, the vanadium catalysts based on complexes with multidentate ligands started drawing considerably more and more interest as their catalytic performance, including thermal stability, can be modified by the use of specific multidentate ligands. The detailed review of the vanadium catalysts, which were employed in the olefin polymerization processes within the last decade, was published by A. M. F. Phillips et al. in Coordination Chemistry Reviews in 2020 [55], and earlier works were summarized by S. Gambarotta in Coordination Chemistry Reviews in 2003 [56]. Here, we summarize some interesting developments of the vanadium complexes for ethylene/1-olefin copolymerization.

Y.S. Li et al. studied the performance of vanadium(III) complexes (Figure 4) bearing bidentate ligands with [O,P], [N,N], and [O,O] donor atoms in ethylene/1-hexene copolymerization [57–59]. Et_2AlCl and $Cl_3CCOOEt$ (ETA) were used as the activator and the reactivator, respectively. The copolymerizations were carried out at 1 atm of ethylene, 1-hexene concentration equal to 0.2 mol/L, and at 25 °C for 5 or 10 min. The complexes **V-1–V-5** showed high activities, 4.08–7.44 kg/(mmol$_V$·h), and they produced copolymers with comonomer content between 2.22–4.20 mol% and with very narrow molecular weight distribution ($M_w/M_n \leq 1.9$). For the complexes with electron donating imino groups, the 1-hexene incorporation ability was shown to increase in line with the decreasing steric hindrance at the 2,6-aryl positions (**V-3** < **V-2**, **V-4** < **V-1**). The presence of the electron withdrawing group in the complex decreased the 1-hexene incorporation (**V-1** > **V-5**). In addition, the properties of the resultant copolymers can be controlled over a wide range by changing the reaction parameters. At 75 °C and at 1-hexene initial concentration of 0.6 mol/L, the **V-3**/Et_2AlCl system was found to produce the copolymer with 1-hexene incorporation as high as 16 mol% [57]. In general, the complexes **V-6–V-12** and **V-13–V-19** exhibited good activities and incorporation ability in the range of 1.08-5.34 kg/(mmol$_V$·h), 2.34–5.21 mol% [58], 0.84–4.56 kg/(mmol$_V$·h), and 2.33–3.75 mol% [59], respectively. Nevertheless, their copolymerization behavior is influenced by the nature of substituents in the ligands. The increasing steric hindrance on the aryloxy moiety of the phenoxy-phosphine ligand was shown to improve the catalytic activity (**V-13** < **V-15** < **V-16**, and **V-17** < **V-18**). However, the increased steric bulk of the ligand in **V-19** slightly lowered the catalytic activity for copolymerization (**V-18** > **V-19**). In addition, the electron-withdrawing effect of the F-substituted phenoxy-phosphine ligand influenced the catalytic performance unfavorably [59]. Moreover, the steric hindrance of the ligand and the electron-withdrawing effect of the ligand was also unfavorable for 1-hexene insertion.

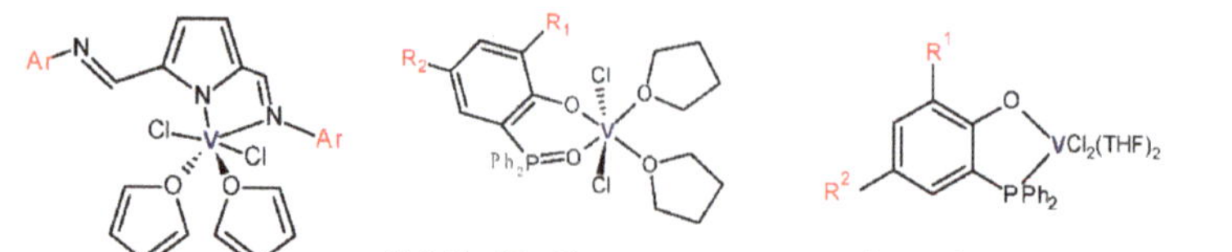

V-1: Ar = C_6H_5
V-2: Ar = 2,6-$Me_2C_6H_3$
V-3: Ar = 2,6-$iPr_2C_6H_3$
V-4: Ar = 2,4,6-$Me_3C_6H_2$
V-5: Ar = C_6F_5 [57]

V-6: $R^1 = R^2$ = H
V-7: R^1 = F, R^2 = H
V-8: R^1 = *t*Bu, R^2 = H
V-9: R^1 = Ph, R^2 = H
V-10: $R^1 = R^2$ = Me
V-11: $R^1 = R^2$ = *t*Bu
V-12: $R^1 = R^2$ = CMe_2Ph [58]

V-13: R^1 = H, R^2 = H
V-14: R^1 = F, R^2 = H
V-15: R^1 = Ph, R^2 = H
V-16: R^1 = *t*Bu, R^2 = H
V-17: R^1 = Me, R^2 = Me
V-18: R^1 = *t*Bu, R^2 = *t*Bu
V-19: R^1 = CMe_2Ph, R^2 = CMe_2Ph [59]

Figure 4. Vanadium complexes with bidentate ligands investigated as precatalysts for ethylene/1-olefin copolymerization [57–59].

The vanadium complexes **V-20** and **V-21** (Figure 5), activated either with $EtAlCl_2$ or with Et_2AlCl, were also examined in ethylene/1-octene copolymerization [37]. The copolymerization tests were carried out in the presence of ETA at 60 °C, for 20 min, under 5 bar ethylene, and in the presence of 2, 5, and 6.5 mL 1-octene. Both these complexes in conjunction with $EtAlCl_2$ exhibited high catalytic activities (66,400–10,400 kg/(mol$_V$·h)) and, taking into account the high pressure of ethylene and low comonomer concentration, they offered acceptable incorporation ability (1.3–3.6 mol%). The produced copolymers were characterized by high molecular weights and low MWD (M_w = 380,000–120,000 g/mol, $M_w/M_n \leq 2$). When the activator was changed for Et_2AlCl, both complexes were less active and they produced copolymers with lower comonomer contents and high dispersity (M_w/M_n = 6.4–7.9).

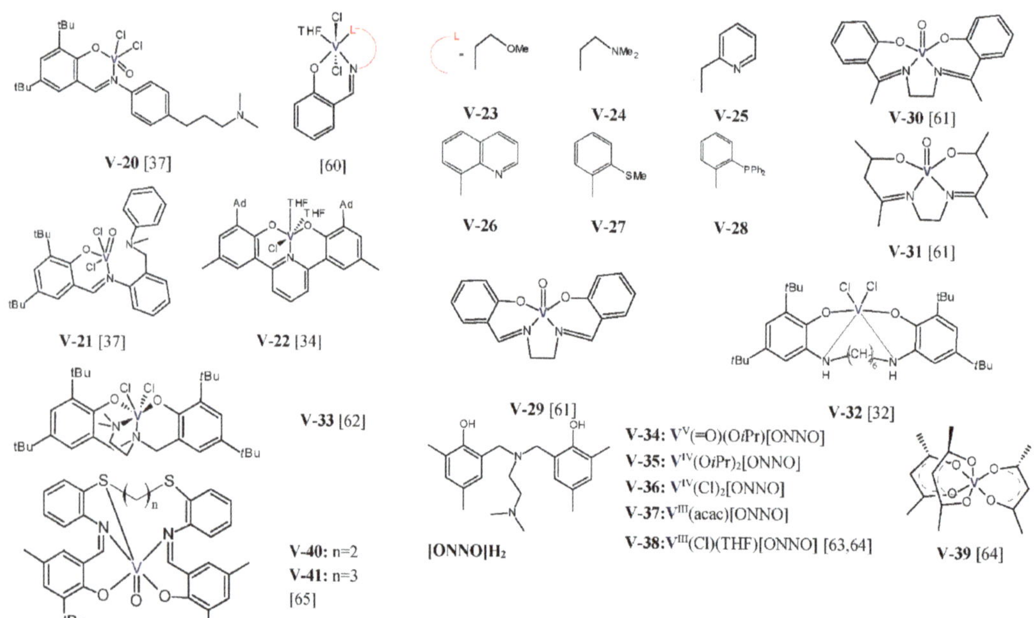

Figure 5. Vanadium complexes investigated as precatalysts for ethylene/1-olefin copolymerization [32,34,37,60–65].

S.R. Golisz [34] reported the use of the vanadium complex **V-22** with the tridentate bis(phenolate) ligand with the pyridinediyl linker, which was analogous to the complexes **Ti-7** and **Zr-7**, in copolymerization of ethylene and 1-octene. **V-22**/MAO exhibited low activity (50 kg/[mol_V·h]) and moderate incorporation ability. The copolymer obtained at high 1-octene/ethylene ratio contained 6.6 mol% of comonomer; still, it was much higher than for the use of group 4 catalysts under the same conditions. J-Q. Wu et al. [60] studied the performance of the vanadium(III) complexes **V-23–V-28** bearing tridentate phenoxy-imine ligands with different pendant donor groups, in copolymerization of ethylene with 1-hexene. At the same copolymerization conditions: 0.27 mol/L 1-hexene, 1 bar ethylene, at 25 °C, with Et_2AlCl as an activator, and ETA as a reactivator, the studied complexes were characterized by different activities, from 0.3 kg/($mmol_V$·h·bar) up to 3.61 kg/($mmol_V$·h·bar), with the figures changing as follows: **V-23** < **V-27** < **V-24** < **V-26** < **V-28** < **V-25**. Hence, the results were dependent both on the pendant donor and on the electronic effect of the backbone of the ligands. All the catalysts produced copolymers with moderate comonomer contents, 2.3–3.2 mol%, with high molecular weights, 41,800–93,800 g/mol, and with unimodal molecular weight distributions (M_w/M_n = 1.9–2.9). The most active complex, **V-25**, produced the copolymer with 3 mol% of 1-hexene and its content in the copolymer increased to 12.9 mol% when the comonomer concentration was raised to 1.35 mol/L. Moreover, enhanced catalytic activity was observed for the catalyst **V-25** with the increasing polymerization temperature up to 50 °C, which is indicative for its good thermal stability.

The oxovanadium(V) complexes (**V-29–V-31**) with tetradentate Schiff base ligands were also used in copolymerization of ethylene with 1-olefins as precatalysts. Those complexes, when activated with $EtAlCl_2$, make it possible to produce poly(ethylene-*co*-1-octene) with the comonomer content from 5.26 to 6.25 mol% and melting points ~112 °C, at the comonomer concentration of 0.40 mol/L. The copolymer with the highest 1-octene content was obtained with the use of **V-29** and it was found to contain isolated comonomer units only [61]. The discussed complexes offered low catalytic activities in copolymerization, from 2.6 kg/(mol_V·h) to 18.1 kg/(mol_V·h), and it grew up in the following line: **V-29** < **V-30** < **V-31**. Higher activities in ethylene/1-octene copolymerization, from 1193.5 to 197.6 kg/(mol_V·h), at the comonomer concentrations changing from 0.20 to 0.73 mol/L, were exhibited by the vanadium catalyst **V-32**/$EtAlCl_2$ which was based on the complex with the tetradentate salan-type ligand. However, no neat copolymer was formed at higher comonomer concentrations, but the mixtures of products were obtained [32].

The copolymerization process of ethylene with 1-octene was also explored for three different comonomer concentrations: 0.19, 0.38, and 0.70 mol/L, with the use of the vanadium(IV) complex **V-33** with the diamino-bis(phenolate) ligand, which was activated by $EtAlCl_2$ [62]. Catalyst activity fell within 350–120 kg/(mol_V·h) and it decreased for the increasing comonomer concentrations. The comonomer incorporation was equal to 4.1 mol% at the lowest comonomer concentration and it increased to 9.1 mol% for 0.70 mol/L. At this comonomer concentration, however, a by-product was formed as a colorless oily liquid in addition to the solid copolymer. It should also be noted that the activator considerably affected the copolymerization process. When the activator was changed for $Al(iBu)_3/Ph_3CB(C_6F_5)_4$, some lowering of the catalytic activity and considerably lower 1-octene incorporation was observed. The obtained copolymer showed a high molecular weight ($M_w \sim 1 \times 10^6$ g/mol) and narrow MWD (M_w/M_n = 1.4) [62].

Lorber et al. employed the vanadium(III-V) complexes **V-34–V-38** bearing similar diamine-bis(phenolate) ligands in the ethylene/1-hexene copolymerization reaction [63,64]. The copolymerizations were conducted in the presence of the $EtAlCl_2$ activator, under 2 bars ethylene, and with 8 mmol 1-hexene. Although **V-38** did not produce any copolymer, the complexes **V-34–V-37** gave the products which contained 3.5–7.4 mol% comonomer, with the activity from 12 to 120 kg/(mol_V·h). The V(III) complex **V-37** exhibited both the highest activity and incorporation ability. When the catalytic properties were compared for **V-37**/$EtAlCl_2$ and for the classical-type system based on vanadium acetylacetonate, **V-39**/$EtAlCl_2$, those catalysts were found to have similar comonomer incorporation abilities: they produced copolymers with 7.4 mol% and 7.5 mol% comonomer. However, the distinguishing feature of the copolymer synthesized with **V-39** was its very high dispersity (M_w/M_n = 225) versus the copolymer produced with **V-37** (M_w/M_n = 1.7). The catalytic performance of **V-37** in copolymerization of ethylene with 1-hexene and ethylene with 1-octene was additionally compared in [64]. It was found that, under the same conditions, the copolymers with higher amounts of comonomer units were created in copolymerization of ethylene with 1-octene (8.0–12.0 mol% vs. 7.4–10.6 mol%).

Vanadium complexes bearing ligands featuring sulfur, nitrogen, and oxygen donors were screened for the copolymerization of ethylene and 1-hexene in the presence of dimethylaluminum chloride as an activator and ETA as reactivator by Homden et al. [65]. Copolymerizations were performed at 25 °C for 30 min under 1 bar of ethylene and with different comonomer concentrations. It was found that **V-40** exhibited the highest incorporation ability (11.3 mol% at 4000 equv of 1-hexene) and the highest activity, 1190 kg/(mol$_V$·h), showed **V-41.** Selected results of ethylene/1-olefin copolymerization with vanadium complexes were displayed in Table 5.

As results from the review of the literature reports on copolymerization of ethylene with 1-olefins show, it is hard to directly compare the performance of the complexes which are used in this process due to the different copolymerization conditions adopted for the experiments, and, in particular, ethylene pressure and 1-olefin concentration values. The studied complexes contained ligands with various denticity and with different donor atoms, and their catalytic performance in copolymerization may be significantly modified, not only by ligand backbone, but also by the type of substituents on the aromatic rings (the effects of their steric and electron donor properties need to be considered). Since the living polymerization catalysts and new coordination polymerization techniques were developed, the group 4 metal complexes made it possible to obtain block copolymers and multiblock copolymers which, so far, could not be produced with the use of the vanadium catalysts. It should also be noted that when the vanadium complexes and the titanium/zirconium complexes with the ligands of the same structure are used in copolymerization, the former usually exhibit higher comonomer incorporation ability and produce more uniform copolymers. However, it should also be stressed that when ethylene/1-octene copolymerization with vanadium complexes activated by $EtAlCl_2$ is conducted with no reactivator, higher comonomer concentrations in the reaction medium may yield mixture of products (copolymerization in the presence of **V-32** and **V-33**). This is probably due to the simultaneous presence of various types of active sites formed by reduction of the initial vanadium compound.

Table 5. Results of ethylene/1-olefin copolymerization with vanadium complexes.

Complex	[V], μmol	Activator	ETA/V	E, bar	Time, min	T, °C	Comonomer		Activity, kg/(mol$_V$·h)	CI, mol%	$M_w \cdot 10^3$, g/mol	M_w/M_n	Ref.
							Type	mol/L					
V-1	0.5	Et_2AlCl	500	1	5	25	1-Hex	0.20	4800	4.20	25.9	1.86	[57]
V-3	0.5	Et_2AlCl	500	1	5	25	1-Hex	0.60	5040	8.80	20.7	1.76	[57]
V-3	0.5	Et_2AlCl	500	1	5	75	1-Hex	0.60	3380	16.1	3.2	1.80	[57]
V-14	1.0	Et_2AlCl	300	1	10	25	1-Hex	0.20	1200	3.44	38.5	2.4	[59]
V-16	1.0	Et_2AlCl	300	1	10	25	1-Hex	0.20	2580	2.92	52.1	1.8	[59]
V-18	1.0	Et_2AlCl	300	1	10	50	1-Hex	0.20	5100	3.45	28.2	1.7	[59]
V-25	1.0	Et_2AlCl	300	1	10	25	1-Hex	0.27	3180	3.00	56.4	2.0	[60]
V-25	1.0	Et_2AlCl	300	1	10	25	1-Hex	1.35	900	12.9	24.5	1.9	[60]
V-33	8.0	$EtAlCl_2$	-	5	30	30	1-Oct	0.19	350	4.1	99.0	2.3	[62]
V-33	8.0	$EtAlCl_2$	-	5	30	30	1-Oct	0.38	155	6.7	-	-	[62]
V-37	10.0	$EtAlCl_2$	-	2	30	ng[1]	1-Hex	1/800 [2]	120	7.4	71.0	1.7	[64]
V-39	10.0	$EtAlCl_2$	-	2	30	ng[1]	1-Hex	1/800 [2]	167	7.5	2882	225.5	[64]

1-Hex = 1-hexene, 1-Oct = 1-octene, E = ethylene, CI = comonomer incorporation, [1] ng—temperature was not given, and [2] catalyst/1-hexene = 1/800.

2.2. Copolymerization of Ethylene with Styrene and Its Derivatives

The properties of polyethylene can be also modified by incorporating units which come from styrene or its derivatives. Depending on the amount of the incorporated styrene units, the products may be from semicrystalline to amorphous [19]. It is hard to copolymerize ethylene with styrene because their reactivity towards most of the catalysts is different, and because of the copolymers chains can be produced together with homopolymers [22,66]. It is for that reason as well as low activity and low styrene incorporation (usually below 1 mol%) that the Ziegler–Natta catalysts failed in the synthesis of ethylene/styrene copolymers [19]. The type CGC homogeneous catalysts turned out undoubtably better in the copolymerization of ethylene and styrene monomers [66–70]. Based on single-site, constrained-geometry catalysts, the INSITE technology was developed for the production of ethylene/styrene interpolymers which are described as "pseudo-random" ones, and contain typically up to 50 mol% of styrene [67,71]. Complexes with the monocyclopentadienyl ligands [72–74] and *ansa*-zirconocenes [75,76] are also used as precursors for ethylene/styrene copolymerization. Their use made it possible to obtain copolymers with various compositions, microstructures and properties, and namely copolymers with isolated styrene units, stereoregular Et-St alternating sequences or block copolymers with isotactic polystyrene blocks [75–78].

Complexes of Group 4 Transition Metals

Early transition metal complexes with chelating ligands have attracted a lot of attention as precatalyst in ethylene/1-olefin polymerization. However, there are not many reports of their use in ethylene/styrene copolymerization. That process employed inter alia titanium bis(phenolate) complexes **Ti-20–Ti-25** (Figure 6) activated with methylaluminoxane [79]. The type of the bridging unit between the phenolate moieties of the ligand was shown to be decisive for incorporation of styrene and catalytic activity. The complexes with the sulfur bridge were observed to be most active and to promote lowest styrene incorporation at the same time. For the reaction conducted at 60 °C and with the styrene/ethylene molar ratio of 5/1, the reported activity and the comonomer content in the copolymer were: 82 kg/(mol_{Ti}·h·mol/L), 6.0 mol% (**Ti-21**); and 109 kg/(mol_{Ti}·h·mol/L), 5.5 mol% (**Ti-20**), respectively. The activity of the complexes with the ethylene bridge amounted to 1 kg/(mol_{Ti}·h·mol/L) and the comonomer incorporation level reached about 35.5 mol% under the same conditions. The sulfoxy-bridged complex, **Ti-22**, gave the styrene incorporation of 10 mol% with the activity of 10 kg/(mol_{Ti}·h·mol/L), and the methylene-bridged complex, **Ti-25**, did not make it possible to incorporate any styrene at all. Such changes in catalytic activity of complexes indicate that the additional coordination of the bridging unit to the titanium center leads to the increased Lewis-acidity and consequently to the increased catalyst activity. On the other hand, the steric hindrance at the active center was also affected by the type of the bridging group and it was responsible for the decreased styrene incorporation. The detailed analysis demonstrated that **Ti-20**/MAO produce random poly(ethylene-*co*-styrene) with broad distribution of molecular weight and the comonomer incorporation [79].

The 1,4-dithiabutanediyl-linked bis(phenolato) titanium complex activated by methylaluminoxane, **Ti-26**/MAO produced exclusively ethylene/styrene copolymers with high activity, up to 1600 kg/(mol_{Ti}·h) [80]. Synthesized copolymers had a broad range of styrene content, from 23 to 68 mol%, and their microstructures were dependent on their compositions. The copolymers with the styrene content of 40 mol% were pseudorandom, and the presence of styrene–styrene homo-sequences (SSS) and isolated ethylene units was observed for the copolymers with higher styrene contents [80]. Synthesis of PE-*block*-PS the diblock copolymers by sequential monomer addition in the presence of various phenoxyimine complexes (**Ti-27–Ti-30**), suitable for living olefin polymerization, i.e., with fluorine substituents, activated by MAO, was reported in [81]. In fact, a blend of the block copolymers and polyethylene was obtained and, at high styrene concentrations, atactic polystyrene was formed as a by-product. It was also found that copolymers synthesized with **Ti-27**/MAO possessed ultra-high molecular weights, with M_n = 540,000–850,000 g/mol for polyethylene blocks and with M_n = 40,000–90,000 g/mol for polystyrene blocks [81].

The copolymerization experiments involving ethylene and styrene derivatives were conducted principally with metallocene, CGC, or scandium catalysts [82–84]. However, the examples described below show that the replacement of styrene with its derivative may induce essential change in the copolymerization behavior of post-metallocene catalysts. The bis(phenoxy-imine) zirconium catalysts, **Zr-22–Zr-24**/MAO, were not able to produce ethylene/styrene copolymers, probably due to the dominant mechanism of the 2,1-insertion, which resulted in the formation of unreactive benzyls [85]. Replacement of styrene with *para-tert*-butylstyrene (TBS) made it possible to conduct the reaction and to produce the copolymer with substantial comonomer incorporation, from 4.9 to 11.3 mol% at a relatively low concentration of comonomer (0.7 mol/L). The copolymers were also characterized by narrow molecular weight distribution (M_w/M_n = 1.86–2.84) and lack of sequential TBS units and TBS-E-TBS triads. Moreover, on moving from ethylene homopolymerization to ethylene/*para-tert*-butylstyrene copolymerization, only a small reduction in the catalyst activity was observed [85].

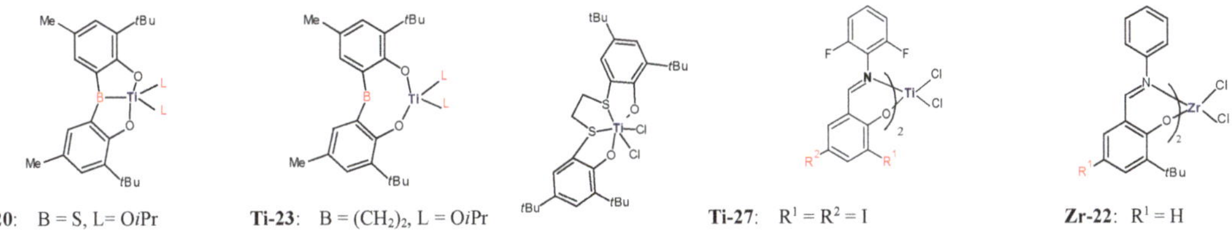

Ti-20: B = S, L = O*i*Pr
Ti-21: B = S, L = Cl
Ti-22: B = S=O, L = O*i*Pr [79]

Ti-23: B = $(CH_2)_2$, L = O*i*Pr
Ti-24: B = $(CH_2)_2$, L = Cl
Ti-25: B = CH_2, L = Cl [79]

Ti-26 [80]

Ti-27: $R^1 = R^2$ = I
Ti-28: R^1 = *t*Bu, R^2 = H
Ti-29: R^1 = $SiMe_3$, R^2 = H
Ti-30: R^1 = $C(CH_3)_2$–C_6F_5, R^2 = H [81]

Zr-22: R^1 = H
Zr-23: R^1 = OMe
Zr-24: R^1 = Cl [85]

Figure 6. Titanium and zirconium complexes investigated as precatalysts for ethylene/styrene copolymerization [79–81,85].

Copolymerization of ethylene with styrene and its derivatives with electron-donating substituents at the para position, *para*-methylstyrene (MS), and TBS, was described in [86]. It was catalyzed by diamine-bis(phenolate) complexes of zirconium (**Zr-6**) and titanium (**Ti-1**) activated either by MMAO or by $Al(iBu)_3/Ph_3CB(C_6F_5)_4$. Copolymerization with **Zr-6**/MMAO was found to produce, exclusively, copolymers with the incorporation level of 2.2–4.6 mol%, whereas other catalytic systems yielded mixtures of copolymers and homopolymers. The catalytic activity of **Zr-6**/MMAO in ethylene/styrene copolymerization at 88.5 mmol comonomer and at 5 bar ethylene was equal to 613.6 kg/(mol_{Zr}·h) and was higher than in ethylene homopolymerization (203.0 kg/[mol_{Zr}·h]); it increased to 985.4 and 1027.2 kg/(mol_{Zr}·h) after changing the comonomer to MS and TBS, respectively. This effect of a styrenic monomer on catalyst activity is consistent with previous findings observed for the catalysts based on **Sc-1**, **Ti-31** and **Zr-25** (Figure 7) [82,83]. Selected results of ethylene/styrene copolymerization were displayed in Table 6.

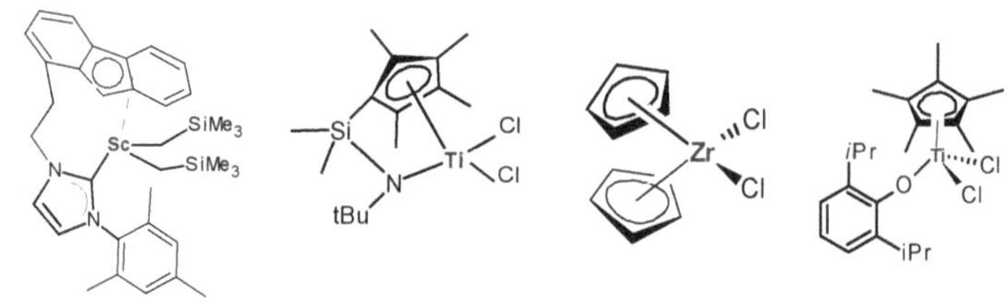

Sc-1 [82] **Ti-31** [83] **Zr-25** [83] **Ti-32** [87]

Figure 7. Complexes investigated as precatalysts for copolymerization of ethylene with alkyl derivatives of styrene or α,ω-alkenols [82,83,87].

Table 6. Ethylene/styrene copolymerization results.

Complex	[M], μmol	Activator	E, bar	Styrene, mol/L	Activity, kg/(mol$_{Mt}$·h)	T_m, °C	χ, %	CI, mol%	Ref.
Ti-21	4	MAO	2.56	1.1	52 [1]	-	-	1.2	[79]
Ti-22	20	MAO	2.56	1.1	10 [1]	-	-	10.0	[79]
Ti-23	20	MAO	2.56	1.1	1 [1]	-	-	35.2	[79]
Ti-26	0.1 [2]	MAO	0.20 [3]	2.2	1600	30.5 [4]	-	55.0	[80]
Ti-26	0.08 [2]	MAO	0.20 [3]	0.55	1300	40.5 [4]	-	40.0	[80]
Ti-1	10	MMAO	5	0.55	28.8	125.0	41.3	0.5	[86]
Zr-6	10	MMAO	5	0.8	475.6	110.8	57.7	3.0	[86]

E = ethylene, CI = comonomer incorporation, [1] activity in kg/(mol$_M$·h·mol/L), [2] mmol/L, [3] mol/L, [4] T_g.

As far as we know, the post-metallocene catalysts based on complexes of group 5 transition metals were not examined in copolymerization of ethylene and styrenic monomers.

2.3. *Copolymerization of Ethylene with α,ω-Alkenols*

Functionalized polyolefins can be synthesized in numerous ways. Polymer post-functionalization is one of the approaches used. That method requires harsh conditions and may lead to undesirable side reactions which eventually change the properties of the polymer in an unwanted manner [20]. The ring-opening metathesis polymerization of functionalized cyclic alkenes and acyclic diene metathesis polycondensation, both followed by hydrogenation, are other approaches for obtaining functionalized polyolefin. However, a multi-step fashion and special monomers needed for those methods are their main drawbacks [20]. An alternative method for the production of functionalized polyolefins is the free-radical copolymerization of olefins and polar vinyl monomers, which is widely applied in industry. This, however, does not provide satisfactory control of the composition and the microstructures of the produced copolymers. It is definitely easier to control the composition and the polar monomer distribution along the polymer backbone in the direct copolymerization of ethylene and polar comonomers promoted by organometallic catalysts. As polar monomers, acrylates, vinyl ethers, vinyl halides, unsaturated carboxylic acids, and esters, mostly methyl esters of those acids, α,ω-alkenols, and others, were used [88,89]. Although late transition-metal catalysts are able to incorporate the unprotected polar vinyl monomers, early transition-metal complexes require the protection of the polar groups to prevent partial or complete deactivation of the catalyst [20,88–90]. One of the groups of polar monomers are unsaturated alcohols with various spacers between the double bond and the functionality, like 5-heksen-1-ol (5H1O), 9-decen-1-ol (9D1O), and 10-undecen-1-ol (10U1O). In order to protect the catalytic active sites against coordination of polar groups and their deactivation, α,ω-alkenols are most frequently reacted with the organoaluminium compounds.

The use of Ziegler–Natta catalysts in the ethylene/α,ω-alkenols copolymerization, alike with other monomers, has limitations because of their low comonomer incorporation ability. More comprehensive and more effective attempts to functionalize polyethylene were carried out in the presence of metallocene catalytic systems. The research in this field was reviewed in detail by M. Atiqullah et al. [91]. Later, copolymerization of ethylene and 4-penten-1-ol as well as 9D1O was shown to be effective by using the half-titanocene catalyst (**Ti-32**) with the additional aryloxy ligand [87,92]. In ethylene/9D1O copolymerization, the incorporation level of the polar monomer was as high as 13.6 mol% [87].

2.3.1. Complexes of Group 4 Transition Metals

Copolymerization of ethylene with α,ω-alkenols and their derivatives involved the use of post-metallocene catalysts based on titanium and zirconium complexes with multidentate ligands of the type [ON], [ONS], [ONNO], and [OSSO] [90,93–97]. Y. Saito et al. verified the ability of the zirconium complex with the [OSSO]-type bis(phenolate) ligand (**Zr-26**, Figure 8), activated by dMAO, to copolymerize ethylene with 5-hexen-1-oxytriisopropylsilane (iPr_3Si-protected 5-hexen-1-ol, 5H1OSi) [93]. That catalytic system was found to be efficient in copolymerization at 25 °C under atmospheric pressure of ethylene and it produced highly random ethylene/5H1OSi copolymers containing from 7.5 to 21.4 mol% 5H1OSi, depending on the comonomer concentrations in the feed. The highest comonomer incorporation was achieved under the most favorable process conditions, i.e., a high amount of comonomer (3.9 mmol) was used and the reaction was conducted over 5 h. The resulting copolymers had M_w in the range of 3800–10,800 g/mol and their dispersity was narrow (M_w/M_n=1.8–2.7). Unfortunately, the catalytic activity of the system decreased drastically compared with its activity in ethylene/1-hexene copolymerization, from 1110 g/($mmol_{Zr}$·h) to 77–470 g/($mmol_{Zr}$·h). That was probably due to the bulkiness of the protecting group in the monomer [93].

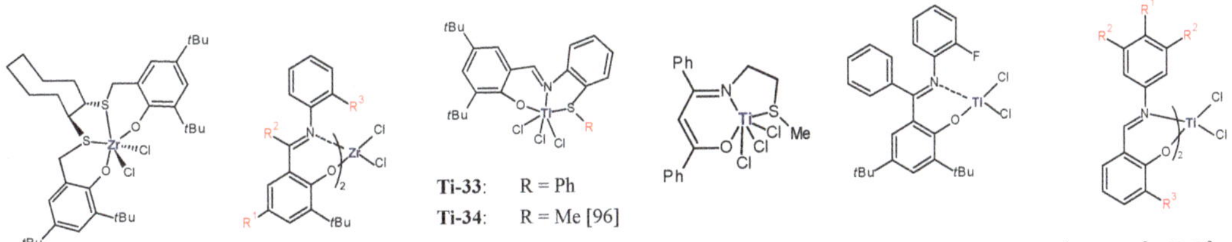

Zr-26 [93]
Zr-27: R¹ = H, R² = H, R³ = H [94]
Zr-28: R¹ = *t*Bu, R² = Ph, R³ = H [94]
Zr-29: R¹ = *t*Bu, R² = Ph, R³ = F [94,95]
Ti-33: R = Ph
Ti-34: R = Me [96]
Ti-35 [96]
Ti-36 [95]
Ti-37: R¹ = *t*Bu, R² = H, R³ = Ph
Ti-38: R¹ = H, R² = H, R³ = *t*Bu
Ti-39: R¹ = H, R² = *t*Bu, R³ = Ph
[90]

Figure 8. Titanium and zirconium complexes investigated as precatalysts for ethylene/α,ω-alkenols copolymerization [90,93–96].

Good catalytic activity and incorporation ability in copolymerization of ethylene with 10-undecen-1-ol and with 9-decen-1-ol showed the zirconium catalysts **Zr-6**/Al(iBu)$_3$/$Ph_3CB(C_6F_5)_4$ [97]. The copolymerizations were carried out at various conditions and the polar monomers were subjected to pretreatment with various alkylaluminium compounds but Al(iBu)$_3$ was found to be the best protecting agent. Ethylene/9-decen-1-ol with the comonomer content as high as 16.4 mol% was obtained at a low comonomer concentration, 0.073 mol/L, under 1 bar of ethylene. Moreover, good comonomer incorporation (up to 5.9 mol% at the same alcohol concentration) was possible at a higher ethylene pressure (3 bar). It is also worth noting that it exhibited much higher activity in copolymerization than in ethylene homopolymerization or ethylene/1-olefin copolymerization which is contrary to most tested catalysts.

Yang et al. employed titanium complexes **Ti-33**, **Ti-34** and **Ti-35** bearing tridentate phenoxy-imine ligands with the additional sulfur donor atom in copolymerization of ethylene and 9-decen-1-ol [96]. Copolymerizations catalyzed by **Ti-33–Ti-35**/MMAO were conducted in the presence of 20 mmol of comonomer pretreated with 1.2 equivalent of

$Al(iBu)_3$ under 1 atm ethylene at 25 °C and at Al/Ti molar ratio of 1000. The catalytic activity of **Ti-33** and **Ti-34** was good, equal to 600 kg/(mol_{Ti}·h) and 700 kg/(mol_{Ti}·h), and the produced copolymers contained 3.5 mol% and 3.4 mol% of the polar comonomer. The less-hindered **Ti-35**, with the strong electron-donating ligand which improved the tolerance of active sites to functional groups, turned out more beneficial for copolymerization. It produced the copolymer with the highest activity, 10,100 kg/(mol_{Ti}·h), and with the highest alcohol content, 11.2 mol%. Moreover, **Ti-35**/MMAO (Al/Ti = 3000) was able to produce the ethylene/9-decen-1-ol copolymer with high activity (1200 kg/[mol_{Ti}·h]) and good comonomer incorporation (3.7 mol%), even when the comonomer was not subjected to pretreatment with $Al(iBu)_3$ [96].

The catalytic performance of the phenoxy-based zirconium complexes, **Zr-27–Zr-29**, activated by MAO, in copolymerization of ethylene and 10-undecen-1-ol was compared by X. Zhang et al. [94]. The reactions were carried out at 25 °C for 5 min under 1 atm ethylene. Comonomer was pretreated with $Al(iBu)_3$ and added in the amount of 2–16 mmol. It was found that the addition of the polar monomer caused only a very slight deactivation of the catalysts and complex **Zr-29** was the most active one. At the α,ω-alcohol concentration equal to 0.08 mol/L, its catalytic activity was equal to $19.2 \cdot 10^3$ kg/(mol_{Zr}·h). The results also showed that the structure of the complex was important for the incorporation of the polar monomer and molecular weights of the copolymers. Both changed in the following order: **Zr-27** < **Zr-28** < **Zr-29**. The complex **Zr-29** gave the copolymers with the highest molecular weights and the highest incorporation of polar comonomer, 3.93 wt% at 8 mmol, which increased to 8.15 wt% (1.4 mol%) at 16 mmol of comonomer in the feed. It should, however, be mentioned that the obtained incorporation level was rather low when the reaction conditions are taken into consideration. The observed changes indicate that the introduction of the phenyl group onto the imine carbon atom and introduction of the fluorine atom at the ortho position of the aniline moiety enhanced the stability of the catalyst [94]. The research on the effect of the properties of the central metal on the catalytic performance of bis(phenoxyketimine)-based complexes (**Zr-29** and **Ti-36**) in ethylene/10-undecen-1-ol copolymerization showed that **Ti-36** was less active than **Zr-29** and its ability to incorporate the comonomer was low, like that for the zirconium complex. Under the same polymerization conditions (1 atm ethylene, 12 mmol α,ω-alkenol, 10 min, 25 °C) they produced copolymers containing 6.2 wt% and 5.6 wt% comonomer, respectively [95]. When the comonomer with a shorter spacer between the polar group and the double bond was employed (5-hexen-1-ol), the catalytic activity and the comonomer content in the copolymer decreased [95].

Mono- and dinuclear zirconium complexes **Zr-14–Zr-17** were tested in ethylene/10-undecen-1-ol and ethylene/5-hexen-1-ol copolymerization over comonomer concentrations ranging from 0.02 to 0.16 mol/L in the presence of MAO as activator and $Al(iBu)_3$ as masking agent [98]. Modest polar comonomer incorporations, from 0.04 mol% to 1.7 mol%, were observed.

Fujita et al. used bis(phenoxy-imine)Ti complexes, inter alia **Ti-37**, **Ti-38**, and **Ti-39**, activated by dMAO in ethylene/5- hexene-1-yl-acetate copolymerization [90]. When 1 mmol comonomer and 1 atm ethylene were applied, these catalysts produced copolymers with low comonomer incorporation, 0.13–0.90 mol%, and high molecular weights, 497,000–273,000 g/mol. It was found that the presence of the phenyl group ortho to the phenoxy-O induces higher comonomer incorporation than the *tert*-Bu group. When a higher amount of 5-hexene-1-yl-acetate was used, it resulted in the higher comonomer contents, up to 3.30 mol% for **Ti-39**/dMAO at 5.25 mmol of the comonomer in the feed.

2.3.2. Complexes of Vanadium

L.P. Lu et al. reported the use of vanadium(III) complexes with bis(imino)pyrrolyl **V-42–V-44**, iminopyrrolyl **V-45**, and β-diketiminate **V-46** ligands (Figure 9) in copolymerization of ethylene with 10-undecen-1-ol [99]. The polar comonomer was subjected to pretreatment with Et_2AlCl as a protecting reagent. The same organoaluminium compound was used as an activator, an ETA was employed as a reactivator. The reactions were conducted at 50 °C for 5 min under 1 atm of ethylene pressure and at the comonomer concentration of 0.1 or 0.5 mol/L. **V-43** was found to offer higher catalytic activity than **V-42**, and the highest activity showed **V-44**. That relation suggests that the large steric hindrance of 2,6-$i Pr_2C_6H_3$ groups and interaction of fluorine atoms with hydroxyls of comonomer more efficiently protect the metal center from coordination of heteroatom and deactivation of catalyst. On the other hand, the ability of all complexes to incorporate the polar comonomer was high (12.1–14.0 mol%) and it was growing as follows: **V-43** < **V-42** < **V-44**, which demonstrates that a bulky 2,6-$i Pr_2C_6H_3$ groups may make it harder for a comonomer molecule to access the active site [99]. The complexes **V-45** and **V-46** were characterized by lower tolerance towards the heteroatoms of comonomer than the bis(imino)pyrrolyl vanadium(III) complexes, but their ability to incorporate the comonomer was equally high as that for **V-44** [99]. Complex **V-46** (Figure 9) was also tested in copolymerization of ethylene with 3-buten-1-ol, 5-hexen-1-ol, and 10-undecen-1-ol. As predicted, decreasing the length of the methylene spacers in the polar comonomer decreased both the catalytic activity and comonomer incorporation [100]. In turn, the comparison of the catalytic properties of vanadium(III) complexes bearing different [N,O] bidentate ligands (**V-46–V-48**) revealed that their capabilities to copolymerize ethylene with 10-undecen-1-ol was only slightly dependent on the ligand structure. The catalytic activities of **V-46–V-48** under the same reaction conditions were 9.72, 8.70, and 11.2 kg/($mmol_V \cdot h$), while the 10-undecen-1-ol incorporation levels in the resultant copolymers were 3.70, 4.00, and 3.40 mol%, respectively [100]. Selected results of ethylene/α,ω-alkenols copolymerization with vanadium and group 4 metal complexes were displayed in Table 7.

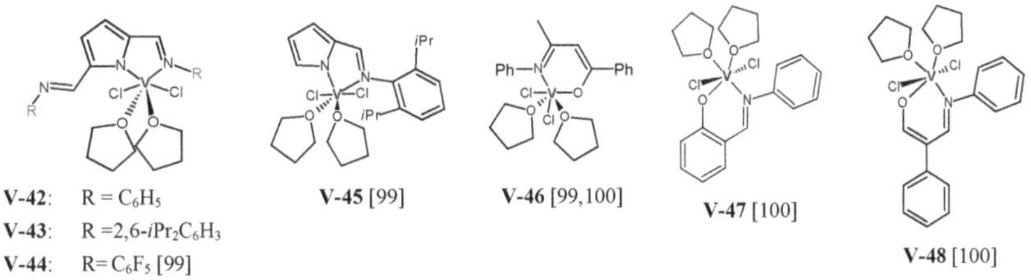

V-42: R = C_6H_5
V-43: R =2,6-$i Pr_2C_6H_3$
V-44: R= C_6F_5 [99]

V-45 [99]

V-46 [99,100]

V-47 [100]

V-48 [100]

Figure 9. Vanadium complexes investigated as precatalysts for ethylene/α,ω-alkenols copolymerization [99,100].

Table 7. Ethylene/α,ω-alkenols copolymerization results.

Complex	[M], μmol	Activator	E, bar	Time, min	Comonomer		Activity, kg/(mol$_M$·h)	CI, mol%	M_w, kg/mol	M_w/M_n	Ref.
					Type	mmol					
Zr-26	2	dMAO	1	60	5H1OSi	2.3	121	7.5	3.8	2.7	[93]
Zr-26	2	dMAO	1	300	5H1OSi	3.9	77	21.4	10.2	1.9	[93]
Zr-6	10.0	Al(*i*Bu)$_3$/B	1	20	9-D-1-ol	6.75	45.3	16.4	2.9	1.6	[97]
Zr-6	10.0	Al(*i*Bu)$_3$/B	3	20	9-D-1-ol	6.75	604.2	2.9	4.7	1.9	[97]
Ti-33	3.5	MMAO	1	10	9-D-1-ol	20.0	600	3.5	940	2.9	[96]
Ti-34	3.5	MMAO	1	10	9-D-1-ol	20.0	700	3.4	115	2.2	[96]
Ti-35	3.5	MMAO	1	1	9-D-1-ol	20.0	10100	11.2	178	2.2	[96]
Zr-29	1.0	MAO	1	5	10-U-1-ol	16.0	13100	1.4	-	-	[94]
V-42	0.5	Et_2AlCl	1	5	10-U-1-ol	5.0	8160	4.2	14.3	1.8	[99]
V-42	0.5	Et_2AlCl	1	5	10-U-1-ol	25.0	2160	13.7	3.8	1.9	[99]

E = ethylene, CI = comonomer incorporation, B = $Ph_3CB(C_6F_5)_4$.

3. Conclusions

The catalytic copolymerization of ethylene with other vinyl monomers like 1-olefins, styrenic and polar monomers, produces materials with unique and interesting set of properties which can fill property gaps not addressed by HDPE. Numerous efficient post-metallocene catalysts based on well-defined complexes of both vanadium and group 4 metals have been developed for ethylene/1-olefin copolymerization. The wide variation of the available structures of the complexes allows the production of copolymers which differ with regard to their composition, chemical composition distribution, molecular weight, and its distribution. It is worth noting that vanadium post-metallocene catalysts often afforded more homogeneous copolymers than their group 4 metal counterparts. Moreover, development of the catalysts, which exhibit living behavior and chain shuttling systems, allowed for the formation of block and multi-block copolymers with various compositions and microstructures. From the work described herein, it is also clear that the successful copolymerization of ethylene with styrene and/or its derivatives was achieved only in the presence of a few titanium and zirconium complexes, while any post-metallocene vanadium complexes were not applicable in that process. As for the copolymerization of ethylene with α,ω-alkenols, complexes of both group 4 and group 5 metals were tested. However, their catalytic activity was usually much lower compared to that for homopolymerization of ethylene, despite the protection of hydroxyl group. Moreover, the reaction conditions for the reported copolymerizations were selected to provide the highest possible incorporation of comonomer. Namely, the process was conducted at atmospheric pressure or lower, which is favorable indeed for higher comonomer contents, but which considerably reduces catalytic activity and molecular weight of the copolymer product. The one exception was the catalytic system based on diamine-bis(phenolate) zirconium complex (**Zr-6**) for which a significant activating effect of the polar comonomer on the catalyst activity was observed.

Author Contributions: Conceptualization, M.B. and J.F.; writing—original draft preparation, M.B. and J.F.; writing—review and editing, M.B.; supervision, M.B. All authors have read and agreed to the published version of the manuscript.

Funding: This research received no external funding.

Conflicts of Interest: The authors declare no conflict of interest.

References

1. Plastics—The Facts 2020. Available online: https://www.plasticseurope.org/pl/resources/publications/4312-plastics-facts-2020 (accessed on 18 October 2021).
2. Yuan, S.F.; Yan, Y.; Solan, G.A.; Ma, Y.; Sun, W.-H. Recent advancements in N-ligated group 4 molecular catalysts for the (co)polymerization of ethylene. *Coord. Chem. Rev.* **2020**, *411*, 213254. [CrossRef]
3. Sauter, D.W.; Taoufik, M.; Boisson, C. Polyolefins, a Success Story. *Polymers* **2017**, *9*, 185. [CrossRef]
4. Kaminsky, W. New polymers by metallocene catalysis. *Macromol. Chem. Phys.* **1996**, *197*, 3907–3945. [CrossRef]
5. Canoa, J.; Kunzb, K. How to synthesize a constrained geometry catalyst (CGC)—A survey. *J. Organomet. Chem.* **2007**, *692*, 4411–4423. [CrossRef]
6. Collins, R.A.; Russell, A.F.; Mountford, P. Group 4 metal complexes for homogeneous olefin polymerisation: A short tutorial review. *Appl. Petrochem. Res.* **2015**, *5*, 153–171. [CrossRef]
7. Baier, M.C.; Zuideveld, M.A.; Mecking, S. Post-metallocenes in the industrial production of polyolefins. *Angew. Chem. Int. Ed.* **2014**, *53*, 9722–9744. [CrossRef]
8. Bryliakov, K.P. Post-metallocene catalysts for olefin polymerization. *Russ. Chem. Rev.* **2007**, *76*, 253–277. [CrossRef]
9. Britovsek, G.J.P.; Gibson, V.C.; Wass, D.F. The search for new-generation olefin polymerization catalysts: Life beyond metallocenes. *Angew. Chem. Int. Ed.* **1999**, *38*, 428–447. [CrossRef]
10. Ronca, S. Polyethylene (Chapter 10). In *Brydson's Plastics Materials*, 8th ed.; Gilbert, M., Ed.; Butterworth-Heinemann: Oxford, UK, 2017; pp. 247–278. [CrossRef]
11. Patel, K.; Chikkali, S.H.; Sivaram, S. Ultrahigh molecular weight polyethylene: Catalysis, structure, properties, processing and applications. *Prog. Polym. Sci.* **2020**, *109*, 101290. [CrossRef]
12. Liu, L.B.; Murakami, N.; Sumita, M.; Miyaska, K. Effect of molecular weight distribution on the structure and mechanical properties of ultradrawn, ultrahigh-molecular-weight polyethylene cast from solution. II. drawability and mechanical properties. *J. Polym. Sci. Part B Polym. Phys.* **1989**, *27*, 2441–2450. [CrossRef]

13. Zucchini, U.; Cecchin, G. Control of molecular-weight distribution in polyolefins synthesized with ziegler-natta catalytic systems. *Adv. Polym. Sci.* **1983**, *51*, 101–154. [CrossRef]
14. Szántó, L.; Feng, Y.; Zhong, F.; Hees, T.; van Ruymbeke, E.; Mülhaupt, R.; Friedrich, C. Ultra-broad molecular weight distribution effects on viscoelastic properties of linear multimodal PE. *J. Rheol.* **2019**, *63*, 773–784. [CrossRef]
15. Sobieraj, M.C.; Rimnac, C.M. Ultra high molecular weight polyethylene: Mechanics, morphology, and clinical behavior. *J. Mech. Behav. Biomed. Mater.* **2009**, *2*, 433–443. [CrossRef] [PubMed]
16. Wang, H.P.; Khariwala, D.U.; Cheung, W.; Chum, S.P.; Hiltner, A.; Baer, E. Characterization of some new olefinic block copolymers. *Macromolecules* **2007**, *40*, 2852–2862. [CrossRef]
17. Zhao, W.; Nomura, K. Design of efficient molecular catalysts for synthesis of cyclic olefin copolymers (coc) by copolymerization of ethylene and α-olefins with norbornene or tetracyclododecene. *Catalysts* **2016**, *6*, 175. [CrossRef]
18. Kaminsky, W. The discovery of metallocene catalysts and their present state of the art. *J. Polym. Sci. Part A Polym. Chem.* **2004**, *42*, 3911–3921. [CrossRef]
19. Rodrigues, A.S.; Carpentier, J.F. Groups 3 and 4 single-site catalysts for styrene–ethylene and styrene-α-olefin copolymerization. *Coord. Chem. Rev.* **2008**, *252*, 2137–2154. [CrossRef]
20. Franssen, N.M.G.; Reek, J.N.H.; de Bruin, B. Synthesis of functional 'polyolefins': State of the art and remaining challenges. *Chem. Soc. Rev.* **2013**, *42*, 5809–5832. [CrossRef] [PubMed]
21. Zhong, S.; Tan, Y.; Zhong, L.; Gao, J.; Liao, H.; Jiang, L.; Gao, H.; Wu, Q. Precision synthesis of ethylene and polar monomer copolymers by palladium-catalyzed living coordination copolymerization. *Macromolecules* **2017**, *50*, 5661–5669. [CrossRef]
22. Galdi, N.; Buonerba, A.; Oliva, L. Olefin–styrene copolymers. *Polymers* **2016**, *8*, 405. [CrossRef]
23. Sastri, V.R. *Commodity Thermoplastics: Polyvinyl Chloride, Polyolefins, and Polystyrene, Chapter 6 [in] Plastics in Medical Devices Properties, Requirements, and Applications*; William Andrew Publishing: Norwich, NY, USA, 2010; pp. 73–119. [CrossRef]
24. Knuuttila, H.; Lehtinen, A.; Nummila-Pakarinen, A. Advanced polyethylene technologies-controlled material properties. *Adv. Polym. Sci.* **2004**, *169*, 13–27. [CrossRef]
25. Chum, P.S.; Swogger, K.W. Olefin polymer technologies—History and recent progress at the dow chemical company. *Prog. Polym. Sci.* **2008**, *33*, 797–819. [CrossRef]
26. Li, J.; Shanks, R.A.; Long, Y. Miscibility and crystallization of metallocene, polyethylene blends with polypropylene. *J. Appl. Polym. Sci.* **2003**, *87*, 1179–1189. [CrossRef]
27. Kim, B.K.; Kim, M.S.; Jeong, H.M.; Kim, K.J.; Jang, J.K. Characterization of ultra low density polyethylenes (PE-ULD). *Angew. Makromol. Chem.* **1992**, *194*, 91–101. [CrossRef]
28. Shamiri, A.; Chakrabarti, M.H.; Jahan, S.; Hussain, M.A.; Kaminsky, W.; Aravind, P.V.; Yehye, W.A. The influence of ziegler-natta and metallocene catalysts on polyolefin structure, properties, and processing ability. *Materials* **2014**, *7*, 5069–5108. [CrossRef]
29. Reybuck, S.E.; Lincoln, A.L.; Ma, S.; Waymouth, R.M. Amine bis(phenolate) zirconium complexes: Influence of ligand structure and cocatalyst on copolymerization behavior. *Macromolecules* **2005**, *38*, 2552–2558. [CrossRef]
30. Białek, M.; Bisz, E. Polypropylene and poly(ethylene-co-1-octene) effective synthesis with diamine-bis(phenolate) complexes: Effect of complex structure on catalyst activity and product microstructure. *J. Polym. Sci. Part A Polym. Chem.* **2017**, *55*, 2467–2476. [CrossRef]
31. Fryga, J.; Białek, M. Effect of AlR_3 (R = Me, Et, iBu) addition on the composition and microstructure of ethylene/1-olefin copolymers made with postmetallocene complexes of group 4 elements. *Polym. J.* **2019**, *51*, 19–29. [CrossRef]
32. Białek, M.; Pochwała, M.; Spaleniak, G. Olefin polymerization and copolymerization by complexes bearing [onno]-type salan ligands: Effect of ligand structure and metal type (titanium, zirconium, and vanadium). *J. Polym. Sci. Part A Polym. Chem.* **2014**, *52*, 2111–2123. [CrossRef]
33. Białek, M.; Czaja, K.; Pietruszka, A. Ethylene/1-olefin copolymerization behaviour of vanadium and titanium complexes bering salen-type ligand. *Polym. Bull.* **2013**, *70*, 1499–1517. [CrossRef]
34. Golisz, S.R.; Bercaw, J.E. Synthesis of early transition metal bisphenolate complexes and their use as olefin polymerization catalysts. *Macromolecules* **2009**, *42*, 8751–8762. [CrossRef]
35. Wang, C.; Ma, Z.; Sun, X.L.; Gao, Y.; Guo, Y.H.; Tang, Y.; Shi, L.P. Synthesis and characterization of titanium(IV) complexes bearing monoanionic [O^-NX] (X = O, S, Se) tridentate ligands and their behaviors in ethylene homo- and copolymerizaton with 1-hexene. *Organometallics* **2006**, *25*, 3259–3266. [CrossRef]
36. Wang, Z.; Peng, A.Q.; Sun, X.L.; Tang, Y. Synthesis and characterization of titanium complexes bearing sulfoxide groups and their catalytic behaviors in ethylene homo- and copolymerization. *Sci. China Chem.* **2014**, *57*, 1144–1149. [CrossRef]
37. Białek, M.; Fryga, J.; Spaleniak, G.; Matsko, M.A.; Hajdasz, N. Ethylene homo- and copolymerization catalyzed by vanadium, zirconium, and titanium complexes having potentially tridentate Schiff base ligands. *J. Cat.* **2021**, *400*, 184–194. [CrossRef]
38. Gao, Y.; Christianson, M.D.; Wang, Y.; Chen, J.; Marshall, S.; Klosin, J.; Lohr, T.L.; Marks, T.J. Unexpected precatalyst σ-ligand effects in phenoxyimine zr-catalyzed ethylene/1-octene copolymerizations. *J. Am. Chem. Soc.* **2019**, *141*, 7822–7830. [CrossRef] [PubMed]
39. Liu, S.; Xing, Y.; Zheng, Q.; Jia, Y.; Li, Z. Synthesis of anthracene-bridged dinuclear phenoxyiminato organotitanium catalysts with enhanced activity, thermal stability, and comonomer incorporation ability toward ethylene (Co)polymerization. *Organometallics* **2020**, *39*, 3268–3274. [CrossRef]

40. Sampson, J.; Choi, G.; Akhtar, M.N.; Jaseer, E.A.; Theravalappil, R.; Al-Muallem, H.A.; Agapie, T. Olefin polymerization by dinuclear zirconium catalysts based on rigid teraryl frameworks: Effects on tacticity and copolymerization behavior. *Organometallics* **2017**, *36*, 1915–1928. [CrossRef]
41. Park, C.Y.; Switzer, J.M.; Speer, J.; Medvedev, G.A.; Caruthers, J.M.; Abu-Omar, M.M. Kinetics of ethylene/1-hexene copolymerization over a single-site hafnium bis(phenolate) catalyst: Insights into insertion complexity and deactivation pathways. *Macromolecules* **2021**, *54*, 4101–4111. [CrossRef]
42. Han, B.; Liu, Y.; Feng, C.; Liu, S.; Li, Z. Development of group 4 metal complexes bearing fused-ring amido-trihydroquinoline ligands with improved high-temperature catalytic performance toward olefin (co)polymerization. *Organometallics* **2021**, *40*, 242–252. [CrossRef]
43. Jun, S.H.; Park, J.H.; Lee, C.S.; Park, S.Y.; Go, M.J.; Lee, J.; Lee, B.Y. Preparation of phosphine-amido hafnium and zirconium complexes for olefin polymerization. *Organometallics* **2013**, *32*, 7357–7365. [CrossRef]
44. Wang, S.; Wang, L.; Zhong, L.; Xu, R.; Wang, X.; Kang, W.; Gao, H. C1-symmetric tert-butyl substituted pyridylamido hafnium complex for ethylene, α-olefin, and styrene polymerizations. *Eur. Polym. J.* **2020**, *131*, 109709. [CrossRef]
45. Kulyabin, P.S.; Uborsky, D.V.; Voskoboynikov, A.Z.; Canich, J.A.M.; Hagadorn, J.R. Pyridylamido hafnium complexes with a silylene bridge: Synthesis and olefin polymerization. *Dalton Trans.* **2020**, *49*, 6693–6702. [CrossRef]
46. Frazier, K.A.; Froese, R.D.; He, Y.; Klosin, J.; Theriault, C.N.; Vosejpka, P.C.; Zhou, Z. Pyridylamido hafnium and zirconium complexes: Synthesis, dynamic behavior, and ethylene/1-octene and propylene polymerization reactions. *Organometallics* **2011**, *30*, 3318–3329. [CrossRef]
47. Furuyama, R.; Mitani, M.; Mohri, J.; Mori, R.; Tanaka, H.; Fujita, T. Ethylene/higher α-olefin copolymerization behavior of fluorinated bis(phenoxy-imine)titanium complexes with methylalumoxane: Synthesis of new polyethylene-based block copolymers. *Macromolecules* **2005**, *38*, 1546–1552. [CrossRef]
48. Hustad, P.D.; Kuhlman, R.L.; Arriola, D.J.; Carnahan, E.M.; Wenzel, T.T. Continuous production of ethylene-based diblock copolymers using coordinative chain transfer polymerization. *Macromolecules* **2007**, *40*, 7061–7064. [CrossRef]
49. Arriola, D.J.; Carnahan, E.M.; Hustad, P.D.; Kuhlman, R.L.; Wenzel, T.T. Catalytic production of olefin block copolymers via chain shuttling polymerization. *Science* **2006**, *312*, 714–719. [CrossRef] [PubMed]
50. Coates, G.W.; Hustad, P.D.; Reinartz, S. Catalysts for the living insertion polymerization of alkenes: Access to new polyolefin architectures using ziegler-natta chemistry. *Angew. Chem. Int. Ed.* **2002**, *41*, 2236–2257. [CrossRef]
51. Mitani, M.; Saito, J.; Ishii, S.; Nakayama, Y.; Makio, H.; Matsukawa, N.; Matsui, S.; Mohri, J.; Furuyama, R.; Terao, H.; et al. FI catalysts: New olefin polymerization catalysts for the creation of value-added polymers. *Chem. Rec.* **2004**, *4*, 137–158. [CrossRef]
52. Doi, Y.; Ueki, S.; Keii, T. "Living" coordination polymerization of propene initiated by the soluble V(acac)$_3$-Al(C_2H_5)Cl system. *Macromolecules* **1979**, *12*, 814–819. [CrossRef]
53. Valente, A.; Mortreux, A.; Visseaux, M.; Zinck, P. Coordinative chain transfer polymerization. *Chem. Rev.* **2013**, *113*, 3836–3857. [CrossRef]
54. Nomura, K.; Zhang, S. Design of vanadium complex catalysts for precise olefin polymerization. *Chem. Rev.* **2011**, *111*, 2342–2362. [CrossRef] [PubMed]
55. Phillips, A.M.F.; Suo, H.; da Silva, M.F.C.G.; Pombeiro, A.J.L.; Sun, W.H. Recent developments in vanadium-catalyzed olefin coordination polymerization. *Coord. Chem. Rev.* **2020**, *416*, 213332. [CrossRef]
56. Gambarotta, S. Vanadium-based Ziegler-/Natta: Challenges, promises, problems. *Coord. Chem. Rev.* **2003**, *237*, 229–243. [CrossRef]
57. Mu, J.S.; Wang, Y.X.; Li, B.X.; Li, Y.S. Synthesis of vanadium(III) complexes bearing iminopyrrolyl ligands and their role as thermal robust ethylene (co)polymerization catalysts. *Dalton Trans.* **2011**, *40*, 3490–3497. [CrossRef] [PubMed]
58. Zhang, S.W.; Lu, L.P.; Long, Y.Y.; Li, Y.S. Ethylene polymerization and ethylene/hexene copolymerization by vanadium(iii) complexes bearing bidentate phenoxy-phosphine oxide ligands. *J. Polym. Sci. Part A Polym. Chem.* **2013**, *51*, 5298–5306. [CrossRef]
59. Zhang, S.W.; Lu, L.P.; Li, B.X.; Li, Y.S. Synthesis, structural characterization, and olefin polymerization behavior of vanadium(iii) complexes bearing bidentate phenoxy-phosphine ligands. *J. Polym. Sci. Part A Polym. Chem.* **2012**, *50*, 4721–4731. [CrossRef]
60. Wu, J.-Q.; Pan, L.; Li, Y.-G.; Liu, S.-R.; Li, Y.-S. Synthesis, structural characterization, and olefin polymerization behavior of vanadium(III) complexes bearing tridentate schiff base ligands. *Organometallics* **2009**, *28*, 1817–1825. [CrossRef]
61. Białek, M.; Leksza, A.; Piechota, A.; Kurzak, K.; Koprek, K. Oxovanadium(IV) complexes with [ONNO]-chelating ligands as catalysts for ethylene homo- and copolymerization. *J. Polym. Res.* **2014**, *21*, 389. [CrossRef]
62. Białek, M.; Bisz, E. Dichlorovanadium(IV) diamine-bis(phenolate) complexes for ethylene (co)polymerization and 1-olefin isospecific polymerization. *J. Cat.* **2018**, *362*, 65–73. [CrossRef]
63. Lorber, C.; Wolff, F.; Choukroun, R.; Vendier, L. Ethylene homo- and copolymerization activity of a series of [ONNO]-type amine bis(phenolate) based vanadium(II–V) catalysts. *Eur. J. Inorg. Chem.* **2005**, *81*, 2850–2859. [CrossRef]
64. Lorber, C. [ONNO]-type amine bis(phenolate)-based vanadium catalysts for ethylene homo- and copolymerization. *Pure Appl. Chem.* **2009**, *81*, 1205–1215. [CrossRef]
65. Homden, D.M.; Redshaw, C.; Hughes, D.L. Vanadium complexes possessing $N_2O_2S_2$-based ligands: Highly active procatalysts for the homopolymerization of ethylene and copolymerization of ethylene/1-hexene. *Inorg. Chem.* **2007**, *46*, 10827–10839. [CrossRef] [PubMed]

66. Guao, N.; Li, L.; Marks, T.J. Bimetallic catalysis for styrene homopolymerization and ethylene-styrene copolymerization. exceptional comonomer selectivity and insertion regiochemistry. *J. Am. Chem. Soc.* **2004**, *126*, 6542–6543. [CrossRef]
67. Stevens, J.C.; Timmers, F.J.; Wilson, D.R.; Schmidt, G.F.; Nickias, P.N.; Rosen, R.K.; Knight, G.W.; Lai, S. Constrained Geometry Addition Polymerization Catalysts, Processes for Their Preparation, Precursors Therefor, Methods of Use, and Novel Polymers Formed Therewith, EP0416815B1. U.S. Patent 6,806,326, 12 September 2006.
68. Arriola, D.J.; Bokota, M.; Campbell, R.E.; Klosin, J.; LaPointe, R.E.; Redwine, O.D.; Shankar, R.B.; Timmers, F.J.; Abboud, K.A. Penultimate effect in ethylene-styrene copolymerization and the discovery of highly active ethylene-styrene catalysts with increased styrene reactivity. *J. Am. Chem. Sci.* **2007**, *129*, 7065–7076. [CrossRef] [PubMed]
69. Sernetz, F.G.; Miilhaupt, R.; Waymouth, R.M. Influence of polymerization conditions on the copolymerization of styrene with ethylene using $Me_2Si(Me_4Cp)$(N-tert-butyl)$TiCl_2$/methylaluminoxane Ziegler-Natta catalysts. *Macromol. Chem. Phys.* **1996**, *197*, 1071–1083. [CrossRef]
70. Sukhova, T.A.; Panin, A.N.; Babkina, O.N.; Bravaya, N.M. Catalytic systems $Me_2SiCp^*N^t$ $BuMX_2/(CPh_3)(B(C_6F_5)_4)$ (M=Ti, X=CH_3, M=Zr, X=iBu) in copolymerization of ethylene with styrene. *J. Polym. Sci. Part A Polym. Chem.* **1999**, *37*, 1083–1093. [CrossRef]
71. Chum, P.S.; Kruper, W.J.; Guest, M.J. Materials properties derived from INSITE metallocene catalysts. *Adv. Mater.* **2000**, *12*, 1759–1767. [CrossRef]
72. Nomura, K. Half-titanocenes containing anionic ancillary donor ligands: Effective catalyst precursors for ethylene/styrene copolymerization. *Catalysts* **2013**, *3*, 157–175. [CrossRef]
73. Longo, H.; Grassi, A.; Oliva, L. Copolymerization of styrene and ethylene in the presence of different syndiospecific catalysts. *Macromol. Chem.* **1991**, *191*, 2387–2396. [CrossRef]
74. Zhang, H.; Nomura, K. Living copolymerization of ethylene with styrene catalyzed by (cyclopentadienyl)(ketimide)titanium(iv) complex-MAO catalyst system. *Macromolecules* **2006**, *39*, 5266–5274. [CrossRef]
75. Caporaso, L.; Izzo, L.; Sisti, I.; Oliva, L. Stereospecific ethylene−styrene block copolymerization with ansa-zirconocene-based catalyst. *Macromolecules* **2002**, *35*, 4866–4870. [CrossRef]
76. Arai, T.; Ohtsu, T.; Suzuki, S. Stereoregular and Bernoullian copolymerization of styrene and ethylene by bridged metallocene catalysts. *Macromol. Rapid Commun.* **1998**, *19*, 327–331. [CrossRef]
77. Ren, J.; Hatfield, G.R. Synthesis and characterization of poly(ethylene-co-styrene). *Macromolecules* **1995**, *28*, 2588–2589. [CrossRef]
78. Venditto, V.; De Tullio, G.; Izzo, L.; Oliva, L. Ethylene-styrene copolymers by ansa-zirconocene- and half-titanocene-based catalysts: Composition, stereoregularity, and crystallinity. *Macromolecules* **1998**, *31*, 4027–4029. [CrossRef]
79. Sernetz, F.G.; Mülhaupt, R.; Fokken, S.; Okuda, J. Copolymerization of ethene with styrene using methylaluminoxane-activated bis(phenolate) complexes. *Macromolecules* **1997**, *30*, 1562–1569. [CrossRef]
80. Capacchione, C.; Proto, A.; Ebeling, H.; Mülhaupt, R.; Okuda, J. Copolymerization of ethylene with styrene catalyzed by a linked bis(phenolato) titanium catalyst. *J. Polym. Sci. Part A Polym. Chem.* **2006**, *44*, 1908–1913. [CrossRef]
81. Weiser, M.-S.; Mülhaupt, R. Formation of polyolefin-*block*-polystyrene block copolymers on phenoxyimine catalysts. *Macromol. Rapid Commun.* **2006**, *27*, 1009–1014. [CrossRef]
82. Li, S.; Wang, M.; Cui, D. Copolymerization of ethylene with styrene catalyzed by a scandium catalyst. *Polym. Chem.* **2018**, *9*, 4757–4763. [CrossRef]
83. Theaker, G.W.; Morton, C.; Scott, P. substituent effects in ethene/styrene copolymerization: Dormant state destabilization? *J. Polym. Sci. Part A Polym. Chem.* **2009**, *47*, 3111–3117. [CrossRef]
84. Guo, N.; Stern, C.L.; Marks, T.J. Bimetallic effects in homopolymerization of styrene and copolymerization of ethylene and styrenic comonomers: Scope, kinetics, and mechanism. *J. Am. Chem. Soc.* **2008**, *130*, 2246–2261. [CrossRef]
85. Theaker, G.W.; Morton, C.; Scott, P. Awakening a dormant catalyst: Salicylaldimine systems for ethene/*tert*-butylstyrene copolymerization. *Dalton Trans.* **2008**, 6883–6885. [CrossRef] [PubMed]
86. Białek, M.; Fryga, J. Homopolymerization of styrenic monomers and their copolymerization with ethylene using group 4 non-metallocene catalysts. *J. Appl. Polym. Sci.* **2020**, *137*, e49349. [CrossRef]
87. Hou, L.; Wang, W.; Sheng, J.; Liu, C. Synthesis and ^{13}C-NMR analysis of an ethylene copolymer with 9-decen-1-ol. *RSC Adv.* **2015**, *5*, 98929–98933. [CrossRef]
88. Dai, S.; Chen, C. Direct synthesis of functionalized high-molecular-weight polyethylene by copolymerization of ethylene with polar monomers. *Angew. Chem. Int. Ed.* **2016**, *55*, 13281–13285. [CrossRef]
89. Na, Y.; Dai, S.; Chen, C. Direct synthesis of polar-functionalized linear low-density polyethylene (LLDPE) and low-density polyethylene (LDPE). *Macromolecules* **2018**, *51*, 4040–4048. [CrossRef]
90. Terao, H.; Ishii, S.; Mitani, M.; Tanaka, H.; Fujita, T. Ethylene/polar monomer copolymerization behavior of bis(phenoxy–imine)ti complexes: Formation of polar monomer copolymers. *J. Am. Chem. Soc.* **2008**, *130*, 17636–17637. [CrossRef]
91. Atiqullah, M.; Tinkl, M.; Pfaendner, R.; Akhtar, M.N.; Hussain, I. Synthesis of functional polyolefins using metallocenes. A comprehensive review. *Polym. Rev.* **2010**, *50*, 178–230. [CrossRef]
92. Wang, W.; Hou, L.; Luo, S.; Zheng, G.; Wang, H. Synthesis and ^{13}C NMR spectroscopy analysis of ethylene copolymer with high content of 4-penten-1-ol. *Macromol. Chem. Phys.* **2013**, *214*, 2245–2249. [CrossRef]
93. Saito, Y.; Nakata, N.; Ishii, A. Copolymerization of ethylene with *i*Pr_3Si-protected 5-hexen-1-ol with an [OSSO]-type bis(phenolato) dichloro zirconium(IV) complex. *Bull. Chem. Soc. Jpn.* **2016**, *89*, 666–670. [CrossRef]

94. Zhang, X.; Chen, S.; Li, H.; Zhang, Z.; Lu, Y.; Wu, C.; Hu, Y. Highly active copolymerization of ethylene with 10-undecen-1-ol using phenoxy-based zirconium/methylaluminoxane catalysts. *J. Polym. Sci. Part A Polym. Chem.* **2005**, *43*, 5944–5952. [CrossRef]
95. Zhang, X.; Chen, S.; Li, H.; Zhang, Z.; Lu, Y.; Wu, C.; Hu, Y. Copolymerizations of ethylene and polar comonomers with bis(phenoxyketimine) group IV complexes: Effects of the central metal properties. *J. Polym. Sci. Part A Polym. Chem.* **2007**, *45*, 59–68. [CrossRef]
96. Yang, X.-H.; Liu, C.-R.; Wang, C.; Sun, X.-L.; Guo, Y.-H.; Wang, X.-K.; Wang, Z.; Xie, Z.; Tang, Y. [O^-NS^R]$TiCl_3$-catalyzed copolymerization of ethylene with functionalized olefins. *Angew. Chem. Int. Ed.* **2009**, *48*, 8099–8102. [CrossRef] [PubMed]
97. Białek, M.; Fryga, J. Effective copolymerization of ethylene with α,ω-alkenols and homopolymerization of α,ω-alkenols catalyzed by aminophenolate zirconium complex. *React. Funct. Polym.* **2019**, *137*, 11–20. [CrossRef]
98. Sampson, J.; Bruening, M.; Akhtar, M.N.; Jaseer, E.A.; Theravalappil, R.; Garcia, N.; Agapie, T. Copolymerization of ethylene and long-chain functional α-olefins by dinuclear zirconium catalysts. *Organometallics* **2021**, *40*, 1854–1858. [CrossRef]
99. Lu, L.P.; Mu, J.S.; Li, Y.S. Copolymerization of ethylene with 10-undecen-1-ol using highly active vanadium(iii) precatalysts bearing bis(imino)pyrrolyl ligands. *Chin. J. Polym. Sci.* **2014**, *32*, 603–608. [CrossRef]
100. Mu, J.-S.; Liu, J.-Y.; Liu, S.-R.; Li, Y.-S. Copolymerizations of ethylene with α-olefin-ω-ols by highly active vanadium(III) catalysts bearing [N,O] bidentate chelated ligands. *Polymer* **2009**, *50*, 5059–5064. [CrossRef]

polymers

MDPI

Article

Influence of Oligopeptide Length and Distribution on Polyisoprene Properties

Chang-Cheng Wang, Rong Zhang, Shiqi Li, Guangsu Huang, Maozhu Tang * and Yun-Xiang Xu *

College of Polymer Science & Engineering, State Key Laboratory of Polymer Materials Engineering, Sichuan University, Chengdu 610065, China; 2019223090113@stu.scu.edu.cn (C.-C.W.); 18980908620@163.com (R.Z.); 18702882636@163.com (S.L.); huangguangsu@scu.edu.cn (G.H.)
* Correspondence: mztang@scu.edu.cn (M.T.); yxxu@scu.edu.cn (Y.-X.X.)

Abstract: The tuning of binding modes of polar groups is the key step to mimicking the structure and properties of natural rubber through the molecular design of synthetic polyisoprenes. Herein, the ordering and binding distances of oligopeptides could be altered systematically by changing their lengths and distribution along the polyisoprene chain, which impose huge impacts on the mechanical properties and chain dynamics of green rubber. In detail, a series of peptide-functionalized polyisoprenes with terminal blocks (B-2A-PIP, B-3A-PIP) or random sequences (R-2A-PIP, R-3A-PIP) are fabricated by using dipeptides (2A) or tripeptides (3A) as crosslinkers to explore the mechanism of terminal interaction on mechanism properties and chain dynamics. B-4A-PIP and R-4A-PIP served as control samples. It is found that the increased oligopeptide length and the block distribution improves the mechanical properties and confine the chain movement by elevate the contents of ordered and compact microstructures, which is indicated by XRD, broadband dielectric spectroscopy (BDS) and consistent with the result of molecular dynamics simulation. New relaxation signals belonging to oligopeptide aggregates are found which showed elevated dielectric strengths upon temperatures increase. Additionally, it also reveals that the binding modes of oligopeptide do not significantly influence the entanglements of polyisoprene.

Keywords: polyisoprene; oligopeptide length; distribution; terminal block

Citation: Wang, C.-C.; Zhang, R.; Li, S.; Huang, G.; Tang, M.; Xu, Y.-X. Influence of Oligopeptide Length and Distribution on Polyisoprene Properties. *Polymers* **2021**, *13*, 4408. https://doi.org/10.3390/polym13244408

Academic Editor: Edina Rusen

Received: 18 November 2021
Accepted: 13 December 2021
Published: 15 December 2021

Publisher's Note: MDPI stays neutral with regard to jurisdictional claims in published maps and institutional affiliations.

Copyright: © 2021 by the authors. Licensee MDPI, Basel, Switzerland. This article is an open access article distributed under the terms and conditions of the Creative Commons Attribution (CC BY) license (https://creativecommons.org/licenses/by/4.0/).

1. Introduction

As an important strategic material, natural rubber (NR) is used in a wide range of livelihood and military applications, such as tires, seals, and aerospace applications. This is due to its excellent comprehensive performance, such as high strength, high toughness, high abrasion resistance, tear resistance, and wet skid resistance [1–3]. However, the natural conditions of the origin greatly limit the quality and yield of natural rubber. Therefore, it is necessary to find alternatives to natural rubber. IR is the most promising synthetic rubber to replace natural rubber, which has similar chemical composition and cis content with NR [4]. Nevertheless, the comprehensive properties of IR are quite inferior to those of NR, such as tensile strength and fatigue resistance [5–7]. Practically, the properties of NR are not only related to the cis content and molecular weight, but also closely related to their terminal structures [8]. In particular, the α-terminal and ω-terminal consist of phospholipids or non-covalently linked proteins, respectively [9]. Different polar terminal structures make distinct contribution on the mechanical properties. However, so far it is relatively challenging to reverse engineer the terminal structures of NR by synthetic approach due to their complicated components. For example, several types of proteins are found in NR and each type is one kind of huge molecule. Therefore, to mimic the terminal structures of NR from the principal level, simplified system is urgently required.

It is generally assumed that the key functions of terminal structures of NR consists of two factors [10]. One is the formation of branching structures which elevated the entanglement of polyisoprene chains, and another one is the dissociation of noncovalent interactions to dissipate energy which are both closely related to the terminal supramolecular interactions. However, although a great deal of research has been conducted focused on the mechanism of supramolecular interactions on comprehensive properties, the support regarding terminal interactions such as NR is rare [11,12]. Moreover, the contributions of terminal groups are generally verified in tensile strength [13], dimensional stability [14], shape memory [15] and so on because of their unique structures. Thus, to construct high performance elastomers, it is valuable to establish the structure–property relationship of terminal interactions on high performance elastomers via precisely tuneable systems.

As a commonly utilized self-assembly tool, oligopeptide has the potential to provide aggregates with tuneable supramolecular interactions and self-sorting properties. If proper terminal structures and oligopeptide structures are selected, precise terminal supramolecular interactions between different polymer chains would occur and their assembly or dissociation could be manipulated. Specifically, the hydrogen bonding numbers of oligopeptide units can be expediently tuned by changing the oligopeptide backbone length [16], which significantly tunes the supramolecular binding strength and impact the elastomer properties. For instance, Jia et al. [17] synthesized a series of star-block copolymers comprising of polyisobutylene stars and oligo(β-alanine) end segments with different length. β-alanine dimer is partially hydrogen-bonded, while the trimer, tetramer and pentamer are fully hydrogen-bonded and form β-sheets. As the oligopeptide length increased, the melting temperatures of crystalline domains raised and eventually increased the rubber elastic moduli.

In addition, the oligopeptides can form aggregates and affect the dynamics of the multiscale units of the polymer. To better understand the structure–property relationships on the molecular unit concepts, the molecular dynamics are necessary. Dielectric spectroscopy is a powerful tool for detecting multi-level molecular units relaxation, which can easily allow to cover 10^{-6}–10^{12} Hz in frequency [18]. In the present work, we use cis-1,4-PI as the bulk material. Owing to the asymmetrical structure, cis-1,4-PI has a dipole moment both parallel and perpendicular to the chain contour [19]. Usually, it presents two relaxation modes under an external electric field, defined as segmental mode (SM) relaxation and normal mode (NM) relaxation, respectively. SM relaxation is caused by the perpendicular component of the dipole, and NM relaxation is caused by the parallel component [20]. Valid information can be obtained from the changes of multi-level molecular units relaxation, and therefore, dielectric spectroscopy will contributes to a better understanding of the structure–property relationships.

In our previous research of mimicking the terminal structure of natural rubber, techniques to prepare terminally oligopeptide functionalized polyisoprene have been developed [21]. However, how to tune the supramolecular binding strength of terminal structures and their principle to maintain terminal interactions need to be further explored. In this paper, besides the previously prepared B-4A-PIP and R-4A-PIP, dipeptide and tripeptide units were grafted onto hydroxyl-functionalized polyisoprenes B-OH-PIP with terminal block and R-OH-PIP with random sequence, producing B-2A-PIP/B-3A-PIP and R-2A-PIP/R-3A-PIP, respectively (Scheme 1). It is found oligopeptide backbone lengths influence the mechanical properties, relaxation processes of elastomers and the self-assembly mode of oligopeptide themselves because of different terminal supramolecular strength and β-sheet aggregation. Furthermore, the terminal block of oligopeptide will self-sort to form stronger terminal interactions regardless of the peptide length, which provide a robust tool to construct the mimicking terminal structures of NR.

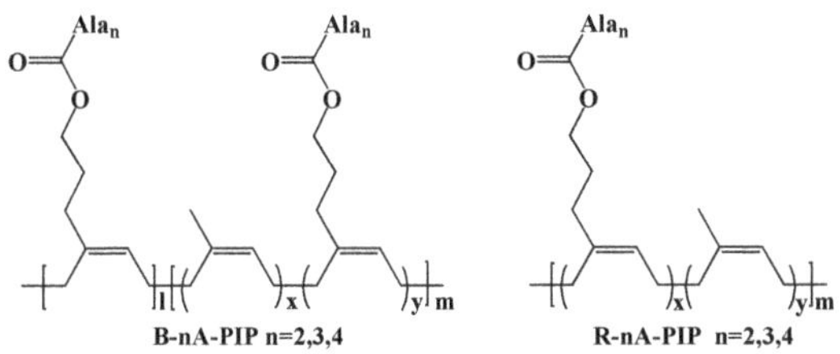

Scheme 1. Chemical structures of oligopeptide functionalized polyisoprene with terminal blocks (B-nA-PIP, n = 2, 3, 4) or random sequences (R-nA-PIP, n = 2, 3, 4).

2. Materials and Methods

2.1. Materials

B-OH-PIP (Mn = 9.2 × 10^5) and R-OH-PIP (Mn = 8.2 × 10^5) were synthesized by the stereoselective coordination polymerization in the glove box according to our previous method [21]. The average number of hydroxyl groups in the terminal block of B-OH-PIP copolymer is approximately forty, while about one hundred hydroxyls are randomly dispersed in the rest polyisoprene chain. For R-OH-PIP, there are about 130 hydroxyl groups randomly dispersed in the polyisoprene chain. Additionally, B-4A-PIP and R-4A-PIP are prepared according to our previous study [21]. Tetrahydrofuran was refluxed over sodium/diphenylketone under nitrogen and then distilled before use. Other chemicals, unless otherwise specified, were purchased from Aldrich (Aldrich, Shanghai, China) and used as received.

2.2. Synthesis of Samples

2.2.1. B-2A-PIP

Step 1: The preparation of 2Ala-OCH2Ph. Boc-2Ala-OCH2Ph was synthesized according to the reported procedure [22]. Boc-2Ala-OCH2Ph (399 mg, 1.14 mmol) was resolved in the dichloromethane (3 mL) for 10 min and then trifluoroacetic acid (TFA, 7 mL) was slowly added into the suspension at the room temperature. The solution was stirred overnight and concentrated by rotary evaporation. The crude product 2Ala-OCH2Ph was then used without further purification. Step 2: The preparation of B-DSC-PIP. N,N′-disuccinimidyl carbonate (76.8 g, 300 mmol) and 4-dimethylaminopyridine (37.0 g, 300 mmol) were resolved in the mixed solution of dry THF (3000 mL) and DMF (1400 mL) in the three-neck bottle while B-OH-PIP (3.0 g) was resolved in dry THF (300 mL) in the flask. Secondly, the latter was slowly added into the mixed solution at 0 °C. The reaction was stirred at the room temperature for 8 h. The mixture was evaporated and poured into acetone to get the precipitate B-DSC-PIP. The crude grey product was collected (2.98 g, 99.3%) and the conversion ratio from hydroxyl group to carbonate group is detected with ^{1}H-NMR, which is 99.4%, as shown in Figure S1. ^{1}H NMR (400 MHz, Chloroform-d) δ: 5.12 (t, 2H), 4.29 (t, 2H), 2.83 (s, 4H), 2.04 (m, 10H), 1.83 (m, 2H), 1.68 (s, 3H). Step3. The synthesis of B-2A-PIP. 2Ala-OCH2Ph (1.14 mmol) and diisopropylethylamine (DIEA, 5.5 mL) were resolved in the mixed solution of THF (150 mL) and water (14 mL) in the three-neck bottle while B-DSC-PIP (1.5 g) was resolved in dry THF (300 mL) in the flask. Secondly, the latter was slowly added into the mixed solution at 0 °C. The reaction was stirred at the room temperature for 8 h. The mixture was evaporated and poured into water to get the precipitate B-2A-PIP. The precipitate was washed with water for three times and the final product was dried in vacuum oven at 45 °C to afford a gray elastomer (1.46 g, 97.3%). The conversion ratio from carbonate group to dipeptide is detected with ^{1}H-NMR, which is 99.6%, as shown in Figure S2. ^{1}H NMR (400 MHz, Chloroform-d) δ: 7.34 (s, 5H), 5.12 (t, 2H), 4.61 (m, 1H), 4.21 (m, 1H), 4.04 (t, 2H), 2.04 (m, 10H), 1.83 (m, 2H), 1.68 (s, 3H).

2.2.2. Synthesis of B-3A-PIP

Step 1: The preparation of 3Ala-OCH2Ph. Boc-3Ala-OCH2Ph was synthesized according to the reported procedures [22]. Boc-3Ala-OCH2Ph (480 mg, 1.14 mmol) was resolved in the dichloromethane (3 mL) for 10 min and then trifluoroacetic acid (TFA, 7 mL) was slowly added into the suspension at the room temperature. The solution was stirred overnight and concentrated by rotary evaporation. The crude product was then used without further purification. Step 2: The synthesis of B-3A-PIP. 3Ala-OCH2Ph (1.14 mmol) and diisopropylethylamine (DIEA, 5.5 mL) were resolved in the mixed solution of THF (150 mL) and water (14 mL) in the three-neck bottle while B-DSC-PIP (1.48 g) was resolved in dry THF (300 mL) in the flask. Secondly, the latter was slowly added into the mixed solution at 0 °C. The reaction was stirred at the room temperature for 8 h. The mixture was evaporated and poured into water to get the precipitate B-3A-PIP. The precipitate was washed with water for three times and the final product was dried in vacuum oven at 45 °C to afford a gray elastomer (1.46 g, 98.6%). The conversion ratio from carbonate group to tripeptide is detected with ^{1}H-NMR, which is 99.6%, as shown in Figure S3. ^{1}H NMR (400 MHz, Chloroform-d) δ: 7.34 (s, 5H), 5.12 (t, 2H), 4.61 (m, 1H), 4.21 (m, 1H), 4.04 (t, 2H), 2.04 (m, 10H), 1.83 (m, 2H), 1.68 (s, 3H).

2.2.3. Synthesis of R-2A-PIP

Step 1: The preparation of R-DSC-PIP. N, N′-disuccinimidyl carbonate (76.8 g, 300 mmol) and 4-dimethylaminopyridine (37.0 g, 300 mmol) were resolved in the mixed solution of dry THF (3000 mL) and DMF (1400 mL) in the three-neck bottle while R-OH-PIP (3.0 g) was resolved in dry THF (300 mL) in the flask. Secondly, the latter was slowly added into the mixed solution at 0 °C. The reaction was stirred at the room temperature for 8 h. The mixture was evaporated and poured into acetone to get the precipitate R-DSC-PIP. The crude grey product was collected (2.96 g, 98.7%) and the conversion ratio from hydroxyl group to carbonate group is detected with ^{1}H-NMR, which is 97.8%, as shown in Figure S4. ^{1}H NMR (400 MHz, Chloroform-d) δ: 5.12 (t, 2H), 4.29 (t, 2H), 2.83 (s, 4H), 2.04 (m, 10H), 1.83 (m, 2H), 1.67 (s, 3H). Step 2: The synthesis of R-2A-PIP. 2Ala-OCH2Ph (1.14 mmol) and diisopropylethylamine (DIEA, 5.5 mL) were resolved in the mixed solution of THF (150 mL) and water (14 mL) in the three-neck bottle while R-DSC-PIP (1.5 g) was resolved in dry THF (300 mL) in the flask. Secondly, the latter was slowly added into the mixed solution at 0 °C. The reaction was stirred at the room temperature for 8 h. The mixture was evaporated and poured into water to get the precipitate R-2A-PIP. The precipitate was washed with water for three times and the final product was dried in vacuum oven at 45 °C to afford a gray elastomer (1.48 g, 98.7%). The conversion ratio from carbonate group to dipeptide is detected with ^{1}H-NMR, which is 98.3%, as shown in Figure S5. ^{1}H NMR (400 MHz, Chloroform-d) δ: 7.34 (s, 5H), 5.12 (t, 2H), 4.61 (m, 1H), 4.21 (m, 1H), 4.04 (t, 2H), 2.04 (m, 10H), 1.83 (m, 2H), 1.67 (s, 3H).

2.2.4. Synthesis of R-3A-PIP

3Ala-OCH2Ph (1.14 mmol) and diisopropylethylamine (DIEA, 5.5 mL) were resolved in the mixed solution of THF (150 mL) and water (14 mL) in the three-neck bottle while R-DSC-PIP (1.5 g) was resolved in dry THF (300 mL) in the flask. Secondly, the latter was slowly added into the mixed solution at 0 °C. The reaction was stirred at the room temperature for 8 h. The mixture was evaporated and poured into water to get the precipitate R-3A-PIP. The precipitate was washed with water for three times and the final product was dried in vacuum oven at 45 °C to afford a gray elastomer (1.43 g, 97.9%). The conversion ratio from carbonate group to tripeptide is detected with ^{1}H-NMR, which is 99.6%, as shown in Figure S6. ^{1}H NMR (400 MHz, Chloroform-d) δ: 7.34 (s, 5H), 5.12 (t, 2H), 4.58 (m, 1H), 4.45 (m, 1H), 4.20 (m, 1H), 4.05 (t, 2H), 2.04 (m, 10H), 1.83 (m, 2H), 1.67 (s, 3H).

2.3. Sample Molding Process

Each product (1.4 g) was resolved in toluene (70 mL) and the solution slowly volatilized in the fume hood to obtain a flat membrane. The product was then moved into the vacuum oven at 40 °C for 12 h to remove the solvent. The species were cut from the membrane according to requirements of different measurements. As a representative, a picture of the sample used for uniaxial tensile testing is shown in Figure S7.

2.4. Characterization

The Bruker ASCEND 400 spectrometer (Bruker, Karlsruhe, Germany) was used to collecting the ^1H NMR spectra, operating at 400 MHz. The samples are resolved in deuterated chloroform solution with TMS as reference. The universal testing machine Instron 5966 (Instron, Boston, MA, USA) was used to measure mechanical properties at 25 °C. The tensile speed is 100 mm/min. The specimen was cut into a dumbbell shaped thin strip with dimension of 35 mm × 2 mm × 1 mm. For each data, the strength was calculated from three measurements and taken from the average value. The Nicolet iS10 (Nicolet, Waltham, MA, USA) was used to measure FTIR spectra in the range of 4000–400 cm^{-1}. The Rigaku X-ray diffractometer (Ultima IV, Akishima-shi, Japan) were used to carry X-ray diffraction experiments with parallel beam optics attachment.

Novocontrol Concept 50 system (Novocontrol GmbH, Montabaur, Germany) was carried on dielectric measurements with Alpha impedance analyzer and Quatro Cryosystem temperature control over the frequency range of 10^{-1}–10^7 Hz. The measured film was about 1 mm thickness and placed between two parallel electrodes with 10 mm diameter. The measured temperatures were from −70 °C to 80 °C at the step of 10 °C. The result were fitted using the Havriliak–Negami function with a conductivity term [23]. In this model, the frequency dependence of the dieletric complex (ε^*) can be described by the equation:

$$\varepsilon^*(\omega) = \varepsilon_\infty + \frac{\Delta\varepsilon}{\left[1 + (i\omega\tau_{HN})^\alpha\right]^\beta} \quad (1)$$

where $\Delta\varepsilon = \varepsilon_s - \varepsilon_\infty$ is the dielectric strength, ε_s and ε_∞ are the relaxed and unrelaxed values of dielectric constant, the parameters α and β ($0 < \alpha, \alpha\beta \leq 1$) define the symmetrical and asymmetrical broadening of the loss peak, and τ_{HN} is the characteristic relaxation time. The relation between τ_{HN} and $\tau_{\max}$ is given by Refs [24,25].

$$\tau_{\max} = \tau_{HN}\left[\sin\frac{\pi\alpha\beta}{2(1+\beta)}\right]^{\frac{1}{\alpha}}\left[\sin\frac{\pi\alpha}{2(1+\beta)}\right]^{-\frac{1}{\alpha}} \quad (2)$$

This characteristic relaxation time ($\tau_{\max}$) obtained from the *HN* equation fit can be correlated with the temperature through the Vogel–Fulcher–Tamman (VFT) equation [26,27]:

$$\tau_{\max} = \tau_0 \exp\left(\frac{B}{T - T_0}\right) \quad (3)$$

where τ_0 and B are empirical parameters and T_0 is the so-called Vogel temperature.

The polymer models were constructed using the polymer builder of Materials Studio package. The GROMACS [28–32] version 2018.3 was used to carry further full-atomistic classical molecular dynamics simulations. The forcefield used for this study was the OPLS all-atom (OPLSAA) force field [33–36], considering its wide applicability in different polymer systems. The timestep used for all simulations was taken as 1 fs. VMD [37] was used to visualize and analyze various properties based on MD trajectories saved every 5000 steps (5 ps). The equilibrated systems have densities of 0.890 ± 0.003 g/cm^3, 0.899 ± 0.003 g/cm^3 and 0.873 ± 0.017 g/cm^3, for B-4A-PIP, B-3A-PIP and B-2A-PIP, respectively.

3. Results and Discussion

3.1. Mechanical Properties and Secondary Structure Analysis

B-OH-PIP and R-OH-PIP were decorated with dipeptides and tripeptides, producing B-2A-PIP/ B-3A-PIP, R-2A-PIP/ R-3A-PIP, respectively. The synthetic routes are shown in Figure S8. B-4A-PIP and R-4A-PIP serve as control samples. All copolymers have similar molecular weights and amounts of polar groups. The tensile strength and strain at break of B-2A-PIP and R-2A-PIP are relatively weak compared to those of B-3A-PIP and R-3A-PIP (Figure 1a), but slightly higher than those of B-OH-PIP and R-OH-PIP (Figure S9), respectively. The limited effect of dipeptide is ascribed to its weak binding strength. Compared to dipeptide counterparts, the mechanical properties of B-3A-PIP and R-3A-PIP are remarkably heightened, as shown in Figure 1b. Moreover, the mechanical properties of B-4A-PIP and R-4A-PIP are further increased (Figure S10a), revealing the significant influence of oligopeptide length on the mechanical properties of polyisoprene. Generally, in the aspect of the effect of peptide distribution, the sample with terminal block showed enhanced yield stress and tensile stress at break and increased length of oligopeptide improved the mechanical properties of elastomers. This indicated increased terminal supramolecular interactions produce certain positive role to mechanical properties of elastomers.

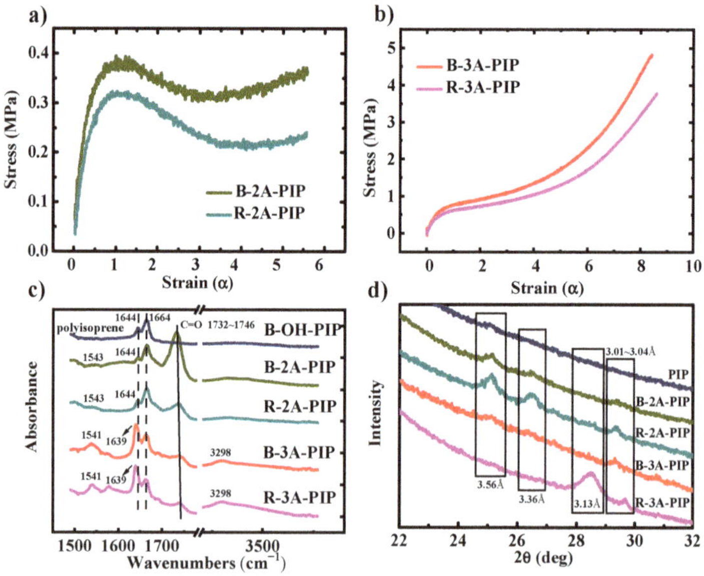

Figure 1. Stress–strain curves as a function of strain of (**a**) B-2A-PIP and R-2A-PIP; (**b**) B-3A-PIP and R-3A-PIP at the tensile rate of 100 mm/min. (**c**) FTIR spectra and (**d**) X-ray diffraction spectra of B-2A-PIP, R-2A-PIP, B-3A-PIP and R-3A-PIP.

Then, FT-IR experiments were further carried out to reveal their difference on the secondary structures of polyisoprene (Figure 1c). Two characteristic bonds at 1644 cm^{-1} and 1664 cm^{-1} are observed for all the samples, attributing to the C=C stretching vibration of polyisoprene [38]. Importantly, obvious β-sheets structures were found at 1531 cm^{-1}, 1632 cm^{-1} and 1706 cm^{-1} in B-4A-PIP and R-4A-PIP (Figure S10b). However, B-2A-PIP and R-2A-PIP give weak amide II absorption at 1543 cm^{-1} and overlapping amide I absorption at 1644 cm^{-1}, corresponding to the random coils [39]. Random coils and extended chain structures of B-3A-PIP and R-3A-PIP are also observed as indicated by the existence of bands at 1541 cm^{-1} (amide II) and at 1639 cm^{-1} (amide I) [39]. The terminal sheet structures are closely related to their corresponding mechanical strength.

Thus, X-ray diffraction experiments were performed to further analyze the structural information of β-sheet (Figure 1d). The diffraction peaks with d spacing of 3.01~3.04 Å are assigned to the hydrogen spacing of β-sheet structures [40,41]. Typically, the augment of hydrogen bonds spacing [42,43] usually indicated weakened binding of hydrogen bonds. Hence, a second reflection around a spacing of 3.13 Å should be attributed to the hydrogen spacing in extended chain structures whose binding strength of hydrogen bonding aggregations is weaker than that of β-sheet structures. Additionally, the reflections at a spacing of 3.56 Å and 3.36 Å should be ascribed to the hydrogen spacing in loosely built β-sheets. In our research, the diffraction peaks of B-4A-PIP assigned to β-sheet and extended chain structures are obvious and sharper than R-4A-PIP (Figure S10c), while the diffraction peak of R-3A-PIP assigned to extended chain structures is also sharp. Additionally, the others are relatively weak. Especially, the diffraction peaks of B-2A-PIP and R-2A-PIP assigned to β-sheets and extended chain structures are both quite weak, while the diffraction peaks assigned to loosely built β-sheets are sharp. It has been reported that the increased oligopeptide length facilitates the formation of ordered and compact structures. However, as the aspect of distribution, the influence of terminal oligopeptide on the secondary structures are not consistent. The oligopeptide aggregates in R-2A-PIP and R-3A-PIP formed more ordered structures than those in B-2A-PIP and B-3A-PIP, mainly reflected in the more pronounced diffraction peaks of loosely built β-sheets and extended chain structures in R-2A-PIP and R-3A-PIP. That is, although the random samples are more ordered, loose aggregates dominate in these ordered structures, while compact aggregates dominate in the block samples (include B-4A-PIP and R-4A-PIP, as shown in Figure S10). Combined with the fact that the block samples have better mechanical properties than the random samples, it seems to indicate that the compact aggregates have a greater effect on the mechanical properties than the loose aggregates.

3.2. Molecular Dynamics-SM Relaxation, NM Relaxation and α′ Relaxation

To explore the influence of terminal oligopeptide and its length on the chain dynamics of copolymers, broadband dielectric spectroscopy (BDS) was conducted. The temperature dependences of dielectric spectroscopy at 6.9 Hz are shown in Figure 2. A low-temperature relaxation just above the glass transition temperature (Tg~−65 °C), which is originated from local motions of the perpendicular dipole moment and assigned as the segmental-mode process (SM) [44,45] (indicated by blue arrows in Figure 2). However, there are no significant differences in the SM relaxation temperatures among these systems. A second broader relaxation peak indicated by black arrows should be assigned to the normal-mode process (NM), which corresponds to the motions of the entire chains [46,47]. The NM relaxation peaks of B-2A-PIP and R-2A-PIP are detected at −30 °C. For the B-3A-PIP sample, its NM relaxation peak shifts toward higher temperature and is detected at 0 °C. Notably, the NM relaxation peaks of R-3A-PIP, B-4A-PIP and R-4A-PIP disappear, maybe due to their stable network in the test condition, as with vulcanized rubber. Moreover, new mode relaxations can also be observed on random copolymers and B-4A-PIP, as indicated by red arrows which are assigned as α′-relaxation. This may be aroused by the terminal interactions.

For more details, quantitative analyses are carried out in the frequency domain. The typical dielectric loss spectra are shown in Figure S11. A clear loss peak emerging around −50 °C (indicated by blue arrows in Figure S11) should be due to SM relaxation [45]. This SM relaxation peak shifts to higher frequency with increasing temperature, which is known as a thermally activated process. The frequency spectra are on further fitted using the Havriliak–Negami function. The temperature dependences of τ_{max} are shown in Figure 3. However, no obvious differences in the SM relaxation speed are observed among these systems. Therefore, it is rational that the terminal oligopeptide and its length have weak influence on the segmental motions.

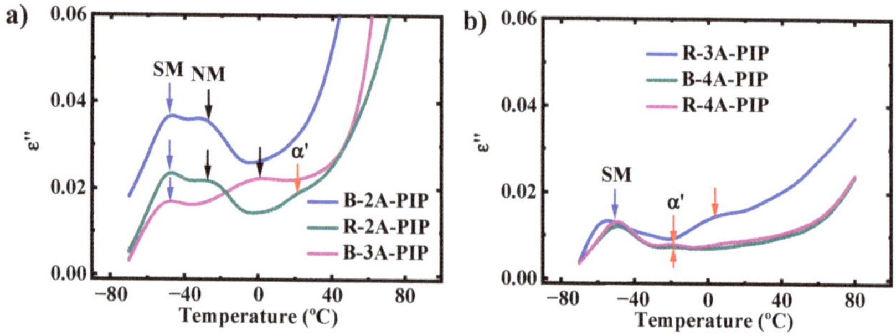

Figure 2. Dielectric loss ε′′ for different copolymers as a function of the temperature at 6.9 Hz ((**a**) B-2A-PIP, R-2A-PIP, B-3A-PIP and (**b**) R-3A-PIP, B-4A-PIP, R-4A-PIP).

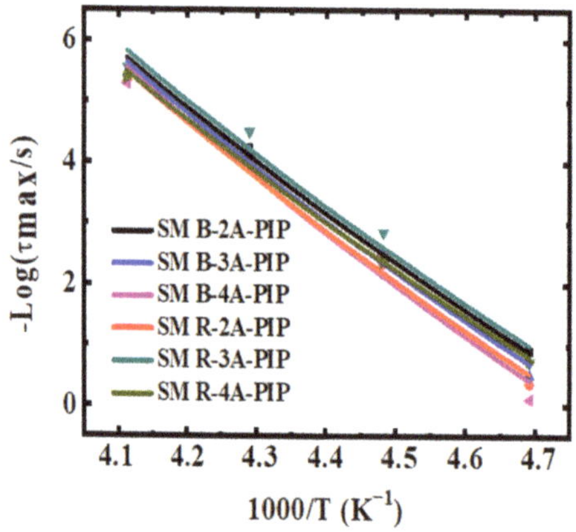

Figure 3. Temperature dependence of τmax for SM relaxation of different samples. Solid lines represent fitting curves.

Quantitative analyses are also carried out for NM relaxations. The typical dielectric loss spectra of B-2A-PIP, R-2A-PIP and B-3A-PIP at higher temperature are shown in Figure 4 and the NM relaxations are indicated by black arrows. The temperature dependence of τ_{max} can also be well described with Vogel–Fulcher–Tamman (VFT) equation. The NM relaxation speed of B-2A-PIP is slower than R-2A-PIP above 254 K, revealing the constraint of terminal blocks on longitudinal chain mobility in the system. Additionally, B-3A-PIP possesses much slower NM relaxation speed than B-2A-PIP. Conclusively, the increased oligopeptide length elevates the temperature of NM relaxation and decreases its speed which may be related to the different stability of β-sheets structures.

Furthermore, while the NM relaxation of B-2A-PIP, R-2A-PIP and B-3A-PIP conform to the VFT equation, the Arrhenius-like equation is used to calculate the apparent activation energy [48,49]:

$$\tau = \tau_0 \exp\left(\frac{E_a}{RT}\right) \tag{4}$$

where E_a is the activation energy and τ_0 is a proportionality constant. As marked in Figure 4d, the E_a of the NM relaxation of B-2A-PIP is 49.4 KJ/mol, while that of B-3A-PIP increases to 71.3 KJ/mol. In fact, the NM process, detectable because of the dipole component parallel to the backbone, corresponds to motions of the entire chain. It indicates that the entire chain relaxation in B-3A-PIP is more severely constrained. The result corresponds well with the X-ray diffraction experiments, which shows the peptide secondary structures in B-3A-PIP are more ordered and compact than those of B-2A-PIP. Therefore, the oligopeptide aggregates in B-3A-PIP are more effective in anchoring the molecular chains, and thus the entire chain relaxation is more severely restricted. In addition, the activation energy of R-2A-PIP is the largest, reaching 100 KJ/mol. This finding confirms that the distribution of oligopeptides has a great influence on chain relaxation. From the secondary structure analysis by X-ray diffraction experiments (Figure 1d), we can find that the peptides in random copolymers tend to form loose aggregates. However, the number of loose aggregates far exceeds that of the more compact aggregates because hydrogen bonds between loose aggregates are more likely to form. Therefore, 2A-PIP possesses more non-covalent crosslinking, which results in the highest NM relaxation activation energy. In addition, compact aggregates are more stable, while the proportion of compact aggregates in R-2A-PIP is relatively small. Therefore, the relaxation time of R-2A-PIP is faster when the temperature is higher than 254 K, as shown in Figure 4d. For R-3A-PIP, B-4A-PIP and R-4A-PIP, the impacts of terminal oligopeptide on the NM relaxation temperature and speed are unclear due to the disappearance of the NM relaxation.

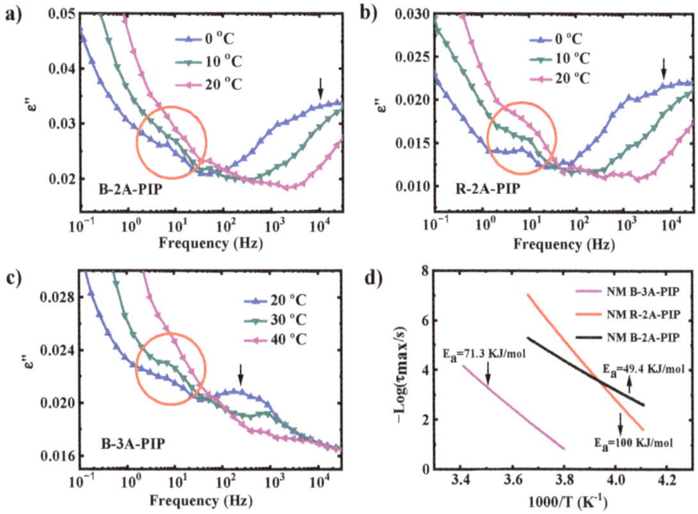

Figure 4. Dielectric loss ε'' as a function of the frequency at for (**a**) B-2A-PIP, (**b**) R-2A-PIP and (**c**) B-3A-PIP. (**d**) The temperature dependence of τmax for NM relaxation of B-2A-PIP, R-2A-PIP and B-3A-PIP.

Interestingly, in the typical dielectric loss spectra of all the samples, there are new peaks in the frequency range of 100–102 Hz, as indicated by red circles in Figures 4 and 5. New relaxation processes are named as B-2A-α' relaxation, R-2A-α' relaxation, B-3A-α' relaxation, R-3A-α' relaxation, B-4A-α' relaxation and R-4A-α' relaxation for B-2A-PIP, R-2A-PIP, B-3A-PIP, R-3A-PIP, B-4A-PIP and R-4A-PIP, respectively. To clarify the origin of α'-relaxation, their dielectric strengths ($\Delta\varepsilon$) are extracted from dielectric spectra and compared with NM relaxation. For polyisoprene, the increased temperature leads to decreased $\Delta\varepsilon$ of NM relaxation [50]. The $\Delta\varepsilon$ of NM relaxation of B-2A-PIP, R-2A-PIP and B-3A-PIP are also decreased with temperature elevation, as shown in Figure S12. Additionally, a slight decrease in $\Delta\varepsilon$ with temperature elevation is characteristic, reflecting the decreasing cooperativity of SM process at higher temperature [51]. However, the $\Delta\varepsilon$ of α'-relaxation exhibits the opposite trend with increasing temperature (Figure 6a,b). Thus, α'-relaxation is neither SM nor NM relaxation. Muller et al. [18] reported that functionalized polybutadiene with a small number of 4-phenyl-1,2,4-triazoline-3,5-dione showed new relaxation mode which is ascribed to the dissociation of the complexes. Inspired by this view, we turn to analyze the secondary structures of alanine oligopeptides and it has been reported that the alanine oligopeptide is arranged in an anti-parallel manner in the secondary structure [52,53]. This anti-parallel arrangement results in a centrosymmetric structure and should not have any dipole moment. Therefore, oligopeptide aggregates exhibit very low dielectric strength. As the temperature rises, the oligopeptide aggregates gradually shift to the dissociated state, which breaks the centrosymmetric structure between oligopeptides. This process increases the dipole moment of the system and, therefore, increases the dielectric strength. Thus, it is reasonable to conclude that α'-relaxation originates from the dissociation of complexed oligopeptides. To further describe the α'-relaxation, we fitted the dielectric spectra with HN equations and extracted the characteristic relaxation time (τ_{max}) in Figure 6c. The temperature dependence of τ_{max} of B-2A-PIP, R-2A-PIP and B-3A-PIP can be well described with Vogel–Fulcher–Tamman (VFT) equation. Compared with NM relaxation, the α'-relaxation speed of B-2A-PIP, R-2A-PIP and B-3A-PIP is slower by several orders of magnitude, all close to 0.1 s. While B-2A-a′ relaxation, R-2A-a′ relaxation and B-3A-a′ relaxation conform to the VFT equation, the Arrhenius-like equation is used to calculate the apparent activation energy (E_a). As shown in Figure 6c, the E_a of B-2A-a′ relaxation and R-2A-a′ relaxation is, respectively, 12.1 KJ/mol and 11.0 KJ/mol, while the activation energy of B-3A-a′ relaxation increases to 13.5 KJ/mol. Strikingly, α'-relaxation peaks of R-3A-PIP, B-4A-PIP and R-4A-PIP almost keep at the same frequency with increasing temperature, which should be ascribed to the restricted mobility of entire chains. The τ_{max} of R-3A-PIP, B-4A-PIP and R-4A-PIP are temperature-independent, which are 0.151 s, 0.143 s and 0.146 s, respectively. Considering the similar activation energy of B-2A, R-2A and B-3A and similar τ_{max} of all six samples, it is speculated that the relaxation of peptide aggregates are closely related to the disentanglement of polyisoprene chains, which are quite independent on the length or distribution of oligopeptides.

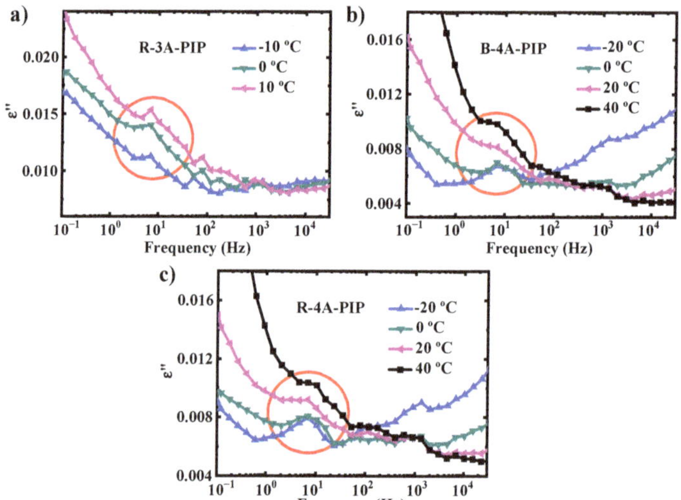

Figure 5. Dielectric loss ε'' as a function of the frequency for (**a**) R-3A-PIP, (**b**) B-4A-PIP and (**c**) R-4A-PIP.

3.3. Classical Molecular Dynamics Simulation

To deeply understand the influence of terminal oligopeptide and its length on the microstructures of copolymers, the molecular dynamics simulation method was carried out. The frame structures of B-2A-PIP, B-3A-PIP and B-4A-PIP are shown in Figure 7. Oligopeptides in block part tend to form nano-sized aggregates, reflecting the influence of terminal oligopeptide on their self-assembly. The structures are further dissected into different layers along the z-axes for better understanding the distribution of oligopeptides and how they interact, as shown in Figure S13–S15. Block oligopeptides are surrounded by abundant random oligopeptides. For B-2A-PIP, the random dipeptides tend to be evenly distributed in all slices. For B-3A-PIP and B-4A-PIP, tripeptides or tetrapeptides increase in the slices with block oligopeptides presented. In particular, one more nano-sized aggregate formed by random tetrapeptides in Figure S14e also increase the number of tetrapeptide chains in that slice. These results can qualitatively support the experimental findings that the microstructures of copolymers become more heterogeneous with increasing oligopeptide length, and may increase the mechanical strength of elastomers. Detailed analyses into structural features of B-4A-PIP shown in Figure 7d,e indicate that block tetrapeptides are likely to form structures with obviously β-sheet feature. In comparison, the interactions between random tetrapeptides can form less ordered β-sheet features.

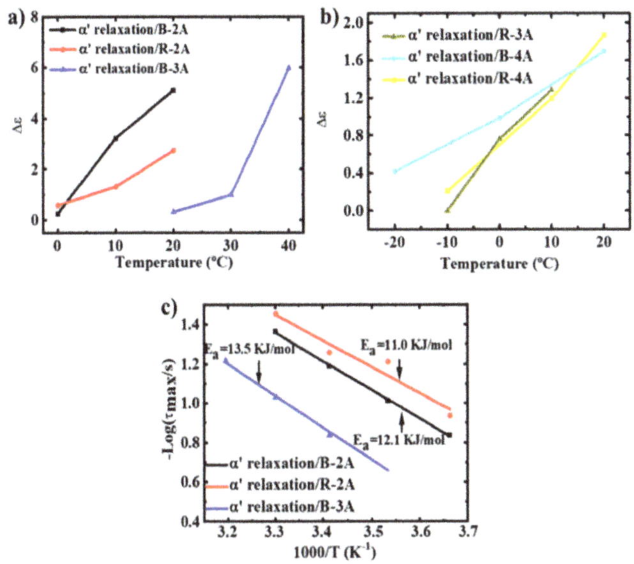

Figure 6. (**a**,**b**) Temperature dependence of Δε for a′ relaxation; (**c**) Arrhenius model for a′ relaxation. Solid line represents fitting curve.

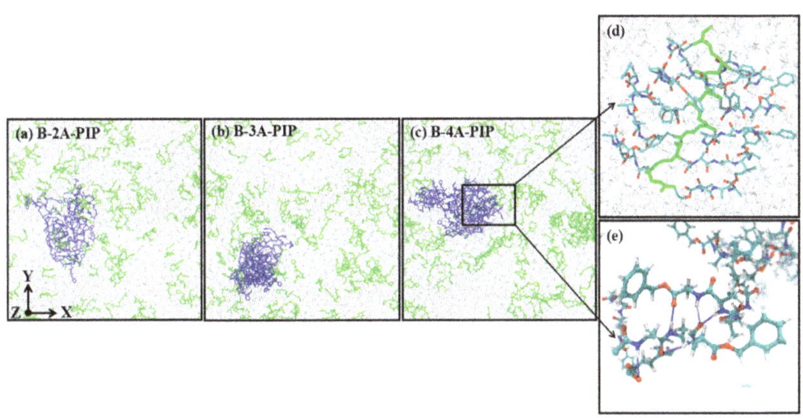

Figure 7. (**a**–**c**) Three different views of frame structures of B-2A-PIP, B-3A-PIP and B-4A-PIP from top with axis denoted figures. A closed black dot indicates that the axis is pointing into the paper. The block segment with 40 repeat units is colored in blue, while the random units are colored in green. PIP units are colored in light grey for clarity. All hydrogen atoms are hidden. (**d**) A part of the block segment of B-4A-PIP. (**e**) A typical random-tetrapeptide interacting units with hydrogen bonds shown in blue dotted lines.

The functions of terminal oligopeptide in copolymers are explored by putting block and random oligopeptides into one or two simulation boxes. Radial distribution functions (RDFs) are used to describe the structures, as shown in Figure 8. Firstly, the interactions between the block and random oligopeptides in one simulation box are analyzed (Figure 8a). In the tetrapeptide system, the peaks formed around 5 Å and 10 Å are sharper than those in the dipeptide and tripeptide systems, as indicated by the black arrows. This suggests that much more ordering structures are formed for tetrapeptides than the others. Increasing the oligopeptide length facilitates more ordering structures. Next, the interactions within block or random oligopeptides in two different simulation boxes are analyzed. In the block system, the curve with similar peak shapes but higher ordinate values are observed for tetrapeptides compared to dipeptides and tripeptides, as indicated by the black arrows. This suggests that increasing the oligopeptide length has a slight influence on the ordered degree of structures while significantly increasing the number of ordered structures in the block system, while in the random system, much sharper and higher peaks of tetrapeptides than the others are observed. It indicates that increasing the oligopeptide length improves both the ordered degree of structures and the number of ordered structures in the random system. Conclusively, the terminal block interaction increased the ordered structures which may well increase the mechanical properties while the increased length of oligopeptide form more, ordered sheet structures in random system. The two factors cause the abnormal difference of β-sheets structures in B-3A-PIP, R-2A-PIP and B-3A-PIP, R-3A-PIP.

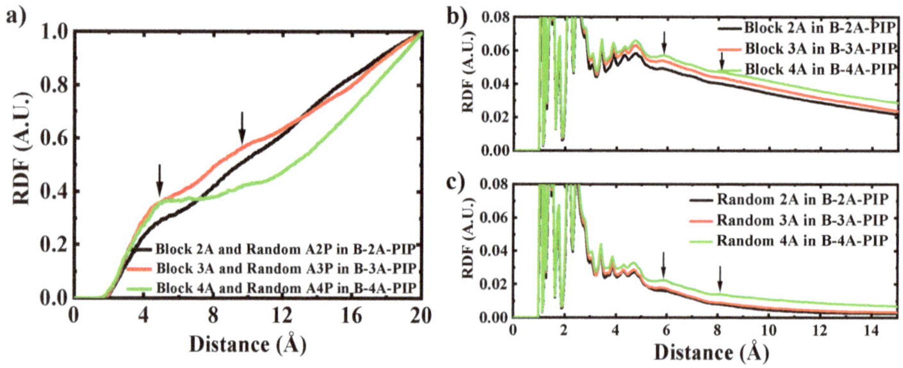

Figure 8. (**a**) The radial distribution functions between block-oligopeptides and random-oligopeptides; (**b**) The radial distribution functions for block-oligopeptides; (**c**) The radial distribution functions for random-oligopeptides.

4. Conclusions

Here, we introduce dipeptide, tripeptide and tetrapeptide into polyisoprene with different hydroxyl distributions, investigating the influence of oligopeptide length and distribution on the mechanical properties, molecular motion and self-assembly behaviors of polyisoprene. The increased oligopeptide length increases the ordered microstructure, which includes compact structure as well as loose structure, and the block distribution of oligopeptides can further increase the proportion of compact aggregates in the ordered microstructure, which is indicated by XRD. Chains dynamics probed by broadband dielectric spectroscopy (BDS) reveals the mechanism. The NM-relaxation and α'-relaxation speed of B-3A-PIP is slower than that of B-2A-PIP, while the E_a of NM-relaxation and α'-relaxation of B-3A-PIP is higher than that of B-2A-PIP, demonstrating that the increased oligopeptide length enhances the restriction effect of oligopeptide aggregates on longitudinal chains movement by increasing the ordered microstructure. In addition, above 254 K, the NM relaxation rate of R-2A-PIP is faster than that of B-2A-PIP due to dissociation of the loose aggregates, demonstrating that there is indeed a larger proportion of compact aggregates in the block samples and, therefore, more severely restricted chain motion in the block samples. However, R-3A-PIP has no significant NM relaxation peak and its α-′ relaxation is temperature-independent. The mechanical properties of B-4A-PIP and R-4A-PIP are further improved compared with B-3A-PIP and R-3A-PIP. B-4A-PIP and R-4A-PIP also have no significant NM relaxation peaks and their α′-relaxations are temperature-independent. In the system with longer oligopeptide length, the mobility of longitudinal chains and oligopeptide complexes are significantly suppressed because of more stable β-sheets structures.

Herein, a new structure–properties relationship regarding terminal interaction in elastomer has been developed. The increased oligopeptide length improves the ordered degree of structures and the number of ordered structures in the matrix which are closely related to the mechanical properties. The binding strength of aggregates derived from the terminal blocks is remarkably improved, which dominantly impacts on the dynamic properties. The mobilities of longitudinal chains and oligopeptide complexes are gradually suppressed. The reinforced robustness of networks ultimately increases the mechanical properties of polyisoprene.

Supplementary Materials: The following are available online at https://www.mdpi.com/article/10.3390/polym13244408/s1, Experimental procedures, analytical data and additional spectra. Figure S1. ^{1}H NMR spectrum of B-DSC-PIP; Figure S2. ^{1}H NMR spectrum of B-2A-PIP; Figure S3. ^{1}H NMR spectrum of B-3A-PIP; Figure S4. ^{1}H NMR spectrum of R-DSC-PIP; Figure S5. ^{1}H NMR spectrum of R-2A-PIP; Figure S6. ^{1}H NMR spectrum of R-3A-PIP; Figure S7. Representative pictures of samples. All samples show the same appearance; Figure S8. Synthesis of B-2A-PIP, R-2A-PIP, B-3A-PIP and R-3A-PIP; Figure S9. Stress–strain curves as a function of strain of PIP, B-OH-PIP and R-OH-PIP; Figure S10. (a) Stress–strain curves as a function of strain; (b) FTIR spectra and (c) X-ray diffraction spectra of B-4A-PIP, R-4A-PIP; Figure S11. Dielectric loss ε'' as a function of the frequency at −60 °C to −30 °C for different copolymers: (a) B-2A-PIP, (b) B-3A-PIP, (c) B-4A-PIP, (d) R-2A-PIP, (e) R-3A-PIP, (f) R-4A-PIP; Figure S12. The temperature dependence of $\Delta\varepsilon$ for NM relaxation of B-2A-PIP (a), R-2A-PIP (b) and B-3A-PIP (c); Figure S13. Three different views of last frame structures of B-2A-PIP from top, side 1 and side 2 with axis denoted figures. A closed black dot indicates such axis is pointing into the paper while the open dot suggests the axis is pointing outwards. The block segment of 2A with 40 repeat units is colored in blue, while the random units are colored in green. PIP units are colored in light grey for clarity. All hydrogen atoms are hidden. We plotted structures for slices along the z-axis. Each slice has a thickness of 2 nm (with Figure S1f slightly greater than 2 nm); Figure S14. Three different views of last frame structures of B-3A-PIP from top, side 1 and side 2 with axis denoted figures. A closed black dot indicates such axis is pointing into the paper while the open dot suggests the axis is pointing outwards. The block segment of 3A with 40 repeat units is colored in blue, while the random units are colored in green. PIP units are colored in light grey for clarity. All hydrogen atoms are hidden. We also plotted structures for slices along the z-axis. Each slice has a thickness of 2 nm (with Figure S1f slightly greater than 2 nm); Figure S15. Three different views of last frame structures of B-4A-PIP from top, side 1 and side 2 with axis denoted figures. A closed black dot indicates such axis is pointing into the paper while the open dot suggests the axis is pointing outwards. The block segment of 4A with 40 repeat units is colored in blue, while the random units are colored in green. PIP units are colored in light grey for clarity. All hydrogen atoms are hidden. We also plotted structures for slices along the z-axis. Each slice has a thickness of 2 nm (with Figure S3f slightly greater than 2 nm).

Author Contributions: Conceptualization, M.T. and Y.-X.X.; data curation, C.-C.W., R.Z., S.L. and M.T.; investigation, C.-C.W.; methodology, C.-C.W., M.T. and Y.-X.X.; project consulting, G.H.; project administration, Y.-X.X.; writing—original draft, C.-C.W. and M.T.; writing—review and editing, Y.-X.X. All authors have read and agreed to the published version of the manuscript.

Funding: The authors gratefully acknowledge the National Natural Science Foundation of China (Grant No. 51973126 and 51333003) for financial support of this research.

Institutional Review Board Statement: Not applicable.

Informed Consent Statement: Not applicable.

Data Availability Statement: The data presented in this study are available on request from the corresponding author.

Conflicts of Interest: The authors declare no conflict of interest.

References

1. Deng, F.; Ito, M.; Noguchi, T.; Wang, L.; Ueki, H.; Niihara, K.; Kim, Y.A.; Endo, M.; Zheng, Q.S. Elucidation of the Reinforcing Mechanism in Carbon Nanotube/Rubber Nanocomposites. *ACS Nano* **2011**, *5*, 3858–3866. [CrossRef] [PubMed]
2. Amnuaypornsri, S.; Sakdapipanich, J.; Tanaka, Y. Green Strength of Natural Rubber: The Origin of the Stress-Strain Behavior of Natural Rubber. *J. Appl. Polym. Sci.* **2010**, *111*, 2127–2133. [CrossRef]
3. Zhu, L.; Tian, X.; Pan, Y.; Chang, T.; Wang, K.; Niu, G.; Zhang, L.; Wang, C.; Han, W. Optimization of Serial Modular Continuous Mixing Process Parameters for Natural Rubber Composites Reinforced by Silica/Carbon Black. *Polymers* **2020**, *12*, 416. [CrossRef] [PubMed]
4. Toki, S.; Che, J.; Rong, L.; Hsiao, B.S.; Amnuaypornsri, S.; Nimpaiboon, A.; Sakdapipanich, J. Entanglements and Networks to Strain-Induced Crystallization and Stress–Strain Relations in Natural Rubber and Synthetic Polyisoprene at Various Temperatures. *Macromolecules* **2013**, *46*, 5238–5248. [CrossRef]
5. Ehabe, E.; Bonfils, F.; Aymard, C.; Akinlabi, A.K.; Sainte, B.J. Modelling of Mooney Viscosity Relaxation in Natural Rubber. *Polym. Test.* **2005**, *24*, 620–627. [CrossRef]
6. Nimpaiboon, A.; Amnuaypornsri, S.; Sakdapipanich, J. Influence of Gel Content on the Physical Properties of Unfilled and Carbon Black Filled Natural Rubber Vulcanizates. *Polym. Test.* **2013**, *32*, 1135–1144. [CrossRef]
7. Toki, S.; Hsiao, B.S.; Amnuaypornsri, S.; Sakdapipanich, J. New Insights into the Relationship between Network Structure and Strain-Induced Crystallization in Un-Vulcanized and Vulcanized Natural Rubber by Synchrotron X-ray Diffraction. *Polymer* **2009**, *50*, 2142–2148. [CrossRef]
8. Tanaka, Y. Structural Characterization of Natural Polyisoprenes: Solve the Mystery of Natural Rubber Based on Structural Study. *Rubber Chem. Technol.* **2001**, *74*, 355–375. [CrossRef]
9. Karino, T.; Ikeda, Y.; Yasuda, Y.; Kohjiya, S.; Shibayama, M.J.B. Nonuniformity in Natural Rubber as Revealed by Small-Angle Neutron Scattering, Small-Angle X-ray Scattering, and Atomic Force Microscopy. *Biomacromolecules* **2007**, *8*, 693–699. [CrossRef]
10. Tanaka, Y.; Tarachiwin, L. Recent Advances in Structural Characterization of Natural Rubber. *Rubber Chem. Technol.* **2009**, *82*, 283–314. [CrossRef]
11. Qin, X.; Wang, J.; Zhang, Y.; Wang, Z.; Li, S.; Zhao, S.; Tan, T.; Liu, J.; Zhang, L. Self-Assembly Strategy for Double Network Elastomer Nanocomposites with Ultralow Energy Consumption and Ultrahigh Wear Resistance. *Adv. Funct. Mater.* **2020**, *30*, 2003429. [CrossRef]
12. Ilyas, M.; Haque, M.A.; Yue, Y.; Kurokawa, T.; Nakajima, T.; Nonoyama, T.; Gong, J.P. Water-Triggered Ductile–Brittle Transition of Anisotropic Lamellar Hydrogels and Effect of Confinement on Polymer Dynamics. *Macromolecules* **2017**, *50*, 8169–8177. [CrossRef]
13. Zhang, Q.; Niu, S.; Wang, L.; Lopez, J.; Chen, S.; Cai, Y.; Du, R.; Liu, Y.; Lai, J.-C.; Liu, L.; et al. An Elastic Autonomous Self-Healing Capacitive Sensor Based on a Dynamic Dual Crosslinked Chemical System. *Adv. Mater.* **2018**, *30*, 1801435. [CrossRef]
14. Lessard, J.J.; Scheutz, G.M.; Sung, S.H.; Lantz, K.A.; Epps, T.H., 3rd; Sumerlin, B.S. Block Copolymer Vitrimers. *J. Am. Chem. Soc.* **2020**, *142*, 283–289. [CrossRef]
15. Ji, F.; Li, J.; Weng, Y.; Ren, J. Synthesis of Pla-Based Thermoplastic Elastomer and Study on Preparation and Properties of Pla-Based Shape Memory Polymers. *Mater. Res. Express* **2019**, *7*, 015315. [CrossRef]
16. Moyassari, A.; Gkourmpis, T.; Hedenqvist, M.S.; Gedde, U.W. Molecular Dynamics Simulations of Short-Chain Branched Bimodal Polyethylene: Topological Characteristics and Mechanical Behavior. *Macromolecules* **2019**, *52*, 807–818. [CrossRef]
17. Scavuzzo, J.; Tomita, S.; Cheng, S.; Liu, H.; Gao, M.; Kennedy, J.P.; Sakurai, S.; Cheng, S.Z.D.; Jia, L. Supramolecular Elastomers: Self-Assembling Star-Blocks of Soft Polyisobutylene and Hard Oligo(B-Alanine) Segments. *Macromolecules* **2015**, *48*, 1077–1086. [CrossRef]
18. Müller, M.; Fischer, E.W.; Kremer, F.; Seidel, U.; Stadler, R. The Molecular Dynamics of Thermoreversible Networks as Studied by Broadband Dielectric Spectroscopy. *Colloid. Polym. Sci.* **1995**, *273*, 38–46. [CrossRef]
19. Nawamawat, K.; Sakdapipanich, J.T.; Ho, C.C. Effect of Deproteinized Methods on the Proteins and Properties of Natural Rubber Latex During Storage. *Macromol. Symp.* **2010**, *288*, 95–103. [CrossRef]
20. Cerveny, S.; Zinck, P.; Terrier, M.; Arrese-Igor, S.; Alegría, A.; Colmenero, J. Dynamics of Amorphous and Semicrystalline 1,4-Trans-Poly(Isoprene) by Dielectric Spectroscopy. *Macromolecules* **2008**, *41*, 8669–8676. [CrossRef]
21. Tang, M.; Zhang, R.; Li, S.; Zeng, J.; Luo, M.; Xu, Y.X.; Huang, G. Towards a Supertough Thermoplastic Polyisoprene Elastomer Based on a Biomimic Strategy. *Angew. Chem. Int. Ed.* **2018**, *57*, 15836–15840. [CrossRef]
22. Loiseau, N.; Gomis, J.-M.; Santolini, J.; Delaforge, M.; André, F. Predicting the Conformational States of Cyclic Tetrapeptides. *Biopolymers* **2003**, *69*, 363–385. [CrossRef]
23. Havriliak, S.; Negami, S. A Complex Plane Representation of Dielectric and Mechanical Relaxation Processes in Some Polymers. *Polymer* **1967**, *8*, 161–210. [CrossRef]
24. Díaz-Calleja, R. Comment on the Maximum in the Loss Permittivity for the Havriliak–Negami Equation. *Macromolecules* **2000**, *33*, 8924. [CrossRef]
25. Liu, J.; Wu, S.; Tang, Z.; Lin, T.; Guo, B.; Huang, G. New Evidence Disclosed for Networking in Natural Rubber by Dielectric Relaxation Spectroscopy. *Soft Matter* **2015**, *11*, 2290–2299. [CrossRef]

26. Ortizserna, P.; Díazcalleja, R.; Sanchis, M.J.; Floudas, G.; Nunes, R.C.; Martins, A.F.; Visconte, L.L. Dynamics of Natural Rubber as a Function of Frequency, Temperature, and Pressure. A Dielectric Spectroscopy Investigation. *Macromolecules* **2010**, *43*, 5094–5102. [CrossRef]
27. Tang, Z.; Zhang, L.; Feng, W.; Guo, B.; Liu, F.; Jia, D. Rational Design of Graphene Surface Chemistry for High-Performance Rubber/Graphene Composites. *Macromolecules* **2015**, *47*, 8663–8673. [CrossRef]
28. Van Der Spoel, D.; Lindahl, E.; Hess, B.; Groenhof, G.; Mark, A.E.; Berendsen, H.J.C. Gromacs: Fast, Flexible, and Free. *J. Comput. Chem.* **2005**, *26*, 1701–1718. [CrossRef]
29. Hess, B.; Kutzner, C.; van, d.S.D.; Lindahl, E.; Hess, B.; Kutzner, C.; van, d.S.D.; Lindahl, E.; Hess, B.; Kutzner, C.; et al. Gromacs 4: Algorithms for Highly Efficient, Load-Balanced, and Scalable Molecular Simulation. *J. Chem. Theory Comput.* **2008**, *4*, 435–447. [CrossRef]
30. Pronk, S.; Pall, S.; Schulz, R.; Larsson, P.; Bjelkmar, P.; Apostolov, R.; Shirts, M.R.; Smith, J.C.; Kasson, P.M.; van der Spoel, D.; et al. Gromacs 4.5: A High-Throughput and Highly Parallel Open Source Molecular Simulation Toolkit. *Bioinformatics* **2013**, *29*, 845–854. [CrossRef]
31. Berendsen, H.J.C.; van der Spoel, D.; van Drunen, R. Gromacs: A Message-Passing Parallel Molecular Dynamics Implementation. *Comput. Phys. Commun.* **1995**, *91*, 43–56. [CrossRef]
32. Lindahl, E.; Hess, B.; van der Spoel, D. Gromacs 3.0: A Package for Molecular Simulation and Trajectory Analysis. *J. Mol. Model.* **2001**, *7*, 306–317. [CrossRef]
33. Jorgensen, W.L.; Maxwell, D.S.; Tirado-Rives, J. Development and Testing of the Opls All-Atom Force Field on Conformational Energetics and Properties of Organic Liquids. *J. Am. Chem. Soc.* **1996**, *118*, 11225–11236. [CrossRef]
34. Rizzo, R.C.; Jorgensen, W.L. Opls All-Atom Model for Amines: Resolution of the Amine Hydration Problem. *J. Am. Chem. Soc.* **1999**, *121*, 4827–4836. [CrossRef]
35. Kaminski, G.A.; Friesner, R.A.; Tirado-Rives, J.; Jorgensen, W.L. Evaluation and Reparametrization of the Opls-Aa Force Field for Proteins Via Comparison with Accurate Quantum Chemical Calculations on Peptides. *J. Phys. Chem. B.* **2001**, *105*, 6474–6487. [CrossRef]
36. Price, M.L.P.; Ostrovsky, D.; Jorgensen, W.L. Gas-Phase and Liquid-State Properties of Esters, Nitriles, and Nitro Compounds with the Opls-Aa Force Field. *J. Comput. Chem.* **2001**, *22*, 1340–1352. [CrossRef]
37. Humphrey, W.; Dalke, A.; Schulten, K. Vmd: Visual Molecular Dynamics. *J. Mol. Graph.* **1996**, *14*, 33–38. [CrossRef]
38. Chen, D.; Shao, H.; Yao, W.; Huang, B. Fourier Transform Infrared Spectral Analysis of Polyisoprene of a Different Microstructure. *Int. J. Polymer. Sci.* **2013**, *2013*, 7866–7873. [CrossRef]
39. Huang, W.; Krishnaji, S.; Tokareva, O.R.; Kaplan, D.; Cebe, P. Influence of Water on Protein Transitions: Morphology and Secondary Structure. *Macromolecules* **2014**, *47*, 8107–8114. [CrossRef]
40. Fawcett, J.K.; Camerman, N.; Camerman, A. The Structure of the Tripeptide L -Alanyl- L -Alanyl- L -Alanine. *Acta Crystallogr. Sect. B: Struct. Sci.* **2010**, *31*, 658–665. [CrossRef]
41. Asakura, T.; Yazawa, K.; Horiguchi, K.; Suzuki, F.; Nishiyama, Y.; Nishimura, K.; Kaji, H. Difference in the Structures of Alanine Tri- and Tetra-Peptides with Antiparallel B-Sheet Assessed by X-ray Diffraction, Solid-State Nmr and Chemical Shift Calculations by Gipaw. *Biopolymers* **2014**, *101*, 13–20. [CrossRef]
42. Takusagawa, F.; Fumagalli, A.; Koetzle, T.F.; Shore, S.G.; Schmitkons, T.; Fratini, A.V.; Morse, K.W.; Wei, C.-Y.; Bau, R. Neutron and X-ray Diffraction Studies of Tris(Methyldiphenylphosphine)[Tetrahydroborato(1-)]Copper, $Cu[P(C_6h_5)_2ch_3]_3(Bh_4)$. The First Accurate Characterization of an Unsupported Metal-Hydrogen-Boron Bridge Bond. *J. Am. Chem. Soc.* **1981**, *103*, 5165–5171. [CrossRef]
43. Ratajczak-Sitarz, M.; Katrusiak, A. Coupling of Molecular Orientation with the Hydrogen-Bond Dimensions and H-Sites in Carboxylic Acids. *J. Mol. Struct.* **2011**, *995*, 29–34. [CrossRef]
44. Kona, F.R.; Buac, D.; A, M.B. Molecular Dynamics in Linear and Multi-Armed Star Polymers of Cis-Polyisoprene as Studied by Broadband Dielectric Spectroscopy. *Macromolecules* **1991**, *23*, 1826–1830.
45. Boese, D.; Kremer, F.; Fetters, L.J. Further Investigation on the Molecular Dynamics in Linear and Multiarmed Star Polymers of Cis -Polyisoprene Studied by Dielectric Spectroscopy. *Polymer* **1990**, *31*, 1831–1837. [CrossRef]
46. Adachi, K.; Kotaka, T. Influence of Entanglement on the Dielectric Normal Mode Process of Cis-Polyisoprene. *Macromolecules* **1984**, *17*, 120–122. [CrossRef]
47. Adachi, K.; Kotaka, T. Dielectric Normal Mode Process in Undiluted Cis-Polyisoprene. *Macromolecules* **1985**, *18*, 466–472. [CrossRef]
48. Lewis, C.L.; Stewart, K.; Anthamatten, M. The Influence of Hydrogen Bonding Side-Groups on Viscoelastic Behavior of Linear and Network Polymers. *Macromolecules* **2014**, *47*, 729–740. [CrossRef]
49. Mijović, J.; Ristić, S.; Kenny, J. Dynamics of Six Generations of Pamam Dendrimers as Studied by Dielectric Relaxation Spectroscopy. *Macromolecules* **2007**, *40*, 5212–5221. [CrossRef]
50. Boese, D.; Kremer, F. Molecular Dynamics in Bulk Cis-Polyisoprene as Studied by Dielectric Spectroscopy. *Macromolecules* **1990**, *23*, 829–835. [CrossRef]
51. Schönhals, A.; Kremer, F.; Schlosser, E. Scaling of the A Relaxation in Low-Molecular Weight Glass-Forming Liquids and Polymers. *Phys. Rev. Lett.* **1991**, *67*, 999–1002. [CrossRef]

52. Asakura, T.; Nishimura, A.; Aoki, A.; Naito, A. Packing Structure of Antiparallel Β-Sheet Polyalanine Region in a Sequential Model Peptide of Nephila Clavipes Dragline Silk Studied Using 13c Solid-State Nmr and Md Simulation. *Biomacromolecules* **2019**, *20*, 3884–3894. [CrossRef]
53. Fawcett, J.K.; Camerman, N.; Camerman, A. The Structure of the Tripeptide L-Alanyl-L-Alanyl-L-Alanine. *Acta Cryst. B* **1975**, *31*, 658–665. [CrossRef]

polymers

MDPI

Article

A Novel Approach of Bioesters Synthesis through Different Technologies by Highlighting the Lowest Energetic Consumption One

Simona Popa [1], Andra Tamas [1], Vasile Simulescu [2], Dorin Jurcau [3], Sorina Boran [1,*] and Giannin Mosoarca [1,*]

[1] Faculty of Industrial Chemistry and Environmental Engineering, Politehnica University Timisoara, Bd. V. Parvan, No. 6, 300223 Timisoara, Romania; simona.popa@upt.ro (S.P.); andra.tamas@upt.ro (A.T.)
[2] Faculty of Chemistry-Biology-Geography, West University of Timisoara, Bv. Vasile Parvan, No. 4, 300223 Timisoara, Romania; vasile.simulescu@e-uvt.ro
[3] Elkim Special SRL, Str. Constantin cel Mare, No. 21, 300265 Timisoara, Romania; lk-eldus@rdstm.ro
* Correspondence: sorina.boran@upt.ro (S.B.); giannin.mosoarca@upt.ro (G.M.); Tel.: +40-256-404228 (S.B.); +40-256-404185 (G.M.)

Abstract: Fatty acids esters have a wide application as bioplasticizers and biolubricants in different industries, obtained mainly in classic batch reactors, through an equilibrium complex reaction, that involves high temperatures, long reaction times, vigorously stirring, and much energy consumption. To overcome these shortcomings, we synthesized a series of fatty acid esters (soybean oil fatty acids being the acid components with various hydroxyl compounds) through novel low energy consumption technologies using a bubble column reactor, a microwave field reactor and for comparison meaning, a classic batch reactor. The obtained bioesters physicochemical properties were similar to one another, a good concordance among their rheological properties was obtained, but the energetic consumption is lower when using the bubble column or the microwave reactors instead of the classical batch reactor.

Keywords: soybean oil fatty acids; bubble column reactor; microwave reactor; lowest energetic consumption; color; rheology

Citation: Popa, S.; Tamas, A.; Simulescu, V.; Jurcau, D.; Boran, S.; Mosoarca, G. A Novel Approach of Bioesters Synthesis through Different Technologies by Highlighting the Lowest Energetic Consumption One. *Polymers* **2021**, *13*, 4190. https://doi.org/10.3390/polym13234190

Academic Editor: Edina Rusen

Received: 9 November 2021
Accepted: 28 November 2021
Published: 30 November 2021

Publisher's Note: MDPI stays neutral with regard to jurisdictional claims in published maps and institutional affiliations.

Copyright: © 2021 by the authors. Licensee MDPI, Basel, Switzerland. This article is an open access article distributed under the terms and conditions of the Creative Commons Attribution (CC BY) license (https://creativecommons.org/licenses/by/4.0/).

1. Introduction

Environmental protection efforts in the industry is focused on reducing wastes by recycling some of these materials [1] and replacing the raw materials of petroleum-based products used in different industries with organic fluids [2,3]. At the same time, modern technologies should be friendly with the environment [4–6], obviously in terms of energy and economic efficiency. Industrial soybean oil is often used, by further processing, as an ingredient for paints, plastics, fibers, detergents, cosmetics and lubricants that show similar viscosity to commercial lubricants [7]. Natural solvent and for example, clear liquid soy derived from soybean oil as methyl esters could serve as a green alternative to synthetic solvents. In addition, due to their low evaporation rate and longer contact time, maintained with a target material, they are recommended as natural adjuvant and surfactant in order to increase crop yields while lowering input costs by improving contact of spray droplets on plant surfaces and more effectively penetrating waxy surfaces. Soy esters present some benefits as safe to handle and store, low toxicity compared to other common substances. Vegetable oils, containing non-toxic and ecofriendly fatty acids, are successfully used by esterification and transesterification syntheses in biodiesel production [8–10], which is one of the main important biodegradable products [2,11–15]. Color properties of certain substances is often used in describing their properties, either organic or inorganic ones, as well as in food field [16–18]. As known, rheological behavior is important because it provides information regarding flow and storage of relevant materials under operation conditions [19] especially when the products are used in paint industry. Rheology is used in many researches, as for the analysis of engine oil lubricants [16–24], hydrogels [25,26], different polymer—plasticizer systems [27,28], heterocycles [29–31], nanofluids [32,33] collagen solutions [34], cyclodextrin nanospongies [35] and other. Energy efficiency represents another important issue in today's technology, in order to find and implement low cost processes [36,37]. At present, the esterification processes in classical batch technology require much energy consumption, because of the necessity of high temperatures (up to 220–250 °C), long reaction time, and vigorously stirring. That is why researches are made to find new energy-saving technological methods [12,13,37].

At present, the microwave heating is gaining more and more influence in technological processes due to its energy economy and environmental advantages, being used in different fields [38–44].

To overcome these shortcomings, the present paper refers to the synthesis of a series of fatty acid esters via two modern technologies that do not deal with solvent extraction of the azeotrope namely a bubble column reactor [11,45] and a microwave field reactor. In this reactor the process time is short, microwave heating being a widely accepted tool for synthetic chemists [26,46]. For the sake of comparison, the synthesis was performed in a classic batch reactor as well. To the best of our knowledge, there is no reported work on a comparison between energy consumption in the various esterification reactors, or a comparative rheological study of the obtained products. Therefore, this paper investigated these aspects, revealing that the bubble column reactor and the microwave field reactor are energy-saving technological methods for the bioester synthesis. All products properties are similar, regardless the synthesis method.

2. Materials and Methods

The soybean oil fatty acids were received from Baichim SRL Bucharest. The organic alcohols n-propanol, n-butanol and n-pentanol, and the catalyst p-toluene-sulfonic acid were purchased from Fluka Honeywell (Charlotte, NC, USA). The fatty acids (R-COOH) from hydrolyzed soybean oil have a typical composition containing 11% palmitic acid, 4% stearic acid, 25% oleic acid, 50% linoleic acid and 9% linolenic acid. The physicochemical properties of fatty acids from used soybean oil are: viscous liquid without mechanical impurities; yellow color; 0.89 g/cm^3 density at 20 °C; 14–16 °C melting point; 193.4 mg KOH/g acid number; −1.458 refraction index at 20 °C.

The main esterification reactions were performed in a bubble column reactor (B), in a microwave reactor (M) and in a classic batch reactor (C), using soybean oil fatty acids as the acid component and three hydroxyl-compounds: n-propanol (1), n-butanol (2), and n-pentanol (3), with the 1:2 mole fraction between the fatty acids from soybean oil and the organic alcohols respectively. The catalyst p-toluene-sulfonic acid was used in proportion of 0.4%. The obtained bioesters with n-propanol are B1, M1 and C1, with n-butanol are B2, M2, C2, and n-pentanol are B3, M3, C3 respectively.

A cylinder glass column with an internal diameter of 0.03 m and a height of 0.3 m provided with a heating mantle was used as bubble column reactor. Agitation was achieved by bubbling nitrogen through a nozzle at the base of the column. The energetic efficiency calculation of the bubble column reactor was developed upon esterification reaction among benzoic acid and propylene glycol, using different reaction conditions [47]. Considering the best conditions achieved in terms of energy consumption (the nozzle of 0.6 mm, argon pressure of 123.6 Pa and with no filling material) esterifications in the bubble column reactor were carried out. Esters of soybean oil fatty acids with the organic alcohols were synthesized in two steps. The first step took place in a flask fitted with a thermometer and a water reflux cooler, where the preheating of the reaction mixture was carried out for one hour under continuous stirring, at 60 °C on an electric stove of 5 kW. Then the product was transferred directly into the column reactor, where the synthesis was carried out at the reflux temperature, to give the products B1, B2 and B3. The reaction conditions and times are presented in Table 1.

Table 1. The reaction conditions for soybean esters production in different technologies.

	Bubble Column			Microwave Field			Classic Reactor		
Bioester	B1	B2	B3	M1	M2	M3	C1	C2	C3
Reaction time, h	3	4	4	2	3	3	9	9	10
Temperature, °C	100	105	110	110	115	120	120	125	130

Fatty acid esters, namely M1, M2 and M3 were obtained in a chemical reactor with a microwave heating oven (model DB-001, China Doble Best, China) provided with a reflux cooler, in a single step in the presence of the catalyst—p-toluene sulfonic acid at the reflux temperature (Table 1). The characteristics of the chemical reactor with microwave (M) heating are: microwave power 0~800 W; microwave frequency 50 MHz 2450+; Shaking magnetic stirrer.

In the classic technology, the reactor (Model ELN9.1, Carl Roth GmbH + Co. KG, Karlsruhe, Germany) was heated with an electric stove of 5 kW. The synthesis was performed in a solution esterification process, using p-toluen sulfonic acid as catalyst. The water azeotrope was extracted with toluene. The esterification was carried out in a single step according to the reaction conditions presented in Table 1, with the formation of products C1, C2 and C3.

In all cases, the esterification was monitored by periodic determination of the acid number over the whole synthesis, and the process was considered to be completed when acid index (IA) was below 1 mg KOH/g. The synthesized esters were purified by neutralization with 10% aqueous sodium carbonate solutions, washed with demineralized water to neutral pH, then vacuum distillation and decolourisation with activated charcoal and filtration. The purified compounds were then subjected to specific analysis.

The physicochemical properties of the bioesters were determined by using standardized techniques: density—SR EN ISO 12185-03, refractive index—SR 7573-95, acidity index—SR EN ISO 660:2009, iodine value—SR EN ISO 3961:2013, color—visual, rheological study was performed using a Brookfield CAP2000+L viscometer (AMETEK GmbH/B.U. Brookfield, Dresden, Germany), according to ASTM D445, temperature range 5–70 °C.

The calculated data are the mean of three independent replicates. Before running the one-way ANOVA analysis, Equal Variances tests (Multiple comparisons and Levene's methods) were performed to verify that the samples have equal variances. The 95% confidence level was adopted and the Tukey pairwise comparisons was applied to establish the significant differences between samples. Minitab 19 software (Minitab, LLC, USA) was utilized to perform the required calculations.

Thermo-gravimetric (TG/DTG) analyses were performed with NETZSCH STA 449F1 STA449F1A-0220-M (NETZSCH-Gerätebau GmbH, Germany)—approximately 3–7 mg of sample was heated in an Al_2O_3 crucible, with 5 °C/min, in nitrogen atmosphere, within the range 20–600 °C. An ion trap mass spectrometer ITQ 1100 coupled with Gas Chromatograph Trace 1310 (Thermo Fisher Scientific, Waltham, MA, USA) was used for qualitative analysis of soybean fatty acids bioesters. MS parameters were set as following: transfer line temperature at 310 °C, source temperature at 170 °C and scan range between 30 and 350 amu. The reaction product structure was established based on the m/z ratio. Fourier Transform Infrared (FT-IR) spectra of the samples were obtained in attenuated total reflectance (ATR) mode on a Bruker Vertex 70 (Bruker Daltonik GmbH, Bremen, Germany) spectrometer equipped with a Platinum ATR, Bruker Diamond Type A225/Q. Spectra were collected in the range 4000–400 cm^{-1} with a resolution of 4 cm^{-1} and with 40 scans/min. A MINOLTA CM 3220d spectrophotometer (Konica Minolta Sensing Europe B.V., Nieuwegein, The Netherlands) was used for the colorimetric analysis of the applied pigment in films in the following conditions: the CIE D65 illuminant (natural day light) and the standard 10° observer function.

3. Results and Discussion

In the present study, three esterification methods were used for obtaining of bioesters from soybean fatty acids with different alcohols. Raw materials are well mixed and esterification reactions can be completed within 3 to 10 h depending on esterification method. The aim was to determine which synthesis method is more energy efficient. As it was expected, microwave and bubble column methods are promising routes with lower energetic consumption.

3.1. Energetic Efficiency Comparison

Taking into account the properties of the heating devices used in the three technologies of this research, the calculated energy consumption of the process developed in the bubble column reactor was approximately 18,000 kJ for one synthesis of 300 g bioester. In the reactor with a microwave field the energy consumption for the same esterification process was approximately 8700 kJ, and in the classical reactor 144,000 kJ. As can be seen, for the same bioester production, the use of the classical reactor requires a much more energy consumption than in the case of the other two technologies, which are also environmental protective ones.

3.2. Bioesters Analysis

In order to verify the properties of the esters of soybean oil fatty acids with various alcohols synthesized via the technologies mentioned above, the purified products were characterized by physicochemical analyzes specific to this class of substances (Table 2 and Figure 1).

Table 2. The properties of esters via the technologies: bubble-column reactor (B1–B3), microwave heated reactor (M1–M3) and classical reactor (C1–C3).

Sample	Aspect	Color	Acid Index (mg KOH/g)	Refractive Index at 20 °C	Density at 25 °C (g/cm^3)	Iodine Value (g I_2/100 g)
B1	viscous, opalescent	orange	<1	1.4562 ± 0.0578	0.8981 ± 0.0271	120 ± 3
B2	gelatinous, opalescent	orange	<1	1.4577 ± 0.0581	0.8926 ± 0.0268	121 ± 4
B3	gelatinous	red-brown	<1	1.4591 ± 0.0584	0.9122 ± 0.0272	121 ± 4
M1	viscous	orange	<1	1.4499 ± 0.0575	0.8673 ± 0.0261	121 ± 3
M2	viscous	orange	<1	1.4598 ± 0.0587	0.8835 ± 0.0264	121 ± 5
M3	viscous	orange	<1	1.4611 ± 0.0584	0.9254 ± 0.0278	121 ± 4
C1	viscous, opalescent	orange	<1	1.4590 ± 0.0581	0.8810 ± 0.0263	120 ± 3
C2	viscous, opalescent	orange	<1	1.4593 ± 0.0579	0.8820 ± 0.0265	121 ± 5
C3	viscous, opalescent	orange	<1	1.4597 ± 0.0584	0.8815 ± 0.0264	121 ± 3

All products showed viscous, opalescent aspect, with a light brown-orange color. The unsaturated degree of oils being appreciated by the iodine index, the iodine values of the synthesized bioesters are in the range of 120–140 g I_2/100 g, according to the one of the soybean oil. As known, acidity is an indication of the presence of free fatty acids that are to be limited due to the formation of soaps, which may lead to the emulsions formation. The values of the acid index indicate a low content of free fatty acids. The values of the refractive index and density at 20 °C, presents a slightly variation. The dynamic viscosity increases with the number of carbon atoms brought by the alcoholic rest. Similar results were reported in the literature [8].

The results from Figure 1 show that there was no statistically significant difference of this parameter between esters, obtained with n-propanol (B1, M1, C1), synthetized in bubble column reactor, microwave reactor and classic batch reactor. The same conclusion can be drawn by observing the results concerning the esters obtained with n-butanol (B2, M2, C2) and n-pentanol (B3, M3, C3) respectively.

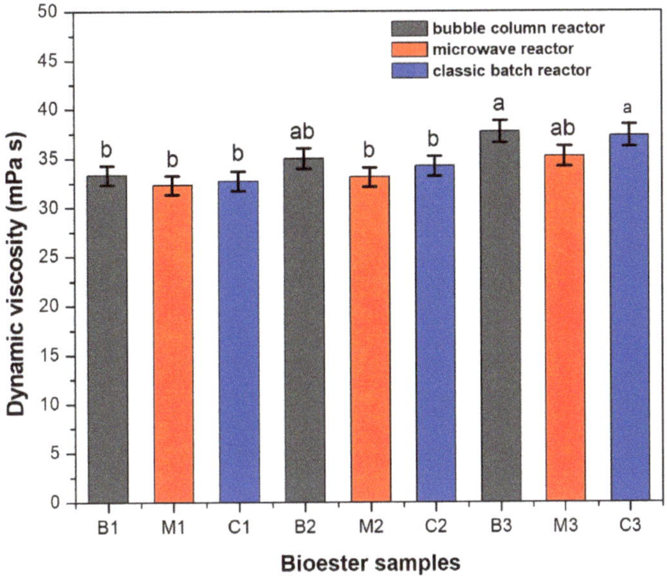

Figure 1. Dynamic viscosity at 25 °C, for the synthetized bioesters. Values are expressed as means of three independent replicates and error bars represent the standard deviation. Columns denoted by different letters indicated significant ($p < 0.05$) differences among different synthesis conditions.

All that suggested the formation of similar esters despite the different reactors used for their synthesis.

3.3. TG/DTG Analysis

Thermo-gravimetric analysis provides essential information on the relative thermal stability of the analyzed compounds, the content of water or other volatile ingredients of synthetic or natural materials [48]. From thermo-analytical curves, recorded for the synthesized esters in the range of 25 and 600 °C in nitrogen atmosphere, the inflection points and the total weight loss at 600 °C (Table 3) indicate that the method used does not affect the physicochemical properties of the obtained esters.

Table 3. The inflection points in the TG analysis.

Sample	T_1 the Inflection Points (°C)	T_2 the Inflection Points (°C)	Total Weight Loss at 200 °C (%)	Total Weight Loss at 600 °C (%)
B1	267	319	3.38	97.55
B2	274	330	3.21	97.61
B3	285	335	3.26	98.66
M1	263	315	2.96	98.77
M2	275	329	3.43	98.91
M3	282	333	3.13	97.64
C1	269	321	3.36	99.11
C2	277	330	2.98	98.98
C3	286	342	3.02	97.93

All of the obtained bioesters present low weight loss until 200 °C (lower than 3.5%). Therefore, the products exhibited good thermal stability below 200 °C so they may be used in technologies where such property is required, for example as natural adjuvant and surfactant in increasing crop yields with lowering costs. Above 300 °C, the weight loss is more important as the temperature increases.

For all studied bioesters the total weight loss appears around 600 °C, and the percentage of the residual mass remaining is insignificant. Three decomposition steps were observed. The first step corresponds to the water loss, are max. 3.5% (Table 3).

The observed mass loss values are similar and are not affected by the nature of the synthesis method. The second decomposition step (between 210 °C and 350 °C) and the third step (between 350 °C and 550 °C) were associated with the destruction of esteric group, the C-C and the C-H bonds. The decomposition process presents two inflection points at T_1 and T_2, associated with the highest decomposition rates.

Nitrogen atmosphere thermal stability analysis of methyl and ethyl esters of soybean oil [49] showed that they have a lower thermal stability compared to the samples synthesized by us by the three obtaining methods.

Comparable results were also obtained for erythritol tetra myristate and erythritol tetra laurate esters [50] or lubricants of the type gallate ester oils, respectively [51].

In contrast the polyester amides series [52] and the esters obtained by transesterification of palm oil-based methyl ester to trimethylolpropane esters [53] show better thermal stability compared to our samples.

Figure 2 presents an example of the degradation process of the B3 product, in nitrogen atmosphere.

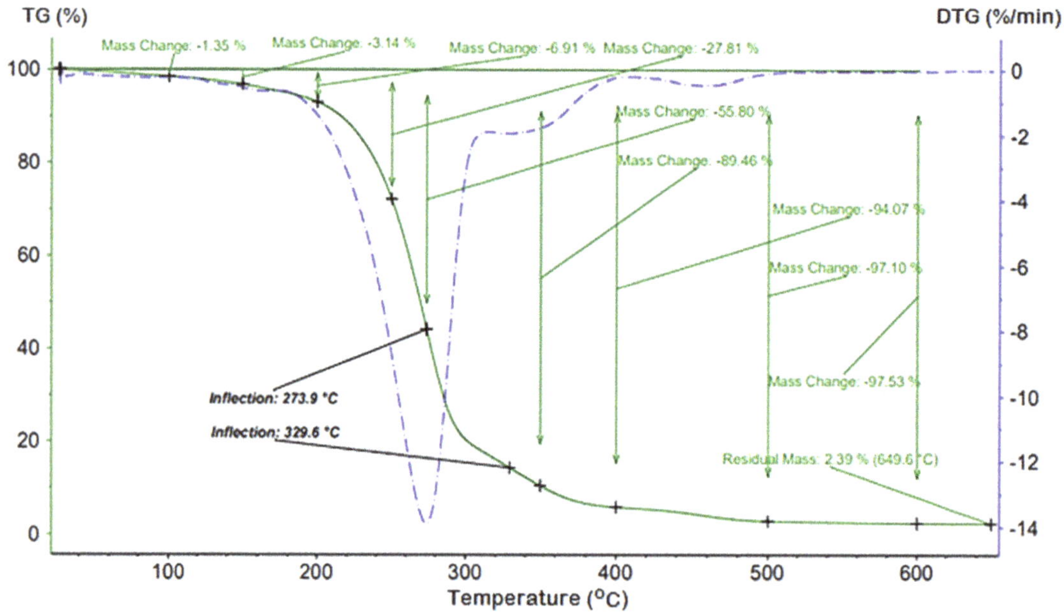

Figure 2. TG/DTG curves of bioester B3, in nitrogen atmosphere.

3.4. GS-MS Analysis

GS-MS spectra (Figures 3–5) are also presented for the B3 bioester. The chromatogram indicates the presence of two main reaction products. The major compound was separated in the analysis conditions at the retention time of 19.15 min. Its peak had the highest intensity from the whole GC chromatogram. The second reaction product, pentyl palmitate, was eluted in the GC chromatogram at the retention time of 17.70 min (Figure 3).

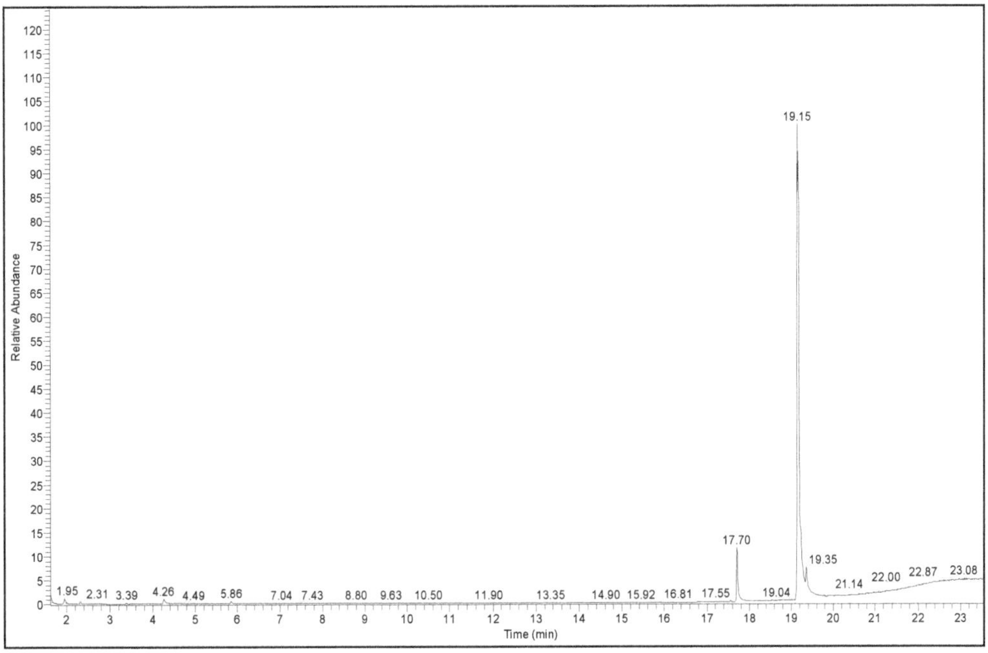

Figure 3. The chromatogram of the reaction mixture resulted from the esterification reaction in the case of B3 bioester.

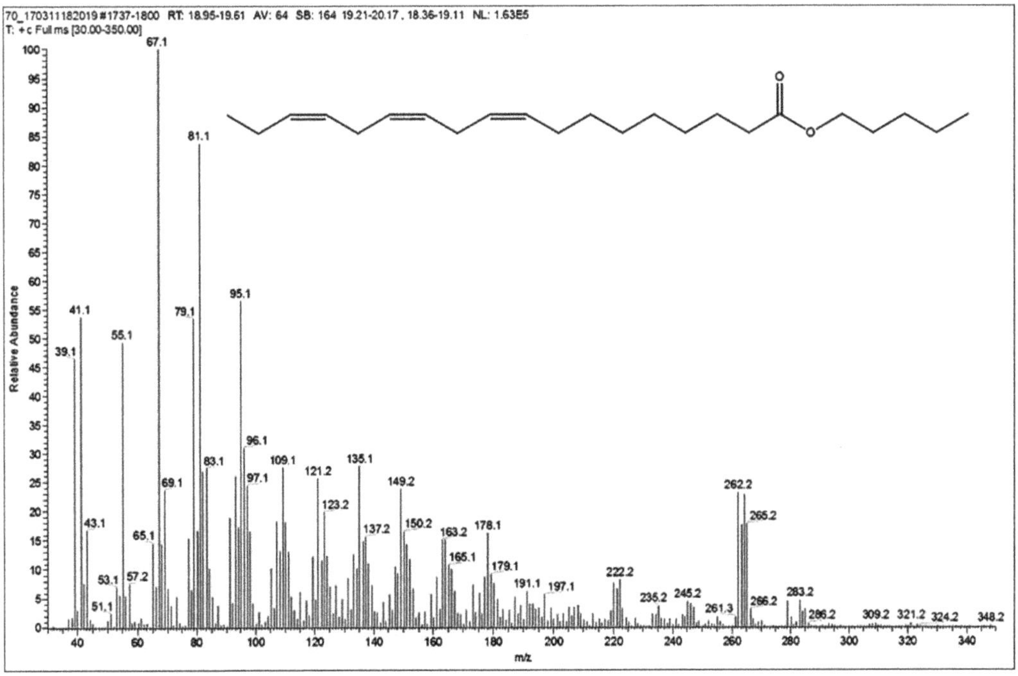

Figure 4. The mass spectrum of the ester, corresponding to the peak separated on the GC column at 19.15 min retention time (M = 348 g/mol), in the case of B3 bioester.

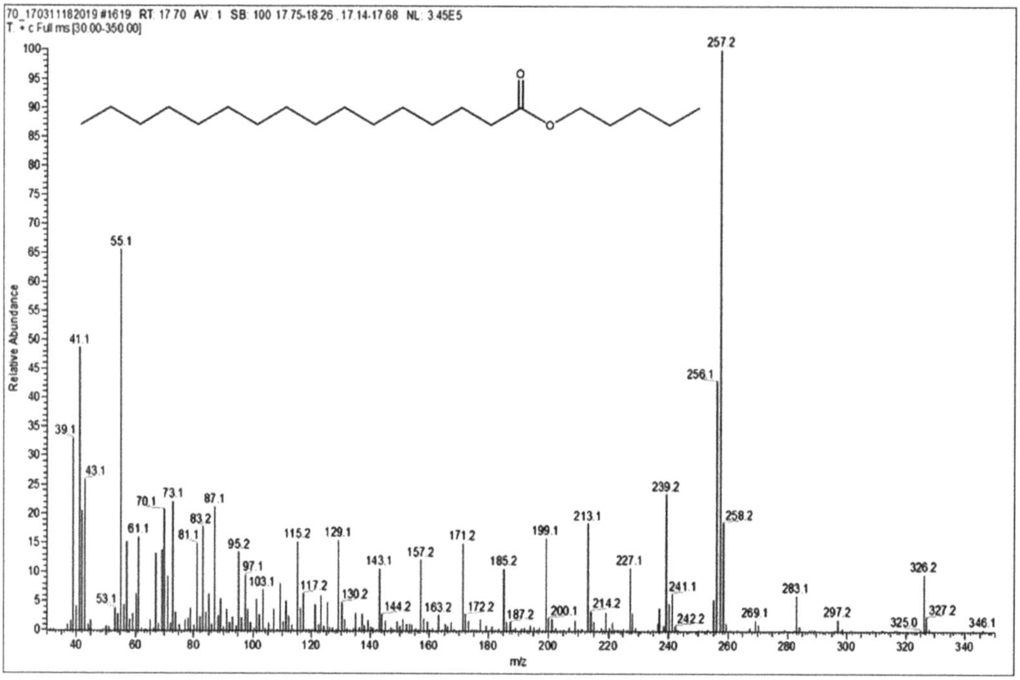

Figure 5. The mass spectrum of the ester, corresponding to the peak separated on the GC column at 17.70 min retention time (M = 326 g/mol), in the case of B3 bioester.

In the Figure 4, the molecular peak at m/z 348, as expected for the main product, was observed with very low intensity, mainly due to its fragmentation. From the MS spectrum it is obvious that the structure of the analyzed compound that the structure is most probably (9Z,12Z,15Z)-octadeca-9,12,15-trienoate. This is proved because the first fragmentation of this compound is expected to occur at carboxyl group, resulting in the signal observed at 262.2 m/z. Therefore, this signal belongs to the fragment remained after losing thepenthoxy group.

Furthermore, on the region of lower mass values, several differences of around 14–15 m/z units were observed. This proved that CH_3 and/or CH_2 fragments are present, either from penthyl or from octadeca-9,12,15-trienoate or even from both.

The MS spectrum of the compound which eluted first (see Figure 3) at 17.7 min, but with a lower intensity, is presented in Figure 5. From this MS spectrum (Figure 5) the molecular peak at 326 m/z can be easily observed, with a better intensity. Moreover, in the same mass spectrum presented in Figure 5, the signals observed at 257.2 and 239.2 m/z indicated the presence of palmitate fragments. Nevertheless, the peak observed at 257.2 m/z showed also the highest intensity from the whole MS spectrum of pentyl-palmitate.

This proved that palmitate fragment had a higher ionization in the used ion source, in comparison with the other fragments formed during MS analysis. The signal obtained at 87.1 m/z proved the presence of penthoxy fragment, as for the previous analyzed ester. Moreover, in this region of the mass spectrum, were also observed more signals, at 71 m/z, 70 m/z and 69 m/z respectively. All of those peaks belong to pentyl fragments.

3.5. FTIR Analysis

The FTIR is often used in polymer analysis [54,55]. The FTIR spectrum of the bioester B2 (Figure 6) reveal peaks at 2926 cm^{-1} and 2855 cm^{-1}. Those peaks can be attributed to the stretching of C-H bonds. At 1736 cm^{-1} the peaks are attributed to the stretching of C=O, typical of esters spectra [56]. The =C–H and C=C bands appear at 3012 cm^{-1} and 1657 cm^{-1}. The peak at 1465 cm^{-1} correspond to the asymmetric stretching of -CH_3 present in the biodiester. This peak is absent in soy oil spectrum. The peak at 1358 cm^{-1} was attributed to the O-CH_2 group and the peak at 1173 cm^{-1} is corresponding to the stretching of O-CH_3. Nevertheless, the peak at 1049 cm^{-1} was to attributed in plane deformation vibration of =C-H bond and the peaks at 926 cm^{-1} and at 725 cm^{-1} are attributed to the C–H wagging bond vibration. All of these proved also the absence of alcohol impurities.

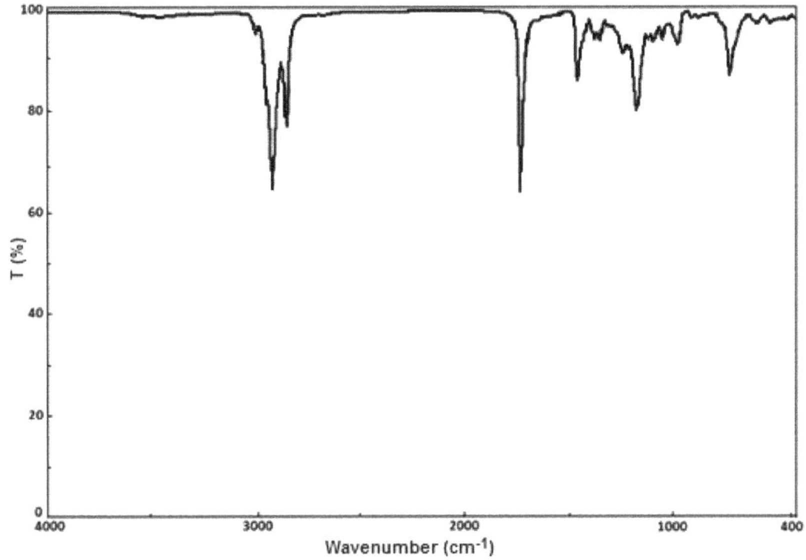

Figure 6. FT-IR analysis of the bioester B3.

3.6. Color Study

For the bioester B3 color study was also performed. The ester was introduced in different concentrations (0.1–2%) in an acrylic resin—that is often used in film industry [57,58]. The film was then deposited on a cellulosic support (wood). Color properties reveal that the reflectance increased with the ester concentration (Figure 7) according to other studies [59,60].

For all dried films, color *CIEL*a*b** parameters were determined: lightness *(L*)*, redness *(a*)*, yellowness *(b*)*, chroma or saturation *(C*)* and hue angle *(h*)* respectively.

The *CIEL*a*b** color properties (Figure 8) reveal that the ester acrylic composition is in the light yellow-olive domain, i.e., the a^* parameter is in the light green domain and the b^* parameter is in the yellow domain. As expected, the film darkness increasing with the ester concentration. The same result were presented in scientific literature [18,60,61].

The total color difference ΔE^*_{ab} may be calculated with the Equation (1) [17,62,63], results being presented in Table 4.

$$\Delta E^*_{ab} = \sqrt{(\Delta L^*)^2 + (\Delta a^*)^2 + (\Delta b^*)^2}, \quad (1)$$

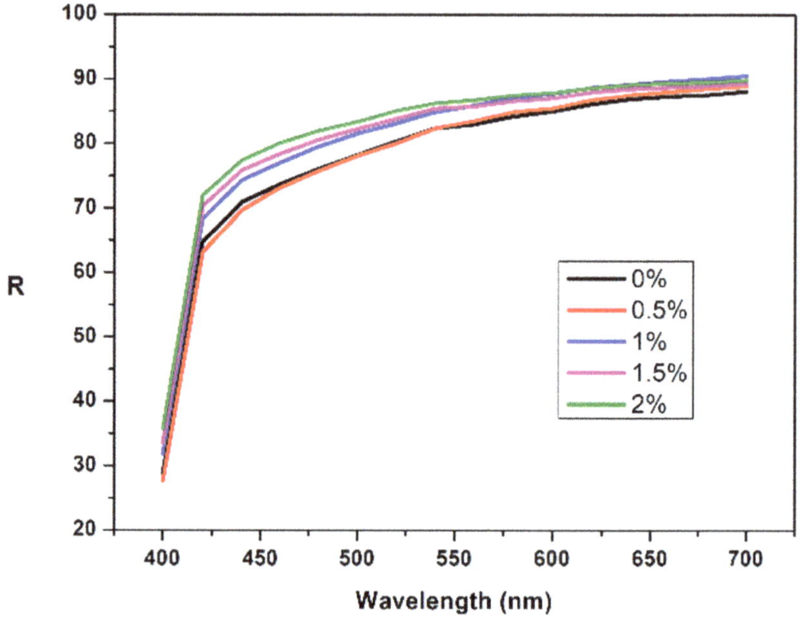

Figure 7. Reflectance spectra of the bioester B3 of the ester acrylic film at different ester concentration.

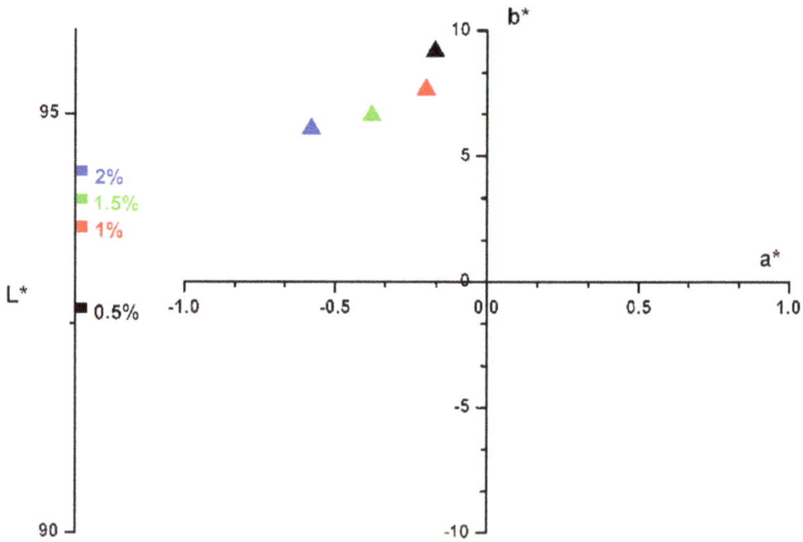

Figure 8. *CIEL*a*b** properties of the bioester B3 of the ester acrylic film at different ester concentration.

Table 4. Color differences of the ester acrylic film at different ester concentration.

Concentration, %	0.5	1.0	1.5	2.0
ΔL^*	0.09	1.06	1.39	1.73
Δa^*	0.06	0.03	−0.15	−0.35
Δb^*	0.86	−0.65	−1.69	−2.22
ΔE^*_{ab}	0.86	1.24	2.19	2.83

3.7. Rheology Comparative Study of the Bioesters

An important parameter that can remarkably influence the rheological properties of all fluids is temperature. Rheology measurements for all esters obtained in this research reveal a non-Newtonian behavior at the temperatures where their viscosity was measured (25–70 °C) and at different share rates (1333–13,333 s^{-1}) [64]. The characteristic equation of Ostwald de Waele model (Equation (2) may be used for interpretation. All bioesters submitted to the rheology tests present a decrease of viscosity when increasing temperature. The same behavior was reported in other papers [18,65,66]. Although the allure of the curves is similar, the viscosity is higher as the number of the carbon atoms increases, exemplified for esters B1–B3 in Figure 9.

$$\eta_a = K\gamma^{n-1}, \quad (2)$$

where: *K*—the index of consistency, Pa·sn; *n*—the flow behavior index.

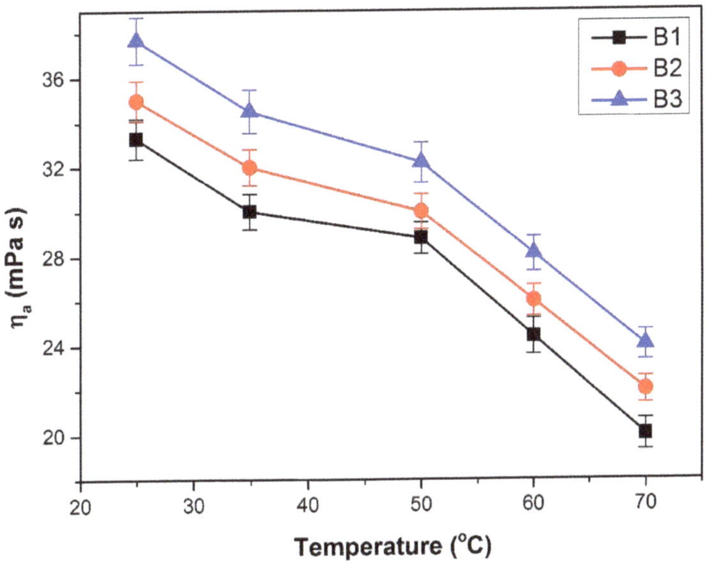

Figure 9. The dependence of the apparent viscosity on temperature for bioesters (B1–B3).

The pseudoplastic flow behavior is proved both by the decrease of the apparent viscosity with the increase of shear rate (shear-thinning behavior) [67], as well as by the sub-unit values of the flow index *n* (Table 5). The decreasing of the apparent viscosity with increasing temperature is also evidenced by the variation of the consistency index *k* correlated with the value of the flow index *n*.

Power law model R^2 values indicate that *k* and *n* values are a good fit. It was observed an increase of consistency coefficient (*k*) with the increase of carbon atoms number in alkyl group from alcohol. As expected, the increase in temperature reduces the consistency coefficient (*k*) and increase the flow behavior index (*n*) and at the same temperature, the decrease of consistency coefficient (*k*) is more evident for ester with long alkyl chain in alchool (B3). The modification of k with temperature revel some influence of Brownian movement. Regardless of the temperature and of the obtaining method, the apparent viscosities of the prepared with n-propyl alcohol (B1, M1, C1), at the same shear rate are very close to one another. For the esters prepared with *n*-butyl alcohol (B2, M2, C2) or *n*-pentanol (B3, M3, C3) the viscosities became more different, their values depending on the technology used (Figure 10).

Table 5. Rheological equations for all bioesters.

Temperature, °C	η_a=k $\dot{\gamma}^{n-1}$	R^2	n	η_a=k $\dot{\gamma}^{n-1}$	R^2	n	η_a=k $\dot{\gamma}^{n-1}$	R^2	n
		B1			M1			C1	
25	η_a= 0.1898 $\dot{\gamma}^{-0.234}$	0.9191	0.766	η_a= 0.1624 $\dot{\gamma}^{-0.214}$	0.8437	0.786	η_a= 0.1670 $\dot{\gamma}^{-0.216}$	0.8283	0.784
35	η_a= 0.1502 $\dot{\gamma}^{-0.218}$	0.9343	0.782	η_a= 0.1594 $\dot{\gamma}^{-0.221}$	0.9298	0.779	η_a= 0.1614 $\dot{\gamma}^{-0.221}$	0.9269	0.779
50	η_a= 0.1644 $\dot{\gamma}^{-0.240}$	0.9836	0.760	η_a= 0.1452 $\dot{\gamma}^{-0.220}$	0.9511	0.780	η_a= 0.1390 $\dot{\gamma}^{-0.214}$	0.9587	0.786
60	η_a= 0.1030 $\dot{\gamma}^{-0.198}$	0.9740	0.802	η_a= 0.0990 $\dot{\gamma}^{-0.188}$	0.9477	0.812	η_a= 0.0936 $\dot{\gamma}^{-0.179}$	0.9758	0.821
70	η_a= 0.0580 $\dot{\gamma}^{-0.147}$	0.9658	0.853	η_a= 0.0550 $\dot{\gamma}^{-0.133}$	0.9107	0.867	η_a= 0.0559 $\dot{\gamma}^{-0.133}$	0.9542	0.867
		B2			M2			C2	
25	η_a= 0.1944 $\dot{\gamma}^{-0.229}$	0.8779	0.771	η_a= 0.1932 $\dot{\gamma}^{-0.238}$	0.8952	0.762	η_a= 0.2029 $\dot{\gamma}^{-0.239}$	0.8879	0.761
35	η_a= 0.1515 $\dot{\gamma}^{-0.211}$	0.9299	0.789	η_a= 0.1520 $\dot{\gamma}^{-0.220}$	0.9025	0.780	η_a= 0.1700 $\dot{\gamma}^{-0.228}$	0.9348	0.772
50	η_a= 0.1400 $\dot{\gamma}^{-0.211}$	0.9546	0.789	η_a= 0.1490 $\dot{\gamma}^{-0.227}$	0.9594	0.773	η_a= 0.1484 $\dot{\gamma}^{-0.222}$	0.9486	0.778
60	η_a= 0.0957 $\dot{\gamma}^{-0.179}$	0.9668	0.821	η_a= 0.0951 $\dot{\gamma}^{-0.189}$	0.9627	0.811	η_a= 0.0968 $\dot{\gamma}^{-0.185}$	0.9504	0.815
70	η_a= 0.0565 $\dot{\gamma}^{-0.130}$	0.9227	0.870	η_a= 0.0608 $\dot{\gamma}^{-0.152}$	0.9326	0.848	η_a= 0.0613 $\dot{\gamma}^{-0.147}$	0.9351	0.853
		B3			M3			C3	
25	η_a= 0.1995 $\dot{\gamma}^{-0.223}$	0.8926	0.777	η_a= 0.2680 $\dot{\gamma}^{-0.275}$	0.938	0.725	η_a= 0.2686 $\dot{\gamma}^{-0.268}$	0.9327	0.732
35	η_a= 0.1482 $\dot{\gamma}^{-0.198}$	0.9258	0.802	η_a= 0.1560 $\dot{\gamma}^{-0.225}$	0.9111	0.775	η_a= 0.1653 $\dot{\gamma}^{-0.223}$	0.9338	0.777
50	η_a= 0.1394 $\dot{\gamma}^{-0.200}$	0.9502	0.800	η_a= 0.1493 $\dot{\gamma}^{-0.230}$	0.9578	0.770	η_a= 0.1375 $\dot{\gamma}^{-0.211}$	0.9633	0.789
60	η_a= 0.0950 $\dot{\gamma}^{-0.166}$	0.9520	0.834	η_a= 0.0956 $\dot{\gamma}^{-0.193}$	0.9557	0.807	η_a= 0.0981 $\dot{\gamma}^{-0.184}$	0.9595	0.816
70	η_a= 0.0575 $\dot{\gamma}^{-0.120}$	0.9321	0.880	η_a= 0.0646 $\dot{\gamma}^{-0.161}$	0.9281	0.839	η_a= 0.0584 $\dot{\gamma}^{-0.138}$	0.9284	0.862

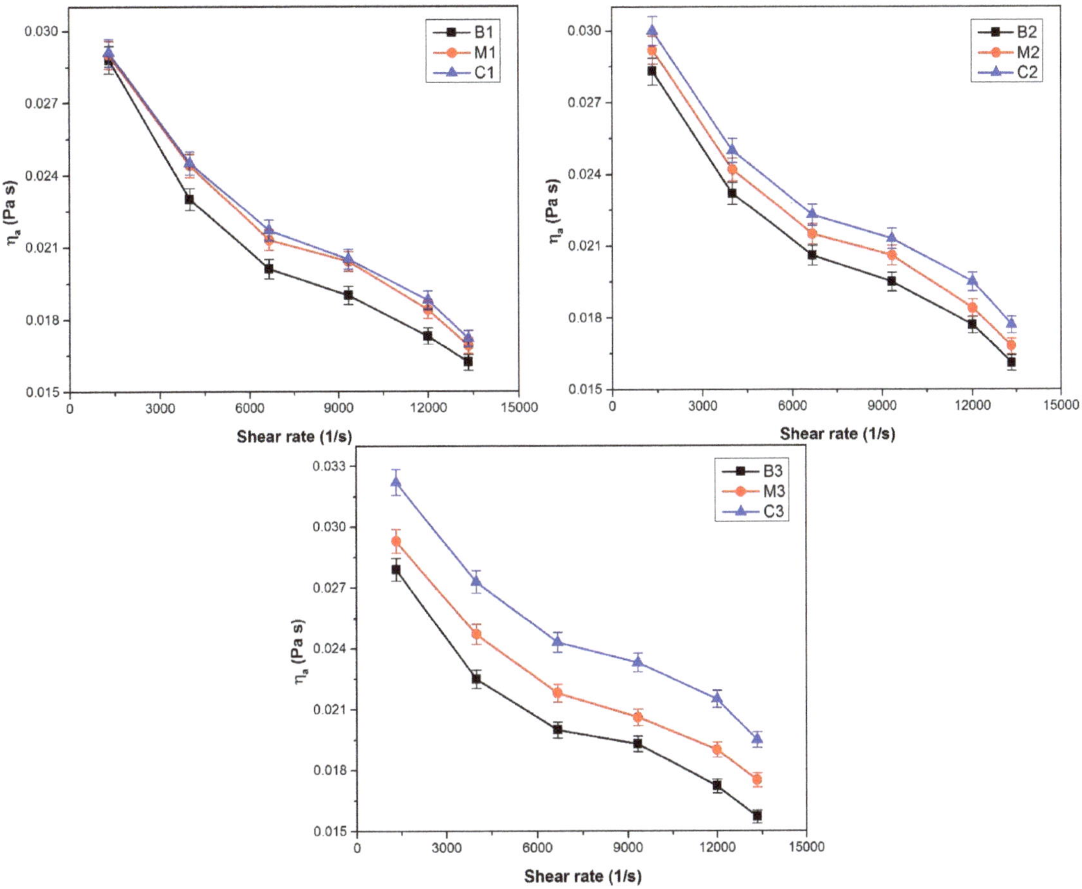

Figure 10. The dependence of the esters apparent viscosity on the share rate.

They can act as thickening agents. According to rheological results, addition of bioester B2 as ingredient for paints, plastics, fibers, detergents, cosmetics and lubricants, is able to modify easily the rheological data and thicken the formulations. The temperature increase leads to micro drops mobility intensification, which influences the activation energy of the system. The phenomenon may be explained by an Arrhenius type equation (Equation (3)):

$$\eta_a = A' \cdot \exp^{Ea/RT}, \quad (3)$$

where: *Ea* is the activation energy of viscous flow, (J/mol); *R* is the gas general constant, (J/mol·K); *T* is absolute temperature and A' represents the material constant, (Pa·s).

The dependence ln η_a = f(*1/T*) was graphically represented, as obtained from the logarithmic form of Equation (3), for apparent viscosity values corresponding to the three chosen values of the shear rate. Particular expressions of Equation (3) as well as the values of the activation energy are presented in Table 6. The activation energy decreases with the increasing of the shear rate, due to the increase of the turbulence and its effect on the linearization tendency of the molecules, considering the reduction of the degree of association of the molecules as well. In both cases, the effort to move molecules is diminished. The activation energy of the bioesters varies in opposite direction with their viscosity when increasing the number of the carbon atoms in alcohols (Figure 10 and Table 6).

Table 6. Particular types of the Arrhenius equation for all esters.

$\dot{\gamma}$	$\eta_a = A' \times 10^3 \cdot \exp^{Ea/RT}$	E_a (kJ/mol)	$\eta_a = A' \times 10^3 \cdot \exp^{Ea/RT}$	E_a (kJ/mol)	$\eta_a = A' \times 10^3 \cdot \exp^{Ea/RT}$	E_a (kJ/mol)
	B1		M1		C1	
4000	$\eta_a = 0.66 \cdot \exp^{1130.7/T}$	9.4	$\eta_a = 0.90 \cdot \exp^{1045.1/T}$	8.7	$\eta_a = 0.83 \cdot \exp^{1074.8/T}$	8.9
9333	$\eta_a = 1.19 \cdot \exp^{886.2/T}$	7.4	$\eta_a = 1.92 \cdot \exp^{754.8/T}$	6.3	$\eta_a = 1.62 \cdot \exp^{809.2/T}$	6.7
13,333	$\eta_a = 1.84 \cdot \exp^{696.6/T}$	5.8	$\eta_a = 2.50 \cdot \exp^{610.8/T}$	5.1	$\eta_a = 3.22 \cdot \exp^{535.6/T}$	4.4
	B2		M2		C2	
4000	$\eta_a = 0.97 \cdot \exp^{1030.8/T}$	8.6	$\eta_a = 0.82 \cdot \exp^{1057.4/T}$	8.8	$\eta_a = 0.77 \cdot \exp^{1093.4/T}$	9.1
9333	$\eta_a = 1.98 \cdot \exp^{757.8/T}$	6.3	$\eta_a = 1.19 \cdot \exp^{891.9/T}$	7.4	$\eta_a = 1.52 \cdot \exp^{813.0/T}$	6.9
13,333	$\eta_a = 3.18 \cdot \exp^{550.5/T}$	4.6	$\eta_a = 2.04 \cdot \exp^{659.4/T}$	5.5	$\eta_a = 2.35 \cdot \exp^{629.0/T}$	5.2
	B3		M3		C3	
4000	$\eta_a = 1.38 \cdot \exp^{947.4/T}$	7.9	$\eta_a = 0.66 \cdot \exp^{1127.9/T}$	9.4	$\eta_a = 0.87 \cdot \exp^{1059.9/T}$	8.8
9333	$\eta_a = 2.52 \cdot \exp^{710.8/T}$	5.9	$\eta_a = 1.24 \cdot \exp^{872.6/T}$	7.2	$\eta_a = 1.91 \cdot \exp^{759.2/T}$	6.3
13,333	$\eta_a = 3.88 \cdot \exp^{519.8/T}$	4.3	$\eta_a = 1.68 \cdot \exp^{711.9/T}$	5.9	$\eta_a = 3.00 \cdot \exp^{559.1/T}$	4.6

4. Conclusions

The synthesis of a series bioesters, using soybean oil fatty acids as the acid component, through three different technologies (in a bubble column reactor, in a reactor heated in a microwave field, and in a classic batch reactor) was performed. Energetic evaluation of the processes pointed out that the processes in the microwave field and in the bubble column reactor are more energetic efficient then in the classic batch reactor. The physicochemical and thermal properties of all esters were determined, and they present similar properties, regardless of the used synthesis routes. Rheological comparative study shows a pseudo-plastic behavior for all esters. Equations of dependence of share stress on share rate, and of apparent viscosity on 1/T are proposed. Activation energy was determined for all samples and revealed an opposite variation with the bioesters viscosity when increasing the number of the carbon atoms in the alcohols from the constitution of esters. It should be noted that the results obtained herein can contribute to the development of new applications containing esters or to the synthesis of biopolymers using the low energy consumption and environmental friendly technologies.

Author Contributions: Conceptualization, S.P., D.J., S.B. and G.M.; methodology, S.P., A.T., V.S., D.J., S.B. and G.M.; software, S.P., S.B. and G.M.; validation, S.P., A.T., V.S., D.J., S.B. and G.M.; formal analysis, S.P., A.T., V.S., D.J., S.B. and G.M.; investigation, S.P., S.B. and G.M.; resources, S.P., S.B. and G.M.; data curation, S.P.; writing—original draft preparation, S.P., D.J. and S.B.; writing—review and editing, S.P., V.S. and G.M.; visualization, S.P. and S.B.; supervision, S.P.; project administration, S.P and D.J.; funding acquisition, S.P and D.J. All authors have read and agreed to the published version of the manuscript.

Funding: The research was supported by the Romanian Ministry of European Funds, Managing Authority for the Increasing Economic Competitiveness Operational Program, grant POS CCE Grant No. PO102418 12/5124/22.05.2014, SMIS 50328, New energetic efficient technologies for some polyesteric copolymer synthesis.

Institutional Review Board Statement: Not applicable.

Informed Consent Statement: Not applicable.

Data Availability Statement: All the experimental data obtained are presented, in the form of table and/or figure, in the article.

Conflicts of Interest: The authors declare no conflict of interest.

References

1. Mosoarca, G.; Negrea, A. Studies regarding the effects of settling tanks sludge recycling on organic matter concentration from treated water. *J. Environ. Prot. Ecol.* **2012**, *13*, 198–202.
2. Syamsuddin, Y.; Murat, M.N.; Hameed, B.H. Synthesis of fatty acid methyl ester from the transesterification of high- and low-acid-content crude palm oil (*Elaeis guineensis*) and karanj oil (*Pongamia pinnata*) over a calcium-lanthanum-aluminum mixed-oxides catalyst. *Bioresour. Technol.* **2016**, *214*, 248–252. [CrossRef]
3. Toscano, G.; Duca, D. Renewable energy content of fatty acid methyl esters (fame) and glycerol. *J. Agric. Eng.* **2009**, *40*, 47–53. [CrossRef]
4. Amani, H.; Asif, M.; Hameed, B.H. Transesterification of waste cooking palm oil and palm oil to fatty acid methyl ester using cesium-modified silica catalyst. *J. Taiwan Inst. Chem.* **2016**, *58*, 226–234. [CrossRef]
5. Stacy, C.J.; Melick, C.A.; Cairncross, R.A. Esterification of free fatty acids to fatty acid alkyl esters in a bubble column reactor for use as biodiesel. *Fuel Process. Technol.* **2014**, *124*, 70–77. [CrossRef]
6. Ye, W.; Gao, Y.; Ding, H.; Liu, M.; Liu, S.; Han, X.; Qi, J. Kinetics of transesterification of palm oil under conventional heating and microwave irradiation, using CaO as heterogeneous catalyst. *Fuel* **2016**, *180*, 574–579. [CrossRef]
7. Oh, J.; Yang, S.; Kim, C.; Choi, I.; Kim, J.H. Synthesis of biolubricants using sulfated zirconia catalysts. *Appl. Catal. A-Gen.* **2013**, *455*, 164–171. [CrossRef]
8. Hamadou, B.; Djomdi; Falama, R.Z.; Delattre, C.; Pierre, G.; Dubessay, P.; Michaud, P. Influence of Physicochemical Characteristics of Neem Seeds (Azadirachta indica A. Juss) on Biodiesel Production. *Biomolecules* **2020**, *10*, 616. [CrossRef]
9. Sanchez, N.; Encinar, J.M.; Nogales, S.; González, J.F. Biodiesel Production from Castor oil by Two-Step Catalytic Transesterification: Optimization of the Process and Economic Assessment. *Catalysts* **2019**, *9*, 864. [CrossRef]
10. Ramkumar, S.; Kirubakaran, V. Biodiesel from vegetable oil as alternate fuel for C.I engine and feasibility study of thermal cracking: A critical review. *Energy Convers. Manag.* **2016**, *118*, 155–169. [CrossRef]
11. Cursaru, D.; Neagu, M.; Bogatu, L. Investigations on the oxidation stability of biodiesel synthesized from different vegetable oils. *Rev. Chim.* **2013**, *64*, 438–441.
12. Dos Santos, P.R.S.; Voll, F.A.P.; Ramos, L.P.; Corazzaa, M.L. Esterification of fatty acids with supercritical ethanol in a continuous tubular reactor. *J. Supercrit. Fluids* **2017**, *126*, 25–36. [CrossRef]
13. Gude, V.G.; Patil, P.; Martinez-Guerra, E.; Deng, S.; Nirmalakhandan, N. Microwave energy potential for biodiesel production. *Sustain. Chem. Process.* **2013**, *1*, 5. [CrossRef]
14. Ilia, G.; Simulescu, V.; Mak, C.A.; Crasmareanu, E. The use of transesterification method for obtaining phosphorus-containing polymers. *Adv. Polym. Technol.* **2014**, *33*, 21437. [CrossRef]
15. Monroe, E.; Somnath, S.; Carlson, J.S.; Eckles, T.P.; Liu, F.; Varman, A.M.; George, A.; Davis, R.W. Superior performance biodiesel from biomass-derived fusel alcohol and low grade oils: Fatty acid fusel esters (FAFE). *Fuel* **2020**, *268*, 117408. [CrossRef]
16. Popa, S.; Milea, M.S.; Boran, S.; Nitu, S.V.; Mosoarca, G.; Vancea, C.; Lazau, R.I. Rapid adulteration detection of cold pressed oils with their refined versions by UV-VIS spectroscopy. *Sci. Rep.* **2020**, *10*, 16100. [CrossRef]
17. Radulescu, M.E.; Visa, A.; Milea, M.S.; Lazau, R.I.; Popa, S.; Funar-Timofei, S. Synthesis, spectral characterization, and theoretical investigations of a new azo-stilbene dye for acrylic resins. *J. Mol. Struct.* **2020**, *1217*, 128380. [CrossRef]
18. Wozniak, M.; Kowalska, M.; Tavernier, S.; Zbikowska, A. Enzymatically Modified Fats Applied in Emulsions Stabilized by Polysaccharides. *Biomolecules* **2021**, *11*, 49. [CrossRef]
19. Umerov, S.O.; Dulina, I.O.; Ragulya, A.V. Rheology of plasticized polymer solutions. *J. Silicate Based Compos. Mater.* **2015**, *67*, 119–125. [CrossRef]
20. Ahmed, W.A.; Yarmo, A.; Salih, N.; Derawi, M.D.; Yusop, M.R.; Salimon, J. Synthesis and lubricity properties analysis of branched dicarboxylate esters based lubricant. *Malaysian J. Anal. Sci.* **2015**, *19*, 106–117.
21. Attia, N.K.; El-Mekkawi, S.A.; Elardy, O.A.; Abdelkader, E.A. Chemical and rheological assessment of produced biolubricants from different vegetable oils. *Fuel* **2020**, *271*, 117578. [CrossRef]
22. Dardan, E.; Afrand, M.; Isfahani, A.H.M. Effect of suspending hybrid nano-additives on rheological behavior of engine oil and pumping power. *Appl. Therm. Eng.* **2016**, *109*, 524–534. [CrossRef]
23. Nogales-Delgado, S.; Encinar, J.M.; Cortes, A.G. High oleic safflower oil as a feedstock for stable biodiesel and biolubricant production. *Ind. Crops Prod.* **2021**, *170*, 113701. [CrossRef]
24. Owuna, F.J.; Dabai, M.U.; Sokoto, M.A.; Dangoggo, S.M.; Bagudo, B.U.; Birnin-Yauri, U.A.; Hassan, L.G.; Sada, I.; Abubakar, A.L.; Jibrin, M.S. Chemical modification of vegetable oils for the production of biolubricants using trimethylolpropane: A review. *Egyp. J. Pet.* **2020**, *29*, 75–82. [CrossRef]
25. Petrova, V.A.; Elokhovskiy, V.Y.; Raik, S.V.; Poshina, D.N.; Romanov, D.P.; Skorik, Y.A. Alginate Gel Reinforcement with Chitin Nanowhiskers Modulates Rheological Properties and Drug Release Profile. *Biomolecules* **2019**, *9*, 291. [CrossRef] [PubMed]
26. Vulpe, R.; Le Cerf, D.; Dulong, V.; Popa, M.; Peptu, C.; Verestiuc, L.; Picton, L. Rheological study of in-situ crosslinkable hydrogels based on hyaluronanic acid, collagen and sericin. *Mat. Sci. Eng. C-Mater.* **2016**, *69*, 388–397. [CrossRef]
27. Helminen, A.; Kylma, J.; Tuominen, J.; Seppala, J.V. Effect of structure modification on rheological properties of biodegradable poly(ester-urethane). *Polym. Eng. Sci.* **2000**, *40*, 1655–1662. [CrossRef]
28. Staroszczyk, H.; Fiedorowicz, M.; Opalinska-Piskorz, J.; Tylingo, R. Rheology of potato starch chemically modified with microwave-assisted reactions. *LWT-Food Sci. Technol.* **2013**, *53*, 249–254. [CrossRef]

29. Simulescu, V.; Crasmareanu, E.; Ilia, G. Synthesis, properties and structures of phosphorus-nitrogen heterocycles. *Heterocycles* **2011**, *83*, 275–291. [CrossRef]
30. Wong, C.P.; Berry, G.C. Rheological studies on concentrated solutions of heterocyclic polymers. *Polymer* **1979**, *20*, 229–240. [CrossRef]
31. Yin, C.Q.; Dong, J.; Zhang, Z.X.; Zhang, Q.H. Rheological behaviour of polyamic acid spinning solutions containing heterocycle. *Mater. Res. Innov.* **2014**, *18*, S4-803–S4-807. [CrossRef]
32. Sharma, A.K.; Tiwari, A.K.; Dixit, A.R. Rheological behaviour of nanofluids: A review. *Renew. Sust. Energ. Rev.* **2016**, *53*, 779–791. [CrossRef]
33. Shu, R.; Gan, Y.; Lv, H.; Tan, D. Preparation and rheological behavior of ethylene glycol-based TiO_2 nanofluids. *Colloid Surf. A Physicochem. Eng. Asp.* **2016**, *509*, 86–90. [CrossRef]
34. Popa, S.; Boran, S.; Simulescu, V. Collagen films obtained from collagen solutions characterized by rheology. *Mater. Plast.* **2017**, *54*, 359–361. [CrossRef]
35. Hoti, G.; Caldera, F.; Cecone, C.; Rubin Pedrazzo, A.; Anceschi, A.; Appleton, S.L.; Khazaei Monfared, Y.; Trotta, F. Effect of the Cross-Linking Density on the Swelling and Rheological Behavior of Ester-Bridged β-Cyclodextrin Nanosponges. *Materials* **2021**, *14*, 478. [CrossRef]
36. Bokhari, A.; Yusup, S.; Chuah, L.F.; Kamil, R.N.M. Relative efficiency of esterified rubber seed oil in a hydrodynamic cavitation reactor and purification via distillation column. *Chem. Eng. Trans.* **2016**, *52*, 775–780.
37. Kuzmina, J.S.; Director, L.B.; Shevchenko, A.L.; Zaichenko, V.M. Energy efficiency analysis of reactor for torrefaction of biomass with direct heating. *J. Phys. Conf. Ser.* **2016**, *774*, 012138. [CrossRef]
38. Encinar, J.M.; Gonzalez, J.F.; Martinez, G.; Sanchez, N.; Pardal, A. Soybean oil transesterification by the use of a microwave flow system. *Fuel* **2012**, *95*, 386–393. [CrossRef]
39. Kazimierowicz, J.; Zieliński, M.; Dębowski, M. Influence of the Heating Method on the Efficiency of Biomethane Production from Expired Food Products. *Fermentation* **2021**, *7*, 12. [CrossRef]
40. Yu, B.-S.; Hung, W.-H.; Fang, J.-N.; Yu, Y.-T. Synthesis of Zn-Saponite Using a Microwave Circulating Reflux Method under Atmospheric Pressure. *Minerals* **2020**, *10*, 45. [CrossRef]
41. Wardhono, E.Y.; Wahyudi, H.; Agustina, S.; Oudet, F.; Pinem, M.P.; Clausse, D.; Saleh, K.; Guénin, E. Ultrasonic Irradiation Coupled with Microwave Treatment for Eco-friendly Process of Isolating Bacterial Cellulose Nanocrystals. *Nanomaterials* **2018**, *8*, 859. [CrossRef] [PubMed]
42. Zieliński, M.; Dębowski, M.; Kazimierowicz, J. Microwave Radiation Influence on Dairy Waste Anaerobic Digestion in a Multi-Section Hybrid Anaerobic Reactor (M-SHAR). *Processes* **2021**, *9*, 1772. [CrossRef]
43. Zimmermann, K. Microwave Technologies: An Emerging Tool for Inactivation of Biohazardous Material in Developing Countries. *Recycling* **2018**, *3*, 34. [CrossRef]
44. Zulqarnain; Ayoub, M.; Yusoff, M.H.M.; Nazir, M.H.; Zahid, I.; Ameen, M.; Sher, F.; Floresyona, D.; Budi Nursanto, E. A Comprehensive Review on Oil Extraction and Biodiesel Production Technologies. *Sustainability* **2021**, *13*, 788. [CrossRef]
45. Popa, S.; Mosoarca, G.; Macarie, L.; Plesu, N.; Ilia, G.; Tara-Lunga-Mihali, M. Copolymerization of butyl acrylate with methyl methacrylate in a bubble column reactor and the use of copolymer in corrosion protection. *Polym. Bull.* **2021**. [CrossRef]
46. Dange, P.N.; Rathod, V.K. Equilibrium and thermodynamic parameters for heterogeneous esterification of butyric acid with methanol under microwave irradiation. *Resour. Efficient. Technol.* **2017**, *3*, 64–70. [CrossRef]
47. Popa, S.; Boran, S. Energetic efficiency calculation for a new experimental reactor. *Mater. Plast.* **2016**, *53*, 410–413.
48. Winkler, H.; Vorwerg, W.; Rihm, R. Thermal and mechanical properties of fatty acid starch esters. *Carbohydr. Polym.* **2014**, *102*, 941–949. [CrossRef]
49. Paul, A.K.; Achar, S.K.; Dasari, S.R.; Borugadda, V.B.; Goud, V.V. Analysis of thermal, oxidative and cold flow properties of methyl and ethyl esters prepared from soybean and mustard oils. *J. Therm. Anal. Calorim.* **2017**, *130*, 1501–1511. [CrossRef]
50. Sari, A.; Karaipekli, A.; Eroglu, R.; Biçer, A. Erythritol Tetra Myristate and Erythritol Tetra Laurate as Novel Phase Change Materials for Low Temperature Thermal Energy Storage. *Energ. Source Part A* **2013**, *35*, 1285–1295. [CrossRef]
51. Hu, C.; Ai, J.; Ma, L.; Wen, P.; Fan, M.; Zhou, F.; Liu, W. Ester Oils Prepared from Fully Renewable Resources and Their Lubricant Base Oil Properties. *ACS Omega* **2021**, *6*, 16343–16355. [CrossRef]
52. Nakayama, Y.; Watanabe, K.; Tanaka, R.; Shiono, T.; Kawasaki, N.; Yamano, N.; Nakayama, A. Synthesis, Properties, and Biodegradation of Sequential Poly(Ester Amide)s Containing γ-Aminobutyric Acid. *Int. J. Mol. Sci.* **2020**, *21*, 3674. [CrossRef]
53. Aziz, N.A.M.; Yunus, R.; Hamid, H.A.; Ghassan, A.A.K.; Omar, R.; Rashid, U.; Abbas, Z. An acceleration of microwave-assisted trans-esterification of palm oil-based methyl ester into trimethylolpropane ester. *Sci. Rep.* **2020**, *10*, 19652–19668. [CrossRef]
54. Veronesi, F.; Guarini, G.; Corozzi, A.; Raimondo, M. Evaluation of the Durability of Slippery, Liquid-Infused Porous Surfaces in Different Aggressive Environments: Influence of the Chemical-Physical Properties of Lubricants. *Coatings* **2021**, *11*, 1170. [CrossRef]
55. Zhang, X.; Li, G.; Chen, Y.; Wang, K.; Yang, E. The Synthesis of Associative Copolymers with Both Amphoteric and Hydrophobic Groups and the Effect of the Degree of Association on the Instability of Emulsions. *Polymers* **2021**, *13*, 4041. [PubMed]
56. Soares, I.P.; Rezende, T.F.; Silva, R.C.; Castro, E.V.R.; Fortes, I.C.P. Multivariate calibration by variable selection for blends of raw soybean oil/biodiesel from different sources using Fourier Transform Infrared spectroscopy (FTIR) spectra data. *Energy Fuels* **2008**, *22*, 2079–2083. [CrossRef]

57. Duan, Y.; Huo, Y.; Duan, L. Preparation of acrylic resins modified with epoxy resins and their behaviors as binders of waterborne printing ink on plastic film. *Colloids Surf. A Physicochem. Eng. Asp.* **2017**, *535*, 225–231. [CrossRef]
58. Jiao, C.; Sun, L.; Shao, Q.; Song, J.; Hu, Q.; Naik, N.; Guo, Z. Advances in Waterborne Acrylic Resins: Synthesis Principle, Modification Strategies, and Their Applications. *ACS Omega* **2021**, *6*, 2443–2449. [CrossRef]
59. Hajipour, A.; Shams-Nateri, A. Expanding the color gamut of inkjet textile printing during color matching. *Color Res. Appl.* **2021**, *46*, 1218–1226. [CrossRef]
60. Popa, S.; Radulescu-Grad, M.E.; Perdivara, A.; Mosoarca, G. Aspects regarding colour fastness and adsorption studies of a new azo-stilbene dye for acrylic resins. *Sci. Rep.* **2021**, *11*, 5889. [CrossRef]
61. Golshan, M.; Amani, F.; Salami-Kalajahi, M. Photophysical and reflectance properties of perylene-3,4,9,10-tetracarboxylic diimide (PTCDI)/rhodamine 6 G hybrid for application in cold paints. *Prog. Org. Coat.* **2021**, *157*, 106308. [CrossRef]
62. Che, J. Relative analysis of the dependency of spectrophotometric data on temperature using textiles, ceramics, plastics, paints, and printed materials. *Color Res. Appl.* **2018**, *43*, 922–927. [CrossRef]
63. Kowalska, M.; Turek, P.; Zbikowska, A.; Babut, M.; Szakiel, J. The Quality of Emulsions with New Synthetized Lipids Stabilized by Xanthan Gum. *Biomolecules* **2021**, *11*, 213. [CrossRef] [PubMed]
64. Boran, S.; Tamas, A.; Mosoarca, G. Soybean bioester obtained in a bubble column esterification reactor—A rheological study. *Studia UBB Chemia* **2019**, *64*, 31–36. [CrossRef]
65. Ladie, R.; Cosentino, C.; Tagliaro, I.; Antonini, C.; Bianchini, G.; Bertini, S. Supramolecular Structuring of Hyaluronan-Lactose-Modified Chitosan Matrix: Towards High-Performance Biopolymers with Excellent Biodegradation. *Biomolecules* **2021**, *11*, 389. [CrossRef]
66. Sienkiewicz, A.; Czub, P. Rheological Analysis of the Synthesis of High-Molecular-Weight Epoxy Resins from Modified Soybean Oil and Bisphenol A or BPA-Based Epoxy Resins. *Materials* **2021**, *14*, 6770. [CrossRef]
67. Ahmad, M.N.; Ishak, M.R.; Taha, M.M.; Mustapha, F.; Leman, Z. Rheological and Morphological Properties of Oil Palm Fiber-Reinforced Thermoplastic Composites for Fused Deposition Modeling (FDM). *Polymers* **2021**, *13*, 3739. [CrossRef]

polymers

MDPI

Article

Strippable Polymeric Nanocomposites Comprising "Green" Chelates, for the Removal of Heavy Metals and Radionuclides

Gabriela Toader [1,†], Daniela Pulpea [1,†], Traian Rotariu [1,*], Aurel Diacon [2], Edina Rusen [2,*], Andreea Moldovan [1], Alice Podaru [1], Raluca Ginghină [3], Florentina Alexe [3], Ovidiu Iorga [3], Sorina Aurora Bajenaru [3], Mihai Ungureanu [1], Florin Dîrloman [1], Bogdan Pulpea [1] and Lucia Leonat [4]

1 Military Technical Academy "Ferdinand I", 39–49 George Coşbuc Boulevard, 050141 Bucharest, Romania; nitagabriela.t@gmail.com (G.T.); pulpea.daniela@gmail.com (D.P.); andreea.voicu89@gmail.com (A.M.); podaru.alice04@gmail.com (A.P.); mihai.ungureanu@mta.ro (M.U.); florin.dirloman@mta.ro (F.D.); pulpea.b@gmail.com (B.P.)
2 Faculty of Applied Chemistry and Materials Science, University 'Politehnica' of Bucharest, 1–7 Gh. Polizu Street, 011061 Bucharest, Romania; aurel.diacon@upb.ro
3 Research and Innovation Center for CBRN Defense and Ecology, 225 Soseaua Oltenitei, 041327 Bucharest, Romania; ginghinaraluca@gmail.com (R.G.); crinaalexe@yahoo.com (F.A.); iorga_ovidiu@yahoo.com (O.I.); aurorabaj@yahoo.com (S.A.B.)
4 National Institute of Materials Physics, 077125 Măgurele, Romania; lucialeonat@gmail.com
* Correspondence: traian.rotariu@mta.ro (T.R.); edina.rusen@upb.ro (E.R.)
† Contributed equally to this study (co-first authors).

Abstract: The issue of heavy metal and radionuclide contamination is still causing a great deal of concern worldwide for environmental protection and industrial sites remediation. Various techniques have been developed for surface decontamination aiming for high decontamination factors (DF) and minimal environmental impact, but strippable polymeric nanocomposite coatings are some of the best candidates in this area. In this study, novel strippable coatings for heavy metal and radionuclides decontamination were developed based on the film-forming ability of polyvinyl alcohol, with the remarkable metal retention capacity of bentonite nanoclay, together with the chelating ability of sodium alginate and with "new-generation" "green" complexing agents: iminodisuccinic acid (IDS) and 2-phosphonobutane-1,2,4-tricarboxylic acid (PBTC). These environmentally friendly water-based decontamination solutions are capable of generating strippable polymeric films with optimized mechanical and thermal properties while exhibiting high decontamination efficiency (DF ≈ 95–98% for heavy metals tested on glass surface and DF ≈ 91–97% for radionuclides ^{241}Am, ^{90}Sr-Y and ^{137}Cs on metal, painted metal, plastic, and glass surfaces).

Keywords: "green" chelates; complexing agents; nanocomposite; strippable coating; surface decontamination; heavy metal; radioactive material; radionuclide

Citation: Toader, G.; Pulpea, D.; Rotariu, T.; Diacon, A.; Rusen, E.; Moldovan, A.; Podaru, A.; Ginghină, R.; Alexe, F.; Iorga, O.; et al. Strippable Polymeric Nanocomposites Comprising "Green" Chelates, for the Removal of Heavy Metals and Radionuclides. *Polymers* **2021**, *13*, 4194. https://doi.org/10.3390/polym13234194

Academic Editor: Alexander B. Morgan

Received: 21 October 2021
Accepted: 26 November 2021
Published: 30 November 2021

Publisher's Note: MDPI stays neutral with regard to jurisdictional claims in published maps and institutional affiliations.

Copyright: © 2021 by the authors. Licensee MDPI, Basel, Switzerland. This article is an open access article distributed under the terms and conditions of the Creative Commons Attribution (CC BY) license (https://creativecommons.org/licenses/by/4.0/).

1. Introduction

Over the last decades, many ecosystems have been altered by human activities, causing the contamination of the environment. Heavy metals, as well as radioactive materials, have been extensively used in industrial applications, medicine, military activity, or various research fields. Despite strict loyalty to all laboratory safety procedures, it is still very possible to encounter heavy metal or radionuclide contamination. During the use of materials containing heavy metals or radioactive metals, various surfaces, such as concrete, steel, glass, rubber, plastic materials, or painted surfaces, from a laboratory, a shooting range [1,2], an industrial or a nuclear facility [3–5], can be contaminated with these hazardous materials.

"Toxic metals", including "heavy metals" or "radioactive metals", are compounds that pose severe environmental problems, negatively affecting the health and the safety of humans at the same time. In very low concentrations, some heavy metals are necessary to support life, but at higher concentrations, they become poisonous due to their bioaccumulation. Occupational exposure to lead is one of the most widespread overexposures. High potential exposures sources include firing ranges, car batteries or pigment industries. Another prevalent heavy metal is mercury. Typical sources of mercury exposure include mining and refining of gold and silver ores. Another category of "toxic metals" is represented by radioactive metals. Radioactive metals are natural or synthetic isotopes of natural non-radioactive metals that can release alpha (α), beta (β), and gamma (γ) radiation [6] In certain circumstances, these metals can be useful for humans, being employed for cancer treatment, material engineering, or for power generation. Uranium is one of the most valuable radioactive material of the modern world. It is the main raw material for nuclear bombs and nuclear power plants. Cesium and strontium are high-yield fission products that are present in significant amounts in fuel pond waters and reprocessing stream liquors [7]. Radioactive contamination can occur as a result of working with radioisotopes, an accident, or even a terrorist attack [8]. A nuclear explosion is followed by the production of a considerable quantity of radioactive cesium isotope ^{137}Cs (1.6 times greater than ^{90}Sr) [3]. ^{137}Cs has long-term consequences due to relatively long half-life [9]. ^{90}Sr is utilized in medicine and industry, but it generates significant concerns regarding the fallout from nuclear weapons or nuclear accidents. The probability of ^{90}Sr being released as a part of a nuclear reactor accident is lower than the one of ^{137}Cs because it is much less volatile, but ^{90}Sr is probably the most hazardous element of the radioactive effect from a nuclear weapon.

A major problem with exploiting radioactive metal is represented by wastes. Once released into the environment they will result in possibly catastrophic effects that may last long periods and cause malignant illnesses. While "high-level radioactive waste (HLW)" mainly comes from spent fuel from commercial or research reactors, reprocessing of spent fuel, nuclear weapons, or propulsions industry, "low-level radioactive waste (LLW)" comes from hospitals and industry, as well as the nuclear fuel cycle [6]. External contamination occurs when a heavy metal or a radioactive material, in the form of dust, powder, or liquid, encounters an object or a person. External contamination can become internal if the hazardous material enters their bodies through ingestion, inhalation, or skin. Unfortunately, surface contamination with heavy metals and radioactive metals can, very often, become airborne or can be easily transferred by contact [10].

Considering the significant number of environmental incidents caused by heavy metals or radioactive materials, an overabundance of formulations for decontamination have been developed, and described in the literature, to successfully address different types of contamination scenarios. Efficient decontamination techniques are essential for minimizing occupational exposures, facilitating waste management, restricting the potential accidental release of hazardous materials, and allowing the reuse of some of the components from nuclear reactors, industrial installations, laboratory equipment, shooting ranges, etc. Well-weighted decisions must be taken when choosing between passing the entire system through the decontamination process or replacing just the contaminated equipment/component [11]. Several in situ chemical decontamination technologies, which can be applied for the removal of these hazardous materials, have been developed: wiping with textiles wetted with a decontamination solution [12]; wet vacuum treatment (when the vacuum cleaner is charged with a decontamination solution) [13]; electrochemical decontamination with use of external electrode [14]; foam decontamination [15]; decontamination by etching pastes and gels [16]; decontamination by removable polymer coatings [5,17]; and decontamination by sorbents [16]. A major problem in chemical decontamination is the production of a high volume of secondary waste that needs additional treatment for radionuclides removal. Moreover, to increase the decontamination factor and accelerate the decontamination process, chemical methods involve the use of concentrated acid solutions and temperatures up to 70–90 °C, which could endanger the health and safety of the workers that are manipulating these corrosive and toxic materials. Electrochemical decontamination is limited by the size of the bath in which the contaminated object must be immersed and could not be used on an industrial scale [12]. Using strippable coatings has the advantage of higher efficiency with simpler equipment, fewer chemical reagents, and less waste volume. Among the above-mentioned techniques, decontamination by removable polymer coatings is described in the literature as being the most rapid and cost-efficient technique [16]. Koryakovskiy et al. explained that the application of pastes and gels to the surface takes about 20 to 30 s/m^2, while their removal takes additional 45 to 90 s/m^2 or even more to wash the residues; still less productive is an electric brush having a treatment rate of about 8 to 10 min/m^2 or electrochemical methods, which are expensive and difficult to manage [16].

It is very important to reduce the work time in the contaminated areas because prolonged exposure to these hazardous metals can lead to serious health and safety problems. Moreover, finding a method to remove and fix the contaminants (in a polymeric matrix for example), efficiently with the shortest possible personnel exposure time, is imperative. In this context, removable polymer coatings fulfill most of the requirements listed above.

This method of decontamination was used for the first time on radioactively hazardous facilities of the naval forces, on nuclear submarines. Since then, polyvinyl alcohol-based decontamination formulations were used on a large scale. These polymeric compositions were employed for Chernobyl accident response activities and displayed the highest decontamination degrees for most of the contaminated materials [16].

However, despite the positive findings, certain problems remain regarding the removable polymeric coatings. One of them is related to their chemical composition because some of the film-forming decontamination solutions contain volatile solvents or corrosive components. In addition, they contain "old-generation" chelating agents which are not biodegradable, thus leading to supplementary disposal issues. Another problem is related to their viscosity control, which also needs improvement, because the available commercial solutions have low viscosities and poor adhesion to smooth surfaces and cause serious problems for the decontamination of vertical surfaces, flowing down by gravity. As already mentioned, decontamination of surfaces with strippable coatings is a technique that has been extensively studied in the last decade, due to its multiple advantages, but especially because it is a decontamination method that generates a considerably smaller amount of post-treatment waste [10]. Even though polyvinyl alcohol (PVA), employed in nearly all these types of strippable coatings, is a biodegradable polymer (in specific circumstances), it is still a synthetic polymer and requires special conditions for biodegradation [18]. Moreover, PVA has gained attention lately, being among the major pollutants of industrial wastewater in the textile industry [19]. Therefore, the environmental issues include not only the well-known contaminating agents but also the polymeric materials due to the problems arising from their subsequent disposal. Thus, finding solutions to improve the biodegradability of these decontamination solutions should become a main concern for an environmentally responsible collective attitude. Sodium alginate (SA) could represent an adequate candidate for decontamination applications due to its unique set of properties. SA is a biodegradable hydrophilic linear polysaccharide obtained from marine brown algae [20]. The major advantage of alginates is represented by their liquid-gel behavior in aqueous solutions, due to the ion exchange phenomenon that occurs between the sodium ions in alginate and other divalent ions (calcium ions especially), leading to a gel structure with higher viscosity. In the presence of divalent ions, the G-blocks of alginate participate at the formation of the intermolecular crosslinking, thus gaining enhanced mechanical properties [21]. Depending on the concentration of divalent ions in the system, the crosslinking process can be temporary or permanent. At lower concentrations of Ca ions, the temporary association of the chains can occur, leading to viscous, thixotropic solutions, while at higher Ca ions concentrations precipitation or gelation will occur due to permanent crosslinking phenomena [21]. In the literature, numerous papers have shown that the gel-forming kinetics have a significant influence on its functional properties involving porosity, swelling behavior, stability, biodegradability, gel strength, or biocompatibility [21,22]. Polysaccharides, including cellulose, chitosan, pectin, or alginate, possess the ability to produce films/coatings [23]. Numerous studies have revealed that some films obtained by employing alginate displayed improved barrier and mechanical properties [23,24].

In the actual context, considering all the shortcomings of the existing decontamination methods detailed above, this work describes a novel approach towards obtaining novel water-based film-forming decontamination solutions, containing "green" chelates, for an ecological tactic of efficiently removing heavy metals and radionuclides. These innovative biodegradable solutions can generate resistant and easy-peelable polymeric nanocomposite coatings, due to their specially designed composition, comprising a polymeric blend (polyvinyl alcohol and sodium alginate), with excellent film-forming abilities conjoined with the reinforcing effect brought by bentonite nanoclay. Once these aqueous solutions are applied on the contaminated surface, the complexation of the contaminants occurs, together with their entrapment in the polymer-clay system. After solvent evaporation, they form resistant continuous films that can be easily removed from the surface by simply being peeled off, thus ensuring fast and efficient decontamination. These modern decontamination solutions generate a considerably lower volume of post-decontamination wastes than the traditional methods. The material resulting after decontamination can be easily compacted and temporarily stored in a special small container, until it can be further disposed of as hazardous waste. Another purpose of this study was to evaluate the effect obtained by reducing the PVA amount employed in the strippable coatings through the introduction of SA, an eco-friendlier alternative. Thus, using different amounts of SA, we were able to reduce the PVA concentration in the nanocomposite films while ensuring the same performance: homogenous film-forming ability, ease of film removal, thermal, and mechanical resistance. Another advantage brought by these alginate-based solutions consists of the ability of SA to bind divalent heavy metal ions or radioactive metals, which ensures higher decontamination degrees in this case. Moreover, these novel alginate–based decontamination solutions have higher viscosities that increase in the presence of divalent ions, due to the crosslinking phenomenon, thus allowing the use of these decontamination solutions also on vertical surfaces. Decontamination can possibly occur through two distinct paths: chemical interaction (complexation by "green" chelating agents [25–27] or SA), and physical interaction (adsorption by bentonite nanoclay). These two mechanisms, in conjunction with alginate crosslinking [28], enhance the overall efficiency of the decontamination solutions.

The novelty of this work consists of an innovative way of combining the ability of polyvinyl alcohol to produce films, with the remarkable adsorption capacity of bentonite nanoclay, together with the chelating ability of alginate and "new-generation" "green" complexing agents: iminodisuccinic acid (IDS) and 2-phosphonobutane-1,2,4-tricarboxylic acid (PBTC), to obtain powerful, versatile, and environmentally-friendly water-based solutions for surface decontamination of heavy metal or radioactive metals. Both "green" chelating agents, PBTC and IDS, are well-known in literature for being efficient in chelating metals in aquatic environments [29]. In comparison with the classical complexing agent EDTA, these new generation "green" complexation agents are considered eco-friendly materials, due to their biodegradability. From our knowledge, we report here for the first time, a decontamination solution, containing two "green" chelates: PBTC and IDS, utilized together with sodium alginate as a complexing agent and, at the same time, as an integrant part of eco-friendly peelable polymeric coatings, for the efficient removal heavy metals or radioactive metals from contaminated surfaces. This paper comprises structural, thermal, and mechanical characterization of the newly synthesized nanocomposites through various analytical techniques and decontamination tests on various types of surfaces (glass, metal, painted metal, plastic, and textile sample from the CBRN individual protection equipment) contaminated with heavy metals and three types of radioactive solutions: alpha (^{241}Am), beta (^{90}(Sr−Y)) and gamma (^{137}Cs).

2. Experimental

2.1. Materials

Reagent: iminodisuccinic acid (**IDS**—BAYPURE® CX 100 solid G (>78% Iminodisuccinic acid Na_4 salt, <15% Aspartic acid, Na_2 salt, <5% Fumaric acid Na2 salt, <0.7% Hydroxysuccinic acid, Na_2 salt, <0.5% Maleic acid Na_2 salt and <4% Water)—Lanxess, Cologne, Germany), 2-phosphonobutane-1,2,4-tricarboxylic acid (**PBTC**—BAYHIBIT® AM (40.0–42.5 % PBTC-Na_4 content in water)—Lanxess, Cologne, Germany), poly(vinyl alcohol) (**PVA** with 98–99% hydrolysis degree, DP $\approx$ 1700–1800, Mw $\approx$ 115000 Da—Loba Chemie, Mumbai, India), nano-clay hydrophilic bentonite (**BT**, Sigma–Aldrich, St. Louis, MO, USA), Glycerol $\geq$ 99% (**GLY**—Sigma–Aldrich, St. Louis, MO, USA); Sodium alginate (Special Ingredients®, **SA**, Garlenda, Savona, Italy). Metal solutions: caesium sulphate (Cs_2SO_4—99.99% trace metal basis, Sigma–Aldrich—0.005 M, 0.05 M and 0.5 M Cs_2SO_4 aqueous solutions); lead standard for AAS (1000 mg/L $\pm$ 4 mg/L Pb in nitric acid, Sigma–Aldrich); strontium standard for AAS (1000 mg/L $\pm$ 4 mg/L Sr in nitric acid, Sigma–Aldrich); cobalt standard for AAS (1000 mg/L $\pm$ 4 mg/L Co in nitric acid, Sigma–Aldrich). Radioactive solutions: ^{137}Cs; ^{241}Am; ^{90}(Sr-Y). Tested surfaces: stainless steel sheets (10 mm $\times$ 10 mm $\times$ 0.4 mm classical 18/8 stainless steel), galvanized metal sheets (10 mm $\times$ 10 mm $\times$ 0.4 mm), brass sheets (10 mm $\times$ 10 mm $\times$ 0.5 mm) and cooper sheets (10 mm $\times$ 10 mm $\times$ 0.2 mm), metal (100 mm $\times$ 100 mm $\times$ 0.5 mm, MIL-46100 military grade steel), painted metal (100 mm $\times$ 100 mm $\times$ 0.5 mm, paint based on urethane modified resin), plastic (100 mm $\times$ 100 mm $\times$ 1 mm, polycarbonate), glass (100 mm $\times$ 100 mm $\times$ 5 mm) and BC/SP2 (Romanian Army Individual Protective Equipment (IPE) material sample (100 mm $\times$ 100 mm $\times$ 0.5 mm)).

2.2. Methods

2.2.1. Synthesis of the Decontamination Solutions

The water-based decontamination solutions were obtained according to the procedure described below. The correlation between the composition of the decontamination solutions and the sample IDs is summarized in Table 1. Every decontamination solution contains water, 5% PVA, 1% BT nano-clay, 2.5% glycerol, and different sodium alginate concentration (0%, 0.25%, 0.5%, 0.75%, or 1%), indicated in Table 1. Each of the last two samples (GD-3-PBTC and GD-3-IDS), contained 1% chelating agent (CA). The aqueous decontamination solutions were obtained through the following steps: the first one consisted in the dissolution of the chelating agent in water (for GD-3-PBTC and GD-3-IDS samples), followed by the dispersion of bentonite by ultrasonication. The next step consisted in the addition of the alginate water solution, followed by the dissolution of PVA. The last step consisted in the addition of glycerol. The solutions were stored at 2–5 °C until they were employed for decontamination tests. All the decontamination solutions containing SA exhibited a liquid-gel behavior in the presence of divalent ions, changing from a relatively low viscosity solution to a gel structure. They were allowed to dry on the target surface, forming thin films, which were subsequently removed by being peeled off. The nanocomposite films were subjected to different analytical investigations, described in Characterization Section 2.3.

2.2.2. Nanocomposite Films Preparation

To evaluate the properties of the nanocomposite films obtained from the decontamination solutions, various analytic techniques were employed (described in the Characterization section). The films subjected to analysis were obtained through casting method. Decontamination solutions were placed in rectangular glass molds (12 cm $\times$ 12 cm $\times$ 2 cm), and they were allowed to dry on a plane horizontal surface, at room temperature and 50–55 relative humidity. Usually, the drying time for these film-forming materials is below 24 h, depending on the thickness of the film and the environmental conditions.

Table 1. Composition of the gel-forming decontamination solutions.

Sample ID	SA (Weight %) *	CA (Weight %) *
GD-0	0	0
GD-1	0.25	0
GD-2	0.5	0
GD-3	0.75	0
GD-4	1	0
GD-3-PBTC	0.75	1
GD-3-IDS	0.75	1

* % calculated from the total mass of the decontamination solution.

2.2.3. Controlled Contamination for SEM-EDS Analysis

Non-radioactive cesium was employed for evaluating in-depth and surface contamination on three different types of metallic surfaces. For this purpose, Cs_2SO_4 solutions were employed for the contamination of stainless-steel surfaces (square-shaped metallic coupons: 1 cm × 1 cm × 0.05 cm) with 3 types of finishing: mirror-finish (SSMF), grinded-finish (SSGF), and etched-finish (SSEF)—immersed in royal water). To reproduce in-depth contamination [30], stainless steel coupons were immersed for 30 min of in 0.005 M Cs_2SO_4 aqueous solution followed by a thermal treatment at 700 °C, in a furnace, for 2 h. To simulate superficial contamination, three different concentrations of Cs_2SO_4 aqueous solutions were employed: 0.005 M, 0.05 M, and 0.5 M. The metallic coupons were placed in Petri dishes and the contamination solution was added until they were completely covered by liquid. Then, the coupons were maintained at 40 °C until the complete evaporation of water. For the decontamination tests, metallic coupon was covered with decontamination solution. Once the drying process (complete evaporation of water, at 25 °C, approximately 8 h) ended, the strippable coatings were easily peeled off from the metallic coupons. Both, polymeric films containing the contaminant and decontaminated surfaces were subsequently subjected to SEM/EDS (Billerica, MA, USA) analysis. This investigation offered preliminary qualitative information about the decontamination process. To quantitatively evaluate the decontamination efficiency, decontamination tests described below were performed.

2.2.4. Decontamination Tests

Decontamination tests aimed to evaluate the removal efficacy of herein reported decontamination solutions. For this purpose, controlled contamination with heavy metals and radionuclides was performed prior to the decontamination step. For the evaluation of the decontamination degree, two distinct analytic investigations were performed: (1) atomic absorption spectrometry and (2) alpha, beta, and gamma radiation measurements.

2.2.5. Evaluation of the Decontamination Efficacy through Atomic Absorption Spectrometry (AAS) Technique

For the controlled contamination with heavy metal, a standard solution of lead was employed. Standard solutions containing Sr and Co were also utilized for controlled contamination (as simulants for their analogous radionuclides). To obtain surfaces with similar contamination degrees, 1 mL from each metal standard solution (Pb, Sr, and Co) was placed in a distinct glass Petri dish (φ = 50 mm) and they were allowed to dry. Then, the decontamination solutions (8 mL) were applied on the contaminated surfaces. After being completely dried, they were easily peeled. Triplicate experiments were performed for a better accuracy. Decontamination efficacy was evaluated by comparing the remnant metal concentration from the surface with the initial concentration of contaminant. For this analysis, after removing the strippable coatings, the Petri dishes were washed three times with 5 mL of distilled water and the collected solutions (3 × 5 mL) constituted the sample that were subjected to AAS investigation, to assess the remnant metal concentration on the decontaminated surface. The decontamination factor was calculated with the following formula: DF = $100 \cdot (C_0 - C_f)/C_0$ [4,30], where DF is the decontamination factor, C_0 is the initial metal concentration, and C_f is the final concentration, reflecting the residual contamination.

In parallel, the concentration of "toxic metal" found in the polymeric nanocomposite film was also calculated using the AAS technique. For this purpose, each of the polymeric nanocomposite films obtained was dissolved in 50 mL of distilled water, sonicated, and then centrifugated and filtered to obtain a clear aqueous solution that was also subjected to AAS analysis. Since these samples required intermediary operations before being ready for AAS analysis, we can use the concentrations obtained only to demonstrate the presence of the "toxic metal" in the exfoliated film, and we can presume that the lacking amount of contaminant remained fixed in the bentonite eliminated after centrifugation and filtration steps. The differences between the values obtained for "toxic concentration" occur due to a certain degree of uncertainty evaluated internally, for the performance of sample preparation procedures for AAS analysis. This consists in a cumulative uncertainty regarding the measuring instruments, the human factor, and the analysis equipment. Losses also occur in the process of extraction, filtration, and dilution. To minimize the uncertainty, triplicate analyses were performed with the same sample, reporting the measured average value. The data collected from these decontamination tests are useful for a better understanding of the metallic contaminant removal process. As already mentioned above, these experiments were repeated three times and the average value was reported.

2.2.6. Evaluation of the Decontamination Efficacy through Radiation Measurement

Three standard radioactive solutions, ^{241}Am for alpha radiation, ^{90}(Sr-Y) for beta radiation, and ^{137}Cs for gamma radiation, were employed for radioactive controlled contamination of 5 types of surfaces: metal, painted metal, plastic, glass, and BC/SP2. The radioactive controlled contamination was performed according to NATO standard AEP-58 [4]. Every tested surface measured a 10 cm^2 area and a specific quantity of each solution was uniformly dispersed on the surfaces to reach a value between 30 ÷ 80 Bq/cm^2 for medium contamination level and 300 ÷ 800 Bq/cm^2 for high contamination level. The activity of the contaminated surface was measured at 10 min after the deposition of the radioactive solution. After measuring the initial activity (A_i) of the surfaces, 10 mL of decontamination solution was poured over the targeted area. The decontaminating solution was allowed to cure and dry. After the completion of this step (after 24 h), the coatings were peeled off and the final activity (A_f) of the surface was measured again immediately after the removal of the nanocomposite film. The decontamination factor (DF) was calculated considering the initial activity of the contaminated surfaces and the final activity of the decontaminated surface, with the aid of the following formula (equivalent to the one utilized for heavy metals): DF = $100\,(A_i - A_f)/A_i$, where A_i represents the initial contamination (Bq/cm^2) and A_f represents the residual contamination (Bq/cm^2), measured after the removal of the nanocomposite film. DF was determined in accordance with NATO standard AEP-58 [4]. The radioactive activity investigations were carried out with Berthold dose/dose rate monitor. The measurements for each type of sample were repeated three times and the average value was reported.

2.3. Characterization

FT-IR spectra were obtained with the aid of a Perkin Elmer Spectrum Two with a Pike Miracle™ ATR modulus, at 4 cm^{-1} resolution, 550 to 4000 cm^{-1}. Thermogravimetric analysis (TGA) was performed on a Netzsch TG 209 F3 Tarsus instrument (Selb, Germany) under nitrogen atmosphere, a flow rate of 20 mL/min, on samples weighting approximately 4 mg, on a temperature ramp starting from ambient temperature up to 700 °C with 10 °C/min heating rate. A 710 Titan 2 universal strength testing machine, equipped with a 3000 N force cell was employed for tensile tests, performed according to ISO 37: 2011(E). The dumbbell-shaped specimens were obtained from the strippable nanocomposite films using a cutting mold device that had 75 mm overall length. The test area implied a length of the narrow section of about 25 ± 1 mm. The variation of the length and force was continuously monitored with an accuracy of ±0.2% at a speed of 8.33 mm/s. Five specimens from each sample were analyzed. The mean values obtained for each material were plotted in a comparative stress–strain graph. To evaluate thermo-mechanical and viscoelastic properties of the nanocomposite coatings, measurements were conducted on a Discovery 850 DMA-TA Instruments, in single cantilever-bending mode. Experiments were run on samples (10 × 30 × 0.5 mm^3 size), at a frequency of 1 Hz, and a temperature ramp starting from −80 °C to 200 °C with a heating rate of 5 °C/min. SEM-EDS analysis was performed with the aid of a Zeiss Gemini 500 microscope coupled with an XFlash 6 EDS detector from Bruker (Billerica, MA, USA). All data from the EDS were analyzed using the ESPRIT Software (version ESPRIT 2, Billerica, MA, USA). For the evaluation of the efficiency of heavy metal removal we employed an atomic absorption spectrometer, PerkinElmer AAnalyst™ 800 (Waltham, MA, USA) high-performance with WinLab32 software (Perkin Elmer, Waltham, MA, USA) for AA. The determinations were performed by subjecting samples to electrothermal atomization in a graphite furnace. The radionuclides decontamination efficiency was determined by measuring the activity of the targeted surfaces (contaminated with ^{241}Am, ^{90} (Sr-Y), and ^{137}Cs), before and after decontamination, using Berthold L123 dose/dose rate monitor with specific detectors for alpha, beta, and gamma radiation.

3. Results and Discussion

The decontamination process with strippable coatings involves the following three steps, presented in Scheme 1: (1) The decontamination solution is applied (poured or sprayed) on the surface contaminated with the "toxic metal" (heavy metal or radioactive metal). For exemplification, Scheme 1 describes contamination with cesium, because ^{137}Cs is one of the radionuclides investigated in this study, but it could be replaced with any other metallic contaminant. The decontamination reaction occurs at the interface between the aqueous decontamination solution and the contaminated metallic surface. If this solution maintains its liquid state, contaminant ions are continuously extracted from the moistened metallic surface into the decontamination solution through complexation and adsorption mechanisms. As the solvent (water) is lost or the polymers undergo crosslinking, the metallic contaminants are entrapped inside the matrix of the polymeric nanocomposite. (2) After stage (1) is complete (less than 24 h), the dry nanocomposite film containing the contaminant can be easily peeled off from the metallic surface. The strong interactions between the components of these materials ensure the formation of a compact polymeric nanocomposite film which maintains its integrity when it is detached from the surface. (3) When the peeling process ends, the metallic surface is considered successfully decontaminated. The strippable coating containing the "toxic metals" should be compacted and sealed in a small container towards final disposal. According to legislation requirements, these materials should be disposed or incinerated as low-level waste [31], but their greatest advantage is represented by the small volume of material generated post-decontamination.

The schematical illustration (Scheme 1) presents only the main mechanisms of action of the active components from our decontamination solution: complexation (by chelating agents) and physical interactions (specific interactions between nano-clay and contaminant cations). Bentonite clay possesses five types of adsorption sites [32]: basal surface site (planar site), interlayer site, hydrated interlayer site, edge site, and the frayed edge site (FES) and their existence allows metal contaminants adsorption through distinct routes [32–34], part of them drawn in Scheme 1. The chelating effect of SA is also a factor which influences decontamination performances, but since it did not bring any outstanding improvements, it was intentionally omitted from this scheme, for the ease of visualization.

The nanocomposite strippable coatings were obtained through casting method, as described in Methods Section 2.2. Figure 1 illustrates the casting process. As can be seen in Figure 1a, the decontamination solution can be easily transferred on the contaminated surface. Depending on the surface type and position, these solutions can be simply poured onto the surface, applied with a roller or a brush, or they can be applied by spraying technique. After being deposited on the targeted area, the decontamination process begins. The decontamination solution is allowed to dry and after the complete evaporation of the solvent (water), the nanocomposite film can be peeled off (Figure 1b), and thus the surface is decontaminated. These decontamination formulations are customizable, because the active ingredients of these decontamination solutions can be selected depending on the targeted contaminating agent (chemical, biological, radiological, or nuclear contaminants). In this study, "green" chelating agents and sodium alginate were employed, and the aim was "toxic metals" (heavy metals and radioactive metals) removal.

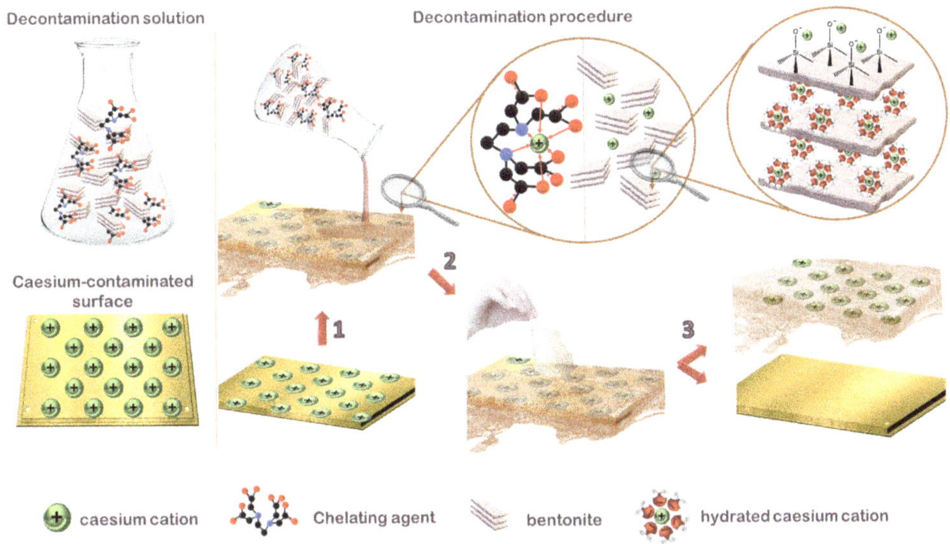

Scheme 1. Schematic illustration of decontamination process: (**1**) application of the decontamination solution on the contaminated surface; (**2**) decontamination process (exemplified by complexation or adsorption mechanisms); (**3**) removal of the contaminants entrapped into the nanocomposite film.

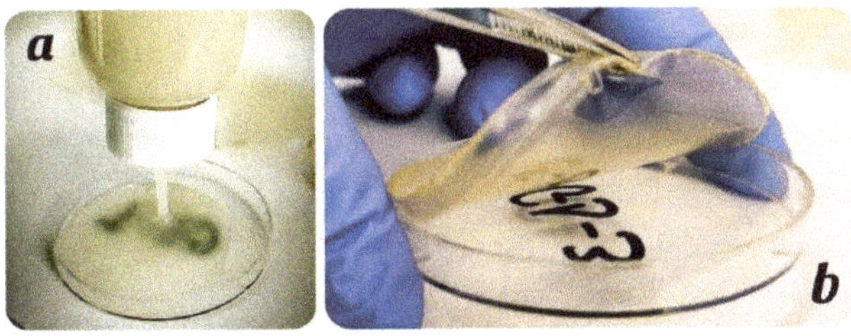

Figure 1. Casting process: (**a**) decontamination solution is transferred on the contaminated surface; (**b**) the surface is decontaminated after film removal.

The composition of the decontamination solutions and the interactions that occur between the components have a significant influence on the performances of the strippable coatings obtained. Thus, various decontamination formulations comprising different SA concentrations (Table 1) were developed to establish which composition of the polymeric matrix is adequate for this type of application.

The FT-IR spectra, illustrated in Figure 2, display the variation of the characteristic peaks of the decontamination strippable coatings with the increase in SA concentration.

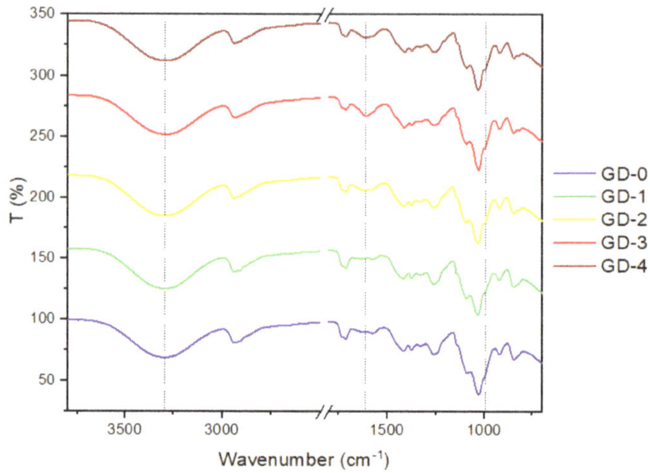

Figure 2. FT-IR spectra of decontamination strippable coatings.

The O–H stretching vibrations at 3293 cm^{-1} were slightly shifted to 3288 cm^{-1}, probably due to the supplementary hydrogen bonds brought by increasing the amount of SA added in the decontamination solutions [35]. By increasing the concentration of SA, the two peaks associated with the asymmetric O–C–O stretching vibrations at 1630 cm^{-1} and 1575 cm^{-1} merged into a singular peak visible at 1610 cm^{-1}, confirming that the interaction between SA and the other components of the decontamination films occurs also trough the carboxylic groups. The intensity of the characteristic peak for bentonite (attributed to Si–O stretching vibrations visible at 1029 cm^{-1}) decreased with the addition of SA. A clearer indication of the interaction between the Si-OH group from the nanoclay and SA can be highlighted through a more visible peak, at 994 cm^{-1}, also given by the Si–O stretching vibrations. The specific peak for mannuronic acid sequence of SA was visible at 820 cm^{-1}, being slightly shifted in comparison with pure SA, due to the interactions established between SA and the other components, mainly hydrogen bonds formation. The peak at 850 is due to Si–O–Si vibrations, also slightly shifted in comparison with pure bentonite spectra.

The thermal and mechanical properties of the decontamination strippable coatings obtained by employing different concentrations of SA were investigated to establish the optimal composition for these strippable coatings designed for surface decontamination. Figure 3 illustrates the results of the TGA analysis performed on the synthesized materials for the assessment of their thermal stability.

As can it be observed in Figure 3a, all the strippable films exhibited a first weight-loss of approximately 8% (up to approximately 140 °C), probably due to the loss of the residual water in their composition. Additionally, the results confirm the thermal stability of the polymeric nanocomposite films up to 140–150 °C, after which a second weight-loss stage is registered varying from 15% (GD-0, GD-1, and GD-2) to 7% (GD-3 and GD-4). Therefore, the addition of SA led to a smaller weight-loss for the material in this temperature range. As it can be noticed in Figure 3b, the peak at 220 °C, assigned to the third weight loss step, is slightly shifted to higher temperatures in the case of GD-2, GD-3, and GD-4 samples; thus, the increase in SA concentration leads to a slightly increased thermal stability for the samples, due to the higher number of hydrogen bonds established between SA and PVA [36].

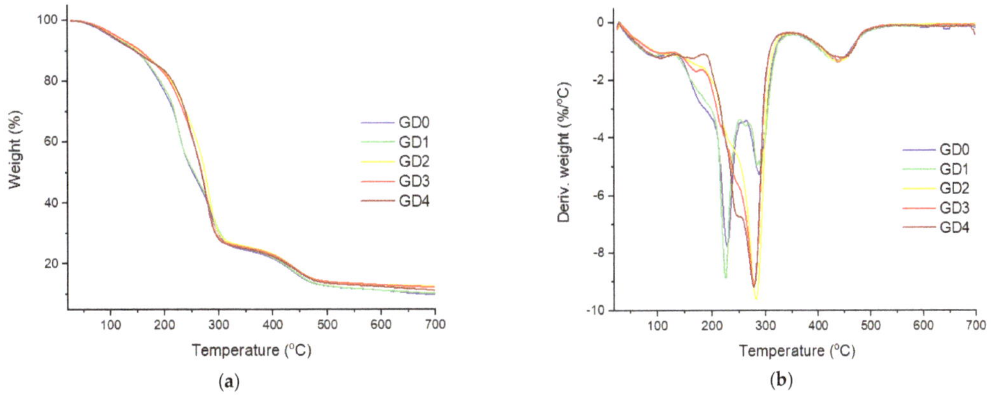

Figure 3. TGA (**a**) and DTG (**b**) plots of decontamination strippable coatings.

Table 2 summarizes the results obtained from tensile tests. According to these results, the strippable films containing SA possess a higher elasticity. Figure 4 describes the correlation between the SA concentration and the mechanical properties of the nanocomposite strippable films. The value of the elastic modulus ($\lambda = \sigma/\varepsilon$) increases with the addition of SA, up to a concentration of approximately 0.8% SA, and then starts to decrease. Considering these results, an optimum amount of SA (approximatively 0.7–0.8%) permits the reduction in PVA concentration, thus affording decontamination strippable coatings with a lower environmental and superior mechanical and thermal properties.

Table 2. Mechanical properties nanocomposite films.

Sample ID	Measured Values		Calculated Values	
	L_b (mm)	F_m (N)	$\sigma = TS_b$ (MPa)	$\varepsilon = E_b$ (%)
GD-0	182.1 ± 26.6	23.2 ± 3.5	3.6 ± 0.5	363.8 ± 53.1
GD-1	135.6 ± 20.3	28.5 ± 2.7	11.5 ± 1.3	271.2 ± 40.6
GD-2	55.9 ± 25.4	22.3 ± 1.9	8.4 ± 1.1	111.8 ± 50.8
GD-3	39.0 ± 6.5	33.6 ± 0.7	9.9 ± 1.0	82.1 ± 12.9
GD-4	29.5 ± 5.5	18.7 ± 5.9	6.9 ± 1.6	58.9 ± 5.5

F_m is the maximum force recorded. Tensile stress, σ, was calculated, according to ISO37:2011(E), taking into account **W** (the average width of the narrow part of the specimen, mm) and **t** (the average thickness, mm): $\sigma = F_m/(W \cdot t)$. Tensile strain, ε, was calculated, according to ISO37:2011(E), taking into account $\mathbf{L_b}$ (length at breaking, mm) and $\mathbf{L_0}$ (the initial length, mm): $\varepsilon = 100 \cdot (L_b - L_0)/L_0$. TS_b is the tensile strength at break (tensile stress recorded at the moment of rupture). E_b stands for elongation at break. The average values for each sample were reported.

Thus, based on the above-mentioned analyses, it can be observed that GD-3 possess remarkable thermal properties and adequate mechanical properties for this type of application. Therefore, the solution containing 0.75% Na-Alginate (GD-3) was chosen as the basic polymeric matrix for the decontamination solutions containing also complexing agents. Thus, starting from this point, the following tests were performed only on the blank solution (GD-0), the solution with optimal level of alginate (GD-3), and the solutions comprising both alginate and chelating agents (GD-3-PBTC and GD-3-IDS).

The investigation of viscoelastic behavior is important for a proper design of the strippable coatings. Considering the results obtained from the tensile tests, which concluded that GD-3 displayed the best mechanical characteristics, four types of polymeric composites were selected to be subjected to DMA analysis: GD-0, GD-3, GD-3-PBTC, and GD-3-IDS. Figure 5 illustrates the comparative graph between these four types of decontamination polymeric composite films.

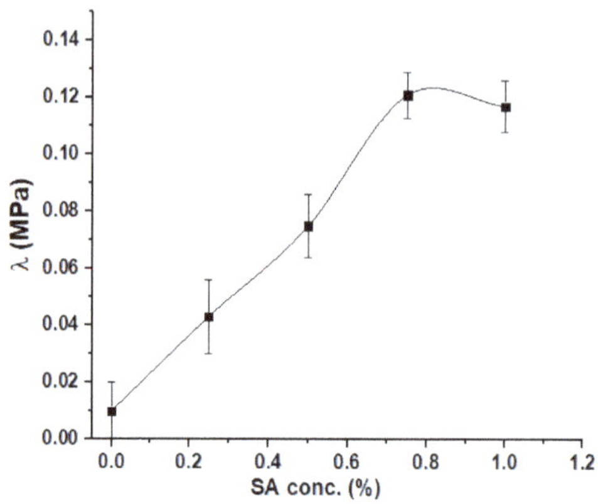

Figure 4. Correlation between the SA concentration and the mechanical properties of the films.

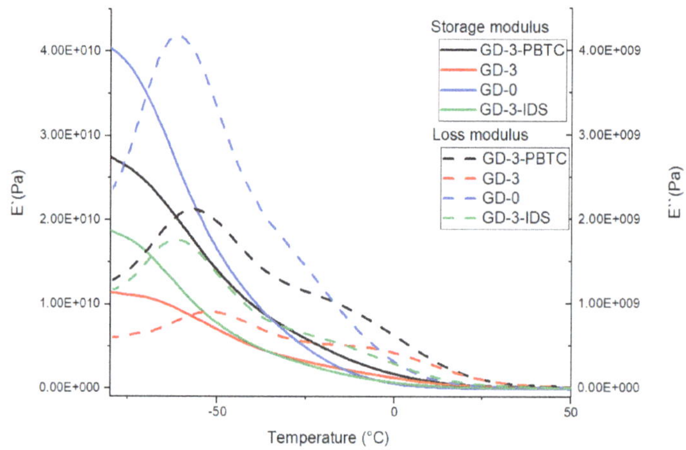

Figure 5. Loss and storage modulus of strippable coatings.

DMA results revealed distinct viscoelastic behavior of the cured strippable coatings depending on the chelating agents employed in the decontamination solutions. An adequate viscoelasticity is important for the peeling process of these coatings because their components should facilitate the final removal of the polymeric film from the decontaminated surface, while ensuring a high decontamination performance. When a stress is applied to a viscoelastic material [37] such as our decontamination coatings, during the peeling process, some parts of the long polymer chain change their position and when the stress is taken away, namely when the coating is completely peeled off, the accumulated back stresses will bring the polymeric nanocomposite film to its original shape, because, theoretically, the polymer chains will return to their initial position. When a stress is applied, the strippable coatings possessing higher viscoelasticity (higher storage and loss modulus values) will undergo a more significant molecular rearrangement. Whereas elasticity (described by the variation of storage modulus) is related to bond stretching, viscosity (described by the variation of loss modulus) is related to the diffusion of molecules inside an amorphous material [38]. From Figure 5, we can conclude that storage and loss modulus decrease in the following order: GD-0 > GD-3-PBTC > GD-3-IDS > GD-3, probably due to the different number of hydrogen bonds established inside the polymeric nanocomposite and depending on different strengths of these hydrogen bonds. The formed hydrogen-bonds network is responsible for the mechanical and structural properties of a material. Therefore, according to the values obtained for storage and loss modulus, it can be noticed that the strippable coatings without alginate and chelating agent exhibited the highest viscoelasticity, the one with alginate displayed the lowest viscoelasticity, while the ones containing IDS and PBTC display improved qualities compared to GD-3 [39].

The thermal properties of the polymeric nanocomposite films employed for decontamination were determined using the DMA technique because it can detect molecular relaxations, such as glass transition temperatures (Tg), with a higher accuracy than DSC or DTA [40,41]. According to the tan delta (δ) plots (Figure 6), the strippable coatings displayed an apparent double glass transition temperature (Table 3). This property was predictable because neat PVA (the polymer matrix employed in these decontamination nanocomposite films) shows typical behavior of a semi-crystalline polymer with two transition regions as well: $T_{g1} \approx$ 30–60 °C and $T_{g2} \approx$ 80–150 °C [39,42]. Our PVA nanocomposite peelable films possess significantly lower glass transition values due to their higher flexibility given by glycerol and the other additives. Therefore, the lower temperature amorphous glass transition, T_{g1}, varies from −61.8 °C to −51.6 °C and could be associated with the glass-rubber transition of the amorphous phase of the polyvinyl alcohol composites while the higher amorphous glass transitions, T_{g2}, varies from 5.7 °C to 14.6 °C and could be assigned to the relaxation of the PVA crystalline domains, representing the temperature range where amorphous, rubbery, and crystalline domains coexist [39,42,43]. Alginate and the chelating agents employed in these decontamination coatings also influence glass transition due to the interactions with the polymeric chain. The hydrogen bonds established between the hydroxyl groups of PVA and the other components of the strippable coating, influence the polymer chains mobility; therefore, both glass transition temperatures will be different for each strippable coating, depending on their composition, as can be observed in Table 3.

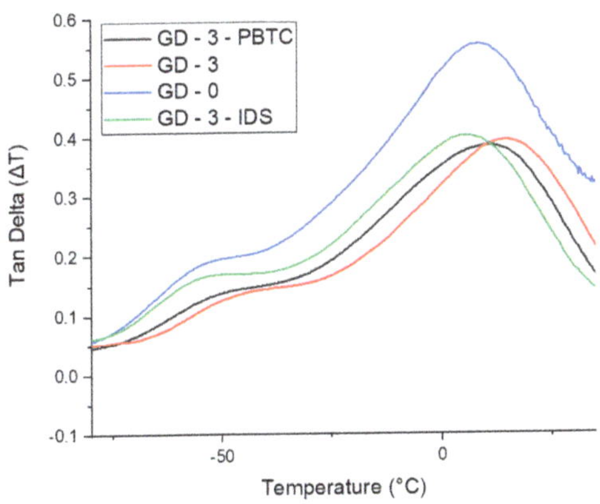

Figure 6. Tan Delta of strippable coatings.

Table 3. Characteristic temperatures * for the polymeric films.

Name	T_{g1} (°C)	T_{g2} (°C)
GD-3	−51.6	14.6
GD-3-IDS	−61.8	5.7
GD-3-PBTC	−56.7	10.3
GD-0	−61.5	8.7

* Tg_1 and Tg_2 correspond to the maximum value of each of the two tan delta peaks.

After characterizing the peelable coatings employed for decontamination, the next objectives consisted of the evaluation of the decontamination efficiency. This goal was accomplished through various analytical techniques: SEM-EDS, AAS, and radiation measurements.

The morphological evaluation of the targeted surfaces, prior- and post-decontamination, offered a preview on the decontamination efficiency through the qualitative information offered by this technique. Since decontamination performances also depend on the type of surface that requires decontamination, SEM/EDS analyses were carried out to observe the contaminant behavior on three types of stainless-steel surfaces: mirror-finish (SSMF), grinded-finish (SSGF), and etched-finish (SSEF). SEM analyses are made in a controlled environment, which cannot be contaminated with radioactive materials; thus, for these analyses, we employed ^{133}Cs as simulant for ^{137}Cs. For the controlled contamination of the above-described surfaces, a 0.005 M Cs_2SO_4 aqueous solution was utilized. The dispersion-pattern of the contaminant on each surface depends on the roughness of the employed material. As shown in Figure 7a, when applied to a mirror-finish metallic surface, contaminant droplets tend to clump in one place. The molecules of these aqueous droplets are held together by strong cohesive forces that are intermolecular attractive forces, which leads to the formation of large crystals that exhibit low-adhesion forces towards the mirror-finish metallic surface; therefore, they were easily removed using strippable coatings method. When surface roughness increases, the contaminant droplets tend to spread much more along the axes of the cracks: linear direction on grinded-finish stainless steel (Figure 7b), and dendritic direction on etched-finished stainless steel (Figure 7c), thus penetrating the interstices of the material.

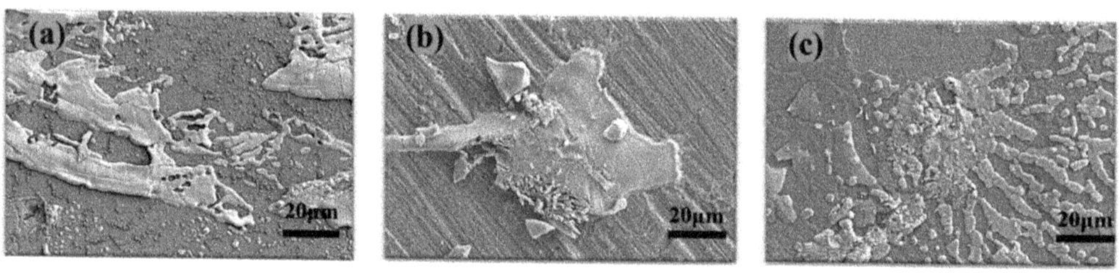

Figure 7. SEM images of cesium surface contamination patterns on different types of surfaces (before decontamination): (**a**) mirror-finish stainless steel, (**b**) grinded-finish stainless steel, and (**c**) etched-finish stainless steel.

For the controlled contamination process, we tried to reproduce cesium contamination at levels representative of those found in the nuclear industry, namely we started with 5 mM (C1) of Cs_2SO_4 aqueous solution, a similar concentration with those mentioned in the literature [7,44]. After being subjected to the controlled contamination process, stainless-steel samples were further analyzed by the EDS technique, the mean values, and standard deviations for the determined and calculated values are reported in Table 4. The results obtained through SEM-EDS technique for the metallic samples subjected to controlled "artificial in-depth contamination", with C1 at 700 °C, did not render major differences in terms of decontamination factors obtained, since cesium could not be detected on the tested surface after decontamination. The only difference, when compared to superficial contamination with C1, consisted of the aspect of the contaminated surface after being subjected to the thermal treatment and the visible solid residues entrapped in the polymeric films (Figure S2 Supplementary Material). Since this attempt led to quite similar results with the samples subjected to "superficial" contamination, only the results obtained for superficial contamination are worth to be further detailed.

As expected, the relative atomic concentration (at.%) [45] was higher on mirror-finish surfaces where the droplets tended to agglomerate, and was lower in the cases of grinded-finish and etched-finish surfaces, according to the increasing number of surface imperfections, as, in these cases, the contaminant spread on the higher surface while also entering the pores and cracks of the metallic sheets, leading, at the same time, to in-depth contamination.

Table 4. EDS analysis of the control sample.

No.	Type of Surface	Cs (at. %) *	Cs (wt.%) *	Cs Abs. Error (%) **
1	SSMF	15.23	58.48	2.42
2	SSGF	10.12	40.22	0.83
3	SSEF	3.74	15.74	0.3

* Detailed in Supplementary Material in Figures S3–S5 and Tables S2–S4, respectively. The values reported for wt% and at% were calculated using the Bruker ESPRIT QUANTAX software of the scanning electron microscope.
** Abs. Errors (%) were calculate by Bruker ESPRIT QUANTAX software at 3 sigma for a single determination.

After the controlled contamination step, the decontamination solutions were poured over the contaminated surfaces and allowed to dry. All four types of decontamination solutions (GD-0, GD-3, GD-3-PBTC, and GD-3-IDS) were tested on mirror-finish, grinded-finish, and etched-finish stainless steel samples. Figure 8 illustrates the stainless-steel surfaces after the peeling process. Even though some polymer traces, probably resulting from the peeling process, are slightly observable, cesium crystals were not visible anymore. Thus, this information can serve as a screening method for the preliminary performance evaluation of the strippable films, thus confirming that all the four strippable coatings can be considered efficient for this type of surfaces, because the amount of residual cesium was undetectable after decontamination. Even if supplementary investigations were performed to prove and quantify the decontamination efficiency, SEM-EDS analysis offered, however, a reasonable preview of decontaminated surfaces.

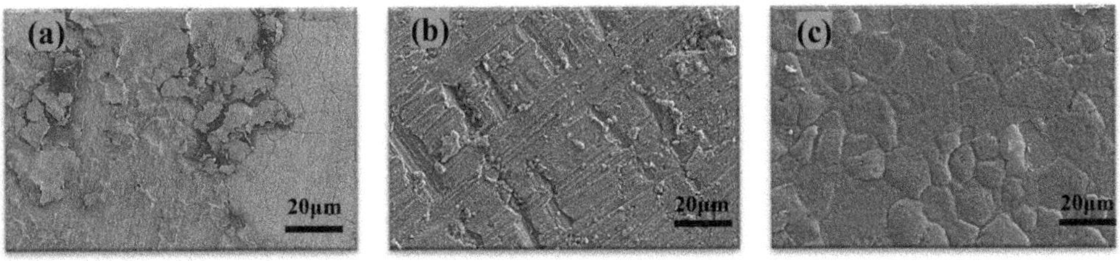

Figure 8. SEM images of surfaces (after decontamination): (**a**) mirror-finish stainless steel, (**b**) grinded-finish stainless steel and, (**c**) etched-finish stainless steel. (These surfaces were contaminated with 0.005 M Cs_2SO_4 (C1) solution and decontaminated with GD-3-IDS.)

SEM-EDS analysis was also performed at higher contamination levels compared to those found in the literature, for a rough estimation of the upper limit of Cs retention in the polymeric coatings. For this purpose, two supplementary concentrated Cs_2SO_4 aqueous solutions (0.05 M (**C2**) and 0.5 M (**C3**)) were employed to find a correlation between the amount of Cs on the metallic surface and the upper limit of Cs retention in the polymeric coatings. The behavior of these strippable coatings on other types of metallic surfaces was also evaluated using galvanized metal (GM), brass (BS), and cooper substrates (CS). These three types of metallic surfaces (GM, BS, and CS) were contaminated with 0.005 M (**C1**), 0.05 M (**C2**), and 0.5 M (**C3**) Cs_2SO_4 aqueous solutions. Contamination and decontamination were performed according to the same procedure (described for **C1**) for all samples, but different amounts of residual cesium were still detected on some of the metallic surfaces, as can be observed in Table 5. Successful decontamination results were obtained for the metallic coupons contaminated with the lowest concentration of Cs (**C1**). For **C1**, after peeling the strippable decontamination coatings, undetectable amounts of residual cesium or a maximum of 0.66 Cs (at. %) for SC2-PBTC were measured for GMS. When the level of contamination was increased, using 0.05 M (**C2**) and 0.5 M (**C3**) Cs_2SO_4, decontamination efficiency seemed to decrease. From Table 5, it can be noticed that after decontamination, residual Cs levels increased, according to the contamination level (C3 > C2 > C1), but the residual levels are also influenced by the type of the surface and the decontamination solution employed. Even so, for the usual contamination levels (**C1**) found in the literature for the nuclear industry, GD-3-IDS displayed the best decontamination performances, for all the tested surfaces; because in this case, the residual cesium was undetectable after decontamination. Figure 9 showed that employing different types of metallic surfaces led to Cs_2SO_4 crystals with distinct shapes, depending on the interaction with these metals. Therefore, on brass surfaces, the crystals have an elongated hexagonal shape, and on galvanized metal and copper surfaces, the crystals have a dendritic shape. When comparing the SEM images from Figure 9 with the ones illustrated in Figure 8, corelated with the data from Table 5, it can be noticed that as expected, the polymeric nanocomposite films become less efficient at higher levels of contamination (concentrations higher than 0.005 M). Thus, if this extreme situation were to be encountered in real contamination circumstances, several repeated decontamination procedures would be required for the complete removal of the contaminant. Another solution could be to increase the concentration of the chelating agent, but this might induce modifications in the properties of the strippable nanocomposite film, which is not desirable because it might reduce its performance.

Table 5. EDS—comparative evaluation of the remnant contaminant after decontamination (means not detectable).

				GD-3-IDS					
Concentration		**C1**			**C2**			**C3**	
Surface	**GM**	**BS**	**CS**	**GM**	**BS**	**CS**	**GM**	**BS**	**CS**
Cs (at. %) experimental	-	-	-	2.97 ± 0.36	1.29 ± 0.21	3.01 ± 0.15	9.47 ± 0.46	4.96 ± 0.36	5.22 ± 0.35
Cs (wt.%) calculated	-	-	-	10.83	8.37	6.09	37.23	28.98	30.06
				GD-3-PBTC					
Concentration		C1			C2			C3	
Surface	GM	BS	CS	GM	BS	CS	GM	BS	CS
Cs (at. %) experimental	0.66 ± 0.12	-	-	2.46 ± 0.28	2.67 ± 0.19	3.95 ± 0.18	4.79 ± 0.37	6.45 ± 0.34	9.86 ± 0.42
Cs (wt.%) calculated	2	-	-	8.97	9.29	9.09	18.25	19.68	40.6

It is useful to reiterate that the SEM-EDS method was employed only as screening method, the data obtained being useful only as preliminary results and complementary information for further investigations. Supplementary analyses with higher accuracy, which offer the possibility of better quantifying the contamination/decontamination levels, calculated according to AEP-58 NATO standard, were further employed (AAS technique and measurement of the activity of the radioactive materials) and are described in the following section of the study.

Decontamination Tests

Heavy metal uptake capacity of the strippable coating was evaluated by applying the decontamination solutions on glass surfaces contaminated with lead (Pb), strontium (Sr), and cobalt (Co), employed in this experiment as simulants for their radioactive analogues, were also subjected to the same procedure of analysis. Atomic adsorption spectrometry (AAS) was utilized to assess the concentration of "toxic metals" before and after decontamination. The data obtained from AAS analysis were utilized to calculate the decontamination efficacy of each solution, for every metal. The decontamination factor, $DF = 100\,(C_0 - C_f)/C_0$ (DF is the decontamination factor, C_0 is the initial metal concentration, and C_f is the final concentration) was calculated. Figure 10 illustrates a comparative plot comprising the results obtained for glass surfaces contaminated with Pb, Sr, and Co.

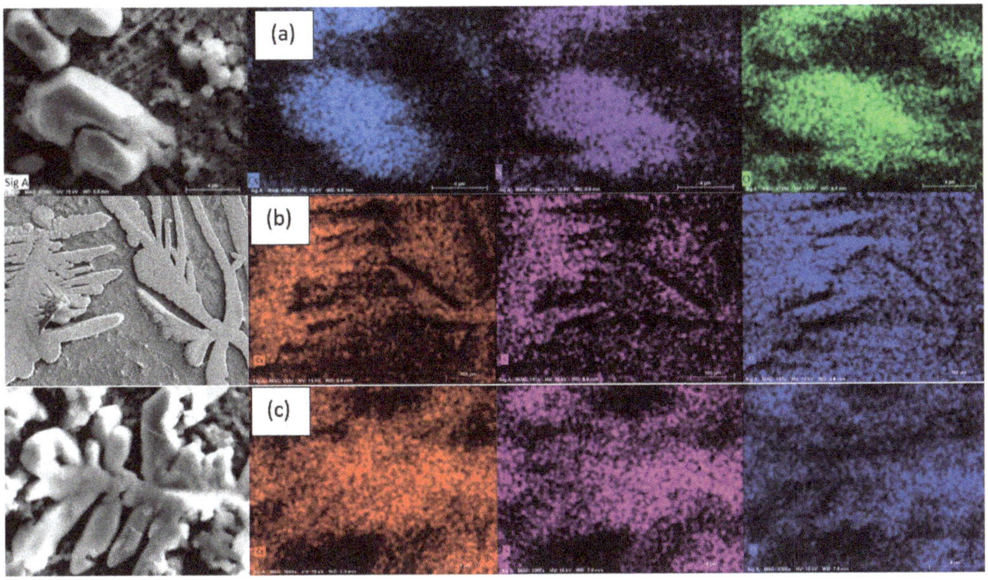

Figure 9. Exemplification of SEM images after decontamination and EDS elemental mapping of Cs, S, and O (the metallic surfaces: (**a**) brass surface (BS), (**b**) galvanized metal surface (GM), and (**c**) copper surface (CS) contaminated with 0.05 M Cs_2SO_4 (C2) solution and decontaminated with GD-3-IDS.

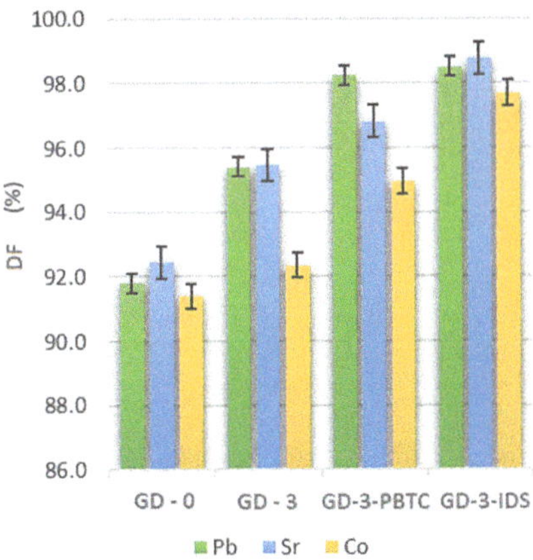

Figure 10. Efficacy of the strippable coatings for toxic metals removal illustrated (DF calculus based on the concentration of metal before and after decontamination).

For each metal tested, different decontamination degrees were reached, depending on the intensity of the interaction between the targeted metal and the surface on which it was deposited and the interaction with the components of the decontamination solutions. As it can be observed in Figure 10, the reference sample (Bk0), exhibited lower decontamination efficiency for all the three metals tested, because, in this case, metal uptake involved only the physical interactions established between the nanoclay adsorbent and the contaminant.

The metallic contaminant was entrapped and fixed in the matrix of the nanocomposite, and it was removed along with the exfoliation of the dry nanocomposite film. The influence of SA on the decontamination efficiency can be deduced from the values obtained and presented in Figure 10 with Alg label attributed to GD-3 decontamination solution. The chelating ability of sodium alginate significantly improved the decontamination performances, obtaining higher DF for all the metals tested. The "green chelates", IDS and PBTC had a significant positive influence on the decontamination efficacy of the solutions, confirmed by the results presented in Figure 10 (Alg-PBTC refers to GD-3-PBTC and Alg-IDS refers to GD-3-IDS). These results showed that IDS led to higher decontamination factors (as anticipated from the preliminary data obtained from SEM-EDS analysis). Therefore, these experiments proved the influence of each of the components on the performances of the decontamination solutions, showing that GD-3-IDS, especially, can be successfully employed for the removal of heavy metals from the contaminated surfaces.

Figure 11 offers information about the presence and abundance of the metals in the nanocomposite coatings after decontamination. However, because the process of obtaining clear AAS solutions required intermediary steps in this case (described in Methods Section 2.2), we cannot retrieve the expected concentration of metal in the values obtained, probably because a part of the metallic contaminant was fixed by bentonite. Even so, most of the contamination can be retrieved in Figure 11, while the results summarized in Figure 10 showed that on the decontaminated surface the toxic metals were only found as traces.

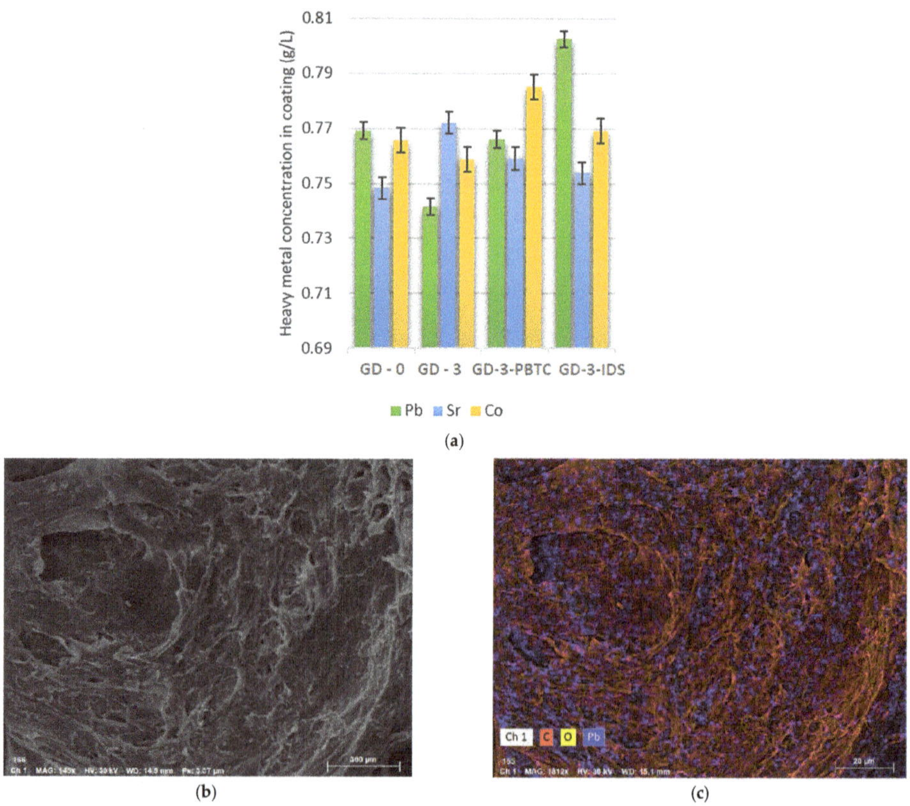

Figure 11. (**a**) Concentration of toxic metals found in the peeled nanocomposite film after decontamination; (**b**) SEM image of the nanocomposite film after decontamination of lead; (**c**) EDS image displaying C, O, and Pb elements present in the film obtained after decontamination of lead.

The most relevant tests in this study are represented by the decontamination investigations performed on radioactive materials. For this purpose, we used five different types of surfaces (metal-M, painted metal-PM, plastic-P, glass-G, and CBRN protective material—BC/SP2) and three radioactive solutions to generate alpha (^{241}Am), beta (^{90}(Sr-Y)), and gamma (^{137}Cs) contamination. The decontamination solution employed for these experiments was GD-3-IDS, due to its remarkable efficiency, sustained by the previous results obtained in this study. The decontamination method herein described is in accordance with AEP-58 NATO standard. This military standard also establishes the minimal requirements for a new decontamination method to be considered efficient: the decontamination factor must have a value higher than 90% [4]. The results depicted in Figure 12 demonstrate that this criterion was successfully reached by GD-3-IDS decontamination solution. As can be seen, all the decontamination factors obtained indicated values above 90%, thus demonstrating that GD-3-IDS polymeric nanocomposite coatings efficiently removed radioactive contamination.

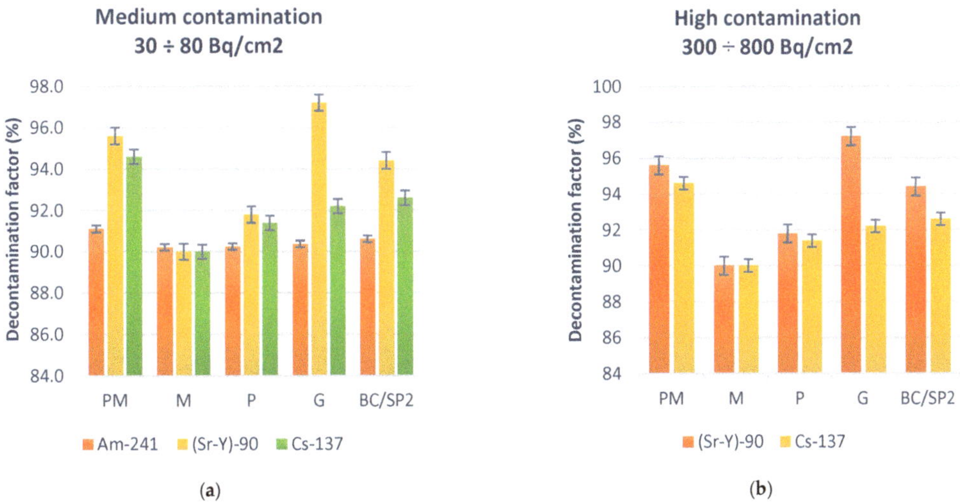

Figure 12. Decontamination factors obtained for GD-3-IDS decontamination solution in two distinct scenarios: (**a**) medium contamination and (**b**) high contamination (PM—painted metal; M—metal; P—plastic; G—glass; and BC/SP2—CBRN protective material).

In the first scenario described in Figure 12 implied a medium contamination level of 30 ÷ 80 Bq/cm^2 while the second scenario implied a high contamination level of 300 ÷ 800 Bq/cm^2. ^{241}Am was tested only in medium contamination scenario, due to safety reasons. From Figure 12, it can be noted that decontamination efficiency of GD-3-IDS varies with the type of targeted surface, probably due to the different adherence of the polymeric matrix to each substrate and with the type of radionuclide employed for contamination. This can be explained by the different interactions established between the contaminant and the contaminated surface, and between the contaminant and the components of the decontamination solution (GD-3-IDS).

The advanced performances of the decontamination solutions developed in this study are sustained especially by the high contamination scenario, because even at this high contamination level, GD-3-IDS still managed to reach the 90%DF imposed by NATO standards. Therefore, even if the values obtained for DF were lower in high contamination than in the case of medium contamination, the strippable nanocomposite films can still be considered efficient in removing ^{241}Am, ^{90}(Sr-Y), and ^{137}Cs contaminants.

4. Conclusions

Novel environmentally friendly surface decontamination solutions were successfully prepared based on PVA/SA/GLY/BT/IDS or PBTC aqueous mixtures. After deposition and during drying the solutions were capable to entrap heavy metals and radionuclide contaminants using physical and chemical processes by generating a composite polymeric film with good wetting capability on a large variety of solid surfaces displaying peeling-off capacity.

Chemical, mechanical, and thermal characterization of the polymeric films were performed using FTIR, tensile tests, TGA, and DMA techniques showing the influence of each component and allowing further optimization and selection of the formula for decontamination performance tests.

Controlled contaminations were performed on various surfaces metals (with different finishes), painted metal, plastic, glass, and CBRN protective material, according to NATO standard AEP-58 and using typical concentrations (or above) for contaminated sites found in the literature.

The decontamination effectiveness was first evaluated in a qualitative manner using SEM-EDS techniques followed by a quantitative approach employing AAS and surface activity measurements on live radioactive agents. The influence of SA, IDS, PBTC concentrations, and surface type over the DF was also emphasized. The presence of sodium alginate, and especially of chelating agents IDS/PBTC, decisively improves the decontamination factor.

Solution GD-3-IDS containing 5% PVA, 1% BT nano-clay, 2.5% glycerol, 0.75% sodium alginate, and 1% IDS chelating agent showed the best results as having decontamination factors that overpassed the DF imposed by NATO standards [4], DF being $\geq$90% for all surfaces tested and for the highest contaminations.

Supplementary Materials: The following are available online at https://www.mdpi.com/article/10.3390/polym13234194/s1. Figure S1—Different types of "polishing" for the stainless-steel coupons employed for decontamination tests prior to contamination step; Figure S2—Contamination (a) immersion in Cs_2SO_4 aqueous solution (C1 = 0.005 M); (b) "in-depth" contamination at 700 °C; (c) metallic coupons after in-depth decontamination; and Decontamination processes (d) applying one of the decontamination solutions; (e) peeled film; (f) different types of metallic samples prepared for SEM-EDS analysis after decontamination; Table S1—AAS instrumental parameters; Figure S3—EDS spectrum for SSMF contaminated with Cs; Table S2—Values obtained from EDS analysis for SSMF contaminated with Cs; Figure S4—EDS spectrum for SSGF contaminated with Cs; Table S3—Values obtained from EDS analysis for SSGF contaminated with Cs; Figure S5—EDS spectrum for SSEF contaminated with Cs; Table S4—Values obtained from EDS analysis for SSEF contaminated with Cs.

Author Contributions: Conceptualization, G.T., D.P., A.D., T.R., E.R. and R.G.; methodology, G.T., D.P., L.L. and T.R.; software, A.M., D.P., B.P., F.D., O.I. and M.U.; validation, G.T., D.P., A.D., T.R. and E.R.; formal analysis, G.T., D.P., A.D., A.M., R.G., A.P., A.M., B.P., L.L., F.A., S.A.B. and O.I.; investigation, G.T., D.P., A.D., A.M., R.G., F.A., S.A.B., L.L., F.D., A.P., A.M. and B.P.; resources, G.T., T.R., A.M., A.D., L.L. and R.G.; data curation, G.T., D.P., F.D., M.U., O.I., A.P., A.M., B.P. and F.A.; writing—original draft preparation, G.T., D.P., T.R., A.D., E.R. and R.G.; writing—review and editing, G.T., D.P., T.R., A.M., A.D., E.R., A.P., R.G., F.A., O.I., S.A.B., B.P., M.U., F.D. and L.L.; visualization, G.T., D.P., R.G., A.M. and F.D.; supervision, G.T., A.D., T.R., L.L. and E.R.; project administration, G.T., A.M., R.G., A.D., L.L. and T.R.; funding acquisition, G.T., A.M., R.G., L.L., A.D. and T.R. All authors have read and agreed to the published version of the manuscript.

Funding: This work was supported by a grant of the Ministry of Research, Innovation and Digitization, CNCS/CCCDI—UEFISCDI through the National Project PN-III-P2–2.1-PTE-2019–0400, ctr. No. 49PTE/2020. L.L. would like to acknowledge the financial support of the Romanian Ministry of Research and Innovation and UEFISCDI through the Project PN-III-P2-2.1-PED-2019-1411, and Core Program PN19-03 (contract no. 21 N/08.02.2019).

Institutional Review Board Statement: Not applicable.

Informed Consent Statement: Not applicable.

Data Availability Statement: The data presented in this study are available on request from the corresponding author.

Acknowledgments: A.D. gratefully acknowledges financial support from the Competitiveness Operational Program 2014–2020, Action 1.1.3: Creating synergies with RDI actions of the EU's HORIZON 2020 framework program and other international RDI programs, MySMIS Code 108792, Acronym project "UPB4H", financed by contract: 250/11.05.2020. Authors are also grateful to Lanxess for donating the complexing agent BAYPURE® CX 100 and BAYHIBIT®. The article processing charges were supported by the Military Technical Academy "Ferdinand I".

Conflicts of Interest: The authors declare no conflict of interest.

References

1. Toader, G.; Stănescu, P.-O.; Zecheru, T.; Rotariu, T.; El-Ghayoury, A.; Teodorescu, M. Water-based strippable coatings containing bentonite clay for heavy metal surface decontamination. *Arab. J. Chem.* **2019**, *12*, 4026–4034. [CrossRef]
2. Pulpea, D.; Rotariu, T.; Toader, G.; Pulpea, G.B.; Neculae, V.; Teodorescu, M. Decontamination of radioactive hazardous materials by using novel biodegradable strippable coatings and new generation complexing agents. *Chemosphere* **2020**, *258*, 127227. [CrossRef]
3. Whicker, F.W.; Eisenbud, M.; Gesell, T. *Environmental Radioactivity from Natural, Industrial and Military Sources*, 4th ed.; Academic Press: Cambridge, MA, USA, 1997.
4. AEP. *AEP-58. Combined Operational Characteristics, Technical Specifications, Test Procedures And Evaluation Criteria For Chemical, Biological, Radiological And Nuclear Decontamination Equipment*; NATO Standard: Brussels, Belgium, 2013.
5. Gray, H.N.; Jorgensen, B.; McClaugherty, D.L.; Kippenberger, A. Smart Polymeric Coatings for Surface Decontamination. *Ind. Eng. Chem. Res.* **2001**, *40*, 3540–3546. [CrossRef]
6. Ying, L.; Li, N. Chapter 8. In *Membrane-Based Separations in Metallurgy*; Jiang, L.Y., Li, N., Eds.; Elsevier: Amsterdam, The Netherlands, 2017; pp. 391–402.
7. Lang, A.R.; Engelberg, D.; Walther, C.; Weiss, M.; Bosco, H.; Jenkins, A.; Livens, F.R.; Law, G.T.W. Cesium and Strontium Contamination of Nuclear Plant Stainless Steel: Implications for Decommissioning and Waste Minimization. *ACS Omega* **2019**, *4*, 14420–14429. [CrossRef]
8. Aono, H.; Takahashi, R.; Itagaki, Y.; Johan, E.; Matsue, N. Cs immobilization using the formation of the glassy phase by the heat-treatment of natural mordenite. *J. Nucl. Mater.* **2018**, *508*, 20–25. [CrossRef]
9. Claverie, M.; Garcia, J.; Prevost, T.; Brendlé, J.; Limousy, L. Inorganic and Hybrid (Organic–Inorganic) Lamellar Materials for Heavy metals and Radionuclides Capture in Energy Wastes Management—A Review. *Materials* **2019**, *12*, 1399. [CrossRef]
10. Yang, H.-M.; Park, C.W.; Lee, K.-W. Polymeric coatings for surface decontamination and ecofriendly volume reduction of radioactive waste after use. *Prog. Nucl. Energy* **2018**, *104*, 67–74. [CrossRef]
11. Kinnunen, P. *ANTIOXI—Decontamination Techniques for Activity Removal in Nuclear Environments*; VTT Technical Research Centre of Finland: Espoo, Finland, 2008.
12. NEA. *Decontamination Techniques Used in Decommissioning Activities*; Nuclear Energy Agency: Paris, France, 1999.
13. International-Atomic-Energy-Agency. Innovative and Adaptive Technologies in Decommissioning of Nuclear Facilities. In *Waste Technology Section*; International-Atomic-Energy-Agency: Vienna, Austria, 2008; Volume IAEA-TECDOC-1602.
14. Grebennikova, T.; Jones, A.N.; Sharrad, C.A. Electrochemical decontamination of irradiated nuclear graphite from corrosion and fission products using molten salt. *Energy Environ. Sci.* **2021**, *14*, 5501–5512. [CrossRef]
15. Yoon, I.-H.; Yoon, S.B.; Sihn, Y.; Choi, M.-S.; Jung, C.-H.; Choi, W.-K. Stabilizing decontamination foam using surface-modified silica nanoparticles containing chemical reagent: Foam stability, structures, and dispersion properties. *RSC Adv.* **2021**, *11*, 1841–1849. [CrossRef]
16. Koryakovskiy, Y.S.; Doilnitsyn, V.A.; Akatov, A.A. Improving the efficiency of fixed radionuclides' removal by chemical decontamination of surfaces in situ. *Nucl. Energy Technol.* **2019**, *5*, 155–161. [CrossRef]
17. He, Z.-Y.; Li, Y.-T.; Zhang, Q.-P.; Li, Y.-J.; Liu, D.-L.; Xiao, Z.-Q.; Zhang, S.-F.; Zhou, Y.-L.; Luo, D.L. Study on the Influencing Factors in the Process of Surface Strippable Decontaminant. *Coatings* **2020**, *10*, 649. [CrossRef]
18. Wu, H.F.; Yue, L.Z.; Jiang, S.L.; Lu, Y.Q.; Wu, Y.X.; Wan, Z.Y. Biodegradation of polyvinyl alcohol by different dominant degrading bacterial strains in a baffled anaerobic bioreactor. *Water Sci. Technol.* **2019**, *79*, 2005–2012. [CrossRef]
19. Halima, N.B. Poly(vinyl alcohol): Review of its promising applications and insights into biodegradation. *RSC Adv.* **2016**, *6*, 39823–39832. [CrossRef]
20. Baldevraj, R.S.M.; Jagadish, R.S. Incorporation of chemical antimicrobial agents into polymeric films for food packaging. In *Multifunctional and Nanoreinforced Polymers for Food Packaging*; Woodhead Publishing: Sawston, UK, 2011; pp. 386–390.
21. George, M.; Abraham, T.E. Polyionic hydrocolloids for the intestinal delivery of protein drugs: Alginate and chitosan–A review. *J. Control Release* **2006**, *114*, 1–14. [CrossRef]
22. Lei, Q.; Huang, Z.Y.; Pan, J.Z.; Bao, J.Q.; Xun, Q.N. Effect of Sodium Alginate on the Properties of Composite Protein Films. *Appl. Mech. Mater.* **2014**, *541–542*, 49–56. [CrossRef]

23. Castro-Yobal, M.A. Evaluation of physicochemical properties of film-based alginate for food packing applications. *e-Polymers* **2021**, *21*, 82–95. [CrossRef]
24. Li, J.; He, J.; Huang, Y.; Li, D.; Chen, X. Improving surface and mechanical properties of alginate films by using ethanol as a co-solvent during external gelation. *Carbohydr. Polym.* **2015**, *123*, 208–216. [CrossRef]
25. Begum, Z.A.; Rahman, I.M.M.; Tate, Y.; Sawai, H.; Maki, T.; Hasegawa, H. Remediation of toxic metal contaminated soil by washing with biodegradable aminopolycarboxylate chelants. *Chemosphere* **2012**, *87*, 1161–1170. [CrossRef]
26. Sahu, A.; Rane, N.V.; Lodaya, B.G.; Pandit, A.B. Green synthesis and kinetic study of eco-friendly chelating agent by hydrothermal process for remediation of heavy metals. *Indian Chem. Eng.* **2021**, 1–16. [CrossRef]
27. Nowack, B. Chelating agents and the environment. *Environ. Pollut.* **2008**, *153*, 1–2. [CrossRef]
28. Bialik-Wąs, K.; Królicka, E.; Malina, D. Impact of the Type of Crosslinking Agents on the Properties of Modified Sodium Alginate/Poly(vinyl Alcohol) Hydrogels. *Molecules* **2021**, *26*, 2381. [CrossRef]
29. Knepper, T.P. Synthetic chelating agents and compounds exhibiting complexing properties in the aquatic environment. *TrAC Trends Anal. Chem.* **2003**, *22*, 708–724. [CrossRef]
30. Toader, G.; Rotariu, T.; Pulpea, D.; Moldovan, A.; Podaru, A.; Gavrila, A.M.; Alexandru, M.; Diacon, A.; Ginghina, R.; Iorga, O.; et al. Polymeric blends designed for surface decontamination. *UPB Sci. Bull. Ser. B Chem. Mater. Sci.* **2021**, *83*, 73–86.
31. Hanley, K. Evaluating the Surface Protection and Decontamination Efficiency of DeconGelTM 1101 toward Cs-137 Spilled on Biological, Salt-Covered, Rusty, Wet and Solid Painted Surfaces. Master's Thesis, Oregon State University, Corvallis, OR, USA, 2010.
32. Lee, J.; Park, S.-M.; Jeon, E.-K.; Baek, K. Selective and irreversible adsorption mechanism of cesium on illite. *Appl. Geochem.* **2017**, *85*, 188–193. [CrossRef]
33. El-Zahhar, A.A. Sorption of cesium from aqueous solutions using polymer supported bentonite. *J. Radioanal. Nucl. Chem.* **2012**, *295*, 1693–1701. [CrossRef]
34. Chikkamath, S.; Patel, M.A.; Kar, A.S.; Tomar, B.S.; Manjanna, J. Sorption and surface complexation modeling of 137Cs on Fe(II)-montmorillonite clay mineral relevant to nuclear waste disposal. *Radiochim. Acta* **2020**, *109*, 73–83. [CrossRef]
35. Knop, S.; Jansen, T.L.C.; Lindner, J.; Vöhringer, P. On the nature of OH-stretching vibrations in hydrogen-bonded chains: Pump frequency dependent vibrational lifetime. *Phys. Chem. Chem. Phys.* **2011**, *13*, 4641–4650. [CrossRef]
36. Çaykara, T.; Demirci, S. Preparation and Characterization of Blend Films of Poly(Vinyl Alcohol) and Sodium Alginate. *J. Macromol. Sci. Part A* **2006**, *43*, 1113–1121. [CrossRef]
37. Bucknall, C.B.; Buckley, C.P.; McCrum, N.G. Viscoelasticity. In *Principles of Polymer Engineering*; Oxford University Press: Oxford, UK, 1997; pp. 117–183.
38. Meyers, M.A.; Chawla, K.K. Elasticity and Viscoelasticity. In *Mechanical Behavior of Materials*; Cambridge University Press: Oxford, UK, 2008; pp. 71–155.
39. Boyer, R.F. An apparent double glass transition in semicrystalline polymers. *J. Macromol. Sci. Part B* **1973**, *8*, 503–537. [CrossRef]
40. Hatakeyama, T.; Quinn, F.X. Differential Thermal Analysis and Differential Scanning Calorimetry. In *Thermal Analysis. Fundamentals and Applications to Polymer Science*; Hatakeyama, T., Ed.; John and Wiley and Sons: Hoboken, NJ, USA, 1999; pp. 5–24.
41. Fong, R.J.; Robertson, A.; Mallon, P.E.; Thompson, R.L. The Impact of Plasticizer and Degree of Hydrolysis on Free Volume of Poly(vinyl alcohol) Films. *Polymers* **2018**, *10*, 1036. [CrossRef]
42. Mohsin, M.; Hossin, A.; Haik, Y. Thermal and mechanical properties of poly(vinyl alcohol) plasticized with glycerol. *J. Appl. Polym. Sci.* **2011**, *122*, 3102–3109. [CrossRef]
43. Park, J.-S.; Ruckenstein, E. Thermal and dynamic mechanical analysis of PVA/MC blend hydrogels. *Polymer* **2001**, *42*, 4271–4280. [CrossRef]
44. Lang, A.; Engelberg, D.; Smith, N.T.; Trivedi, D.; Horsfall, O.; Banford, A.; Martin, P.A.; Coffey, P.; Bower, W.R.; Walther, C.; et al. Analysis of contaminated nuclear plant steel by laser-induced breakdown spectroscopy. *J. Hazard. Mater.* **2018**, *345*, 114–122. [CrossRef] [PubMed]
45. Kolbesen, B.O. *Analytical Techniques for Semiconductor Materials and Process Characterization 6 (ALTECH 2009)*; The Electrochemical Society: Pennington, NJ, USA, 2009; p. 1999.

polymers

MDPI

Article

Removal of Trithiocarbonyl End Group of RAFT-Polymerized Poly(stearyl acrylate) and Effect of the End Group on Thermal and Structural Properties

Eri Oishi [1], Masumi Takamura [2] and Tatsuhiro Takahashi [1,*]

[1] Department of Organic Materials Science, Graduated School of Organic Materials Science, Yamagata University, 4-3-16 Jonan, Yonezawa 992-8510, Yamagata, Japan; tky45776@st.yamagata-u.ac.jp
[2] Open Innovation Platform, Yamagata University, 4-3-16 Jonan, Yonezawa 992-8510, Yamagata, Japan; masumi_takamura@yz.yamagata-u.ac.jp
* Correspondence: effort@yz.yamagata-u.ac.jp

Abstract: The effect of a long alkyl end group on the thermal and structural properties of RAFT (reversible addition-fragmentation chain transfer)-polymerized poly(stearyl acrylate) (PSA) was investigated. RAFT-polymerized PSA was prepared using 2-cyano-2-[(dodecylsulfanylthiocarbonyl) sulfanyl] propane (CDTP) with long alkyl group as a chain transfer agent and azobisisobutyronitrile (AIBN) as an initiator. The RAFT polymerization resulted in the polymerized structure having trithiocarbonyl (TTC) at one end and isobutyronitrile at the other end. RAFT-polymerized PSA was prepared with two different molecular weights. The TTC end group was replaced by isobutyronitrile using radical reaction with AIBN through optimization of the conditions, which resulted in isobutyronitrile at both ends. The effect of the end group on the thermal and structural properties was investigated using differential scanning calorimetry and X-ray diffraction, and the results indicated that the long alkyl group from TTC lowers the melting point and semi-crystalline structure in the case of low molecular weight PSA.

Keywords: RAFT polymerization; poly(stearyl acrylate); poly(octadecyl acrylate); semi-crystalline polymer; side chain crystalline polymer; end group effect

Citation: Oishi, E.; Takamura, M.; Takahashi, T. Removal of Trithiocarbonyl End Group of RAFT-Polymerized Poly(stearyl acrylate) and Effect of the End Group on Thermal and Structural Properties. *Polymers* **2021**, *13*, 4169. https://doi.org/10.3390/polym13234169

Academic Editor: Edina Rusen

Received: 23 October 2021
Accepted: 26 November 2021
Published: 28 November 2021

Publisher's Note: MDPI stays neutral with regard to jurisdictional claims in published maps and institutional affiliations.

Copyright: © 2021 by the authors. Licensee MDPI, Basel, Switzerland. This article is an open access article distributed under the terms and conditions of the Creative Commons Attribution (CC BY) license (https://creativecommons.org/licenses/by/4.0/).

1. Introduction

End group modification of polymers has received considerable attention, especially for polymers with specific functional groups as side chains, such as polyacrylate and polyacrylamide, with an aim to further improve functionalization toward various applications [1–4]. There are two end groups for linear polymers without branching structures, and the influence of the end groups is significant when the molecular weight of the polymer is lower. Reports related to polymers with narrow molecular weight distribution produced by living radical polymerization has recently increased in a drive to enhance the effect of end groups with reproducibility and uniformity.

Although there are various methods for living radical polymerization, reversible addition-fragmentation chain transfer (RAFT) polymerization has been frequently utilized for various monomers, such as conjugated vinyl monomers (i.e., acrylate) and non-conjugated monomers (i.e., vinyl acetate) by optimization of the chain transfer agents (CTAs) [5–9]. There have been many previous reports on RAFT-polymerized polymers with trithiocarbonyl (TTC) groups as the one end group, which was modified in various ways, including by using nucleophile agents [10–13], oxidation agents [14,15], and protonation [10,13,16].

Among the various end modifications, the most frequently utilized method is the use of radical initiators [13,17–20]. The chemical reaction proceeds with the dozens of times the initiator (based on moles) for polymerization is added and reacted, which results in cleavage of the C-S bond between the polymer end and the TTC group, and the further addition reaction of the initiator radical. The end group, after the modification with this method, is determined by the chemical structure of the applied radical initiators. Therefore, various radical initiators have been utilized for the modification of end groups produced by RAFT polymerization, such as azo radical initiators [17–20] and peroxide radical initiators [18,20].

The polymer crystallinity of acrylate polymers, either amorphous or semi-crystalline, is determined by the number of alkyl groups in the side chain. An acrylate polymer with longer alkyl groups of more than 6 carbons as side chains is semi-crystalline and has strong hydrophobic properties and a low glass transition temperature. Among these polymers, poly(stearyl acrylate) (PSA) has been utilized for applications such as the surface modification of polyethylene [21], control of polymer deterioration [22], and as a positive temperature coefficient (PCT) thermistor [23]. The crystalline region forms as the side chains line up in parallel for a semi-crystalline polymer with a side chain, which is different from that of a semi-crystalline main chain polymer, such as polyethylene. The melting point of the semi-crystalline polymer increases with the molecular weight; however, there have been no reports on the effect of end groups on the thermal properties and crystalline structure of such polymers.

In this paper, PSA, a semi-crystalline polymer with side chains, was synthesized through RAFT polymerization. The complete removal of the one end group of CTA was examined under various conditions for optimization. The effect of the end group on the crystalline properties was investigated. 2-Cyano-2-[(dodecylsulfanylthiocarbonyl) sulfanyl] propane (CDTP), a trithiocarbonyl (TTC) type CTA with a long alkyl chain of 12 carbons, was used as the CTA for the RAFT polymerization. Azobisisobutyronitrile (AIBN) was used as a radical initiator to remove the TTC end groups. The exact amount of TTC removed was carefully analyzed using nuclear magnetic resonance spectroscopy (NMR), elemental analysis, and matrix assisted laser deposition/ionization time of flight mass spectrometry (MALDI-TOF-MS). The effect of TTC removal on the crystallinity was examined using differential scanning calorimetry (DSC) together with analysis of the degree of crystallinity from X-ray diffraction measurements. PSAs with two different molecular weights were synthesized and the influence of the end group on the molecular weight was also examined.

2. Materials and Methods

2.1. Materials

Stearyl acrylate (SA) was obtained from Tokyo Chemical Industry (Tokyo, Japan), and the inhibitor was removed by recrystallization in methanol. CDTP was obtained from Fujifilm Wako Pure Chemical Co. (Osaka, Japan). AIBN and dehydrated toluene were obtained from Kanto Chemical Co., Inc. (Tokyo, Japan). AIBN was purified by recrystallization in ethanol.

2.2. Synthesis of PSA by RAFT Polymerization

Scheme 1 shows the synthesis route for RAFT-polymerized PSA with CDTP as the CTA, and Table 1 shows the properties of the polymerized PSA. The details of synthesis are as follows.

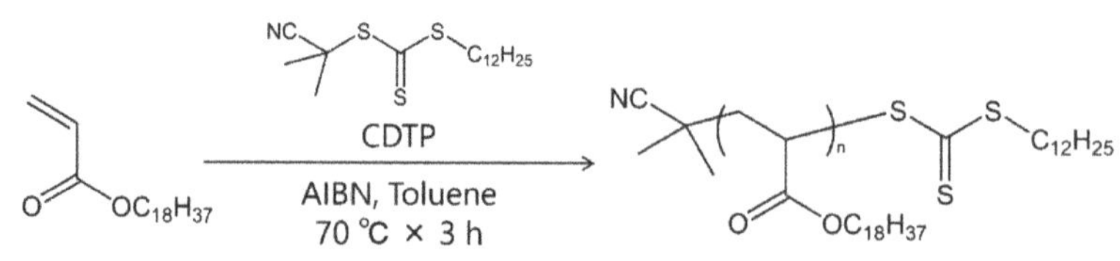

Scheme 1. Synthesis of PSA by RAFT polymerization using CDTP as chain transfer agent.

Table 1. RAFT polymerization of SA with CDTP as CTA.

Polymer	$[SA]_0/[CDTP]_0/[AIBN]_0$ [a]	Conv. [%] [b]	M_n(theo.) [c]	M_n(SEC) [d]	M_w(SEC) [d]	M_w/M_n [d]
PSA4.9k	45/3/1	85	4500	4900	5200	1.05
PSA17k	300/3/1	78	26,000	16,700	18,500	1.11

[a] Solution polymerization with toluene (2M). [b] Calculated by ^{1}H-NMR. [c] Theoretical Mn calculated with $[SA]_0$, $[CDTP]_0$, and monomer conversion. [d] PSt-calibrated SEC values.

PSA4.9k: SA (2.94 g, 9.07 mmol), CDTP (210.4 mg, 0.61 mmol), and recrystallized AIBN (34.6 mg, 0.21 mmol) were dissolved in 4.5 mL of anhydrous toluene in a 30 mL Schlenk flask. Three freeze-pump-thaw cycles were performed, after which the flask was filled with nitrogen and removed from the vacuum line. The solution was stirred at 70 °C for 3 h and the reaction was then quenched. The reacted solution was reprecipitated in methanol, followed by filtration. The polymer was dried under vacuum after three precipitation/filtration cycles to obtain the product as a yellow powder.

PSA17k: SA (0.98 g, 3.01 mmol), CDTP (10.8 mg, 0.03 mmol), and recrystallized AIBN (1.6 mg, 0.01 mmol) were dissolved in 1.5 mL of anhydrous toluene in a 30 mL Schlenk flask. Three freeze-pump-thaw cycles were performed, after which the flask was filled with nitrogen gas and removed from the vacuum line. The solution was stirred at 70 °C for 3 h, and the reaction was then quenched. Purification of the polymer was the same as that for PSA4.9k, to obtain the product as a pale-yellow powder.

2.3. Removal of TTC End Group from RAFT Polymerized PSA and Modifying Isobuthironitrile End Group

Scheme 2 shows removal of TTC and modifying isobuthironitrile end group reactions. The synthesized PSA was dissolved in toluene (100 mg/mL), and AIBN was added to the solution in a PSA:AIBN ratio of 1:30. Three freeze-pump-thaw cycles were performed, after which the flask was filled with nitrogen gas. The reaction of TTC end group removal from PSA4.9k was conducted with various temperatures and reaction times: 90 °C for 2 h, 80 °C for 2.5 h (same condition as [17]), 70 °C for 16 h, and 65 °C for 24 h. The reaction of TTC removal from PSA17k was performed at 70 °C for 16 h. The reacted solution was precipitated in acetone, followed by filtration, and drying under vacuum, to obtain each product after the reactions.

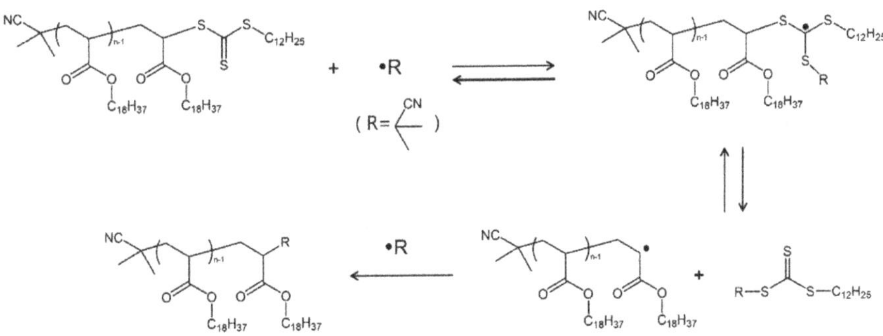

Scheme 2. Removal of TTC end group and modifying isobutyronitrile end group by conventional radical reaction with AIBN as radical resource.

2.4. Polymer Characterization

^{1}H-NMR (JNM-EC500, JEOL, Tokyo, Japan; 500 MHz) was used to determine monomer conversion, the number average of molecular weight, and the TTC removal rate. $CDCl_3$ was used as the solvent for NMR analysis, and the chemical shift was calibrated using residual $CHCl_3$ (at 7.26 ppm) as the internal standard. Size exclusion chromatography (SEC;

GULLIVER 1500, JASCO, Tokyo, Japan) analysis was performed using a chromatograph equipped with a pump, an absorbance detector (UV, λ = 254 nm), and three polystyrene gel columns, based on a conventional calibration curve using polystyrene standards. The eluent was tetrahydrofuran (THF) at a flow rate of 1.0 mL/min at 40 °C. Elemental analysis (EA; CHN and S) was performed with a CHNS/O Elemental Analyzer 2400 II (Perkin-Elmer, Waltham, MA, USA) calibrated using stearic acid, sulfanilamide, sulfathiazole, 1,3-diphenylthiourea (Kishida Chemical Co., Ltd., Elemental analysis standard, Japan) and cystine (Perkin-Elmer, Waltham, MA, USA). Matrix-assisted laser desorption ionization time-of flight mass spectrometry (MALDI-TOF MS) analysis was performed on a JMS-3000 Linear TOF spectrometer (JEOL Ltd., Japan) at an acceleration voltage of 20 kV in the positive linear mode. External mass calibration was performed using a PSt standard (Mn = 5000). Trans-2-[3-(4-tert-buthylphenyl)-2-methyl-2-propanylidene]malononitrile (DCTB) was used as the matrix and sodium trifluoroacetate as the cationization agent. The polymer (10 mg/mL), matrix (20 mg/mL), and cationization agent (1 mg/mL) were dissolved in THF, and these solutions were mixed at a volume ratio of 1:5:1. 1 μL of the mixed solution was deposited on a MALDI sample plate, and the spots were dried in air at room temperature. Thermogravimetric analysis (TGA) was conducted using a TG-DTA8122 with Smart loader analyzer (Rigaku, Japan) in a nitrogen atmosphere, with calibration using indium. All samples were weighed (8–9 mg) and then heated to 500 °C at 5 °C/min. Differential scanning calorimetry (DSC) analysis was conducted using a calorimeter (DSC Q200, TA Instruments, Tokyo, Japan) with a refrigerated cooling system (RCS, TA Instruments Japan) that was calibrated with sapphire for calibration of the cell resistance and thermal capacity, and with indium for the cell constant and temperature calibration. All samples were weighed (2–3 mg), heated to 80 °C at 5 °C/min followed by holding at 80 °C for 10 min, and then cooled to −30 °C at 5 °C/min followed by holding at −30 °C for 10 min. The heating and cooling cycle was conducted 2 times. All values reported in this work were taken from the first cooling and second heating cycles. X-ray diffraction (XRD) measurements were performed at the BL03XU beamline, Spring-8, Japan.

3. Results and Discussion

3.1. Optimization for TTC End Group Removal and Modification with the Isobutyronitrile End Group by Radical Reaction with AIBN

Table 2 shows the various conditions for removal of the TTC end group from RAFT-polymerized PSA and then modification with the isobutyronitrile end group. The TTC removal rate was calculated from ^{1}H-NMR measurements. Figure 1 shows ^{1}H-NMR spectra for the polymers reacted under various conditions (Entries 1–4) and the PSA4.9k precursor. The peaks of focus are *b′* and *h*, where peak *h* is S-CH_2- in TTC, peak *b′* is S-CH- the SA unit next to TTC, and the integral ratio of *h*:*b′* = 2:1 for the PSA 4.9k precursor. Therefore, if all TTC was removed, then the peaks of *b′* and *h* should disappear. Perrier et al. [17] reported the successful removal of the TTC end group from RAFT-polymerized poly(methyl methacrylate) (PMMA) and modification with various end groups by radical reaction with azo-type initiators in a initiator:PMMA ratio of 20:1 at 80 °C for 2.5 h. Therefore, 30 eq. AIBN was taken in the present experiment under the condition of 80 °C for 2.5 h (Table 2 and Figure 1, Entry 3) to obtain PSA without TTC. Although the AIBN decomposition rate at 80 °C for 2.5 h was 72.2%, the TTC end groups were not removed. We then attempted the procedure using a higher temperature of 90 °C for 3 h (Table 3 and Figure 1, Entry 4), by which the AIBN decomposition rate was 99.5%; therefore, more AIBN was decomposed under this condition. However, the TTC removal reaction did not occur.

Table 2. TTC removing reaction from RAFT-polymerized PSA.

Entry	Precursor	Removal Condition Using Precursor				
		$[AIBN]_0/[PSA]_0$	Temp. [°C]	Time [h]	AIBN Decomposition Rate [%] [a]	TTC Removal [b]
1	PSA4.9k	30/1	65 °C	24 h	81.1	32%
2	PSA4.9k	30/1	70 °C	16 h	88.9	88%
3	PSA4.9k	30/1	80 °C	2.5 h	72.2	none
4	PSA4.9k	30/1	90 °C	3 h	99.5	none
5	PSA17k	30/1	70 °C	16 h	88.9	N/A

[a] Calculated data. [b] Calculated by ^{1}H-NMR in Figure 1.

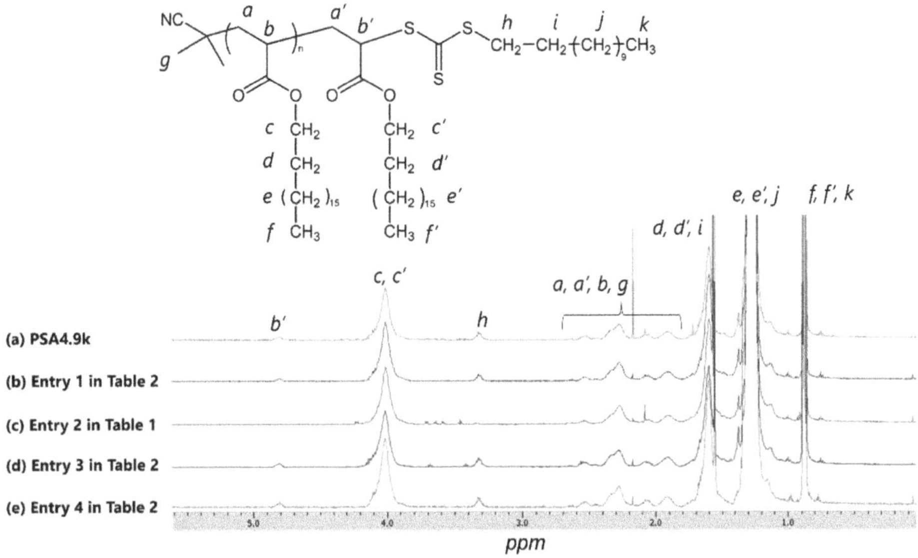

Figure 1. ^{1}H-NMR spectra for (**a**) RAFT-polymerized PSA 4.9k in Table 1; (**b–e**) TTC removal from PSA 4.9k conducted under various conditions.

The time-decomposition rate for AIBN at 65–90 °C was calculated and plotted (Figure 2). At 90 °C, the 1 h decomposition rate was over 80%, which indicates that radicals are immediately generated, although radical deactivation is also very fast. At 80 °C, the 1 h later decomposition rate was 40%; the decomposition rate was less than that at 90 °C. Although it should be very fast, there was concern that radical deactivation would occur before the radicals reacted with TTC. In addition, the radical concentration was very high initially, but rapidly decreased with time.

Entries 1 and 2 in Table 2 show the TTC removal reaction conducted at 65 °C for 24 h and 70 °C for 16 h, respectively. These AIBN decomposition rates per unit time were lower than those at 90 °C and 80 °C; however, radical generation was sustainable for a long time. Entry 2 in Figure 1 shows that the *b′* peak almost disappeared and the h peak was decreased. The reason why only the h peak remained was that the removed TTC (the structure is the same as CDTP) would be included in purified PSA after the TTC removal reaction (Entry 2). The TTC removal rate was calculated from the ratio of (c+c′) and *b′* integrals from ^{1}H-NMR measurements. The TTC removal rate in Entry 2 was 88%, which demonstrates that the condition of sustainable radical generation for a long time is necessary for the TTC removal reaction by AIBN with the RAFT-polymerized PSA. The TTC removal reaction was then conducted at 65 °C for 24 h (Entry 1); this condition continued to generate radicals for a longer time than Entry 2. However, Entry 1 in Figure 1 shows TTC remained (i.e., peak *b′* and *h*) more so than Entry 2, and the TTC removal rate was only 32%. Radicals continued to be gradually generated at 65 °C for 24 h; however, the concentration of radicals per unit time was not sufficient to remove the TTC end group from RAFT-polymerized PSA. The removal of TTC from PSA17k was performed at 70 °C for 16 h only (Entry 5). PSA17k has higher molecular weight, so that the difference before and after reaction could not be determined because the proton peaks of b' and h were very weak. Therefore, TTC removal from PSA17k was conducted using the most suitable condition for TTC removal from PSA4.9k.

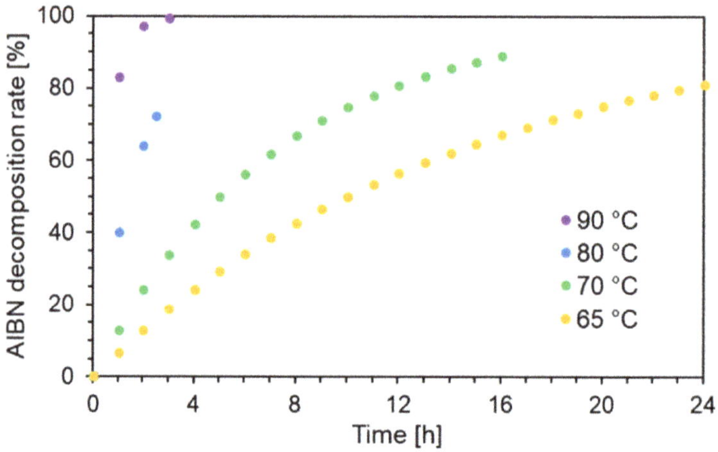

Figure 2. Calculated time-decomposition rate for AIBN.

3.2. Polymer Characterization

3.2.1. Size Exclusion Chromatography (SEC)

Figure 3 shows SEC chromatograms before and after the TTC removal reaction from PSA4.9k and PSA17k, and Table 3 shows the number average molecular weight (M_n), weight average molecular weight (M_w), and polydispersity index (PDI; M_w/M_n) for each sample. These results show that the molecular weight and PDI were almost the same before and after the TTC removal reaction. Side reactions have often been reported, where 2 polymer radicals are coupled during removal of the RAFT-polymerized end group by radical reaction using radical initiators before the generated radical and polymer radical were coupled [18]. Our results shows that the coupling of two polymer radicals did not occur during the TTC removal reaction at 70 °C for 16 h, and the polymer main chain was not damaged from excessive amounts of isobutyronitrile radicals generated by AIBN.

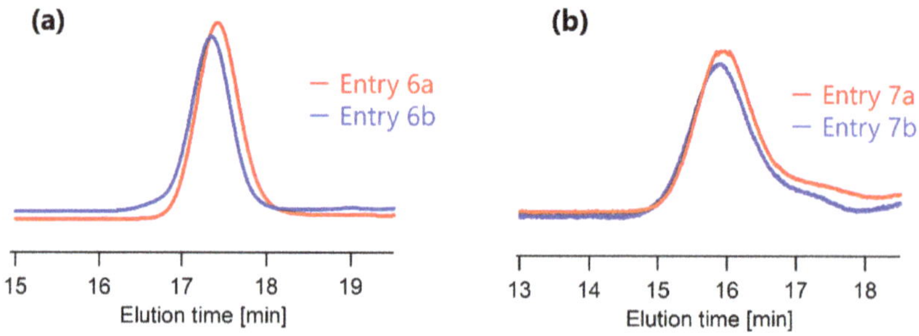

Figure 3. SEC chromatograms for before and after TTC removal reaction from (**a**) PSA4.9k (Entry 6a and 6b in Table 3) and (**b**) PSA17k (Entry 7a and 7b in Table 3).

Table 3. SEC value for PSAs.

Entry	TTC Removal Rate [%] [a]	M_n (GPC) [b]	M_w (GPC) [b]	M_w/M_n [b]	Remarks
6a	0	4900	5200	1.05	Same as PSA4.9k
6b	88	5400	5600	1.04	Entry 2
7a	0	16,700	18,500	1.11	PSA17k
7b	100	17,900	19,700	1.11	Entry 5

[a] Calculated by ^{1}H-NMR. [b] PSt-calibrated SEC values.

3.2.2. Elemental Analysis

Table 4 shows the results of elemental analysis. The theoretical mass percentage of Entries 6a and 7a were calculated from the SEC results. Entries 6b and 6c were calculated for the TTC removal rate using NMR, and Entry 7b was calculated from the estimation of the TTC removal rate as 100%. The experimental mass percentage of S for Entry 6a is slightly less than theoretical. The TTC end groups may have been removed during polymer synthesis or purification. Entry 6b is that for TTC removed from Entry 6a, the experimental mass percentage of S is significant, even if the N value is the same as the theoretical. This also confirmed that the TTC removed by the radical reaction remained with the polymer, because the *b′* peak disappeared, whereas the *h* peaks remained in the ^{1}H-NMR spectra (Figure 1). Therefore, a purification method to completely wash the removed TTC from the polymer is an issue for future work. In Entry 7b, S was not detected (or was only trace amounts below the detection limit), which indicates that TTC was completely removed. However, the mass of N was half the theoretical value, and although TTC was successfully removed, it is possible that the isobutyronitrile group derived from AIBN could not be introduced.

Table 4. Results of elemental analysis for PSAs.

Entry	TTC Removal Rate [%] [a]	C [%] (Theo. [b])	H [%] (Theo. [b])	N [%] (Theo. [b])	S [%] (Theo. [b])	Remarks
6a	0	76.2 (76.4)	12.4 (12.2)	0.3 (0.3)	1.7 (1.9)	PSA4.9k in Table 1
6b	88	77.0 (76.8)	12.2 (12.2)	0.4 (0.4)	1.2 (1.3)	Entry 2 in Table 2
6c	32	76.7 (77.4)	12.2 (12.3)	0.6 (0.6)	0.4 (0.2)	Entry 1 in Table 2
7a	0	77.6 (77.3)	12.4 (12.4)	0.1 (0.2)	0.5 (0.5)	PSA17k in Table 1
7b	100	77.4 (77.7)	12.4 (12.4)	0.1 (0.2)	0.0 (0.0)	Entry 5 in Table 2

[a] Calculated by ^{1}H-NMR. [b] Calculated by SEC value and TTC removal rate by ^{1}H-NMR.

3.2.3. MALDI-TOF MS Analysis

Analysis using MALDI-TOF MS was performed to identify the end structure of PSA though the removal of TTC with an excess amount of AIBN. Figure 4 shows the MALDI-TOS MS spectra and Table 5 summarizes the various PSA chemical structures, which are speculated from the mass peaks detected by MALDI-TOF MS. The MALDI-TOS MS spec trum of the PSA structure that was synthesized by RAFT polymerization, shown in Figure 4a, revealed only single mass peaks (m/z = 3938.2, n = 11 with Na^+), which was in agreement with the mass of PSA with TTS as the one end group and isobutyronitrile as the other (P1 in Table 5). Figure 4b shows the MALDI-TOF MS spectrum for 32% TTC removed PSA, the mass peak of which indicated isobutyronitrile at both ends of PSA (P2 in Table 5, m/z = 4070.9, n = 12 with K^+), in addition to that of the $[P1+Na]^+$ structure. The largely increased mass peak for isobutyronitrile at both ends $[P2+K]^+$ was shown for the 88% TTC removed PSA (Figure 4c), which was prepared by optimization of the conditions for removal by AIBN, although the mass peak for TTC $[P1+Na]^+$ as one end was not detected. In addition, two newly generated mass peaks (m/z = 4309.5, 4324.5) were detected, which was in good agreement with the PSA structures (shown as P3 in Table 5) with cationated ions of Na and K, respectively. There have been former reports regarding the structural change by laser effects or oxidation during MALDI measurements [24,25]; however, no peaks related with these changes were detected in the present study. From these results, it is clearly demonstrated that the isobutyronitrile group from AIBN was possibly introduced through the end group modification of PSA to remove the TTC end group due to the CTA at 70 °C for 16 h with an excess amount of AIBN.

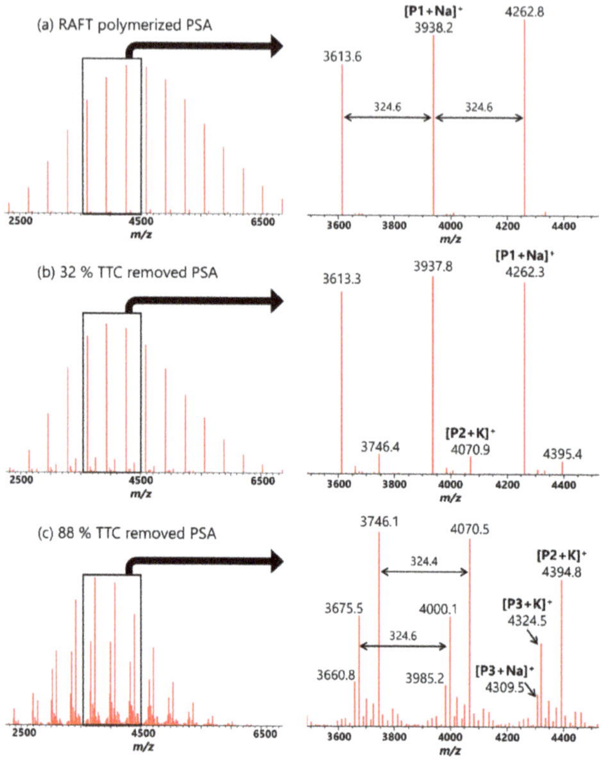

Figure 4. MALDI-TOF MS spectra for (**a**) RAFT polymerized PSA (PSA4.9k in Table 1), (**b**) 32% TTC removed PSA (entry 1 in Table 2, entry 6c in Table 4), and (**c**) 88% TTC removed PSA (entry 2 in Table 2, entry 6b in Table 4). The TTC removal rate was calculated from the ^{1}H-NMR results.

Table 5. Polymer structures corresponding to the peaks in Figure 4.

Polymer	Structure	n	Cationization	*m*/*z*	
				Theoretical	Experimental
P1	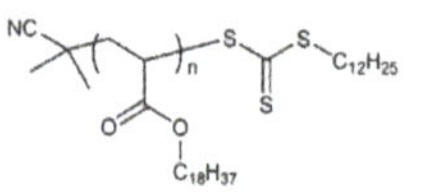	11	Na	3938.6	3938.2
		12	Na	4263.1	4262.3
P2	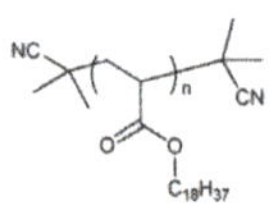	12	K	4069.8	4070.9
		13	K	4394.3	4394.8
P3	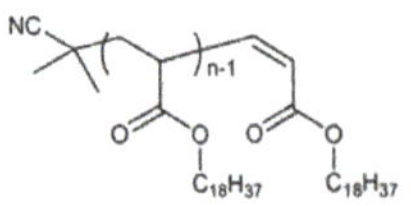	13	Na	4309.1	4309.5
		13	K	4325.2	4324.5

3.2.4. TGA

Figure 5 shows TGA results for entries 6a, 6b, 7a, and 7c in Tables 3 and 4. All samples were not decomposed in the range of room temperature to 200 °C. The weight loss of entries 6a and 6b with smaller molecular weights began at around 250 °C, and then all samples were rapidly pyrolyzed at 350 °C.

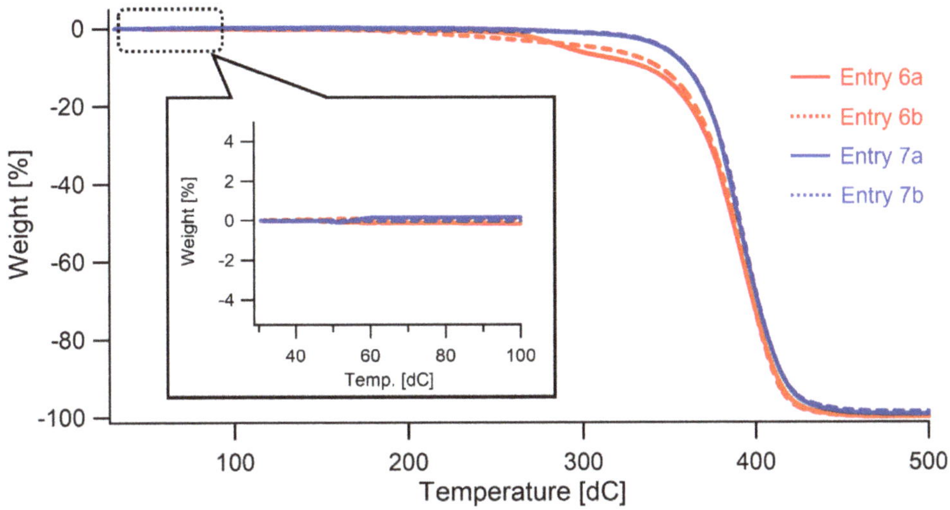

Figure 5. TGA curves corresponding to entries 6a, 6b, 7a, and 7b in Tables 3 and 4 to determine the decomposition temperature of PSAs.

3.2.5. DSC Analysis

Figure 6 shows two DSC curves: PSA with TTC (Entry 6a in Table 4) and PSA with the isobutyronitrile group (Entry 6b in Table 4) after modification with the respective end groups. An increase of molecular weight causes the melting point to be higher [26,27], and the crystallization temperature is dependent on the cooling speed [28]. Although these two samples have almost the same molecular weight, with the only differences being the end group of either TTC or isobutyronitrile, both the melting point and the crystallization point at the peaks were unexpectedly shifted to higher temperature by ca. 3 °C. This result suggests that the end group, long alkyl (TTC) lowers both the melting and crystallization temperatures.

Here, let us discuss the melting point temperature of the end group itself, low molecular weight CDTP, because CDTP has a long C12 alkyl chain that exhibits crystalline characteristics (melting point around 10 °C). On the other hand, the monomer of PSA, i.e., SA, also has a long C18 alkyl chain that exhibits crystalline characteristics (melting temperature around 30 °C). This melting point temperature difference of the two low molecular weight compounds is interpreted as the longer alkyl (C18, SA, monomer of PSA) has a higher melting point than that of the shorter alkyl (C12, CDTP, end group); therefore, the C12 alkyl melts and crystallizes at lower temperature. The other possible interpretation is that the end group of CDTP exists in an amorphous region instead of a crystalline region. Whichever the case, it is considered that at least the end of CDTP lowers the melting point temperature of PSA and makes the crystallite size smaller.

Figure 7 shows DSC curves of PSA, which has higher molecular weight, before (Entry 7a) and after (Entry 7b) the modification. A comparison of Figure 7 with Figure 6 shows that the modification makes no difference to the melting and crystallization characteristics for the larger molecular weight PSA. The volume percentage of the CDTP end group differs largely between PSA4.9k and PSA17k. In the case of PSA17k, the volume percentage of the CDTP end group is very small (ca. 2%), while that of PSA4.9k is large (ca. 7%). The effect of end group modification thus becomes negligible for PSA17k.

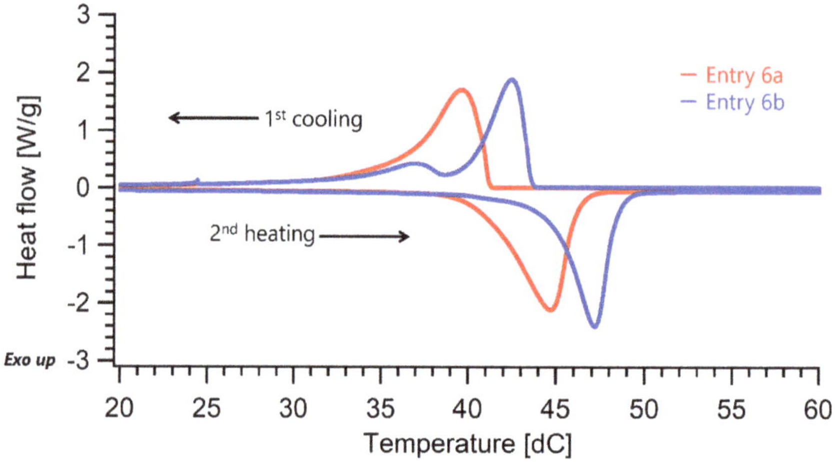

Figure 6. DSC curves before (red line, Entry 6a) and after TTC removal from PSA4.9k (blue line, Entry 6b). The heating/cooling rate was 5 °C/min.

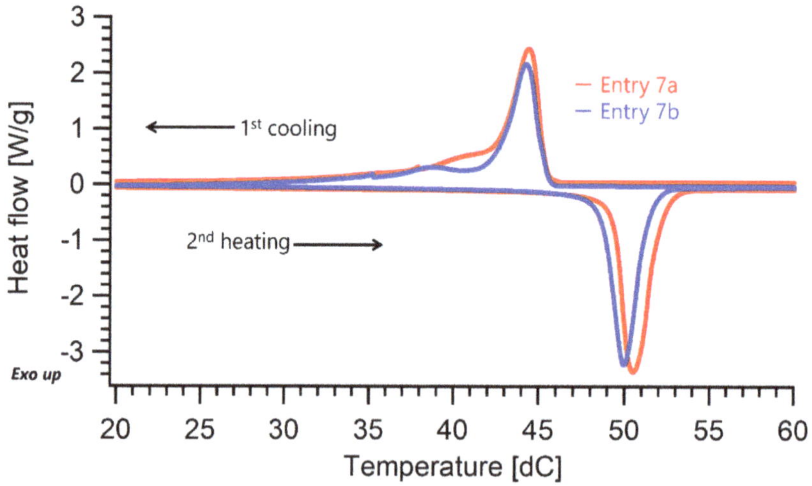

Figure 7. DSC curves before (red line, Entry 7a) and after TC removal from PSA17k (blue line, Entry 7b). The heating/cooling rate was 5 °C/min.

Here, let us consider the crystallite size through careful observation of the DSC curves in Figures 6 and 7. The common point between PSA4.9k (Figure 6) and PSA17k (Figure 7) is that there are two crystallization peaks, small and large, after the removal of the end long alkyl chain, as shown in Figure 6 (6b) and Figure 7 (7b). Figure 7 (7a) shows that the lower peak is not clear and should be considered as a shoulder. DSC characteristics for a larger size crystallite indicate crystallization at higher temperature, which suggests that the higher and larger peak comes from the formation of larger size crystallites (42.4 °C in Figure 6 (6b) and 44.2 °C in Figure 7 (7b)). On the other hand, the lower and smaller peak (or shoulder) is related to the formation of smaller size crystallites (36.9 °C in Figure 6 (6b) and 38.6 °C in Figure 7 (7b)). PSA prior to modification exhibited only a single broad peak without a smaller peak or shoulder, which indicates that the end group causes a wider size distribution of crystallites, so that the peak distribution widens down to the lower temperature region. There should be no conflict regarding the interpretation before and after the modification in terms of the crystallite size and DSC curves.

Let us discuss the degree of TTC removal in detail. The removal of TTC causes the crystallite size to become larger. The second lower peak is relatively large in Entry 6b than in Entry 7b. Assuming residual TTC of ca. 12% and the volume percentage of TTC because of lower molecular weight into consideration, it can be considered that the remaining TTC causes smaller crystallite size with a larger distribution, which results in widening of the peak down to lower temperature. The DSC curve of Entry 7a shows a shoulder at lower temperature, which suggests the amount of TTC in Entry 7a is smaller than that in Entry 6a, so that the content of middle size crystallites would be less because there is less probability to disturb the crystallite growth compared with that of Entry 6a.

3.2.6. XRD

Figure 8 shows the XRD patterns for PSA4.9k and PSA17k, before and after TTC removal. All the diffraction patterns were almost identical with the peak at 20° 2θ, which indicates the crystalline structure is the same because the crystalline structure of PSA is not affected by the end group. The degree of crystallinity is calculated to be ca. 50% for Entries 6a, 6b, 7a, and 7b. The crystallinity was almost the same, even if the molecular weight of PSA was different.

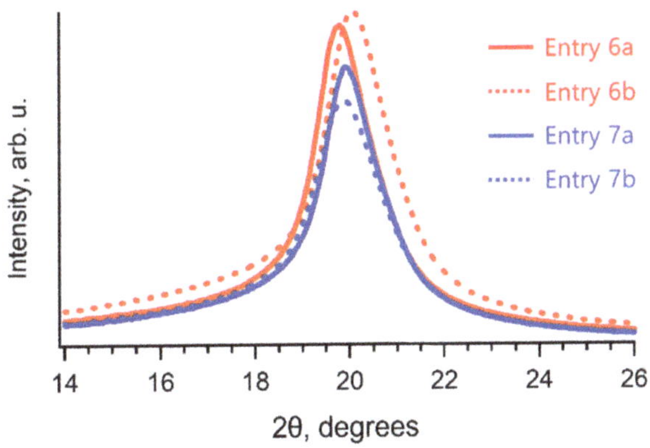

Figure 8. X-ray diffraction (XRD) patterns. Entry 6a (red solid line) is for PSA4.9k before TTC removal, and Entry 6b (red dashed line) is that after TTC removal. Entry 7a (blue solid line) is for PSA17k before TTC removal and Entry 7b (blue dashed line) is that after TTC removal.

4. Conclusions

PSA was prepared by RAFT polymerization using CDTP as the CTA to examine the effect of the end group on the thermal and structural properties. The long alkyl C12 end group of PSA polymerized by the RAFT process was removed by a radical reaction using AIBN. The reaction condition was optimized, especially with respect to the temperature, which resulted in the successful removal of TTC, which was characterized by ^{1}H-NMR, elemental analysis, and SEC. The polymer structures before and after removal of TTC were determined by MALDI-TOF MS spectra. We demonstrated removal TTC and modifying isobutyronitrile end group were succeeded by using AIBN. The thermal and structural properties were characterized using TGA, DSC, and XRD, which provided evidence that the long alkyl end groups of TTC lower the melting point and crystallization peak temperatures by 3 °C, which in turn affects the formation of different crystallite sizes with identical crystalline structure and degree of crystallinity.

Author Contributions: E.O. performed the measurements, analyzed experimental data, and wrote the manuscript. T.T. supervised the entire project. M.T. participated in comprehensive discussion and provided advice and suggestions. All authors have read and agreed to the published version of the manuscript.

Funding: This research received no external funding.

Data Availability Statement: The raw data presented in this study are available on request from the corresponding author.

Acknowledgments: The authors thank Shotaro Nishitsuji, Yamagata University for conducting the XRD measurements at Spring-8, Go Matsuba, Yamagata University for the useful discussion on semi-crystalline polymer properties, and Hideharu Mori, Yamagata University for also useful discussion on polymer synthesis, end group modification, and MALDI-TOF MS analysis.

Conflicts of Interest: The authors declare no conflict of interest.

References

1. Xia, Y.; Nicholas, A.D.; Burke, N.A.D.; Stolver, H.D.H. End Group Effect on the Thermal Response of Narrow-Disperse Poly(N-isopropylacrylamide) Prepared by Atom Transfer Radical Polymerization. *Macromolecules* **2006**, *39*, 2275–2283. [CrossRef]
2. Kujawa, P.; Segui, F.; Shaban, S.; Diab, C.; Okada, Y.; Tanaka, F.; Winnik, F.M. Impact of End-Group Association and Main-Chain Hydration on the Thermosensitive Properties of Hydrophobically Modified Telechelic Poly(N-isopropylacrylamides) in Water. *Macromolecules* **2006**, *39*, 341–348. [CrossRef]
3. Roth, P.J.; Jochum, F.D.; Forst, F.R.; Zentel, R.; Theato, P. Influence of End Groups on the Stimulus-Responsive Behavior of Poly[oligo(ethylene glycol) methacrylate] in Water. *Macromolecules* **2010**, *43*, 4638–4645. [CrossRef]
4. Jochum, F.D.; Borg, L.; Roth, P.J.; Theato, P. Thermo- and Light-Responsive Polymers Containing Photoswitchable Azobenzene End Groups. *Macromolecules* **2009**, *42*, 7854–7862. [CrossRef]
5. Chiefari, J.; Chong, Y.K.; Ercole, F.; Krstina, J.; Jeffery, J.; Le, T.P.T.; Mayadunne, R.T.A.; Meijs, G.F.; Moad, C.L.; Moad, G.; et al. Living Free-Radical Polymerization by Reversible Addition-Fragmentation Chain Transfer: The RAFT Process. *Macromolecules* **1998**, *31*, 5559–5562. [CrossRef]
6. Moad, G.; Rizzardo, E.; Thang, S.H. Living Radical Polymerization by the RAFT Process. *Aust. J. Chem.* **2005**, *58*, 379–410. [CrossRef]
7. Moad, G.; Rizzardo, E.; Thang, S.H. Living Radical Polymerization by the RAFT Process—A Second Update. *Aust. J. Chem.* **2009**, *62*, 1402–1472. [CrossRef]
8. Moad, G.; Rizzardo, E.; Thang, S.H. Living Radical Polymerization by the RAFT Process—A Third Update. *Aust. J. Chem.* **2012**, *65*, 985–1076. [CrossRef]
9. Keddie, D.J.; Moad, G.; Rizzardo, E.; Thang, S.H. RAFT Agent Design and Synthesis. *Macromolecules* **2012**, *45*, 5321–5342. [CrossRef]
10. Moad, G.; Chong, Y.K.; Postma, A.; Rizzardo, E.; Thang, S.H. Advances in RAFT polymerization: The synthesis of polymers with defined end-groups. *Polymer* **2005**, *46*, 8458–8468. [CrossRef]
11. Harrisson, S. Radical-Catalyzed Oxidation of Thiols by Trithiocarnate and Dithioester RAFT Agents: Implications for the Preparation of Polymers with Terminal Thiol Functionality. *Macromolecules* **2009**, *42*, 897–898. [CrossRef]
12. Yu, B.; Chan, J.W.; Hoyle, C.E.; Lowe, A.B. Sequential Thiol-Ene/Thiol-Ene and Thiol-Ene/Thiol-Yne Reactions as a Route to Well-Defined Mono and Bis End-Functionalized Poly(N-isoprolacrylamide). *J. Polym. Sci. A Polym. Chem.* **2009**, *47*, 3544–3557. [CrossRef]
13. Postma, A.; Davis, T.P.; Evans, R.A.; Li, G.; Moad, G.; O'Shea, M.S. Synthesis of Well-Defined Polystyrene with Primary Amine End Groups through the Use of Phthalimid-Functional RAFT Agents. *Macromolecules* **2006**, *39*, 5293–5306. [CrossRef]
14. Dietrich, M.; Glassner, M.; Gruendling, T.; Schmid, C.; Falkenhagen, J.; Barner-Kowollik, C. Facile conversion of RAFT polymers into hydroxyl functional polymers: A detailed investigation of variable monomer and RAFT agent combinations. *Polym. Chem.* **2010**, *1*, 634–644. [CrossRef]
15. Gruendling, T.; Pickford, R.; Guilhaus, M.; Barner-Kowollik, C. Degradation of RAFT Polymers in a Cyclic Ether Studied via High Resolution ESI-MS: Implications for Synthesis, Storage, and End-Group Modification. *J. Polym. Sci. A Polym. Chem.* **2008**, *46*, 7447–7461. [CrossRef]
16. Chong, Y.K.; Moad, G.; Rizzardo, E.; Thang, S.H. Thiocarbonylthio End Group Removal from RAFT-Synthesized Polymers by Radical-Induced Reduction. *Macromolecules* **2007**, *40*, 4446–4455. [CrossRef]
17. Perrier, S.; Takolpuckdee, P.; Mars, C.A. Reversible Addition-Fragmentation Chain Transfer Polymerization: End Group Modification for Functionalized Polymers and Chain Transfer Agent Recovery. *Macromolecules* **2005**, *38*, 2033–2036. [CrossRef]
18. Chen, M.; Moad, G.; Rizzardo, E. Thiocarbonylthio End Group Removal from RAFT-Synthesized Polymers by a Radical-Induced Process. *J. Polym. Sci. A Polym. Chem.* **2009**, *47*, 6704–6714. [CrossRef]
19. Willcock, H.; O'Reilly, R.K. End group removal and modification of RAFT polymers. *Polym. Chem.* **2010**, *1*, 149–157. [CrossRef]

20. Moad, G.; Rizzardo, E.; Thang, S.H. End-functional polymers, thiocarbonylthio group removal/transformation and reversible addition-fragmentation-chain transfer (RAFT) polymerization. *Polym. Int.* **2011**, *60*, 9–25. [CrossRef]
21. Miho, Y.; Hirai, S.; Nakano, R.; Sekiguchi, H.; Yao, S. Modification of polyethylene using side-chain crystalline block copolymer and evaluation of hydrophilicity. *Polym. J.* **2018**, *50*, 439–445. [CrossRef]
22. Wang, S.; Kesava, S.V.; Gomez, E.D.; Robertson, M.L. Sustainable Thermoplastic Elestmers Derived from Fatty Acids. *Macromolecules* **2013**, *46*, 7202–7212. [CrossRef]
23. Yokota, T.; Inoue, Y.; Terakawa, Y.; Reeder, J.; Kaltenbrunner, M.; Ware, T.; Yang, K.; Mabuchi, K.; Murakawa, T.; Sekino, M.; et al. Ultraflexible, large-area, physiological temperature sensors for multipoint measurements. *Proc. Natl. Acad. Sci. USA* **2015**, *112*, 14533–14538. [CrossRef]
24. Favier, A.; Ladaviere, C.; Charreyre, M.T.; Pichot, C. MALDI-TOF MS Investigation of the RAFT Polymerization of a Water-Soluble Acrylamide Derivative. *Macromolecules* **2004**, *37*, 2026–2034. [CrossRef]
25. Xu, J.; He, J.; Fan, D.; Wang, X.; Yang, Y. Aminolysis of Polymers with Thiocarbonylthio Termini Prepared by RAFT Polymerization: The Difference between Polystyrene and Polymethacrylates. *Macromolecules* **2006**, *39*, 8616–8624. [CrossRef]
26. Jeon, J.; Lee, H.B.-R.; Bao, Z. Flexible Wireless Temperature Sensors Based on Ni Microparticle-Filled Binary Polymer Composites. *Adv. Mater.* **2013**, *25*, 850–855. [CrossRef] [PubMed]
27. Kim, B.; Mather, P.T. Amphiphilic Telechelics Incorporating Polyhedral Oligosilsesquioxane: 1. Synthesis and Characterization. *Macromolecules* **2002**, *35*, 8378–8384. [CrossRef]
28. Schick, C. Differential scanning calorimetry (DSC) of semicrystalline polymers. *Anal. Bioanal. Chem.* **2009**, *395*, 1589–1611. [CrossRef]

polymers

MDPI

Article

Eco–Friendly Peelable Active Nanocomposite Films Designed for Biological and Chemical Warfare Agents Decontamination

Gabriela Toader [1,†], Aurel Diacon [2,†], Traian Rotariu [1,*], Mioara Alexandru [3,*], Edina Rusen [2], Raluca Elena Ginghină [4], Florentina Alexe [4], Ramona Oncioiu [4], Florina Lucica Zorila [3], Alice Podaru [1], Andreea Elena Moldovan [1], Daniela Pulpea [1], Ana Mihaela Gavrilă [5], Tanta Verona Iordache [5] and Raluca Șomoghi [5,6]

1 Military Technical Academy "Ferdinand I", 39-49 George Cosbuc Boulevard, 050141 Bucharest, Romania; gabriela.toader@mta.ro (G.T.); podaru.alice04@gmail.com (A.P.); andreea.voicu89@gmail.com (A.E.M.); pulpea.daniela@gmail.com (D.P.)
2 Faculty of Applied Chemistry and Materials Science, University 'Politehnica' of Bucharest, 1-7 Gh. Polizu Street, 011061 Bucharest, Romania; aurel.diacon@upb.ro (A.D.); edina.rusen@upb.ro (E.R.)
3 Microbiology Laboratory of Horia Hulubei National Institute for R&D in Physics and Nuclear Engineering, 30 Reactorului St., 077125 Bucharest-Magurele, Romania; florina.zorila@nipne.ro
4 Research and Innovation Center for CBRN Defense and Ecology, 225 Soseaua Oltenitei, 041327 Bucharest, Romania; raluca.ginghina@nbce.ro (R.E.G.); florentina.alexe@nbce.ro (F.A.); ramona.oncioiu@nbce.ro (R.O.)
5 National Institute of Research and Development for Chemistry and Petrochemistry, 202 Splaiul Independentei, 060041 Bucharest, Romania; ana.gavrila@icechim.ro (A.M.G.); tanta-verona.iordache@icechim.ro (T.V.I.); r.somoghi@gmail.com (R.Ș.)
6 Petroleum-Gas University of Ploiesti, 100680 Ploiesti, Romania
* Correspondence: traian.rotariu@mta.ro (T.R.); mioara.alexandru@nipne.ro (M.A.)
† Co-first authors, equally contributed to this study.

Citation: Toader, G.; Diacon, A.; Rotariu, T.; Alexandru, M.; Rusen, E.; Ginghină, R.E.; Alexe, F.; Oncioiu, R.; Zorila, F.L.; Podaru, A.; et al. Eco–Friendly Peelable Active Nanocomposite Films Designed for Biological and Chemical Warfare Agents Decontamination. *Polymers* **2021**, *13*, 3999. https://doi.org/10.3390/polym13223999

Academic Editor: Eduardo Guzmán

Received: 26 October 2021
Accepted: 12 November 2021
Published: 19 November 2021

Publisher's Note: MDPI stays neutral with regard to jurisdictional claims in published maps and institutional affiliations.

Copyright: © 2021 by the authors. Licensee MDPI, Basel, Switzerland. This article is an open access article distributed under the terms and conditions of the Creative Commons Attribution (CC BY) license (https://creativecommons.org/licenses/by/4.0/).

Abstract: In the context of imminent threats concerning biological and chemical warfare agents, the aim of this study was the development of a new method for biological and chemical decontamination, employing non-toxic, film-forming, water-based biodegradable solutions, using a nano sized reagent together with bentonite as trapping agents for the biological and chemical contaminants. Bentonite-supported nanoparticles of Cu, TiO_2, and Ag were successfully synthesized and dispersed in a polyvinyl alcohol (**PVA**)/glycerol (**GLY**) aqueous solution. The decontamination effectiveness of the proposed solutions was evaluated by qualitative and quantitative analytical techniques on various micro-organisms, with sulfur mustard (**HD**) and dimethyl methylphosphonate (**DMMP**) as contaminants. The results indicate that the peelable active nanocomposite films can be successfully used on contaminated surfaces to neutralize and entrap the hazardous materials and their degradation products. Mechanical and thermal characterization of the polymeric films was also performed to validate the decontamination solution's potential as peelable-film generating materials. The removal efficacy from the contaminated surfaces for the tested micro-organisms varied between 93% and 97%, while for the chemical agent **HD**, the highest decontamination factor obtained was 90.89%. **DMMP** was almost completely removed from the contaminated surfaces, and a decontamination factor of 99.97% was obtained.

Keywords: eco-friendly; nanoparticles; chemical warfare agents; sulfur mustard (**HD**); dimethyl methylphosphonate (**DMMP**); biological agents; nanocomposite films; decontamination; antimicrobial materials; peeling; active coatings

1. Introduction

The most recent global crisis caused by the SARS-CoV-2 pandemic demonstrated that it is vital to be prepared for emerging sanitary, biological, chemical, or environmental hazards. Decontamination has always represented a major challenge, but especially now, in the current situation of the COVID-19 crisis, developing efficient methods for the neutralization and the removal of the contaminants should become a priority. Although chemical and biological weapons are forbidden by the Chemical Weapons Convention [1] and Biological and Toxin Weapons Convention [2], certain states are still suspected to currently possess chemical weapons, as well as biological weapons. Moreover, even if a virus, like SARS-CoV-2, is not on the list of biological weapons, it is still important to develop new methods for biohazards management. Considering the large number of nosocomial infections with pathogens resistant to classical methods of treatment and decontamination (with antibiotics and disinfectants), in recent years, the emphasis has been on the development of new, versatile products with antimicrobial properties. Another important threat is represented by the main ingredients used for the synthesis of chemical warfare agents (**CWA**): sulfur mustard, sarin, soman, tabun, Vx, etc., which are also found in the manufacturing processes of various chemical and pharmaceutical industries (chlorine, phosgene, and cyanides). The use of chemical and biological agents from military stockpiles or biological civilian applications drives the critical need to improve decontamination capabilities worldwide. Therefore, decontamination plays a crucial role in defense against biological and chemical warfare agents (BCWA). After a chemical or biological attack, decontamination is vital. This complex process converts hazardous materials into products that can be safely handled. The methods that are typically applied are nucleophilic reactions or oxidations [3]. Toxic chemicals or micro-organisms must be eliminated by the application of efficient decontamination methods as quickly as possible in order to be able to resume routine activities. For military purposes, decontamination is undertaken to restore the combat effectiveness of equipment and personnel as rapidly as possible [4]. Most current decontamination procedures are labor- and resource-intensive, require excessive amounts of water, are corrosive and/or toxic, and are not considered environmentally safe [4–7]. Current research and development are focused on developing decontamination systems that would overcome these limitations and effectively decontaminate a broad spectrum of chemical and biological agents (CB agents) from all surfaces and materials [6,8–11]. There is no single technology that will be applicable in all situations and all types of contaminations because the nature and extent of contamination are different in different places. Surface decontamination is very difficult to achieve as contaminants can be located within the pores and cracks of materials, which makes their removal more challenging [12]. Depending on the type of contaminating agent, many decontamination methods for surfaces can be found in the literature: aspiration [13], abrasion of the surface layer [14], rinsing with water or with solvents [15], foams [15,16], gels [16], polymeric coatings [17–19], etc. The main disadvantages and limitations of the existing surface-decontamination solutions are: most of them are corrosive and/or toxic, affecting the decontaminated substrate and also exposing the user to hazardous materials; most of the existing decontamination methods are not considered environmentally safe because they require excessive amounts of reactants, solvents, or water, generating enormous quantities of post-decontamination waste, which requires subsequent decontamination; the decontamination systems that require large amounts of water do not represent a feasible solution because water can often be difficult to find (for example, on the battlefield [4]), and the disposal of this contaminated water will also further represent a cumbersome problem. Recent trends in BCWA decontamination technologies involve the use of materials capable of neutralizing contaminants under atmospheric conditions via hydrolysis and/or oxidation routes, under mild conditions of the reaction. Besides the classical decontamination methods, the use of modern methods that imply using polymeric films/coatings seems to bring multiple advantages for BCWA removal [11,17,20]. In comparison with the traditional techniques, these decontamination methods usually consist of applying a smaller amount of material (containing the active ingredients) onto the contaminated surface, thus resulting in a coating that will entrap and neutralize the contaminant and can be easily removed and compactly stored at the end of the decontamination process. This new decontamination method can be found referred to in that literature as "stripping/peelable coating techniques" [5,11,17,18,20]. This technique applies to a wide range of contaminants and surfaces/materials. Thus, polymeric peelable films represent a modern and versatile method for surface decontamination [6,10,14,21,22]. Various polymers can be listed as film-forming polymers: acrylates [17,21,23,24], silicones [25], vinyl polymers [12,26,27], polyurea [28–32], alginate derivates [17,33–35]. Film-forming materials are already commercially available as products comprising paint-like polymeric mixtures that can be applied by spray-on or roll-on/brush-on techniques, and they can form peelable coatings for decontamination/decommissioning purposes: CBI Polymers, New York, USA—DeconGel™ 1108, Instacote Inc Protective Coatings, Erie, USA—InstaCote™ CC Wet/CC Strip, or Bartlett Nuclear Inc., Plymouth, USA—StripCoat TLC Free™. Although these products do generate a smaller volume of secondary waste and they ensure reasonable DFs, their main disadvantage is represented by the toxicity and corrosivity of some of the active ingredients comprised in these commercial formulations.

The development of nanotechnology in various fields has experienced exponential growth over the last decade. In the biological and chemical decontamination fields, formulations based on nanoparticles and metal oxide nanoparticles have attracted a tremendous interest due to their remarkable properties. Small particle sizes and high specific surface areas bring multiple advantages and unique physicochemical properties that facilitate the adsorption and degradation of toxic compounds [36,37]. Advances in the preparation of metallic NPs and metallic oxides like ZnO, MgO, CaO, CeO_2, ZrO_2, TiO_2, etc., have led to the development of a new class of antimicrobial materials and decontaminants for chemical warfare agents with a high stability under harsh process conditions [8,37]. Highly ionic metallic NPs (e.g., Cu-NPs, Ag-NPs) are of particular interest due to their numerous reactive surface sites with atypical crystal morphologies. Ag and Cu nanoparticles immobilized on metal-oxide substrate have been demonstrated to neutralize viruses, bacteria, and fungi [38]. Nano-scaled copper particles (Cu-NPs) have many applications in industry, such as in gas sensors, high-temperature superconductors, solar cells, and other applicatoin. Copper ions have demonstrated antimicrobial activity against a wide range of micro-organisms (*Staphylococcus aureus*, *Salmonella enteric*, *Campylobacter jejuni*, *Escherichia coli*, and *Listeria monocytogenes*) [39]. The antibacterial effect exhibited against bacterial cell functions can occur through various mechanisms, depending on the physicochemical properties of NPs and the type of interactions between bacterial cells (e.g., adhesion to s Gram-negative bacterial cell wall due to electrostatic interaction [39]). These interactions lead to a disruption of the integrity of the bacterial membrane and finally cause the death of the micro-organism. Copper NPs possess better properties in comparison with other expensive metals with antimicrobial activity, such as silver and gold [40]. Silver nanoparticles (Ag-NPs) are known to neutralize both bacteria and viruses through metal-ion binding. In 2003, during the first SARS outbreak, Al_2O_3-supported Ag was investigated for the neutralization of SARS coronavirus, *E. coli* (bacterium), and *Debaryomyces polymorfus* (fungi). After only five minutes of exposure to the Ag nanoparticles, the three pathogens were inactivated successfully. The mechanism was not investigated, but it is assumed that catalytic oxidation is responsible and not metal poisoning (Au and Cu inactivate bacteria, viruses, and fungi only under aerobic conditions) [38].

Metal NPs and metal oxide NPs are also efficient for the decontamination of chemical warfare agents [41,42]. Sulfur mustard (**HD**) can be decontaminated through dichlorination, oxidation, or hydrolysis mechanisms, thus being converted into non-toxic products [41]. From all the materials used for the chemical degradation of **HD**, it was demonstrated that nano-oxides can adsorb and degrade sulfur mustard to thiodiglycol and divinyl sulfide at room temperature. A disadvantage of this decontamination method is that requires several hours for full degradation [42]. There are also studies regarding metal-organic frameworks (MOFs) constructed from metal ions or clusters and multifunctional organic linkers through self-assembly, which have been reported as perfect candidates for chemical and biological decontamination. The most well-known is Cu-BTC MOF, and it was also demonstrated to be capable of hydrolyzing **HD** and nerve agents under ambient conditions via its coordination of water molecules, which have an important practical value [43,44]. Silver nanoparticles (Ag-NPs) encapsulated in MOF were reported as efficient decontaminants for **HD** [43].

While the abovementioned BCWA decontamination methods offer promising possibilities, they also possess a series of disadvantages, such as high production costs, laborious production processes, toxicity and corrosivity of some of the active ingredients, generation of a large amount of post-decontamination waste, unsatisfying decontamination degrees, etc.

The novelty of this paper consists in the development of a new method of biological and chemical decontamination by employing non-toxic, film-forming, water-based biodegradable solutions using both neutralization and adsorption mechanisms for the removal of the contaminant from a surface by employing a nanosized reagent together with bentonite as trapping agents for BCWA contaminants. Once they are applied to the contaminated surface, the neutralization of the contaminants occurs, followed by their entrapment in the polymer-clay system. After drying, these solutions form strippable films that can be easily removed from the surface. Decontamination tests herein reported confirmed the antimicrobial activity of the decontamination solutions (DF $\geq$ 93%) and the successful neutralization and removal of chemical agents: up to 90% decontamination efficiency for **HD** and over 99% decontamination efficiency for dimethyl methylphosphonate. Therefore, this study reveals that BC contaminants were successfully neutralized and entrapped in the polymer matrix, demonstrating that this novel ecological approach towards obtaining innovative peelable active nanocomposite films for the removal of biological and chemical agents from contaminated surfaces could represent a powerful environmentally responsible tool for decontamination applications in the future.

2. Materials and Methods

2.1. Materials

Poly(vinyl alcohol) (**PVA**, 86.7–88.7% hydrolysis degree, Mw $\approx$ 130,000 Da, DP $\approx$ 2700, Sigma–Aldrich), hydrophilic bentonite (**BT**, Nanomer® PGV, Sigma–Aldrich, St. Louis, MO, USA), anhydrous glycerol (**GLY**, Sigma–Aldrich, St. Louis, MO, USA), copper(II) acetate monohydrate ($\geq$98%, Sigma Aldrich, St. Louis, MO, USA), titanium(IV) oxide (nanopowder, <100 nm particle size, 99.5% trace metals basis, Sigma Aldrich, St. Louis, MO, USA), Triton™ X-100 solution (Sigma Aldrich, St. Louis, MO, USA), ascorbic acid ($\geq$99%, Sigma Aldrich, St. Louis, MO, USA), Silver nitrate ($\geq$99.0%, Sigma Aldrich, St. Louis, MO, USA), tri-sodium citrate dihydrate (Sigma Aldrich, St. Louis, MO, USA), and sodium borohydride ($\geq$99.0%, Sigma Aldrich, St. Louis, MO, USA) were used as received. For the chemical decontamination tests, real chemical warfare agents (**CWA**) were used: bis(2-chloroethyl) sulfide (**HD,** sulfur mustard, purity: 95%, own synthesis), together with a chemical warfare simulant: dimethyl methylphosphonate (**DMMP**, as simulant for nerve agents, $\geq$97%, Sigma Aldrich). All the tests involving the decontamination of the toxic agents utilized in this study were performed at the Research and Innovation Center for CBRN Defense and Ecology in the 'Chemical Analysis Laboratory' from Bucharest, the only OPCW-designated laboratory in Romania.

2.2. Methods

2.2.1. Preparation of Decontamination Solutions

Solutions free of metal nanoparticles and bentonite were initially prepared to serve as reference points (Table 1). The BCWA decontamination solutions based on bentonite-supported metal nanoparticles were obtained as follows: the metallic salts were dissolved in water (according to Table 2), various amounts of bentonite and TiO_2 were dispersed in these solutions (continuous magnetic stirring, 800 rpm), and the nanoparticle precursor and bentonite (or bentonite and TiO_2) were kept in contact under stirring for 24 h. After this, the obtained dispersions were sonicated for 30 min while the corresponding reducing agents (according to Tables 2 and 3) were added. **PVA** was introduced next, and the dispersions were maintained at 95 °C under vigorous stirring until the complete dissolution of the polymer. Finally, the glycerol was added last, while the dispersions were allowed to cool down.

Table 1. Composition of the reference samples.

Sample	PVA (wt.%)	GLY (wt.%)	BT (wt.%)
SD1	10	-	-
SD2	10	5	-
SD3	10	5	1

Table 2. Composition of the solutions containing CuNPs.

Sample	PVA (wt.%)	GLY (wt.%)	BT (wt.%)	Cu $(CH_3COO)_2$ • H_2O (wt.%)	TiO_2 (wt.%)	*Dispersing Agent* Triton X100 (wt.%)	*Reducing Agent* Ascorbic Acid (wt.%)
SD4	10	5	0.5	0.04	-	-	0.143
SD5	10	5	1	0.04	-	-	0.143
SD6	10	5	1.5	0.04	-	-	0.143
SD7	10	5	0.5	0.04	1	0.25	0.143
SD8	10	5	1	0.04	1	0.25	0.143
SD9	10	5	1.5	0.04	1	0.25	0.143

Table 3. Composition of the solutions containing AgNPs.

Sample	PVA (wt.%)	GLY (wt.%)	BT (wt.%)	$AgNO_3$ (wt.%)	*Reducing Agent 1* Trisodium Citrate Dihydrate (wt.%)	*Reducing Agent 2* $NaBH_4$ (wt.%)
SD10	10	5	0.5	0.04	0.012	0.01
SD11	10	5	1	0.04	0.012	0.01
SD12	10	5	1.5	0.04	0.012	0.01

2.2.2. Preparation of the Nanocomposite Films

The obtained solutions were further used for decontamination tests, but they were also employed for obtaining square-shaped, thin nanocomposite films that were useful for characterization through different analytic procedures. To obtain the nanocomposite films by the casting method, approximately 100 mL of each decontamination solution was introduced in a square (12 cm × 12 cm) glass mold, placed on a perfectly flat surface, and allowed to dry (at 25 °C, 50–55% relative humidity). Afterwards, the films obtained were peeled and employed for further investigation.

2.3. *Decontamination Tests*

2.3.1. Biological Decontamination Tests

The biological decontamination tests involved the characterization of the antimicrobial activity of the decontamination solutions, followed by the evaluation of the efficacy of the strippable nanocomposites for the removal of biological contaminants from the targeted surfaces. For the characterization of antimicrobial activity of these substances, three of the most used methods appropriate for this type of sample were applied: minimal inhibitory concentration (MIC), minimal bactericidal concentration (MBC) and time-kill test. The MIC value is defined as the lowest concentration of the antimicrobial agent that inhibits the visible growth of the micro-organisms tested. This is usually expressed in mg/mL or mg/L. The MBC is defined as the lowest concentration of antibacterial agent needed to kill 99.9% of the final inoculum after incubation for 24 h under a standardized set of conditions. Time-kill assay is the most appropriate method for determining bactericidal effect. It is a good method for obtaining information about the interaction between the antimicrobial agent and the microbial strain. The time-kill test reveals a time-dependent or a concentration-dependent antimicrobial effect [45].

Minimal Inhibitory Concentration

The antimicrobial activity of the decontamination solutions was evaluated against *Staphylococcus aureus* (ATCC 6538) as a model for Gram-positive bacteria and *Escherichia coli* (ATCC 8739) and *Pseudomonas aeruginosa* (ATCC 9027) as a model for Gram-negative bacteria. *S. aureus*, *E. coli*, and *Ps. aeruginosa* were chosen, considered standard micro-organisms for testing the antimicrobial properties of newly synthesized products [46]. After cultivation overnight in Muller Hinton broth (MHb) (Merck) at 37 °C with stirring (200 rpm), the bacterial strains were harvested. Portions of suspension were harvested by centrifugation and resuspended in phosphate buffer saline (PBS) (Sigma-Aldrich, St. Louis, MO, USA). The suspensions were adjusted to approximately 10^6 CFU/mL [47]. Minimum inhibitory concentrations (MIC) were established for each solution by the broth microdilution method [48,49]. Two-fold serial dilutions of each solution were performed in MHb in duplicate. Negative and positive controls were associated [50]. The inhibitory effect of the substances was evaluated starting from 50% concentration (the samples of substances were diluted 1:1 with MHb). A total of 10 µL of the micro-organism suspensions (~10^4 CFU) was added in each well corresponding to the testing samples and controls. Because the solutions are turbid themselves and the bacterial growth is difficult to discern, at the end of the incubation period, 10 µL of resazurin 0.1% was added to each well. After 2 h of incubation with resazurin, the plates were read.

Minimal Bactericidal Concentration

MBC was determined after broth microdilution by subculturing the content of each well that did not show any visible signs of growth on the surface of non-selective agar plates (Muller-Hinton agar, MHa, Merck, Darmstadt, Germany). This allowed for the determination of the number of surviving cells (CFU/mL) after 24 h of incubation at 37 °C [45].

Time-Kill Test

Portions of more concentrated bacterial strain suspensions (10^7 CFU/mL), prepared as previously described, were treated with the studied substances at 2× MIC concentration for solutions with MIC values established and 50% concentration for the others and kept in direct contact for 2 h and 24 h, respectively, at 37 °C. At each established time, portions of bacterial cultures were serial diluted in PBS and then plated on Muller Hinton agar medium (MHa). After incubation at 37 °C for 24 h, bacterial survival was evaluated [51].

Efficacy of Biological Contaminant Removal from the Targeted Surfaces

The efficacy of micro-organism removal from the surface was determined by calculating the decontamination factor achieved by polymeric nanocomposite films after exfoliation. Sterile surfaces (sterile Petri dishes) were contaminated with known concentrations of micro-organisms. Two micro-organisms were applied separately: *Staphylococcus aureus* (ATCC 6538) and *Escherichia coli* (ATCC 8739). The contaminated plates were dried, and then the polymer solutions were applied. The plates were kept under airflow at room temperature. After 24 h, the formed polymeric films were exfoliated. After this, the culture media was applied by spreading on the decontaminated surfaces of Petri dishes. The plates were incubated at 37 °C for 24 h. After the incubation period, the number of CFUs was counted. DF was calculated by considering the initial contamination and the number of CFUs counted after the incubation period.

2.3.2. Chemical Warfare Agent Decontamination Tests

Chemical decontamination using one real warfare agent, sulfur mustard (Yperite or **HD**), and one simulant for nerve agents, dimethyl methylphosphonate (**DMMP**), was performed as follows: firstly, controlled contamination was performed, followed by the application of the decontamination solution on the contaminated surface (after 5 min from contamination); following the film-curing process (20–24 h), the nanocomposite films containing the degradation products of **HD** and **DMMP** were peeled off and subjected to extraction in DCM, followed by GC-MS analysis.

In the first step (controlled contamination, 10 $g_{(toxic)}/m^2$), some metallic probes, measuring 10 cm^2, were contaminated with 7.87 μL (10 mg) of **HD** (ρ = 1.27 g/cm^3) and other metallic probes, also measuring 10 cm^2, were contaminated with 8.61 μL (10 mg) of **DMMP** (ρ = 1.145 g/cm^3). After 5 min of toxic-metallic probe direct contact, approximately 1.5 mL of decontamination solution was poured over the contaminated area, making sure the surface was completely covered by the liquid. Once the decontamination solution was placed on the contaminated surface, the active ingredients set up the degradation of the toxic agent while being adsorbed by bentonite nanoclay and entrapped in the polymeric matrix of the nanocomposite. After approximately 20 h, the polymeric nanocomposite coating containing the entrapped contaminants could be easily peeled off. Both the decontaminated surface and the film obtained were subjected to extraction in 10 mL of dichloromethane (DCM). The organic extracts were dried over sodium sulfate, filtered on 45 μm a Sartorius filter, and analyzed by the GC-MS technique. To evaluate the decontamination efficiency, the decontamination factor (DF) was calculated by considering the initial concentration of contaminant and the residual toxic found on the metallic probe after decontamination.

$$DF = 100 \times (C_0 - C_f)/C_0 \quad (1)$$

where DF is the decontamination factor, C_0 is the initial toxic concentration, and C_f is the final concentration, reflecting the residual contamination. Measurements were repeated in triplicate, and the average values obtained were reported.

2.4. *Characterization*

To acquire quality imaging of the samples, a high-resolution transmission electron microscope (HRTEM), type TECNAI F30 G2STWIN, Fei Company, Oregon, USA, was used at 300 kV acceleration voltage and with a resolution of 1 Å. The correlation between dynamic viscosity and shear gradient of the decontamination solutions was studied to establish a model flow profile of the polymeric solution with superior film-forming characteristics. Rheological tests were performed on a Rheotest 2.1 device (Rheotest Medingen GmbH, Ottendorf-Okrilla, Germany) with coaxial cylinders at room temperature (25 °C) to determine the behavior of the solutions. The amount of solvent evaporated in time at different temperatures (25 °C, 30 °C, and 35 °C) was used to investigate the drying profile of the nanocomposite films. An ATS 120 Axis Thermobalance was used to measure the evaporation rate of 4 mL of sample for evaluation of the film-formation process. Promas software calculated the evaporation rate by weighting the sample every 150 s. FT-IR spectra were obtained using a Perkin Elmer Spectrum Two (Perkin Elmer, Waltham, MA, USA) with a Pike MiracleTM ATR modulus and a 4 cm^{-1} resolution, from 550 to 4000 cm^{-1}. To investigate the mechanical proprieties, polymeric films were obtained by casting method and then cut in a dumbbell shape with 75 mm overall length and a narrow section of about 25 ± 1 mm, which were subsequently subjected to tensile tests on a 710 Titan 2 universal strength-testing machine equipped with a 3000 N force cell, according to ISO 37: 2011(E). The test involves continuous observation of the length and force variation with an accuracy of ±0.2% at a speed of 8.33 mm/s. To compare the results, the mean values of each sample were plotted in a stress/strain graph. Five specimens from each sample were subjected to tensile tests. Samples weighing approximately 25–30 mg were subjected to thermal tests, heated from 30 °C to 450 °C with a constant heating rate of 5 °C/min on a DTA OZM 551 Ex Differential Thermal Analysis System equipped with Meavy dedicated software. GC-MS investigations were performed on a GC Thermo Scientific Trace 1310 (Thermo Fisher Scientific, Waltham, MA, USA) gas chromatograph coupled with a TSQ 9000 triple quadrupole mass spectrometer (MS/MS) (Thermo Fisher Scientific, Waltham, MA, USA) using a TR5MS GOLD capillary column (5% phenyl 95% dimethylpolysiloxane). The injection mode used was splitless with an injector temperature of 250 °C and helium as carrier gas (1.5 mL/min). The temperature program started from 40 °C, up to 300 °C, with a rate of 10 °C/minute. Electron impact ionization (EI) mode (mass range between 40 and 650 amu) was used. The compounds were identified based on the interpretation of MS/EI fragmentation.

3. Results and Discussion

This first step of this study consisted of the synthesis of bentonite-supported metal and metal oxide nanoparticles that are suitable for decontamination applications, capable of reacting with chemical warfare agents to form non-toxic products while also neutralizing biologic agents due to their anti-bacterial properties. Thus, nanosized hydrophilic bentonite was used as support for the metal/metal oxides nanoparticles during their generation. In the decontamination process, bentonite can act as an efficient adsorbent for the contaminants, facilitating their deactivation induced by the presence of the nanoparticles. The bentonite-supported metal/metal oxide nanoparticles were dispersed in an aqueous solution of polyvinyl alcohol (**PVA**), a biodegradable polymeric matrix, which plays an essential role in holding together (and thus "binding") all the components of the decontamination solution (including the entrapped contaminant). The excellent film-forming capacity of **PVA** ensures the formation of peelable films from these decontamination solutions, which facilitates the efficient removal of contaminants from different types of surfaces.

TEM analysis was employed as the first characterization tool to confirm the generation of Cu and Ag nanoparticles in the decontamination solutions (Figure 1). From the images presented in Figure 1A–D, it can be noticed that for the samples containing 1% bentonite (SD5), the CuNPs diameter is around 20 nm, while for the SD6 (containing 1.5% bentonite), the particle size is around 40–60 nm. In the case of SD4 (0.5% bentonite), the presence of independent nanoparticles outside the clay structure was not visible by TEM. Thus, as the concentration of bentonite increased in the samples, larger nanoparticles were observed. Similar behavior was evidenced for the AgNPs (samples SD10, SD11, and SD12). The particle size of the AgNPs increased at a higher clay content. An explanation for the particle size modification can be attributed to the adjustment of the growth process due to the presence of clay, which causes smaller particles to destabilize, due to the clay's inherent electrostatic charge [52], and promotes growth. In the case of the samples containing TiO_2, no visible effect on the particle size was observed; the CuNPs' particle size varied, depending on the bentonite content.

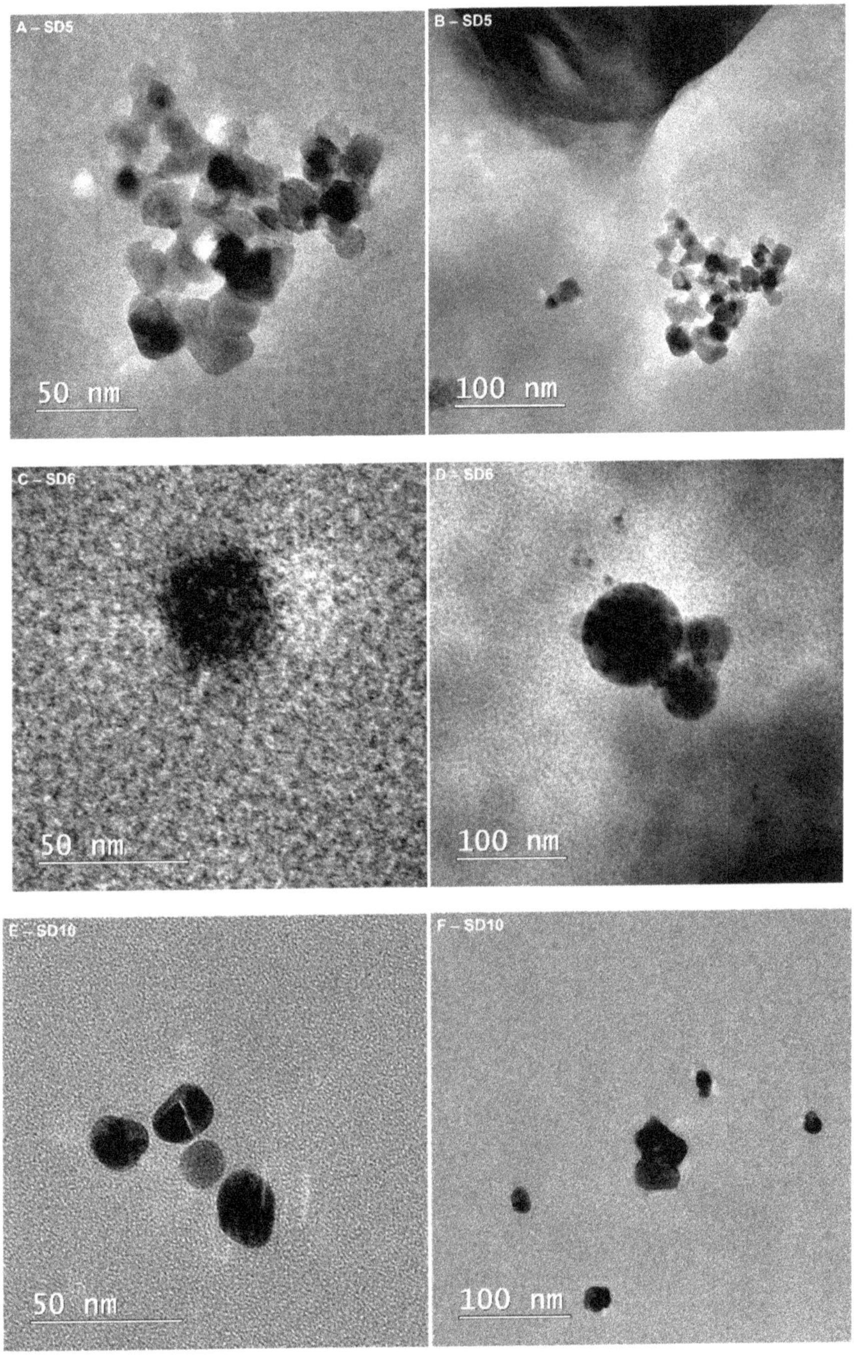

Figure 1. *Cont.*

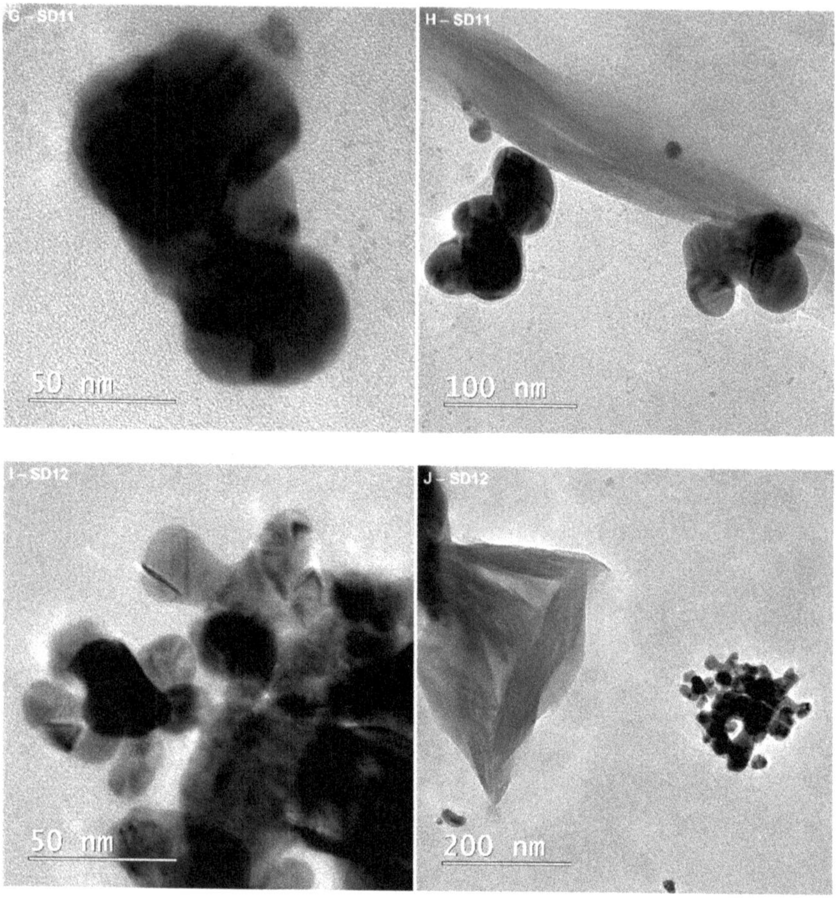

Figure 1. TEM image of bentonite-supported metal nanoparticles employed in the decontamination solutions: (**A**,**B**) SD5-mag 100,000×, mag 50,000×; (**C**,**D**) SD6-mag 180,000×, mag 50,000×; (**E**,**F**) SD10-mag 180,000×, mag 50,000×; (**G**,**H**) SD11-mag 180,000×, mag 60,000×; (**I**,**J**) SD12-mag 180,000×, mag 29,000×.

An important parameter that influences both the polymer solution deposition procedure and the type of surface that will undergo decontamination is represented by the dynamic viscosity of the solution. Thus, the solutions presenting a higher viscosity are suitable for application by brush technique, whereas less viscous solutions can be deposited by spray technique. A higher viscosity can also affect the mobility of the molecules influencing the rate of adsorption of the biological compounds, as well as decrease the capacity of the solution to enter the pores and cracks of the surface. Thus, the influence of each component on the viscosity of the solutions was evaluated, and the results are illustrated in Figure 2. Polyvinyl alcohol (**PVA**) has an essential role in film formation (SD1), while the addition of glycerol (SD2) improves the elasticity of the peelable films and bentonite aids in the improvement of the biological/chemical-agent retention inside the film by complexation and adsorption processes (SD3). The effect of glycerol and bentonite on the viscosity of the solution can be explained by the physical interactions that occur between the polymer and these components, such as the formation of hydrogen bonds. Considering these aspects, all the components employed can form hydrogen bonds with water or **PVA**, thus connecting the macromolecules through physical interaction. Consequently, there is an apparent increase in the molecular weight of the polymer, which leads to an increase in the viscosity of the decontamination solution. Additionally, the variation of the dynamic viscosity of the decontamination solutions containing copper and silver nanoparticles was investigated. Compared with the control sample (SD3), it can be noticed that for all the solutions, similar dynamic viscosity values were obtained, due to the low concentration of nanoparticles.

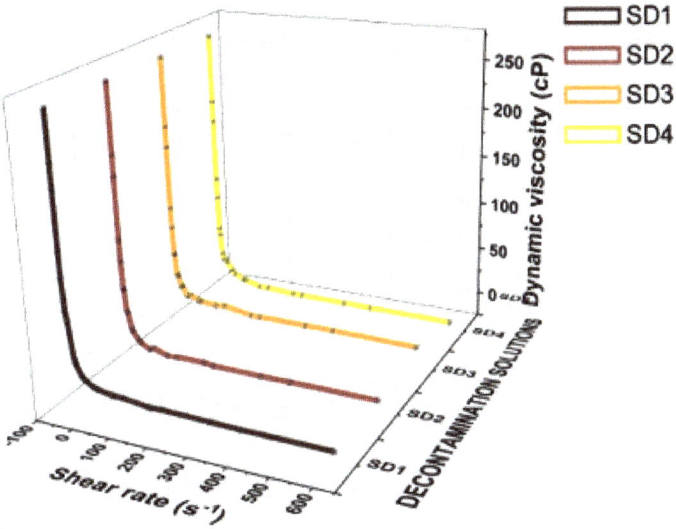

(a)

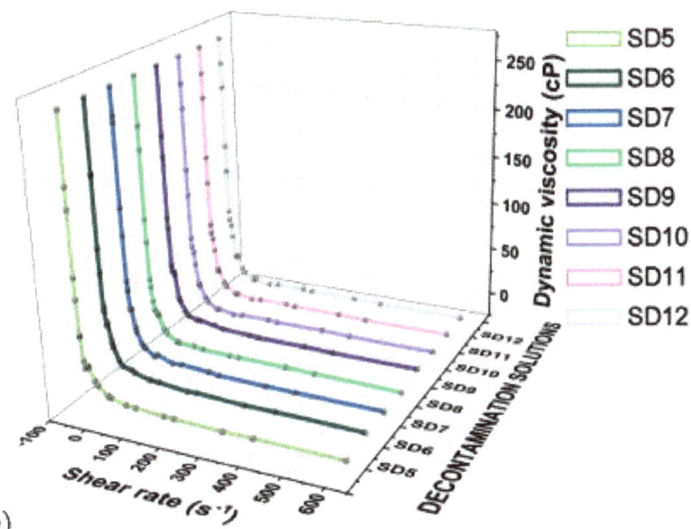

(b)

Figure 2. Variation of the dynamic viscosity with shear rate for the decontamination solutions, (**a**) SD1, SD2, SD3, SD4 and (**b**) SD5, SD6, SD7, SD8, SD9, SD10, SD11, SD12.

In order to determine the time required for solvent evaporation from the decontamination solution and film formation, determination of the evaporation rate is essential. The capacity of solvent evaporation (water in this case), after solution deposition, depends on several factors: temperature, humidity, type of surface, etc. but also on the viscosity and nature of solution constituents. To ascertain these values, parameters such as sample surface and quantity of solution were kept constant. The evaporation parameters are presented in Table S1. As the temperature is increased, the solvent evaporation rate is more predominantly influenced by the chemical composition of the solution, the interaction between the components, and solution viscosity. The measurements taken at temperatures between 25 and 40 °C showed that the solution's components influence the evaporation rate, which is due to the intermolecular interaction between the components and water molecules; thus, the stronger the interaction, the longer the interval required for drying.

FT-IR spectroscopy was employed to highlight the formed hydrogen bond between the polymer and glycerol, respectively, as well as the presence of bentonite in the polymer films. The results are presented in Figure 3 and detailed in Figure S1. The broad peaks around 3300 cm^{-1} can be attributed to hydroxyl groups from the **PVA** chain, while the values in the range of 1095–1085 cm^{-1} indicate the presence of hydrogen bonds formed between **PVA** and glycerol. The weak bands at 3028 and 2919 cm^{-1} can be assigned to C–H stretching vibrations. The strong peak highlighted around 1031 cm^{-1} has a double meaning: it appears due to the strong vibrations of the C–O bonds of a primary alcohol, and at the same time, it also indicates the presence of Si–O bonds due to bentonite clay. At the same time, an increase in absorbance is observed with the addition of bentonite.

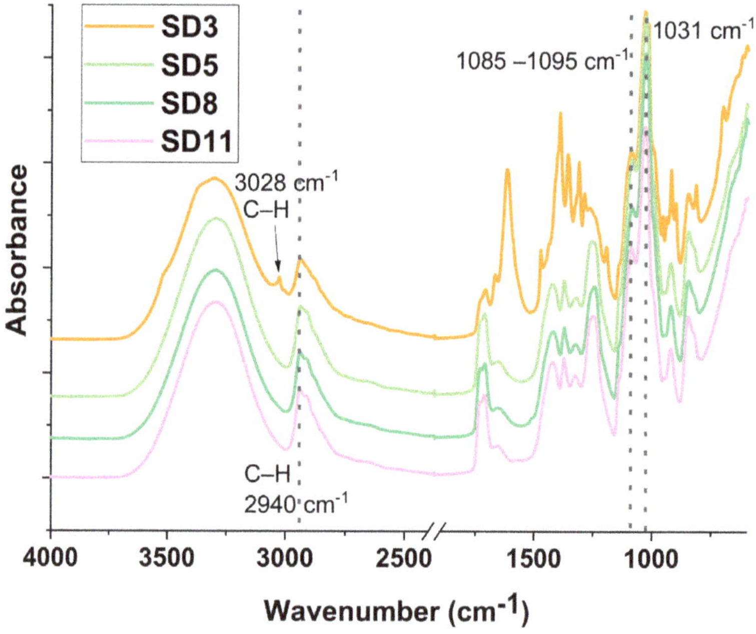

Figure 3. FT-IR spectra of the polymer films (SD3, SD5, SD8, and SD11).

The mechanical properties of polymer films are very important for the peeling process and efficient decontamination of surfaces, which requires a high resistance but also a certain degree of elasticity. Figure 4 illustrates comparative stress-strain plots for all the nanocomposites obtained. As can be observed, the addition of glycerol led to higher stress-strain values. In the absence of glycerol, films containing only polyvinyl alcohol (SD1) are more resistant, but they are much too rigid and brittle to be used for surface decontamination. Based on the results of the tensile strength tests, it can be stated that each component modifies the mechanical properties of the films. Moreover, this aspect must be considered when formulating decontamination solutions to obtain the desired characteristics. Thus, to obtain films easily exfoliated from the surface of interest while avoiding fracture of the composite material, careful selection of components and their ratio is required. The exact values of the mechanical parameters are given in the Table S2. As can be observed, the nanocomposites containing metallic nanoparticles also displayed good mechanical properties. The polymeric films maintained their integrity after the completion of peeling. Thus, when these materials were subjected to low stretching forces (typical for a peeling process) the nanocomposite film had enough mechanical resistance and did not break. The mechanical resistance of the nanocomposites employed for surface decontamination is afforded by the synergistic effect between the reinforcing nanoclay, the polymer matrix, and the glycerol (acting as plasticizer).

The differential thermal analysis presented in Figure 5 allowed the evaluation of the thermal characteristics of the polymer nanocomposite films. The thermal behavior of the film containing only **PVA** (**SD1**) is slightly different than that of the films containing glycerol. Thus, the peak situated at approximately 236 °C can be attributed to the melting of the crystalline regions/domains of **PVA**. Comparing SD2 and SD3, very small differences in terms of thermal transitions can be noticed. The first characteristic signal is endothermic for a temperature range between 70 and 150 °C (Figure 5A), which can be attributed to the evaporation of water trapped between the polymer chains. The shift of this peak to slightly higher values in the case of SD3 can be explained by the increased interaction due to the presence of bentonite. The second signal between 175 and 260 °C, also endothermic, can be attributed to the melting of the polymer (T_m), while the signals after 270 °C can be attributed to the polymer degradation process. Similarly, the presence of bentonite (SD3), copper nanoparticles (SD5, SD8), or silver nanoparticles (SD11) in the composition of the polymer films leads to a slight modification of the characteristic temperature response; nevertheless, the responses of all samples are within the abovementioned temperature intervals. The decrease in melting temperature could be attributed to an increase in the thermal conductivity and polymer chain mobility due to the presence of the metallic nanoparticles (Table S3).

3.1. Decontamination Tests

The decontamination tests were performed to prove and evaluating the efficacy of this BCWA decontamination method. Biological decontamination tests were performed first, using *E. coli*, *Ps. aeruginosa* and *S. aureus* (as simulants for biological agents), and they were followed by chemical decontamination tests run on one real chemical warfare blistering agent, **HD**, and one simulant for neuroparalytic agents, **DMMP**. Figure 6 illustrates the steps taken to perform the decontamination using the herein-reported film-forming solutions.

The method for biological or chemical decontamination consists of the utilization of the synthesized eco-friendly active solutions (containing bentonite-supported nanoparticles) for the degradation/neutralization and entrapment of toxic agents, followed by the exfoliation of the formed film, which contains the degradation products resulting from the neutralization of the targeted hazardous materials.

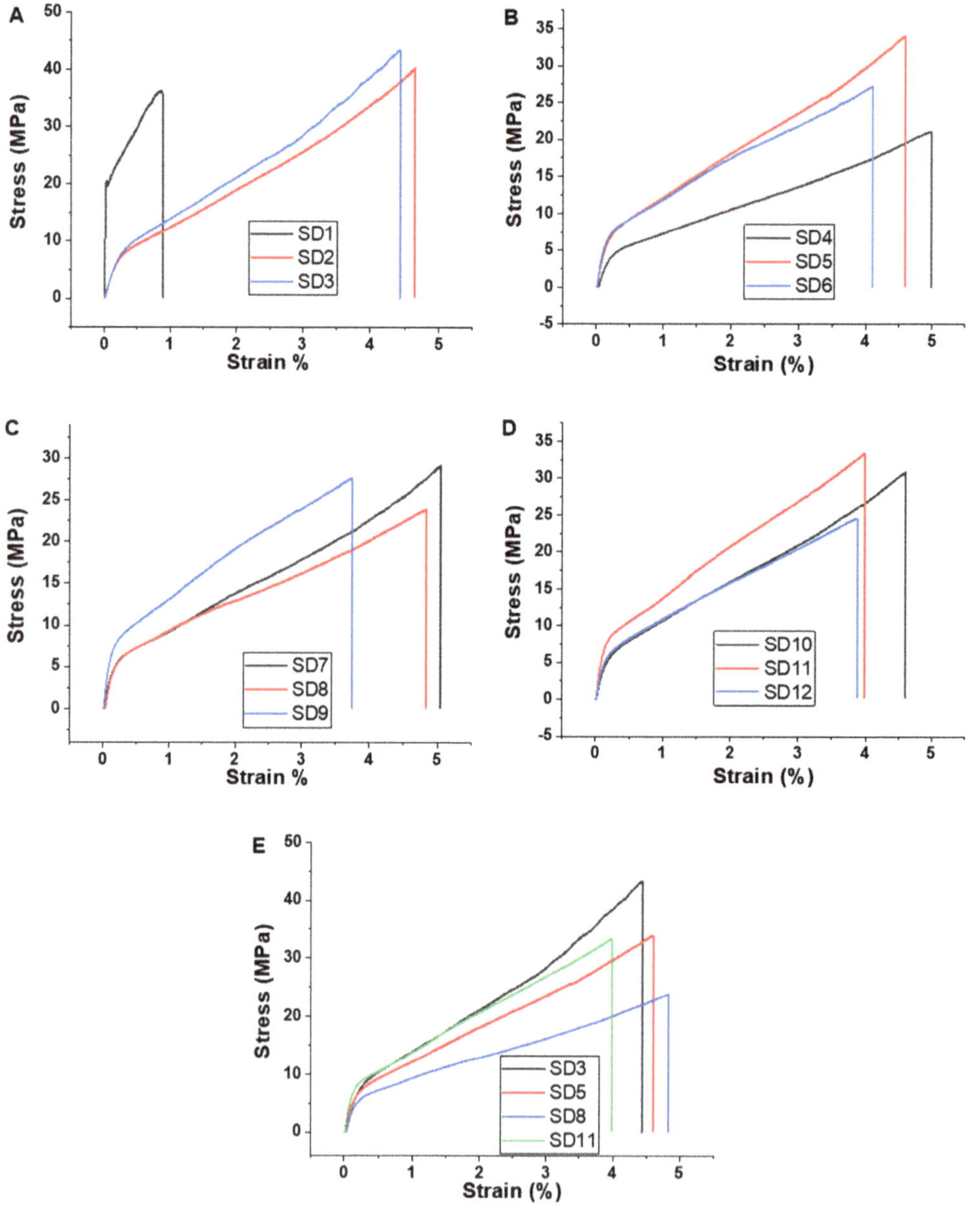

Figure 4. Tensile tests results for the polymer films: (**A**) SD1, SD2, SD3; (**B**) SD4, SD5, SD6; (**C**) SD7, SD8, SD9; (**D**) SD10, SD11, SD12; (**E**) SD3, SD5, SD8, SD11.

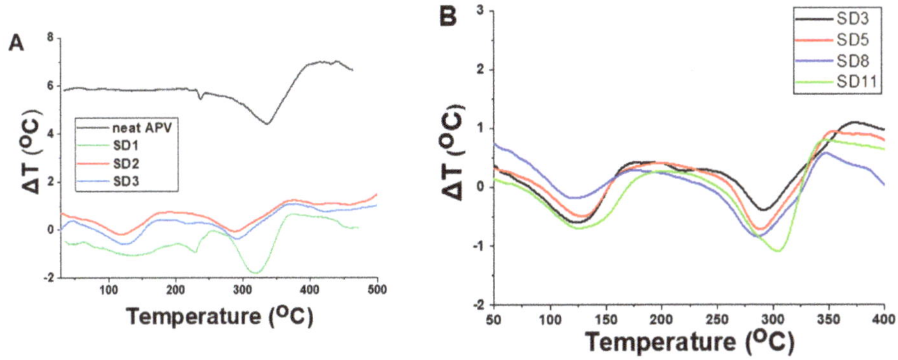

Figure 5. DTA thermograms for the polymer films (**A**) neat APV, SD1, SD2, SD3 and (**B**) SD3, SD5, SD8, SD11.

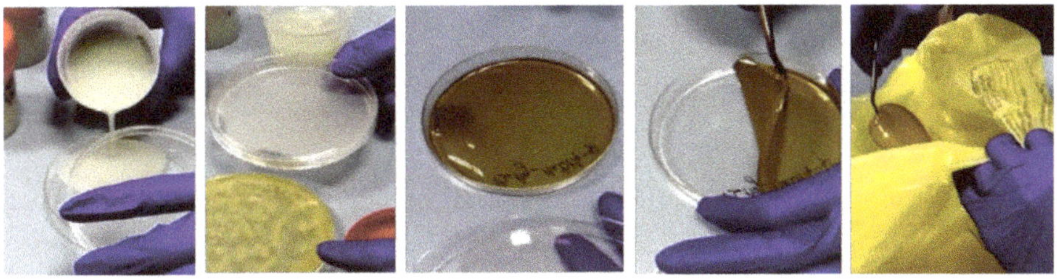

(a) (b) (c) (d) (e)

Figure 6. Decontamination using eco-friendly active nanocomposite peelable coatings: (**a**) decontamination solution, (**b**) decontamination solution is allowed to neutralize the contaminant, (**c**) dried peelable film, (**d**) peeling process and (**e**) decontamination waste.

3.2. Biological Decontamination Tests

The results obtained of decontamination tests performed on biological contaminants are further detailed. An antimicrobial activity assay generated the MIC and MBC values of the decontamination solutions, displayed in Table 4 and Figure 7A,B. Solutions SD4, SD5, and SD6 revealed the lowest antimicrobial activity against bacterial strains used in the test. Solutions SD7, SD8, and SD9 showed low antimicrobial activity against Gram-negative bacteria (*E. coli*, *Ps. aeruginosa*) and pronounced activity against Gram-positive bacteria (*S. aureus*), MBC being established in this case. Solutions SD10, SD11, and SD12 showed stronger antimicrobial activity against Gram-negative bacteria, with MBC values established, but lower than those presented for Gram-negative bacteria for solutions SD7, SD8, and SD9.

Table 4. Minimal inhibitory concentration (MIC) and Minimal bactericidal concentration (MBC).

Sample/ Micro-Organism	*E. coli*		*Ps. aeruginosa*		*S. aureus*	
	MIC (%)	MBC (%)	MIC (%)	MBC (%)	MIC (%)	MBC (%)
SD4	>50	–	50	–	50	–
SD5	>50 *	–	50	–	50	–
SD6	>50 *	–	50	–	25	–
SD7	>50 *	–	25	–	1.56	25
SD8	50	–	25	–	0.78	12.5
SD9	25	–	25	–	0.39	25
SD10	6.25	25	0.39	3.125	3.125	–
SD11	12.5	25	0.78	12.5	6.25	–
SD12	12.5	25	0.39	6.25	6.25	–
SD3 (BK)	–	–	–	–	–	–

* Under the test conditions, the antimicrobial activity of the solution against the *E. coli* micro-organism could not be highlighted.

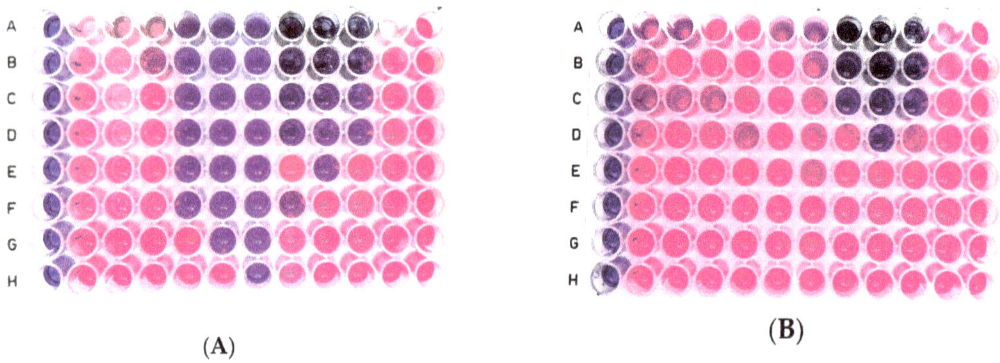

Figure 7. ((**A**) *S. aureus* ATCC 6538) MIC determination against *S. aureus* ATCC 6538 observed from broth microdilution assay using MH broth and resazurin (columns from left to right: NC, SD4, SD5, SD6, SD7, SD8, SD9, SD11, SD10, SD12, SD3 (Bk), PC). and (**B**) *E. coli* ATCC 8739.observed from broth microdilution assay using MH broth and resazurin (columns from left to right: NC, SD4, SD5, SD6, SD7, SD8, SD9, SD11, SD10, SD12, SD3 (Bk), PC).

The inhibition of the bacterial strain growth could be explained by specific interactions of nanoparticles with the cell envelope of micro-organisms [53]. When nanoparticles are small enough, they can penetrate membrane pores. Nanoparticles that can enter the cell membrane interact with bacterial enzymes, damaging the cell [54]. Some nanoparticles interact electrostatically with the bacterial membrane, and reactive oxygen species are generated, leading to the disruption of the membrane and DNA damages [55].

A time-kill assay was performed on *E. coli* and *S. aureus*. After the proposed contact times (2 h and 24 h) between strains and bentonite-supported nanoparticle solutions, bacterial growth was evaluated. In the case of the time-kill assay, it is observed that after 2 h of contact, all the decontamination solutions showed activity against both micro-organisms. After periods longer than 2 h of contact, in the case of *E. coli* strains, an increase in the number of CFU/mL was observed. Most likely, in the case of this bacteria, nanoparticles adhered to the surface of the cells or penetrated inside the membranes and were blocked. It is known that nanoparticles containing Cu, Zn, and Ti ions bind to negatively charged membranes (such as *E. coli*) [56]. This would explain the survival rate and resumption of the growth and division cycle. Considering that after 2 h of contact, an accelerated increase in the number of CFU/mL was observed, only the values recorded at time t1 (2 h of direct contact) were represented graphically (the results are illustrated in Table 5, Figures 8 and S3), highlighting their antimicrobial activity during the short contact period. In the case of S. aureus, the activity is more pronounced, with observed antimicrobial activity even after 24 h in the case of SD7, SD8, SD9, SD10, SD11, and SD12 (results illustrated in Figure S4). The substrate solution (BK) has a slight activity on *E. coli* but no activity on *S. aureus*.

Table 5. Bacterial cell population decrease (%) after 2 h of contact with decontamination solutions.

Micro-Organism/Sample	SD4	SD5	SD6	SD7	SD8	SD9	SD10	SD11	SD12	SD3 (BK)
E. coli	96.25	98.65	97.53	98.1	97.23	93.25	95.43	95.43	80.25	47.5
S. aureus	87.2	93.7	86.6	90.4	94.6	95.8	96.1	97	97.35	−75

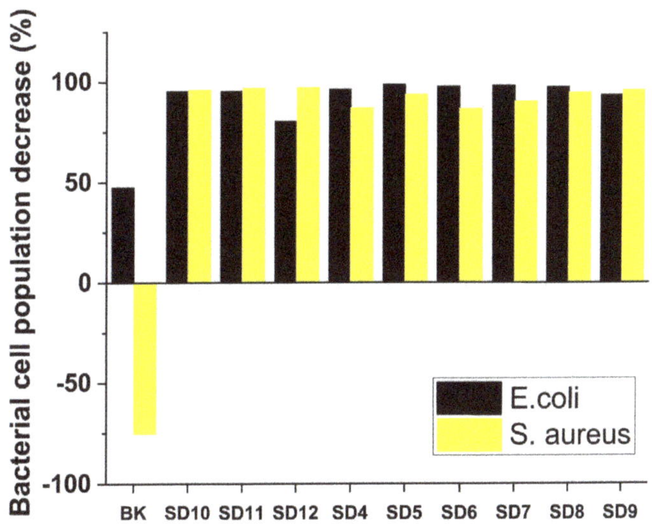

Figure 8. Bacterial cell population decrease (%) after 2 h of contact with decontamination solutions.

Rutala et al. [57] showed that the use of soap and water can sometimes be less efficient due to their lower microbial reduction capacity (≤80% reduction, in comparison with a phenolic disinfectant, which offers 94–95% reduction) and also due to the possibility of contamination of the soap solution. However, a few hours later, the bacterial count was nearly back to the pretreatment level [57]. In an Ayliffe et al. study [57,58], bacterial contamination of soap and water without a disinfectant increased from 10 CFU/mL to 34,000 CFU/mL after cleaning a hospital ward. If the soap solution or the mop are reused, contamination will, in fact, be transferred from one room to another.

The use of strippable coatings offers the advantage of avoiding these re-contamination incidents like the ones described above because on each contaminated surface, a new coating is formed, and after removal, it can be compacted and sealed in small containers dedicated to biological waste. As a conventional substitute for the classical soap and water, disinfectants significantly improved microbial removal when a conventional string mop was used (95% vs. 68%) [57], but using a microfiber mop instead of the conventional mop could also prevent the possibility of transferring microbes from room to room if a new microfiber pad is used in each room. By comparing the classical decontamination methods with the advantages of the strippable coating method, coupled with the DF values obtained for our decontamination solutions, we can affirm that this new method, based on peelable films, ensures sufficiently high values of microbial reduction while bringing the advantage of consisting of eco-friendly materials.

To evaluate the efficacy of biological contaminant removal from the targeted surfaces, controlled contamination of Petri dishes with *E. coli* and *S. aureus* was performed, followed by the addition of the decontamination solution. At the end of the curing process, the obtained nanocomposite coatings were easily peed off (Figure 6), and the decontaminated surface was further investigated to evaluate the decontamination efficiency.

The surfaces of Petri dishes were contaminated with portions of 10 μL suspension of *E. coli* (5 × 103 CFU/Petri dish) and *S. aureus* (7 × 103 CFU/Petri dish). Following the application of the decontamination solutions and removal of the peelable films, the number of residual micro-organisms on the targeted surface was assessed by cultivation in culture media (MHa). The effectiveness of the biological decontamination can be expressed utilizing the decontamination factor (DF). The decontamination factor can be calculated by the following equation:

$$DF = 100 \times (C_i - C_f)/C_i \quad (2)$$

where C_i represents the contamination level before applying the decontamination solution and C_f reflects the residual contamination [6]. Table 6 presents the DFs obtained for *E. coli* and *S. aureus*, and to facilitate comparison, Figure 9 summarizes all these values.

Table 6. Efficacy of removal of *E. coli* and *S. aureus* strains from surfaces.

Micro-Organism/Sample	SD4	SD5	SD6	SD7	SD8	SD9	SD10	SD11	SD12	SD3 (BK)
E. coli	95	94	94.6	95.8	95.4	94.2	95.8	95	94	93.6
S. aureus	95.9	95.6	95.4	96.3	96.4	96.1	97.1	96.9	96.6	95.6

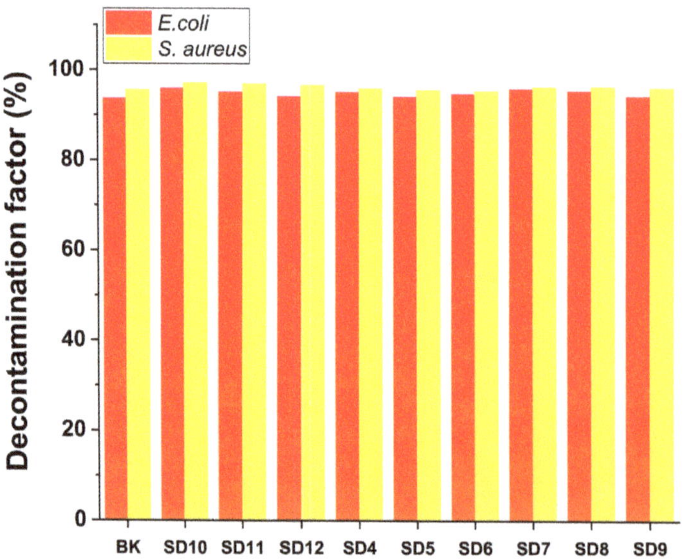

Figure 9. Biological decontamination efficacy.

The efficacy of removal for the tested micro-organisms varies (93% < DF < 97%). Therefore, we can affirm that these polymeric decontamination solutions represent a useful tool for biological decontamination of surfaces. Basically, biological decontamination occurred through two mechanisms: the first one consists of the entrapment of micro-organisms in the polymeric matrix of the nanocomposite due to the excellent adsorptive properties of bentonite nanoclay; and the second one consists of the active inhibition of the activity of micro-organisms with the aid of the antimicrobial effect of the bentonite-supported Ag, Cu, and TiO_2 nanoparticles present in the decontamination solutions. Therefore, even if some of the decontamination solutions did not show remarkable antimicrobial activity, they can still be successfully used for decontamination as they have a great potential for entrapping and sealing the biological contaminants inside the polymeric matrix of the nanocomposite film obtained following the evaporation of the solvent (water). The increased stability of the peelable films herein reported could ensure minimization of risks associated with biological contamination, ensuring immediate decontamination by first covering and then capturing the contaminant inside the polymeric film. The peeled nanocomposite films containing the entrapped contaminant can be further subjected to analysis for the identification and evaluation of the concentration of the contaminant.

3.3. Chemical Decontamination Tests

Chemical decontamination tests followed the biological decontamination tests. Since the biological decontamination tests showed that the solutions based on bentonite-supported silver nanoparticles displayed the best results, we employed only these solutions (SD10, SD11, and SD12) for the tests performed on real chemical warfare agents. This choice was also influenced by safety concerns, as working with real warfare agents imposes higher risks and requires specially trained personal. Thus, we tried to limit the number of experiments by employing only these three decontamination solutions, and we maintained the relevant steps for the decontamination procedure in order to obtain accurate information. For the same reasons, we also tested a simulant. Chemical decontamination of metallic surfaces measuring 10 cm^2 was accomplished in three stages: the first one consisted of the controlled contamination of the metallic coupons with **HD** and **DMMP** (10 mg/10 cm^2), respectively; the second one consisted of applying the decontamination solution on the contaminated surface and allowing it to neutralize the toxin and to form the film by evaporation of the solvent at room temperature (25 °C); the last step consisted of DCM extraction of the decontaminated surface and of the peeled film. The results obtained for **HD** and **DMMP** are presented in Table 7, Figure 10. Some relevant chromatograms were selected and are shown in Figures S6–S9.

Decontamination factor was calculated according to the following equation, also described in the *Methods* section: $DF = 100 \times (C_0 - C_f)/C_0$, where DF is the decontamination factor, C_0 is the initial toxic concentration found on the tested metallic surface, and C_f is the final concentration found on the decontaminated metallic surface, reflecting the residual contamination (according to the area of the characteristic peak of toxin). The values obtained are illustrated in Figure 11.

Table 7. Evaluation of chemical decontamination efficacy with the aid of GC-MS results.

Sample ID	DF (%)	Residual Contamination (%)	Residual Contamination (mg/10 cm^2)
S0_HD	0	100	10
HD_SD10_M	78.28	21.72	2.1722
HD_SD11_M	90.89	9.11	0.9111
HD_SD12_ M	66.73	33.27	3.3273
HD_SD10_P	N/A	21.02	2.1018
HD_SD11_P	N/A	14.58	1.4579
HD_SD12_P	N/A	36.47	3.6470
S0_DMMP	0	100	10
DMMP_SD10_M	99.97	0.03	0.0026
DMMP_SD11_M	99.98	0.02	0.0018
DMMP_SD12_ M	99.96	0.04	0.0036
DMMP_SD10_P	N/A	3.00	0.2619
DMMP_SD11_P	N/A	5.79	0.5063
DMMP_SD12_P	N/A	7.53	0.6584

HD—sulfur mustard; S0_HD—sulfur mustard blank; **DMMP**—dimethyl methylphosphonate; S0_**DMMP**—dimethyl methylphosphonate blank; M—samples extracted in DCM from the metalic surfaces after decontamination; P—samples extracted in DCM from the nanocomposite film after decontamination.

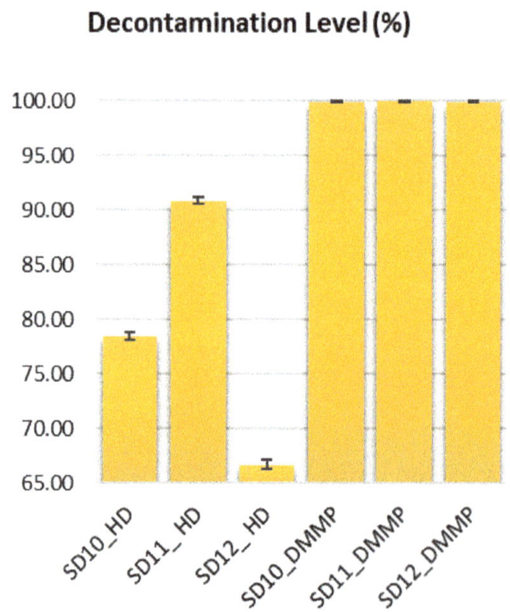

Figure 10. Chemical decontamination efficacy: decontamination factors.

As it can be noticed in Table 7 and Figure 10, the SD11 decontamination solution achieved the highest decontamination factor for **HD**. In the case of **DMMP**, employed as simulant for nerve chemical agents, the decontamination factors obtained were much higher, with all decontamination solutions (SD10, SD11, and SD12) being highly efficient. It is well known that **HD** is more difficult to decontaminate due to its chemical structure, as it tends to establish stronger interactions with the metallic substrate on which it is deposited. Even so, SD11 managed to efficiently remove more than 90% from the contaminated surface. The other decontamination solutions were not so efficient, probably due to their composition. We can presume that SD10 was not able to entrap the same amount of toxin, probably due to the lower content of bentonite, which was reflected in a higher residual contamination ≈ 21.72%. On the other hand, even if SD12 had more bentonite and theoretically greater adsorptive capacity, having a slightly higher viscosity and less NP active centers (bigger NPs and lower specific surface) with decreased mobility in a more viscous media, this could have led to much lower DF values.

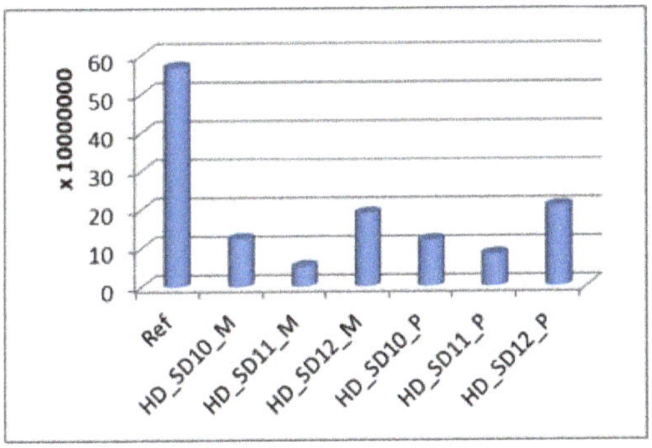

(**a**) Sulfur mustard

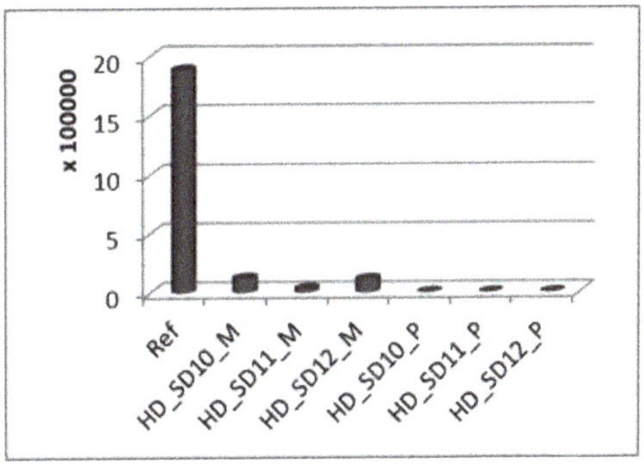

(**b**) 1,4—Dithiane

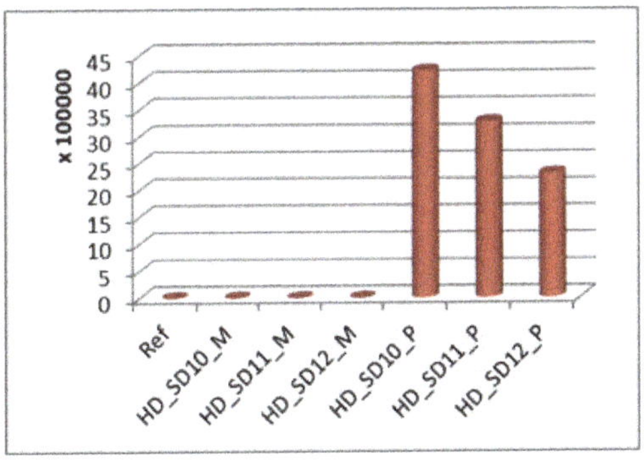

(**c**) Thiodiglycol (TDG)

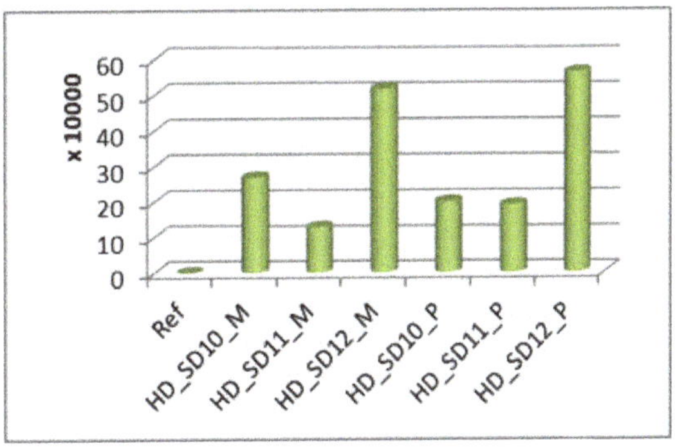

(**d**) Ethanol.2-((2-chloroethyl)thio)1-acetate

Figure 11. *Cont.*

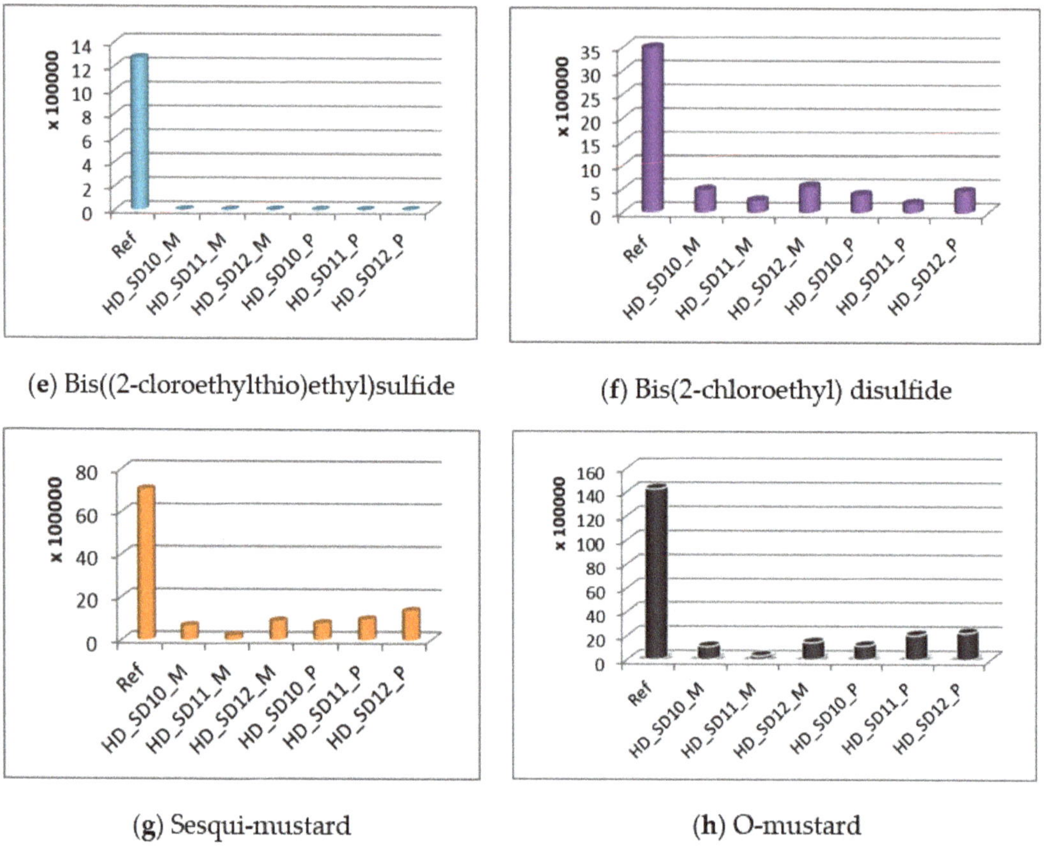

Figure 11. Degradation products of **HD** tracked during the decontamination process (*y*-axis values correspond to the values obtained for the area of the specific peak (counts) of each analyte): (**a**) sulfur mustard, (**b**) 1,4—dithiane, (**c**) thiodiglycol, (**d**) ethanol.2-((2-chloroethyl)thio)1-acetate, (**e**) Bis((2-cloroethylthio)ethyl)sulfide, (**f**) Bis(2-chloroethyl) disulfide, (**g**) sesqui-mustard and (**h**) O-mustard.

The achieved **HD** decontamination efficiency was 90.89% for SD11. This experiment demonstrates the functionality of the polymeric films used for the capture and removal of the toxic agent and also a small amount of waste generation. A comparative assessment between the result obtained and SD11 polymer films on **HD**-contaminated metallic surfaces (DF ≈ 90.89%) and conventional decontamination products, which are commercially available (bleach (full strength)- DF ≈ 86% or hydrogen peroxide- DF ≈ 71% for HD-contaminated metallic surfaces) [59], shows a clear improvement in terms of decontamination efficiency, afforded by the strippable coatings herein reported (SD11).

An ideal achievement is a 100% efficient decontamination, but in the case of operational decontamination, this percentage is relatively difficult to obtain for yperite, as the decontamination process depends on a multitude of factors, such as decontamination time, ambient temperature, the contact time elapsed between the contaminated surface and toxic agent until the application of the decontamination solution, and last but not least, the type of material that has been contaminated. In this regard, the U.S. Environmental Protection Agency (EPA) conducted some studies covering aspects of mustard decontamination using commercial decontamination solutions, wherein it concluded that depending on the factors listed above, decontamination efficiency can vary in practice between 37% and max. 95%

when applying the decontamination product only once [59,60]. Thus, we can affirm that SD11 polymeric films present unique perspectives in their operational use for removing **HD** contaminant.

In general, **DMMP** is much easier to decontaminate because of the weaker interactions it establishes with the metallic surfaces on which it is deposited in decontamination tests. Even so, it still requires an adequate decontaminating agent. All the decontamination solutions managed to reach DF values greater than 99.96%. These high values of DF were obtained due to the remarkable adsorption capacity of the materials employed for decontamination and the compatibility of the components of the decontamination solutions with **DMMP** but also due to the weak interactions between **DMMP** and the metallic substrate.

The last step in the evaluation of chemical decontamination efficacy consisted of the investigation of the degradation products of sulfur mustard. The **HD** solution employed for controlled contamination had a purity of 95%. Thus, 5% of the solution contained by-products of the synthesis of **HD** and small amounts of degradation products (Table S5). The solution utilized for controlled contamination, together with the degradation products of **HD**, was also tracked during the decontamination process because part of these synthesis by-products from the **HD** initial solution is also part of the blister agent class, possessing a higher blistering action than neat **HD**. Thus, even if they are found in a small concentration in the initial solution, the higher toxicity of these compounds imposes the necessity of examining their degradation. The results are illustrated in Figure 11 and Table S5.

In Figure 11, it can be noticed that the decontamination solution does not just entrap the toxic, but it also actively decomposes **HD** (and the other initial components of the contaminating solution) into less toxic compounds. These results offer clear evidence of the ability of the decontamination solutions to efficiently neutralize the toxic agent. The decontamination is performed by two pathways: the chemical degradation of the toxins, which is possible with the aid of the active components, as well as the entrapment and sealing-off of the degradation products and the toxic compounds that were only partially degraded. As can be seen in Figure 11, **HD** was only partially degraded, as **HD** can still be detected on the surface and in the nanocomposite films after DCM extraction. sesqui-mustard and O–mustard, which are well-known for their higher toxicity, both present in the initial contaminating solution were also partially degraded. In comparison with **HD**, sesqui–mustard, and O-mustard, the compound 1,4—dithiane was not visible in the polymeric film after decontamination. Thiodiglycol, the hydrolysis product of **HD**, was found in significant quantities in the samples obtained from DCM extraction of the polymeric film after decontamination, thus offering evidence of the high capacity of these decontamination solutions to hydrolyze **HD**.

Based on the chemical decontamination tests, it can be concluded that these novel water-based decontamination solutions are a useful and versatile tool for the neutralization and removal of chemical warfare agents, ensuring high decontamination levels.

4. Conclusions

This study proposes new decontamination solutions consisting of innovative, ecological, peelable active nanocomposite films specially designed for biological and chemical warfare agents. These film-forming decontamination solutions are water-based solutions obtained from eco-friendly materials. Bentonite-supported nanoparticles (Cu, TiO_2, and Ag) were successfully synthesized in aqueous solution and were employed in the decontamination formulations as active agents facilitating the neutralization of the hazardous materials. The unicity of these formulations consists of their environmentally responsible composition and high capacity to entrap and neutralize BCWA contaminants.

Particle-size control of the synthesized nanoparticles was accomplished by employing three different concentrations of bentonite nanoclay, which also served as adsorbent in the decontamination solutions, trapping the contaminants that diffuse in the polymeric composite network until the end of the drying process. Bentonite-supported silver nanoparticles displayed high antimicrobial activity and had a positive effect on the degradation process of the chemical warfare agent sulfur mustard, as well as **DMMP**, a nerve agent simulant. TEM analyses confirmed the nanometric dimensions of the obtained metallic particles. The decontamination formulations were further prepared based on these active ingredients and a water-soluble polymer, APV. Their viscosity was evaluated, revealing only minor differences between them due to the low concentration of nanoparticles and nanoclay (up to 1.5%). Viscosity influences the application method (spraying vs. brushing), but it also influences the motion of the active ingredients towards the contaminants within the polymeric matrix. The evaporation rate of each decontamination solution was evaluated to assess the necessary time for obtaining the peelable nanocomposite films. Chemical, mechanical, and thermal characterizations of the polymeric nanocomposite films were performed using FT-IR, tensile tests, DTA, and DMA techniques, showing the influence of each component on the final properties of the polymeric nanocomposite designed for BCWA decontamination. The decontamination effectiveness was first evaluated by qualitative and quantitative approaches, employing specific analytic tools for each type of contaminant. The influence of the concentration of bentonite nanoclay, and subsequently, the influence of the nature and size of the synthesized nanoparticles over the decontamination efficiency were also emphasized. The presence of nanoparticles led to higher decontamination factors. The solutions containing Ag-NPs displayed more antimicrobial activity. Copper nanoparticles displayed less antimicrobial activity, but this aspect was improved by the addition of TiO_2 nanoparticles. The efficacy of removal for the tested micro-organisms varies (93% < DF < 97%), thus confirming that these polymeric decontamination solutions represent a useful tool for biological decontamination of surfaces. The decontamination solution containing 1% bentonite nanoclay and Ag-NPs (**SD11**) displayed the best results for **HD** decontamination (DF $\geq$ 90.89%). In contrast, **DMMP** was almost completely removed from the contaminated surfaces, displaying a decontamination factor of DF $\approx$ 99.97% $\pm$ 0.01.

In conclusion, the eco-friendly, peelable active nanocomposite films designed for biological and chemical warfare agent decontamination can be successfully used on contaminated surfaces, reducing the risk of spreading bio-contaminants or chemical agents by neutralizing and entrapping the hazardous materials and their degradation products into the polymer nanocomposite matrix. In comparison with classical decontamination methods, employing ecological peelable coatings brings multiple advantages: significantly lower consumption of water and reagents, significantly lower amount of post-decontamination waste, ease of application, eco-friendly components, and high decontamination factors for both biological and chemical agents.

Supplementary Materials: The following are available online at https://www.mdpi.com/article/10.3390/polym13223999/s1, Table S1. Evaporation rate of the decontamination solutions; Figure S1. FT-IR spectra of the polymer nanocomposite films; Figure S2. Tensile tests results; Table S2. Tensile tests results; Figure S3. Cell population after 2 h in contact with the decontamination solutions for *E. coli; Table S3. Antimicrobial activity of the decontamination solutions against* E. coli; Figure S4. Cell population after 24 h in contact with the decontamination solutions for *E. coli; Table S4. Antimicrobial activity of the decontamination solutions against* S. aureus; Figure S5. Mass spectra of **HD** (up—from analysis; down—NIST database) RT—10.71 min; Figure S6. Chromatograms multigraph; Figure S7. Total Ions Chromatograms overlap for all samples, representing **HD** at the RT (retention time): 10.70; Figure S8. Chromatograms overlap for reference sample and samples extracted from the metallic surface after decontamination; Figure S9. Chromatograms overlap for reference sample and samples extracted from the nanocomposite film after decontamination; Table S5. Degradation products of **HD** monitorized during the decontamination process; Figure S10. Degradation products of Yperite; Figure S11. Mass spectra of **HD** (up—from analysis; down—NIST database) RT—5.93 min; Figure S12. Chromatograms overlap for all samples; Figure S13. Chromatograms overlap for all samples; Figure S14. Chromatograms overlap for reference sample and samples extracted from the metallic surface after decontamination; Figure S15. Chromatograms overlap for reference sample and samples extracted from the nanocomposite film after decontamination.

Author Contributions: Conceptualization, G.T., A.D., T.R., M.A., E.R. and R.E.G.; methodology, G.T., A.D., T.R. and M.A.; software, A.E.M., D.P., A.M.G. and R.Ș.; validation, G.T., A.D., T.R., M.A., T.V.I. and E.R.; formal analysis, G.T., A.D., M.A., R.E.G., F.A., R.O., F.L.Z., A.P., A.E.M., D.P., A.M.G. and R.Ș.; investigation, G.T., A.D., M.A., R.E.G., F.A., R.O., F.L.Z., A.P., A.E.M., D.P., T.V.I. and R.Ș.; resources, G.T., T.R., M.A., A.D., T.V.I. and R.E.G.; data curation, G.T., A.D., M.A., F.A., R.O., F.L.Z., A.P., A.E.M., D.P. and R.Ș.; writing—original draft preparation, G.T., A.D., T.R. and M.A.; writing—review and editing, G.T., A.D., T.R., M.A., E.R., R.E.G., F.A., R.O., F.L.Z., A.P., A.E.M., D.P., A.M.G., T.V.I. and R.Ș.; visualization, G.T., A.D., M.A., E.R., D.P.; supervision, G.T., A.D., T.R., M.A. and E.R.; project administration, G.T., A.E.M., R.E.G., A.D. and T.R.; funding acquisition, G.T., A.E.M., R.E.G., M.A., A.D. and T.R. All authors have read and agreed to the published version of the manuscript.

Funding: The authors are thankful for the financial support provided by the Executive Agency for Higher Education, Research, Development, and Innovation Funding (UEFISCDI), the Ministry of Education of Romania, through the National Projects PN-III-P2-2.1-PED-2019-4222 ctr. no. 427PED/2020 and PN-III-P2-2.1-PTE-2019-0400 ctr. no. 49PTE/2020.

Institutional Review Board Statement: Not applicable.

Informed Consent Statement: Not applicable.

Data Availability Statement: The data presented in this study are available on request from the corresponding author.

Acknowledgments: Aurel Diacon gratefully acknowledges financial support from the Competitiveness Operational Program 2014–2020, Action 1.1.3: Creating synergies with RDI actions of the EU's HORIZON 2020 framework program and other international RDI programs, MySMIS Code 108792, Acronym project "UPB4H", financed by contract: 250/11.05.2020.

Conflicts of Interest: The authors declare no conflict of interest.

References

1. OPCW. The Convention on the Prohibition of the Development, Production, Stockpiling and Use of Chemical Weapons and on Their Destruction (the Chemical Weapons Convention or CWC). Available online: https://www.opcw.org/ (accessed on 29 April 1997).
2. UN. The Convention on the Prohibition of the Development, Production and Stockpiling of Bacteriological (Biological) and Toxin Weapons and on Their Destruction. Available online: https://www.un.org/disarmament/biological-weapons/ (accessed on 26 March 1975).
3. Wagner, G.W.; Yang, Y.-C. Rapid Nucleophilic/Oxidative Decontamination of Chemical Warfare Agents. *Ind. Eng. Chem. Res.* **2002**, *41*, 1925–1928. [CrossRef]
4. Wartell, M.A.; Kleinman, M.T.; Huey, B.M.; Duffy, L.M. *Strategies to Protect the Health of Deployed, U.S. Forces: Force Protection and Decontamination*; National Academic Press: Washington, DC, USA, 1999.
5. EPA. Evaluation of the Curing Times of Strippable Coatings and Gels as Used for Radiological Decontamination. In *EPA Report*; U.S. Environmental Protection Agency: Washington, DC, USA, 2014.
6. Kumar, V.; Goel, R.; Chawla, R.; Silambarasan, M.; Sharma, R.K. Chemical, biological, radiological, and nuclear decontamination: Recent trends and future perspective. *J. Pharm. Bioallied Sci.* **2010**, *2*, 220–238. [CrossRef] [PubMed]
7. Zorila, F.L.; Ionescu, C.; Craciun, L.S.; Zorila, B. Atomic force microscopy study of morphological modifications induced by different decontamination treatments on *Escherichia coli*. *Ultramicroscopy* **2017**, *182*, 226–232. [CrossRef]
8. Dent, E.; Espina-Benitez, M.B.; Pitault, I.; Pollet, T.; Blaha, D.; Bolzinger, M.A.; Rodriguez-Nava, V.; Briancon, S. Metal oxide nanoparticles for the decontamination of toxic chemical and biological compounds. *Int. J. Pharm.* **2020**, *583*, 119373. [CrossRef] [PubMed]
9. Gephart, R.T., 3rd; Coneski, P.N.; Wynne, J.H. Decontamination of chemical-warfare agent simulants by polymer surfaces doped with the singlet oxygen generator zinc octaphenoxyphthalocyanine. *ACS Appl. Mater. Interfaces* **2013**, *5*, 10191–10200. [CrossRef] [PubMed]
10. Gray, H.N.; Bergbreiter, D.E. Applications of polymeric smart materials to environmental problems. *Environ. Health Perspect.* **1997**, *105*, 55–63.
11. Gray, H.N.; Jorgensen, B.; Mc Claugherty, D.L.; Kippenberger, A. Smart Polymeric Coatings for Surface Decontamination. *Ind. Eng. Chem. Res.* **2001**, *40*, 3540–3546. [CrossRef]
12. Toader, G.; Stănescu, P.O.; Zecheru, T.; Rotariu, T.; El-Ghayoury, A.; Teodorescu, M. Water-based strippable coatings containing bentonite clay for heavy metal surface decontamination. *Arab. J. Chem.* **2019**, *12*, 4026–4034. [CrossRef]
13. Almeida, C.C.; Garcia, R.H.L.; Cambises, P.B.S.; Da Silva, T.M.; Paiva, J.E.; Carneiro, J.C.G.G.; Rodrigues, D.L. Radiation Protection Procedures for the Dismantling and Decontamination of Nuclear Facility. In Proceedings of the International Nuclear Atlantic Conference, Recife, Brazil, 24–29 November 2013.

14. Bossart, S.; Blair, D.M. Decontamination Technologies for Facility Reuse. In Proceedings of the WM'03 Conference, Tucson, AZ, USA, 23–27 February 2003; pp. 1–13.
15. Kohli, R. Chapter 1—Removal of Surface Contaminants Using Ionic Liquids. In *Developments in Surface Contamination and Cleaning*; Elsevier: Amsterdam, The Netherlands, 2013; pp. 1–63.
16. Sutton, M.; Burastero, S.R.; Perkins, J.; Chiarappa-Zucca, M.L.; Andresen, B.D. Alpha-Aminobenzyl-Alpha, Alpha-Diphosphoric Acid Selective Chelation of Beryllium. U.S. Patent 11,731,421, 1 November 2007.
17. Yang, H.-M.; Park, C.W.; Lee, K.-W. Polymeric coatings for surface decontamination and ecofriendly volume reduction of radioactive waste after use. *Prog. Nucl. Energy* **2018**, *104*, 67–74. [CrossRef]
18. Pulpea, D.; Rotariu, T.; Toader, G.; Pulpea, G.B.; Neculae, V.; Teodorescu, M. Decontamination of radioactive hazardous materials by using novel biodegradable strippable coatings and new generation complexing agents. *Chemosphere* **2020**, *258*, 127227. [CrossRef]
19. Toader, G.; Rotariu, T.; Moldovan, E.A.; Esanu, S.R.; Pulpea, D.; Podaru, D.; Dirloman, F.; Iordache, T.V.; Gavrila, A.M.; Istrate, M. Biodegradable Polymeric Nanocomposites, with Film-Forming Properties, Intended for Surface Decontamination and Process for Obtaining and Using Them. Rumanian Patent OSIM A00340, 16 June 2021.
20. Toader, G.; Rotariu, T.; Pulpea, D.; Moldovan, A.; Podaru, A.; Gavrila, A.M.; Alexandru, M.; Diacon, A.; Ginghina, R.; Iorga, O.; et al. Polymeric blends designed for surface decontamination. *UPB Sci. Bull. Series B* **2021**, *83*, 73–86.
21. He, Z.; Li, Y.; Xiao, Z.; Jiang, H.; Zhou, Y.; Luo, D. Synthesis and Preparation of (Acrylic Copolymer) Ternary System Peelable Sealing Decontamination Material. *Polymers* **2020**, *12*, 1556. [CrossRef]
22. Koryakovskiy, Y.S.; Doilnitsyn, V.A.; Akatov, A.A. Improving the efficiency of fixed radionuclides' removal by chemical decontamination of surfaces in situ. *Nucl. Energy Technol.* **2019**, *5*, 155–161. [CrossRef]
23. Bromberg, L.; Creasy, W.R.; McGarvey, D.J.; Wilusz, E.; Hatton, T.A. Nucleophilic Polymers and Gels in Hydrolytic Degradation of Chemical Warfare Agents. *ACS Appl. Mater. Interfaces* **2015**, *7*, 22001–22011. [CrossRef]
24. Teodorescu, M.; Lingu, A.; Stanescu, P.O. Preparation and Properties of Novel Slow-Release NPK Agrochemical Formulations Based on Poly(acrylic acid) Hydrogels and Liquid Fertilizers. *Ind. Eng. Chem. Res.* **2009**, *48*, 6527–6534. [CrossRef]
25. Klykken, P.; Servinski, M.; Thomas, X. Silicone Film-Forming Technologies for Health Care Applications. *Dow Corning* **2004**, *1*, 1–8.
26. Ben Halima, N. Poly(vinyl alcohol): Review of its promising applications and insights into biodegradation. *RSC Adv.* **2016**, *6*, 39823–39832. [CrossRef]
27. Moshin, M. Thermal and mechanical properties of poly(vinyl alcohol) plasticized with glycerol. *J. Appl. Polym. Sci.* **2011**, *122*, 3102–3109.
28. Casalini, R.; Bogoslovov, R.; Qadri, S.B.; Roland, C.M. Nanofiller reinforcement of elastomeric polyurea. *Polymer* **2012**, *53*, 1282–1287. [CrossRef]
29. Li, Y.; Chen, C.; Hou, H.; Cheng, Y.; Gao, H.; Zhang, P.; Liu, T. The Influence of Spraying Strategy on the Dynamic Response of Polyurea-Coated Metal Plates to Localized Air Blast Loading: Experimental Investigations. *Polymers* **2019**, *11*, 1888. [CrossRef]
30. Toader, G.; Rusen, E.; Teodorescu, M.; Diacon, A.; Stanescu, P.O.; Damian, C.; Rotariu, T.; Rotariu, A. New polyurea MWCNTs nanocomposite films with enhanced mechanical properties. *J. Appl. Polym. Sci.* **2017**, *134*, 45061. [CrossRef]
31. Toader, G.; Rusen, E.; Teodorescu, M.; Diacon, A.; Stanescu, P.O.; Rotariu, T.; Rotariu, A. Novel polyurea polymers with enhanced mechanical properties. *J. Appl. Polym. Sci.* **2016**, *133*, 43967. [CrossRef]
32. Toader, G.; Diacon, A.; Rusen, E.; Rizea, F.; Teodorescu, M.; Stanescu, P.O.; Damian, C.; Trana, E.; Rotariu, A.; Bucur, F.; et al. A Facile Synthesis Route of Hybrid Polyurea-Polyurethane-MWCNTs Nanocomposite Coatings for Ballistic Protection and Experimental Testing in Dynamic Regime. *Polymers* **2021**, *13*, 1618. [CrossRef] [PubMed]
33. Castro-Yobal, M.A.; Contreras-Oliva, A.; Saucedo-Rivalcoba, V.; Rivera-Armenta, J.L.; Salinas-Ruiz, J.; Herrera-Corredor, A.; Hernández-Ramírez, G. Evaluation of physicochemical properties of film-based alginate for food packing applications. *e-Polymers* **2021**, *21*, 82–95. [CrossRef]
34. Lei, Q.; Huang, Z.Y.; Pan, J.Z.; Bao, J.Q.; Xun, Q.N. Effect of Sodium Alginate on the Properties of Composite Protein Films. *Appl. Mech. Mater.* **2014**, *541*, 49–56. [CrossRef]
35. Li, J.; He, J.; Huang, Y.; Li, D.; Chen, X. Improving surface and mechanical properties of alginate films by using ethanol as a co-solvent during external gelation. *Carbohydr. Polym.* **2015**, *123*, 208–216. [CrossRef] [PubMed]
36. Salerno, A.; Pitault, I.; Devers, T.; Pelletier, J.; Briançon, S. Model-based optimization of parameters for degradation reaction of an organophosphorus pesticide, paraoxon, using CeO(2) nanoparticles in water media. *Environ. Toxicol. Pharmacol.* **2017**, *53*, 18–28. [CrossRef]
37. Wagner, G.W.; Koper, O.B.; Lucas, E.; Decker, S.; Klabunde, K.J. Reactions of VX, GD, and HD with Nanosize CaO: Autocatalytic Dehydrohalogenation of HD. *J. Phys. Chem. B* **2000**, *104*, 5118–5123. [CrossRef]
38. He, H.; Dong, X.; Yang, M.; Yang, Q.; Duan, S.; Yu, Y.; Han, J.; Zhang, C.; Chen, L.; Yang, X. Catalytic inactivation of SARS coronavirus, Escherichia coli and yeast on solid surface. *Catal. Commun.* **2004**, *5*, 170–172. [CrossRef] [PubMed]
39. Gyawali, R.; Ibrahim, S.; Abu Hasfa, S.H.; Smqadri, S.Q.; Haik, Y. Antimicrobial Activity of Copper Alone and in Combination with Lactic Acid against Escherichia coli O157:H7 in Laboratory Medium and on the Surface of Lettuce and Tomatoes. *J. Pathog.* **2011**, *2011*, 650968. [CrossRef] [PubMed]

40. Usman, M.S.; El Zowalaty, M.E.; Shameli, K.; Zainuddin, N.; Salama, M.; Ibrahim, N.A. Synthesis, characterization, and antimicrobial properties of copper nanoparticles. *Int. J. Nanomed.* **2013**, *8*, 4467–4479.
41. Singh, V.V.; Jurado-Sánchez, B.; Sattayasamitsathit, S.; Orozco, J.; Li, J.; Galarnyk, M.; Fedorak, Y.; Wang, J. Multifunctional Silver-Exchanged Zeolite Micromotors for Catalytic Detoxification of Chemical and Biological Threats. *Adv. Funct. Mater.* **2015**, *25*, 2147–2155. [CrossRef]
42. Wagner, G.W. Decontamination of Chemical Warfare Agents with Nanosize Metal Oxides. In *Defense Applications of Nanomaterials*; American Chemical Sociaty: Washington, DC, USA, 2005; pp. 139–152.
43. Roy, A.; Srivastava, A.K.; Singh, B.; Mahato, T.H.; Shah, D.; Halve, A.K. Degradation of sulfur mustard and 2-chloroethyl ethyl sulfide on Cu–BTC metal organic framework. *Microporous Mesoporous Mater.* **2012**, *162*, 207–212. [CrossRef]
44. Roy, A.; Srivastava, A.K.; Singh, B.; Shah, D.; Mahato, T.H.; Gutch, P.K.; Halve, A.K. Degradation of sarin, DEClP and DECNP over Cu-BTC metal organic framework. *J. Porous Mater.* **2013**, *20*, 1103–1109. [CrossRef]
45. Balouiri, M.; Sadiki, M.; Ibnsouda, S.K. Methods for in vitro evaluating antimicrobial activity: A review. *J. Pharm. Anal.* **2016**, *6*, 71–79. [CrossRef] [PubMed]
46. Borges, A.; Abreu, A.C.; Ferreira, C.; Savedra, M.J.; Simões, L.C.; Simões, M. Antibacterial activity and mode of action of selected glucosinolate hydrolysis products against bacterial pathogens. *J. Food Sci. Technol.* **2015**, *52*, 4737–4748. [CrossRef]
47. Cui, Y.; Oh, Y.J.; Lim, J.; Youn, M.; Lee, I.; Pak, H.K.; Jo, W.; Park, S. AFM study of the differential inhibitory effects of the green tea polyphenol (-)-epigallocatechin-3-gallate (EGCG) against Gram-positive and Gram-negative bacteria. *Food Microbiol.* **2012**, *29*, 80–87. [CrossRef] [PubMed]
48. CLSI. Methods for Dilution Antimicrobial Susceptibility Tests for Bacteria That Grow Aerobically. In *CLSI Document M07-A9*; CLSI: Wayne, PA, USA, 2012.
49. De Brucker, K.; Delattin, N.; Robijns, S.; Steenackers, H.; Verstraeten, N.; Landuyt, B.; Luyten, W.; Scoofs, L.; Dovgan, B.; Fröhlich, M.; et al. Derivatives of the mouse cathelicidin-related antimicrobial peptide (CRAMP) inhibit fungal and bacterial biofilm formation. *Antimicrob. Agents Chemother.* **2014**, *58*, 5395–5404. [CrossRef]
50. Wiegand, I.; Hilpert, K.; Hancock, R.E. Agar and broth dilution methods to determine the minimal inhibitory concentration (MIC) of antimicrobial substances. *Nat. Protoc.* **2008**, *3*, 163–175. [CrossRef]
51. Konaté, K.; Mavoungou, J.F.; Lepengué, A.N.; Aworet-Samseny, R.R.; Hilou, A.; Souza, A.; Dicko, M.H.; M'Batchi, B. Antibacterial activity against β- lactamase producing Methicillin and Ampicillin-resistants Staphylococcus aureus: Fractional Inhibitory Concentration Index (FICI) determination. *Ann. Clin. Microbiol. Antimicrob.* **2012**, *11*, 18. [CrossRef]
52. Zhou, D.; Abdel-Fattah, A.I.; Keller, A.A. Clay Particles Destabilize Engineered Nanoparticles in Aqueous Environments. *Environ. Sci. Technol.* **2012**, *46*, 7520–7526. [CrossRef]
53. Lesniak, A.; Salvati, A.; Santos-Martinez, M.J.; Radomski, M.W.; Dawson, K.A.; Åberg, C. Nanoparticle Adhesion to the Cell Membrane and Its Effect on Nanoparticle Uptake Efficiency. *J. Am. Chem. Soc.* **2013**, *135*, 1438–1444. [CrossRef] [PubMed]
54. Behzadi, S.; Serpooshan, V.; Tao, W.; Hamaly, M.A.; Alkawareek, M.Y.; Dreaden, E.C.; Brown, D.; Alkilany, A.M.; Farokhzad, O.C.; Mahmoudi, M. Cellular uptake of nanoparticles: Journey inside the cell. *Chem. Soc. Rev.* **2017**, *46*, 4218–4244. [CrossRef] [PubMed]
55. Petersen, E.J.; Reipa, V.; Watson, S.S.; Stanley, D.L.; Rabb, S.A.; Nelson, B.C. DNA damaging potential of photoactivated p25 titanium dioxide nanoparticles. *Chem. Res. Toxicol.* **2014**, *27*, 1877–1884. [CrossRef] [PubMed]
56. Wang, L.; Hu, C.; Shao, L. The antimicrobial activity of nanoparticles: Present situation and prospects for the future. *Int. J. Nanomed.* **2017**, *12*, 1227–1249. [CrossRef] [PubMed]
57. Rutala, W.A.; Weber, D.A. *Guideline for Disinfection and Sterilization in Healthcare Facilities*; CDPC: Atlanta, GA, USA, 2008.
58. Ayliffe, G.A.; Collins, B.J.; Lowbury, E.J.L.; Babb, J.R.; Lilly, H.A. Ward floors and other surfaces as reservoirs of hospital infection. *J. Hyg.* **1967**, *65*, 515–536. [CrossRef]
59. Oudejans, L. *Decontamination of Agent Yellow, a Lewisite and Sulfur Mustard Mixture*; U.S. Environmental Protection Agency: Washington, DC, USA, 2015.
60. Oudejans, L. *Decontamination of Sulfur Mustard and Thickened Sulfur Mustard Using Chlorine Dioxide Fumigation*; U.S. Environmental Protection Agency: Washington, DC, USA, 2011.

polymers

MDPI

Article

Novel Polyurethanes Based on Recycled Polyethylene Terephthalate: Synthesis, Characterization, and Formulation of Binders for Environmentally Responsible Rocket Propellants

Florin Marian Dîrloman [1,†], Gabriela Toader [1,†], Traian Rotariu [1,*], Tudor Viorel Ţigănescu [1,2], Raluca Elena Ginghină [3], Răzvan Petre [3], Florentina Alexe [3], Mihai Ionuţ Ungureanu [1], Edina Rusen [4], Aurel Diacon [4], Adi Ghebaur [4], Monica Duldner [5], Alina Elena Coman [5] and Robert Ţincu [6]

1 Military Technical Academy "Ferdinand I", 39-49 George Cosbuc Boulevard, 050141 Bucharest, Romania; florin.dirloman@mta.ro (F.M.D.); gabriela.toader@mta.ro (G.T.); viorel.tiganescu@mta.ro (T.V.Ţ.); mihai.ungureanu@mta.ro (M.I.U.)
2 Military Equipment and Technologies Research Agency, 16 Aeroportului Street, Clinceni, 077025 Ilfov, Romania
3 Research and Innovation Center for CBRN Defense and Ecology, 225 Soseaua Oltenitei, 041327 Bucharest, Romania; raluca.ginghina@nbce.ro (R.E.G.); razvan.petre@nbce.ro (R.P.); florentina.alexe@nbce.ro (F.A.)
4 Faculty of Applied Chemistry and Materials Science, University Politehnica of Bucharest, 1-7 Gh. Polizu Street, 011061 Bucharest, Romania; edina.rusen@upb.ro (E.R.); aurel.diacon@upb.ro (A.D.); adi.ghebaur@upb.ro (A.G.)
5 National Institute of Research and Development for Chemistry and Petrochemistry, 202 Splaiul Independentei, 060041 Bucharest, Romania; monica.duldner@icechim.ro (M.D.); alina-elena.coman@icechim.ro (A.E.C.)
6 Center for Organic Chemistry "C.D. Nenitescu" of the Romanian Academy, 202B Splaiul Independentei, 060023 Bucharest, Romania; tincurobert@gmail.com
* Correspondence: traian.rotariu@mta.ro
† Co-first authors, equally contributed to this work.

Citation: Dîrloman, F.M.; Toader, G.; Rotariu, T.; Ţigănescu, T.V.; Ginghină, R.E.; Petre, R.; Alexe, F.; Ungureanu, M.I.; Rusen, E.; Diacon, A.; et al. Novel Polyurethanes Based on Recycled Polyethylene Terephthalate: Synthesis, Characterization, and Formulation of Binders for Environmentally Responsible Rocket Propellants. *Polymers* **2021**, *13*, 3828. https://doi.org/10.3390/polym13213828

Academic Editor: Sándor Kéki

Received: 30 September 2021
Accepted: 28 October 2021
Published: 5 November 2021

Publisher's Note: MDPI stays neutral with regard to jurisdictional claims in published maps and institutional affiliations.

Copyright: © 2021 by the authors. Licensee MDPI, Basel, Switzerland. This article is an open access article distributed under the terms and conditions of the Creative Commons Attribution (CC BY) license (https://creativecommons.org/licenses/by/4.0/).

Abstract: Novel polyurethane-based binders, specifically designed for environmentally responsible rocket propellant composites, were obtained by employing the polyester-polyols that resulted from the degradation of polyethylene terephthalate waste. A new class of "greener" rocket propellants, comprising polyurethanes (based on recycled PET) as the binder, phase stabilized ammonium nitrate (PSAN) as the eco-friendly oxidizer, and triethylene glycol dinitrate (TEGDN) as the energetic plasticizer, together with aluminum as fuel and Fe_2O_3 as the catalyst, is herein reported. The components of the energetic mixtures were investigated (individually and as composite materials) through specific analytical tools: ^{1}H-NMR, FT-IR, SEM-EDX, DTA and TGA, tensile and compression tests, DMA, and micro-CT. Moreover, the feasibility of this innovative solution is sustained by the ballistic performances exhibited by these composite materials in a subscale rocket motor, proving that these new formulations are suitable for rocket propellant applications.

Keywords: polyethylene terephthalate recycling; polyurethane; binder; polymeric composite; rocket propellant; mechanical properties; subscale rocket motor

1. Introduction

Solid rocket propellants are a particular class of energetic materials that are developed to ensure the propulsion of spacecrafts, missiles, and rockets toward a target. The main energetic transformation is represented by steady combustion in an enclosed environment, specifically designed to release hot gases at high speeds to obtain the appropriate propulsion thrust [1]. The main aspects that influence the combustion behavior are the constituent elements and the configuration of the grains. Structurally, solid rocket composite propellants are typically based on a heterogeneous combination of distinct compounds that serve as fuel, oxidizer, burn-rate modifiers, and binder [1,2]. Usually, the applicability of polyurethane or polyurea matrices lies in the field of thermal insulations, hydro-isolations, adhesives, and ballistic protection [3–6]. In the case of energetic materials, they have a dual purpose, as a binding agent and as an organic fuel [1,2]. These polymeric matrices are used for binding other solid components of the rocket propellants, to protect them against environmental agents and to confer a certain geometry and mechanical strength. In state-of-the-art composite rocket propellants, ammonium perchlorate (AP) is the preferred oxidizer, almost without exception, due to its outstanding properties [1,2,7,8]. However, in the last decade, environmental impact issues have urged research activities for a replacement with "green" oxidizers, such as phase-stabilized ammonium nitrate (PSAN) or ammonium dinitramide (ADN) [2,9–11]. A common practice in rocket propellant technology consists in adding metallic additives to increase the heat output, the density, and the specific impulse of the energetic material. Aluminum fine powder, magnesium, and boron are the main candidates as fuels in rocket propellant formulations. Together with the oxidizers and additives, the solid fuel is introduced in the polymeric matrix to ensure homogeneity, protection, and proper mechanical characteristics of the energetic mixture [2,8,12–16].

Historically, the beginning of the use of polymers in the development of solid rocket propellants is marked by the introduction of nitrocellulose (NC). NC was mainly employed in homogeneous (colloidal) rocket propellants, providing the necessary structural integrity to the propellants so that they could be molded into different geometric configurations, according to the type of launching system [1,2,17]. In heterogeneous propellants, polysulfide polymers were initially used as the binding agent. The drawback of this polymer was the high incompatibility with metallic fuels, causing safety problems occasionally leading to autoignition during long-term storage [2,18]. In this context, the necessity of developing a new class of binders became imperative. The polybutadiene chain was found to be a more suitable binder for the energetic composites, due to its high elasticity and low glass temperature. The copolymer of butadiene and acrylic acid (PBAA) was the first one to be used from this class [1,2,18,19]. The low viscosity of these binders allowed significant solid loading ratios of oxidizer and metallic fuel in the energetic composites. However, these materials exhibited poor mechanical properties. This aspect was overcome using a polybutadiene acrylonitrile copolymer (PBAN). The low cost of production, the low viscosity, and the impulse increase made PBAN, at that time, the ideal candidate as a rocket propellant binder [1,2,18–21]. However, the high temperature required to cure these propellants led to the development of other types of binders, such as carboxyl-terminated polybutadiene (CTPB). CTPB-based composites displayed significantly enhanced mechanical properties, especially at lower temperatures, in comparison with PBAA or PBAN binders, without affecting the specific impulse, density, or solid loading ratios [1,2,18,19]. The curative agents used for these types of polymers employed as binders are usually based on di- or tri-functional epoxides or aziridines [18]. Modern solid rocket propellants use hydroxyl-terminated polybutadiene (HTPB), an inert binder from the same class as CTPB, which ensures the optimal combination of thermodynamic and mechanical properties [7,8,18,20]. The HTPB binders exhibit superior elongation capacity at low temperatures and better ageing properties in contrast with CTPB [18,19]. The curing process of HTPB is quicker due to the use of isocyanates [19,22,23]. Despite the advantages of HTPB binders, the problem of using this type of inert binder in a composite rocket propellant is represented by the necessity of adding an extra amount of oxidizer to balance the oxygen score and achieve the appropriate performances. Consequently, it is desirable to develop more versatile polymers with substantial energetic character. The integration of azido- or nitro-functional groups on the polymeric chain is conducted for a significant performance enhancement with lower amounts of energetic materials. Among the polymers employed as energetic binders in solid rocket propellants, two worth mentioning are polyglycidyl nitrates (PGN) and glycidyl azido polymers (GAP) [10,18]. GAP are considered excellent energetic polymers due to their unique structure, which ensures a high heat of formation. These polymers are recommended for rocket propellant composites that comprise eco-friendly oxidizers with weaker energetic characteristics, such as ammonium nitrate, to balance the energy loss caused by replacing classical oxidizers such as ammonium perchlorate [10].

In recent times, the pollution of ecosystems due to the inappropriate disposal of PET waste is widespread. Moreover, the high level of carbon dioxide emissions released into the air because of their combustion has led to the application of preventive measures. As a countermeasure to these ecological issues, interest in the development of materials based on recycled PET has achieved substantial growth [24]. As a result, in December 2015, the European Commission adopted a plan based on a circular economy strategy that will be applied in the context of reusing plastic wastes [24–27]. For obtaining eco-friendly polyols, with extensive applicability, PET waste can be recycled by undergoing a controlled degradation procedure that aims to generate oligomers that can be subsequently utilized for specific applications [27–30].

In this paper, a new group of flexible polyurethanes was synthesized from polyols obtained through the recycling of PET waste. A commercial polyol and a commercial aromatic polyisocyanate were also employed for the development of polyurethane formulations. These polyurethanes were subsequently used as binders in various rocket propellant composites, also comprising an eco-friendly oxidizer. The polyurethane networks were softened with an energetic plasticizer (triethylene glycol dinitrate, TEGDN) to boost the exothermic decomposition of these rocket propellants and to obtain composites with appropriate mechanical behavior suitable for this type of application. The synthesized eco-friendly composites can be successfully employed as rocket propellants, an aspect that was demonstrated through the subscale rocket motor experimental investigations that we performed.

Therefore, the novelty of this work consists in the innovative path of demonstrating the applicability of novel synthesized polyurethanes as binders for new environmentally friendly rocket propellants, herein reported. These original energetic mixtures, comprising newly synthesized components such as polyurethanes (from recycled PET) as the binder, PSAN as the eco-friendly oxidizer, TEGDN as the energetic plasticizer, together with aluminum for fuel and Fe_2O_3 as the catalyst, achieved remarkable performances as rocket propellants, demonstrable by the results obtained through the measurements performed using the subscale rocket motor.

Furthermore, the experimental data revealed that these new "green" binders could successfully replace classical binders from existing rocket propellants, since they possess comparable performances (similar thermal stability, up to 300 °C [31], similar mechanical properties [32]), but multiple advantages in comparison with HTPB binders [33], emphasized in the experimental section. The new rocket propellant composites, besides their environmentally friendly character, displayed better performances: up to a 30% improvement of the specific volume (in comparison with HTPB-based rocked propellant from consecrated aviation missiles). The heat of combustion (≈1000 cal/g) and T_g values (−60 °C to −32 °C) were comparable with nitrocellulose double base propellants (≈1020 cal/g and −35.5 °C) [34,35], which ensure that they maintain their performances even at lower environmental temperatures. The polyols obtained from PET degradation, the polyurethanes, and the composite rocket propellant formulations were investigated through specific analytical tools. Moreover, the feasibility of these novel energetic composites was demonstrated through the evaluation of the combustion rate and maximum pressure, proving that these new materials are suitable for rocket propellant applications.

Polyurethane networks, mainly those based on HTPB, are currently considered state-of-the-art propellant binder systems, being extensively used for composite solid propellants. Unfortunately, as already described above, they possess several disadvantages, among which the drawbacks of their inadequate life-cycle assessment can also be mentioned. In contrast, the materials developed in this study are especially designed to be environmentally friendly, while ensuring improved performances as new "green" alternatives for rocket propellants.

2. Materials and Methods

2.1. Materials

2.1.1. Materials for the Catalytic Degradation of Polyethylene Terephthalate (PET)

PET (Mn ≈ 40,000–70,000 Da, melting range: 254–260 °C) was obtained from post-consumer bottles cut into square shaped flakes (5 mm × 5 mm). The PET flakes were washed with distilled water and dried at 100 °C for 6 h before being introduced into the degradation reactor. Poly(ethylene glycol) of an average Mn≈ 570–630 Da (PEG600, Sigma Aldrich, St. Louis, MO, USA), titanium(IV) isopropoxide (TTIP, 97%, Sigma Aldrich, St. Louis, MO, USA), adipic acid (AA, 99%, Sigma Aldrich, St. Louis, MO, USA), sebacic acid (SA, 99%, Sigma Aldrich, St. Louis, MO, USA), and tin(II) 2-ethylhexanoate ($C_{16}H_{30}O_4Sn$, 92.5–100%, Sigma Aldrich, St. Louis, MO, USA) were used as received.

2.1.2. Materials for Polyurethane Synthesis

The two types of polyester-polyols (named RP1 and RP2 in this paper, according to the compositions described below in 'Methods' section) that resulted from PET degradation were vacuum dried for 24 h at 50 °C and were subsequently used for polyurethane synthesis. Setathane D1160, hydroxyl content 5.4%, (SET, Allnex, Brussels, Belgium) was vacuum dried for 24 h at 50 °C before being employed in polyurethane synthesis. Diphenylmethane–4,4′–diisocyanate, -NCO content 31.5% (MDI, technical product Desmodur® 44V20L, Covestro, Leverkusen, Germany), was used as received. The energetic plasticizer, triethylene glycol dinitrate (TEGDN), was synthesized and purified in the Military Technical Academy (MTA) according to a procedure found in the literature [36].

2.1.3. Materials for Composite Rocket Propellants Based on Polyurethane Binders

The above-mentioned synthesized polyurethanes were employed as binder for the new rocket propellant composite formulations. The "green" oxidizer, phase-stabilized ammonium nitrate (PSAN), was prepared in MTA: ammonium nitrate (AN, min. 98%, Honeywell Fluka™, Seelze, Germany) and potassium nitrate (KN, 99%, ACROS Organics™, Fair Lawn, NJ, USA) were co-crystallized from aqueous solution, according to the procedure described in the literature [37]. As metallic fuel, two types of aluminum powder, with an average particle size <5 μm (99.5%, Sigma Aldrich, St. Louis, MO, USA) and 100 μm (99.5%, Sigma Aldrich, St. Louis, MO, USA), were used as received. As catalyst, iron oxide (99.9%, powder, Fe_2O_3, Sigma Aldrich, St. Louis, MO, USA) was employed as received.

2.2. Methods

2.2.1. Polyester-Polyols Synthesis through the Catalytic Degradation of Polyethylene Terephthalate (PET)

The polyester-polyols (RP1 and RP2) necessary for the synthesis of the binder were obtained following two main steps. In the first step, 1 mol of PET, 1.8 mol of PEG600, and tin(II) 2-ethylhexanoate (0.5% from PET wt.) were introduced in a four-neck round-bottom flask (with inlets fitted for a mechanical stirrer, a thermometer, a reflux condenser, and a nitrogen gas purging line), and they were maintained under an inert atmosphere, at total reflux, with continuous stirring, for 3 h at 190 °C. The reaction mixture was then cooled below 100 °C. In the second step were added 0.3 mol of PEG600, 0.5 mol of AS, 0.5 mol of AA, and TIPT (0.5% from PET wt.), and the reaction mixture was maintained under inert atmosphere, at total reflux, with continuous stirring, for another 4–6 h at 190–200 °C. The main possible reaction products obtained through this two-step catalytic degradation are illustrated in Scheme 1. RP1 and RP2 were obtained according to the procedure described, with the following observations: RP1 was obtained by employing both AA and SA (0.5 mol of AS, 0.5 mol of AA), while RP2 was obtained by employing only AA (0.75 mol of AA).

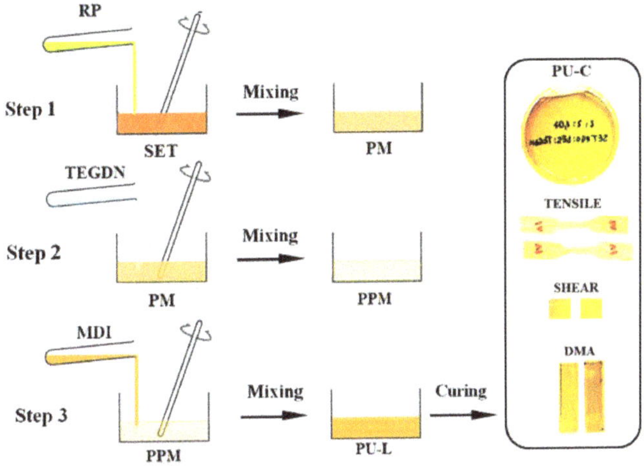

Scheme 1. Schematic presentation of the synthesis and curing of the polyurethane binders (PM: polyols mixture; PPM: plasticizer-polyols mixture; PU-L: polyurethane liquid state; PU-C: polyurethane cured state).

2.2.2. Synthesis of Polyurethane Binder

Starting from the reaction products described, a series of miscible blends was prepared by mixing (Scheme 1, Step 1 and Step 2) the newly synthesized polyester-polyols (RP1 or RP2) with the commercial polyol (SET) and with the synthesized energetic plasticizer (TEGDN). Therefore, RP1, RP2, SET, and TEGDN were mixed in different molar ratios, as detailed in Table 1 (Scheme 1, Step 1 and Step 2), before polyurethanes synthesis.

Table 1. The composition of the investigated polyols.

Sample	Composition	$-OH^{SET}/-OH^{RP}$ Ratio [1]	TEGDN (wt%) [2]
S50	SET	-	-
SRP1	SET:RP1	0.75:0.25	-
SRP1T1	SET:RP1 + TEGDN	0.75:0.25	15%
SRP1T2	SET:RP1 + TEGDN	0.75:0.25	30%
SRP1	SET:RP2	0.75:0.25	-
SRP1T1	SET:RP2 + TEGDN	0.75:0.25	15%
SRP1T2	SET:RP2 + TEGDN	0.75:0.25	30%

[1] -OH molar ratios for SET and RP blends. [2] TEGDN added to SET:RP blends, as weight percentage calculated from the total weight of the binder.

To investigate the possibility of utilizing these newly obtained polyurethanes as binders in rocket propellant composites, polyurethane formulations were obtained by following two working strategies: the first one employing only a commercial polyester-polyol (SET) and the second one consisted in the addition of the polyester-polyols that resulted from the catalytic degradation of PET (RP1 and RP2) to the commercial polyol (SET), as described in Table 1. Thus, the polyurethanes were obtained by reacting the commercial polyol (SET) and the recycled polyols (RP1 and RP2) with the curing agent Desmodur® 44V20L (MDI, diphenylmethane-4,4′-diisocyanate) by adjusting the -NCO/-OH molar ratios, according to the compositions detailed in Table 2. The energetic plasticizer, TEGDN, was added in two different proportions: 15 wt.% and 30 wt.% of the total binder content, according to Tables 1 and 2. The polyurethane synthesis process occurred in one simple step: MDI was added to the polyol blends (Tables 1 and 2) and the mixture was stirred vigorously (except for binders that also included energetic plasticizer (TEGDN), which was introduced into the polyol blends before the addition of the curing agent). The mixing process was carried out at 50 °C to improve processability. The resulting compounds were poured into glass molds, and they were allowed to cure at 60 °C in a vacuum oven. The polyurethane films were exfoliated from the glass molds after the completion of the curing process and were subjected to specific investigations, as described in Results section.

Table 2. Composition of the polyurethane binders.

Sample	Composition	-NCO/-OH Ratio	TEGDN (wt%)
PU_1	DES:SET	1:1	-
PU_2		3:2	-
PU_3		2:1	-
PU_1T1	DES:SET+TEGDN	1:1	15%
PU_2T1		3:2	15%
PU_3T1		2:1	15%
PU_1T2		1:1	30%
PU_2T2		3:2	30%
PU_3T2		2:1	30%
PU_11	DES:SET: RP1	1:0.75:0.25	-
PU_21		3:1.5:0.5	-
PU_31		2:0.75:0.25	-
PU_11T1	DES:SET:RP1+TEGDN	1:0.75:0.25	15%
PU_21T1		3:1.5:0.5	15%
PU_31T1		2:0.75:0.25	15%
PU_11T2		1:0.75:0.25	30%
PU_21T2		3:1.5:0.5	30%
PU_31T2		2:0.75:0.25	30%
PU_12	DES:SET:RP2	1:0.75:0.25	-
PU_22		3:1.5:0.5	-
PU_32		2:0.75:0.25	-
PU_12T1	DES:SET:RP2+TEGDN	1:0.75:0.25	15%
PU_22T1		3:1.5:0.5	15%
PU_32T1		2:0.75:0.25	15%
PU_12T2		1:0.75:0.25	30%
PU_22T2		3:1.5:0.5	30%
PU_32T2		2:0.75:0.25	30%

2.2.3. Rocket Propellant Composite Formulations

The new composite formulations consisted in a solid–liquid mixture, in which the liquid component, represented by the novel polyurethane binders, incorporated the solid components, comprising an oxidizer (PSAN, two granulometric sizes), a metallic fuel (two granulometric sizes of aluminum powder), and catalyst (iron oxide). To provide adequate mechanical strength for these rocket propellant composites and to facilitate the mixing process, the oxidizer chosen for this application used bi-granular (200 µm and 50 µm) mixture. To ensure a good homogeneity of the composite rocket propellant, the liquid and the solid phases were mixed separately, as follows: the liquid mixture was prepared by simply mixing the polyols with the curing agent (MDI); separately, the solid mixture containing the oxidant (PSAN), the metallic fuel (Al), and the catalyst (Fe_2O_3) was sieved and sorted by granule size, then dried. Then, the liquid and the solid phases were thoroughly mixed and introduced in cylindrical molds with 45 mm diameter and allowed to cure at 60 °C in a vacuum oven. The described procedure can be better understood in conjunction with the illustrations presented in Scheme 2.

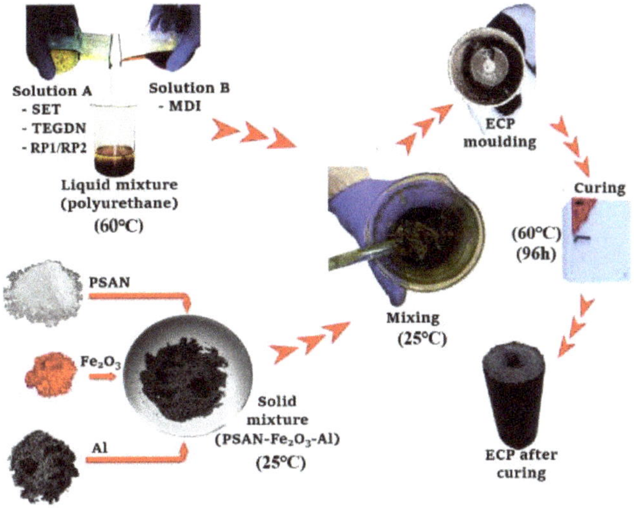

Scheme 2. Schematic representation of the rocket propellant composite fabrication process.

The novel composite rocket propellant formulations were produced in small batches (25 g) and pressed at 10 bar (as required by the characteristic tests necessary for the evaluation of the performances of solid rocket propellants: evaluation of combustion behavior, evaluation of the mechanical properties, and morphological analysis) [38]. Subsequently, the samples were cured under vacuum for 96 h at 60 °C. The binders employed in this processing step were selected according to their homogeneity and their thermal and mechanical properties. Therefore, only the polyurethanes with molar ratio -NCO/-OH of 2:1 plasticized with 30 wt.% TEGDN were eligible to be used as binders for the rocket propellant formulations. Table 3 provides information regarding the composition of the new eco-friendly composite propellants (ECP) obtained in this study. All the formulations contained 57 wt.% PSAN with 200 μm granulation and 16 wt.% of PSAN with 50 μm granulation, as well as 1 wt.% of burn rate regulator Fe_2O_3.

Table 3. Formulations for the rocket propellant composite.

Sample	Components [wt.%]					
	PU_3	PU_3T2	PU_31	PU_31T2	Al <5 μm	Al 100 μm
ECP_A5	15				12	
ECP_A100	15					12
ECP_B5		15			12	
ECP_B100		15				12
ECP_C5			15		12	
ECP_C100			15			12
ECP_D5				15	12	
ECP_D100				15		12

2.2.4. Characterization

The hydroxyl number of the polyols (L_{OH}) was determined according to ASTM-D4274-21 [39]. The acid number (I_A) was determined according to ASTM D4662-20 [40]. The number-average molecular weight (M_n) was quantified according to the "end -groups" method, using the acid number and hydroxyl number, without removing the free glycols, by using the following formula: $Mn = (2 \cdot 56.1 \times 1000)/(I_A + I_{OH})$ [41,42]. The FT-IR spectra of the synthesized polyols were recorded using a Spectrum Two FT-IR Spectrometer

(PerkinElmer, MA, USA). The parameters used for the FT-IR ATR determinations were number of scans: 32; resolution: 4 cm^{-1}; spectral range: 600–4000 cm^{-1}. ^{1}H-NMR spectra of the polyols were acquired with a Gemini Varian 300 MHz 300 MHz spectrometer (Varian Inc., Palo Alto, CA, USA). To conduct the ^{1}H-NMR analyses, the samples were dissolved in 0.5 mL of deuterated chloroform ($CDCl_3$). The viscosity of the samples was measured at 25 °C on a Rheotest 2.1 (Germany) device with coaxial cylinders. A shear rate varying from 1 to 1400 s^{-1} was applied for all samples. The thermal properties of the synthesized polyurethanes were investigated using DTA and TGA analyses. DTA OZM 551 Ex Differential Thermal Analysis System (OZM Research, Hrochův Týnec, Czech Republic), with Meavy dedicated software, was used for the evaluation of the influence of each component on the thermal properties of these new materials. Each test was performed on 35 mg of sample heated between 25 °C and 300 °C with a 5 °C/min heating rate, according to STANAG 4515 [43]. The thermogravimetric analysis (TGA) of the synthesized polyurethanes was executed using a Netzsch TG 209 F3 Tarsus instrument (NETZSCH, Selb, Germany). The experiments were performed at a heating rate of 10 °C/min under nitrogen flow, where samples of approximately 4 mg were heated from ambient temperature 25 °C to 700 °C. The morphology of the polyurethane films and the distribution of the main elements have been investigated by means of SEM-EDX (scanning electron microscopy) using a Tescan Vega II LMU SEM instrument (TESCAN, Brno, Czechia) at 10 keV acceleration voltage, under high vacuum. To ensure good data acquisition, the samples were sputter-coated with gold. The distribution of the metallic particles inside the polyurethane matrix of the rocket propellant composite was studied through micro-CT (μCT) method. Thus, the SkyScan micro-CT was attached to the previously mentioned Tescan Vega II LMU SEM instrument. Samples were examined as follows: exposure time, 4 s per projection at electron beam currents of 100 nA; accelerating voltage, 30 KeV; step size, 1°; scanning time, 24 min. The reconstructed images were obtained using the NRecon program by using float-point data values for internal calculations during reconstruction, which allow the operator to define the density window as a range of reconstructed values. The reconstruction results were visualized with the aid of the program DataViewer® 2D/3D Micro-CT Slice Visualization (Micro Photonics Inc., Allentown, PA, USA). Tensile tests were carried out by using a 710 Titan 2 Universal Strength Tester by James H. Heal and Co. Ltd. testing machine (Richmond Works Halifax, UK). The tests were carried out according to the international standard ISO 37: 2011(E) [44]. Polyurethane samples were prepared by cutting tensile specimens (standard dumbbell: 5 mm width of the narrow parallel part, 100 mm total length, the distance between gauge marks 20 mm). Tests were performed at a rate of extension of 500 mm/min, starting from 50 mm jaw separation (plain jaw faces). For each polyurethane film, five tensile tests were performed, and the average value was reported. The low strain rate compression tests of the composite propellant were performed on an Instron 2519-107 Universal Test Machine (Instron, Norwood, MA, 02062-2643, USA). Cylindrical specimens, with a diameter of 20 mm and a length of 20 mm, were tested at a compression rate of 50 mm/min. Each test was repeated three times. Dynamic mechanical analysis (DMA) tests were carried out on a Discovery DMA 850 apparatus from TA Instruments (New Castle, DE, USA). Samples were analyzed in single cantilever mode, frequency 1 Hz, on a temperature ramp between −100 °C and 50 °C, using a heating rate of 5 °C/min maintained by cooling with liquid nitrogen. On the same instrument, 3-point bending clamp was installed to evaluate the sample sheets during a bending process consisting of successive stress, amplitude varying from 1 to 100 μm, 1 Hz, initial preload force 0.01 N, at 25 °C. The measurements of the frequency-dependent shear modulus were performed on the DMA 850 instrument with shear-sandwich clamps, at frequencies varying from 0.01 Hz to 10 Hz, 20 μm amplitude at 25 °C. For the single cantilever temperature ramp and 3 point bending analyses performed on DMA 850, the specimens were cut from the propellant cylinders in rectangular shapes with the following dimensions: 60 mm × 12.5 mm × 3 mm; for the shear-sandwich set-up, they were cut in square shapes of 10 mm × 10 mm × 1.5 mm. The experimental tests regarding the burning mechanics of the new environmentally friendly rocket propellants were conducted on a subscale rocket motor TRM35 (OZM Research, Hrochův Týnec, Czech Republic). The specimens were designed as cylinders with inner hole with the following dimensions: 40 mm × 12 mm × 45 mm (outer diameter × inner diameter × length).

3. Results and Discussion

The degradation of PET through glycolysis involved the use of PEG600, employed in the first step, followed by the addition of AA and SA in the second step. The AA and SA were introduced in the second step of the reaction to ensure the consumption of the unreacted PEG600, while also serving as chain extenders and maintaining the low viscosity of the final reaction products mixture. Another purpose was to ensure a high concentration of OH-ending groups for the synthesis of the polyurethane binders. Scheme 3 illustrates only the main, hypothetical reaction products that resulted from PET, because it is difficult to depict all the combinations that may arrive from this situation.

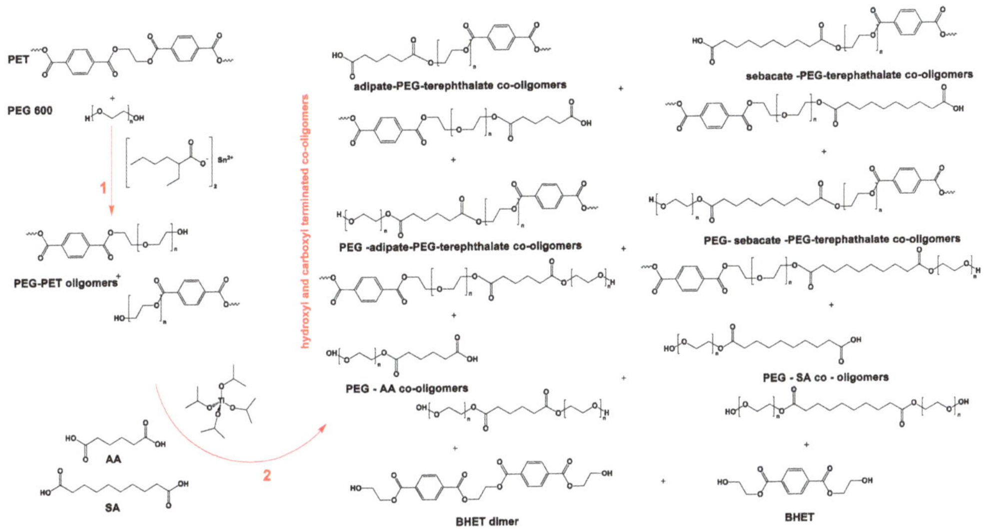

Scheme 3. Hypothetical scheme comprising the main probable degradation products obtained through the two-step catalytic degradation of PET.

Before proceeding with the synthesis of the polyurethane binders, a thorough analysis of the synthesized polyols was imperative. Thus, the products obtained through the two-step catalytic degradation of PET, together with the blends obtained with the commercial polyol and with the energetic plasticizer, were investigated through various analytical techniques; ^{1}H-NMR and FT-IR were employed for the evaluation of the chemical structure of the polyester-polyol, while the titration methods offered information about the hydroxyl number and acid number. Rheological properties were also investigated, because the viscosity of the polyol blends directly influences the polyurethane synthesis (ease of mixing the components, reaction rate, homogeneity of the obtained materials) and the preparation of the final composite formulation.

Figure 1 illustrates the ^{1}H-NMR spectra of the polyols that resulted from the degradation of PET (RP1 and RP2). In both cases, the peaks labeled with 1 can be assigned to the protons from the aromatic rings (δ_H = 8.05 ppm, singlet) from the terephthalates moieties derived from PET, and the peaks labeled with 2 and 3, respectively, and 5 and 6 can be attributed to the protons from methylene groups adjacent to -COO groups (δ_H = 4.44 ppm, triplet and δ_H = 4.16 ppm, triplet, respectively, and δ_H = 2.26 ppm, multiplet and δ_H = 1.59 ppm, multiplet), while peak 4 (δ_H = 3.8 ppm, triplet) can belong to the protons adjacent to -OH groups, and peak 7 (δ_H = 1.24 ppm) can be attributed to the protons from methylene groups adjacent to other methylene groups. The intense peak from 3.59 ppm can be attributed to PEG fragments. Figures S1–S4 from the "Supplementary Material" file illustrate the ^{1}H-NMR spectra of the polyol blends that resulted from mixing RP1 and RP2, respectively, with SET and with SET+TEGDN. In comparison with the plots obtained for the neat polyester-polyols derived from PET (RP1 and RP2), the plots from the "Supplementary Material" file contain two additional peaks in the 5.4–5.5 ppm region, specific for the tertiary proton of the -CH2CHCH2- backbone of SET (this commercial polyol is mainly based on castor oil) and one new peak at δ_H = 4.61 ppm, specific for the protons adjacent to nitro groups in TEGDN.

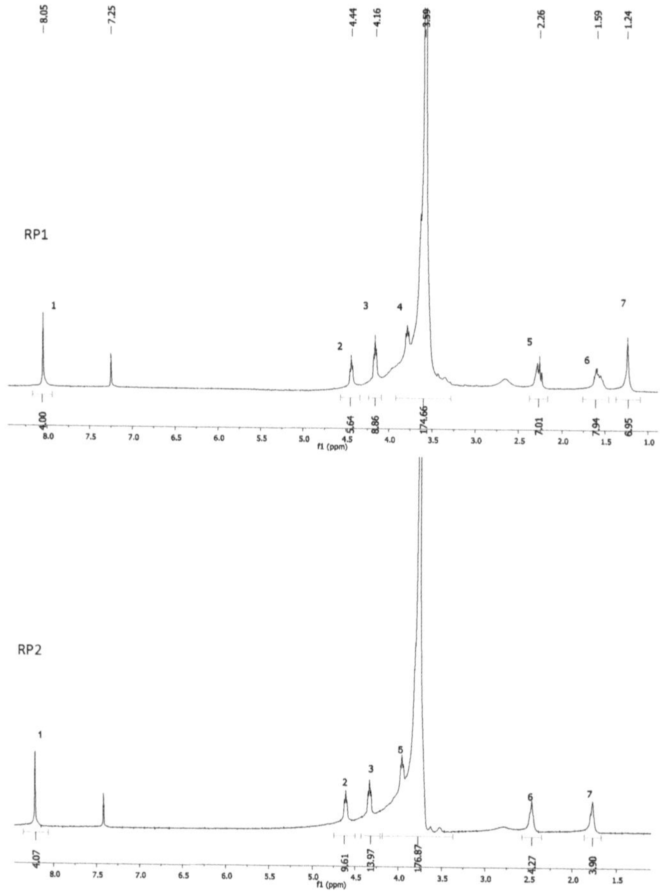

Figure 1. ^{1}H-NMR spectra of the polyol mixtures that resulted from the degradation of PET.

In all the polyol mixtures (polyester-polyols obtained from the degradation of PET (RP1 and RP2) and polyol mixtures (SRP1 and SRP2, SRP1T1 and SRP1T2) (see details on composition in Table 1), FT-IR spectra confirmed the presence of free hydroxyl groups by the presence of a peak at the 3667–3670 cm^{-1} region. The presence of intermolecular bonded hydroxyls was also visible through a broad peak situated in the 3475–3490 cm^{-1} region. The intense peak that can be observed around 1740 cm^{-1} can be attributed to C=O stretching from ester groups. C-OH stretching vibrations are visible in the 1097–1100 cm^{-1} region. The peaks from 1269 cm^{-1} and 1252 cm^{-1} can be assigned to C-O bonds, while 2866 cm^{-1} and 2945 cm^{-1} can be assigned to C-H bonds. C-H bending can be evidenced through the peaks present at 1454 cm^{-1} and 1350 cm^{-1}. For the samples containing TEGDN (SRP1T1 and SRP1T2), three additional peaks can be observed at 1626 cm^{-1}, 1277 cm^{-1}, and 855 cm^{-1}, which represent the -O-NO_2 stretching vibrations. The data can be observed in Figure 2. The RP2 mixture spectra are presented in Figure S5 in the "Supplementary Material" file.

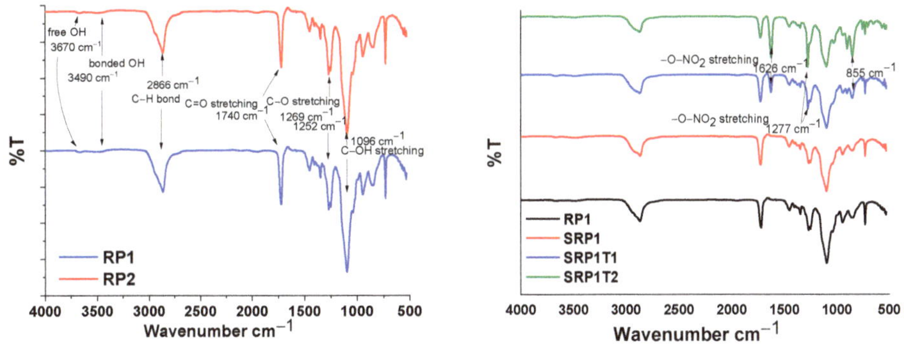

Figure 2. FT-IR spectra of RP1 and RP2 polyols and mixtures SRP1, SRP1T1, and SRP1T2.

The number-average molecular weight (M_n) of the compounds that resulted from the degradation of PET was estimated based on the experimental values obtained from the hydroxyl number and the acid number. The results are presented in Table 4. As can be observed, when the ratio of glycol/PET (RP1) was higher, the numerical-average molecular weight (M_n) obtained for the glycolyzed products was lower. Since these products will be included in the synthesis of the polyurethane binders designed for energetic mixtures, the results presented in Table 4 are useful for further calculations (e.g., calculation of the amount of MDI necessary for the synthesis of polyurethane binders).

Table 4. Assessment of I_{OH} and I_A for the polyester-polyols that resulted from the degradation of PET, through titration methods.

Sample	Properties			
	I_{OH} mgKOH/g	I_A mgKOH/g	M_n, calc. *	Description
RP1	30.86	1.47	3470	Pale yellow, viscous liquid
RP2	33.36	1.08	3257	Light brown, viscous liquid

* Mn calc. = (2 × 56.1 × 1000)/(I_A + I_{OH}) [41,42].

An important aspect for this type of application (polyurethane-based binders for rocket propellants) needs to be underlined: when the PET degradation reaction is completed, the free glycols should not be separated. Thus, they offer the possibility to maintain a low viscosity for the glycolyzed products mixture, which is an important factor for the synthesis of polyurethane binders. The presence of these unreacted glycols led to lower viscosities of the RP1 and RP2, also demonstrated through rheological measurements illustrated in Figure 3a–d. The neat commercial polyol (SET) exhibited lower dynamic viscosities than the polyol mixtures that resulted from the degradation of PET (RP1 and RP2). The addition of the energetic plasticizer to the polyester-polyol blends can significantly improve the performance of these rocket propellants. Therefore, analyzing its influence on the physical and chemical properties of these energetic composites is of great importance.

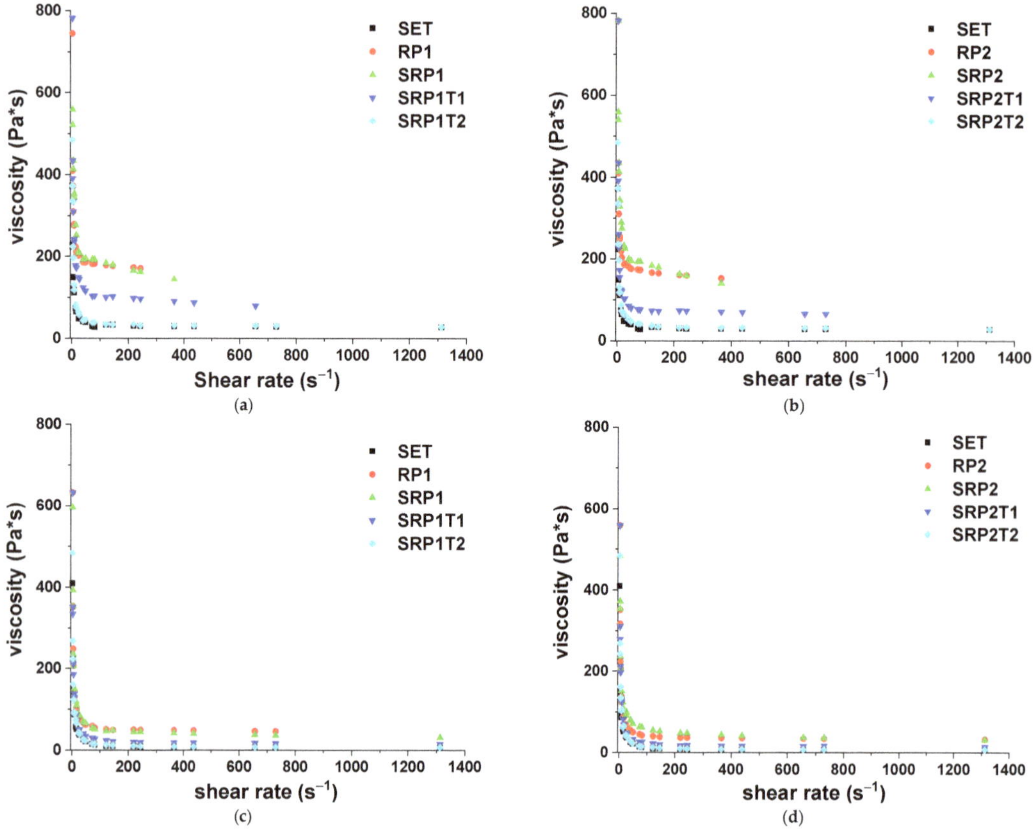

Figure 3. The effects of SET and TEGDN upon the viscosity of polyester-polyols.

The plots illustrated in Figure 3a–d describe the behavior of the polyol mixtures at 25 °C (Figure 3a,b) and 50 °C (Figure 3c,d), with the increase of the shear rate. As can be observed, RP1 and RP2 do not seem to exceed some of the tolerable limits [1,2] for this type of application, their viscosities allowing them to be easily mixed with the solid components of the rocket propellants. Moreover, the addition of SET and TEGDN led to even lower viscosities, facilitating the synthesis of homogenous binders. The viscosity of the mixtures SRP1 and SRP2, which resulted from the addition of the commercial polyol (SET) to the recycled polyols (RP1 and RP2), was slightly increased in comparison to neat RP1 and RP2. However, this aspect does not represent a problem for this type of application, because the addition of TEGDN results in a decrease of the polyol mixture's viscosity. It was observed that the viscosity of the polyol blends diminished with the extent of TEGDN added: the dynamic viscosity of SRP1T1 was higher than for SRP1T2. The polyol mixtures based on RP2 exhibited the same behavior: the dynamic viscosity of SRP2T1>SRP2T2. A lower viscosity of the polyol mixture will facilitate the mixing process of the rocket propellant composite, at least in terms of rheological properties.

To obtain specimens with a certain geometric shape, specific for each investigation method employed for this study, after following the three steps of polyurethane synthesis illustrated in Scheme 1, the freshly prepared liquid polyurethanes were poured into glass vessels and placed in a vacuum oven. During the curing process, the time was recorded, and the process of surface hardening was examined periodically. According to the observations summarized in Figure S6 from the "Supplementary Material" file, we can conclude that the curing time of polyurethane samples is inversely proportional to the -NCO/-OH ratios. More precisely, when the concentration of MDI increased, the curing time decreased. All three types of SET-MDI films successfully incorporated a large amount of TEGDN, while in the case of the polyurethanes based on the polyols that resulted from PET, some problems were identified for PU_11T1, PU_11T2, PU_12T1, and PU_12T2, where the reaction seemed incomplete. In the case of PU_22T2 and PU_32T2, the surfaces had an irregular aspect. Moreover, the color changed and the 'air-bubble' trapping that was observed in the case of PU_2, PU_3, PU_21, PU_31, PU_22, and PU_32, was diminished as we increased the amount of TEGDN added to the polyol blends. A brief description of the cured samples and curing time (T_C) for each formulation is depicted in Figure S6 in the "Supplementary Material" file.

The visual inspection of the cured polyurethane films (illustrated in Figure S6 in the "Supplementary Material" file) offered evidence about the potential polyol blends that can serve as candidates for the polyurethane binders. Based on their visual aspect, for a detailed investigation, polyurethane films having 3:2 and 2:1 -NCO/-OH ratios, with and without TEGDN, were chosen to be subjected to SEM-EDX analysis. These examinations were performed only for RP-based polyurethanes, because the aim of this study consisted in developing eco-friendly binders for "green" rocket propellants. The morphological results provided by SEM analysis can be found in Figure S7 in the "Supplementary Material" file. The visual inspection of the polyurethane films revealed that the surfaces of the samples based on RP2 (samples PU_22T2 and PU_32T2) were irregular. Therefore, an elemental mapping of carbon, nitrogen, and oxygen, with the aid of an energy dispersive X-rays (EDX) technique, was necessary for a better understanding of the interaction between the elements. According to EDX images, displayed in Figure S8, all the polyurethane formulations based on RP1 showed good homogeneity. The same can be observed for PU_22 and PU_32 samples. However, in the case of PU_22T2 and PU_32T2, there is an obvious lack of homogeneity, probably induced by the incompatibility of TEGDN with the polyol blends containing RP2. Therefore, for the next steps of our study, based on the information obtained through SEM-EDX analysis, we decided to employ only the polyol formulations containing RP1.

The thermal behavior of polyurethane films (with and without the energetic plasticizer, TEGDN) was investigated by TGA and DTA analytic techniques. Due to safety reasons, the rocket propellant composite formulations were analyzed only by DTA means. Differential thermal analysis (DTA) gives complementary information about the thermal properties of the synthesized materials. Typical DTA thermograms for the polyurethane films, based on 3:2 and 2:1 -NCO/-OH group molar ratios, are shown in Figure 4a,b. Figure 4c displays a comparative plot between neat TEGDN and the rocket propellant composites obtained in this study. There is only one exothermic peak of great significance for both films and composites. The decomposition temperature attributed to this peak varies in the interval of 190–200 °C; detailed values, for each formulation, are summarized in Table S3 from the "Supplementary Material file" file. Based on the DTA analysis, this high exothermic peak, present in all thermograms of the formulations containing TEGDN (Figure 4c), corresponds to the decomposition temperature of this energetic plasticizer [36]. After evaluating the thermograms illustrated in Figure 4 together with the values summarized in Table S3 from the "Supplementary Material file" file, it can be observed that the higher compatibility between RP1 and TEGDN (confirmed by SEM-EDX) led to a more efficient energetic binder, which can be demonstrated by the existence of a higher and sharper peak of decomposition for the rocket propellant ECP_D5 (comprising a polyurethane binder based on RP1, molar ratio 2:1, sample PU_31T2). The positive influence of RP1 on the thermal performances of the synthesized materials can be observed in all the thermograms containing RP1 and TEGDN (Figure 4a–c), samples PU_31T2 and ECP_D5. Therefore, we can conclude that the optimal polyurethane binder, in terms of thermal performances of the rocket propellant, is represented by the polyurethane based on RP1 (sample PU_31T2), which seems to enhance the exothermic efficiency of the energetic formulation.

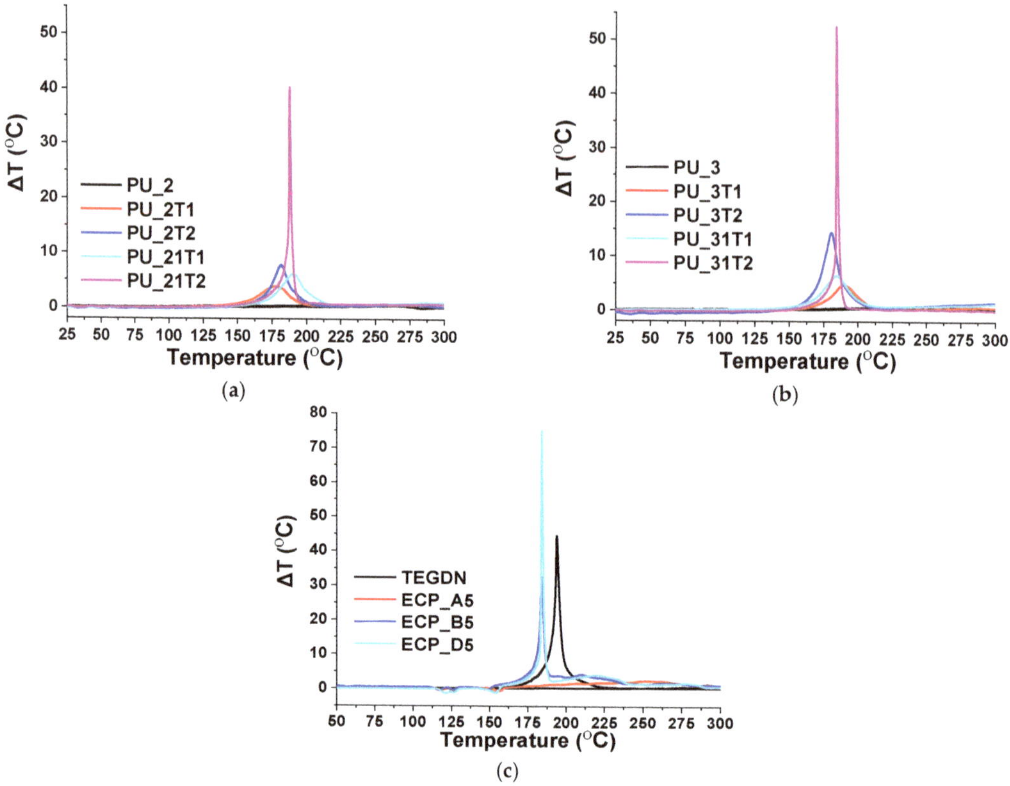

Figure 4. DTA thermograms for polyurethane films: (**a**) 3:2 -NCO/-OH group molar ratio; (**b**) 2:1 -NCO/-OH group molar ratio; (c) neat TEGDN and composite rocket propellants.

Moreover, the sharpness of the characteristic exothermic peak gives indications about reaction kinetics, being related to the rate of reaction. It is commonly known that the energetic materials undergo fast exothermic degradation transformations, which also occur with our rocket propellant composites, a necessary characteristic for obtaining efficient propulsion. This positive effect was not present when PU_3T2 was employed in the synthesis of the rocket propellants (sample ECP_B5). As can be observed in Figure 4, the peak height of the composite ECP_D5 (based on polyurethane containing TEGDN, sample PU_31T2) is higher than the peak height of neat TEGDN. Therefore, this gives clear evidence of the positive effect that RP1 has on the energetic formulation, enhancing the exothermic decomposition process, which can be considered beneficial and desirable for this type of application. The samples containing only the commercial polyol (SET), presented here for comparison, namely the samples named here PU_3 and ECP_A5, displayed lower values in terms of thermal performances. Therefore, we can conclude that the most efficient polyurethane formulation that can be employed as an energetic binder for the rocket propellants involved both commercial (SET) and recycled (RP1) polyols together with the energetic plasticizer (TEGDN, 30 wt.%). Namely, the optimal polyurethane binder was the sample coded with PU_31T2. Since in the DTA analysis the polyurethane films based on a 2:1 -NCO/-OH molar ratio (sample PU_31T2) displayed the best values for these rocket propellants, thermogravimetric analysis was performed on the samples based on SET (samples PU_3 and PU_3T2) and SET+RP1 (samples PU_31 and PU_31T2), to obtain complementary information regarding the thermal properties of the polyurethane films. Figure 5 displays TGA (a) and DTG (b) curves for the new synthesized polyurethane films. As can be observed in Figure 5 and Table S4 from the "Supplementary Material" file, the decomposition process of the polyurethane films based on SET (sample PU_3) began slightly earlier than in the case of the polyurethane films obtained with SET+RP1 (sample PU_31). This trend was not similar for the samples containing TEGDN, because the decomposition process for sample PU_31T2 began earlier due to better compatibility and miscibility of RP1 with the energetic plasticizer (as already confirmed by SEM and DTA analyses). The peaks displayed in the DTG curve offer information about the succession of the degradation phases of the polyurethane films. The first peak, situated around 190 °C, can be attributed to the decomposition of TEGDN embedded in the polyurethane matrix, while the peaks that appeared starting at 300 °C can be assigned to the degradation of the polyurethane matrices.

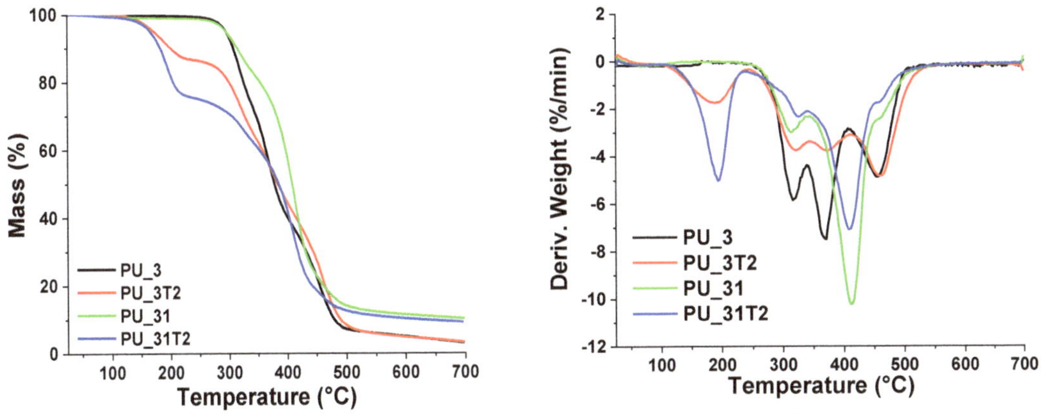

Figure 5. (**a**) TGA and (**b**) DTG profiles of the polyurethanes.

Table S5 from the "Supplementary Material" file summarizes the weight loss percentage, at certain temperatures, for the polyurethane binders. As can be observed, those containing TEGDN, samples PU_3T2 and PU_31T2, displayed earlier a significant weight loss than the polyurethane matrices without the energetic plasticizer (samples PU_3 and PU_31). The mechanical properties of the polyurethane matrix directly influence the mechanical behavior of the propellant grains. Thus, the commercial and recycled polyol-based polyurethanes (with and without TEGDN) were subjected to tensile stress. The comparative tensile stress–strain profiles of PU_3, PU_3T2, PU_31, and PU_31T2 are depicted in Figure 6, while in Figure S9 from the "Supplementary Material" file can be found the stress–strain curves of each specimen of the polyurethane formulation. As can be observed, the maximum values of tensile stress and strain were reached by the polyurethanes based on the commercial polyol, samples PU_3 and PU_3T2. Even though, theoretically, samples PU_31 and PU_3 should display similar mechanical behavior (both are based on Setathane (SET)), tensile tests revealed lower mechanical resistance for PU_31 polyurethanes. The difference between the mechanical properties of these two analogue matrices (samples PU_3 and PU_31) appeared due to the introduction of RP1 during synthesis. This polyol blend (SRP1) was obtained by mixing SET (commercial polyol) with RP1 (polyol that resulted from the degradation of PET). Besides the long aliphatic chains of SET (which is a branched, castor oil-based polyol [45]), it contains multiple shorter chains belonging to RP1, which led to a decrease of the overall mechanical resistance of these polymeric films. The introduction of an energetic plasticizer in the polyurethane network, which is compatible with the other reactants employed for the synthesis (RP1, SET, and MDI), is imperative for enhancing the performance characteristics of the correspondent rocket propellant. As can be seen from Figure 6, the addition of TEGDN led to a higher elongation of the tested specimens and thus an increase of strain values, but the stress values were slightly decreased. The Young's modulus followed the same trend, decreasing with the addition of RP1 and TEGDN. Explicit values for the Young's modulus, maximum tensile stress, and strain of the synthesized polyurethane films are presented in Table 5. The mechanical properties of the rocket propellant composites are not influenced only by the nature and concentration of the polyurethane binder, but it also depends on the properties and concentration of the solid components (oxidizer, metallic fuel). Thus, to investigate the influence of the metallic fuel granule size on the mechanical properties of the newly synthesized composite rocket propellants, different propellant formulations were subjected to tensile tests (Table 6) and compression (Figure 7 and Table 7).

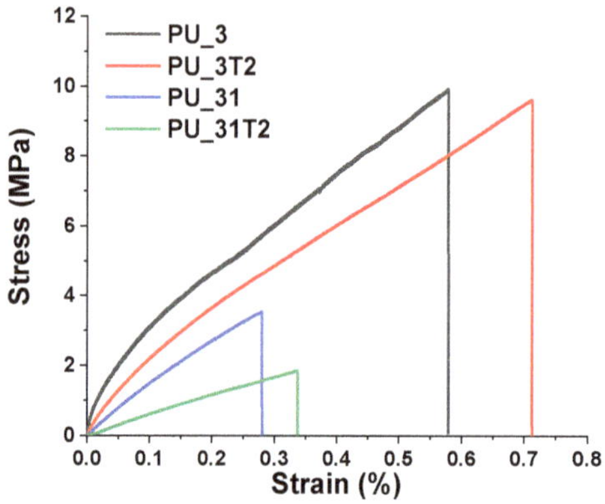

Figure 6. Tensile stress–strain plots of polyurethane films.

Table 5. Tensile characteristics of polyurethane films.

Sample	Young's Modulus (MPa)	Maximum Tensile Stress (MPa)	Maximum Tensile Strain	Deformation Energy (Tensile Toughness) (J m^{-3})
PU_3	18.01 ± 2.06	10.19 ± 1.54	0.56 ± 0.05	3.95 ± 0.82
PU_31	13.27 ± 1.86	9.22 ± 1.29	0.69 ± 0.07	3.16 ± 0.66
PU_3T2	12.79 ± 0.76	3.30 ± 0.31	0.26 ± 0.02	0.45 ± 0.08
PU_31T2	5.92 ± 1.78	1.97 ± 0.39	0.34 ± 0.03	0.32 ± 0.06

Table 6. Tensile properties for the novel composite rocket propellants.

Sample	Young's Modulus (MPa)	Tensile Stress (MPa)	Tensile Strain (%)
ECP_B5	140.6 ± 0.10	1.3 ± 0.11	0.01
ECP_C5	106.3 ± 0.21	1.1 ± 0.20	0.01
ECP_D5	44.9 ± 0.06	0.4 ± 0.07	0.01

Compression deformation at the breaking point of the composite propellants, as a function of the metallic fuel (aluminum) particle size, is illustrated in Figure 7. During the tests, the percentage of binder used was the same for all samples. Hardness appears to be directly influenced by particle size and binder type used. The most rigid sample is ECP_A5, due to PU_3 and small particles (<5 μm), while the most elastic is ECP_D100, based on PU31_T2 and large particles (100 μm), as shown in Table 7 for a strain value around 0.2%. Comparing formulations with similar grain sizes, the stress–strain differences appear due to the use of RP1 and TEGDN. Thus, before adding TEGDN, ECP_A5 and ECP_A100 showed a compressive stress resistance four times higher than that of ECP_B5 and ECP_B100, respectively. Similar behavior was distinguished for ECP_C5, ECP_C100, ECP_D5, and

ECP_D100, based on RP1. The use of RP1 decreases the resistance only twice compared to those that did not contain it. As anticipated, the addition of the energetic plasticizer, TEGDN, has a significant impact on the compression behavior of these materials.

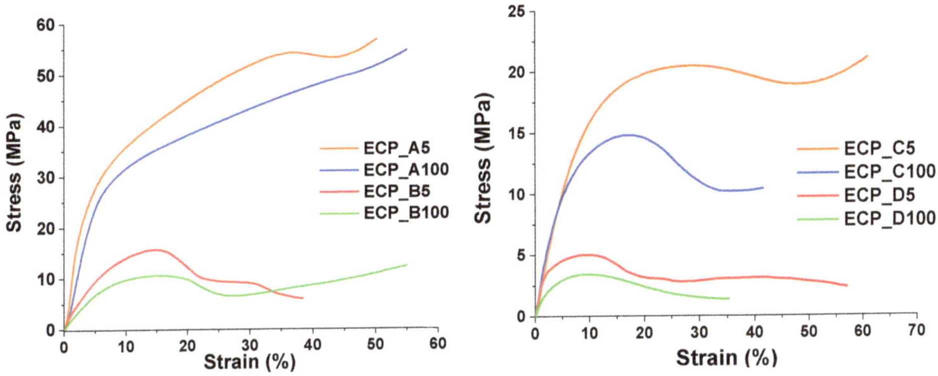

Figure 7. Compressive stress–strain profiles for composite rocket propellants.

Table 7. Compressive strain–stress values of the composite rocket propellants.

Sample	Compressive Strain (%)	Compressive Stress (MPa)
ECP_A5	0.22	0.90 ± 0.13
ECP_A100	0.23	0.52 ± 0.03
ECP_B5	0.22	0.75 ± 0.03
ECP_B100	0.26	0.41 ± 0.01
ECP_C5	0.29	0.49 ± 0.09
ECP_C100	0.23	0.66 ± 0.07
ECP_D5	0.22	0.61 ± 0.04
ECP_D100	0.25	0.38 ± 0.03

Since the formulations based on small particle granulation exhibited higher compression resistance, the tensile tests were performed only for this category (aluminum particles size < 5 μm). The composite tensile test results are shown in Figure S10, except for ECP_A5, because its mechanical resistance exhibited the limits of traction force of the instrument employed for this analysis. As can be observed in Table 7, the stiffness of the rocket propellant specimens varied with the modification of the grain sizes and their composition. Specifically, the rocket propellants based on RP1 displayed poorer mechanical resistance than those based only on SET. On the other hand, the composite propellants containing smaller grain sizes (samples ECP_B5, ECP_C5, and ECP_D5) of the solid components displayed better mechanical resistance, due to a better arrangement of these grains inside the polymeric matrix. The results indicate that the mechanical characteristics of the new materials are closely related to their internal structure, homogeneity of the solids dispersion in the binder matrix, particles sizes, and good interaction of the organic binder with the grains of the inorganic oxidizer and metallic fuel.

The mechanical behavior of these binders, such as elasticity, viscosity, hardness, brittleness, or rigidity, under a variety of stresses (for example pressure or traction) influences the mechanical properties of the rocket propellant composites. Even if the polyurethane matrices based on the commercial polyol (PU_3 and PU_3T2) displayed higher mechanical resistance than those containing the polyols from PET degradation (PU_31 and PU_31T2), we can conclude that these recycled materials possess a mechanical resistance suitable to be employed as binders in rocket propellants. Moreover, the selection of the polymeric matrix of the binder must take into consideration the advantages brought by the recycled polyols through their outstanding thermal properties. Thus, a compromise can be made, and despite their poorer mechanical resistance (which is situated within some tolerable values), the binders based on recycled PET can enhance the thermal performances of the rocket propellant and contribute to a less polluted environment. Moreover, its higher flexibility will ensure resilience, a better shock resistance, which is very important for maintaining the geometry of the propellant, preventing crack formation and adhesion failure.

DMA analysis was aimed at investigating the thermo-mechanical properties of the synthesized polyurethanes and polymeric composites in a single cantilever bending mode (on temperature ramp from −100 °C to +50 °C or +100 °C), three-point bending mode (ambient temperature, +25 °C), and shear-sandwich configuration (ambient temperature, +25 °C). The rocket propellant composites were tested only up to +50 °C due to safety reasons [46], as they are energetic materials. Figure 8 illustrates a comparative plot of the storage modulus, loss modulus, and tan delta obtained in a single cantilever bending mode for the polyurethanes (Figure 8) based on polyols obtained from the degradation of PET (with and without TEGDN, namely PU_31 and PU_31T2). The DMA results for the corresponding polymeric composites are presented in Figure 8d–f: based on commercial polyol, ECP_B5; based on polyols recycled from PET, ECP_D5. According to STANAG 4540 [47], the maximum value of the loss modulus E″ peak of the composites for rocket propellants can be considered their glass transition temperature. Therefore, according to this standard for energetic materials, we can assume that the glass transition temperatures of our samples correspond to the maximum of the loss modulus E″ peak. Thus, PU_31 (the polyurethane based on polyols recycled from PET) displayed a glass transition temperature at around −25 °C, while its homologous polyurethane containing TEGDN (PU_31T2) displayed a lower T_g (−53 °C), due to its plasticizing effect. The rocket propellant composite (comprising the polyurethane based on polyols recycled from PET), ECP_D5, displayed a glass transition temperature at approximately −32 °C. In comparison with the neat polyurethanes, this transition is also influenced by the interactions that occur between the solid components of the rocket propellant composite and the polymeric matrix of the binder. The structure and the properties of the rocket propellant composites are influenced by the composition and ratio of the blended polyols employed for the synthesis of the binder. Thus, it can be observed that the composite containing only SET, ECP_B5, displayed two glass transition temperatures, due to a stronger segregation [48] of the hard and the soft segments of the polyurethane employed in this type of composite. In this case, the polymeric chains of SET (castor-oil-based polyol), possessing considerably higher lengths, will generate larger zones where soft segments will merge. Thus, the structuring of these polyurethanes will have a particular microscopic aspect, consisting of congregated hard segments (consisting mostly of aromatic regions originating from MDI, Desmodur®44V20L), which will look similar to "isles areas" in the soft polymeric matrix (consisting of long aliphatic chains originating from SET). This behavior can also be observed in SEM-EDX images illustrated in Figure S7. On the other hand, the recycled polyol blends did not lead to the same behavior as the polyurethanes, probably because in this case, soft segments are more homogenous and intercalated with hard segments due to the shorter length of the diols employed for the synthesis of these binders. Thus, the rocket propellant composites that contain binders based on polyurethanes obtained from recycled polyol blends displayed only one T_g point (−32 °C).

Many properties of viscoelastic materials are dependent on time (frequency). The viscoelastic nature of polyurethane binders employed in this study requires consideration of their creep behavior during the design process of the rocket propellant composites. This variation is illustrated in Figure 9a–f, for the shear storage modulus, shear loss modulus, and δ, based on data collected in the shear-sandwich DMA set-up. As can be observed in Figure 9, the storage modulus presented a significant increase in the sample containing the commercial polyol and TEGDN (PU_3T2), indicating that this plasticized polyurethane has a great potential for storing the energy during a loading cycle [49].

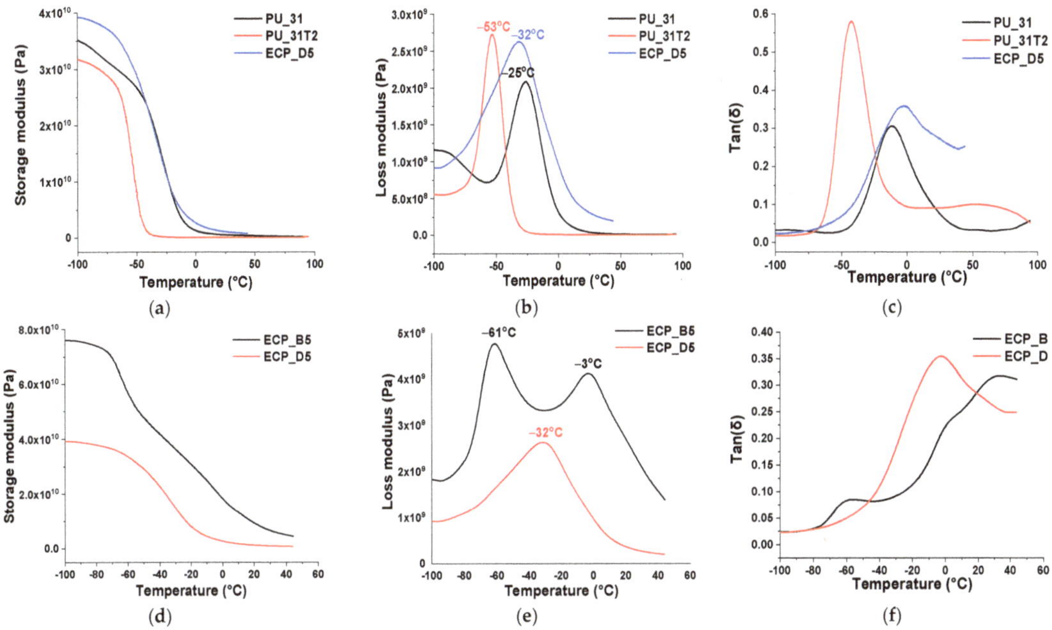

Figure 8. Dynamic mechanical analysis for polyurethanes and composite propellants. (**a**,**d**) Storage modulus for polyurethane films and for the corresponding composites; (**b**,**e**) Loss modulus for polyurethane films and for the corresponding composites; (**c**,**f**) Tan delta plots for polyurethane films and for the corresponding composites.

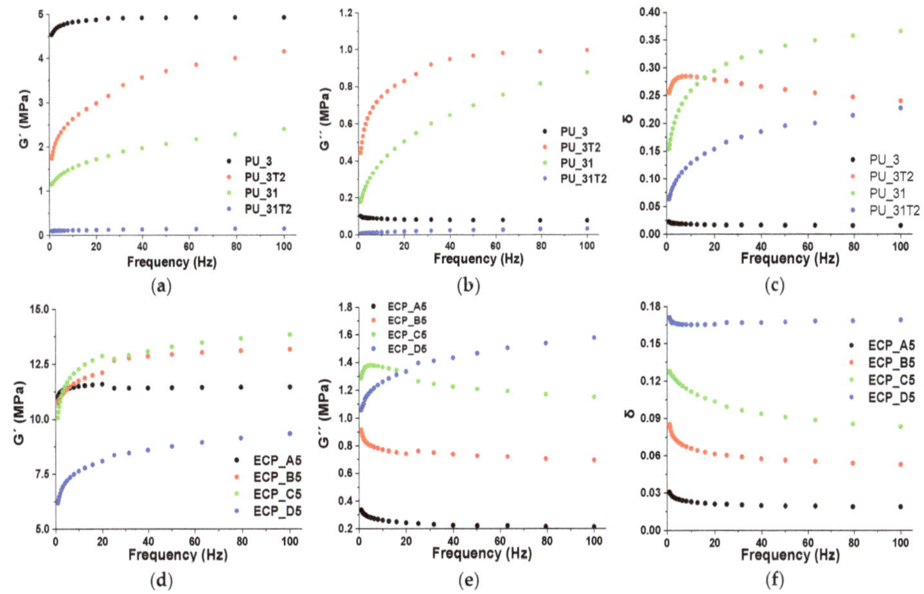

Figure 9. Shear storage modulus (**a**,**d**), shear loss modulus (**b**,**e**), and tan delta (**c**,**f**) profiles vs. frequency for synthesized polyurethanes and for the corresponding composite propellants.

Consequently, this parameter can be related to the shape recovery of this polymer during loading. Similar behavior was observed for PU_31, the polyurethane based mainly on polyols derived from PET. Therefore, we can sustain that these materials also possess a remarkable capacity for storing energy. This ability is important for the binders employed in the rocket propellant composites because, when the energetic charge/loading encounters a shear stress situation, the polyurethane binder absorbs the energy, thus preventing the accidental initiation of the energetic composite. The loss modulus, which characterizes the ability of the polyurethane binder to dissipate this energy through the internal molecular motions of the polymeric chains, displayed higher values for the PU_3T2 and PU_31 samples. Therefore, tracking the changes of the shear modulus as a function of frequency offers evidence about the "damping" behavior of these polyurethanes by describing the dissipation of mechanical energy through internal motion (loss modulus, tan delta). Comparing these mechanical properties of the synthesized polyurethanes with those obtained for the corresponding rocket propellant composites (Figure 9d–f), we can affirm that the composite based off the recycled polyol blends and TEGDN (ECP_D5) displayed the most significant increase in terms of storage and loss modulus. Thus, we can presume that this rocket propellant composite displayed better ability to store and disperse mechanical energy. Tan delta (δ) represents the ratio of the loss to the storage modulus, and it is often called damping, being a measure of the capacity of the dissipation of energy in a material under cyclic loadings. All the composites designed for use as rocket propellant exhibited similar frequency-dependent damping behaviors. These polyurethane-based composites displayed slightly decreased values for tan delta at higher frequencies. This behavior can be explained by the gradual slippage exhibited by the main backbone chain of the polyurethane binders while shear stress was applied. Their damping capacity reflects their ability to dissipate mechanical energy. Therefore, we can conclude that they will be able to reduce the vibration amplitude significantly. The damping properties of the solid components from these rocket propellants conjoined with the remarkable damping properties of the employed polyurethane matrices (Figure 9c), which led to high-performance energetic composites, are suitable to be safely used as rocket propellants.

The rocket propellant composites (based on the synthesized polyols) were also analyzed using a three-point bending clamp. This is a typical flexural test that uses three identically sized cylindrical rollers to bend the sample. The three-point bending mode was employed as a method of analysis of the rocket propellant composites because, according to literature data, this type of deformation is usually considered "cleaner" than either the single/dual cantilever or tension modes, since clamping effects are eliminated [50]. For each sample, the oscillation amplitude was incrementally increased from 1 μm to 500 μm. As can be observed in Figure 10, the composites based on recycled polyols and TEGDN (ECP_D5) displayed higher flexibility, being the only rocket propellant composite that did not break during the bending test. The other energetic composites broke earlier because they were more rigid. An appropriate flexibility of these materials ensures the ease of charging the rocket motor with the composite propellants and guarantees that these energetic composites will maintain their integrity when the rocket motor is subjected to external mechanical stress.

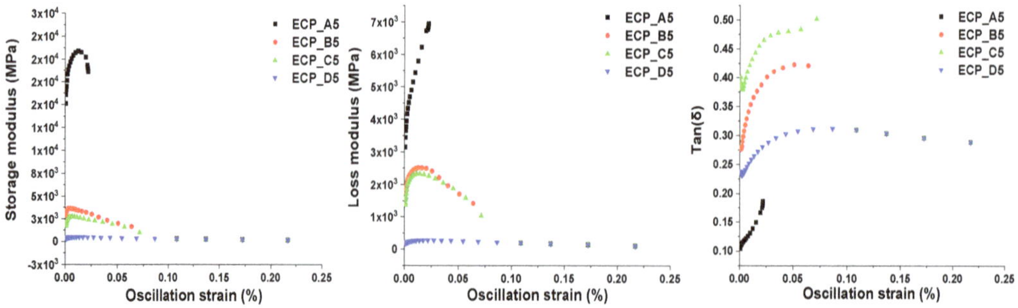

Figure 10. Bending behavior of the composite rocket propellants.

The highly aqueous solubility of PSAN employed in these rocket propellant composites allowed its extraction from the polymeric matrix and the assessment of the inner morphology of the energetic composites pressed in a cylindrical geometry. For this goal, small transversal slices of ECP were cut and placed in an aqueous solution until the oxidizer particles completely dissolved and migrated from the polyurethane matrix into the water solution. The samples were recovered after 6 h and dried at 50 °C for 12 h. The morphology of the resulted specimens was evaluated through the SEM-EDX technique. Figure 11 shows the SEM-EDX morphological images of the "oxidizer free" composite samples, investigated at different scales. According to these results, the voids that can be observed in these images correspond to the spaces that were previously occupied by the grains of the oxidizer. The dissolution of the oxidizer in water leads to a porous polyurethane matrix (still containing the metallic component). This experiment allows us to evaluate the way that the oxidizer solid grains are dispersed inside the polyurethane matrix, this being an important aspect for the investigation of the burning rate behavior inside the motor rocket chamber. Furthermore, based on the EDX analysis depicted in Figure 11, the distribution of aluminum and iron oxide particles inside the polyurethane matrix can be evaluated. Figure S12 from the "Supplementary Material" file presents the SEM-EDX morphological analysis of the surface for the composite formulations with oxidizer and both types of polyurethanes (based on commercial polyol, PU_3, and based on polyols obtained from PET degradation, PU_31, 2:1 molar ratio -NCO/-OH). Figure S13 and Table S1 illustrate the elemental composition of the "oxidizer free" composite samples. For comparison, Figures S14 and S15, and Table S2 respectively illustrate a morphological characterization of a pressed solid mixture of an oxidant and metallic fuel, without a binder. In the absence of the binder, the mechanical resistance of this pressed mixture is almost inexistent, being a very friable structure.

Based on this SEM-EDX evaluation of the rocket propellant composites, we can conclude that the solid grains of the oxidizer and metallic fuel are uniformly distributed inside the polyurethane binder matrix, which ensures a good cohesion between all the solid components, conferring proper explosive and mechanical characteristics.

As mentioned above, our composites, being specifically designed to be employed as rocket propellants, require a uniform dispersion of the solid components inside the polymeric matrix to ensure a constant burning rate. Thus, for obtaining complementary information to the SEM-EDX analyses presented above, the synthesized composites were also subjected to μCT analysis to evaluate the 3D distribution of the solid components inside the rocket propellant composites. As can be observed in Figure 12, all the composites displayed a homogenous three-dimensional distribution of the solid components in the polyurethane matrix. There are some slightly notable differences between the four rocket propellant composites illustrated in Figure 12, due to the addition of TEGDN and also due to the addition of the recycled polyol, which led to a less dense network of hard nanodomains inside the polyurethane matrix than the commercial polyol.

The evaluation of the ballistic performances of these new rocket propellants based on PSAN, Al, and polyurethane matrices synthesized from polyester-polyols derived from PET degradation cannot be complete without confirming its applicability in a firing set-up. For this reason, a series of static experimental firings was performed with the ECP_D5 composite on a subscale rocket motor (SRM) TRM-35 to determine its performance characteristics. The experimental system was designed to investigate the burning behavior of the composite propellant for small-grain rockets before introducing them into a large-scale launching system. Figure 13 shows the experimental set-up used in the test. The propellant grain was wet pressed and cured in a cylindrical geometry with the inner bore, with obstructed front and rear ends, providing a neutral burning [2] from the inner and outer surface of the cylinder, as can be seen in detail in Figure 13. A structural view of the subscale rocket motor stand burner and the rocket propellant cylindrical composites is shown in Figures S16 and S17 from the "Supplementary Material" file. The ignition was ensured by a small amount of pyrotechnic composition, presented in Figure S18. To ensure the pressurization of the combustion chamber and the propellant grain ignition, the nozzle was equipped with an aluminum membrane that breaks after ignition. Figure S19 provides a view of the eco-friendly rocket propellant at the convergent section nozzle and the shape of the flame generated during an outdoor combustion session.

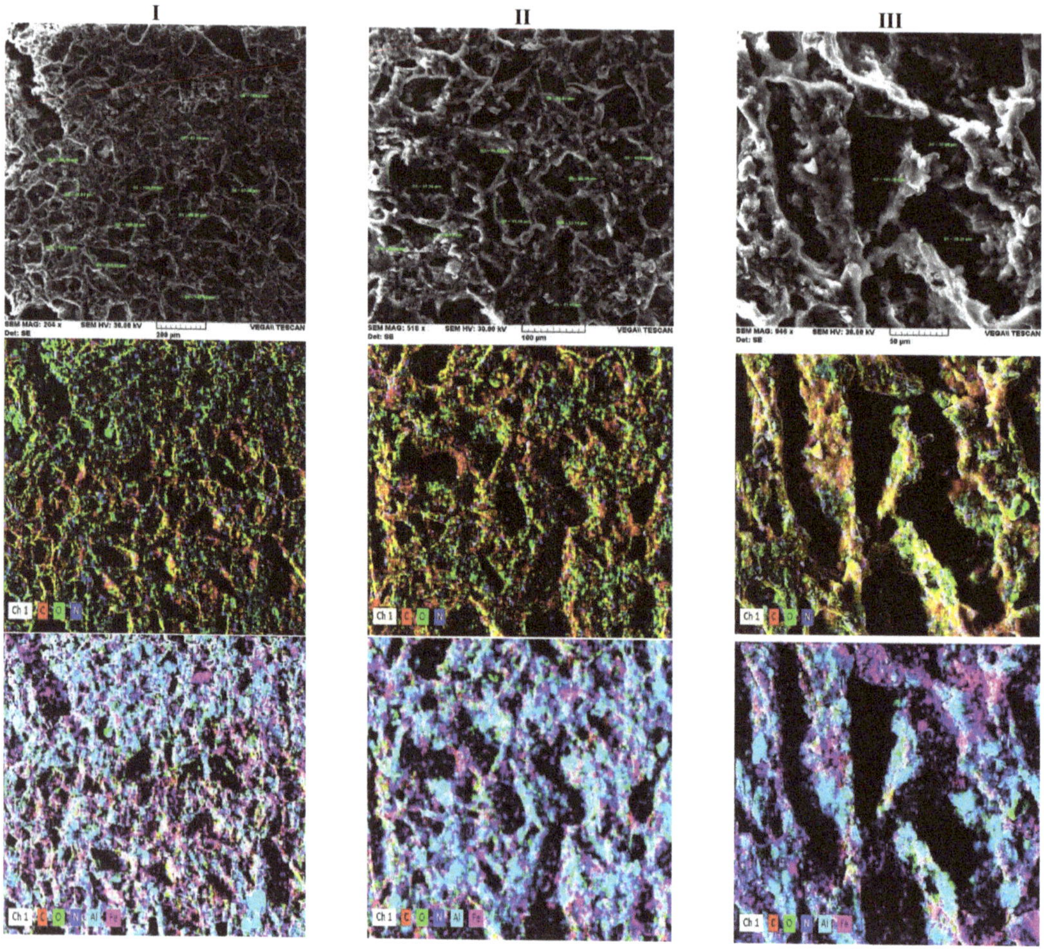

Figure 11. SEM and EDX analysis of the porous ECP_D5 formulation (after oxidizer extraction) at different scales: (**I**) 200 μm; (**II**) 100 μm; (**III**) 50 μm.

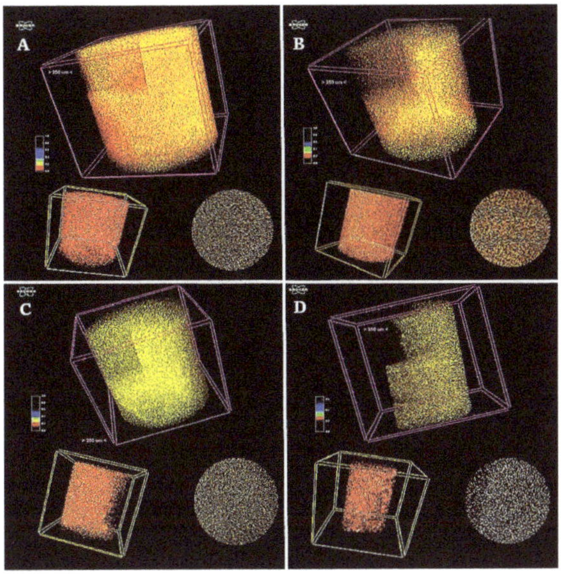

Figure 12. Micro-CT images (250 μm voxel size) of composite formulations: (**A**) ECP_A5; (**B**) ECP_B5; (**C**) ECP_C5; (**D**) ECP_D5.

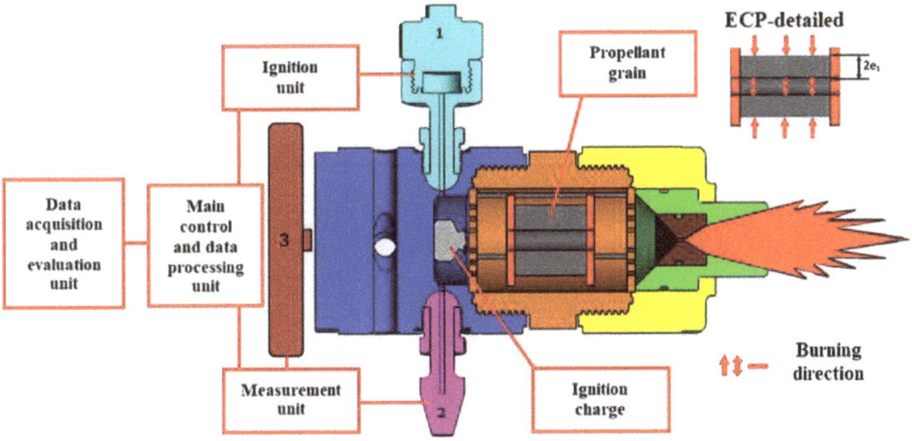

Figure 13. Subscale rocket motor set-up: (**1**) ignition; (**2**) pressure transducer; (**3**) thrust transducer (e_1: burning propellant thickness).

The burning characteristics for the environmentally responsible rocket propellant obtained in this study are illustrated in Figure 14. Although the aspect of the burning profile pressure vs. time does not perfectly match the theoretical burning profiles of rocket propellant, due to the low scale of the experiment, we were able to emphasize that it provides promising results for this type of application. Thus, the average value of the pressure $(\overline{p_{th}})$, calculated with Equation (1), was 39.23 bars, which indicates good premises for this type of propellant, being in accordance with the values presented in the literature [50]. An average combustion rate $(\overline{u_{th}})$ of 2.78 mm/s was calculated using Equation (2), which is in the same range with the burning rate of ammonium perchlorate-based composite propellants [51]. The burning profile of the sample indicates a progressive burning at the beginning and then a regressive burning, probably due to the erosion of the propellant grain at the obstructed front/end surfaces and the transition towards a spherical geometry.

$$\overline{p_{th}} = \frac{1}{t_h - t_z}\int_{t_z}^{t_h} p(t)dt \quad (1)$$

$$\overline{u_{th}}(\overline{p_{th}}) = \frac{e_1}{t_h - t_z} \quad (2)$$

where: e_1 is the burning thickness, as presented in Figure 13 and $(t_h - t_z) = t_b$ is the burning time interval.

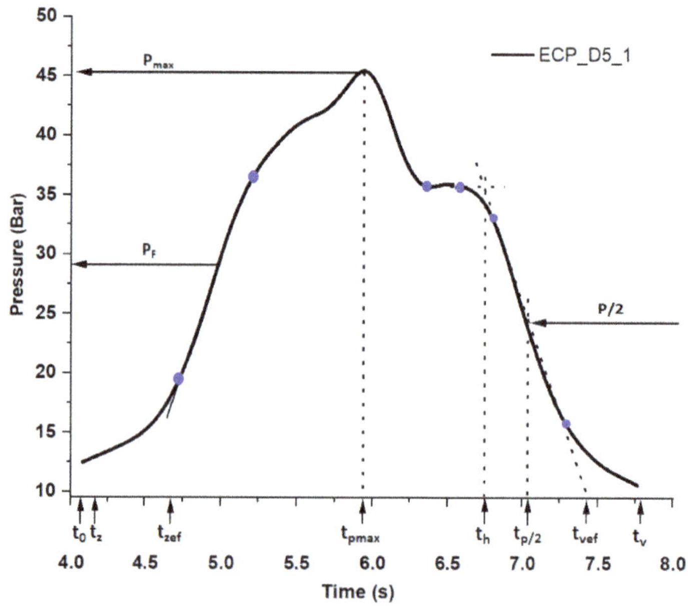

where: t_0—ignition time, defined by electric or mechanic impulse.
t_z—beginning of burning, defined by level of pressure.
t_{zef}—effective beginning of burning, defined by tangent line (2 points).
t_{pmax}—maximum pressure time, defined by P_{max}.
t_h—end of burning, defined by junction of 2 tangent lines (4 points).
$t_{p/2}$—half pressure end, defined by half value of average pressure.
t_{vef}—effective efflux time, defined by tangent line (2 points).
t_{vef}—end of efflux time, defined by level of pressure.

Figure 14. The burning rate profile and characteristics for the developed propellant.

4. Conclusions

Novel polyurethane binders and their applications in future environmentally responsible composite rocket propellants were investigated. To highlight the advantages brought by this new "green" approach, in comparison with state-of-the-art HTPB binders (extensively utilized nowadays in this field), we further summarized the most important achievements of this study. The polyurethane-based binders were synthesized using polyester-polyols obtained from catalytic degradation of recycled PET, commercial polyols, and the energetic plasticizer TEGDN. To demonstrate that the polyurethanes are suitable for this type of application, they were subjected to structural and rheological characterization (^{1}H-NMR, FT-IR, and viscosity analysis), while the new composite propellants were subjected to morphological, thermal, and mechanical characterizations using various analytical techniques (SEM-EDX, DTA, TGA, DMA, μCT, and tensile and compression tests). Moreover, the composite propellant formulations developed were analyzed in terms of ballistic performances by real firing tests in a subscale rocket motor.

^{1}H-NMR and FT-IR characterizations confirmed that the synthesized polyester-polyol has the appropriate chemical structure, while their viscosity follows the requirements imposed in the development of propellants. SEM-EDX and μCT analyses proved the homogenous dispersions of the solid load inside most of the composites. The uniformity of the oxidizers and fuel distribution inside the polymeric matrix was also indicated by the continuous combustion of the propellant grain, thus improving the energetic performances of these composites. TGA measurements indicated that the presence of TEGDN lowered the decomposition temperature of the polyurethanes, an aspect also confirmed by DTA investigations. However, the polyurethanes possess good thermal stability (up to about 300 °C for PU_3 and PU_31 and up to 190 °C for PU_3T2 and PU_31T2). Even if the combustion process of the rocket propellants begins at lower temperatures than the neat polyurethane binders, they still can be safely utilized, each composite possessing decomposition onset temperatures above 165 °C.

DMA profile of the polyurethanes from recycled PET plasticized with TEGDN displayed a very low glass transition temperature (−53 °C), while for the propellant based on it, the T_g was slightly higher due to the presence of solid loading. DMA analysis also demonstrated that the polyurethanes and propellant specimens based on polyester-polyols synthesized from PET waste possess a good capacity for absorbing and dissipating energy. Tensile and compressive test results also showed that the developed specimens (polyurethanes and propellants) have an acceptable mechanical behavior, in accordance with the minimal requirements for solid rocket propellant binders. Thus, in comparison with the existing HTPB binders present in available rocket propellants, despite their poorer mechanical resistance (which is situated within some tolerable values), the binders based on recycled PET waste can enhance the thermal performances of the rocket propellant and contribute to a less polluted environment. Additionally, the higher flexibility will ensure resilience, a superior shock resistance necessary for the prevention of crack formation and adhesion failure, thus insuring the geometric stability of the propellant grain.

Small-scale real firing testing of the novel solid composite propellants indicated ballistic performances in accordance with those exhibited by state-of-the-art solid propellants (adequate values for pressure and combustion rate, improved energetic performances). We can conclude that the extensive study herein reported offers a comprehensive image of the possibility to replace the existing binders, such as HTPB, with binders derived from PET, while maintaining high performance standards, thus substantially minimizing the environmental impact of the rocket propellants. This ecological approach could be integrated into the life cycle assessment of environmentally responsible rocket propellants that should be developed in the future. By developing future "greener" rocket propellants based on polyols synthesized from PET waste and "clean" oxidizers, a great contribution could be added to the circular economy process and to the global effort to protect the environment and human health.

Supplementary Materials: The following are available online at https://www.mdpi.com/article/10.3390/polym13213828/s1. Included are NMR spectra of SRP1, SRP2, SRP1T2, and SRP2T2 (Figures S1–S4), FT-IR of polyurethanes based on RP2 (Figure S5), images of polyurethane binders and curing time (Figure S6), SEM images of polyurethane formulations based on RP1 and RP2 (with and without TEGDN) (Figure S7), EDX mapping of the polyurethane formulations from S7 (Figure S8), tensile stress–strain plots for PU formulations (Figure S9), tensile stress–strain plots for composite propellant formulations (Figure S10), compression test plots for composite formulations (Figure S11), SEM-EDX images of composite propellant formulations (Figure S12), EDX spectra for oxidizer-free and binder-free formulations (Figure S14 and S15), stand burner equipped with subscale rocket motor TRM-35 (Figure S16), structural configuration of ECP_D5 (Figure S17), pyrotechnic composition for propellant ignition (Figure S18), flame configuration of ECP_D5 during combustion

(Figure S19), behavior of PUs during manual bending (Figure S20), structural configurations of composite propellants for mechanical analysis (Figure S21), weight and atomic composition of free oxidizer composite formulations (Table S1), weight and atomic composition of free binder solid mixture (Table S2), thermal characteristics for polyurethane films, composite propellant formulations, and energetic plasticizer (Table S3), thermal properties of synthesized polyurethanes (Table S4), the decomposition process of the polyurethanes (weight loss versus temperature) (Table S5), and heat of combustion, specific volume and Tg for our new composite propellants in comparison with the existing commercial formulations (Table S6).

Author Contributions: Conceptualization, F.M.D., T.R., E.R. and G.T.; methodology, T.R., A.D. and E.R.; software: R.E.G., F.A. and M.I.U.; validation, F.M.D., T.R., G.T., T.V.Ţ., R.P., E.R., A.G. and A.D.; formal analysis, F.M.D., G.T., R.E.G. and M.I.U., F.A.; investigation, F.M.D., G.T., A.G., R.E.G., A.E.C., M.D., F.A., M.I.U., R.P., E.R., A.D. and R.Ţ.; resources, R.P., A.D. and A.G.; data curation, F.M.D., G.T., R.P., R.E.G., F.A., M.D., A.E.C., M.I.U. and R.Ţ.; writing—original draft preparation, F.M.D. and G.T.; writing—review and editing, F.M.D., G.T., T.R., A.D. and E.R.; visualization, F.M.D., G.T., R.E.G., T.V.Ţ., A.D., R.P., A.G., M.D., A.E.C. and R.Ţ.; supervision, T.R., G.T., E.R. and T.V.Ţ.; project administration, T.R. and A.D.; funding acquisition, T.R. and A.D. All authors have read and agreed to the published version of the manuscript.

Funding: This work was supported (funded) by the Romanian Ministry of Education and Scientific Research (UEFISCDI) under the Complex Projects Realized in Consortium Program, project no. 70PCCDI/2018.

Institutional Review Board Statement: Not applicable. This study did not involve humans or animals.

Informed Consent Statement: Not applicable. This study did not involve humans.

Acknowledgments: Aurel Diacon gratefully acknowledges the financial support from the Competitiveness Operational Program 2014–2020, Action 1.1.3: Creating synergies with RDI actions of the EU's HORIZON 2020 framework program and other international RDI programs, MySMIS Code 108792, Acronym project "UPB4H", financed by contract: 250/11.05.2020.

Conflicts of Interest: The authors declare no conflict of interest.

References

1. Agrawal, J.P. *High Energy Materials: Propellants, Explosives and Pyrotechnics*; John Wiley & Sons: Weinheim, Germany, 2010; pp. 209–316.
2. Kubota, N. *Propellants and Explosives: Thermochemical Aspects of Combustion*; John Wiley & Sons: Weinheim, Germany, 2015.
3. Bucur, F.; Trana, E.; Rotariu, A. Numerical and Experimental Study on the Locally Blast Loaded Polyurea Coated Steel Plates. *Mater. Plast.* **2019**, *56*, 492–499. [CrossRef]
4. Toader, G.; Diacon, A.; Rusen, E.; Rizea, F.; Teodorescu, M.; Stanescu, P.; Damian, C.; Rotariu, A.; Trana, E.; Bucur, F.; et al. A Facile Synthesis Route of Hybrid Polyurea-Polyurethane-MWCNTs Nanocomposite Coatings for Ballistic Protection and Experimental Testing in Dynamic Regime. *Polymers* **2021**, *13*, 1618. [CrossRef]
5. Toader, G.; Rusen, E.; Teodorescu, M.; Diacon, A.; Stanescu, P.O.; Rotariu, T.; Rotariu, A. Novel polyurea polymers with enhanced mechanical properties. *J. Appl. Polym. Sci.* **2016**, *133*. [CrossRef]
6. Voicu, A.E.; Rotariu, T.; Teodorescu, M.; Zecheru, T.; Tiganescu, T.V.; Orban, O. pH sensitive polymeric binders for energetic materials. *Mater. Plast.* **2017**, *54*, 103. [CrossRef]
7. Chaturvedi, S.; Dave, P.N. Solid propellants: AP/HTPB composite propellants. *Arab. J. Chem.* **2019**, *12*, 2061–2068. [CrossRef]
8. Pang, W.-Q.; DeLuca, L.T.; Fan, X.-Z.; Glotov, O.G.; Wang, K.; Qin, Z.; Zhao, F.-Q. Combustion behavior of AP/HTPB/Al composite propellant containing hydroborate iron compound. *Combust. Flame* **2020**, *220*, 157–167. [CrossRef]
9. Jos, J.; Mathew, S. Ammonium Nitrate as an Eco–Friendly Oxidizer for Composite Solid Propellants: Promises and Challenges. *Crit. Rev. Solid State Mater. Sci.* **2017**, *42*, 470–498. [CrossRef]
10. Judge, M.D.; Lessard, P. An Advanced GAP/AN/TAGN Propellant. Part I: Ballistic Properties. *Propellants, Explos. Pyrotech.* **2007**, *32*, 175–181. [CrossRef]
11. Larsson, A.; Wingborg, N. Green propellants based on ammonium dinitramide (ADN). *Adv. Spacecr. Technol.* **2011**, 139–156. [CrossRef]
12. Ao, W.; Fan, Z.; Liu, L.; An, Y.; Ren, J.; Zhao, M.; Liu, P.; Li, L.K. Agglomeration and combustion characteristics of solid composite propellants containing aluminum-based alloys. *Combust. Flame* **2020**, *220*, 288–297. [CrossRef]
13. Davis, A. Solid propellants: The combustion of particles of metal ingredients. *Combust. Flame* **1963**, *7*, 359–367.

14. DeLuca, L.T. Overview of Al-based nanoenergetic ingredients for solid rocket propulsion. *Def. Technol.* **2018**, *14*, 357–365. [CrossRef]
15. Li, L.-B.; Chen, X.; Musa, O.; Zhou, C.-S.; Zhu, M. The effect of pressure and oxygen concentration on the ignition and combustion of aluminum–magnesium fuel-rich propellant. *Aerosp. Sci. Technol.* **2018**, *76*, 394–401. [CrossRef]
16. TRANĂ, E.; Rotariu, A.N.; Rotariu, T.; Pulpea, B.G.; Moldoveanu, C.E.; Bucur, F.; Matache, L.C.; Gozin, M. Experimental Study on Aluminum Foils Use in Blast Enhancement Application. *Proc. Rom. Acad.-Ser. A Math. Phys. Tech. Sci.* **2019**, *20*, 275–282.
17. Li, Y.; Yang, W.; Ying, S. Preparation and Characteristics of Foamed NC-Based Propellants. *Propellants Explos. Pyrotech.* **2014**, *39*, 677–683. [CrossRef]
18. Ang, H.G.; Pisharath, S. *Energetic Polymers: Binders and Plasticizers for Enhancing Performance*; John Wiley & Sons: Weinheim, Germany, 2012.
19. Mastrolia, E.J.; Klager, K. Solid Propellants Based on Polybutadiene Binders. In *Ultratrace Metal Analysis in Biological Sciences and Environment*; American Chemical Society (ACS): Washington, DC, USA, 1969; pp. 122–164.
20. Gołofit, T.; Zysk, K. Thermal decomposition properties and compatibility of CL-20 with binders HTPB, PBAN, GAP and polyNIMMO. *J. Therm. Anal. Calorim.* **2015**, *119*, 1931–1939. [CrossRef]
21. Sell, T.; Vyazovkin, S.; Wight, C.A. Thermal decomposition kinetics of PBAN-binder and composite solid rocket propellants. *Combust. Flame* **1999**, *119*, 174–181. [CrossRef]
22. Ducruet, N.; Delmotte, L.; Schrodj, G.; Stankiewicz, F.; Desgardin, N.; Vallat, M.-F.; Haidar, B. Evaluation of hydroxyl terminated polybutadiene-isophorone diisocyanate gel formation during crosslinking process. *J. Appl. Polym. Sci.* **2013**, *128*, 436–443. [CrossRef]
23. Sekkar, V.; Bhagawan, S.; Prabhakaran, N.; Rao, M.R.; Ninan, K. Polyurethanes based on hydroxyl terminated polybutadiene: Modelling of network parameters and correlation with mechanical properties. *Polymers* **2000**, *41*, 6773–6786. [CrossRef]
24. Achilias, D. *Material Recycling: Trends and Perspectives*; BoD–Books on Demand: 2012; IntechOpen Limited: London, UK, 2012. [CrossRef]
25. Chilton, T.; Burnley, S.; Nesaratnam, S. A life cycle assessment of the closed-loop recycling and thermal recovery of post-consumer PET. *Resour. Conserv. Recycl.* **2010**, *54*, 1241–1249. [CrossRef]
26. Gravagnuolo, A.; Angrisano, M.; Girard, L.F. Circular Economy Strategies in Eight Historic Port Cities: Criteria and Indicators Towards a Circular City Assessment Framework. *Sustainability* **2019**, *11*, 3512. [CrossRef]
27. Vermeșan, H.; Mangău, A.; Tiuc, A.-E. Perspectives of Circular Economy in Romanian Space. *Sustainability* **2020**, *12*, 6819. [CrossRef]
28. Pacheco-Torgal, F. *Introduction to the Use of Recycled Plastics in Eco-Efficient Concrete*; Woodhead Publishing: Sawston, UK, 2019; pp. 1–8.
29. Park, S.-H.; Alammar, A.; Fulop, Z.; Pulido, B.A.; Nunes, S.P.; Szekely, G. Hydrophobic thin film composite nanofiltration membranes derived solely from sustainable sources. *Green Chem.* **2021**, *23*, 1175–1184. [CrossRef]
30. Sadler, J.C.; Wallace, S. Microbial synthesis of vanillin from waste poly(ethylene terephthalate). *Green Chem.* **2021**, *23*, 4665–4672. [CrossRef] [PubMed]
31. Sinha, Y.K.; Sridhar, B.T.N.; Santhosh, M. Thermal decomposition study of HTPB solid fuel in the presence of activated charcoal and paraffin. *J. Therm. Anal. Calorim.* **2014**, *119*, 557–565. [CrossRef]
32. Wang, Z.; Qiang, H.; Wang, T.; Wang, G. Tensile behaviors of thermal aged HTPB propellant at low temperatures under dynamic loading. *Mech. Time-Depend. Mater.* **2020**, *24*, 141–159. [CrossRef]
33. Mahottamananda, S.N.; Kadiresh, P.N.; Jayapal, S.N.M. A study on thermal stability and combustion performance of hydroxyl-terminated polybutadiene-paraffin blended fuel. *Energy Sources Part A Recover. Util. Environ. Eff.* **2020**, 1–12. [CrossRef]
34. Paraschiv, T.; Tiganescu, T.V.; Iorga, G.O.; Ginghina, R.E.; Grigoroiu, O.C. Experimental and Theoretical Study on Three Combustion Models for the Determination of the Performance Parameters of Nitrocellulose—Based Propellants. *Rev. Chim.* **2020**, *71*, 87–97. [CrossRef]
35. Cegła, M.; Borkowski, J.; Zmywaczyk, J.; Koniorczyk, P. Dynamic mechanical analysis of double base rocket propellants. *Bull. Mil. Univ. Technol.* **2016**, *65*, 47–56. [CrossRef]
36. Urbansky, T. *Chemistry and Technology of Explosives*; Pergamon Press Book: Oxford, UK, 1965; pp. 154–156.
37. Dîrloman, F.M.; Rotariu, T.; Țigănescu, T.V.; Toader, G.; Zecheru, T.; Iordache, T.V. Phase stabilized ammonium nitrate for future green rocket propellants. *UPB Sci. Bull.* **2021**, *83*, 87–100.
38. NATO. NPFC-MIL-STD-2100-"Propellant, Solid, Characterization of (Except Gun Propellant)". 1979. Available online: http://everyspec.com/MIL-STD/MIL-STD-2000-2999/MIL-STD-2100_10191/ (accessed on 25 June 2021).
39. ASTM-D4274-21. *Standard Test Methods for Testing Polyurethane Raw Materials: Determination of Hydroxyl Numbers of Polyols*; revision ed.; ASTM International: Pennsylvania, PA, USA, 2021.
40. ASTM D4662-20. *Standard Test Methods for Polyurethane Raw Materials: Determination of Acid and Alkalinity Numbers of Polyols*; ASTM International: Pennsylvania, PA, USA, 2020.
41. Duldner, M.-M.; Bartha, E.; Capitanu, S.; Nica, S.; Coman, A.E.; Tincu, R.; Sarbu, A.; Filip, P.I.; Apostol, S.; Gare, S. Attempts to Upcycle PET Wastes into Bio-based Long-lasting Insulating Materials. *Rev. Chim.* **2019**, *70*, 2301–2307. [CrossRef]
42. Goethals, E.J. *Telechelic Polymers*; CRC Press: Boca Raton, FL, USA, 1988; Volume 416.

43. STANAG 4515(2). *Explosives, Thermal Analysis Using Differential Thermal Analysis (DTA), Differential Scanning Calorimetry (DSC), Heat Flow Calorimetry (HFC) and Thermogravimetric Analysis (TGA)*; NATO Standardization Office: Brussels, Belgium, 2015.
44. ISO 37:2011. *Rubber, Vulcanized or Thermoplastic—Determination of Tensile Stress-Strain Properties*; International Organization for Standardization: Geneva, Switzerland, 2011; p. 27.
45. SETATHANE®D 1150. 2021. Available online: https://allnex.com/en/product/010faff2-5d57-4c45-bf7d-429e151cc23f/setathane-d-1150 (accessed on 14 September 2021).
46. Keizers, H.L.J.; Brouwer, G.R.; Weijl, J.; Weterings, F. Evaluation of rocket motor safelife based on condition monitoring and ageing modelling. In *AVT-089 Specialist Meeting: Advances in Rocket Propellant Performance. Life and Disposal for Improved System Performance and Reduced Costs*; RTO Meeting Proceedings 91; RTO: Aalborg, Denmark, 2002.
47. NATO, STANAG 4540. *Explosives, Procedures for Dynamic Mechanical Analysis (DMA) and Determination of Glass Transition Temperature*; NATO Standardization Office: Brussels, Belgium, 2002.
48. Lee, J.; Kim, S.; Oh, K. Bio-Based Polyurethane Foams with Castor Oil Based Multifunctional Polyols for Improved Compressive Properties. *Polymers* **2021**, *13*, 576. [CrossRef] [PubMed]
49. Remišová, E.; Zatkaliková, V.; Schlosser, F. Study of Rheological Properties of Bituminous Binders in Middle and High Temperatures. *Civ. Environ. Eng.* **2016**, *12*, 13–20. [CrossRef]
50. Lee-Sullivan, P.; Dykeman, D. Guidelines for performing storage modulus measurements using the TA Instruments DMA 2980 three-point bend mode. *Polym. Test.* **2000**, *19*, 155–164. [CrossRef]
51. Nakka, R.A. Richard Nakka's Experimental Rocketry Web Site. 1997. Available online: http://www.nakka-rocketry.net/ (accessed on 30 June 2021).

polymers

MDPI

Article

Biosynthesis of Poly-ß-Hydroxybutyrate (PHB) from Different Bacterial Strains Grown on Alternative Cheap Carbon Sources

Sherif M. El-Kadi [1], Mohssen Elbagory [2,3], Hassan A. H. EL-Zawawy [4], Hossam F. A. EL-Shaer [5], Adel A. Shoukry [4], Sahar El-Nahrawy [3], Alaa El-Dein Omara [3,*] and Dina Fathi Ismail Ali [6]

1 Agricultural Microbiology Department, Faculty of Agriculture, Damietta University, Damietta 22052, Egypt; sherifelkadi@du.edu.eg
2 Department of Biology, Faculty of Science and Arts, King Khalid University, Mohail Assir 61321, Saudi Arabia; mhmohammad@kku.edu.sa
3 Department of Microbiology, Soils, Water and Environment Research Institute, Agricultural Research Center, Giza 12112, Egypt; sahar.elnahrawy@yahoo.com
4 Botany Department (Microbiology), Faculty of Agriculture, Al-Azhar University, Cairo 11651, Egypt; zawawy.hassan@gmail.com (H.A.H.E.-Z.); alshokry17@azhar.edu.eg (A.A.S.)
5 Botany Department (Genetics), Faculty of Agriculture, Al-Azhar University, Cairo 11651, Egypt; hosamelshaer805@azhar.edu.eg
6 Agricultural Microbiology Department, Faculty of Agriculture, Mansoura University, Mansoura 35516, Egypt; dfali@mans.edu.eg
* Correspondence: alaa.ahmed@arc.sci.eg or alaa.omara@yahoo.com

Citation: El-Kadi, S.M.; Elbagory, M.; EL-Zawawy, H.A.H.; EL-Shaer, H.F.A.; Shoukry, A.A.; El-Nahrawy, S.; Omara, A.E.-D.; Ali, D.F.I. Biosynthesis of Poly-ß-Hydroxybutyrate (PHB) from Different Bacterial Strains Grown on Alternative Cheap Carbon Sources. *Polymers* **2021**, *13*, 3801. https://doi.org/10.3390/polym13213801

Academic Editor: Edina Rusen

Received: 4 October 2021
Accepted: 29 October 2021
Published: 3 November 2021

Publisher's Note: MDPI stays neutral with regard to jurisdictional claims in published maps and institutional affiliations.

Copyright: © 2021 by the authors. Licensee MDPI, Basel, Switzerland. This article is an open access article distributed under the terms and conditions of the Creative Commons Attribution (CC BY) license (https://creativecommons.org/licenses/by/4.0/).

Abstract: Thirty bacterial isolates were tested on three different media for Poly-ß-hydroxybutyrate (PHB) production. The best bacterial isolates for producing PHB were screened and identified based on molecular biology; then, using three different alternative carbon sources (dried whey, sugar beet molasses and date molasses), physical properties were evaluated by Infrared (IR) spectrometry and Gas chromatography–mass spectrometry (GC-MS/MS) analysis. Our results showed that the best isolates identified based on molecular biology were *Bacillus paramycoides* MCCC 1A04098, *Azotobacter salinestris* NBRC 102611 and *Brevundimonas naejangsanensis* BIO-TAS2-2. The addition of sugar beet molasses to the medium of *A. salinestris* increased the cell dry weight (CDW), PHB concentration, PHB% and conversion coefficient (4.97 g/L, 1.56 g/L, 31.38% and 23.92%, respectively). The correlation coefficient values between PHB g/L and CDW g/L varied between very strong and moderate positive correlation. IR of the produced PHB from *B. paramycoides* and *A. salinestris* showed similar bands which confirmed the presence of PHB; however, *B. naejangsanensis* showed weak bands, indicating lower PHB concentration. The chemical composition obtained showed that the GC-MS of the PHB extracted represents 2, 4-ditert-butylphenol for *B. paramycoides* and isopropyl ester of 2-butenoic acid for both of *A. salinestris* and *Brevundimonas naejangsanensis*. Therefore, PHB produced by microorganisms can be considered a biodegradable polyester, and represents a promising technique for the development of eco-friendly and fully biodegradable plastics.

Keywords: agro-industrial waste; GC-MS/MS spectrometry; IR spectrometry; PHB

1. Introduction

Since petroleum-based plastics have become a major cause of environmental pollution, degradable plastics have attracted a great deal of attention worldwide for their unique properties, such as thermoplasticity. However, production of these vital plastics is limited due to their high cost of production. Hence, studies looking for bacterial strains with better yields and improving their culture conditions are of the greatest importance in order to reduce the costs of PHB production [1]. The slow growth of the PHB market since the 1980s is due to the high cost of production for the amount of PHB produced. Currently, only about 1% of plastic materials on the market are biodegradable, and with the increase in environmental pollution, there is a need to increase the production of PHB [2]. PHB and different biodegradable polyesters are promising tools for the development of eco-friendly and fully biodegradable plastics. PHBs are a class of biopolymers produced by different microbial species. Unlike plastics derived from petrochemicals, PHB is biodegradable and biocompatible in nature. PHBs have advanced applications in the medical sector (surgical pins and sutures, wound dressings, bone replacements), the packaging industry (packaging films, bags, cosmetics containers, shampoo bottles), and agriculture (plastic mulch, as a carrier for long-term doses of herbicides, fungicides and insecticides) [3–6]. PHB is produced using various raw materials such as agricultural and food industry wastes, which contributes to its economic feasibility. Research is currently concerned with reducing the cost of production [7]. The use of food industrial byproducts in the production of PHB is considered one of the growing areas of modern biotechnology; these products can be used as a nutrient-rich medium at no additional cost. Whey and molasses are very important byproducts that are rich in nutrients and which bacteria can convert into PHB [8]. The use of molasses as a potential low-cost substrate for PHB biopolymer production has been studied by [9]. The use of molasses, which is a cheap substrate available in Egypt, may help reduce the cost of producing such biopolyesters [3]. Beet molasses has been shown to be an excellent starting material for PHB production by *Azotobacter vinelandii*. The substrate cost for PHB production from beet molasses in fed-batch culture was one-third of that using glucose [10]. The results obtained by [11] indicate that by using *Bacillus subtilis* RS1, agro-industry wastes such as sugarcane molasses can be used as an inexpensive substrate for PHB production. Date molasses and date seed oil have also been used as the sole carbon source for PHB biosynthesis [12]. Processes are currently underway to improve PHB production via genetic modification of some strains in order to develop desirable and suitable properties for some diverse applications [7]. Pavan [13] aimed to conduct an economic assessment of PHB production process using molasses. The alternative sources showed equally low production costs (USD 4.28 per kg) and better economic indicators among the evaluated scenarios. In recent years, using renewable raw materials for PHB production as degradable bioplastics has been a good method of environment consideration. Low-cost carbon substrates have the potential to reduce PHB production cost and increase sustainability [14]. According to [11], the global PHB market is expected to reach USD 93.5 million by 2021. In the Middle East and Arab countries, Date palm (*Pheonix dactylifera* L.) is an important commercial crop. Egypt is one of the largest date producers among Arab and world countries. Egyptian date production in 2020 was 1.9 million tons, which represented 21.2% of world production. Date soft types are highly perishable and easily attacked by microbial enzymes; thus, date and date molasses are considered good materials for fermentation processes [15–17]. The main reason for the low prevalence of bioplastic products is the high cost of production.

Therefore, this study aims to produce PHB using efficient PHB-producing bacteria on different available alternative carbon sources (whey, sugar beet molasses and date molasses) and to study their properties by infrared (IR) spectroscopy and gas chromatography GC-MS/MS mass spectrometry analysis.

2. Materials and Methods

2.1. Bacterial Isolation

Four different soil samples were obtained from Damietta and Dakahlia Governorates (New Damietta city, Kafer El-Batekh, Kafer Saad, and Kalabshou), Egypt, were used for bacterial isolation. From these soil samples, 30 bacterial isolates were randomly isolated on three different selective media (10 for each medium): medium (A) [18], medium (B) [19] and medium (C) [20], for PHB production. These isolates were identified in the Agricultural Microbiology Department, Faculty of Agriculture, Damietta University, Damietta, Egypt. Morphological and physiological characteristics of bacterial isolates such as shape, Gram stain, spore stain, capsule stain and motility were microscopically studied, as were indole and catalase tests, starch hydrolysis and casein hydrolysis [21–23].

Isolates obtained from each of the three previously-mentioned media (10 randomly picked from each medium) were maintained on slants of Nutrient agar medium [24], Modified Ashby's medium [25], and Nutrient-rich medium [26], incubated at 5 °C, and sub-cultured monthly.

2.2. *Efficacy of PHB-Producing Isolates*

2.2.1. Cultivation Media

A cultivation process was implemented so that isolates from each medium could be cultivated on the same isolation media for PHB production. Carbon and nitrogen sources for each medium were sterilized separately and were aseptically added before inoculation and after cooling of each medium.

For medium (A) [18], 20 g/L glucose (carbon source) and 2.0 g/L ammonium acetate (nitrogen source) were added to the sterilized medium. The pH was adjusted to 7. For to medium (B), 10 g/L fructose was added as a carbon source [19] and 1.2 g/L ammonium acetate (nitrogen source) was added to the sterilized medium; the pH was adjusted to 7.2. Finally, 30 g/L sucrose was used as a carbon source for medium (C) [20], and 1 g/L ammonium sulfate was added to the sterilized medium. The pH was adjusted to 6.8.

2.2.2. Cultivation System

Batch cultivation on conical flasks was used in this experiment. Twenty ml of each medium was prepared from maintenance cultures to be used as an inoculum. Incubation was performed in an incubator shaker at 200 rpm at 30 °C for 48 h. The inoculum was aseptically transferred to 50 mL of the same medium and incubated under the same conditions for the same period mentioned above [27]. All experiments were conducted in three replications.

2.2.3. Biomass Determination

A 50 mL sample of each culture was centrifuged at 5000× *g* for 10 min. Cell pellets were washed twice with sterile distilled water. The precipitated biomass was dried at 100 °C for 24 h or until a constant weight was reached. This value was used for determination of cell dry weight (CDW) [28].

2.3. *PHB Determination*

Ten ml of chloroform was added to the dried cell biomass obtained from the previous centrifuged cultures and incubated at 70 °C for 10 min, then centrifuged at 5000× *g* for 10 min. The resulting solution was separated and collected, and the chloroform was allowed to evaporate in order to obtain PHBs for further assessment. Then, the precipitate was dried at 100 °C for 24 h or until a constant weight was reached. This value was used for PHB–free cell dry weight determination [19]. The difference in weight between the previous two values was considered as PHB production (g/50 mL); this value was calculated by [29,30], as g/L:

$$\text{PHB (\%)} = \text{PHB weight (g/L)} \times 100/\text{total cell dry weight (g/L)}$$

$$\text{Conversion coefficient (\%)} = \text{PHB weight (g/L)} \times 100/\text{utilized sugar weight (g/L)}$$

2.4. *Determination of Utilized Sugars and Nitrogen*

Total carbohydrates were determined as glucose [31]. The total sugars remaining in the culture broth were estimated, and by subtracting this value from the amount of total sugars in the medium composition at the beginning of the experiment, the consumed sugars or utilized sugars were calculated. Total nitrogen was measured according to the Kjeldahl digestion method [32].

2.5. The Relationship between PHB Production, CDW and Utilized Sugar (g/L)

The correlation coefficient values (r) was calculated according the equation given in [33]. The correlation between PHB g/L and CDW g/L was calculated where X is the value of PHB g/L, $\overline{X}$ is the arithmetical mean of PHB g/L, Y is the value of CDW g/L, and $\overline{Y}$ is the arithmetical mean of the CDW g/L values. The value of CDW was replaced with the value of utilized sugar when the correlation coefficient was calculated between PHB g/L and utilized sugar.

$$r = \frac{\Sigma(X - \overline{X})(Y - \overline{Y})}{\sqrt{\Sigma(X - \overline{X})^2}\sqrt{\Sigma(Y - \overline{Y})^2}}$$

If the value of the correlation coefficient (r) = +1, there is a perfect positive correlation; if (r) = −1, a perfect negative correlation; if r = 0, there is no correlation. If the value of the correlation coefficient is between r = 0.0–0.2, 0.2–0.4, 0.4–0.6, 0.6–0.8 or 0.8–1.0, this means a very weak correlation, weak correlation, moderate correlation, strong correlation or very strong correlation, respectively [33].

2.6. Identification of Bacterial Isolates Based on Molecular Biology

The highest PHB-producing isolates were inoculated in suitable media and incubated overnight at 37 °C. The resulting bacterial suspension was pelleted at 10,000 rpm for 5 min and genomic DNA was extracted using a Pure Link genomic DNA mini kit. PCR amplification of 16s rDNA was performed using the isolated DNA. The PCR conditions were set as follows: initial denaturation at 95 °C for 5 min, followed by 25 cycles of denaturation at 95 °C for 40 s, annealing at 55 °C for 2 min and primer extension at 72 °C for 1 min, ending with the final elongation step at 72 °C for 7 min. PCR products were gel purified and sent for sequencing with 16S rRNA primer. The obtained sequences were trimmed to obtain a sequence with ladder 1–1.5 Kb DNA. Then, the sequences were BLAST search analyzed on the (www.ncbi.nlm.nih.gov, accessed on 1 September 2021) to identify the isolate [34].

2.7. PHB Production Using Cheap Alternative Carbon Sources

Three different alternative carbon sources were used; dried whey and date molasses were purchased from Damietta City market, Damietta City, Egypt, and sugar beet molasses was purchased from the Sugar Beet Factory Kalabsho, Belkas, Dakahlia Governorate, Egypt. Molasses was pretreated according to the method in [35]. Total sugars and total nitrogen of all sources were determined. The values of carbon and nitrogen were considered during preparation of the cultivation media which already contain these sources.

2.8. The Physical Properties of PHB Polymer

IR was used to confirm the formation of PHB using a MATTISON 5000 FTIR spectrometer at Mansoura University, Faculty of Pharmacy, Chemistry Department, Spectral Analysis Unit. A 50 mm mortar and pestle were used. Hydraulic presses were performed using a PIKE die press. A SPECAC press instrument was used for preparation of the KBr sample discs of investigated drugs, pressing under a 7 mm die at a pressure of 2 tons for 1 min. A SHIMADZU CORPORATION balance BL-220H (Kyoto, Japan) was utilized throughout the work. IR spectra of the PHB polymer were determined at 2 cm^{-1} spectral resolution by using a MATTSON 5000 FTIR spectrometer. An excitation wavelength at 3900 nm was provided and the laser power at the sample position was typically 500 nm. Spectra were obtained with a spectral resolution of 4 cm^{-1}, and 1024 [36].

For GC-MS/MS analysis, the chemical composition was obtained after extraction and derivatized using BSTFA reagent. The derivatization product MTBSTFA-HB was analyzed on a Finnigan GCQ (Finnigan MAT, Austin, TX, USA), which is an ion trap MS with external ionization. A 30 m fused silica capillary column DB-5MS, I.D. 0.25 mm, film thickness 0.25 mm (J&W Scientific, Folsom, CA, USA) was used for separation. The samples were injected in splitless mode. The injection temperature was 250 °C, temperature of ion source 200 °C and transfer line 275 °C. Helium velocity was 40 cm s^{-1}. The oven temperature, initially at 60 °C, was gradually increased (20 °C min^{-1}) to 250 °C and held at this temperature for 15.5 min. The MS/MS detection consisted of a sequence of three steps: (i) isolation of the "parent" ion of the detected compounds (immediate rejection of other ions except the "parent" ion); (ii) fragmentation of the "parent" ion; and (iii) analysis of the fragmentation products. These steps were realized using GCQ MS/MS software, version 2.0. The MS/MS measurements were performed at a collision energy of 0.8 V for parent ions of m/z 275 (for the analyte) and m/z 331 (for the internal standard). The product spectra in the mass range of m/z 100–331 were scanned. The retention time of the analyte was 8.34 ± 0.04 min and of the internal standard 10.2 ± 0.04 min [37].

2.9. Statistical Analysis

Data obtained throughout this study were analyzed by computer-assisted one-way ANOVA, using the software package stat graphics version 5.0 (costat). Least significance differences (LSDs) were calculated at level of significance $p < 0.05$ [38].

3. Results

3.1. Morphological and Physiological Characteristics of the Bacterial Isolates

Thirty bacterial isolates (10 isolates for each medium) were randomly isolated on three different selective media, medium (A), medium (B), and medium (C). Some of morphological and physiological characteristics of bacterial isolates were obtained. In addition, 10 bacterial isolates which were isolated on medium (A) were bacilli, Gram positive, spore positive, capsule positive and negative, and motile, and the indole test, catalase test, starch hydrolysis and casein hydrolysis were positive (Table S1, Supplementary Materials). From these characteristics, all isolates seemed to be *Bacillus* spp. [21]. Finally, 10 bacterial isolates which were isolated on medium (B) were diplococcic, Gram negative, non-spore, capsule positive, and motile and non-motile, and the indole test, catalase test, starch hydrolysis and casein hydrolysis were positive (Table S2, Supplementary Materials). From these characteristics, all isolates seemed to be *Azotobacter* spp. [21].

Bacterial isolates which were isolated on medium (C) were bacilli, cocci and short rods. Gram stain, spore stain, and capsule were positive and negative. Motile and none motile, indole test, catalase test, starch hydrolysis and casein hydrolysis were positive (Table S3, Supplementary Materials).

3.2. Screening the Best PHB-Producing Bacterial Isolates

The results presented in Table 1 show that all tested isolates cultivated on medium (A) produced PHB. The CDW varied from 0.98 to 2.99 g/L. PHB production ranged from 0.10 to 0.62 g/L. PHB% ranged between 6.84 and 27.55%. Sugar utilization ranged from 14.32 g/L to 18.51 g/L. The values of the conversion coefficient ranged between 0.62% and 3.90%. Significant differences were observed between bacterial isolates in most cases. Generally, the best isolate which yielded the highest PHB production with reasonable sugar utilization was No. A6. Therefore, this isolate was chosen and used in the subsequent experiments. Isolates No. A1, A2 and A6 also produced more than 20% PHB.

In addition, all tested isolates grown on medium (B) also produced PHB (Table 1). The highest values of cell dry weight, PHB production, PHB%, sugar utilization and conversion coefficient were 4.00 g/L, 0.93 g/L, 28.51%, 20.93 g/L, and 6.17%, respectively. However, the lowest values of the above parameters were 2.56 g/L, 0.73 g/L, 19.50%, 15.07 g/L, and 2.86%, respectively. It was observed that there were significant differences between the bacterial isolates; thus, the best isolate with the highest value of PHB (0.93 g/L) was No. P3. This isolate was therefore chosen for identification based on molecular biology and used in the experiments on alternative carbon sources. PHB was produced from all isolates grown on medium (C) (Table 1). The CDW values varied from 0.79 to 3.1 g/L. PHB production ranged from 0.08 to 0.73 g/L, and PHB% ranged between 6.04 and 23.67%. Sugar utilization ranged from 6.09 g/L to 8.17. The values of the conversion coefficient ranged between 1.00% and 8.93%. Generally, the best isolate which yielded the highest PHB production (0.73 g/L) with reasonable sugar utilization was No. W8. It was noted that there were significant differences between the bacterial isolates. Therefore, this isolate was chosen and used in the subsequent experiments.

Table 1. Screening of different bacterial isolates to produce PHB on different media.

Isolate Number	C.D.W (g/L)	PHB (g/L)	PHB (%)	Utilized Sugar (g/L)	Conversion Coefficient (%)
			Medium A		
A1	2.99 a	0.61 b	20.40 c	17.03 c	3.58 b
A2	2.52 b	0.51 b	20.23 b	15.36 d	3.32 b
A3	1.27 f	0.11 de	8.66 h	16.00 d	0.68 e
A4	1.69 d	0.15 de	8.87 h	14.32 e	1.04 d
A5	1.17 fg	0.13 de	11.11 g	18.51 a	0.70 e
A6	2.25 c	0.62 a	27.55 a	15.88 d	3.90 a
A7	0.98 h	0.17 d	17.34 e	16.85 c	1.00 d
A8	1.46 e	0.10 e	6.84 i	15.94 d	0.62 e
A9	1.03 gh	0.12 de	11.65 f	17.77 b	0.67 e
A10	1.58 de	0.29 c	18.35 d	15.80 d	1.83 b
LSD 0.05	0.145	0.042	0.404	0.486	0.140
			Medium B		
P1	3.61 b	0.85 b	23.54 c	15.34 e	5.54 b
P2	3.93 a	0.83 b	21.11 d	19.58 c	4.23 de
P3	3.47 bc	0.93 a	26.80 ab	15.07 e	6.17 a
P4	3.30 c	0.83 b	25.14 bc	15.19 e	5.46 b
P5	4.00 a	0.78 bc	19.50 d	15.44 e	5.05 c
P6	3.62 b	0.78 bc	21.54 d	17.32 d	4.50 d
P7	2.78 d	0.74 c	26.62 ab	19.17 c	3.86 e
P8	2.56 e	0.73 c	28.51 a	20.93 b	3.48 f
P9	3.88 a	0.82 b	21.12 d	19.40 c	4.22 de
P10	3.40 bc	0.84 b	24.75 bc	29.33 a	2.86 g
LSD 0.05	0.165	0.053	1.571	0.455	0.365
			Medium C		
W1	1.37 d	0.10 ef	6.6 d	7.94 a	1.25 ef
W2	2.04 b	0.28 b	13.72 b	6.82 e	4.10 b
W3	1.85 c	0.17 d	9.37 c	7.71 b	2.20 d
W4	1.15 e	0.17 d	15.40 b	7.39 c	2.30 d
W5	1.36 d	0.21 c	14.73 b	7.10 d	2.95 c
W6	1.98 b	0.16 d	8.25 cd	6.09 f	2.62 cd
W7	1.82 c	0.13 de	7.68 cd	7.65 b	1.69 e
W8	3.10 a	0.73 a	23.67 a	8.17 a	8.93 a
W9	0.79 f	0.11 ef	14.79 b	7.44 c	1.47 ef
W10	1.33 d	0.08 f	6.04 d	7.96 a	1.00 f
LSD 0.05	0.121	0.032	1.748	0.203	0.433

C.D.W: cell dry weight; PHB: Poly-ß-hydroxybutyrate; A1–10: symbol for unknown bacterial isolate grown on medium (A) [15]; P1–10: symbol for unknown bacterial isolate grown on medium (B) [16]; W1–10: symbol for unknown bacterial isolate grown on medium (C) [17]. Means followed by different letters indicate significant differences between treatments according to Duncan's test ($p < 0.05$).

3.3. *Identification of the Best Bacterial Isolates Based on Molecular Method*

Isolates No. A6, P3 and W8 (which were the best isolates for PHB production in the previous experiments) were identified according to the Polymerase Chain Reaction (PCR) method by Sigma Company (Cairo, Egypt) [34]. Isolate No. A6 was identified by the

16S ribosomal RNA gene and the partial sequence Query1 GAGCTTGCTCTTATGAAGT-TAGCGGCGGACGGGTGAGTAACACGTGGGTAACCTGCCCAT 60.

The sequences of this isolate were accessed through a database (www.ncbi.nlm.nih.gov, accessed on 1 September 2021) using the accession number, and it was found to belong to the *Bacillus paramycoides* strain MCCC 1A04098 (Figure 1A). Isolate No. P3 was identified using the same method, by the partial sequence Query 1 AATACCCTGCAGTCTTGACGT-TACCGGCAGAATAAGCACCGGCTAACTTCGTGCCAGCAG 60. It was identified as the *Azotobacter salinestris* strain NBRC 102611 (Figure 1B). Isolate No. W8 was also identified using the same method, by the partial sequence Query 1 GAATAACTCAGGGAAACTTGT-GCTAATACCGAATGTGCCCTTTGGGGGAAAGATTTATCG 60. It was identified as *Brevundimonas naejangsanensis* strain BIO-TAS2-2 (Figure 1C).

3.4. PHB Production from Cheap Alternative Carbon Sources

The values of carbon and nitrogen were considered during preparation of the cultivation media, which already contained these sources. In dried whey, sugar beet molasses and date molasses, the carbon contents were 70%, 51%, and 37%, respectively, and the nitrogen contents were 15%, 13%, and 14.1%, respectively. Table 2 shows the effect on PHB production from *B. paramycoides* strain MCCC 1A04098 of replacing glucose in medium (A) with the three cheap alternative carbon sources (dried whey, treated sugar beet molasses, and treated date molasses). The addition of treated sugar beet molasses was the most economically suitable, as it increased PHB production (0.98 g/L) and the conversion coefficient (6.5%). Furthermore, the utilized sugar (15.11 g/L) was moderate and slightly lower than the control value (15.88 g/L). On the other hand, dried whey yielded the lowest PHB production (0.51 g/L), even lower than the control. The values of treated date molasses showed the same pattern as treated sugar beet molasses, except for sugar utilization which was the highest among all carbon sources (16.40 g/L). It was observed that there were significant differences between bacterial strains. In the present study, the maximum production of PHB and CDW by *B. paramycoides* was 0.98 g/L and 2.84 g/L from sugar beet molasses, respectively.

A

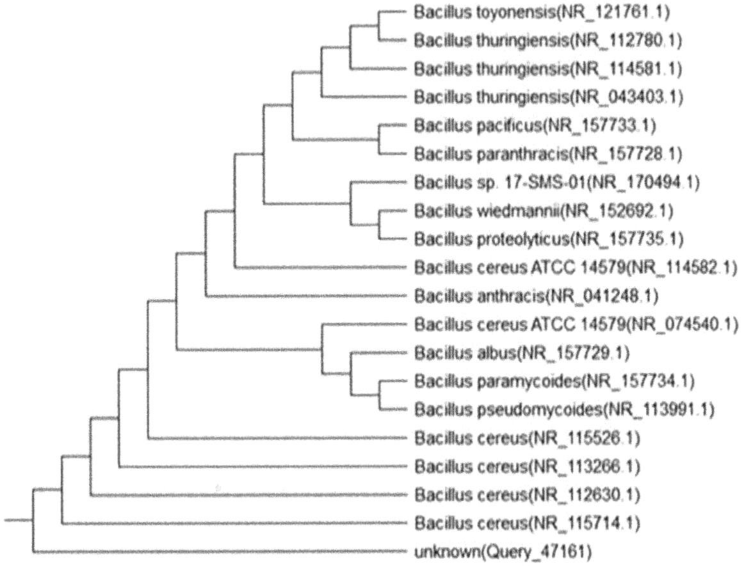

Figure 1. *Cont.*

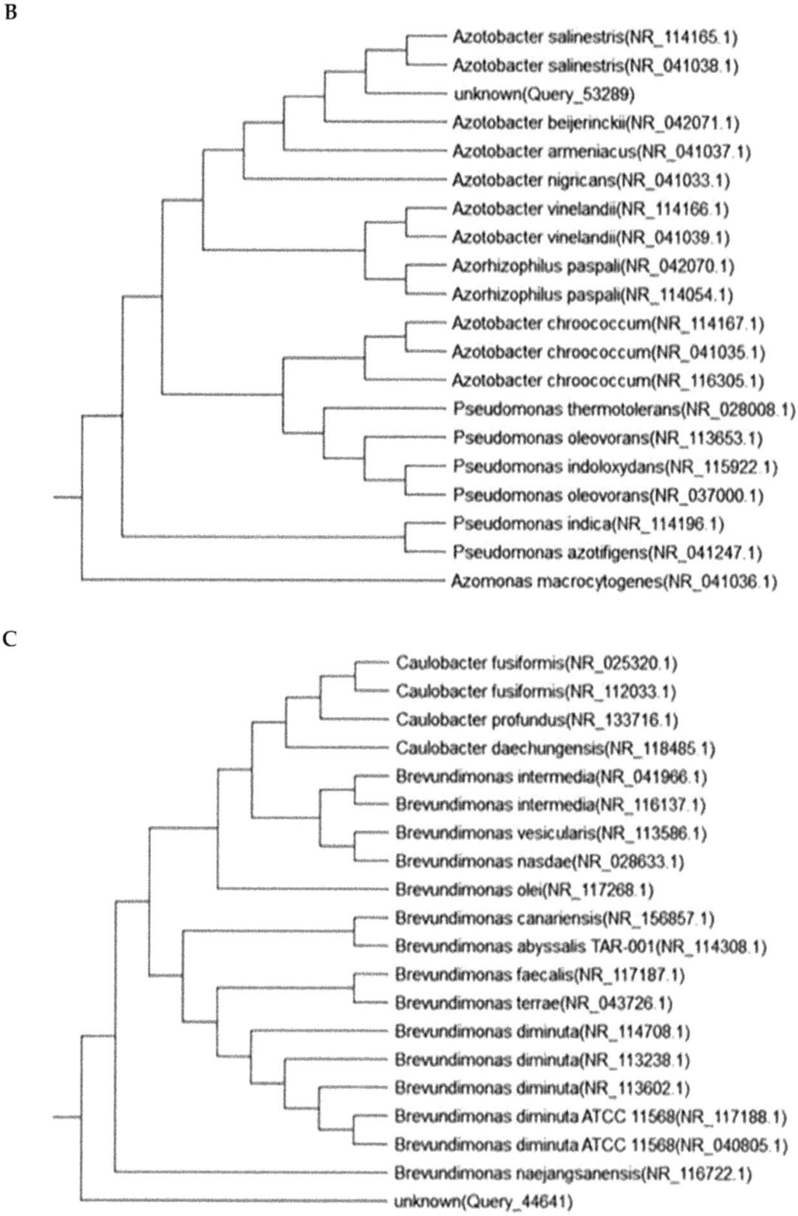

Figure 1. Genotype tree of *Bacillus paramycoides* MCCC 1A04098 (**A**), *Azotobacter salinestris* NBRC 102611 (**B**) and *Brevundimonas naejangsanensis* BIO-TAS2-2 (**C**).

Table 2. Effect of cheap alternatives carbon sources on PHB production by different strains of *Bacillus paramycoides*, *Azotobacter salinestris* and *Brevundimonas naejangsanensis*.

Alternatives Carbon Sources	C.D.W (g/L)	PHB (g/L)	PHB (%)	Utilized Sugar (g/L)	Conversion Coefficient (%)
			***Bacillus paramycoides*strain MCCC 1A04098**		
Dried whey	1.56 ^{d}	0.51 ^{d}	33.81 ^{a}	13.74 ^{b}	3.71 ^{c}
sugar beet molasses	2.84 ^{a}	0.98 ^{a}	34.50 ^{a}	15.11 ^{ab}	6.50 ^{a}
Date molasses	2.67 ^{b}	0.88 ^{b}	32.86 ^{a}	16.40 ^{a}	5.36 ^{b}
(Control, glucose)	(2.25 ^{c})	(0.62 ^{c})	(27.55 ^{b})	(15.88 ^{a})	(3.90 ^{c})
LSD 0.05	0.123	0.079	4.499	1.391	0.804
			***Azotobacter salinestris* strain NBRC 102611**		
Dried whey	2.47 ^{d}	0.66 ^{d}	26.51 ^{bc}	5.15 ^{d}	12.81 ^{c}
Sugar beet molasses	4.97 ^{a}	1.56 ^{a}	31.39 ^{a}	6.52 ^{c}	23.92 ^{a}
Date molasses	4.61 ^{b}	1.36 ^{b}	29.45 ^{ab}	6.83 ^{b}	19.91 ^{b}
(Control, fructose)	(3.10 ^{c})	(0.73 ^{c})	(23.67 ^{c})	(8.17 ^{a})	(8.93 ^{d})
LSD 0.05	0.227	0.047	2.951	0.220	0.984
			***Brevundimonas naejangsanensis* strain BIO-TAS2-2**		
Dried whey	1.63 ^{d}	0.64 ^{d}	39.27 ^{b}	11.59 ^{d}	5.51 ^{b}
Sugar beet molasses	3.57 ^{a}	1.36 ^{b}	40.36 ^{b}	16.71 ^{a}	9.73 ^{a}
Date molasses	3.19 ^{c}	1.50 ^{a}	47.14 ^{a}	16.03 ^{b}	9.35 ^{a}
(Control, sucrose)	(3.47 ^{b})	(0.93 ^{c})	(26.80 ^{c})	(15.07 ^{c})	(6.17 ^{b})
LSD 0.05	0.161	0.069	3.537	0.284	0.657

Means followed by different letters indicate significant differences between treatments according to Duncan's test ($p < 0.05$).

Fructose was replaced with alternative carbon sources in medium (B) for PHB production using the *A. salinestris* strain NBRC 102611. Table 2 showed that the use of treated beet molasses yielded the highest PHB production (1.56 g/L) and highest conversion coefficient (23.92%), with moderate sugar utilization (6.52 g/L) in comparison with the control values. Dried whey yielded the lowest PHB production (0.66 g/L). It was observed that there were significant differences between the bacterial isolates.

Sucrose was replaced with alternative carbon sources in medium (C) for PHB production from the *Brevundimonas naejangsanensis* strain BIO-TAS2-2. Table 2 showed that the use of treated date molasses yielded the highest PHB production (1.5 g/L) and the highest conversion coefficient (9.35%), with moderate sugar utilization (16.03 g/L) in comparison with the control values. Dried whey yielded the lowest PHB production (0.64 g/L). It was observed that there were significant differences between the bacterial isolates.

3.5. The Correlation Coefficient Value between PHB Production, CDW and Utilized Sugar (g/L) of All Tested Isolates and Strains

The correlation coefficient was calculated according the equation in [33]. Statistical analysis indicated there was a very strong positive correlation ($r = 0.987$) between PHB g/L and CDW g/L in the *A. salinestris* strain and a strong positive correlation ($r = 0.457$) in the isolates grown on medium (B). A very weak negative correlation ($r = -0.161$) between PHB g/L and utilized sugar g/L in the isolates grown on medium (B) was observed, with the highest value between them being $r = 0.897$ (a very strong positive correlation) in the case of *B. naejangsanensis* (Table 3).

Table 3. The correlation analysis between PHB, dry cell weight and utilized sugar.

Strains or Isolates Obtained From	Correlation Coefficient Analysis (r Value)	
	Correlation between PHB g/L and CDW (g/L)	Correlation between PHB g/L and Utilized Sugar (g/L)
Medium (A)	0.901	−0.012
Bacillus paramycoides	0.958	0.527
Medium (B)	0.457	−0.161
Azotobacter salinestris	0.987	0.051
Medium (C)	0.840	0.239
Brevundimonas naejangsanensis	0.724	0.897

Medium (A) [15]; Medium (B) [16]; Medium (C) [17].

3.6. *The Physical Properties of PHB Polymer*

3.6.1. Fourier Transform Infrared Spectroscopy (FTIR)

Poly ß-hydroxybutyrate (PHB) Polymer extracted from *B. paramycoides* was characterized using IR spectra in the range of 600–4000 cm^{-1}, as displayed in Figure 2a. The IR spectra of PHB produced show obvious peaks at 1262 and 1725 cm^{-1}, corresponding to specific rotations around carbon atoms specific to –CH and C=O stretching of the ester group present in the molecular chain [36]. Strong peaks at 1034 and 1097 indicate the presence of C-O stretching. Additionally, the absorption bands at 2926 and 2963 cm^{-1} are due to C–H stretching vibrations of methyl and methylene groups, which confirms PHB formation. PHB extracted from *A. salinestris*, as shown in Figure 2b, is similar to that extracted from *B. paramycoides*. Furthermore, the IR spectra of the extract from *A. salinestris* display similar bands at 1032, 1098 1262, 1725, 2925 and 2963 cm^{-1} which confirm the presence of PHB, as shown in Figure 2b. On the other hand, the IR spectra of PHB extracted from *Brevundimonas naejangsanensis* illustrate weak bands at 1264 and 1748 cm^{-1}, indicating lower PHB concentration in comparison with the other obtained extracts, as shown in Figure 2c. In addition, all peaks at 3400 to 3600 cm^{-1} correspond to specific amino groups.

3.6.2. Gas Chromatography–Mass Spectrometry (GC-MS) Analysis

Chemical composition was obtained after extraction and derivatized using BSTFA reagent. Figure 3a shows a peak at retention time 15.91 min, which represents Polyhydroxybutyrate (PHB) from *B. paramycoides*. The molecular ion peak in the mass spectrum at m/z 104 agrees well with the molecular formula for PHB. In addition, Figure 3b shows a peak at 20.08 min, which represents 2, 4-ditert-butylphenol. The area under the peak denotes the PHB content, which was determined to be 4.85%.

Figure 4a shows the GC-MS chromatogram of the PHB extracted from *A. salinestris*; the peak at retention time 16.07 min represents the isopropyl ester of 2-butenoic acid (Figure 4b), confirming the present of Polyhydroxybutyrate (PHB). The area under the peak denotes the PHB content, which was determined to be 10.73% (Figure 4b). Figure 5a shows the GC-MS chromatogram of the PHB extracted from *Brevundimonas naejangsanensis*, showing a peak at retention time 16.07 min which represents the isopropyl ester of 2-butenoic acid (Figure 5b), confirming the presence of Polyhydroxybutyrate (PHB). The area under the peak denoting the PHB content was determined to be 0.85%, with very low mass spectrum matching.

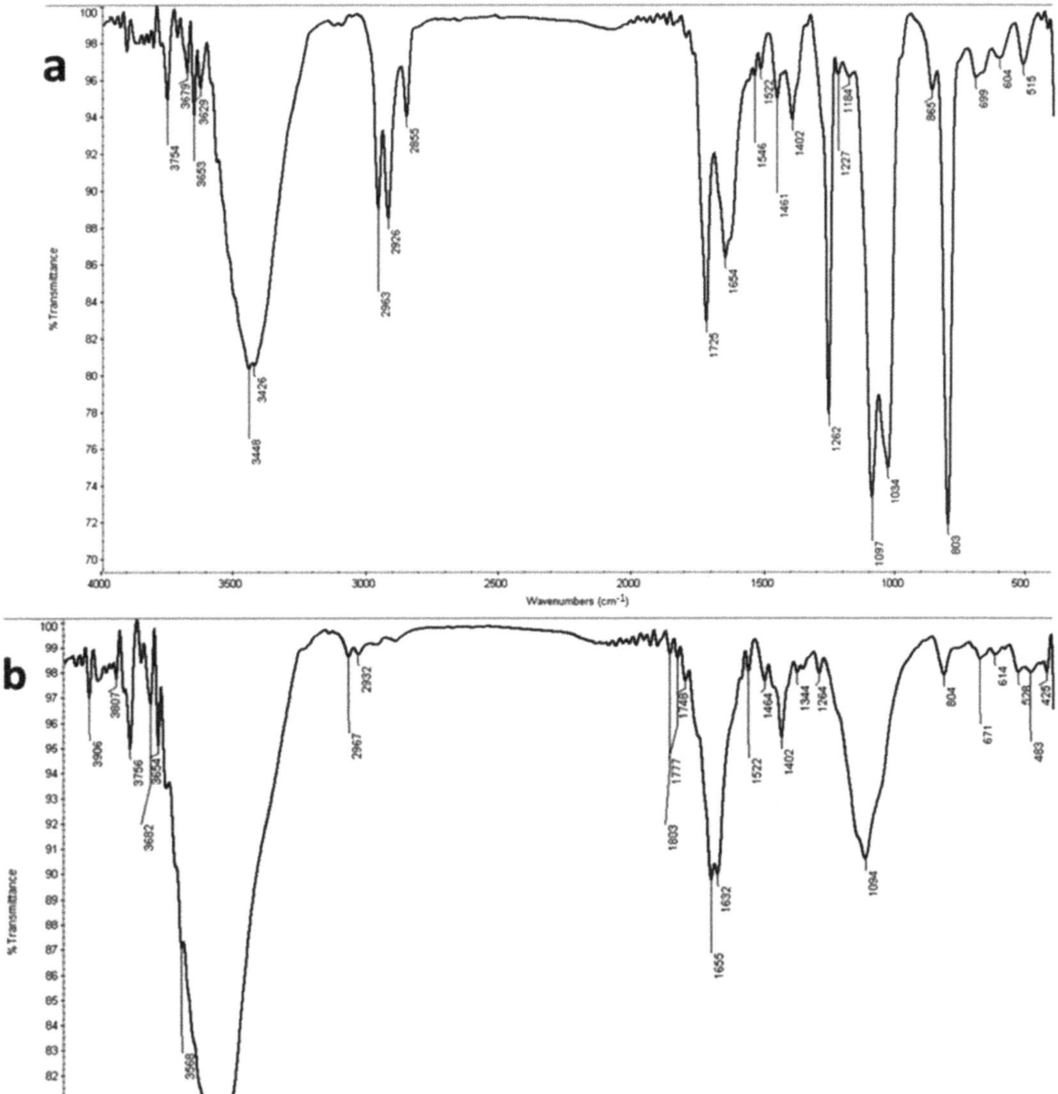

Figure 2. *Cont.*

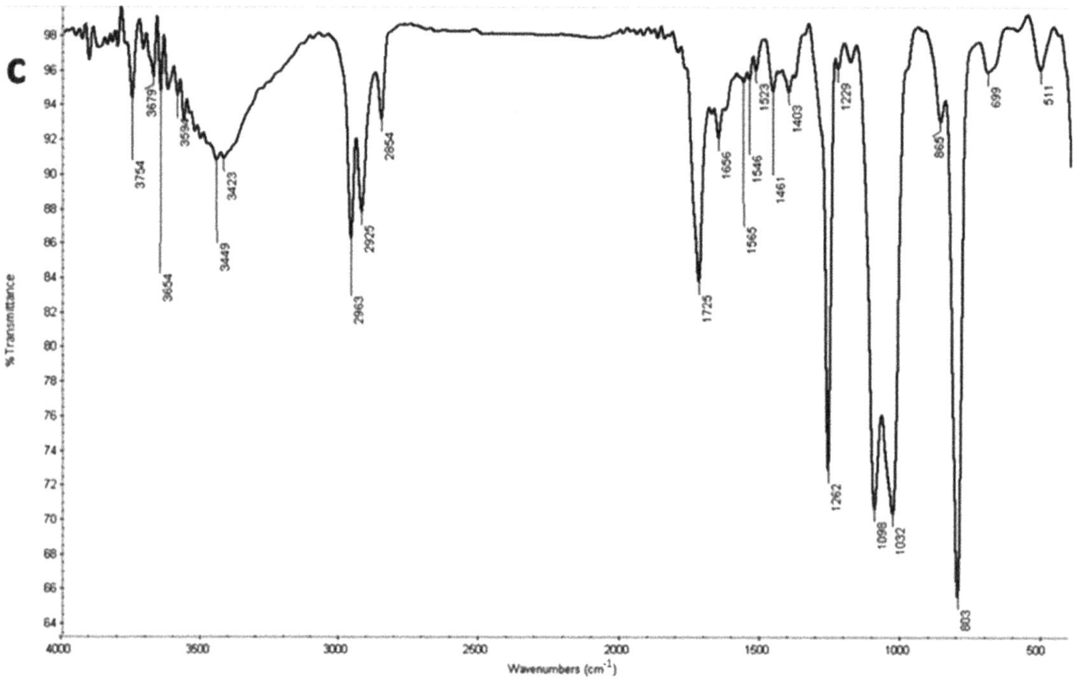

Figure 2. IR spectrum of PHB extracted from *B. paramycoides* (**a**), *A. salinestris* (**b**), and *Brevundimonas naejangsanensis* (**c**).

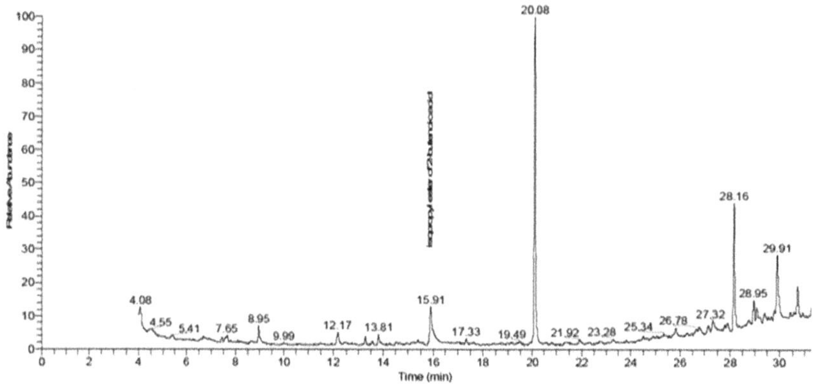

(a)

Figure 3. *Cont.*

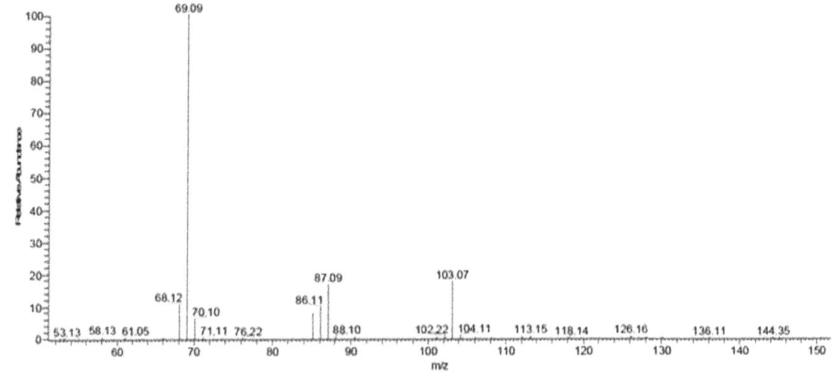

(b)

Figure 3. (**a**) Total GC-MS chromatogram of PHB extracted from *Bacillus paramycoides*. (**b**): The mass spectrum of isopropyl ester of 2-butenoic acid of PHB extracted from *Bacillus paramycoides*.

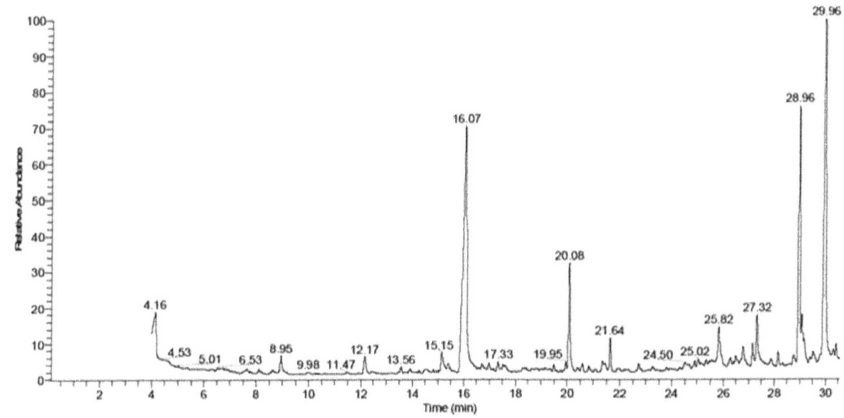

(a)

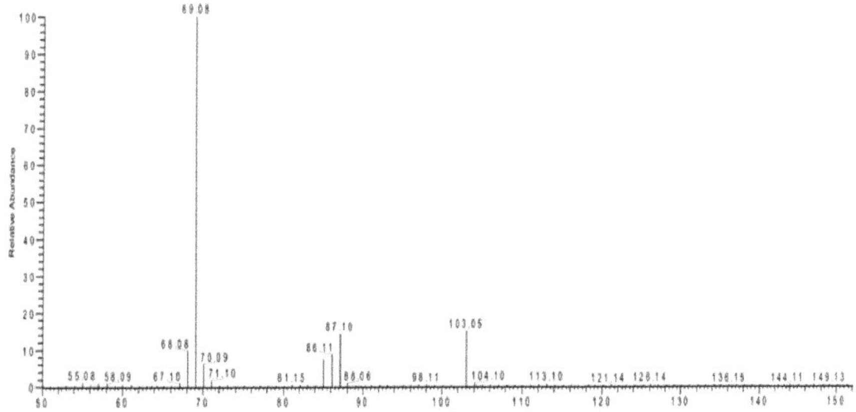

(b)

Figure 4. (**a**): Total GC-MS chromatogram of PHB extracted from *Azotobacter salinestris*. (**b**): The mass spectrum of isopropyl ester of 2-butenoic acid of PHB extracted from *Azotobacter salinestris*.

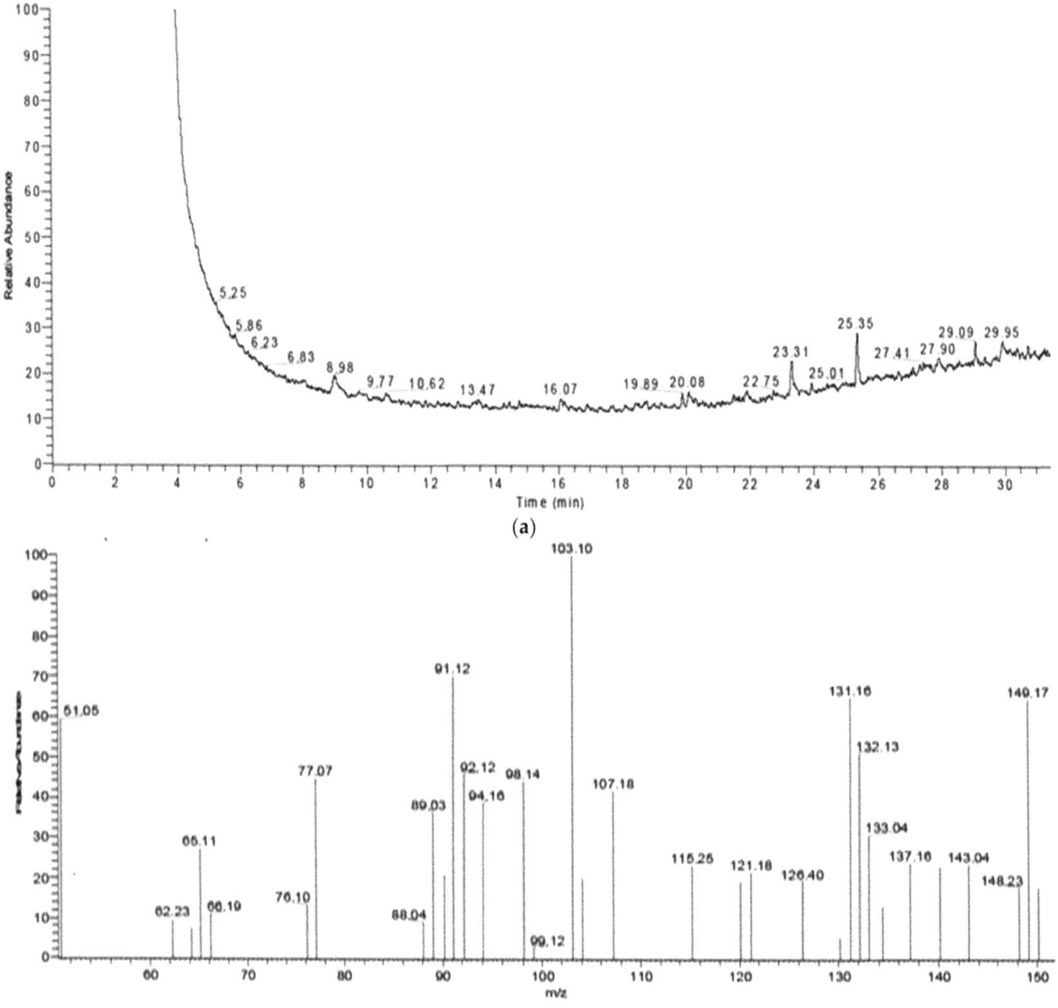

(a)

(b)

Figure 5. (**a**): Total GC-MS chromatogram of PHB extracted from *Brevundimonas naejangsanensis*. (**b**): The mass spectrum of isopropyl ester of 2-butenoic acid of PHB extracted from *Brevundimonas naejangsanensis*.

4. Discussion

4.1. Screening the Best PHB-Producing Bacterial Isolates

Significant differences were observed between the bacterial isolates in most cases. Generally, the best isolate which yielded the highest PHB production with reasonable sugar utilization was No. A6 (Table 1). Isolates No. A1, A2 and A5 produced more than 20% PHB. These results were similar to the results obtained by [39], who isolated three bacteria from the roots of beans (*Vicia faba*) grown in Qalyubia Governorate, Egypt to produce PHB using the flask culture technique. These isolates collected more than 20% of PHB, and were identified as *Pseudomonas fluorescens* S48, *B. megaterium* 7A and *B. megaterium* UBF19. Four isolates (A1, A2, A6 and A10) were higher than those found by [18], who reported that the highest value of PHB in *B. megaterium* was 0.27 g/L. Furthermore, all of these isolates were higher than [40], who produced PHB by batch fermentation of *B. subtilis* from glucose and reached only 0.077 g/L.

In addition, all tested isolates grown on medium (B) produced PHB (Table 1). It was observed that there were significant differences between bacterial isolates. Thus, the best isolate with the highest value of PHB (0.93 g/L) was No. P3. Multiple previous studies [19,41] produced PHB from *A. chroococcum* in amounts ranging from 2.0 to 3.0 g/L, higher than in the current study; however, [10] produced only about 1.0 g/L from *A. vinelandii*, lower than this study. In [42], the production of PHB was studied using the shaking culture method with *A. chroococcum* resulting in a PHB% of 46.80%. On the other hand, PHB was produced from all isolates grown on medium (C) in the present study (Table 1).

4.2. Identification of the Best Bacterial Isolates Based on Molecular Method

Isolates No. A6, P3 and W8 were identified according to the Polymerase Chain Reaction (PCR) method. The sequence of the A6 isolate was accessed using the accession number and it was found to belong to the *Bacillus paramycoides* strain MCCC 1A04098 (Figure 1A). Although several reports have produced PHB from *Bacillus* sp., such as *B. amyloliquefaciens*, *B. aryabhattai*, *B. brevis*, *B. cereus*, *B. circulans*, *B. coagulans*, *B. firmus*, *B. laterosporus*, *B. licheniformis*, *B. macerans*, *B. megaterium*, *B. sphaericus*, *B. subtilis*, and *B. thuringiensis* [1,14,43], and a few researchers [44–48] have used *B. mycoides* for producing PHB, to the best of our knowledge only this study has produced PHB from *B. paramycoides*.

In addition, the isolate No. P3 was identified as *Azotobacter salinestris* strain NBRC 102611 (Figure 1B). Three species of the genus *Azotobacter* sp. have commonly appeared in the literature on the production of PHB, including *A. chroococcum* [19,28,41,42,49], *A. vinelandii* [50–53], and *A. beijerinckii* [54–56]; however, only one used *A. salinestris* for PHB production [57] before this study. Isolate No. W8 was identified as the *Brevundimonas naejangsanensis* strain BIO-TAS2-2 (Figure 1C). Only one study has used *Brevundimonas naejangsanensis* for PHB production [58] prior to the current study. This study is also the first to report the newly isolated bacterial strain designated as *Bacillus paramycoides* strain MCCC 1A04098 as potential source for PHB production.

4.3. PHB Production from Cheap Alternative Carbon Sources

Table 2 shows the effect of replacing glucose in medium (A) with three cheap alternative carbon sources, namely dried whey, treated sugar beet molasses, and treated date molasses, on PHB production from *B. paramycoides* strain MCCC 1A04098. Generally, in order to decrease the cost of PHB and increase the cell dry weight, PHB concentration, PHB%, and conversion coefficient, the addition of treated sugar beet molasses in the medium of *B. paramycoides* is recommended.

Bacillus sp. is capable of producing hydrolytic enzymes, which can degrade agro-industrial and other wastes that can be utilized as carbon sources for PHB production [43]. These results were similar to [46], where the highest PHB yield was observed in *B. mycoides* DFC1 (1.28 g/L) using wheat starch, and lower than [47], where a maximum dry cell weight of 4.35 g/L was obtained. In addition, these results were higher than those obtained in [59], where the maximum PHB% obtained by *B. subtilis* was 13.02% ± 1.67%. As for the attempts that have been made to reduce the cost of production, there have been many attempts, such as [1], who reported that PHB production from *B. aryabhattai* showed good polymer accumulation in basal medium with glycerol after 48 h of incubation and accumulated 1.79 g/L. Hence, glycerol may be a better option as it is cheaper than glucose and can be obtained as a byproduct from industries such as biodiesel production. In the same trend, the current study was better than [3], who studied the effect of different carbon sources on PHB production by *B. megaterium*. The highest production of PHB was observed with cane molasses as the sole carbon source (40.8, mg/L). The same results were obtained by [60], who reported that sugar beet molasses could be a suitable substrate for the production of PHB with *B. megaterium*. Cell dry weight was 16.7 g/L and PHB production was

0.6 g/L with batch cultivation. Sugarcane molasses was also used as a sole carbon source by [11,61,62], using *B. cereus* and *B. subtilis* to obtain a PHB% of 57.5%, 49.9% and 44.7%, respectively.

Fructose was replaced with alternative carbon sources in medium (B) for PHB production from *A. salinestris* strain NBRC 102611 (Table 2). In [52] it was found that there are more than of 300 bacterial species that can produce PHB, of which only a few have been used on a commercial scale. Molasses and whey have been reported to be excellent substrates for PHB production [35,57,63]. Similar results were obtained by [60], who reported that the use of sugar beet molasses in PHB production led to a two- to three-fold increase in biomass production compared to the control experiment. In [10] it was reported that the results obtained with beet molasses were the best compared with other sugars, especially unrefined sugar; in addition, it was proven that it is not only a source of carbon, but also contains some other nutrients that may interfere with the synthesis of PHB. Thus, to decrease the cost of PHB, the addition of sugar beet molasses in the medium of *A. salinestris* to increase cell dry weight, PHB concentration, PHB% and conversion coefficient appears very promising.

Sucrose was replaced with alternative carbon sources in medium (C) for PHB production from *Brevundimonas naejangsanensis* strain BIO-TAS2-2 (Table 2). Both the current study and [12] used date molasses as a novel carbon source for *B. naejangsanensis* and *Cupriavidus necator* for PHB production. The current findings were homogeneous with [58], who tested the ability of *B. vesicularis* to accumulate PHB and found that with acid-hydrolyzed wood (sawdust) as the substrate, PHB production was improved. In order to reduce the production cost by 50% and make it as economical as possible, more than 96% of sugars must be consumed, and the cells must contain more than 90% PHB with a yield not less than 64% of the DCW.

4.4. The Correlation Coefficient Value between PHB Production, CDW and Utilized Sugar (g/L) of All Tested Isolates and Strains

Multiple studies [64,65] have found a positive correlation between PHB production and CDW with r = 0.999, which was similar to our results where r = 0.987 (Table 3). The available studies on the correlation between PHB production and CDW were insufficient, with little information available; [64–66] published that there was a positive correlation between PHB and CDW. On the contrary, Aslim [67], reported there was a low correlation between CDW and the PHB contents of *Lactobacillus* cultures. Thus, there is a need for further study on this point. In this context, [68] reported that the correlation between PHB synthesis and nitrogen fixation was little discussed in purple non-sulfur bacteria species. In keeping with this trend, [65,69] studied the correlation between PHB production, pH, and viscosity, finding that it was decreased or that no correlation could be observed. However, there are no studies dealing with the correlation between PHB production and the amount of sugar consumed, which is the most important economic factor affecting polymer production.

4.5. The Physical Properties of PHB Polymer

4.5.1. Fourier Transform Infrared Spectroscopy (FTIR)

Poly ß-hydroxybutyrate (PHB) Polymer extracted from the studied strains was characterized using IR spectra in the range of 600–4000 cm^{-1}, as displayed in Figure 2. The results were obtained by [14], which observed that 1624–1724 cm^{-1} was associated with the C=O stretching of the ester carbonyl bond. The bond at 1529 cm^{-1} was characteristic of the stretching and deformation vibration of the C-H group, and those at 2960 and 3277 cm^{-1} were characteristic of the stretching and deformation vibrations of the terminal OH groups. The PHB polymer functional groups were confirmed as C=O groups by FT-IR spectroscopy. In addition, analysis of the spectrum of PHB was observed at 3400, 1639, 3018, 2978, 2842, 1216, 1044 and 669–765 δ, respectively, for the (-OH) broad peak, (C=C) double bond, (≡C-H) acetylenic bond, (-CH3) methyl group, (-S-H) thiol weak adsorption, (≡C-O-C≡) ether, sulfoxide and (F-Cl) haloalkanes [41]. The same results were obtained by [70], who showed that the bands of PHB between 3300 and 3700 cm^{-1} were ascribed to O-H groups in the phenolic and aliphatic structures. The band at 2923 cm^{-1} was assigned to C-H stretching in aromatic methoxyl groups, or in methyl and methylene groups of the side chain. The band at 1645 cm^{-1} was attributed to unconjugated C=O stretching. The bands centered at 1508 and 1593 cm^{-1} were associated with aromatic structural vibrations. The bands between 1400 and 1450 cm^{-1} originated from C-H deformation coupled with the vibration from the aromatic ring. The bands at 1217 and 1128 cm^{-1} indicated guaiacyl and syringyl groups, respectively. Moreover, Rathika [11], reported that, the broad transmittance peak at 3281 cm^{-1} could be ascribed to the stretching of O–H groups. Peaks observed at 2924 and 2852 cm^{-1} were associated with the C-H stretching of bonds of methyl (CH3) and methylene (CH2) groups. The intense absorption peak at 1722 cm^{-1} was the characteristic carbonyl (C=O) stretching of ester groups in the extracted PHAs.

The 1H NMR spectrum showed the expected resonances for PHB as demonstrated by the methine group (-CH-) between 5.22 and 5.28 ppm, a methylene group (-CH2-) between 2.45 and 2.62 ppm, and the methyl group (-CH3) between 1.26 and 1.28 ppm, as in standard PHB. Scans of 13C NMR showed peaks at 169.13, 67.62, 40.81 and 19.76 ppm, which represent the carbonyl carbon (-C-), ester (-O-CH-) group, methylene (-CH2-) and the methyl (-CH3) groups, as shown in the standard. These results confirm the material as a homopolymer of 3-hydroxybutyrates, i.e., poly-3-hydroxybutyrate [1].

4.5.2. Gas Chromatography–Mass Spectrometry (GC-MS) Analysis

Our observation in Figures 3–5 were in agreement with [11], who detected PHAs extracted from *B. subtilis* RS1 cultured and processed in a medium of sugarcane molasses using GC–MS and revealed significant peaks at retention times 20.9, 23.1, and 23.8 min, corresponding to methyl esters of pentadecanoic acid and hexadecanoic acid, respectively. Furthermore, the results revealed that hexadecanoic acid methyl ester is the predominant monomer of PHA produced from *B. subtilis* RS1. In addition, Ramya [14] reported that the bioplastics produced by *B. cereus* RBL6 were characterized by GC–MS. *B. cereus* RBL6 was reported to produce methyl-3-hydroxy hexadecanoic acid, which belongs to the monomer chains of the medium-chain-length PHA class.

5. Conclusions

Our study aimed to use cheap and available alternative carbon sources (whey, sugar beet molasses and date molasses) to produce PHB using *Bacillus paramycoides*, *Azotobacter salinestris* and *Brevundimonas naejangsanensis*. The results showed that the addition of sugar beet molasses in the medium of *A. salinestris* increased the cell dry weight (CDW), PHB concentration, PHB% and conversion coefficient. In addition, statistical analysis indicated that the correlation coefficient values between PHB g/L and CDW g/L varied between r = 0.987 and r = 0.457 (very strong positive correlation and moderate positive correlation); however, these values were r = −0.161 and r = 0.897 when the correlation was calculated between PHB g/L and utilized sugar g/L. In addition, the IR of the produced PHB from *B. paramycoides* and *A. salinestris* showed similar bands, which confirmed the presence of PHB; however, *B. naejangsanensis* showed weak bands. The chemical composition obtained by GC-MS of the PHB extract represented 2, 4-ditert-butylphenol for *B. paramycoides*, and isopropyl ester of 2-butenoic acid for both of *A. salinestris* and *Brevundimonas naejangsanensis*.

Supplementary Materials: The following are available online at https://www.mdpi.com/article/10.3390/polym13213801/s1: Table S1. Morphological and physiological characteristics of bacterial isolates on medium (A); Table S2. Morphological and physiological characteristics of bacterial isolates on medium (B); Table S3. Morphological and physiological characteristics of bacterial isolates on medium (C).

Author Contributions: Conceptualization, S.M.E.-K., A.E.-D.O., M.E. and D.F.I.A.; methodology, S.M.E.-K., A.E.-D.O., S.E.-N., A.A.S., H.F.A.E.-S., H.A.H.E.-Z. and D.F.I.A.; software, S.M.E.-K., A.E.-D.O., H.A.H.E.-Z. and D.F.I.A.; validation, S.M.E.-K., A.E.-D.O., M.E. and D.F.I.A.; formal analysis, S.M.E.-K., A.E.-D.O., H.F.A.E.-S., H.A.H.E.-Z. and D.F.I.A.; investigation, S.M.E.-K., A.E.-D.O., H.A.H.E.-Z. and D.F.I.A.; resources, S.M.E.-K., A.E.-D.O., H.F.A.E.-S. and H.A.H.E.-Z. and D.F.I.A.; data curation, S.M.E.-K., A.E.-D.O., S.E.-N., M.E. and D.F.I.A.; writing—original draft preparation, S.M.E.-K., A.E.-D.O. and D.F.I.A.; writing—review and editing, S.M.E.-K., A.E.-D.O. and D.F.I.A.; visualization, S.M.E.-K., A.E.-D.O. and D.F.I.A.; supervision, S.M.E.-K., funding acquisition, M.E. All authors have read and agreed to the published version of the manuscript.

Funding: This research received no external funding.

Institutional Review Board Statement: Not applicable.

Informed Consent Statement: Not applicable.

Data Availability Statement: The data that supports the findings of this study are contained within the article or supplementary material and available from the corresponding author upon reasonable request.

Acknowledgments: The authors extend their appreciation to the Deanship of Scientific Research at King Khalid University for funding this work through the Research Groups Project under grant number R.G.P. 1/112/42. All the authors are grateful for the support provided by Faculty of Agriculture, Damietta University, Faculty of Agriculture, Mansoura University and Soils, Water and Environment Research Institute (SWERI), Agriculture Research Center (ARC), Egypt.

Conflicts of Interest: The authors declare no conflict of interest.

References

1. Pillai, A.B.; Kumar, A.J.; Thulasi, K.; Kumarapillai, H. Evaluation of short-chain-length polyhydroxyalkanoate accumulation in Bacillus aryabhattai. *Braz. J. Microbiol.* **2017**, *48*, 451–460. [CrossRef]
2. Zheng, Y.; Chen, J.; Ma, Y.; Chen, G. Engineering biosynthesis of polyhydroxyalkanoates (PHA) for diversity and cost reduc-tion. *Metab. Eng.* **2020**, *58*, 82–93. [CrossRef] [PubMed]
3. Gouda, M.K.; Swellam, A.E.; Omar, S.H. Production of PHB by a Bacillus megaterium strain using sugarcane molasses and corn steep liquor as sole carbon and nitrogen sources. *Microbiol. Res.* **2001**, *156*, 201–207. [CrossRef] [PubMed]
4. Rasheed, T.; Adeel, A.H.; Fahmeeda, K.; Farooq, S.; Muhammad, B.; Hafiz, M.N.I. Carbon nanotubes assisted analytical detection–Sensing/delivery cues for environmental and biomedical monitoring. *TrAC Trends Anal. Chem.* **2020**, 116066. [CrossRef]
5. Altaf, F.; Muhammad, N.; Zaib, J.; Tahir, A.; Muhammad, A.; Muhammad, B.; Tayyaba, N.; Farooq, S. Synthesis and characterization of PVA/starch hydrogel membranes incorporating essential oils aimed to be used in wound dressing applications. *J. Polym. Environ.* **2021**, *29*, 156–174. [CrossRef]
6. Khan, A.; Muhammad, N.; Farooq, S.; Zaib, J.; Tayyaba, N.; Ofaira, A.; Tazien, R.; Naseem, I. Metal organic frameworks derived sustainable polyvinyl alcohol/starch nanocomposite films as robust materials for packaging applications. *Polymers* **2021**, *13*, 2307. [CrossRef]
7. Sirohi, R.; Pandey, J.P.; Gaur, V.K.; Gnansounou, E.; Sindhu, R. Critical overview of biomass feedstocks as sustainable substrates for the production of polyhydroxybutyrate (PHB). *Bioresour. Technol.* **2020**, *311*, 123536. [CrossRef]
8. Carlozzi, P.; Touloupakis, E.; Lorenzo, T.D.; Giovannelli, A.; Seggiani, M.; Cinelli, P.; Lazzeri, A. Whey and molasses as inexpensive raw materials for parallel production of biohydrogen and polyesters via a two-stage bioprocess: New routes to-wards a circular bioeconomy. *J. Biotechnol.* **2019**, *303*, 37–45. [CrossRef]
9. Khardenavis, A.A.; Kumar, V.A.N.; Chakrabarti, M.S. Utilization of molasses spent wash for production of bioplastics by waste activated sludge. *Waste Manag.* **2009**, *29*, 2558–2565. [CrossRef] [PubMed]
10. Page, W.J. Production of polyhydroxyalkanoates by Azotobacter vinelandii UWD in beet molasses culture. *FEMS Microbiol. Lett.* **1992**, *103*, 149–157. [CrossRef]
11. Rathika, R.; Janaki, V.; Shanthi, K.; Kamala-Kannan, S. Bioconversion of agro-industrial effluents for polyhydroxyalkanoates production using Bacillus subtilis RS1. *Int. J. Environ. Sci. Technol.* **2018**, *16*, 5725–5734. [CrossRef]
12. Purama, R.K.; Al-Sabahi, J.N.; Sudesh, K. Evaluation of date seed oil and date molasses as novel carbon sources for the pro-duction of poly(3Hydroxybutyrate-co-3Hydroxyhexanoate) by Cupriavidus necator H16 Re 2058/pCB113. *Ind. Crop. Prod.* **2018**, *119*, 83–92. [CrossRef]
13. Pavan, F.A.; Junqueira, T.L.; Watanabe, M.D.B.; Bonomi, A.; Quines, L.K.; Schmidell, W.; de Aragao, G.M.F. Economic anal-ysis of polyhydroxybutyrate production by Cupriavidus necator using different routes for product recovery. *Biochem. Eng. J.* **2019**, *146*, 97–104. [CrossRef]
14. Ramya, R.; Sangeetha Devi, R.; Manikandan, A.; Rajesh Kannan, V. Standardization of biopolymer production from sea-weed associative bacteria. *Int. J. Biol. Macromol.* **2017**, *102*, 550–564.

15. Bekheet, S.A.; El-Sharabasy, S.F. Date palm status and perspective in Egypt. In *Date Palm Genetic Resources and Utilization*; Springer: Berlin/Heidelberg, Germany, 2015; pp. 75–123.
16. El-Dengawy, E.F.A.; Samaan, A.; El-Shobaky, M.A.; El-Kadi, S.M.; Saleh, M.A.A. Evaluation of rutability, quality and micro-bial load in Hayani date palm fruits during cold storage as affected by applying some safe postharvest treatments. *J. Plant Prod. Mansoura Univ.* **2018**, *9*, 805–813.
17. Echegaray, N.; Pateiro, M.; Gullón, B.; Amarowicz, R.; Misihairabgwi, J.M.; Lorenzo, J.M. Phoenix dactylifera products in human health—A review. *Trends Food Sci. Technol.* **2020**, *105*, 238–250. [CrossRef]
18. Aslim, B.; Yuksekdag, Z.N.; Beyatli, Y. Determination of PHB growth quantities of certain Bacillus species isolated from soil. *Yurkish Elctronic J. Biotechnol.* **2002**, *52*, 24–30.
19. Pozo, C.; Martinez-Toledo, M.V.; Rodelas, B.; Gonzalez-Lopez, J. Effects of culture conditions on the production of polyhydroxyalkanoates by Azotobacter chroococcum H23 in media containing a high concentration of alpechin (wastewater from olive oil mills) as primary carbon source. *J. Biotechnol.* **2002**, *97*, 125–131. [CrossRef]
20. Wang, F.; Lee, S.Y. Poly(3-Hydroxybutyrate) Production with High Productivity and High Polymer Content by a Fed-Batch Culture of Alcaligenes latus under Nitrogen Limitation. *Appl. Environ. Microbiol.* **1997**, *63*, 3703–3706. [CrossRef]
21. Vos, P.; Garrity, G.; Jones, D.; Krieg, N.R.; Ludwig, W.; Rainey, F.A.; Schleifer, K.H.; Whitman, W.B. *Bergey's Manual of Systematic Bacteriology*, 2nd ed.; Springer: New York, NY, USA, 2009; Volume 3.
22. Akhter, M.; Hossain, S.J.; Hossain, S.; Datta, R.K. Isolation and characterization of salinity tolerant Azotobacter sp. *Greener J. Biol. Sci.* **2012**, *2*, 43–51.
23. Bhaduri, J.; Kundu, P.; Mitra, D.; Roy, S.K. Isolation and characterization of nitrogen fixing bacteria Azotobacter sp. from tea field soil of dooars and Darjeeling region of north Bengal, India. *Int. J. Eng. Sci. Invent.* **2016**, *5*, 46–51.
24. Difco. *Difco Manual of Dehydrated Culture Media and Reagents for Microbiological and Clinical Laboratory Procedures*, 9th ed.; Difco Laboratories Incorporated: Detroit, MI, USA, 1997; Volume 451.
25. Abd-El-Malek, Y.; Ishac, Y.Z. Evaluation of Methods Used in Counting Azotobacters. *J. Appl. Bacteriol.* **1968**, *31*, 267–275. [CrossRef] [PubMed]
26. Du, G.; Chen, J.; Yu, J.; Lun, S. Continuous production of poly-3-hydroxybutyrate by Ralstonia eutropha in a two-stage culture system. *J. Biotechnol.* **2001**, *88*, 59–65. [CrossRef]
27. Khanna, S.; Srivastava, A.K. Optimization of nutrient feed concentration and addition time for production of poly (hydroxybutyrate). *Enzym. Microb. Technol.* **2006**, *39*, 1145–1151. [CrossRef]
28. Kim, B.S. Production of poly(3-hydroxybutyrate) from inexpensive substrates. *Enzym. Microb. Technol.* **2000**, *27*, 774–777. [CrossRef]
29. Lee, S.Y.; Chol, J.I. Polyhydroxyalkanoates biodegradable polymer. In *Manual of Industrial Microbiology and Biotechnology*, 2nd ed.; Demain, A.L., Davies, J.E., Eds.; ASM Press: Washington, DC, USA, 1998; pp. 616–627.
30. Wang, Y.J.; Hua, F.L.; Tsang, Y.F.; Chan, S.Y.; Sin, S.N.; Chua, H.; Yu, P.H.F.; Ren, N.Q. Synthesis of PHAs from waster un-der various C:N ratios. *Bioresour. Technol.* **2007**, *98*, 1690–1693. [CrossRef]
31. Dubois, M.; Gilles, K.A.; Hamilton, J.K.; Rebers, P.A.; Smith, F. Colorimetric Method for Determination of Sugars and Related Substances. *Anal. Chem.* **1956**, *28*, 350–356. [CrossRef]
32. Baethgen, W.E.; Alley, M.M. A manual colorimetric procedure for measuring ammonium nitrogen in soil and plant Kjeldahl digests. *Commun. Soil Sci. Plant Anal.* **1989**, *20*, 961–969. [CrossRef]
33. Asuero, A.G.; Sayagoand, A.; González, A.G. The correlation coefficient: An overview. *Crit. Rev. Anal. Chem.* **2006**, *36*, 41–59. [CrossRef]
34. Huggett, J.F.; Novak, T.; Garson, J.A. Differential susceptibility of PCR reactions to inhibitors: An important and unrecog-nized phenomenon. *BMC Res. Notes* **2008**, *1*, 70. [CrossRef] [PubMed]
35. Lazaridou, A.; Roukas, T.; Biliaderis, C.G.; Vaikousi, H. Characterization of pullulan produced from beet molasses by Aureobasidium pullulans in a stirred tank reactor under varying agitation. *Enzyme Microbial Technol.* **2002**, *31*, 122–132. [CrossRef]
36. Sindhu, R.; Ammu, B.; Binod, P.; Deepthi, S.K.; Ramachandran, K.B.; Soccol, C.R.; Pandey, A. Production and characteriza-tion of poly-3-hydroxybutyrate from crude glycerol by Bacillus sphaericus NII 0838 and improving its thermal properties by blending with other polymers. *Braz. Arch. Biol. Technol.* **2011**, *54*, 783–794. [CrossRef]
37. Elhottova, D.; Triska, J.; Petersen, S.O.; Santruckova, H. Analysis of poly-b-hydroxybutyrate in environmental samples by GC-MS/MS. *Fresenius J. Anal. Chem.* **2000**, *367*, 157–164. [CrossRef] [PubMed]
38. Ing, S.J. CoStat statistical software. *Int. J. Bio-Med. Comput.* **1987**, *20*, 227–228. [CrossRef]
39. Gamal, R.F.; Abd El-Hady, H.M.; Hassan, E.A.; El-Tayeb, T.S.; Aboutaleb, K.A. Production of Polyhydroxyalkanoate (PHAs) and copolymer [P (HB-co-HV)] by soil bacterial isolates in batch and two-stage batch cultures. *Egypt. J. Microbiol.* **2001**, *46*, 109–123.
40. Yadav, J.; Balabantaray, S.; Patra, N. Statistical optimization of fermentation conditions for the improved production of poly-β-hydroxybutyrate from Bacillus subtilis. *Chem. Eng. Commun.* **2017**, *204*, 1122–1128. [CrossRef]
41. Saha, S.P.; Patra, A.; Paul, A.K. Studies on intracellular degradation of polyhydroxyalkanoic acid–polyethylene glycol co-polymer accumulated by Azotobacter chroococcum MAL-201. *J. Biotechnol.* **2007**, *132*, 325–333. [CrossRef] [PubMed]
42. El-Sawah, M.; Kassem, M.; Selim, A.; Ashour, E.H.; El-Kadi, S. Production of Some Biodegradable Polymers by Some Bacterial Isolates. *J. Agric. Chem. Biotechnol.* **2008**, *33*, 6053–6064. [CrossRef]
43. Mohanrasu, K.; Rao, R.G.R.; Dinesh, G.; Zhang, K.; Prakash, G.S.; Song, D.-P.; Muniyasamy, S.; Pugazhendhi, A.; Jeyakanthan, J.; Arun, A. Optimization of media components and culture conditions for polyhydroxyalkanoates production by Bacillus megaterium. *Fuel* **2020**, *271*, 117522. [CrossRef]

44. Thakur, P.S.; Borah, B.; Baruah, S.D.; Nigam, J.N. Growth associated production of poly-3-hydroxybutyrate by Bacillus my-coides. *Folia Microbiol.* **2001**, *46*, 488–494. [CrossRef] [PubMed]
45. Borah, B.; Thakur, P.; Nigam, J. The influence of nutritional and environmental conditions on the accumulation of poly-beta-hydroxybutyrate in Bacillus mycoides RLJ B-017. *J. Appl. Microbiol.* **2002**, *92*, 776–783. [CrossRef] [PubMed]
46. Aarthi, N.; Ramana, K.V. Identification and characterization of polyhydroxybutyrate producing Bacillus cereus and Bacillus mycoides strains. *Int. J. Environ. Sci.* **2011**, *1*, 744–756.
47. Narayanan, A.; Ramana, K.V. Polyhydroxybutyrate production in Bacillus mycoides DFC1 using response surface optimiza-tion for physico-chemical process parameters. *3 Biotech* **2012**, *2*, 287–296. [CrossRef]
48. Narayanan, A.; Kumar, V.A.S.; Ramana, K.V. Production and Characterization of Poly (3-Hydroxybutyrate-co-3-Hydroxyvalerate) from Bacillus mycoides DFC1 Using Rice Husk Hydrolyzate. *Waste Biomass-Valorization* **2013**, *5*, 109–118. [CrossRef]
49. Savenkova, L.; Gercberga, Z.; Kizhlo, Z.; Stegantseva, E. Effect of phosphate supply and aeration on poly-b-hydroxybutyrate production in Azotobacter chroococcum. *Process. Biochem.* **1999**, *34*, 109–114. [CrossRef]
50. Page, W.J.; Cornish, A. Growth of Azotobacter vinelandii UWD in fish peptone medium and simplified extraction of poly-b-hydroxybutyrate. *Appl. Environ. Microbiol.* **1993**, *59*, 4236–4244. [CrossRef]
51. Yan, Y.B.; Wu, Q.; Zhang, R.Q. Dynamic accumulation and degradation of poly (3-hydroxyalkanoate)s in living cells of Azo-tobacter vinelandii UWD characterized by 13C NMR. *FEMS Microbiol. Lett.* **2000**, *193*, 269–273. [CrossRef]
52. Garcia, A.; Segura, D.; Espin, G.; Galindo, E.; Castillo, T.; Pena, C. High production of poly-b-hydroxybutyrate (PHB) by an Azotobacter vinelandii mutant altered in PHB regulation using a fed-batch fermentation process. *Biochem. Eng. J.* **2014**, *82*, 117–123. [CrossRef]
53. Smith, M.J.; Francis, M.B. A Designed A. vinelandii–S. elongates co culture for chemical photoproduction from air, water, phosphate and trace metals. *ACS Synth. Biol.* **2016**, *5*, 955–961. [CrossRef] [PubMed]
54. Senior, P.J.; Dawes, E.A. The regulation of poly-b-hydroxybutyrate metabolism in Azotobacter beijerinckii. *Biochem. J.* **1973**, *134*, 225–238. [CrossRef]
55. Ward, A.C.; Rowley, B.I.; Dawes, E.A. Effect of oxygen and nitrogen limitation on poly-b-hydroxybutyrate biosynthesis in ammonium growth Azotobacter beijerinckii. *J. Gen. Microbiol.* **1977**, *102*, 61–68. [CrossRef]
56. Carter, B.S.; Dawes, E.A. Effect of oxygen concentration and growth rate on glucose metabolism, poly-b-hydroxybutyrate biosynthesis and respiration of Azotobacter beijerinckii. *J. Gen. Microbiol.* **1979**, *110*, 393–400. [CrossRef]
57. Silva, J.A.; Tobella, L.M.; Becerra, J.; Godoy, F.; Martínez, M.A. Biosynthesis of poly-β-hydroxyalkanoate by Brevundimonas vesicularis LMG P-23615 and Sphingopyxis macrogoltabida LMG 17324 using acid-hydrolyzed sawdust as carbon source. *J. Biosci. Bioeng.* **2007**, *103*, 542–546. [CrossRef] [PubMed]
58. Lin, Y.-Y.; Chen, P.T. Development of polyhydroxybutyrate biosynthesis in Bacillus subtilis with combination of PHB-associated genes derived from Ralstonia eutropha and Bacillus megaterium. *J. Taiwan Inst. Chem. Eng.* **2017**, *79*, 110–115. [CrossRef]
59. Schmid, M.; Song, H.; Raschbauer, M.; Emerstorfer, F.; Omann, M.; Stelzer, F.; Neureiter, M. Utilization of desugarized sugar beet molasses for the production of poly(3-hydroxybutyrate) by halophilic Bacillus megaterium uyuni S29. *Process. Biochem.* **2019**, *86*, 9–15. [CrossRef]
60. Sharma, P.; Bajaj, B.K. Production and characterization of poly-3-hydroxybutyrate from Bacillus cereus PS 10. *Int. J. Biol. Macromol.* **2015**, *81*, 241–248. [CrossRef]
61. Singh, G.; Kumari, A.; Mittal, A.; Goel, V.; Yadav, A.; Aggarwal, N.K. Cost Effective Production of Poly-β-Hydroxybutyrate by Bacillus subtilis NG05 Using Sugar Industry Waste Water. *J. Polym. Environ.* **2012**, *21*, 441–449. [CrossRef]
62. Page, W.J.; Bhanthumnavin, N.; Manchak, J.; Ruman, M. Production of poly (13-hydroxybutyrate-13-hydroxyvalerate) copoly-mer from sugars by Azotobacter salinestris. *Appl. Microbiol. Biotechnol.* **1997**, *48*, 88. [CrossRef]
63. Pais, J.; Serafim, L.S.; Freitas, F.; Reis, M.A. Conversion of cheese whey into poly(3-hydroxybutyrate-co-3-hydroxyvalerate) by Haloferax mediterranei. *New Biotechnol.* **2016**, *33*, 224–230. [CrossRef]
64. Sun, W.; Cao, J.; Teng, K.; Meighen, E. Biosynthesis of poly-3-hydroxybutyrate in the luminescent bacterium, Vibrio harveyi, and regulation by the lux autoinducer, N-(3-hydroxybutanoyl)homoserine lactone. *J. Biol. Chem.* **1994**, *269*, 20785–20790. [CrossRef]
65. Kucuk, C. Production of poly-β-hydroxybutyrate by root nodule bacteria of Phaseolis vulgaris. *Jokull J.* **2016**, *66*, 2018–2224.
66. Mercan, N.; Aslim, B.; Ksekdaú, Z.N.Y.; Beyatli, Y. Production of poly-b-Hydroxybutyrate (PHB) by some Rhizobium bacteria. *Turk. J. Biol.* **2002**, *26*, 215–219.
67. Aslim, B.F.; Caliskan, Y.; Gunduz, B.U. Poly-β-hydroxybutyrate production by lactic acid bacteria. *FEMS Microbiol. Lett.* **1998**, *159*, 293–297. [CrossRef]
68. Wu, S.C.; Liou, S.Z.; Lee, C.M. Correlation between bio-hydrogen production and polyhydroxybutyrate (PHB) synthesis by Rhodopseudomonas palustris WP3-5. *Bioresour. Technol.* **2012**, *113*, 44–50. [CrossRef]
69. Mahishi, L.; Tripathi, G.; Rawal, S. Poly(3-hydroxybutyrate) (PHB) synthesis by recombinant Escherichia coli harbouring Streptomyces aureofaciens PHB biosynthesis genes: Effect of various carbon and nitrogen sources. *Microbiol. Res.* **2003**, *158*, 19–27. [CrossRef] [PubMed]
70. Liu, D.; Yan, X.; Si, M.; Deng, X.; Min, X.; Shi, Y.; Chai, L. Bioconversion of lignin into bioplastics by *Pandoraea* sp. B-6: Molecular mechanism. *Environ. Sci. Pollut. Res.* **2019**, *26*, 2761–2770. [CrossRef]

polymers

MDPI

Article

Synthesis and Study of Morphology and Biocompatibility of Xanthan Gum/Titanium Dioxide-Based Polyurethane Elastomers

Shazia Naheed [1,*], Muhammad Shahid [1], Rashida Zahoor [1], Zumaira Siddique [1], Nasir Rasool [1], Sajjad Haider [2,*] and Shaukat Khan [3]

1 Department of Chemistry, Government College University, Faisalabad 38030, Pakistan; alchemist019021@gmail.com (M.S.); rashidazahoor786@gmail.com (R.Z.); zumairasiddique@gmail.com (Z.S.); nasirrasool@gcuf.edu.pk (N.R.)
2 Chemical Engineering Department, College of Engineering, King Saud University, P.O. Box 800, Riyadh 11421, Saudi Arabia
3 School of Chemical Engineering, Yeungnam University, 280-Daehak-Ro, Gyeongsan 712-749, Korea; shaukat85@yu.ac.kr
* Correspondence: shazianaheed@gcuf.edu.pk (S.N.); shaider@ksu.edu.sa (S.H.); Tel.: +92-(312)-9655311 (S.N.)

Abstract: A series of xanthan gum/titanium dioxide-based polyurethane elastomers were synthesized through the prepolymer method by the step growth polymerization. In the present work, xanthan gum was used as a bioactive material, with TiO_2 as a nanofiller. The structural characterization of newly prepared polyurethane samples was carried out with the help of Fourier Transform Infrared Spectroscopy. Thermogravimetric Analysis gave us the information about the thermal stability. Differential Scanning Calorimetry directs the thermal changes in the polyurethane samples. The Atomic Force Microscopy technique revealed that the degree of micro-phase separation increases by augmenting the % age of TiO_2, which was further confirmed by X-Ray Diffraction results. XRD confirmed the crystallinity of the final sample at about $2\theta = 20°$. Antimicrobial activity determined through the Disc Diffusion Method, and the results indicated that the synthesized polyurethane have antimicrobial activity. The water absorption capability of the polyurethane samples showed that these polymer samples are hydrophilic in nature.

Keywords: polyurethanes; xanthan gum; titanium dioxide; atomic force microscopy; X-Ray diffraction

Citation: Naheed, S.; Shahid, M.; Zahoor, R.; Siddique, Z.; Rasool, N.; Haider, S.; Khan, S. Synthesis and Study of Morphology and Biocompatibility of Xanthan Gum/Titanium Dioxide-Based Polyurethane Elastomers. *Polymers* **2021**, *13*, 3416. https://doi.org/10.3390/polym13193416

Academic Editor: Edina Rusen

Received: 30 August 2021
Accepted: 27 September 2021
Published: 5 October 2021

Publisher's Note: MDPI stays neutral with regard to jurisdictional claims in published maps and institutional affiliations.

Copyright: © 2021 by the authors. Licensee MDPI, Basel, Switzerland. This article is an open access article distributed under the terms and conditions of the Creative Commons Attribution (CC BY) license (https://creativecommons.org/licenses/by/4.0/).

1. Introduction

Polyurethanes have become the 6th most used group of polymers in the last few decades and have gained more importance due to multiple use in different fields, such as coatings, adhesives, furniture, and foams [1,2]. Polymeric composites are currently at an important crossroad, where research is shifted toward more sustainable bio-based materials. In the last few years, biomaterials and biocomposites have gained extraordinary attention. Biocomposites have additional benefits, apart from their eco-friendly nature [3]. Polyurethanes are often applied as biomaterials because they have good mechanical properties and low adsorption of biomolecules [4].

In recent years, numerous polymeric materials have utilized for the manufacturing of different medical devices that interact with blood, body liquids, and tissues [5]. The soft, as well as hard, segment of polyurethanes exhibited elastomeric properties [6]. Basically, the polyurethanes are incorporated in medical applications due to their toughness, cost effectiveness, and durability [7–9].

By comparing them with additives, they have gained much more attention due to ease of processing and, specifically, the countless architectural variety for optimizing their stages. The characteristics of the final polyurethane could be modified and controlled by the addition of additives [10]. They have a basic skeleton for lightweight materials, which is necessary for the transport industry. Reducing average weight decreases fuel usage, as well as greenhouse gases. Moreover, most of them are recyclable, bioactive, and biodegradable, depending upon the ingredients [11]. The physical properties of a polyurethane, such as density, tensile strength, high water abrasion, etc., distinguish its functioning [12]. They have admirable properties, such as hardness, tensile, compressive, impact resistance, etc. In general, polyurethanes are used in many different applications. These properties are tightly correlated with the biphasic nature of the segmented polyurethanes in the hard and soft phase. This, in turn, depends upon the chemical nature and composition of both phases. Flexible segment polyether and polyester-based polyurethanes (PUs) are susceptible to degradation under hydrolytic and oxidation environments [13].

The mechanical characterization of microcapsules of polyurethanes exhibits increased mechanical strength by a variable quantity of chain extenders, i.e., 1,4-butanediol (BDO) [14]. In the modern era, the standard of living could be enhanced with the employment of polymeric biomaterials. Generally, such biomaterials are simulated/artificial polymers, which are used in artificial organs, implants, dentistry, abrasion bandages, and drug delivery systems [15]. However, investigation on the degradation, morphology, and thermal and mechanical behavior is crucial to determine the end use of these materials.

In various extra and intracorporeal devices, biomedical polyurethanes are extensively applied. Typical illustrations include pacemaker leads insulation, indwelling tubes, heart pump tubes, and balloons of angioplasty [16]. Aliphatic diisocyanates, such as hexamethylene diisocyante (HDI), or cycloaliphatic diisocyanates, such as 4,4′-methylenebis(cyclohexyl isocyanate) ($H_{12}MDI$) and isophorone diisocyanate (IPDI), have bid superior stability over the aromatic isocyanate. The aliphatic diisocyanates show improved phase separation compared to corresponding aromatic diisocyanates. They also show improved phase separation behavior over the corresponding aromatic diisocyanates [3]. Today, extensive studies have been accomplished on polyurethane due to their inimitable possessions, as well as to illustrate their different behaviors [17].

By supplementing the diols of lower molecular weight as chain extender, Barikani and Hepburn reported that thermal stability of polyurethanes was enhanced [18]. Polyurethane biocompatible organic polymers, with an alginate nucleus and chitosan shell, used as nanoparticles, were prepared for increasing encapsulation efficiency, more regular insulin liberation, and better insulin accessibility [19]. Today, polysaccharides extracted from plants, such as guar gum or pectin, and extracted from algae, such as alginate, are replaced by some bacterial exopolysaccharides, such as xanthan gum and gellan gum [20]. *Xanthomonas campestris* produces the xanthan gum, which is basically heteropolysaccharide. Xanthan gum has extensive applications in industries, such as food, oil, pharmaceuticals, etc., due to its rheological properties, i.e., pseudo plasticity, high viscosity, etc. [1].

Biopolymer collection has been turned into a noteworthy natural issue today because of expanding use of biomaterials for pharmaceutical applications. Reprocessing of polymeric wastes by recycling them has been extensively applied. Now, there is a dire need to synthesize environmentally friendly biodegradable polymer [15].

Biocompatibility is an index which is a basic requirement for a polymeric material to be appropriate as an ideal biomaterial, which, in turn, depends upon degradable nature, cytotoxicity, and the other mechanical properties [21].

A few biomaterials out of bulk commodity polymers have been found promising. The novel polymeric materials of nanocomposites of nylon 6/clay has led to emanation having unanimity of incomparable characteristics [22]. For structural development, the amalgamation of layered nanofillers can radically influence the blends' and polymers' microphase morphology by acting as templates [23]. In order to promote biodegradability and biocompatibility, current studies have marked the porosity worthy of acritical stuff, more distinctively, porosities distribution and structure [24].

In this study, the xanthan gum/TiO_2-based polyurethanes (XTPUs) were prepared by augmentation of weight (% age) of TiO_2, by the prepolymer method. The microstructure biocompatibility of the nanocomposites was investigated. It was hoped that the introduction of nano TiO_2 could not only improve the physical properties and biocompatibility of PU but also inhibit the growth of bacteria even when nanofiller TiO_2 were used in low concentrations and were embedded in the polyurethane (PU) matrix.

2. Experimental

2.1. Materials and Synthesis

Analytical grade chemicals were used in this research work, provided by Sigma Chemical Co. (St. Louis, MO, USA), including Hydroxyl-terminated polybutadiene (HTPB, Mn = 3000 g/mol); Isophorone Diisocyanate (IPDI), 1,4-butanediol (BDO) was used as chain extender, titanium dioxide (TiO_2) was used as a nanofiller, and the commercial xanthan gum (XG) was used as bioactive material. These were used to synthesize polyurethanes' elastomers by the step growth synthesis method. Before the use of these chemicals, for removal of moisture, they were dried at 80 °C in an electrical oven. During the drying process, they were placed in a vacuum for 24 h.

The prepolymer synthesis was done, according to a reported method [25]. Firstly, TiO_2 (% by weight) and XG (% by weight) were dispersed and mechanically stirred in one mole of HTPB for 2 h at 100 °C until it completely dispersed in four-necked specially-designed apparatus equipped with round bottom flask, heating oil bath, magnetic stirrer, reflux condenser, nitrogen inlet, and dropping funnel at 100 °C. Then, it was reacted with two moles of IPDI for 2 h at 80–100 °C, in order to obtain isocyanate terminated (NCO) polyurethane prepolymer. The stirring continued until NCO-terminated prepolymer was synthesized. Titration with n-butylamine (ASTM D 2572-80) was conducted to obtain the NCO contents of the polymer. The synthesis of prepolymer was confirmed using FT-IR spectroscopy [26].

2.2. Synthesis of Final Polymer

The final xanthan gum/TiO_2-based polyurethane (XTPU) polymer was obtained by stirring, for 30 min, the polyurethanes prepolymer with the chain extender 1,4-butanediol (1.2 mol). On the appearance of homogeneity and completion of dispersion of the chain extender in the reaction mixture, the polymer, in liquid form, was poured on the Teflon plate in order to develop a sheet having thickness of almost 2–3 mm. Then, at 100 °C for 24–48 h, the circulating hot air oven was used to cure the synthesized polymer. Before testing, i.e., characterization by various techniques, the cured sheets was stored at 25 °C for one week in order to attain almost 40% humidity. A series of six polymer samples were prepared by keeping constant the weight % age of XG, i.e., 1% and by varying weight % age of TiO_2 from 0% to 5%, as shown in Table 1. Figure 1 presents the schematic demonstration of the chemical route for synthesis of the polymer.

Table 1. General formulation of polyurethanes.

Sample Code	IPDI (mole)	HTPB (mole)	BDO (mole)	TiO_2 %	Xanthan Gum %
XTPU-1	2	0.8	1.2	0	0
XTPU-2	2	0.8	1.2	1	1
XTPU-3	2	0.8	1.2	2	1
XTPU-4	2	0.8	1.2	3	1
XTPU-5	2	0.8	1.2	4	1
XTPU-6	2	0.8	1.2	5	1

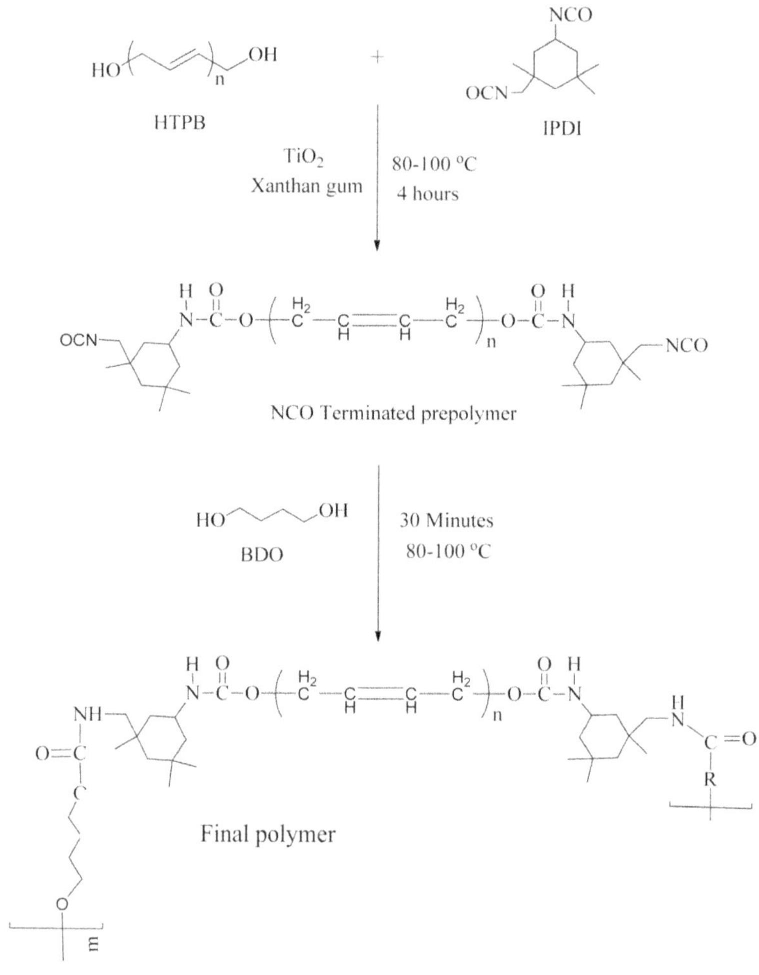

Figure 1. Synthesis of polyurethane.

3. Characterization

3.1. Fourier Transform Infrared Spectroscopy (FT-IR)

Structural characterization of synthesized polymer was done by BRUKER TENOSR II FT-IR spectrometer (Bruker, Billerica, MA, USA). The infrared spectra recorded after 15-s interludes, by using 8 scans over a resolutions of 2 cm^{-1}. The KBr beam splitter and DTGS detector were provided to the spectrometer. Data was collected, processed, and presented by thermo scientific spectroscopy software against a time-dependent series. The mechanism of reaction, as well as crosslinking behavior, was interpreted by FT-IR [27].

3.2. Atomic Force Microscopy (AFM)

AFM (CP-II, Veeco, Newport Beach, CA, USA) was used to assess surface morphology of prepared samples. A phosphorus-doped silicon-integrated pyramidal tip was used as support, in order to acquire images in tapping mode with a triangular cantilever (force constant of 20–80 N/m). Simultaneously, the images of topography and phase separation were recorded. The root-mean square average of the surface roughness was calculated within the given area as the standard deviation of all heights. Image Pro Plus 4.5 software

(Media Cybernetics, Rockville, MD, USA) was used to measure the average hard and soft domain size from the phase images [28].

3.3. *Thermogravimetric (TGA) and Differential Scanning Calorimeter (DSC)*

Thermogravimetric and Differential Scanning Calorimeter analysis were accomplished using an SDT Q600 V20.9 Build 20 (TA Instruments, Newcastle, DE, USA) under a dry nitrogen flow. The temperature range was set at 0–600 °C, and tamp (heating) rate was retained at 10 °C/min. The samples for TGA and DSC were primed on a glass substrate by spin coating of the mixed solution. Then, it was cured at various temperature steps. Aluminium sample pans were utilized to seal the almost 4 mg of each sample. Then, the prepared materials were analyzed under dry nitrogen by DSC over a heating rate of 10 °C/min from 0 to 500 °C. The glass transition temperature (Tg) crystallization temperature (Tc) and melting temperature (Tm) of the prepared polymers were investigated with the help of differential scanning calorimeter SDT Q600 V20.9 Build 20 [29].

3.4. *Antimicrobial Activity*

The capacity of a substance to have contact with the human body tissues without affecting the human body is called biocompatibility. The samples were subjected to assess the antimicrobial activity. The inhibition studies were done by actively growing bacterial cells. First of all, a nutrient agar media of 1000 mL was prepared, and 150 mL was poured in separate 150-mL flasks. The autoclave was used for 15 min at 120 °C by putting in agar medium nutrients flasks. Later on, it was cooled and bacteria, i.e., *Escherichia coli* (gram-negative) and *Macrococcus* (Gram-positive), were added to about 15 µL in the above two flasks. The sterile petri plates were used, and almost 20 mL of each agar nutrient medium was transferred in these plates, while room temperature was maintained. For 24 h, the samples were incubated at 37 °C, and they were then placed in petri dishes. The zone of inhibition, in which growth of bacteria was inhibited, was calculated by the diffusion active compound in the surrounding of the sample [30,31].

3.5. *X-Ray Diffractometry (XRD)*

In X-ray diffractometry, X-rays were used to irradiate XTPU polymers, and the scattering pattern was noted. Basically, the scattered radiations intensity noted as a function of the scattering angle θ. The X-rays scattering is due to electron density difference. A small angle scattering of X-Rays is used to investigate the microstructures in the range of tens to thousands angstrom (Å). It is basically a phase segregation in polymers. Meanwhile, X-ray scattering of a wide angle was also applied to calculate the crystallinity of the polymer samples at atomic level. The different reflections of XRD were the result of crystalline behavior of the polymer samples, which was calculated by using Bragg's law.

$$2d \sin \theta = n\lambda, \quad (1)$$

where d is the basically distance between crystalline planes, θ is the angle of X-ray beam which it makes with the planes, n is an integer, and λ is the wavelength. The dispersion range (2θ) of 0–70° was used to record the relative intensity [32].

4. Results and Discussion

4.1. *FT-IR Analysis*

Figure 2 represents the FT-IR spectra of 1,4-butanediol, Isophrone Diisocyanate, HTPB, Xanthan Gum, NCO-terminated Prepolymer, and XTPU final Polymer. Figure 2a shows the FT-IR spectrum of 1, 4-butanediol and indicated that, due to the OH group, a very strong peak appeared at 3400 cm^{-1}. Due to the CH_2 group the peak appeared at 2900 cm^{-1}. In Figure 2b, the FT-IR spectrum of IPDI showed the anti-symmetric stretching peak of the CH_2 group at 2939.52 cm^{-1} and symmetric stretching peak of the CH_2 group at 2862.0 cm^{-1} was also observed. In the spectrum of IPDI, due to the NCO group, a sharp peak at

2250.71 cm^{-1} was also observed. In Figure 2c, the FT-IR spectrum of HTPB where the peak of hydroxyl (OH) appeared at 3736.12 cm^{-1} due to stretching vibration is shown. There appeared a peak due to anti-symmetric stretching vibration of the CH_2 group, observed at 2945.30 cm^{-1}. Figure 2d shows the most significant bands for the xanthan gum in the range of 4000–500 cm^{-1}. It includes an axial deformation of OH at 3300–3450 cm^{-1} and an axial deformation of C-H at 2855–2926 cm^{-1}, which may due to absorption of symmetrical and asymmetrical stretching of -CH_3 or may be due to -CH_2-groups. There was also an aldehydic (-CHO) peak at 1710–1730 cm^{-1}. An axial deformation of C-O of enols was observed at 1530–1650 cm^{-1}. An axial deformation of C-O at 1045–1150 cm^{-1} was also observed. In Figure 2e, the NCO-terminated prepolymer clearly showed that peaks due to the OH group diminished. The peak due to the NH group appeared at 3325 cm^{-1}. The NCO group's intensity was lower to some extent, which means that isocyanate groups reacted completely. The formation of the prepolymer was confirmed by the appearance of peak of the NH group at 3325 cm^{-1} and supported its proposed structure. The anti-symmetric peak of the CH_2 group was seen at 2945.30 cm^{-1}. The stretching peak of –C=O was observed at 1724.36 cm^{-1}. Figure 2f shows, by extending prepolymer with 1,4-BDO, that the FT-IR spectra showed a very strong peak at about 1707 cm^{-1}, which was assigned to C=O stretching of urethane. Peaks corresponding to the absorption of NH, C=O, and C=O were observed at 3325 cm^{-1}, 1707 cm^{-1} (non-hydrogen bonded), 1643 cm^{-1} (hydrogen bonded), and 1225 cm^{-1}, respectively, which indicate the new synthesized product being in the urethane group. The observed N-H bending vibrations at 1598 cm^{-1}, C-O-C stretching absorption band corresponding to linkage between OH and NCO groups to form urethane bond in the range 1057–1130 cm^{-1}, also provide strong evidence for the formation of XTPU.

The FT-IR spectra of XTPU 1 to 6 with varying weight % age of TiO_2 and constant weight % age, i.e., 1% xanthan gum, are shown in Figure 3. All the spectra confirm the formation of urethane linkage in the final XTPU polymer samples. In Figure 3, the XTPU-1 spectra showed the formation of urethane linkage NH at the peak 3750.17 cm^{-1}, by the disappearing peak of NCO at 2156 cm^{-1}. It showed the symmetric and asymmetric peak at 2840 cm^{-1} of the –CH_2 group and 2913.45 cm^{-1}, respectively. The peak of –C=O group appeared at 1697.82 cm^{-1}. FT-IR spectra of XTPU-2 is shown in Figure 3. This spectra showed the formation of urethane linkage NH at the peak 3750 cm^{-1}. It also displayed the symmetric and a-symmetric peak of the –CH_2 group at 2800 cm^{-1} and 2912.59 cm^{-1}. The –C=O group peak appeared at 1698.02 cm^{-1}. The peak of NCO disappeared at 2160 cm^{-1}. The FT-IR spectra of XTPU-3 is shown in Figure 3. This spectrum characterizes the peak of –C=O group at 1698.02 cm^{-1}. The peak of N-H forms at 3750.13 cm^{-1}. It also gives the symmetric and a-symmetric peak at 2850 cm^{-1} and 2911.38 cm^{-1}. Figure 3 shows the FT-IR spectra of XTPU-4. This spectrum characterizes the peak of –C=O group at 1698.02 cm^{-1}. The peak of N-H forms at 3853.72 cm^{-1}. It also gives the symmetric and a-symmetric peak at 2843.05 cm^{-1} and 2913.05 cm^{-1}. In Figure 3, the FT-IR spectra of XTPU-5 shows the formation of N-H at 3760 cm^{-1}. It shows the formation of –C=O peak at 1698.05 cm^{-1}. It also shows the a-symmetric and symmetric peaks of CH_2 group at 2950 cm^{-1} and 2840 cm^{-1}. Furthermore, it gives information about the disappearing peak of the NCO group at 2300 cm^{-1}. Figure 3 also shows FT-IR spectra of XTPU-6 and confirmed the formation of N-H at 3753 cm^{-1}. Moreover, it illustrates the formation of –C=O peak at 1697.97 cm^{-1}. It shows the a-symmetric and symmetric peaks of –CH_2– group at 2912.56 cm^{-1} and 2843 cm^{-1}. It also gives information about the disappearing peak of the NCO group at 2300 cm^{-1}.

4.2. Evaluation of Antimicrobial Activity

The gram positive, as well as gram negative, bacteria and fungi were used for antimicrobial activity, so they can be classified as evaluation of antibacterial activity and evaluation of antifungal activity.

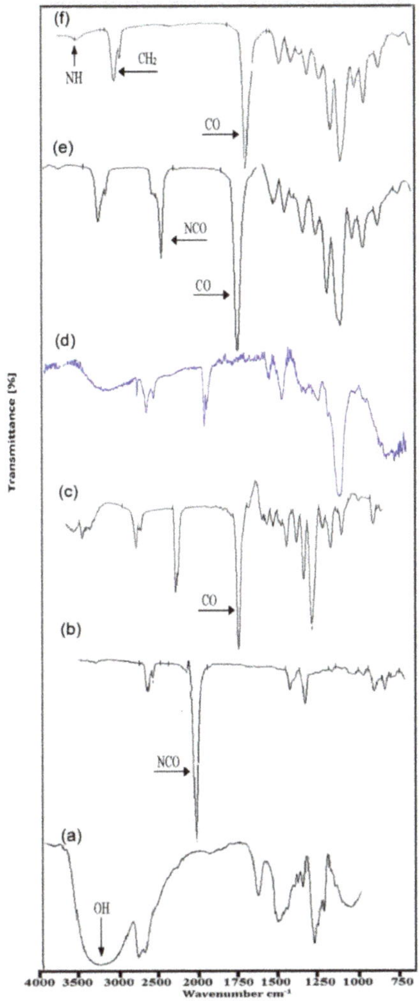

Figure 2. FT-IR spectra of (**a**) 1,4-Butanediol (BDO), (**b**) Isophrone diisocyanate, (**c**) Hydroxyl Terminated Polybutadiene (HTPB), (**d**) Xanthan Gum (XG), (**e**) NCO-terminated Prepolymer, and (**f**) XTPU final Polymer.

4.2.1. Evaluation of Antibacterial Activity

Biocompatibility is necessary for any material to be in contact with living tissues without resulting any harm to living body. Most polyurethane polymers are biocompatible, and their biocompatibility was evaluated by antibacterial activity. Antibacterial analysis was carried out through the disc diffusion technique, and *Escherichia coli* and *Macrococcus* (a gram negative and gram positive, respectively) strains of bacteria were used for this purpose. On agar plates, the bacterial cultures were spread, and punched samples of polyurethane polymer of 5 mm diameter samples were applied over the plates. This complete setup was carried out at 37 °C in an incubator overnight. The results for biocompatibility with *E. coli* and *Macrococcus* are presented in Figure 4. The resistance of polyurethane samples against *E. coli* the gram negative and *Macrococcus* the gram positive bacteria are shown in the inhibition zone. It was found that antibacterial ability or biodegradability of a polymer depends upon the concentration of TiO_2, along with content of xanthan gum used. We may increase the quality by changing the composition. It was observed that samples based on BDO show less antibacterial activity against *E. coli*. Antibacterial activity was influenced by bacterial strain used as given in results as reported in literature [30].

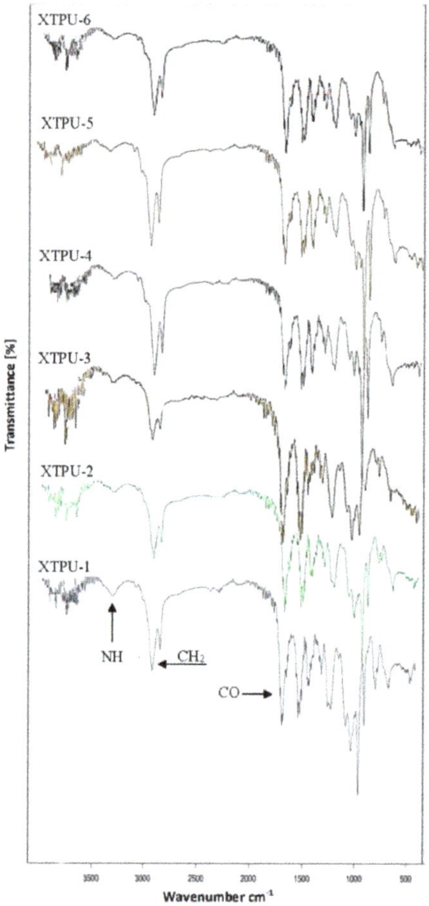

Figure 3. FT-IR spectrum of polyurethane samples XTPUs having 0–5% TiO_2, respectively.

4.2.2. Evaluation of Antifungal Activity

Antifungal activity was evaluated against *Aspergillus flavus* using disc diffusion method. Vogel's media was used for fungal culture growth. The selected polyurethane polymer samples were punched in disc of 5 mm diameter and applied over the place incubated at 37 °C overnight in incubator. Antifungal activity of 6 polyurethane samples with a variable content of Titanium dioxide and 1% xanthan gum was evaluated. The observed antifungal activity is shown in Figure 5. A clear zone of inhibition was observed on the plates of XTPU-5 and XTP-6; however, other polymer samples inhibited fungal growth at the area under the sample disc. The results exhibited that antifungal or biodegradability of polymer depends upon the content of TiO_2 used. We may increase the quality by changing the composition. Moreover, the zone of inhibition can be increased by increasing the weight % age of TiO_2.

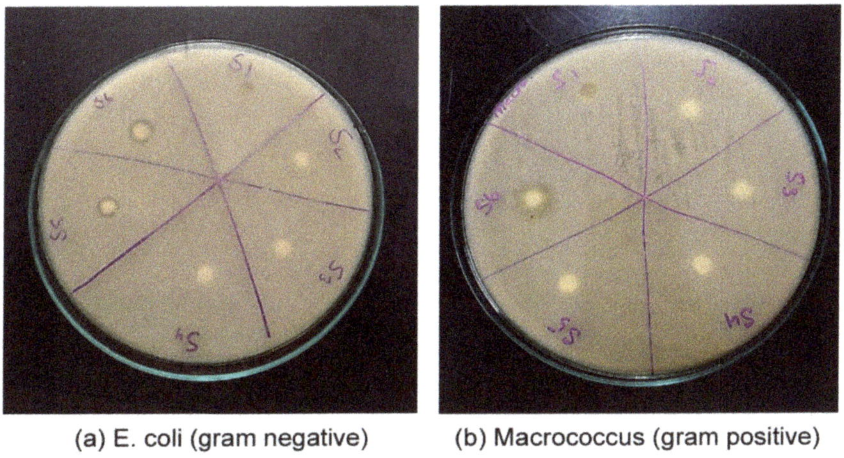

Figure 4. Antibacterial assay of samples against *E. coli* and *Macrococcus*.

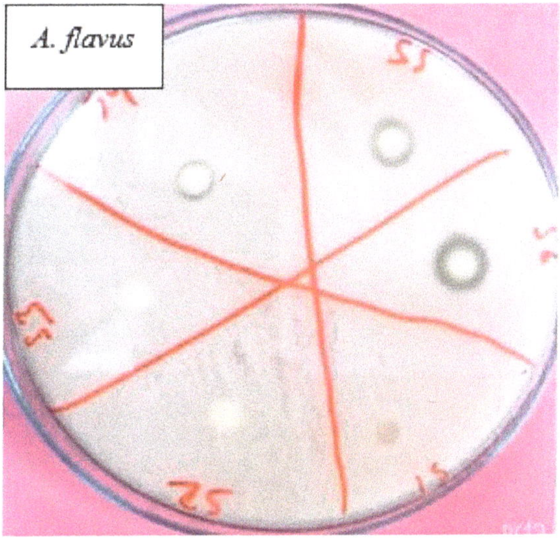

Figure 5. Antifungal assay of polyurethane samples against *Aspergillus flavus*.

4.3. Thermal Analysis

The thermal analyses of xanthan gum, titanium dioxide-based polyurethanes were performed by Thermogravimetric analysis (TGA) and Differential Scanning Calorimetry (DSC). These analyses are further explained below.

4.3.1. Thermogravimetric Analysis (TGA)

The thermal properties of xanthan gum, titanium dioxide-based polyurethanes (XTPU) were analyzed by using the TGA technique. The TGA thermograms of XTPU 1 to 6 are shown in Figure 6. The thermograms of TGA of all XTPU samples showed thermal stability and thermal degradation behavior. In an inert atmosphere, the thermal analysis data revealed that all of the XTPU polymer samples are thermally stable up to 140–225 °C. All the polymer samples faced a 10% loss of weight in the range of 305–355 °C. These samples faced a 20% loss of weight in the range of 372–410 °C. They also faced a 50% loss of weight in the range of 423–463 °C, and maximum decomposition observed in the samples occurred at 460–495 °C. The comparative results showed that the chain extender, i.e., BDO, and bioactive material, i.e., xanthan gum, play a vital role. The results illustrated that poly (ethylene glycol adipate)-based polyurethane [33] is thermally less stable compared to xanthan gum, TiO_2-based polyurethanes with BDO as a chain extender. The values of temperature showed that pure polymers have lower degradation temperature, i.e., 460 °C, while, at the same time, the polymers having 1% XG, 5% of TiO_2 have higher degradation temperature, i.e., 495 °C, showing degradation temperature variation about 35 °C. Results revealed that the percentage mole ratio of xanthan gum and TiO_2 enhances the degradation temperature of polyurethane more than pure polyurethane, i.e., XTPU-1. The chain extender, i.e., BDO, and xanthan gum increase the decomposition temperature and thermal stability of XTPU, as shown in Table 2.

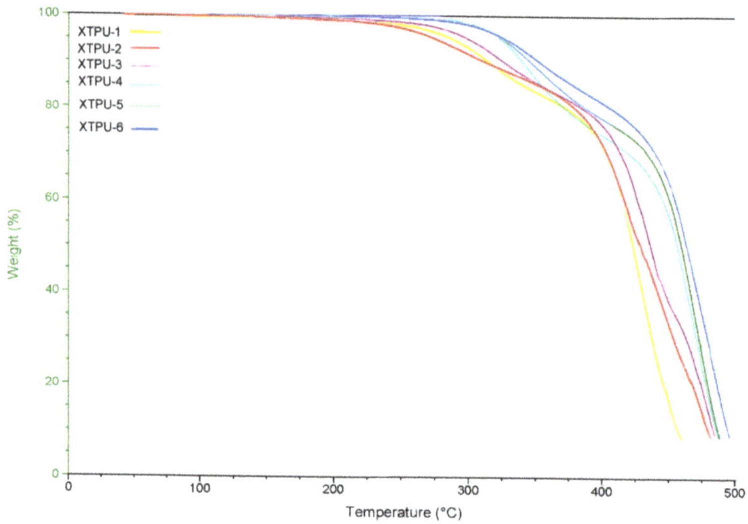

Figure 6. TGA thermogram of XTPUs at 500 °C.

Table 2. Thermal stability data of the XTPU samples based on TGA.

Sample Code	IPDI (mole)	HTPB (mole)	BDO (mole)	TiO_2 %	X.G %	T_0 (°C)	T_{20} (°C)	T_{50} (°C)	T_{80} (°C)	T_{max} (°C)
XTPU-1	2	0.8	1.2	0	0	140	372	423	448	460
XTPU-2	2	0.8	1.2	1	1	150	380	430	470	479
XTPU-3	2	0.8	1.2	2	1	162	385	435	475	485
XTPU-4	2	0.8	1.2	3	1	190	372	455	479	487
XTPU-5	2	0.8	1.2	4	1	210	390	457	482	489
XTPU-6	2	0.8	1.2	5	1	225	405	467	485	495

Temperature at which 0%, 20%, 50%, and 80% weight losses obtained from TGA. Maximum decomposition temperature obtained from TGA.

4.3.2. Differential Scanning Calorimetry (DSC) Study

The effect of chain extenders length, XG, and TiO_2-based polyurethanes samples were also studied by DSC measurements. The XTPU thermograms are shown in Figure 7. In the XTPU 1 to 6 samples, it is clear by DSC analysis that thermal changes took place when temperature was increased from 0 to 500 °C. It is clear from the data of Table 3 and from DSC thermograms that polymer degradation started in the range of 140–225 °C, which is also in accordance with TGA thermograms. The polymers samples showed glass transition temperature (Tg) in between 178–193 °C, crystallization temperature in the range 368–373 °C, melting temperature (Tm) in the range of 430–450 °C, and decomposition/degradation temperature in the range of 460–495 °C [34]. Here, it has been observed that the melting transition was found to be high near the decomposition temperature. This causes an arrangment of the chains and facilitates a greater interaction between chains, increasing the miscibility between the rigid and soft segment, as seen in TM and Tc, for all materials. The glass transition temperature observed at these values is relative to the rigid segment. The value of the Tg of the polyurethane sometimes also depends on the type of nanofiller (TiO_2) employed in the synthesis. This value is indicative of the soft and rigid segment mixing degee. The higher value of Tg usually represents the higher miscibilty or campatibity degree of rigid-flexible segments [35]. The results revealed that presence of a nanofiller, i.e., TiO_2, and its bonding between xanthan gum and diisocyanate, i.e., IPDI, increases the stability and crystalline behavior of the synthesized polymer. So, the crystillinity may be observed because of nanofiller. The most probable reason for this behavior in the nanocomposite is reduction in mobility of chains of urethanes, which reduces the process of degradation, as reported in literature [36]. This can also be related to the formation of longer crosslinked microdomains of rigid segments or the structure with a greater degree of organization. This trend agrees with the results obtained in the literature.

Table 3. Thermal stability data of the XTPU samples based on DSC.

Sample Code	IPDI (mole)	HTPB (mole)	BDO (mole)	TiO_2 %	X.G %	T_g (°C)	T_c (°C)	T_m (°C)	T_d (°C)
XTPU-1	2	0.8	1.2	0	0	178	368	430	460
XTPU-2	2	0.8	1.2	1	1	180	365	435	479
XTPU-3	2	0.8	1.2	2	1	183	370	435	485
XTPU-4	2	0.8	1.2	3	1	185	370	440	487
XTPU-5	2	0.8	1.2	4	1	191	371	447	489
XTPU-6	2	0.8	1.2	5	1	193	373	450	495

T_g is Glass transition temperature; T_c is Crystallization temperature; T_m is melting temperature; T_d is degradation/decomposition temperature.

4.4. X-ray Diffraction Study

The crystalline behavior of XTPU samples was calculated by using the crystalline peak intensity of respective samples. The Debye-Scherer (powder) method, applying Bragg's relation, was used to estimate the d-spacing of various XTPUs [37].

In XTPU samples, relative contents, structure regularity, and thermodynamic incompatibility affect the phase separation of soft, as well as hard, segment. A well oriented crystallinity was perceived at $2\theta = 20°$, as reported in literature [38]. The phase separation of the soft and hard segment in XTPUs is attributed to crystallinity of sample, and it was revealed by X-ray diffraction studies. So, crystalline behavior improved as % age of TiO_2 in the final XTPU increased. Finally, in the current study, we can attribute the crystallinity to the soft segment, and increasing percentage of TiO_2 did not show any appreciable change in polymer structure. The chemical cross-linking of xanthan gum restricts the soft segment melting. So, it can be determined that only XTPU-6, having 1% xanthan gum and 5% TiO_2, showed higher crystallinity, which has been associated with phase separation at $2\theta = 20°$, as shown in Figure 8.

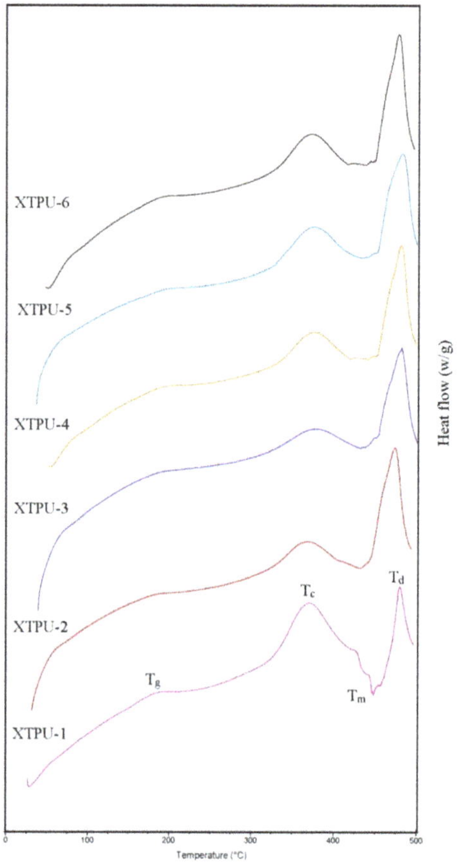

Figure 7. DSC curves of XTPU-1, 2, 3, 4, 5, and 6 at 500 °C.

4.5. Atomic Force Microscopy (AFM)

The change in surface morphology of XTPU samples was mainly investigated by AFM. It provides important information relating surface characterization. In the current study, the final morphology of XTPUs films is visualized in Figure 9. In these phase images, the darker regions correspond to soft segments, while lighter domain or crystalline regions correspond to hard segments [2]. The XTPUs can be visualized for the microphase structures as a roughness gradient of 2.5 µm, 158 nm, and 300 nm for XTPU-1, XTPU-3, and XTPU-6 respectively. The complete and better dispersion of nanofiller, also with biomaterial, was also observed within samples. These results showed (the circled region) homogeneous structure of polymer and pattern of distribution of xanthan gum with TiO_2 in the polyurethane polymer. AFM images of XTPUs have shown that hard segments are completely dispersed in the soft segment matrix. Moreover, it was inferred that, by increasing the content of TiO_2 from 0% to 5% in 6 polymer samples from XTPU-1 to XTPU-6, the order in pattern crystalline character and stability increases and cracking decreases in the polymer, according to reported literature [25]. Morover, in regard to polyurethane composition, it was also reported that hydrogen bonding was present in urethane groups, which may be the cause of phase segregation. The higher molecular weight of macrodiol and xanthan gum provided ordered and compact arrangments of soft and hard phases. However, various factors can be reponsible for phase segregation in PU, such as the diminsions of soft and hard segments, the polarity of groups, the chemical nature, and molecular weight [37,39].

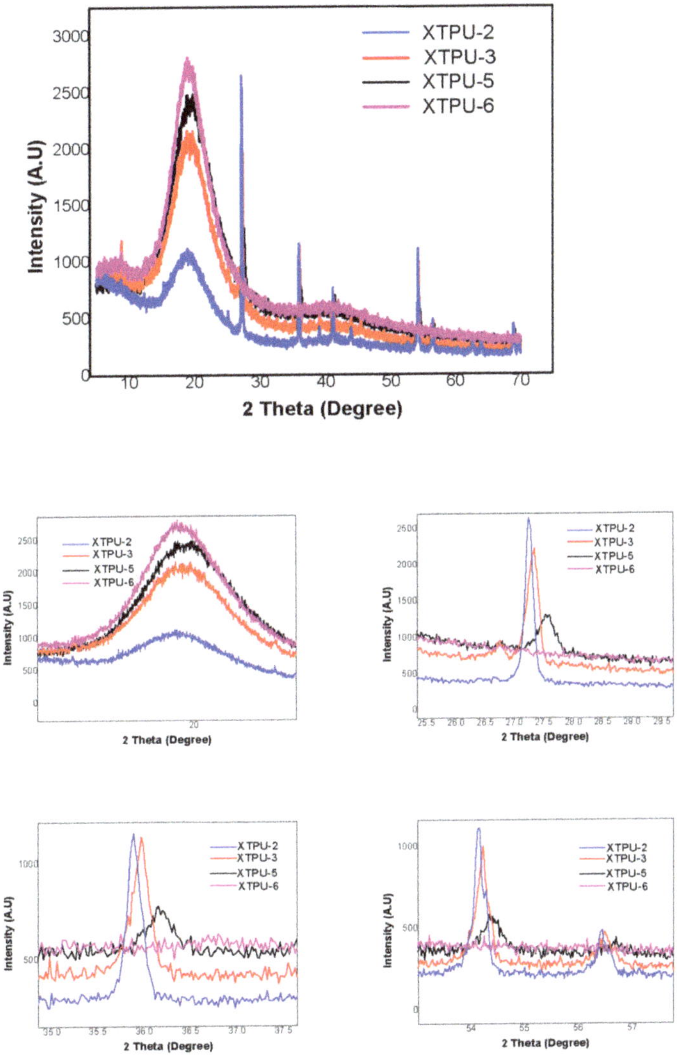

Figure 8. X-ray diffractograms of XTPU-2, XTPU-3, XTPU-5, and XTPU-6 with varying titanium dioxide % age.

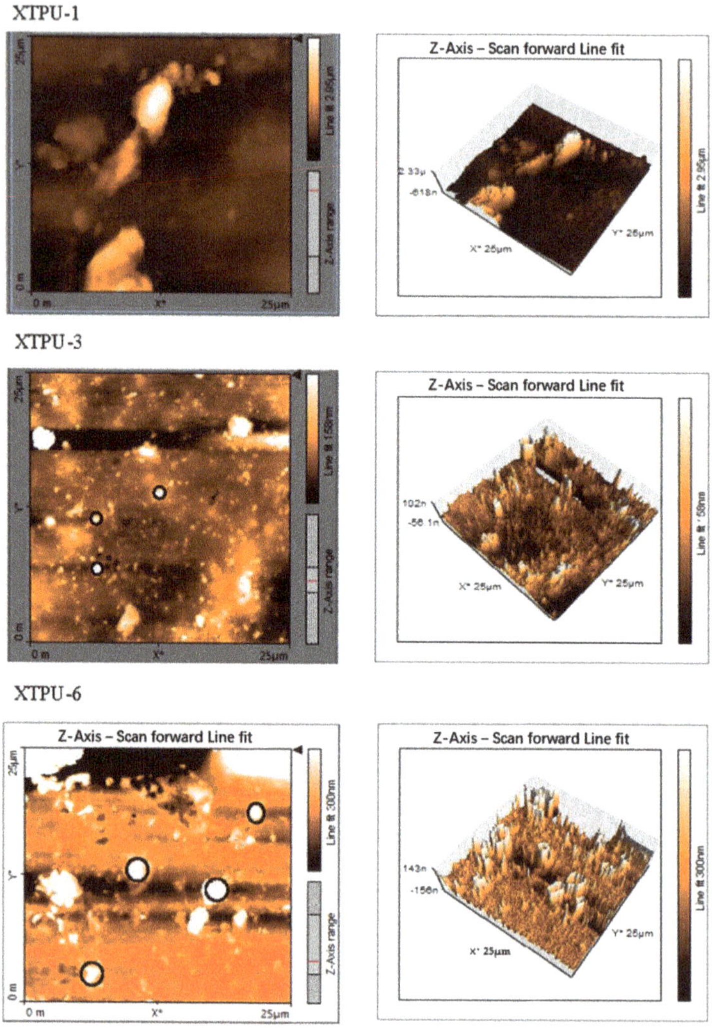

Figure 9. Three-dimensional AFM images of XTPU-1, XTPU-3, and XTPU-6 with 0%, 2%, and 5% titanium dioxide, respectively.

4.6. Evaluation of Water Absorption

The water absorption test and hydrophobicity/hydrophilicity of polyurethane samples are used for evaluation of biodegradability of synthesized polymer. The degree of swelling of polymer components and their affinity to water is one of the indices of decomposition rate of polymeric materials under the influence of environmental factors. Due to presence of soft segment HTPB and XG into the back bone, the prepared polymers are hydrolytic resistant and can be utilized to perform indoor and outdoor environments. So, water absorption capability can be used to determine the hydrolytic degradation of them. The investigation of the hydrolytic stability of XTPUs has shown an increase of the water absorption. Such an increase may be due to the presence of free polar fragments (OH groups), which determine the hydrophilic properties of the polymer. Probably, a part of XG hydrophilic hydroxyls involved in the formation of intermolecular bonds with the polar groups of XTPUs remains unengaged [40]. Water absorption as a function of time and type of samples is summarized in Table 4. Water absorption was calculated by the following formula:

$$\text{Water absorption} (\%) = \frac{m_w - m_d}{m_d} \times 100 \quad (2)$$

where m_d and m_w are the masses of dry and wet XTPU samples, accordingly [38]. There was a remarkable water absorption indicated with passage of time in these prepared samples of polyurethanes. The results noticeably indicated that the polymer samples are hydrophilic in nature [41].

Table 4. Percentage of water absorption.

Sr. No	Samples Code	Temperature (°C)	% Age of Water Absorption				
			Day 1	Day 2	Day 3	Day 4	Day 5
01	XTPU-1	37	0.94	0.95	0.98	1.10	1.21
02	XTPU-2	37	0.95	0.96	0.99	1.10	1.23
03	XTPU-3	37	0.95	0.98	1.10	1.21	1.31
04	XTPU-4	37	1.10	1.20	1.3	1.5	1.61
05	XTPU-5	37	1.80	1.85	1.90	1.95	1.99
06	XTPU-6	37	1.96	1.98	1.99	2.0	2.1

5. Conclusions

Xanthan gum/Titanium dioxide-based polyurethane elastomers with concordant crystallinity and hydrophilicity were synthezied. The reactants were HTPB, IPDI, and Titanium dioxide that produced prepolymer, extended with BDO. Xanthan gum was used as a bioactive material. Molecular characterization of the prepared elastomers was carried out by FT-IR spectroscopy, and the appearance of NH peaks with disappearance of NCO and OH peaks indicates that the proposed structure of polyurethane has been accomplished. The AFM technique revealed that the degree of micro-phase separation increases with augmenting % age of TiO_2, which was further confirmed by XRD results. It was revealed by XRD that the crystalline behavior of the synthesized samples due to amorphous region of macrodiol. The % age of TiO_2 influence on TGA and DSC values gave the information about the thermal stability and thermal changes in the polyurethane samples. The higher value of DSC is due to segments' mobility in the hard blocks' microregions (hard domain) and their destruction. Such changes may be caused by three-dimensional XG molecules, which form steric hindrence during polymer formation. Antimicrobial activity determined through the Disc Diffusion Method and the results indicated that the synthesized polyurethane had antimicrobial activity against the *Macrococcus*, *E. coli*, and *Aspergillus flavus*.

Author Contributions: Conceptualization, S.N. and S.H.; methodology, S.N. and M.S.; formal analysis, S.N. and N.R.; investigation, M.S. and S.N.; data curation, M.S., S.N. and N.R.; writing—original draft preparation, M.S.; writing—review and editing, M.S., S.N., Z.S. and R.Z.; supervision, S.N.; project administration, S.N., N.R. and S.K. All authors have read and agreed to the published version of the manuscript.

Funding: The authors sincerely appreciate funding from Researchers Supporting Project number (RSP-2021/399), King Saud University, Riyadh, Saudi Arabia.

Institutional Review Board Statement: Not applicable.

Informed Consent Statement: Not applicable.

Data Availability Statement: The data presented in this study are available on request from the corresponding author.

Conflicts of Interest: The authors declare no conflict of interest.

References

1. Zhang, Z.-G.; Chen, H.-Z. Xanthan production on polyurethane foam and its enhancement by air pressure pulsation. *Appl. Biochem. Biotech.* **2010**, *162*, 2244–2258. [CrossRef]
2. Klinedinst, D.B.; Yilgör, E.; Yilgör, I.; Beyer, F.L.; Wilkes, G.L. Structure–property behavior of segmented polyurethaneurea copolymers based on an ethylene-butylene soft segment. *Polymer* **2005**, *46*, 10191–10201. [CrossRef]
3. Hottle, T.A.; Bilec, M.M.; Landis, A.E. Sustainability assessments of bio-based polymers. *Polym. Degrad. Stab.* **2013**, *98*, 1898–1907. [CrossRef]
4. Huang, S.L.; Chao, M.S.; Ruaan, R.C.; Lai, J.Y. Microphase separated structure and protein adsorption of polyurethanes with butadiene soft segment. *Eur. Polym. J.* **2000**, *36*, 285–294. [CrossRef]
5. Ishihara, K.; Iwasaki, Y. Biocompatible elastomers composed of segmented polyurethane and 2-methacryloyloxyethyl phosphorylcholine polymer. *Polym. Adv. Technol.* **2000**, *11*, 626–634. [CrossRef]
6. Zia, K.M.; Zuber, M.; Bhatti, I.A.; Barikani, M.; Sheikh, M.A. Evaluation of biocompatibility and mechanical behavior of polyurethane elastomers based on chitin/1, 4-butane diol blends. *Int. J. Biol. Macromol.* **2009**, *44*, 18–22. [CrossRef] [PubMed]
7. Middleton, J.C.; Tipton, A.J. Synthetic biodegradable polymers as orthopedic devices. *Biomaterials* **2000**, *21*, 2335–2346. [CrossRef]
8. Wang, Y.; Hong, Q.; Chen, Y.; Lian, X.; Xiong, Y. Surface properties of polyurethanes modified by bioactive polysaccharide-based polyelectrolyte multilayers. *Colloids Surfaces B Biointerfaces* **2012**, *100*, 77–83. [CrossRef] [PubMed]
9. Zhou, B.; Hu, Y.; Li, J.; Li, B. Chitosan/phosvitin antibacterial films fabricated via layer-by-layer deposition. *Int. J. Boil. Macromol.* **2014**, *64*, 402–408. [CrossRef]
10. Szycher, M. *Szycher's Handbook of Polyurethanes*, 2nd ed.; CRC Press: Boca Raton, FL, USA, 2012.
11. Gaidukova, G.; Ivdre, A.; Fridrihsone, A.; Verovkins, A.; Cabulis, U.; Gaidukovs, S. Polyurethane rigid foams obtained from polyols containing bio-based and recycled components and functional additives. *Ind. Crop. Prod.* **2017**, *102*, 133–143. [CrossRef]
12. Kuan, H.; Ma, C.-C.M.; Chang, W.-P.; Yuen, S.-M.; Wu, H.-H.; Lee, T.-M. Synthesis, thermal, mechanical and rheological properties of multiwall carbon nanotube/waterborne polyurethane nanocomposite. *Compos. Sci. Technol.* **2005**, *65*, 1703–1710. [CrossRef]
13. Zia, K.M.; Bhatti, H.N.; Bhatti, I.A. Methods for polyurethane and polyurethane composites, recycling and recovery: A review. React. *Funct. Polym.* **2007**, *67*, 675–692. [CrossRef]
14. Hu, Y.; Liu, C.; Shang, Q.; Zhou, Y. Synthesis and characterization of novel renewable castor oil-based UV-curable polyfunctional polyurethane acrylate. *J. Coatings Technol. Res.* **2017**, *15*, 77–85. [CrossRef]
15. Nair, L.S.; Laurencin, C.T. Biodegradable polymers as biomaterials. *Prog. Polym. Sci.* **2007**, *32*, 762–798. [CrossRef]
16. Gorna, K.; Gogolewski, S. Molecular stability, mechanical properties, surface characteristics and sterility of biodegradable polyurethanes treated with low-temperature plasma. *Polym. Degrad. Stab.* **2003**, *79*, 475–485. [CrossRef]
17. Barikani, M.; Hepburn, C. The relative thermal stability of polyurethane elastomers. II: Influence of polyol-diisocyanate molar block ratios with a single and mixed diisocyanate system. *Cell. Polym.* **1987**, *6*, 29–36.
18. Tant, M.R.; McManus, H.L.N.; Rogers, M.E. *High-Temperature Properties and Applications of Polymeric Materials*; ACS Publications: Washington, DC, USA, 1995; pp. 1–20. [CrossRef]
19. Bhattacharyya, A.; Mukherjee, D.; Mishra, R.; Kundu, P. Preparation of polyurethane–alginate/chitosan core shell nanoparticles for the purpose of oral insulin delivery. *Eur. Polym. J.* **2017**, *92*, 294–313. [CrossRef]
20. Freitas, F.; Alves, V.D.; Reis, M.A. Advances in bacterial exopolysaccharides: From production to biotechnological applications. *Trends Biotech.* **2011**, *29*, 388–398. [CrossRef] [PubMed]
21. Gultekin, G.; Atalay-Oral, C.; Erkal, S.; Sahin, F.; Karastova, D.; Tantekin-Ersolmaz, S.B.; Guner, F.S. Fatty acid-based polyurethane films for wound dressing applications. *J. Mater. Sci. Mater. Electron.* **2008**, *20*, 421–431. [CrossRef] [PubMed]
22. Usuki, A.; Kojima, Y.; Kawasumi, M.; Okada, A.; Fukushima, Y.; Kurauchi, T.; Kamigaito, O. Synthesis of nylon 6-clay hybrid. *J. Mater. Res.* **1993**, *8*, 1179–1184. [CrossRef]
23. Yurekli, K.; Karim, A.; Amis, E.J.; Krishnamoorti, R. Influence of Layered Silicates on the Phase-Separated Morphology of PS−PVME Blends. *Macromolecules* **2003**, *36*, 7256–7267. [CrossRef]
24. Jabbari, E.; Khakpour, M. Morphology of and release behavior from porous polyurethane microspheres. *Biomaterials* **2000**, *21*, 2073–2079. [CrossRef]
25. Naheed, S.; Zuber, M.; Barikani, M.; Salman, M. Molecular engineering and morphology of polyurethane elastomers containing various molecular weight of macrodiol. *Mater. Sci. Eng. B* **2020**, *264*, 114960. [CrossRef]
26. *Standard Test Method for Isocyanate Groups in Urethane Materials or Prepolymers*; ASTM D2572-80; ASTM: West Conshohocken, PA, USA, 2010.
27. Yilgör, I.; Yilgor, E.; Das, S.; Wilkes, G.L. Time-dependent morphology development in segmented polyetherurea copolymers based on aromatic diisocyanates. *J. Polym. Sci. Part B Polym. Phys.* **2009**, *47*, 471–483. [CrossRef]
28. Hsu, S.-H.; Tseng, H.-J.; Lin, Y.-C. The biocompatibility and antibacterial properties of waterborne polyurethane-silver nanocomposites. *Biomaterials* **2010**, *31*, 6796–6808. [CrossRef]
29. Yilgor, I.; Yilgor, E.; Guler, I.G.; Ward, T.C.; Wilkes, G.L. FTIR investigation of the influence of diisocyanate symmetry on the morphology development in model segmented polyurethanes. *Polymer* **2006**, *47*, 4105–4114. [CrossRef]
30. Naz, F.; Zuber, M.; Zia, K.M.; Salman, M.; Chakraborty, J.; Nath, I.; Verpoort, F. Synthesis and characterization of chitosan-based waterborne polyurethane for textile finishes. *Carbohydr. Polym.* **2018**, *200*, 54–62. [CrossRef]

31. Tabasum, S.; Zuber, M.; Jamil, T.; Shahid, M.; Hussain, R. Antimicrobial and pilling evaluation of the modified cellulosic fabrics using polyurethane acrylate copolymers. *Int. J. Biol. Macromol.* **2013**, *56*, 99–105. [CrossRef]
32. Barrioni, B.R.; de Carvalho, S.M.; Oréfice, R.L.; de Oliveira, A.A.R.; Pereira, M.D.M. Synthesis and characterization of biodegradable polyurethane films based on HDI with hydrolyzable crosslinked bonds and a homogeneous structure for biomedical applications. *Mater. Sci. Eng. C* **2015**, *52*, 22–30. [CrossRef]
33. Azzam, R.A.; Mohamed, S.K.; Tol, R.; Everaert, V.; Reynaers, H.; Goderis, B. Synthesis and thermo-mechanical characterization of high performance polyurethane elastomers based on heterocyclic and aromatic diamine chain extenders. *Polym. Degrad. Stab.* **2007**, *92*, 1316–1325. [CrossRef]
34. Malakpour, S.E.; Rostamizadeh, H. Synthesis of novel polyurethanes with Fluorescein Linkages. *Iran. Polym. J.* **1999**, *8*, 099V.
35. Travinskaya, T.V.; Brykova, A.N.; Savelyev, Y.V.; Babkina, N.V.; Shtompel, V.I. (Bio) degradable ionomeric polyurethanes based on xanthan: Synthesis, properties, and structure. *Inter. J. Polym Sci.* **2017**, *103*, 903–908. [CrossRef]
36. Chen, M.; Zhou, D.-L.; Chen, Y.; Zhu, P.-X. Analyses of structures for a synthetic leather made of polyurethane and microfiber. *J. Appl. Polym. Sci.* **2006**, *103*, 903–908. [CrossRef]
37. Naheed, S.; Zuber, M.; Salman, M.; Rasool, N.; Siddique, Z.; Shaik, M.; Sharaf, M.; Abdelgawad, A.; Sekou, D.; Awwad, E. Impact of Macrodiols on the Morphological Behavior of H12MDI/HDO-Based Polyurethane Elastomer. *Polymers* **2021**, *13*, 2060. [CrossRef]
38. Yang, C.-H.; Liu, F.-J.; Liu, Y.-P.; Liao, W.-T. Hybrids of colloidal silica and waterborne polyurethane. *J. Colloid Interface Sci.* **2006**, *302*, 123–132. [CrossRef]
39. da Silva, V.D.; dos Santos, L.M.; Subda, S.M.; Ligabue, R.; Seferin, M.; Carone, C.L.P.; Einloft, S. Synthesis and characterization of polyurethane/titanium dioxide nanocomposites obtained by in situ polymerization. *Polym. Bull.* **2013**, *70*, 1819–1833. [CrossRef]
40. Asensio, M.; Costa, V.; Nohales, A.; Bianchi, O.; Gómez, A.C.M. Tunable Structure and Properties of Segmented Thermoplastic Polyurethanes as a Function of Flexible Segment. *Polymers* **2019**, *11*, 1910. [CrossRef]
41. Naheed, S.; Zuber, M.; Barikani, M. Synthesis and thermo-mechanical investigation of macrodiol-based shape memory polyurethane elastomers. *Int. J. Mater. Res.* **2017**, *108*, 515–522. [CrossRef]

polymers

MDPI

Article

cis-1,4 Selective Coordination Polymerization of 1,3-Butadiene and Copolymerization with Polar 2-(4-Methoxyphenyl)-1,3-butadiene by Acenaphthene-Based α-Diimine Cobalt Complexes Featuring Intra-Ligand π-π Stacking Interactions

Beibei Wang [1,2], Heng Liu [3,*], Tao Tang [1,2] and Xuequan Zhang [3]

1 Changchun Institute of Applied Chemistry, Chinese Academy of Sciences, 5625 Renmin Street, Changchun 130022, China; bbwang@ciac.ac.cn (B.W.); ttang@ciac.ac.cn (T.T.)
2 Changchun Institute of Applied Chemistry, University of Science and Technology of China, Hefei 230026, China
3 Key Laboratory of Rubber-Plastics, Ministry of Education/Shandong Provincial Key Laboratory of Rubber-Plastics, Qingdao University of Science & Technology, Qingdao 266042, China; xqzhang@qust.edu.cn
* Correspondence: hengliu@ciac.ac.cn; Tel.: +86-431-85262305

Abstract: Highly *cis*-1,4 selective (up to 98%) coordination–insertion polymerization of 1,3-butadiene (BD) has been achieved herein using acenaphthene-based α-diimine cobalt complexes. Due to the presence of intra-ligand π-π stacking interactions, the complexes revealed high thermostability, affording polybutadiene products in high yields. Moreover, all of the obtained polymers possessed a relatively narrow molecular weight distribution as well as high molecular weight (up to 92.2×10^4 Dalton). The molecular weights of the resultant polybutadienes could be finely tuned by varying polymerization parameters, including temperature, Al/Co ratio, etc. Moreover, the copolymerization of butadiene with polar monomer 2-(4-methoxyphenyl)-1,3-butadiene (2-MOPB) was also successfully realized to produce a type of polar *cis*-1,4 polybutadiene (*cis*-1,4 content: up to 98.1%) with a range of 2-MOPB content (0.46–1.83%). Water contact angle measurements indicated that the insertion of a polar monomer into a polymer chain could significantly improve the polymer's surface property.

Keywords: *cis*-1,4 selective; 1,3-butadiene; α-diimine cobalt complexes; copolymerization; π-π stacking interactions

Citation: Wang, B.; Liu, H.; Tang, T.; Zhang, X. *cis*-1,4 Selective Coordination Polymerization of 1,3-Butadiene and Copolymerization with Polar 2-(4-Methoxyphenyl)-1,3-butadiene by Acenaphthene-Based α-Diimine Cobalt Complexes Featuring Intra-Ligand π-π Stacking Interactions. *Polymers* **2021**, *13*, 3329. https://doi.org/10.3390/polym13193329

Academic Editor: Edina Rusen

Received: 26 August 2021
Accepted: 22 September 2021
Published: 29 September 2021

Publisher's Note: MDPI stays neutral with regard to jurisdictional claims in published maps and institutional affiliations.

Copyright: © 2021 by the authors. Licensee MDPI, Basel, Switzerland. This article is an open access article distributed under the terms and conditions of the Creative Commons Attribution (CC BY) license (https://creativecommons.org/licenses/by/4.0/).

1. Introduction

Hydrocarbon polymers, which mainly include polyolefins and polydienes, have been widely used in modern human life because of their excellent chemical and physical properties in combination with their low cost, superior processability, and good recyclability [1–3]. However, due to the lack of polar functional groups, most hydrocarbon polymers exhibit very poor surface properties, such as low adhesive properties, low compatibility with polar fillers, etc., which greatly limits their applications in a broader scope [4–6]. Copolymerization with polar monomers to access functionalized hydrocarbon polymers is the most direct and economic strategy to solve the above issue; nevertheless, most catalytic systems have extremely high oxophilic active species that are able to be decomposed by heteroatoms (such as N, O, S, or other polar groups) due to their strong and irreversible coordination to the metal center. In 1995, Brookhart et al. disclosed a family of α-diimine late transition metal-based catalytic systems [7,8]. Owing to the low oxophilic nature of the metal centers and the properties of α-diimine coordination ligand that are able to be regulated exactly, a catalytic system that shows a high heteroatom tolerance and that can directly catalyze the copolymerization of olefins with polar comonomers, has created a new field aiming to access functionalized polyolefins, and colossal advances have been made this field in the past few years [9–12]. Similar to polar polyolefin materials, functionalized polydienes elastomers are also of great significance because the incorporated functional groups can greatly improve the compatibility between the polydienes matrix and reinforce polar fillers, such as carbon black, silica, etc., which therefore produces well-performing elastomers with good modulus, tensile strength, and other dynamic mechanical properties [13]. However, research n copolymerization with polar copolymers in this field o is very limited [14,15]. In the present research, a series copolymerizations of 1,3-butadiene with polar comonomer 2-(4-methoxyphenyl)-1,3-butadiene (2-MOPB) is conducted, and the corresponding catalytic performances and microstructures of the resulting copolymers are discussed.

Despite of the inherent advantage of late-transition metal complexes for the direct preparing of polar copolymers, they still currently suffer from some shortages, and one of them is the thermal stability of the active species. Currently, the specific high-temperature decomposition mechanism for late-transition metal catalysts is still unclear, yet one widely accepted explanation ascribes it to the C-H activation that occurs between the metal center and alkyl groups located on the ortho-position of the *N*-aryl fragment after it rotates to the coordination plane [16]. Therefore, suppressing *N*-aryl rotations is believed to be one of the most efficient strategies to enhance the thermal stability of these complexes, and in the past few years, various intra-ligand covalent and non-covalent interactions, e.g., hydrogen bonding, π,π-stacking, etc., have been introduced to ligand design, aiming to fasten the *N*-aryl moiety and to subsequently prohibit its rotation [17–19]. In the present research, ortho-substituted dibenzhydryl groups were introduced to the *N*-aryl moiety of α-diimine cobalt complexes, and surprisingly, it was found that one of the aryl groups in dibenzhydryl could form π,π-interaction with the acenaphthenyl backbone, which subsequently endows the complexes with excellent thermal stability. Detailed research discussing these findings will be presented in the following sections.

2. Materials and Methods

2.1. General Considerations and Materials

All reactions were conducted under a dry and oxygen-free argon atmosphere using Schlenk techniques or under a nitrogen atmosphere in an MBraun glovebox (MBraun, Garching bei München, Germany). Toluene for polymerization was dried with sodium and was distilled under the nitrogen atmosphere. Dichloromethane and diethyl ether were also dried with calcium hydride and were distilled under the nitrogen atmosphere. EASC (Aladdin, Shanghai, China) was purchased and then diluted in a 2 M toluene solution. Catalysts were dissolved in toluene and were added to the polymerization vessel with the injector. Other commercially available reagents were purchased and were used without purification. Ligands **L1–L4** were synthesized following the procedures reported upon in the literature [17]. The compound 2-MOPB was also prepared according to previous reports [11].

NMR spectra of the organic compounds were recorded on a Bruker AV400 (Bruker Optics, Ettlingen, Germany) (FT, 400 MHz for ^{1}H; 100 MHz for ^{13}C) spectrometer at room temperature with tetramethylsilane (TMS; $\delta = 0$ ppm) as an internal reference, and then the NMR spectra of copolymers were measured at a room temperature using a Varian Unity 400 MHz spectrometer (Varian, Inc., Palo Alto, CA, USA). Polymer characteristics were investigated on a TOSOH HLC-8220 GPC instrument (Tosoh Asia Pte. Ltd., Tokyo, Japan) at 40 °C with polystyrene as a standard. Tetrahydrofuran (THF) was used as the eluent at a flow rate of 1.0 mL min^{-1}. The polymer microstructure analysis was performed using a Nicolet Is10 FTIR (Thermo fisher, Waltham, MA, USA). Copolymer water contact angles were measured using a JY-PHB contact angle meter (YouTe, Chengde, China).

The crystal of **Co1** was obtained by hexane diffusion to its saturated dichloromethane solution, while the crystal of **L2** precipitated from its saturated dichloromethane/hexane mixed solution at room temperature. All of the crystal performances took place in the nitrogen atmosphere, and data collection took place at a low temperature on a Bruker SMART APEX diffractometer (Bruker Optics, Germany) with a CCD area detector using graphite monochromated Mo Kα radiation (λ = 0.71073 Å). The raw frame data were processed using SAINT (Bruker Optics, Germany) and SADABS (Bruker Optics, Germany) to yield the reflection data file. The crystal class characteristics and unit cell parameters were performed with the SMART program package. The structures were determined using the SHELXTL program. The hydrogen atoms were located at the calculated positions and were included in the structure calculation without further refinement of the parameters. CCDC numbers for **Co1** and **L1**: 2095433-2095434.

2.2. *Typical Procedure for Butadiene Homopolymerization and Copolymerization*

Polymerizations of 1,3-butadiene were performed by introducing butadiene (0.1 g/mL in toluene) and EASC solution into 20 mL glass flasks equipped with a magnetic stirrer to mix the solution. At the setting polymerization temperature, a cobalt complex toluene solution was injected. Then, the polymerization was initiated and performed for a designated amount of time. After the polymerization was finished, acidified MeOH (HCl, 10 wt%) and antioxidant were added, and the polymer was collected and purified by pouring ethanol (150 mL) into a flask to precipitate. The polymer was dried at 40 °C in a vacuum oven to a constant weight. For the copolymerization of 2-MOPB and butadiene, the same procedure was performed except for that the mixed monomer of the evaluated amount of 2-MOPB and butadiene was added to the glass flasks for polymerization.

2.3. *Synthesis of Complexes (**Co1–Co4**)*

2.3.1. Synthesis of Complex **Co1**

To a mixture of **L1** (240 mg, 0.32 mmol) and $CoCl_2$ (42 mg, 0.32 mmol), dichloromethane (20 mL) was added at room temperature. Then, the mixture was stirred overnight at room temperature. The solvent was removed until 2 mL were left. An amount of 50 mL hexane was added while the mixture was being stirred. The mixture was filtered, and the residue was washed with hexane (3 × 10 mL) and was dried under vacuum; the pure target product (197 mg, 0.22 mmol) was collected as a brown solid at a 69.9% yield. FTIR (KBr, cm^{-1}): 3057(w), 3024(w), 2958(w), 2920(w), 2853(w), 1652($v_{C=N}$, s), 1466(m), 1450(m), 1383(w), 1355(w), 1317(w), 1204(w), 1157(w), 1095(w), 1025(w), 862(m). MS (ESI): m/z calcd. for $C_{56}H_{48}Cl_2CoN_2$, $[M]^+$ 878.83, found $[M\text{-}Cl]^+$ 842.3.

2.3.2. Synthesis of Complex **Co2**

The procedure was similar to the synthesis of **Co2**. Yield: 215 mg, 0.23 mmol, (yield: 73.0%). FTIR (KBr, cm^{-1}): 3058(w), 3025(w), 2959(w), 2920(w), 2853(w), 1647($v_{C=N}$, s), 1469(m), 1452(m), 1383(w), 1357(w), 1290(w), 1174(w), 1089(w), 1031(w), 1016 (w), 862(m). MS (ESI): m/z calcd. For $C_{54}H_{42}Cl_4CoN_2$, $[M]^+$ 919.67, found $[M\text{-}Cl]^+$ 884.2.

2.3.3. Synthesis of Complex **Co3**

The procedure was similar to the synthesis of **Co3**. Yield: 149 mg, 0.16 mmol, (yield: 51.2%). FTIR (KBr, cm^{-1}): 3059(w), 3026(w), 2959(w), 2921(w), 2852(w), 1648($v_{C=N}$, s), 1452(m), 1440 (m), 1383(w), 1204(w), 1177(w), 1092(w), 1033(w), 863(m). MS (ESI): m/z calcd. for $C_{56}H_{48}Cl_2CoN_2O_2$, $[M]^+$ 910.83, found $[M\text{-}Cl]^+$ 874.2.

2.3.4. Synthesis of Complex **Co4**

The procedure was similar to the synthesis of **Co4**. Yield: 235 mg, 0.25 mmol, (yield: 77.3%). FTIR (KBr, cm^{-1}): 3054(w), 3025(w), 2956(w), 2920(w), 2854(w), 1650($v_{C=N}$, s), 1451(m), 1437(m), 1383(w), 1364(w), 1318(w), 1203(w), 1076(w), 1033(w), 862(m). MS (ESI): m/z calcd. for $C_{62}H_{48}Cl_2CoN_2$, $[M]^+$ 950.9, found $[M\text{-}Cl]^+$ 915.3.

3. Results

3.1. *Synthesis and Characterization of the Cobalt Complexes*

Ligands **L1–L4** were synthesized by the condensation reaction between ortho-substituted aniline and acenaphthene according to the reported method [20,21]. Their corresponding cobalt complexes **Co1–Co4** could subsequently be accessed in high yields by directly treating **L1–L4** with a stoichiometric amount of $CoCl_2$ (Scheme 1) in dichloromethane and with subsequent precipitation by diethyl ether. All of the complexes **Co1–Co4** were well characterized by FTIR and mass spectra. In the FTIR spectra, the stretching vibration absorptions of the C=N groups in complexes **Co1–Co4** appeared at 1647.9–1652.7 cm^{-1}, which was shifted slightly compared with their counterparts in the free ligands, indicating good coordination between the α-diimine nitrogen atoms and the metal cobalt. Single crystal structures of ligand **L2** and complex **Co1** were further characterized by single crystal X-ray crystallographic analysis, and their structures are shown in Figures 1 and 2. As expected, a distorted tetrahedron geometry was observed for complex **Co1**, and the bond angles and bond distances in **L2** and **Co1** are typical for previously reported α-diimine Co(II) complexes [22–24].

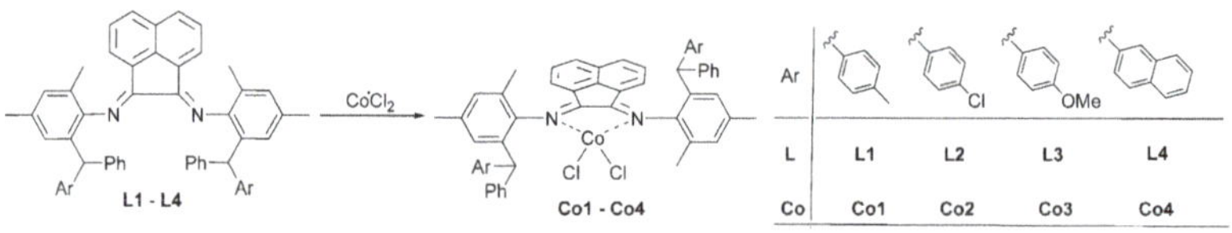

Scheme 1. Synthetic procedure for cobalt complexes **Co1–Co4**.

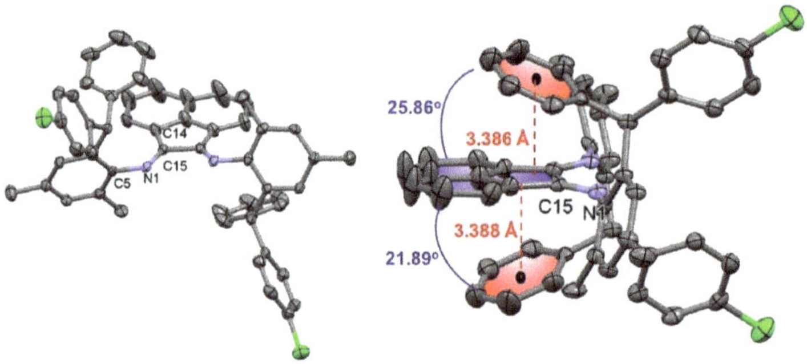

Figure 1. ORTEP drawing of complex L2 with thermal displacement parameters at 50% probability. Bond lengths (Å): N1-C15, 1.268(5), N1-C5, 1.420(5). Bond angles (deg): C5-N1-C15, 122.97(3).

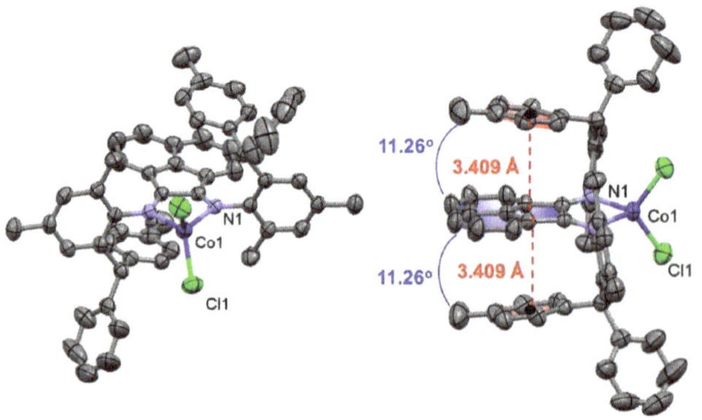

Figure 2. ORTEP drawing of complex Co1 with thermal displacement parameters at 50% probability. Bond lengths (Å): Co1-Cl1, 2.202(8), Co1-N1, 2.075(2), N1-C1, 1.279(3) N1-C8, 1.442(3). Bond angles (deg): Cl1-Co1-Cl1, 114.69(5), N1-Co1-Cl1, 117.29(6), N1-Co1-N1, 81.87(11), C1-N1-Co1, 110.81(16), C1-N1-C8, 121.26(19), C8-N1-Co1, 127.79(14).

For the ligand **L2**, both imine groups revealed *E,E*-configurations, which were different from some of the previous literature, which stated that *E,Z*-configurations were adopted [25–28]. Such a difference was probably due to the intra-ligand π,π-interactions that were found between the acenaphthenyl plane and the phenyl rings on the dibenzhydryl groups, which favored the *E,E*-configuration as the more thermodynamically stable state. As seen from the illustrative structure shown in Figure 1 (right), the two phenyl rings on the dibenzhydryl groups (colored into red) connected to the ortho-position of the aniline fragments were nearly parallel with the acenaphthenyl plane (colored into blue), revealing dihedral angles of only 25.85° and 21.89°, respectively. The distances from the center of the two phenyl rings to the acenaphthenyl plane are also very short, with values of 3.386 Å and 3.388 Å, respectively. The nearly parallel structure as well as the short distances clearly indicate the presence of intra-ligand π,π-interactions.

Such a π,π-interaction can be also clearly observed in the complex **Co1**. Similar to **L2**, the distances from the centers of the two phenyl rings (colored into red) on the dibenzhydryl groups to the acenaphthenyl plane (colored into blue) are also very short, producing values of 3.408 Å (due to the symmetric structure, the two distances are identical) (Figure 2). The dihedral angles between the two phenyls and the acenaphthenyl plane are even smaller, producing values of only 11.26°. Such a smaller dihedral angle indicates a stronger π,π-interaction in the complex than the one in the ligand. This result makes sense when considering the electrons flowing from the acenaphthenyl group to the imine groups after coordination with the cobalt metal, which renders acenaphthenyl more electron-deficient and therefore produced a stronger interaction with the electron-rich phenyl rings on the dibenzhydryl substituents [18,19].

3.2. 1,3-Butadiene Homopolymerization

Cobalt complexes have previously been explored as efficient precatalysts for 1,3-diene polymerization [29–32]; nevertheless, most of them suffer from poor thermostabilities, i.e., at high temperatures (generally higher than 50 °C), the active species deactivated fast and produced polybutadiene products at rather low yields. Until now, no relative literature has put forward a specific reason for this; however, inspired by late-transition metal-mediated ethylene polymerization, the active species could be deactivated by C-H activation between the metal center and the ortho-substituted alkyl groups on the aniline fragments [16]. Suppressing such C-H activation through the rotation restrain of *N*-aryls has previously been proven as an effective strategy to enhance the thermostabilities of late-transition metal-based catalysts [17,33]. For the present cobalt complexes, due to the presence of intra-ligand π,π-interactions, such *N*-aryl rotations and thereby C-H activations could be also suppressed, resulting in the anticipation of improved thermostabilities. Based on the above consideration, the present cobalt complexes were subsequently explored for 1,3-butadiene polymerization at various temperatures.

All of the 1,3-butadiene homopolymerizations employed ethylaluminum sesquichloride (EASC) as a cocatalyst because in our previous reports, EASC was determined to be one of the most efficient cocatalysts. Before exploring the thermal catalytic performances, the cocatalyst EASC/Co ratio was first optimized. As per the data shown in Table 1, with the increment of EASC loading, gradually increasing catalytic activities were witnessed for **Co1** mediated systems, which was consistent with previous reports. Moreover, monotonously decreasing molecular weights of the resultant polybutadienes were revealed, indicating the facilitated polybutadienyl chain transfer reactions from the propagating cobalt active species to aluminum alkyl compounds as well as increased activation and formation of catalytic species when the EASC equivalents increased. At EASC/Co = 300, the resulting polymer possesses a similar molecular weight to a commercial polybutadiene counterpart that has a Mooney viscosity of approximately 45; therefore, this condition was applied to subsequent studies. It is worthy of noting that the *cis*-1,4- contents of the polybutadienes almost remained completely unchanged when varying the Al/Co ratios, indicating their subtle influences on the regioselectivity of the active species.

Table 1. Polymerization of 1,3-butadiene by cobalt complexes Co1–Co4.

Entry [a]	Cat.	EASC/[Co]	T_{rxn} (°C)	Yield (g)	Activity [b]	Microstructure [c]			M_n [d]	M_w/M_n [d]
						trans-1,4 %	cis-1,4 %	1,2 %		
1	**Co1**	50	30	0.33	3.3	2.4	95.3	2.3	57.2	2.17
2	**Co1**	100	30	0.42	4.2	1.6	95.8	2.6	35.7	2.04
3	**Co1**	300	30	0.62	6.2	1.4	94.1	4.5	13.4	2.06
4	**Co1**	600	30	0.81	8.1	1.2	95.8	3.0	6.8	2.21
5	**Co1**	300	0	0.06	0.6	2.0	97.0	1.0	77.2	1.86
6	**Co1**	300	50	0.64	6.4	4.6	93.6	1.8	19.2	8.56
7	**Co1**	300	70	— [f]	—	—	—	—	—	—
8	**Co2**	300	0	0.04	0.4	2.8	95.7	1.5	92.2	1.64
9	**Co2**	300	30	0.83	8.3	1.5	94.9	3.6	35.0	2.18
10	**Co2**	300	50	0.45	4.5	4.6	95.4	1.7	2.9	4.43
11	**Co2**	300	70	— [f]	—	—	—	—	—	—
12	**Co3**	300	0	0.02	0.2	2.0	96.8	1.2	91.8	1.77
13	**Co3**	300	30	0.38	3.8	1.3	96.2	2.5	41.0	1.84
14	**Co3**	300	50	0.33	3.3	2.9	95.3	1.5	3.2	2.45
15	**Co3**	300	70	— [f]	—	—	—	—	—	—
16	**Co4**	300	0	0.17	1.7	1.0	98.1	0.9	36.2	2.07
17	**Co4**	300	30	0.83	8.3	4.3	91.4	4.3	7.6	1.88
18	**Co4**	300	50	0.83	8.3	5.9	90.0	4.1	7.8	6.22
19	**Co4**	300	70	0.78	7.8	9.3	87.4	3.3	5.7	8.94
20	**Co4**	300	90	0.70	7.0	9.5	86.9	3.6	2.3	3.86
21 [e]	**Co4**	300	50	0.83	99.6	4.3	85.3	13.0	3.7	1.54

[a] General conditions: solution, toluene = 8.3 mL, (BD) = 1.85 M, Co complex = 5 μmol. (BD):(Co) = 3000, polymerization time = 2 h. [b] 10^4 g of PBD (mol of Co)$^{-1}$ h^{-1}. [c] Microstructure determined by ^{1}H NMR in $CDCl_3$ and IR. [d] M_n: 10^4 g mol^{-1}, M_n and M_w/M_n determined using GPC. [e] Polymerization time was 10 min. [f] Not detected.

The thermal stabilities of the complexes **Co1–Co4** were next investigated systematically by conducting polymerization at various temperatures. As per the data summarized in Table 2, their thermal catalytic performances were highly dependent on the ligand structures. For complex **Co1**, its catalytic activity increased dramatically from 0.6 × $10^{4.1}$ to 6.4 × 10^4 g PBD (mol of Co)$^{-1}$ h^{-1} when the temperature was elevated from 0 °C to 50 °C; nevertheless, further temperature increases led to a completely deactivated system, suggesting the poor stability of the corresponding active species. Similar results were observed for complexes **Co2** and **Co3**: their highest catalytic activities were achieved at 30 °C; however, elevating the temperature resulted in quickly deactivated systems, and at 70 °C, no polymer products were obtained. In contrast to **Co1–Co3**, outstanding high catalytic efficiency as well as high thermostability were revealed from **Co4**-mediated polymerizations. As per the data shown in Table 1 and Figure 3, all of the 1,3-butadiene monomers were converted to polymer products when conducting the polymerization at 30 °C with a catalytic activity of 8.3 × 10^4 g PBD (mol of Co)$^{-1}$ h^{-1}; increasing the temperature to 50 °C had little influence on the polymer yields, indicating the much better thermostability of the active species than the other three analogs. To our satisfaction, **Co4** also afforded high polymer yields at high temperatures between 70 and 90 °C, at which point, most of the late transition metal complexes had been deactivated. These outstanding catalytic efficiencies of **Co4** at elevated temperatures were probably due to the enhanced π,π-interaction between the electron-rich naphthyl group with an acenaphthenyl backbone that could suppress the rotation of the *N*-aryls at a high temperature; for the other three complexes, however, the π,π-interaction interaction was comparatively weaker, and the stacked structure would have decomposed at high temperatures, which therefore could not guarantee the thermostability of the active species. Another thing worthy of note is that quite a high reaction rate was revealed from the **Co4**-mediated systems; when implementing polymerization at 50 °C, the monomer could be fully converted in only 10 min, demonstrating high catalytic activity of 99.6 × 10^4 g PBD (mol of Co)$^{-1}$ h^{-1}.

Table 2. The copolymerization of 2-MOPB with BD using **Co4**.

Entry [a]	Cat.	[2-MOPB]/[Co]	Yield (g)	Activity [b]	Inc. (mol %)	Microstructure [c]			M_n [d]	M_w/M_n [d]
						trans-1,4 %	*cis*-1,4 %	1,2 %		
1	Co4	5:1	0.58	5.8	0.46	10.1	85.5	4.4	28.4	2.23
2	Co4	10:1	0.46	4.6	0.53	11.4	85.4	3.2	17.4	2.07
3	Co4	20:1	0.44	4.4	0.66	6.9	90.2	2.8	12.9	2.02
4	Co4	50:1	0.40	4.0	1.83	9.4	88.4	2.2	5.4	1.97

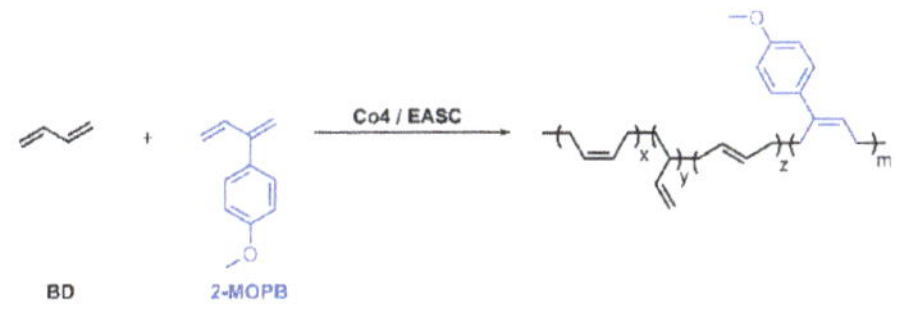

[a] General conditions: solution, toluene = 8.3 mL, (BD) = 1.85 M, Co complexes = 5 μmol. (BD) + (2-MOPB):(Co) = 3000, EASC: (Co) = 300, polymerization time = 2 h, polymerization temperature = 20 °C. [b] 10^4 g of PBD (mol of Co)$^{-1}$ h^{-1}. [c] Microstructure is determined by ^{1}H NMR in $CDCl_3$ and IR. [d] M_n: 10^4 g mol^{-1}, M_n and M_w/M_n determined using GPC.

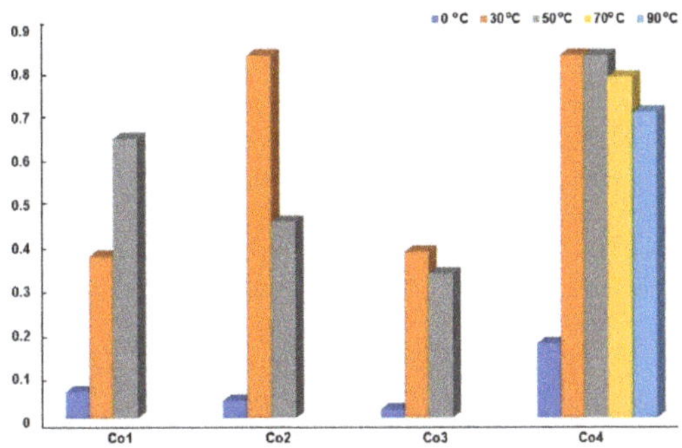

Figure 3. Influences of polymerization temperature on polymer yields.

Temperature also imposed a significant influence on the microstructures and molecular weights of the resultant polymers. For **Co4**-mediated polymerizations, a monotonous *cis*-1,4- content decrease of from 98.1% to 86.9% occurred when the temperature increased from 0 °C to 90 °C. Such a decrease can be ascribed to the facilitated *anti–syn* isomerization of the terminal π-allylic propagating unit through the π–σ-rearrangement at high temperatures, leading to decreased *cis*-1,4- content and simultaneously increased *trans*-1,4- units (Figure 4). Regarding the molecular weight of the resultant polymers, monotonously decreased M_ns due to facilitated chain transfers at high temperatures were revealed.

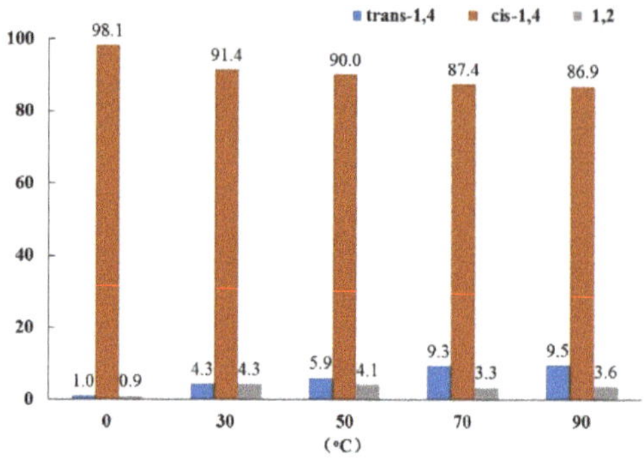

Figure 4. The microstructures of the polybutadiene obtained at different temperatures with **Co4**.

3.3. *Copolymerization of 2-(4-Methoxyphenyl)-1,3-butadiene with 1,3-Butadiene*

Considering the high catalytic efficiency of complex **Co4**, the copolymerization of 1,3-butadiene with polar comonomer 2-(4-methoxyphenyl)-1,3-butadiene (2-MOPB) was conducted next to improve the surface properties of polybutadiene elastomers, and the results were summarized in Table 2. It was determined that although **Co4** could successfully catalyze the copolymerizations, the polar comonomer incorporation rate was very low, even when 50 equivalents of 2-MOPB to **Co4** were introduced. Under these conditions, the incorporated contents are shown to be in the range of 0.46–1.83%. The ^{1}H and ^{13}C NMR spectra of poly(2-MOPB-*co*-BD) are shown in Figure 5. Due to the relatively low incorporation of comonomer in the polymer, it can be reasonably predicted that the 2-MOPB unification was always isolated in the polymer chain. For the ^{1}H NMR of the resultant polar copolymer, the resonances at δ 2.22 and 2.49 ppm were assigned to the newly generated methylene groups directly bonding to the double bond, which corresponded to the resonances at δ 26.97 and 29.70 ppm in the ^{13}C NMR spectrum, respectively, while the resonances at δ 5.58 ppm corresponded to the double bond groups generated after 1,4 insertion. Moreover, ^{1}H NMR produced the singlet at 3.79 ppm, which can be ascribed to the methoxy group that can also be observed at 55.25 ppm in the ^{13}C NMR spectrum. The signals at 6.84 ppm corresponded to the aryl protons of 2-MOPB. This is consistent with previous reports [11]. When it comes to the other aryl hydrogens (H_3), they may be overlapped by the *d*-chloroform resonances. Further analysis on the ^{13}C NMR spectrum revealed that the signals at δ 161.20, 127.42, and 113.55 ppm corresponded to the aryl carbons of 2-MOPB. Moreover, some of the other groups were hardly assigned due to the low incorporation. According to these results, the copolymerization of 2-MOPB and butadiene was achieved successfully. Due to the poisoning of the active species by the methoxyl group in the comonomer, gradually decreased polymer yields as well as the molecular weights of the resultant polybutadienes were demonstrated. The static water contact angle (WCA) was tested for the copolymers to determine the changes in the surface properties. As per the results shown in Figure 6, the WCA of P(2-MOPB-*co*-BD) decreased from 95.6° to 89.1° when the 2-MOPB content increased from 0.46% to 1.83%, which was much lower than the value of 102.9° in PBD, which is indicative of the much-improved surface properties.

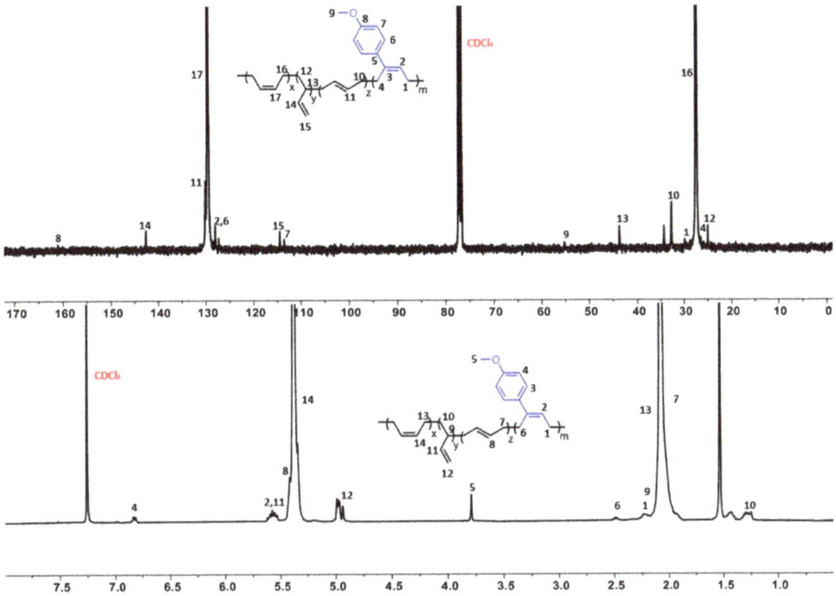

Figure 5. The ^{1}H and ^{13}C NMR spectra of poly(2-MOPB-*co*-BD) from entry 4, Table 2.

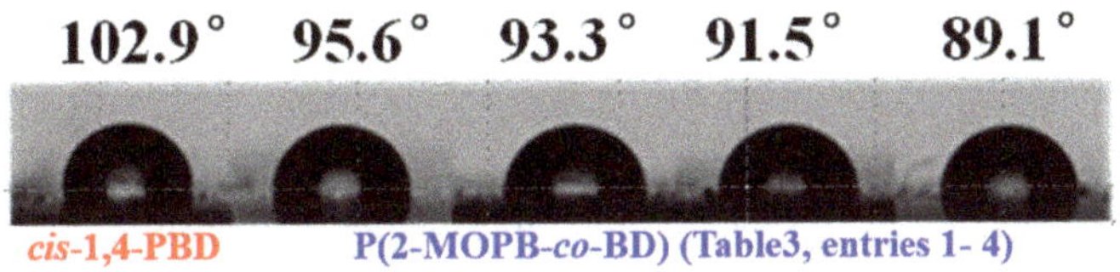

Figure 6. Contact angle characteristic of poly(2-MOPB-*co*-BD).

4. Conclusions

A new series of acenaphthene-based α-diimine cobalt complexes were synthesized that can polymerize 1,3-butadiene to access highly *cis*-1,4 selective (up to 98%) polybutadiene in all conditions. Due to the formed intra-ligand π-π stacking interactions, the complexes performed high catalytic activities at elevated temperature, and for complex Co4, the activities were maintained at 7.0×10^4 g PBD (mol of Co)$^{-1}$ h^{-1}, which had not been reported before. Furthermore, all of the obtained polymers possessed a relatively narrow molecular weight distribution as well as high molecular weight (up to 92.2×10^4 Dalton). The complexes were also able to promote the copolymerization of 1,3-butadiene with polar 2-MOPB, directly producing functionalized polymers, although their incorporation content was limited. After introducing polar 2-MOPB into the backbone, the surface properties of the polymers significantly improved, and greatly decreased water contact angles were revealed.

Author Contributions: B.W. performed the experiments and analyzed the data. H.L. and B.W. contributed to the analysis of the results and the writing of the manuscript. T.T. and X.Z. contributed to the discussion section and revised the paper. H.L. and X.Z. provided the funds. All authors have read and agreed to the published version of the manuscript.

Funding: We are grateful for the financial support from the National Natural Sciences Foundation of China (21801236, U1862206).

Institutional Review Board Statement: Not applicable.

Informed Consent Statement: Not applicable.

Data Availability Statement: The data presented in this study are available on request from the corresponding author.

Conflicts of Interest: The authors declare no conflict of interest.

References

1. Wilson, D. Recent advances in the neodymium catalysed polymerisation of 1,3-dienes. *Makromol. Chem. Symp.* **1993**, *66*, 273. [CrossRef]
2. Friebe, L.; Nuyken, O.; Obrecht, W. Neodymium-Based Ziegler/Natta Catalysts and their Application in Diene Polymerization. In *Neodymium Based Ziegler Catalysts—Fundamental Chemistry*; Nuyken, O., Ed.; Springer: Berlin/Heidelberg, Germany, 2006; pp. 1–154.
3. Ricci, G.; Sommazzi, A.; Masi, F.; Ricci, M.; Boglia, A.; Leone, G. Well-defined transition metal complexes with phosphorus and nitrogen ligands for 1,3-dienes polymerization. *Coord. Chem. Rev.* **2010**, *254*, 661–676. [CrossRef]
4. Boffa, L.S.; Novak, B. Copolymerization of Polar Monomers with Olefins Using Transition-Metal Complexes. *Chem. Rev.* **2000**, *100*, 1479–1494. [CrossRef] [PubMed]
5. Amin, S.B.; Marks, T.J. Versatile Pathways for In Situ Polyolefin Functionalization with Heteroatoms: Catalytic Chain Transfer. *Angew. Chem. Int. Ed.* **2008**, *47*, 2006–2025. [CrossRef]
6. Chen, E.Y.-X. Coordination Polymerization of Polar Vinyl Monomers by Single-Site Metal Catalysts. *Chem. Rev.* **2009**, *109*, 5157–5214. [CrossRef] [PubMed]
7. Johnson, L.K.; Killian, C.M.; Brookhart, M. New Pd(II)- and Ni(II)-Based Catalysts for Polymerization of Ethylene and .alpha.-Olefins. *J. Am. Chem. Soc.* **1995**, *117*, 6414–6415. [CrossRef]
8. Johnson, L.K.; Mecking, S.; Brookhart, M. Copolymerization of Ethylene and Propylene with Functionalized Vinyl Monomers by Palladium(II) Catalysts. *J. Am. Chem. Soc.* **1996**, *118*, 267–268. [CrossRef]
9. Wang, T.; Wu, C.; Cui, D. cis-1,4 Selective Copolymerization of Butadiene and Functionalized α-Olefins via Polar Group Activation Mechanism. *Macromolecules* **2020**, *53*, 6380–6386. [CrossRef]
10. Leicht, H.; Göttker-Schnetmann, I.; Mecking, S. Stereoselective Copolymerization of Butadiene and Functionalized 1,3-Dienes with Neodymium-Based Catalysts. *Macromolecules* **2017**, *50*, 8464–8468. [CrossRef]
11. Yao, C.; Liu, N.; Long, S.; Wu, C.; Cui, D. Highly cis-1,4-selective coordination polymerization of polar 2-(4-methoxyphenyl)-1,3-butadiene and copolymerization with isoprene using a β-diketiminato yttrium bis(alkyl) complex. *Polym. Chem.* **2015**, *7*, 1264–1270. [CrossRef]
12. Shi, Z.; Guo, F.; Meng, R.; Jiang, L.; Li, Y. Stereoselective copolymerization of amino-functionalized styrene with butadiene using a half-sandwich scandium complex. *Polym. Chem.* **2016**, *7*, 7365–7369. [CrossRef]
13. Buonerba, A.; Cuomo, C.; Speranza, V.; Grassi, A. Crystalline Syndiotactic Polystyrene as Reinforcing Agent of cis-1,4-Polybutadiene Rubber. *Macromolecules* **2010**, *43*, 367–374. [CrossRef]
14. Gong, D.; Tang, F.; Xu, Y.; Hu, Z.; Luo, W. Cobalt catalysed controlled copolymerization: An efficient approach to bifunctional polyisoprene with enhanced properties. *Polym. Chem.* **2021**, *12*, 1653–1660. [CrossRef]
15. Xu, Y.; Zhao, J.; Gan, Q.; Ying, W.; Hu, Z.; Tang, F.; Luo, W.; Luo, Y.; Jian, Z.; Gong, D. Synthesis and properties investigation of hydroxyl functionalized polyisoprene prepared by cobalt catalyzed co-polymerization of isoprene and hydroxylmyrcene. *Polym. Chem.* **2020**, *11*, 2034–2043. [CrossRef]
16. Tempel, D.J.; Johnson, L.K.; Huff, R.L.; White, P.S.; Brookhart, M. Mechanistic Studies of Pd(II)−α-Diimine-Catalyzed Olefin Polymerizations1. *J. Am. Chem. Soc.* **2000**, *122*, 6686–6700. [CrossRef]
17. Wang, X.; Dong, B.; Yang, Q.; Liu, H.; Hu, Y.; Zhang, X. Boosting the Thermal Stability of α-Diimine Palladium Complexes in Norbornene Polymerization from Construction of Intraligand Hydrogen Bonding and Simultaneous Increasing Axial/Equatorial Bulkiness. *Inorg. Chem.* **2021**, *60*, 2347–2361. [CrossRef] [PubMed]
18. Gong, Y.; Li, S.; Tan, C.; Kong, W.; Xu, G.; Zhang, S.; Liu, B.; Dai, S. π–π interaction effect in insertion polymerization with α-Diimine palladium systems. *J. Catal.* **2019**, *378*, 184–191. [CrossRef]
19. Xiao, Z.; Zheng, H.; Du, C.; Zhong, L.; Liao, H.; Gao, J.; Gao, H.; Wu, Q. Enhancement on Alternating Copolymerization of Carbon Monoxide and Styrene by Dibenzobarrelene-Based α-Diimine Palladium Catalysts. *Macromolecules* **2018**, *51*, 9110–9121. [CrossRef]
20. Du, C.; Zhong, L.; Gao, J.; Zhong, S.; Liao, H.; Gao, H.; Wu, Q. Living (co)polymerization of ethylene and bio-based furfuryl acrylate using dibenzobarrelene derived α-diimine palladium catalysts. *Polym. Chem.* **2019**, *10*, 2029–2038. [CrossRef]
21. Wang, F.; Tanaka, R.; Li, Q.; Nakayama, Y.; Shiono, T. Chain-Walking Polymerization of Linear Internal Octenes Catalyzed by α-Diimine Nickel Complexes. *Organometallics* **2018**, *37*, 1358–1367. [CrossRef]
22. Rosa, V.; Gonzalez, P.J.; Avilés, T.; Gomes, P.T.; Welter, R.; Rizzi, A.C.; Passeggi, M.C.G.; Brondino, C.D. Synthesis, Solid-State Structures, and EPR Spectroscopic Studies on Polycrystalline and Single-Crystal Samples of α-Diimine Cobalt(II) Complexes. *Eur. J. Inorg. Chem.* **2006**, *2006*, 4761–4769. [CrossRef]
23. Chen, L.; Li, Y.; Yue, S.; Ling, J.; Ni, X.; Shen, Z. Chemoselective RAFT Polymerization of a Trivinyl Monomer Derived from Carbon Dioxide and 1,3-Butadiene: From Linear to Hyperbranched. *Macromolecules* **2017**, *50*, 9598–9606. [CrossRef]

24. Rosa, V.; Carabineiro, S.A.; Avilés, T.; Gomes, P.T.; Welter, R.; Campos, J.M.; Ribeiro, M.D.R. Synthesis, characterisation and solid state structures of α-diimine cobalt(II) complexes: Ethylene polymerisation tests. *J. Organomet. Chem.* **2008**, *693*, 769–775. [CrossRef]
25. Chakraborty, S.; Chattopadhyay, G.; Saha, C. ChemInform Abstract: A Tandem Reduction—Oxidation Protocol for the Conversion of 1-Keto-1,2,3,4-tetrahydrocarbazoles to Carbazoles via Tosylhydrazones Through Microwave Assistance: Efficient Synthesis of Glycozoline, Clausenalene, Glycozolicine, and Deoxycarbazomycin B and the Total Synthesis of Murrayafoline A. *J. Heterocycl. Chem.* **2013**, *44*, 91–98.
26. Li, S.; Zhao, Y.; Dai, S. Synthesis of polyethylene thermoplastic elastomer by using robust α–diimine Ni(II) catalysts with abundant t Bu substituents. *J. Appl. Polym. Sci.* **2021**, *59*, 638–645. [CrossRef]
27. Peng, D.; Chen, C. Photoresponsive Palladium and Nickel Catalysts for Ethylene Polymerization and Copolymerization. *Angew. Chem. Int. Ed.* **2021**. [CrossRef]
28. Rosa, V.; Avilés, T.; Aullón, G.; Covelo, B.; Lodeiro, C. A New Bis(1-naphthylimino)acenaphthene Compound and Its Pd(II) and Zn(II) Complexes: Synthesis, Characterization, Solid-State Structures and Density Functional Theory Studies on the syn and anti Isomers. *Inorg. Chem.* **2008**, *47*, 7734–7744. [CrossRef]
29. Fang, L.; Zhao, W.-P.; Han, C.; Zhang, C.-Y.; Liu, H.; Hu, Y.-M.; Zhang, X.-Q. 1,3-Butadiene Polymerizations Catalyzed by Cobalt and Iron Dichloride Complexes Bearing Pyrazolylimine Ligands. *Chin. J. Polym. Sci.* **2018**, *37*, 462–470. [CrossRef]
30. Fang, L.; Zhao, W.; Han, C.; Liu, H.; Hu, Y.; Zhang, X. Isoprene Polymerization with Pyrazolylimine Cobalt(II) Complexes: Manipulation of 3,4-Selectivities by Ligand Design and Use of Triphenylphosphine. *Eur. J. Inorg. Chem.* **2019**, *2019*, 609–616. [CrossRef]
31. Liu, H.; Zhuang, R.; Dong, B.; Wang, F.; Hu, Y.-M.; Zhang, X.-Q. Mono- and Binuclear Cobalt(II) Complexes Supported by Quinoline-2-imidate Ligands: Synthesis, Characterization, and 1,3-Butadiene Polymerization. *Chin. J. Polym. Sci.* **2018**, *36*, 943–952. [CrossRef]
32. Liu, H.; Yang, S.-Z.; Wang, F.; Bai, C.-X.; Hu, Y.-M.; Zhang, X.-Q. Polymerization of 1,3-butadiene catalyzed by cobalt(II) and nickel(II) complexes bearing pyridine-2-imidate ligands. *Chin. J. Polym. Sci.* **2016**, *34*, 1060–1069. [CrossRef]
33. Wang, X.; Dong, B.; Yang, Q.; Liu, H.; Zhang, C.; Zhang, X. α-Diimine nickel complexes bearing axially bulky terphenyl and equatorially bulky dibenzobarrelene groups: Synthesis, characterization and olefin polymerization studies. *Polym. Chem.* **2020**, *11*, 6783–6793. [CrossRef]

polymers

MDPI

Article

Preparation of Emulsifier-Free Styrene–Acrylic Emulsion via Reverse Iodine Transfer Polymerization

Tao Huang and Shuling Gong *

College of Chemistry and Molecular Sciences, Wuhan University, Wuhan 430072, China; 2017102030109@whu.edu.cn
* Correspondence: gongsl@whu.edu.cn

Abstract: Styrene–acrylic emulsions containing hydroxyl functional monomer unit's component are widely used for maintenance coating. In this paper, a stable emulsifier-free styrene–acrylic emulsion with solid content over 43% could be obtained in 210 min via reverse iodine transfer polymerization (RITP). By adding a mixture of methacrylic acid (MAA) and poly(ethylene glycol)methyl ether methacrylate (PEGMA) into a system containing a high content of hydroxyl functional monomer component (19.4 wt.% of the total monomer mass), styrene (St) could be copolymerized with methyl methacrylate (MMA); the modified film exhibited good hardness properties, good adhesive properties, and low water absorption. An increase in the amount of PEGMA decreased the glass transition temperature (T_g). When 1.4 times the reference amount of initiator was added, the highest molecular weight M_n could reach 40,000 g.·mol^{-1} with 0.25 times the reference amount of iodine in the emulsion. The largest tensile strength of the dried emulsion film over 5.5 MPa endowed the material with good mechanical properties. Living polymerization was proven by the kinetics of RITP emulsion and chain extension reaction. TEM micrographs manifest the emulsification of the seed random copolymer. This paper may provide a potential methodology for preparing polymer materials with excellent mechanical properties.

Keywords: styrene–acrylic emulsion; cooperated; reverse iodine transfer polymerization; polymeric emulsifier

Citation: Huang, T.; Gong, S. Preparation of Emulsifier-Free Styrene–Acrylic Emulsion via Reverse Iodine Transfer Polymerization. *Polymers* **2021**, *13*, 3348. https://doi.org/10.3390/polym13193348

Academic Editors: Edina Rusen and Eduardo Guzmán

Received: 6 August 2021
Accepted: 26 September 2021
Published: 29 September 2021

Publisher's Note: MDPI stays neutral with regard to jurisdictional claims in published maps and institutional affiliations.

Copyright: © 2021 by the authors. Licensee MDPI, Basel, Switzerland. This article is an open access article distributed under the terms and conditions of the Creative Commons Attribution (CC BY) license (https://creativecommons.org/licenses/by/4.0/).

1. Introduction

Styrene–acrylic emulsions are widely used as industrial maintenance coating for the acrylate unit's resistance to photodegradation and the styrene unit's resistance to hydrolysis [1,2]. Furthermore, styrene–acrylic emulsions are also used for preparing nanoparticles [3–9], which are applied in the treatment of bacterial infections [10,11] or encapsulation medicine [12,13]. The optionality of the monomer endows emulsion polymers with some special properties. In emulsion polymerization, methacrylic Acid (MAA), methacrylic acid-β-hydroxyethyl ester (HEMA), methyl methacrylate (MMA), styrene (St), *n*-butyl acrylate (BA), and *n*-butyl methacrylate (BMA) are widely used monomers [14,15]. For example, the carboxyl group from MAA units or acrylic acid (AA) units endows the polymer with adhesive properties [16]; the hydroxyl group from methacrylic acid (MAA) units or methacrylic acid-β-hydroxyethyl ester (HEMA) can be crosslinked with amino resin [17], whereby the modified film with the crosslinked structure exhibits good mechanical properties [18]. MMA or St are used as hard monomers, which can increase the glass transition temperature of the polymer [1]. BA is used as a soft monomer [19]. Thus, styrene–acrylic emulsion polymers with functional groups such as carboxyl or hydroxyl in the pendant group have wide application prospects.

Styrene–acrylic emulsions can be prepared via emulsion polymerization with or without an emulsifier. Emulsion polymerization with a nonpolymeric emulsifier is conducted via the polymerization of weak water-soluble monomers within the emulsion of a nonpolymeric emulsifier [11]. In emulsifier-free polymerization, an initiator [20] or water-soluble monomer [21] can act as a polymeric emulsifier. HEMA is a water-soluble monomer; however, polymers with HEMA units may not be stably dispersed in water easily. A high content of HEMA units or hydroxyl groups in a polymer favors the crosslinking reaction between HEMA units and amino resin [22], and this can guarantee high crosslinking density that enhances mechanical properties [18]. However, there exists difficulty in preparing an acrylic emulsion polymer with a high content of HEMA for the low stability of the emulsion, and the low stability is derived from crosslinked polymer led by the crosslinking, which is caused by transesterification among the pendant hydroxyl groups of the polymer chain [23,24] and the interpolymer complex. Our group identified coagulum over 0.5 wt.% and flocculation when an excessive amount of HEMA ($\geq$25 wt.% total monomer mass) was added in the polymerization [25]. For this reason, the content of HEMA should not be too high when preparing a high-solid-content acrylic emulsion to restrain the crosslinked polymer. Thus, it is necessary to study a synthetic methodology to prepare a stable emulsion with HEMA units in the polymer chain. An interpolymer complex is formed in acidic conditions because of the hydrogen bonding between the hydroxyl and carboxyl groups in monomer units [26,27], and the complex decreases the stability of the emulsion. For this reason, the ammonium salt of MAA [28] was used as a polymeric surfactant to restrain the formation of the interpolymer complex in our previous paper, and a stable emulsion with a high content of hydroxyl functional monomer HEMA (19.4 wt.% of the total monomer mass) units was prepared; the highest M_n of the polymer could reach 34,100 g·mol^{-1}, but the content of MAA in the total monomer mass was 12.3 wt.%. Furthermore, the emulsion viscosity with St units in the polymer chain was high, and the fluidity of the emulsion was not good. The carboxyl group in MAA units may corrode metal when ammonia solution is volatilized; hence, choosing a polymeric emulsifier that does not corrode metal and can stabilize an emulsion with St units in the polymer chain is necessary. Nonionic polymeric emulsifier PEGMA was added to stabilize the emulsion with St units [29–31], and the copolymer particles were uniform. In terms of the chemical structure, the group of poly(ethylene glycol) ethyl ether in PEGMA does not corrode metal. On the basis of the above understanding, PEGMA was chosen to study the preparation of a styrene–acrylic emulsion in this paper.

Emulsion polymerization can be conducted via controlled radical polymerization (CRP) [32]. In CRP, the controlled mechanism is based on the reversible deactivation of growing radicals [32,33]. An emulsion copolymer with PEGMA units and St units in the polymer chain can be prepared via CRP such as nitroxide-mediated polymerization (NMP) [21] and reversible addition fragmentation chain transfer polymerization (RAFT) [34–36]. The monomer benzyl methacrylate (BnMA) is analogous to St in chemical structure, and an emulsion copolymer containing PEGMA units and BnMA units can be prepared via photo-controlled iodine-mediated green reversible deactivation radical polymerization (RDRP) [37]. In our previous paper, an emulsion copolymer containing HEMA units and MAA units was prepared via RITP, and the emulsion was stable. Compared with NMP polymerization and RAFT polymerization, RITP does not require complicated chemicals to regulate the polymerization, and the chain transfer angents are synthesized in situ during the polymerization [38]. On the other hand, only a few of the chain transfer agents in RAFT are commercially available, and disadvantages of the polymer prepared by RAFT polymerization include its odor and color [39]. Iodine is commercially available, and there is no odor in the dried emulsion film. Thus, iodine transfer polymerization was chosen. Furthermore, PEGMA can be copolymerized with HEMA [24] or a monomer mixture comprising MAA and St [40], forming a stable emulsion. Therefore, it is probable that an emulsion copolymer containing PEGMA units, HEMA units, MAA units, and St units may be prepared via RITP emulsion polymerization.

In this paper, the copolymerization of St and an acrylate-based monomer (Figure 1) was studied via RITP emulsion polymerization. The research was conducted by changing the mass ratio of methyl methacrylate/St, the mass ratio of PEGMA/MAA, the amount of PEGMA, and the amount of iodine. Kinetics experiments of the random copolymerization and chain extension reaction with St units and BA units proved the living polymerization of the chain. The emulsion polymer was characterized by monomer conversion, viscosity, particle size, molecular weight, T_g, particle morphology, and Fourier-transform infrared (FTIR) spectra. The protective properties of the modified film were measured by pencil hardness rank, adhesive property, toluene absorption, and water resistance. The mechanical properties of the dried emulsion film or modified film were measured by tensile experiment.

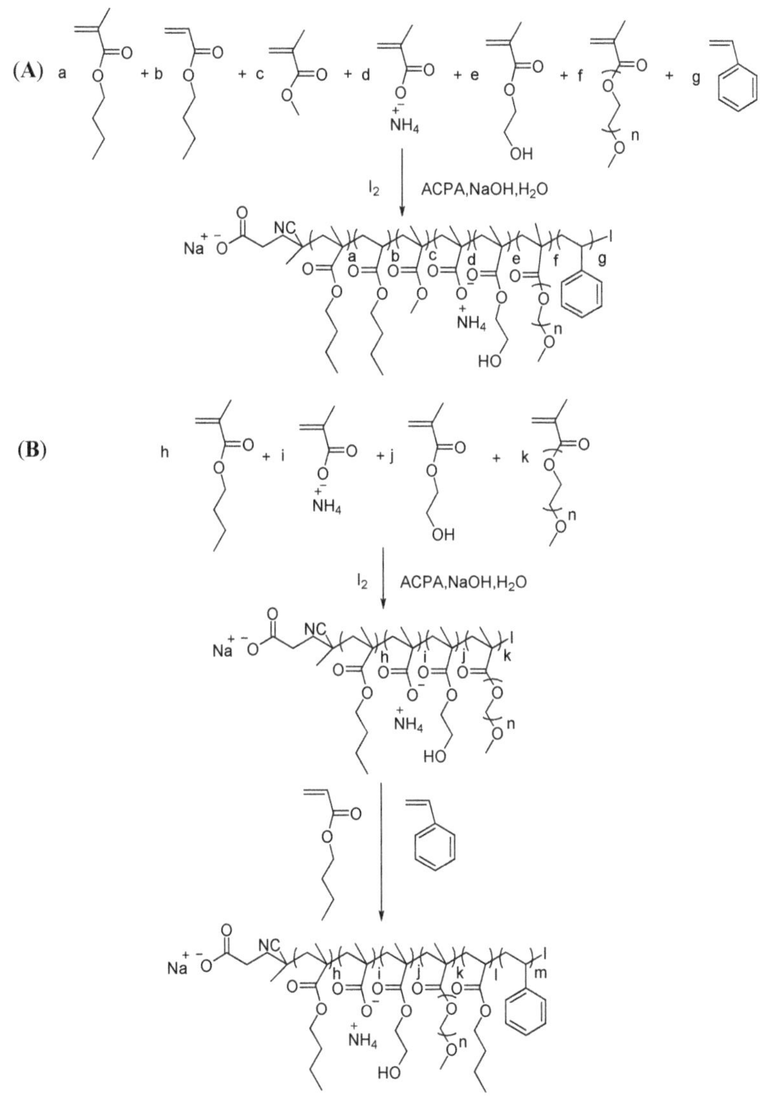

Figure 1. (**A**) Synthetic route of styrene–acrylic emulsion copolymer via RITP and (**B**) chain extension reaction via RITP.

2. Materials and Methods

2.1. Materials

Ammonia solution (25–28 wt.%), *n*-butyl acrylate (BA), methyl methacrylate (MMA), styrene (St), *n*-butyl methacrylate (BMA), methacrylic acid (MAA), *p*-toluene sulfonic acid (TsOH), and sodium hydroxide (NaOH) were purchased from Sinopharm Chemical Reagent Co., Ltd. (Shanghai, China), as analytical reagent grade (AR). Methacrylic acid-β-hydroxyethyl ester (HEMA) was purchased from Tianjin Institute of Chemical Reagents, AR. Poly (ethylene glycol) methyl ether methacrylate (PEGMA, average molecular weight = 475 $g \cdot mol^{-1}$) was purchased from Shanghai Aladdin Biochemical Technology Co., Ltd (Shanghai, China). 4,4′–Azobis (4-cyanovaleric acid) (ACPA, 98%, AR), containing ca. 20% water, was purchased from Energy Chemical. *N*, *N*-Dimethylethanolamine (DMEA) was purchased from Tianjin Kemiou Chemical Reagent Co., Ltd. (Tianjin, China), AR. Hexamethylolmethymelamine (HMMM, MF Resin) was provided by H. J. Unkel Co., Ltd. (Zhuhai, China). The monomers BA, MMA, St, and BMA were extracted by washing four times with 10 wt.% aqueous sodium hydroxide solution in a separatory funnel, followed by washing with deionized water four times in a separatory funnel. Other materials were used as received.

2.2. *Methods*

2.2.1. Emulsifier-Free Copolymerization of Acrylate Monomers

In a typical example illustrated in Figure 1A, BA (1.027 g, 8.01 mmol), BMA (6.162 g, 40.33 mmol), MMA (0.822 g, 8.21 mmol), St (1.233 g, 11.83 mmol), HEMA (2.618 g, 20.12 mmol), 9.138 g of deionized water, and 273 mg (1.074 mmol) of I_2 were added to a flat-bottom flask (100 mL), and the mixture was stirred by a magnetic stirrer for 15 min. Thereafter, a neutralized MAA solution comprising MAA (1.664 g, 19.33 mmol), deionized water (4.141 g), and ammonia solution (1.49 g, 21.87–24.50 mmol NH_3) was added to the flask and stirred for 15 min. Next, 3.461 g (1.235 mmol) of ACPA 10 wt.% solution (2.25 g ACPA, 642 mg NaOH, and 19.608 g H_2O) was added. The reaction system was deoxygenated by bubbling with high-purity nitrogen for 25 min at room temperature while stirring. Afterward, the ingredients were placed under the atmosphere of high-purity nitrogen provided by a balloon, and the mixture was heated to 80 °C while stirring; the total heating time was 210 min. The reaction was ceased by exposure to air. The total theoretical monomer mass content by weight was 40 wt.%. The monomer conversion and the solid content were determined by gravimetric analysis.

When the mass ratio of the monomer or the mass of the I_2 was changed, the procedures were as described above.

2.2.2. Chain Extension Reaction with BA and St

Firstly, BMA (6.162 g, 40.33 mmol), HEMA (2.618 g, 20.12 mmol), deionized water (9.892 g), and I_2 (164 mg, 0.644 mmol) were added to a flat-bottom flask (100 mL), and the mixture was stirred by a magnetic stirrer for 15 min. Thereafter, a PEGMA solution comprising PEGMA (0.832 g, 1.75 mmol) and H_2O (2.27 g) was added into the flask and stirred for 15 min; then, a neutralized MAA solution comprising MAA (0.832 g, 9.66 mmol), deionized water (3.80 g), and ammonia solution (0.708 g, 10.39–11.64 mmol NH_3) was added to the flask and stirred for 10 min. Afterward, 3.461 g (1.235 mmol) of ACPA 10 wt.% solution was added. The reaction system was deoxygenated by bubbling with high-purity nitrogen for 25 min at room temperature while stirring. Next, the ingredients were placed under the atmosphere of high-purity nitrogen provided by a balloon, and the mixture was heated to 80 °C while stirring. The sample was removed by a syringe under a nitrogen atmosphere after heating for 130 min (the first stage). Thereafter, the nitrogen gas-saturated mixture comprising BA (1.027 g, 8.01 mmol) and St (2.054 g, 19.72 mmol) was immediately added to the flask, and polymerization continued for 90 min (the second stage).

When the mixture comprising BA and HEMA was added at the second stage, the procedures were as described above. The reaction time of the first stage was 168 min, and that of the second stage was 33 min.

2.2.3. Modification of the Emulsion Polymer

Firstly, the emulsion (2.5 g) and deionized water (2.08 g) were added to a round flask (25 mL) at room temperature, and then stirred for 5 min. Afterward, HMMM (0.64 g) was added while stirring for 10 min. TsOH solution (85 μL, 10 wt.%) was added dropwise to the mixture, and the mixture was stirred for 1 h.

To measure the hardness, adhesive properties, and water resistance, the ingredients of the modified film were coated on a clean tinplate and coverslip, and these samples were placed in room temperature for 2 h to evaporate water naturally. Thereafter, the tinplate and coverslip were heated at 80 °C for 120 min and then 150 °C for 40 min. After heating, the tinplate and coverslip were cooled to room temperature via natural cooling.

To measure tensile strength, the ingredients of the modified film were coated on a horizontal tetrafluoroethylene plate, and the samples were placed in room temperature for 2 h to evaporate water naturally. The plate was heated at 80 °C for 120 min and then 150 °C for 40 min. Afterward, the modified film could be torn off the plate.

2.2.4. Preparation of Dried Emulsion Films

To measure tensile strength, the emulsion mixture was poured onto a horizontal tetrafluoroethylene plate, and the emulsion was dried under the radiation of an infrared lamp for at least 24 h; then, the film could be torn off the plate. At least three samples of the layer could be obtained per piece of emulsion film.

2.3. Characterizations

2.3.1. Viscosity

The emulsion viscosity was measured using a DV-79 digital viscometer (Shanghai Ni Run Intelligent Technology Co., Ltd., Shanghai, China) with a rotor (E-type rotor, F-type rotor, or G-type rotor) at 25 °C. The rotational rate of the rotor was 75 or 750 rpm when the viscosity was in the corresponding measurement range of the rotor type and rotational rate.

2.3.2. Monomer Conversion

The determination of monomer conversion was as follows:

$$\text{Conversion\%} = \text{m (Dried emulsion)}/(\text{m (Emulsion solution)} \times \text{w (Total monomer)}) \quad (1)$$

where w (Total monomer) is the total mass content of the monomer in weight.

2.3.3. Particle Diameter

After adding deionized water to the emulsion according to the volume, the emulsion sample was diluted 1000-fold. Then, the particle size diameter and the particle size distribution of the diluted emulsion sample were measured using a Zetasizer Nano ZS laser particle sizer, Malvern Instruments Ltd. (Shanghai, China) at 25 °C; the Malvern Zetasizer ZS device that we used worked at 173° in backscattering mode.

2.3.4. Molecular Weight

The emulsion sample was further neutralized using DMEA, and the excess DMEA could be removed via rotary evaporation under reduced pressure. Afterward, the sample was dispersed in THF (20 mg/mL), and the mixture was filtered using an organic phase filter head. Then, 50 μL of the filtered sample was injected into the device. The number average molecular weight (M_n), the weight average molecular weight (M_w), and the index of the molecular weight distribution ($Đ$) of the polymer were measured using a gel permeation chromatograph (GPC) from US WATERS Corporation (Milford, MA, USA). The GPC device was equipped with a Waters-2414 Refractive Index Detector, and the eluent THF flowed at a rate of 1.0 mL/·min at 25 °C. The material of the GPC columns was styrene–divinylbenzene copolymer. The GPC columns used were WATERS Styragel

HR1 + Styragel HR3 + Styragel HR4. The calibrated material in the chromatographic column was polymethyl methacrylate.

2.3.5. Glass Transition Temperature

T_g was tested using a TA Q20 DSC Instrument (TA Instruments, New Castle, DE, USA), and the measurement was conducted under nitrogen atmosphere. The measuring temperature range was −15 °C to 120 °C, and the heating/cooling rate was 10 °C/min.

2.3.6. Fourier-Transform Infrared Spectroscopy

The sample was prepared by drying the emulsion under the radiation of an infrared lamp. The infrared spectrum of the sample was tested using a Thermo Nicolet Fourier-transform infrared spectroscope (FTIR) (Thermo Fisher, Waltham, MA, USA), and the device was equipped with a diamond from Smiths Detection for attenuated total reflectance (ATR). The scanning range was 4000–500 cm^{-1}.

2.3.7. Transmission Electron Microscopy

The 1000-fold diluted sample was added to the copper mesh, and then stained with 3% phosphotungstic acid (PTA) solution. After evaporation of water in the mesh, the mesh was observed using a JEM-2100 transmission electron microscope from JEOL Corporation (Tokyo, Japan), with an accelerating voltage of 220 kV. The sample morphology could be observed after vacuuming.

2.3.8. Hardness Rank of the Modified Film

The hardness of the modified film was tested by sliding pencil lead on the film surface. The procedures and the hardness rank were conducted according to China National Standard GB/T 6739-2006/ISO 15184:1998.

First, the pencil was pressed down on the film surface at an angle of 45° by hand. The pencil was shifted at a rate of 0.5–1 mm/s and for at least 7 mm. If the length of the marking was longer than 3 mm, a pencil with lower hardness was used until no marking longer than 3 mm existed. When a marking shorter than 3 mm existed, the pencil hardness rank was regarded as the hardness rank of the film. This measurement was conducted at a temperature of 23 ± 2 °C. The measurement was conducted at least three times.

2.3.9. Adhesive Property Rank of the Modified Film

The adhesive property was evaluated by the affected area of cuts on the surface of the modified film or dried emulsion film adhered to the tinplate piece. The procedures and the evaluation standard were conducted according to China National Standard GB/T 9286-1998 equivalent ISO 2409:1992.

Six cuts were made in each direction of the lattice pattern, and the lattice was made by scratching the knife under the surface of the film with the use of ruler. The spacing of the cuts in each direction were equal to 1 mm (0–60 μm thick film) or 2 mm (60–120 μm thick film). The tape was removed on the surface of the cuts by grasping the free end and pulling it off in 0.5–1.0 s at about 60°. When none of the lattice squares were detached, classification was regarded as rank 0. When 0–5%, 5–15%, 15–35%, and 35–65% of the area was affected, classification was regarded as 1, 2, 3, and 4, respectively. Any degree of flaking that could not be classified as 4 was classified as 5. This measurement was conducted at a temperature of 23 ± 2 °C. The measurement was conducted at least three times.

2.3.10. Water Resistance of the Modified Film

The water resistance of the modified film was measured by the change in the film immersed in boiling water. The test was based on a comparison with a modified film not immersed in boiling water. The procedures and the evaluation standard were conducted according to China National Standard GB/T 1733-1993.

Water resistance of the films was determined by immersing three sample pieces into boiling water for 2 h, and then the immersed part was dried by filter paper. Next, the appearance of these three sample pieces in terms of gloss, whitening, wrinkling, etc. was compared with those not immersed in boiling water.

2.3.11. Water Absorption of the Modified Film

The modified film was weighed. Then, the same piece of modified film sample was immersed in deionized water for 24 h. After 24 h, the sample immersed in water was weighed. The determination of water absorption was as follows:

$$\text{Water absorption\%} = 100\% \times \text{m (Dried film)}/(\text{m (Film immersed in water)}). \quad (2)$$

2.3.12. Toluene Absorption of the Modified Film

The modified film was immersed in toluene for 24 h. The determination of toluene absorption was as follows:

$$\text{Toluene absorption\%} = 100\% \times \text{m (Dried film)}/(\text{m (Film immersed in Toluene)}) \quad (3)$$

2.3.13. Tensile Strength

Tensile strength of the dried emulsion film or modified film was measured using an Electronic Universal Testing Machine from MTS SYSTEMS (China) Co., Ltd. (Shanghai, China) with SANS-Power Test software, and the measurement was conducted at 10 mm·min^{-1} at 20 ± 2 °C. The shape of the film was rectangular in general. The length of the film was longer than 12 mm, and the width ranged from 5–12 mm.

3. Results and Discussion

3.1. Participation of Styrene in Emulsifier-Free Polymerization

As shown in Table 1, when all hard monomer MMA was replaced by St, the emulsion was not stable throughout the reaction when only the ammonium salt of MAA was used as a polymeric emulsifier in the emulsion system. In Tables 1 and 2, the total mass of MMA and St was constant, while the mass ratio of MMA/St was changed; the mass of other ingredients was constant. When hard monomer MMA participated in the polymerization, the emulsion was stable, the monomer conversion was higher than 95%, and the solid content was over 40 wt.%. The monomer conversion was not changed obviously for the four mass ratios. The emulsion was stored and remained stable for 6 months. The emulsion polymer had a measured M_n range from 23,000 to 30,000 g·mol^{-1}, which is analogous to the range in our previous paper. The measured M_n was higher than the theoretical M_n. Iodine is hydrolyzed in water [41], and this hydrolysis decreases the amount of iodine that is involved in the synthesis of the chain transfer agent in situ [42–44]. Furthermore, ammonia solution facilitates hydrolysis. Thus, the deviation of measured M_n from the theoretical M_n was caused by the hydrolysis of I_2 in water. The measured M_n increased with the increase in MMA. The index of the molecular weight distribution decreased overall with the increase in MMA. According to the work of Tonnar et al. [38,45], the index of the molecular weight distribution with iodine in the polymerization was smaller than that without iodine, and the index with iodine in the polymerization ranged from 1.40 to 2.20 in most cases. In comparison with typical controlled radical emulsion polymerization such as RAFT polymerization [8,36] or atom transfer radical polymerization (ATRP) [46], the index of the molecular weight distribution in RITP was relatively large. However, the diameter of the particle (d_p) was not increased with the increase in MMA, and the d_p ranged from 350–430 nm. When the mass ratio of MMA to St was 4:6, the viscosity value of emulsion was 2350 mPa·s at 750 rpm. This viscosity was larger than that (525 mPa·s at 750 rpm) in our previous paper with the mass ratio of BA to MMA kept at 1:2 [28]. More MMA led to an overall decrease in the emulsion viscosity. Side products are unavoidable because chain transfer occurs due to collision between two active centers attached to polymer chains in polymerization [47] or emulsion polymerization [48–50]. Experimentally, an emulsion with lower viscosity can improve the diffusion of the added hydrophobic monomer or monomer mixture, and this may ensure that random copolymerization or block copolymerization with St monomer units is conducted more fluently. The results in Table 1 show that, when all MMA was replaced by St, the emulsion was not conducted fluently with only the ammonium salt of MAA used as a polymeric emulsifier, whereas when both St and MMA were present, the emulsion was stable and of low viscosity.

Table 1. Results of emulsion with St units in polymer chain.

m(MMA)/m(St)	Conversion (%)	$M_{n,th}$ [a] (10^3 g·mol^{-1})	$M_{n,GPC}$ (10^3 g·mol^{-1})	Đ	d_p(nm) [b]/PDI	Viscosity (mPa·s)/Rotor Type/Rotor Rate (rpm)	Reaction Phenomena
0/10	-	-	-	-	-	-	Opalescent, a lot of gel.
4/6	99.5	7.34	18.6	1.89	359/0.139	2350/F/750	Opalescent, high viscosity, no gel.
6/4	>99.5	7.38	24.1	1.70	429/0.040	639/F/750	Opalescent with weak blue color.
8/2	96.8	7.15	27.5	1.56	409/0.042	664/F/750	Opalescent with weak blue color.
10/0	>99.5	7.38	30.2	1.62	357/0.066	525/F/750	Opalescent with weak blue color.

[a] $M_{n,th}$ = (mass of monomer) × (monomer conversion)/(2 × $n_{I2,initial}$) + M_{AI}, in which M_{AI} = 275.02 g·mol^{-1}. [b] d_p: particle size diameter; PDI: polydispersity index of particle size diameter. Conditions: MAA solution was neutralized by 1.39 g of ammonia solution; m(BA)/[m(MMA) + m(St)] = 1/2; m(MMA) + m(St) = 2.054 g; n(MAA)/n(HEMA)/n(BMA)/n(BA)/n(ACPA)/n(I_2) = 18/18.74/40.36/7.46/1.15/1; the total mass of ingredients without ACPA solution and I_2 was maintained at 30.08 g in theory.

Table 2. Properties of the cured film synthesized by emulsion with St units in polymer chain.

m(MMA)/m(St)	Pencil Hardness of the Dried Emulsion Film	Pencil Hardness of the Modified Film	Adhesive Property of the Dried Emulsion Film	Adhesive Property of the Modified Film	Toluene Absorption (Wt.%)	Water Absorption (Wt.%)	Water Resistance in Boiled Water
4/6	1 H	2 H	1	0	12.2	5.70	Whitening
6/4	1 H	2 H	1	0	12.0	5.42	Translucent
8/2	1 H	2 H	1	0	11.4	4.47	Whitening
10/0	-	2 H	-	0	16.9	3.82	Translucent

Conditions: MAA solution was neutralized by ammonia solution; n(MAA) /n(HEMA)/n(BMA)/n(BA)/n(ACPA)/n(I_2) = 18/18.74/40.36/7.46/1.15/1; m(BA)/[m(MMA) + m(St)] = 1/2; m(MMA) + m(St) = 2.054 g; ammonia solution (1.39 g); the total mass of ingredients without ACPA solution and I_2 was maintained at 30.08 g in theory.

Before modification by MF resin, the pencil hardness rank of the three kinds of dried emulsion film was 1 H in room temperature, and their adhesive property rank was 1. Furthermore, the dried film was translucent, and it could be dispersed in water when the mixture was stirred. The cured film surface became hard with a pencil hardness rank of 2 after the modification in Table 2. The reason is that a crosslinked structure polymer was prepared via the reaction of hydroxy functional acrylics with MF resin [22], and this crosslinked structure limited the movement of the segment [18], thus leading to the hardness of the polymer film being higher than that of the dried emulsion film. Furthermore, the adhesiveness rank of the cured film was 0, suggesting that the adhesive property of the modified film was better than that of the dried emulsion film. Toluene absorption at the mass ratio of 10/0 was higher than that for the other three mass ratios; hence, the toluene resistance of the cured film was increased when St was added. The cured film was not dissolved in toluene, as polymers with a crosslinked structure only swell in some kinds of organic solvent [51]. Water absorption decreased when the mass ratio of MMA/St increased. The cured film was translucent after immersing in boiled water when the monomer mass ratio was 6/4 or 10/0, indicating water resistance. The best results in terms of good pencil hardness, good adhesive property, low toluene absorption, low water absorption, and good water resistance of the modified film were obtained when the mass ratio of MMA/St was 6/4.

Styrene is a hydrophobic monomer, and the phenyl structure of St units in the polymer promotes chain rigidity. Poly (ethylene oxide) (PEO) in the pendant group of a polymeric emulsifier can form a dense protective layer around the surface of the latex particles [29,52] prepared via conventional radical emulsion polymerization. Some reports have proposed that hydrophilic polymeric emulsifiers such as poly (ethylene glycol) ethyl ether methacrylate (PEG-EEA) can be copolymerized with St to prepare a stable emulsion [29,30,53] via conventional radical polymerization. Furthermore, MAA can cooperate with PEGMA to stabilize an emulsion with St units [40] via conventional radical polymerization. Considering the above properties of polymers containing St monomer units, a high-solid-content styrene–acrylic emulsion may be prepared by adding PEGMA to an RITP system. As shown in Table 3, when there was no PEGMA in the polymerization system, the emulsion was of high viscosity and poor fluidity. In Tables 3 and 4, the total mass of PEGMA and MAA was constant, while the mass ratio of PEGMA/MAA was changed; no MMA was added, and St was added; the mass of other ingredients was constant. Increasing the mass content of PEGMA resulted in lower monomer conversion, lower molecular weight, higher index of molecular weight distribution (*Đ*), lower diameter, lower viscosity, and good fluidity. The dense protective layer derived from PEGMA on the latex particles may hinder the access of hydrophobic monomers such as St, BA, and BMA, which may not be beneficial for the increase in particle size. The steric effects depend upon the size of the substituents [54], whereby larger substituents in the PEGMA may lead to a larger steric effect; thus, propagating chain radicals with PEGMA leads to a larger steric effect than that with ammonium salt of MAA, resulting in lower reactivity, a lower propagating reaction rate, and lower monomer consumption. Therefore, an increase in PEGMA may also lead to a lower molecular weight and lower monomer conversion. Furthermore, the dense protective layer may decrease the frictional effect between particles, which may decrease the viscosity of the emulsion. The emulsion was stable when the mass ratio of PEGMA/MAA was 5/5, indicating that half the referenced mass amount of MAA could be used to prepare a stable emulsion with analogous M_n range and low viscosity. However, when the mass ratio of PEGMA/MAA was 7/3, the stability of the emulsion was not good, and a little white precipitate existed in the emulsion. This phenomenon indicated that enough MAA is needed to guarantee higher monomer conversion and maintain the stability of the emulsion with St units in the polymer chain. Furthermore, a high mass content of PEGMA may not guarantee high monomer conversion, high molecular weight, and low polydispersity index in this polymerization system. The above results show that the cooperation of PEGMA and MAA facilitated stabilization of the styrene–acrylic emulsion, and all MMA monomers could be replaced by St.

Table 3. Results of emulsion with PEGMA units in polymer chain.

m(PEGMA)/ *m*(MAA)	Conversion (%)	$M_{n,th}$ (10^3 g·mol^{-1})	$M_{n,GPC}$ (10^3 g·mol^{-1})	*Đ*	d_p(nm)/PDI	Viscosity (mPa·s)/Rotor Type/Rotor Rate (rpm)	Reaction Phenomena
0/10	-	-	-	-	-	-	Opalescent, high viscosity, and poor fluidity.
3/7	>99.5	7.14	27.4	1.64	336/0.187	1210/G/750	Opalescent, no gel.
5/5	94.6	6.61	23.3	1.70	271/0.201	151/F/750	Opalescent with weak blue color.
7/3	83.0	5.70	21.0	1.79	220/0.037	38.2/F/750	Opalescent with weak blue color, little white precipitate.

Conditions: *m*(MAA) + *m*(PEGMA) = 1.664 g; no MMA in the emulsion polymerization system; *n*(St)/*n*(HEMA)/*n*(BMA)/*n*(BA)/*n*(ACPA)/*n*(I_2) = 18.36/18.74/40.36/7.46/1.15/1; *m*(BA)/*m*(St) = 1/2; MAA solution was neutralized by ammonia solution; the total mass of ingredients without ACPA solution and I_2 was maintained at 30.08 g in theory.

Table 4. Properties of the cured film synthesized by emulsion with PEGMA units in polymer chain.

m(PEGMA)/m(MAA)	Pencil Hardness Rank	Adhesion Property Rank	Water Absorption (Wt.%)
3/7	2 H	1	5.70
5/5	2 H	1	4.69
7/3	2 H	4	2.57

Conditions: m(MAA) + m (PEGMA) = 1.664 g; no MMA in the polymerization system; n(St)/n(HEMA)/n(BMA)/n(BA)/n(ACPA)/n(I_2) = 18.36/18.74/40.36/7.46/1.15/1; m(BA)/m(St) = 1/2; MAA solution was neutralized by ammonia solution; the total mass of ingredients without ACPA Solution and I_2 was maintained at 30.08 g in theory.

From the perspective of the chemical structure of PEGMA, no groups in PEGMA monomer units exist that can react with MF resin. This property is different from that of MAA units or HEMA units. Thus, it is necessary to research the influence of the mass ratio of MAA/PEGMA on the properties of the cured film. The pencil hardness rank of the cured film was not changed by the mass ratio of MAA/PEGMA, as shown in Table 4. The adhesive property worsened when the mass ratio of PEGMA/MAA was 7/3. The water absorption decreased with the increase in mass ratio, which may be because less MAA decreased the hydrophilicity of the chain segment in the crosslinked film when no MMA participated in polymerization. The best results in terms of good pencil hardness, good adhesive property, and low water absorption were obtained when the mass amount of PEGMA was equal to that of MAA.

From the perspective of the chemical structure of PEGMA, it can be used as a polymeric emulsifier. However, the emulsification ability of PEGMA in this emulsion polymerization was unknown. As shown in Table 5, the emulsion was stable when the polymeric emulsifier consisted of neutralized MAA, and all St monomer units were replaced by MMA. However, the emulsion was not stable when all MAA was replaced by PEGMA, and there existed flocculation that could not be dispersed in water or THF. These phenomena suggest that the emulsification ability of neutralized MAA units in the polymer chain was stronger than that of PEGMA, and the emulsion was not stable when all neutralized MAA was replaced by PEGMA. The hydrophilic/lipophilic balance (HLB) value of the ammonium salt of MAA is 21.25, while that of PEGMA is 9.68. Thus, the lipophilic property of PEGMA is stronger than that of neutralized MAA. However, the steric hindrance due to the side group of PEGMA is larger than that of neutralized MAA, which may result in a lower reactivity of propagating radical with PEGMA in the chain than that of the neutralized MAA. In summary, the emulsion ability of neutralized MAA throughout the polymerization period was stronger than that of PEGMA. St is more hydrophobic than MMA; hence, the emulsion polymerization with St units must be conducted with the addition of neutralized MAA.

Table 5. Results of emulsion with MMA units in polymer chain.

m(MAA)/ m(PEGMA)	Reaction Time(min)	Conversion (%)	$M_{n,GPC}$ (10^3 g·mol^{-1})	Đ	d_p(nm)/PDI	Viscosity (mPa·s)/Rotor Type/Rotor Rate (rpm)	Reaction Phenomena
10/0	210	>99.5	30.2	1.62	357/0.066	525/F/750	Milky white with weak blue color.
0/10	320	-	-	-	-	-	Pale yellow color, a lot of flocculation

Conditions: MAA solution was neutralized by ammonia solution; m(MAA) + m(PEGMA) = 1.664 g; n(HEMA)/n(BMA)/n(BA)/n(ACPA)/n(I_2) = 18.74/40.36/7.46/1.15/1; m(BA)/m(MMA) = 1/2; no St in the emulsion; the total mass of ingredients without ACPA solution and I_2 was maintained at 30.08 g in theory.

As shown in Table 3, when the mass ratio of PEGMA/MAA was 3/7, the viscosity was much higher than that of the others. As mentioned in Table 1, changing the mass ratio of MMA/St could tune the viscosity when no PEGMA units existed in the polymer chain. Similarly, changing the mass ratio of MMA/St could tune the viscosity when MAA and PEGMA were used as polymeric emulsifiers. This experiment was done, and the results are shown in Table 6. In Tables 6 and 7, the total mass of MMA and St was constant, while the mass ratio of MMA/St was changed; the mass of other ingredients was constant, and the monomer conversion was higher than 95%. The monomer conversion decreased when the mass ratio was increased from 0/10 to 8/2. The measured molecular weight decreased when the mass ratio of MMA/St was increased from 0/10 to 4/6, while the measured molecular weight was reduced when the mass ratio of MMA/St was increased from 4/6 to 8/2. The measured molecular weight ranged between 21,000 and 28,000 g·mol^{-1}, and the index of molecular weight distribution ranged between 1.55 and 1.87. The largest d_p existed when the mass ratio of MMA/St was 8/2, and d_p at the six mass ratios ranged from 330 to 430 nm. When no MMA or no St took part in the polymerization, the viscosity was higher than 1100 mPa·s. The smallest viscosity existed when the mass ratio of MMA/St was 4/6. The viscosity of the emulsion with no St units in the polymer chain was larger than that with both St and MMA units in the polymer chain. This indicates that a styrene–acrylic polymer with analogous molecular weight and lower viscosity could be prepared when MAA was combined with PEGMA, and the emulsion with some MMA replaced by St was stable when PEGMA was added to the system.

Table 6. Results of emulsion with MMA and St units in polymer chain.

m(MMA)/*m*(St)	Conversion (%)	$M_{n,th}$ (10^3 g·mol^{-1})	$M_{n,GPC}$ (10^3 g·mol^{-1})	Đ	d_p(nm)/PDI	Viscosity (mPa·s)/Rotor Type/Rotor Rate (rpm)	Solid Content (wt.%)
0/10	>99.5	7.14	27.4	1.64	336/0.187	1210/G/750	45.8
2/8	99.5	7.10	22.7	1.75	340/0.050	454/F/750	41.8
4/6	98.8	7.06	21.3	1.85	333/0.097	46.9/E/750	41.5
6/4	97.6	6.97	22.8	1.73	377/0.062	501/F/750	41.0
8/2	96.2	6.87	27.3	1.57	454/0.249	415/F/750	40.4
10/0	>99.5	7.14	26.5	1.65	430/0.061	1270/G/750	42.6

Conditions: *m*(MMA) + *m*(St) = 2.054 g, *m*(MAA) + *m*(PEGMA) = 1.664 g, and *m*(PEGMA)/*m*(MAA) = 3/7; *n*(MAA)/*n*(PEGMA)/*n*(HEMA)/*n*(BMA)/*n*(BA)/*n*(ACPA)/*n*(I_2) = 12.60/0.98/18.74/40.36/7.46/1.15/1; ammonia solution (1.10 g); *m*(BA)/[*m*(St) + *m*(MMA)] = 1/2; the total mass of ingredients without ACPA solution and I_2 was maintained at 30.08 g in theory.

Table 7. Properties of the cured film synthesized by emulsion with MMA and St units in polymer chain.

m(MMA)/*m*(St)	Pencil Hardness Rank	Adhesive Property Rank	Toluene Absorption (Wt. %)	Water Absorption (Wt. %)	Water Resistance in Boiled Water
0/10	2 H	1	16.0	5.70	Whitening
4/6	1 H	0	13.7	4.05	Whitening
6/4	2 H	0	13.8	2.98	Whitening
8/2	2 H	0	12.8	3.67	Translucent
10/0	3 H	0	12.4	2.28	Translucent

Conditions: *m*(MMA) + *m*(St)= 2.054 g, *m*(MAA) + *m*(PEGMA) = 1.664 g, and *m*(PEGMA)/*m*(MAA) = 3/7; *n*(MAA)/*n*(PEGMA)/*n*(HEMA)/*n*(BMA)/*n*(BA)/*n*(ACPA)/*n*(I_2) = 12.60/0.98/18.74/40.36/7.46/1.15/1; *m*(BA)/[*m*(St) + *m*(MMA)] = 1/2; ammonia solution (1.10 g); the total mass of ingredients without ACPA solution and I_2 was maintained at 30.08 g in theory.

In Table 7, the properties of the cured film synthesized by emulsion with MMA units and St units in the polymer chain are shown. Here, the pencil hardness of the cured films was at rank 2 in most cases, and the adhesive property was good for all six mass ratios of MMA/St. The toluene absorption of the cured film ranged between 12.3 wt.% and 16.1 wt.%, and the difference was overall minimal. The water absorption of the cured film ranged between 2.20 wt.% and 5.75 wt.%. The best results in terms of good pencil hardness, good adhesive property, low toluene absorption, low water absorption, and good water resistance of the modified film were obtained when the mass ratio of MMA/St was 10/0.

In summary, PEGMA could be combined with MAA in the reaction system to stabilize a styrene–acrylic emulsion with HEMA units in the polymer chain, and an emulsion polymer with analogous molecular weight range and relative low viscosity was synthesized. Moreover, the cured film exhibited a good hardness property, good adhesive property, low toluene absorption, low water absorption, and good water resistance.

3.2. Influence of the Amount of PEGMA on Emulsifier-Free Polymerization

In our previous paper, MAA accounting for at least 8.6 wt.% of the total monomer mass was added to stabilize the emulsion; however, the viscosity of the polymerization upon changing the amount of MAA was higher than 1200 mPa·s for most of the experiment [28]. As mentioned before, the viscosity of the emulsion without MMA units in the polymer chain could be decreased when the proportion of PEGMA was increased. Furthermore, the emulsion with St units in the polymer chain was stable and of relatively low viscosity for the combination of MAA and PEGMA. Therefore, it is necessary to study the influence of the amount of PEGMA on the emulsion when both MMA and St are involved in the polymerization.

In Tables 8 and 9, the mass of PEGDMA was changed, and the mass ratio of MMA/St was constant; the mass of other ingredients was constant, and the total mass content was constant. Monomer conversion was over 98%, and the solid content of the emulsion could reach 45 wt.%, as shown in Table 8. The monomer conversion did not change obviously with the change in mass ratio. The measured molecular weight decreased when the mass ratio PEGMA/MAA was increased from 1/7 to 5/7. The reason may be that PEGMA could be used as an emulsifier and monomer, and the polymeric emulsifier favored the generation of oligomers, whereas the generated oligomers increased the index of molecular weight distribution (*Đ*) and the PDI; *Đ* or PDI was increased at this mass ratio. The molecular weight was increased when the mass ratio was increased from 5/7 to 12/7. The reason may be that PEGMA was used as a monomer, and the concentration of monomers increased with the increase in PEGMA amount, while the increase in monomer concentration could increase the average degree of emulsion polymerization [55]. The measured molecular weight ranged between 18,000 and 22,000 g·mol^{-1}, and the index of molecular weight distribution ranged between 1.70 and 1.91. Larger-diameter particles existed when the mass ratio of PEGMA/MAA was 5/7 or 7/7. The viscosity ranged from 520 mPa·s to 770 mPa·s in most cases. When the mass ratio of PEGMA/MAA was between 5/7 and 12/7, the viscosity was increased with the increase in PEGMA, but the tendency was opposite for the molecular weight distribution index and the diameter. The emulsion was stable throughout the reaction and remained stable for 6 months.

The T_g of the emulsion polymer decreased with the increase in PEGMA, as shown in Figure 2. According to the Fox equation of T_g [14], T_g of a random copolymer can be changed as a function of the weight fractions of the monomer unit and T_g of the component monomer, whereby more of the soft monomer unit leads to a decrease in the polymer T_g. The T_g of the PEGMA homopolymer (M_n of PEGMA monomer is 475 g·mol^{-1}) is −62.8 °C [56]; hence, PEGMA is a soft monomer. Thus, a greater amount of PEGMA would lead to a decrease in the T_g of a random polymer. Moreover, the T_g at all six mass ratios was in the range of room temperature applied for acrylic resin, indicating that the emulsion has a potential application in coating.

Table 8. Influence PEGMA on emulsion with St units and MMA units in polymer chain.

m(PEGMA)/ m(MAA)	Conversion (%)	$M_{n,th}$ (10^3 g·mol^{-1})	$M_{n,GPC}$ (10^3 g·mol^{-1})	*Đ*	d_p(nm)/PDI	Viscosity (mPa·s)/Rotor Type/Rotor Rate (rpm)	Solid Content (wt.%)
1/7	>99.5	6.98	21.9	1.80	342/0.073	594/F/750	42.7
3/7	98.8	7.20	21.3	1.85	333/0.097	46.9/E/750	41.5
5/7	>99.5	7.29	19.0	1.90	521/0.134	530/F/750	44.4
7/7	>99.5	7.45	21.0	1.84	507/0.176	552/F/750	44.9
9/7	>99.5	7.60	21.3	1.78	442/0.054	620/F/750	45.1
12/7	99.4	7.78	21.9	1.72	439/0.157	767/F/750	46.1

Conditions: m(MMA) + m(St) = 2.054 g and m(MMA)/m(St) = 4/6; m(MAA) = 1.165 g; n(MAA)/n(HEMA)/n(MMA)/n(St)/n(BMA)/n(BA)/n(ACPA)/n(I_2) = 12.60/18.74/7.64/11.02/40.36/7.46/1.15/1; ammonia solution (1.10 g); m(BA)/[m(St) + m(MMA)] = 1/2; the total mass of ingredients without ACPA solution and I_2 was maintained at 30.08 g in theory.

Table 9. Properties of the cured film with different amounts of PEGMA.

m(PEGMA)/*m*(MAA)	Pencil Hardness Rank	Adhesive Property Rank	Water Absorption (wt. %)
1/7	2 H	0	3.52
3/7	1 H	0	4.05
5/7	2 H	0	5.03
7/7	2 H	0	7.85
9/7	2 H	0	4.41
12/7	2 H	0	2.89

Conditions: *m*(MMA) + *m*(St) = 2.054 g and *m*(MMA)/*m*(St) = 4/6; *m*(MAA) = 1.165 g; *n*(MAA)/*n*(HEMA)/ *n*(MMA)/*n*(St)/*n*(BMA)/*n*(BA)/*n*(ACPA)/*n*(I_2) = 12.60/18.74/7.64/11.02/40.36/7.46/1.15/1; *m*(BA)/[*m*(St) + *m*(MMA)] = 1/2; ammonia solution (1.10 g); the total mass of ingredients without ACPA solution and I_2 was maintained at 30.08 g in theory.

As illustrated in Table 9, the pencil hardness rank of the cured film was at 2 H in most cases, and the adhesive property was good for all six mass ratios of PEGMA/MAA. The hardness and the adhesive property were not influenced by the mass ratio of PEGMA/MAA. Water absorption ranged between 2.80 wt.% and 7.90 wt.%. Water absorption was increased between the mass ratios of 1/7 and 7/7. The reason may be that, when the amount of MAA was constant and the amount of PEGMA was less than MAA, more PEGMA was located on the surface of the film, as PEGMA units are hydrophilic. PEGMA is a soft monomer, and PEGMA units favor the movement of chain segments in the crosslinking reaction, thereby allowing more polymer to react with amino resin. This may have enhanced the crosslinking reaction and decreased the hydrophilicity of the modified film. Thus, water absorption was decreased between the mass ratios of 7/7 and 12/7.

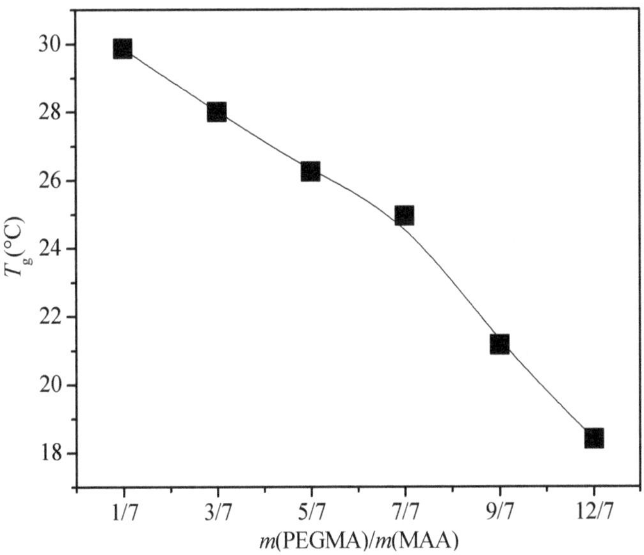

Figure 2. T_g of the emulsion polymer with different mass ratios of PEGMA/MAA.

The tensile strength and the elongation of dried emulsion polymer films are shown in Figure 3. In Figure 3A, the dried emulsion polymer film with T_g = 29.9 °C exhibited the largest maximum tensile strength (over 5.0 MPa). The maximum tensile strength of the dried emulsion polymer film was increased when the T_g of the emulsion polymer was increased, but the trend was opposite for the elongation at break. Below T_g, there is insufficient energy for whole segments of the polymer chains to move; hence, the polymer film is stiff and deformation is resisted [57]. When the film is elongated at a temperature lower than T_g, a higher T_g of the tested polymer leads to higher energy, enabling whole segments to move and more outside force to elongate the polymer film. Thus, the maximum tensile strength of the dried emulsion film increased with the increase in T_g when the tested temperature was below T_g. The largest maximum tensile strength (5.39 MPa) in the polymer with T_g = 29.9 °C was higher than that of the polyacrylate polymer with T_g = 43.9 °C (2.98 MPa) in our previous paper, indicating that the styrene–acrylic emulsion has potential application in preparing materials with mechanical properties.

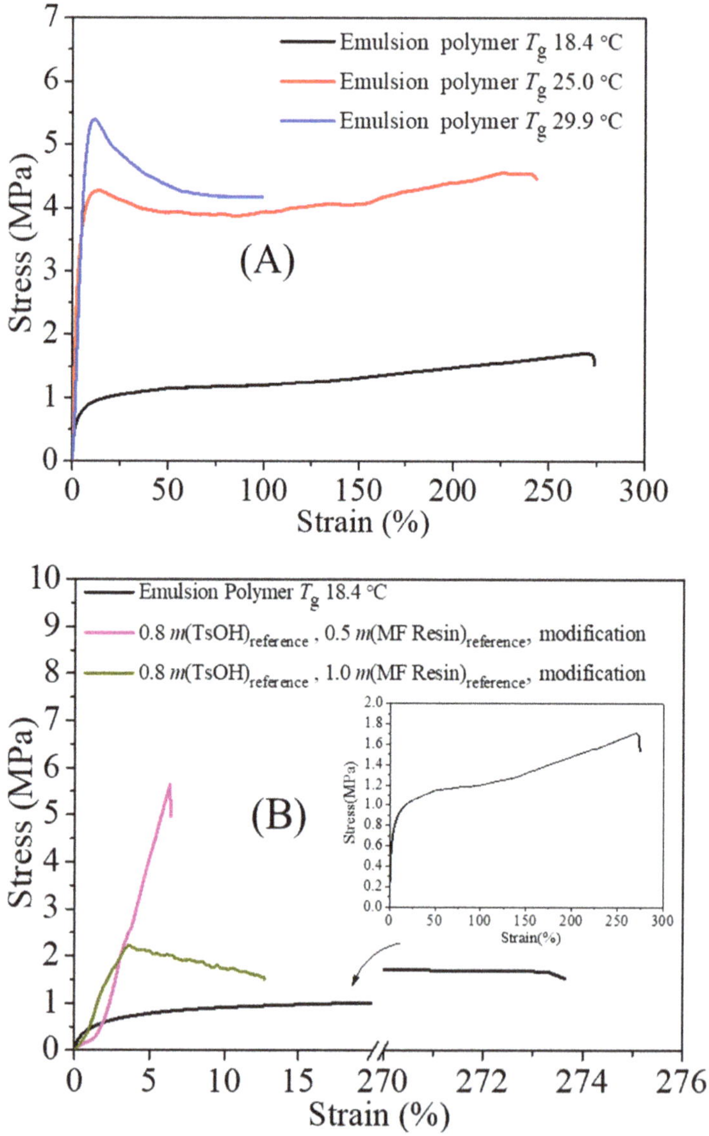

Figure 3. Stress–strain curves of (**A**) dried emulsion polymer film, and (**B**) the film modified via reaction of emulsion polymer with MF resin.

The maximum tensile strength of the modified film was higher than that of the film unmodified, as shown in Figure 3B. When half the mass of the reference MF resin was used, the maximum tensile strength of the modified film was more than 5.5 MPa. These results indicate that the tensile strength property of the emulsion film could be improved via modification of the polymer with HEMA units. Hydroxyl groups from HEMA units in the polymer chain react with MF resins to form a crosslinking structure [22], which restricts the motion of the polymer chains [18]; therefore, the strength of the modified film was higher than that of the emulsion film without modification.

In conclusion, the best results in terms of measured molecular weight, index of molecular weight distribution, particle size, viscosity, solid content, adhesive property rank, pencil hardness rank, and maximum tensile strength were obtained when the mass ratio of PEGMA/MAA was 7/7.

3.3. Influence of Iodine on Copolymerization

In Table 10, no MMA was added, hard monomer St was added, and the mass ratio of PEGMA/MAA was 1/1; the mass of iodine was changed. As shown for runs 1a to 5a in Table 10, the molar ratio of ACPA/I_2 was 1.15 when 1.0 times the mass amount of ACPA was added, and the measured M_n increased overall with the decrease in iodine, while the highest measured M_n was no more than 30,000 g·mol^{-1}. However, the increment of M_n was not significant. The monomer conversion was no more than 97%, and the solid content was no more than 41 wt.%. In Tonnar's work [38], the molar ratio of ACPA/I_2 was 1.6, and the measured M_n increased obviously with the decrease in I_2 in the presence of ACPA, while the highest measured M_n was 47,000 g·mol^{-1}. This molar ratio of ACPA/I_2 could be used to prepare a polymer with a measured M_n of more than 30,000 g·mol^{-1}. When 1.4 times the mass amount of ACPA was added, the molar ratio of ACPA/I_2 was 1.61, the measured M_n was increased from 19,400 g·mol^{-1} to 32,900 g·mol^{-1} (as shown for runs 1b to 5b in Table 10), and the largest measured M_n was more than 30,000 g·mol^{-1}. As shown for runs 3a to 3b or runs 5a to 5b, the measured M_n for 1.4 times the reference mass amount of ACPA was higher than that for 1.0 times the reference mass amount of ACPA; the reason for this phenomenon is unknown. The monomer conversion was more than 98%, and the solid content was over 42 wt.%. The monomer conversion in Table 10 did not change obviously overall when the iodine amount was increased. The monomer conversion, diameter, and solid content for 1.4 times the reference mass amount of ACPA were higher than those for 1.0 times the reference mass amount of ACPA.

Table 10. Influence of iodine on emulsion with St units in polymer chain.

Run	$m(I_2)/m(I_2)_0$	Conversion (%)	$M_{n,th}$ (10^3 g·mol^{-1})	$M_{n,GPC}$ (10^3 g·mol^{-1})	$Đ$	d_p(nm)/PDI	Solid Content (wt.%)
1a $^{\alpha}$	1/1	94.6	6.75	23.3	1.70	271/0.201	39.7
1b $^{\beta}$	1/1	>99.5	6.98	19.4	1.97	287/0.052	42.6
2a	4/5	96.4	8.35	24.8	1.71	295/0.194	40.4
3a	3/5	94.6	10.84	24.8	1.63	282/0.107	39.8
3b	3/5	98.5	11.27	32.1	1.45	333/0.096	42.2
4a	2/5	93.5	15.94	26.0	1.58	382/0.181	39.3
5a	1/4	90.4	24.50	27.1	1.50	324/0.093	38.0
5b	1/4	>99.5	27.08	32.9	1.42	373/0.255	43.7

$^{\alpha}$ $m(I_2)_0$ = 0.273 g; m(ACPA) = m(ACPA)$_0$ = 0.346 g. Conditions: m(PEGMA)/m(MAA) = 1/1 and m(MAA) + m(PEGMA) = 1.664 g; no MMA in the polymerization system; n(St)/n(HEMA)/n(BMA)/n(BA)/n(ACPA)/$n(I_2)_0$ = 18.36/18.74/40.36/7.46/1.15/1; m(BA)/m(St) = 1/2; ammonia solution (0.70 g); the total mass of ingredients without ACPA solution and I_2 was maintained at 30.08 g in theory. $^{\beta}$ $m(I_2)_0$ = 0.273 g; m(ACPA) = 1.4m(ACPA)$_0$ = 0.485 g. Conditions: m(PEGMA)/m(MAA) = 1/1 and m(MAA) + m(PEGMA) = 1.664 g; no MMA in the polymerization system; n(St)/n(HEMA)/n(BMA)/n(BA)/n(ACPA)/$n(I_2)_0$ = 18.36/18.74/40.36/7.46/1.61/1; m(BA)/m(St) = 1/2; ammonia solution (0.70 g); the total mass of ingredients without ACPA solution and I_2 was maintained at 30.08 g in theory.

The hard monomer St is usually combined with MMA in radical polymerization for determination of the copolymerization parameters based on kinetic data and quantum-chemical considerations [58]. Thus, we investigated whether the above molecular weight tendency in Table 10 existed in the styrene–methyl methacrylate-based styrene–acrylic emulsion polymer. In Table 11, the mass ratio of MMA/St was 4/6, the mass ratio of PEGMA/MAA was 3/7, and the mass of iodine was changed. As shown for runs 1b to 3b in Table 11, the monomer conversion was not changed obviously with the increase in the iodine amount, but the measured M_n for 1.4 times the mass amount of ACPA increased obviously with the decrease in iodine, and the highest measured M_n could reach 40,000 g·mol^{-1}. This molecular weight tendency could have led to some changes in the mechanical property of the polymer film. Thus, it was necessary to measure the tensile strength of the dried emulsion film with different measured M_n. The maximum tensile strength was increased with the increase in polymer M_n, and the largest maximum tensile strength was more than 5.5 MPa, as shown in Figure 4. Polymer chains with a high molecular weight become large and are, hence, entangled [18]; thus, a higher molecular weight promotes entanglements, which can act as junction points and govern the material's mechanical response [59]. Thus, a polymer with a high molecular weight exhibits high strength, including tensile strength. The elongation at break of the dried emulsion film with M_n 40,700 g·mol^{-1} was more than 100%, indicating flexibility of the polymer film. Therefore, the polymer with the highest M_n over 40,000 g·mol^{-1} has some significance, and this polymerization methodology may provide potential application for preparing styrene–acrylic emulsions used in materials with excellent mechanical properties.

Table 11. Influence of iodine on emulsion with MMA and St units in polymer chain.

Run	$m(I_2)/m(I_2)_0$	Conversion (%)	$M_{n,th}$ (10^3 g·mol^{-1})	$M_{n,GPC}$ (10^3 g·mol^{-1})	Đ	d_p(nm)/PDI	Solid Content (wt.%)
1a $^{\alpha}$	1/1	98.8	7.06	21.3	1.85	333/0.097	41.5
1b $^{\beta}$	1/1	>99.5	7.14	21.9	1.81	327/0.088	43.9
2b	3/5	93.3	10.94	25.2	1.66	407/0.074	39.9
3b	1/4	>99.5	27.73	40.7	1.27	403/0.132	43.6

$^{\alpha}$ $m(I_2)_0$ = 0.273 g; m(ACPA) = m(ACPA)$_0$ = 0.346 g. Conditions: m(MMA)/m(St) = 4/6 and m(MMA) + m(St) = 2.054 g; m(BA)/[m(MMA) + m(St)] = 1/2; m(PEGMA)/m(MAA) = 3/7; m(MAA) + m(PEGMA) = 1.664 g; n(MAA)/n(PEGMA)/n(HEMA)/n(BMA)/n(BA)/n(ACPA)/$n(I_2)_0$ = 12.60/0.98/18.74/40.36/7.46/1.15/1; ammonia solution (1.10 g); the total mass of ingredients without ACPA solution and I_2 was maintained at 30.08 g in theory. $^{\beta}$ $m(I_2)_0$ = 0.273 g; m(ACPA) = 1.4m(ACPA)$_0$ = 0.485 g. Conditions: m(MMA)/m(St) = 4/6 and m(MMA) + m(St) = 2.054 g; m(BA)/[m(MMA) + m(St)] = 1/2; m(PEGMA)/m(MAA) = 3/7; m(MAA) + m(PEGMA) = 1.664 g; n(MAA)/n(PEGMA)/n(HEMA)/n(BMA)/n(BA)/n(ACPA)/$n(I_2)_0$ = 12.60/0.98/18.74/40.36/7.46/1.15/1; ammonia solution (1.10 g); the total mass of ingredients without ACPA solution and I_2 was kept at 30.08 g in theory.

The above result indicates that a styrene–acrylic emulsion polymer with a relatively higher molecular weight could be prepared by reducing the I_2 amount when 1.4 times the mass amount of initiator was used, and the dried emulsion film could exhibit a larger maximum tensile strength at higher molecular weight. Therefore, the polymer film has potential application in materials with excellent mechanical properties.

3.4. Kineticks in RITP Emulsion Copolymerization

The monomer conversion was increased with time after 40 min, as shown in Figure 5. In the first 40 min of reaction time, the mixture was brown, indicating that iodine was not completely consumed in the induction period.

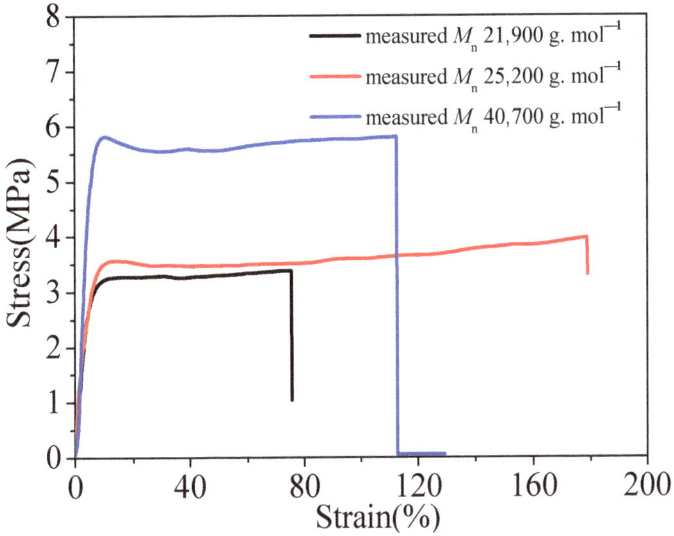

Figure 4. Stress–strain curves of emulsion film with different measured molecular weights.

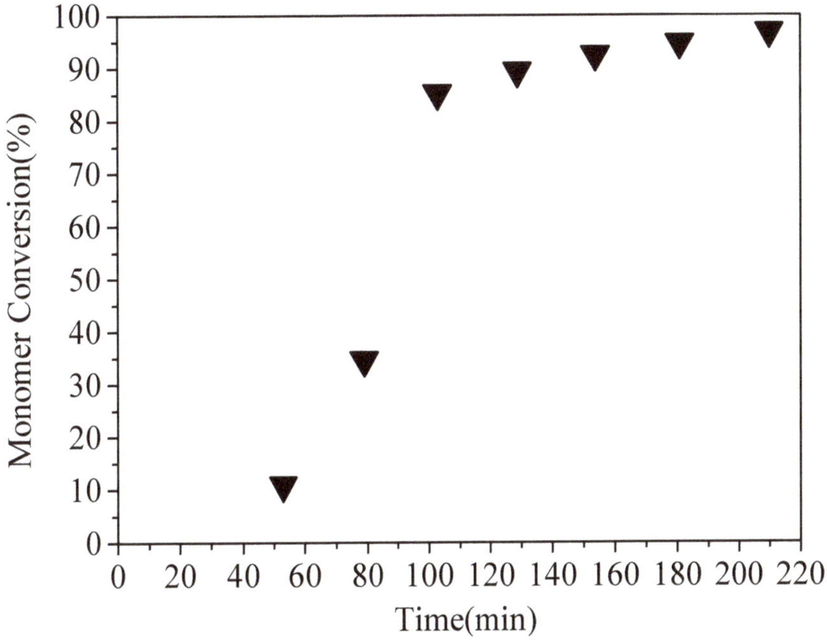

Figure 5. Evolution of monomer conversion versus time for RITP emulsion copolymerization. Conditions: m(PEGMA)/m(MAA) = 1/1 and m(MAA) + m(PEGMA) = 1.664 g; m(BA)/m(St) = 1/2; n(St)/n(HEMA)/n(BMA)/n(BA)/n(ACPA)/$n(I_2)$ = 18.36/18.74/40.36/7.46/1.15/1, ammonia solution (0.70 g); $m(I_2)$ = 0.273 g; m(ACPA) = 0.346 g; no MMA in the polymerization system; the total mass of ingredients without ACPA Solution and I_2 was maintained at 30.08 g in theory; the reaction time was 210 min.

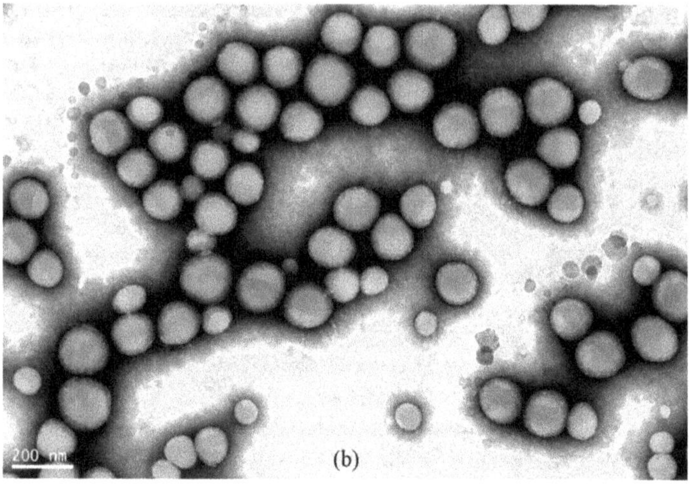

Figure 7. TEM micrograph of (**a**) the seed polymer and (**b**) the block copolymer described in Table 12.

We investigated whether the random copolymer can be used as macro-chain transfer agent or macro-emulsifier in the chain extension reaction when BA and HEMA were added in at second polymerization stage. As shown in Table 13, the emulsion of the random copolymer poly (PEGMA-*co*-MAA^a-*co*-BMA-*co*-MMA-St) was stable and milky white with a weak blue color. When the monomer mixture comprising BA and HEMA was added at the second polymerization stage, the emulsion was not stable, and a lot of coagulation existed after 33 min of reaction time. The pH value of the withdrawn sample ranged between 6.5 and 7, and there were no nonionic MAA monomers that participated in random copolymerization with HEMA, thereby restraining the formation of an interpolymer complex. Crosslinking caused by transesterification among the pendent hydroxyl groups can lead to crosslinked polymers [23,24], and these crosslinked polymers may lead to instability of the emulsion. Thus, the emulsion was not stable, and a large amount of gel existed after 33 min. The instability of the emulsion in the chain extension reaction period indicated that HEMA could not be copolymerized fluently in the chain extension reaction of this emulsion polymerization system.

Table 13. Chain extension reaction with BA and HEMA.

Type	Stage Time (min)	Conversion (%)	Solid Content (wt.%)	Emulsion Appearance
Seed Polymer Poly(PEGMA-*co*-MAA^a-*co*-BMA-*co*-MMA-St)	168	96.2	38.5	Opalescent with weak blue color.
Block Copolymer Poly(PEGMA-*co*-MAA^a-*co*-BMA-*co*-MMA-St)-*b*-Poly(PEGMA-*co*-MAA^a-*co*-BMA-*co*-MMA-St-*co*-BA-*co*-HEMA)	33	-	-	Opalescent, a lot of gel.

MAA^a was neutralized by ammonia solution. Conditions: m(PEGMA)/m(MAA) = 12/7 and m(MAA) + m(PEGMA) = 3.1611 g; m(BA)/ [m(St) + m(MMA)] = 1/2; ammonia solution (1.10 g); n(PEGMA)/n(MAA)/n(St)/n(HEMA)/n(BMA)/n(BA)/n(MMA)/n(ACPA)/n(I_2) = 6.52/21/18.38/31.22/67.28/12.44/12.74/1.92/1; the total mass of ingredients without ACPA solution and I_2 was maintained at 30.08 g in theory.

In conclusion, the living polymerization of the random copolymer chain was proven by the chain extension reaction containing BA and St in the emulsion. Moreover, the copolymer prepared via the chain extension reaction exhibited a higher measured M_n, lower $Đ$, larger particle size, and higher solid content than the random copolymer at the first polymerization stage. The TEM results indicate the increase in particle size and the regular shape of the micelle. However, the chain extension reaction with BA and HEMA was not successful; thus, there exists some limitation to the chain extension reaction of the seed random copolymer with other kinds of monomer.

3.6. Infrared Spectra of Polymer

The FTIR spectra of the copolymer are illustrated in Figure 8. The wide absorption peak at 3430 cm^{-1} was caused by O–H stretching vibration. The wide absorption peak at 3224 cm^{-1} was derived from N–H stretching vibration of NH_4^+ located in the ammonium salt of MAA units. The two peaks at 2955 and 2868 cm^{-1} were ascribed to C–H stretching vibrations of $-CH_3$ and $-CH_2$ groups, respectively. The two peaks at 1451 and 1383 cm^{-1} were caused by C–H bending vibrations of $-CH_3$ and $-CH_2$ groups, respectively. The strong absorption peak at 1723 cm^{-1} was derived from the stretching vibration of the carbonyl ester C=O. The peak at 1544 cm^{-1} was caused by the asymmetric stretching vibration of carbonyl anion COO^-. The bands of 1240 and 1068 cm^{-1} were derived from asymmetric stretching vibration and symmetric stretching vibration of the ester group C–O–C, respectively. The strong absorption peak at 1150 cm^{-1} was ascribed to the bending vibration of the ether group C–O–C from PEGMA units and that of C–O–H from HEMA units. With the increase in mass ratio of PEGMA/MAA, the peak intensity at 1544 cm^{-1} was deceased, indicating a decrease in MAA content in the polymer chain. The FTIR spectra indicate the influence of the mass ratio of PEGMA/MAA on the signal changes representing polymeric groups.

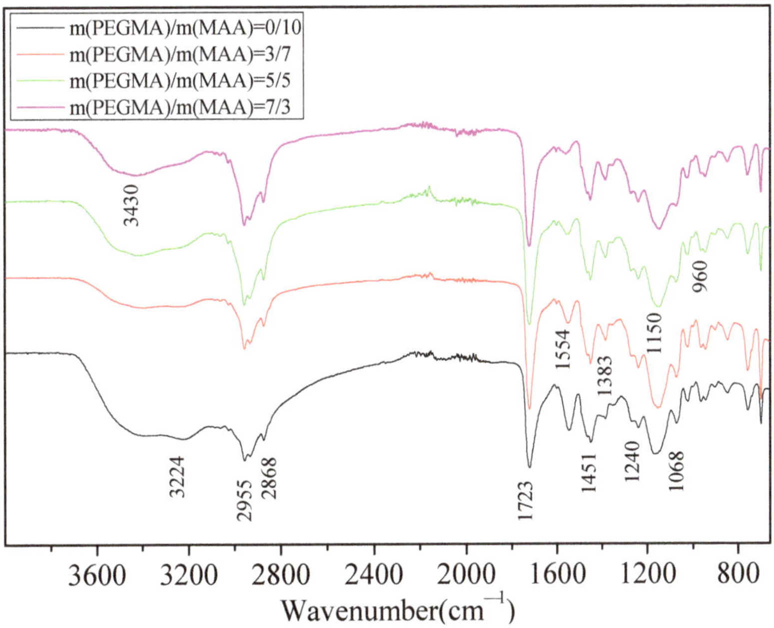

Figure 8. FTIR spectra of copolymer with different mass ratios of PEGMA/MAA. Polymerization conditions: m(MAA) + m(PEGMA) = 1.664 g; no MMA in the polymerization system; the MAA solution was neutralized by ammonia solution; m(BA)/m(St) = 1/2; n(St)/n(HEMA)/n(BMA)/n(BA)/n(ACPA)/$n(I_2)$ = 18.36/18.74/40.36/7.46/1.15/1; the total mass of ingredients without ACPA solution and I_2 was maintained at 30.08 g in theory.

4. Conclusions

Emulsifier-free styrene–acrylic emulsions prepared via RITP were studied in this paper. When the ammonium salt of MAA was used as the only kind of polymeric surfactant and St was added into the polymerization system containing HEMA monomer component, the emulsion did not flow fluently and was sticky. Thus, it was necessary to explore a method to prepare stable styrene–acrylic emulsions with a high content of HEMA monomer component. In this paper, a stable styrene–acrylic emulsion with low viscosity and an analogous M_n range to previous work (from 23,000 to 30,000 g·mol^{-1}) could be prepared when PEGMA was combined with the ammonium salt of MAA. The acrylate-based polymer emulsion with HEMA units in the polymer chain was not stable when only PEGMA was used as a polymeric emulsifier. St is more hydrophobic than MMA; hence, the polymerization containing St monomer units was conducted in combination with the ammonium salt of MAA and PEGMA. The influence of PEGMA and I_2 on the emulsion was studied. With the increase in PEGMA amount, the T_g of the polymer decreased; the largest maximum tensile strength (5.39 MPa) in the polymer with T_g = 29.9 °C was larger than that of the polyacrylate polymer with T_g = 43.9 °C (2.98 MPa) in our previous paper. When polymerization was conducted in 1.4 times the reference amount of initiator, M_n was obviously increased with the decrease in I_2, and the highest M_n of the polymer with HEMA units (40,700 g·mol^{-1}) was larger than that in our previous paper (32,700 g·mol^{-1}) when the mass ratio of BA/BMA was 1/6, while the largest maximum tensile strength of the dried styrene–acrylic emulsion polymer film with the highest M_n was more than 5.5 MPa. The living polymerization of the random copolymer chain was proven by a kinetics experiment and chain extension reaction; the uniformly shaped particles and the increase in particle diameter according to TEM indicated that the random copolymer can serve as a seed emulsifier in the polymerization. The novelty lies in that the polymerization was conducted with a high solid content (ranging from 40 wt.% to 46 wt.%) and a high content of hydroxyl monomer component, while St took part in the polymerization; a moderate M_n range (20,000–41,000 g·mol^{-1}) was achieved in a short period of time (<4 h), and a stable emulsion with moderate viscosity (ranging from 100 mPa·s to 700 mPa·s) was successfully prepared. The polymerization was conducted in water, and no costly nonpolymeric emulsifier or organic solvent was used. Furthermore, the amount of MAA could be reduced when MAA was combined with PEGMA, and this strengthened the stability of the emulsion when ammonium hydroxide used for neutralizing MAA was volatilized quickly in hot weather. This protocol represents an environmentally friendly system tailored for the direct preparation of an emulsion used for maintenance coating. In summary, the work in this article provides a synthetic method for preparing high-solid-content styrene–acrylic emulsions with HEMA units in the polymer chain, and the synthetic method may have application in preparing coatings or polymer materials with excellent mechanical properties.

Author Contributions: Conceptualization, T.H. and S.G.; methodology, T.H. and S.G.; validation, T.H.; formal analysis, T.H.; investigation, T.H.; writing—original draft preparation, T.H.; writing—review and editing, T.H.; visualization, T.H.; supervision, S.G.; funding acquisition, S.G. Both authors have read and agreed to the published version of the manuscript.

Funding: This research received no external funding.

Institutional Review Board Statement: Not applicable.

Informed Consent Statement: Not applicable.

Conflicts of Interest: The authors declare no conflict of interest.

References

1. Jones, F.N.; Nichols, M.E.; Pappas, S.P. Latexes. In *Organic Coatings: Science and Technology*, 4th ed.; John Wiley & Sons: Hoboken, NJ, USA, 2017; p. 133.
2. Jiao, C.; Sun, L.; Shao, Q.; Song, J.; Hu, Q.; Naik, N.; Guo, Z. Advances in Waterborne Acrylic Resins: Synthesis Principle, Modification Strategies, and Their Applications. *ACS Omega* **2021**, *6*, 2443–2449. [CrossRef] [PubMed]
3. Sajjadi, S.; Jahanzad, F. Nanoparticle formation by highly diffusion-controlled emulsion polymerisation. *Chem. Eng. Sci.* **2006**, *61*, 3001–3008. [CrossRef]

4. Schmid, A.; Scherl, P.; Armes, S.P.; Leite, C.A.P.; Galembeck, F. Synthesis and Characterization of Film-Forming Colloidal Nanocomposite Particles Prepared via Surfactant-Free Aqueous Emulsion Copolymerization. *Macromolecules* **2009**, *42*, 3721–3728. [CrossRef]
5. Xu, G.; Lei, X.; Hu, J.; Pi, P.; Yang, Z. Preparation of Raspberry-Like Nanocomposite Microspheres Based on Soap-Free Cationic Latex and Nano-SiO_2 Particles. *Polym-Plast. Technol. Eng.* **2013**, *52*, 510–513. [CrossRef]
6. Bladé, T.; Malosse, L.; Duguet, E.; Lansalot, M.; Bourgeat-Lami, E.; Ravaine, S. Synthesis of nanoscaled poly(styrene-co-n-butyl acrylate)/silica particles with dumbbell- and snowman-like morphologies by emulsion polymerization. *Polym. Chem.* **2014**, *5*, 5609–5616. [CrossRef]
7. Guimarães, T.R.; Khan, M.; Kuchel, R.P.; Morrow, I.C.; Minami, H.; Moad, G.; Perrier, S.; Zetterlund, P.B. Nano-Engineered Multiblock Copolymer Nanoparticles via Reversible Addition–Fragmentation Chain Transfer Emulsion Polymerization. *Macromolecules* **2019**, *52*, 2965–2974. [CrossRef]
8. Deane, O.J.; Musa, O.M.; Fernyhough, A.; Armes, S.P. Synthesis and Characterization of Waterborne Pyrrolidone-Functional Diblock Copolymer Nanoparticles Prepared via Surfactant-free RAFT Emulsion Polymerization. *Macromolecules* **2020**, *53*, 1422–1434. [CrossRef]
9. Sundberg, D. Structured, Composite Nanoparticles from Emulsion Polymerization–Morphological Possibilities. *Biomacromolecules* **2020**, *21*, 4388–4395. [CrossRef]
10. Garay-Jimenez, J.C.; Gergeres, D.; Young, A.; Lim, D.V.; Turos, E. Physical properties and biological activity of poly(butyl acrylate–styrene) nanoparticle emulsions prepared with conventional and polymerizable surfactants. *Nanotechnol. Biol. Med.* **2009**, *5*, 443–451. [CrossRef]
11. Turos, E.; Garay-Jimenez, J.C.; Sona, A.J. An evaluation of non-ionic surfactants on the cytotoxicity and activity of poly(butyl acrylate/styrene) nanoparticle emulsions. *J. Nanopart Res.* **2019**, *21*, 165. [CrossRef]
12. Kakran, M.; Antipina, M.N. Emulsion-based techniques for encapsulation in biomedicine, food and personal care. *Curr. Opin. Pharmacol.* **2014**, *18*, 47–55. [CrossRef] [PubMed]
13. Rajaee, A.; Farzi, G. Encapsulation of paclitaxel in ultra-fine nanoparticles of acrylic/styrene terpolymer for controlled release. *Colloid Polym. Sci.* **2016**, *294*, 95–105. [CrossRef]
14. Jones, F.N.; Nichols, M.E.; Pappas, S.P. Polymerization and Film Formation. In *Organic Coatings: Science and Technology*, 4th ed.; John Wiley & Sons: Hoboken, NJ, USA, 2017; p. 12.
15. Barandiaran, M.J.; Cal, J.C.; Asua, J.M. Emulsion Polymerization. In *Polymer Reaction Engineering*; Blackwell Publishing Ltd.: Oxford, UK, 2007; p. 233.
16. Wang, T.; Canetta, E.; Weerakkody, T.G.; Keddie, J.L.; Rivas, U. pH Dependence of the Properties of Waterborne Pressure-Sensitive Adhesives Containing Acrylic Acid. *ACS Appl. Mater. Interfaces* **2009**, *1*, 631–639. [CrossRef] [PubMed]
17. Jones, F.N.; Nichols, M.E.; Pappas, S.P. Latexes. In *Organic Coatings: Science and Technology*, 4th ed.; John Wiley & Sons: Hoboken, NJ, USA, 2017; p. 126.
18. Balani, K.; Verma, V.; Agarwal, A.; Narayan, R. Physical, Thermal, and Mechanical Properties of Polymers. In *Biosurfaces*; John Wiley & Sons: Hoboken, NJ, USA, 2014; p. 337.
19. Barandiaran, M.J.; Cal, J.C.; Asua, J.M. Emulsion Polymerization. In *Polymer Reaction Engineering*; Blackwell Publishing Ltd.: Oxford, UK, 2007; p. 238.
20. Darabi, A.; Rezaee Shirin-Abadi, A.; Avar, S.; Cunningham, M.F. Surfactant-free emulsion copolymerization of styrene and methyl methacrylate for preparation of water-redispersible polymeric powders. *J. Polym. Sci. Part A Polym. Chem.* **2018**, *56*, 2376–2381. [CrossRef]
21. Darabi, A.; Cunningham, M.F. Preparation of Poly(poly(ethylene glycol)methyl ether methacrylate-co-styrene)-b-poly(2-(diethylamino)ethyl methacrylate-co-acrylonitrile) by nitroxide-mediated polymerisation in water. *Polymer* **2017**, *115*, 255–260. [CrossRef]
22. Jones, F.N.; Nichols, M.E.; Pappas, S.P. Amino Resins. In *Organic Coatings: Science and Technology*, 4th ed.; John Wiley & Sons: Hoboken, NJ, USA, 2017; p. 156.
23. Robinson, K.L.; Khan, M.A.; de Paz Báñez, M.V.; Wang, X.S.; Armes, S.P. Controlled Polymerization of 2-Hydroxyethyl Methacrylate by ATRP at Ambient Temperature. *Macromolecules* **2001**, *34*, 3155–3158. [CrossRef]
24. Ouchi, M.; Yoda, H.; Terashima, T.; Sawamoto, M. Aqueous metal-catalyzed living radical polymerization: Highly active water-assisted catalysis. *Polym. J.* **2012**, *44*, 51–58. [CrossRef]
25. Zhang, C.-Y.; Zhu, Z.-W.; Gong, S.-L. Synthesis of stable high hydroxyl content self-emulsifying waterborne polyacrylate emulsion. *J. Appl. Polym. Sci.* **2017**, *134*, 44844. [CrossRef]
26. Mun, G.A.; Khutoryanskiy, V.V.; Akhmetkalieva, G.T.; Shmakov, S.N.; Dubolazov, A.V.; Nurkeeva, Z.S.; Park, K. Interpolymer complexes of poly(acrylic acid) with poly(2-hydroxyethyl acrylate) in aqueous solutions. *Colloid Polym. Sci.* **2004**, *283*, 174–181. [CrossRef]
27. Mun, G.A.; Nurkeeva, Z.S.; Beissegul, A.B.; Dubolazov, A.V.; Urkimbaeva, P.I.; Park, K.; Khutoryanskiy, V.V. Temperature-Responsive Water-Soluble Copolymers Based on 2-Hydroxyethyl Acrylate and Butyl Acrylate. *Macromol. Chem. Phys.* **2007**, *208*, 979–987. [CrossRef]
28. Huang, T.; Yuan, Q.-X.; Gong, S.-L. Emulsifier-Free Acrylate-Based Emulsion Prepared by Reverse Iodine Transfer Polymerization. *Polymers* **2020**, *12*, 730. [CrossRef]

29. Ottewill, R.H. Stabilization of polymer colloid dispersions. In *Emulsion Polymerization and Emulsion Polymers*; Lovell, P.A., El-Aasser, M.S., Eds.; John Wiley & Sons: Chichester, UK, 1997; p. 105.
30. Tuncel, A. Emulsion copolymerization of styrene and poly(ethylene glycol) ethyl ether methacrylate. *Polymer* **2000**, *41*, 1257–1267. [CrossRef]
31. Hong, C.K.; Hwang, M.J.; Ryu, D.W.; Moon, H. Preparation of copolymer particles by emulsion polymerization using a polymerizable amphiphilic macromonomer. *Colloids Surf. A Physicochem. Eng. Asp.* **2008**, *331*, 250–256. [CrossRef]
32. Zetterlund, P.B.; Thickett, S.C.; Perrier, S.; Bourgeat-Lami, E.; Lansalot, M. Controlled/Living Radical Polymerization in Dispersed Systems: An Update. *Chem. Rev.* **2015**, *115*, 9745–9800. [CrossRef] [PubMed]
33. Ravve, A. Free-Radical Chain-Growth Polymerization. In *Principles of Polymer Chemistry*, 3rd ed.; Springer: New York, NY, USA, 2012; p. 115.
34. Rieger, J.; Stoffelbach, F.; Bui, C.; Alaimo, D.; Jérôme, C.; Charleux, B. Amphiphilic Poly(ethylene oxide) Macromolecular RAFT Agent as a Stabilizer and Control Agent in ab Initio Batch Emulsion Polymerization. *Macromolecules* **2008**, *41*, 4065–4068. [CrossRef]
35. Zhang, X.; Boissé, S.; Zhang, W.; Beaunier, P.; D'Agosto, F.; Rieger, J.; Charleux, B. Well-Defined Amphiphilic Block Copolymers and Nano-objects Formed in Situ via RAFT-Mediated Aqueous Emulsion Polymerization. *Macromolecules* **2011**, *44*, 4149–4158. [CrossRef]
36. Truong, N.P.; Dussert, M.V.; Whittaker, M.R.; Quinn, J.F.; Davis, T.P. Rapid synthesis of ultrahigh molecular weight and low polydispersity polystyrene diblock copolymers by RAFT-mediated emulsion polymerization. *Polym. Chem.* **2015**, *6*, 3865–3874. [CrossRef]
37. Ni, Y.; Tian, C.; Zhang, L.; Cheng, Z.; Zhu, X. Photocontrolled Iodine-Mediated Green Reversible-Deactivation Radical Polymerization of Methacrylates: Effect of Water in the Polymerization System. *ACS Macro Lett.* **2019**, *8*, 1419–1425. [CrossRef]
38. Tonnar, J.; Lacroix-Desmazes, P.; Boutevin, B. Living Radical Ab Initio Emulsion Polymerization of n-Butyl Acrylate by Reverse Iodine Transfer Polymerization. In *Controlled/Living Radical Polymerization: From Synthesis to Materials*; Matyjasjzewski, K., Ed.; American Chemical Society: Washington, DC, USA, 2006; Volume 944, pp. 604–619.
39. Koltzenburg, S.; Maskos, M.; Nuyken, O. Radical Polymerization. In *Polymer Chemistry*; Springer: Berlin/Heidelberg, Germany, 2017; p. 241.
40. Tuncel, A.; Serpen, E. Emulsion copolymerization of styrene and methacrylic acid in the presence of a polyethylene oxide based-polymerizable stabilizer with a shorter chain length. *Colloid Polym. Sci.* **2001**, *279*, 240–251. [CrossRef]
41. Nagy, K.; Körtvélyesi, T.; Nagypál, I. Iodine Hydrolysis Equilibrium. *J. Solut. Chem.* **2003**, *32*, 385–393. [CrossRef]
42. Tonnar, J.; Lacroix-Desmazes, P.; Boutevin, B. Controlled Radical Ab Initio Emulsion Polymerization of n-Butyl Acrylate by Reverse Iodine Transfer Polymerization (RITP): Effect of the Hydrolytic Disproportionation of Iodine. *Macromol. Rapid Commun.* **2006**, *27*, 1733–1738. [CrossRef]
43. Tonnar, J.; Lacroix-Desmazes, P.; Boutevin, B. Living Radical ab Initio Emulsion Polymerization of n-Butyl Acrylate by Reverse Iodine Transfer Polymerization (RITP): Use of Persulfate as Both Initiator and Oxidant. *Macromolecules* **2007**, *40*, 6076–6081. [CrossRef]
44. Lacroix-Desmazes, P.; Tonnar, J.; Boutevin, B. Reverse Iodine Transfer Polymerization (RITP) in Emulsion. *Macromol. Symp.* **2007**, *248*, 150–157. [CrossRef]
45. Tonnar, J.; Lacroix-Desmazes, P.; Boutevin, B. Controlled Radical Polymerization of Styrene by Reverse Iodine Transfer Polymerization (RITP) in Miniemulsion: Use of Hydrogen Peroxide as Oxidant. *Macromolecules* **2007**, *40*, 186–190. [CrossRef]
46. Lorandi, F.; Wang, Y.; Fantin, M.; Matyjaszewski, K. Ab Initio Emulsion Atom-Transfer Radical Polymerization. *Angew. Chem. Int. Ed.* **2018**, *57*, 8270–8274. [CrossRef] [PubMed]
47. Cowie, J.M.G.; Arrighi, V. Free-Radical Addition Polymerization. In *Polymers: Chemistry and Physics of Modern Materials*, 3rd ed.; CRC Press: Boca Raton, FL, USA, 2007; pp. 67–70.
48. McCord, E.F.; Shaw, W.H.; Hutchinson, R.A. Short-chain branching structures in ethylene copolymers prepared by high-pressure free-radical polymerization: An NMR analysis. *Macromolecules* **1997**, *30*, 246–256. [CrossRef]
49. Plessis, C.; Arzamendi, G.; Leiza, J.R.; Schoonbrood, H.A.S.; Charmot, D.; Asua, J.M. Seeded Semibatch Emulsion Polymerization of n-Butyl Acrylate. Kinetics and Structural Properties. *Macromolecules* **2000**, *33*, 5041–5047. [CrossRef]
50. Asua, J.M. Emulsion polymerization: From fundamental mechanisms to process developments. *J. Polym. Sci. Part A Polym. Chem.* **2004**, *42*, 1025–1041. [CrossRef]
51. Cowie, J.M.G.; Arrighi, V. Polymers in Solution. In *Polymers: Chemistry and Physics of Modern Materials*, 3rd ed.; CRC Press: Boca Raton, FL, USA, 2007; p. 197.
52. Brown, R.; Stützel, B.; Sauer, T. Steric stabilization by grafting and copolymerization of water-soluble oligomers and polymers. *Macromol. Chem. Phys.* **1995**, *196*, 2047–2064. [CrossRef]
53. Ottewill, R.H.; Satgurunathan, R.; Waite, F.A.; Westby, M.J. Non-ionic polystyrene latices in aqueous media. *Br. Polym. J.* **1987**, *19*, 435–440. [CrossRef]
54. Ravve, A. Free-Radical Chain-Growth Polymerization. In *Principles of Polymer Chemistry*, 3rd ed.; Springer: New York, NY, USA, 2012; p. 84.
55. Barandiaran, M.J.; Cal, J.C.; Asua, J.M. Emulsion Polymerization. In *Polymer Reaction Engineering*; Blackwell Publishing Ltd.: Oxford, UK, 2007; p. 250.

56. Szabó, Á.; Szarka, G.; Iván, B. Synthesis of Poly(poly(ethylene glycol) methacrylate)–Polyisobutylene ABA Block Copolymers by the Combination of Quasiliving Carbocationic and Atom Transfer Radical Polymerizations. *Macromol. Rapid Commun.* **2015**, *36*, 238–248. [CrossRef] [PubMed]
57. Ravve, A. Physical Properties and Physical Chemistry of Polymers. In *Principles of Polymer Chemistry*, 3rd ed.; Springer: New York, NY, USA, 2012; p. 22.
58. Ravve, A. Common Chain-Growth Polymers. In *Principles of Polymer Chemistry*, 3rd ed.; Springer: New York, NY, USA, 2012; p. 370.
59. Jones, F.N.; Nichols, M.E.; Pappas, S.P. Polymerization and Film Formation. In *Organic Coatings: Science and Technology*, 4th ed.; John Wiley & Sons: Hoboken, NJ, USA, 2017; p. 9.
60. Tonnar, J.; Lacroix-Desmazes, P. One-pot surfactant-free functional latexes by controlled radical polymerization in ab initio emulsion. *Soft Matter* **2008**, *4*, 1255–1260. [CrossRef] [PubMed]
61. Lim, J.; Yeap, S.P.; Che, H.X.; Low, S.C. Characterization of magnetic nanoparticle by dynamic light scattering. *Nanoscale Res. Lett.* **2013**, *8*, 381. [CrossRef]
62. Bell, N.C.; Minelli, C.; Tompkins, J.; Stevens, M.M.; Shard, A.G. Emerging Techniques for Submicrometer Particle Sizing Applied to Stöber Silica. *Langmuir* **2012**, *28*, 10860–10872. [CrossRef] [PubMed]
63. Kumar, N.; Kumbhat, S. Characterization Tools for Nanomaterials. In *Essentials in Nanoscience and Nanotechnology*; John Wiley & Sons: Hoboken, NJ, USA, 2016; pp. 77–148.

polymers MDPI

Article

Molecular Brushes with a Polyimide Backbone and Poly(ε-Caprolactone) Side Chains by the Combination of ATRP, ROP, and CuAAC

Anna V. Kashina *, Tamara K. Meleshko, Natalia N. Bogorad, Viktor K. Lavrentyev and Alexander V. Yakimansky

Institute of Macromolecular Compounds, Russian Academy of Sciences, Bolshoy pr. 31, 199004 Saint Petersburg, Russia; meleshko@hq.macro.ru (T.K.M.); bogorad@hq.macro.ru (N.N.B.); lavrentev1949@mail.ru (V.K.L.); yakimansky@yahoo.com (A.V.Y.)

* Correspondence: kashina.anna@mail.ru

Abstract: An approach to the synthesis of the novel molecular brushes with a polyimide (PI) backbone and poly(ε-caprolactone) (PCL) side chains was developed. To obtain such copolymers, a combination of various synthesis methods was used, including polycondensation, atom transfer radical polymerization (ATRP), ring opening polymerization (ROP), and Cu (I)-catalyzed azide-alkyne Huisgen cycloaddition (CuAAC). ATRP of 2-hydroxyethyl methacrylate (HEMA) on PI macroinitiator followed by ROP of ε-caprolactone (CL) provided a "brush on brush" structure PI-*g*-(PHEMA-*g*-PCL). For the synthesis of PI-*g*-PCL two synthetic routes combining ROP and CuAAC were compared: (1) polymer-analogous transformations of a multicenter PI macroinitiator with an initiating hydroxyl group separated from the main chain by a triazole ring followed by ROP of CL, or (2) a separate synthesis of macromonomers with the desirable functional groups (polyimide with azide groups and PCL with terminal alkyne groups), followed by a click reaction. Results showed that the first approach allows to obtain graft copolymers with a PI backbone and relatively short PCL side chains. While the implementation of the second approach leads to a more significant increase in the molecular weight, but unreacted linear PCL remains in the system. Obtained macroinitiators and copolymers were characterized using ^{1}H NMR and IR spectroscopy, their molecular weight characteristics were determined by SEC with triple detection. TGA and DSC were used to determine their thermal properties. X-ray scattering data showed that the introduction of a polyimide block into the polycaprolactone matrix did not change the degree of crystallinity of PCL.

Keywords: molecular brushes; polyimide; poly(ε-caprolactone); ATRP; ROP; CuAAC

Citation: Kashina, A.V.; Meleshko, T.K.; Bogorad, N.N.; Lavrentyev, V.K.; Yakimansky, A.V. Molecular Brushes with a Polyimide Backbone and Poly(ε-Caprolactone) Side Chains by the Combination of ATRP, ROP, and CuAAC. *Polymers* **2021**, *13*, 3312. https://doi.org/10.3390/polym13193312

Academic Editor: Edina Rusen

Received: 27 August 2021
Accepted: 22 September 2021
Published: 28 September 2021

Publisher's Note: MDPI stays neutral with regard to jurisdictional claims in published maps and institutional affiliations.

Copyright: © 2021 by the authors. Licensee MDPI, Basel, Switzerland. This article is an open access article distributed under the terms and conditions of the Creative Commons Attribution (CC BY) license (https://creativecommons.org/licenses/by/4.0/).

1. Introduction

Polyimides (PI) are an essential class of polymers with a set of valuable properties. In addition to heat, thermal, radiation, and chemical resistance, good film-forming properties, high mechanical strength, and excellent insulating properties, polyimides demonstrate good biocompatibility in vitro and in vivo, as well as low toxicity [1–3]. Multicomponent copolymers, in which PI blocks are combined with blocks of biocompatible aliphatic polymers, have great potential for the development of materials for tissue engineering.

Although atom transfer radical polymerization (ATRP) has been successfully used to synthesize many types of copolymers, it remains a challenge to synthesize copolymers of different architectures using monomers that polymerize through fundamentally different mechanisms, such as ATRP and ring opening polymerization (ROP). To solve this problem, one can use initiators containing two types of functional groups capable of initiating each of these processes in parallel and independently [4,5]. Second approach is introduction of a functional group that initiates the polymerization of the second monomer at the stage of initiation or termination of the first monomer polymerization (post-modification) [6].

Thus, in [7], graft copolymers with poly(2-hydroxyethyl methacrylate) (PHEMA) backbone and block copolymer side chains containing blocks of poly(ε-caprolactone) PCL and poly(butyl acrylate) (PBA) were synthesized. Copolymers of PCL with poly(octadecyl methacrylate) (POMA) and poly(N,N-dimethylamino-2-ethyl methacrylate) (PDMAEMA) PCL-*b*-PODMA-*b*-PDMAEMA or PCL-*b*-(PODMA-*co*-PDMAEMA) were obtained in [5], in which the PODMA block had a "pseudo-brush" structure due to long aliphatic tails. Another example of a successful combination of the different methods of controlled polymerization is the synthesis of a three-armed star-shaped polymer with arms of different structures, for which different methods were used [4]. Thus, the PCL arm was obtained using ROP, the polystyrene (PS) arm was obtained by the NMP, and the poly(tert-butyl acrylate) (PTBA) was obtained by the ATRP.

The ROP is effectively combined with other CRP methods in the synthesis of various copolymers. Thus, in [8], the authors describe the preparation of triblock copolymers with a central PCL block and peripheral PDMAEMA blocks using a combination of ROP with RAFT. Synthesized by ROP PCL with hydroxyl end groups was functionalized using 4-cyanopentanoic acid dithionaphthalenoate (CPADN) and used as a macro-RAFT agent for polymerization of DMAEMA to obtain the targeted triblock copolymers. An efficient method for the synthesis of linear di- and triblock copolymers with PCL blocks with narrow molecular weight distribution is to carry out ROP on mono- or difunctional polymer initiators. ROP of CL on the corresponding polymer initiators with hydroxyl chain ends provides copolymers in which PCL blocks are covalently attached to a block of polydimethylsiloxane [9], polyisobutylene [10], and poly(ethylene oxide) [11].

Grafting of polyester chains with narrow molecular weight distribution, in particular, PCL for the production of molecular brushes is also of great interest. The method of click chemistry is often used to graft PCL chains [12,13]. For example, in [12], the surface modification of the nanodispersed cellulose systems by grafting PCL chains using click-chemistry is described. However, to obtain molecular brushes with polyester side chains, ROP is easier to perform [14]. Therefore, numerous articles are devoted to the synthesis of molecular brushes with homopolymer PCL side chains or block copolymer side chains with a PCL block [12,13]. Carrying out polymerization of CL on multifunctional polymer initiators with varying numbers and positions of initiating groups leads to the production of PCL blocks with narrow molecular weight distribution, which are usually grafted to the carbochain backbone. To our knowledge, there are no data on the introduction of PCL blocks into the side chains of molecular brushes with a polyarylene backbone.

At the same time, works on combining aromatic PIs with aliphatic polyesters, in particular with PCL, have been going on for more than two decades. With simple mixing of homopolymer PI and PCL, for example, through a common solvent, phase separation of polymers is observed at the macroscopic level [15]. A more uniform phase system of PI and PCL is obtained by synthesizing chemically bonded PI and PCL blocks [15–20]. Until recently, the covalent bonding of these polymers was carried out either by stepwise polycondensation, leading to the production of linear alternative segmented (or multisegmented) block copolymers [18,19], or by cross-linking between PI and PCL chains in a polymer mixture or between PI and PCL layers in a multilayer coating [16,17,20]. A valuable feature of such copolymers is microphase separation [17], which gives them new interesting, for example, membrane properties. Thus, based on multisegmented linear block copolymers containing segments of polyurethanimide and PCL, effective diffusion membranes have been developed for the separation of binary mixtures of organic liquids [17]. However, the preparation of such segmented block copolymers is very laborious. Another method for combining PI and PCL by preparing block copolymers with sequentially attached PI and PCL blocks was proposed in [21]. In [22] the synthesis of linear block copolyimide with PCL and PI blocks based on 3,3′-dioxybenzidine was proposed using ROP of CL on a difunctional polyimide initiator. The obtained copolyimide was intended to improve the dispersion of carbon nanotubes in low-boiling organic solvents.

It should be noted that the introduction of PCL blocks into multiblock copolymers is of considerable interest from the point of view of further applications of these copolymers, since PCL blocks are capable of undergoing alkaline [23] and plasma [24] etching and biodegradation. The synthesis of new initiators and ROP of monomers and macromonomers containing functional groups on such initiators is a promising strategy for the preparation of macromolecules of complex architecture. In recent years, due to the development of the ROP method, it has been successfully used to obtain copolymers of different chemical structures and architecture.

Earlier, our research group reported the synthesis of triblock copolymers based on polyimide with external blocks of poly(ε-caprolactone) (PCL) [25], as well as grafted pentablock copolymers of linear-brush topology with external PCL and PMMA blocks and an internal brush-type block PI-*g*-PMMA [26]. This work is devoted to the development of methods for the synthesis of previously undescribed molecular brushes with a PI backbone and PCL side chains. To obtain such copolymers, a combination of various synthesis methods was used, including polycondensation, ATRP, ROP, and Cu (I) catalyzed azide-alkyne Huisgen cycloaddition (CuAAC) [27].

2. Materials and Methods

2.1. Materials

N-methyl-2-pyrrolidone (N-MP, 98%, Aldrich, St. Louis, MO, USA), toluene (analytical grade) methylene chloride, chloroform, THF, DMF (all reagent grade, Vekton, Voronezh, Russia), trimethylamine (≥99%, Aldrich, Overijse, Belgium), and pyridine (99%, Acros Organics, NJ, USA) were purified through standard techniques. 3,3′,4,4′-(1,3-diphenoxybenzene)tetracarboxylic dianhydride (99%, Ambinter Stock Screening Collections, China), 2,4-diaminophenol dihydrochloride (98%, Lancaster, Eastgate, White Lund, Morecambe, England), and potassium iodide (≥99.5%, Aldrich, St. Louis, MO, USA) were dried at elevated temperature in a vacuum prior to synthesis. 2-Hydroxyethyl methacrylate (HEMA) (99%, Acros Organics, UK) and ε-caprolactone (99%, Aldrich, St. Louis, MO, USA) were distilled twice under vacuum. Copper(I) chloride (99%, Aldrich, St. Louis, MO, USA) was stirred in glacial acetic acid overnight, filtered, and washed with argon-purged methanol (reagent grade, Vekton, Russia). Sn(II) 2-ethyl hexanoate ($Sn(EH)_2$) (~95%, Aldrich, Japan) was distilled under vacuum. 2-Bromoisobutyryl bromide (98%, Aldrich, St. Louis, MO, USA), 2,2′-bipyridine (bpy) (≥99%, Aldrich, St. Louis, MO, USA), sodium azide (NaN_3) (≥99.5%, Aldrich, St. Louis, MO, USA), propargyl alcohol (99%, Aldrich, Steinheim, Germany), N,N,N′,N″,N″-pentamethyldiethylenetriamine (PMDETA) (99%, Aldrich, Steinheim, Germany), copper(I) bromide (CuBr) (98%, Aldrich, St. Louis, MO, USA), and LiBr (≥99%, Aldrich, St. Louis, MO, USA) were used without additional purification.

2.2. Synthesis of PI-g-PHEMA as Branched Multicenter Macroinitiator with Hydroxyl Groups in Each Repeating Unit of the Side Chains of the Grafted Copolyimide

The polyimide multifunctional macroinitiator with initiating groups in almost every monomer unit was obtained by polycondensation of dianhydride of 3,3′,4,4′-(1,3-diphenoxybenzene)-tetracarboxylic acid and 2,4-diaminophenol, its phenol groups being further modified by 2-bromoisobutyryl bromide. The initial PI1 and the related PI2 multifunctional macroinitiator (Figure 1, see Result and Discussion section) were synthesized using procedures described in our previous publications [28–30].

To obtain a brush-type macroinitiator with hydroxyl groups in the side chains, graft copolyimide PI-*g*-PHEMA was synthesized by polymerization of 2-hydroxyethyl methacrylate (HEMA) on PI2 macroinitiator (Figure 1) by the ATRP. A typical synthesis procedure was as follows. A weighed portion of PI2 (0.075 g, 0.11 mmol per initiating group) and 2,2′-bipyridine (0.052 g, 0.33 mmol) was placed into a 25 mL Schlenk flask equipped with a magnetic stirrer, then DMF (16.5 mL, 0.21 mol) and HEMA (3.65 mL, 0.03 mol) were added with the syringe. The solvent, monomer, and syringes were purged with argon. The flask was sealed with a rubber septum, and the mixture was stirred until the powder was completely dissolved. Then, three freeze−pump−thaw cycles were carried out (evacuation for 15 min), after which the flask was filled with argon. After opening the septum in a stream of argon, CuCl (0.011 g, 0.11 mmol) was added to the reaction mixture, after which the flask was closed again with the septum. Three more freeze–pump–thaw cycles (evacuation for 15 min) of the reaction mixture were carried out. The flask was filled with argon and thermostated in an oil bath, and placed on a magnetic stirrer with a temperature controller, at 30 °C.

After a given reaction time, the reaction mixture was quickly cooled to room temperature and diluted two times with THF. To remove copper salts from the mixture, reaction solution was passed through a column filled with Al_2O_3, then concentrated using a rotary evaporator, and the polymerization product was precipitated into a water/methanol mixture with a volume ratio of 1/6. The filtered powder was dried under vacuum at 50 °C.

2.3. Synthesis of Graft Copolymers PI-g-(PHEMA-g-PCL)

The synthesis of the copolymer was carried out by ROP of CL in bulk on PI-*g*-PHEMA. A typical synthesis procedure is as follows: 0.05 g (0.23 mmol per initiating group) of PI-*g*-PHEMA was added to a 25-mL Schlenk flask equipped with a magnetic stirrer, and three freeze–pump–thaw cycles were carried out (evacuation for 15 min), after which the flask was filled with argon. Then, 1.2 mL (0.011 mol) of ε-caprolactone was introduced in an argon flow and thermostated at 130 °C using an oil bath. After complete dissolution of the macroinitiator, 0.05 mL (0.15 mmol) of $Sn(Oct)_2$ was introduced into the reaction in an argon flow. Upon completion of polymerization, the reaction mixture was cooled to room temperature and diluted with methylene chloride. The resulting solution was passed through a silica gel column to purify the product from catalyst and monomer impurities. The solution was then concentrated on a rotary evaporator and the product was precipitated into cooled petroleum ether. The polymer was dried under vacuum at 30 °C.

2.4. Introduction of Azide Groups into Polyimide

The functionalization of the polyimide with azide groups was carried out as follows. A weighed portion of PI2 macroinitiator (0.5 g, 0.66 mmol per initiating group) (Figure 1) and DMF (14.4 mL, 0.19 mol) was placed in a 25-mL round-bottom flask equipped with a magnetic stirrer, and the mixture was stirred until the powder was completely dissolved. Then NaN_3 (0.214 g, 3.3 mmol) was added to the flask and purged with argon for 30 min. The mixture was stirred on a magnetic stirrer at room temperature overnight. The reacted polymer was precipitated into methanol. The filtered powder was dried under vacuum at 50 °C.

2.5. Introduction of Distant Hydroxyl Groups to Polyimide

To obtain PI4 initiator (Figure 8, see Result and Discussion section), an azide-alkyne cycloaddition was carried out between the azide groups of PI3 (Figure 8) and the acetylene groups of propargyl alcohol. The typical synthesis was as follows. A sample of PI3 macroinitiator (0.25 g, 0.32 mmol per initiating group) (Figure 8) was placed in a 25 mL Schlenk flask equipped with a magnetic stirrer. Propargyl alcohol (92.9 μL, 1.6 mmol) and PMDETA (66.7 μL, 0.32 mmol) were added, and then DMF (4 mL, 0.052 mol) was added using a syringe. The solvent and syringes were purged with argon. The flask was sealed with a rubber septum, and the mixture was stirred until the powder was completely dissolved. Then, three freeze–pump–thaw cycles (evacuation for 15 min) were carried out, after which the flask was filled with argon. After opening the septum, CuBr (0.0457 g, 0.32 mmol) was added to the reaction mixture in an argon flow, after which the flask was closed again with the septum, three more freeze–pump–thaw cycles (evacuation for 15 min) of the reaction mixture were carried out, the flask was filled with argon and thermostated in an oil bath, placed on a magnetic stirrer with a temperature regulator at 50 °C overnight. Then the reaction mixture was cooled to room temperature and the product was precipitated into methanol. Residual copper was disposed of by changing the precipitant and reprecipitation from DMF. The filtered powder was dried under vacuum at 50 °C.

2.6. Synthesis of Linear Homopolymer PCL with Terminal Alkyne Groups

The synthesis was carried out by ROP of CL in a solution in toluene using propargyl alcohol as an initiator. A typical synthesis procedure is as follows: Propargyl alcohol (51 μL, 0.89 mmol), ε-caprolactone (1.97 mL, 17.8 mmol), and toluene (2.52 mL, 23.7 mmol) were added to a 10 mL Schlenk flask equipped with a magnetic stirrer. Then three freeze−pump−thaw cycles (evacuation for 15 min) were carried out, after which the flask was filled with argon. Then, 0.08 mL (0.25 mmol) of $Sn(Oct)_2$ was introduced in an argon flow into the reaction, after which the flask was closed again with a septum, three more freeze−pump−thaw cycles (evacuation for 15 min) of the reaction mixture were carried out, the flask was filled with argon and thermostated in the oil bath, placed on a magnetic stirrer with a temperature controller, at 100 °C for a specified time (3 h) and at 80 °C overnight. Upon completion of polymerization, the reaction mixture was cooled to room temperature and diluted with methylene chloride. The resulting solution was passed through a silica gel column to purify the product from catalyst and monomer impurities. The solution was then concentrated on a rotary evaporator and the product was precipitated into cooled petroleum ether. The polymer was dried under vacuum at 30 °C.

2.7. Synthesis of Grafted Copolyimides PI-g-PCL

To obtain the targeted copolymers, two approaches were used. The first approach consisted in ROP of CL on a macroinitiator with distant hydroxyl groups ("graft from"). A typical experiment was as follows. A 10 mL Schlenk flask equipped with a magnetic stirrer was charged with 0.05 g (0.054 mmol per initiating group) of PI4 macroinitiator (Figure 8), sealed with a rubber septum, and then 3 mL (26.9 mmol) of ε-caprolactone was introduced in an argon flow. The mixture was thermostated at 130 °C in an oil bath. After complete dissolution of the initiator, 0.123 mL (0.38 mmol) of $Sn(Oct)_2$ was introduced into the flask in an argon flow, and the reaction mixture was thermostated for a preset time. The molar ratio of PI4/CL was 1/500. The amount of $Sn(Oct)_2$ was 5 wt.% in relation to the monomer. Upon completion of polymerization, the reaction mixture was rapidly cooled to room temperature and diluted with methylene chloride. The resulting solution was passed through a silica gel column to purify the product from catalyst and monomer impurities. The solution was then concentrated using a rotary evaporator and the product was precipitated into cooled petroleum ether. The polymer was dried under vacuum at 30 °C.

The second approach was to carry out an azide-alkyne cycloaddition between the azide groups PI3 macroinitiator (Figure 8) and the alkyne groups of linear PCL. A typical reaction was carried out as follows. A weighed portion of PI3 macroinitiator (0.05 g, 0.064 mmol per initiating group) (Figure 8), PCL (0.447 g, 3.9 mmol per initiating group), and PMDETA (13 μL, 0.064 mmol) was placed in a 25-mL Schlenk flask equipped with a magnetic stirrer, then DMF (3.7 mL, 47.8 mmol) was added using a syringe. The solvent and syringes were purged with argon. The flask was sealed with a rubber septum, and the mixture was stirred until the powder was completely dissolved. Then, three freeze−pump−thaw cycles (evacuation for 15 min) were carried out, after which the flask was filled with argon. After opening the septum, CuBr (0.009 g, 0.064 mmol) was added to the reaction mixture in an argon flow, after which the flask was closed again with the septum, three more freeze−pump−thaw cycles (evacuation for 15 min) of the reaction mixture were carried out, the flask was filled with argon and thermostated in an oil bath placed on a magnetic stirrer with a temperature regulator, at 50 °C overnight. After that the reaction mixture was quickly cooled to room temperature and diluted two times with THF. To remove copper salts from the mixture, it was passed through a column filled with Al_2O_3, then concentrated using a rotary evaporator, and precipitated into cooled petroleum ether. The filtered powder was dried under vacuum at 30 °C.

2.8. Methods

^{1}H NMR spectra were recorded on Bruker AC-400 (400.1 MHz), with the signals of the solvent (DMSO-d_6, $CDCl_3$) as the reference.

IR spectra were recorded using a Shimadzu IR Affinity-1S spectrophotometer in the ATR mode (multiple attenuated total internal reflection) with a 4 cm^{-1} resolution and 30 scans.

Molecular weights and dispersities (Đ) of the samples were estimated by size exclusion chromatography (SEC) on Agilent-1260 Infinity fitted with differential refractive index (RID), light scattering (LS), and viscometry (VS) detectors equipped with two Agilent PLgel MIXED-C columns (7.5 × 300 mm, 5 μm), 1 × 43 PLgel 5 M guard column (50 × 7.5 mm), and autosampler. The analysis was carried out at 50 °C using 0.1 M LiBr in DMF as eluent at a flow rate of 1.0 mL min^{-1}. The salt was added to suppress the aggregation of macromolecules. For triple detection analysis the system was calibrated using PMMA standard with $M_p = 2.16 \times 10^3$ g mol^{-1} with dn/dc=0.052. Before analysis, all samples were passed through a 0.45 μm Nylon filters. SEC data were analyzed using Agilent GPC/SEC Software version 1.2.

Thermogravimetric analysis (TGA) of obtained copolymers was performed using a TG 209 F1 Libra analyzer (NETZSCH, Germany) in the temperature range of 25–800 °C and heating rate of 10 deg min^{-1} under nitrogen. The weight of samples was 2–3 mg.

Melting temperatures and enthalpies of synthesized copolymers were determined by differential scanning calorimetry (DSC) using a DSC 204 F1 Phoenix device (NETZSCH, Germany) at a heating rate of 10 deg min^{-1} in the range from −30 to +120 °C under nitrogen. The weight of samples was 2.5–3.5 mg.

The wide-angle X-ray scattering (WAXS) was carried out on an upgraded DRON 2.0 diffractometer (Saint-Petersburg, Russia) with CuKα radiation with a 0.154 nm wavelength.

3. Results and Discussion

PCL blocks were introduced into copolyimides by two fundamentally different ways. In the first case, to obtain grafted copolyimides, a precursor molecular brush with poly(2-hydroxyethyl methacrylate) (PHEMA) side chains was synthesized. Hydroxyl groups located in each repeating unit of the side chains of such brushes were used as initiators for carrying out ROP of CL. As a result the "brush on brush" structures PI-*g*-(PHEMA-*g*-PCL) were obtained. The second approach was aimed at obtaining graft copolymers with PCL side chains directly attached to the PI backbone. For this, a combination of ROP and CuAAC was used in two different variations.

3.1. Synthesis of Multifunctional Polyimide Initiators

The synthesis of multifunctional polyimide initiators with initiating OH-groups PI1 and 2-bromoisobutyrate groups PI2 in each repeating unit (Figure 1) based on soluble polyimide obtained by polycondensation of 4,4′-(1,3-phenylene-dioxy)bisphthalic anhydride and 2,4-diaminophenol was carried out according to a previously developed method [29]. The synthesis of polyimide was carried out according to the standard two-stage scheme with the preparation of polyamido acid (PAA) at the first stage and its subsequent thermal imidization in solution. During the synthesis of PAA, the equimolar ratio of monomers was strictly observed. The presence of a signal at 10.3 ppm in the ^{1}H NMR spectrum of the product of thermal imidization of PAA indicated the presence of phenolic groups in PI1 (Figure 2). To obtain initiator PI2, samples of initiator PI1 were used as precursors, and treated with 2-bromoisobutyryl bromide under conditions that ensure almost complete esterification of the hydroxyl groups of PI1 [29]. The degree of functionalization by initiating 2-bromo-isobutyrate groups of PI2 initiator intended to perform ATRP, was 97 ÷ 99 mol%

according to the ^{1}H NMR spectroscopy data, i.e., virtually every repeating unit contained an initiating group.

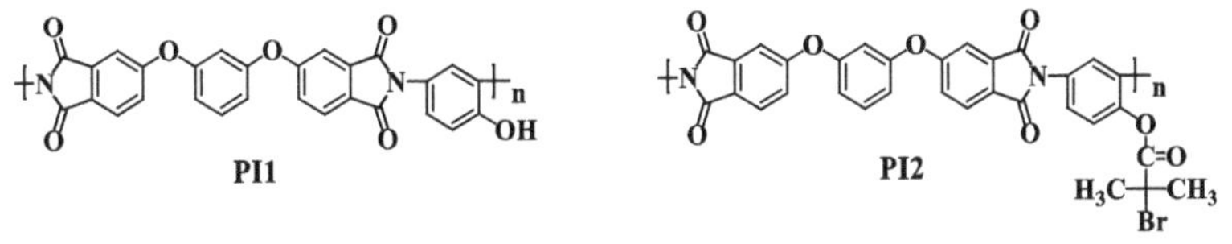

Figure 1. Multifunctional polyimide macroinitiators.

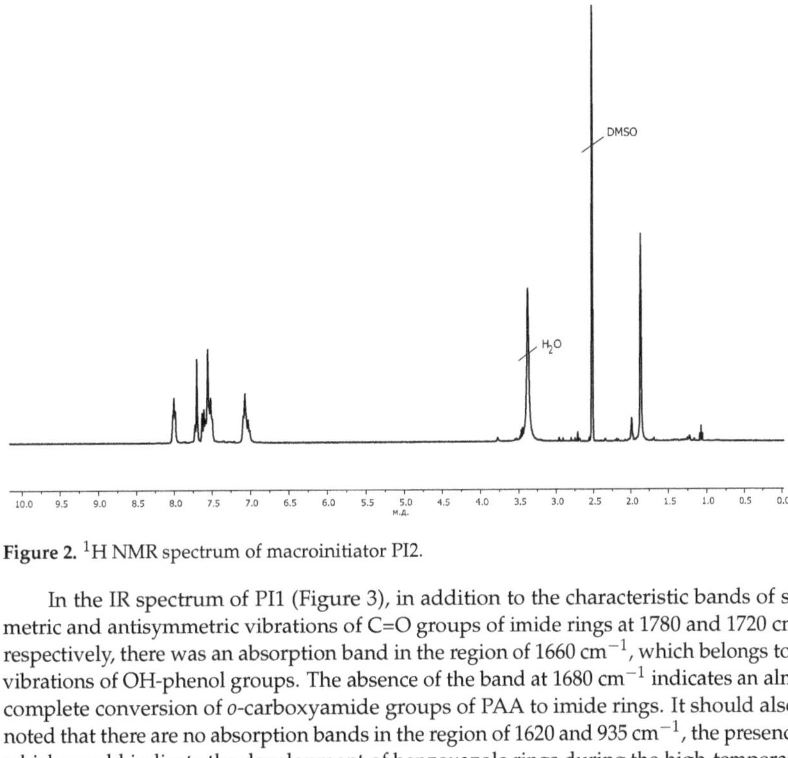

Figure 2. ^{1}H NMR spectrum of macroinitiator PI2.

In the IR spectrum of PI1 (Figure 3), in addition to the characteristic bands of symmetric and antisymmetric vibrations of C=O groups of imide rings at 1780 and 1720 cm^{-1}, respectively, there was an absorption band in the region of 1660 cm^{-1}, which belongs to the vibrations of OH-phenol groups. The absence of the band at 1680 cm^{-1} indicates an almost complete conversion of *o*-carboxyamide groups of PAA to imide rings. It should also be noted that there are no absorption bands in the region of 1620 and 935 cm^{-1}, the presence of which would indicate the development of benzoxazole rings during the high-temperature reaction with the participation of phenol and imide groups [31].

It should be noted that in an attempt to obtain initiator similar to PI1 but with a hydroxymethyl group in each repeating unit, during the thermal imidization of PAA based on (3,5-diaminophenyl)methyl alcohol an irreversible gelation rapidly develops at 170–180 °C in the N-MP solution which leads to a product insoluble in amide solvents. Apparently, interlink and/or interchain interactions of a more mobile and reactive (than phenol group) hydroxymethyl group leads to the formation of insoluble supramolecular structures due to hydrogen bonds.

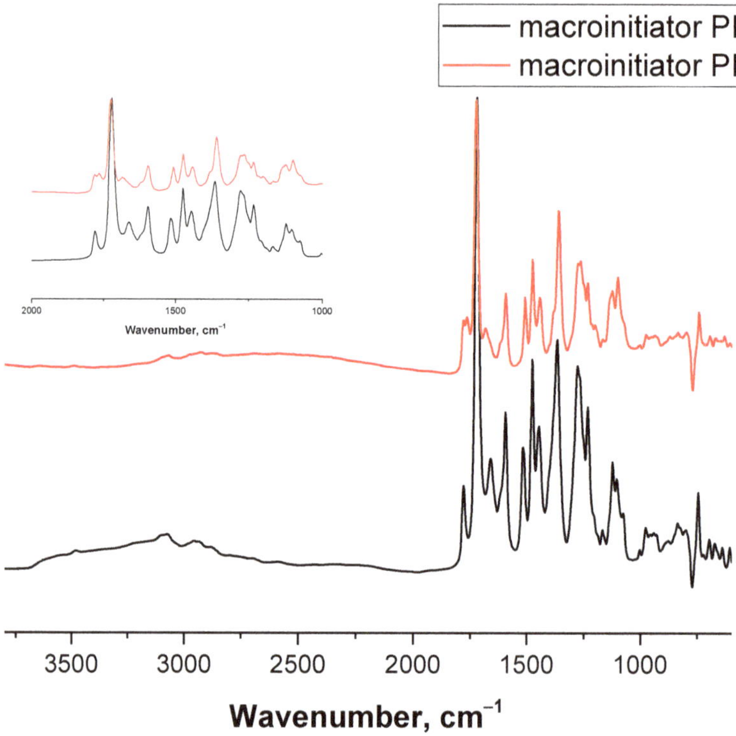

Figure 3. IR spectrum of macroinitiator PI1.

3.2. Synthesis of Brush-on-Brush Type Copolyimides

Previously, we reported [26], that polymerization of CL on PI macroinitiator PI2 (Figure 4) with phenol-type initiation groups practically does not proceed and the main polymerization product is homopolymer PCL. So, the introduction of PCL blocks into the side chains was carried out by performing ROP at the hydroxyethyl groups of the side chains of the polyimide brush with poly(2-hydroxyethyl methacrylate) side chains PI-*g*-PHEMA, which was used as a branched multifunctional macroinitiator. To obtain it, ATRP of 2-hydroxyethyl methacrylate was carried out on macroinitiator PI2. Table 1 shows the ATRP conditions and the monomer conversion values determined gravimetrically.

The formation of the copolymer PI-*g*-PHEMA was confirmed by the shift of the signals of the methyl protons of 2-bromo-isobutyrate groups from the 1.9 ppm to the 1.5 ppm in the ^{1}H NMR spectrum of the ATRP products (Figure 5) and the appearance of signals of methylene protons of the –CH_2–OH group at the 3.8 ppm while maintaining the signals of aromatic protons in the range of 6.0–8.5 ppm, related to the polyimide backbone. The intensities of the peaks from the PI backbone decreased due to its lower content in the final molecular brush.

In the IR spectrum (Figure 6) of the HEMA polymerization product, in comparison with the spectrum of the macroinitiator, absorption bands of OH groups at 3500 cm^{-1}, of methylene groups at 2950 cm^{-1} belonging to the PHEMA chains appeared, and the band at 1765 cm^{-1} disappeared, which refers to the vibrations of an ester group with an electronegative substituent (2-Br-substituent) in the α-position.

The structure of the obtained copolymers was proved by the complex use of ^{1}H NMR spectroscopy and SEC. In the ^{1}H NMR spectrum (Figure 5) of the CL polymerization product, signals of methylene protons –OCH_2– and –$COCH_2$– groups of PCL chains were observed at 4.1 and 2.3 ppm, respectively.

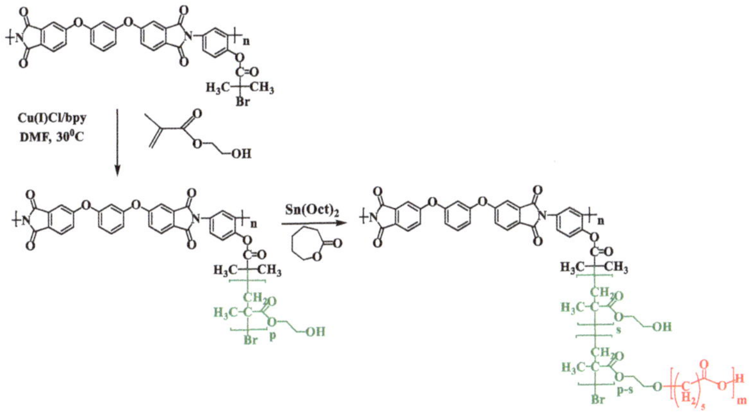

Figure 4. Synthesis of PI-*g*-PHEMA and PI-*g*-(PHEMA-*g*-PCL).

Table 1. Synthesis of PI-*g*-PHEMA by ATRP on macroinitiator PI2 in DMF solution.

NO	PI2/bpy/CuCl/HEMA *	[HEMA],%	T, °C	t, h	α, %
1	1/3/1/50	12	30	1	13.3
2	1/3/1/50	12	30	2	19.4
3	1/3/1/50	12	30	20	gel

*–molar ratio of the components in the reaction system, α–conversion of the monomer.

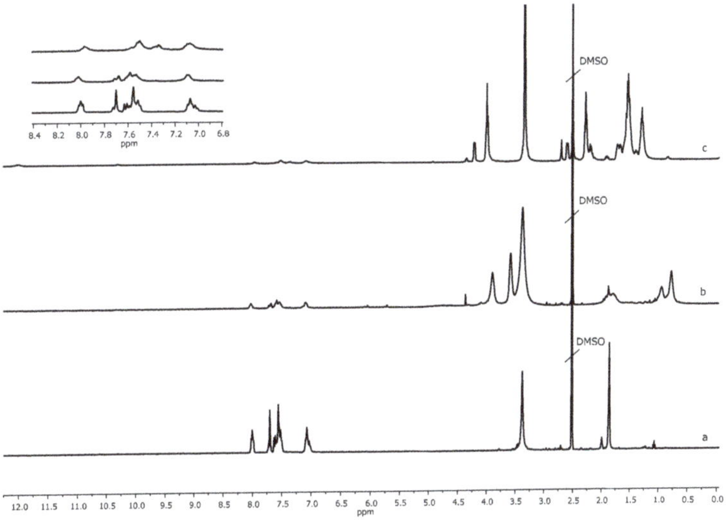

Figure 5. ^{1}H NMR spectra of macroinitiator PI2 (**a**), PI-*g*-PHEMA (**b**) and PI-*g*-(PHEMA-*g*-PCL) (**c**).

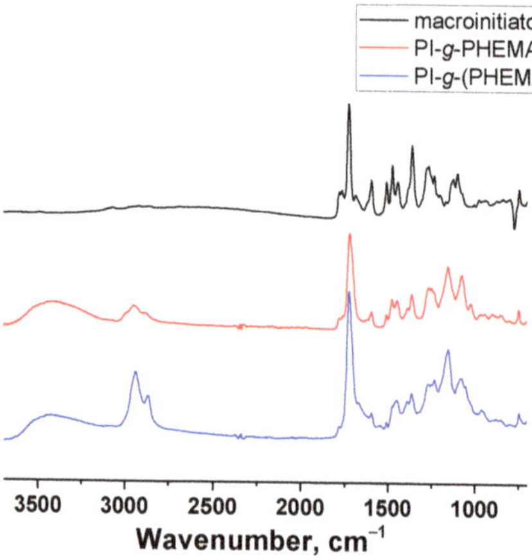

Figure 6. IR spectra of macroinitiator PI2, PI-*g*-PHEMA and PI-*g*-(PHEMA-*g*-PCL).

Sequential offset of the MWD curves in the macroinitiator PI2, PI-*g*-PHEMA and PI-*g*-(PHEMA-*g*-PCL) series to the high MM values (Figure 7) and the presence of only one peak for PI-*g*-(PHEMA-*g*-PCL) indicates the grafting of PCL side chains to the PHEMA chains. According to the MW data corresponding to this series of polymers (Table 2), when passing from PI to the "brush on brush" structure, the MW increases by almost four times. A significant decrease in the value of the increment of the refractive index dn/dc in the series obviously reflects an increase in the ratio of aliphatic units in the copolymer.

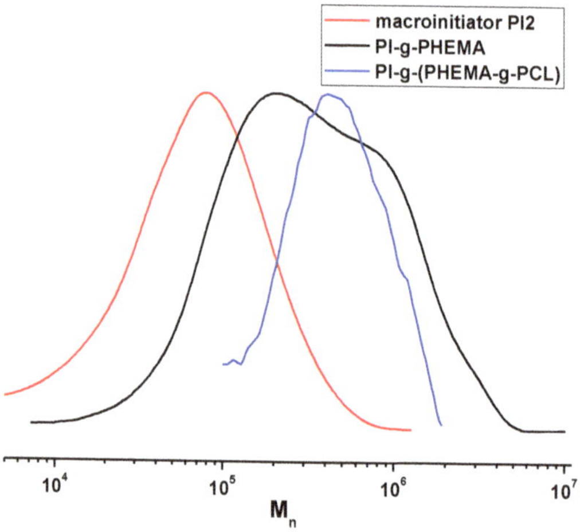

Figure 7. MWD curves of macroinitiator PI2, PI-*g*-PHEMA, and PI-*g*-(PHEMA-*g*-PCL).

Table 2. MW characteristics of macroinitiator PI2, PI-*g*-PHEMA and PI-*g*-(PHEMA-*g*-PCL).

NO	Sample	$M_n \times 10^{-3}$	Đ	dn/dc, mL/g
1	PI2 macroinitiator	60	2.8	0.148
2	PI-*g*-PHEMA	180	3.9	0.087
3	PI-*g*-(PHEMA-*g*-PCL)	220	2.5	0.058

3.3. *Synthesis of Molecular Brushes PI-g-PCL with a Polyimide (PI) Backbone and Polycaprolactone (PCL) Side Chains Using a Combination of ROP and CuAAC*

To obtain the targeted copolymers PI-*g*-PCL, the following synthesis scheme was proposed, which includes two routes (Figure 8).

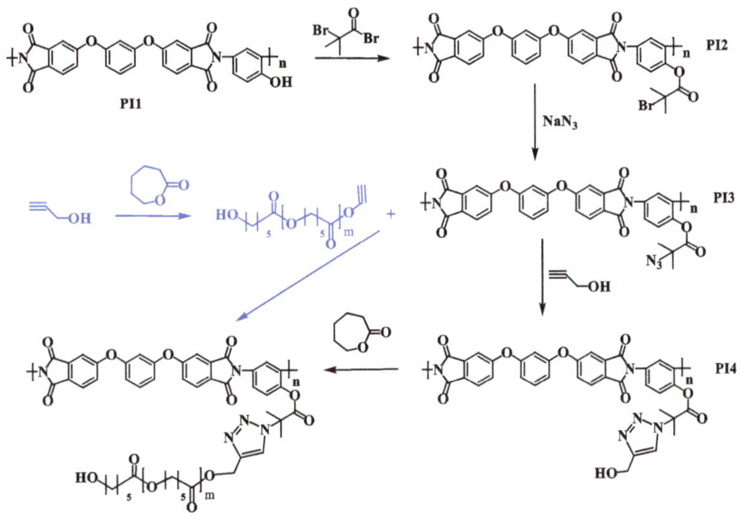

Figure 8. Routes for the synthesis of grafted copolyimides with PCL side chains by ROP and CuAAC.

3.3.1. Synthesis of PI with Azide Groups

At the first stage, samples of prepolymer PI3 (Figure 8) with azide functional groups were obtained. For this, polyimide macroinitiator PI2 (Figure 8) with 2-bromo-isobutyrate groups was modified under the action of sodium azide NaN_3. Azidating agent was taken in 4 ÷ 10-fold excess. The reaction proceeded at room temperature for two hours. According to the ^{1}H NMR spectra (Figure 9), functionalization with sodium azide proceeded completely, as a result of which the signal at 1.9 ppm shifted to the 1.35 ppm. The degree of functionalization was determined by the content of 2-bromo-isobutyrate groups in the initial polyimide macroinitiator, which was 60%.

Table 3 presents the molecular weight characteristics of the initial PI2 macroinitiator with 2-bromo-isobutyrate groups and the obtained PI3 macroinitiators with azide groups.

During the isolation of the azidation product, the polymer is apparently fractionated, which leads to a decrease in the dispersity and MW of the obtained macroinitiators PI3.

Table 3. MW characteristics of PI macroinitiators.

№	Sample	In/NaN_3 Ratio	$M_n \times 10^{-3}$	Đ
1	Macroinitiator PI2	-	25	2.7
2	Macroinitiator PI3	1/10	22	1.3
3		1/5	17.5	1.2
4		1/5	20	1.2

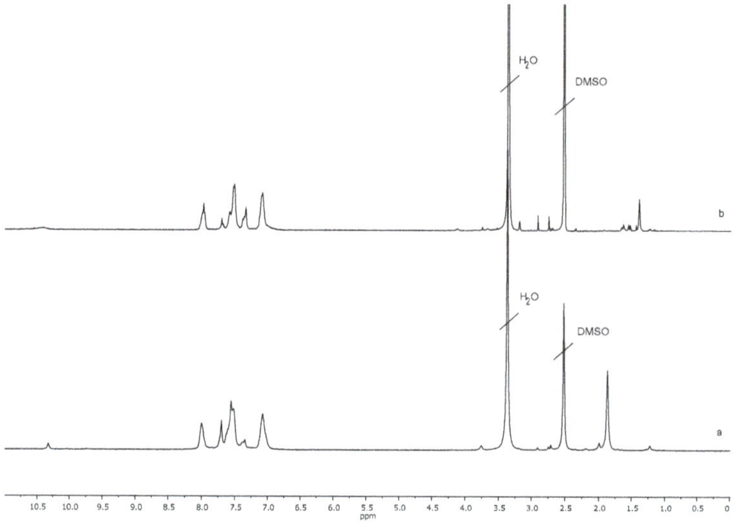

Figure 9. ^{1}H NMR spectra of macroinitiator PI2 (**a**) and macroinitiator PI3 (**b**).

3.3.2. Synthesis of PI with Hydroxyl Side Groups Attached to the PI Backbone through the 1,2,3-Triazole Linker

To implement the first route of the synthesis, the obtained prepolymer PI3 (Figure 8) was introduced into reaction with the propargyl alcohol. As a result of a click reaction, initiating ROP hydroxyl groups separated from the benzene ring of the backbone by a triazole ring (macroinitiator PI4, Figure 8) were introduced into the initiator. The reaction was carried out at a 50 °C for 2 hours in an N-MP using a 1.5-fold excess of propargyl alcohol and a CuCl/PMDETA catalytic system. Figure 10 shows the IR spectra of macroinitiators PI3 and PI4 (Figure 8). In the IR spectrum of the macroinitiator PI4, in addition to the characteristic absorption bands of imide rings (at 1370 cm^{-1} and a doublet at 1776 cm^{-1} and 1717 cm^{-1}), there is a peak at 2300 cm^{-1}, which refers to the stretching vibrations of the ethynyl group (–C≡CH) (Figure 10).

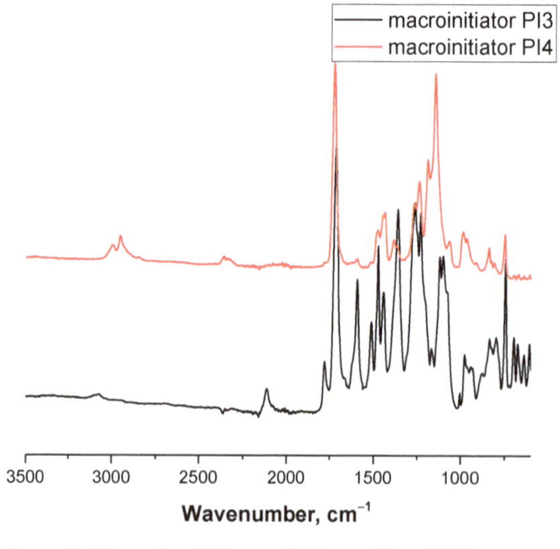

Figure 10. IR spectra of PI macroinitiators PI3 and PI4.

The resulting polymer PI4, according to the SEC data, had the following characteristics: $M_n = 34 \times 10^{-3}$, $Đ = 1.2$.

3.3.3. Synthesis of Molecular Brushes with a PI Backbone and PCL Side Chains by Polymerization of ε-Caprolactone on a Multicenter Macroinitiator PI4

The resulting macroinitiator PI4 (Figure 8) was used to carry out ROP of CL in bulk (initiator/monomer molar ratio = 1/1000) at 130 °C in an inert atmosphere in the presence of tin (II) octanoate as a catalyst. The ^{1}H NMR spectrum of the obtained polymer (Figure 11) contains signals of methylene protons of the –OCH_2– and –$COCH_2$– groups at 4.0–4.1 ppm and 2.2–2.3 ppm and the middle methylene groups of the PCL side chains at 1.6–1.7 ppm and 1.35–1.45 ppm. In the spectrum of the polymerization product, signals of aromatic protons are not visible, due to the fact that in chloroform the polyimide is completely shielded by the PCL side chains.

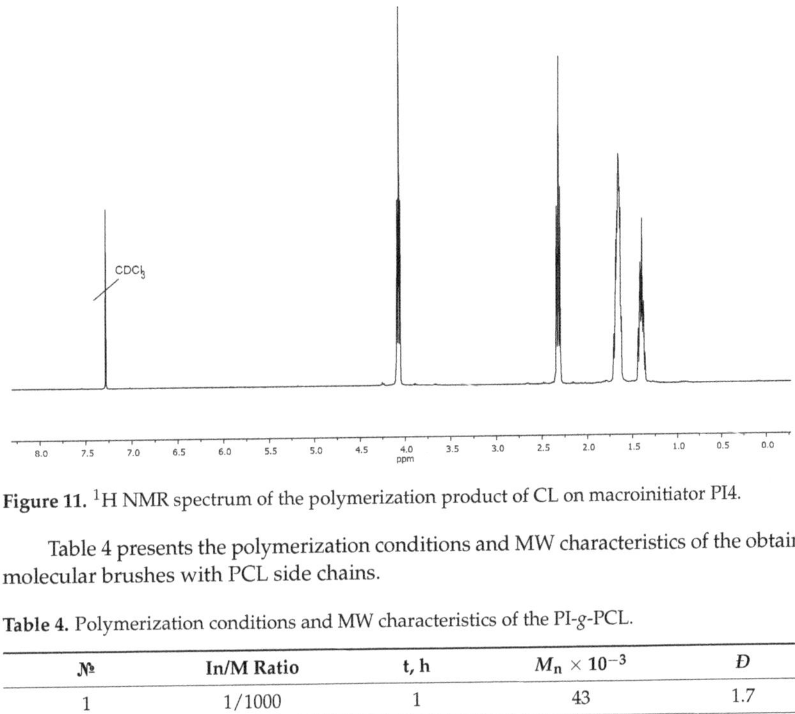

Figure 11. ^{1}H NMR spectrum of the polymerization product of CL on macroinitiator PI4.

Table 4 presents the polymerization conditions and MW characteristics of the obtained molecular brushes with PCL side chains.

Table 4. Polymerization conditions and MW characteristics of the PI-g-PCL.

№	In/M Ratio	t, h	$M_n \times 10^{-3}$	Đ
1	1/1000	1	43	1.7
2	1/500	2	52	1.5

Therefore, this approach allows to obtain graft copolymers with a PI backbone and relatively short PCL side chains, which is illustrated by the MWDs in Figure 12.

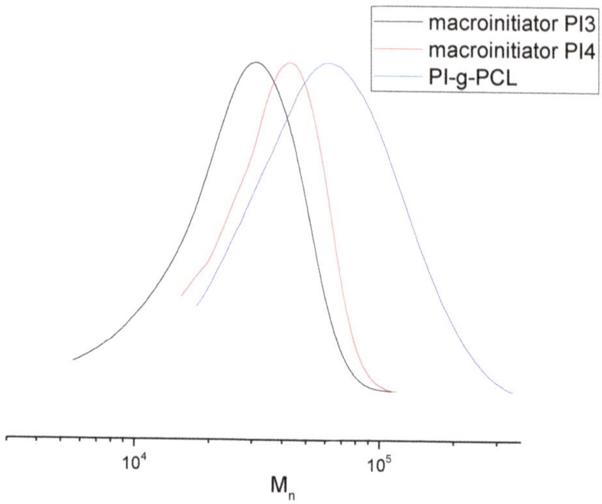

Figure 12. MWD curves of macroinitiators PI3 and PI4, and grafted copolyimide PI-*g*-PCL **5**.

3.3.4. Synthesis of Molecular Brushes with a PI Backbone and PCL Side Chains by CuAAC of Prepolymers

To implement the second route, a linear PCL (Figure 8) with a terminal triple bond was synthesized. Polymerization was carried out in bulk at 100 °C using propargyl alcohol as an initiator and tin (II) octanoate as a catalyst. The ^{1}H NMR spectrum of the polymerization product (Figure 13) contains all characteristic signals of PCL: signals of methylene protons of the $-OCH_2-$ and $-COCH_2-$ groups at the 4.0–4.1 ppm and 2.2–2.3 ppm and the middle methylene groups of the PCL chains at 1.6–1.7 ppm and 1.35–1.45 ppm. Moreover, clear signal at the 4.3 ppm, related to the protons of the $-C{\equiv}CH$ end groups is visible in the ^{1}H NMR spectrum (Figure 13).

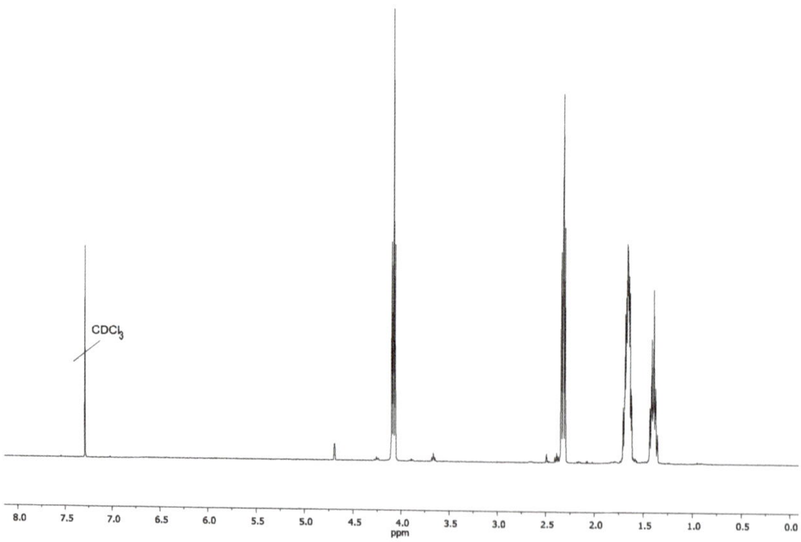

Figure 13. ^{1}H NMR spectrum of linear PCL with pendant $-C{\equiv}CH$ groups.

Table 5 presents the polymerization conditions and MW characteristics of the obtained linear PCL with the –C≡CH end groups.

Table 5. Polymerization conditions and MW characteristics of the linear PCL.

№	Solvent	In/M Ratio	$M_n \times 10^{-3}$	Đ
1	-	1/50	13.7	1.3
2	toluene	1/20	4.5	1.6

Azide-alkyne cycloaddition was carried out between linear PCL and macroinitiator PI3 (Figure 8), as a result of which the targeted copolyimides PI-*g*-PCL were also obtained. The click-reaction was carried out in an N–MP at 70 °C for 24 h, using CuCl/PMDETA as the catalytic complex. The molar ratio of the reagents was PI3/PCL/CuCl/PMDETA = 1/2/1/1.

Table 6 presents the MW characteristics of the products of click-reaction between linear PCL and macroinitiator with azide groups PI3. It can be seen that the implementation of the second approach leads to a more significant increase in molecular weight (Table 6), but unreacted linear PCL remains in the system. It is found in the form of a second peak in the chromatogram in the lower molecular weight region and its MW completely coincides with the mass of the previously analyzed linear PCL with alkyne end groups.

Table 6. MW characteristics of the click-reaction products.

Sample	Peak	$M_n \times 10^{-3}$	Đ
1	1	171	2.2
	2	13	1.5

3.4. TGA, DSC and X-ray Study of the Synthesized Copolymers

Table 7 and Figure 14 presents the data on the thermal stability of the PI macroinitiator, PCL macromonomer, and linear and grafted copolyimides based on it. All copolymers lost most of the weight before 400 °C. But they are mostly stable till 275 °C, while linear PCL start to decompose already at 210 °C. The residual mass depends on the content of PI block in the copolymer. While linear PCL retains less than 1% of mass after 800 °C, molecular brush with long PCL side chains (PI-*g*-PCL 1)–2%, and molecular brush with short PCL side chains (PI-*g*-PCL 2)–15%.

Melting temperatures (T_m) and enthalpy (ΔH_m) of copolymers determined by the DSC method are presented in Table 7. The DSC curves of synthesized copolymers are presented with the endothermic melting peaks (Figure 15). All copolymers melt in the range from 52 to 58 °C and the melting enthalpy increases with the increase of PI content in the copolymer.

Table 7. Thermal properties of synthesized copolymers.

Polymer	TGA			DSC	
	τ_5 (°C)	τ_{10} (°C)	Mass Residua, %	T_m, °C	ΔH_m, J/g
PI macroinitiator	232.9	375.0	53.83	-	-
PCL	210.5	259.6	0.82	54.3	−87.49
PCL-*b*-PI-*b*-PCL	284.6	311.7	7.08	58.1	−65.54
PI-*g*-PCL 1	274.3	296.9	2.12	55.8	−70.93
PI-*g*-PCL 2	279.1	303.7	14.81	52.3	−68.39

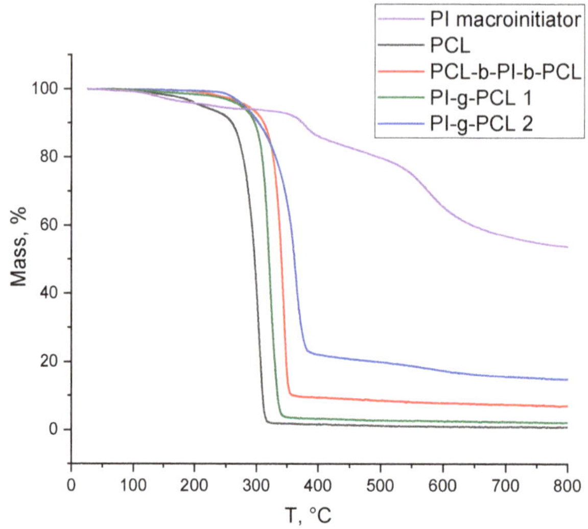

Figure 14. TGA curves for the synthesized macroinitiators, macromonomer, and copolymers.

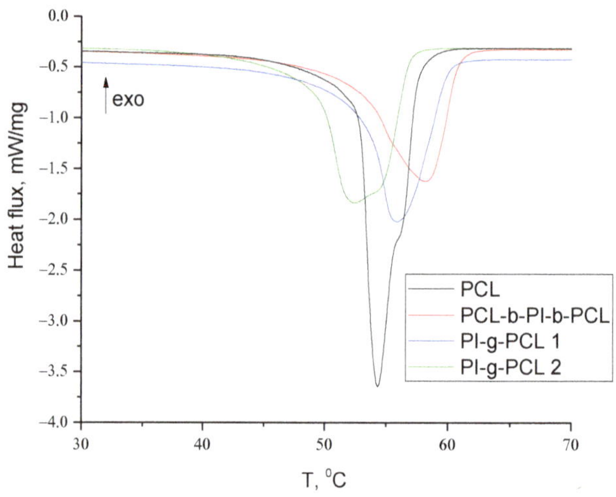

Figure 15. DSC curves for the synthesized macromonomer and copolymers.

Comparison of diffractograms of linear PCL with films of its linear and grafted copolymers showed that they are practically symbate (Figure 16). Consequently, the introduction of a polyimide block into the polycaprolactone matrix practically did not lead to a change in the degree of crystallinity of PCL. Apparently, the length of the PI blocks between PCL blocks is not enough for the PI to form the amorphous areas in the copolymer, so they're ordered inside PCL crystalline structure.

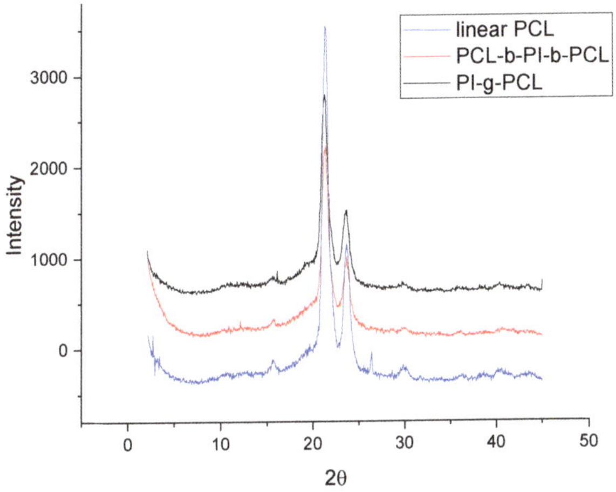

Figure 16. Diffractograms of linear PCL, triblock-copolymer PCL-*b*-PI-*b*-PCL, and grafted copolyimide PI-*g*-PCL.

4. Conclusions

Multicomponent copolymers, in which PI blocks are combined with blocks of biocompatible aliphatic polymers, have great potential for the development of materials for tissue engineering. The introduction of PCL blocks into multiblock copolymers is of considerable interest from the point of view of further applications of these copolymers since PCL blocks are capable of undergoing alkaline and plasma etching and biodegradation. The synthesis of new initiators and the ROP of monomers and macromonomers containing functional groups on such initiators is a promising strategy for the preparation of macromolecules of complex architecture.

In this work we suggested an approach to the synthesis of the novel polymer molecular brushes with a polyimide (PI) backbone and poly(ε-caprolactone) (PCL) side chains. Using combination of polycondensation, ATRP, ROP, and CuAAC – targeted grafted copolyimides PI-*g*-PCL were synthesized as well as more complex "brush on brush" structures PI-*g*-(PHEMA-*g*-PCL). Comparison of different combinations of ROP and CuAAC showed that polymer-analogous transformations of a multicenter PI macroinitiator with an initiating hydroxyl group separated from the main chain by a triazole ring and carrying out ROP on it allows obtaining graft copolymers with a PI backbone and relatively shorter PCL side chains. While a separate synthesis of macromonomers with the desirable functional groups (polyimide with azide groups and PCL with terminal alkyne groups), followed by a click reaction leads to a more significant increase in molecular weight, but unreacted linear PCL remains in the system. X-ray scattering data showed that the introduction of a polyimide block into the polycaprolactone matrix practically did not change the degree of crystallinity of PCL.

Author Contributions: Conceptualization, T.K.M. and A.V.Y.; methodology, A.V.K. and T.K.M.; validation, A.V.K.; formal analysis, A.V.K.; investigation, A.V.K., N.N.B. and V.K.L.; writing—original draft preparation, A.V.K.; writing—review and editing, A.V.Y.; visualization, A.V.K.; supervision, T.K.M. and A.V.Y.; funding acquisition, A.V.Y. All authors have read and agreed to the published version of the manuscript.

Funding: We acknowledge funding from the Ministry of Science and Higher Education of the Russian Federation (Agreement No. 075-15-2020-794).

Institutional Review Board Statement: Not applicable.

Informed Consent Statement: Not applicable.

Data Availability Statement: The data presented in this study are available on request from the corresponding author.

Conflicts of Interest: The authors declare no conflict of interest.

References

1. Constantin, C.P.; Aflori, M.; Damian, R.F.; Rusu, R.D. Biocompatibility of Polyimides: A Mini-Review. *Materials* **2019**, *12*, 3166. [CrossRef]
2. Herth, E.; Guerchouche, K.; Rousseau, L.; Calvet, L.E.; Loyez, C. A biocompatible and flexible polyimide for wireless sensors. *Microsyst. Technol.* **2017**, *23*, 5921–5929. [CrossRef]
3. Starr, P.; Agrawal, C.M.; Bailey, S. Biocompatibility of common polyimides with human endothelial cells for a cardiovascular microsensor. *J. Biomed. Mater. Res. Part. A* **2016**, *104*, 406–412. [CrossRef]
4. Tunca, U.; Ozyurek, Z.; Erdogan, T.; Hizal, G. Novel miktofunctional initiator for the preparation of an ABC-type miktoarm star polymer via a combination of controlled polymerization techniques. *J. Polym. Sci. Part. A Polym. Chem.* **2004**, *42*, 4228–4236. [CrossRef]
5. Jakubowski, W.; Lutz, J.-F.; Slomkowski, S.; Matyjaszewski, K. Block and random copolymers as surfactants for dispersion polymerization. I. Synthesis via atom transfer radical polymerization and ring-opening polymerization. *J. Polym. Sci. Part. A Polym. Chem.* **2005**, *43*, 1498–1510. [CrossRef]
6. Zhu, F.; Ngai, T.; Xie, Z.; Wu, C. Self-Assembly Assisted Coupling of End Functional Block Copolymers. *Macromolecules* **2003**, *36*, 7405–7408. [CrossRef]
7. Lee, H.; Jakubowski, W.; Matyjaszewski, K.; Yu, S.; Sheiko, S.S. Cylindrical Core−Shell Brushes Prepared by a Combination of ROP and ATRP. *Macromolecules* **2006**, *39*, 4983–4989. [CrossRef]
8. Zhu, C.; Jung, S.; Luo, S.; Meng, F.; Zhu, X.; Park, T.G.; Zhong, Z. Co-delivery of siRNA and paclitaxel into cancer cells by biodegradable cationic micelles based on PDMAEMA–PCL–PDMAEMA triblock copolymers. *Biomaterials* **2010**, *31*, 2408–2416. [CrossRef] [PubMed]
9. Azemar, F.; Faÿ, F.; Réhel, K.; Linossier, I. Control of hydration and degradation properties of triblock copolymers polycaprolactone- b -polydimethylsiloxane- b -polycaprolactone. *J. Appl. Polym. Sci.* **2014**, *131*. [CrossRef]
10. Castano, M.; Alvarez, A.; Becker, M.L.; Puskas, J.E. Synthesis of polyisobutylene-polycaprolactone block copolymers using enzyme catalysis. *Express Polym. Lett.* **2016**, *10*, 693–700. [CrossRef]
11. Hu, Y.; Xie, J.; Tong, Y.W.; Wang, C.-H. Effect of PEG conformation and particle size on the cellular uptake efficiency of nanoparticles with the HepG2 cells. *J. Control. Release* **2007**, *118*, 7–17. [CrossRef] [PubMed]
12. Benkaddour, A.; Jradi, K.; Robert, S.; Daneault, C. Grafting of Polycaprolactone on Oxidized Nanocelluloses by Click Chemistry. *Nanomaterials* **2013**, *3*, 141–157. [CrossRef] [PubMed]
13. Zhao, P.; Yan, Y.; Feng, X.; Liu, L.; Wang, C.; Chen, Y. Highly efficient synthesis of polymer brushes with PEO and PCL as side chains via click chemistry. *Polymer* **2012**, *53*, 1992–2000. [CrossRef]
14. Du, J.-Z.; Tang, L.-Y.; Song, W.-J.; Shi, Y.; Wang, J. Evaluation of Polymeric Micelles from Brush Polymer with Poly(ε-caprolactone)-b -Poly (ethylene glycol) Side Chains as Drug Carrier. *Biomacromolecules* **2009**, *10*, 2169–2174. [CrossRef]
15. Liu, T.; Ozisik, R.; Siegel, R.W. Phase separation and surface morphology of spin-coated films of polyetherimide/polycaprolactone immiscible polymer blends. *Thin Solid Films* **2007**, *515*, 2965–2973. [CrossRef]
16. Park, M.-H.; Jang, W.; Yang, S.-J.; Shul, Y.; Han, H. Synthesis and characterization of new functional poly (urethane-imide) crosslinked networks. *J. Appl. Polym. Sci.* **2006**, *100*, 113–123. [CrossRef]
17. Gnanasekaran, D.; Reddy, B.S. A facile synthesis of mixed soft-segmented poly(urethane-imide)-polyhedral oligomeric silsesquioxone hybrid nanocomposites and study of their structure-transport properties. *Polym. Int.* **2014**, *63*, 507–513. [CrossRef]
18. Jonquières, A.; Roizard, D.; Lochon, P. Polymer design for pervaporation membranes: Influence of the soft segment size of block copolymers (polyurethaneimides or polyureaimides) on their pervaporation features. *J. Memb. Sci.* **1996**, *118*, 73–84. [CrossRef]
19. Jonquières, A.; Roizard, D.; Cuny, J.; Vicherat, A.; Lochon, P. Polarity measurements in block copolymers (polyurethaneimides) and correlation with their pervaporation features. *J. Appl. Polym. Sci.* **1995**, *56*, 1567–1579. [CrossRef]
20. Zakharchenko, S.; Sperling, E.; Ionov, L. Fully Biodegradable Self-Rolled Polymer Tubes: A Candidate for Tissue Engineering Scaffolds. *Biomacromolecules* **2011**, *12*, 2211–2215. [CrossRef] [PubMed]
21. Ju, J.; Wang, Q.; Wang, T.; Wang, C. Low dielectric, nanoporous fluorinated polyimide films prepared from PCL-PI-PCL triblock copolymer using retro-Diels–Alder reaction. *J. Colloid Interface Sci.* **2013**, *404*, 36–41. [CrossRef]
22. Liu, C.; Liu, B.; Chan-Park, M.B. Synthesis of polycaprolactone-polyimide-polycaprolactone triblock copolymers via a 2-step sequential copolymerization and their application as carbon nanotube dispersants. *Polym. Chem.* **2017**, *8*, 674–681. [CrossRef]
23. Gupta, D.; Singh, A.K.; Kar, N.; Dravid, A.; Bellare, J. Modelling and optimization of NaOH-etched 3-D printed PCL for enhanced cellular attachment and growth with minimal loss of mechanical strength. *Mater. Sci. Eng. C* **2019**, *98*, 602–611. [CrossRef] [PubMed]
24. Ko, Y.M.; Myung, S.W.; Kim, B.H. O2/Ar plasma treatment for enhancing the biocompatibility of hydroxyapatite nanopowder and polycaprolactone composite film. *J. Nanosci. Nanotechnol.* **2015**, *15*, 6048–6052. [CrossRef] [PubMed]

25. Meleshko, T.K.; Kashina, A.V.; Saprykina, N.N.; Kostyuk, S.V.; Vasilenko, I.V.; Nikishev, P.A.; Yakimanskii, A.V. Synthesis and morphology of polycaprolactone–block-polyimide–block-polycaprolactone triblock copolymers for film separation membranes. *Russ. J. Appl. Chem.* **2017**, *90*, 602–612. [CrossRef]
26. Kashina, A.V.; Meleshko, T.K.; Bogorad, N.N.; Bezrukova, M.A.; Yakimanskii, A.V. Synthesis of Pentablock Copolymers of the Mixed Linear-Brush Topology by Controlled Radical Polymerization and Ring-Opening Polymerization Reactions. *Polym. Sci.-Ser. C* **2019**, *61*, 174–185. [CrossRef]
27. Danese, M.; Bon, M.; Piccini, G.; Passerone, D. The reaction mechanism of the azide–alkyne Huisgen cycloaddition. *Phys. Chem. Chem. Phys.* **2019**, *21*, 19281–19287. [CrossRef]
28. Meleshko, T.K.; Il'gach, D.M.; Bogorad, N.N.; Kukarkina, N.V.; Yakimansky, A.V. Synthesis of graft copolyimides via controlled radical polymerization of methacrylates with a polyimide macroinitiator. *Polym. Sci. Ser. B* **2014**, *56*, 118–126. [CrossRef]
29. Meleshko, T.K.; Il'gach, D.M.; Bogorad, N.N.; Kukarkina, N.V.; Vlasova, E.N.; Dobrodumov, A.V.; Malakhova, I.I.; Gorshkov, N.I.; Krasikov, V.D.; Yakimanskii, A.V. Synthesis of multicentered polyimide initiators for the preparation of regular graft copolymers via controlled radical polymerization. *Polym. Sci. Ser. B* **2010**, *52*, 589–599. [CrossRef]
30. Yakimansky, A.V.; Meleshko, T.K.; Ilgach, D.M.; Bauman, M.A.; Anan'eva, T.D.; Klapshina, L.G.; Lermontova, S.A.; Balalaeva, I.V.; Douglas, W.E. Novel regular polyimide-graft-(polymethacrylic acid) brushes: Synthesis and possible applications as nanocontainers of cyanoporphyrazine agents for photodynamic therapy. *J. Polym. Sci. Part. A Polym. Chem.* **2013**, *51*, 4267–4281. [CrossRef]
31. Rusakova, O.Y.; Kostina, Y.V.; Rodionov, A.S.; Bondarenko, G.N.; Alent'ev, A.Y.; Meleshko, T.K.; Kukarkina, N.V.; Yakimanskii, A.V. Study of the mechanism of the thermochemical reaction of polyimides with hydroxyl groups via vibrational-spectroscopy and quantum-chemistry methods. *Polym. Sci. Ser. A* **2011**, *53*, 791–799. [CrossRef]

polymers

MDPI

Article

The Influence of Substituents in Phosphazene Catalyst-Flame Retardant on the Thermochemistry of Benzoxazine Curing

Natalia V. Bornosuz [1], Roman F. Korotkov [1], Alexander A. Kolenchenko [1], Alexey V. Shapagin [2], Alexey V. Orlov [1], Irina Yu. Gorbunova [1], Vyacheslav V. Kireev [1] and Igor S. Sirotin [1,*]

1 Faculty of Petrochemistry and Polymer Materials, Mendeleev University of Chemical Technology, 125047 Moscow, Russia; bornosuz@muctr.ru (N.V.B.); ro.korotkov@muctr.ru (R.F.K.); kolenchenkoalex@muctr.ru (A.A.K.); alexeyorlovvladimirovich3829@gmail.com (A.V.O.); igorbunova@muctr.ru (I.Y.G.); kireev@muctr.ru (V.V.K.)
2 Frumkin Institute of Physical Chemistry and Electrochemistry Russian Academy of Sciences (IPCE RAS), 31, Bld. 4 Leninsky Prospect, 119071 Moscow, Russia; shapagin@mail.ru
* Correspondence: isirotin@muctr.ru; Tel.: +7-(499)-978-91-98

Abstract: This work is devoted to the influence of phosphazene modifiers with different substituents on the curing process, thermal properties and flammability of benzoxazine resin. Novel catalysts with m-toluidine substituents were introduced. The catalytic activity of studied phosphazene compounds decreased in the row: hexachlorocyclotriphosphazene (HCP) > tetra m-toluidine substituted phosphazene PN-mt (4) > hexa m-toluidine substituted phosphazene PN-mt (6) > hexaphenoxy-cyclotriphosphazene (HPP), where HPP is totally inactive. Two types of catalysis: basic and acid were proposed. A brief study of resulting properties of polybenzoxazines was presented. The addition of any studied modifier caused the decrease of glass transition temperature and thermal stability of polymers. The morphology of cured compositions was characterized by matrix-dispersion phase structure. All phosphazene containing polybenzoxazines demonstrated the improved flame resistance.

Keywords: benzoxazines cure; phosphazenes; flame retardant; catalysis; thermal analysis

Citation: Bornosuz, N.V.; Korotkov, R.F.; Kolenchenko, A.A.; Shapagin, A.V.; Orlov, A.V.; Gorbunova, I.Y.; Kireev, V.V.; Sirotin, I.S. The Influence of Substituents in Phosphazene Catalyst-Flame Retardant on the Thermochemistry of Benzoxazine Curing. *Polymers* **2021**, *13*, 3111. https://doi.org/10.3390/polym13183111

Academic Editor: Paul Joseph

Received: 13 August 2021
Accepted: 2 September 2021
Published: 15 September 2021

Publisher's Note: MDPI stays neutral with regard to jurisdictional claims in published maps and institutional affiliations.

Copyright: © 2021 by the authors. Licensee MDPI, Basel, Switzerland. This article is an open access article distributed under the terms and conditions of the Creative Commons Attribution (CC BY) license (https://creativecommons.org/licenses/by/4.0/).

1. Introduction

Polymer composite materials (PCM) have become ingrained in our everyday lives. They have the advantages of low specific gravity combined with high strength, moisture and chemical resistance, radio transparency, excellent dielectric properties, durability, etc. Because of these properties PCM came into common use in automotive and shipbuilding industries, aircraft, sports, medical applications, and many other fields. However, technological progress and economic development have helped scientists improve the quality of PCM by modifying binder and filler. As one of the main disadvantages of both polymers and composites based on them is their high flammability due to the organic nature of the matrix, which limits wider application of PMC, coke-forming phenolic or polybenzoxazine matrices [1] or epoxy and polyester binders modified with flame retardants are usually used in transport interior industry [2,3]. However, the introduction, for instance, of phosphorus-containing flame retardants, as a rule, leads to a significant decrease in mechanical properties and heat resistance of the polymer due to the absence of compatibility between matrix and flame retardant resulting in a two-phase system.

Our object of study is benzoxazine binders for PCM. It is a novel class of monomers commercialized in 2008. For now, many corporations such as Huntsman Advanced Materials and Henkel have branded benzoxazine binders and prepregs. Benzoxazine monomers have the feature of thermal self-curing (Figure 1) that could be proceeded by different mechanisms proposed in the literature [4]. Polybenzoxazines are characterized by a number of beneficial properties, including high mechanicals, dimensional stability and heat resistance, near-zero curing shrinkage, and low moisture absorption [5,6]. Some polybenzoxazines are completely nonflammable, for instance, ones based on 4,4′-diaminodiphenylmethane (P-d) and 3,3′-dichloro-4,4′-diaminodiphenylmethane (P-q) with phenol [7]. However, most of them have just reduced flammability reaching V-1/V-2 category by UL-94 tests. Another common problem of benzoxazine is high curing temperature, and therefor low energy efficiency, that limits their use in different technologies. The main way to solve this problem is to use the catalysts [8–12] such as phenols, strong acids [9], carboxylic acids and Lewis acids for the homopolymerization of benzoxazines. The base catalysts such as amines [13], imidazoles [14], and organophosphorus compounds, which exhibit a weaker catalyzing effect in comparison with acidic ones, are also applicable.

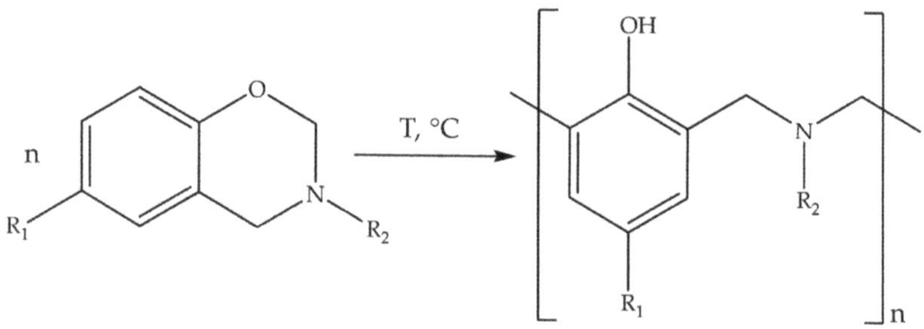

Figure 1. Formation of polybenzoxazine network by thermal curing process.

Thus, these are two problems of benzoxazine that our scientific group are to cope with: flammability and catalysis. Pursuing this complex goal phosphazene-based compounds may become widespread.

Phosphazenes are a unique class of heterocyclic compounds, discovered by Liebig, Wöhler, and Rose in 1834. The unique features of phosphazenes are their outstanding flexible molecular design and a wide range of homologous cyclo- and linear phosphazenes applied in many different fields. A lot of functionalized epoxy resins, benzoxazines, and amine-based curing agents with phosphazene core are proposed in the literature [6,15–20]. Cyclophosphazenes, owing to their flexible design, have the possibility of replacing chlorine atoms in hexachlorocyclotriphosphazenes with practically any substituents that in fact determine the properties of the resultant compound. They also may form numerous complexes. That is why scientists often use phosphazenes as a base for catalysts, extractants, and curing agents. Moreover, due to their inorganic nature they take on market as flame retardants [3]. The synergistic effect of phosphorus and nitrogen atoms enclosed in an unsaturated skeleton contributes to enhanced flame resistance [21,22] in the process of thermooxidative destruction (combustion). The mechanism of the synergistic effect in phosphazenes is not fully understood; however, it is known that the organophosphorus molecules are efficient radical scavengers and flame quenching materials, and combustion processes are essentially exothermic free-radical reactions, so the existence of radical stabilizers impedes combustion by the quenching mechanism. On the other hand, the nitrogen-containing moieties release an inert gaseous byproduct to form a highly porous char that provides thermal insulation and prevents the combustion from spreading [23].

Phosphazene-based retardants as other retardants can be either additive or reactive. The most frequently additive modifiers contribute to the reduction of mechanical properties of resulting polymers forming a two-phase system that could be defective. Therefore, miscible or reactive flame retardants are preferable.

Our previous work [24] was devoted to a novel catalyst-flame retardant based on phosphazene for benzoxazine monomer based on bisphenol A, m-toluidine, and paraphormaldehyde (BA-mt). The efficiency of hexakis–(3–methylphenylamino)cyclotriphosphazene (PN-mt) as a catalyst and flame retardant was proved. In this work we intend to continue our study of this type of phosphazene catalyst-flame retardant, disclose the influence patterns of substituents in hexachlorocyclotriphosphazenes on the curing process, and give a brief study of thermal properties and flammability of the resulting polymer.

The following modifiers for BA-mt matrix were chosen: hexachlorocyclotriphosphazenes, Tetrakis–(3–methylphenylamino) dichlorocyclotriphosphazene, and hexakis–(3–methylphenylamino) cyclotriphosphazene and hexaphenoxycyclotriphosphazene. The first and the last are well-known compounds: an initial reagent and additive in flame retardant, respectively. They are expected to be the extreme points: the most reactive and nonreactive, respectively. The second and the third compounds, our novel modifiers, differed in the degree of m-toluidine substitution, so we expect the reduction of reactivity in the above mentioned list.

2. Materials and Methods

2.1. Starting Materials

Hexachlorocyclotriphosphazene (HCP) was purchased from Shandong Chuangyu Chemical Co., Ltd., Shandong, China. HCP was recrystallized from hexane before use. M-toluidine (Acros Organics, Geel, Belgium) was distilled twice under vacuum before use. Bisphenol A (4,4′–(Propane–2,2–diyl)diphenol) was supplied by by Kazanorgsintez PJSC, Kazan, Russia and used without further purification. Phenol was distilled before use. Excipients were used without preliminary purification. Solvents were purified according to known methods, and their physical characteristics corresponded to literature data [25].

2.2. Synthesis of Benzoxazine Monomer Based on Bisphenol A, M-Toluidine, and Paraphormaldehyde (BA-mt)

Benzoxazine BA-mt (bis(3–(m–tolyl)–3,4–dihydro–2H–1,3–benzoxazine) is a well-known commercialized monomer. It was chosen as the matrix for compositions to continue our study of a novel catalyst, revealed in our previous work [24]. It was synthesized using the method reported in the literature [26]. The product yield was 95%. Completely cured BA-mt has a glass transition temperature of 217 °C for samples cured under following conditions: 2 h 180 °C, 4 h 200 °C, and 1 h 210 °C.

2.3. Synthesis of Hexakis–(3–methylphenylamino)cyclotriphosphazene (PN-mt (6))

PN-mt (6) was obtained according to the method reported in our previous work [24]. A 250 mL round-bottom flask equipped with a magnetic stirrer and a reflux condenser was charged with 10 g (0.0287 mol) of HCP, dissolved in advance in 50 mL of 1,4-dioxane, 24.60 g (0.230 mol) of meta-toluidine, and 20.32 g (0.230 mol) triethylamine. The reaction mixture was refluxed under intensive stirring for 8 h. Further the solvent and excess of triethylamine and meta-toluidine were distilled off using a vacuum rotary evaporator. At first the remaining white mass with a yellowish tinge was repeatedly washed with distilled water. The product was twice precipitated from ethanol into water and, finally, dried in a vacuum until a constant mass was achieved. The final product was a white fine powder. The yield was 16.51 g (82%).

2.4. Synthesis of Tetrakis–(3–methylphenylamino)dichlorocyclotriphosphazene (PN-mt (4))

PN-mt (4) was obtained according to the following procedure. A 250 mL round-bottom flask equipped with a magnetic stirrer and a reflux condenser was charged with 10 g (0.0287 mol) of HCP, previously dissolved in 50 mL of 1,4-dioxane, 12.30 g (0.115 mol) of meta-toluidine and 11.61 g (0.115 mol) triethylamine. The reaction mixture was refluxed under intensive stirring for 6 h. Then, the solvent and excess of triethylamine and meta-toluidine were distilled off using a vacuum rotary evaporator. At first the remaining white mass with a yellowish tinge was repeatedly washed with distilled water. The product was twice reprecipitated from ethanol into water and, finally, dried in a vacuum until a constant mass was achieved. The final product was a white fine powder. The yield was 16.12 g (75%).

2.5. Synthesis of Hexaphenoxycyclotriphosphazene (HPP)

HPP was obtained according to the following procedure. A 200 mL round-bottom flask equipped with a reflux condenser and an overhead stirrer was charged with 10.00 g (0.0287 mol) HCP, 17.55 g (0.1868 mol) phenol, and 47.56 g (0.3446 mol) K_2CO_3, then 125 mL of acetonitrile was added. The mixture was boiled for 14 h. At the end of the process, the hot solution was filtered from salts and the solvent was distilled off. After that, the dry product was dissolved in 125 mL of toluene and washed three times with 10% sodium hydroxide solution. Then, it was washed with distilled water until neutral pH. The resulting solution was dried to remove traces of water with sodium sulfate for 1 h with intensive stirring. The resulting suspension was filtered, and the solvent was distilled off to constant weight. As a result, 16.32 g of white crystals of hexaphenoxycyclotriphosphazene were obtained. The yield was 82%.

2.6. Composition Preparation

In order to determine the effect of substituents in phosphazene compounds on the curing process of benzoxazine monomer BA-mt, the following formulations presented in Table 1 were prepared. The variation of modifier's concentration in the mixtures was determined by several applied purposes: to achieve catalysis enough for compete cure under set curing conditions, to enhance flame resistance, and to maintain heat resistance at an appropriate level.

Table 1. Formulations of mixtures in parts per hundred resin (phr).

Formulation Index	Catalyst, Phr	Catalyst
BA-mt [1]	0	–
HCP-5	5	
HCP-10	10	HCP [2]
HCP-15	15	
PN-mt (4)-5	5	
PN-mt (4)-10	10	PN-mt (4) [3]
PN-mt (4)-15	15	
PN-mt (6)-5	5	
PN-mt (6)-10	10	PN-mt (6) [4]
PN-mt (6)-15	15	
HPP-5	5	
HPP-10	10	HPP [5]
HPP-15	15	

[1] BA-mt—bis(3—(m—tolyl)—3,4-dihydro—2H—1,3-benzoxazine, [2] HCP-Hexachlorocyclotriphosphazene, [3] PN-mt(4)—Tetrakis-(3-methylphenylamino)dichlorocyclotriphosphazene, [4] PN-mt(6)—Hexakis-(3-methylphenylamino)cyclotriphosphazene, [5] HPP—Hexaphenoxycyclotriphosphazene.

Compositions with HCP were prepared by dissolving of calculated amount of HCP in BA-mt melt at 100 °C for 10 min. The other compositions were prepared by dissolving the calculated amount of catalyst in 1,4-dioxane at 80 °C followed by addition of the obtained catalyst solution to the calculated amount of BA-mt. The mixtures were stirred at 80 °C for 10 min to achieve a homogeneous transparent solution. Subsequent degassing of the systems was performed at 120 °C for 30 min at a residual pressure of 1.0 kPa. Obtained compositions were either used as received for curing study or cured at 180 °C for 6 h for further thermal analysis and microscopy. Samples cured at 160 °C − 2 h + 180 °C − 2 h + 200 °C − 2 h + 210 °C − 1 h were utilized for flammability tests.

2.7. Measurements

The ^{1}H, ^{31}P, and ^{13}C NMR spectra were obtained in DMSO-d6 solutions with a Bruker AV-400 spectrometer (Bruker Corporation, Bremen, Germany) operating at 400, 162, and 100 MHz, respectively. The signals, due to the deuterated solvents, were used as internal references. The chemical shifts of the signals were calculated relative to the signals of tetramethylsilane (^1H, ^{13}C) and phosphoric acid (^{31}P), which were used as references. The spectra were processed with the help of the MestReNova Lab software package (version 12.0.4, MESTRELAB RESEARCH, S.L, Santiago de Compostela, Spain).

Differential scanning calorimeter DSC 214 Polyma (Netzsch, Selb, Germany) was used for monitoring thermal effects [27]. The temperature characteristics of the curing and glass transition temperatures of cured samples were determined according to ISO 11357-5:1999 [28] and ISO 11357-2:1999 [29], respectively. The heating rate for all measurements was 10 °C/min. All tests were performed in the temperature range 50–300 °C in a nitrogen atmosphere at a rate flow of 40 mL/min. The weight of the samples ranged from 5 to 10 mg. For data processing, Proteus Thermal Analysis version 8.0.2 software (Netzsch, Selb, Germany) was used.

The simultaneous thermal analyzer STA 449 F3 Jupiter (Netzsch, Selb, Germany) was used to evaluate the thermal stability of cured resin samples. The heating rate for all measurements was 20 °C/min. All tests were performed in the temperature range 50–1000 °C in a helium atmosphere at a flow rate of 100 mL/min. The weight of the samples ranged from 10 to 20 mg. For data processing, Proteus Thermal Analysis version 8.0.2. software (Netzsch, Selb, Germany) was used.

Flammability tests were carried out in accordance with Vertical Burning Test UL-94 (ASTM D3801–20a, Northbrook, IL, USA) [30] for 5 samples. The dimensions of the samples were 127 mm × 12.7 mm × 2 mm.

The morphology of cracks was investigated on an FEI Quanta 650 scanning electron microscope (Thermo Fisher Scientific, Waltham, MA, USA) in the secondary electron mode at an accelerating voltage of 10 kV. The phase structure was developed by plasma etching of a low-frequency oxygen discharge at a universal vacuum station (Edwards Coating System E306A, England). Sample preparation consisted of vacuum thermal sputtering of gold onto the crack surface (Edwards Coating System E306A, Burgess Hill, Great Britain).

3. Results and Discussions

3.1. Synthesis and Characteristics of Products

Structures of all compounds used in compositions are presented in Figure 2.

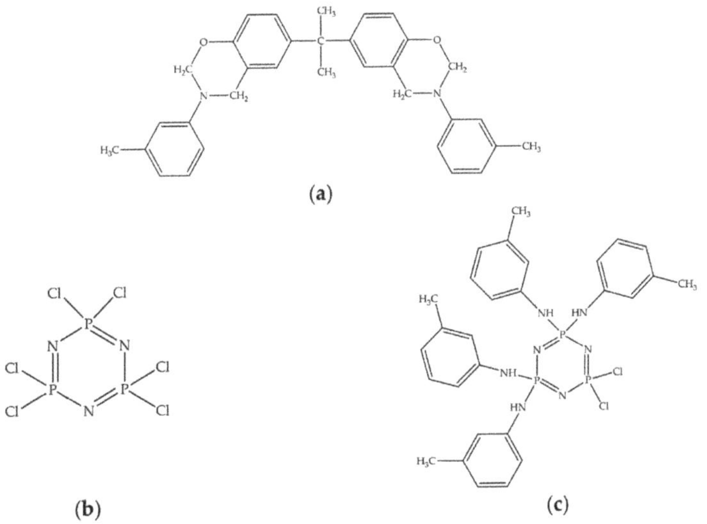

Figure 2. *Cont.*

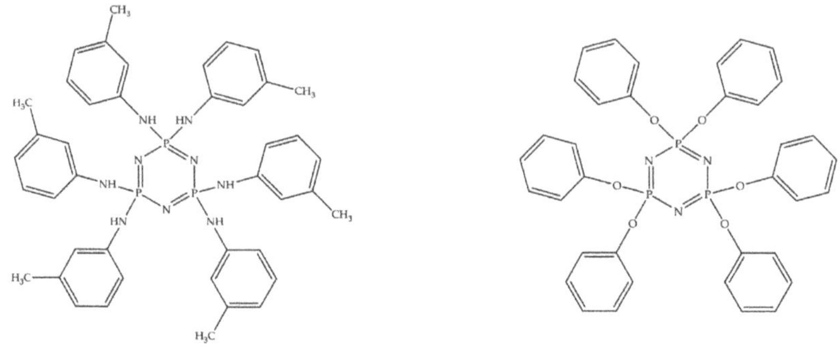

(d) (e)

Figure 2. Structures of matrix BA-mt (**a**) and the modifiers: HCP (**b**), PN-mt(4) (**c**), PN-mt(6) (**d**), HPP (**e**).

Benzoxazine monomer BA-mt chosen for our catalysis study was obtained according to the method described in [26]. Monomer was characterized by ^{1}H (Figure S1 in Supplementary Materials), ^{13}C (Figure S2 in Supplementary Materials) NMR spectroscopy, DSC (Figure S3 in Supplementary Materials) and TGA (Figure S4 in Supplementary Materials). ^{1}H NMR spectrum corresponds the structure of synthetized BA-mt. Signals at the area δ_H = 4.67 ppm (Ar–CH_2–N) and δ_H = 5.42 ppm (O–CH_2–N) correspond to methylene groups in the oxazine ring. The broadened signal at 4.25–4.5 ppm corresponds to the methylene protons of oligomeric compounds. The presence of oligomerized monomer in the product is either confirmed by a slight broadening of the DSC peak corresponded to the benzoxazine curing. Other thermal characteristics are T_{peak} = 233.3 °C and ΔH = 320.1 J/g. Thermal resistance of the cured matrix was evaluated by TGA. The temperature of 5% mass loss was 358.5 °C and residual mass at 1000 °C was 30.38%.

The modifier HPP obtained according to Section 2.5 was characterized by ^{1}H (Figure S5 in Supplementary Materials), ^{13}C (Figure S6 in Supplementary Materials), and ^{31}P (Figure S7 in Supplementary Materials) NMR spectroscopy, DSC (Figure S8 in Supplementary Materials) and TGA (Figure S9 in Supplementary Materials). The ^{1}H-NMR spectrum of the product shows proton signals in the region of 6.85–7.40 ppm that correspond to signals of phenoxy substituents. In the area of 2.5 ppm there is a signal of residual protons of DMSO-d6. The ^{13}C-NMR spectrum contains signals of carbon atoms of benzene rings of phenoxy substituents in the ranges 120.48–129.90 ppm and 149.92 ppm. In the region 39.13–39.97 ppm carbon signal of DMSO-d6 is present. The ^{31}P-NMR spectrum contains signals of phosphorus atoms in HPP (8.55 ppm). DSC exotherm detected the melting temperature of 112.6 °C. Destruction of the compound was evaluated by TGA revealing the temperature of 5% mass loss being 324.8 °C and residual mass at 1000 °C − 4%.

The modifier PN-mt(4) obtained according to Section 2.4 was characterized by ^{1}H (Figure S10 in Supplementary Materials), ^{13}C (Figure S11 in Supplementary Materials), and ^{31}P (Figure S12 in Supplementary Materials) NMR spectroscopy, DSC (Figure S13 in Supplementary Materials) and TGA (Figure S14 in Supplementary Materials). The ^{1}H-NMR spectrum of the product shows proton signals in the range of 2.12–2.24 ppm which correspond to the signals of the methyl groups of m-toluidine, 7.02–7.08 ppm and 8.02–8.05 ppm the protons of the benzene ring of m-toluidine, as well as 6.64–6.67 ppm corresponding signals of protons of secondary amine groups attached to phosphorus atoms. The ratio of protons of the secondary amine groups to methyl protons of m-toluidine is 1:3 as calculated. The ^{13}C-NMR spectrum of the product contains signals from the carbon atoms of the methyl groups of m-toluidine in the regions of 21.19–21.28 ppm, 115.02–141.37 ppm atoms of benzene rings. In the range 38.85−40.11 ppm signals of carbons ^{13}C DMSO-d6 are present. The ^{31}P-NMR spectrum of the product contains signals of phosphorus atoms corresponding to the degrees of substitution: three- (geminal)—in the regions of 22–23, 6.5–7.5, and −0.5–0.5 ppm, and tetra (geminal) −21–22 and (−1.5)–(−1) ppm. Such a specific nature of substitution, which differs from the most common non-geminal one, requires a separate study. DSC exotherm detected the melting temperature of 195.2 °C. Destruction of the compound was evaluated by TGA revealing the temperature of 5% mass loss being 261.9 °C and residual mass at 1000 °C − 35%.

The modifier PN-mt(6) obtained according to Section 2.3 was characterized by ^{1}H (Figure S15 in Supplementary Materials), ^{13}C (Figure S16 in Supplementary Materials), and ^{31}P (Figure S17 in Supplementary Materials) NMR spectroscopy, DSC (Figure S18 in Supplementary Materials) and TGA (Figure S19 in Supplementary Materials). The 1H-NMR spectrum of the product shows proton signals in the range of 2.29 ppm which correspond to the signals of the methyl groups of m-toluidine, and 6.66–7.32 ppm—to the protons of the benzene ring of m-toluidine. It was considered that the signal of the proton of the amine group was superimposed on the signal of one of the protons of the aromatic ring, as reported in the article [24]. The absence of a similar overlap of signals for PN-mt(4) could be explained by the fact that, in the case of incomplete substitution of chlorine atoms in the initial HCP, the fraction of intermolecular hydrogen bonding increased (between the molecules of PN-mt(4) and the solvent, DMSO-d6). Therefore, the protons of the secondary -NH- groups in PN-mt (4) were more deshielded and shifted to the downfield compared with PN-mt(6). The ^{13}C-NMR spectrum of the product contains signals from the carbon atoms of the methyl groups of m-toluidine in the region of 21.39 ppm, signals in the region of 114.87–142.56 ppm. correspond to the atoms of the benzene rings. In the region 39.08–39.91 ppm signals of carbon ^{13}C DMSO-d6 are present. The ^{31}P-NMR spectrum of the product contains a singlet signal in the region of 2.39 ppm, which indicates the complete replacement of chlorine atoms in the initial HCP by meta-toluidine radicals. DSC exotherm detected the melting temperature of 242.7 °C. Destruction of the compound was evaluated by TGA revealing the temperature of 5% mass loss being 266.7 °C and residual mass at 1000 °C − 30.74%.

Recrystallized HCP was characterized by ^{31}P NMR- spectrum as well as by DSC and TGA presented in Supplementary Figures S20, S21, and S22, respectively. The ^{31}P-NMR spectrum contains singlet with δ_P = 19.9 ppm of phosphorus atom. DSC exotherm detected the melting temperature of 112.7 °C. According to the TG analysis the sublimation of HCP is observed [31].

3.2. *The Influence of Phosphazene Compounds on the Curing Process*

The influence of chosen modifiers on the curing process was studied by DSC method. The key points of DSC scans are presented in Table 2. All DSC curves are presented in Figure 3.

Table 2. Characteristic temperatures of DSC curves of uncured samples.

Formulation Index	T_{onset}, °C	T_{peak}, °C	T_{end}, °C	ΔH, J/g
BA-mt	224.6	233.3	241.8	320.1
HCP-5	121.7 (142.1) *	146.1 (189.9) *	167.5 (227.6) *	208.6
HCP-10	122.4 (139.2) *	143.6 (187.1) *	167.9 (225.8) *	215.6
HCP-15	121.8 (134.7) *	138.6 (185.1) *	155.5 (224.8)	225.1
PN-mt(4)-5	196.4	209.0	225.3	280.0
PN-mt(4)-10	190.6	204.9	222.0	277.2
PN-mt(4)-15	186.7	201.6	218.8	269.4
PN-mt(6)-5	203.5	218.0	236.2	312.3
PN-mt(6)-10	200.4	216.0	238.2	301.4
PN-mt(6)-15	196.5	214.0	242.3	285.6
HPP-5	223.4	232.4	243.1	290.6
HPP-10	223.7	233.0	243.1	261.8
HPP-15	224.7	233.5	242.1	254.3

* the second peak according to the peak separation (Figure 4).

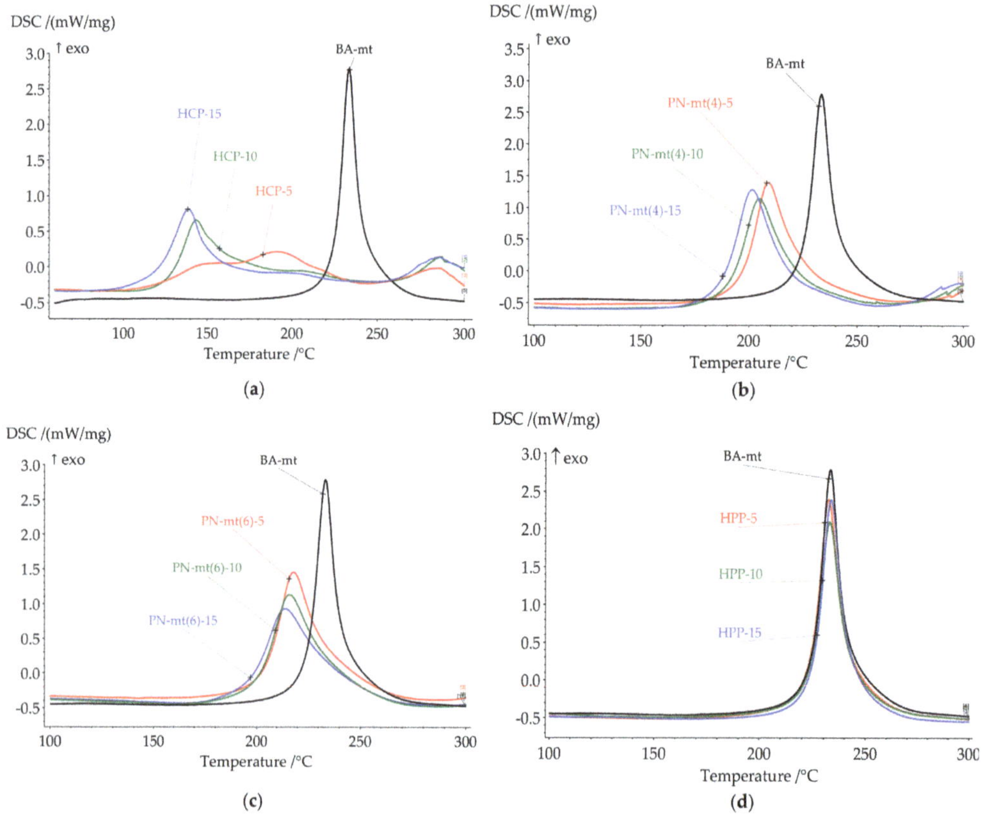

Figure 3. DSC curves of uncured compositions with HCP (**a**), PN-mt(4) (**b**), PN-mt(6) (**c**), HPP (**d**).

HPP almost does not change key temperatures, but influences the heat released. Nevertheless, the lone electron pair of the nitrogen atoms in PN-mt located in the phosphazene core was expected to catalyze benzoxazine ring-opening polymerization (ROP), no catalysis was observed.

The use of HCP as a catalyst of benzoxazine curing significantly decreased the onset temperature of the reaction. The curing process of compositions was characterized by complex process with at least two stages. That is why peak separation was carried out (Figure 4). The ratio between the first and second peaks areas increased in a row, HCP-5, HCP-10, and HCP-15 being 19:81, 72:28, and 81:19. We suppose that oligomeric compounds presented in BA-mt resin react by nucleophilic substitution with HCP (Figure 5). The released HCl catalyzed ROP corresponded to the first peak (Figure 6a). With the temperature raise the autocatalytic effect became more significant associated with the increase of hydroxylic group's concentration during ROP. Further curing of composition led to decrease in mobility of molecules and decrease in HCl concentration due to steric hindrance or the fixation of HCl by tertiary nitrogen atoms in the formed Mannich bridges. This effect resulted in the second wide peak on DSC curve corresponded to homopolymerization of benzoxazine. With an increase of HCP content, the first stage prevails over the second accompanying by the growth of heat released.

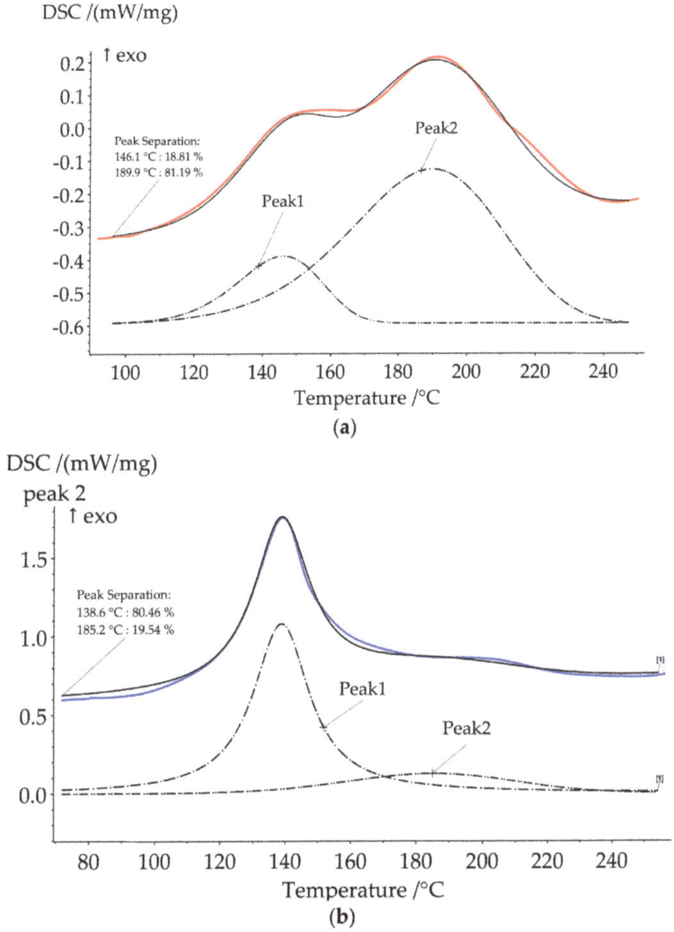

Figure 4. Peak separation for formulations HCP-5 (**a**) and HCP-15 (**b**).

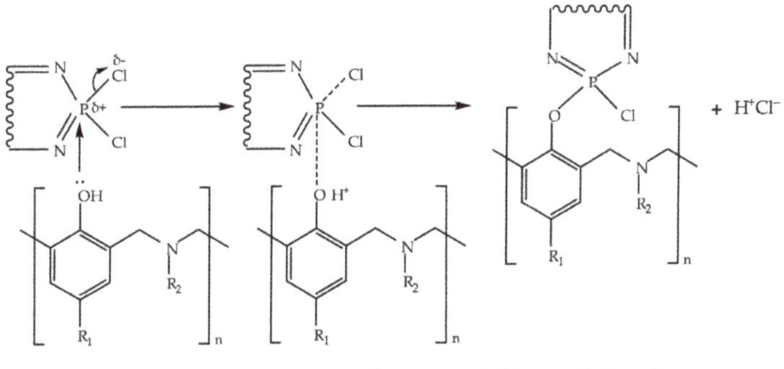

Figure 5. The proposed mechanism of phenolysis of chlorocyclophosphazenes.

The most interesting catalytic effect was observed in compositions with PN-mt(4) and PN-mt(6). The addition of 5 phr PN-mt(4) caused the decrease of the onset temperature to 196.3 °C; the addition of 5 phr PN-mt(6) decreased the onset temperature to 203.5 °C. The further increase of catalyst containment to 15 phr slightly reduces onset temperature (to 186.7 °C in case of 15 phr PN-mt(4) and to 196.5 °C in case of 15 phr PN-mt(6)). Additionally catalysts affect the amount of released heat. In case when we used 5 phr PN-mt(4) released heat decreases to 280.0 J/g; 5 phr PN-mt(6)—released heat decreases to 312.3 J/g. Further increase of PN-mt(4) content did not significantly change the exothermic effect of the polymerization. The unique catalytic effect consisted in the possibility to catalyze benzoxazine ROP by released HCl for PN-mt(4) (Figure 6a) and a nucleophilic attack of the lone electron pair of the nitrogen atoms in PN-mt located in substituents (Figure 6b) on the methylene bridge connecting the oxygen and nitrogen atoms in the benzoxazine ring for PN-mt(4) and PN-mt(6) [32]. The basic catalytic effect was weaker for the benzoxazine polymerization than acidic one. Thus, the use of PN-mt(4) led to the implementation of two different mechanisms of catalysis resulting in a greater catalytic effect than in case of PN-mt(6). Moreover we supposed that compositions with PN-mt(6) were induced by steric hindrance.

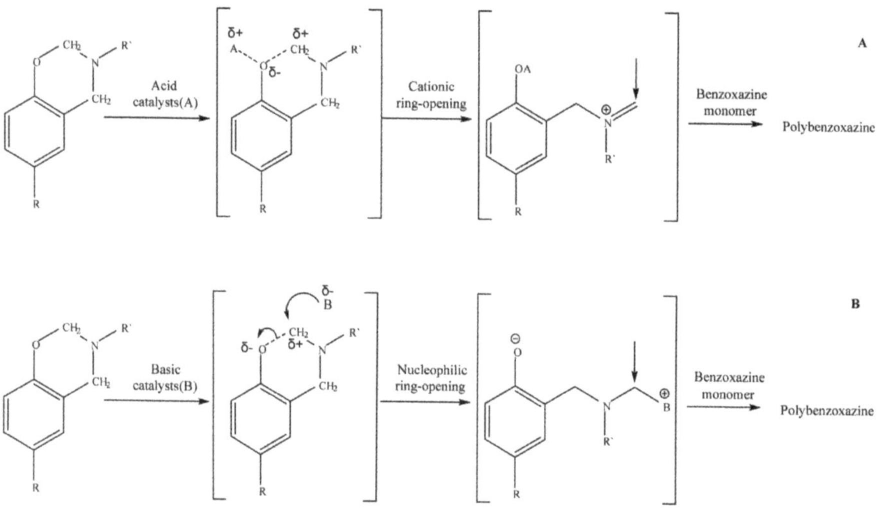

Figure 6. The proposed cationic catalytic mechanism ((**A**)—in case of released HCl) and anionic catalytic mechanism ((**B**)—in case of secondary amine group).

Thus, the catalytic activity of obtained phosphazene compounds on curing process of benzoxazine resin decreased in row: HCP < PN-mt(4) < PN-mt(6) < HPP. This tendency was also confirmed by the fact that acidic catalysis was more effective than basic one. The absence of catalytic effect of HPP proved that there was no catalytic activity of lone electron pair of nitrogen atom in the core of phosphazene ring.

In comparison with literature data, the catalytic effect of HCP was quite explicit. The lowering of the onset by 100 °C was a significant effect, which was better than the catalytic effect of acids and phenols described in [8]. The decrease of peak polymerization temperature for PN-mt(6) and PN-mt(4) was 15 °C and 24 °C respectively. This effect was in good agreement with literature data of catalytic effect of amines, imidazoles, and indole on benzoxazine ring-opening reaction. Obtained phosphazene compounds were secondary amines and the catalytic effect was similar to other secondary amines [33].

3.3. Thermal Analysys of Cured Compositions

All compositions under study were cured according to the modeling curing program 180 °C for 6 h. Thermal analysis of resulting polymers was proceeded by DSC and TGA methods to correlate heat flow effects with mass loss caused by destruction. The key points of DSC and TGA results were presented in Table 3. All DSC curves were presented in Figure 7 and TGA curves were presented in Figure 8.

Table 3. Characteristic temperatures of TGA, DSC and LOI data of cured samples and TGA data of neat modifiers.

Formulation Index	TGA			DSC		LOI **
	T_{onset}, °C	$T_{5\%}$, °C	Residual Mass, %	$T_{g(middle)}$, °C	ΔH_{res}, J/g	
BA-mt	348.5	358.5	30.38	203.3	38.4	29.652
HCP-5	298.7	304.7	28.27	161.4 *	0 *	28.808
HCP-10	299.6	306.1	29.88	193.4 *	0 *	29.452
HCP-15	312.5	319.6	35.22	221.7 *	0 *	31.588
PN-mt(4)-5	307.2	318.3	31.43	158.5	1.0	30.072
PN-mt(4)-10	313.1	320.0	32.51	159.3	0	30.504
PN-mt(4)-15	314.4	320.3	33.49	160.5	0	30.896
PN-mt(6)-5	329.1	329.2	26.49	180.5	9.3	28.096
PN-mt(6)-10	324.8	327.1	27.74	167.4	6.0	28.596
PN-mt(6)-15	322.9	327.0	29.20	160.0	5.0	29.18
HPP-5	345.9	350.1	32.52	195.9	21.7	30.508
HPP-10	342.6	341.9	30.91	186.5	14.4	29.864
HPP-15	336.5	336.5	29.20	176.3	9.1	29.18
HCP	118.5	–	4.56	–	–	
PN-mt(4)	257.6	–	34.82	–	–	
PN-mt(6)	254.1	–	30.74	–	–	
HPP	332.0	–	3.90	–	–	

* DSC results after the second run (Figure S23 in Supplementary Materials). ** Char yield at 1000 °C values were used for calculation.

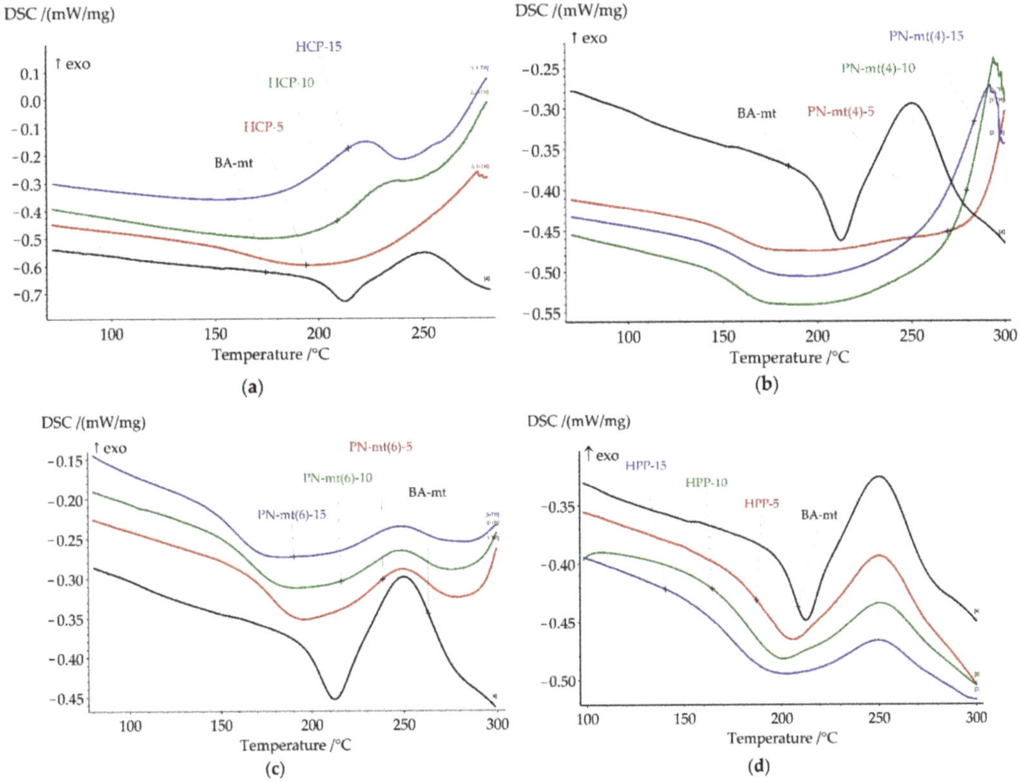

Figure 7. DSC curves of cured compositions with HCP (**a**), PN-mt(4) (**b**), PN-mt(6) (**c**), HPP (**d**) in the scanning mode.

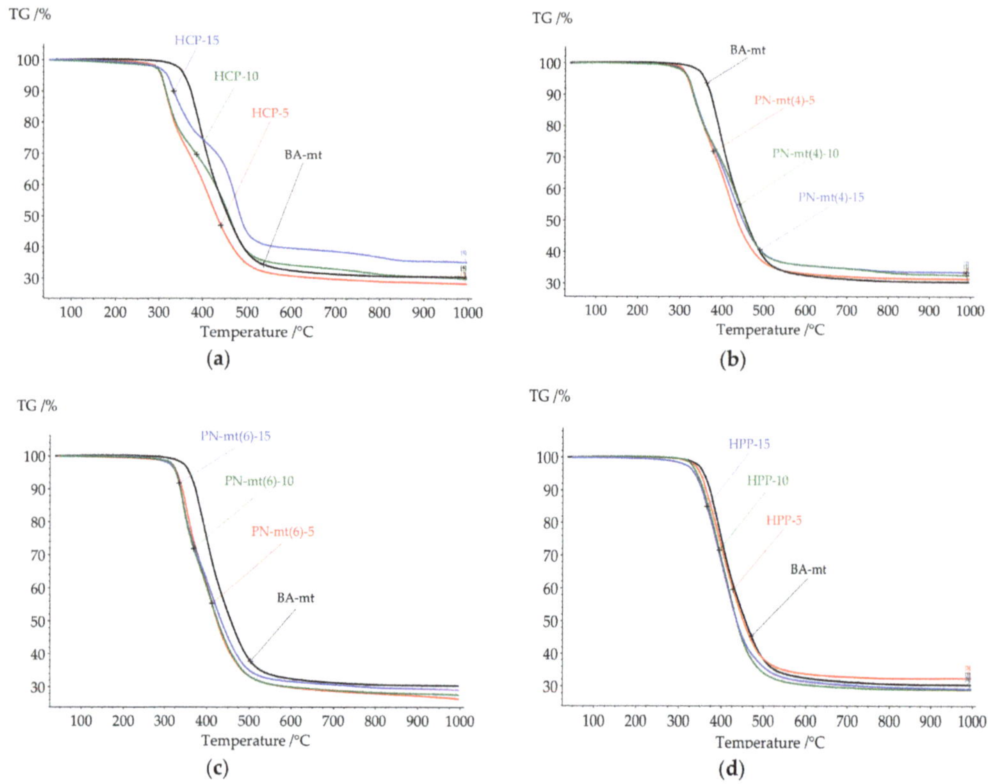

Figure 8. TGA curves of cured compositions with HCP (**a**), PN-mt(4) (**b**), PN-mt(6) (**c**), HPP (**d**) in the scanning mode.

Glass transition temperature (Tg) of the neat BA-mt cured at 180 °C for 6 h was 203.3°C. After glass transition the exothermic peak was observed with enthalpy 38.4 J/g corresponding to the post-curing of benzoxazine resin. This temperature program resulted in incomplete curing for matrix (degree of conversion 88%) which was chosen especially for explicit evaluation of modifiers influence in compositions on the final degree of conversion.

Cured compositions with HCP addition did not demonstrate glass transition until a significant exothermic effect (Figure 7a). According to the DSC analysis the exothermic effect starts for HCP-5 at 215 °C; for HCP-10 at 210 °C; and for HCP-15 at 200 °C. According to the TG analysis the weight loss started at 300 °C so we supposed that exothermic effect was related to the post-curing reaction and the begin of destruction. The second DSC scan confirmed our assumption as no exothermic peaks until 280 °C were observed (Figure S23 in Supplementary Materials).

The addition of PN-mt(4) led to the complete cure, while PN-mt(6) and HPP led to the slight increase of degree of conversion: 98% and 97% respectively for 15 phr compared with neat BA-mt (88%). The addition of 5 phr of HCP and PN-mt(4) dramatically reduced Tg of polymers till 161.4 °C and 158.5 °C, respectively. Further raise of concentration of HCP and PN-mt(4) till 15 phr had different effects. HCP caused the great increase of Tg to 221.7 °C, while composition PN-mt(4)-15 reached only 160.5 °C. PN-mt(6) and HPP influenced in another way. They monotonously reduced Tg from 180.5 °C and 195.9 °C for 5 phr to 160.0 °C and 176.3 °C for 15 phr, respectively.

The outstanding dependence of glass transition temperature for HCP could be related to the formation of rigid polymer network due to high reactivity of HCP detected by DSC

(Figure 3a). One of the indirect evidences for proposed curing mechanism of phenolysis (Figure 5) was the rust on the steel casting mold which was used for HCP curing.

The other modifiers reduce glass transition temperature. This effect could be caused by incomplete compatibility of resin and catalyst. Another reason was the significant size of the modifier molecules, which was comparable to nanoparticles with a diameter of 1–2 nm. It resulted in a change in the spatial regularity of the formed polybenzoxazine network and a decrease in the number of hydrogen bonds between hydrogen atoms in phenolic groups and nitrogen in polymerized benzoxazines. The decrease of glass transition temperature upon addition of PN-mt(4) could also be explained by the fact that the nucleophilic substitution reaction was incomplete and chlorine atoms of PN-mt(4) were not completely replaced by benzoxazine polymeric matrix. Based on this assumption PN-mt(4) participated in the reaction as monofunctional additive disrupted the organized benzoxazine network resulted in decreasing of glass transition temperature.

DSC analysis revealed that for compositions with HCP, PN-mt(4) and PN-mt(6) at the end of a measurement an exothermic peak was observed that indicated that samples lost their thermal stability. To study degradation process TGA measurements in inert atmosphere were carried out.

Thermal stability of the studied compositions had the same tendencies as the glass transition temperature. The addition of HCP and PN-mt(4) dramatically reduced the onset of thermal destruction followed by a slight increase of values in the row from 5 to 15 phr. PN-mt(6) and HPP influenced in another way. They monotonously reduced onset of the destruction.

It was proposed in literature that the thermal decomposition of polybenzoxazines occurs stepwise [34,35]. At the first stage of the destruction, aromatic compounds were formed (benzene, derivatives of phenol, aniline). On the second step low-molecular compounds (hydrocarbons, carbon dioxide, aliphatic amines, etc.) were formed, followed by carbonization.

Cured compositions with HCP, PN-mt4, and PN-mt6 demonstrated similar thermal characteristics. The destruction of cured compositions endured two stages of decomposition. Whereas in systems with HPP only one stage was observed.

3.4. Morphology of Resulting Polybenzoxazines

It was important for the flame retardant additive and catalysts to be dispersed or miscible in the polymer matrix for improved properties. The morphology of cured compositions was studied by scanning electron microscopy (SEM). The structure of cured neat BA-mt (Figure 9a) revealed homogeneous globular structure. Cured compositions with additives was characterized by matrix-dispersion phase structure (Figure 9b–i). The matrix consists of globular phase with the size of 20–100 nm. The dispersed phase was irregularly distributed in the matrix and consists of globules with the size of 130–350 nm. In some cases, the globules of dispersed phase formed associates. This phase structure was typical for all systems under study regardless of reactivity of additives and their concentration. We supposed that forming of dispersed phase structure occurred during polymerization which led to deterioration in thermodynamic compatibility of phases. As a result the heterogeneous phase structure was formed. For the system HPP-15 one could see minor amount of the second phase, so mostly this system was compatible, and its additive influenced greatly on Tg. It is well known that there are two main factors that influence Tg. The first one is compatibility of matrix and modifier. If the system is compatible its glass transition temperature would shift to the glass transition temperature of modifier. The second factor is the influence of modifier on the network density due to copolymerization of matrix and modifier. Thus glass transition temperature of systems with HCP, PN-mt(4), and PN-mt(6) was deendent on the predominance of one of these factors.

3.5. Flammability of Cured Compositions

Flammability of cured compositions was estimated by UL-94 standard. The results of burning tests are shown in Table 4. The cured neat BA-mt achieved only V-1 flammability category of the UL-94 standard. The use of obtained additives in amount of 5 phr does not reduce the total burning time of the specimens except for the use of HCP. This effect could be explained by low concentration of the additive and also by the introduction of a phosphazene ring to the polymeric matrix in case of using HCP. For all used flame retardant agents, the total time of burning decreased with the increase of additive concentration due to an increase of the phosphazene containment. It is also important to note that no afterglow, either of burning particles or drops, was observed during burning tests. The HPP concentration more than 10 phr allows to achieve the V-0 flammability category. The HPP is a well-known commercialized flame retardant which efficiency was proved by many works [22,36,37]. HCP affects the flammability of benzoxazine resin at a content of 5 phr and more. The concentration of 10 phr HCP and more resulted in plastic with V-0 flammability category. We supposed that the great flame-retardant properties of HCP were associated with introducing the phosphazene ring to polymeric matrix and high phosphorous content due to low molecular mass of HCP (versus the other additives). We supposed that the flame retardant effect of PN-mt(4) and PN-mt(6) will be similar to the HPP. However, the total time of burning for compositions containing PN-mt(4) and PN-mt(6) is higher than for compositions with HPP. PN-mt(4) contained less phenyl residues resulted in better flammability of compositions with PN-mt(4). Another possible explanation is that PN-mt(4) and PN-mt(6) affected the polymerization process and hence could affect the structure of the matrix. The composition with 15 phr PN-mt(6) achieved the V-0 flammability category but due to the above-mentioned reasons compositions with PN-mt(4) reaches only V-1 flammability category.

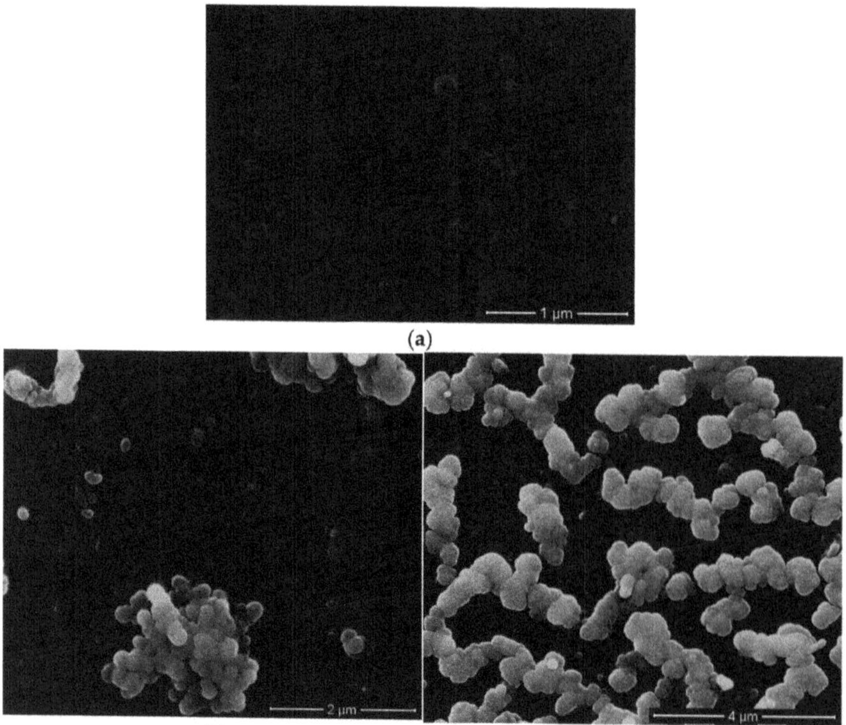

(a)

(b) (c)

Figure 9. *Cont.*

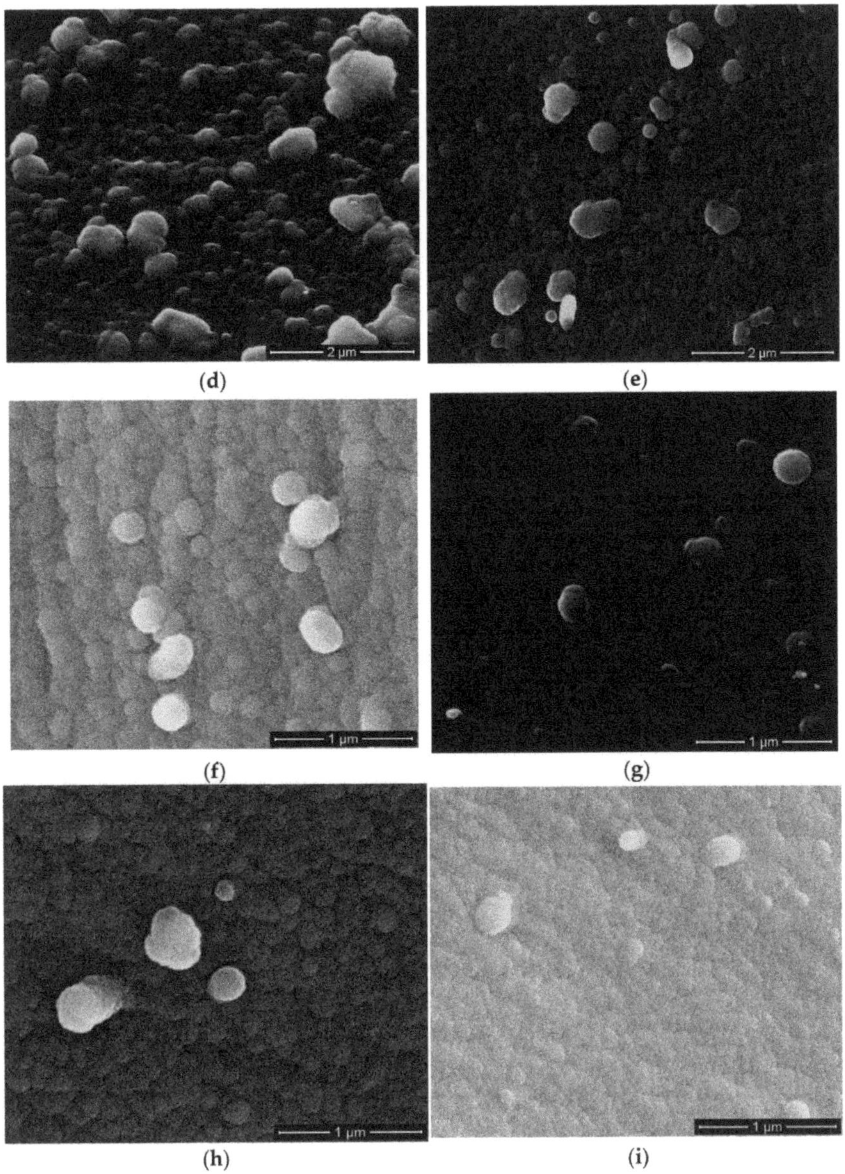

Figure 9. Morphology of cured BA-mt (**a**) and compositions HCP-5 (**b**), HCP-15 (**c**), PN-mt(4)-5 (**d**), PN-mt(4)-15 (**e**), PN-mt(6)-5 (**f**), PN-mt(6)-15 (**g**), HPP-5 (**h**), HPP-15 (**i**).

Table 4. Flammability test results according to UL-94 standard.

Composition	Flame Application	Specimen № 1	Specimen № 2	Specimen № 3	Specimen № 4	Specimen № 5	ΣT, S	Afterglow, S	Drops	Category
BA-mt	t_1, s t_2, s	24 3	10 4	3 21	12 30	22 2	131	no	no	V-1
HCP-5	t_1, s t_2, s	20 7	17 2	8 0	5 13	21 2	95	no	no	V-1
HCP-10	t_1, s t_2, s	0 0	0 8	0 6	1 0	0 5	20	no	no	V-0
HCP-15	t_1, s t_2, s	3 0	0 3	0 0	0 0	0 2	8	no	no	V-0
PN-mt(4)-5	t_1, s t_2, s	24 8	23 6	20 0	25 1	19 5	131	no	no	V-1
PN-mt(4)-10	t_1, s t_2, s	17 10	25 5	7 15	9 6	26 7	127	no	no	V-1
PN-mt(4)-15	t_1, s t_2, s	15 4	12 2	8 0	7 11	11 2	72	no	no	V-1
PN-mt(6)-5	t_1, s t_2, s	12 14	21 2	24 10	17 9	20 6	135	no	no	V-1
PN-mt(6)-10	t_1, s t_2, s	20 2	29 0	25 4	16 3	19 4	122	no	no	V-1
PN-mt(6)-15	t_1, s t_2, s	7 2	1 5	1 8	2 9	4 0	39	no	no	V-0
HPP-5	t_1, s t_2, s	0 12	9 18	0 20	19 27	4 25	134	no	no	V-1
HPP-10	t_1, s t_2, s	1 1	1 2	0 1	1 7	0 1	15	no	no	V-0
HPP-15	t_1, s t_2, s	0 0	1 0	0 0	0 2	1 0	4	no	no	V-0

According to calculationos by the Van Krevelen–Hovtyzer equation LOI all polymer compositions could be related to nonflammable category as LOI > 28 [38]. So, we can conclude that all presented additives improve the flame resistance of benzoxazine resin due to an increase of phenyl residues content and to the phosphazene structure that possesses the synergistic effect of phosphorus and nitrogen introduced into the mixture.

The addition of phosphazene compounds also affect the flammability of benzoxazine resin. It was shown that 5 phr of all catalysts except HCP did not decrease the burning time of plastic specimens. HCP, PN-mt(6), and HPP allowed to reach the V-0 flammability category when 15 phr was added. However, the addition of novel additive PN-mt(4) to benzoxazine resin did not reach the V-0 category but decreased the total burning time from 131 s (0 phr) to 71 s (15 phr).

4. Conclusions

The influence of modifiers with substituents in the phosphazene core including novel flame retardant agent–catalyst PN-mt(4) on the curing process, thermal stability, morphology, and flammability of benzoxazine resin BA-mt was studied in this work. The catalytic activity of phosphazene compounds on the curing process of benzoxazine resin decreased in the row: HCP > PN-mt(4) > PN-mt(6) > HPP where HPP was totally inactive. HCP, PN-mt(4), and PN-mt(6) catalysts decreased the onset polymerization temperature and residual enthalpy for cured samples. The two catalysis mechanisms were proposed. For chlorine-containing catalysts HCl released during phenolysis catalyze the ROP by acidic mechanism proposed in our work. For m-toluidine containing catalysts the lone electron pair of the nitrogen atoms located in substituents catalyze ROP by basic mechanism. The addition of minor amounts of modifiers overall caused the decrease of the glass temperature and the thermal stability of the cured resin. However, the addition of 15 phr HCP resulted in the rigid polymer network due to multifunctionality of modifier and hence increased the glass transition temperature. The morphology of cured modified compositions was characterized by matrix-dispersion phase structure. The use of phosphazene catalysts affects the flammability of benzoxazine resin. Incorporation of HCP, PN-mt(6), and HPP allowed to reach the V-0 flammability category when 15 phr was added.

Supplementary Materials: The following are available online at https://www.mdpi.com/article/10.3390/polym13183111/s1, Figure S1: 1H NMR spectrum of benzoxazine BA-mt. Figure S2: 13C NMR spectrum of benzoxazine BA-mt. Figure S3. DSC curve of BA-mt. Figure S4. TGA curve of BA-mt. Figure S5. 1H NMR spectrum of HPP. Figure S6. 13C NMR spectrum of HPP. Figure S7. 31P NMR spectrum of HPP. Figure S8. DSC curve of HPP. Figure S9. TGA curve of HPP. Figure S10. 1H NMR spectrum of PN-mt(4). Figure S11. 13C NMR spectrum of PN-mt(4). Figure S12. 31P NMR spectrum of PN-mt(4). Figure S13. DSC curve of PN-mt(4). Figure S14. TGA curve of PN-mt(4). Figure S15. 1H NMR spectrum of PN-mt(6). Figure S16. 13C NMR spectrum of PN-mt(6). Figure S17. 31P NMR spectrum of PN-mt(6). Figure S18. DSC curve of PN-mt(6). Figure S19. TGA curve of PN-mt(6). Figure S20. 31P NMR spectrum of HCP. Figure S21. DSC curve of HCP. Figure S22. TGA curve of HCP. Figure S23. DSC curves of the second DSC run of cured compositions with HCP.

Author Contributions: Data curation, N.V.B., R.F.K., A.A.K. and A.V.S.; Formal analysis, A.V.O.; Supervision, I.Y.G. and I.S.S.; Writing—original draft, N.V.B., R.F.K. and A.V.O.; Writing—review and editing, V.V.K. and I.S.S. All authors have read and agreed to the published version of the manuscript.

Funding: This work was supported by the Mendeleev University of Chemical Technology (№ -2020-029).

Institutional Review Board Statement: Not applicable.

Informed Consent Statement: Not applicable.

Data Availability Statement: The data presented in this study are available on request from the corresponding author.

Conflicts of Interest: The authors declare no conflict of interest.

References

1. Machado, I.; Shaer, C.; Hurdle, K.; Calado, V.; Ishida, H. Towards the development of green flame retardancy by polybenzoxazines. *Prog. Polym. Sci.* **2021**, *121*, 101435. [CrossRef]
2. Morgan, A.B. 23—Flame retardant fiber-reinforced composites. In *Handbook of Fire Resistant Textiles*; Kilinc, F.S., Ed.; Woodhead Publishing Series in Textiles; Woodhead Publishing: Philadelphia, PA, USA, 2013; pp. 623–652, ISBN 978-0-85709-123-9.
3. Levchik, S.V.; Weil, E.D. A review of recent progress in phosphorus-based flame retardants. *J. Fire Sci.* **2006**, *24*, 345–364. [CrossRef]
4. Lochab, B.; Monisha, M.; Amarnath, N.; Sharma, P.; Mukherjee, S.; Ishida, H. Review on the Accelerated and low-temperature polymerization of benzoxazine resins: Addition polymerizable sustainable polymers. *Polymers* **2021**, *13*, 1260. [CrossRef] [PubMed]
5. Kiskan, B.; Ghosh, N.N.; Yagci, Y. Polybenzoxazine-based composites as high-performance materials: Polybenzoxazine-based composites. *Polym. Int.* **2011**, *60*, 167–177. [CrossRef]
6. Sirotin, I.S.; Sarychev, I.A.; Vorobyeva, V.V.; Kuzmich, A.A.; Bornosuz, N.V.; Onuchin, D.V.; Gorbunova, I.Y.; Kireev, V.V. Synthesis of phosphazene-containing, bisphenol a-based benzoxazines and properties of corresponding polybenzoxazines. *Polymers* **2020**, *12*, 1225. [CrossRef] [PubMed]
7. Petrakova, V.; Kireev, V.; Onuchin, D.; Sarychev, I.; Shutov, V.; Kuzmich, A.; Bornosuz, N.; Gorlov, M.; Pavlov, N.; Shapagin, A.; et al. Benzoxazine monomers and polymers based on 3,3′-dichloro-4,4′-diaminodiphenylmethane: Synthesis and characterization. *Polymers* **2020**, *13*, 1421. [CrossRef]
8. Ishida, H.; Rodriguez, Y. Catalyzing the curing reaction of a new benzoxazine-based phenolic resin. *J. Appl. Polym. Sci.* **1995**, *58*, 1751–1760. [CrossRef]
9. Dunkers, J.; Ishida, H. Reaction of benzoxazine-based phenolic resins with strong and weak carboxylic acids and phenols as catalysts. *J. Polym. Sci. Part A Polym. Chem.* **1999**, *37*, 1913–1921. [CrossRef]
10. Liu, C.; Shen, D.; Sebastián, R.M.; Marquet, J.; Schönfeld, R. Mechanistic studies on ring-opening polymerization of benzoxazines: A mechanistically based catalyst design. *Macromolecules* **2011**, *44*, 4616–4622. [CrossRef]
11. Chutayothin, P.; Ishida, H. Cationic ring-opening polymerization of 1,3-benzoxazines: Mechanistic study using model compounds. *Macromolecules* **2010**, *43*, 4562–4572. [CrossRef]
12. Liu, C.; Shen, D.; Sebastián, R.M.; Marquet, J.; Schönfeld, R. Catalyst effects on the ring-opening polymerization of 1,3-benzoxazine and on the polymer structure. *Polymer* **2013**, *54*, 2873–2878. [CrossRef]

13. Sun, J.; Wei, W.; Xu, Y.; Qu, J.; Liu, X.; Endo, T. A curing system of benzoxazine with amine: Reactivity, reaction mechanism and material properties. *RSC Adv.* **2015**, *5*, 19048–19057. [CrossRef]
14. Yue, J.; He, L.; Zhao, P.; Gu, Y. Engineering benzoxazine/epoxy/imidazole blends with controllable microphase structures for toughness improvement. *ACS Appl. Polym. Mater.* **2020**, *2*, 3458–3464. [CrossRef]
15. Sarychev, I.A.; Sirotin, I.S.; Borisov, R.S.; Mu, J.; Sokolskaya, I.B.; Bilichenko, J.V.; Filatov, S.N.; Kireev, V.V. Synthesis of resorcinol-based phosphazene-containing epoxy oligomers. *Polymers* **2019**, *11*, 614. [CrossRef]
16. Kireev, V.V.; Bilichenko, Y.V.; Borisov, R.S.; Mu, J.; Kuznetsov, D.A.; Eroshenko, A.V.; Filatov, S.N.; Sirotin, I.S. Synthesis of bisphenol a based phosphazene-containing epoxy resin with reduced viscosity. *Polymers* **2019**, *11*, 1914. [CrossRef]
17. Terekhov, I.V.; Filatov, S.N.; Chistyakov, E.M.; Borisov, R.S.; Kireev, V.V. Synthesis of oligomeric epoxycyclotriphosphazenes and their properties as reactive flame-retardants for epoxy resins. *Phosphorus Sulfur Silicon Relat. Elem.* **2017**, *192*, 544–554. [CrossRef]
18. Tan, Z.-W.; Wu, X.; Zhang, M.; Qiu, J.-J.; Liu, C.-M. Synthesis and properties of main-chain oligomeric benzoxazine precursor containing cyclotriphosphazene units. *High Perform. Polym.* **2014**, *26*, 906–913. [CrossRef]
19. Yang, G.; Wu, W.-H.; Wang, Y.-H.; Jiao, Y.-H.; Lu, L.-Y.; Qu, H.-Q.; Qin, X.-Y. Synthesis of a novel phosphazene-based flame retardant with active amine groups and its application in reducing the fire hazard of epoxy resin. *J. Hazard. Mater.* **2019**, *366*, 78–87. [CrossRef] [PubMed]
20. Amarnath, N.; Appavoo, D.; Lochab, B. Eco-friendly halogen-free flame retardant cardanol polyphosphazene polybenzoxazine networks. *ACS Sustain. Chem. Eng.* **2018**, *6*, 389–402. [CrossRef]
21. Leu, T.-S.; Wang, C.-S. Synergistic effect of a phosphorus–nitrogen flame retardant on engineering plastics. *J. Appl. Polym. Sci.* **2004**, *92*, 410–417. [CrossRef]
22. Pan, M.; Huang, R.; Wang, T.; Huang, D.; Mu, J.; Zhang, C. Preparation and properties of epoxy resin composites containing hexaphenoxycyclotriphosphazene. *High Perform. Polym.* **2014**, *26*, 114–121. [CrossRef]
23. Sun, J.; Wang, X.; Wu, D. Novel spirocyclic phosphazene-based epoxy resin for halogen-free fire resistance: Synthesis, curing behaviors, and flammability characteristics. *ACS Appl. Mater. Interfaces* **2012**, *4*, 4047–4061. [CrossRef]
24. Bornosuz, N.V.; Gorbunova, I.Y.; Kireev, V.V.; Bilichenko, Y.V.; Chursova, L.V.; Svistunov, Y.S.; Onuchin, D.V.; Shutov, V.V.; Petrakova, V.V.; Kolenchenko, A.A.; et al. Synthesis and application of arylaminophosphazene as a flame retardant and catalyst for the polymerization of benzoxazines. *Polymers* **2020**, *13*, 263. [CrossRef]
25. Riddick, J.A.; Bunger, W.B.; Sakano, T.K. *Organic Solvents: Physical Properties and Methods of Purification*, 4th ed.; John Wiley: New York, NY, USA, 1986.
26. Aizawa, T.; Hirai, Y.; Numata, S. Method for Producing Benzoxazine Resin. U.S. Patent US7041772B2, 9 May 2006.
27. Vyazovkin, S.; Burnham, A.K.; Criado, J.M.; Pérez-Maqueda, L.A.; Popescu, C.; Sbirrazzuoli, N. ICTAC kinetics committee recommendations for performing kinetic computations on thermal analysis data. *Thermochim. Acta* **2011**, *520*, 1–19. [CrossRef]
28. 14:00–17:00 ISO 11357-5:1999. Available online: https://www.iso.org/cms/render/live/en/sites/isoorg/contents/data/standard/02/71/27143.html (accessed on 14 January 2021).
29. 14:00–17:00 ISO 11357-2:1999. Available online: https://www.iso.org/cms/render/live/en/sites/isoorg/contents/data/standard/02/55/25545.html (accessed on 14 January 2021).
30. Horner, A. *Aircraft Materials Fire Test Handbook*; United States Federal Aviation Administration: Washington, DC, USA, 2000.
31. Rhili, K.; Chergui, S.; El Douhaibi, A.S.; Siaj, M. Hexachlorocyclotriphosphazene functionalized graphene oxide as a highly efficient flame retardant. *ACS Omega* **2021**, *6*, 6252–6260. [CrossRef] [PubMed]
32. Liu, C.; Chen, Q.-Y. Chapter 2—Catalytic accelerated polymerization of benzoxazines and their mechanistic considerations. In *Advanced and Emerging Polybenzoxazine Science and Technology*; Ishida, H., Froimowicz, P., Eds.; Elsevier: Amsterdam, The Netherlands, 2017; pp. 9–21, ISBN 978-0-12-804170-3.
33. Wang, H.; Zhu, R.; Yang, P.; Gu, Y. A study on the chain propagation of benzoxazine. *Polym. Chem.* **2016**, *7*, 860–866. [CrossRef]
34. Ran, Q.; Gu, Y.; Ishida, H. Chapter 11—Thermal degradation mechanism of polybenzoxazines. In *Advanced and Emerging Polybenzoxazine Science and Technology*; Ishida, H., Froimowicz, P., Eds.; Elsevier: Amsterdam, The Netherlands, 2017; pp. 171–204, ISBN 978-0-12-804170-3.
35. Hemvichian, K.; Ishida, H. Thermal decomposition processes in aromatic amine-based polybenzoxazines investigated by TGA and GC–MS. *Polymer* **2002**, *43*, 4391–4402. [CrossRef]
36. Höhne, C.-C.; Wendel, R.; Käbisch, B.; Anders, T.; Henning, F.; Kroke, E. Hexaphenoxycyclotriphosphazene as FR for CFR Anionic PA6 via T-RTM: A study of mechanical and thermal properties. *Fire Mater.* **2017**, *41*, 291–306. [CrossRef]
37. Feng, H.; Qian, L.; Lu, L. Synergistic effect of polyimide charring agent and hexaphenoxycyclotriphosphazene on improving fire safety of polycarbonate: High graphitization to strengthen the char layer. *Polym. Adv. Technol.* **2021**, *32*, 1135–1149. [CrossRef]
38. Van Krevelen, D.W.; te Nijenhuis, K. *Properties of Polymers: Their Correlation with Chemical Structure, Their Numerical Estimation and Prediction from Additive Group Contributions*; Elsevier: Amsterdam, The Netherlands, 2009; ISBN 978-0-08-091510-4.

polymers

MDPI

Article

Eight-Fold Interpenetrating Diamondoid Coordination Polymers for Sensing Volatile Organic Compounds and Metal Ions

Venkatesan Lakshmanan [1], Yi-Ting Lai [1], Xiang-Kai Yang [1], Manivannan Govindaraj [1], Chia-Her Lin [2,*] and Jhy-Der Chen [1,*]

1 Department of Chemistry, Chung-Yuan Christian University, Chung Li 32023, Taiwan; venkatesanflower95@gmail.com (V.L.); e0912583362@gmail.com (Y.-T.L.); xiangkaishulin@gmail.com (X.-K.Y.); manivannanjent@gmail.com (M.G.)

2 Department of Chemistry, National Taiwan Normal University, Taipei 11677, Taiwan

* Correspondence: chiaher@ntnu.edu.tw (C.-H.L.); jdchen@cycu.edu.tw (J.-D.C.); Tel.: +886-3-265-3351 (J.-D.C.)

Abstract: Reactions of divalent metal salts with 4,4-oxybis(N-(pyridine-4-yl)-benzamide), **L**, and naphthalene-1,4-dicarboxylic acid (1,4-H_2NDC) in various solvents gave [Zn(**L**)(1,4-NDC)·$H_2O]_n$, **1**, [Cd(**L**)(1,4-NDC)(H_2O)·MeOH$]_n$, **2**, and [Co(**L**)(1,4-NDC)($H_2O)_{0.5}$·MeOH$]_n$, **3**, which have been structurally characterized. Complexes **1**–**3** show eight-fold interpenetrating frameworks with the **dia** topology, which exhibit porosities substantiated by CO_2 adsorption, whereas **1** and **2** manifest stability in aqueous environments and show high selectivity toward sensing of mesitylene molecules and Fe^{3+} ions with low detection limits and good reusability up to five cycles.

Keywords: coordination polymer; crystal structure analysis; entanglement; luminescence

Citation: Lakshmanan, V.; Lai, Y.-T.; Yang, X.-K.; Govindaraj, M.; Lin, C.-H.; Chen, J.-D. Eight-Fold Interpenetrating Diamondoid Coordination Polymers for Sensing Volatile Organic Compounds and Metal Ions. *Polymers* **2021**, *13*, 3018. https://doi.org/10.3390/polym13183018

Academic Editor: Piotr Dobrzynski

Received: 2 August 2021
Accepted: 5 September 2021
Published: 7 September 2021

Publisher's Note: MDPI stays neutral with regard to jurisdictional claims in published maps and institutional affiliations.

Copyright: © 2021 by the authors. Licensee MDPI, Basel, Switzerland. This article is an open access article distributed under the terms and conditions of the Creative Commons Attribution (CC BY) license (https://creativecommons.org/licenses/by/4.0/).

1. Introduction

Coordination polymers (CPs) have shown potential applications in luminescent sensing, gas storage and separation, drug delivery, catalysis, ion exchange and magnetism [1–5]. Metal ions are linked together by organic ligands to form extended frameworks, which are often porous and comprised into infinite arrays. In addition, polycarboxylates play a vital role in the organization of diverse CPs with differential dimensionalities, which show multiple potential coordination sites that are highly active to coordinate with metal centers. Linear and angular types of polycarboxylate ligands have been used extensively in the preparation of CPs in a mixed system [6,7]. Entanglement involving interpenetration, polycatenation and self-catenation is a very interesting phenomenon, which is also a major factor contributing to the diversities of CPs due to their aesthetic and intriguing topological interests [8]. Despite many such interesting CPs being reported, the manipulation of the structural diversity is still a challenge.

Considerable effort has been invested into understanding the entanglement of CPs containing bis-pyridyl-bis-amide (bpba) ligands [9–13]. Two highly interpenetrated Zn(II) and Cd(II) CPs with eight- and nine-fold interpenetrating diamondoid modes, respectively, have been reported by using the long flexible *N*,*N*′-di(4-pyridyl)adipoamide ligand and the short rigid 1,4-benzenedicarboxylic acid [13]. Construction of a high-fold interpenetrating diamondoid framework with linear bidentate ligand may thus be ascribed to the length and flexibility of the spacer ligands and the geometrical arrangements of the metal centers that may modulate pore size and shape in a single network. While the entanglement may reduce the solvent accessible volume required for the molecules/metal ions adsorptions, the modulated pore size and shape may enhance the selectivity. Moreover, the tunable porosity of luminescent CPs has a great impact on the sensitivity and selectivity for the detection of volatile organic compounds (VOCs) and metal ions that are harmful to the environment and public health [14].

In this study, three CPs constructed from the angular ligand 4,4-oxybis(N-(pyridine-4-yl)-benzamide (**L**), 1,4-H_2NDC and divalent metal salts, namely, [Zn(**L**)(1,4-NDC)·H_2O]$_n$, **1**, [Cd(**L**)(1,4-NDC)(H_2O)·MeOH]$_n$, **2**, and [Co(**L**)(1,4-NDC)(H_2O)$_{0.5}$·MeOH]$_n$, **3**, are reported, which show eight-fold interpenetrating diamondoid networks. Complexes **1** and **2** exhibit different emission intensity enhancement for distinct VOCs, and show a high sensitivity toward detection of Fe^{3+} ion in aqueous systems. The synthesis, structural characterization and the detections of VOCs and metal ions of these CPs form the subject of this report.

2. Materials and Methods

2.1. General Procedures

Elemental analyses of (C, H, N) were performed on a PE 2400 series II CHNS/O (PerkinElmer Instruments, Shelton, CT, USA) or an Elementar Vario EL III analyzer (Elementar Analysensysteme GmbH, Hanau, German). Infrared spectra were obtained from a JASCO FT/IR-460 Plus spectrometer with pressed KBr pellets (JASCO, Easton, MD, USA). Thermal gravimetric analyses (TGA) were carried out on an SII Nano Technology Inc. TG/DTA 6200 over the temperature range of 30 to 900 °C at a heating rate of 10 °C min^{-1} under N_2 (SEIKO Instruments Inc., Chiba, Japan). UV–vis spectrum was performed on a UV-2450 spectrophotometer (Dongguan Hongcheng Optical Products Co., Dongguan, China). Emission spectra for the solid samples were determined with a Hitachi F-4500 fluorescence spectrophotometer (Hitachi, Tokyo, Japan). Gas sorption measurements were conducted using a Micromeritics ASAP 2020 system (Micromeritics Instruments Co., Norcross, GA, USA). Powder X-ray diffraction patterns were carried out with a Bruker D8-Focus Bragg-Brentano X-ray powder diffractometer equipped with a CuKα sealed tube (λ_α = 1.54178 Å) at 40 kV and 40 mA (Bruker Corporation, Karlsruhe, Germany).

2.2. Materials

The reagent Zn(OAc)$_2$·2H_2O was purchased from SHOWA Co. (Saitama, Japan), Cd(OAc)$_2$·2H_2O from Fisher Scientific Co. (Hampton, NY, USA), Co(OAc)$_2$·4H_2O from J. T. Baker Co. (Phillipsburg, NJ, USA), and naphthalene-1,4-dicarboxylic acid (1,4-H_2NDC) from Alfa Aesar Co. (Ward Hill, MA, USA). The ligand 4,4′-oxybis(N-(pyridine-4-yl)-benzamide) (**L**) was prepared according to a published procedure with some modification [15].

2.3. Preparations

2.3.1. [Zn(**L**)(1,4-NDC)·H_2O]$_n$, **1**

Zn(OAc)$_2$·2H_2O (0.022 g, 0.10 mmol), **L** (0.041 g, 0.10 mmol) and 1,4-H_2NDC (0.022 g, 0.10 mmol) in 8 mL H_2O and 2 mL methanol were sealed in a 23 mL Teflon-lined flask inside a stainless steel autoclave, which was heated at 120 °C for 2 days, and then the reaction system was cooled down at a rate of 2 °C per hour until the room temperature was reached. Colorless crystals were obtained, which were purified with methanol. Yield: 0.041 g (58%), using Zn(II) as the limited reagent. Anal. calcd. for $C_{36}H_{26}N_4O_8Zn$ (MW = 707.98): C, 61.07; N, 7.91; H, 3.70%. Found: C, 61.22; N, 8.07; H, 4.01%. FT-IR (cm^{-1}): 3237(m, N-H), 3168(m, N-H), 3070(w), 1686(s), 1596(s), 1512(s), 1490(s), 1429(m), 1332(m), 1236(m), 1210(m), 1170(s).

2.3.2. [Cd(**L**)(1,4-NDC)(H_2O)·MeOH]$_n$, **2**

The procedures for **2** were similar to **1** except Cd(OAc)$_2$·2H_2O (0.013 g, 0.05 mmol), **L** (0.021 g, 0.05 mmol) and 1,4-H_2NDC (0.011 g, 0.05 mmol) in 2 mL H_2O and 8 mL methanol were used. Colorless crystals were obtained. Yield: 0.023 g (58%). Anal. calcd. for $C_{37}H_{30}CdN_4O_9$ (MW = 787.05): C, 56.34; N,7.11; H, 3.80%. Found: C, 55.69; N, 7.05; H, 3.44%. FT-IR (cm^{-1}): 3168(m), 3068(m), 1664(s), 1595(s), 1552(s), 1494(s), 1426(s), 1364(s), 1334(s), 1289(m), 1235(s), 1211(m), 1169(s), 1100(m), 1016(m), 836(m), 760(m), 583(m).

2.3.3. [Co(**L**)(1,4-NDC)$(H_2O)_{0.5}$·MeOH]$_n$, **3**

Complex **3** was prepared according to the procedures for **1** except $Co(OAc)_2 \cdot 4H_2O$ (0.025 g, 0.10 mmol), **L** (0.041 g, 0.10 mmol) and 1,4-H_2NDC (0.022 g, 0.10 mmol) were used. Purple crystals were obtained. Yield: 0.036 g (49%). Anal. calcd for $C_{37}H_{29}CoN_4O_{8.5}$ (MW = 724.57): C, 61.33; N, 7.73; H, 4.03%. Found: C, 60.50; N, 7.56; H, 3.90%. FT-IR (cm^{-1}): 3230(m), 3158(m), 3069(m), 1683(m), 1595(s), 1510(s), 1494(s), 1427(s), 1363(s), 1330(s), 1274(m), 1235(s), 1208(m), 1170(m), 1098(m), 1022(m), 838(m), 759(w), 584(w). Figure S1 shows the FT-IR spectra of complexes **1–3**.

2.4. *X-ray Crystallography*

A Bruker AXS SMART APEX II CCD diffractometer, which was equipped with graphite-monochromated MoKα (λ_α = 0.71073 Å) radiation, was used to collect the diffraction data for complexes **1–3** [16], and the data were reduced by using the well-established computational procedures. The structure factors were treated with Lorentz and polarization corrections and an empirical absorption correction based on "multi-scan" [17]. The Direct or Patterson method was applied to locate the positions of some of the heavier atoms, and the remaining atoms were identified in several alternating difference Fourier maps and least-square refinements. Hydrogen atoms were added by using the HADD command in SHELXTL. The coordinated water molecule, O(8), of complex **2** is disordered such that two orientations of the oxygen atom can be found, and the occupancy of each orientation was set to be 0.5 and were refined isotropically. Moreover, the occupancy of O(8) of **3** was refined to be 0.5. Table 1 lists the basic information regarding the crystal parameters and structure refinement.

Table 1. Crystal data for complexes **1–3**.

Complex	1	2	3
Formula	$C_{36}H_{26}ZnN_4O_8$	$C_{37}H_{30}CdN_4O_9$	$C_{37}H_{29}CoN_4O_{8.5}$
Formula weight	707.98	787.05	724.57
Crystal system	monoclinic	monoclinic	monoclinic
Space group	$P2_1/c$	$P2_1/n$	$P2_1/n$
a, Å	9.78070(1)	8.5591(4)	8.5240(6)
b, Å	19.3224(2)	21.5741(11)	21.5176(16)
c, Å	17.9006(2)	19.0086(9)	18.1376(13)
α, °	90	90	90
β, °	103.1124(8)	98.065(3)	96.850(2)
γ, °	90	90	90
V, Å^3	3294.77(6)	3475.3(3)	3303.0(4)
Z	4	4	4
D_{calc}, mg/m^3	1.427	1.504	1.457
F(000)	1456	1600	1496
μ (MoK_α), mm^{-1}	0.805	0.690	0.582
Range (2θ) for data collection, deg.	$3.14 \leq 2\theta \leq 52.00$	$2.87 \leq 2\theta \leq 52.00$	$4.41 \leq 2\theta \leq 50.12$
Independent reflections	6474 [R(int) = 0.0398]	6845 [R(int) = 0.0633]	5855 [R(int) = 0.1880]
Data/restraints/parameters	6474/0/437	6845/1/496	5855/0/487
Quality-of-fit indicator [a]	1.034	1.076	1.027
Final R indices [I > 2σ(I)] [b,c]	R1 = 0.0618, wR2 = 0.1761	R1 = 0.0648, wR2 = 0.1609	R1 = 0.0655, wR2 = 0.1477
R indices (all data)	R1 = 0.0871 wR2 = 0.1939	R1 = 0.0935 wR2 = 0.1751	R1 = 0.1986 wR2 = 0.2248

[a] quality-of-fit = $[\Sigma w(|F_o^2| - |F_c^2|)^2/(N_{observed} - N_{parameters})]^{1/2}$. [b] $R_1 = \Sigma||F_o| - |F_c||/\Sigma|F_o|$. [c] $wR_2 = [\Sigma w(F_o^2 - F_c^2)^2/\Sigma(F_o^2)^2]^{1/2}$. $w = 1/[\sigma^2(F_o^2) + (ap)^2 + (bp)]$, $p = [\max(F_o^2 \text{ or } 0) + 2(F_c^2)]/3$. **1**, a = 0.1067, b = 3.1664; **2**, a = 0.0754, b = 7.0332; **3**, a = 0.0849, b = 3.8585.

3. Results and Discussion

3.1. Crystal Structures of 1–3

Complexes **1–3** are isostructural and their single crystals conform to the monoclinic system. While the crystal structure of **1** was solved in the space group $P2_1/c$, those of **2** and **3** were in $P2_1/n$. Their asymmetric units are mainly composed of one divalent metal ion [Zn(II), **1**; Cd(II), **2** and Co(II), **3**], one **L** and one 1,4-NDC^{2-} ligands, and there is also one cocrystallized water molecule in **1**, one coordinated water and one cocrystallized MeOH molecule in **2**, and a half coordinated water and one cocrystallized MeOH molecule in **3**, respectively. Figure 1a–c depict the drawings showing the coordination environments about the metal ions of complexes **1–3**. The Zn(II) ions of **1** is four-coordinated by two pyridyl nitrogen atoms [Zn-N = 2.024(3) Å and 2.037(3) Å] from two **L** ligands and two carboxylate oxygen atoms [Zn-O = 1.905(5) and 1.940(3) Å] from two 1,4-NDC^{2-} ligands, forming a distorted tetrahedral geometry, while the Cd(II) ion of **2** is seven-coordinated by two pyridyl nitrogen atoms [Cd-N = 2.272(5) and 2.297(5) Å] from two **L** ligands, four carboxylate oxygen atoms [Cd-O = 2.248(5)–2.577(4) Å] from two 1,4-NDC^{2-} ligands and one oxygen atom [Cd-O = 2.416(13) Å] of the coordinated water molecule, resulting in a distorted pentagonal bipyramidal geometry. As shown in the X-ray crystallographic section, the occupancy of the coordinated water of **3** was refined to be 0.5; six and five coordination are thus possible for the Co(II) ion. The six-coordinated Co(II) ion is supported by two pyridyl nitrogen atoms [Co-N = 2.069(5) and 2.088(6) Å] from two **L** ligands, three carboxylate oxygen atoms [Co-O = 1.913(10)–2.302(5) Å] from two 1,4-NDC^{2-} ligands and one oxygen atom [Co-O = 2.238(9) Å] of the coordinated water molecule, revealing a distorted octahedral geometry, while the removal of the O(8) atom results in a five-coordinated Co(II) ion and forms a distorted square pyramidal geometry.

Considering the divalent metal ions as four-connection nodes and the ligands as linkers, the structures of **1–3** can be simplified as 3D nets with the (6^6)-**dia** topology (Figure 1d) revealing the 8-fold [4 + 4] interpenetration (Figure 1e), determined using ToposPro [18]. Noticeably, the diamondoid Zn(II) and Cd(II) CPs constructed from the flexible *N,N′*-di(4-pyridyl)adipoamide and 1,4-benzenedicarboxylic acid show eight- and nine-fold interpenetrating modes, respectively [13], indicating the nature of the metal center alters the degree of interpenetration. In marked contrast, complexes **1** and **2** with the Zn(II) and Cd(II) ions, respectively, are both eight-fold interpenetrated, demonstrating that the degree of interpenetration of the diamondoid CPs in mixed systems is also subject to the changes of the shape and flexibility of the polycarboxylate and bpba ligands.

3.2. Ligand Conformations and Bonding Modes

For the bpba ligand, the two C=O (or N-H) groups that are on the same and opposite directions define the *cis* and *trans* conformations, respectively [19]. Accordingly, all of the **L** ligands in **1–3** adopt the *cis* conformation. A schematic drawing defining the C-O-C (θ) angle and dihedral angle (d) is shown in Figure 2, and the angles were calculated and listed in Table 2. Clearly, although all of the **L** ligands in **1–3** adopt the same *cis* conformation, they differ in the dihedral angles, presumably due to the different metal identity.

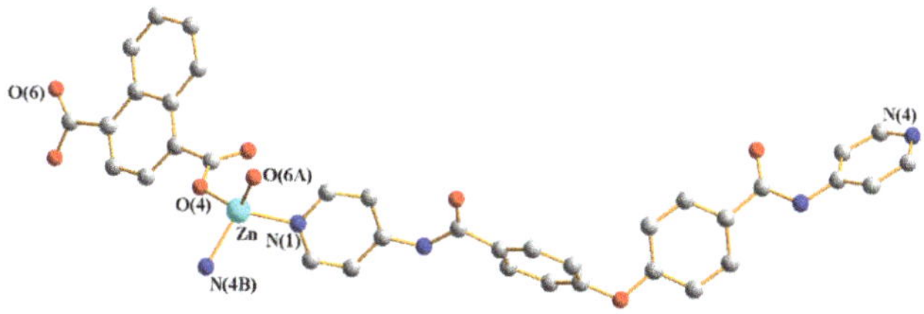

(a)

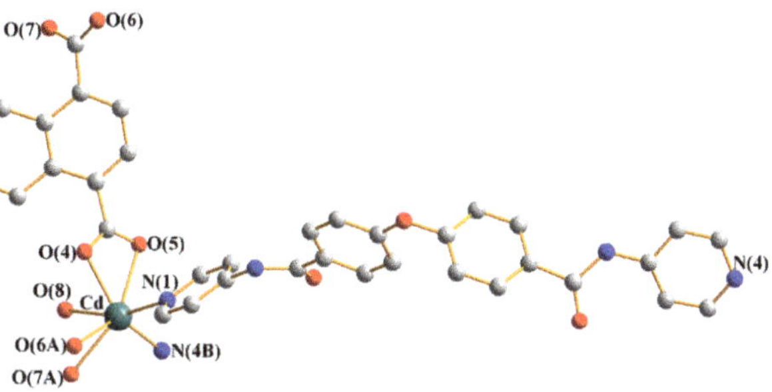

(b)

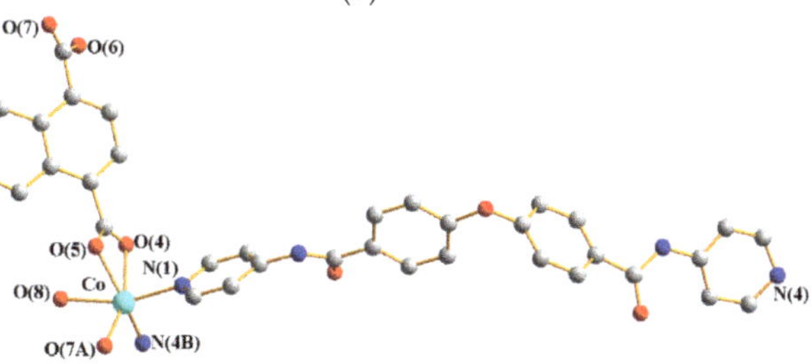

(c)

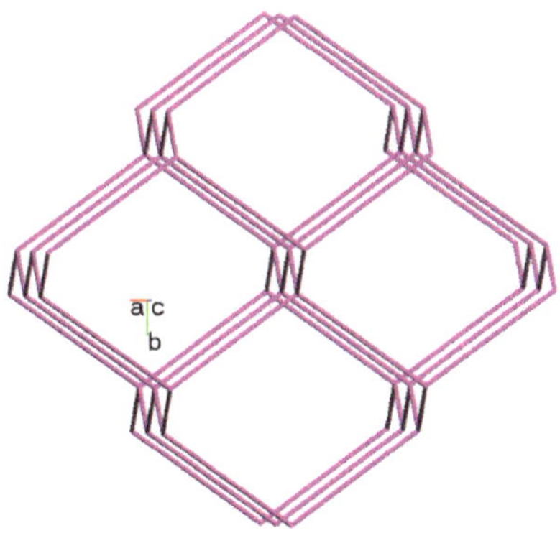

(d)

Figure 1. *Cont.*

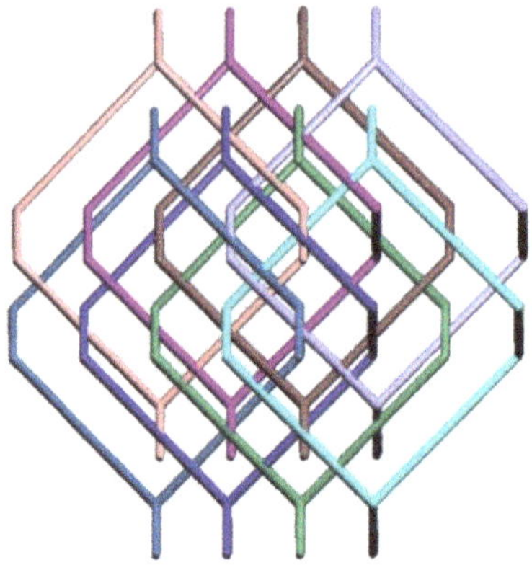

(e)

Figure 1. Coordination environments about the (**a**) Zn(II) ion in **1**. Symmetry transformations used to generate equivalent atoms: (A) x, −y + 5/2, z − 1/2; (B) x − 2, −y + 3/2, z − 1/2; (**b**) Cd(II) ion in **2**. Symmetry transformations used to generate equivalent atoms: (A) x + 1/2, −y + 3/2, z + 1/2; (B) x − 3/2, −y + 1/2, z + 1/2; and (**c**) Co(II) ion in **3**. Symmetry transformations used to generate equivalent atoms: (A) x − 3/2, −y + 1/2, z + 1/2; (B) x + 1/2, −y + 3/2, z + 1/2. (**d**) A drawing showing a 3D framework with the **dia** topology. (**e**) A schematic view of the eight-fold [4 + 4] interpenetration.

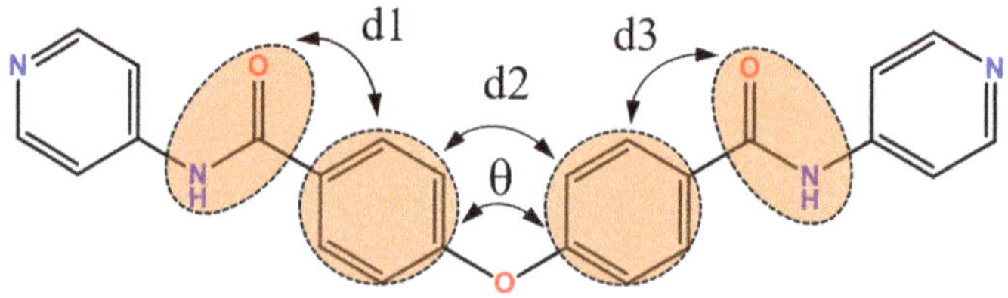

Figure 2. A drawing defining the C-O-C (θ) and dihedral angles (d) for **L**.

Table 2. Selected angles of the **L** ligands in complexes **1–3**.

Complex	C-O-C Angle (°)	Dihedral Angle (°)		
		d1	d2	d3
1	117.0	30.49	74.52	21.12
2	116.9	25.51	87.63	35.97
3	116.9	28.54	88.70	34.02

Moreover, the 1,4-NDC^{2-} ligand of **1** bridge two Zn(II) ions through two carboxylate oxygen atoms from two carboxylate groups to reveal the coordination mode I, while those of **2** and **3** chelate and bridge two metal ions through four carboxylate oxygen atoms, coordination mode II, and three carboxylate oxygen atoms, coordination mode III, respectively (Figure 3). Noticeably, while the Zn(II), Cd(II) and Co(II) metal centers of **1–3** adopt the distorted tetrahedral, pentagonal bipyramidal and octahedral (square pyramidal) geometries, respectively, giving rise to different coordination modes for the 1,4-NDC^{2-} ligands; their structural topologies are not subject to the metal identities.

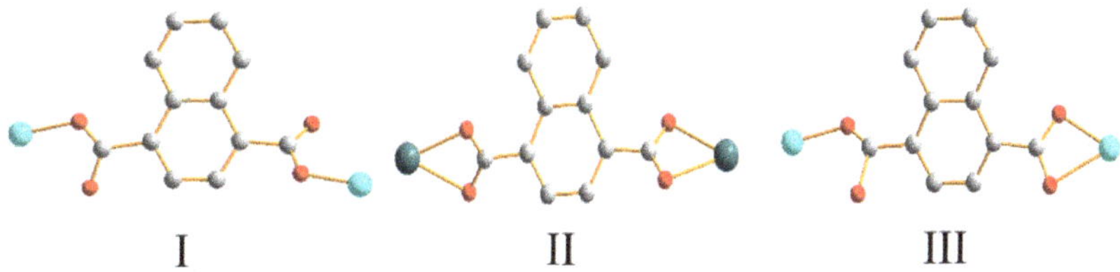

Figure 3. Various coordination modes of 1,4-NDC^{2-} ligands in **1–3**.

3.3. PXRD Patterns and Thermal Analysis

As shown in Figures S2–S4, the experimental PXRD patterns of complexes **1–3** are in consistency with their corresponding simulated ones, demonstrating the purities of the bulk samples. On the other hand, their thermal gravimetric analysis (TGA) curves performed in a nitrogen atmosphere from 30 to 800 °C (Figure S5) display two-step weight losses involving removal of cocrystallized solvents and the coordinated ligands were found. From Table 3, it can be shown that the starting temperature for the removal of the organic ligands of **3**, 300 °C, is higher than those of **1–2**, probably indicating that the framework of **3** is stronger than those of **1–2** upon heating.

Table 3. Thermal properties of **1–3**.

Complex.	Weight Loss of Solvents °C (found/calc), %	Weight Loss of Ligands °C (found/calc), %
1	~120 (2.76/2.54)	240–800 (84.89/88.21)
2	~140 (3.33/4.07)	250–800 (80.83/81.90)
3	~145 (3.25/4.42)	300–800 (87.34/87.64)

3.4. Chemical Stability

To estimate the chemical stability, the PXRD patterns of complexes **1–3** in water, as well as their dehydration and rehydration products, were measured. Complexes **1–3** were first immersed into water for seven days and then dried under vacuum. Figures S6–S8 show that only complexes **1** and **2** are stable in water. Moreover, dehydration (heated at 150 °C for 1 h) and rehydration (immersed into water for 1 h) of complexes **1** and **2** gave PXRD patterns comparable to the simulated ones, indicating the stability upon solvent removal and adsorption. The stabilities of **1** and **2** in acidic and basic solutions at 50 °C for 24 h were further investigated. As shown in Figures S9 and S10, the PXRD patterns of **1** in pH 2–12 and **2** in pH 4–10 are almost identical to the simulated ones, suggesting the better acid and base resistances of **1** than **2**.

3.5. Gas Sorption

The low-pressure gas sorption properties at 273 K and 298 K were examined for complexes **1** and **2** and confirmed by CO_2 gas absorption and desorption isotherms, resulting in a type II isotherm behavior [20,21]. As illustrated in Figures S11–S13, the experimental amount of adsorbed CO_2 gas in the sample at 273 K is higher than that at 298 K. While desolvated samples of **1** and **2** adsorbed 1.40565 and 1.55795 mmol/g of CO_2 at 273 K, they adsorbed 0.83070 and 1.05312 mmol/g at 298 K, respectively. The Brunauer–Emmett–Teller (BET) surface areas of **1** and **2** evaluated from the CO_2 gas uptake are 411.3 and 302.4 $m^2\ g^{-1}$ and the Langmuir surface areas are 482.6 and 330.5 $m^2\ g^{-1}$ at 273 K, with total pore volumes of 0.05 and 0.06 $cm^3\ g^{-1}$, respectively. The pore size distribution and sorption/desorption hysteresis are presumably related to very narrow pore sizes. Calculations using nonlocal density functional theory based on CO_2 sorption data at 273 K suggest that the pore sizes for **1** and **2** are similar and are around 1.06 nm, respectively, indicating microporous materials. The PXRD patterns of the samples of **1** and **2** measured after the experiments match quite well with those of the simulated ones, indicating permanent porous features of **1** and **2** (Figure S13c,d).

3.6. Photoluminescence Properties

An ample number of studies have shown that CPs with d^{10} metal centers exhibit great sensing properties [22–26]. Therefore, the solid-state emission spectra of the water-stable **1** and **2**, as well as 1,4-H_2NDC and **L** ligands, were investigated at room temperature and the results are shown in Figure S14 and Table 4. The emission band of **L** appears at 468 nm upon excitation at 376 nm, while 1,4-H_2NDC exhibits an intense emission band at 480 nm upon two excitations at 280 and 370 nm, which may be attributed to intra-ligand $\pi^* \rightarrow \pi$ or $\pi^* \rightarrow n$ transition.

Table 4. Luminescent properties of **L**, 1,4-H_2NDC, **1** and **1a**, **2** and **2a**.

Compound	Excitation λ_{ex} (nm)	Emission λ_{em} (nm)	Compound	Excitation λ_{ex} (nm)	Emission λ_{em} (nm)
L	376	468	1,4-H_2NDC	280, 370	480
1	350	390	**1a**	350	390
2	350	395	**2a**	350	395

As-synthesized complexes **1** and **2** may contain water molecules in the pores, they were first activated by immersion into methanol for two days and heated at 80 °C for 1 h to give **1a** and **2a**, respectively. It can be seen (Figure S15) that complexes **1a** and **2a** show identical emission wavelengths to those of original **1** and **2**. Moreover, the PXRD patterns (Figure S16) of **1a** and **2a** match quite well with those of **1** and **2**, respectively, indicating no framework changes and suggesting ligand-based emissions with little contribution from the cocrystallized solvents. The different emission wavelengths in **1** and **2** are probably due to the distinction in the metal identity and the coordination mode of the 1,4-NDC^{2-} ligand. Zn(II) and Cd(II) atoms are more difficult to undergo oxidation and reduction reactions, due to their closed-shell electronic configurations; therefore, the emissions of complexes **1** and **2**, as well as **1a** and **2a**, are probably due to ligand-to-ligand charge transfer (LLCT), while metal-to-ligand charge transfer (MLCT) is also possible.

3.7. Detection of Volatile Organic Compounds (VOCs)

The CO_2 gas adsorption studies confirm the permanent pores in the eight-fold interpenetrating CPs **1** and **2**, which provide a unique opportunity to investigate the role of the entangled bpba-CPs in the sensing of abundant VOCs [26–32] and metal ions [33]. The activated complexes of **1** and **2**, **1a** and **2a**, were exposed to various VOCs by immersing grounded samples (30 mg) into 5 mL different organic solvents, which were then sonicated for 10 min and collected by centrifugation and then air dried. The emission spectra of the VOC-absorbed complexes upon excitation at 350 nm were measured at room temperature (Figure 4a), revealing that the emission intensities of VOC-absorbed **1a** and **2a** reached the maximum for the mesitylene molecules. The PXRD patterns shown in Figure S16 for the mesitylene absorbed **1a** and **2a** indicate that they maintain the identical structural frameworks. The luminescence enhancement, $[(I - I_o)/I_o]$ (I_o = luminescence intensity of **1a** or **2a**; I = luminescence intensity of mesitylene-loaded **1a** or **2a**) due to the adsorption of mesitylene are calculated as 1.91 and 1.65 for **1a** and **2a**, respectively. The organic uptakes were estimated from the weight-losses in TGA curves of VOC-loaded **1a** and **2a** (Figure S17), showing that the uptakes of mesitylene molecules per unit cell are 2.53 in **1a** and 2.97 in **2a**, respectively. The emission wavelengths of mesitylene-absorbed **1a** and **2a** shows red-shifts of 13 and 11 nm, respectively, compared with solvent-free **1a** and **2a** (Figures S18a and S19a), indicating the possible interligand couplings [34]. The mesitylene molecules may have the strongest C-H$\cdots\pi$ interactions with **L** linkers among other exposed solvents, inducing the largest bathochromic shift and the intensity enhancement [35].

Recently, it has been shown that C-H···π interaction can be modulated to enhance the luminescence [36].

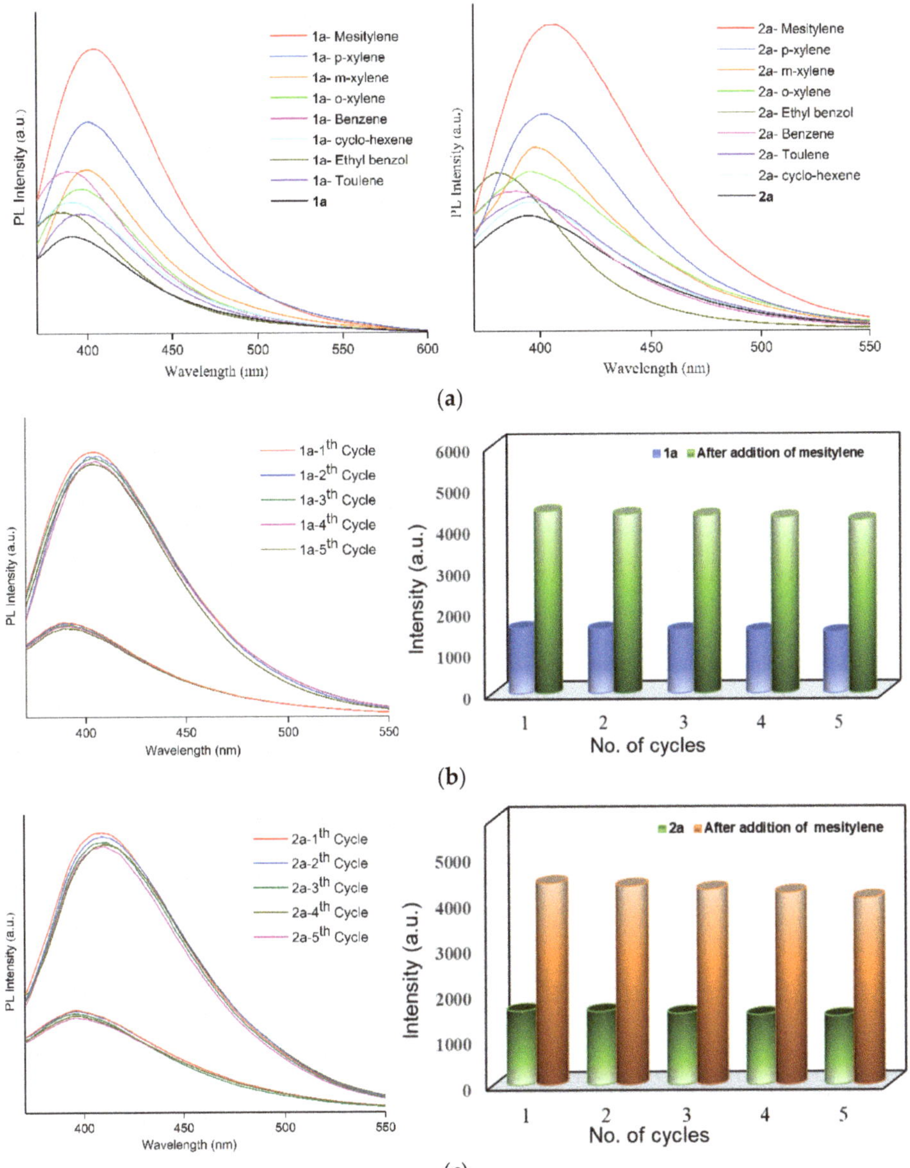

Figure 4. (**a**) Emission spectra of activated **1** (left) and **2** (right) loaded with different VOC's. Emission spectra showing the recyclability of (**b**) **1a** and (**c**) **2a** exposed with mesitylene for five repeated cycles.

To estimate the selective sensing performances of **1a** and **2a** toward mesitylene, luminescence titration experiments were performed by the addition of the increasing mesitylene (0–5 mL) to **1a** and **2a**. The volume of mesitylene and the emission intensity show linear dependence of R^2 = 0.983 and 0.985 (Figures S18b,c and S19b,c). Table S1 shows the intensity enhancements of several reported complexes due to the interactions with the specific organic solvents, showing comparable luminescence enhancements for **1a** and **2a** in the detection of mesitylene molecules. Further, to determine the recyclability of mesitylene adsorption in complexes **1a** and **2a**, their luminescence sensing abilities were explored for five regeneration cycles. The PXRD patterns in Figure S16 indicate that the main diffraction peaks of **1a** and **2a** are quite identical with those of the mesitylene-treated samples, while Figure 4b,c display their similar emission spectra and intensities, respectively, suggesting that their frameworks remain rigid for five cycles.

3.8. Detection of Metal Ions

In order to explore the potential application of **1** and **2** in luminescent sensing of metal cations, 30 mg samples of **1** and **2**, respectively, were immersed into 10 mL aqueous solutions of nitrate salts $M(NO_3)_x$ (M = Al^{3+}, Cd^{2+}, Cu^{2+}, Mg^{2+}, Ni^{2+}, Zn^{2+}, Na^+, K^+, Cr^{3+} and Fe^{3+}) with a concentration of 1×10^{-3} M. After 1 h, the solids were filtered and then the solid-state emission spectra were measured at room temperature. As shown in Figure S20, remarkable luminescence quenching of about 95% for **1** and 92% for **2** were found in the detection of Fe^{3+} ions. To further explore the quenching effect of Fe^{3+} ions, sensing dependence of luminescence intensity on the concentration of Fe^{3+} was investigated by immersing finely grounded samples (30 mg) of **1** and **2** into Fe^{3+} aqueous solutions with various concentrations (0.005–0.15 mM) for 1 h. As shown in Figure 5b,c (left), the emission intensities were getting lower and almost completely quenched upon increasing the concentration of Fe^{3+}. Quantitatively, the quenching capacity of Fe^{3+} ion can be rationalized by the Stern–Volmer equation: $I_0/I = 1 + K_{SV} \times [Q]$, where [Q] is the concentration of Fe^{3+}, K_{SV} is the quenching constant, and I_0 and I are the emission intensities in the absence and presence of Fe^{3+}, respectively [37]. As demonstrated in Figure 5b,c (right), the titration curves for Fe^{3+} ions in **1** and **2** are virtually linear at low concentrations, which gave the linear correlation coefficient (R^2) of 0.997 for **1** and 0.975 for **2**, respectively, while the S-V curves at higher concentrations became nonlinear, which can be described by concurrent dynamic and static quenching or the self-absorption process. The Stern−Volmer constant (K_{SV}) values are 6.895×10^5 for **1** and 9.940×10^5 M^{-1} for **2**, respectively, in the presence of Fe^{3+}.

Furthermore, the detection limits were calculated according to the standard equation $3\sigma/k$, where σ is the standard deviation from the blank measurements and k is the absolute value of the calibration curve at lower concentration [38], giving 2.35 and 1.01 μM for **1** and **2**, respectively. Noticeably, complexes **1** and **2** show comparative K_{SV} with some previously reported complexes toward the detection of Fe^{3+} ions in aqueous phase solution (Table S2). The recyclability test showed no significant differences in the PXRD patterns (Figure S21) and the luminescence intensities (Figure S22) for five regeneration cycles, indicating the reusability and demonstrating that the quenching effect was not due to the structural decomposition of **1** and **2** upon the interactions with the Fe^{3+} ions.

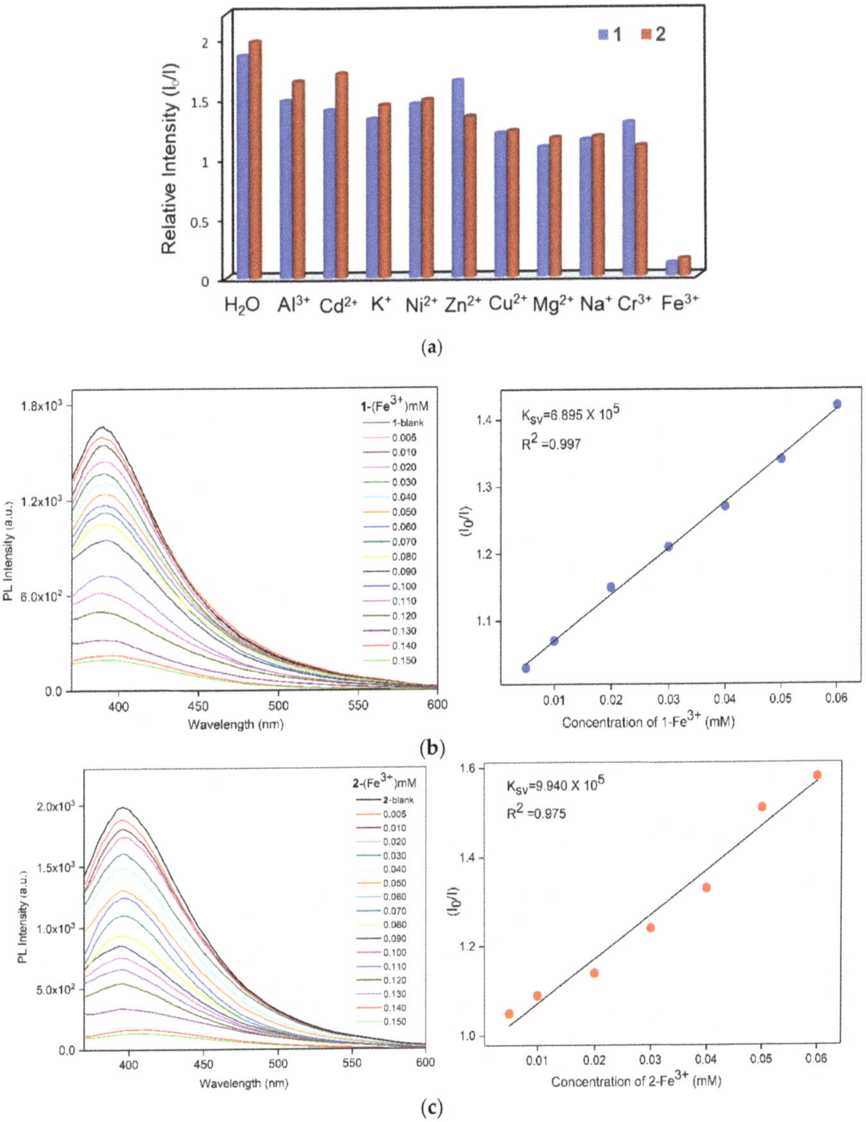

Figure 5. (**a**) Bar diagrams showing the relative emission intensities of **1** and **2** in various metal ions. Emission spectra of the activated (**b**) **1** and (**c**) **2** at various concentrations of Fe^{3+} (left), and the Stern–Volmer (sv) plot of I_0/I versus Fe^{3+} concentration (right).

Several mechanisms for luminescence quenching such as framework collapse, cation exchange, and interactions between the incoming metal ion and the organic linker that result in competitive absorption of the excited energies of the sensed metal ion and the host CPs, have been suggested [39]. As mentioned above, by comparing the PXRD patterns and luminescence intensities, the framework collapse can be ruled out. The interactions between Fe^{3+} ions and complexes **1** and **2** through the amide carbonyl oxygen atoms may thus lead to the luminescence quenching [40]. The UV-vis absorption spectrum of Fe^{3+} in aqueous solution and the corresponding excitation and emission spectra of complexes **1** and **2** are shown in Figure S23. Partial overlaps between the absorption spectrum of Fe^{3+} ion and the excitation spectra of complexes **1** and **2** are observed, indicating that the excitation energies of **1** and **2** can be partially absorbed by the Fe^{3+} ions, and the luminescence quenching can most probably be ascribed to competitive energy absorption [39].

4. Conclusions

Three eight-fold interpenetrated 3D CPs with the (6^6)-**dia** topology have been successfully prepared by using angular **L** and the linear dicarboxylic acid 1,4-H_2NDC. The Zn(II), Cd(II) and Co(II) metal centers of **1–3** adopt the distorted tetrahedral, pentagonal bipyramidal and octahedral/square pyramidal geometries, respectively, resulting in different coordination modes for the 1,4-NDC^{2-} ligands and indicating that the structural topology is not subject to the metal geometry. Their structural stabilities are assessed particularly in the aqueous environments, revealing that complexes **1** and **2** with close-shell Zn(II) and Cd(II) ions are much more stable than **3** with the open-shell Co(II) ion. The surface areas and pore structures of **1** and **2** were observed via CO_2 gas sorption measurements and the activated samples of **1** and **2** explored by emission spectra and PXRD patterns demonstrate efficient detection of mesitylene and Fe^{3+} ions with high selectivity and recyclability. This investigation oversees the sensing properties of the highly entangled bpba-based CPs toward VOCs and metal ions.

Supplementary Materials: The following are available online at https://www.mdpi.com/article/10.3390/polym13183018/s1. FT-IR spectra for **1–3** (Figure S1). PXRD patterns for **1–3** (Figures S2–S4). TGA curves for **1–3** (Figure S5). CO_2 sorption isotherms of **1** and **2** (Figures S11–S13). Emission spectra for **1** and **2** (Figures S14 and S15). Sensing properties of reported compounds toward various aromatic VOCs (Table S1). Sensing properties of reported compounds toward Fe^{3+} ions (Table S2). Crystallographic data for **1–3** have been deposited with the Cambridge Crystallographic Data Centre, CCDC No. 2100027-2100029.

Author Contributions: Investigation, V.L. and Y.-T.L.; data curation, X.-K.Y. and M.G.; supervision, C.-H.L. and J.-D.C. All authors have read and agreed to the published version of the manuscript.

Funding: This research was funded by the Ministry of Science and Technology, Taiwan: MOST 109-2113-M-033-009.

Acknowledgments: We are grateful to the Ministry of Science and Technology of the Republic of China for their support.

Conflicts of Interest: The authors declare no conflict of interest.

References

1. Lustig, W.P.; Mukherjee, S.; Rudd, N.D.; Desai, A.V.; Li, J.; Ghosh, S.K. Metal-organic frameworks: Functional luminescent and photonic materials for sensing applications. *Chem. Soc. Rev.* **2017**, *46*, 3242–3285. [CrossRef]
2. Li, B.; Wen, H.-M.; Cui, Y.; Zhou, W.; Qian, G.; Chen, B. Emerging Multifunctional Metal-Organic Framework Materials. *Adv. Mater.* **2016**, *28*, 8819–8860. [CrossRef] [PubMed]
3. Pamela, B.S.; Chen, H.-T.; Lin, C.-H. De novo synthesis and particle size control of iron metal organic framework for diclofenac drug delivery. *Microporous Mesoporous Mater.* **2020**, *309*, 110495.
4. Ma, L.; Abney, C.; Lin, W. Enantioselective catalysis with homochiral metal–organic frameworks. *Chem. Soc. Rev.* **2009**, *38*, 1248–1256. [CrossRef] [PubMed]
5. Zhou, Y.-C.; Dong, W.-W.; Jiang, M.-Y.; Wu, Y.-P.; Li, D.-S.; Tian, Z.-F.; Zhao, J. A new 3D 8-fold interpenetrating 6^6-dia topological Co-MOF: Syntheses, crystal structure, magnetic properties and electrocatalytic hydrogen evolution reaction. *J. Solid-State Chem.* **2019**, *279*, 120929. [CrossRef]
6. Du, M.; Jiang, X.-J.; Zhao, X.-J. Direction of unusual mixed-ligand metal–organic frameworks: A new type of 3-D polythreading involving 1-D and 2-D structural motifs and a 2-fold interpenetrating porous network. *Chem. Commun.* **2005**, *44*, 5521–5523. [CrossRef]
7. Yang, J.; Ma, J.-F.; Liu, Y.-Y.; Ma, J.-C.; Batten, S.R. A series of Lead (II) Complexes with $\pi-\pi$ Stackings: Structural Diversities by Varying the Ligands. *Cryst. Growth Des.* **2009**, *9*, 1894–1911. [CrossRef]
8. Carlucci, L.; Ciani, G.; Proserpio, D.M. Polycatenation, polythreading and polyknotting in coordination network chemistry. *Coord. Chem. Rev.* **2003**, *246*, 247–289. [CrossRef]

9. Hsu, C.-H.; Huang, W.-C.; Yang, X.-K.; Yang, C.-T.; Chhetri, P.M.; Chen, J.-D. Entanglement and irreversible structural transformation in Co (II) Coordination Polymers Based on Isomeric Bis-pyridyl-bis-amide Ligands. *Cryst. Growth Des.* **2019**, *19*, 1728–1737. [CrossRef]
10. Wu, T.-T.; Hsu, W.; Yang, X.-K.; He, H.-Y.; Chen, J.-D. Entanglement in Co (II) coordination networks: Polycatenation from single net to 2-fold and 3-fold interpenetrated nets. *CrystEngComm* **2015**, *17*, 916–924. [CrossRef]
11. Liu, Y.-F.; Hu, J.-H.; Lee, W.-T.; Yang, X.-K.; Chen, J.-D. Structural Transformations of Cobalt(II) Coordination Polymers Constructed from *N,N'*-Di-3-pyridyladipoamide and Tetracarboxylic Acids: Disentanglement upon Water Coordination. *Cryst. Growth Des.* **2020**, *20*, 7211–7218. [CrossRef]
12. Hsu, Y.-F.; Lin, C.-H.; Chen, J.-D.; Wang, J.-C. A Novel Interpenetrating Diamondoid Network from Self-Assembly of *N,N'*-Di(4-pyridyl)adipoamide and Copper Sulfate: An Unusual 12-Fold, [6 + 6] Mode. *Cryst. Growth Des.* **2008**, *8*, 1094–1096. [CrossRef]
13. Cheng, J.-J.; Chang, Y.-T.; Wu, C.-J.; Hsu, Y.-F.; Lin, C.-H.; Proserpio, D.M.; Chen, J.-D. Highly interpenetrated diamondoid nets of Zn(II) and Cd(II) coordination networks from mixed ligands. *CrystEngComm* **2012**, *14*, 537–543. [CrossRef]
14. Kumar, P.; Deep, A.; Kim, K.-H.; Brown, R.J. Coordination polymers: Opportunities and challenges for monitoring volatile organic compounds. *Prog. Polym. Sci.* **2015**, *45*, 102–118. [CrossRef]
15. Ganguly, S.; Mukherjee, S.; Dastidar, P. Single-Crystal-to-Single-Crystal Breathing and Guest Exchange in Co(II) Metal–Organic Frameworks. *Cryst. Growth Des.* **2016**, *16*, 5247–5259. [CrossRef]
16. Bruker AXS Inc. *APEX2, V2008.6, SADABS V2008/1, SAINT V7.60A, SHELXTL V6.14*; Bruker AXS Inc.: Madison, WI, USA, 2008.
17. Sheldrick, G.M. A short history of SHELX. *Acta Crystallogr.* **2008**, *A64*, 112–122. [CrossRef] [PubMed]
18. Blatov, V.A.; Shevchenko, A.P.; Proserpio, D.M. Applied Topological Analysis of Crystal Structures with the Program Package ToposPro. *Cryst. Growth Des.* **2014**, *14*, 3576–3586. [CrossRef]
19. Thapa, K.B.; Chen, J.-D. Crystal engineering of coordination polymers containing flexible bis-pyridyl-bis-amide ligands. *CrystEngComm* **2015**, *17*, 4611–4626. [CrossRef]
20. Tseng, T.W.; Lee, L.W.T.; Luo, T.; Chien, P.H.; Liu, Y.H.; Lee, S.L.; Wang, C.M.; Lu, K.L. Gate-opening upon CO_2 adsorption on a metal–organic framework that mimics a natural stimuli-response system. *Dalton Trans.* **2017**, *46*, 14728–14732. [CrossRef]
21. Lee, C.-H.; Huang, H.-Y.; Liu, Y.-H.; Luo, T.-T.; Lee, G.-H.; Peng, S.-M.; Jiang, J.-C.; Chao, I.; Lu, K.-L. Cooperative Effect of Unsheltered Amide Groups on CO_2 Adsorption Inside Open-Ended Channels of a Zinc(II)−Organic Framework. *Inorg. Chem.* **2013**, *52*, 3962–3968. [CrossRef]
22. Yan, Y.; Peng, D.; Ma, J.-F.; Kan, W.-Q.; Liu, B.; Yang, J. A Series of Metal–Organic Frameworks Based on Different Salicylic Derivatives and 1,1'-(1,4- Butanediyl)bis(imidazole) Ligand: Syntheses, Structures, and Luminescent Properties. *Cryst. Growth Des.* **2011**, *11*, 5540–5553.
23. Niu, D.; Yang, J.; Guo, J.; Kan, W.-Q.; Song, S.-Y.; Du, P.; Ma, J.-F. Syntheses, Structures, and Photoluminescent Properties of 12 New Metal–Organic Frameworks Constructed by a Flexible Dicarboxylate and Various N-Donor Ligands. *Cryst. Growth Des.* **2012**, *12*, 2397–2410. [CrossRef]
24. Sun, D.; Yan, Z.-H.; Blatov, V.A.; Wang, L.; Sun, D.-F. Syntheses, Topological Structures, and Photoluminescences of Six New Zn(II) Coordination Polymers Based on Mixed Tripodal Imidazole Ligand and Varied Polycarboxylates. *Cryst. Growth Des.* **2013**, *13*, 1277–1289. [CrossRef]
25. Yang, J.-X.; Xin, Z.; Qin, Y.-Y.; Yao, Y.-G. N-Donor Auxiliary Ligand Influence on the Coordination Mode Variations of V- Shaped Triazole Dicarboxylic Acid Ligand Affording Seven New Luminescent Zn(II) Compounds with Variable Structural Motifs. *Cryst. Growth Des.* **2020**, *20*, 6366–6381. [CrossRef]
26. Gong, Y.; Li, J.; Qin, J.B.; Wu, T.; Cao, R.; Li, J.H. Metal(II) Coordination Polymers Derived from Bis-pyridyl-bis-amide Ligands and Carboxylates: Syntheses, Topological Structures, and Photoluminescence Properties. *Cryst. Growth Des.* **2011**, *11*, 1662–1674. [CrossRef]
27. Das, P.; Mandal, S.K. A dual-functionalized, luminescent and highly crystalline covalent organic framework: Molecular decoding strategies for VOCs and ultrafast TNP sensing. *J. Mater. Chem. A.* **2018**, *6*, 16246–16256. [CrossRef]
28. Parmar, B.; Bisht, K.K.; Rachuri, Y.; Suresh, E. Zn(ii)/Cd(ii) based mixed ligand coordination polymers as fluorosensors for aqueous phase detection of hazardous pollutants. *Inorg. Chem. Front.* **2020**, *7*, 1082–1107. [CrossRef]
29. Chang, Y.-T.; Lo, S.-H.; Lin, C.-H.; Hung, L.-I.; Wang, S.-L. Indium Phosphite-Based Porous Solids Exhibiting Organic Sensing and a Facile Route to Superhydrophobicity. *Chem. Eur. J.* **2018**, *24*, 12474–12479. [CrossRef]
30. Rani, P.; Gauri; Husain, A.; Bhasin, K.K.; Kumar, K. A Doubly Interpenetrated Cu(II) Metal–Organic Framework for Selective Molecular Recognition of Nitroaromatics. *Cryst. Growth Des.* **2020**, *20*, 7141–7151. [CrossRef]
31. Chen, J.-K.; Yang, S.-M.; Li, B.-H.; Lin, C.-H.; Lee, S. Fluorescence Quenching Investigation of Methyl Red Adsorption on Aluminum-Based Metal–Organic Frameworks. *Langmuir* **2018**, *34*, 1441–1446. [CrossRef]
32. Das, D.; Biradha, K. Luminescent Coordination Polymers of Naphthalene Based Diamide with Rigid and Flexible Dicarboxylates: Sensing of Nitro Explosives, Fe (III) Ion, and Dyes. *Cryst. Growth Des.* **2018**, *18*, 3683–3692. [CrossRef]
33. Shi, Y.-S.; Qiang, Y.; Zhang, J.-W.; Cui, G.-H. Four dual-functional luminescent Zn(ii)-MOFs based on 1,2,4,5-benzenetetracarboxylic acid with pyridylbenzimidazole ligands for detection of iron(iii) ions and acetylacetone. *CrystEngComm* **2021**, *23*, 1604–1615. [CrossRef]

34. Zhang, M.; Feng, G.; Song, Z.; Zhou, Y.-P.; Chao, H.-Y.; Yuan, D.; Tan, T.T.Y.; Guo, Z.; Hu, Z.; Tang, B.Z.; et al. Two-Dimensional Metal–Organic Framework with Wide Channels and Responsive Turn-On Fluorescence for the Chemical Sensing of Volatile Organic Compounds. *J. Am. Chem. Soc.* **2014**, *136*, 7241–7244. [CrossRef] [PubMed]
35. Zhang, J.; Wang, J.; Long, S.; Peh, S.B.; Dong, J.; Wang, Y.; Karmakar, A.; Yuan, Y.D.; Cheng, Y.; Zhao, D. Luminescent Metal-Organic Frameworks for the Detection and Discrimination of o-Xylene from Xylene Isomers. *Inorg. Chem.* **2018**, *57*, 13631–13639. [CrossRef]
36. Liu, M.; Wei, Y.; Ou, Q.; Yu, P.; Wang, G.; Duan, Y.; Geng, H.; Peng, Q.; Shuai, Z.; Liao, Y. Molecular Design Strategy for Simultaneously Strong Luminescence and High Mobility: Multichannel CH-π Interaction. *J. Phys. Chem. Lett.* **2021**, *12*, 938–946. [CrossRef] [PubMed]
37. Ge, F.-Y.; Sun, G.-H.; Lei, M.; Ren, S.-S.; Zheng, H.-G. Four New Luminescent Metal–Organic Frameworks as Multifunctional Sensors for Detecting Fe^{3+}, $Cr_2O_7^{2-}$ and Nitromethane. *Cryst. Growth Des.* **2020**, *20*, 1898–1904. [CrossRef]
38. Zhang, S.-H.; Zhang, S.-Y.; Li, J.-R.; Huang, Z.-Q.; Jing, Y.; Yue, K.-F.; Wang, Y.-Y. Rational synthesis of an ultra-stable Zn(ii) coordination polymer based on a new tripodal pyrazole ligand for the highly sensitive and selective detection of Fe^{3+} and $Cr_2O_7^{2-}$ in aqueous media. *Dalton Trans.* **2020**, *49*, 11201–11208. [CrossRef] [PubMed]
39. Pamei, M.; Puzari, A. Luminescent transition metal–organic frameworks: An emerging sensor for detecting biologically essential metal ions. *Nano-Struct. Nano-Objects* **2019**, *19*, 100364. [CrossRef]
40. Yang, X.-K.; Lee, W.-T.; Hu, J.-H.; Chen, J.-D. Zn(II) and Cd(II) coordination polymers with a new angular bis-pyridyl-bis-amide: Synthesis, structures and sensing properties. *CrystEngComm* **2021**, *23*, 4486–4493. [CrossRef]

polymers

MDPI

Article

Experimental Study on High-Performers Quaternary Copolymer Based on Host–Guest Effect

Tao Xu, Jincheng Mao *, Yang Zhang *, Xiaojiang Yang, Chong Lin, Anqi Du, Heng Zhang, Quan Zhang and Jinhua Mao *

State Key Laboratory of Oil and Gas Reservoir Geology and Exploitation, Southwest Petroleum University, Chengdu 610500, China; 15827742134@163.com (T.X.); xjyang@swpu.edu.cn (X.Y.); 201999010016@swpu.edu.cn (C.L.); 13880147377@163.com (A.D.); 201811000155@stu.swpu.edu.cn (H.Z.); zhangsanfengdbsy@163.com (Q.Z.)

* Correspondence: jcmao@swpu.edu.cn (J.M.); 201521000202@stu.swpu.edu.cn or yangzhang10234@163.com (Y.Z.); 201921000706@stu.swpu.edu.cn or 18328085244@163.com (J.M.); Tel.: +86-28-8303-3546 (J.M. & Y.Z. & J.M.)

Abstract: A quaternary polymer (HGP) was prepared by the free-radical polymerization of acrylamide, acrylic acid, maleic anhydride functionalized *β*-cyclodextrin (MAH-*β*-CD), and N-(3-methacrylamidopropyl)-N, N-dimethylnaphthalen-1-aminium chloride (NAP). It was found that host–guest behavior occurred most effectively at a molar rate of NAP and CD with 1:1, which exhibited better solubility than hydrophobically associative polymer. Moreover, the as-prepared polymer has superior salt tolerance, shear resistance, and viscoelasticity due to host–guest strategy. More importantly, the HGP solution simulates the distribution of formation water in the Bohai SZ1-1 oilfield has good rheological properties at 120 °C. All results show that the proposed polymer could be a competitive candidate in oilfield applications such as fracturing fluids, displacement fluids, and drilling fluids.

Keywords: host–guest strategy; hydrophobically associating water-soluble polymers; rheological behaviors; salt tolerance; *β*-cyclodextrin

Citation: Xu, T.; Mao, J.; Zhang, Y.; Yang, X.; Lin, C.; Du, A.; Zhang, H.; Zhang, Q.; Mao, J. Experimental Study on High-Performers Quaternary Copolymer Based on Host–Guest Effect. *Polymers* **2021**, *13*, 2972. https://doi.org/10.3390/polym13172972

Academic Editor: Edina Rusen

Received: 29 May 2021
Accepted: 27 August 2021
Published: 1 September 2021

Publisher's Note: MDPI stays neutral with regard to jurisdictional claims in published maps and institutional affiliations.

Copyright: © 2021 by the authors. Licensee MDPI, Basel, Switzerland. This article is an open access article distributed under the terms and conditions of the Creative Commons Attribution (CC BY) license (https://creativecommons.org/licenses/by/4.0/).

1. Introduction

Water-soluble polymers have become increasingly important in oilfield applications such as hydraulic fracturing, enhanced oil recovery, and drilling [1]. As the petroleum industry taps deeper reservoirs, there is an urgent requirement that chemicals should maintain performances in harsh reservoirs with high temperatures and salinities [2,3]. Polyacrylamide and partially hydrolyzed polyacrylamide (HPAM) are widely used in polymer flooding and hydraulic fracturing [4]. However, HPAM with poor heat resistance and salt tolerance cannot withstand the harsh reservoir [5–8].

In addition, scholars have paid increased attention in recent decades to hydrophobically associating water-soluble polymers (HAWSPs), which have good thickening and salt resistance. Hydrophobic association between the hydrophobic groups of HAWSP increases the hydrodynamic volume and increases viscosity significantly [9,10]. Compared with conventional water-soluble polymers, HAWSPs have poor solubility due to the presentence of hydrophobic groups, which one of the factors restricting its application in oil fields.

The host–guest effect has attracted attention in recent years because it can improve the performance of polymers. *β*-cyclodextrin (*β*-CD) is the most used in this field [11–14]. *β*-CD can form supramolecular structures with guest chains by hydrogen bonding and van der Waals forces [15,16]. There is a network structure in such a polymer solution caused by the inclusion of the *β*-CD and the guest such as phenyls, long-chain alkyls, adamantyl, and azobenzene [17–21]. Weickenmeier et al. obtained a polymer solution with an apparent supramolecular network structure attributed to the host–guest effect between *β*-CD polymers and hydrophobically modified polymers [22]. Wei Bing et al. grafted a β-CD and an adamantane ring structure on the side chain of the polymer; the superior properties make this novel polymer promising in enhancing oil recovery [23]. Zou CJ et al. prepared a new supramolecular retarded acid system by β-CD and 2-Phosphonobutane-1, 2, 4-Tricarboxylic Acid, which can decrease the corrosion rate with formations [24]. Due to the spatial position of the host–guest molecules on the polymer chain and the difference in their molecular activation energy, the rheological behavior of the polymer solution system is significantly different. According to previous studies, grafting of β-CD groups in a HAWSP can produce host–guest interactions, improving properties such as solubility and temperature, salt, and shear resistance [25,26].

Inspired by the above, a quaternary polymer (HGP) was designed by copolymerizing acrylamide (AM), acrylic acid (AA), and two kinds of monomer: *N*-(3-methacrylamidopropyl)-*N,N*-dimethylnaphthalen-1-aminium chloride (NAP) and modified maleic anhydride (MAH-β-CD). NAP and MAH-β-CD were selected as functional monomers for the following reasons: (1) the hydrophobic cationic monomer NAP can form hydrophobic microdomains in an aqueous solution due to hydrophobic association, which significantly increases the viscoelasticity. (2) Due to the poor solubility of hydrophobically associating polymers, the introduction of cyclodextrin can speed up the dissolution of the polymer. At the same time, NAP and MAH-β-CD have host–guest effects in the solution. It is reported that the envelope constants between cyclodextrin and naphthalene are 4.67×10^3 $molL^{-1}$ (Figure 1) [27]. (3) In addition, NAP and MAH-β-CD are rigid groups, which not only increase the rigidity of the backbone of the polymer but also improve the temperature resistance of the polymer.

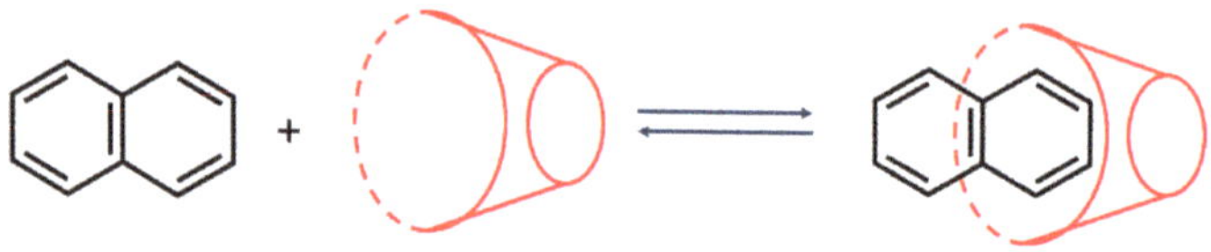

Figure 1. Cyclodextrin and naphthalene have host–guest effects.

HGP embraces good solubility, thickening performance, shear tolerance, and salt and temperature resistance due to the host–guest effect.

2. Experimental Section

2.1. Materials

β-cyclodextrin(β-CD, BR 99%), maleic anhydride (MAH, AR 99%), acetone (AR 98%), chloroform Dimethylamino propyl methacrylamide (DMAPMA, AR 98%), 1-Chloromethyl naphthalene (AR 98%), Anhydrous methanol (AR 99.5%) and acetonitrile (AR 99.5%), Sodium carbonate (AR 99.5%), Ethyl ether, Acrylamide(AM, AR 98%), Acrylic acid (AA, AR 99.5%), 2,2′-azobis (2-methylpropionamidine) dihydrochloride (V50, AR 98%), sodium hydroxide, sodium chloride (NaCl, AR 99%), potassium chloride (KCl, AR 99%), Calcium chloride anhydrous ($CaCl_2$, AR 99%), and magnesium chloride ($MgCl_2$, AR 99%) were purchased from Chengdu Kelong Chemical Reagents Corp. (Chengdu, China). HPAM (Mw = 16 to 18 $\times 10^6$) was purchased from the Hengju Oil Field Chemical Reagents Co, Ltd. (Beijing, China). Deionized (DI) water was obtained from a water purification system.

2.1.1. Synthesis of MAH-β-CD

The activity of various hydroxyl groups in cyclodextrin is different. According to nucleophilic properties, the reaction activity order of the three hydroxyl groups is C6 > C2 > C3. The steric hindrance of the 6-hydroxyl group is small, and the reaction activity is higher under acidic conditions. Therefore, the electrophilic reagent attacks the 6-hydroxyl group, and the reagent with lower reaction activity has the highest selectivity on the 6-primary hydroxyl group. In addition, through references [28,29], it was found that when the molar ratio of 1:1 for the CD and MAH, the yield was higher. A mixture of β-CD (5.68 g, 0.005 mol),

MAH (4.90 g, 0.005 mol), and anhydrous *N,N*-dimethyl formamide (30 mL) was stirred at 80 °C for 10 h. Then, the reaction mixture was cooled to room temperature, and a large amount of chloroform was added to precipitate the compound. The precipitate was filtered and washed three times with a large amount of acetone. Vacuum drying the precipitate at 75 °C for 24 h resulted in a 62% yield of a yellow powder (MAH-β-CD) (Scheme 1).

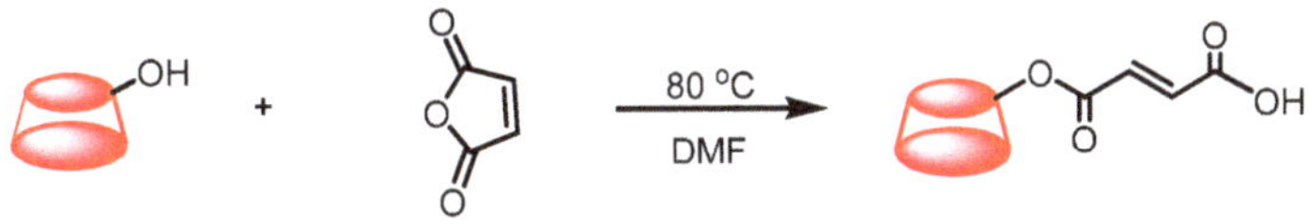

Scheme 1. Synthesis route of maleic anhydride cyclodextrin (MAH-β-CD).

2.1.2. Synthesis of NAP

A mixture of (10.215 g, 0.06 mol) DMAPMA and (10.598 g, 0.06 mol) 1-(chloromethyl) naphthalene was added to 60 mL anhydrous methanol and 60 mL acetonitrile, and sodium carbonate (2.11 g) was added. The system was protected by argon and reacted at 60 °C for 48 h at reflux. After the reaction, the solid was filtered, and methanol and acetonitrile were removed at 55 °C by rotary evaporator. The concentrated substance was added to diethyl ether; there white powder was produced, named NAP (yield 91%) (Scheme 2).

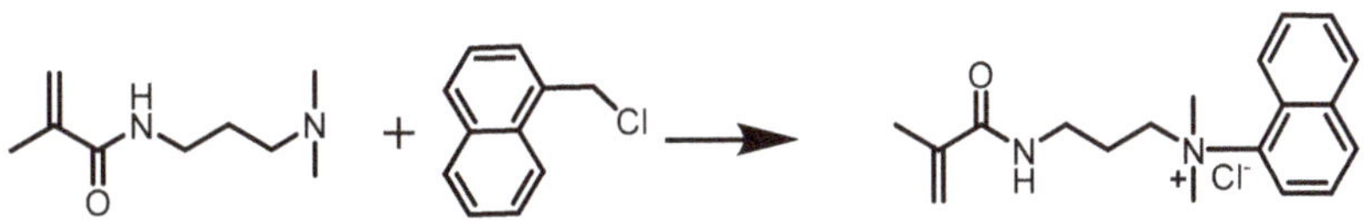

Scheme 2. Synthesis route of NAP.

2.1.3. Synthesis of Copolymer

The appropriate amounts of AM (10 g, 0.14 mol), AA (2.5 g, 0.03 mol), MAH-β-CD (0.56 g, 4.54×10^{-4} mol), and NAP (0.15 g, 4.54×10^{-4} mol) were stirred into deionized water under a nitrogen atmosphere for 30 min, and the total monomer concentration was maintained at 30 wt.%. Then, the pH was adjusted to 7.0–8.0 with sodium hydroxide. After complete dissolution, V50 solution was added, the content of which was 0.035 wt.% of the total mass of the monomer. At last, the solution was under a UV light fixture (T5 8-W UVB) at 25 °C for 8 h (yield 89–92%). The product was purified three times by precipitation with ethanol dried at 25 °C. In this study, three copolymers, HGP, HP-T, and GP-T, were prepared, in which the molar ratio of MAH-β-CD to NAP in the three copolymers was 1:1, 1:0, and 0:1, respectively. The synthesis route is shown in Scheme 3.

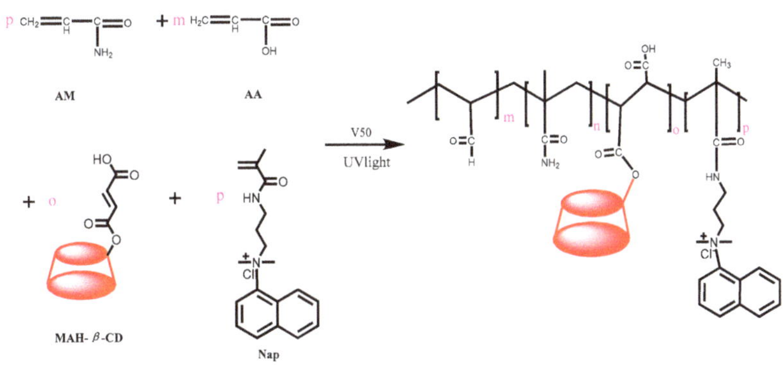

Scheme 3. Synthesis route of copolymer.

2.2. Characterization

The ^{1}H NMR spectra of NAP, HGP, and MAH-β-CD were measured on a Bruker AVANCE III HD 400 spectrometer (Bruker BioSpin, Switzerland). The Fourier Transform Infrared Spectroscopy (FT-IR) of HGP was measured with Nicolet 6700 spectrophotometer.

2.3. Effect of β-CD and MAH-β-CD on NAP CMC

The influence of β-CD and MAH-β-CD concentration on the surface tension of NAP was measured by KRUSS DSA30S tensiometer at 25 °C. The measurements were repeated three times independently, and the error was represented by the error strip in the figure.

2.4. Effect on the Viscosity of the Functional Monomer

To investigate the effect of the molar ratio of the MAH-β-CD and NAP on the viscosity of polymer solutions, the polymer with different molar ratio of MAH-β-CD to NAP was prepared. The measurement was repeated three times independently.

2.5. Conductivity

In order to explore the impact of the host–guest effect on copolymers' solubility, the DDS-307$^+$ conductivity meter (Chengdu Century Ark Technology Co., Ltd., Chengdu, China) was used to measure the conductivity of HGP, GP-T, and HP-T, and all experiments were conducted at 25 °C. The measurement was repeated three times independently.

2.6. Thickening Performance

The viscosities of the copolymer solutions with various concentrations were measured by HAAKE RS600 rheometer (Thermo Fisher Scientific, 81 Wyman Street, Waltham, MA, USA) at 170 s^{-1} and 25 °C. The measurement was repeated three times independently.

2.7. Microstructure Analysis

The microstructure of the HGP, HP-T and GP-T solutions was observed by environmental scanning electron microscope (ESEM; Quanta 450, Hillsboro, OR, USA). All samples were frozen at −50 °C using liquid nitrogen, and the frozen surfaces of the samples were observed with the ESEM operating at an accelerating voltage of 20 kV.

2.8. Viscoelasticity

The viscoelasticity of copolymers solutions was measured by an Anton Paar rheometer (MCR302) with CP50-1-SN30644 plate fixture (diameter = 0.099 mm) at 25 °C. The measurement was repeated three times independently.

2.9. Shear Tolerance

The shear tolerances of the copolymer solutions (0.6 wt.%) were measured by a continuous shear from 7.31 s^{-1} to 1000 s^{-1} at 25 °C for 30 min by the HAAKE MAR III RS 600 rheometer. The shear recoveries of shear rates at 170 s^{-1} and 510 s^{-1} were measured under the same conditions. The measurement was repeated three times independently. The thermal shear of HGP, HP-T, and GP-T solutions was investigated. The temperature was raised from 30 °C to 120 °C over a period of 120 min at a constant shear rate of 170 s^{-1}.

2.10. Salt Tolerance

To investigate the salt tolerance of the copolymer solutions HGP, HP-T, GP-T, and HPAM, the viscosity of polymers (0.6 wt.%) in different concentrations of sodium chloride was measured. Similarly, the salt tolerance of the copolymers was investigated in calcium chloride solution and magnesium chloride solution, respectively. The viscosity was measured by the HAAKE rheometer at 170 s^{-1} and 25 °C.

3. Results and Discussion

3.1. Characterization

The HGP structure was confirmed by ^1H NMR, as shown in Figure 2c. The proton signals at 4.70 ppm were assigned to the solvent proton (D_2O). Proton signals of 3.0 ppm were assigned to $–N^+–CH_3$, and 5.56 to 6.24 ppm were associated with the proton peak of naphthalene. A 0.88 ppm proton signal could be assigned to the $-CH_3$ proton in the polymer backbone. The proton signals of 1.603 ppm and 2.143 ppm were attributed to the $-CH_2-CH-$ in the polymer backbone. The two protons of the main chain of the MAH-β-CD were around 2.45 ppm (-CH-) and at 1.70 ppm ($-CH_2-$). The proton signal at 4.998 ppm was associated with the β-CD glucose unit (O-CH-O). The mass signal was considered to be the glucose proton units C-CH-OH and $C-CH_2-OH$ at 3.681 ppm. The 3.272 ppm proton signal was derived from the C-CH-O of the glucose proton unit, which corresponds to Figure 2a,b; the experimental results verified the identity of the synthesized polymer to the target.

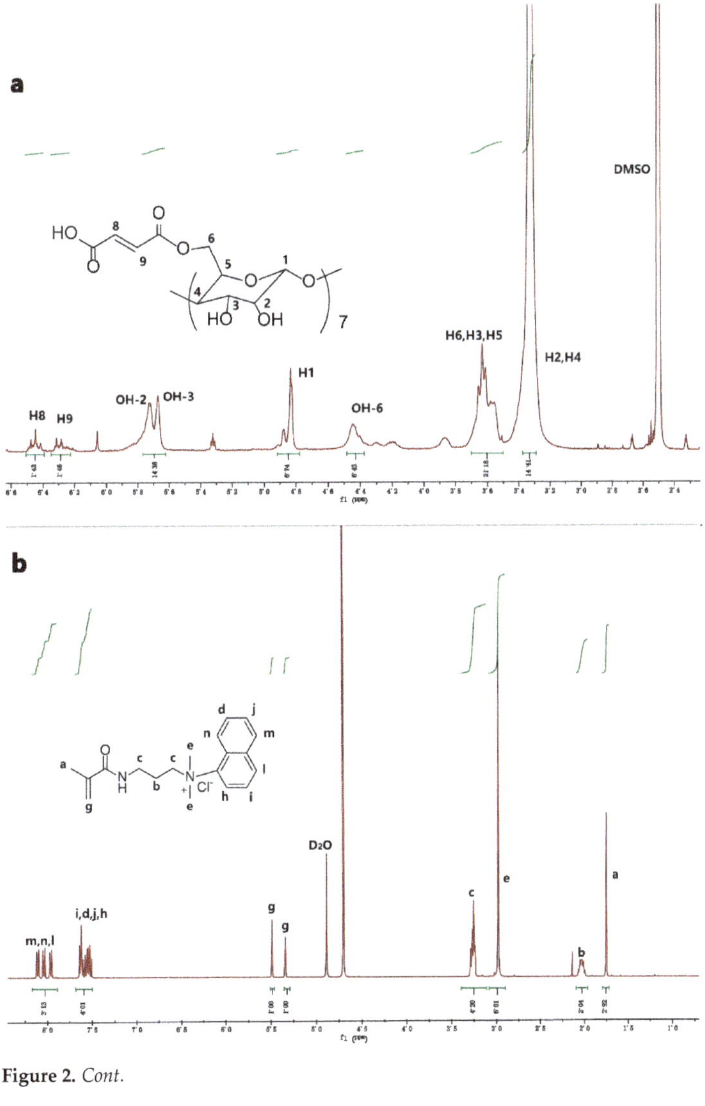

Figure 2. *Cont.*

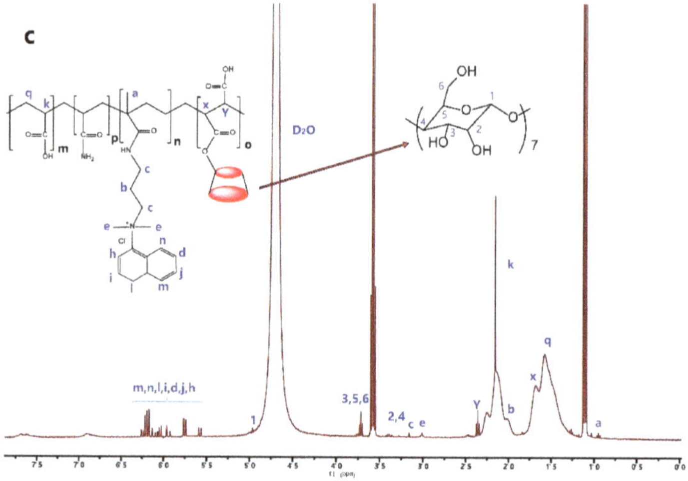

Figure 2. ^{1}H NMR spectrum: (**a**) MAH-β-CD, DMSO; (**b**) NAP, D_2O; (**c**) HGP, D_2O.

HGP was characterized and confirmed by the FT-IR spectrum shown in Figure 3. The absorption bands at 3455 cm^{-1} and 1673 cm^{-1} were due to the stretching vibration of N-H and C=O in the acrylamide groups (–$CONH_2$). The band at 3098 cm^{-1} was attributed to the C-H stretching vibration of arene hydrocarbons. The band at 2927 cm^{-1} was due to the stretching vibration peak of -CH_2-. The peak at 1446 cm^{-1} was assigned to the stretching vibration of the arene C-C skeleton. The 1415.4 cm^{-1} peak was attributed to the in-plane bending vibration peak of amide C-N and N-H. The bands observed at 1336 cm^{-1} was due to the vibration of -CH_2-. The band at 844 cm^{-1} was the in-plane and out-plane bending vibration of arene C-H. Furthermore, the stretching vibration of C-O-C referred to the band at 938.00 cm^{-1} which corresponded to the skeleton vibrations of β-CD. The FT-IR spectra confirmed that NAP and MAH-β-CD had been successfully introduced into the polymer.

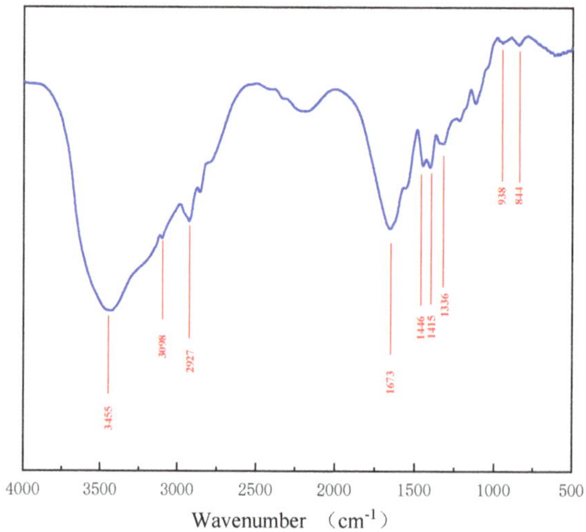

Figure 3. The FTIR spectrum of HGP.

3.2. Effect of β-CD and MAH-β-CD on NAP CMC

The critical micelle concentration (CMC) of NAP was measured at 25 °C. As shown in Figure 4a, with the increase in concentration, the surface tension of the NAP solution decreases continuously. When the concentration is 3.54×10^{-4} mol/L, the surface tension drops to the minimum and rarely changes with the increase in concentration, indicating that the CMC of NAP is about 3.54×10^{-4} mol/L. Figure 4b shows the relationship between the surface tension of NAP (3.54×10^{-4} mol/L) and the concentration of cyclodextrin (β-CD or MAH-β-CD) at 25 °C. Clearly, with the increase in CD/NAP, the system surface tension increases continuously. When CD:NAP ≈ 1, the surface tension reaches the maximum and tends to be stable. It can be seen that MAH-β-CD, like the β-CD, has a similar binding effect on hydrophobic guest NAP, and an inclusion ratio of CD:NAP ≈ 1.

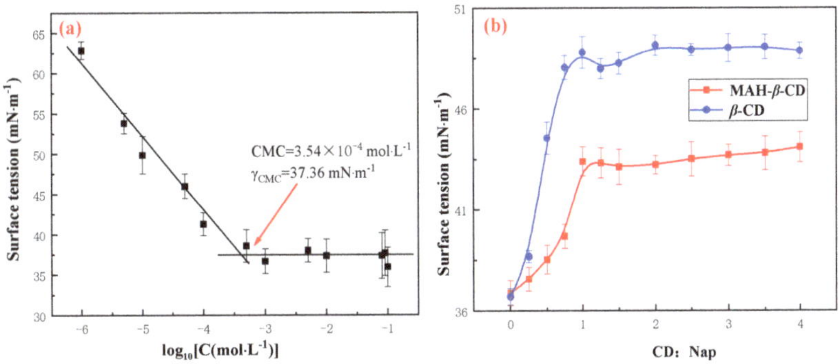

Figure 4. (**a**) Surface tension of NAP at 25 °C; (**b**) relationship between CD:NAP and NAP surface tension at 25 °C.

3.3. Effect on the Viscosity of Host–Guest Monomer Ratio

It is well known that monomer content has a marked effect on polymer viscosity [30]. As shown in Figure 5, with the increase in the molar ratio of the host–guest monomer, the viscosity of the solution increased first and then decreased. When the molar ratio of MAH-β-CD to NAP was 1:1, the solution viscosity reached the maximum 194 mPa·s, corresponding to the effect of CD on NAP CMC. The highest effective host–guest structure in the solution occurred when the molar ratio was 1:1, resulting in a significant increase in viscosity.

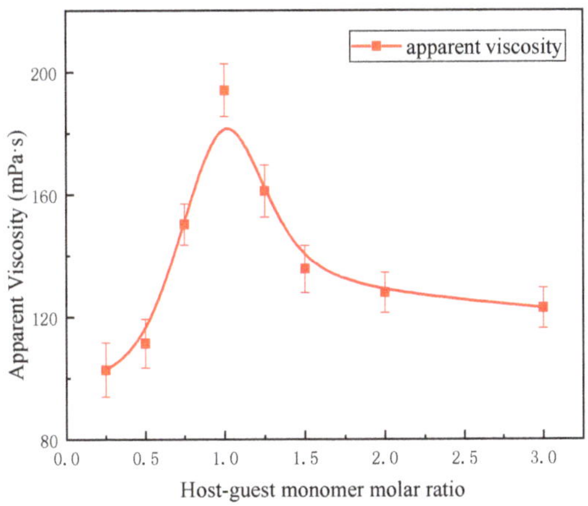

Figure 5. The 0.6 wt.% HGP viscosity at various monomer molar ratios with a shear rate of 170 s^{-1} at 25 °C.

3.4. Conductivity

It is important for application in oilfields that polymers dissolve in water immediately. When the polymer is completely dissolved, the conductivity hardly changes over time. As shown in Figure 6, the conductivity of HGP, GP-T, and HP-T gradually stabilized over time. The conductivity of HGP and HP-T tends to stabilize quickly, at 4.2 min and 5.1 min, respectively. However, the GP-T was relatively slow and did not become stable until 8.0 min. β-CD can wrap the hydrophobic groups to form an inclusion compound due to its unique structures, including hydrophobic inner cavity and hydrophilic outer cavity. Thus, the hydrophobic association between the hydrophobic groups in the polymer was shielded, accelerating the diffusion rate of the polymer molecules to water, and reducing the dissolution time of the polymer. From this perspective, polymers using the host–guest strategy have more promising application prospects than hydrophobic associating polymers.

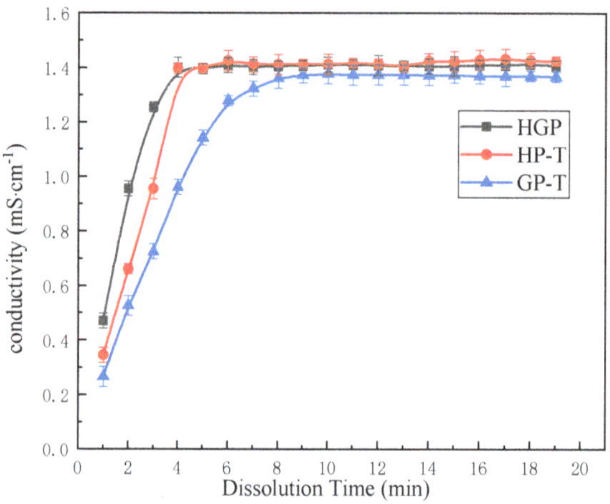

Figure 6. Conductivity changes in the dissolution process of copolymers at 25 °C.

3.5. Thickening Performance

The viscosity of the polymer solution increases as the concentration increases. As shown in Figure 7, the four polymers exhibited good proportional linear thickening properties. When the concentration increased from 1000 mg/L to 3500 mg/L, the growth rate in viscosity of the HGP was the highest (168%), and HP-T had the lowest liquid viscosity growth rate (92.5%). This could be interpreted as follows: at low concentrations, the system thickening effect was not obvious, but when the concentration was more than 1000 mg/L, the inclusion effect became significant, and the host–guest recognition between the β-CD and NAP on the polymer formed a cross-linked 3D network structure. In addition, the hydrophobic association also contributes to viscosity, as shown in Figure 8. As a result, the viscosity of HGP rose sharply.

3.6. Microstructure Analysis

ESEM was used to investigate the reasons for the increase in polymer viscosity. In Figure 9, microcosmic network structures in HGP are much denser than those of GP-T and HP-T. As shown in Figure 9a, there is a noticeable multilayer 3D structure containing pores with different sizes in the HGP solution as a result of the host–guest effect, which could not only support the polymer chain but also encapsulate a lot of water. In other words, the host–guest effect can increase the apparent viscosity and deformation resistance.

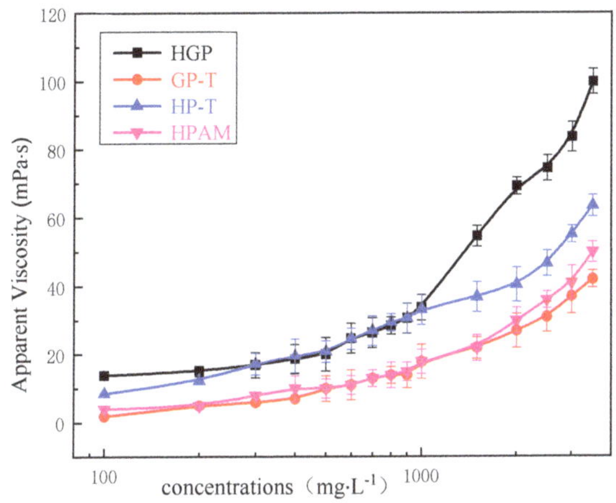

Figure 7. Apparent viscosity as a function of copolymer concentrations for HGP, GP-T, HP-T, and HPAM at shear rate of 170 s^{-1} at 25 °C.

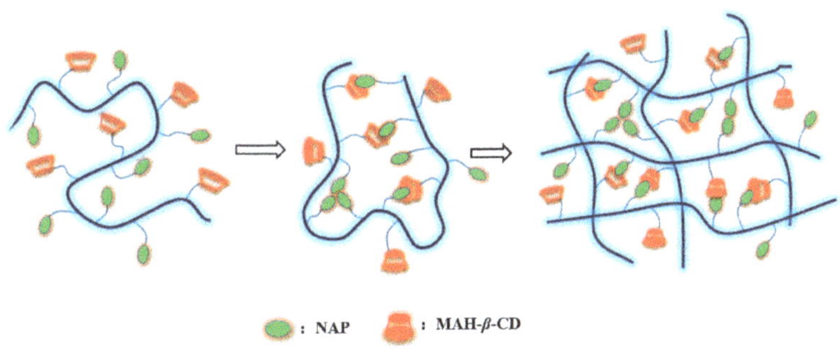

Figure 8. The thickening mechanism of HGP.

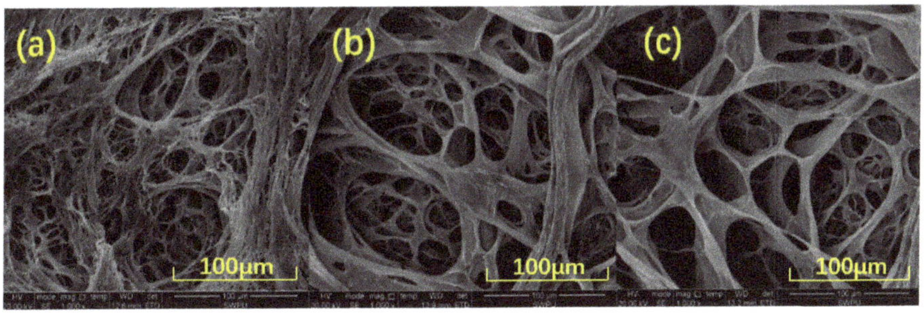

Figure 9. SEM morphologies of 0.6 wt.% GP-T, HP-T, and HGP in DI water (1000×). (**a**) HGP, (**b**) HP-T, (**c**). GP-T.

3.7. Viscoelasticity

The change in storage modulus (G'), loss modulus (G''), and intersection (G_c) depended on the frequency [4]. Figure 10 shows that for HGP, HP-T, and GP-T, at a low frequency where the G' was less than G'', the viscous modulus played a dominant role. The storage and loss moduli increased with the increase in frequency. As the scanning frequency further increased above the characteristic frequency where G' and G'' crossed each other (G_c), the G' was more significant than the G'', indicating that the G' played a dominant role. The relaxation time t_c corresponding to the intersection G_c can be used to describe the solution viscoelasticity. A longer t_c verifies that the solution contributes more to elastic efficiency. Table 1 shows the relaxation times of different solutions according to the calculation results; the t_c of HGP, GP-T, and HP-T are about 2.68, 0.451, and 1.3103 s, respectively, suggesting that the host–guest structure can endow the polymer solution with superior viscoelasticity.

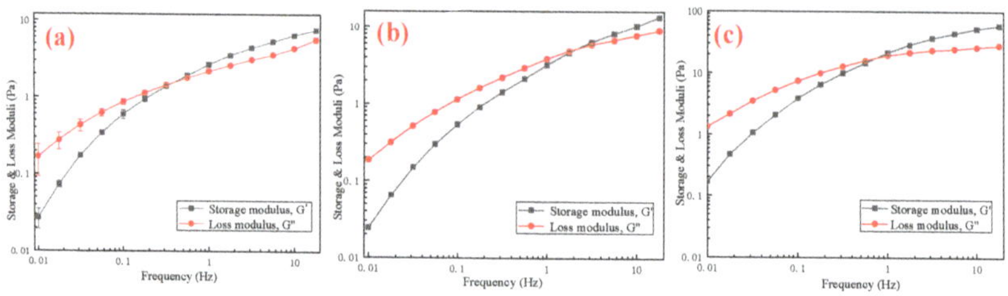

Figure 10. G' and G'' varying as a function of angular frequency for three polymers. (**a**) HGP, (**b**) GP-T, (**c**) HP-T.

Table 1. Crossover frequencies and longest relaxation times of three polymer solutions.

	Polymer		
	HGP	GP-T	HP-T
Crossover frequency (Hz)	0.372	2.20	0.767
Relaxation time (s)	2.688	0.451	1.303

3.8. Shear Tolerance

As shown in Figure 11, the viscosity decreased as the shear rate increased, indicating that those polymer solutions are typical pseudoplastic fluids. With the increase in shear rate, the forces between groups and molecular chains were destroyed, including hydrogen bonds, van der Waals forces, and hydrophobic associations. The flow resistance of the system dropped dramatically, resulting in a decrease in apparent viscosity. In addition, the viscosities of HGP, GP-T, and HP-T were 87.8 mPa·s, 47.5 mPa·s, and 53.2 mPa·s, respectively, at a shear rate of 1000 s^{-1}. A similar situation can also be observed in Figure 12. When the shear rate alternated between 170 s^{-1} and 510 s^{-1}, the HGP solution showed the best shear recovery performance, indicating that the HGP solution had good shear resistance properties caused by the interaction between the β-CD and the hydrophobic group. When the shear occurred, the host–guest interaction was destroyed, producing many free host and guest groups at the cross-section. While shear rate decreased, the host–guest effect reformed, leading to the recovery in the viscosity of polymer solutions.

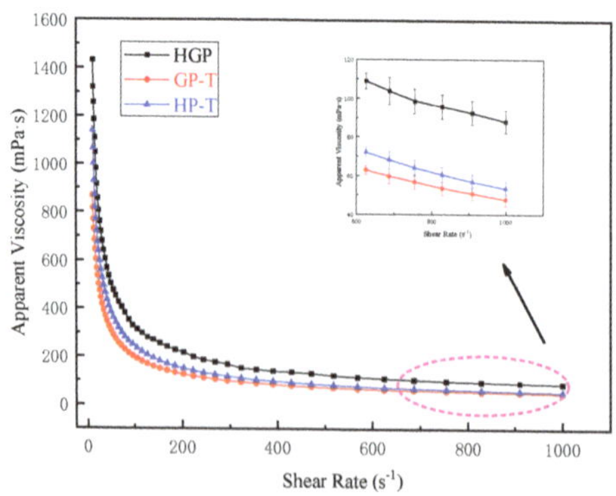

Figure 11. Apparent viscosity relative to shear rate of the HGP, GP-T, and HP-T at 0.6 wt.%.

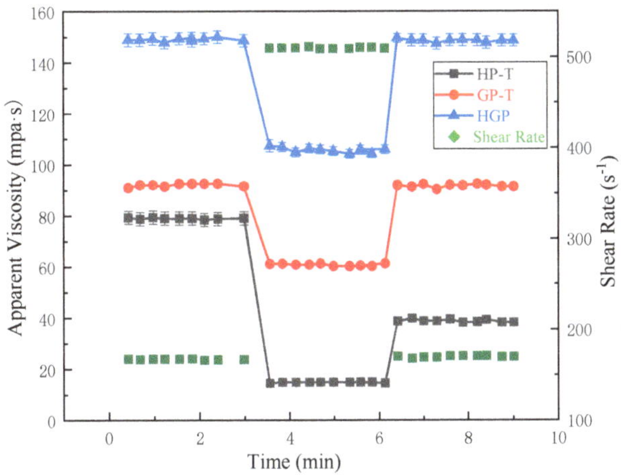

Figure 12. The HGP, GP-T, and HP-T apparent viscosity relative to alternating shear rate.

The polymer will experience shear and temperature increase in the reservoir. Thus, it is necessary to evaluate the behaviors of polymers in formation water. In the experiment, formation water of the SZ1-1 oilfield in the Bohai Sea was used to prepare 0.6 wt.% polymer solution at 25 °C; the formation water composition is shown in Table 2.

Table 2. Formation water ion composition.

Composition	Na^+/K^+	Ca^{2+}	Mg^{2+}	CO_3^{2-}	HCO_3^-	SO_4^{2-}	Cl^-
Content (mg·L^{-1})	3091.64	276.17	158.68	14.21	311.48	85.29	5436.34

Figure 13 shows that the viscosity of the polymer solution decreased as the temperature increased. From 30 °C to 120 °C, the HGP viscosity maintains at approximately 90 mPa·s, significantly greater than HP-T (51 mPa·s) and GP-T (30 mPa·s), indicating that the HGP has greater temperature resistance than HP-T and GP-T. β-CD encapsulating the hydrophobic chain was an exothermic reaction [24,31], leading to the dissociation of the β-CD inclusion complex with increasing temperature due to the Brownian motion. In addition, the weakening of the hydrophobic association resulted in a considerable reduction in viscosity. However, the conformation of the polymer at high temperature determines its hydrodynamic volume, which determines the viscosity. The conformation is mainly affected by the rigidity of the polymer and the forces between groups and molecular chains. Although rigidity of the polymer worsens the flexibility of the molecular chain, it can make the molecular chain more extended [32]. Therefore, increasing the rigidity of the polymer can effectively increase the temperature resistance, and the introduction of the hydrophobic group can also change the conformation of the polymer at high temperatures [33]. Thus, the temperature resistance of the polymer solutions at high temperatures was HGP > GP-T > HP-T.

3.9. Salt Tolerance

Inorganic salt has a negative influence on the polymer solution viscosity, especially the divalent ones, such as Ca^{2+} and Mg^{2+} [34]. As shown in Figure 14a, the viscosity change of the HGP solution was more pronounced than those of the HP-T and GP-T solutions. When the NaCl concentration was less than 20,000 mg/L, the viscosity of the HGP solution decreased as the NaCl concentration increased. This can be interpreted as sodium ions shielding the carboxylate ions on the polymer, which reduces electrostatic repulsion between the anions, causing the polymer backbone to curl, resulting in the reduction in viscosity at the initial stage.

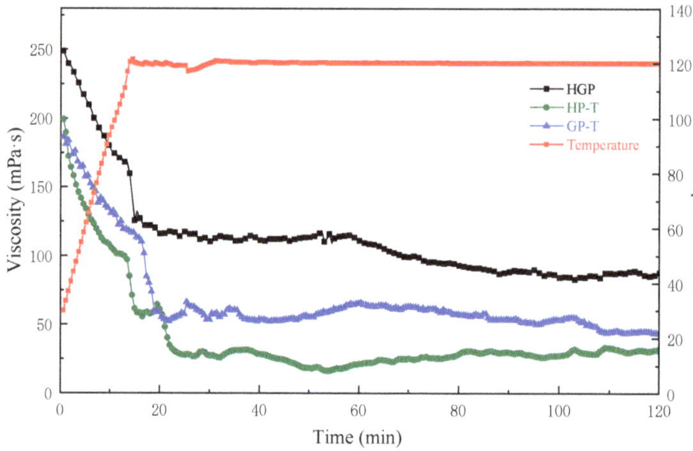

Figure 13. Rheological properties of copolymers HGP, HP-T, and GP-T at 120 °C.

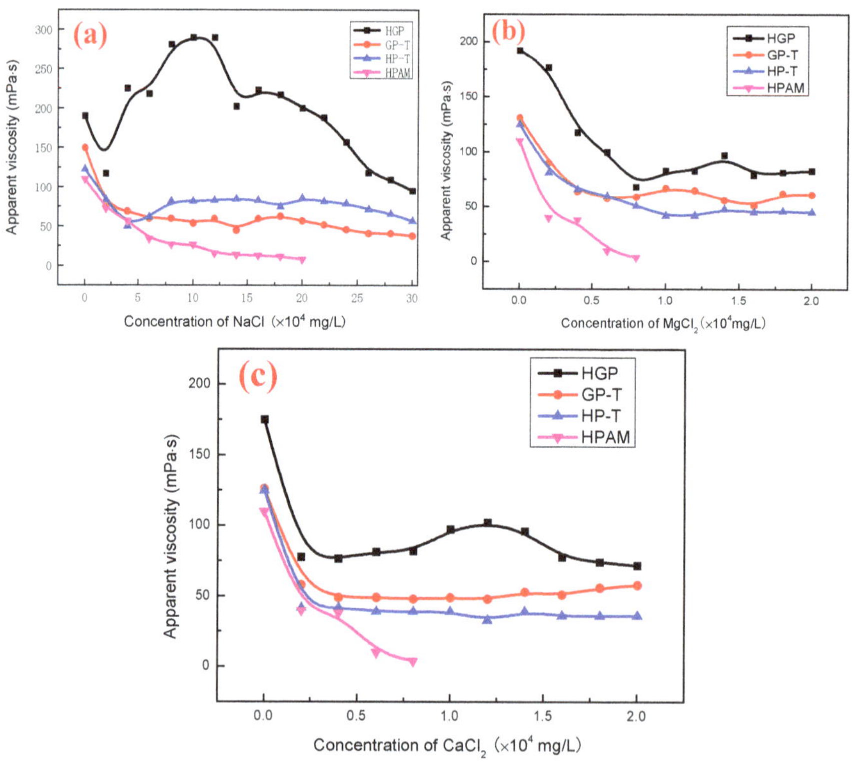

Figure 14. Effects of inorganic salt concentrations on the viscosities of HGP, GP-T, HP-T, and HPAM, 0.6 wt.%. (**a**) NaCl, (**b**) $MgCl_2$, (**c**) $CaCl_2$.

When the content of NaCl increases from 2×10^4 mg/L to 12×10^4 mg/L, the polarity of the solution increases, which strengthens the hydrophobic association and the host–guest effect. Macroscopically, the viscosity of HGP increased from 117 mPa·s to 290 mPa·s. Further, when the NaCl concentration reached 30×10^4 mg/L, the viscosity lowered from 290 mPa·s to 89 mPa·s. This can be clarified as when the salt concentration further increases, the electrostatic shielding effect was greater than the host–guest interaction and hydrophobic association, resulting in the polymer backbone crimp and a sharp decrease in viscosity. For HPAM, the electrostatic shielding effect of salt on carboxylate ions was significant, and the polymer backbone crimped seriously, so the viscosity of HPAM is almost the same as that of water [35].

Similarly, the effect of $CaCl_2$ and $MgCl_2$ on the viscosity of the polymer solution was also investigated, as shown in Figure 14b,c. With the increase in concentrations of $CaCl_2$ and $MgCl_2$, the viscosity of the HGP was higher than that of HP-T and GP-T, indicating that the host–guest effect between NAP and β-CD can improve the salt tolerance of the polymer. When the brine concentration reached 2×10^4 mg/L, the viscosity of HGP in $CaCl_2$ and $MgCl_2$ decreased from the initial viscosity to 71.6 and 82.5, respectively. However, when the concentration of inorganic salts reached 7000 mg/L, the viscosity of the HPAM solution was reduced almost to zero.

4. Conclusions

In this study, a water-soluble polymer, HGP, was synthesized that combined the host–guest effect of β-cyclodextrin and NAP. From the experiments, six conclusions were drawn:

(1) HGP with good properties was synthesized by free-radical polymerization. NAP, MAH-β-CD, and HGP were characterized by ^{1}H NMR spectroscopy.
(2) HGP using the host–guest strategy encompasses a better solubility than HAWSPs.
(3) The highest effectiveness of host–guest structures can be formed in an aqueous solution when the molar ratio of β-CD to NAP is 1:1.
(4) When the concentration increased from 1000 mg/L to 3500 mg/L, the growth rate in viscosity of the HGP was 168%. The microscopic reason was that microcosmic network structures in HGP are much denser than those of GP-T and HP-T.
(5) The host–guest effect between NAP and MAH-β-CD groups improves the performance of HGP, including the shear, viscoelasticity, and salt tolerance.
(6) For the polymer solutions prepared by the formation water of the SZ1-1 oilfield in the Bohai Sea, HGP maintains the viscosity of 90 mPa·s at 120 °C, 170 s^{-1}, which was higher than HP-T and GP-T.

Therefore, HGP has good application prospects for drilling, hydraulic fracturing, and enhanced oil recovery.

Author Contributions: Formal analysis, T.X. and Y.Z.; Funding acquisition, X.Y. and C.L.; Investigation, A.D., Q.Z. and J.M. (Jinhua Mao); Software, H.Z.; Writing—original draft, T.X.; Writing—review and editing, T.X. and J.M. (Jincheng Mao). All authors have read and agreed to the published version of the manuscript.

Funding: This research was funded by National High Technology Research & Development Program (Grant Nos. 2016ZX05053, 2016ZX05014-005-007), China Postdoctoral Science Foundation (Grant Nos. 2019M650250), and the National Natural Science Foundation of China.

Institutional Review Board Statement: Not applicable.

Informed Consent Statement: Not applicable.

Data Availability Statement: The data presented in this study are available on request from the corresponding author.

Acknowledgments: The research is partly supported by the National High Technology Research & Development Program (2016ZX05053, 2016ZX05014-005-007), China Postdoctoral Science Foundation

(2019M650250), and the National Natural Science Foundation of China (41902303). We thank Jincheng Mao, Xiaojiang Yang, and Chong Lin for providing the helpful of Funding.

Conflicts of Interest: The authors declare no conflict of interest.

References

1. Wever, D.A.Z.; Picchioni, F.; Broekhuis, A. Polymers for enhanced oil recovery: A paradigm for structure–property relationship in aqueous solution. *Prog. Polym. Sci.* **2011**, *36*, 1558–1628. [CrossRef]
2. Kulawardana, E.U.; Koh, H.; Kim, D.H.; Liyanage, P.J.; Upamali, K.A.; Huh, C.; Weerasooriya, U.; Pope, G.A. Rheology and Transport of Improved EOR Polymers under Harsh Reservoir Conditions. *SPE Improv. Oil Recovery Symp.* **2012**. [CrossRef]
3. Bock, J.; Valint, P.L.; Pace, S.J. Enhanced Oil Recovery with Hydrophobically Associating Polymers Containing Sulfonate Functionality. U.S. Patent US4702319 A, 27 October 1987.
4. Zhang, Y.; Mao, J.; Zhao, J.; Yang, X.; Xu, T.; Lin, C.; Mao, J.; Tan, H.; Zhang, Z.; Yang, B.; et al. Preparation of a Hydrophobic-Associating Polymer with Ultra-High Salt Resistance Using Synergistic Effect. *Polymers* **2019**, *11*, 626. [CrossRef] [PubMed]
5. Nasr-El-Din, T. Water-soluble hydrophobically associating polymers for improved oil recovery: A literature review. *J. Pet. Sci. Eng.* **1998**, *19*, 265–280.
6. Zhao, T.; Xing, J.; Dong, Z.; Tang, Y.; Pu, W. Synthesis of Polyacrylamide with Superb Salt-Thickening Performance. *Ind. Eng. Chem. Res.* **2015**, *54*, 151019101705006. [CrossRef]
7. Yu, W.; Feng, Y.; Wang, B.; Lu, Z. A novel thermoviscosifying water-soluble polymer: Synthesis and aqueous solution properties. *J. Appl. Polym. Sci.* **2010**, *116*, 3516–3524.
8. Zou, C.; Gu, T.; Xiao, P.; Ge, T.; Wang, M.; Wang, K. Experimental Study of Cucurbit[7]uril Derivatives Modified Acrylamide Polymer for Enhanced Oil Recovery. *Ind. Eng. Chem. Res.* **2014**, *53*, 7570–7578. [CrossRef]
9. Abdel-Alim, A.H.; Hamielec, A.E. Hamielec, Shear degradation of water-soluble polymers. I. Degradation of polyacrylamide in a high: Hear couette viscometer. *J. Appl. Polym. Sci.* **1973**, *17*, 3769–3778. [CrossRef]
10. Basedow, A.M.; Ebert, K.H. Effects of mechanical stress on the reactivity of polymers: Activation of acid hydrolysis of dextran by ultrasound. *Polym. Bull.* **1979**, *1*, 299–306. [CrossRef]
11. Osman, S.K.; Brandl, F.P.; Zayed, G.M.; Teßmar, J.K.; Göpferich, A.M. Cyclodextrin based hydrogels: Inclusion complex formation and micellization of adamantane and cholesterol grafted polymers. *Polymer* **2011**, *52*, 4806–4812. [CrossRef]
12. Wenz and Gerhard, Cyclodextrin Inclusion Polymers Forming Hydrogels. *Adv. Polym. Sci.* **2009**, *222*, 175–203.
13. Topuz, F.; Holtzl, T.; Szekely, G. Scavenging organic micropollutants from water with nanofibrous hypercrosslinked cyclodextrin membranes derived from green resources. *Chem. Eng. J.* **2021**, *419*, 129443. [CrossRef]
14. Cecone, C.; Hoti, G.; Krabicová, I.; Appleton, S.L.; Caldera, F.; Bracco, P.; Zanetti, M.; Trotta, F. Sustainable synthesis of cyclodextrin-based polymers by exploiting natural deep eutectic solvents. *Green Chem.* **2020**, *22*, 5806–5814. [CrossRef]
15. Chen, X.; Liu, J.; Yan, J.-L.; Fang, Y. Host-Guest Interactions between N,N′-Bis(ferrocenylmethylene)-Diaminobutane and Benzenetetracarboxylic Dianhydride Bridged Bis(β-cyclodextrin)s. *Chin. J. Chem.* **2006**, *24*, 1687–1691. [CrossRef]
16. Jing, L.; Yan, J.; Chen, X.; Fang, Y. Studies on the conformational behavior of acenaphthylene-labeled poly(acrylamide- co -acryloyl-6-deoxy-6-amino-β-cyclodextrin). *Colloid Polym. Sci.* **2007**, *285*, 881–889.
17. Li, S.; Zhang, X.; Liang, H.; Wang, X. Supramolecular Hydrogels Obtained by Host-Guest Interactions of Cyclodextrin Dimers with Adamantyl-Containing LCST Copolymers. *Acta Chim. Sin.* **2012**, *70*, 1013–1020. [CrossRef]
18. Huh, K.M.; Cho, Y.W.; Chung, H.; Kwon, I.C.; Jeong, S.Y.; Ooya, T.; Lee, W.K.; Sasaki, S.; Yui, N. Supramolecular hydrogel formation based on inclusion complexation between poly(ethylene glycol)-modified chitosan and alpha-cyclodextrin. *Macromol. Biosci.* **2004**, *4*, 92. [CrossRef] [PubMed]
19. Liu, Y.Y.; Fan, X.; Sun, L. Synthesis and properties of poly(n-isopropylacrylamide)/β-cyclodext rin ionic polymer. *Acta Polym. Sin.* **2005**, *1*, 471–475.
20. Wang, J.; Jiang, M. Studies on self-assembly from lightly-crosslinked polymers via inclusion interaction. *Acta Polym. Sin.* **2007**, *7*, 979–985. [CrossRef]
21. Weickenmeier, M.; Wenz, G.; Huff, J. Association thickener by host guest interaction of a β-cyclodextrin polymer and a polymer with hydrophobic side-groups. *Macromol. Rapid Commun.* **1997**, *18*, 1117–1123. [CrossRef]
22. Charlot, A.; Auzely-Velty, R.; Rinaudo, M. Specific interactions in model charged polysaccharide systems. *J. Phys. Chem. B* **2008**, *107*, 8248–8254. [CrossRef]
23. Pu, W.-F.; Yang, Y.; Wei, B.; Yuan, C. The Potential of a β-Cyclodextrin/Adamantane Modified Copolymer in Enhancing Oil Recovery through Host-Guest Interactions. *Ind. Eng. Chem. Res.* **2016**, *55*, 8679–8689. [CrossRef]
24. Zou, C.; Liao, W.; Zhang, L.; Chen, H. Study on acidizing effect of β-cyclodextrin-PBTCA inclusion compound with sandstone. *J. Pet. Sci. Eng.* **2011**, *77*, 219–225. [CrossRef]
25. Liu, X.; Jiang, W.; Gou, S.; Ye, Z.; Feng, M.; Lai, N.; Liang, L. Synthesis and evaluation of novel water-soluble copolymers based on acrylamide and modular β-cyclodextrin. *Carbohydr. Polym.* **2013**, *96*, 47–56. [CrossRef]
26. Zou, C.; Zhao, P.; Hu, X.; Yan, X.; Zhang, Y.; Wang, X.; Song, R.; Luo, P. β-Cyclodextrin-Functionalized Hydrophobically Associating Acrylamide Copolymer for Enhanced Oil Recovery. *Energy Fuels* **2013**, *27*, 2827–2834. [CrossRef]
27. Huang, G.L.; Zhang, Y.; Ren, F.; Zheng, F.L. Study on the difference of envelope between 1–bromonaphthalene, 2–bromonaphthalene and β-cyclodextrin. *J. Jimei Univ. (Nat. Ed.)* **2003**, *1*, 22–27.

28. He, Y.; Xu, Z.H.; Wu, F.; Qing, D. Synthesis and evaluation of a novel amphiphilic polymer containing β-cyclodextrin. *Polym. Sci.* **2014**, *56*, 822–829. [CrossRef]
29. Yan, J.; Du, L.; Lu, F.; Li, Z.; Yang, J.; Bie, S.; Zhang, J. Synthesis and properties of functionalized β-cyclodextrin copolymer and its metal complexes. *Polym. Bull.* **2006**, *57*, 481–489.
30. Candau, F.; Regalado, E.J.; Selb, J. Recent advances in associating polymers prepared by micellar polymerization. *Macromol. Symp.* **2000**, *150*, 241–249. [CrossRef]
31. Shan, J.; Li, J.N.; Jiang, Z.T. Inclusion reactions of β-cyclodextrin and its derivatives with cinnamaldehyde in Cinnamomum loureirii essential oil. *Eur. Food Res. Technol.* **2010**, *230*, 543–550.
32. Yang, B.; Mao, J.; Zhao, J.; Shao, Y.; Zhang, Y.; Zhang, Z.; Lu, Q. Improving the Thermal Stability of Hydrophobic Associative Polymer Aqueous Solution Using a "Triple-Protection" Strategy. *Polymers* **2019**, *11*, 949. [CrossRef] [PubMed]
33. Zhang, Y.; Mao, J.; Zhao, J.; Yang, X.; Zhang, Z.; Yang, B.; Zhang, W.; Zhang, H. Preparation of a Novel Ultra-High Temperature Low-Damage Fracturing Fluid System Using Dynamic Crosslinking Strategy. *Chem. Eng. J.* **2018**, *354*, 913–921. [CrossRef]
34. Tian, J.; Mao, J.; Zhang, W.; Yang, X.; Lin, C.; Cun, M.; Mao, J.; Zhao, J. Application of a Zwitterionic Hydrophobic Associating Polymer with High Salt and Heat Tolerance in Brine-Based Fracturing Fluid. *Polymers* **2019**, *11*, 2005. [CrossRef] [PubMed]
35. Wang, W.; Yang, H.; Yan, Y.; Wang, M. Study on characterization of heat-resistant and salt-tolerant polymer pama and properties of its salting solution. *Pet. Process. Petrochem.* **2014**, *45*, 31–34.

polymers

Article

Preparation of Bisphenol-A and Polydimethylsiloxane (PDMS) Block Copolycarbonates by Melt Polycondensation: Effects of PDMS Chain Length on Conversion and Miscibility

Zibo Zhou and Guozhang Wu *

Shanghai Key Laboratory of Advanced Polymeric Materials, School of Materials Science & Engineering, East China University of Science & Technology, Shanghai 200237, China; s13867700926@163.com
* Correspondence: wgz@ecust.edu.cn

Abstract: This study aimed to improve polydimethylsiloxane (PDMS) conversion in the preparation of polycarbonate (PC)–polydimethylsiloxane (PDMS) copolymer through melt polycondensation. We examined the transesterification process of PDMS with diphenyl carbonate (DPC) and its copolymerization products with bisphenol-A (BPA) for different chain lengths of PDMS. The key factors affecting PDMS conversion were investigated. Results showed that long-chain PDMS required a higher critical transesterification level (38.6%) to improve miscibility with DPC. During polycondensation, side reactions were more prone to occur when the equilibrium transesterification level of long-chain PDMS was lower. PDMS conversion was also lower when more short-chain PDMS was fed. Increasing the chain length of PDMS also reduced PDMS conversion. Notably, increasing the amount of KOH can significantly improve PDMS conversion throughout the polycondensation stage by increasing the equilibrium transesterification level of long-chain PDMS, thereby inhibiting the occurrence of side reactions.

Keywords: PC-PDMS copolymer; melt polycondensation; miscibility; equilibrium transesterification level; conversion

Citation: Zhou, Z.; Wu, G. Preparation of Bisphenol-A and Polydimethylsiloxane (PDMS) Block Copolycarbonates by Melt Polycondensation: Effects of PDMS Chain Length on Conversion and Miscibility. *Polymers* **2021**, *13*, 2660. https://doi.org/10.3390/polym13162660

Academic Editors: Edina Rusen and Agnieszka Tercjak

Received: 18 July 2021
Accepted: 3 August 2021
Published: 10 August 2021

Publisher's Note: MDPI stays neutral with regard to jurisdictional claims in published maps and institutional affiliations.

Copyright: © 2021 by the authors. Licensee MDPI, Basel, Switzerland. This article is an open access article distributed under the terms and conditions of the Creative Commons Attribution (CC BY) license (https://creativecommons.org/licenses/by/4.0/).

1. Introduction

Polycarbonate (PC) is widely applied in various fields such as electronics, automotive parts, constructions, and medical equipment due to its excellent optical and mechanical properties, including transparency and impact strength [1–3]. However, its utilization still faces challenges because new designs tend to emphasize more on key application areas requiring thin-wall parts, complex geometries, and high-impact products that can withstand very low temperatures with no compromise on flow and processability. Meanwhile, PC shows a V-2 rating in the UL-94 test, meaning it does not meet fire-safety requirements. For this reason, PC modification with other components through copolymerization and melt blending has been extensively studied over the last four decades [4–6].

Polydimethylsiloxane (PDMS) molecules have relatively flexible polymer backbones (or chains) with a low glass-transition temperature, resulting in low temperature ductility, high thermal stability, and versatile functionality. Thus, PDMS is an ideal flame retardant or impact modifier for PC. These disadvantages can be compensated for through copolymerization with PDMS, which generally shows higher polymer-chain flexibility [7,8]. The introduction of PDMS into the PC main chain enables an easy process of PC and improves its low-temperature toughness, weatherability, hydrolytic stability, and flame retardancy [9–15]. Compared with traditional interfacial phosgenation, melt-condensation polymerization is a green process not requiring the use of hazardous phosgene and chlorinated solvent. Thus, this efficient and friendly method can be applied in preparing PC-PDMS copolymer for industrial production.

The preparation of PC-PDMS copolymers by melt polycondensation is greatly hindered by the difficulty of increasing PDMS conversion. PDMS and DPC are immiscible with each other due to the large difference in their solubility parameters (δ_{DPC} = 10.53 cal$^{1/2}$ cm$^{-3/2}$ vs. δ_{PDMS} = 7.3 cal$^{1/2}$ cm$^{-3/2}$) [16,17]. Thus, PDMS cannot be well-dispersed in DPC or PC oligomers in the melt state, thereby suppressing the transesterification reaction of PDMS to some extent [18]. King et al. [19] demonstrated that hydroxyl-terminated PDMS with long chains cannot achieve high conversion through melt polycondensation (only 3–5%), as experimentally observed by Koenig et al. [20] Moreover, the immiscibility between PC and PDMS is a key factor affecting the heat resistance, impact resistance, and optical transparency of PC-PDMS copolymers. Zhang et al. [21] and Zhou et al. [22] reported that PC-PDMS copolymer formed in situ in PC/PDMS blend under the action of transesterification catalyst helps improve the miscibility between the two. Therefore, we consider that PDMS and DPC should be reacted first in the transesterification stage. Afterwards, the end group of PDMS changes, which is beneficial to improving the miscibility with DPC or PC oligomers, and then it reacts with BPA or PC oligomers to obtain the desired products. Thus, the change in miscibility influences PDMS conversion during transesterification.

A series of siloxane equilibrium reactions occur during transesterification under the action of strong bases, including the linear condensation between PDMS and PDMS or the cyclic degradation of PDMS itself, thereby affecting the transesterification ratios of PDMS [23]. The former increases the average chain length of PDMS, and the latter decreases its average chain length. Either raising temperature or lowering pressure can break the equilibrium state and affect the form of PDMS in the melt. GE and Bayer have found that the siloxane chain is severely decomposed by siloxane-chain scission followed by siloxane depolymerization to give cyclic siloxanes during the typical melt process [24,25]. Cyclic siloxanes formed in this process remain in the polymer and have an exceptionally disruptive effect on applications in the electrical/electronic sector [26–28]. Another potential problem regarding PC-PDMS copolymer synthesis, which is relatively neglected often, is the stability of the hydroxyl end groups in PDMS oligomers. Hydroxyl-end groups can backbite the terminal silicon atoms in PDMS oligomers, leading to the formation of cyclic siloxane and the loss of end-group functionality, thereby inhibiting the introduction of PDMS into the PC chain. To improve PDMS conversion, the equilibrium conversion rate of PDMS in the transesterification stage must be increased to promote the miscibility of PDMS with PC oligomers while reducing the residual amount of PDMS and inhibiting the occurrence of side reactions.

Accordingly, the present study initially examined the transesterification process between PDMS and DPC and investigated the degree of transesterification reaction required for different chain lengths of PDMS and DPC to reach a critical miscibility state, as well as the final equilibrium transesterification level. The effects of the feed amount of PDMS and the average chain length on PDMS conversion during melt polycondensation were then investigated. We found that PDMS conversion in the final product depended on the equilibrium conversion rate at the transesterification stage. Finally, a solution strategy was provided to effectively increase PDMS conversion.

2. Materials and Methods

2.1. Materials

Bisphenol-A (BPA; 99.0%), diphenyl carbonate (DPC; 99.0%), and potassium hydroxide (KOH; 95%) were provided by Aladdin. Anhydrous alcohol (CH_3CH_2OH; 99.7%), dichloromethane (CH_2Cl_2; 99.5%), chloroform ($CHCl_3$; 99.0%), and n-hexane (C_6H_{14}; 97%) were purchased from Shanghai Titan. DPC was purified by recrystallization in anhydrous ethanol three times.

Silanol-terminated PDMS with an average chain length of 5.3 and 22.5 was supplied by Shenzhen Ji-Peng Silicone Fluoride Materials Co. (Shenzhen, China) and denoted as $Silanol_{5.3}$ and $Silanol_{22.5}$, respectively. Another silanol-terminated PDMS with an average chain length of 56.2, denoted as $Silanol_{56.2}$, was provided by ThermoFisher Scientific (Shanghai, China). The average chain length of PDMS was determined by ^{1}H-NMR as described below. The chemical structures of BPA, PDMS, and DPC are shown in Figure 1.

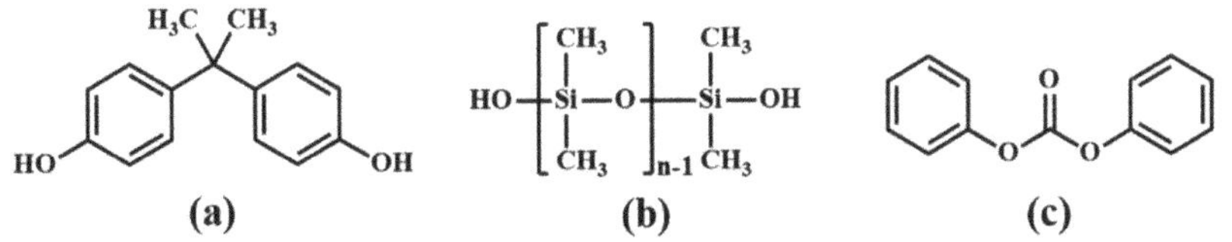

Figure 1. Chemical structures of (**a**) BPA, (**b**) PDMS, and (**c**) DPC.

2.2. Melt-Transesterification Reaction of DPC and PDMS

Equimolar amounts of DPC and PDMS with different average chain lengths were poured into a 100 mL four-neck round-bottom glass flask fitted with a reflux condenser. The flask was evacuated and refilled with dry nitrogen gas three times and then heated in a 180 °C silicone oil bath. The addition of KOH (0.01 mol% vs. PDMS) into the melted mixture was considered the start of the reaction, and a small amount of sample was extracted from the reactor at set intervals. Each sample was sealed and stored in a refrigerator for analysis and direct characterization by nuclear magnetic resonance (NMR) and optical microscopy (OM).

2.3. Synthesis of BPA/PDMS Copolycarbonates

Different amounts of $Silanol_{5.3}$ were introduced into BPA and DPC mixtures. The molar ratio of DPC/BPA was set at 1.06, and the BPA/PDMS molar ratio was changed to 100/0, 99/1, 98/2, 96/4, 94/6, 92/8, and 90/10. The total amount of monomer was 77 mmol. A KOH amount of 100 ppm to diols on a mole basis was used. Whole reactants were placed in a 100 mL three-neck round-bottom flask equipped with a mechanical stirrer and high-vacuum bearing. The temperature was increased to 180 °C with mechanical stirring under N_2 atmosphere. The material was allowed to melt for 10 min, and then the temperature was kept constant for 15 min to make the transesterification reach equilibrium. Then, the temperature was increased to 210 °C, and the pressure was gradually reduced to 120 mmHg. The temperature was held for 45 min to distill off phenol. Subsequently, the temperature was raised to 270 °C over the next 15 min, after which the pressure was reduced to 30 mmHg. After further reaction for 10 min, the final temperature was set at 280 °C, and the pressure was reduced to less than 1 mmHg. This condition was maintained for 10 min to complete the reaction. The synthesized polymer was cooled to room temperature, dissolved in dichloromethane, and precipitated in an appropriate amount of n-hexane accompanied by ultrasonic cleaning for 1 h to remove unreacted PDMS. The final product was collected by vacuum filtration and dried in a vacuum oven at 100 °C for 24 h until their use for characterizations.

A similar procedure was performed using $Silanol_{22.5}$ and $Silanol_{56.2}$ as raw materials. The molar ratio of DPC to BPA was kept at 1.06, and the amount of PDMS was set at 6 mol% BPA.

2.4. Characterizations

ATR-FTIR experiments were performed on a Nicolet 5700 spectrometer with a DTGS detector in the range of 400–4000 cm^{-1} at a resolution of 4 cm^{-1}. The samples were dissolved in dichloromethane at a concentration of 5 wt%, and then 50 μL of solution was spread on a KBr pellet. The solvent was removed by vacuum desiccation at room temperature before ATR-FTIR analysis.

^{1}H, ^{13}C, and ^{29}Si NMR spectroscopic measurements were conducted using a Bruker DMX-600 NMR spectrometer. $CDCl_3$ with TMS served as deuterium solvents. The incorporated ratios, block contents, and conversions of PDMS were designated as N_{PDMS}, W_{PDMS}, and Conversion, respectively. They were measured by

$$N_{PDMS} = \frac{\frac{[a]}{6 \times n_{PDMS}}}{\frac{[a]}{6 \times n_{PDMS}} + \frac{[b]}{6}} \quad (1)$$

$$W_{PDMS} = \frac{74.1 \times [a]}{74.1 \times [a] + 254 \times [b]} \times 100\% \tag{2}$$

$$\text{Conversion}(\%) = \frac{N_{PDMS}}{N_0} \tag{3}$$

where [*a*] represents an integrated value of a methyl group in a dimethylsiloxane moiety observed at around δ −0.06 to 0.3, [*b*] represents an integrated value of a methyl group in a BPA moiety observed at around δ 1.5 to 1.9, and N_0 represents the initial feed amount of PDMS.

The molecular weights and molecular-weight distributions of silanols were determined by gel permeation chromatography (GPC). The measurements were performed using a Waters 1515 HPLC pump equipped with a refractive-index detector (Waters 2414). Tetrahydrofuran (THF) was used as the mobile phase at a flow rate of 1 mL/min at 35 °C. Weight-average molecular weight (M_w) and polydispersity index (PDI) were determined against linear polystyrene standards.

The viscosity-average molecular weight (M_η) of the polycondensation product was determined with an Ubbelohde viscometer at 25 ± 0.5 °C using $CHCl_3$ as the solvent and was calculated by Equations (4) and (5) as follows [29]:

$$[\eta] = \sqrt{2(\eta_{sp} - \ln \eta_r)}/C \tag{4}$$

$$[\eta] = KM_\eta^\alpha \tag{5}$$

where C is the concentration of the solution (always 0.01 g mL^{-1}), η_r is the relative viscosity, and η_{sp} is the specific viscosity. According to ref. [29], the characteristic parameters *K* and α of BPA-PC are 0.0111 mL g^{-1} and 0.82, respectively.

Similarly, the M_η of PDMS can also be determined using an Ubbelohde viscometer. When $CHCl_3$ was replaced with toluene as the solvent, it can be described as follows:

$$[\eta] = 2 \times 10^{-2} M_\eta^{0.66} \tag{6}$$

Color difference (ΔC), which represents the yellowness of the synthesized product, was measured with an ultraviolet spectrophotometer (UV-*vis*; UV1900, Youke Instrument). A $CHCl_3$ solution with a concentration of 0.01 g mL^{-1} was used, and ΔC was determined by:

$$\Delta C = \left(\left| \frac{1}{3} - \frac{T_{445}}{T} \right| + \left| \frac{1}{3} - \frac{T_{555}}{T} \right| + \left| \frac{1}{3} - \frac{T_{600}}{T} \right| \right) \times 100\% \tag{7}$$

where T_{445}, T_{555}, and T_{600} represent the transmittance of the solution relative to $CHCl_3$ at wavelengths of 445, 555, and 600 nm. Furthermore, $T = T_{445} + T_{555} + T_{600}$.

The morphology of transesterification products was observed using an OM system (Olympus CX21). We took a drop of transesterification product onto a glass slide, gently pressed the sample to flow naturally, and used a low-power zoom lens (40 × 10) to observe the changes in the morphology of the two phases.

The glass-transition temperature (T_g) of PC was measured with a differential scanning calorimetry (DSC) system (DSC25, TA Instruments) within 20 °C to 250 °C at a heating rate of 10 °C min^{-1} in N_2 atmosphere. To eliminate the thermal history of the samples, T_g was determined as the inflection point of the transition during the second heating process.

The decomposition temperature was tested using a thermogravimetric analysis (TGA) apparatus (Netzsch STA 409). The samples were heated from 50 °C to 800 °C at a heating rate of 10 °C min^{-1} under a nitrogen flow of 50 mL min^{-1}. The 5% weight-loss temperature was recoded as $T_{d,5\%}$.

3. Results and Discussions

3.1. Molecular Structures of Silanol-Terminated PDMS

Figure 2 shows the ATR-IR spectra of PDMS with different chain lengths. A broad absorption peak appeared at 3000–3700 cm^{-1} due to the stretching vibration of the terminal hydroxyl group. The average chain length of $Silanol_{5.3}$ was shorter with a higher hydroxyl content, resulting in a broader absorption peak. Meanwhile, shorter-chain PDMS oligomers were more likely to form hydrogen bonds between the chains, leading to a redshift of the absorption peak [30].

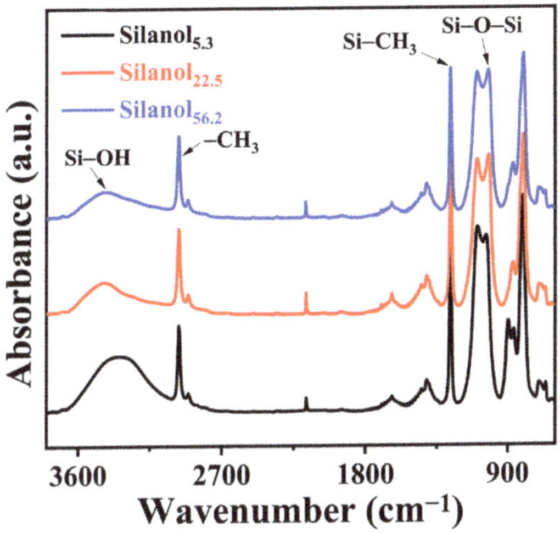

Figure 2. ATR-IR spectra of silanol-terminated PDMS with different block lengths.

Figure 3 shows the GPC curves of PDMS with different chain lengths. From the graph, the molecular weights and molecular-weight distributions of Silanols can be analyzed. We observed that PS standard and PDMS with different chain lengths peaked at 33 min, indicating that this peak was not the main peak of PDMS. The GPC data of the main peak of PDMS with different chain lengths are given in Table 1. $Silanol_{56.2}$ and $Silanol_{22.5}$ with longer chain lengths had wider molecular-weight distribution and a more complex structure than $Silanol_{5.3}$.

Table 1. Basic information of silanol-terminated PDMS with different chain lengths.

Sample ID	Viscosity (cP)	M_η [1] (g/mol)	M_w [2] (g/mol)	PDI [2]	$W_{theo.}$ [3]	$W_{cal.}$ [4]	n_{PDMS} [5]
$Silanol_{5.3}$	30	1700	2200	1.12	8.0	8.3	5.3
$Silanol_{22.5}$	60	2200	2630	1.21	4.0	4.1	22.5
$Silanol_{56.2}$	100	4900	5200	1.23	0.8	0.8	56.2

[1] Calculated with Equation (6); [2] determined by THF-GPC using polystyrene standards (RI detector); [3] theoretical hydroxyl group content of Silanols ($W_{theo.}$); [4] calculated hydroxyl group content of Silanols ($W_{cal.}$) measured by ^{1}H-NMR analysis; [5] calculated by the $W_{cal.}$ obtained from ^{1}H-NMR.

Figure 4 shows the ^{1}H-NMR spectra of PDMS with different average chain lengths. The peaks at δ 2.13 and 2.14 ppm were assigned to the hydrogen atom of terminal hydroxyl groups of $Silanol_{22.5}$ and $Silanol_{56.2}$, respectively. The shorter-chain PDMS had a higher content of hydroxyl groups, which were easily induced by the hydrogen bonding of the terminal hydroxyl group of neighboring molecular chains and moved toward a lower field at δ 3.20 ppm, where the peak widened. The relative values of the absorption peak area of $Si–CH_3$ near 0 ppm on the main chain and Si–OH enabled the calculation of the terminal hydroxyl content of PDMS with different average chain lengths. The average chain length of PDMS obtained was similar to the theoretical chain length.

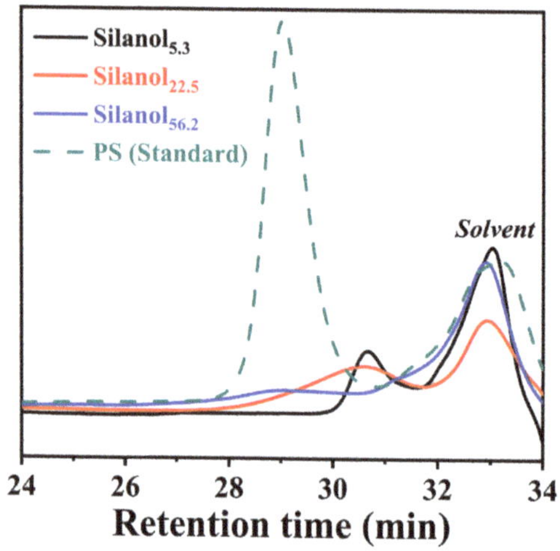

Figure 3. GPC curves of silanol-terminated PDMS with different chain lengths.

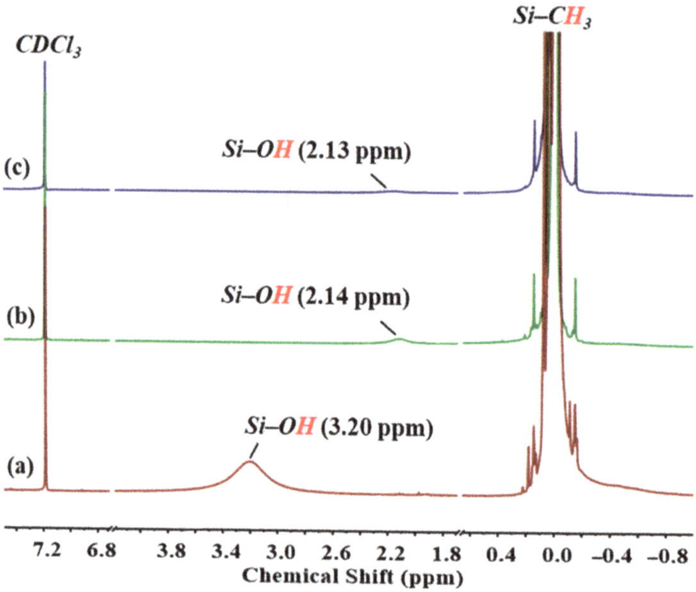

Figure 4. ^{1}H-NMR spectra of (**a**) $Silanol_{5.3}$, (**b**) $Silanol_{22.5}$, and (**c**) $Silanol_{56.2}$.

3.2. Melt-Transesterification Process of DPC and PDMS

Differences may exist in the miscibility of DPC and PDMS with different average chain lengths during the melt-transesterification stage, which may affect the equilibrium transesterification conversion of PDMS. For this reason, we reacted PDMS having different chain lengths with DPC catalyzed by KOH. Samples were extracted from the reactor at set intervals for analyses through ^{1}H-NMR and OM to verify whether the miscibility of the mixture was improved by the transesterification between PDMS and DPC.

Figure 5 gives the OM photographs of transesterification products obtained from DPC and PDMS with different chain lengths reacted at different times. Transesterification temperature was set at 180 °C, and the molar ratio of DPC to PDMS was controlled at 1:1. Figure 5 shows that when the chain length of PDMS was shorter, more DPC was added and more DPC crystals appeared on the OM photographs. However, the results in Figure 5 revealed that the experiments of $Silanol_{5.3}$ had fewer DPC crystals before KOH addition than the other two Silanols with larger average chain lengths. The ^{1}H-NMR spectra of the transesterification products of DPC and PDMS with different average chain lengths before KOH addition are shown in Figure 6. $Silanol_{5.3}$ had already undergone transesterification with DPC, and at this point, 1.8% DPC conversion had been achieved. By the time the reaction reached 3 min, 5.2% DPC was consumed, and the relatively small volume of $Silanol_{5.3}$ more easily dissolved in the remaining DPC due to the change in end group, reflected in the OM photographs as the complete disappearance of DPC crystals. The DPC consumption deduced from ^{1}H-NMR at this moment can be regarded as the critical transesterification level when DPC and PDMS reached mutual miscibility.

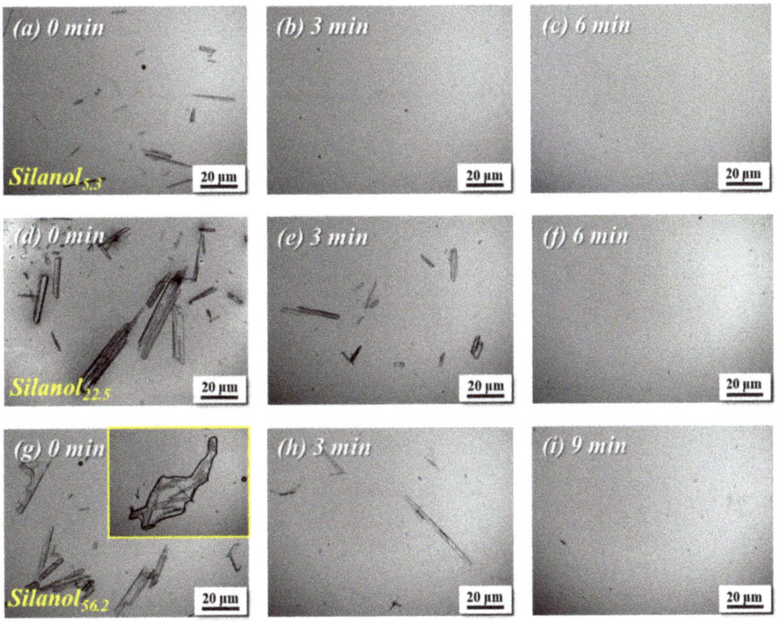

Figure 5. OM photographs of transesterification products of DPC/$Silanol_{5.3}$ reacted for (**a**) 0 min, (**b**) 3 min and (**c**) 6 min, DPC/$Silanol_{22.5}$ reacted for (**d**) 0 min, (**e**) 3 min and (**f**) 6 min and DPC/$Silanol_{56.2}$ reacted for (**g**) 0 min, (**h**) 3 min and (**i**) 9 min at 180 °C.

Table 2 shows the critical transesterification level of DPC and PDMS with different average chain lengths to reach mutual miscibility. With an increased average chain length of PDMS, the critical transesterification level for the complete disappearance of DPC crystals also increased. A conversion rate of 17.0% for DPC was required with $Silanol_{22.5}$ and 38.6% for DPC with $Silanol_{56.2}$ for PDMS to be completely dissolved in DPC. This phenomenon indicated that PDMS with a longer average chain length required a higher critical transesterification level to be soluble in the melt with DPC, also indicating that PDMS with different chain lengths had certain solubility differences.

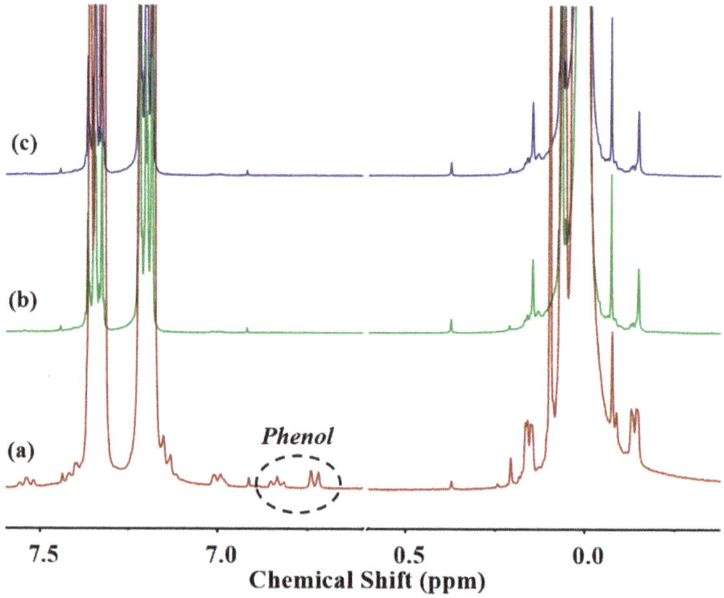

Figure 6. ^{1}H-NMR spectra of first transesterification product of (**a**) $Silanol_{5.3}$, (**b**) $Silanol_{22.5}$, and (**c**) $Silanol_{56.2}$ reacted with DPC under the action of 100 ppm KOH at 180 °C.

Table 2. Critical transesterification level of silanol-terminated PDMS with different chain lengths reacted with DPC to reach two-phase intermiscibility.

Sample ID	Critical Transesterification Level (%)
$Silanol_{5.3}$	5.3
$Silanol_{22.5}$	17.0
$Silanol_{56.2}$	38.6

Figure 7 presents the change curve of the transesterification level of DPC and PDMS with different chain lengths over time. It should be pointed out that under the catalysis of high concentrations of KOH, prolonging the reaction time may cause side effects such as thermal degradation or rearrangement of PDMS, which affect the calculation of the transesterification level. Therefore, we determined a suitable duration of 100 min for the transesterification. Transesterification between $Silanol_{56.2}$ and DPC already reached equilibrium by 100 min, and the conversion of DPC was constant at 38.6%. Conversely, the two sets of experiments on $Silanol_{5.3}$ and $Silanol_{22.5}$ did not yet reach equilibrium, and the transesterification levels of the end product were 51.5% and 45.6%, respectively. This finding explained why long-chain PDMS was more difficult to introduce into the PC backbone, affecting its conversion in the melt-polycondensation process for the preparation of PC-PDMS copolymers.

Figure 8 reveals the ^{29}Si-NMR spectra of pristine PDMS with different chain lengths and their transesterification products with DPC. Figure 8 shows that a new peak appeared at around δ −13.72 ppm (α), which was believed to originate from DPC-PDMS oligomers formed by transesterification. Moreover, according to the literature, some cyclic siloxanes may be present during the preparation of silanols, and the main structures would be D_3, D_4, D_5, and D_6, showing peaks at −9.2, −20, −22.6, and −23.0 ppm in that order [31,32]. Figure 8 shows that PDMS with different chain lengths did not display any peak at the above positions, and no cyclic siloxanes formed in the final transesterification products after reacting with DPC at 180 °C for 100 min. This finding meant that PDMS itself did not suffer from side reactions such as cyclic degradation during the melt-transesterification stage within 100 min. However, the unconverted long-chain PDMS during transesterification may be more susceptible to side reactions at high temperatures and strong bases, reducing PDMS conversion. The experimental results confirmed that improving PDMS conversion in melt polycondensation and suppressing possible side reactions required the selection of PDMS with lower viscosity, shorter chain length, and higher activity as the raw material. This selection would ensure the enhancement in PDMS conversion in the transesterification stage and the elimination of the effect of terminal reactive groups on the side reactions, such as chain tailoring or self-condensation.

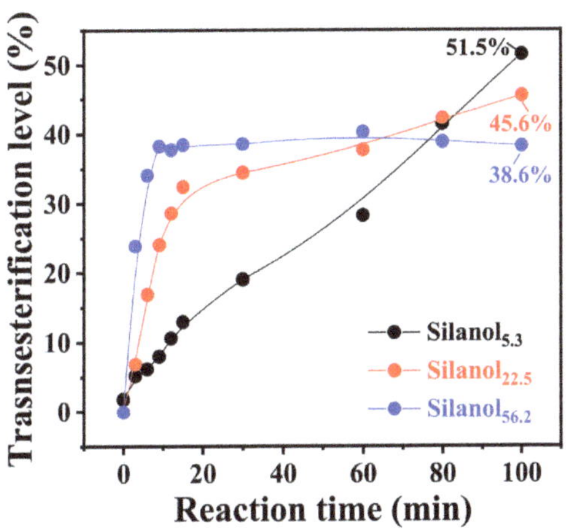

Figure 7. Time dependence of transesterification level of silanols with different average chain lengths reacted with DPC catalyzed by KOH at 180 °C.

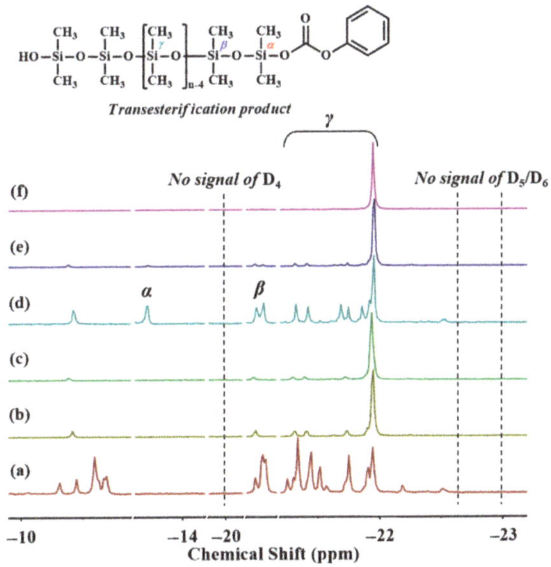

Figure 8. ^{29}Si-NMR spectra of (**a**) $Silanol_{5.3}$, (**b**) $Silanol_{22.5}$, and (**c**) $Silanol_{56.2}$, as well as the final transesterification products of (**d**) $Silanol_{5.3}$, (**e**) $Silanol_{22.5}$, and (**f**) $Silanol_{56.2}$ reacted with DPC for 100 min at 180 °C.

3.3. Influence of Silanol Feeding on Conversion

The study on the transesterification process between DPC and PDMS helped us select $Silanol_{5.3}$, which had a lower viscosity, higher hydroxyl content, and shorter chain length, for further transesterification with DPC and BPA. Due to the difference in reactivity between BPA and $Silanol_{5.3}$, competition existed during the transesterification with DPC, so we expected that the feed amount of $Silanol_{5.3}$ affected the PDMS conversion. Hence, we prepared a series of silanol-based copolycarbonates by varying the silanol feed amount.

The sequence distribution of BPA and $Silanol_{5.3}$ moieties was determined by the chemical shifts at around δ 150.5–151.2 ppm in the ^{13}C-NMR spectrum (Figure 9). The carbonyl carbons split into three peaks corresponding to C_1 (PDMS-PDMS), C_2 (BPA-PDMS), and C_3 (BPA-BPA). The chemical shifts of the three peaks detected at δ 150.8, 151.0, and 151.1 ppm were assigned to the carbon atoms. The central carbon atom of BPA linked within the chain also split into three peaks (C_4, C_5, and C_6) as a result of the copolymerization [33]. Molar ratios of the C_1, C_2, and C_3 sequences of $Silanol_{5.3}$-based copolycarbonates can also be calculated by the integral ratio of the three different types of peaks [34].

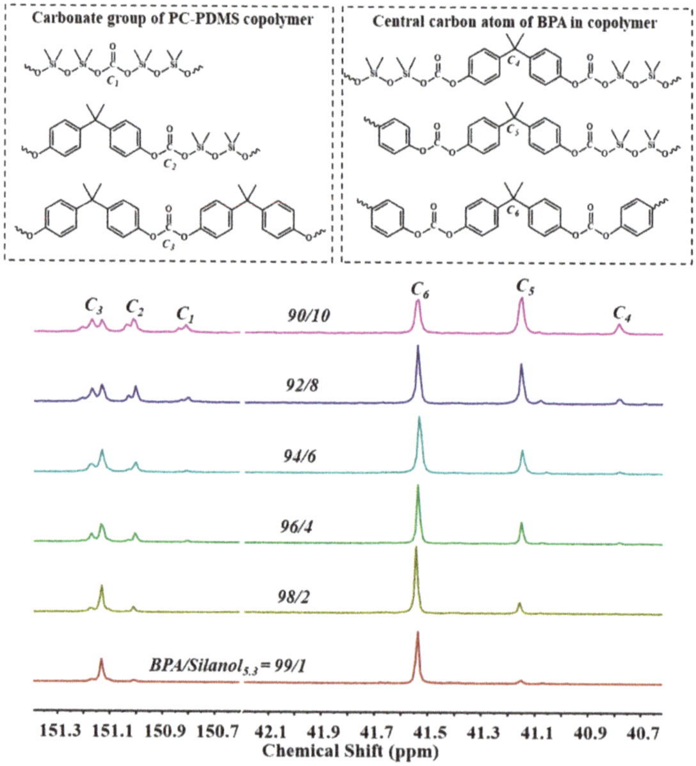

Figure 9. ^{13}C-NMR spectra of PC-PDMS copolymer with different feed ratios of BPA to $Silanol_{5.3}$.

Table 3 summarizes the trends of the sequence distribution of $Silanol_{5.3}$-based copolycarbonates with $Silanol_{5.3}$ feed amount. Table 3 shows that the contents of C_1 and C_2 structures gradually increased with increased $Silanol_{5.3}$ feed amount, whereas the content of C_3 structures gradually decreased. The number-average sequence length of the BPA segment (L_{nBPA}) decreased with increased $Silanol_{5.3}$ feeding, whereas L_{nPDMS} gradually increased. When the molar ratio of BPA/PDMS was 90/10, L_{nPDMS} was equal to 2, meaning that the PDMS segments in the copolymer were primarily in the form of

$Silanol_{5.3}$–O–C(=O)–O–$Silanol_{5.3}$. Changing the feed amount of PDMS affected the distribution of PDMS segments in the copolymer.

Table 3. Microstructure of silanol-terminated PDMS copolycarbonates with different feed ratios of BPA to PDMS.

BPA/PDMS	C_1 [1] (mol%)	C_2 [1] (mol%)	C_3 [1] (mol%)	L_{nBPA} [2]	L_{nPDMS} [2]	B [3]
99/1	-	6.5	93.5	29.9	1	1.03
98/2	-	14.3	85.7	13.0	1	1.08
96/4	4.7	22.4	72.9	7.5	1.4	0.84
94/6	5.0	25.3	69.6	6.5	1.4	0.87
92/8	12.1	29.0	58.9	5.1	1.8	0.74
90/10	16.7	33.5	49.7	4.0	2.0	0.75

[1] Molar content of BPA-BPA, BPA-PDMS, and PDMS-PDMS sequences in the carbonyl carbon region; [2] block length of BPA and PDMS measured by ^{13}C-NMR (Figure S1 in Supplementary Materials); [3] degrees of randomness (B) of copolymers calculated by ^{13}C-NMR (Figure S1).

Table 4 demonstrates the effect of the initial feed ratio of BPA to $Silanol_{5.3}$ on the molecular-structure characteristics of the $Silanol_{5.3}$-based copolycarbonates. BPA and DPC were well miscible at the molecular level, but BPA was less reactive than $Silanol_{5.3}$, and the poor miscibility between $Silanol_{5.3}$ and DPC may need to be improved through transesterification. This phenomenon may affect the performance of the polycondensation product. For instance, the highest M_η of synthesized products was achieved when the molar ratio of DPC to diols was 1:1. With decreased relative content of DPC, the conversion of $Silanol_{5.3}$ was supposed to decrease. However, PDMS conversion increased when the BPA/PDMS feed ratio was 90/10 probably because the short-chain, high-reactivity $Silanol_{5.3}$ fed in higher amounts was more likely to produce the C_1 structure. Consequently, conversion was higher due to the inhibition of cyclic degradation by the adjacent PC segments. The short-chain $Silanol_{5.3}$ was used as a third monomer in the copolymerization with BPA and DPC, and the PDMS conversion decreased with increased PDMS feed amount. With further increased PDMS feed amount, PDMS conversion remained over 65%.

Table 4. Molecular-structure characteristics of silanol copolycarbonates with different feed ratios of BPA and $Silanol_{5.3}$ obtained from ^{1}H-NMR.

BPA/PDMS [1]	DPC/diols [2]	M_η (g/mol)	ΔC (%)	Conversion [3] (%)	T_g [4] (°C)	$T_{d, 5\%}$ [5] (°C)
100/0	1.06	15,900	0.20	-	148.4	424
99/1	1.05	17,500	0.18	85.8	137.7	424
98/2	1.04	18,400	0.22	70.6	131.8	402
96/4	1.02	29,200	0.11	66.0	128.4	400
94/6	1.00	37,800	0.18	68.2	129.5	420
92/8	0.98	26,000	0.31	67.9	109.9	391
90/10	0.96	24,700	0.50	72.6	101.7	385

[1] Based on the initial molar percentages of BPA and PDMS; [2] based on the initial feeding ratios of DPC and diols; [3] calculated by ^{1}H-NMR analysis according to Equation (3); [4] measured by DSC at a heating rate of 10 °C min^{-1} (2nd scan); [5] degradation temperature for 5% weight loss was determined by TGA at a heating rate of 10 °C min^{-1} (with N_2).

3.4. Influence of Chain Length on Conversion

After the investigation of transesterification in Section 3.2, we found differences in the miscibility between DPC and PDMS with different chain lengths, which required improvement through transesterification. The equilibrium transesterification levels of PDMS with different chain lengths were dissimilar, and the residual PDMS oligomers increased the tendency of side reactions to occur in the polycondensation stage. Furthermore, PDMS conversion in the transesterification stage directly determined PDMS conversion in the final products. Therefore, we selected PDMS with different chain lengths as substrates and controlled the same feeding ratio of BPA to PDMS to investigate the effect of the average chain length of PDMS on its conversion.

Figure 10 shows the ^{13}C-NMR spectra of PC-PDMS copolymers prepared from different-chain-length silanols under the same conditions. A longer PDMS block length in copolymers caused the *B* to approach 1, and random copolymers were more easily obtained. Table 5 presents the molecular structure of PC-PDMS copolymers prepared from PDMS with different chain lengths. PDMS conversion significantly decreased when longer-chain PDMS was added (~68% vs. 13%).

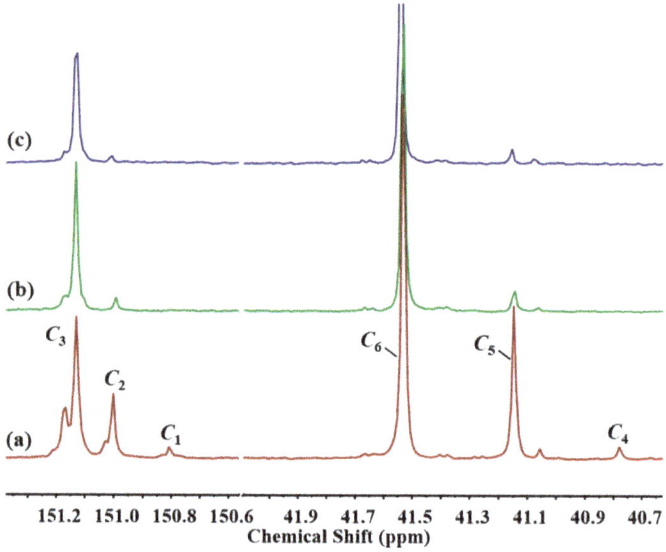

Figure 10. ^{13}C-NMR spectra of PC-PDMS copolymers using (**a**) $Silanol_{5.3}$, (**b**) $Silanol_{22.5}$, and (**c**) $Silanol_{56.2}$ as raw materials with a feeding mole ratio of BPA/PDMS = 94/6.

Table 5. Molecular-structure characteristics of PC-PDMS copolymers with different chain lengths of silanol-terminated PDMS obtained from ^{1}H-NMR and ^{13}C-NMR.

Material Code	BPA/PDMS [1]	PDMS [2] (wt%)	Conversion [2] (%)	L_{nBPA} [3]	L_{nPDMS} [3]	B [3]
$Silanol_{5.3}$	94/6	6.5	68.2	6.5	1.4	0.87
$Silanol_{22.5}$	94/6	6.9	67.7	28.2	1.0	1.04
$Silanol_{56.2}$	94/6	1.4	13.1	50.3	1.0	1.02

[1] Based on the initial molar percentages of BPA and PDMS; [2] determined by ^{1}H-NMR analysis; [3] calculated by ^{13}C-NMR.

To improve the conversion of long-chain PDMS in the polycondensation process and to suppress side reactions such as cyclic degradation, the critical transesterification level of DPC and PDMS must be reached to achieve mutual miscibility. The equilibrium conversion rate of PDMS must also be increased to introduce as much PDMS into the PC oligomers as possible at a lower temperature. Accordingly, we proposed a countermeasure to overcome this problem of low equilibrium conversion of PDMS in the transesterification stage by increasing the amount of KOH used as catalyst (Table 6).

Figure 11 shows the curves of transesterification level of long-chain $Silanol_{56.2}$ and DPC catalyzed by different amounts of KOH as a function of time. With increased KOH amount, the equilibrium transesterification level of long-chain $Silanol_{56.2}$ increased (~49.0% vs. 38.6%), and this idea can be carried over to the preparation of the polycondensation products. Table 7 provides information about the molecular structure of the PC-PDMS copolymers prepared from long-chain $Silanol_{56.2}$ catalyzed by different amounts of KOH. Increased equilibrium conversion at the transesterification stage significantly increased PDMS conversion throughout the polycondensation (90.4% vs. 13.1%).

Table 6. Polymerization of silanol copolycarbonates with different chain lengths of PDMS and BPA as two different diols and DPC as the carbonate source.

Material Code	BPA/PDMS [1]	DPC/diols [2]	M_η [3] (g/mol)	ΔC [4] (%)	T_g [5] (°C)	$T_{d, 5\%}$ [6] (°C)
Pure PC	100/0	1.06	15,900	0.20	148.4	424
$Silanol_{5.3}$	94/6	1.00	37,800	0.18	129.5	420
$Silanol_{22.5}$	94/6	1.00	25,000	0.09	139.7	412
$Silanol_{56.2}$	94/6	1.00	37,000	0.23	150.7	432

[1] Based on the initial feeding ratios of BPA and PDMS; [2] based on the initial molar percentages of DPC and diols; [3] measured by an Ubbelohde viscometer using $CHCl_3$ as the solvent through Equations (4) and (5); [4] determined by UV-*vis* spectrometer through Equation (7); [5] measured by DSC at a heating rate of 10 °C min^{-1} (2nd scan); [6] degradation temperature for 5% weight loss was measured by TGA at a heating rate of 10 °C min^{-1} (with N_2).

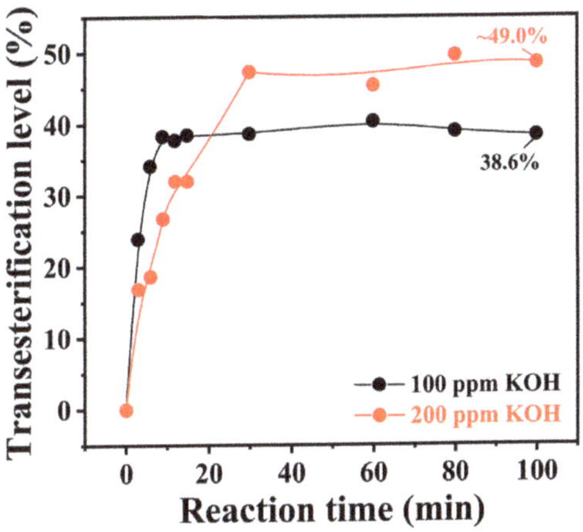

Figure 11. Time dependence of transesterification level of $Silanol_{56.2}$ reacted with DPC under the catalysis of different amounts of KOH at 180 °C.

Table 7. Molecular-structure characteristics of PC-PDMS copolymers prepared by different amounts of KOH obtained from ^{1}H-NMR and ^{13}C-NMR.

Material Code	BPA/PDMS [1]	PDMS [2] (wt%)	Conversion [2] (%)	L_{nBPA} [3]	L_{nPDMS} [3]	B [3]
$Silanol_{56.2}$	94/6	1.4	13.1	50.3	1.0	1.02
$Silanol_{56.2}$ [4]	94/6	9.3	90.4	25.2	1.0	1.04

[1] Based on the initial molar percentages of BPA and PDMS; [2] determined by ^{1}H-NMR analysis; [3] calculated by ^{13}C-NMR; [4] a KOH amount of 200 ppm to diols on a mole basis was used.

4. Conclusions

We investigated the transesterification process between DPC and PDMS with different chain lengths using KOH as catalyst and successfully prepared a series of high-molecular-weight PC-PDMS copolymers with low color difference through melt polycondensation. Transesterification experiments confirmed that with increased chain lengths of PDMS, the difficulty of miscibility between PDMS and DPC increased. Consequently, a higher critical transesterification level was needed to dissolve in DPC than PDMS with lower chain lengths. Furthermore, the conversion rate of $Silanol_{56.2}$ was only 38.6% when the transesterification with DPC reached equilibrium, and unreacted PDMS oligomers were more prone to undergo side reactions in the polycondensation stage. The feed amount of PDMS and its chain length affected the conversion of PDMS throughout melt polycondensation. The short-chain Silanol$_{5.3}$, as the third raw material in the copolymerization with BPA and DPC, was more likely to produce a C_1 structure, and PDMS conversion decreased with increased PDMS fed. With further increased average chain length of PDMS, conversion clearly decreased. With increased KOH amount, the equilibrium transesterification level of Silanol$_{56.2}$ and DPC was promoted, thereby enhancing the conversion rate of long-chain Silanol$_{56.2}$ in the final product (90.4% vs. 13.1%). These results can serve as a reference for the preparation of high-conversion and high-quality PC-PDMS copolymers through melt polycondensation.

Supplementary Materials: The following are available online at https://www.mdpi.com/article/10.3390/polym13162660/s1, Figure S1: Typical ^{13}C-NMR spectrum ($CDCl_3$, 150 MHz) of PC-PDMS copolymer and calculation of L_{nBPA}, L_{nPDMS} and B.

Author Contributions: Writing–Original Draft Preparation, Z.Z.; Writing–Review and Editing, G.W. and Z.Z. Both authors have read and agreed to the published version of the manuscript.

Funding: This research received no external funding.

Institutional Review Board Statement: Not applicable.

Informed Consent Statement: Not applicable.

Data Availability Statement: The data presented in this study are available on request from the corresponding author.

Conflicts of Interest: The authors declare no conflict of interest.

References

1. Bryant, R.G. *Encyclopedia of Polymer Science and Technology*; John Wiley & Sons: New York, NY, USA, 2006.
2. Legrand, D.G.; Bendler, J.T. *Handbook of Polycarbonate Science and Technology*; Marcel Dekker: New York, NY, USA, 1999.
3. Schnell, H. *Chemistry and Physics of Polycarbonates*; Interscience: New York, NY, USA, 1964.
4. Perret, B.; Schartel, B. The effect of different impact modifiers in halogen-free flame retarded polycarbonate blends—II. Fire behaviour. *Polym. Degrad. Stab.* **2009**, *94*, 2194–2203. [CrossRef]
5. Levchik, S.V.; Weil, E.D. Overview of recent developments in the flame retardancy of polycarbonates. *Polym. Int.* **2005**, *54*, 981–998. [CrossRef]
6. Wawrzyn, E.; Schartel, B.; Seefeldt, H.; Karrasch, A.; Jäger, C. What reacts with what in bisphenol A polycarbonate/silicon rubber/bisphenol A bis(diphenyl phosphate) during pyrolysis and fire behavior? *Ind. Eng. Chem. Res.* **2012**, *51*, 1244–1255. [CrossRef]
7. Wolf, M.P.; Salieb-Beugelaar, G.B.; Hunziker, P. PDMS with designer functionalities-Properties, modifications strategies, and applications. *Prog. Polym. Sci.* **2018**, *83*, 97–134. [CrossRef]
8. Kashiwaga, T.; Gilman, J.W.; Grand, A.F.; Wilkie, C.A. Silicon-based flame retardants. In *Fire Retardancy of Polymeric Materials*; CRC Press: New York, NY, USA, 2000.
9. Vaughn, J.H.A. Organopolysiloxane-Polycarbonate Block Copolymers. U.S. Patent 3,189,662, 15 June 1965.
10. Vaughn, J.H.A. Organopolysiloxane Polycarbonate Block Copolymers. U.S. Patent 3,419,634, 31 December 1968.
11. Aoki, Y. Polycarbonate Resin Composition and Moled Article. U.S. Patent 9617422B2, 11 April 2017.
12. Hein, C.L.; Patil, R.; Zhou, B.; Lin, G.Y.; Schmidt, C.; Zoller, D.; Ma, S.L.; Hassman, C. Polycarbonate Copolymers, Articles Formed Therefrom, and Methods of Manufacture. W.O. Patent 2018164706A1, 13 September 2018.
13. Chun, B.K.; Bahn, H.M.; Hwang, Y.Y.; Park, J.J.; Hong, M.H.; Lee, K.J.; Ko, U.; Son, Y.W. Copolycarbonate and Composition Containing Same. W.O. Patent 2016089118A2, 9 June 2016.
14. Kimura, T.; Tando, K.; Oda, K. Polycarbonate-Polydiorganosiloxane Copolymer, Resin Composition of Polycarbonate-Polydiorganosiloxane Copolymer, and Production Method for Resin Composition of Polycarbonate-Polydiorganosiloxane Copolymer. W.O. Patent 2019124556A1, 27 June 2019.
15. Pang, X.; Ge, X.; Ji, J.; Liang, W.; Liu, R.; Chen, X.; Yin, G.; Ge, J. Improving oxygen permeability and thermostability of polycarbonate via copolymerization modification with bio-phenol polysiloxane. *Polymers* **2019**, *11*, 1302. [CrossRef] [PubMed]
16. Ohashi, K.E.; Suzuki, H.; Muraoka, T.; Yoshisato, E.; Nagashima, R. Method of Separating and Recovering Aromatic Carbonate and Production Process. U.S. Patent 20030162989A1, 28 August 2003.
17. Lee, J.N.; Park, C.; Whitesides, G.M. Solvent compatibility of poly (dimethylsiloxane)-based microfluidic devices. *Anal. Chem.* **2003**, *75*, 6544–6554. [CrossRef] [PubMed]
18. Barton, A.F.M. Solubility parameters. *Chem. Rev.* **1975**, *75*, 731–753. [CrossRef]
19. King, J.A., Jr.; McCloskey, P.J.; Davis, G.C. Method for Preparation of Block Copolysiloxanecarbonate. U.S. Patent 5,504,177, 2 April 1996.

20. Koenig, A.; Ebert, W.; Koehler, W. Process for the Preparation of Polysiloxane-Polycarbonate Block Cocondensates. D.E. Patent 19710081A1, 17 September 1998.
21. Zhang, S.M.; Zhang, H.X.; Zhang, W.Y.; Wu, Z.Q.; Chen, F.; Fu, Q. Toughening of polycarbonate through reactive melt blending: Effect of hydroxyl content and viscosity of hydroxyl-terminated polydimethysiloxane. *Chin. J. Polym. Sci.* **2014**, *32*, 823–833. [CrossRef]
22. Zhou, W.J.; Osby, J. Siloxane modification of polycarbonate for superior flow and impact toughness. *Polymer* **2010**, *51*, 1990–1999. [CrossRef]
23. Scott, D.W. Equilibria between linear and cyclic polymers in methylpolysiloxanes. *J. Am. Chem. Soc.* **1946**, *68*, 2294–2298. [CrossRef]
24. Banach, T.E.; Davis, G.C.; Smigelski, P.M.; McCloskey, P.J. Method for Making Siloxane Copolycarbonates. U.S. Patent 6252013B1, 26 June 2001.
25. Konig, A.; Kuhling, S. Process for the Production of Poly(Diorganosiloxane)/Polycarbonate Block Copolymers. U.S. Patent 5,783,651, 21 July 1998.
26. Malmgren-Hansen, B.; Olesen, S.; Pommer, K.; Funch, L.W.; Pedersen, E. *Survey of Chemical Substances in Consumer Products Survey No. 32—2003 Emission and Evalution of Chemical Substances from Selected Electrical and Electronic Products*; The Danish Environmental Protection Agency: Copenhagen, DA, USA, 2003.
27. Woodburm, K.; Drottar, K.; Domoradzki, J.; Durham, J.; Mcnett, D.; Jezowski, R. Determination of the dietary biomagnification of octamethylcyclotetrasiloxane and decamethylcyclopentasiloxane with the rainbow trout (*Oncorhynchus mykiss*). *Chemosphere* **2013**, *93*, 779–788. [CrossRef] [PubMed]
28. Yilgor, I.; Yilgor, E. Thermal stabilities of end groups in hydroxyalkyl terminated polydimethylsiloxane oligomers. *Polym. Bull.* **1998**, *40*, 525–532. [CrossRef]
29. He, M.J.; Chen, W.X.; Dong, X.X. *Polymer Physics*, 3rd ed.; Fundan University Press: Shanghai, China, 2007.
30. Cezard, C.; Rice, C.A.; Suhm, M.A. OH-stretching red shifts in bulky hydrogen-bonded alcohols: Jet spectroscopy and modeling. *J. Phys. Chem. A* **2006**, *110*, 9839–9848. [CrossRef] [PubMed]
31. Vallejo-Montesinos, J.; Villegas, A.; Jacobo-Azuara, A.; Martínez, J.M.; Ramírez, E.; Cervantes, J. Synthesis and Properties in Solution of Gaussian Homo Asymmetric Polysiloxanes with a Bulky Side Group. *J. Inorg. Organomet. P.* **2012**, *22*, 1332–1340. [CrossRef]
32. Demirci, A.; Yamamoto, S.; Matsui, J.; Miyashita, T.; Mitsuishi, M. Facile synthesis of cyclosiloxane-based polymers for hybrid film formation. *Polym. Chem.* **2015**, *6*, 2695–2706. [CrossRef]
33. Williams, E.A.; Cargioli, J.D.; Hobbs, S.Y. The ^{13}C and ^{29}Si nuclear magnetic resonance analysis of bisphenol A polycarbonate-polydimethylsiloxane block copolymers. *Macromolecules* **1977**, *10*, 782–785. [CrossRef]
34. Yamadera, R.; Murano, M. The determination of randomness in copolyesters by high resolution nuclear magnetic resonance. *J. Polym. Sci. Part A Polym. Chem.* **1967**, *5*, 2259–2268. [CrossRef]

polymers

MDPI

Article

Facile Synthesis of Thermoplastic Polyamide Elastomers Based on Amorphous Polyetheramine with Damping Performance

Jie Jiang [1], Qiuyu Tang [1], Xun Pan [2], Jinjin Li [1], Ling Zhao [1,3], Zhenhao Xi [1,*] and Weikang Yuan [1]

1 State Key Laboratory of Chemical Engineering, School of Chemical Engineering, East China University of Science and Technology, Shanghai 200237, China; jjiangecust@gmail.com (J.J.); qiuyu.tang@foxmail.com (Q.T.); lijinjin@ecust.edu.cn (J.L.); zhaoling@ecust.edu.cn (L.Z.); wkyuan@ecust.edu.cn (W.Y.)

2 Flinders Institute for Nanoscale Science and Technology, Flinders University, Sturt Road, Bedford Park, SA 5042, Australia; caroline.pan@flinders.edu.au

3 College of Chemistry and Chemical Engineering, Xinjiang University, Urumqi 830046, China

* Correspondence: zhhxi@ecust.edu.cn

Abstract: Novel thermoplastic polyamide elastomers (TPAEs) consisting of long-chain semicrystalline polyamide 1212 (PA1212) and amorphous polyetheramine were synthesized via one-pot melt polycondensation. The method provides accessible routes to prepare TPAEs with a high tolerance of compatibility between polyamide and polyether oligomers compared with the traditional two-step method. These TPAEs with 10 wt % to 76 wt % of soft content were obtained by reaction of dodecanedioic acid, 1,12-dodecanediamine, and poly(propylene glycol) (PPG) diamine. The structure–property relationships of TPAEs were systematically studied. The chemical structure and the morphologic analyses have revealed that microphase separation occurs in the amorphous region. The TPAEs that have long-chain PPG segments consist of a crystalline polyamide domain, amorphous polyamide-rich domain, and amorphous polyetheramine-rich domain, while the ones containing short-chain PPG segments comprise of a crystalline polyamide domain and miscible amorphous polyamide phase and amorphous polyetheramine phase due to the compatibility between short-chain polyetheramine and amorphous polyamide. These novel TPAEs show good damping performance at low temperature, especially the TPAEs that incorporated 76 wt % and 62 wt % of PPG diamine. The TPAEs exhibit high elastic properties and low residual strain at room temperature. They are lightweight with density between 1.01 and 1.03 g/cm^3. The long-chain TPAEs have well-balanced properties of low density, high elastic return, and high shock-absorbing ability. This work provides a route to expand TPAEs to damping materials with special application for sports equipment used in extremely cold conditions such as ski boots.

Keywords: one-pot method; polyamide; elastomer; PPG diamine; damping property

Citation: Jiang, J.; Tang, Q.; Pan, X.; Li, J.; Zhao, L.; Xi, Z.; Yuan, W. Facile Synthesis of Thermoplastic Polyamide Elastomers Based on Amorphous Polyetheramine with Damping Performance. *Polymers* **2021**, *13*, 2645. https://doi.org/10.3390/polym13162645

Academic Editor: Angels Serra

Received: 7 July 2021
Accepted: 4 August 2021
Published: 9 August 2021

Publisher's Note: MDPI stays neutral with regard to jurisdictional claims in published maps and institutional affiliations.

Copyright: © 2021 by the authors. Licensee MDPI, Basel, Switzerland. This article is an open access article distributed under the terms and conditions of the Creative Commons Attribution (CC BY) license (https://creativecommons.org/licenses/by/4.0/).

1. Introduction

Thermoplastic elastomers (TPEs) are usually multiblock segmented copolymers consisting of thermodynamic incompatible hard blocks and soft blocks, normally leading to a microphase separated structure. The soft blocks are often polyether or aliphatic polyester with low glass transition temperature (T_g), enabling elastomeric properties to the copolymers. Moreover, the hard blocks containing crystalline units form a physical cross-linked site. Chemical covalent links between the two blocks can avoid macro-phase separation. These materials bridge the gap between thermoplastics and elastomers. Thermoplastic polyurethanes (TPUs), thermoplastic polyether ester elastomers (TPEEs), and thermoplastic polyamide elastomers (TPAEs) are the three main types of segmented TPEs. TPEs have received much industrial and commercial interest duo to their good thermomechanical properties, excellent chemical resistance, wide service temperature range, and ease of processing.

TPAEs have been recently developed as a relatively new family of engineering plastics because of their highly efficient energy return, good performance at low temperature, and selective gas permeability. They are widely used in sports equipment, such as ski boots, cleated shoes, and running shoes. Generally, TPAEs are segmented poly(ether-*block*-amide) copolymers composed of polyether as soft segments and polyamide as hard segments. The performance of the material depends not only on the ratio of soft and hard segments but also the nature of a soft block when using a certain polyamide. Currently, the main soft blocks for TPAEs are PEG [1] and PTMG, enabling antistatic property and good flexibility, respectively. Commercially available TPAEs have been mainly developed by Arkema and Evonik under the trade name of PEBAX and VESTAMID, respectively. The copolymers are synthesized by the polycondensation of carboxyl terminated polyamide (PA6, PA11, or PA12 oligomers) and hydroxyl terminated polyether (PTMG or PEG), resulting in ester linkage [2,3], which is the so-called "two-step method". The structure–property relationships of TPAEs have been extensively studied [4–10]. For example, TPAEs consisting of PTMG has been mostly used for footwear [11–14] and gas separation [15,16], while those containing PEG as a flexible phase are employed to selectively separate CO_2 [17–20] and permanent antistatic packaging [21]. It should be noted that PTMG and PEG are a semicrystalline C4 building block and C2 building block, respectively. The C3 building block has been hardly used in the preparation of TPAEs. In our previous work, a linear semicrystalline poly(trimethylene glycol) (PPDO) has been utilized to prepare TPAEs with high elasticity [22]. Nevertheless, for these semicrystalline blocks, the melting of the corresponding segments will cause the variation of storage modulus, which is not conducive to the use of materials in this temperature range [2]. To enrich the TPAE family, poly(propylene glycol) (PPG, C3) is an alternative amorphous polyether with a methyl side group on each repeat unit, which has weak dispersion and a low dipole moment, resulting in the superior elasticity to polyurethane elastomers [23,24]. However, the reactivity of secondary hydroxyl groups in PPG is relatively low, and its poor compatibility with polyamide oligomers prevents the formation of high molecular weight copolymers. Considering the condensation of hydroxyl or amino groups with carboxyl groups, PPG diamine (trade name Jeffamine D) is an alternative C3 building block. It is a non-crystalline aliphatic polyetheramine with low viscosity and low tendency to form a hydrogen bond, which is generally used as a gasoline additive, curing agents of epoxy resins [25,26], unit of epoxy-imine vitrimer [27], precursor of main-chain polybenzoxazines resin [28], or soft segments of Pus [29], providing high toughness, thermal stability, or flexibility. For PPG-based TPAEs, monomer casting PA6 modified with maximum 6 wt % of this polyetheramine shows an obvious decrease in crystallization ability and an elastic deformation behavior [30]. Grymans et al. selected PA46 salt and PPG diamine to prepare PA46-PPG segmented copolymers via condensation followed by solid-phase polycondensation [31]. However, severe macro-phase separation occurs despite adding *m*-cresol as solvent. The solid products are inhomogeneous after post-polymerization when using mole mass 2000 of Jeffamine D. Jo et al. synthesized PA6-PPG copolymers with soft content from 10 wt % to 56 wt % by melt copolymerization [32–34], whose microphase separation structure was systematically investigated. Shibasaki et al. reported a monodisperse poly(N-methyl benzamide) copolymerized with PPG diamine in solution to afford non-hydrogen-bondable block copolymers, which are non-crystalline materials but showing weak tensile modulus [35]. Luo et al. prepared fluorene-based PA-PPG membranes via solution polycondensation with high carbon dioxide permeability [36,37].

It is noteworthy that for the sporting applications of TPAEs, the attenuation of vibration is an important property. The methyl side group in the PPG backbone allows strong energy dissipation by internal friction, enabling the shock-absorbing capacity of the TPAEs when using them as potential damping materials, which has not been extensively investigated. PA1212 is an important semicrystalline engineering plastic, which owns high toughness and strength, low temperature resistance, and low shrinkage [38]. It has close properties with PA12, and the polymerization of PA1212 does not require high temperature and high pressure of PA12, which is mostly used to produce commercial PEBAX. Furthermore, it is not easy to prepare the high molecular weight of TPAEs using the "two-step method" due to the incompatibility between polyamide and polyether oligomer and the difficulties in adjusting the stoichiometric balance of reactive groups [39]. In this work, one-pot melt polycondensation was proposed to firstly synthesize a series of segmented TPAEs with variable compositions of PA1212 block (C12) and PPG block (C3). The TPAEs were prepared directly from monomers of PA1212 and PPG diamine. This procedure is feasible with the advantages of low starting viscosity, ease of stoichiometric balance of acid and amine groups, and high tolerance of compatibility between polyamide and polyether compared with the two-step method. This one-pot strategy is its simplicity, high efficiency, and low cost, especially using commercial PPG diamine (C3) to manufacture advanced materials. The chemical structures of novel TPAEs were studied using nuclear magnetic resonance (NMR) and Fourier transform infrared spectroscopy (FT-IR). The microphase-separated structures were characterized by differential scanning calorimetry (DSC), dynamic thermomechanical analysis (DMA), and X-ray diffraction and confirmed by microscopy technics. Additionally, tensile tests and cyclic tensile tests were performed to evaluate the mechanical properties of the series of new copolymers, indicating that the synthesized TPAEs have high elasticity. The novel TPAEs have the damping properties at low temperature, especially those with high flexible content due to the incorporation of the amorphous PPG segment. The effects of content and length of the PPG segment on microphase separated morphology were systematically investigated. Thermal gravimetric analysis (TGA) was applied to find that the TPAEs have good thermal stability. These novel family of TPAEs are believed to have great potential for low-temperature applications.

2. Experiment Section

2.1. Materials

Dodecanedioic acid (≥99%), deuterated trifluoroacetic acid (TFA-d, 99.5 atom %D), *m*-cresol (≥99%), and 1,1,1,3,3,3-Hexafluoro-2-propanol (HFIP, 99%) were purchased from Shanghai Titan Science and Technical Company (Shanghai, China). Polyetheramine JeffamineD400 (Mn ≈ 400 g·mol^{-1}) and JeffamineD2000 (Mn ≈ 2000 g·mol^{-1}) (PPG diamine) were purchased from Shanghai Aladding Biochemical Technology Co., Ltd. (Shanghai, China), and 1,12-dodecanediamine (≥99%) was supplied by Zibo Guangtong Chemical Company (Shandong, China). Sodium hypophosphite monohydrate ($NaH_2PO_2 \cdot H_2O$, ≥98%) was purchased from Sinopharm Chemical Reagent Co., Ltd. (Shanghai, China). All the reagents were used as received without further purification.

2.2. Characterization

2.2.1. Intrinsic Viscosity

The intrinsic viscosity $[\eta]$ of TPAEs was determined by an Ubbelohde viscometer using *m*-cresol as solvent with a concentration of 0.5 g·dL^{-1} at 25 ± 0.01 °C. $[\eta]$ was calculated by the following Solomon—Ciuta equation of a single-point method (Equation (1)).

$$[\eta] = \sqrt{2(\eta_{sp} - ln\eta_r)}/c \tag{1}$$

where η_{sp} is specific viscosity, η_r is relative viscosity, and c is the TPAE concentration (g·dL^{-1}).

2.2.2. Density Measurement

The density of samples was determined using a water displacement method according to *ASTM D792-13*; the density (ρ) was calculated by Equation (2):

$$[\rho] = \left(\frac{a}{a-b}\right)\rho_{water} \tag{2}$$

where a is the apparent mass of sample in air, b is the apparent mass of sample completely immersed in water, and ρ_{water} is the density of water.

2.2.3. Nuclear Magnetic Resonance (NMR)

The TPAEs samples (15 mg) were dissolved in 0.5 mL of deuterated trifluoroacetic acid (TFA-d). ^{1}H NMR and 2D NMR spectra were recorded on an AVANCE-600 spectrometer (Bruker Ascend, 600 MHz, Germany) at ambient temperature using TFA-d as solvent. The chemical shift of the solvent is 11.6 ppm. Two-dimensional (2D) correlation spectroscopy (^{1}H-^{1}H COSY) using the standard Bruker pulse program cosygpppqf with the following parameters: spectral width SW1 = SW2 = 11,904.7 Hz, acquisition time 0.086 s, relaxation delay 2.0 s; processing, SI = 1024 (f2, f1), WDW = QSINE.

2.2.4. Fourier Transform Infrared Spectroscopy (FT-IR)

The films of TPAE were analyzed on a Nicolet 6700 spectrometer (Thermo Fisher Scientific, USA) equipped with ZnSe crystal using an attenuated total reflectance (ATR) model. Films were prepared by drop-casting from HFIP and kept at room temperature for 48 h followed by drying in vacuum at 60 °C for another 48 h to remove the solvent. The spectra were recorded from 4000 to 650 cm^{-1} with a spectral resolution of 4 cm^{-1}.

2.2.5. Gel Permeation Chromatography (GPC)

GPC was performed using HFIP as eluent at 40 °C on an Agilent PL-GPC50 equipped with a differential refractive index detector. Two PLHFIPgel columns (300 × 7.5 mm, 9 µm) were used. The sample concentration and eluent flow rate were 3 mg·mL^{-1} and 1.0 mL·min^{-1}, respectively. Calibration of the measurements was carried out with poly(methyl methacrylate) standards.

2.2.6. Differential Scanning Calorimetry (DSC)

Thermal properties of TPAEs were characterized on a Q2000 DSC apparatus (TA instrument, USA). The instrument was calibrated with indium before measurements. A sample was heated to 200 °C at a rate of 10 °C·min^{-1} and held for 3 min under nitrogen atmosphere to eliminate any thermal history. Then, the sample was cooled to −80 °C and reheated to 200 °C at a cooling/heating rate of 10 °C·min^{-1}. The second endothermal curve was recorded to analyze the thermal behavior. The X_c crystalline fraction referring to the amount of PA1212 was calculated according to Equation (3):

$$X_c = \Delta H_m / (w_{PA} \Delta H_m^0) \quad (3)$$

where the ideal enthalpy of fusion of pure PA1212 ΔH_m^0 = 292.2 J/g [40].

2.2.7. Wide-Angle X-ray Diffraction (WAXD)

The WAXD measurements were implemented on an X-ray diffractometer (Bruker D8 Advance, Germany) equipped with a Cu K$_\alpha$ radiation source (wavelength λ = 0.1542 nm) at ambient temperature. The diffraction patterns were recorded in the range of 2θ from 5° to 80°. Samples with about 0.5 mm of thickness were prepared by casting from HFIP.

2.2.8. Small-Angle X-ray Scattering (SAXS)

Films of TPAEs casting from HFIP solution with about 0.5 mm in thickness were used for the tests. Samples were subsequently annealed under vacuum at 80 °C for 12 h to avoid the influence of process history on the morphology. SAXS profiles were obtained from a SAXSess mc2 apparatus (Anton Paar, Austria) equipped with a Cu K$_\alpha$ monochromatic radiation (wavelength λ = 0.15406 nm). The measurements were operated at 40 kV and 30 mA at ambient temperature. The scattering intensity (I) was obtained as a function of scattering vector (q).

2.2.9. Atomic Force Microscopy (AFM)

AFM (Bruker ICON, Germany) was used to investigate the morphology of TPAEs in tapping mode with an Al reflective coated silicon probe (RTESPA, cantilever:

T × L × W = 3.4 × 125 × 40 μm^3, 300 kHz, 40 N·m^{-1}). NanoScope Analysis software was employed for the data processing. Samples were dissolved dimethylacetamide (DMAc) to acquire 0.2 wt % of polymer solution. Thin films were obtained by drop casting on the silicon wafer followed by carefully evaporating the solvent above 10 °C of the crystalline temperature of the samples. Then, all the wafers were immediately isothermal treated for 20 min.

2.2.10. Transmission Electron Microscopy (TEM)

The morphology observation of TPAEs samples was performed on a transmission electron microscope (JEM-1400, Japan); 0.5 wt % sample solutions were prepared by dissolving the TPAEs in DMAc. Thin films were fabricated by dropping the solution on the copper mesh followed by carefully evaporating the solvent above 10 °C of the crystalline temperature of the samples. Then, all the samples were immediately isothermal treated for 20 min.

2.2.11. Dynamic Thermomechanical Analysis (DMA)

Dynamic thermomechanical analysis was conducted on a DMA Q800 (TA Instruments, USA) equipped with a liquid nitrogen cooling apparatus. The specimens with a rectangular geometry (30 × 5 × 0.5 mm^3) were prepared by casting from HFIP solution. The measurements were performed in the tension mode at a heating rate of 5 °C·min^{-1} from −100 to 100 °C in nitrogen atmosphere at a frequency of 1 Hz. The storage modulus (E') and loss factor ($tan\ \delta$) were recorded as a function of temperature for each sample. Furthermore, the temperature of the $tan\ \delta$ peak is used to define the glass transition temperature (T_g) of each sample.

2.2.12. Mechanical Properties

Mechanical testing was performed using dumbbell-shaped samples (35 × 2 × 0.5 mm^3), according to the guideline of ISO 527-2), which were prepared by casting from HFIP and cut by a mold. The tensile testing was conducted on an Instron 3367 (USA) universal testing machine at a speed of 50 mm·min^{-1} at room temperature. Cyclic tensile tests were conducted on a CMT6103 testing machine (MTS, USA). The test specimens were repeatedly exposed to consecutive cycles to a constant strain of 100% (or 200%) with a constant speed of 50 mm·min^{-1}. The recovery was measured by observing the residual strain after 10 cycles.

2.3. Synthesis of TPAEs

A typical one-pot synthesis procedure is described as follows. To a 250 mL three-necked flask equipped with a mechanical stirrer, an argon inlet, and a condenser, dodecanedioic acid (17.27 g, 0.075 mol), 1,12-dodecanediamine (7.52 g, 0.0375 mol), Jeffamine D2000 (75 g, 0.0375 mol), and sodium hypophosphite monohydrate (1 wt %) were added. The mixture was slowly heated to 220 °C and kept for 2.5 h in argon flow at a stirring speed of 200 rpm until no water condensation was observed. Afterwards, the vacuum was gradually applied to 100 Pa within 30 min. After 2 h of reaction, the reaction temperature was raised to 240 °C and the vacuum was kept constant at 80–100 Pa for another hour. The resulting product was poured into ice water and dried under reduced pressure.

3. Results and Discussion

3.1. Synthesis and Structure Characterization of TPAEs

The one-pot polycondensation for novel TPAEs was performed using monomers of PA1212 and polyetheramine in the presence of sodium hypophosphite monohydrate, as shown in Scheme 1. The molar amount of DA is equal to the sum of DN and PPG diamine to maintain the stoichiometric balance of functional groups. Dodecanedioic acid (DA) can react with both 1,12-dodecanediamine (DN) and PPG diamine, and the reaction of DA and DN produces the PA1212 segment. Practically, the equimolar of DA and DN forms PA1212 and the excess DA limits the molar weight of the PA1212 oligomer on one hand and links the PPG diamine segment on the other hand. These TPAEs can also be considered as copolyamides due to the amide linkage between polyamide and polyether, thus conferring more resistance to hydrolysis than the most reported TPAEs containing ester linkages. Six TPAEs with varied soft contents were synthesized by adjusting the feed ratio of dodecanedioic acid, 1,12-dodecanediamine, and polyetheramine. The polyetheramine used are long-chain Jeffamine D2000 and short-chain Jeffamine D400 with mole masses of approximately 2000 g·mol^{-1} and 400 g·mol^{-1}, respectively, and the polyamide has a regular-length segment. The sample code was expressed as TPAE-x, where x corresponds to the soft content determined according to the ^{1}H NMR spectra.

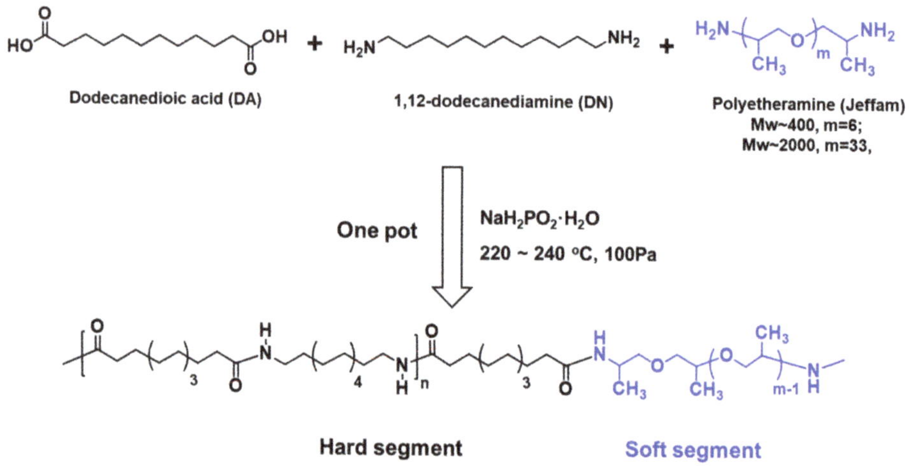

Scheme 1. One-pot synthesis process of TPAEs based on PA1212 and polyetheramine.

The chemical structures of the TPAEs are confirmed by NMR and FT-IR. Figure 1a,b and Figure 2 depict the ^{1}H NMR, ^{13}C NMR, and ^{1}H-^{1}H COSY spectra of TPAE-0.76, respectively. Each peak is assigned to the corresponding hydrogen and carbon in TPAE. The characteristic peaks at δ = 2.75 ppm (H^1) and δ = 3.60 ppm (H^2) are attributed to CH_2 proton next to the carboxyl and nitrogen atom respectively, indicating that the PA1212 segment has formed. The correspondingly CH_2 carbons are found at δ = 33 ppm (C^1) and δ = 42 ppm in the ^{13}C spectrum, respectively. It is noteworthy that the appearance of resonance at δ = 4.39 ppm (H^5) confirms that polyetheramine has been bonded to the PA1212 segment, which agrees with the results of PA6-based copolyetheramide [32]. The corresponding CH carbon appears at δ = 50 ppm. The peaks of CH_2 proton (C^7, C^8, C^9) in the aliphatic chain of PA1212 emerge at 1.31–1.81 ppm. The strong signal at 1.31 ppm is attributed to side CH_3 (C^6) of polyetheramine. The peaks at 3.82 ppm and 3.99 ppm belong to CH_2 and CH of the polyetheramine backbone, respectively. The corresponding carbon resonance can be found in a ^{13}C NMR spectrum. Thus, the soft content can be calculated according to peak areas of H^5, H^2, and H^3 in ^{1}H NMR [5,6], and the mass ratios of soft segment are listed in Table 1. The exact compositions of copolymers are close to theoretical calculation based on the feeding ratio, indicating the synthesized copolymers with expected structure. The slight decreases of soft contents calculated by NMR are because the hydrophilic polyetheramine was brought out with water vapor. It is noted that the peak at 180 ppm suggests that there are more residual carboxyl groups in TPAE than amine groups. ^{1}H-^{1}H COSY can demonstrate coupling relations of protons on adjacent carbon atoms, presenting cross dots on the spectrum. The cross-signals found in the COSY spectrum prove the presence of the protons in the copolymer backbone. The polyetheramine segment has been successfully bonded to the PA1212 segment confirmed by the cross-peaks at 5/6 and 5/3.

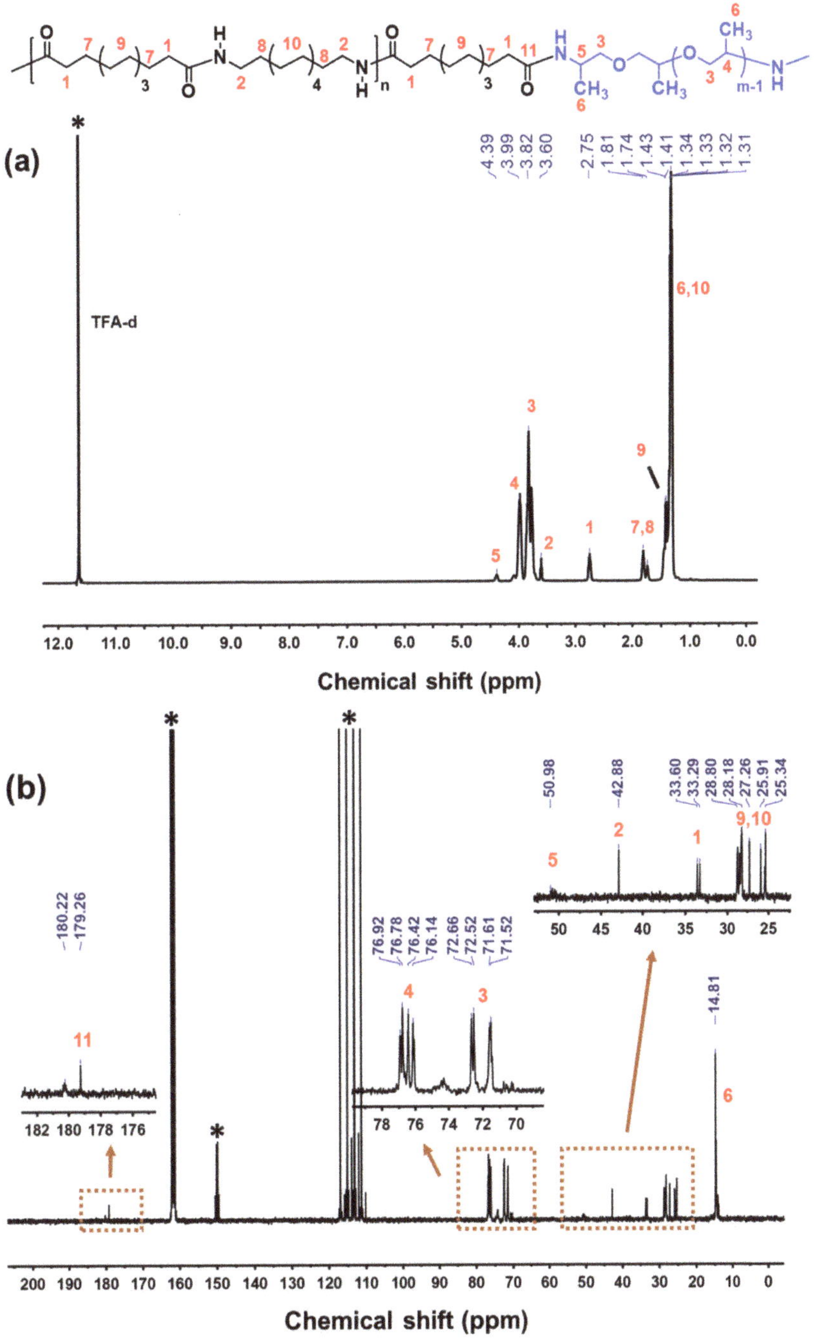

Figure 1. ^{1}H NMR (**a**) and ^{13}C NMR (**b**) spectrum of TPAE-0.76 (* Solvent: TFA-d).

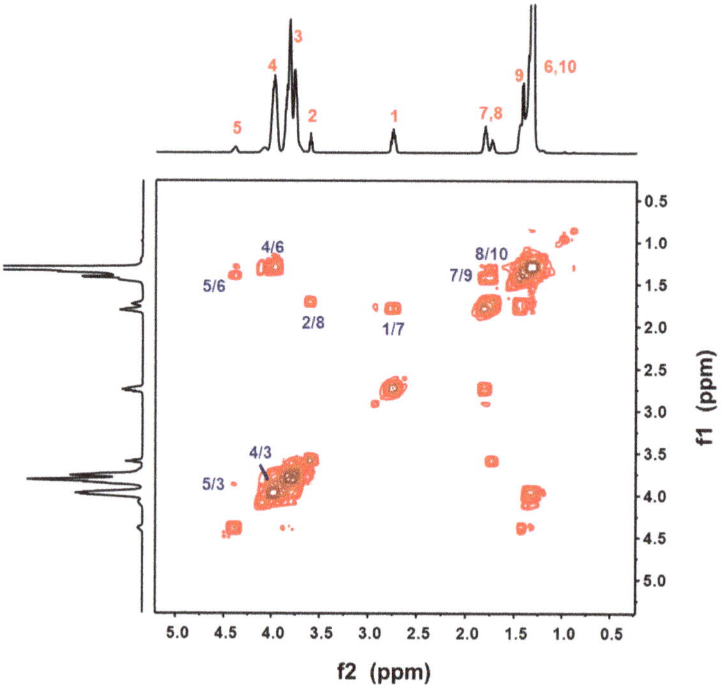

Figure 2. ^1H-^1H COSY spectrum of TPAE-0.76.

Table 1. Compositions, GPC results, [η], density of TPAEs.

Sample Code	Feed Ratio of DA/DN/Jeffam (mol/mol)	MW of Jeffam (g/mol)	Calculated MW of PA1212 Segment (g/mol)	[η] (dL/g)	Density (g/cm^3)	Soft Content in feed	Soft Content by NMR	M_n [a] (×10^4) (g/mol)	M_w [a] (×10^4) (g/mol)	PDI [a]
TPAE-0.76	2/1/1	2000	624	1.26	1.01	0.77	0.76	2.34	5.48	2.34
TPAE-0.62	3/2/1	2000	1018	1.39	1.01	0.66	0.62	3.02	7.87	2.61
TPAE-0.43	8/7/1	2000	2988	1.36	1.02	0.45	0.43	3.82	8.83	2.30
TPAE-0.39	2/1/1	400	624	1.32	1.03	0.40	0.39	3.55	7.56	2.13
TPAE-0.25	3/2/1	400	1018	1.43	1.02	0.27	0.25	4.50	11.09	2.47
TPAE-0.10	8/7/1	400	2988	1.17	1.02	0.12	0.10	3.75	9.22	2.46

[a] Determined by GPC.

The structure analysis of TPAEs by FT-IR is presented in Figure 3. The characteristic peaks at 3308 cm^{-1} (amide A), 1634 cm^{-1} (amide I), and 1536 cm^{-1} (amide II) are attributed to the PA1212 segment [41]. The incorporation of the polyetheramine segment to the PA1212 backbone is confirmed by the distinct absorption band at 1093 cm^{-1} (C-O-C stretching vibration) that became more pronounced with increased soft content, which is ascribed to the polyetheramine segment [41]. The peaks at 2850–2970 cm^{-1} are assigned to the stretching vibration of -C-H-.

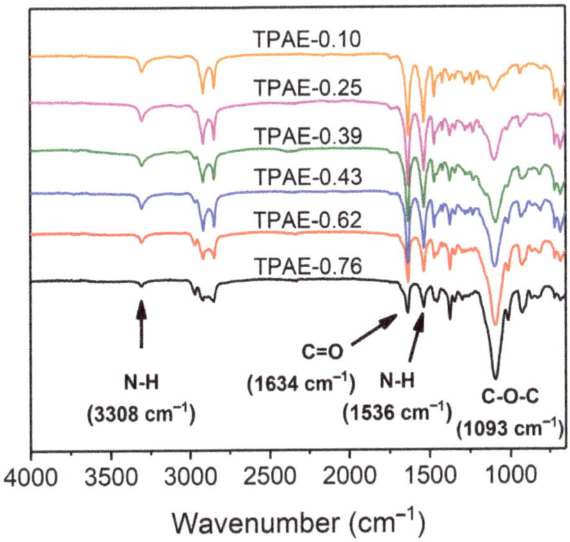

Figure 3. FT-IR spectra of TPAEs.

3.2. *Thermal Characterization*

To gain a deep insight into the phase separation behavior and the crystal structure of the copolymers with the different block length of hard and soft segments, the endothermic behavior of different TPAEs was recorded by DSC testing and illustrated in Figure 4. The corresponding characteristic parameters are listed in Table 2.

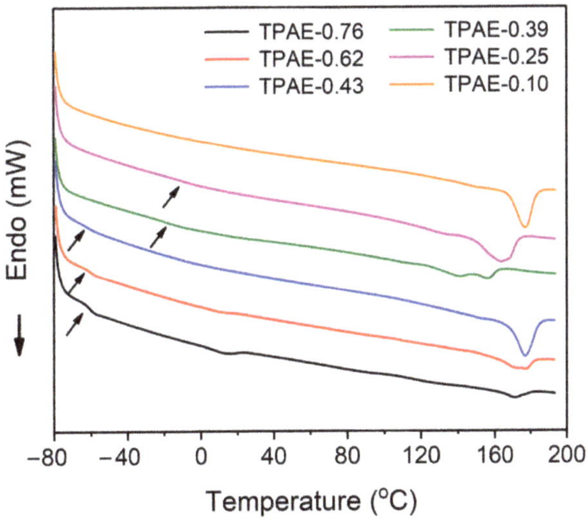

Figure 4. DSC thermograms of different TPAEs.

Table 2. DSC and DMA results of TPAEs.

Sample Code	T_m (°C)	ΔH_m J/g	T_c (°C)	ΔH_c J/g	X_c (%)	T_g (DSC) (°C)	T_g (DMA) (°C)	E' at −25 °C (MPa)	E' at 25 °C (MPa)
TPAE-0.76	170.8	6.65	133.7	6.88	9.5	−60.3	−56.4	54.0	16.3
TPAE-0.62	177.4	16.06	137.9	16.22	14.5	−60.1	−53.6	77.4	30.0
TPAE-0.43	176.9	37.65	147.1	36.67	22.6	−60.8	−51.8	414.4	168.7
TPAE-0.39	156.2	35.00	111.2	35.37	19.6	−19.5	−15.7	2378.1	213.8
TPAE-0.25	163.3	43.48	127.9	43.46	19.8	−10.2	−9.2	2008.5	306.7
TPAE-0.10	176.8	54.03	140.6	55.70	20.5	N.D.	N.D.	1154.1	436.5

N.D. means not determined.

First, the polyetheramine is completely amorphous, as confirmed by the DSC scan (Figure S1, Supplementary Material). It can be observed that each copolymer sample has only one endothermal peak during the heating scan, which is ascribed to the melting behavior of the semicrystalline polyamide segment. This is much different from the double crystalline TPAEs containing PTMG, PEG, or PPDO soft segments with two distinct endothermal peaks. The amorphous nature of PPG may endow high flexibility and elasticity. Furthermore, the copolymers having long-chain Jeffamine D2000 exhibit similar T_g in spite of different compositions (TPAE-0.76, TPAE-0.62, and TPAE-0.43), which suggests that microphase separation occurs in the amorphous region. The T_g values of the three samples are close to −69.5 °C of the pristine soft block (Figure S1, Supplementary Material). Differently, the copolymers containing short-chain Jeffamine D400 show increasing T_g values with increasing polyamide content, which indicates that the flexible polyetheramine segments are compatible with the rigid polyamide segments in the amorphous region [33] (TPAE-0.39 and TPAE-0.25). However, in sample TPAE-0.10, the T_g is not detected, suggesting the high miscibility between polyetheramine and polyamide.

For containing different chain lengths and content of polyetheramine, the melting temperature of TPAEs increases obviously as the hard content increases. It can be found that the segmented polyamide has a lower melting temperature than homo PA1212 (184.2 °C). The melting temperature of copolymers containing short-chain Jeffamine D400 show a relatively large difference to homo PA1212. It is noted that TPAE-0.43 and TPAE-0.39 have the similar hard content but distinct melting points. This is because short-chain Jeffamine D400 is miscible with polyamide and restricts the growth of polyamide crystals, resulting in a reduction of crystalline degree (from 22.6% to 19.6%). Therefore, it is an effective strategy to decrease the processing temperature of the elastomer by using short chains of polyether. Correspondingly, the crystallization temperature exhibits the same trend as melting temperature (Figure S2, Supplementary Material), which also indicates that soft blocks are compatible to the hard blocks in the amorphous region and that short chain polyetheramine is miscible in this amorphous region. Otherwise, the multiple melting endotherms are caused by (1) the melting of crystallites with different lamellae thickness, (2) the melting of different crystal forms, or (3) the crystallization and remelting of imperfect crystallites [33].

3.3. Morphological Characterizations

The wide-angle X-ray diffraction of six TPAEs with different compositions are shown in Figure 5. The copolymer films with non-crystalline soft segments show two main diffraction peaks at around $2\theta = 20.1°$ and $24.0°$ that are attributed to the (100) crystal plane and (010)/(110) crystal plane [40,41], suggesting that the films adopt a triclinic α-crystal phase. The corresponding d spacing values are 0.44 nm and 0.37 nm, which represent the inter-chain distance within the hydrogen bonded-sheet and the inter-sheet distance, respectively [42]. The microphase separation takes place in the amorphous region of novel TPAEs.

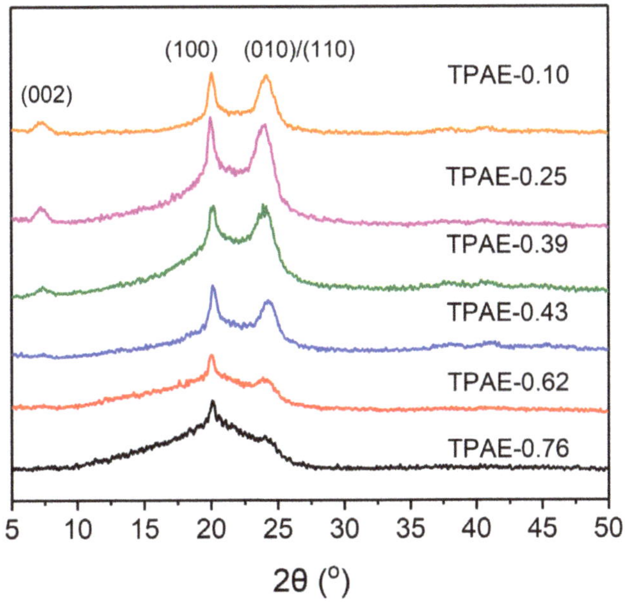

Figure 5. WAXD patterns of the TPAEs at ambient temperature.

Small-angle X-ray scattering was conducted to further explore the microphase separated structure of the TPAEs. Figure 6a presents the Lorentz corrected SAXS profiles (Iq^2 vs. q) of the TPAEs with different compositions. According to Bragg's law, the corresponding long period (L) can be calculated based on the maximum value of the scattering vector (q_{max}), $L = 2\pi/q_{max}$, and the results are illustrated in Figure 6b. Only one broad scattering peak appears in all the TPAEs samples in Figure 6a. The peak indicates the existence of a periodic structure. The intensity peaks of copolymers with Jeffamine D400 (TPAE-0.39, TPAE-0.25, and TPAE-0.10) are similarly broad, while the ones with Jeffamine D2000 (TPAE-0.76, TPAE-0.62, and TPAE-0.43) are broadened as the soft content increases, suggesting the existence of a long periodic structure in copolymers with Jeffamine D2000, and the structure disappears as soft content increases. Furthermore, due to the non-crystallization of soft segments, the scattering peaks are attributed to the repeat distance between the polyamide crystalline domains. The L values decrease from 16.03 nm of TPAE-0.76 to 11.26 nm of TPAE-0.10 with increasing soft content and stay close when copolymers have short-chain Jeffamine D400. The reduction of L in value is owing to the crystallization of polyamide [43,44]. It is proof that short-chain polyetheramine is miscible with polyamide amorphous domain.

The surface topographical characteristic of the TPAEs with different compositions was investigated by AFM at ambient temperature, and the height images are shown in Figure 7. The AFM height image of homo PA1212 is included for comparison. The copolymers display well-ordered crystalline structure and the spherulites of the polyamide domain with a clear boundary can be found except for the high soft content of TPAE-0.76 and TPAE-0.62. It should be noteworthy; however, these two copolymers contain a polyamide crystal phase according to DSC and WAXD results. The slender and rigid polyamide domain is observed, which can be explain by the assumption that the polyamide crystalline domain forms randomly ordered lamellar crystals by the dilute solution casting method [5]. For the other samples, the spherulitic morphology consisting of crystalline polyamide and amorphous polyamide segments are well dispersed in the amorphous polyetheramine. It is believed that the amorphous polyetheramine chains are comprised of the spherulites or filled among the lamellar crystals [41].

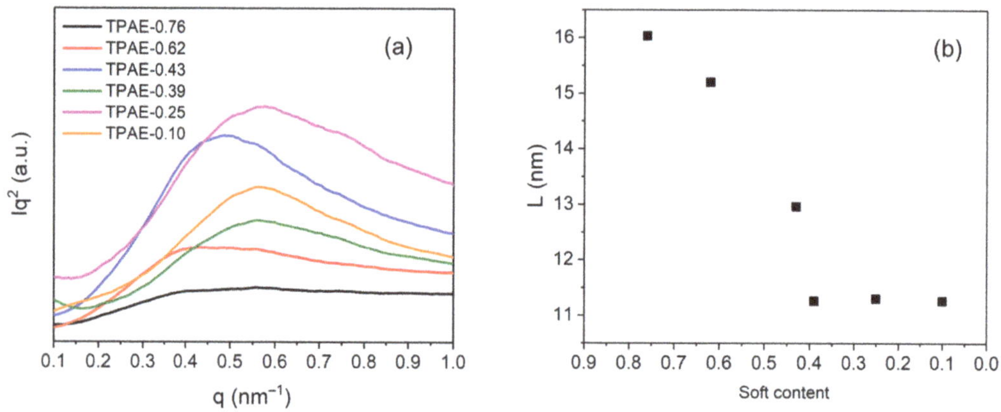

Figure 6. (**a**) Lorentz corrected SAXS profiles of the TPAEs and (**b**) long period vs. soft contents.

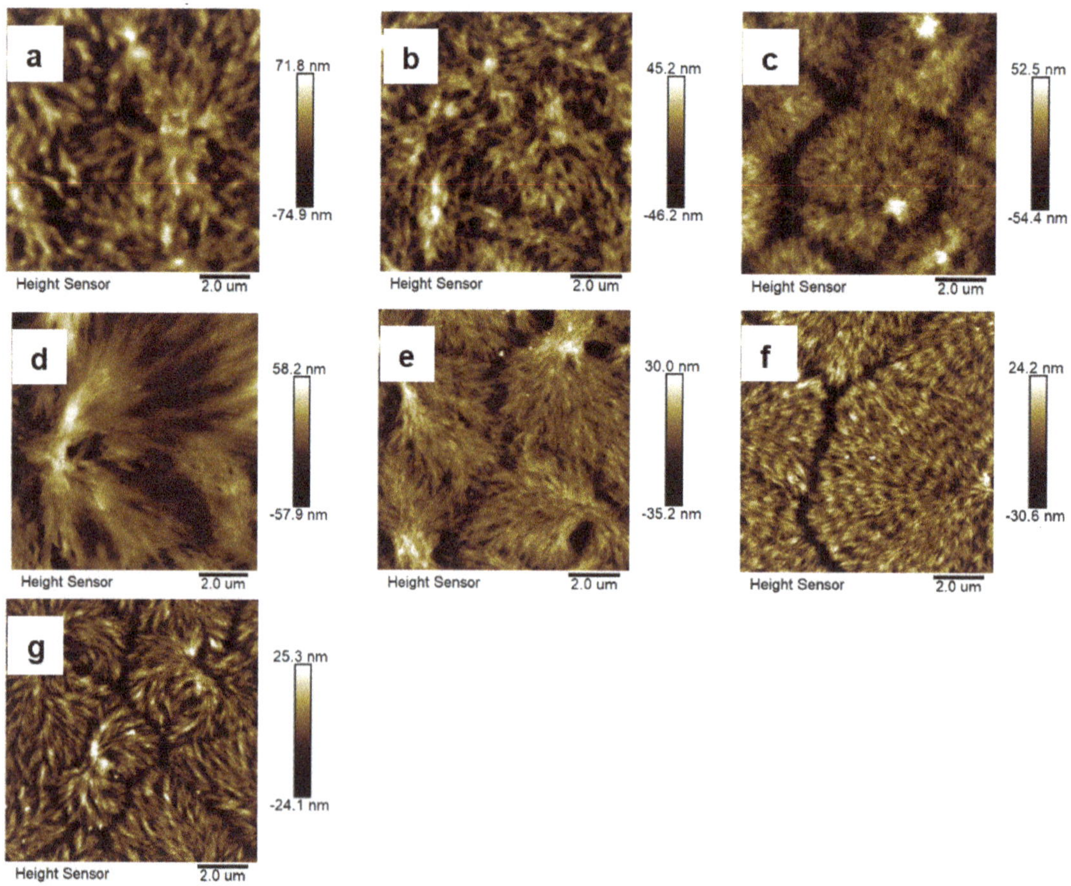

Figure 7. AFM height images of drop-casting TPAEs: (**a**) TPAE-0.76, (**b**) TPAE-0.62, (**c**) TPAE-0.43, (**d**) TPAE-0.39, (**e**) TPAE-0.25, (**f**) TPAE-0.10, and (**g**) homo PA1212.

In addition, the non-crystalline soft segments have a distinguishing effect on the morphology features. It can be found that when using long-chain Jeffamine D2000, the spherulitic structure forms only in TPAE-0.43 with relative high polyamide content, showing packed down-feather-like lamellar topography; for TPAE-0.39 to TPAE-0.10 with short-chain Jeffamine D400, the bundle-like structure with filamentous polyamide crystals turns into similar spherulite with a feather-like structure. Furthermore, with the increasing of polyamide content, the short-chain polyetheramine has a much greater dilution effect on the crystallization of polyamide than the long-chain polyetheramine because of the difference in the compatibility of distinct polyetheramines with polyamide. As a result, the degree of microphase separation of copolymers with long-chain soft segments is higher than that with short chain. The above deductions are further validated from the polarized optical microscopy (POM) images (Figure S3, Supplementary Material). Only very small crystals can be found in TPAEs with low amount of soft content (containing short-chain polyetheramine), and small size and sparse packed spherulites of polyamidecan be observed in TPAE-0.10.

To further investigate the morphological evolution with different compositions of the TPAEs, TEM was performed, and the representative images are shown in Figure 8. The TPAE-0.76 and TPAE-0.62 with high soft content grow long rigid-rod polyamide crystals distributed in a disordered form in amorphous soft domain. The rigid-rod crystals are assembled into a bundle-like structure in TPAE-0.39 as the increasing of polyamide content. The densely packed spherulites with feather-like structure can be seen in TPAE-0.10 with high polyamide content. These results are agreeable with the above deductions by AFM measurements.

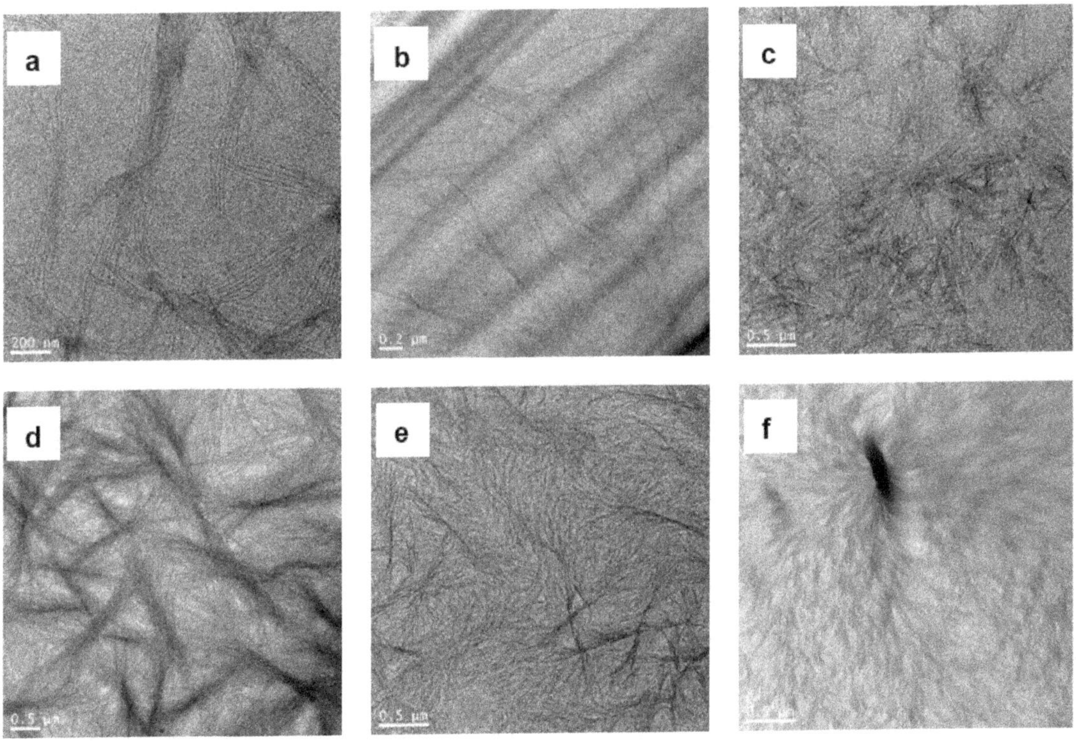

Figure 8. TEM images of (**a**) TPAE-0.76, (**b**) TPAE-0.62, (**c**) TPAE-0.43, (**d**) TPAE-0.39, (**e**) TPAE-0.25, and (**f**) TPAE-0.10.

3.4. Dynamic Thermomechanical Analysis

To analyze the relationships between the microstructure and thermal mechanical properties of elastomers, DMA measurements were carried out on all the synthesized novel TPAEs. Figure 9 illustrates the temperature dependence of the storage modulus (E') and loss factor (*tan δ*) for the TPAEs with variable compositions. The glass transition temperature T_g (peak value of *tan δ*) and E' at −25 °C and 25 °C are listed in Table 2, correspondingly. It can be found that the copolymers have decreased E' with decreasing soft content at low temperature ca. −70 °C and exhibit a plateau of E' in the glassy state. Then, E' drops with the glass transition behavior. The E' values of TPAE-0.76 and TPAE-0.62 at −25 °C are far less than those of the other three samples, which are lower than commercial PEBAX at the similar compositions [5]. That means that these two copolymers with high soft content are well flexible at low temperature. Following the transition, the copolymers exhibit a rubbery state and a low dependence on temperature, providing elastic performance. The E' value at room temperature increases with the decreasing of soft content, which indicates the corresponding stiffness increases. It is worth noting that the E' value of TPAE-0.43 is smaller than that of TPAE-0.39, while the soft contents of two TPAE are very close. The E' of TPAE-0.43 has little change from −25 to 25 °C. To obtain TPAE with high elastic property, in addition to considering increasing the soft content, it is more significant to apply long-chain soft segments.

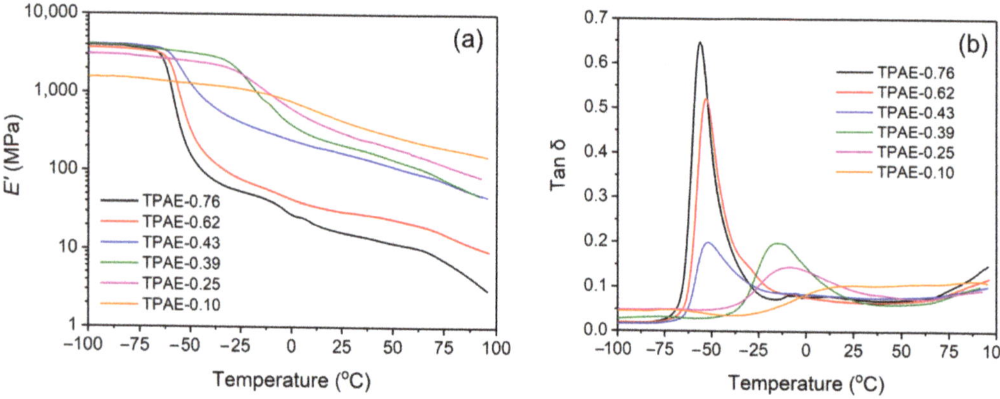

Figure 9. Thermomechanical analysis of TPAEs: (**a**) storage modulus (E') vs. temperature and (**b**) loss factor (*tan δ*) vs. temperature.

Figure 9b shows the thermal transitions of the TPAEs. The peak values of *tan δ* decrease with the increasing of polyamide content, indicating that polyamide crystals restrain the chain mobility of amorphous PPG. This is because the polyamide domain can act as a physical crosslinker in TPAEs [45]. The copolymers containing long-chain soft segments exhibit similar T_g values that are independent of the soft content, whereas those having short-chain polyetheramine show increasing T_g values as the soft content increases. The DMA results are in good agreement with the DSC results. The values of T_g measured by DMA are higher than those by the DSC method, which is because the copolymer chain begins to relax at the temperature before the *tan δ* peak, which is measured by mechanical response [46].

Tan δ is defined using the ratio of loss modulus (E'') and storage modulus (E') and is used as an assessment of the ability of energy dissipation by elastomers. The strain hysteresis caused by the viscoelasticity of copolymers can result in energy absorption. Therefore, TPAEs are widely used as damping materials for attenuating vibration and reducing noise due to their significantly viscoelastic properties. Damping materials should meet the requirement of *tan δ* > 0.3, and the corresponding temperature region is defined as the damping temperature range [47,48]. The *tan δ* peak values of PEBAX are reported to be lower than 0.16, which is indicative of good energy return when using as an elastomer [5]. However, the *tan δ* peak values of the synthesized TPAEs are over 0.14 except for TPAE-0.10. Especially, the *tan δ* peak values of TPAE-0.76 and TPAE-0.62 with high PPG content are distinctly high (>0.5), which indicates that these two TPAEs have excellent shock absorbing capacity. It is because the energy dissipation by internal friction can be promoted since this C3 building block bears a methyl side group in the backbone [48]. As $tan\ \delta > 0.3$, the damping temperature range for TPAE-0.76 and TPAE-0.62 are from −61 to −47 °C (ΔT = 14 °C) and from −58 to −45 °C (ΔT = 13 °C), respectively. It can be found that the synthesized TPAEs have excellent viscoelastic properties and damping performance by introducing a PPG block, which is important for toughness applications such as ski boots and rail gasket in an extremely cold situation. Thus, the novel TPAEs with long-chain semicrystalline polyamide and a high content of amorphous polyetheramine enrich the family of TPAEs for damping materials.

3.5. Mechanical Properties

Figure 10 presents the uniaxial tensile stress–strain curves of the TPAEs with different compositions at room temperature. The typical elastic behavior of the specimen is determined by the lack of a yield point during the deformation except for TPAE-0.10, which has the highest hard content and shows thermoplastic character. The polyetheramine-based TPAEs have good elastic properties according to the manner in which TPAE-0.25 reveals an intermediate stress–strain behavior. The Young' moduli (the slope of initial linear elastic domain) increase with increasing polyamide content due to the increase in interconnectivity within the rigid polyamide domain [49]. Moreover, the tensile strength increases as the rigid content increases, which is independent of the polyetheramine length. The specimens exhibit good elongation at break that increases with the increase of soft content. Based on the testing results, the tensile property of TPAEs highly depends on the soft content instead of the segment length.

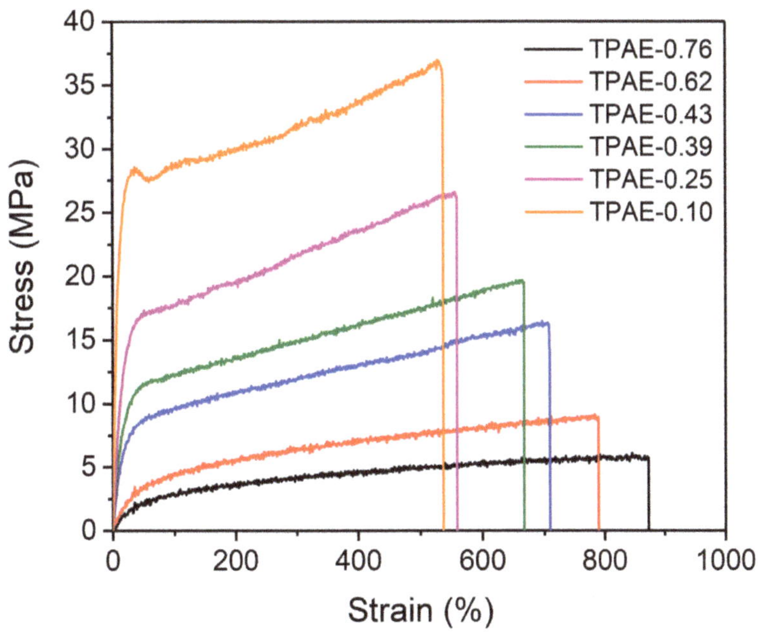

Figure 10. Tensile stress–strain curves of TPAE films.

To further study the elastic properties of the TPAEs, mechanical hysteresis was measured by a cyclic tensile test and the representative stress–strain curves are presented in Figure 11. The residual strain was calculated after 10 cycles of tensile testing. As we expected, the residual strain increases with the decreasing of soft content and increasing of testing deformation (Table S1, Supplementary Material). The novel TPAEs using diamine-terminated PPG present good elastic recovery and flexibility. In fact, the specimens after testing just have a small extent of residual strain after storing at room temperature overnight (Figure S4, Supplementary Material).

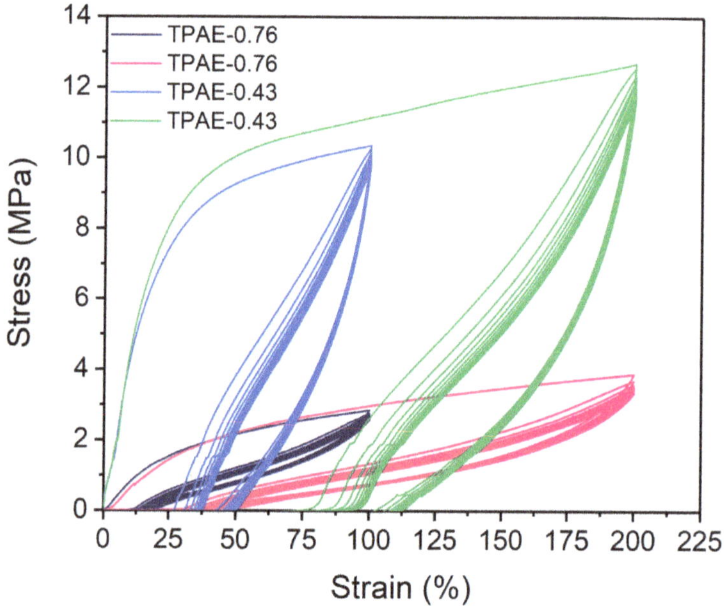

Figure 11. Selected cyclic tensile stress–strain curves of TPAE films.

4. Conclusions

Novel TPAEs consisting of long-chain semicrystalline PA1212 (C12) as the hard segment and amorphous diamine terminated PPG (C3) as the soft segment were successfully synthesized via one-pot melt polycondensation. The chemical structures of the copolymers were characterized by NMR and FT-IR, and the results prove that the diamine terminated PPG segments have been incorporated to PA1212 blocks with expected composition. The molecular weights of the TPAEs were approaching the commercial grade. The microphase-separated morphologies and tensile properties of novel TPAEs have been systematically investigated by DSC, X-ray diffraction, AFM, SEM, and DMA. It has been found that microphase separation occurs in the amorphous region, and short-chain PPG diamine is miscible with the amorphous phase of PA1212. The copolymers adopt the same α-crystal phase as homo PA1212, and the crystallinity of the copolymers (PA1212 segments) varies with the content and segment length of PPG diamine. The microphase separation structure, associated with the macro properties, is tunable according to the contents and length of the flexible PPG diamine. Due to the amorphous PPG segment and long-chain semicrystalline PA1212, the new family of TPAEs shows high elasticity and damping performance, especially at low temperature for TPAE-0.76 and TPAE-0.62. The novel TPAEs show balanced properties of tremendous lightness, elastic return, and attenuation of vibration, suggesting their potential applications such as running shoes, ski boots, and tennis rackets even products used in extreme cold condition such as rail gaskets.

Supplementary Materials: The following are available online at https://www.mdpi.com/article/10.3390/polym13162645/s1, Figure S1: DSC thermogram of Jeffamine D2000, Figure S2: DSC thermogram of TPAEs, Figure S3: POM images of (a) (b) TPAE-0.39, (c) TPAE-0.25 (d) TPAE-0.10, (e) homo PA1212, Figure S4: Photos of specimens before and after cyclic tensile testing for (a) TPAE-0.76 and (b) TPAE-0.43 (stored at room temperature overnight), Figure S5: Thermal gravimetric curves of TPAEs, Table S1: The residual strain of the specimens after cyclic tensile testing, Table S2 Characteristic thermal degradation parameters of TPAEs.

Author Contributions: Conceptualization, J.J. and Z.X.; Formal analysis, J.J., Q.T. and X.P.; Funding acquisition, L.Z. and Z.X.; Investigation, J.J.; Supervision, W.Y.; Writing—original draft, J.J.; Writing—review and editing, X.P., J.L. and Z.X. All authors have read and agreed to the published version of the manuscript.

Funding: This work was financially supported by the National Natural Science Foundation of China (Grant No. 21978089 and No. 21878256), the Fundamental Research Funds for the Central Universities (Grant No. 22221818010), and the 111 Project (Grant No. B20031).

Institutional Review Board Statement: Not applicable.

Informed Consent Statement: Not applicable.

Data Availability Statement: The data presented in this study are available on request from the corresponding author.

Conflicts of Interest: There are no conflict of interest to declare.

References

1. Husken, D.; Feijen, J.; Gaymans, R.J. Hydrophilic segmented block copolymers based on poly(ethylene oxide) and monodisperse amide segments. *J. Polym. Sci. Part A Polym. Chem.* **2007**, *45*, 4522–4535. [CrossRef]
2. Boulares, A.; Tessier, M.; Marechal, E. Synthesis and characterization of poly (copolyethers-block-polyamides) II. Characterization and properties of the multiblock copolymers. *Polymer* **2000**, *41*, 3561–3580. [CrossRef]
3. Boulares, A.; Tessier, M.; MarÉChal, E. Synthesis and Characterization of Poly(Copolyethers-block-Polyamides). I. Structural Study of Polyether Precursors. *J. Macromol. Sci. Part A* **1998**, *35*, 933–953. [CrossRef]
4. Barbi, V.; Funari, S.S.; Gehrke, R.; Scharnagl, N.; Stribeck, N. SAXS and the Gas Transport in Polyether-b lock-polyamide Copolymer Membranes. *Macromolecules* **2003**, *36*, 749–758. [CrossRef]
5. Sheth, J.P.; Xu, J.; Wilkes, G.L. Solid state structure–property behavior of semicrystalline poly (ether-block-amide) PEBAX® thermoplastic elastomers. *Polymer* **2003**, *44*, 743–756. [CrossRef]
6. Bai, L.; Hong, Z.; Wang, D.; Li, J.; Wang, X.; Pan, G.; Li, L.; Li, X. Deformation-induced phase transitions of polyamide 12 in its elastomer segmented copolymers. *Polymer* **2010**, *51*, 5604–5611. [CrossRef]
7. Tavernier, B.; Mewis, J.; Van Puyvelde, P.; Takenaka, M.; Ernst, B.; Hashimoto, T. Effect of thermomechanical history on the crystallization of poly(ether-block-amide). *Polym. Eng. Sci.* **2008**, *48*, 2418–2425. [CrossRef]
8. Sauer, B.B.; McLean, R.S.; Brill, D.J.; Londono, D.J. Morphology and orientation during the deformation of segmented elastomers studied with small-angle X-ray scattering and atomic force microscopy. *J. Polym. Sci. Part B Polym. Phys.* **2002**, *40*, 1727–1740. [CrossRef]
9. Song, Y.; Yamamoto, H.; Nemoto, N. Segmental orientations and deformation mechanism of poly (ether-block-amide) films. *Macromolecules* **2004**, *37*, 6219–6226. [CrossRef]
10. Fakirov, S. *Handbook of Condensation Thermoplastic Elastomers*; WILEY-VCH Verlag GmbH & Co. KGaA: Weinheim, Germany, 2005; Chapter 9; pp. 243–257.
11. Burns, G.T.; Tam, N. Is it the shoes? A simple proposal for regulating footwear in road running. *Br. J. Sports Med.* **2020**, *54*, 439–440. [CrossRef] [PubMed]
12. Fan, D.; Shi, Z.; Li, N.; Qiu, J.; Xing, H.; Jiang, Z.; Li, M.; Tang, T. Novel Method for Preparing a High-Performance Auxetic Foam Directly from Polymer Resin by a One-Pot CO_2 Foaming Process. *ACS Appl. Mater. Interfaces* **2020**, *12*, 48040–48048. [CrossRef]
13. Ma, Z.; Wei, A.; Li, Y.; Shao, L.; Zhang, H.; Xiang, X.; Wang, J.; Ren, Q.; Kang, S.; Dong, D. Lightweight, flexible and highly sensitive segregated microcellular nanocomposite piezoresistive sensors for human motion detection. *Compos. Sci. Technol.* **2021**, *203*, 108571. [CrossRef]
14. Barzegari, M.R.; Hossieny, N.; Jahani, D.; Park, C.B. Characterization of hard-segment crystalline phase of poly(ether-block-amide) (PEBAX®) thermoplastic elastomers in the presence of supercritical CO_2 and its impact on foams. *Polymer* **2017**, *114*, 15–27. [CrossRef]
15. Armstrong, S.; Freeman, B.; Hiltner, A.; Baer, E. Gas permeability of melt-processed poly(ether block amide) copolymers and the effects of orientation. *Polymer* **2012**, *53*, 1383–1392. [CrossRef]
16. Bondar, V.; Freeman, B.D.; Pinnau, I. Gas sorption and characterization of poly (ether-b-amide) segmented block copolymers. *J. Polym. Sci. Part B Polym. Phys.* **1999**, *37*, 2463–2475. [CrossRef]

17. Liu, S.L.; Shao, L.; Chua, M.L.; Lau, C.H.; Wang, H.; Quan, S. Recent progress in the design of advanced PEO-containing membranes for CO2 removal. *Prog. Polym. Sci.* **2013**, *38*, 1089–1120. [CrossRef]
18. Zhao, D.; Ren, J.; Wang, Y.; Qiu, Y.; Li, H.; Hua, K.; Li, X.; Ji, J.; Deng, M. High CO_2 separation performance of Pebax®/CNTs/GTA mixed matrix membranes. *J. Membr. Sci.* **2017**, *521*, 104–113. [CrossRef]
19. Car, A.; Stropnik, C.; Yave, W.; Peinemann, K.-V. PEG modified poly(amide-b-ethylene oxide) membranes for CO_2 separation. *J. Membr. Sci.* **2008**, *307*, 88–95. [CrossRef]
20. Wang, Y.; Li, H.; Dong, G.; Scholes, C.; Chen, V. Effect of Fabrication and Operation Conditions on CO_2 Separation Performance of PEO–PA Block Copolymer Membranes. *Ind. Eng. Chem. Res.* **2015**, *54*, 7273–7283. [CrossRef]
21. Wang, J.; Bao, L.; Zhao, H.; Lei, J. Preparation and characterization of permanently anti-static packaging composites composed of high impact polystyrene and ion-conductive polyamide elastomer. *Compos. Sci. Technol.* **2012**, *72*, 976–981. [CrossRef]
22. Jiang, J.; Tang, Q.; Pan, X.; Xi, Z.; Zhao, L.; Yuan, W. Structure and Morphology of Thermoplastic Polyamide Elastomer Based on Long-Chain Polyamide 1212 and Renewable Poly(trimethylene glycol). *Ind. Eng. Chem. Res.* **2020**, *59*, 17502–17512. [CrossRef]
23. Xu, J.; Chen, W.; Wang, C.; Zheng, M.; Ding, C.; Jiang, W.; Tan, L.; Fu, J. Extremely stretchable, self-healable elastomers with tunable mechanical properties: Synthesis and applications. *Chem. Mater.* **2018**, *30*, 6026–6039. [CrossRef]
24. Wang, D.; Xu, J.; Chen, J.; Hu, P.; Wang, Y.; Jiang, W.; Fu, J. Transparent, Mechanically Strong, Extremely Tough, Self-Recoverable, Healable Supramolecular Elastomers Facilely Fabricated via Dynamic Hard Domains Design for Multifunctional Applications. *Adv. Funct. Mater.* **2019**, *30*, 1907109. [CrossRef]
25. Nohales, A.; Solar, L.; Porcar, I.; Vallo, C.I.; Gómez, C.M. Morphology, flexural, and thermal properties of sepiolite modified epoxy resins with different curing agents. *Eur. Polym. J.* **2006**, *42*, 3093–3101. [CrossRef]
26. Strachota, A.; Kroutilová, I.; Kovářová, J.; Matějka, L. Epoxy networks reinforced with polyhedral oligomeric silsesquioxanes (POSS). Thermomechanical properties. *Macromolecules* **2004**, *37*, 9457–9464. [CrossRef]
27. Mo, R.; Song, L.; Hu, J.; Sheng, X.; Zhang, X. An acid-degradable biobased epoxy-imine adaptable network polymer for the fabrication of responsive structural color film. *Polym. Chem.* **2020**, *11*, 974–981. [CrossRef]
28. Deliballi, Z.; Kiskan, B.; Yagci, Y. Main-chain benzoxazine precursor block copolymers. *Polym. Chem.* **2018**, *9*, 178–183. [CrossRef]
29. Ma, S.; Zhang, H.; Sablong, R.J.; Koning, C.E.; van Benthem, R. t-Butyl-Oxycarbonylated Diamines as Building Blocks for Isocyanate-Free Polyurethane/Urea Dispersions and Coatings. *Macromol. Rapid Commun.* **2018**, *39*, 1800004. [CrossRef]
30. Xu, S.; Ye, L. Synthesis and properties of monomer cast nylon-6-b-polyether amine copolymers with different structures. *RSC Adv.* **2015**, *5*, 32460–32468. [CrossRef]
31. Van Hutten, P.; Walch, E.; Veeken, A.; Gaymans, R. Segmented block copolymers based on polyamide-4, 6 and poly (propylene oxide). *Polymer* **1990**, *31*, 524–529. [CrossRef]
32. Yu, Y.C.; Jo, W.H. Segmented block copolyetheramides based on nylon 6 and polyoxypropylene. I. Synthesis and characterization. *J. Appl. Polym. Sci.* **1994**, *54*, 585–591. [CrossRef]
33. Yu, Y.C.; Jo, W.H. Segmented block copolyetheramides based on nylon 6 and polyoxypropylene. II. Structure and properties. *J. Appl. Polym. Sci.* **1995**, *56*, 895–904. [CrossRef]
34. Yu, Y.C.; Jo, W.H.; Lee, M.S. Segmented block copolyetheramides based on nylon 6 and polyoxypropylene. III. SAXS analysis. *J. Appl. Polym. Sci.* **1997**, *64*, 2155–2163.
35. Mori, T.; Masukawa, S.; Kikkawa, T.; Fujimori, A.; Satoh, A.; Matsumoto, K.; Jikei, M.; Oishi, Y.; Shibasaki, Y. Rapid synthesis and properties of segmented block copolymers based on monodisperse aromatic poly(N-methyl benzamide) and poly(propylene oxide). *RSC Adv.* **2017**, *7*, 33812–33821. [CrossRef]
36. Zhu, T.; Yang, X.; Zhang, Y.; Zheng, Y.; He, X.; Luo, J. Random and block copolymer membranes based on flexible etheric-aliphatic soft segments designed for CO_2/CH_4 separation. *J. Nat. Gas Sci. Eng.* **2018**, *54*, 92–101. [CrossRef]
37. Zhu, T.; Yang, X.; Zheng, Y.; He, X.; Chen, F.; Luo, J. Preparation of poly(ether-block-amide)/poly(amide-co-poly(propylene glycol)) random copolymer blend membranes for CO_2/N_2 separation. *Polym. Eng. Sci.* **2019**, *59*, E14–E23. [CrossRef]
38. Liu, M.; Zhao, Q.; Wang, Y.; Zhang, C.; Mo, Z.; Cao, S. Melting behaviors, isothermal and non-isothermal crystallization kinetics of nylon 1212. *Polymer* **2003**, *44*, 2537–2545. [CrossRef]
39. Yuan, R.; Fan, S.; Wu, D.; Wang, X.; Yu, J.; Chen, L.; Li, F. Facile synthesis of polyamide 6 (PA6)-based thermoplastic elastomers with a well-defined microphase separation structure by melt polymerization. *Polym. Chem.* **2018**, *9*, 1327–1336. [CrossRef]
40. Ren, M.; Mo, Z.; Chen, Q.; Song, J.; Wang, S.; Zhang, H.; Zhao, Q. Crystallization kinetics and morphology of nylon 1212. *Polymer* **2004**, *45*, 3511–3518. [CrossRef]
41. Zhu, P.; Dong, X.; Cao, Y.; Wang, L.; Liu, X.; Wang, Z.; Wang, D. The Brill transition in polyether-b-amide segmented copolymers and composition dependence. *Eur. Polym. J.* **2017**, *93*, 334–346. [CrossRef]
42. Song, J.; Zhang, H.; Ren, M.; Chen, Q.; Sun, X.; Wang, S.; Zhang, H.; Mo, Z. Crystal Transition of Nylon-12,12 under Drawing and Annealing. *Macromol. Rapid Commun.* **2005**, *26*, 487–490. [CrossRef]
43. Yang, J.; Liang, Y.; Luo, J.; Zhao, C.; Han, C.C. Multilength Scale Studies of the Confined Crystallization in Poly(l-lactide)-block-Poly(ethylene glycol) Copolymer. *Macromolecules* **2012**, *45*, 4254–4261. [CrossRef]
44. Cao, Y.; Zhu, P.; Wang, Z.; Zhou, Y.; Chen, H.; Müller, A.J.; Wang, D.; Dong, X. Influence of soft block crystallization on microstructural variation of double crystalline poly(ether-mb-amide) multiblock copolymers. *Polym. Cryst.* **2018**, *1*, 1–11. [CrossRef]

45. Chi, D.; Liu, F.; Na, H.; Chen, J.; Hao, C.; Zhu, J. Poly(neopentyl glycol 2,5-furandicarboxylate): A Promising Hard Segment for the Development of Bio-based Thermoplastic Poly(ether-ester) Elastomer with High Performance. *ACS Sustain. Chem. Eng.* **2018**, *6*, 9893–9902. [CrossRef]
46. Lu, Y.; Larock, R.C. New hybrid latexes from a soybean oil-based waterborne polyurethane and acrylics via emulsion polymerization. *Biomacromolecules* **2007**, *8*, 3108–3114. [CrossRef] [PubMed]
47. Li, Z.; Lu, X.; Tao, G.; Guo, J.; Jiang, H. Damping elastomer with broad temperature range based on irregular networks formed by end-linking of hydroxyl-terminated poly(dimethylsiloxane). *Polym. Eng. Sci.* **2016**, *56*, 97–102. [CrossRef]
48. Zhao, X.; Shou, T.; Liang, R.; Hu, S.; Yu, P.; Zhang, L. Bio-based thermoplastic polyurethane derived from polylactic acid with high-damping performance. *Ind. Crop. Prod.* **2020**, *154*, 112619. [CrossRef]
49. Sijbrandi, N.J.; Kimenai, A.J.; Mes, E.P.C.; Broos, R.; Bar, G.; Rosenthal, M.; Odarchenko, Y.; Ivanov, D.A.; Dijkstra, P.J.; Feijen, J. Synthesis, Morphology, and Properties of Segmented Poly(ether amide)s with Uniform Oxalamide-Based Hard Segments. *Macromolecules* **2012**, *45*, 3948–3961. [CrossRef]

polymers

MDPI

Article

Novel Chemical Architectures Based on Beta-Cyclodextrin Derivatives Covalently Attached on Polymer Spheres

Stefan Bucur [1], Ionel Mangalagiu [1,2], Aurel Diacon [3], Alexandra Mocanu [3], Florica Rizea [3], Raluca Somoghi [4], Adi Ghebaur [3,5], Aurelian Cristian Boscornea [3] and Edina Rusen [3,*]

1 Faculty of Chemistry, Alexandru Ioan Cuza University of Iasi, 11 Carol 1st Bvd, 700506 Iasi, Romania; bucurm.stefan@gmail.com (S.B.); ionelm@uaic.ro (I.M.)
2 Institute of Interdisciplinary Research—CERNESIM Centre, Alexandru Ioan Cuza University of Iasi, 11 Carol I, 700506 Iasi, Romania
3 Faculty of Applied Chemistry and Materials Science, University Politehnica of Bucharest, 1-7 Gh. Polizu Street, 011061 Bucharest, Romania; aurel_diacon@yahoo.com (A.D.); mocanu_alexandra85@yahoo.com (A.M.); flori_rizea@yahoo.com (F.R.); adi.ghebaur@yahoo.com (A.G.); cristian.boscornea@yahoo.com (A.C.B.)
4 National Research and Development Institute for Chemistry and Petrochemistry—ICECHIM, 202 Splaiul Independenţei, 060021 Bucharest, Romania; raluca.somoghi@yahoo.com
5 Advanced Polymer Materials Group, University Politehnica of Bucharest, Gh. Polizu Street, 011061 Bucharest, Romania
* Correspondence: edina.rusen@upb.ro

Citation: Bucur, S.; Mangalagiu, I.; Diacon, A.; Mocanu, A.; Rizea, F.; Somoghi, R.; Ghebaur, A.; Boscornea, A.C.; Rusen, E. Novel Chemical Architectures Based on Beta-Cyclodextrin Derivatives Covalently Attached on Polymer Spheres. *Polymers* **2021**, *13*, 2338. https://doi.org/10.3390/polym13142338

Academic Editor: Andrea Mele

Received: 18 June 2021
Accepted: 14 July 2021
Published: 16 July 2021

Publisher's Note: MDPI stays neutral with regard to jurisdictional claims in published maps and institutional affiliations.

Copyright: © 2021 by the authors. Licensee MDPI, Basel, Switzerland. This article is an open access article distributed under the terms and conditions of the Creative Commons Attribution (CC BY) license (https://creativecommons.org/licenses/by/4.0/).

Abstract: This study presents the synthesis and characterization of polymer derivatives of beta-cyclodextrin (BCD), obtained by chemical grafting onto spherical polymer particles (200 nm) presenting oxirane functional groups at their surface. The polymer spheres were synthesized by emulsion polymerization of styrene (ST) and hydroxyethyl methacrylate (HEMA), followed by the grafting on the surface of glycidyl methacrylate (GMA) by seeded emulsion polymerization. The BCD-polymer derivatives were obtained using two BCD derivatives with hydroxylic (BCD-OH) and amino groups ($BCD-NH_2$). The degree of polymer covalent functionalization using the BCD-OH and $BCD-NH_2$ derivatives were determined to be 4.27 and 19.19 weight %, respectively. The adsorption properties of the materials were evaluated using bisphenol A as a target molecule. The best fit for the adsorption kinetics was Lagergren's model (both for Q_e value and for R^2) together with Weber's intraparticle diffusion model in the case of ST-HEMA-GMA-$BCD-NH_2$. The isothermal adsorption evaluation indicated that both systems follow a Langmuir type behavior and afforded a Q_{max} value of 148.37 mg g^{-1} and 37.09 mg g^{-1} for ST-HEMA-GMA-$BCD-NH_2$ and ST-HEMA-GMA-BCD-OH, respectively. The BCD-modified polymers display a degradation temperature of over 400 °C which can be attributed to the existence of hydrogen bonds and BCD thermal degradation pathway in the presence of the polymers.

Keywords: beta-cyclodextrin; emulsion polymerization; adsorption kinetics; adsorption isotherms; bisphenol A adsorption

1. Introduction

Bisphenol A (BPA) is one of the most abundant chemical synthetic additives produced for the manufacturing of epoxy resins, polysulfones, unsaturated polyesters, polyacrylate resins polycarbonate plastics, and rubber [1,2]. Human exposure to BPA [3,4] or its bisphenol-based substitutes [5] exhibited cytotoxicity, neurotoxicity, endocrine-disrupting effects, reproductive toxicity, uterine cancer, and interference of cellular pathways, capable of mimicking some of the hormones of the human body [1,3–5]. A recent study [6] revealed that the estimated intake of BPA is 30.76 ng/kg per body weight per day, as this compound enters in various environmental media (air, soil, aquatic systems) and the food chain due to improper recycling procedures [1,7]. This value, although in some countries did not exceed the maximum acceptable values for daily intake, is all the more worrying as the first six countries with the highest values belong to the European Union [6]. Thus, it is of tremendous importance on the one hand to develop efficient decontamination procedures and on the other hand to impose stricter laws regarding plastic manufacturing or recycling.

Different methods for BPA removal were directed to biological treatment, membrane adsorption processes, or advanced oxidation [1]. The biological treatment involves the degradation of BPA by immobilization of enzymes such as oxidoreductase or polyphenol oxidases [8] that are capable of oxidizing organic pollutants, such as phenols, organic dyes, or drugs. Advanced oxidation is a method that removes BPA by the generation of highly reactive radicals or the application of photocatalytic treatment to break down the molecules of BPA into less harmful compounds from water, sediment, and soil [1,9–11]. The membrane adsorption technology is based on the selective adsorption capacity of different membranes of targeted pollutants (by chemical or physical interaction) and is sometimes preferred for BPA removal since it does not generate new harmful by-products [12,13].

The drawbacks of these methods are not only related to the potential of generating harmful by-products but also to the difficulty of controlling the necessary parameters for efficient removal of BPA (such as pH and temperature in the case of enzymes or photocatalytic removal) and laborious/expensive synthesis methods for adsorption membranes [2].

Cyclodextrins (CDs) are a class of three-dimensional (3D) cyclic oligosaccharides composed of D-glucose units linked together by α-1,4-glucosidic bonds with hydrophilic surface and hydrophobic internal hollow [14,15]. The amphiphilic, biodegradable, and non-hazardous properties of such compounds, as well as the possibility to design new cyclodextrin derivatives, has led to a wide range of applications for drug delivery systems, antiviral therapy, cosmetics, agriculture, enzymology, catalysis, enantiomers separation, and environmental protection, in which it has been shown they can be excellent candidates for decontamination of aqueous media, air or soil [15–19].

As environmental issues are becoming more important, intense research interest has been dedicated to the removal of pollutants from water using different cyclodextrins-polymer systems that are tailored to eliminate organic or inorganic pollutants based on the adsorption process of the targeted molecule [20–22]. Based on the 3D spatial structure of CDs and on the hydrophilic/hydrophobic building, the enhanced removal of pollutants is based on guest-host interactions between the cyclodextrin derivative and the targeted pollutant. The additional presence of polymer chains can improve the adsorption capacity of the contaminating agent depending on their chemical structure or morphology, since polymer adsorbents can include polymers with a spherical shape, intrinsic porosity, metal/covalent-organic frameworks, as well as hyper-crosslinked polymers [23–25].

Thus, literature data is abundant in different synthesis approaches that have the same goal, to improve the adsorption efficiency of the cyclodextrins-polymers (CDP) systems for wastewater decontamination. For example, beta-cyclodextrin (BCD) was reacted with epichlorohydrin (EPI) and further with trimesoyl chloride to obtain a CDP system that was further embedded by interfacial cross-linking into a nylon microfiltration membrane to create a porous structure with enhanced adsorption capacities of water contaminants [26]. The adsorption efficiency capacity of modified BCD is amazing considering that structurally different contaminants like BPA, methylene blue, and copper can be adsorbed simultaneously by using citric acid-crosslinked-BCD polymers [27]. Recent approaches demonstrated that BCD-based polymers obtained by crosslinking with EPI can be used in municipal wastewater treatment pilot plants to remove several micropollutants, including BPA, with over 80% efficiency [28]. Also, the BCDP adsorption capacities were improved by the presence of nano-adsorbents such as Fe_3O_4, SiO_2, silver nanoparticles, or carbon nanotubes due to their high specific surface area and absence of internal diffusion resistance that enhances the kinetics for the adsorption processes of different contaminating agents such as BPA, organophoshorous insecticides, or p-nitrophenol from water [9,20,29].

Thus, in this work, novel BCD modified polymer particles with spherical morphology were synthesized for a possible decontamination process of wastewater. Utilizing the oxirane functional groups present at the surface of polymer particles, two types of BCD derivatives (with hydroxylic (BCD-OH) and amino groups (BCD-NH_2)—Scheme 1) were chemically grafted to the polymer colloids (Scheme 2). The materials designed aimed to improve the interaction between the polymeric adsorbent, pollutant, and the contaminated media. One of the main goals of this study was to determine the linking capacities of the BCD to the polymer, as well as the maximum complexation capacity of the BCD-modified polymers toward a targeted molecule. Thus, BPA was selected as a model molecule to investigate the adsorption kinetics, complexation mechanism, and isotherms.

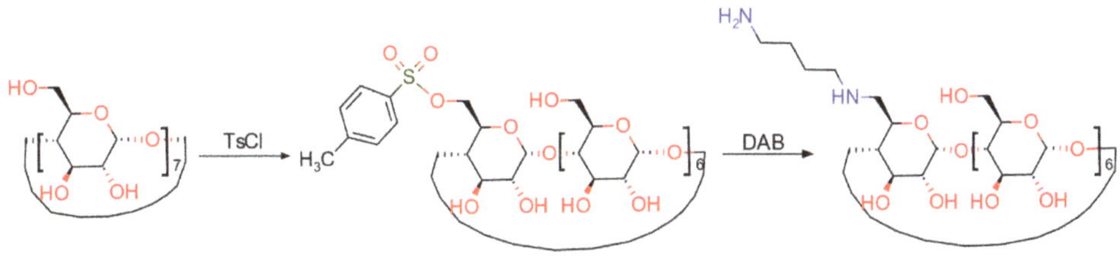

Scheme 1. Synthesis route for BCD-NH_2.

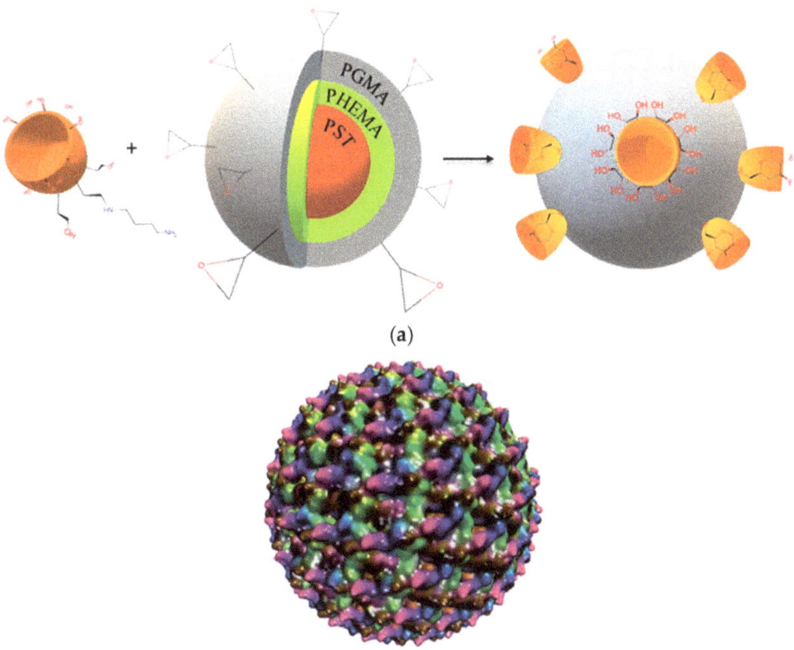

(**a**)

(**b**)

Scheme 2. Chemical synthesis of ST-HEMA-GMA-BCD-NH_2. (**a**) The reaction of BCD-NH_2 with ST-HEMA-GMA; (**b**) ST-HEMA-GMA-BCD-NH_2 complex structure (3D).

2. Materials and Methods

2.1. Materials

Styrene (ST) (Sigma-Aldrich, St. Louis, MO, USA) has been purified through vacuum distillation. 2-Hydroxyethyl methacrylate (HEMA) (Sigma-Aldrich, St. Louis, MO, USA) and glycidyl methacrylate (GMA) (Merck, Darmstadt, Germany) have been purified by passing over short columns of activated basic alumina. Potassium persulfate ($K_2S_2O_8$) (KPS) (Merck, Darmstadt, Germany) has been recrystallized from an ethanol/water mixture and then vacuum-dried. β-cyclodextrin (BCD-OH) (≥95.0%, Wacker Chemie, Munich, Germany) was vacuum-dried before use for 24 h. p-Toluene sulfonyl chloride (Ts-Cl) (reagent grade, ≥98%, Merck, Darmstadt, Germany), sodium hydroxide (reagent grade, ≥98%, pellets (anhydrous), (Sigma-Aldrich, St. Louis, MO, USA), 1,4-Diaminobutane (99%, Sigma-Aldrich, St. Louis, MO, USA), acetone (for analysis, >99%, Chemical Company Iasi), dimethyl sulfoxide (DMSO) (Aldrich-anhydrous, Darmstadt, Germany), pyridine (Aldrich, Darmstadt, Germany), glucose (Aldrich anhydrous, Darmstadt, Germany), phenol (Merck, Darmstadt, Germany), sulfuric acid (Sigma-Aldrich, St. Louis, MO, USA), bisphenol A (Merck, Darmstadt, Germany), ethanol (Chimopar, Bucuresti, România) were used as received.

2.2. *Methods*

2.2.1. ST-HEMA Emulsion Polymerization

A mixture of ST (1.3 mL), respectively, HEMA (0.3 mL) and KPS (25 mg) was added to distilled water (20 mL). The mixture was purged with nitrogen and then maintained for 8 h at 75 °C under continuous stirring. The final dispersion was dialyzed in distilled water for 7 days, using cellulose dialysis membranes (molecular weight cut-off: 12,000–14,000), to remove the unreacted monomers and initiator.

2.2.2. Seeded Emulsion Polymerization of GMA (ST-HEMA-GMA)

To the dialyzed emulsion presented previously was added 0.3 mL GMA and KPS (25 mg). The mixture was purged with nitrogen and then maintained for 8 h at 75 °C under continuous stirring. The final dispersion was dialyzed in distilled water for 7 days, using cellulose dialysis membranes (molecular weight cut-off: 12,000–14,000), to remove the unreacted monomers and initiator.

2.2.3. Synthesis of Diamino Butane Monosubstituted BCD (BCD-NH_2)

The Ts-BCD was synthesized by a method similar to that reported by Brady et al. [30] (For the NMR spectra and Maldi see Supporting info Figures S1 and S2). The procedure for obtaining BCD-NH_2 was as follows: 5.5 g Ts-BCD (4.266 mmol) were dissolved in 166 mL of 1,4-diaminobutane (DAB), slowly warmed up to 70 °C, and kept at this temperature for 24 h. At the end of reaction time, the solvent was vacuum distilled and the solid resulted was dissolved in a minimum amount of water. This syrup was added dropwise into 150–200 mL of acetone and precipitates a white solid. At least two acetone precipitations are needed to obtain a white powder solid, otherwise the product seems oily. The final product was obtained by drying the sample in a vacuum oven at 40 °C for 2 days with a 54% yield. **^{1}H-NMR** (DMSO-d6, 500 MHz): δ 1.405 (m, H-8, H-9), 2.5–2.562 (s, NH_2, NH, DMSO), 2.688 (m, H-7), 2.882–2.860 (m, H-10), 3.346–3.304 (m, H-2, H-4), 3.627–3.551 (m, H-3, H-5, H-6), 4.474 (s, 6-OH), 4.823 (s, H-1), 5.736 (br, 2-OH and 3-OH) and **^{13}C NMR** (DMSO-d6, 125 MHz): δ 26.94 (C-9), 29.33 (C-8), 40.69 (C-10), 49.05 (C-7), 49.40 (C-6), 59.93 (C-6*), 72.05 (C-5), 72.43 (C-2), 73.07 (C-3), 81.55 (C-4), 101.96 (C-1). (see Supplementary information Figures S3–S6)).

MALDI: calc. for $C_{46}H_{80}N_2O_{34}$, M is 1204.45 Da (monoisotopic mass); the simple $[M]^+$ or $[M+H]^+$ were not identified but the mass 1215 corresponds to a complex adduct $[2M+Na]^{2+}$.

2.2.4. The Reaction of BCD-OH with ST-HEMA-GMA

The ST-HEMA-GMA emulsion was dried in an oven at 70 °C on a glass plate. A solid powder was obtained after the water evaporation. 0.5 g ST-HEMA-GMA powder was dispersed in 10 mL DMSO and heated to 80 °C. After 30 min, 0.6 g BCD-OH was added to the mixture together with 0.01 mL pyridine, as catalysis. The reaction was kept to the temperature for 12 h and precipitated into hot water. The obtained white powder was filtered and dried in the oven.

2.2.5. The Reaction of BCD-NH_2 with ST-HEMA-GMA

0.5 g ST-HEMA-GMA powder was dispersed in 10 mL DMSO and heated to 80 °C. After 30 min, 0.7 g BCD-NH_2 was added to the mixture together with 0.01 mL pyridine, as catalysis. The reaction was kept to the temperature for 12 h and precipitated into hot water. The obtained white powder was filtered and dried in the oven.

2.3. Characterization

The morphology of the polymer nanoparticles was evaluated using transmission electron microscopy (TEM) (Tecnai G2 F20 TWIN Cryo-TEM, FEI Company), at 300 kV acceleration voltage, at a 1 Å resolution.

FTIR spectra were recorded on a Bruker VERTEX 70 spectrometer using 32 scans with a resolution of 4 cm^{-1} in 4000–600 cm^{-1} region. The samples were analyzed using the attenuated total reflection (ATR) technique.

The UV-Vis spectra were recorded using a V-550 Able Jasco spectrophotometer, using a bandwidth of 1 nm, and a scanning speed of 1000 nm min^{-1}. The BCD content was quantified by determining the reducing sugars of the polymer using concentrated H_2SO_4 acidolysis and phenol colorimetric analysis [31]. The procedure employed for the determination of BCD content and the calibration curve are presented in the Supporting Info.

The thermogravimetric analyses (TGA) were performed using a Netzsch TG 209 F3 Tarsus equipment considering the next parameters: nitrogen atmosphere flow rate 20 mL min^{-1}; samples mass: ~3 mg; temperature range: room temperature −700 °C; heating rate: 10 °C min^{-1} in an alumina crucible.

NMR experiments were carried out on Bruker Avance III 500, ^1H NMR spectra were recorded at 500 MHz using the solvent peaks as internal references and ^{13}C NMR spectra were recorded at 125 MHz. All NMR experiments were conducted according to the literature, and they can be found in the supplementary material.

Matrix-assisted laser desorption ionization time of flight (MALDI-TOF) mass spectrometry experiments were carried out on Shimadzu AXIMA Performance, operated in high-resolution reflectron mode using α-Cyano-4-hydroxycinnamic acid as matrix.

The fluorescence spectra have been registered using a FP-6500 Able Jasco spectrofluorometer.

3. Results and Discussion

The first stage of our study consisted in the synthesis of polymeric particles presenting the capacity for the chemical attachment of BCD derivatives (BCD-OH and BCD-NH_2, Scheme 1). The spherical shape was selected due to its high specific surface and high surface-to-volume ratio. The presence of an oxirane functional group at the surface of the polymer particles also permits the grafting of both beta-cyclodextrin derivatives (BCD-OH and BCD-NH_2). Therefore, the synthesis strategy involved the emulsion polymerization of ST-HEMA system [32], followed by the seeded polymerization of GMA on the surface of the polymer particles [33]. As a result, an oxirane functional group capable of reacting with both hydroxyl and amino functional groups will be present on the surface of the polymer particles after the seeded polymerization. Figure 1 presents TEM images of the polymer particles at different stages during the synthesis route: ST-HEMA (after the emulsion polymerization), ST-HEMA-GMA (after the seeded polymerization), and ST-HEMA-GMA-BCD-NH_2 (after BCD-NH_2 grafting to the polymeric particle).

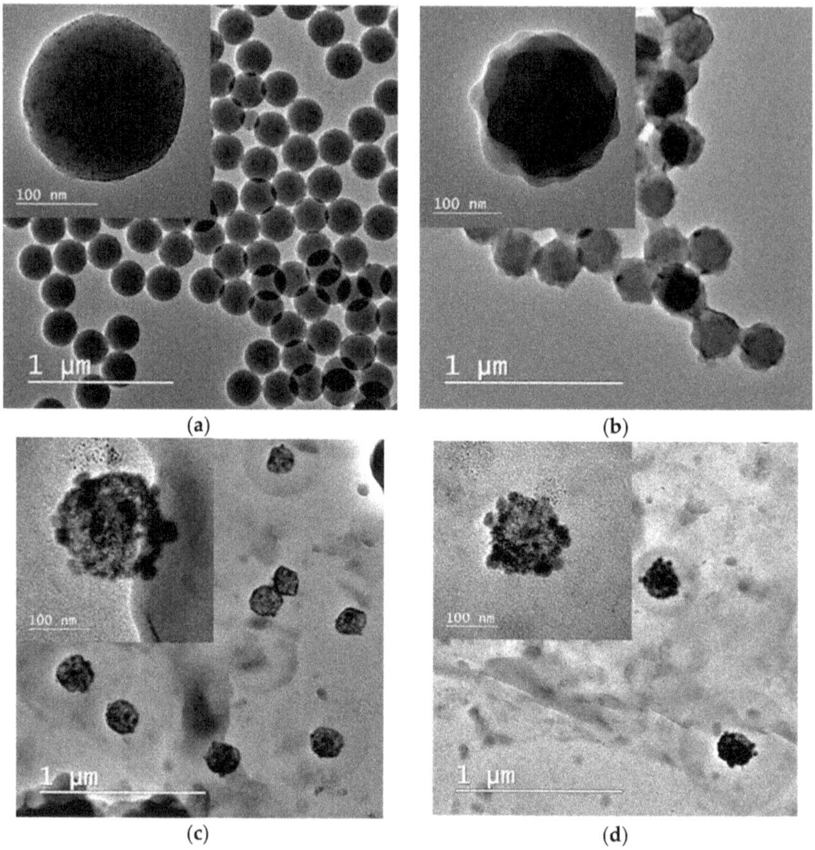

Figure 1. TEM images of the samples: (**a**) ST-HEMA; (**b**) ST-HEMA-GMA; (**c**) ST-HEMA-GMA-BCD-NH_2 and (**d**) ST-HEMA-GMA-BCD-OH.

From the analysis of Figure 1, it can be observed that the ST-HEMA polymer particles are around 200 nm with a monodisperse size distribution. The average particle size and the size distribution are increased after the GMA seeded polymerization. Thus, the average particle size increases to around 220–230 nm, and a core-shell structuring can be noticed (inset detail Figure 1b). In the case of BCD-NH_2 and BCD-OH grafting on the surface of the polymer particles, aggregates with an average dimension of 20 nm [34] can be observed deposited on the polymer particles after the seeded polymerization (darker spheres around the ST-HEMA-GMA Figure 1c,d).

To sustain the chemical grafting of the BCD derivative through the oxirane group reaction, the materials were analyzed by FT-IR spectroscopy (Figure 2). Thus, the specific vibration of oxirane 907 cm^{-1} decreased while the intensity of the signal at 3420 cm^{-1}, specific for -OH vibration increased. Moreover, the appearance of C-O-C vibration signal at 1033 cm^{-1} and 3266 cm^{-1} signal specific for NH vibration confirm the chemical attachment of the BCD derivatives to the polymeric particles. The signal at 1026 cm^{-1} specific for the vibration of primary alcohol functional groups can also be identified in the ST-HEMA-GMA-BCD-OH and ST-HEMA-GMA-BCD-NH_2 samples. In addition, the C−N bond that is formed during BCD-NH_2 grafting to the polymer particles can be highlighted in the BCD-NH_2 and ST-HEMA-GMA-BCD-NH_2 samples at 1143 cm^{-1}.

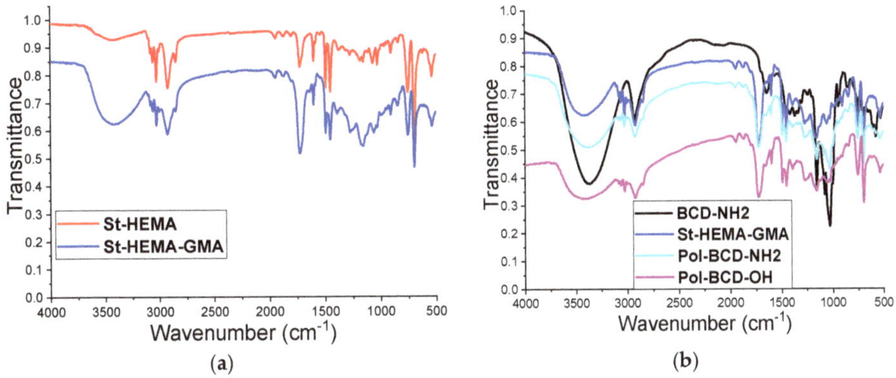

Figure 2. FT-IR spectra for ST-HEMA, BCD-NH_2, ST-HEMA-GMA-BCD-NH_2, ST-HEMA-GMA-BCD-OH, ST-HEMA-GMA, ST-HEMA. (**a**)ST-HEMA; ST-HEMA-GMA; (**b**) BCD-NH_2; ST-HEMA-GMA; Pol-BCD-NH_2; Pol-BCD-OH.

After the BCD attachment to the surface of the polymer particles, the degree of grafting was determined by NMR, TGA, and UV-Vis spectroscopy (see Supporting info Figure S7—calibration curve and determination method). Thus, BCD-OH content was determined at 4.27%, while in the case of BCD-NH_2 the content was significantly higher at 19.19%. The difference in BCD content in the final materials can be explained by the higher reactivity of the oxirane group towards the amino than hydroxyl groups.

The organic pollutants present in water, such as bisphenol A, can be removed relatively easily and at a low cost by adsorption processes that can exploit the presence of BCD cage structure immobilized on polymeric supports [35]. In this study, we have determined the quantity of bisphenol A adsorbed per gram of material using ST-HEMA-GMA-BCD materials (Q_e (mg bisphenol A/g polymer)). Thus, using a calibration curve for fluorescence intensity depending on the bisphenol A concentration (Supporting info Figure S8), the amount adsorbed onto the BCD modified polymer spheres was determined using Equation (1):

$$Q_e = \frac{(c_0 - c_e) \times V}{m} \quad (1)$$

where V is the solution volume (mL), c_0 (mg L^{-1}) and c_e (mg L^{-1}) are the initial and final solution concentrations of bisphenol A and m is the mass of BCD modified polymer particles (mg).

After 240 min of interaction at 25 °C, the adsorption capacity Q_e values obtained were 9.51 and 12.57 mg/g for ST-HEMA-GMA-BCD-OH and ST-HEMA-GMA-BCD-NH_2, respectively. The higher Q_e value obtained for the polymers modified using the BCD-NH_2 can be related to the higher BCD grafting degree. It is easily observed that the large difference in grafting efficiency is in contrast with the relatively small difference between the adsorption capacity at equilibrium. This can be explained by the formation of hydrogen bonds between the hydroxyl groups of BCD-OH and bisphenol A [20]. Nevertheless, the higher bisphenol A adsorption characteristics of ST-HEMA-GMA-BCD-NH_2 are evident. Consequently, a kinetic analysis of the adsorption process was performed for ST-HEMA-GMA-BCD-NH_2 (Figure 3).

Figure 3 illustrates the adsorption of bisphenol A onto the ST-HEMA-GMA-BCD-NH_2 as a function of time in contact. It can be noted that the absorption rate is fast up to 60 min which is followed by a slight decrease up to 200 min, followed by an equilibrium tendency of the system in the 200–300 min. The first stage can be explained by a high bisphenol A concentration in solution which diffuses to the polymer particles surface. The second stage, from 60 to 200 min, represents intermediary behavior during which the adsorption rate decreases as the internal diffusion resistance increases, which is finally followed by the equilibrium characteristics after 200 min. Several mathematical models were analyzed to determine the adsorption efficiency and the mechanism that it follows: Lagergren's pseudo-first-order kinetic model (Equation (2)), Ho's pseudo-second-order model (Equation (3)), and Weber's intra diffusion model (Equation (4)):

$$\ln(Q_e - Q_t) = \ln Q_e - k_1 t \quad (2)$$

$$\frac{t}{Q_t} = \frac{1}{k_2 Q_e^2} + \frac{t}{Q_e} \quad (3)$$

$$Q_t = K_p \sqrt{t} \quad (4)$$

where Q_e (mg g^{-1}) and Q_t (mg g^{-1}) are the amounts of bisphenol A adsorbed per unit mass of ST-HEMA-GMA-BCD-NH_2 at equilibrium and t (min), respectively. k_1 is the pseudo-first-order adsorption rate constant (min^{-1}), k_2 is the pseudo-second-order adsorption rate constant (g (mg min)$^{-1}$), and K_p is the intraparticle diffusion constant (mg g^{-1} $min^{-0.5}$).

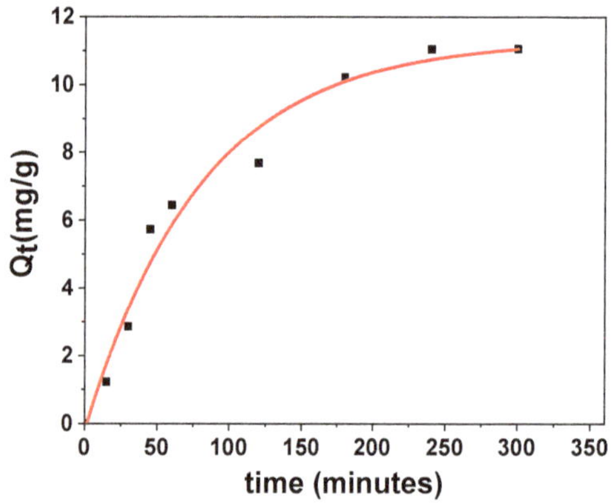

Figure 3. Adsorption kinetic curve of ST-HEMA-GMA-BCD-NH_2 at pH 5 and 25 °C.

The kinetic parameters of Lagergren's pseudo-first-order, Ho's pseudo-second-order, and Weber's intra particles diffusion models equations were calculated by slope-intercept of the linear fitting plots of ln(Q_e-Q_t) versus t, t/Q_t versus t, and Q_t versus t, respectively (Figure 4).

Comparing the experimental data with the results from the mathematical models for the adsorption process, the best fit was Lagergren's (both for Q_e value and for R^2) together with Weber's intraparticles diffusion model for the step that controls the mass transfer (Figure 3 and Table 1). Analyzing the values for K_p of Weber's model, this corresponds to a rapid adsorption process in the first stage of contacting [36].

Table 1. Kinetic parameters for the adsorption of bisphenol A by ST-HEMA-GMA-BCD-NH_2.

	Lagergren's Pseudo-First-Order			**Ho's Pseudo-Second-Order Model**			**Weber's Intraparticles Diffusion Model**	
$Q_{e, exp}$ (mg g^{-1})	**$k_1 \times 10^3$ (min^{-1})**	**Q_e (mg g^{-1})**	**R^2**	**$k_2 \times 10^3$ (g mg^{-1} min^{-1})**	**Q_e (mg g^{-1})**	**R^2**	**K_p (mg g^{-1} $min^{-0.5}$)**	**R^2**
12.57	4.7	12.36	0.933	2.1	8.79	0.9054	0.50873	0.9882

The study of the adsorption isotherms offers information on the interaction between the adsorbate and the adsorbent and allows the determination of the adsorption capacity of the adsorbent, which is an important parameter for system evaluation. The most intensively used isotherm adsorption model are Langmuir (Equation (5)) and Freundlich (Equation (6)).

$$\frac{1}{Q_e} = \frac{1}{Q_{max}} + \frac{1}{Q_m K_L} \times \frac{1}{c_e} \quad (5)$$

$$\ln Q_e = \ln K_F + \frac{1}{n} \ln c_e \quad (6)$$

where Q_e (mg g^{-1}) indicates the amount of adsorbate at equilibrium; Q_{max} (mg g^{-1}) indicates the maximum amount of adsorbate at equilibrium; c_e (mg L^{-1}) is the equilibrium concentration of the adsorbate in the solution; K_L (L mg^{-1}) and K_F (L mg^{-1}) are the Langmuir and Freundlich constants, respectively; and n is the heterogeneity factor.

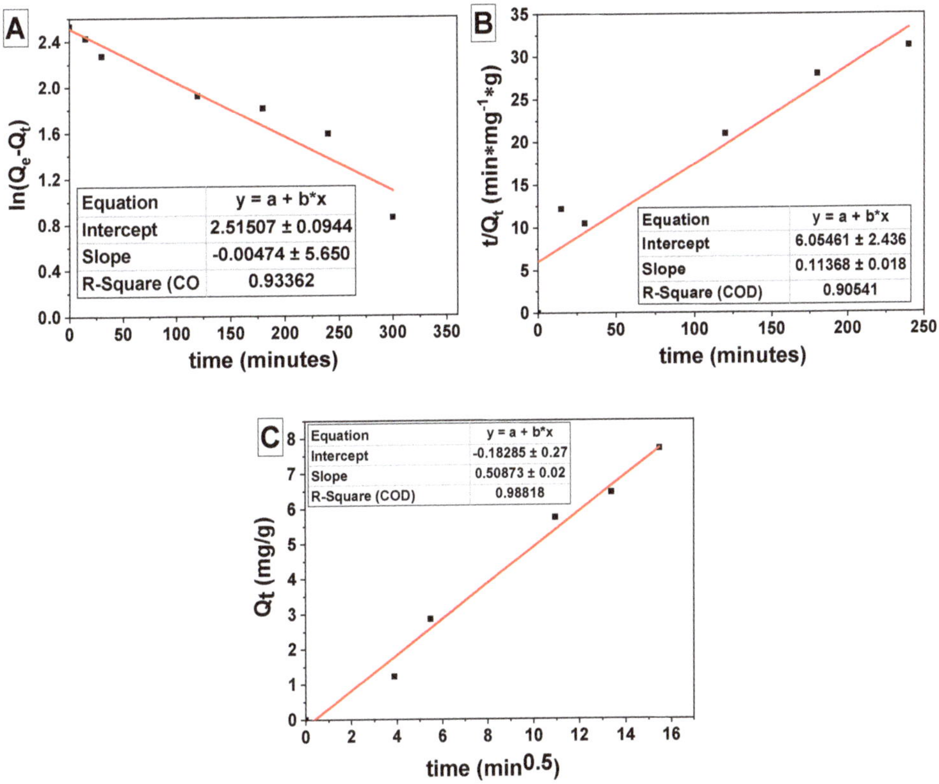

Figure 4. The linear plots of Lagergren's pseudo-first-order (**A**), Ho's pseudo-second-order (**B**), and Weber's intraparticles diffusion models (**C**).

The isothermal adsorption at the equilibrium of bisphenol A by ST-HEMA-GMA-BCD-NH_2 and ST-HEMA-GMA-BCD-OH are represented in (Supporting info Figure S9). From this, it can be noted that the adsorption capacity increases with the increase of adsorbate and reaches saturation as the bisphenol A concentration exceeds 45 mg/L and 40 mg/L in the case of ST-HEMA-GMA-BCD-NH_2 and ST-HEMA-GMA-BCD-OH, respectively. The linearization of the two selected isotherms Equation (5), Equation (6), and the parameters calculated are presented in Figure 5 and Table 2.

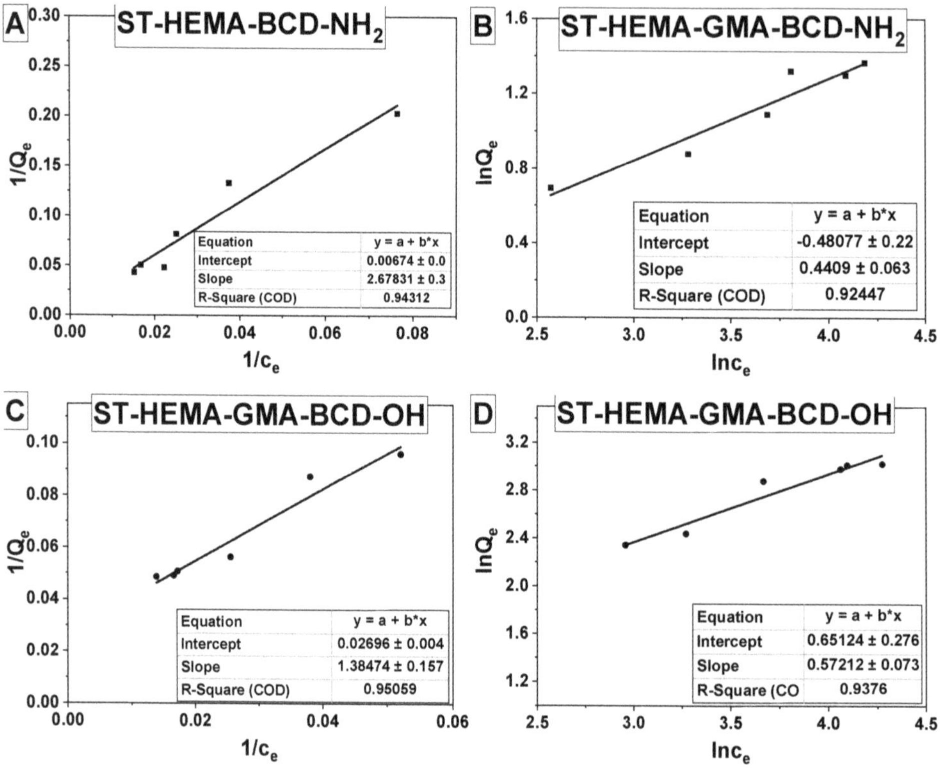

Figure 5. (**A**,**C**) Langmuir plots illustrating the linear dependences of $1/Q_e$ on $1/c_e$, and (**B**,**D**) Freundlich plot illustrating the linear dependences of $\ln Q_e$ on $\ln c_e$ for adsorption of bisphenol A by ST-HEMA-BCD-NH_2 and ST-HEMA-BCD-OH, respectively.

Table 2. Isotherm parameters for the adsorption of bisphenol A by ST-HEMA-GMA-BCD-NH_2 and ST-HEMA-GMA-BCD-OH.

Absorbent	Adsorbate	Langmuir			Freundlich		
		K_L/ (L mg)$^{-1}$	Q_{max}/ (mg g^{-1})	R^2	K_F/ (L mg^{-1})	n	R^2
ST-HEMA-GMA-BCD-NH_2	Bisphenol A	0.0025	148.37	0.9431	0.6183	2.268	0.9244
ST-HEMA-GMA-BCD-OH	Bisphenol A	0.0195	37.09	0.9505	1.9179	1.7478	0.9376

The comparison of the Langmuir and Freundlich isotherm parameters as listed in Table 2 indicates better linearity in the case of the Langmuir plots, suggesting that the adsorption of bisphenol A by ST-HEMA-GMA-BCD-NH_2 and ST-HEMA-GMA-BCD-OH can be regarded as a monolayer adsorption process. This means that the adsorption sites located on the surface of the polymer particles were homogeneously distributed and an equivalent adsorption force is displayed. Additionally, it could be evaluated from the Langmuir equation that the Q_{max} for bisphenol A was 148.37 mg g^{-1} and 37.09 mg g^{-1} for ST-HEMA-GMA-BCD-NH_2 and ST-HEMA-GMA-BCD-OH, respectively. (Table 3) For a cross-linked polymer, three main adsorption mechanisms were proposed which involve (1) host–guest interactions in the polymers cavities, (2) interactions in the pores of the polymeric network, and (3) interactions on the surface (physical sorption) [37]. The difference between the adsorption characteristics of the two polymers (Q_{max}) can be correlated to the amount of BCD grafted to the polymer particles [38]. Thus, due to its higher reactivity, BCD-NH_2 afforded a larger substitution degree which is reflected by the enhanced Q_{max} value.

Table 3. Comparison of adsorption capacities of different polymer adsorbents for bisphenol A.

Adsorbent	Phenolic Pollutants	Q_{max} (mg g^{-1})	Reference
Hyper-crosslinked β-CD porous polymer	Bisphenol A	278	[24]
Graphene oxide-β-CD nanocomposites	Bisphenol A	373.4	[39]
BCD polymer functionalized Fe_3O_4 magnetic nanoparticles	Bisphenol A	74.63	[20]
BCD grafted cellulose beads	Bisphenol A	30.77	[40]
(BCD+epichlorohydrin) polyBCD	Bisphenol A	84	[41]
BCD-Functionalized Mesoporous Magnetic Clusters	Bisphenol A	52.7	[42]
BCD-poly(glycidyl methacrylate)-SiO_2- nanoparticles	Bisphenol A	22.48	[43]
Diatomite cross-linked BCD polymers	Bisphenol A	83.57	[44]
ST-HEMA-GMA-BCD-NH_2	Bisphenol A	148.37	This study
ST-HEMA-GMA-BCD-OH	Bisphenol A	37.09	This study

Comparing the values for Q_{max} for the adsorption of bisphenol A using different adsorbents (Table 3), it can be noted that the values obtained for our materials are reasonably high. The increased efficiency is due to the high ratio between the specific surface to the mass of the polymer particles and the high reactivity of the surface functional groups (oxirane). These characteristics permit the synthesis of BCD functionalized polymer particles with a good weight ratio (BCD to polymer support) and a proper distribution of BCD on the surface of the polymer particles. Therefore, the possibility of guest–host interaction is facilitated from the steric point of view (see Supplementary Info Figure S10). In addition to this interaction, there is also the possibility for adsorption processes in the pores of the cross-linked polymer particles, respectively physical interaction on the surface of the materials [20]. To confirm the interaction between the polymeric structures modified with BCD and bisphenol A (BPA) FTIR analysis was performed on the materials after the adsorption evaluation. From the spectra (see Supplementary Materials Figure S11), the appearance of a novel signal at 1150 cm^{-1} corresponding to the C-O vibration from BPA [45] can be observed, confirming the presence of BPA on the surface of the polymer particles.

The thermal resistance is one of the most important properties of polymers and their derivatives, in our case the polymer particles modified with BCD-OH and BCD-NH_2. Figure 6 presents the TGA and DTG curves for the synthesized polymers and starting materials.

Comparing the temperatures corresponding for a 10% weight loss (Table 4) it can be noted that BCD-OH has the highest thermal resistance, followed by ST-HEMA-GMA-BCD-OH and ST-HEMA-BCD-NH_2, possibly due to the hydrogen bonds formed between the -OH groups. The ST-HEMA and ST-HEMA-GMA display lower thermal stability. The TGA analysis also establishes the temperature where the maximum weight loss takes place. In this case, the T_{max} indicated that the most stable materials were ST-HEMA-GMA-BCD-OH and ST-HEMA-GMA-BCD-NH_2, which display stability up to 400 °C.

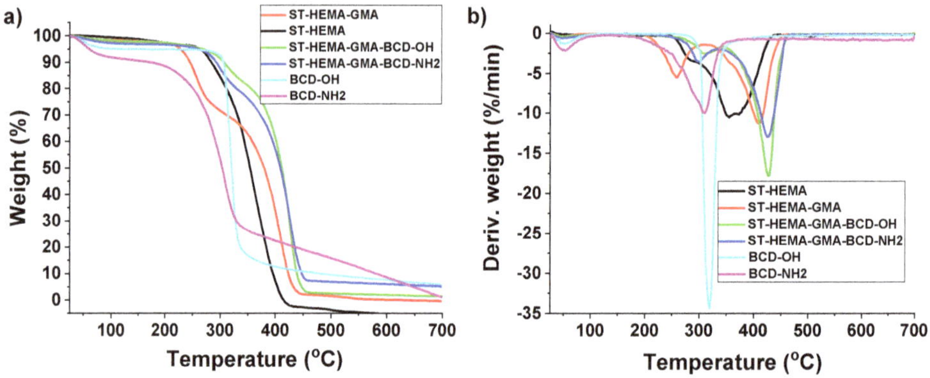

Figure 6. (**a**) TGA curves for BCD-NH_2, BCD-OH, ST-HEMA-GMA-BCD-NH_2, ST-HEMA-GMA-BCD-OH, ST-HEMA-GMA, ST-HEMA and (**b**) DTG curves for BCD-NH_2, BCD-OH, ST-HEMA-GMA-BCD-NH_2, ST-HEMA-GMA-BCD-OH, ST-HEMA-GMA, ST-HEMA.

Table 4. Thermal properties of BCD-NH_2, BCD-OH, ST-HEMA-GMA-BCD-NH_2, ST-HEMA-GMA-BCD-OH, ST-HEMA-GMA, ST-HEMA.

Sample	$T_{10\%}$, [°C]	T_{max}, [°C]
ST-HEMA	283	283; 355; 373
ST-HEMA-GMA	244	257; 409
BCD-NH_2	172	310
BCD-OH	306	318
ST-HEMA-GMA-BCD-OH	303	313; 425
ST-HEMA-GMA-BCD-NH_2	291	299; 424

$T_{10\%}$ = decomposition onset temperature (measured at 10% weight loss); T_{max} = the maximum decomposition temperature, corresponding to the maximum of DTG peak (the first derivative of the thermogravimetric curve).

4. Conclusions

A monoamine derivative of BCD was synthesized, characterized by NMR and mass spectroscopy. The BCD derivative was chemically grafted to ST-HEMA-GMA particles obtained by seeded emulsion polymerization. The GMA was used during the seeded polymerization due to its high reactivity towards -OH and NH_2 functional groups. The chemical attachment of the BCD derivatives (BCD-OH and BCD-NH_2) to the surface of the polymer particles was qualitatively confirmed by FT-IR spectroscopy and quantitatively by acidolysis and phenol colorimetric analysis. The degree of polymer functionalization using the BCD-OH and BCD-NH_2 derivatives were 4.27 and 19.19 weight %.

The adsorption properties of the materials were evaluated using bisphenol A as target molecule. The results from the mathematical models for the adsorption kinetics process indicated that the best fit was Lagergren's model (both for Q_e value and for R^2), together with Weber's intraparticles diffusion model for the step that controls the mass transfer in the case of ST-HEMA-GMA-BCD-NH_2. The isothermal adsorption evaluation indicated that both systems follow a Langmuir type behavior and afforded a Q_{max} value of 148.37 mg g^{-1} and 37.09 mg g^{-1} for ST-HEMA-GMA-BCD-NH_2 and ST-HEMA-GMA-BCD-OH, respectively. Due to the higher reactivity of BCD-NH_2, a larger substitution degree of the polymer particles was obtained, which is reflected by the enhanced Q_{max} value.

The thermogravimetric analysis of the materials indicated that the functionalization of the polymer particles afforded an increase of the thermal resistance. Thus, the BCD modified polymers present a degradation temperature of over 400 °C, which can be attributed to the hydrogen bonds and BCD thermal degradation pathway in the presence of the polymers.

Supplementary Materials: NMR and Maldi characterization Figures S1–S6; UV-Vis and fluorescence determinations procedures and calibration curves (Figure S7 and S8); adsorption isotherms (Figure S9), 3D rendering of bisphenol A insertion in BCD-NH_2 (Figure S10) and FTIR spectra of the polymers modified with BCD after BPA adsorption study (Figure S11) are available online at https://www.mdpi.com/article/10.3390/polym13142338/s1.

Author Contributions: Conceptualization, E.R.; methodology, F.R.; software S.B.; validation, E.R., I.M. and A.D.; formal analysis, A.G.; investigation, R.S.; resources, A.M.; data curation, A.D.; writing—original draft preparation, E.R.; writing—review and editing, E.R.; visualization, A.D.; supervision, E.R.; project administration, I.M.; funding acquisition, A.M., A.C.B. All authors have read and agreed to the published version of the manuscript.

Funding: This research received no external funding.

Acknowledgments: Authors are thankful to the POSCCE-O 2.2.1, SMIS-CSNR 13984-901, No. 257/28.09.2010 Project, CERNESIM, for the NMR experiments. Raluca Șomoghi acknowledges financial support by Ministry of Research and Inovation, Nucleu Programme, Project PN.19.23.01.01 Smart-Bi. This work was supported by a grant of the Ministry of Research, Innovation and Digitization, CNCS/CCCDI—UEFISCDI, project number PN-III-P1-1.1-TE-2019-1387 (contract number TE141/2020), within PNCDI III.

Conflicts of Interest: The authors declare no conflict of interest.

References

1. Ohore, O.E.; Zhang, S. Endocrine disrupting effects of bisphenol A exposure and recent advances on its removal by water treatment systems. A review. *Sci. Afr.* **2019**, *5*, e00135. [CrossRef]
2. Bolong, N.; Ismail, A.F.; Salim, M.R.; Rana, D.; Matsuura, T.; Tabe-Mohammadi, A. Negatively charged polyethersulfone hollow fiber nanofiltration membrane for the removal of bisphenol A from wastewater. *Sep. Purif. Technol.* **2010**, *73*, 92–99. [CrossRef]
3. Liu, J.; Zhang, L.; Lu, G.; Jiang, R.; Yan, Z.; Li, Y. Occurrence, toxicity and ecological risk of Bisphenol A analogues in aquatic environment—A review. *Ecotoxicol. Environ. Saf.* **2021**, *208*, 111481. [CrossRef]
4. Wyżga, B.; Połeć, K.; Olechowska, K.; Hąc-Wydro, K. The impact of toxic bisphenols on model human erythrocyte membranes. *Colloids Surf. B Biointerfaces* **2020**, *186*, 110670. [CrossRef] [PubMed]
5. Zhang, H.; Ding, T.; Luo, X.; Li, J. Toxic effect of fluorene-9-bisphenol to green algae Chlorella vulgaris and its metabolic fate. *Ecotoxicol. Environ. Saf.* **2021**, *216*, 112158. [CrossRef] [PubMed]
6. Huang, R.-p.; Liu, Z.-h.; Yuan, S.-f.; Yin, H.; Dang, Z.; Wu, P.-x. Worldwide human daily intakes of bisphenol A (BPA) estimated from global urinary concentration data (2000–2016) and its risk analysis. *Environ. Pollut.* **2017**, *230*, 143–152. [CrossRef] [PubMed]
7. Sharma, P.; Chadha, P. Bisphenol A induced toxicity in blood cells of freshwater fish Channa punctatus after acute exposure. *Saudi J. Biol. Sci.* **2021**. [CrossRef]
8. Zdarta, J.; Staszak, M.; Jankowska, K.; Kaźmierczak, K.; Degórska, O.; Nguyen, L.N.; Kijeńska-Gawrońska, E.; Pinelo, M.; Jesionowski, T. The response surface methodology for optimization of tyrosinase immobilization onto electrospun polycaprolactone–chitosan fibers for use in bisphenol A removal. *Int. J. Biol. Macromol.* **2020**, *165*, 2049–2059. [CrossRef]
9. Fang, Z.; Hu, Y.; Cheng, J.; Chen, Y. Continuous removal of trace bisphenol A from water by high efficacy TiO_2 nanotube pillared graphene-based macrostructures in a photocatalytically fluidized bed. *Chem. Eng. J.* **2019**, *372*, 581–589. [CrossRef]
10. Yang, S.; Wu, P.; Liu, J.; Chen, M.; Ahmed, Z.; Zhu, N. Efficient removal of bisphenol A by superoxide radical and singlet oxygen generated from peroxymonosulfate activated with Fe0-montmorillonite. *Chem. Eng. J.* **2018**, *350*, 484–495. [CrossRef]
11. Zhen, J.; Zhang, S.; Zhuang, X.; Ahmad, S.; Lee, T.; Si, H.; Cao, C.; Ni, S.-Q. Sulfate radicals based heterogeneous peroxymonosulfate system catalyzed by CuO-Fe3O4-Biochar nanocomposite for bisphenol A degradation. *J. Water Process Eng.* **2021**, *41*, 102078. [CrossRef]
12. Zahari, A.M.; Shuo, C.W.; Sathishkumar, P.; Yusoff, A.R.M.; Gu, F.L.; Buang, N.A.; Lau, W.-J.; Gohari, R.J.; Yusop, Z. A reusable electrospun PVDF-PVP-MnO_2 nanocomposite membrane for bisphenol A removal from drinking water. *J. Environ. Chem. Eng.* **2018**, *6*, 5801–5811. [CrossRef]
13. Nasseri, S.; Ebrahimi, S.; Abtahi, M.; Saeedi, R. Synthesis and characterization of polysulfone/graphene oxide nano-composite membranes for removal of bisphenol A from water. *J. Environ. Manag.* **2018**, *205*, 174–182. [CrossRef] [PubMed]
14. Liu, Z.; Ye, L.; Xi, J.; Wang, J.; Feng, Z.-g. Cyclodextrin polymers: Structure, synthesis, and use as drug carriers. *Prog. Polym. Sci.* **2021**, *118*, 101408. [CrossRef]
15. Luo, Q.; He, L.; Wang, X.; Huang, H.; Wang, X.; Sang, S.; Huang, X. Cyclodextrin derivatives used for the separation of boron and the removal of organic pollutants. *Sci. Total Environ.* **2020**, *749*, 141487. [CrossRef] [PubMed]

16. Gaálová, J.; Michel, M.; Bourassi, M.; Ladewig, B.P.; Kasal, P.; Jindřich, J.; Izák, P. Nafion membranes modified by cationic cyclodextrin derivatives for enantioselective separation. *Sep. Purif. Technol.* **2021**, *266*, 118538. [CrossRef]
17. Omtvedt, L.A.; Dalheim, M.Ø.; Nielsen, T.T.; Larsen, K.L.; Strand, B.L.; Aachmann, F.L. Efficient Grafting of Cyclodextrin to Alginate and Performance of the Hydrogel for Release of Model Drug. *Sci. Rep.* **2019**, *9*, 9325. [CrossRef]
18. Jicsinszky, L.; Martina, K.; Cravotto, G. Cyclodextrins in the antiviral therapy. *J. Drug Deliv. Sci. Technol.* **2021**, *64*, 102589. [CrossRef] [PubMed]
19. Morillo, E.; Madrid, F.; Lara-Moreno, A.; Villaverde, J. Soil bioremediation by cyclodextrins. A review. *Int. J. Pharm.* **2020**, *591*, 119943. [CrossRef]
20. Gong, T.; Zhou, Y.; Sun, L.; Liang, W.; Yang, J.; Shuang, S.; Dong, C. Effective adsorption of phenolic pollutants from water using β-cyclodextrin polymer functionalized Fe_3O_4 magnetic nanoparticles. *RSC Adv.* **2016**, *6*, 80955–80963. [CrossRef]
21. Mpatani, F.M.; Aryee, A.A.; Kani, A.N.; Guo, Q.; Dovi, E.; Qu, L.; Li, Z.; Han, R. Uptake of micropollutant-bisphenol A, methylene blue and neutral red onto a novel bagasse-β-cyclodextrin polymer by adsorption process. *Chemosphere* **2020**, *259*, 127439. [CrossRef]
22. Zhou, Y.; Cheng, G.; Chen, K.; Lu, J.; Lei, J.; Pu, S. Adsorptive removal of bisphenol A, chloroxylenol, and carbamazepine from water using a novel β-cyclodextrin polymer. *Ecotoxicol. Environ. Saf.* **2019**, *170*, 278–285. [CrossRef]
23. Xiao, G.; Fu, L.; Li, A. Enhanced adsorption of bisphenol A from water by acetylaniline modified hyper-cross-linked polymeric adsorbent: Effect of the cross-linked bridge. *Chem. Eng. J.* **2012**, *191*, 171–176. [CrossRef]
24. Li, X.; Zhou, M.; Jia, J.; Ma, J.; Jia, Q. Design of a hyper-crosslinked β-cyclodextrin porous polymer for highly efficient removal toward bisphenol a from water. *Sep. Purif. Technol.* **2018**, *195*, 130–137. [CrossRef]
25. Raza, S.; Wen, H.; Peng, Y.; Zhang, J.; Li, X.; Liu, C. Fabrication of SiO_2 modified biobased hydrolyzed hollow polymer particles and their applications as a removal of methyl orange dye and bisphenol-A. *Eur. Polym. J.* **2021**, *144*, 110199. [CrossRef]
26. Wang, Z.; Guo, S.; Zhang, B.; Fang, J.; Zhu, L. Interfacially crosslinked β-cyclodextrin polymer composite porous membranes for fast removal of organic micropollutants from water by flow-through adsorption. *J. Hazard. Mater.* **2020**, *384*, 121187. [CrossRef] [PubMed]
27. Huang, W.; Hu, Y.; Li, Y.; Zhou, Y.; Niu, D.; Lei, Z.; Zhang, Z. Citric acid-crosslinked β-cyclodextrin for simultaneous removal of bisphenol A, methylene blue and copper: The roles of cavity and surface functional groups. *J. Taiwan Inst. Chem. Eng.* **2018**, *82*, 189–197. [CrossRef]
28. Fenyvesi, É.; Barkács, K.; Gruiz, K.; Varga, E.; Kenyeres, I.; Záray, G.; Szente, L. Removal of hazardous micropollutants from treated wastewater using cyclodextrin bead polymer—A pilot demonstration case. *J. Hazard. Mater.* **2020**, *383*, 121181. [CrossRef]
29. Tizro, N.; Moniri, E.; Saeb, K.; Panahi, H.A.; Ardakani, S.S. Preparation and application of grafted β-cyclodextrin/thermo-sensitive polymer onto modified Fe_3O_4@SiO_2 nano-particles for fenitrothion elimination from aqueous solution. *Microchem. J.* **2019**, *145*, 59–67. [CrossRef]
30. Brady, B.; Lynam, N.; O'Sullivan, T.; Ahern, C.; Darcy, R. β-Cyclodextrin-6 A-(4-methylbenzenesulfonate). *Org. Synth.* **2000**, *77*, 220–224. [CrossRef]
31. DuBois, M.; Gilles, K.A.; Hamilton, J.K.; Rebers, P.A.; Smith, F. Colorimetric Method for Determination of Sugars and Related Substances. *Anal. Chem.* **1956**, *28*, 350–356. [CrossRef]
32. Rusen, E.; Mocanu, A.; Marculescu, B.; Somoghi, R.; Butac, L.; Miculescu, F.; Cotrut, C.; Antoniac, I.; Cincu, C. Obtaining complex structures starting from monodisperse poly(styrene-co-2-hydroxyethylmethacrylate) spheres. *Colloids Surf. A Physicochem. Eng. Asp.* **2011**, *375*, 35–41. [CrossRef]
33. Rusen, E.; Mocanu, A.; Marculescu, B. Obtaining of monodisperse particles through soap-free and seeded polymerization, respectively, through polymerization in the presence of C60. *Colloid Polym. Sci.* **2010**, *288*, 769–776. [CrossRef]
34. Ryzhakov, A.; Do Thi, T.; Stappaerts, J.; Bertoletti, L.; Kimpe, K.; Sá Couto, A.R.; Saokham, P.; Van den Mooter, G.; Augustijns, P.; Somsen, G.W.; et al. Self-Assembly of Cyclodextrins and Their Complexes in Aqueous Solutions. *J. Pharm. Sci.* **2016**, *105*, 2556–2569. [CrossRef]
35. Simões, S.M.N.; Rey-Rico, A.; Concheiro, A.; Alvarez-Lorenzo, C. Supramolecular cyclodextrin-based drug nanocarriers. *Chem. Commun.* **2015**, *51*, 6275–6289. [CrossRef]
36. Wu, F.-C.; Tseng, R.-L.; Juang, R.-S. Initial behavior of intraparticle diffusion model used in the description of adsorption kinetics. *Chem. Eng. J.* **2009**, *153*, 1–8. [CrossRef]
37. Ye, H.; Zhao, B.; Zhou, Y.; Du, J.; Huang, M. Recent advances in adsorbents for the removal of phthalate esters from water: Material, modification, and application. *Chem. Eng. J.* **2021**, *409*, 128127. [CrossRef]
38. Zhou, Z.; Xiao, Y.; Hatton, T.A.; Chung, T.-S. Effects of spacer arm length and benzoation on enantioseparation performance of β-cyclodextrin functionalized cellulose membranes. *J. Membr. Sci.* **2009**, *339*, 21–27. [CrossRef]
39. Gupta, V.K.; Agarwal, S.; Sadegh, H.; Ali, G.A.M.; Bharti, A.K.; Hamdy Makhlouf, A.S. Facile route synthesis of novel graphene oxide-β-cyclodextrin nanocomposite and its application as adsorbent for removal of toxic bisphenol A from the aqueous phase. *J. Mol. Liq.* **2017**, *237*, 466–472. [CrossRef]
40. Lin, Q.; Wu, Y.; Jiang, X.; Lin, F.; Liu, X.; Lu, B. Removal of bisphenol A from aqueous solution via host-guest interactions based on beta-cyclodextrin grafted cellulose bead. *Int. J. Biol. Macromol.* **2019**, *140*, 1–9. [CrossRef]
41. Kitaoka, M.; Hayashi, K. Adsorption of Bisphenol A by Cross-Linked β-Cyclodextrin Polymer. *J. Incl. Phenom. Macrocycl. Chem.* **2002**, *44*, 429–431. [CrossRef]

42. Lee, J.H.; Kwak, S.-Y. Rapid adsorption of bisphenol A from wastewater by β-cyclodextrin-functionalized mesoporous magnetic clusters. *Appl. Surf. Sci.* **2019**, *467–468*, 178–184. [CrossRef]
43. Wang, N.; Zhou, L.; Guo, J.; Ye, Q.; Lin, J.-M.; Yuan, J. Adsorption of environmental pollutants using magnetic hybrid nanoparticles modified with β-cyclodextrin. *Appl. Surf. Sci.* **2014**, *305*, 267–273. [CrossRef]
44. Shi, S.; Ocampo-Pérez, R.; Lv, J.; Liu, Q.; Nan, F.; Liu, X.; Xie, S.; Feng, J. Diatomite cross-linked β-Cyclodextrin polymers: A novel vision of diatomite adsorbent for the removal of bisphenol A. *Environ. Technol. Innov.* **2021**, *23*, 101602. [CrossRef]
45. Ullah, R.; Ahmad, I.; Zheng, Y. Fourier Transform Infrared Spectroscopy of "Bisphenol A". *J. Spectrosc.* **2016**, *2016*, 2073613. [CrossRef]

polymers

MDPI

Article

The Effect of Titanium Tetra-Butoxide Catalyst on the Olefin Polymerization

Mohammed S. Alsuhybani and Eid M. Alosime *

King Abdulaziz City for Science and Technology (KACST), P.O. Box 6086, Riyadh 11442, Saudi Arabia; sohybani@kacst.edu.sa
* Correspondence: alosimi@kacst.edu.sa

Abstract: The purpose of this study was to assess the ability of titanium Ti(IV) alkyloxy compounds supported by organic polymer polyvinyl chloride (PVC) to polymerize ethylene by feeding triethylaluminium (TEA) as a cocatalyst. Additionally, the impacts of the molar ratio of [Al]/[Ti] on the catalytic activities in ethylene's polymerization and of the comonomer through utilization of diverse quantities of comonomers on a similar or identical activity were studied. The optimal molar ratio of [Al]/[Ti] was 773:1, and the prepared catalyst had an initial activity of up to 2.3 kg PE/mol Ti. h. when the copolymer was incorporated with 64 mmol of 1-octene. The average molecular weight (M_w) of the copolymer produced with the catalysts was between 97 kg/mol and 326 kg/mol. A significant decrease in the M_w was observed, and PDI broadened with increasing concentration of 1-hexene because of the comonomer's stronger chain transfer capacity. The quick deactivation of titanium butoxide $Ti(OBu)_4$ on the polymers was found to be associated with increasing oxidation when supported by the catalyst. The presence of Ti(III) after reduction with the aluminum alkyls cleaves the carbon-chlorine bonds of the polymer, producing an inactive polymeric Ti(IV) complex. The results show that synergistic effects play an important role in enhancing the observed rate of reaction, as illustrated by evidence from scanning electron microscopy (SEM). The diffusion of cocatalysts within catalytic precursor particles may also explain the progression of cobweb structures in the polymer particles.

Keywords: polyethylene; olefin polymerization; Ziegler–Natta; polyvinyl chloride

Citation: Alsuhybani, M.S.; Alosime, E.M. The Effect of Titanium Tetra-Butoxide Catalyst on the Olefin Polymerization. *Polymers* **2021**, *13*, 2109. https://doi.org/10.3390/polym13132109

Academic Editor: Edina Rusen

Received: 16 June 2021
Accepted: 25 June 2021
Published: 26 June 2021

Publisher's Note: MDPI stays neutral with regard to jurisdictional claims in published maps and institutional affiliations.

Copyright: © 2021 by the authors. Licensee MDPI, Basel, Switzerland. This article is an open access article distributed under the terms and conditions of the Creative Commons Attribution (CC BY) license (https://creativecommons.org/licenses/by/4.0/).

1. Introduction

Ziegler-Natta (ZN) catalysts are the most commonly used catalysts in the olefin polymerization industry [1]. They are mainly composed of transition metal compounds, such as titanium, chromium, and vanadium precursors [2], and they are considered the best option for olefin polymerization industries because of their high productivity and good morphology control [3]. They undergo activation through the use of either an activator or a cocatalyst that activates inactive sites [4]. The most commonly used cocatalysts are alkylaluminium based, such as triethylaluminium (TEA) and tri-octyl aluminum [5,6].

$MgCl_2$ in combination with either $TiCl_4$ or $TiCl_3$ enhances the effectiveness of ZN catalysts and TEA cocatalysts [7,8]. In the synthesis of a novel chromium SiO_2/MgO-based ZN catalyst using water-soluble magnesium sources, the Cr-Ti catalysts have been reported to increase the polymerization activity and can generate polyethylene with a favorable hydrogen response [9]. Kinetic investigations of ethylene polymerization have demonstrated that two types of active sites, $TiCl_4$ and Cp_2ZrCl_2, are formed when $TiCl_4$ and zirconocene (Cp_2ZrCl_2) catalysts are anchored with a $MgCl_2(THF)_2$ support and then activated using TEA and methylaluminoxane [10]. Furthermore, silicon dioxide (SiO_2) has also been utilized as a support [11].

A $MgCl_2/SiO_2$ bisupport utilizes magnesium acetate as a source of Mg. When $MgCl_2/SiO_2$ reacts with $TiCl_4$ and $VOCl_3$ under different reaction sequences, ZN hybrid titanium/vanadium catalysts are formed [12]. The effectiveness of the Ti/V hybrid catalysts lies between that of the MgTi/Si and MgV/Si catalysts. However, the Ti/V hybrid catalysts, which result from a coreaction with $TiCl_4$ and $VOCl_3$, show increased activity compared with the Ti/V hybrid catalysts prepared using a two-step mechanism [13].

The polymerization rate is influenced by factors such as the concentration of the active center, the propagation rate constant, and the monomer concentration [14]. The concentration of the monomer changes gradually as one moves from the surface toward the core of the polymer/catalyst particles [15,16]. The concentration of the monomer is assumed to remain even across the entire polymer medium, whereas the propagation rate constant highly depends on the stereospecificity of the active centers [17]. The increased concentration of the active center promotes the activity of the ZN catalysts [18]. In addition, an appropriate amount of diethylaluminium chloride is added to the ZN catalyst system. Here, alongside TEA as the cocatalyst, the catalytic activity increases [19].

The functional groups within polymer structures play a crucial role in promoting the formation of either chemical bonds or interactions between the polymers and catalysts [20]. However, polymers such as atactic poly(propylene), natural rubber, and polyvinyl chloride are characterized by low-surface free energies and a lack of functional grouping. Therefore, they cannot interact with ZN catalysts unless functional groups are introduced [21]. Catalytic systems containing chlorine either as part of the support or as a cocatalyst exhibit enhanced activity; therefore, to allow for the study of their activities, PP, NR, and PVC are chlorinated before being subjected to the heterogenization of the ZN catalyst [22].

Recently, it has been discovered that the use of Ti(IV) alkoxide complexes—with 1,2- and 1,4-diolate ligands activated by a binary activator {Bu_2Mg + Et_2AlCl} and as a catalyst during the polymerization process of ethylene—results in the formation of an ultrahigh molecular weight polyethylene (UHMWPE) [23]. Ti(IV) complexes with diolate ligands are also very efficient in the copolymerization of ethylene and α-olefins. However, the effectiveness of the various diolate complexes differs based on the size of the chelate ring and the amount of aluminum chloride used [23].

In the current paper, we present our experiments, which are aimed at further understanding ZN catalysts and the processes for olefin polymerization. The titanium tetrabutoxide $Ti(OBu)_4$ compound was used as a compound-containing oxygen to determine the impact on the $TiCl_4$ catalyst, which was supported using PVC polymeric material. The latter was treated by a Grignard compound, and TEA ($Al(C_2H_5)_3$) was used as a cocatalyst. The utilization of a PVC-based polymeric support in preparing the catalyst provided important benefits compared with contemporary methods that use polymerization catalysts supported by magnesium chloride ($MgCl_2$) and SiO_2. Moreover, the PVC-based polymeric supports (or particles) for preparing the catalysts have shortened the dehydration duration and steps, allowing for lower temperatures compared with the polymerization catalysts supported by $MgCl_2$ and SiO_2. Therefore, PVC-based polymeric supports (or particles) are more suitable for synthesizing catalysts because they simplify the synthesis process, leading to a significant reduction in the cost of preparing catalysts. Additionally, the cost of PVC-based support tends to be considerably lower compared with polymerization catalysts that are supported by $MgCl_2$ and SiO_2. Similarly, the catalyst substantially utilizes the lowest levels of catalyst precursors in preparing the catalyst compared with the polymerization catalysts supported by $MgCl_2$ and SiO_2. The catalytic activities of $Ti(OBu)_4$ catalysts, for both ethylene homopolymerization and its copolymerization, ethylene, and 1-octene and ethylene and 1-hexene, were assessed.

2. Experiment

2.1. Materials

Ethylene gas was supplied from Abdullah Hashem Co., Saudi Arabia, with a purity of 99.95%. Here, *n*-hexane served as a polymerization medium in the heterogenous phase and was purchased from BDH® with a purity of 99%. In addition, 1-octane was purchased from Ried Dehean, and 1-hexene was obtained from Advanced Engineering, UK, and dried over 5 Å molecular sieves before use. Butylmagnesium chloride (BuMgCl) was purchased from Aldrich Chemical (2 M in THF). Titanium(IV) butoxide ($Ti(OBu)_4$) with 100% purity was purchased from Akzo Chemie America. PVC spheres with an average particle size of 50 μm were used (supplied by SABIC, Riyadh, Saudi Arabia). All support and catalyst synthesis and characterizations were performed under inert gas.

2.2. *Synthesis of PVC/BuMgCl/Ti(OBu)$_4$·TiCl$_4$ Catalyst*

The PVC/BuMgCl/$Ti(OBu)_4 \cdot TiCl_4$ catalyst was synthesized using a three-neck round-bottom flask equipped with a condenser and stirrer. Here, 5.0 g of PVC was added into the flask, and the flask containing the PVC was heated to 90 °C using an oil bath for 30 min. Then the flask was evacuated at 40 mbar pressure for 30 min. The flask and its contents were purged with dried nitrogen three times to ensure that the flask did not contain any oxygen, and the PVC was slurried using 30 mL of n-hexane. Afterwards, 1.5 mL of 2 M BuMgCl was added to the slurry, and the resultant mixture was stirred for 5 min at 60 °C. After this, 0.2 mL of 1 M $Ti(OBu)_4$ was added to the previous slurry, and the resulting mixture was stirred for 60 min at 60 °C. Once it was swept using nitrogen for a period of 60 min, the resultant precursor was then vacuum dried at 110 °C for 30 min. The obtained powder was then mixed with 2 mL of $TiCl_4$ (1 M) followed by stirring for 1 h. The mixture was brown before the participate distilled out; the hexane was then removed by cannula. The mixture was washed with 30 mL n-hexane three times and then dried and purged with nitrogen. A solid brown spherical catalyst particle was obtained. The contents of Ti and Mg were determined as 0.000379 and 0.00507 mmol respectively using inductively coupled plasma spectroscopy.

2.3. *Ethylene Homopolymerization*

Ethylene polymerization was tested at five different molar ratios of cocatalyst TEA to Ti [Al]/[Ti]: 309, 618, 773, 927, and 1236 mmol, to which 0.02 g of the catalyst was added before pressurization with ethylene.

Ethylene homopolymerizations were carried out in a 2 L glass jacketed reactor. The reactor was purged with a high-vacuum pump with nitrogen at 130 °C for 1 h to ensure that all moisture and oxygen were removed. Afterwards, 400 mL of n-hexane was added into the reactor. The reactor was stirred and heated to a reaction temperature of 88 °C. Ethylene was then introduced, followed by the TEA cocatalyst, which was left in for 5 min. The PVC/BuMgCl/$Ti(OBu)_4 \cdot TiCl_4$ catalyst was then fed into the reactor, followed by ethylene gas to maintain a specific pressure of 7 bar for 1 h of the polymerization reaction. The reactor was maintained and controlled at the desired reaction temperature throughout the polymerization process. Upon completion, the ethylene flow was stopped, and the reactor pressure was gradually lowered through venting. Ethylene consumption was automatically recorded using a flowmeter. The reactor was opened, and the polymer product was collected, filtered, and washed with petroleum ether, methanol, and chloric acid before being dried overnight in a vacuum oven at 70 °C.

2.4. *Ethylene/1-Octene and Ethylene/1-Hexene Copolymerization*

Copolymerizations of ethylene/1-octene (32, 48, 64, 96, and 127 mmol of 1-octene) and ethylene/1-hexene (40, 60, 68, 80, and 160 mmol of 1-hexene) were carried out in n-hexane in a 2 L glass jacketed reactor equipped with a mechanical stirrer. The reactor was purged with a high-vacuum pump with nitrogen at 130 °C for 1 h to ensure that all moisture and oxygen were removed. The reactor was charged with 400 mL of 1-hexene. Next, the ethylene gas feed was started, followed by equilibration at the desired polymerization temperature. After 15 min, a solution of TEA cocatalysts in n-hexane was added and stirred for 5 min, and an n-hexane solution (1 mL) of the titanium complex PVC/BuMgCl/$Ti(OBu)_4 \cdot TiCl_4$ was then added to the reactor with vigorous stirring to initiate polymerization. During the polymerization process, the mixture of ethylene copolymer was fed to the reactor continuously to maintain atmospheric pressure. Upon completion of the polymerization process, the ethylene flow was stopped, and the reactor pressure was slowly released through venting. Ethylene consumption was automatically recorded using a flowmeter. The reactor was opened, and the polymer product was collected, filtered, and washed with petroleum ether, methanol, and chloric acid before being dried overnight in a vacuum oven at 70 °C.

2.5. Characterization

The polymer features were investigated through differential scanning calorimetry (DSC), thermogravimetric analysis (TGA), gel permeation chromatography (GPC), and scanning electron microscopy (SEM). DSC measurements of the polymer samples were carried out using a differential scanning calorimeter (DSC-8500, PerkinElmer, Shelton, CT, USA). The sample weight was approximately 10 mg, and the employed heating rate was 10 °C/min. The sample was heated to 200 °C and allowed to rest for 2 min to remove the thermal history; it was then cooled down to 25 °C at 10 °C/min and finally heated to 200 °C at 10 °C/min to record the second heating curve and melting temperature (T_m) because the first heating cycle is affected by the mechanical and thermal history of sample preparation. The enthalpy of fusion (ΔH_f) of each sample was also calculated using the DSC curve.

The thermal behavior of the mixing catalyst/cocatalyst molar ratio of the polymerization was evaluated using a thermogravimetric analyzer (TGA 1, Perkin Elmer, Shelton, CT, USA). Each sample was heated from room temperature to 700 °C at a rate of 10 °C/min. Thermal decomposition temperatures and weight loss were determined from the weight–temperature curve. The TGA test was performed under a nitrogen atmosphere.

The average molecular weight (M_w) and polydispersity index (M_w/M_n) were measured using a gel permeation chromatograph (GPC 2000, Waters Corp., Milford, MA, USA) with standard polystyrene as a reference and 1,2,4-trichlorobenzene (TCB) as the solvent. Polymers were dissolved at 150 °C in TCB in a concentration of 1.0 mg/mL before injection into the chromatograph.

The morphological characteristics of the catalyst particles incorporated during polymerization and copolymerization were studied by SEM (JSM-IT300, Jeol, Japan).

In this experiment, high-temperature ^{13}C NMR (Bruker Avance 400, Bruker Biospin Karlsruhe, Germany) was used to determine the incorporation of comonomers into respective polymers. About 100 mg of the sample was then equipped into an NMR tube (5 mm) with the organic solvent 1,4-dichlorobenzene-d_4 with a sample concentration of approximately 15 mg/mL. This was then scanned using Varian Inova-400 MHz at 110 °C at 100.62 MHz having a delay of 3 s for 10 h. The internal reference for the polymer chain was considered to be the carbon backbone at 30 ppm. Subsequently, 1-octene and 1-hexene content were introduced and their contents calculated taking into consideration the normalized integration area within the ^{13}C NMR spectrum as outlined in the study by Seger and Maciel [24].

3. Results and Discussion

3.1. Ethylene Homopolymerization and Characterization of Homopolymers

The use of the catalyst system PVC/BuMgCl/$Ti(OBu)_4 \cdot TiCl_4$ in the polymerization of a single type of monomer to form a homopolymer was examined at various stimulus ratios to determine the best [Al]/[Ti] ratio resulting in high polymer productiveness. The catalyst molar ratio's power to affect the activity of the catalyst and monomer inclusion as a function of the [Al]/[Ti] molar ratio in the polymerization process is summarized in Table 1. With only the [Al]/[Ti] molar ratio varying during the procedure, the catalyst amount was maintained at a constant 0.02 g. The optimum cocatalyst amount was influenced by the low [Al]/[Ti] ratio value. Considering the optimal conditions observed under the previous conditions (Figure 1), the molar ratio [Al]/[Ti] corresponds to 773, which represents a very high [Al]/[Ti] ratio. In addition, ethylene polymerization over PVC/BuMgCl/$Ti(OBu)_4 \cdot TiCl_4$ is more economical than traditional heterogenous catalysts: as the activities increase, less catalyst will be required to produce the same amount of polymer in the polymerization reactor. As a consequence, there will be less residual catalyst in the product, thereby reducing the cost [25]. The catalytic activity decreased as the [Al]/[Ti] ratio increased from 927 to 1236, presumably because the prereduction influence of the alkyl aluminum reagent decreased the valence of the active site to an appropriate level and a somewhat higher dosage of cocatalyst brought about the deactivation from overreduction [13]. However, the necessity of having optimized processing conditions has decreased, which allows for a significant increase in the versatility of the process, making the commercialization of new polyolefin materials possible.

Table 1. Influence of [Al]/[Ti] molar ratios on catalyst activity in ethylene polymerization.

[Al]/[Ti] Ratio	Catalyst Activity (g PE/g Cat. h)
309	700
618	800
773	1100
927	1050
1236	550

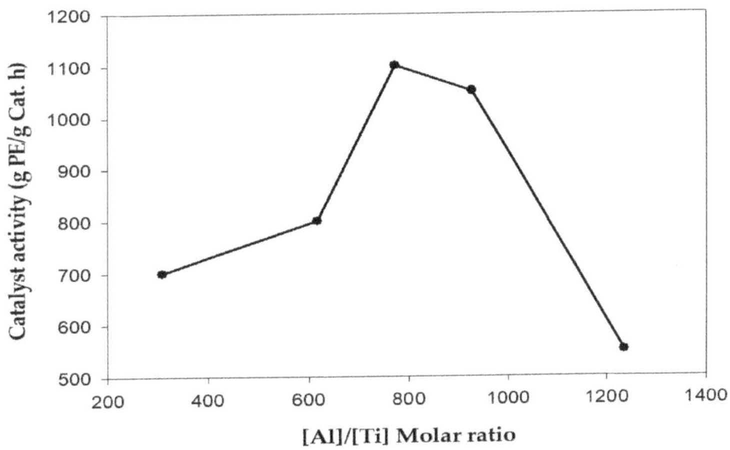

Figure 1. Variation of catalyst activity as a function of [Al]/[Ti] molar ratio in ethylene polymerization.

In general, the TEA is known to alkylate the titanium species prior to the respective incoming alkene molecule coordination to neighboring vacant sites in the titanium. Chain elongation is thus produced by alkyl migration on the organic ligand thus leaving another vacant site for insertion of another molecule. The outlined insertion process is repetitive leading to the creation of a polymer chain. Eventually, the polymer chain is then hewed from the respective catalyst through the process of beta-hydrogen elimination [8]. An increase in the [Al]/[Ti] ratio for a small amount of catalyst explains the increase in catalyst activity. This reduces the number of activated sites, hence inducing the polymerization centers and leading to a decrease in polymerization activities, as stated in the kinetic model proposed by Hongrui et al. [26]. The importance of the presence of a cocatalyst to stimulate the active centers is its ability to polymerize, as in the following reactions (Schemes 1 and 2):

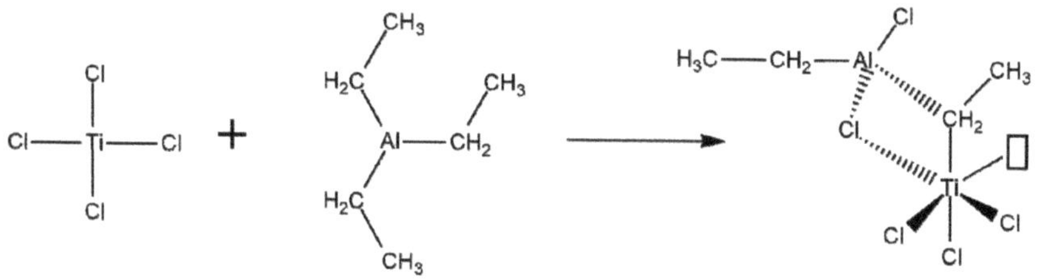

(1)

Titanum tetrachloride $TiCl_4$ Triethylaluminium $Al(C_2H_5)_3$ $TiCl_3R \bullet AlR_2Cl$

Scheme 1. Chemical reaction of cocatalyst in Ziegler-Natta catalyzed olefin polymerization.

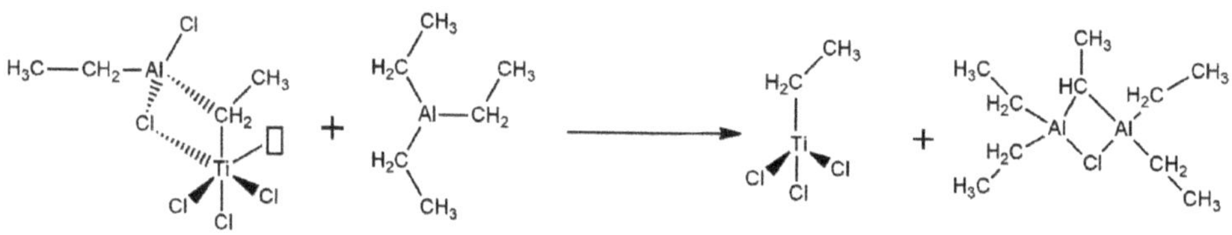

(2)

$TiCl_3R . AlR_2Cl$ Triethylaluminium $Al(C_2H_5)_3$ $TiCl_3R$ Al_2R_5Cl

Scheme 2. Ethylene polymerization reaction with heterogeneous Ti-based Ziegler-Natta catalyst, where R is Et.

A fraction of the catalyst material will not be fully prepolymerized if its residence time in the prepolymerization reactor is too short. Indeed, too short a residence time can result in catalyst deactivation or the formation of undesired fines in the principal reactor. The prepolymerization in the dilute slurry phase of the batch operation introduces extra costs because of its discontinuity and requires separation of the inert hydrocarbon and the prepolymer material. Lack of catalyst activity with increasing concentrations of aluminum results from the increase in the cocatalyst. The presence of TEA leads to adsorption of the effective centers and competition between monomer molecules for adsorption onto these centers. As a result, the decrease in active centers does not favor polymerization. Furthermore, large quantities of cocatalyst reduce the oxidation state of the titanium atom in the active center. Consequently, there will be a weakening of activities in the active center, producing an ineffective center.

A systematic increase in the average velocity profiles for polymerization over time is the result of the slower process of alkylation. The rate profiles expressed as grams of polymers produced per hour and per gram of catalyst are presented in Figure 2. An [Al]/[Ti] ratio of 618 or lower was used for catalysis preparation, indicating a constant rise in activity over time. However, the rate profiles from the [Al]/[Ti] ratios of 773 and 927 show a rapid increase in the polymerization rate from the beginning to the end, reaching a peak and maintaining the status to the end. The opposite was seen for the [Al]/[Ti] ratio of 1236, in which the kinetic curve indicates that the maximum point was acquired within a short time. The change is proof that formation and active site deactivation are two independent processes [7]. The best productivity in the operation of the catalyst was achieved with a molar ratio of 773. Nevertheless, at a high yield (>2 $g{\cdot}g^{-1}$), the catalyst portions, despite being small, likely drift toward the surface of the polymer particles [27]. Similar results were reported by Zheng et al. [15], who investigated the effect of [Al]/[Ti] catalytic activity and found that catalytic activity was high, as shown by the yield activity of Ti species. They considered TEA the strongest agent for alkylating titanium chloride because of the balance of the two active sites changed as the relative proportion of metal species was altered.

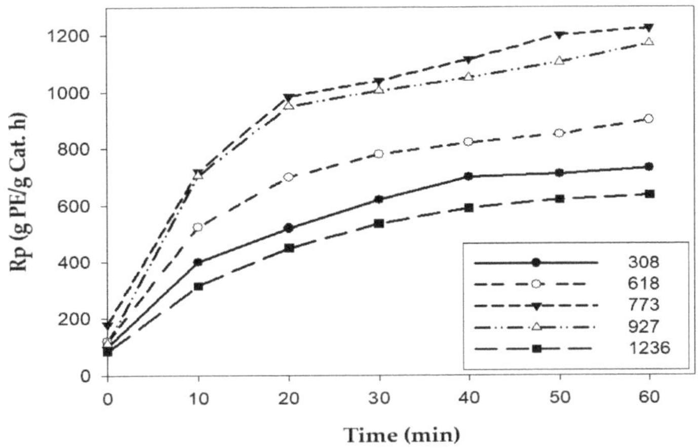

Figure 2. Corresponding rate profiles of the ethylene homopolymerization carried out using various [Al]/[Ti] molar ratios.

Figure 3 shows the SEM results for ethylene polymerization using various molar ratios of [Al]/[Ti]. (Details regarding the presence of all constituent atoms of the catalysts according to the analysis of EDS images are given in the Supplementary Materials, Figures S1–S5). It is evident that the catalyst particles remain in a good spherical shape and that the nodal structure blossoms. The increase in cocatalyst at a [Al]/[Ti] ratio of 1236 showed that the particles of the polymer differed greatly in size and were less porous and "fused." The sample shows many fibrils. In addition, the current study found that the time necessary to attain maximum polymerization activity determined the structure of the polymer particles, in line with the observations of Nooijen [28]. The time required to reach the maximum polymerization activity is contingent on the rate of reaction between the catalyst precursor and the cocatalyst, which consequently depends on the type of cocatalyst. For instance, suppose it takes a long time to attain the maximum polymerization activity; the polymer particles will tend to show open cobweb structures [29]. The gradual increases in catalytic activity are the reason for the development of cobweb structures. This is because the increase in activity can cause the polymer to crack, leading to the formation of open structures with fibrils [28,29], as indicated in Figure 3c.

(a)

(b)

Figure 3. *Cont.*

(**c**)

Figure 3. SEM result of polymer and catalyst particle in various [Al]/[Ti] molar ratios: (**a**) 309, (**b**) 773, and (**c**) 1236.

The data acquired from the GPC offer an estimate of the polymer M_w, indicating that polymerization was affected by the [Al]/[Ti] ratio (Figure 4). An increase in the dosage of the cocatalysts is shown, which would contribute to the decrease in M_w. The strong chain effect of alkyl metal salts can lower the M_w through the transfer of the nascent polymer chain of the active sites [30].

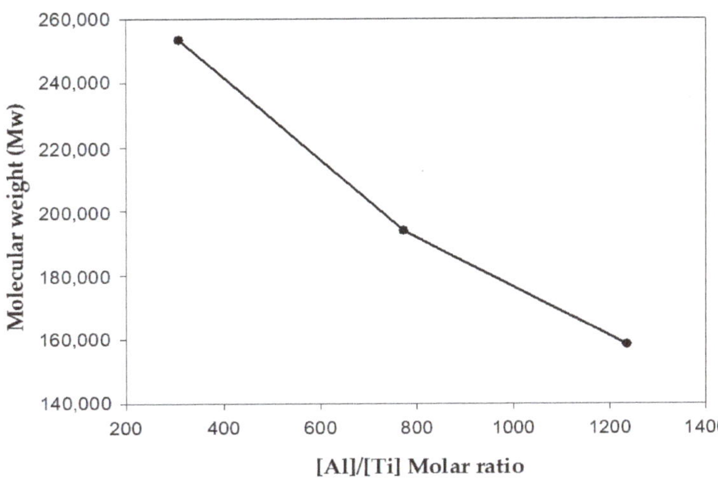

Figure 4. The relationship between the [Al]/[Ti] molar ratio and the molecular weight of polyethylene.

To further investigate the influence of the cocatalyst/catalyst [Al]/[Ti] molar ratio on the thermal stability of homopolymerization, characterization of the obtained polymers was carried out through the use of TGA and DSC. Figure 5 indicates the TGA thermograms for five molar ratios of cocatalyst TEA to Ti [Al]/[Ti]: 309, 618, 773, 927, and 1236 mol. Table 2 shows the T_{max} of decomposition. The catalytic effects for 618, 927, and 1236 are higher than that for 309. The most active catalysis was observed for 773, with the earlier stages having a low decomposition process. Thus, a certain level of thermal stabilization is produced by the presence of catalysts, which dramatically accelerate the decomposition process at a certain temperature. The T_m values were determined by assigning the maximum endotherm peaks. Figure 6 and Table 3 represent the DSC thermogram with the melting points and ΔH_f, whose increase depends on the [Al]/[Ti] molar ratios; it increases until the ratio reaches 773 moles, and then it begins to decrease. The limitations in the chain branches of the homopolymers were identifiable in the current study [12]. As noted by Smith et al. [31], slow polymerization processes relative to industrial-scale polymerization rates like in the case of slurry polymerization that is performed at room temperature, the growing chains' local temperature is anticipated to be relatively less than the dissolution temperature of a given linear polyethylene within the medium of reaction which is at roughly 90 °C in heptane. Subsequently, considering the growing chain, a maximum of 36 CH_2 units of the given macromolecule could be considered dissolved in the respective surrounding medium prior to commencement of crystallization. What is more, as noted by the authors, the outlined length was relatively below the expected distance of 140 CH_2 units that are required in the overall formation of various entanglements within the polyethylene melts. As such, the resultant formed macromolecule is highly linear. The use of the required amount of cocatalysts is key to boosting the operation of the catalyst. The cocatalyst can reduce the alkylate metal transition, which initiates the first polymer chain through the elimination of impurities [32]. Active sites can be deactivated because of overreduction in the chain transfer reaction, which is key for catalytic activities [13].

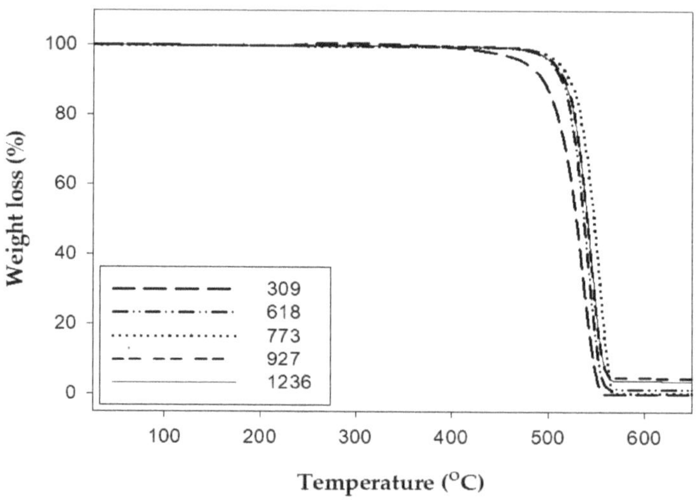

Figure 5. TGA thermograms of the ethylene homopolymerization carried out using various [Al]/[Ti] molar ratios.

Table 2. Temperature for the maximum rate of decomposition obtained from the TGA thermograms of polyethylene with various [Al]/[Ti] molar ratios.

[Al]/[Ti] Ratio (mol/mol)	T_{max} (°C)
309	536.81
618	540.86
773	553.71
927	543.29
1236	544.95

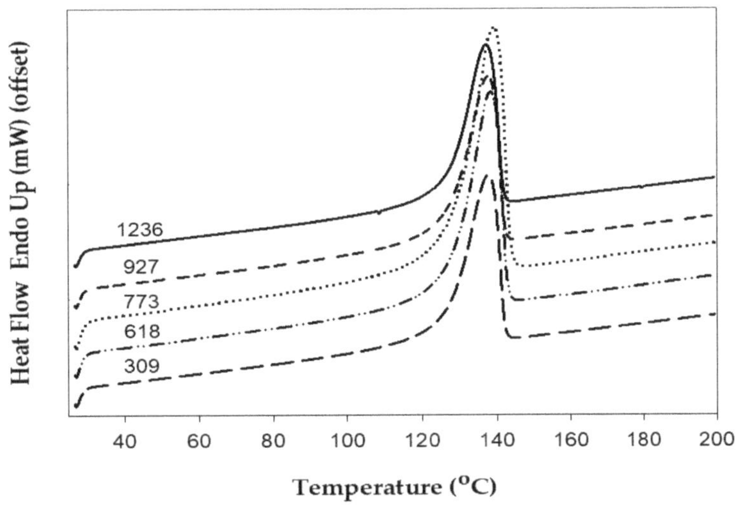

Figure 6. DSC second heating curves for polyethylene with various [Al]/[Ti] molar ratios.

Table 3. Melting temperature (T_m) and enthalpy of fusion (ΔH_f) obtained from the DSC of polyethylene with various [Al]/[Ti] molar ratios.

[Al]/[Ti] ratio	T_m (°C)	ΔH_f (J/g)
309	137.83	172.9197
618	138.4	262.5728
773	139.32	316.2398
927	137.72	179.0029
1236	137.25	203.4728

3.2. Ethylene/1-Octene and Ethylene/1-Hexene Copolymerizations and Characterization of the Copolymers

The ethylene/1-octene and ethylene/1-hexene copolymerization behaviors were assessed after the behaviors of ethylene homopolymerization were compared. The appropriate dosage of the TEA cocatalyst in homopolymerization with the $TiCl_4$ catalyst was [Al]/[Ti] = 773. A similar TEA cocatalyst dosage was utilized in both ethylene/1-hexane and ethylene/1-octene copolymerization. Nevertheless, the copolymers of ethylene with α-olefins had numerous benefits compared with other homopolymers. These benefits include larger process capacity, lower viscosity, and higher flexibility; hence, an investigation of the performance of the copolymerization of the aforementioned PVC/BuMgCl/Ti(OBu)$_4$·$TiCl_4$

catalyst was important. Additional quantities of α-olefin were a determining factor in ethylene/α-olefin copolymerization because the addition of α-olefin can significantly impact its polymer microstructure and considerably affect the behavior of catalyst polymerization. Various quantities of 1-hexene and 1-octene were added in the consequent polymerization of the catalyst in the current study, and their influence on catalytic properties is shown in Table 4, which also shows noticeably enhanced activity with the PVC/BuMgCl/$Ti(OBu)_4 \cdot TiCl_4$ catalyst. There was an initial increase in the catalytic activities before a gradual decrease as the concentration of the comonomer increased. For example, the amount of 1-hexene increased from 40 to 160 mmol, while the 1-octene quantity increased from 32 to 127 mmol, and optimum activity was attained at 64 and 60 mmol, respectively. These results show that the copolymerization of 1-hexene, 1-octene, and ethylene was strongly activated by the comonomer. Tuskaev et al. [23], Hongrui et al. [26], and Muñoz-Escalona et al. [33] have studied the notable preliminary enhancement in activity by α-olefin, which is known as the comonomer effect. These studies were grouped or categorized according to physical and chemical effects. Similar to the hydrogen effect, inserting ethylene into the Ti–H bond leads to the production of the $Ti–C_2H_5$ complex, which remains a dormant site via the β-agostic effect. $Ti–C_2H_5$ complex sites are not active for polymerization. On the other hand, inserting 1-octene or 1-hexene resulted in the formation of active sites of $Ti–C_6H_{13}$, $Ti–C_8H_{17}$, and $Ti–C_6H_{13}$. Additionally, the comonomer unit causes a donor effect, improving the activities of the catalyst. Conversely, the physical effect causes a decrease in the polymer's crystallinity as a result of the insertion of the comonomer, leading to accelerated monomer diffusion. Nevertheless, a comonomer will have significant and diverse impacts on the reactions based on the types of catalyst and the conditions of the polymerization [34]. However, the mechanism for the comonomer effect in ethylene/α-olefin copolymerization over ZN catalysts remains unclear, warranting more studies that can offer a better description. In the present study, the process initiated by the $Ti(OBu)_4$ catalyst illustrates the features found in olefin polymerization.

Table 4. Copolymer activities of different comonomers and polymer properties.

Sample	Amount of Comonomer (mmol)	Activity (g PE/mmol Cat. h)	T_m [a] (°C)	ΔH_f [a] (J/g)	Mw [b] ($\times 10^5$ g/mol)	PDI [b]	Comonomer Incorporation [mol%] [c]
1-octene	32	1750	135	164.1	1.97	5.8	3.26
	48	2200	135	190.9	-	-	-
	64	2300	133	144.9	1.77	4.9	4.45
	96	2250	133	150.5	-	-	-
	127	2050	132	157.9	1.86	5.2	4.06
1-hexene	40	1500	134	151.3	2.47	7.3	1.34
	60	2000	133	144.2	-	-	-
	68	1900	132	143.2	3.26	4.9	4.87
	80	1650	138	161.3	-	-	-
	160	1450	136	147.5	2.31	6.8	-

Other polymerization conditions: cocatalyst [Al]/[Ti] molar ratio of 773:1. [a] T_m and ΔH_f calculated by DSC thermograms; [b] M_w and PDI estimated by GPC in TCB; [c] omonomer incorporated into copolymers estimated by high temperature ^{13}C NMR and calculated according to the literature methods [24]. "-" Not determined.

Table 4 shows a summary of the characterization of DSC. Figures 7 and 8 show the respective second heating curves for ethylene/1-octene and ethylene/1-hexene. The findings reveal a reduction in the ΔH_f and the T_m of the copolymers because of the increased concentration of 1-hexene and 1-octene, as shown in Table 4; this occurred because of the insertion of a larger amount of comonomers in the mainstay of the polymer chain, which caused an increase in the concentration of comonomers. Based on the T_m values, the catalyst displayed lower sensitivity to 1-octene as the amount of 1-octene increased. The catalyst showed a greater sensitivity to 1-hexene as a result of the increased concentration of comonomer due to the introduction of the 1-hexene comonomer. The long chains caused an increase in T_m in addition to that caused by 80 and 160 mmol of 1-hexene because the configuration of chains became stiffer, leading to the appearance of additional extra small peaks in the melting line because of crystal gaps [35]. Furthermore, UHMWPE is believed to exhibit sharper incipient melting, which shows that the crystallinity difference between chains is smaller [35].

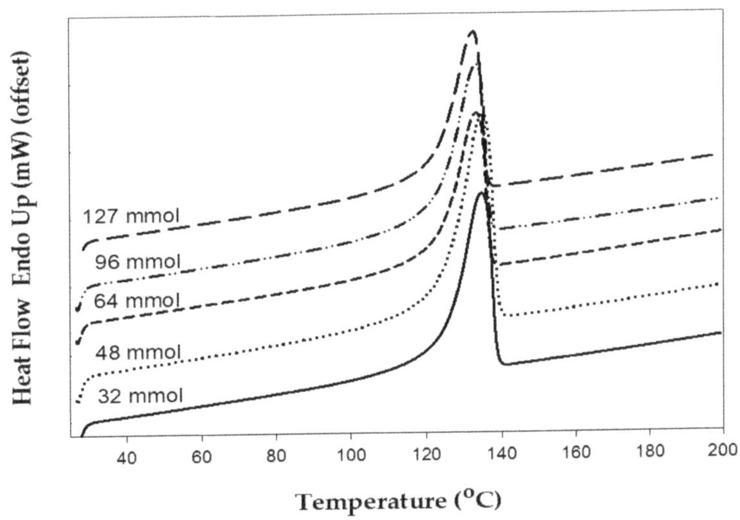

Figure 7. DSC second heating curve of the copolymer with various 1-octene concentrations.

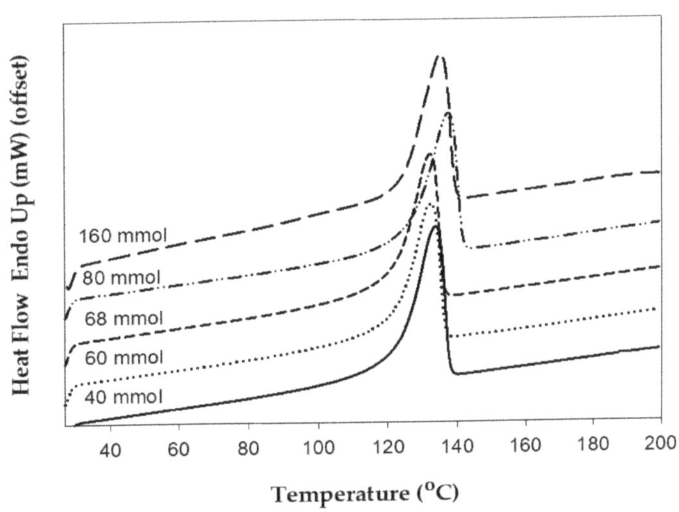

Figure 8. DSC second heating curve of the copolymer with various 1-hexene concentrations.

The present research utilized GPC to characterize the resultant copolymers. The findings reveal that the average M_w was reduced, except in the case of 68 mmol 1-hexene. There was an increase in the concentration of the 1-hexene to 160 mmol (up from 40 mmol), while the concentration of 1-octene increased to 127 mmol (up from 32 mmol). These findings are consistent with those of Yang et al. [36], who found that the strength of comonomer impacts tends to be greater in active centers, causing the production of a lower

M_w polymer compared with those that produce higher M_w polymers; this then results in a substantial reduction of the average M_w of the entire polymer while its molecular weight distribution (MWD) is broadened. Furthermore, 1-hexene broadened the MWD more than PVC/BuMgCl/Ti(OBu)$_4$·TiCl$_4$ catalysts did, and the polymerization of the 64 mmol of ethylene/1-octene demonstrated that the catalyst was highly balanced. The catalyst had the highest activity among the comonomers, while its PE products displayed the lowest MWD. Therefore, the lower sensitivity to 1-hexene for the catalyst is associated with the existing supports and the technique used for their preparation. In this case, the PVC-based polymeric support was the main support in the catalyst system. Thus, future studies should investigate the support used in catalyst systems.

SEM analysis of the comonomer concentrations was conducted to determine if there was a higher activity in ethylene/1-octene copolymerization. The results indicate that the catalyst particles had a satisfactory shape, showing increasing comonomer concentrations, as can be seen in Figure 9. (Details of the existence of all constituent atoms of the catalysts by analysis of EDS images are given in the Supplementary Materials, Figures S6–S8.) Moreover, a spherical copolymer was effectively disseminated or dispersed because it was replicated in the catalyst morphology. The copolymer particles grew gradually with increasing the 1-octene concentration, mainly because the catalyst particles in the preliminary polymerization phase break down into sub-particles and primary particles for easier separation. The primary catalyst particles initiate 1-octene/ethylene copolymerization, resulting in the formation of copolymer primary particles with sizes of 4–23 μm. Generally, these findings show that a PVC-based polymeric support has the capacity to allow for underground crystallization on the catalyst surface. The determining factor of the process of polymerization is the 1-octene concentration in n-hexane. These findings reveal that the preparation of the adduct solution containing a higher 1-octene concentration yields higher viscosity, preventing the dispersion of 1-octene, leading to its accumulation on external surfaces [37].

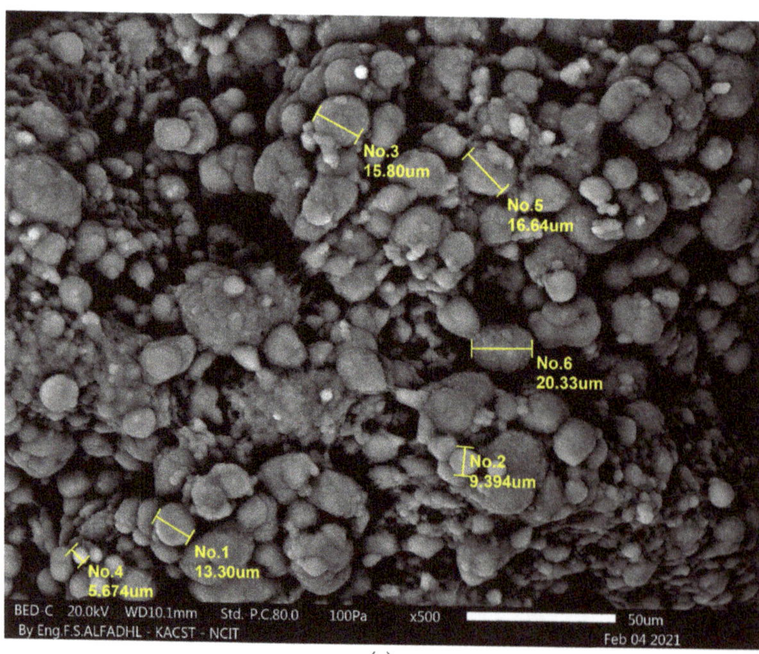

(a)

Figure 9. *Cont.*

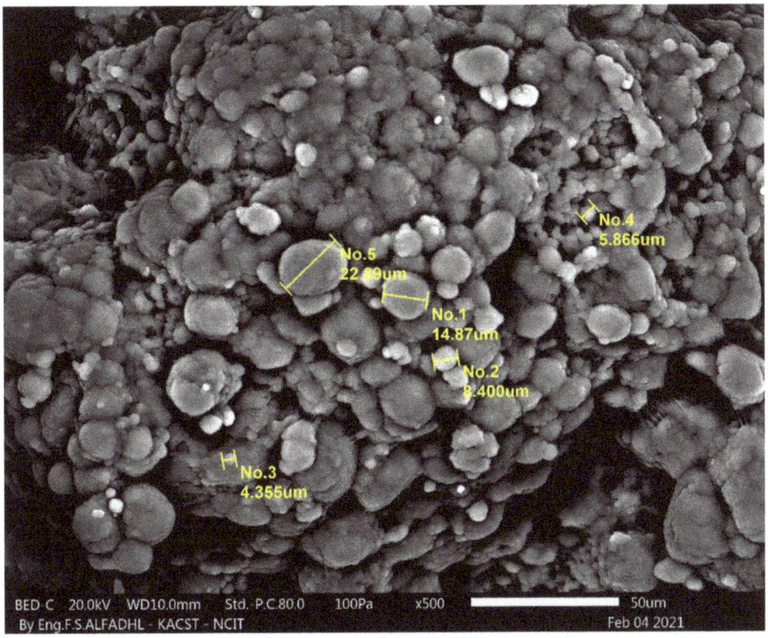

(b)

(c)

Figure 9. SEM images of the copolymer particles polymerized with catalyst using various 1-octene concentrations: (**a**) 32 mmol, (**b**) 64 mmol, and (**c**) 127 mmol.

The contents of 1-octene together with 1-hexene relative to their incorporation of the copolymers as obtained from the catalyst were investigated using high-temperature ^{13}C NMR. In this particular study, the copolymers chosen to be analyzed for incorporation were copolymers synthesized using 32, 64, and 127 mmol of 1-octene and 40 and 68 mmol of 1-hexene. The results of the analysis are outlined in Table 4 and Figure S9 illustrates the ^{13}C NMR spectra within the Supplementary Materials. In this reaction, it can be asserted that too small quantities of 1-octene and 1-hexene may not affect the overall copolymerization behavior of given catalysts including copolymer's microstructures while too large quantities of the comonomer may equally not be of benefit for the resultant activity. Consequently, the results of the study affirmed that having similar amounts of titanium species within a catalyst makes butoxide ligand beneficial for the overall insertion of 1-octene. The comonomer incorporation ability of respective hybrid ZN catalysts is affected prominently by titanium modification. As such, considering the initial stage, the molar ratio of 773 may be key in promoting the incorporation of 1-octene from 3.26% to 4.45% especially for Ti catalyst thus having 36.5% greater incorporation. There would be a pronounced difference in the 1-hexene incorporation thus there is the importance of controlling the comonomer content within the polymer.

4. Conclusions

Ti(IV) complexes that contain halogen atom alternatives such as Cl and are supported on PVC as a result of Grignard compound activation produce catalysts with better integration of 1-hexene and 1-octene. They also cause stronger activation of ethylene polymerization. An active catalyst was produced in the slurry homopolymerization of ethylene with the ZN catalyst, which was created using $Ti(OBu)_4$. To obtain the highest activity, the optimum ratio of the cocatalyst TEA to Ti [Al]/[Ti] was 773:1, which resulted in higher activity. The higher the [Al]/[Ti] ratio, the smaller the molecular weight, suggesting that a higher [Al]/[Ti] ratio is associated with reduced M_w. These findings illustrate that 1-octene integration attained the highest measured activity of 2.3 kg of PE/mol of cat·h when fused with a 64 mmol copolymer. The increase in the concentration of 1-octene was found to be associated with a reduction in the catalytic activity in $PVC/BuMgCl/Ti(OBu)_4 \cdot TiCl_4$. Both the ethylene/1-hexene copolymers and the catalyst demonstrated considerably lower activities. Therefore, the embodied catalyst has a polyethylene of higher M_w, extensive MWD, and a negative comonomer impact with a consumption rate that is moderately or substantially unstable. Based on the SEM findings, the activated titanium sites played an important role in decreasing the crystallinity of the polymeric PVC. They also supported higher metal-active species on the inner surfaces of the support available to the comonomer in the slurry stage, which can be greatly increased with a rise in reaction activity. Moreover, the comonomer caused donor effects, leading to increased activity. From the standpoint of the physical impact, the generated copolymers caused a decrease in crystallinity, leading to the acceleration of monomer diffusion in the ethylene, which was assumed to be impassable in the crystallized polyethylene. The results and conclusions acquired in this investigation can be used as a guide to optimizing the synthesis of catalysts. Further, the experimental results of this study confirm theoretical studies that indicate the performance of active sites of ZN active sites having butoxide as a ligand of the titanium.

Supplementary Materials: The following are available online at https://www.mdpi.com/article/10.3390/polym13132109/s1. Figure S1: SEM-EDS images of the ethylene polymerization carried out using [Al]/[Ti] molar ratio at 309. Figure S2: SEM-EDS images of the ethylene polymerization carried out using [Al]/[Ti] molar ratio at 618 mmol. Figure S3: SEM-EDS images of the ethylene polymerization carried out using [Al]/[Ti] molar ratio at 773 mmol. Figure S4: SEM-EDS images of the ethylene polymerization carried out using [Al]/[Ti] molar ratio at 927 mmol. Figure S5: SEM-EDS images of the ethylene polymerization carried out using [Al]/[Ti] molar ratio at 1236 mmol. Figure S6: SEM-Eds images of the ethylene/1-octene copolymerization carried out using 1-octene concentration at 32 mmol. Figure S7: SEM-Eds images of the ethylene/1-octene copolymerization carried out using 1-octene concentration at 64 mmol. Figure S8: SEM-Eds images of the ethylene/1-octene copolymerization carried out using 1-octene concentration at 127 mmol. Figure S9: High Temperature 13C NMR spectra of the copolymers: (a) 32 mmol of 1-octene, (b) 64 mmol of 1-octene, (c) 127 mmol of 1-octene, (d) 40 mmol of 1-hexene, and (e) 68 mmol of 1-hexene.

Author Contributions: Mohammed S. Alsuhybani assisted with experiment design, performed the experiments and the data analysis. Eid M. Alosime assisted with analyzed and interpreted of data, wrote and critical reviewed/edited of the paper. All authors have read and agreed to the published version of the manuscript.

Funding: This research received no external funding.

Institutional Review Board Statement: Not applicable.

Informed Consent Statement: Not applicable.

Data Availability Statement: The data presented in this study are available on request from the corresponding author.

Acknowledgments: The authors are thankful to King Abdulaziz, City for Science and Technology for providing all the necessary lab facilities and equipment and to King Saud University for assisting with the polymerization process. In addition, we would like to thank Eng. Fadhl Alfadhl for doing the SEM image and Eng. Raid Binkhnin for assistance with the TGA and DSC testing.

Conflicts of Interest: The authors declare no conflict of interest.

References

1. D'Amore, M.; Thushara, K.; Piovano, A.; Causa, M.; Bordiga, S.; Groppo, E. Surface investigation and morphological analysis of structurally disordered $MgCl_2$ and $MgCl_2/TiCl_4$ Ziegler–Natta catalysts. *ACS Catal.* **2016**, *6*, 5786–5796. [CrossRef]
2. Zhou, Y.; He, X.; Fu, T.; Zhao, N.; Liu, B. Novel Cr/Ti bimetallic polyethylene catalysts synthesized through introduction of chromium species into the ($SiO_2/MgR_2/MgCl_2$)· $TiCl_x$ Ziegler-Natta catalyst. *J. Organomet. Chem.* **2019**, *888*, 1–15. [CrossRef]
3. Ribour, D.; Spitz, R.; Monteil, V. Modifications of the active sites distribution in the Ziegler Natta polymerization of propylene using Lewis acids. *J. Polym. Sci. Part. A Polym. Chem.* **2010**, *48*, 2631–2635. [CrossRef]
4. D'Amore, M.; Credendino, R.; Budzelaar, P.H.; Causá, M.; Busico, V. A periodic hybrid DFT approach (including dispersion) to $MgCl_2$-supported Ziegler–Natta catalysts–1: $TiCl_4$ adsorption on $MgCl_2$ crystal surfaces. *J. Catal.* **2012**, *286*, 103–110. [CrossRef]
5. Pongchan, T.; Praserthdam, P.; Jongsomjit, B. Gas-phase polymerization of ethylene over Ti-based Ziegler–Natta catalysts prepared from different magnesium sources. *Mater. Today Chem.* **2020**, *18*, 100366. [CrossRef]
6. Pongchan, T.; Praserthdam, P.; Jongsomjit, B. Temperature effect on propylene polymerization behavior over Ziegler-Natta catalyst with different cocatalyst systems. *Mater. Res. Express* **2020**, *7*, 025309. [CrossRef]
7. Philippaerts, A.; Ensinck, R.; Baulu, N.; Cordier, A.; Woike, K.; Berthoud, R.; Cremer, G.D.; Severn, J.R. Influence of the particle size of the $MgCl_2$ support on the performance of Ziegler catalysts in the polymerization of ethylene to ultra-high molecular weight polyethylene and the resulting polymer properties. *J. Polym. Sci. Part. A Polym. Chem.* **2017**, *55*, 2679–2690. [CrossRef]
8. Mandal, B.M. *Fundamentals of Polymerization*; World Scientific: Hackensack, NJ, USA, 2013.
9. Wang, J.; Cheng, R.; He, X.; Liu, Z.; Zhao, N.; Liu, B. Introduction of chromium species into the ($SiO_2/MgO/MgCl_2$) $TiCl_x$ Zieglere-Natta catalyst for better catalytic performance. *J. Organomet. Chem.* **2015**, *798*, 299–310. [CrossRef]
10. Ochędzan-Siodłak, W.; Nowakowska, M. Titanium catalyst ($TiCl_4$) supported on $MgCl_2$ (THF)$(AlEt_2Cl)_{0.34}$ for ethylene polymerization. *Polimery* **2007**, *52*, 184–189. [CrossRef]
11. Zhao, N.; Cheng, R.; He, X.; Liu, Z.; Liu, B.; Zhan, R.; Gao, Y.; Zou, E.; Wang, S. Novel SiO_2-Supported Silyl-Chromate (Cr)/Imido-Vanadium (V) Bimetallic Catalysts Producing Polyethylene and Ethylene/1-Hexene Copolymers with Bimodal Molecular-Weight Distribution. *Macromol. Chem. Phys.* **2014**, *215*, 1434–1445. [CrossRef]
12. Fu, T.; Liu, Z.; Cheng, R.; He, X.; Tian, Z.; Liu, B. Ethylene Polymerization over $MgCl_2/SiO_2$ Bi-Supported Ziegler–Natta Hybrid Titanium/Vanadium Catalysts. *Macromol. Chem. Phys.* **2017**, *218*, 1700027. [CrossRef]
13. Kashiwa, N. The discovery and progress of $MgCl_2$-supported $TiCl_4$ catalysts. *J. Polym. Sci. Part. A Polym. Chem.* **2004**, *42*, 1–8. [CrossRef]
14. Keii, T.; Terano, M.; Kimura, K.; Ishii, K. A kinetic argument for a quasi-living polymerization of propene with a $MgCl_2$-supported catalyst. *Makromol. Chem. Rapid Commun.* **1987**, *8*, 583–587. [CrossRef]
15. Zheng, W.; He, A.; Liu, C.; Shao, H.; Wang, R. The influences of alkylaluminium as cocatalyst on butene-1 polymerization with $MgCl_2$-supported $TiCl_4$ Ziegler-Natta catalysts. *Polymer* **2020**, *210*, 122998. [CrossRef]
16. Zheng, W.-P.; Ma, Y.-P.; Du, D.-L.; He, A.-H.; Shao, H.-F.; Liu, C.-G. Polymerization Kinetics of Propylene with the $MgCl_2$-Supported Ziegler-Natta Catalysts—Active Centers with Different Tacticity and Fragmentation of the Catalyst. *Chin. J. Polym. Sci.* **2020**, *1*, 11. [CrossRef]
17. Matsuoka, H.; Liu, B.; Nakatani, H.; Terano, M. Variation in the Isospecific Active Sites of Internal Donor-Free $MgCl_2$-Supported Ziegler Catalysts: Effect of External Electron Donors. *Macromol. Rapid Commun.* **2001**, *22*, 326–328. [CrossRef]
18. Chadwick, J.C. Polyolefins-Catalyst and Process Innovations and their Impact on Polymer Properties. *Macromol. React. Eng.* **2009**, *3*, 428–432. [CrossRef]
19. Hu, J.; Han, B.; Shen, X.-R.; Fu, Z.-S.; Fan, Z.-Q. Probing the roles of diethylaluminum chloride in propylene polymerization with $MgCl_2$-supported Ziegler-Natta catalysts. *Chin. J. Polym. Sci.* **2013**, *31*, 583–590. [CrossRef]

20. Jericó, S.; Schuchardt, U.; Joekes, I.; Kaminsky, W.; Noll, A. Chlorinated organic polymers as supports for Ziegler-Natta catalysts. *J. Mol. Catal. A Chem.* **1995**, *99*, 167–173. [CrossRef]
21. Whitehurst, D. Catalysis by heterogenized transition-metal complexes. *Chem. Inform.* **1980**, *44-49*, 44–49.
22. Pino, P.; Mülhaupt, R. Stereospecific polymerization of propylene: An outlook 25 years after its discovery. *Angew. Chem. Int. Ed. Engl.* **1980**, *19*, 857–875. [CrossRef]
23. Tuskaev, V.A.; Gagieva, S.C.; Kurmaev, D.A.; Melnikova, E.K.; Zubkevich, S.V.; Buzin, M.I.; Nikiforova, G.G.; Vasil'ev, V.G.; Saracheno, D.; Bogdanov, V.S. Olefin polymerization behavior of titanium (IV) alkoxo complexes with fluorinated diolate ligands: The impact of the chelate ring size and the nature of organoaluminum compounds. *Appl. Organomet. Chem.* **2020**, *34*, 5933. [CrossRef]
24. Seger, M.; Maciel, G. Quantitative 13C NMR analysis of sequence distributions in poly (ethylene-co-1-hexene). *Anal. Chem.* **2004**, *76*, 5734–5747. [CrossRef] [PubMed]
25. Galli, P.; Vecellio, G. Technology: Driving force behind innovation and growth of polyolefins. *Prog. Polym. Sci.* **2016**, *26*, 1287–1336. [CrossRef]
26. Hongrui, Y.; Letian, Z.; Dandan, Z.; Zhisheng, F.; Zhiqiang, F. Effects of alkylaluminum as cocatalyst on the active center distribution of 1-hexene polymerization with $MgCl_2$-supported Ziegler Natta catalysts. *Catal. Commun.* **2015**, *62*, 104–106.
27. McKenna, T.F.; Di Martino, A.; Weickert, G.; Soares, J.B. Particle growth during the polymerisation of olefins on supported catalysts, 1–nascent polymer structures. *Macromol. React. Eng.* **2010**, *4*, 40–64. [CrossRef]
28. Nooijen, G. On the importance of diffusion of cocatalyst molecules through heterogeneous Ziegler/Natta catalysts. *Eur. Polym. J.* **1994**, *30*, 11–15. [CrossRef]
29. Graff, R.; Kortleve, G.; Vonk, C.G. On the size of the primary particles in ziegler catalysts. *J. Polym. Sci. Part. B Polym. Lett.* **1970**, *8*, 735–739. [CrossRef]
30. Tregubov, A.A.; Zakharov, V.A.; Mikenas, T.B. Supported titanium-magnesium catalysts for ethylene polymerization: A comparative study of catalysts containing isolated and clustered titanium ions in different oxidation states. *J. Polym. Sci. Part. A Polym. Chem.* **2010**, *47*, 6362–6372. [CrossRef]
31. Smith, P.; Chanzy, H.; Rotzinger, B. Drawing of virgin ultrahigh molecular weight polyethylene: An alternative route to high strength fibres. *Polym. Commun.* **1985**, *26*, 258–260.
32. Andoni, A.; Chadwick, C.; Niemantsverdriet, J.; Thüne, A. Preparation method for well-defined crystallites of $MgCl_2$-supported Ziegler-Natta catalysts and their observation by AFM and SEM. *Macromol. Rapid Commun.* **2007**, *28*, 1466–1471. [CrossRef]
33. Muñoz-Escalona, A.; García, H.; Albornoz, A. Homo and copolymerization of ethylene with highly active catalysts based on $TiCl_4$ and grignard compounds. *J. Appl. Polym. Sci.* **2010**, *34*, 977–988. [CrossRef]
34. Białek, M.; Czaja, K.; Pietruszk, A. Ethylene/1-olefin copolymerization behaviour of vanadium and titanium complexes bearing salen-type ligand. *Polym. Bull.* **2013**, *70*, 1499–1517. [CrossRef]
35. Geçim, G.; Erkoç, E. Gas phase polymerization of ethylene towards UHMWPE. *Turk. J. Chem.* **2020**, *44*, 695–711. [CrossRef] [PubMed]
36. Yang, H.; Huang, B.; Fu, Z.; Fan, Z. Ethylene/1-hexene copolymerization with supported Ziegler–Natta catalysts prepared by immobilizing $TiCl_3$ (OAr) onto $MgCl_2$. *J. Appl. Polym. Sci.* **2015**, *132*, 1–9. [CrossRef]
37. Zhou, Y.; Zhang, R.; Ren, H.; He, X.; Li, B.; Zhao, N.; Liu, B. Ethylene polymerization over novel organic magnesium based V/Ti bimetallic Ziegler-Natta Catalysts. *J. Organomet. Chem.* **2020**, *908*, 121066. [CrossRef]

polymers

MDPI

Article

A Facile Synthesis Route of Hybrid Polyurea-Polyurethane-MWCNTs Nanocomposite Coatings for Ballistic Protection and Experimental Testing in Dynamic Regime

Gabriela Toader [1], Aurel Diacon [2,*], Edina Rusen [2], Florica Rizea [2], Mircea Teodorescu [2], Paul O. Stanescu [2], Celina Damian [2], Adrian Rotariu [1], Eugen Trana [1], Florina Bucur [1] and Raluca Ginghina [3]

1 Faculty of Weapon Systems Engineering and Mechatronics, Military Technical Academy, 39-49 George Cosbuc Boulevard, 050141 Bucharest, Romania; nitagabriela.t@gmail.com (G.T.); adrian.rotariu@mta.ro (A.R.); eugen.trana@mta.ro (E.T.); florina.bucur@mta.ro (F.B.)

2 Department of Bioresources and Polymer Science, Faculty of Applied Chemistry and Materials Science, University Politehnica of Bucharest, 1-7 Gh. Polizu Street, 011061 Bucharest, Romania; edina.rusen@upb.ro (E.R.); flori_rizea@yahoo.com (F.R.); mircea.teodorescu@upb.ro (M.T.); paul_stanescu@yahoo.com (P.O.S.); celina.damian@yahoo.com (C.D.)

3 Scientific Research Center for CBRN Defense and Ecology, 225 Soseaua Oltenitei, 041327 Bucharest, Romania; ginghinaraluca@gmail.com

* Correspondence: aurel.diacon@upb.ro

Citation: Toader, G.; Diacon, A.; Rusen, E.; Rizea, F.; Teodorescu, M.; Stanescu, P.O.; Damian, C.; Rotariu, A.; Trana, E.; Bucur, F.; et al. A Facile Synthesis Route of Hybrid Polyurea-Polyurethane-MWCNTs Nanocomposite Coatings for Ballistic Protection and Experimental Testing in Dynamic Regime. *Polymers* **2021**, *13*, 1618. https://doi.org/10.3390/polym13101618

Academic Editor: Andrew B. Lowe

Received: 20 April 2021
Accepted: 14 May 2021
Published: 17 May 2021

Publisher's Note: MDPI stays neutral with regard to jurisdictional claims in published maps and institutional affiliations.

Copyright: © 2021 by the authors. Licensee MDPI, Basel, Switzerland. This article is an open access article distributed under the terms and conditions of the Creative Commons Attribution (CC BY) license (https://creativecommons.org/licenses/by/4.0/).

Abstract: This study describes a simple, practical, inexpensive, improved, and efficient novel method for obtaining polyurea-polyurethane-multiwall carbon nanotubes (MWCNTs) nanocomposites with enhanced mechanical properties, and their experimental testing in a dynamic regime. SEM and micro-CT investigations validated the homogeneity of the nanocomposite films and uniform dispersion of the nanofiller inside the polymeric matrix. The experimental measurements (TGA, DSC, DMA, and tensile tests) revealed improved thermal and mechanical properties of these new materials. To demonstrate that these nanocomposites are suitable for ballistic protection, impact tests were performed on aluminum plates coated with the polyurea-polyurethane MWCNTs nanocomposites, using a Hopkinson bar set-up. The experimental testing in the dynamic regime of the polyurea-polyurethane-coated aluminum plates confirmed that the nanocomposite layers allow the metal plate to maintain its integrity at a maximum force value that is almost 200% higher than for the uncoated metallic specimens.

Keywords: polyurea-polyurethane; nanocomposite; ballistic protection; coatings; mechanical properties; multiwall carbon nanotubes (MWCNTs); Hopkinson bar

1. Introduction

Nowadays, war migrates from the battlefields towards less predictable areas. All existing ammunition types and improvised explosive devices represent serious threats. The survival of defense, public order, or national security crews, and the vehicles that they use during a mission, is conditioned by several factors, including ballistic protection. Specific categories of threat require different levels of ballistic protection, regulated by the standards valid in each state.

New effective solutions for ballistic protection are offered by the evolution of composite materials, the synthesis of new materials, and the selection and correlation of performant materials [1]. These materials must be resistant to the action of a shock wave or the impact of a projectile [2–5], but they must also be lightweight so they do not overload the armor of a trooper or vehicle [6]. Composites reinforced with aramid fibers are described in the literature as one of the materials with the highest ratios between weight (mass) and level of impact protection, but their cost is considered a disadvantage [6,7]. Another alternative to steel armor is the use of an aluminum substrate coated with composite materials [8]. Numerous studies have shown that the rigidity and hardness of a material can be improved using nanoparticles [7,9–12].

To obtain superior mechanical properties for a particular type of polymeric matrix, it is necessary to take into account the size, shape, and chemical properties of the nanometric filler particles [13]. Multiwall carbon nanotubes (MWCNTs) have been extensively used in nanocomposites due to their outstanding properties: high Young's modulus, stiffness, flexibility, and conductivity. Additionally, only a 1% MWCNTs content can lead to an increase of up to 36–42% of the modulus of elasticity for composite material [14–16]. Even though carbon nanotubes have great potential as nanofillers, a major concern related to the performance of MWCNTs-based nanocomposites is represented by the difficulty of achieving their homogenous dispersion. The dispersability of carbon nanotubes in the polymer matrix is conditioned by their chemical and physical compatibility. Therefore, in some cases, the functionalization of MWCNTs can become mandatory for their compatibilization with the organic substrate.

Polyurea has been widely reported as being an appropriate polymeric matrix for impulsive loading ballistic protection [17–22], due to its embedded reinforcing nanoscale hard domains, uniformly dispersed and chemically linked inside its soft elastic nanodomains [9,23]. Therefore, many researchers have investigated polyurea for their distinctive synergistic properties. Li et al. [24] employed Jeffamine® D2000 and two different types of isocyanates, isophorone diisocyanate and hexamethylene diisocyanate, to obtain self-healing elastomers. Even though they possess this self-healing property, these elastomers exhibit insufficient mechanical resistance (maximum tensile stress of approximately 3.5 MPa) for impulsive loading applications. Therefore, numerous studies [17,25,26] confirmed the beneficial contribution of polyurea coatings applied to metallic surfaces. Bai et al. [27] obtained a polyurea with improved mechanical resistance by employing diphenylmethane diisocynante and Unilink 4200 diamine, reaching a maximum true stress of approximately 13 MPa (in the quasi-static regime, at low strain rates) and 27 MPa under a dynamic regime. Li et al. [28] investigated the response of stainless steel plates coated with a commercial polyurea (LINE XS-350) to impulsive loadings. This commercial polyurea displayed a maximum quasi-static true stress of only 22.4 MPa. The use of polyurea coatings affords a better response of the coated metal sheet at the action of a shock wave or the impact of a projectile by suffering lower deformations than the neat metallic plates. Therefore, in terms of ballistic protection, the efficacy of polyurea is already validated by an important volume of data available in the literature. Polyurea-MWCNTs systems [29–31] were also demonstrated in multiple studies as good candidates for impulsive loading applications. However, one major inconvenience is that the chemical modification of the nanofiller sometimes involves significant costs that may not prove to be economically viable.

In many cases, a good dispersion of nanoparticles in the polymer matrix can be obtained only by prior functionalization. This process can sometimes be difficult to perform and expensive, as the synthesis process involves several steps. To simplify the manufacturing process and to reduce the production costs for polyurea nanocomposites designed for ballistic protection, a commercially available MWCNT product was used. This product consists of a concentrate of MWCNTs pre-dispersed in a polyester-polyol system, which ensures a good dispersion in the polyurea matrix. This simplified synthesis method could be more advantageous for coating extended areas specific for ballistic protection applications. Thus, the metal substrates can be protected against cracking/failure by being coated with this high-performance polyurea nanocomposite by simply spraying the premixed reactants onto the targeted surface.

Therefore, taking into consideration the unique set of properties of polyurea-polyurethane and the multiple advantages of using pre-dispersed MWCNTs, this study intended to provide a novel approach towards a new facile synthesis route for obtaining high-performing polyurea-polyurethane nanocomposite coatings. This study also comprises a section on the experimental testing in a dynamic regime in order to prove the real contribution of this type of nanocomposite coating on the deformation mitigation of metallic plates for assessing the suitability of the materials' design in relation to ballistic protection applications. To improve the response of the metallic structures to impulsive loadings, we herein decided to employ our previously synthesized polyurea matrix [17], which was obtained from poly (propylene glycol) bis(2-aminopropyl ether) with Mn $\approx$ 2000 Da, diphenylmethane-4,4′-diisocyanate, and 4′-diaminodiphenylmethane as chain extender. This polyurea possesses superior mechanical resistance, displaying a maximum quasi-static true stress of 33.76 MPa. Its unique set of properties established the premises for obtaining a performant nanocomposite with superior mechanical resistance (up to 40.84 MPa maximum quasi-static true stress) for ballistic protection applications. To the extent of our knowledge, this is the first paper that proposes an improved route for obtaining performant polyurea-polyurethane nanocomposites with pre-dispersed MWCNTs, conjoined with the evaluation of the mitigation effect brought by these performant materials and using the Hopkinson bar method. The novelty of this paper consists of both the straightforwardness of the polyurea coating fabrication method and the experimental set-up approach for the evaluation of the behavior of these materials at impact with a projectile.

2. Materials and Methods

2.1. Materials

Poly (propylene glycol) bis(2-aminopropyl ether) − Mn $\approx$ 2000 Da (PPG2000, Sigma Aldrich, St. Louis, MO, USA), 4,4′-diaminodiphenylmethane (DADPHM, Sigma Aldrich, St. Louis, MO, USA), and diphenylmethane-4,4′-diisocyanate (MDI, technical product Desmodur® 44V20L, Covestro, Leverkusen, Germany) were pre-dispersed in multiwalled carbon nanotubes in a polyester polyol based resin (MWCNTs, Graphistrength® CPU2-30, Arkema, Colombes, France; Graphistrength® CPU2-30 is a MultiWall Carbon Nanotubes (MWCNT), at a concentrate that is used as an additive for polyurethane-based materials, coatings or adhesives. It contained 30 wt% of MWCNT, perfectly dispersed in a polyester polyol. Typical final MWCNT loadings in the final compounds were in the range 0.1 to 2 wt%, depending on the host matrix characteristics, targeted performances, processing methods, and conditions. Acetone (Sigma Aldrich, St. Louis, MO, USA), was used as received.

2.2. Methods

2.2.1. Preparation of Polyurea-Polyurethane Nanocomposite Films

Five distinct types of polyurea-polyurethane nanocomposites were synthesized (Table 1) to obtain polymeric films with different mechanical properties.

Table 1. The composition of polyurea nanocomposite films.

Sample	MWCNTs (wt. %)
PU (blank)	0
PU-NC1	0.05
PU-NC2	0.1
PU-NC3	0.2
PU-NC4	0.3

For the synthesis of the blank sample, the diamines (PPG2000 and DADPHM) and the isocyanate (MDI) were solubilized separately in acetone, obtaining two solutions: Solution **A** (containing the two components with amino functional groups, PPG2000: DADPHM molar ratio 1:2) and Solution **B** (containing the component with isocyanate functional groups). The molar ratio between the isocyanate and the amino (primary amine) groups was maintained for all reactions at 1:1. Solutions **A** and **B** were vigorously stirred for 10 to

15 s at room temperature, and then the mixture was transferred on a Petri dish placed on a perfectly horizontal surface. The reaction mixture was maintained at 20 °C and relative humidity of between 50% and 55%, until the reaction was complete. After 24 h, a polyurea film with a thickness of about 0.4 mm was obtained.

Polyurea-polyurethane-based composites were obtained using the same procedure, according to the compositions described in Table 1. The only difference was the introduction of carbon nanotubes in solution **A**, thus obtaining Solution **C**. To obtain the best possible dispersion of MWCNT in the polyurea matrix, solution **C** was subjected to the action of ultrasound for one hour, before being mixed with solution **B**. For a better understanding of the behavior of the synthesized polymeric films on impact with a projectile, samples consisting of aluminum plates coated with polyurea-polyurethane nanocomposites (described in Table 1) were prepared by a casting technique (Scheme 1).

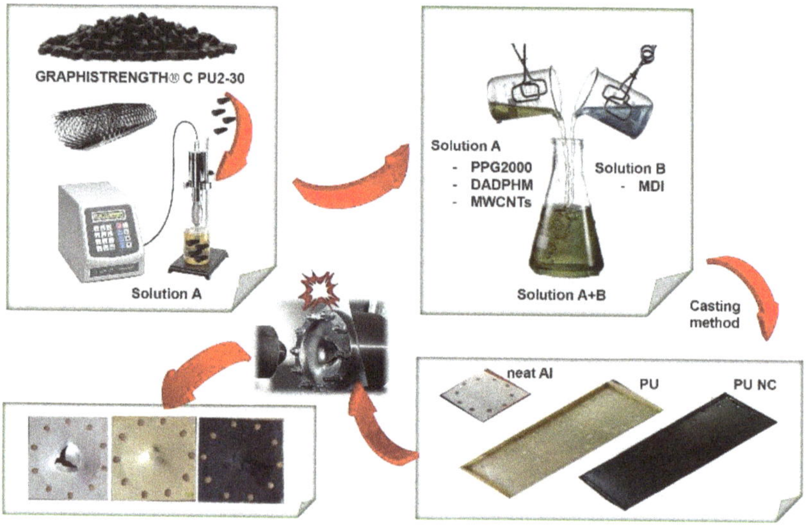

Scheme 1. Representation of the fabrication steps for the polyurea-polyurethane nanocomposite films.

The nanocomposites, obtained as described above, contain a hybrid polymeric matrix consisting of both polyurea and polyurethane zones. The polyurethane regions surround the MWCNTs, since these carbon nanotubes were pre-dispersed in a polyester polyol matrix before being part of the polyurea matrix. Even though these polyurethane areas are present in the nanocomposite, their concentration is neglectable in comparison with the polyurea matrix.

2.2.2. Characterization

The morphology of the polyurea-polyurethane-MWCNTs nanocomposite films was investigated by SEM (scanning electron microscopy) using a Tescan Vega II LMU SEM instrument (TESCAN, Brno, Czechia) at 10 keV acceleration voltage. The distribution of the MWCNTs inside the polyurea-polyurethane matrix was examined via micro-CT technique. The SkyScan micro-CT attachment allowed for converting the Tescan Vega II LMU SEM to an X-ray microtomograph for non-destructive imaging and for measuring of the object's internal microstructure of specimens. Analysis parameters: Exposure time—4 s per projection at electron beam currents of 100 nA; accelerating voltage—30 KeV; step size—1°; scanning time—24 min. Reconstruction was performed by the NRecon program which used float-point data values for internal calculations during reconstruction, and afterward allowed the operator to define the density window as a range of the reconstructed values.

The full set of reconstruction results was visualized by the program DataViewer® 2D/3D Micro-CT Slice Visualization (Micro Photonics Inc., Allentown, PA, USA). A thermogravimetric analysis (TGA) of the synthesized nanocomposites was performed on a Thermal Analysis Q500 instrument (TA Instruments, New Castle, DE, USA). Samples of about 2 mg were heated under nitrogen flow, with 10 °C/min, from 25 °C to 700 °C. The glass transition temperatures (T_g) of the polyurea-polyurethane nanocomposites were established using differential scanning calorimetry (DSC). All the samples of around 10 mg were analyzed using a NETZSCH DSC 204 F1 Phoenix instrument (NETZSCH, Selb, Germany), under nitrogen flow, at 10 °C/min heating rate, in two heating/cooling cycles, between −80 and 200 °C. Stress–strain curves were obtained using an Instron 3382 testing machine (Instron, Norwood, MA, USA). The samples were prepared for the tensile tests by cutting the nanocomposite films in a rectangular shape, at standard dimensions for tensile specimen, with 5 mm width and 100 mm length. For each specimen, the rate of the extension was set at 500 mm/min, and the separation of the initial jaws was set at 50 mm (plain jaw faces). For each type of nanocomposite film, five tensile tests were carried out and the average of the measured values and the standard deviation for each point was registered. For a more accurate approach of the interpretation of these tests, the values of true stress (σ_T) and true strain (ε_T) were employed for determining Young's modulus values, which were calculated according to the mathematical model described in [32]. The definition of true stress (σ_T) states that this σ_T signifies the instantaneous applied load divided by the instantaneous cross-sectional area. True stress is related to engineering stress (σ) through the following equation: $\sigma_T = \sigma\ (1 + \varepsilon)$. The definition of true strain (ε_T) states that ε_T signifies the rate of instantaneous increase in the instantaneous gauge length. The true strain is related to engineering strain (ε) by $\varepsilon_T = \ln\ (1 + \varepsilon)$. A comparative multigraph containing all the true stress/true strain values characteristic for each synthesized material was plotted to evaluate the influence of the nanofiller on their mechanical properties. This multigraph was designed to show only the curves with the closest parameters to the mean values from each set of specimens. The dynamic and mechanical behavior of the samples was evaluated in single cantilever bending mode between −80 and 200 °C, with a controlled heating rate of 5 °C/min, using a TRITEC 2000 B-Dynamic mechanical analysis (DMA, Martignat, France) instrument.

For the evaluation of the behavior of polyurea-polyurethane nanocomposite films in a dynamic regime, a series of experimental measurements were performed on aluminum metal plates coated with polyurea and polyurea-polyurethane nanocomposites. The measuring instruments used were an ultra-high-speed video camera PHOTRON (Photron, Tokyo, Japan) and a PCB force transducer (PCB Piezotronics, Depew, NY, USA).

The samples used for these tests consisted of an aluminum metal plate with a thickness of 0.5 mm and a free diameter of 100 mm, on which a layer of polyurea-polyurethane nanocomposite of approximately 1 mm was previously deposited. Each sample was fixed (Figure 1b) and orientated, with the uncoated side towards the projectile launching direction. The experiments were carried out using a Hopkinson bar air propulsion system (Figure 1c), utilizing a spherical head projectile (Figure 1a). The impact strength obtained at the impact between the projectile and the samples was measured using a piezoelectric sensor connected to the mounting bracket of the tested aluminum plate. These experiments were performed on three types of materials: neat aluminum plate, polyurea coated aluminum plate (PU), and polyurea-polyurethane-MWCNTs nanocomposite-coated aluminum plates (PU-NC2, PU-NC3, and PU-NC4).

Figure 1. Hopkinson bar set-up for experimental testing of the polyurea-polyurethane-MWCNTs nanocomposite-coated aluminum plates in dynamic regime: (**a**) Spherical head projectile; (**b**) sample before the experiment (coated on the backside with polyurea (PU) or polyurea-polyurethane-MWCNTs nanocomposite (PU-NC)); (**c**) image captured during the experiment, at the moment of the impact of the projectile with the sample.

3. Results and Discussion

Polyurea and MWCNT nanocomposites designed for ballistic protection applications must meet certain performance criteria related to their thermal and mechanical strength. Materials exposed to the impact with a projectile or at the action of a shock wave can suffer both considerable deformation (or failure of the structure) and thermal degradation. Thus, polyurea coatings have the role of reducing these devastating effects.

The first step in our study consisted in the morphology characterization of the nanocomposites using SEM analysis. In Figure 2 SEM images of the PU-NC4 sample are presented. The nanocomposite consists of a continuous film; however, it can be noticed that, in some areas (highlighted with a white ring), there is a lack of homogeneity. The explanation could be due to the MWCNTs' aggregation in the polymer matrices and their escape at the surface of the polymer films, respectively.

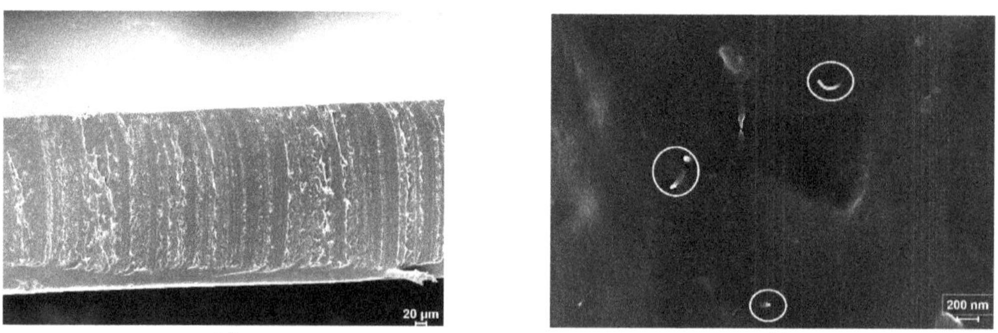

(**a**) (**b**)

Figure 2. SEM analysis of the polyurea-polyurethane-MWCNTs nanocomposite films: (**a**) in cross-section; (**b**) from the top.

To strengthen the hypothesis about the aggregation of the MWCNTs in sample PU-NC4, the second step of this study involved an investigation of the distribution of the MWCNTs inside the polyurea-polyurethane matrix through a micro-CT technique. CTVol® Micro-CT Surface Rendering Software allowed a realistic visualization of the samples (Figures 3a–d and 4) and displayed a homogenous spatial distribution of the nanofiller inside the polymeric matrix. Reconstruction of the collected images was performed by the NRecon reconstruction software (local version), and the full set of reconstruction results were visualized via DataViewer® 2D/3D Micro-CT Slice Visualization(Micro Photonics Inc., Allentown, PA, USA). Representative 2D cross-section slices of each reconstructed sample were illustrated in Figure 3e–h. As can be observed from Figure 3, the MWCNTs are well dispersed inside the polyurea-polyurethane matrix. Moreover, the increase of MWCNTs concentration is well distinguishable in the four micro-tomographs.

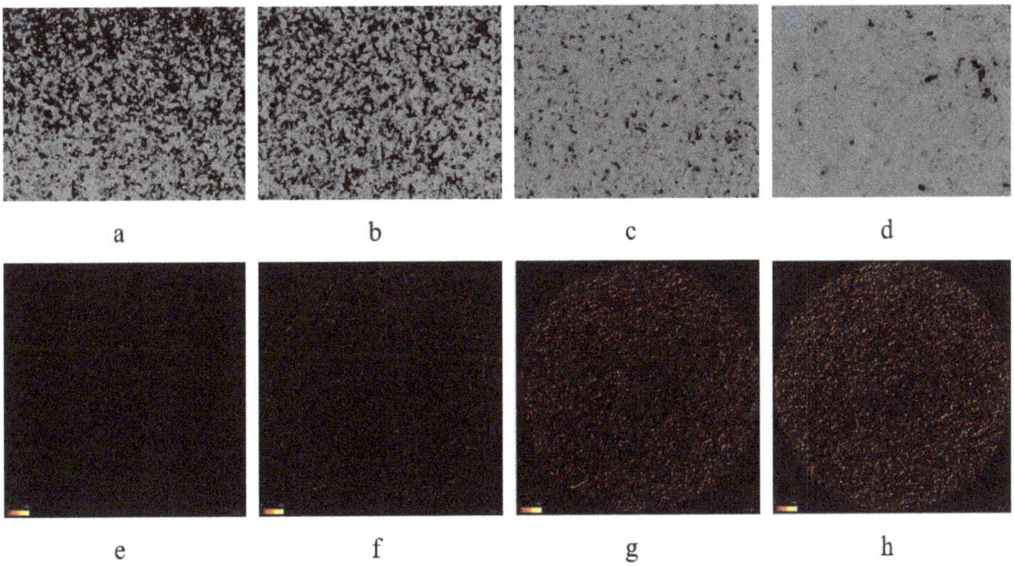

Figure 3. Micro-CT images (70 μm voxel size) of polyurea-polyurethane-MWCNTs nanocomposites comprising: (**a**,**e**) 0.05 wt.% MWCNTs, (**b**,**f**) 0.1 wt.% MWCNTs, (**c**,**g**) 0.2 wt.% MWCNTs, (**d**,**h**) 0.3 wt.% MWCNTs.

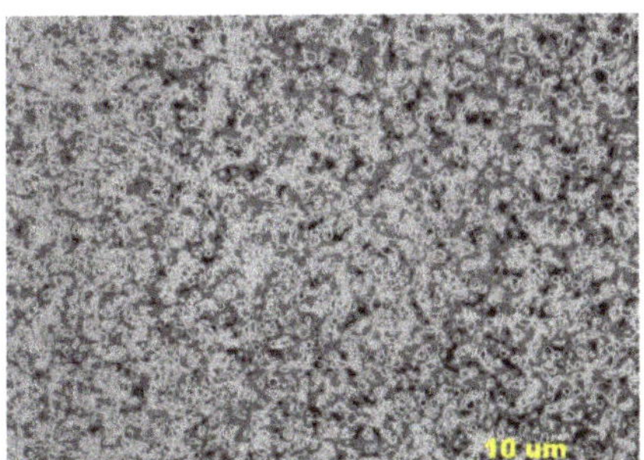

Figure 4. Micro-CT image (10 μm voxel size) of PU-NC4 sample.

Even if the nanocomposite samples containing 0.3 wt.% MWCNTs (PU-NC4) exhibited an apparent macro-homogeneity, there are regions where these nanoparticles tend to agglomerate, as can be observed in Figure 4.

Thermogravimetric analysis showed that the synthesized polyurea-polyurethane-MWCNTs films displayed a high thermal resistance, as can be observed in Figure 5 and Tables 2 and 3.

According to the TGA analysis, it appears that the introduction of carbon nanotubes into polyurea films had a slightly positive effect on thermal resistance, as their degradation process began later compared to the control sample (PU). However, as the concentration of carbon nanotubes increased, this effect decreased. The temperature corresponding to the maximum decomposition rate was also higher in the case of nanocomposites. Starting at 354 °C (in the case of PU samples) and at 366 °C, respectively (in the case of the nanocomposite samples), a significant weight loss process could be observed due to the decomposition process of the polymeric matrix (Figure 5).

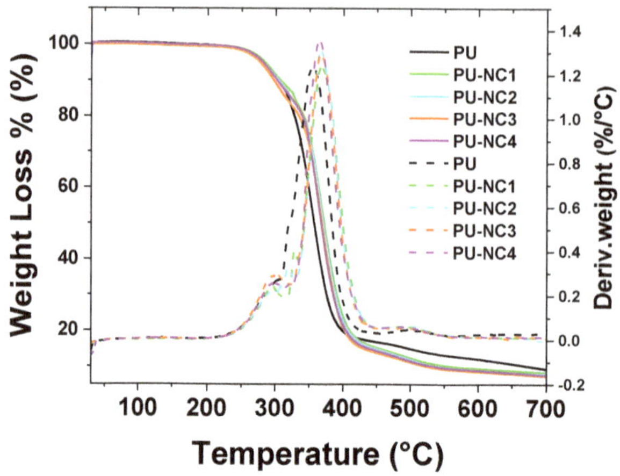

Figure 5. TGA and DTG analyses of the polyurea-polyurethane-MWCNTs nanocomposites.

Table 2. Thermal properties of polyurea-polyurethane-MWCNTs nanocomposites.

Sample	$T_{10\%}$, [°C]	T_{max}, [°C]
PU	303.0	354.0
PU-NC1	307.0	367.0
PU-NC2	304.5	366.5
PU-NC3	303.5	366.5
PU-NC4	302.0	364.0

$T_{10\%}$ = decomposition onset temperature (measured at 10% weight loss); T_{max} = the maximum decomposition temperature, corresponding to the maximum of DTG peak (the first derivative of the thermogravimetric curve).

Table 3. The decomposition process of the polyurea-polyurethane-MWCNTs nanocomposites.

Sample	Weight Loss [%]								
	200 °C	250 °C	300 °C	350 °C	400 °C	450 °C	500 °C	550 °C	600 °C
PU	0.2	1.3	9.1	41.4	80.2	83.3	85.1	86.9	87.9
PU-NC1	0.2	1.1	8.5	25.7	75.9	84.8	87.7	89.7	90.6
PU-NC2	0.7	1.7	9.0	26.2	77.1	85.2	88.1	90.2	91.1
PU-NC3	0.6	1.8	9.8	28.7	79.1	86.3	88.9	90.7	91.5
PU-NC4	0.2	1.3	9.4	28.1	78.5	85.7	88.6	90.4	91.2

Another important aspect related to the performance of polyurea-based materials designed for ballistic protection is related to their glass transition temperature. At lower temperatures, but still above the value of the glass transition temperature, polyurea tends to pass into the glass phase due to the energy dissipation phenomenon generated during a deformation caused by the action of a shock wave or impact with a projectile [17,20,31]. As can be observed in Figure 6, the synthesized materials exhibited two glass transition temperatures, each of them corresponding to the flexible and rigid nanodomains of the polyurea nanocomposite, respectively. Table 4 summarizes the two T_g values obtained for the synthesized materials. It can be noticed that the values obtained for the glass transition temperatures associated with the flexible nanodomains (T_{g1}) were quite similar to the blank sample (PU) because there are no significant differences between the aliphatic chains situated at a considerable distance from the rigid nanodomains, but there is a visible difference between the T_{g2} values obtained for the reference sample (PU) and the nanocomposite samples (PU-NC) because the carbon nanofiller joins the rigid nanodomains, thus reducing the mobility of the polymeric chains situated in their proximity.

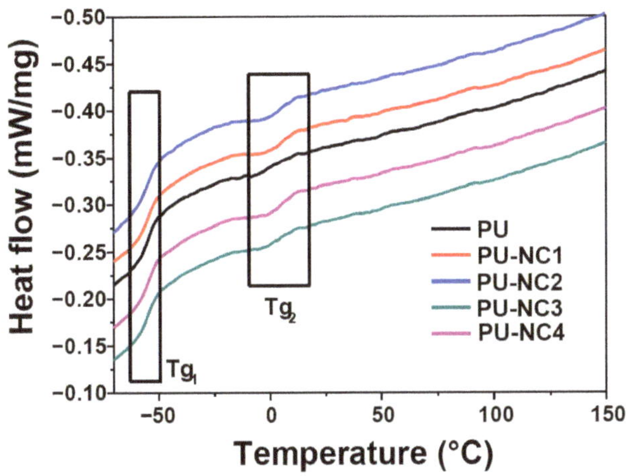

Figure 6. DSC curves obtained for polyurea-polyurethane-MWCNTs nanocomposites.

Table 4. Glass transition temperature obtained for polyurea-polyurethane-MWCNTs nanocomposites (Tg1–glass transition corresponding to the flexible nanodomains; Tg2–glass transition corresponding to the hard nanodomains).

Sample	Tg1, °C	Tg2, °C
PU	−54.0	6.0
PU-NC1	−54.5	9.5
PU-NC2	−55.5	10.0
PU-NC3	−55.5	10.0
PU-NC4	−55.5	10.0

To evaluate the behavior of polyurea-polyurethane films under the mechanical stress action, the samples were subjected to tensile tests. Table 5 and Figure 7 illustrate the obtained results. As can be noticed, the introduction of carbon nanotubes into the polyurea matrix led to an increase of Young's modulus, which means that they have become more rigid than the reference sample.

Although these materials were stiffer than the control sample, they showed higher deformation energy, which suggests that these materials can dissipate more energy. Since these materials were specially designed for ballistic protection, their high capacity to absorb and dissipate energy represents a remarkable advantage.

The pre-dispersed MWCNTs achieved a good dispersion in the polymer matrix until it reached a concentration of 0.2% MWCNTs but, at higher concentrations, the nanofiller tends to agglomerate. This situation leads to a decrease in the tensile strength of the samples containing 0.3% MWCNTs (PU-NC4). Since PU-NC3 displayed the highest value of tensile true strain (σ_T) and the highest deformation energy (Figure 7 and Table 5), we can affirm that these maximum values indicate the concentration of 0.2% MWCNTs as representing the optimal composition (Figure 8) for this type of polyurea nanocomposite. The results are also sustained by SEM and Micro-CT analyses through the aggregation of the MWCNTs in the case of PU-NC4.

Table 5. Mechanical properties of polyurea-polyurethane- MWCNTs nanocomposites.

Sample	MWCNTs Conc. (%)	Young's Modulus * (MPa)	Maximum Tensile True Stress (σ_T) (MPa)	Deformation Energy (Tensile Toughness) ($J \cdot m^{-3}$)
PU	0	48.65 ± 1.02	33.76 ± 0.42	11.31 ± 0.29
PU-NC1	0.05	70.11 ± 0.98	34.48 ± 0.62	12.90 ± 0.13
PU-NC2	0.1	79.22 ± 1.28	35.87 ± 0.55	13.57 ± 0.21
PU-NC3	0.2	81.54 ± 0.83	40.84 ± 0.42	16.53 ± 0.18
PU-NC4	0.3	43.87 ± 1.44	33.91 ± 0.58	12.90 ± 0.43

* Young's Modulus was calculated according to the mathematical model described by Xue and Hutchinson [32].

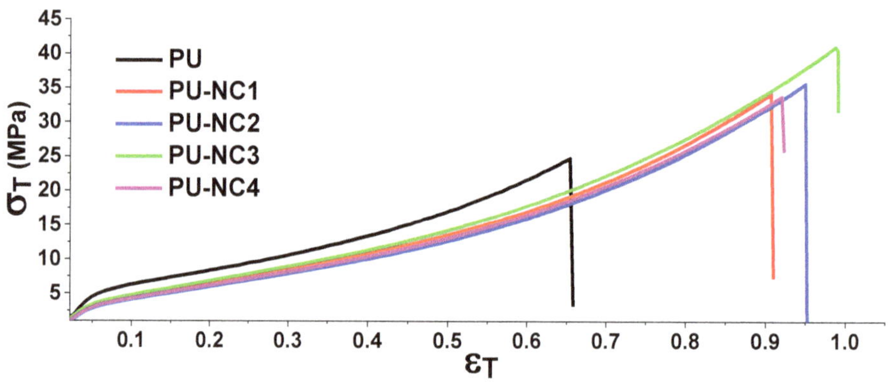

Figure 7. True stress–true strain plots for the synthesized polyurea-polyurethane nanocomposite films.

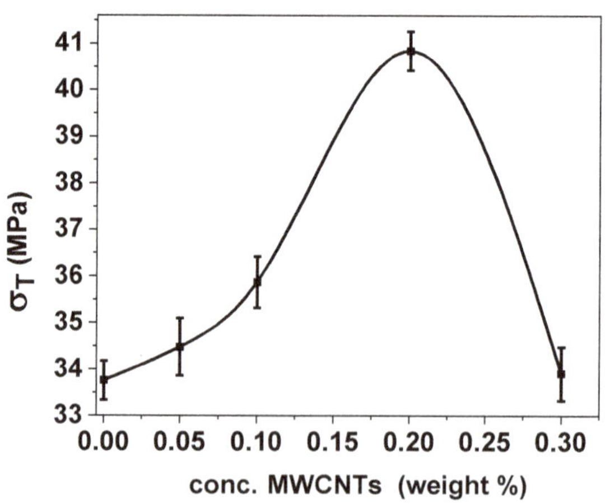

Figure 8. Variation of the maximum of true stress, depending on the concentration of the nanofiller.

The impact strength of a material is an index of its toughness. When a load is applied to a polymer, a part of the energy is dissipated throughout the polymer mass, while another part is stored in the material and will be released after the load is removed. The

DMA analysis aimed to follow the evolution of storage (E′) and loss (E″) modulus of the synthesized nanocomposites to evaluate their potential to be used as coatings for the structures that need improved ballistic protection. Figure 9 illustrates a comparative plot of the E′ and E″ values for the blank sample (PU), and for the nanocomposite which displayed the best tensile test results (PU-NC3).

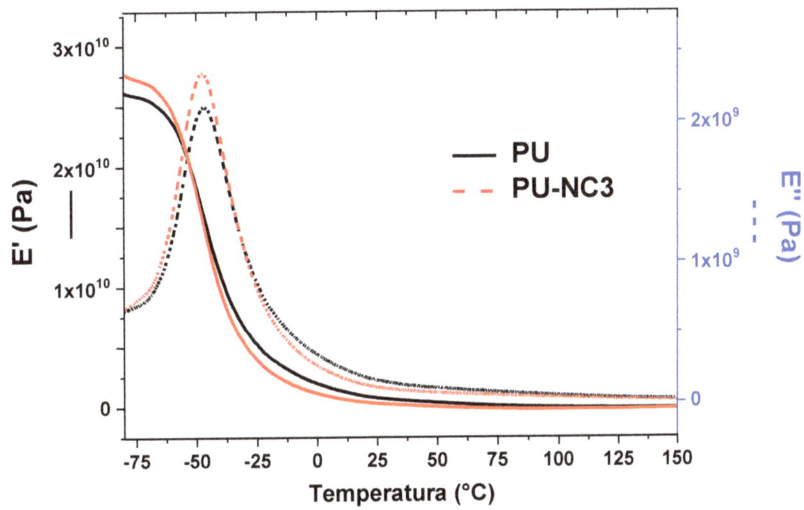

Figure 9. Comparative E′ and E″ plot between polyurea and polyurea-polyurethane nanocomposites.

Both E′ and E″ were higher in the case of PU-NC3, showing that this nanocomposite possesses the ability to store a significant amount of energy, and it can also dissipate a higher quantity of energy in comparison with the neat polyurea (PU). The peaks corresponding to the loss modulus can allow the estimation of the glass transition temperature at around −50 °C, indicating that this value obtained through DMA technique is following DSC analysis.

The characterization of these materials specially designed for ballistic protection would not be complete without experimental measurements which aim to evaluate their behavior at the impact with a projectile or the action of a shock wave. Therefore, for evaluating the efficiency of the obtained materials, round-headed projectiles were employed for impacting aluminum plates coated with the synthesized polyurea-polyurethane nanocomposites on the backside.

For the experimental testing in the dynamic regime, we used a Hopkinson bar set-up, with every sample being fixed and aligned for axial symmetry. An air gun, which allows the variation of pressure, launches the striker (the spherical head projectile in our case) towards the direction of the center of the tested specimen. Thus, the sample is rapidly impacted by the compressive stress wave generated. At impact, the samples undergo rapid and permanent deformations or fractures. Tables 6 and 7 and Figure 10 display the results obtained with the Hopkinson bar tests. As can be noticed in Table 7, at an initial projectile acceleration pressure of 0.2 bars, all the samples underwent deformations, but none of them suffered fissures. Starting with 0.3 bars, small cracks appeared in the aluminum plate, but the samples covered with polyurea, even if they displayed higher deformations, did not suffer fractures, meaning that neat polyurea already offers significant improvement. At a higher initial projectile acceleration pressure of 0.4 bars, the aluminum plate underwent a higher fracture, but the polyurea-coated specimens did not exhibit fissures. Only at 0.5 bars of initial projectile acceleration pressure could a small fissure be observed on the aluminum plate coated with polyurea, but the polyurea film remained unbroken due to its higher elasticity, and the specimens coated with the nanocomposite suffered only low deformations and no fractures. At a higher initial projectile acceleration pressure of 0.6 bars, polyurea films PU and the nanocomposite containing the lowest MWNCTs concentration PU-NC2 started to crack. At 0.6 bars, the aluminum plate coated with PU-NC3 suffered low fractures, but this polymeric film resisted up to 0.8 bars. Therefore, we can affirm that, only at values higher than 0.8 bars for the initial projectile acceleration pressure, did both layers (metallic layer and polymeric nanocomposite coating) undergo visible fractures; at lower values, the metallic layer was the only one that cracked. In the case of PU-NC4 specimens, the metallic layer exhibited low fissures at 0.7 bars, and both layers (metallic plate and polymeric nanocomposite coating) were fractured at values higher than 0.8 bars. The increase in the resistance of the aluminum plates covered with PU and PU-NC can also be highlighted by the recordings made with the help of the pressure transducer (Table 6). From the values displayed in Table 6, it can be noticed that, for the tests performed on the uncoated metallic plates, the maximum value of the force was 2.6 kN at the pressure of 0.3 bar. This characteristic displayed a decreasing trend for the tests performed at higher pressures, and could be associated with the loss of the integrity of the plate (fracturing phenomenon). On the other hand, for the specimens coated with polyurea, the values measured with the pressure transducer were almost double for the maximum force, more precisely 5 kN for PU-NC2 at 0.5 bars. For higher initial projectile acceleration pressure, all specimens, including the ones coated with the nanocomposite, followed the same trend: after reaching a maximum value of the force, there was a decrease, a phenomenon associated with the loss of the integrity of the metal plates.

Table 6. Maximum force values obtained during the impact tests with Hopkinson bar.

Pressure (Bar)	Aluminum Plate (Uncoated) Maximum Force (kN)	Aluminum Plate (Coated with PU) Maximum Force (kN)	Aluminum Plate (Coated with PU-NC2) Maximum Force (kN)	Aluminum Plate (Coated with PU-NC3) Maximum Force (kN)	Aluminum Plate (Coated with PU-NC4) Maximum Force (kN)
0.2	2.0	2.0	2.1	-	-
0.3	2.6	3.0	2.8	2.3	3.0
0.4	2.5	3.6	3.8	3.8	4.0
0.5	1.8	4.1	4.3	5.0	4.4
0.6	-	3.6	3.9	4.1	4.8
0.7	-	-	-	-	4.0

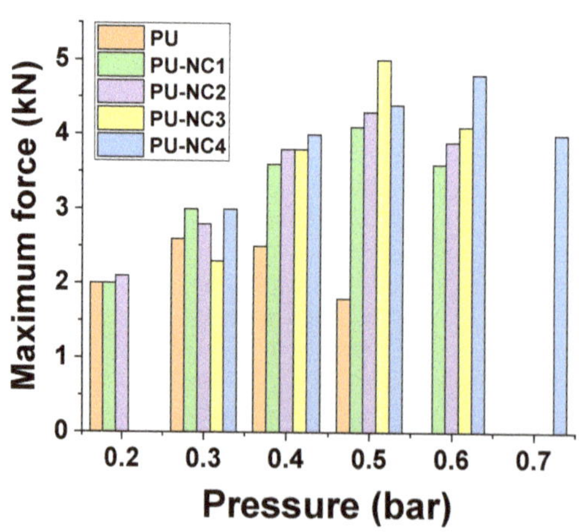

Figure 10. Maximum force values obtained during the impact tests with Hopkinson bar.

Table 7. Images of the tested specimens, captured after the impact tests with Hopkinson bar (pressure in bar and pattern of deformation).

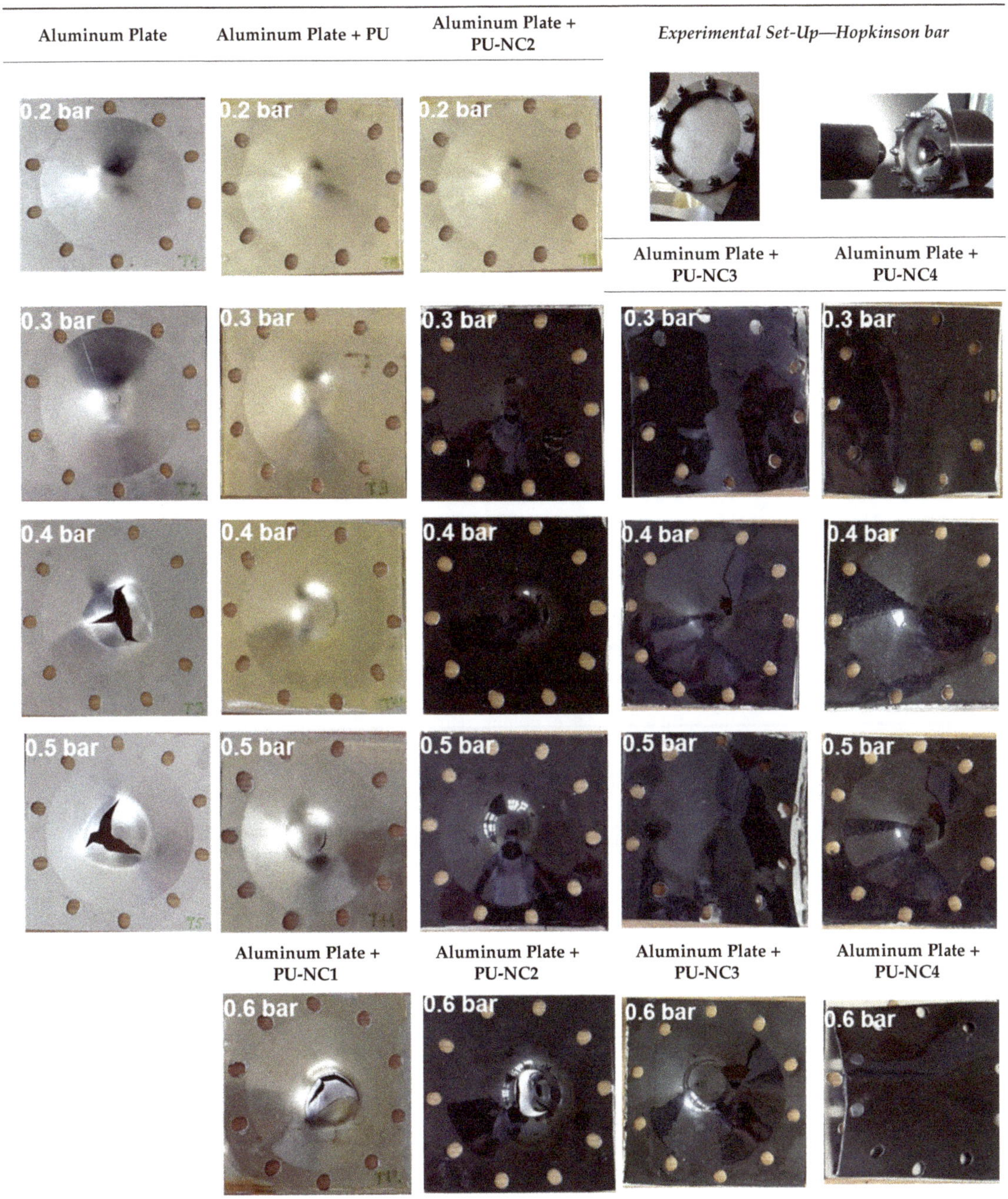

Table 7. *Cont.*

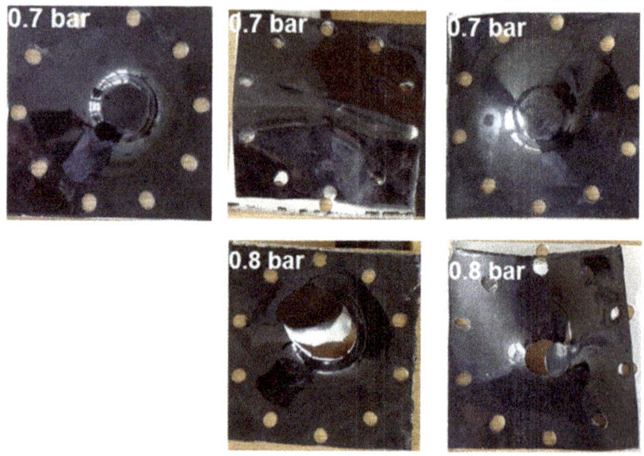

4. Conclusions

Polyurea-polyurethane and MWCNTs nanocomposites were obtained to be used for ballistic protection applications through a facile synthesis approach involving MWCNTs pre-dispersed in a polyester polyol-based resin. To demonstrate that these materials are suitable for this type of application, they were subjected to thermal and mechanical characterization using different analysis techniques (SEM, μCT, TGA, DSC, DMA, tensile tests), and they were also subjected to ballistic tests using a Hopkinson bar system. SEM and micro-CT analyses confirmed the homogenous dispersion of the MWCNTs inside the polyurea-polyurethane matrix for the samples containing 0.05%, 0.1%, and 0.2% MWCNTs. In the nanocomposites containing 0.3% MWCNTs, it can be noticed that, in some regions, there is a lack of homogeneity due to the tendency to form aggregates at higher concentrations of this nanofiller. TGA showed that the nanocomposite films have good thermal stability (up to about 300 °C). The presence of MWCNTs delayed the onset of the decomposition process of polyurea-polyurethane films, and the maximum degradation temperature was about 10 °C higher than that of the reference polyurea sample (PU). DSC curves displayed two glass transition temperatures due to the coexistence of the two segregated nanodomains from the componence of polyurea matrix (flexible and rigid nanodomains). According to tensile test results, it turned out that, from all the synthesized materials, the polyurea-polyurethane nanocomposite film containing 0.2% MWCNTs is the optimal option for ballistic protection applications, since it possesses the highest deformation energy. DMA analysis also demonstrated that PU-NC3 samples had a remarkable capacity for absorbing and dissipating energy. Experimental testing in a dynamic regime of the polyurea-coated aluminum plates showed that the polymeric layer allows the metal plate to maintain its integrity at an acceleration pressure value that is almost three times higher than the one for the uncoated metallic specimen. Synthesized nanocomposites possess unique properties that recommend them to be used in the modernization of ballistic protection equipment and devices. Thus, military vehicles or bulletproof vests could be safer, but also lighter and less expensive.

We can conclude that the synthesized polyurea-polyurethane-MWCNTs nanocomposites possessing this exceptional combination of properties and advantages are suitable for employment as complementary materials for ballistic protection.

Author Contributions: Conceptualization, G.T., A.D. and E.R.; methodology, E.R., M.T. and A.R.; software, R.G.; validation, G.T., A.D., E.R., F.R., M.T. and E.T.; formal analysis, G.T., A.D., E.R. and R.G.; investigation, G.T., A.D., F.R., M.T., P.O.S., C.D., A.R., E.T., F.B. and R.G.; resources, F.R. and E.T.; data curation, G.T., A.D., P.O.S., C.D., A.R., F.B. and R.G.; writing—original draft preparation, G.T., A.D. and E.R.; writing—review and editing, G.T., A.D. and E.R.; visualization, G.T., A.D., E.R., F.R., P.O.S., C.D., E.T. and F.B.; supervision, A.D., E.R., M.T., and A.R.; project administration, E.R. and A.R.; All authors have read and agreed to the published version of the manuscript.

Funding: This work was partially supported by the National Authority for Scientific Research from the Ministry of Education, Research and Youth of Romania through the National Project PN-III-P2-2.1-PTE-2019-0400 Ctr. No. 49PTE/2020. The authors would also like to thank for the financial support provided by the National Authority for Scientific Research from the Ministry of Education, Research and Youth of Romania through the National Project PN-II-PT-PCCA-2013 No. 278/2014.

Institutional Review Board Statement: Not applicable.

Informed Consent Statement: Not applicable.

Conflicts of Interest: The authors declare no conflict of interest.

References

1. Parveen, S.; Rana, S.; Fangueiro, R. A Review on Nanomaterial Dispersion, Microstructure, and Mechanical Properties of Carbon Nanotube and Nanofiber Reinforced Cementitious Composites. *J. Nanomater.* **2013**, *2013*, 710175. [CrossRef]
2. Goga, D.; Ţigănescu, T.; Pulpea, B.; Moldoveanu, C.; Rotaru, C. A Quantitative Method of Comparative Assessment of Primers Ignition Performances. *Adv. Mil. Technol.* **2017**, *12*, 217–228. [CrossRef]
3. Pulpea, G.B.; Voicu, A.; Matache, L.; Haller, L.; Mandache-Dodoiu, A.D.; Ginghină, R. Numerical Simulation of Interior Ballistics for Pyrotechnics Systems. *J. MTA Rev.* **2017**, *27*, 39–44.
4. Gavrila, A.M.; Iordache, T.V.; Lazau, C.; Rotariu, T.; Cernica, I.; Stroescu, H.; Stoica, M.; Orha, C.; Bandas, C.E.; Sarbu, A. Biomimetic Sensitive Elements for 2,4,6-Trinitrotoluene Tested on Multi-Layered Sensors. *Coatings* **2020**, *10*, 273. [CrossRef]
5. Voicu, A.E.; Rotariu, T.; Teodorescu, M.; Zecheru, T.; Tiganescu, T.V.; Orban, O. pH sensitive polymeric binders for energetic materials. *Mater. Plast.* **2017**, *54*, 103. [CrossRef]
6. Da Silva, J.E.L., Jr.; Paciornik, S.; d'Almeida, J.R.M. Evaluation of the effect of the ballistic damaged area on the residual impact strength and tensile stiffness of glass-fabric composite materials. *Compos. Struct.* **2004**, *64*, 123–127. [CrossRef]
7. Ávila, A.F.; Neto, A.S.; Nascimento, H., Jr. Hybrid nanocomposites for mid-range ballistic protection. *Int. J. Impact Eng.* **2011**, *38*, 669–676. [CrossRef]
8. Findik, F.; Tarim, N. Ballistic impact efficiency of polymer composites. *Compos. Struct.* **2003**, *61*, 187–192. [CrossRef]
9. Casalini, R.; Bogoslovov, R.; Qadri, S.B.; Roland, C.M. Nanofiller reinforcement of elastomeric polyurea. *Polymer* **2012**, *53*, 1282–1287. [CrossRef]
10. Bogue, R. Nanocomposites: A review of technology and applications. *Assem. Autom.* **2011**, *31*, 106–112. [CrossRef]
11. Bhattacharya, M. Polymer Nanocomposites—A Comparison between Carbon Nanotubes, Graphene, and Clay as Nanofillers. *Materials* **2016**, *9*, 262. [CrossRef] [PubMed]
12. Ho, M.-W.; Lam, C.-K.; Lau, K.-T.; Ng, D.H.L.; Hui, D. Mechanical properties of epoxy-based composites using nanoclays. *Compos. Struct.* **2006**, *75*, 415–421. [CrossRef]
13. Crosby, A.J.; Lee, J.Y. Polymer Nanocomposites: The "Nano" Effect on Mechanical Properties. *Polym. Rev.* **2007**, *47*, 217–229. [CrossRef]
14. Hsiao, C.-C.; Lin, T.S.; Cheng, L.Y.; Ma, C.-C.M.; Yang, A.C.M. The Nanomechanical Properties of Polystyrene Thin Films Embedded with Surface-grafted Multiwalled Carbon Nanotubes. *Macromolecules* **2005**, *38*, 4811–4818. [CrossRef]
15. Kumar, V.; Lee, G.; Monika; Choi, J.; Lee, D.-J. Studies on composites based on HTV and RTV silicone rubber and carbon nanotubes for sensors and actuators. *Polymer* **2020**, *190*, 122221. [CrossRef]
16. Kumar, V.; Alam, M.N.; Manikkavel, A.; Choi, J.; Lee, D.-J. Investigation of silicone rubber composites reinforced with carbon nanotube, nanographite, their hybrid, and applications for flexible devices. *J. Vinyl Addit. Technol.* **2020**. [CrossRef]
17. Toader, G.; Rusen, E.; Teodorescu, M.; Diacon, A.; Stanescu, P.O.; Rotariu, T.; Rotariu, A. Novel polyurea polymers with enhanced mechanical properties. *J. Appl. Polym. Sci.* **2016**, *133*. [CrossRef]
18. Raman, S.N.; Ngo, T.; Lu, J.; Mendis, P. Experimental investigation on the tensile behavior of polyurea at high strain rates. *Mater. Des.* **2013**, *50*, 124–129. [CrossRef]
19. Mohotti, D.; Ngo, T.; Raman, S.N.; Ali, M.; Mendis, P. Plastic deformation of polyurea coated composite aluminium plates subjected to low velocity impact. *Mater. Des.* **2014**, *56*, 696–713. [CrossRef]
20. Grujicic, M.; Pandurangan, B.; He, T.; Cheeseman, B.A.; Yen, C.F.; Randow, C.L. Computational investigation of impact energy absorption capability of polyurea coatings via deformation-induced glass transition. *Mater. Sci. Eng. A* **2010**, *527*, 7741–7751. [CrossRef]

21. Fragiadakis, D.; Gamache, R.; Bogoslovov, R.B.; Roland, C.M. Segmental dynamics of polyurea: Effect of stoichiometry. *Polymer* **2010**, *51*, 178–184. [CrossRef]
22. Grujicic, M.; He, T.; Pandurangan, B.; Svingala, F.R.; Settles, G.S.; Hargather, M.J. Experimental Characterization and Material-Model Development for Microphase-Segregated Polyurea: An Overview. *J. Mate. Eng. Perform.* **2012**, *21*, 2–16. [CrossRef]
23. Ackland, K.; Anderson, C.; Ngo, T.D. Deformation of polyurea-coated steel plates under localised blast loading. *Int. J. Impact Eng.* **2013**, *51*, 13–22. [CrossRef]
24. Li, T.; Zheng, T.; Han, J.; Liu, Z.; Guo, Z.-X.; Zhuang, Z.; Xu, J.; Guo, B.-H. Effects of Diisocyanate Structure and Disulfide Chain Extender on Hard Segmental Packing and Self-Healing Property of Polyurea Elastomers. *Polymers* **2019**, *11*, 838. [CrossRef] [PubMed]
25. Amini, M.R.; Isaacs, J.; Nemat-Nasser, S. Investigation of effect of polyurea on response of steel plates to impulsive loads in direct pressure-pulse experiments. *Mech. Mater.* **2010**, *42*, 628–639. [CrossRef]
26. Roland, C.M.; Fragiadakis, D.; Gamache, R.M.; Casalini, R. Factors influencing the ballistic impact resistance of elastomer-coated metal substrates. *Philos. Mag.* **2013**, *93*, 468–477. [CrossRef]
27. Bai, Y.; Liu, C.; Huang, G.; Li, W.; Feng, S. A Hyper-Viscoelastic Constitutive Model for Polyurea under Uniaxial Compressive Loading. *Polymers* **2016**, *8*, 133. [CrossRef]
28. Li, Y.; Chen, C.; Hou, H.; Cheng, Y.; Gao, H.; Zhang, P.; Liu, T. The Influence of Spraying Strategy on the Dynamic Response of Polyurea-Coated Metal Plates to Localized Air Blast Loading: Experimental Investigations. *Polymers* **2019**, *11*, 1888. [CrossRef]
29. Petre, R.; Zecheru, T.; Petrea, N.; Ginghina, R.; Sandu, S.; Muresan, M.; Matache, L.C.; Sava, A.C.; Neatu, F. Synthesis and Mechanical Properties of Polyurea-Based Hybrid Composites for Ballistic Individual Protection. *Mater. Plast.* **2018**, *55*, 315. [CrossRef]
30. Sahoo, N.G.; Rana, S.; Cho, J.W.; Li, L.; Chan, S.H. Polymer nanocomposites based on functionalized carbon nanotubes. *Prog. Polym. Sci.* **2010**, *35*, 837–867. [CrossRef]
31. Toader, G.; Rusen, E.; Teodorescu, M.; Diacon, A.; Stanescu, P.O.; Damian, C.; Rotariu, T.; Rotariu, A. New polyurea MWCNTs nanocomposite films with enhanced mechanical properties. *J. Appl. Polym. Sci.* **2017**, *134*, 45061. [CrossRef]
32. Xue, Z.; Hutchinson, J.W. Neck retardation and enhanced energy absorption in metal–elastomer bilayers. *Mech. Mater.* **2007**, *39*, 473–487. [CrossRef]

polymers

MDPI

Article

Kaolin-Enhanced Superabsorbent Composites: Synthesis, Characterization and Swelling Behaviors

Mengna Chen [1], Xuelong Chen [2], Caiyan Zhang [1], Baozheng Cui [1], Zewen Li [1], Dongyu Zhao [1] and Zhe Wang [1,*]

[1] Department of Polymer Materials and Engineering, School of Chemistry and Materials Science, Heilongjiang University, Harbin 150080, China; Mengna999456@163.com (M.C.); zhangcaiyan061995@163.com (C.Z.); 2201459@s.hlju.edu.com (B.C.); lizewen145@126.com (Z.L.); zhaodyu@aliyun.com (D.Z.)

[2] School of Material Science and Engineering, Nanyang Technological University, 50 Nanyang Avenue, Singapore 639798, Singapore; xchen014@e.ntu.edu.sg

* Correspondence: zhe.wang@msn.com

Abstract: One type of low-cost and eco-friendly organic-inorganic superabsorbent composite (SAPC) was synthesized by free radical polymerization of acrylic acid (AA), starch (ST), sodium alginate (SA) and kaolin (KL) in aqueous solution. The structure and morphology of the SAPC were characterized by Fourier transform infrared spectrometer (FT-IR), scanning electron microscope (SEM), X-ray diffraction (XRD) and thermogravimetric analysis (TGA). The influence of different reaction conditions on water absorption of SAPC, i.e., SA and KL contents, AA neutralization degree (ND), potassium persulfate (KPS) and N, N′-methylenebisacrylamide (MBA) loading were systematically studied. Under the optimal synthesis conditions, very high water absorption of 1200 g/g was achieved. The swelling kinetic mechanism of SAPC was studied by pseudo-second order swelling kinetics model and Ritger-Peppas model. The performances of SAPC under different environments were tested and results revealed that this new SAPC had excellent swelling capacity, high water retention, good salt tolerance in monovalent salt solution (NaCl solution) and good pH tolerance between 4 and 10.

Keywords: sodium alginate; starch; kaolin; free radical polymerization; superabsorbent composites; characterization; swelling behavior

Citation: Chen, M.; Chen, X.; Zhang, C.; Cui, B.; Li, Z.; Zhao, D.; Wang, Z. Kaolin-Enhanced Superabsorbent Composites: Synthesis, Characterization and Swelling Behaviors. *Polymers* **2021**, *13*, 1204. https://doi.org/10.3390/polym13081204

Academic Editor: Edina Rusen

Received: 22 March 2021
Accepted: 2 April 2021
Published: 8 April 2021

Publisher's Note: MDPI stays neutral with regard to jurisdictional claims in published maps and institutional affiliations.

Copyright: © 2021 by the authors. Licensee MDPI, Basel, Switzerland. This article is an open access article distributed under the terms and conditions of the Creative Commons Attribution (CC BY) license (https://creativecommons.org/licenses/by/4.0/).

1. Introduction

Superabsorbent polymers (SAPs) are one type of three-dimensional (3D), cross-linked hydrophilic materials that can absorb and hold a large amount of water within a certain period of time [1–4]. Before the emergence of SAP, natural biomass materials, such as cotton, have been used for traditional absorbents [5]. These materials absorb water through capillarity, yet with low water absorption capacity, and cannot retain water for a long time, making them unsuitable for many life and industry applications. By contrast, SAP mainly absorbs water through chemical adsorption, with efficient water absorption and large water retention capacity [6]. Owing to excellent performance, SAP has been widely used in diversified fields, such as personal care, industry, chemicals, forestry, agriculture, wastewater treatment and drug-delivery systems [7–9].

Although with wide applications, SAP still has several key issues that need to be addressed, such as high synthesis costs, potential toxicity and serious environmental influences [10,11]. To reduce production costs and improve performance, many inorganic clay minerals (montmorillonite, kaolin, bentonite and attapulgite) and natural polymeric materials such as polysaccharides (starch, sodium alginate, cellulose, chitosan and their derivatives) have been used to synthesize environmentally friendly organic-inorganic superabsorbent composites (SAPCs) [12–14].

Kaolin is a hydrous aluminosilicate with layered structure, which has advantages of large specific surface area, large cation exchange capacity and low cost [15,16]. The addition of kaolin can not only significantly reduce the production cost, but is also able to improve water absorption, gel strength, thermal stability and mechanical properties of the final products [17,18].

Starch is a natural polysaccharide, mainly composed of amylose and amylopectin [19]. The former possesses a linear chain composed of α-1,4-linked glucose unites, while the latter has a highly branched chain of α-1,4-linked glucose units interlinked by α-1,6-linked bonds [20]. Sodium alginate, an abundant natural polysaccharide extracted from various species of brown seaweed, is composed of β-1,4-linked D-mannuronic acid and α-1,4-linked L-guluronic acid with various proportions, which can be arranged into different sequences in the polymeric backbone [21,22]. Taking advantages of excellent properties, such as biocompatibility, biodegradability, low cost and low toxicity, starch and sodium alginate have become ideal backbones for the synthesis of SAP [23]. By now, graft polymerization of vinyl monomers such as acrylamide, acrylonitrile, and acrylic acid with starch and sodium alginate, respectively, has been reported. Erdener Karadağ et al. synthesized novel composite sorbent AAm/MA hydrogels containing starch and kaolin used for water sorption and dye uptake [24]. Jihuai Wu et al. fabricated one type of starch-graft-acrylamide/kaolin superabsorbent composite with water absorbency 4000 times higher than its own weight [25]. Linhui Zhu et al. studied the adsorption behaviors of sodium alginate graft poly(acrylic acid-co-2-acrylamide-2-methyl propane sulfonic acid)/kaolin hydrogel composite towards dyes [26]. Sodium alginate graft poly(acrylic acid-co-acrylamide)/kaolin composite hydrogel has also been prepared by Yaoji Tang et al [27]. However, there are few reports on the simultaneous introduction of starch and sodium alginate into the system to graft polymerize with vinyl monomer. It is reported that the use of a polysaccharide mixture in SAP synthesis shows better mechanical stability and water retention capacity compared to a single polysaccharide [28]. The addition of sodium alginate can improve pH sensitivity of starch-based SAP [29]. In addition, the preparation route of SAP from sodium alginate nowadays is attractive as to make full use of marine resources and reduce manufacture cost and pollution. Nevertheless, the viscosity of sodium alginate solution is extremely high. During the synthesis of alginate-based SAP, the viscosity of the reaction system often sharply increases due to the addition of sodium alginate, which restricts the movement of reactants and reduces the monomer conversion rate. It is hypothesized that adding a certain amount of starch can reduce the content of sodium alginate and viscosity of the system.

Based on these reasons, a novel organic-inorganic SAPC was synthesized by free radical copolymerization with acrylic acid, starch, sodium alginate and kaolin. The structure of the SAPC was characterized by Fourier transform infrared spectrometer, scanning electron microscope, X-ray diffraction and thermogravimetric analysis. The influence of different reaction conditions on water absorption of SAPC was studied. A pseudo-second order kinetics model and Ritger-Peppas model also illustrated the swelling kinetic mechanism of SAPC. The performance of this new SAPC was tested. Results showed that the SAPC had excellent swelling capacity, salt tolerance, water retention and pH sensitivity.

2. Experimental

2.1. Materials

Starch (ST, food grade) was purchased from Shandong Lejiake Food Co., Ltd. (Dezhou, Shandong, China). Sodium alginate (SA, analytical grade) was from Shanghai Yuanye Biotechnology Co., Ltd. (Shanghai, China). Acrylic acid (AA, analytical grade) was purchased from Tianjin Zhiyuan Chemical Reagent Co., Ltd. (Tianjin, China), kaolin (KL, chemically pure) was purchased from Shanghai Fengxian Fengcheng Reagent Factory (Shanghai, China), and potassium persulfate (KPS, analytical grade) was purchased from, Tianjin Komiou Chemical Reagent Co., Ltd. (Tianjin, China). N, N′-methylenebisacrylamide (MBA, chemically pure) was obtained from Tianjin Guangfu Research Institute of Fine

Chemical (Tianjin, China). Sodium hydroxide (NaOH, analytical grade) was supplied by Tianjin Tianli Chemical Reagent Co., Ltd. (Tianjin, China). All solutions were prepared with distilled water.

2.2. Synthesis of SAPC

The ST (3 g) was dissolved in distilled water (40 mL) and then placed in a 500 mL four-port flask with a condenser, a nitrogen tube and a mechanical stirrer. The solution was heated to 85 °C and continuously stirred for 1 h. Then, the solution of SA was added to the flask and continued to stir until temperature dropped to 50 °C, and then KPS was added to generate radicals. After the temperature reached 70 °C and stirring for 15 min, a mixture containing KL, MBA and partially neutralized AA (20 mL AA, neutralized AA with 25% (*w*/*w*) NaOH solution under ice bath conditions) was added to the above system, followed by reaction for 3 h. A nitrogen purge was used throughout the process. The product was dried in a vacuum oven at 60 °C until the weight stabilized. Afterwards, the product was milled, and the size of all powder products used for testing was about 80 mesh.

For comparison, the synthesis of neat SAP was carried out in the same way except without KL. The structure of ST and SA is shown in Scheme 1, while the mechanistic pathway for synthesis of SAPC is shown in Scheme 2.

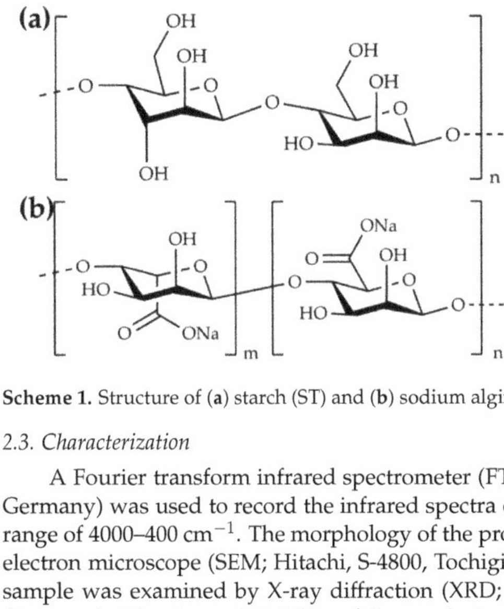

Scheme 1. Structure of (**a**) starch (ST) and (**b**) sodium alginate (SA).

2.3. Characterization

A Fourier transform infrared spectrometer (FT-IR; Bruker, Equinox 55, Mannheim, Germany) was used to record the infrared spectra of raw materials and products in the range of 4000–400 cm^{-1}. The morphology of the products was observed with a scanning electron microscope (SEM; Hitachi, S-4800, Tochigi, Japan). The crystal structure of the sample was examined by X-ray diffraction (XRD; Bruker, D8 ADVANCE, Mannheim, Germany). The thermal stability of the products was analyzed by thermogravimetric analysis (TGA; Mettler, TGA/DSCI, Greifensee, Switzerland) from 35 to 800 °C with heating rate 10 °C min^{-1} under nitrogen atmosphere.

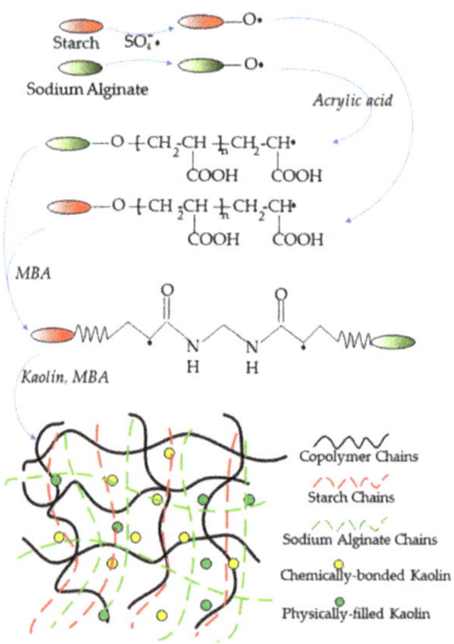

Scheme 2. The mechanistic pathway for synthesis of superabsorbent composite (SAPC).

2.4. *Performance Test of SAPC*

2.4.1. Swelling Ratios

To determine the swelling ratio, dry composites (0.1 ± 0.001 g) were immersed in 500 mL of solutions (distilled water, various salt solutions and pH solutions) for 4 h at room temperature to achieve equilibrium. The swollen product was filtered through a 100-mesh screen to remove excess moisture and then weighed. Each specimen was tested thrice, and the average value was used. The water absorbent capacity was calculated using the following Equation (1):

$$Q_{eq} = \frac{(m_1 - m_0)}{m_0} \tag{1}$$

where Q_{eq} is the water absorbent capacity and m_0 and m_1 are the mass of the dried and swollen products, respectively.

2.4.2. Swelling Kinetics

The pre-weighed dry product (0.1 ± 0.001 g) was added into distilled water (500 mL), and swelled for a period of time (5, 10, 15, 30, 45, 60, 120, 180 and 240 min). The water absorption at time t was measured by the method described in Section 2.4.1.

2.4.3. Swelling in Salt Solutions

NaCl and $FeCl_3$ were used to prepare salt solutions with different concentrations (0.02–0.1 mol/L). The pre-weighed dry product (0.1 ± 0.001 g) was dispersed in various salt solutions (500 mL), and water absorption was measured by the method described in Section 2.4.1.

2.4.4. Water Retention Measurement

The fully swollen product (80 g) was placed in petri dishes at various temperatures (25, 45 and 60 °C), and then weighted every 2 h. According to Equation (2), water retention of the product was calculated:

$$W_r = \frac{(M_t - M_0)}{(M_1 - M_0)} \times 100\% \tag{2}$$

where W_r is the water retention rate of product, M_t is the mass of the swollen product at time t, M_0 is the mass of the dry product and M_1 is the initial weight of the fully swollen products.

2.4.5. Swelling Measurement in Buffer Solutions

The buffer solutions (hydrochloric acid and sodium hydroxide solutions) with pH values from 2 to 12 were prepared to test the pH sensitivity of the SAPC. A pH-meter (PHS-3C, accuracy is $\pm$ 0.01) provided by Shanghai INESA Scientific Instrument Co., Ltd. (Shanghai, China) was used to detect the pH value of the solution. Then, the dry product (0.1 $\pm$ 0.001 g) was dispersed in various buffer solutions (500 mL), and water absorption was calculated according to Equation (1).

3. Results and Discussion

3.1. *FT-IR Analysis*

The FT-IR spectra of the ST, SA, SAP, SAPC and KL are shown in Figure 1. The FT-IR spectrum of ST is shown in Figure 1a, peaks at 1158, 1081 and 1005 cm^{-1} were the stretching vibration of the C-O-C bond [30,31]. After polymerization, these peaks shifted and weakened. As shown in the spectra of SA (Figure 1b), the strong bands at 1610 and 1410 cm^{-1} were due to the asymmetric and symmetric stretching vibrations of -COO$^-$ groups, respectively [32]. The peak at 1032 cm^{-1} was related to the stretching vibration of C-OH [33], which moved to 1030 cm^{-1} and 1036 cm^{-1}, respectively, after polymerization, and the intensity of this peak weakened significantly. Figure 1c shows the FT-IR spectrum of SAP, the absorption peak at 3469 cm^{-1} was attributed to the stretching vibration of -OH, and three new peaks appeared at 1711, 1567 and 1409 cm^{-1}. The peak at 1711 cm^{-1} may be related to the ester group formed during graft polymerization. Also, the new absorption at 1567 and 1409 cm^{-1} could be ascribed to the asymmetric and symmetric stretching of -COO$^-$, respectively. These results showed that both ST and SA were involved in the grafting reaction. In the infrared spectrum of KL (Figure 1e), the absorption peaks of -OH were observed at 3696-3621 cm^{-1} [25]. As revealed in Figure 1d,e, the -OH peak shifted from 3469 cm^{-1} to 3457 cm^{-1}, and a characteristic Si-O peak was observed at 471 cm^{-1}, which indicated that -OH on KL was also involved in the reaction.

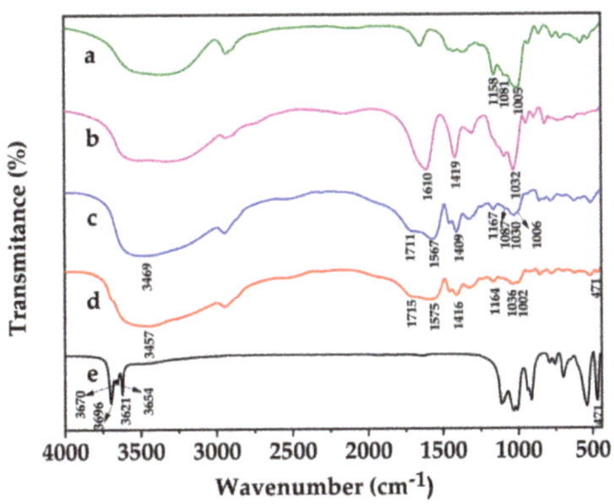

Figure 1. FT-IR spectra of (**a**) starch (ST), (**b**) sodium alginate (SA), (**c**) superabsorbent polymer (SAP), (**d**) superabsorbent composite (SAPC) (4 wt. % KL) and (**e**) kaolin (KL).

3.2. *Morphology*

The appearance of SAPC at different states is exhibited in Figure 2a,b. The dried SAPC was a light-yellow solid, while the swollen SAPC was a transparent hydrogel. The volume of dried SAPC was significantly smaller than fully swollen SAPC.

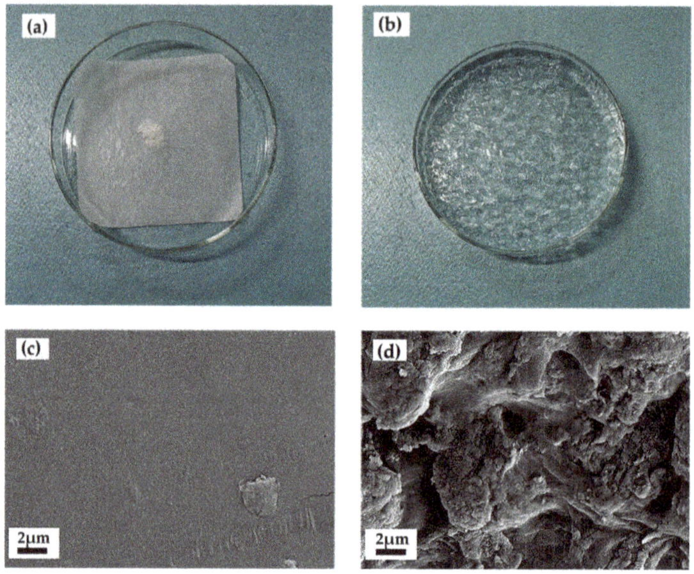

Figure 2. Digital images of (**a**) dried SAPC (80 mesh) and (**b**) swollen SAPC. SEM of (**c**) SAP and (**d**) SAPC (4 wt. % KL); the scaling bars are 2 μm.

SEM of dried SAP and SAPC are shown in Figure 2c,d. The surface of SAP was compact and smooth, making it not conducive to the liquid entering into the polymer network. By contrast, after addition of KL, the surface of SAPC displayed a more rough and loose morphology with porous structure.

3.3. XRD Analysis

The XRD patterns of KL, SAPC and SAP are shown in Figure 3. The strong peak of KL at $2\theta = 12.59°$ had an interplanar distance of $d = 0.702$ nm. This peak shifted to $2\theta = 12.44°$ (0.711 nm) in SAPC due to the intercalation of KL by polymer network. No obvious peaks were observed for SAP due to the amorphous structure. In contrast, SAPC showed the typical crystallite reflections associated with KL, which indicated that KL was uniformly dispersed in the polymer matrix, and an amorphous superabsorbent composite was synthesized.

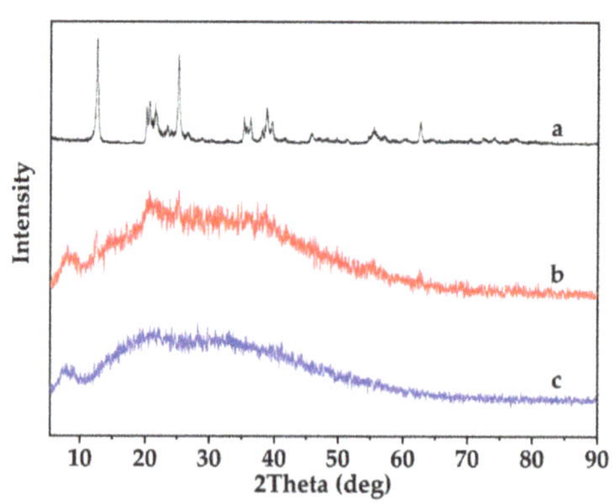

Figure 3. X-ray diffraction XRD patterns of (**a**) KL, (**b**) SAPC (4 wt. % KL) and (**c**) SAP.

3.4. TGA Analysis

The TGA curves of SAP and SAPC (4 wt. % KL) are shown in Figure 4. In the range from 35 to 203 °C, the minor weight loss of SAP and SAPC was mainly ascribed to the evaporation of adsorbed moisture and bound water. The weight loss of SAP and SAPC at 203–343 °C were related to the dehydration of saccharide rings and breaking of the C-O-C bond in ST and SA chain. Also, the weight loss from 343 °C to 416 °C corresponded to the decomposition of the carboxyl groups of the copolymers. The major mass loss of SAP and SAPC occurred in the temperature range of 416-480 °C and 416-495 °C, which was due to the decomposition of the 3D network structure. It is noted between 343 and 495 °C that the decomposition rate of SAP was obviously higher than that of SAPC. Additionally, the weight residual of SAPC was 49.5%, 6.4% higher than that of SAP. These facts seem to indicate improved thermal stability of SAPC compared to that of SAP, probably owing to the heat dissipation impedance effect of KL.

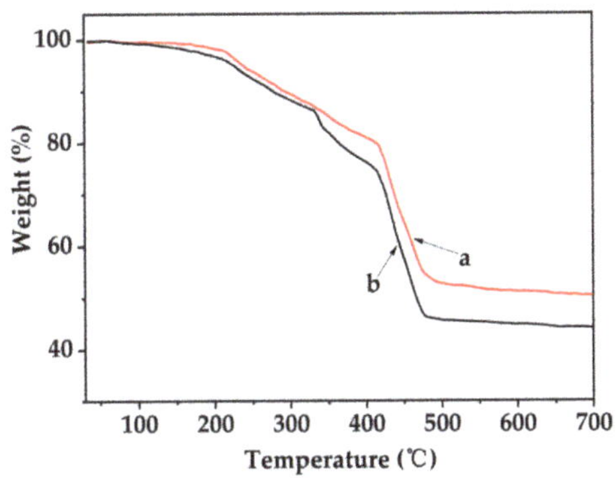

Figure 4. Thermal gravimetry analysis (TGA) curves of (**a**) SAP and (**b**) SAPC (4 wt. % KL).

3.5. Effects of Reaction Conditions on SAPC Performance

3.5.1. Effect of SA Content

The effect of SA content on water absorption of SAPC is shown in Figure 5a. Water absorption increased from 736 to 933 g/g, with the SA content changing from 10 wt. % to 15 wt. %, and then decreased to 763 g/g, with SA content further increasing to 20 wt. %. When the content of SA, which act as the basic skeleton, was too low, AA self-polymerized, resulting in the inability to form an effective 3D network to absorb water. With much higher SA content, sharply increased viscosity resulted in the system, restricting the monomer movement and reducing the conversion rate. Thus, a less developed 3D polymer network and suppressed water absorption capacity could be expected.

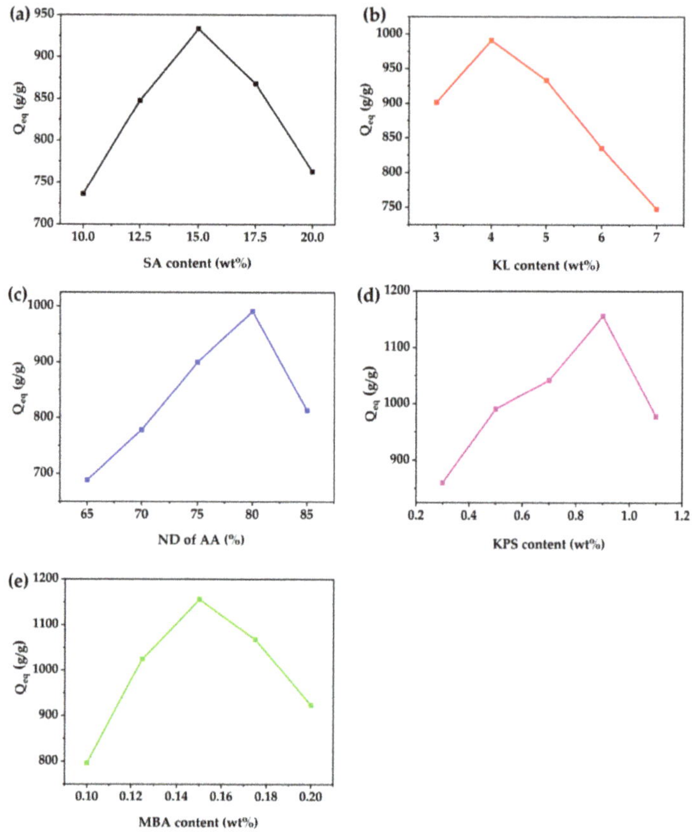

Figure 5. Effects of (**a**) SA content (ST = 15 wt %, KL = 5 wt %, neutralization degree (ND) of acrylic acid (AA) = 80%, potassium persulfate (KPS) = 0.5 wt %, N, N′-methylenebisacrylamide (MBA) = 0.15 wt %), (**b**) KL content (ST = 15 wt %, SA = 15 wt %, ND of AA = 80%, KPS = 0.5 wt %, MBA = 0.15 wt %), (**c**) ND of AA (ST = 15 wt %, SA = 15 wt %, KL = 4 wt %, KPS = 0.5 wt %, MBA = 0.15 wt %), (**d**) KPS content (ST = 15 wt %, SA = 15 wt %, KL = 4 wt %, ND of AA = 80%, MBA = 0.15 wt %) and (**e**) MBA content (ST = 15 wt %, SA = 15 wt %, KL = 4 wt %, ND of AA = 80%, KPS = 0.9 wt %) on the water absorption of SAPC. These contents are based on the weight percentage of raw materials and AA (20 mL). Q_{eq} represents the water absorption of swelling for 4 h.

3.5.2. Effect of KL Content

Figure 5b depicts the effect of KL content on water absorption. Water absorption of SAPC first increased and then decreased with the increase of KL content from 3 wt. % to 7 wt. %. KL with a large number of hydrophilic groups could be chemically cross-linked with a polymer chain, forming a cross-linked polymer network with KL particles as the additional cross-linking points. The introduction of appropriate KL particles weakens the hydrogen-bonding interaction between carboxyl groups in the polymer and reduces the physical entanglement of the grafted polymer network chain [34,35]. Nevertheless, with excessive KL (>4 wt. %), the excess KL particles increased the degree of cross-linking, leading to a decrease of the penetration space of water molecules. In addition, the extra kaolin particles were also filled into the polymer network space in the form of physical filling. Yet the water absorption rate of KL itself was low, as a result, the water absorption of SAPC inevitably decreased. Therefore, appropriate KL loading is critical to obtaining SAPC with high water absorption capacity.

3.5.3. Effect of Neutralization Degree of AA

The water absorptance capacity of SAPC was closely related to the neutralization degree (ND) of AA (Figure 5c). AA has higher reaction activity and polymerization rate than sodium acrylate [36]. Water absorption of SAPC increased as ND increased from 65% to 80%. When ND was lower, the polymerization reaction completed rapidly within a short time, resulting in a highly cross-linked network structure with low swelling capacity. With the increase of ND, the polymerization rate decreased correspondingly. In addition, the increase of -COO^- groups and Na^+ content in the polymer network increased both the repulsive force between the anions on the polymer chains and the osmotic pressure difference between the inside and outside of the polymer network, which was conducive to the entry of water or other small molecules [37–39]. However, when ND was higher than 80%, water absorption dropped significantly, attributed to the reaction of -COO^- groups with excess Na^+, resulting in weakening of the repulsive force.

3.5.4. Effect of KPS Content

Water absorption was significantly affected by the initiator concentration and the average kinetic chain length. With low KPS content (0.3–0.9 wt. %), an integral polymer network could not be formed due to the few grafting points and a large number of unreacted monomers in the reaction system. With too much KPS (0.9–1.1 wt. %), the excessive free radicals terminated the propagating chains earlier, shortening the average kinetic chain length. In both scenarios, low water absorption capacity resulted (Figure 5d).

3.5.5. Effect of MBA Content

The effect of the MBA content on water absorption of SAPC is shown in Figure 5e. Water absorption of SAPC increased first and then decreased with the increase of MBA content from 0.1 to 0.2 wt. %, and a maximum value of 1156 g/g was achieved with 0.15 wt. % MBA loading. Lower or higher MBA content resulted in reduced water absorption. The water absorption capacity of resin was closely related to its spatial network structure. When the content of MBA was not sufficient, cross-linking density was low and a complete spatial network structure could be established, leading to low water absorption and poor mechanical properties of SAPC. However, a higher concentration of MBA produced a denser cross-linked structure, which made it difficult for liquid molecules to enter.

The optimal reaction conditions of SAPC were when the masses of ST, SA, KL, KPS and MBA were 15 wt. %, 15 wt. %, 4 wt. %, 0.9 wt. % and 0.15 wt. % of AA, respectively. The neutralization degree of AA was 80%. In this work, SAP and SAPC were synthesized under the optimal conditions for further tests.

3.6. Performance Tests of SAPC

3.6.1. Swelling Kinetics of SAPC in Distilled water

The pseudo-second order swelling kinetics model (Equation (3)) and the Ritger-Peppas model (Equation (4)) were used to analyze the experimental swelling data to evaluate the swelling behavior of SAPC, with results shown in Figure 6a,b [40–42]:

$$\frac{t}{q_t} = \frac{1}{(k_2 q_e^2)} + \frac{t}{q_e} \tag{3}$$

$$F = \frac{q_t}{q_e} = \mathrm{k}t^n \tag{4}$$

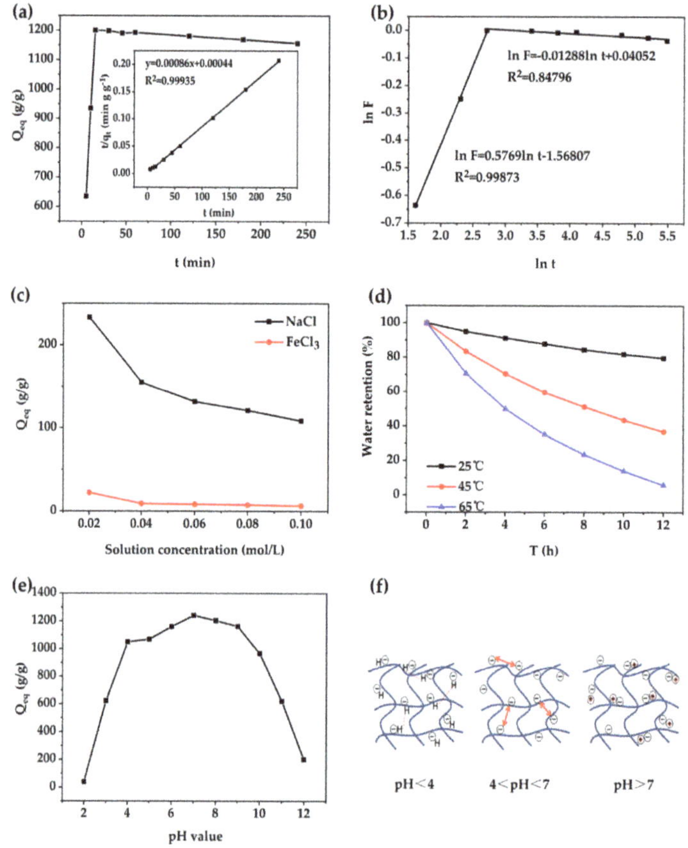

Figure 6. (**a**,**b**) Swelling kinetics of SAPC (4 wt. % KL), (**c**) Swelling capacity in different salt solutions, (**d**) Water retention at different temperatures, (**e**) The swelling capacity of SAPC (4 wt. % KL) in different pH solutions and (**f**) the swelling mechanism at various pH values.

To calculate n and k, take the natural logarithm of Equation (4):

$$ln\,F = ln\,(\frac{q_t}{q_e}) = ln\,k + nln\,t \tag{5}$$

where t is absorption time and k_2 is the rate constant of the pseudo-second order model. q_t and q_e correspond to water absorption of SAPC at time t and at equilibrium, respectively. F denotes fractional uptake at time t. k and n are the characteristic constants of the polymer and the diffusion index, respectively.

From Figure 6a, water absorption of SAPC increased rapidly in the first 15 min, reached swelling equilibrium at 15 min, and then stabilized after 15 min. The maximum water absorption was 1200 g/g. It was also found that the relationship curve between t and t/q_t was linear with R^2 value (0.99935) close to 1, indicating that the swelling behavior of SAPC could be fitted by the pseudo-second order kinetics model. Since this model is based on chemical adsorption assumptions, it suggests that the chemisorption was the main way of the water absorption process.

Based on the polymer chain relaxation rate and the relative diffusion rate of water into the polymer network, the water diffusion mechanism can be classified into five types. That is, pseudo-Fickian diffusion ($n < 0.5$), Fickian diffusion ($n = 0.5$), non-Fickian diffusion

($0.5 < n < 1.0$), Case II transport diffusion ($n = 1$) and relaxed diffusion ($n > 1$) [43–45]. These equations were applied to the initial stages of swelling. According to Figure 6b, within 0–15 min, $0.5 < n < 1$, indicating that the water diffusion mechanism was consistent with the non-Fickian diffusion mechanism, the diffusion and relaxation were considered to be isochronally effective. After 15 min, $n < 0.5$, the main reason of swelling was the diffusion of water molecules in the polymer network.

3.6.2. Swelling Behavior in Salt Solutions

The influences of cation and saline type on water absorption were studied using $FeCl_3$ and NaCl, with results shown in Figure 6c. For the same saline solution, as cation concentration increased, water absorption of SAPC decreased. This result was due to the charge screening effect of cations, which reduced the osmotic pressure difference between the polymer network and the external saline solution and decreased the repulsive force between the -COO^- groups of the polymer chain. Comparing the two absorption curves, it was found that water absorption of SAPC was lower in the multivalent cation solution. This is attributed to the complexation between carboxylate anions of chains and multivalent cations, resulting in an increase of the network cross-linking density and a drastic decrease in water absorption [46,47].

3.6.3. Water Retention Properties

The water retention of fully swollen SAPC at different temperatures, i.e., 25, 45 and 60 °C, was studied, and the results are shown in Figure 6d. As can be seen, water retention of SAPC decreased with time, and the water retention curve at 25 °C was gentler than that at 45 and 60 °C. At 25 °C, the water retention rate was about 80% after 12 h. However, the water retention rate of SAPC at 45 °C and 60 °C still reached about 37% and 6% after 12 h, respectively. The above results showed that SAPC had excellent water retention ability and thus broad application prospects.

3.6.4. Swelling Behavior in Buffer Solutions

As a new superabsorbent composite, the swelling capacity in different pH solutions has great influence on its application in various fields. Hence, the swelling behavior of SAPC in different buffer solutions was studied, with results shown in Figure 6e and the mechanism illustrated in Figure 6f. At low pH values (pH ≤ 4), due to the higher concentration of H^+, most of the -COO^- groups were transformed into -COOH groups, which weakened the electrostatic repulsion between -COO^- groups and strengthened the hydrogen-bonding interaction between the -COOH groups of polymer chains [48]. Therefore, water absorption was low. With an increase of pH value ($4 \leq$ pH ≤ 7), water absorption of SAPC increased, owing to gradual ionization of -COOH groups, which resulted in an increase of electrostatic repulsion between -COO^- groups. At a high pH value (pH ≥ 7), as Na^+ concentration increased in the solution, water absorption decreased correspondingly. The decrease was due to the charge shielding effect of excessive Na^+ on the -COO^- groups, which weakened the electrostatic repulsion force and reduced the osmotic pressure difference between the inside and outside of the polymer network. Generally, this type of SAPC had excellent pH tolerance between 4 and 10.

4. Conclusions

To summarize, a series of SAPC was successfully prepared by free radical polymerization of SA, ST, AA and KL in aqueous solution. The optimal reaction conditions of SAPC were achieved when the mass of ST, SA, KL, KPS and MBA were 15 wt. %, 15 wt. %, 4 wt. %, 0.9 wt. % and 0.15 wt. % of AA, respectively, with a neutralization degree of AA 80%. With optimized reactants concentrations, a maximum water absorption of 1200 g/g was obtained. FT-IR spectra confirmed the success of the polymerization reaction. SEM images revealed the rough and porous surface of SAPC which was conducive to the liquid entering into the polymer network. XRD patterns proved that KL was uniformly dispersed in the polymer matrix. TGA spectra indicated that the addition of KL improved the thermal stability of SAPC. The swelling kinetic of SAPC was studied, showing that the swelling behavior of SAPC followed the pseudo-second order kinetic model and the non-Fickian diffusion model. Along with excellent water absorption capacity, the fabricated SAPC also had excellent water retention property, good salt tolerance in monovalent salt solution and pH tolerance between 4 and 10, making it promising in many applications.

Author Contributions: Conceptualization, M.C. and Z.W.; methodology, M.C. and Z.W.; validation, M.C. and Z.W.; formal analysis, M.C. and Z.W.; investigation, M.C., C.Z., B.C.; resources, M.C., Z.L., Z.W.; data curation, M.C., C.Z., B.C.; writing—original draft preparation, M.C. and Z.W.; writing—review and editing, M.C., X.C., Z.W.; visualization, M.C. and Z.W.; supervision, D.Z. and Z.W.; project administration, Z.W.; funding acquisition, M.C. and Z.W. All authors have read and agreed to the published version of the manuscript.

Funding: This work was supported by grants from the National Natural Science Foundation of China (contract grant number 31270608) and the Heilongjiang Educational Committee (contract grant number 1511385).

Institutional Review Board Statement: Not applicable.

Informed Consent Statement: Not applicable.

Data Availability Statement: Data availability upon request.

Conflicts of Interest: The authors declare no conflict of interests.

References

1. Fang, S.; Wang, G.; Xing, R.; Chen, X.; Liu, S.; Qin, Y.; Li, K.; Wang, X.; Li, R.; Li, P. Synthesis of superabsorbent polymers based on chitosan derivative graft acrylic acid-co-acrylamide and its property testing. *Int. J. Biol. Macromol.* **2019**, *132*, 575–584. [CrossRef] [PubMed]
2. Adair, A.; Kaesaman, A.; Klinpituksa, P. Superabsorbent materials derived from hydroxyethyl cellulose and bentonite: Preparation, characterization and swelling capacities. *Polym. Test.* **2017**, *64*, 321–329. [CrossRef]
3. Mu, Y.; Du, D.; Yang, R.; Xu, Z. Preparation and performance of poly(acrylic acid-methacrylic acid)/montmorillonite microporous superabsorbent nanocomposite. *Mater. Lett.* **2015**, *142*, 94–96. [CrossRef]
4. Alam, M.N.; Christopher, L.P. Natural Cellulose-Chitosan Cross-Linked Superabsorbent Hydrogels with Superior Swelling Properties. *ACS Sustain. Chem. Eng.* **2018**, *6*, 8736–8742. [CrossRef]
5. Wan, T.; Zhou, Z.; Huang, R.; Zou, C.; Xu, M.; Cheng, W.; Li, R. Synthesis and swelling properties of microcrystal muscovite composite superabsorbent. *Appl. Clay Sci.* **2014**, *101*, 199–204. [CrossRef]
6. Cheng, S.; Liu, X.; Zhen, J.; Lei, Z. Preparation of superabsorbent resin with fast water absorption rate based on hydroxymethyl cellulose sodium and its application. *Carbohydr. Polym.* **2019**, *225*, 115214. [CrossRef]
7. Zhang, Q.; Wang, Z.; Zhang, C.; Aluko, R.E.; Yuan, J.; Ju, X.; He, R. Structural and functional characterization of rice starch-based superabsorbent polymer materials. *Int. J. Biol. Macromol.* **2020**, *153*, 1291–1298. [CrossRef]
8. Sawut, A.; Yimit, M.; Sun, W.; Nurulla, I. Photopolymerisation and characterization of maleylatedcellulose-g-poly(acrylic acid) superabsorbent polymer. *Carbohydr. Polym.* **2014**, *101*, 231–239. [CrossRef]
9. Essawy, H.A.; Ghazy, M.B.M.; Abd El-Hai, F.; Mohamed, M.F. Superabsorbent hydrogels via graft polymerization of acrylic acid from chitosan-cellulose hybrid and their potential in controlled release of soil nutrients. *Int. J. Biol. Macromol.* **2016**, *89*, 144–151. [CrossRef]
10. Nagarpita, M.V.; Roy, P.; Shruthi, S.B.; Sailaja, R.R.N. Synthesis and swelling characteristics of chitosan and CMC grafted sodium acrylate-co-acrylamide using modified nanoclay and examining its efficacy for removal of dyes. *Int. J. Biol. Macromol.* **2017**, *102*, 1226–1240. [CrossRef]
11. Chen, Y.; Tang, H.; Liu, Y.; Tan, H. Preparation and study on the volume phase transition properties of novel carboxymethyl chitosan grafted polyampholyte superabsorbent polymers. *J. Taiwan Inst. Chem. Eng.* **2016**, *59*, 569–577. [CrossRef]
12. Álvarez-Castillo, E.; Bengoechea, C.; Guerrero, A. Composites from by-products of the food industry for the development of superabsorbent biomaterials. *Food Bioprod. Process.* **2020**, *119*, 296–305. [CrossRef]
13. Fang, S.; Wang, G.; Li, P.; Xing, R.; Liu, S.; Qin, Y.; Yu, H.; Chen, X.; Li, K. Synthesis of chitosan derivative graft acrylic acid superabsorbent polymers and its application as water retaining agent. *Int. J. Biol. Macromol.* **2018**, *115*, 754–761. [CrossRef]
14. Yadav, M.; Rhee, K.Y. Superabsorbent nanocomposite (alginate-g-PAMPS/MMT): Synthesis, characterization and swelling behavior. *Carbohydr. Polym.* **2012**, *90*, 165–173. [CrossRef]
15. Ge, X.; Chang, M.; Jiang, W.; Zhang, B.; Xing, R.; Bulin, C. Selective location of kaolin and effects of maleic anhydride in kaolin/poly(ε-caprolactone)/poly(lactic acid) composites. *Appl. Clay Sci.* **2020**, *189*, 105524. [CrossRef]

16. Pourjavadi, A.; Ayyari, M.; Amini-Fazl, M.S. Taguchi optimized synthesis of collagen-g-poly(acrylic acid)/kaolin composite superabsorbent hydrogel. *Eur. Polym. J.* **2008**, *44*, 1209–1216. [CrossRef]
17. Bao, Y.; Ma, J.; Sun, Y. Swelling behaviors of organic/inorganic composites based on various cellulose derivatives and inorganic particles. *Carbohydr. Polym.* **2012**, *88*, 589–595. [CrossRef]
18. Zhang, J.; Chen, H.; Wang, A. Study on superabsorbent composite. III. Swelling behaviors of polyacrylamide/attapulgite composite based on acidified attapulgite and organo-attapulgite. *Eur. Polym. J.* **2005**, *41*, 2434–2442. [CrossRef]
19. Shalvir, A.; Liu, Q.; Abdekhodaie, M.J.; Wu, X.Y. Novel modified starch-xanthan gum hydrogels for controlled drug delivery: Synthesis and characterization. *Carbohydr. Polym.* **2010**, *79*, 898–907. [CrossRef]
20. Dragan, E.S.; Apopei, D.F. Synthesis and swelling behavior of pH-sensitive semi-interpenetrating polymer network composite hydrogels based on native and modified potatoes starch as potential sorbent for cationic dyes. *Chem. Eng. J.* **2011**, *178*, 252–263. [CrossRef]
21. Phang, Y.; Chee, S.; Lee, C.; Teh, Y. Thermal and microbial degradation of alginate-based superabsorbent polymer. *Polym. Degrad. Stabil.* **2011**, *96*, 1653–1661. [CrossRef]
22. Pereira, R.; Carvalho, A.; Vaz, D.C.; Gil, M.H.; Mendes, A.; Bártolo, P. Development of novel alginate based hydrogel films for wound healing applications. *Int. J. Biol. Macromol.* **2013**, *52*, 221–230. [CrossRef] [PubMed]
23. Olad, A.; Doustdar, F.; Gharekhani, H. Fabrication and characterization of a starch-based superabsorbent hydrogel composite reinforced with cellulose nanocrystals from potato peel waste. *Colloid Surf. A-Physicochem. Eng. Asp.* **2020**, *601*, 124962. [CrossRef]
24. Karadağ, E.; Topag, F.; Kundakci, S.; Üzüm, Ö.B. Novel composite sorbent AAm/MA hydrogels containing starch and kaolin for water sorption and dye uptake. *Bull. Mater. Sci.* **2014**, *37*, 1637–1646. [CrossRef]
25. Wu, J.; Wei, Y.; Lin, J.; Lin, S. Study on starch-graft-acrylamide/mineral powder superabsorbent composite. *Polymer.* **2003**, *44*, 6513–6520. [CrossRef]
26. Zhu, L.; Guan, C.; Zhou, B.; Zhang, Z.; Yang, R.; Tang, Y.; Yang, J. Adsorption of Dyes onto Sodium Alginate Graft Poly(Acrylic Acid-*co*-2-Acrylamide-2-Methyl Propane Sulfonic Acid)/ Kaolin Hydrogel Composite. *Polym. Polym. Compos.* **2017**, *25*, 627–634. [CrossRef]
27. Tang, Y.; Wang, Q.; Zhou, B.; Ma, D.; Ma, Z.; Zhu, L. Synthesis of Sodium Alginate Graft Poly(Acrylic Acid-Co-Acrylamide)/ Kaolin Composite Hydrogel and the Study on Its Sorption of Rhodamine B. *Polym. Polym. Compos.* **2015**, *23*, 467–474. [CrossRef]
28. Sorour, M.; El-Sayed, M.; El Moneem, N.A.; Talaat, H.; Shaalan, H.; El Marsafy, S. Free radical grafting kinetics of acrylamide onto a blend of starch/chitosan/alginate. *Carbohydr. Polym.* **2013**, *98*, 460–464. [CrossRef]
29. Lee, J.; Park, S.; Roh, H.; Oh, S.; Kim, S.; Kim, M.; Kim, D.; Park, J. Preparation and Characterization of Superabsorbent Polymers Based on Starch Aldehydes and Carboxymethyl Cellulose. *Polymers* **2018**, *10*, 605. [CrossRef]
30. Lan, G.; Zhang, M.; Liu, Y.; Qiu, H.; Xue, S.; Zhang, T.; Xu, Q. Synthesis and Swelling Behavior of Super-Absorbent Soluble Starch-g-poly(AM-co-NaAMC$_{14}$S) Through Graft Copolymerization and Hydrolysis. *Starch/Stärke* **2019**, *71*, 1800272. [CrossRef]
31. Qiao, D.; Yu, L.; Bao, X.; Zhang, B.; Jiang, F. Understanding the microstructure and absorption rate of starch-based superabsorbent polymers prepared under high starch concentration. *Carbohydr. Polym.* **2017**, *175*, 141–148. [CrossRef]
32. Gharekhani, H.; Olad, A.; Mirmohseni, A.; Bybordi, A. Superabsorbent hydrogel made of NaAlg-g-poly(AA-co-AAm) and rice husk ash: Synthesis, characterization, and swelling kinetic studies. *Carbohydr. Polym.* **2017**, *168*, 1–13. [CrossRef]
33. Wang, W.; Wang, A. Synthesis and swelling properties of pH-sensitive semi-IPN superabsorbent hydrogels based on sodium alginate-g-poly(sodium acrylate) and polyvinylpyrrolidone. *Carbohydr. Polym.* **2010**, *80*, 1028–1036. [CrossRef]
34. Shi, X.; Wang, W.; Kang, Y.; Wang, A. Enhanced Swelling Properties of a Novel Sodium Alginate-Based Superabsorbent Composites: NaAlg-g-poly(NaA-co-St)/APT. *J. Appl. Polym. Sci.* **2012**, *125*, 1822–1832. [CrossRef]
35. Feng, E.; Ma, G.; Wu, Y.; Wang, H.; Lei, Z. Preparation and properties of organic-inorganic composite superabsorbent based on xanthan gum and loess. *Carbohydr. Polym.* **2014**, *111*, 463–468. [CrossRef]
36. Zhang, J.; Zhang, F. Recycling waste polyethylene film for amphoteric superabsorbent resin Synthesis. *Chem. Eng. J.* **2018**, *331*, 169–176. [CrossRef]
37. Shi, X.; Wang, W.; Wang, A. Effect of surfactant on porosity and swelling behaviors of guar gum-g-poly(sodium acrylate-co-styrene)/attapulgite superabsorbent hydrogels. *Colloid Surf. B Biointerfaces* **2011**, *88*, 279–286. [CrossRef] [PubMed]
38. Rashidzadeh, A.; Olad, A. Slow-released NPK fertilizer encapsulated by NaAlg-g-poly(AA-co-AAm)/MMT superabsorbent nanocomposite. *Carbohydr. Polym.* **2014**, *114*, 269–278. [CrossRef]
39. Behrouzi, M.; Moghadam, P.N. Synthesis of a new superabsorbent copolymer based on acrylic acid grafted onto carboxymethyl tragacanth. *Carbohydr. Polym.* **2018**, *202*, 227–235. [CrossRef]
40. Zhang, J.; Zhang, F. A new approach for blending waste plastics processing: Superabsorbent resin synthesis. *J. Clean Prod.* **2018**, *197*, 501–510. [CrossRef]
41. Motamedi, E.; Motesharezedeh, B.; Shirinfekr, A.; Samar, S.M. Synthesis and swelling behavior of environmentally friendly starch-based superabsorbent hydrogels reinforced with natural char nano/micro particles. *J. Environ. Chem. Eng.* **2020**, *8*, 103583. [CrossRef]
42. Diao, M.; Li, Q.; Xiao, H.; Duan, N.; Xu, J. Synthesis and adsorption properties of superabsorbent hydrogel and peanut hull composite. *J. Environ. Chem. Eng.* **2014**, *2*, 1558–1567. [CrossRef]
43. Ahmed, E.M.; Aggor, F.S.; Awad, A.M.; El-Aref, A.T. An innovative method for preparation of nanometal hydroxide superabsorbent Hydrogel. *Carbohydr. Polym.* **2013**, *91*, 693–698. [CrossRef]

44. Dalaran, M.; Emik, S.; Güçlü, G.; İyim, T.B.; Özgümüş, S. Study on a novel polyampholyte nanocomposite superabsorbent hydrogels: Synthesis, characterization and investigation of removal of indigo carmine from aqueous solution. *Desalination* **2011**, *279*, 170–182. [CrossRef]
45. Olad, A.; Doustdar, F.; Gharekhani, H. Starch-based semi-IPN hydrogel nanocomposite integrated with clinoptilolite: Preparation and swelling kinetic study. *Carbohydr. Polym.* **2018**, *200*, 516–528. [CrossRef]
46. Pourjavadi, A.; Ghasemzadeh, H.; Soleyman, R. Synthesis, Characterization, and Swelling Behavior of Alginate-g-Poly(sodium acrylate)/Kaolin Superabsorbent Hydrogel Composites. *J. Appl. Polym. Sci.* **2007**, *105*, 2631–2639. [CrossRef]
47. Ma, G.; Yang, Q.; Ran, F.; Dong, Z.; Lei, Z. High performance and low cost composite superabsorbent based on polyaspartic acid and palygorskite clay. *Appl. Clay Sci.* **2015**, *118*, 21–28. [CrossRef]
48. He, G.; Ke, W.; Chen, X.; Kong, Y.; Zheng, H.; Yin, Y.; Cai, W. Preparation and properties of quaternary ammonium chitosan-g-poly(acrylic acid-co-acrylamide) superabsorbent hydrogels. *React. Funct. Polym.* **2017**, *111*, 14–21. [CrossRef]

polymers

MDPI

Article

Photophysical, Thermal and Structural Properties of Thiophene and Benzodithiophene-Based Copolymers Synthesized by Direct Arylation Polycondensation Method

Newayemedhin A. Tegegne [1,*], Zelalem Abdissa [2] and Wendimagegn Mammo [2]

1 Department of Physics, Addis Ababa University, Addis Ababa P.O. Box 1176, Ethiopia
2 Department of Chemistry, Addis Ababa University, Addis Ababa P.O. Box 1176, Ethiopia; zelalem2ab@yahoo.com (Z.A.); wendimagegn.mammo@aau.edu.et (W.M.)
* Correspondence: newaye.medhin@aau.edu.et

Abstract: Three low-band-gap copolymers based on isoindigo acceptor units were designed and successfully synthesized by direct arylation polycondensation method. Two of them were benzodithiophene (BDT)-isoindigo copolymers (**PBDTI-OD** and **PBDTI-DT**) with 2-octlydodecyl (OD) and 2-decyltetradecyl (DT) substituted isoindigo units, respectively. Thiophene donor and DT-substituted isoindigo acceptor units were copolymerized to synthesize **PTI-DT**. The copolymers have a broad absorption range that extends to over 760 nm with a band gap ≈1.5 eV. The photophysical property studies showed that the BDT-based copolymers have non-polar ground states. Their emission exhibited the population of the intramolecular charge transfer (ICT) state in polar solvents and tightly bound excitonic state in non-polar solvents due to self-aggregation. On the contrary, the emission from the thiophene-based copolymers was only from the tightly bound excitonic state. The thermal decomposition temperature of the copolymers was above 380 °C. The X-ray diffraction pattern of the three copolymers showed a halo due to $\pi - \pi$ stacking. A second, sharper peak was observed in the BDT-based copolymer with a longer side chain on the isoindigo unit (**PBDTI-DT**), and the thiophene-based copolymers with **PTI-DT**, exhibiting a better structural order.

Keywords: intramolecular charge transfer; copolymers; $\pi - \pi$ stacking; direct arylation polycondensation; and excitonic state

Citation: Tegegne, N.A.; Abdissa, Z.; Mammo, W. Photophysical, Thermal and Structural Properties of Thiophene and Benzodithiophene-Based Copolymers Synthesized by Direct Arylation Polycondensation Method. *Polymers* **2021**, *13*, 1151. https://doi.org/10.3390/polym13071151

Academic Editor: Edina Rusen

Received: 28 February 2021
Accepted: 22 March 2021
Published: 4 April 2021

Publisher's Note: MDPI stays neutral with regard to jurisdictional claims in published maps and institutional affiliations.

Copyright: © 2021 by the authors. Licensee MDPI, Basel, Switzerland. This article is an open access article distributed under the terms and conditions of the Creative Commons Attribution (CC BY) license (https://creativecommons.org/licenses/by/4.0/).

1. Introduction

The introduction of π-conjuageted polymers to the fast-growing solar cell technology has brought forth new features of flexibility and solution processabiltiy in the so-called organic solar cells (OSCs) devices. OSCs have shown a leap in efficiency in the last five years to above 18% in single-junction [1], and over 17% in multi-junction [2], devices. This tremendous improvement is mainly attributed to the introduction of new donor polymers and non-fullerene acceptors that can harvest the broad solar spectrum with appropriate energetics in the active layer. A vast number of polymers have been synthesized and characterized to enrich the structure-property relation recipe for efficient OSCs. Alternating donor–acceptor (D–A) copolymers are commonly synthesized to obtain low-band-gap polymers with good solar harvest that extends to the near-infra-red region. Some important parameters that need to be addressed during the synthesis of the D–A copolymers include the HOMO level, which should be lower-lying, with the LUMO lying in the region for efficient exciton dissociation.

The synthesis of low-band-gap polymers using a simple and economically viable technique has gained the attention of researchers due to its potential for industrialization of the OSC technology. The direct arylation polycondensation method (DAP) has notably shortened the synthesis route compared to the commonly used Suzuki and Stille cross-coupling techniques. In this reaction, the unsubstituted arylene monomers are directly coupled with dihaloarylene units with the help of transition metals—usually palladium acetate. The by-product at each of the synthesis steps is, therefore, drastically reduced, making the DAP method more environmentally friendly and economically viable. In addition, the DAP method made the synthesis of polymers that were difficult to prepare with the traditional Suzuki and Stille cross-coupling techniques possible [3]. The mass production of OSCs will need an enormous amount of donor and acceptor organic materials. An easy and fast synthesis method such as DAP is undoubtedly important in the realization of the commercialization of OSC technology. Despite its many interesting merits, the method has a drawback: its selectivity between C–H bonds for some monomers is poor. Consequently, the method is more commonly used to synthesize homopolymers like poly-3-hexylthiophene (P3HT) and benzodithiophene (BDT) [4–6]. However, D–A copolymers synthesized by DAP method are limited. Recently, our group and others have reported the successful synthesis of D–A copolymers using the DAP method [7–10].

The choice of D and A units in the synthesis of D–A copolymers determines their electrical, optical and structural properties, which that will subsequently determine the performance of the OSC that the materials are used for. Isoindigo is a widely used acceptor unit due to its good electron-withdrawing properties. The fused ring benzo[1,2-b:4,5-b′]dithiophene (BDTs) and thiophene units are the most commonly used donor units in the synthesis of D–A copolymers that produced high power conversion efficeny (PCE) of over 11% in OSCs [11]. These polymers are usually contain aloxy or alkyl side chains to enhance their solubility. A basic understanding of the structure–property relationships of the polymers is vital for the optimization of their structure for higher PCE of OSCs. The optical properties, intra/inter-chain interactions in the polymers, and their morphologies in the films are some of the parameters that play important roles in the photogeneration of charges in OSCs. In addition, their thermal properties determine their lifetimes in harsh environment conditions, and also their application in flexible OSC devices, which are normally processed at temperatures below 150 °C.

In this work, we report the synthesis of three D–A copolymers based on an isoindigo acceptor unit. The two copolymers have alkyl-substituted BDT (**PBDTI-OD** and **PBDTI-DT**) donor units, while in the third copolymer (**PTI-DT**), an alkyl-substituted thiophene unit is used. The structures of the BDT-based copolymers were systematically tailored by increasing the side chain length in the isoindigo units from 2-octyldodecyl (OD) in **PBDTI-OD** to 2-decyltetradecyl (DT) in **PBDTI-DT**, while keeping the backbone structure the same. The third copolymer, **PTI-DT**, was synthesized to study the effect of electron-donating groups by changing the donor units from BDT to thiophene. The structural differences in the three copolymers allowed us to investigate the effect of backbone structure on their photo-physics, thermal and structural properties. We found two relaxation channels in the BDT-based copolymers, while a one-channel relaxation was found in **PTI-DT**. The copolymers showed excellent thermal stability with decomposition temperatures (T^D = 5% mass loss) above 380 °C. The X-ray diffraction patterns of the copolymers suggested that a better structural ordering was found in the thiophene-based copolymer.

2. Materials and Methods

2.1. Materials and Synthesis

Three copolymers based on an isoindigo acceptor were designed and synthesized using a direct heteroarylation polymerization (DAP) method for the first time. The two copolymers were synthesized by the copolymerization of a 4,8-bis(decyl)benzo[1,2-b;4,5-b′]dithiophene (BDT) (**1**) donor unit and an isoindigo acceptor unit. N-Alkylation of the isoindigo moiety was achieved with 2-octyldecylbromide and 2-decyltetradecylbromide to obtain compounds **2** and **3** which were used to synthesize **PBDTI-OD** and **PBDTI-DT**, respectively, as shown in Scheme 1. In the synthesis of **PBDTI-OD/PBDTI-D**, a 1:1 ratio of BDT (**1**) (0.5 mmol) and isoindigo (**2/3**) (0.5 mmol) was copolymerized using palladium acetate, (0.01 mmol), pivalic acid (0.15 mmol) and potassium carbonate (1.3 mmol) as the catalyst, acidic additive and base, respectively. The polymerization reaction took place in an inert atmosphere at a temperature of 100 °C using anhydrous dimethylacetamide (DMAc) for **PBDTI-DT** and a mixture of DMAc and toluene as the reaction solvents for **PBDTI-OD**. The heating was stopped after 7 and 19 h for **PBDTI-OD** and **PBDTI-DT**, respectively, and the solutions were left to cool down to room temperature. The copolymers were precipitated in methanol and filtered. The crude copolymers were dissolved in chloroform (CF) and washed with aqueous solution of ethylene-diamine tatraacetic acid (EDTA) (pH = 8) to remove trace metals, followed by 0.1 N HCl and distilled water. **PBDTI-OD** was Soxhlet-extracted with methanol, hexane, diethylether and chloroform while **PBDTI-DT** was extracted with methanol, acetone, ethylacetate, hexane, diethylether and chloroform. The CF extract of **PBDTI-OD** was concentrated and precipitated in acetone and filtered. After drying the polymer at 40 °C in a vacuum oven for 24 h, the solid obtained was further stirred in hexane overnight and filtered to obtain 0.36 g blue-black copolymer (56%). Similarly, the CF extract of **PBDTI-DT** was concentrated precipitated in methanol, filtered, dried in an oven at 40 °C for 24 h to afford 0.32 g dark brown solid (43%).

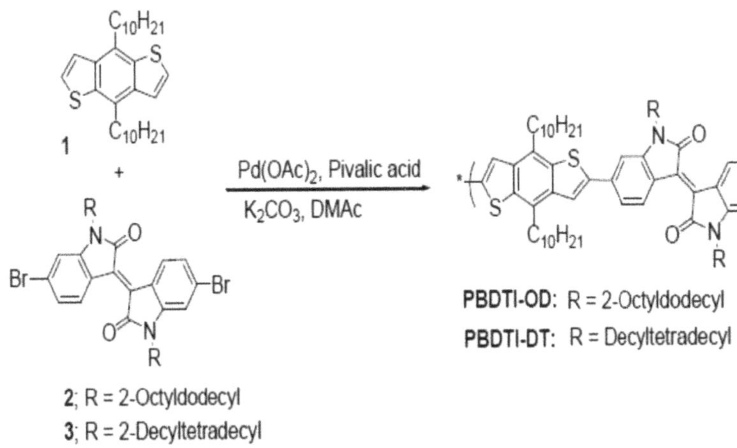

Scheme 1. Syntheses scheme of **PBDTI-OD** and **PBDTI-DT**.

PTI-DT was synthesized by copolymerizing 3-octylthiophene (**4**, 0.5 mmol) with isoindigo (**3**), as shown in Scheme 2. The polymerization was done in anhydrous DMAc for 3.5 h. Following the same procedure, the crude polymer was extracted. **PTI-DT** was Soxhlet-extracted in methanol, acetone, ethyl acetate and CF, which afforded 0.37 g (66%) dark-brown solid after drying in an oven at 40 °C overnight. Note here that the long side chains attached to both the donor and acceptor units of the copolymers increased their solubilities in chlorinated solvents like CF and 1,2-dichlorobenzene (oDCB). Comparing the three copolymers, **PBDTI-OD** was found to be less soluble in these chlorinated solvents like CF and oDCB. On the other hand, introduction of the long side chains might create steric hindrance that might twist the backbones of the copolymers. However, BDT has a planar structure that is expected to make the backbones of **PBDTI-OD** and **PBDTI-DT** less twisted and have a longer conjugation length than the thiophene-based copolymer, **PTI-DT** [12,13].

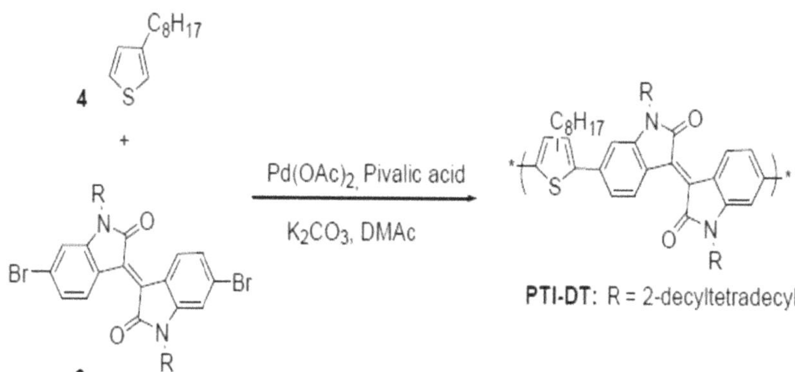

Scheme 2. Synthesis scheme of **PTI-DT**.

The molecular weights of the copolymers weree determined by size exclusion chromatography was done using a Waters Alliance GPVC2000 instrument with a refractive index detector. The columns used were Waters Stragel® HMW 6EX2 and Waters Stragel® HT 6EX1 and the measurements were taken in 1,2 4-trichlorobenzene at a temperature of 135 °C. The relative molecular mass was measured against polystyrene standards. The average molecular weights (M_w) of **PBDTI-OD**, **PBDTI-DT** and **PTI-DT** were found to be 53,461, 71,234 and 23,574 g/mol, respectively. The high M_ws of the copolymers are due to the long alkyl side chains appended both in the donor and acceptor units. The calculated poly-dispersity indices (PDI = M_w/M_n) of the three copolymers show that **PTI-DT** has a more uniform molecular weight distribution. The molecular weight and yield data are summarized in Table 1.

Table 1. Molecular weight and yield data of **PBDTI-OD**, **PBDTI-DT** and **PTI-DT**.

Copolymer	Yield (%)	M_n (g/mol)	M_w (g/mol)	PDI
PBDTI-OD	56.25	16,932	53,461	3.2
PBDTI-DT	43.71	17,974	71,234	4.0
PTI-DT	66.07	12,537	23,574	1.9

2.2. Absorption and Photoluminescence Spectroscopy

The absorption spectra of **PBDTI-OD**, **PBDTI-DT** and **PTI-DT** in both solutions and spin-coated films were recorded using a Perkin Elmer Lambda 19, UV/Vis/NIR spectrophotometer. The thin films were spin-coated from a CF solution of concentration 10 mg/mL at a spin speed of 1000 rpm and the films were annealed at 85 °C for 10 min. Similarity, their photoluminescence (PL) spectra were recorded both in solutions and spin-coated films using JY Horiba, 'Fluoromax'-4 spectrofluorometer. The absorption and PL spectra in solutions were taken in CF, o-DCB, and cyclohexane (Chex) with polarity indices of 4.1, 2.7 and 0.2, respectively. The PL spectra of the copolymers in solutions and films were taken at excitations matching their longer wavelength electronic transition. The solutions for PL measurement absorb only 0.05 at the maxima to avoid self absorption.

2.3. X-ray Diffraction

The X-ray diffraction (XRD) patterns of the copolymers were recorded for for 2θ values from 2 to 40° in drop-casted films on a glass substrates using Shimadzu Scientific Instruments X-ray diffractometer 7000S with a copper target, Cu-Kα-radiation with

40 kV × 40 mA power. The interlayer spacing between the planes (d) was calculated using Bragg's Equation (1).

$$n\lambda = 2dsin(\theta), \quad (1)$$

where n, λ and θ are order of diffraction (1), wavelength of the X-ray (0.154 nm) and angle of diffraction, respectively.

2.4. Thermogravimetric Analysis

The thermal properties of the copolymers were investigated using DTG-60H thermogravimetric analyser in platinum crucibles under nitrogen atmosphere flowing at a rate of 50 mL/min.

2.5. Electrochemistry

Square wave voltammetry (SWV) was done to measure the frontier molecular orbitals of the copolyers in a three-electrode setup. Platinium electrodes were used both as working and counter electrodes, while Ag/AgCl was used as a reference electrode, which was calibrated against ferocene/ferrocynium ion (Fc/Fc^+). The redox potential of Fc/Fc^+ was found to be 0.44 eV versus the reference electrode, taking its energy level to be 4.8 eV below the vacumm. The copolymer films were drop-casted on the Pt disc working electrode, and a 0.1 M solution of tetrabutylammonium hexafluorophosphate (Bu_4NPF_6) in anyhdrous acetonitrile served as a supporting electrolyte. The higherst occupied molecular orbital (HOMO) and lowest unoccupied molecular orbital (LUMO) energy levels were determined from the onsets of oxidation and reduction potentials, respectively, using Equation (2) [14].

$$E_{HOMO/LUMO} = e(E_{ox/red} + 4.4)eV \quad (2)$$

3. Results and Discussion

3.1. Optical and Electrochemical Properties of PBDTI-OD, PBDTI-DT and PTI-DT

The absorption spectra of **PBDTI-OD, PBDTI-DT** and **PTI-DT** were taken both in CF solutions and as thin films, as shown in Figure 1a,b. The spectra of **PBDTI-OD** and **PBDTI-DT** in CF solutions are characterized by one dominant transition peaking at 631 and 628 nm, respectively, and two peaks below 450 nm. On the other hand, the absorption spectrum of **PTI-DT** is quite different from the BDT-based copolymers where a dominant one transition in solution that peaks at 615 nm and a modest peak at 450 nm are observed. The two-energy-band profile exhibited in **PBDTI-OD, PBDTI-DT** and **PTI-DT** is common in push–pull copolymers. The low energy band is due to the $S_0 \rightarrow S_1$ transition, usually with intra-molecular charge transfer (ICT) characteristics due to the charge transfer between the donor and acceptor moieties in the copolymers. On the other hand, the high energy band is due to the $\pi - \pi^*$ transition, which is common in the π-conjugated polymers due to the alternating single and double bonds in their backbone [15]. The presumably assigned ICT state transition in **PBDTI-OD** and **PBDTI-DT** is red-shifted by more than 11 nm with respect to **PTI-DT**, suggesting a better electron-donating property of the BDT moiety [16]. This could also be due to the long alkyl side chains on the thiophene and isoindigo units of **PTI-DT** that can sterically intract, leading to a twisting in the backbone of the copolymer. Computaion analusis of a similar BDT-based copolymer showed that the copolymers are likely to adopt a planar structure [16] which facilitates ICT in **PBDTI-OD** and **PBDTI-DT** chains.

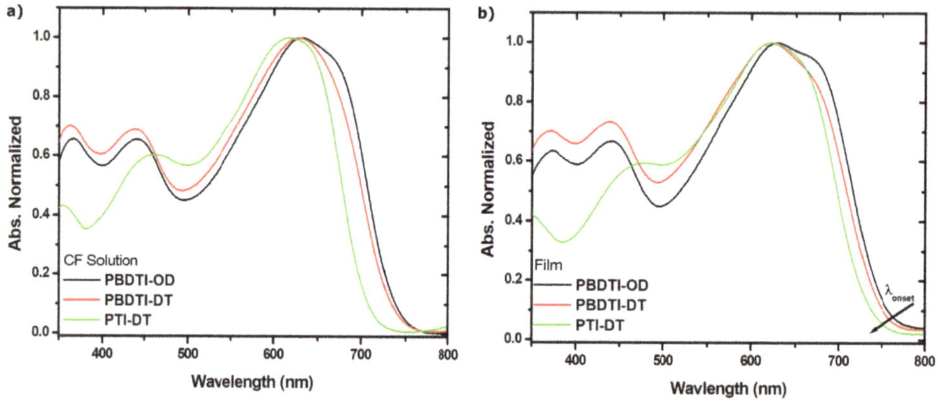

Figure 1. Absorption spectra of **PBDTI-OD** (black), **PBDTI-DT** (red) and **PTI-DT** (green) in (**a**) CF solutions and (**b**) thin films.

The absorption spectra of thin films of **PBDTI-OD**, **PBDTI-DT** and **PTI-DT** were found to be red-shifted by 30, 25 and 40 nm from their solution spectra, respectively. The red-shift in absorbance upon solidification shows a better interchain interaction in the films due to $\pi - \pi$ stacking, which is beneficial for charge transport when the copolymers are used in OSCs. The larger red-shift in absorbance of **PTI-DT** upon solidification indicates that a different intra- and intermolecular interaction from the DBT-based copolymers exists due to the thiophene donor unit. One of the reasons for this could be that **PTI-DT** is twisted in the absence of strong intermolecular interaction in dilute solution. However, the stronger inter-molecular force in the solid films pushes the copolymer into a planar conformation, resulting in a siginicant red-shift in absorption onset [17,18]. Similarly, the smaller absorption onset shift in the films of **PBDTI-DT** compared to **PBDTI-OD** shows the more planar structure of **PBDTI-OD** due to the shorter side chain on the isoindigo unit [19]. The optical band gaps (E_g) of the three copolymers were calculated from the onsets of absorption spectra of the films and were found to be 1.53, 1.54 and 1.56 eV for **PBDTI-OD**, **PBDTI-DT**, and **PTI-DT**, respectively. The copolymers have a similar bandgap, that is suitable to harvest substantial solar irradiation in the high solar flux region [20]. The backbone structures of the copolymers have negligible effect on their band gaps. The optical properties of the copolymers are summarized in Table 2.

Table 2. Electrochemical and Optical Properties of **PBDTI-OT**, **PBDTI-DT** and **PTI-DT**.

Name	λ_{max}Soln (nm)		λ_{max}Film (nm)		[1] Shift (nm)		[2] $E_g{}^{Opt}$	E_{HOMO}	E_{LUMO}
	λ_{max1}	λ_{max2}	λ_{max1}	λ_{max2}	$\Delta\lambda_{max1}$	$\Delta\lambda_{max1}$	eV	eV	eV
PBDTI-OD	366, 440	631	374, 440	631	8, 0	0	1.53	−5.5	−3.9
PBDTI-OT	363, 437	628	370, 438	623	7, 4	−5	1.54	−5.7	−3.8
PTI-DT	—, 457	615	—, 461	623	-, 4	8	1.56	−5.6	−3.7

[1] Abs peak shift from solution to film, [2] $E_g = 1200/\lambda_{onset}$.

The PL spectra of the thin films were recorded by exciting at their absorption maxima in the longer wavelength region and were found to be Stokes-shifted by 179, 189 and 185 nm in **PBDTI-OD**, **PBDTI-DT** and **PTI-DT**, respectively (see Figure 2). Since an efficeint overalp between the absorption and emission spectra of the copolymers is needed for Förster resonance energy transfer (FRET), the large Stokes shift in the copolymers will inhibit interchain excitation energy transfer [21]. However, the large Stokes shift also shows that multiple processes have taken place to stabilize the first excited state. Since ICT formation in solid films is not efficient, the emission from the first excited state ($S_0 \leftarrow S_1$) is ascribed to tightly bound intrachain exciton relaxation. The strong $\pi - \pi$ interaction upon solidification of the copolymers enables an efficient interchain charge transfer interaction that dominates over intrachain charge transfer (i.e., ICT population). Our work on three similar copolymers with two and three thiophene donor units, coupled with an isoindgo acceptor unit, showed similar results [7,8].

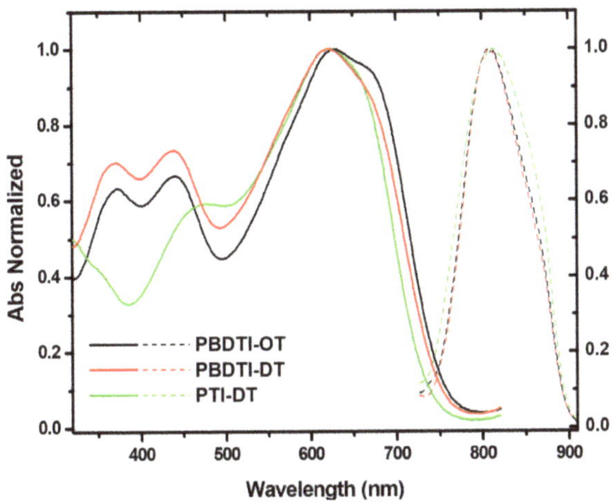

Figure 2. Absorption (solid line) and PL (broken line) spectra of **PBDTI-OD** (black), **PBDTI-DT** (red) and **PTI-DT** (green).

The HOMO and LUMO energy levels of the copolymers were determined using SWV from the onsets of their oxidation and reduction potentials, as shown in Figure 3. The HOMO and LUMO energy levels of the copolymers are summarized in Table 2. The difference in the HOMO levels between the copolymers with the same donor unit (**PBDTI-OD** and **PBDTI-DT**) was 0.2 eV, whereas the difference is only 0.1 eV between **PBDTI-DT** and **PTI-DT** (with the same acceptor unit). The highest LUMO level was found in **PBDTI-OD**, followed by **PBDTI-OD** and **PTI-DT**. The π-electron density in the copolymers is not expected to be significantly affected by the alkyl side chains [22]. The reason for the difference in HOMO levels between **PBDTI-OD** and **PBDTI-DT** should, therefore, be due to the effect of the length of the alkyl side chains on the reorganization of the copolymers in the drop-casted films [8,19]. This is expected due to the relatively lower solubility of **PBDTI-OD**.

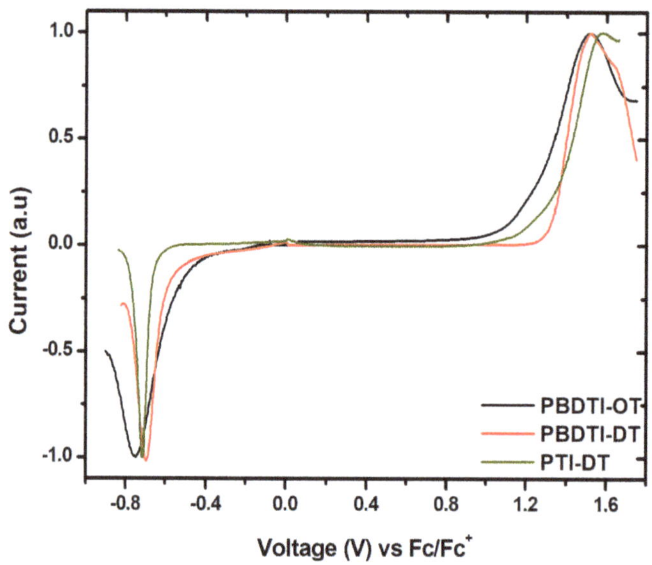

Figure 3. Square wave voltammograms of **PBDTI-OD** (black), **PBDTI-DT** (red) and **PTI-DT** (dark yellow).

The difference in the energetics of **PBDTI-DT** and **PTI-DT** is significantly dependent on the electron-donating strength of BDT vesus thiophene units, respectively. The stronger electron-donating property of BDT would enhance the ICT in **PBDTI-DT**, which will stabilize its LUMO level compared to **PTI-DT**, in which the donor unit is a thiophene [23]. Note here that the reorganization of the copolymers in the casted films also plays a role. The medium-lying HOMO levels of the copolymers are beneficial for attaining high open-circuit voltage when they are used in OSCs as donor materials, since $V_{OC} = e(E_{LUMO}^{Acceptor} - E_{HOMO}^{Donor})$, where e is the elementary charge [24]. The LUMO offset of the copolymers with a fullerene acceptor should be around 0.3 eV for efficient exciton disscociation at the donor/acceptor interface [24,25]. **PBDTI-OD**, **PBDTI-DT** and **PTI-DT** have LUMO offsets of 0.1, 0.2 and 0.3 eV, respectively with respect to the commonly used fullerene acceptor, $PC_{71}BM$ (LUMO = 4.0 eV). The LUMO offsets in the BDT-based copolymers are below the recommended value, which is detrimental to exciton dissociation in fullerene-based OSCs, while efficient exciton dissociation is expected in **PTI-DT**-based OSCs. Hence, **PBDTI-OD** and **PBDTI-DT** might work better when blended with non-fullerene acceptors.

3.2. Photophysical Properties of the Copolymers

To further understand the photophysics of the copolymers, their absorption and PL spectra were recorded in CF, oDCB and Chex solution in a concentration range of 125–5.7 μg/mL. The absorbances of all the copolymers in the three solvents showed negligible shift with concentration gradient, which indicates the absence of aggregation in the solutions. The absorption spectra of the copolymers in the three solvents of different polarity are almost similar, with a slight difference in the apperance of a shoulder when the poor solvent (Chex) was used. However, the low-energy absorption peaks of the copolymers remained the same with increasing solvent polarity, as shown in Figure 4a–c (solid line). This confirms that the ground states (S_0) of the copolymers are non-polar. The PL spectra of **PBDTI-OD** showed a red-shift of 18 nm as the solvent polarity increased from 2.7 (oDCB) to 4.2 (CF). The Stokes shift of **PBDTI-OD** in Chex is higher than both oDCB and CF, despite its low polarity index of 0.2. Similarly, a 29 nm red-shift in PL of **PBDTI-DT** is found with increasing solvent polairty from 2.7 (oDCB) to 4.2 (CF). An ever larger Stokes shift in was observed when Chex was used as a solvent. The PL spectra of **PTI-DT** showed negligible shifts in the three solvents.

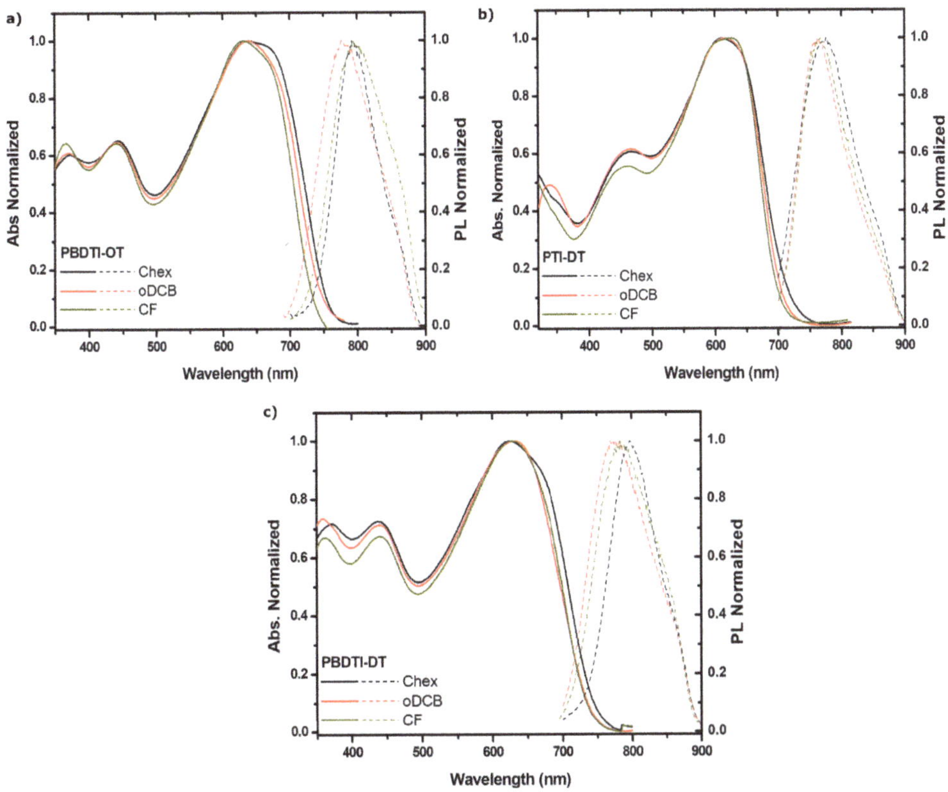

Figure 4. Absorption (solid line) and PL (broken line) of (**a**) **PBDTI-OT**, (**b**) **PBDTI-DT** and (**c**) **PTI-DT** in Chex (black), oDCB (red) and CF (dark yellow) solutions.

The red-shift in PL of **PBDTI-OD** and **PBDTI-DT** with increasing solvent polarity while their absorption remained the same confirms the higher dipole moments of their first excited states (S_1). This is consistent with our assumption that the longer wavelength region transitions have ICT characterers. Hence, their electronic structure changes from the non-polar $D - A$ to the dipolar $D^+ = A-$ configuration [15]. The larger Stokes shift observed in Chex solutions of **PBDTI-OD** and **PBDTI-DT** shows the population of a different state than ICT. Consequently, the electrons in the first excited state of the copolymers have two channels; one is the ICT state and the other is the tightly bound excitonic state. The latter is populated when there is a significant interchain interaction either in films or solutions. A poor solvent like Chex is expected to form aggregates that induce interchain interaction in the solutions, resulting in red-shifted PL. Thus, the emission in the non-polar solvent could be explained by the tightly bound interchain exciton model instead of the commonly used ICT model [26,27]. However, the Uv-Vis absorbances of **PBDTI-OD** and **PBDTI-DT** taken in solutions of similar concentration, as the PL measurements confirmed the absence of aggregation. Therefore, the aggregation effect observed in the Chex solutions should be a self-aggregation, causing the copolymer chains to form tight coils. As a result, the excited electrons could be transferred to the other part of the chain acting as interchain excitons [7,28]. Consequently, the PL spectra were red-shifted despite the low polarity of the solvent.

The photophysics of **PTI-DT** is quite different in that both its ground and exited states are not influenced by the polarities of the solvents. The similar Stokes shift, both in polar (CF and oDCB) and the non-polar (Chex) solvents, shows that self-aggregation of the copolymer chain exists in all solvents. **PTI-DT** showed a large absorption onset shift as well as a peak shift, while the BDT-based copolymers showed no peak shift upon solidification (see Table 2). This can be attributed to the twisting of the thiophene-based copolymer in dilute solution in the absence of a strong intermolecular interaction. The twisting in the backbone of **PTI-DT** favours the self-aggregation effect in any solvent. The ICT population from the first excited state of **PTI-DT** is dominated by the excitonic state, making the emitting state insensitive to polarity of the environment.

In summary, the BDT-based copolymers have a bi-relaxation channel that can be modulated with the solvents, while **PTI-DT** has a quasi-one-relaxation channel. The population of the ICT and tightly bound excitonic states were found in the BDT-based copolymers, while only the latter was found in **PTI-DT**. The results confirm the importance of donor units in the synthesis of D–A copolymers for efficient ICT processes.

3.3. Thermal Properties of PBDTI-OT, PBDTI-DT and PTI-DT

The thermal properties of **PBDTI-OT**, **PBDTI-DT** and **PTI-DT** were studied using TGA, as shown in Figure 5. The three copolymers have agood thermal stabilities, with decomposition temperatures (T^D = 5% weight loss) above 380 °C. The thermal degradation in these materials follows two steps: one around 380 °C and the other above 550 °C. The first degradation accounts for the breaking up of the side chains. The shorter side chain in the isoindigo unit of **PBDTI-OT** slightly increased its thermal stability compared to **PBDTI-DT**, despite its lower molecular weight. The second thermal degradation is due to the decomposition of the backbone structures of the copolymers. The slight increase in the degradation temperature in **PBDTI-OT** also shows the strong bonding in its backbone structure compared to **PBDTI-DT** and **PTI-DT**. In general, the thermal stabilities of the copolymers make them applicable to flexible OSCs, which are normally processed at temeratures below 150 °C.

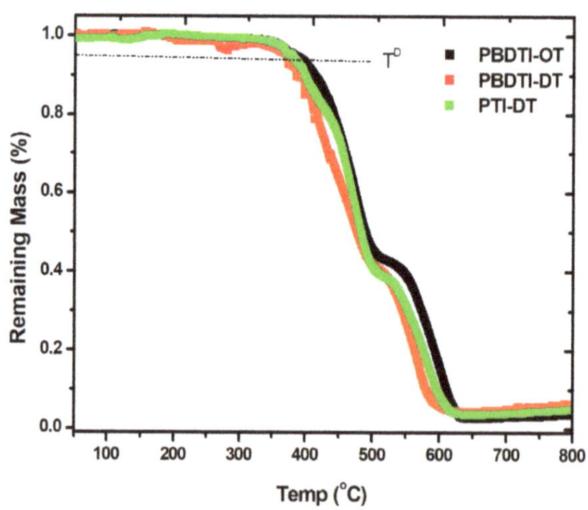

Figure 5. TGA measurement of **PBDTI-OT** (black), **PBDTI-DT** (red) and **PTI-DT** (green).

3.4. Structural Properties of PBDTI-OT, PBDTI-DT and PTI-DT Films

The structures of the copolymers were studied by powder XRD in drop-casted thin films, and the results are shown in Figure 6. The XRD traces of the three copolymers exhibit a broad peak (010) centered around $2\theta = 22°$ due to the $\pi - \pi$ stacking.

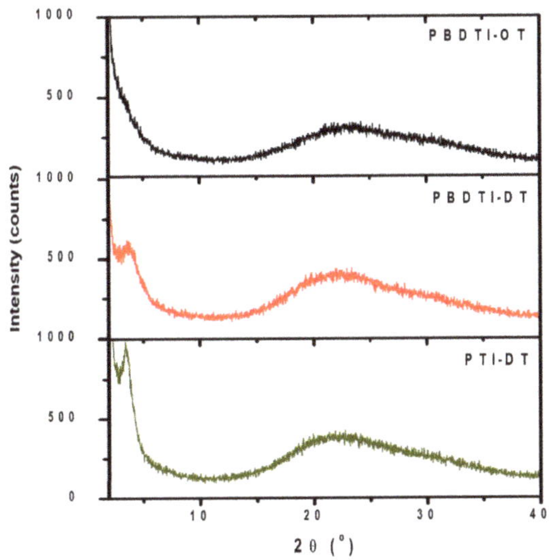

Figure 6. XRD traces **PBDTI-OT** (black), **PBDTI-DT** (red) and **PTI-DT** (dark yellow) in drop-casted films.

The calculated $\pi - \pi$ stacking distances using Bragg's equation Equation (1) are 0.38, 0.41 and 0.41 nm in **PBDTI-OT**, **PBDTI-DT** and **PTI-DT** films, respectively, confirming the presence of strong intermolecular interaction in the copolymer backbones [29]. **PBDT-DT** and **PTI-DT** exhibit a (100) diffraction peak at $2\theta = 3.82°$ and $3.60°$, corresponding to lammelar spacings of 2.3 and 2.5 nm, respectively. The inter-lammelar spacing in **PTI-DT** (2.5 nm) is higher than the well-studied thiophene-based homopolymer, poly-3-hexylthiophne (P3HT) (d = 1.63) [30]. This is due to the longer side chains both in the thiophene and isoindigo units of **PTI-DT** compared to P3HT. Note here that the (100) plane diffraction peak is sharper in **PTI-DT**, indicating its slightly higher crystallinity over the BDT-based copolymer. On the other hand, the BDT-based copolymer with a shorter side chain on the isoindigo unit (**PBDT-OT**) is fully amorphous. The higher crystallity of thiophene-based copolymer is beneficial for improved charge transport.

4. Conclusions

Three copolymers with BDT (**PBDTI-OD** and **PBDTI-DT**) and thiophene (**PTI-DT**) donor units and isoindigo acceptor units were designed and synthesized using the DAP method. The molecular weights of the BDT-based copolymers were more than two-fold higher than that of **PTI-DT**. The copolymers absorb in a wide range from from 300 to above 760 nm in thin films, making them low-band-gap polymers with $E_g \approx 1.5$ eV. The photophysical properties of the copolymers were studied in both films and solutions. Solvent polarity-dependent spectroscopic studies on **PBDTI-OD** and **PBDTI-DT** showed that the ground state is not sensitive to polarity of the environment, while the excited state is. This led us to conclude that the ground states of the BDT-based copolymers are non-polar. The PL spectra of the BDT-based copolymers were red-shifted with increasing solvent polarity of chlorinated solvents, confirming the dipolar characteristics of their first excited states. Therefore, the first excited state has ICT characteristics. However, the Stokes shift in **PBDTI-OD** and **PBDTI-DT** in the non-polar solvent, Chex, was not blue-shifted, as expected from its low polarity index. Self-aggregation of the copolymers in the poor solvent was found to favour the population of a tightly bound excitonic state, thereby inhibiting the generation of ICT state. The absorption and PL spectra of the thiophene-based copolymer were not influenced by solvent polarity. This led us to conclude that self-aggregation in the thiophene-based copolymer is induced despite the solvent polarity. Therefore, ICT state population in **PTI-DT** is not efficient. The copolymers showed excellent thermal stability, with a decomposition temperature above 380 °C. The XRD patterns of the three copolymers showed a halo from (010) plane due to $\pi - \pi$ stacking. A second diffraction peak was observed in **PBDTI-DT** and **PTI-DT**, confirming a better crystallinity in these copolymers.

Author Contributions: Z.A. and W.M. synthesized the copolymer and wrote the synthesis part of the manuscript. N.A.T. characterized the polymers and part of wrote the manuscript. All authors have read and agreed to the published version of the manuscript.

Funding: This research was funded by UNESCO and the international Development research Center, Ottawa, Canada, grant number 4500406703.

Institutional Review Board Statement: Not applicable.

Informed Consent Statement: Not applicable.

Data Availability Statement: Data will be available based on request.

Acknowledgments: This work was carried out with the aid of a grant from UNESCO and the international Development research Center, Ottawa, Canada. The views expressed herein do not necessarily represent those of UNESCO, IDRC or its Board of Governors. Z.A. and W.M. would like to acknowledge the financial support from the international Science Program (ISP), Uppsala, Sweden.

Conflicts of Interest: The authors declare no conflict of interest.

References

1. Liu, Q.; Jiang, Y.; Jin, K.; Qin, J.; Xu, J.; Li, W.; Xiong, J.; Liu, J.; Xiao, Z.; Sun, K.; et al. 18% Efficiency organic solar cells. *Sci. Bull.* **2020**, *65*, 272–275. [CrossRef]
2. Meng, L.; Zhang, Y.; Wan, X.; Li, C.; Zhang, X.; Wang, Y.; Ke, X.; Xiao, Z.; Ding, L.; Xia, R.; et al. Organic and solution-processed tandem solar cells with 17.3% efficiency. *Science* **2018**, *361*, 1094–1098. [CrossRef] [PubMed]
3. Mercier, L.G.; Leclerc, M. Direct (hetero) arylation: A new tool for polymer chemists. *Acc. Chem. Res.* **2013**, *46*, 1597–1605. [CrossRef]
4. Saito, H.; Chen, J.; Kuwabara, J.; Yasuda, T.; Kanbara, T. Facile one-pot access to π-conjugated polymers via sequential bromination/direct arylation polycondensation. *Polym. Chem.* **2017**, *8*, 3006–3012. [CrossRef]
5. Rudenko, A.E.; Thompson, B.C. Influence of the carboxylic acid additive structure on the properties of poly (3-hexylthiophene) prepared via direct arylation polymerization (DArP). *Macromolecules* **2015**, *48*, 569–575. [CrossRef]
6. Cicha, C.L.; Gockel, S.N.; Helmin, A.J.; Wilcox, W.D.; Janzen, D.E.; Pappenfus, T.M. Benzodithiophene homopolymers via direct (hetero) arylation polymerization. *Polym. Bull.* **2018**, *75*, 5667–5675. [CrossRef]
7. Tegegne, N.A.; Abdissa, Z.; Mammo, W.; Andersson, M.R.; Schlettwein, D.; Schwoerer, H. Ultrafast excited state dynamics of a bithiophene-isoindigo copolymer obtained by direct arylation polycondensation and its application in indium tin oxide-free solar cells. *J. Polym. Sci. Part B Polym. Phys.* **2018**, *56*, 1475–1483. [CrossRef]
8. Tegegne, N.A.; Abdissa, Z.; Mammo, W.; Uchiyama, T.; Okada-Shudo, Y.; Galeotti, F.; Porzio, W.; Andersson, M.R.; Schlettwein, D.; Vohra, V.; et al. Effect of Alkyl Side Chain Length on Intra-and Intermolecular Interactions of Terthiophene–Isoindigo Copolymers. *J. Phys. Chem. C* **2020**, *124*, 9644–9655. [CrossRef]
9. Tegegne, N.A.; Wendimu, H.; Abdissa, Z.; Mammo, W.; Andersson, M.R.; Hone, F.G.; Andoshee, D.M.; Olaoye, O.; Bosman, G. Light-induced degradation of a push–pull copolymer for ITO-free organic solar cell application. *J. Mater. Sci. Mater. Electron.* **2020**, *31*, 21303–21315. [CrossRef]
10. Zimmermann, D.; Sprau, C.; Schröder, J.; Gregoriou, V.G.; Avgeropoulos, A.; Chochos, C.L.; Colsmann, A.; Janietz, S.; Krüger, H. Synthesis of D-π-A-π type benzodithiophene-quinoxaline copolymers by direct arylation and their application in organic solar cells. *J. Polym. Sci. Part A Polym. Chem.* **2018**, *56*, 1457–1467. [CrossRef]
11. Ye, L.; Zhang, S.; Zhao, W.; Yao, H.; Hou, J. Highly efficient 2D-conjugated benzodithiophene-based photovoltaic polymer with linear alkylthio side chain. *Chem. Mater.* **2014**, *26*, 3603–3605. [CrossRef]
12. Liu, Y.; Wan, X.; Wang, F.; Zhou, J.; Long, G.; Tian, J.; Chen, Y. High-performance solar cells using a solution-processed small molecule containing benzodithiophene unit. *Adv. Mater.* **2011**, *23*, 5387–5391. [CrossRef]
13. Li, M.; Ni, W.; Wan, X.; Zhang, Q.; Kan, B.; Chen, Y. Benzo [1, 2-b: 4, 5-b′] dithiophene (BDT)-based small molecules for solution processed organic solar cells. *J. Mater. Chem. A* **2015**, *3*, 4765–4776. [CrossRef]

14. Eckhardt, H.; Shacklette, L.; Jen, K.; Elsenbaumer, R. The electronic and electrochemical properties of poly (phenylene vinylenes) and poly (thienylene vinylenes): An experimental and theoretical study. *J. Chem. Phys.* **1989**, *91*, 1303–1315. [CrossRef]
15. Jespersen, K.G.; Beenken, W.J.; Zaushitsyn, Y.; Yartsev, A.; Andersson, M.; Pullerits, T.; Sundström, V. The electronic states of polyfluorene copolymers with alternating donor-acceptor units. *J. Chem. Phys.* **2004**, *121*, 12613–12617. [CrossRef] [PubMed]
16. Ma, Z.; Wang, E.; Jarvid, M.E.; Henriksson, P.; Inganäs, O.; Zhang, F.; Andersson, M.R. Synthesis and characterization of benzodithiophene–isoindigo polymers for solar cells. *J. Mater. Chem.* **2012**, *22*, 2306–2314. [CrossRef]
17. Wang, B.; Zhang, J.; Tam, H.L.; Wu, B.; Zhang, W.; Chan, M.S.; Pan, F.; Yu, G.; Zhu, F.; Wong, M.S. Impact of alkyl side chains on the photovoltaic and charge mobility properties of naphthodithiophene–benzothiadiazole copolymers. *Polym. Chem.* **2014**, *5*, 836–843. [CrossRef]
18. Osaka, I.; Saito, M.; Koganezawa, T.; Takimiya, K. Thiophene–thiazolothiazole copolymers: Significant impact of side chain composition on backbone orientation and solar cell performances. *Adv. Mater.* **2014**, *26*, 331–338. [CrossRef]
19. Zhang, S.; Ye, L.; Zhao, W.; Liu, D.; Yao, H.; Hou, J. Side chain selection for designing highly efficient photovoltaic polymers with 2D-conjugated structure. *Macromolecules* **2014**, *47*, 4653–4659. [CrossRef]
20. Scharber, M.C.; Mühlbacher, D.; Koppe, M.; Denk, P.; Waldauf, C.; Heeger, A.J.; Brabec, C.J. Design rules for donors in bulk-heterojunction solar cells—Towards 10% energy-conversion efficiency. *Adv. Mater.* **2006**, *18*, 789–794. [CrossRef]
21. Yu, Y.; Yang, B.; Yuan, Y.; Zhang, H. Novel side-chain alternative copolymer combined FRET and DRET with large pseudo-Stokes shift and polarity-sensitive fluorescence behavior. *J. Mater. Chem. C* **2019**, *36*, 11285–11292. [CrossRef]
22. Tessarolo, M.; Gedefaw, D.; Bolognesi, M.; Liscio, F.; Henriksson, P.; Zhuang, W.; Milita, S.; Muccini, M.; Wang, E.; Seri, M.; et al. Structural tuning of quinoxaline-benzodithiophene copolymers via alkyl side chain manipulation: Synthesis, characterization and photovoltaic properties. *J. Mater. Chem. A* **2014**, *2*, 11162–11170. [CrossRef]
23. Zhang, A.; Xiao, C.; Meng, D.; Wang, Q.; Zhang, X.; Hu, W.; Zhan, X.; Wang, Z.; Janssen, R.A.; Li, W. Conjugated polymers with deep LUMO levels for field-effect transistors and polymer–polymer solar cells. *J. Mater. Chem. C* **2015**, *3*, 8255–8261. [CrossRef]
24. Brabec, C.J.; Cravino, A.; Meissner, D.; Sariciftci, N.S.; Fromherz, T.; Rispens, M.T.; Sanchez, L.; Hummelen, J.C. Origin of the open circuit voltage of plastic solar cells. *Adv. Funct. Mater.* **2001**, *11*, 374–380. [CrossRef]
25. Mola, G.T.; Abera, N. Correlation between LUMO offset of donor/acceptor molecules to an open circuit voltage in bulk heterojunction solar cell. *Phys. B Condens. Matter* **2014**, *445*, 56–59. [CrossRef]
26. Sowmiya, M.; Tiwari, A.K.; Saha, S.K. Study on intramolecular charge transfer fluorescence properties of trans-4-[4′-(N, N′-dimethylamino) styryl] pyridine: Effect of solvent and pH. *J. Photochem. Photobiol. A Chem.* **2011**, *218*, 76–86. [CrossRef]
27. Belletête, M.; Bouchard, J.; Leclerc, M.; Durocher, G. Photophysics and solvent-induced aggregation of 2, 7-carbazole-based conjugated polymers. *Macromolecules* **2005**, *38*, 880–887. [CrossRef]
28. Rolczynski, B.S.; Szarko, J.M.; Son, H.J.; Liang, Y.; Yu, L.; Chen, L.X. Ultrafast intramolecular exciton splitting dynamics in isolated low-band-gap polymers and their implications in photovoltaic materials design. *J. Am. Chem. Soc.* **2012**, *134*, 4142–4152. [CrossRef]
29. Choi, M.H.; Song, K.W.; Moon, D.K. Alkylidenefluorene–isoindigo copolymers with an optimized molecular conformation for spacer manipulation, π–π stacking and their application in efficient photovoltaic devices. *Polym. Chem.* **2015**, *6*, 2636–2646. [CrossRef]
30. Vanlaeke, P.; Swinnen, A.; Haeldermans, I.; Vanhoyland, G.; Aernouts, T.; Cheyns, D.; Deibel, C.; D'Haen, J.; Heremans, P.; Poortmans, J.; et al. P3HT/PCBM bulk heterojunction solar cells: Relation between morphology and electro-optical characteristics. *Sol. Energy Mater. Sol. Cells* **2006**, *90*, 2150–2158. [CrossRef]

MDPI
St. Alban-Anlage 66
4052 Basel
Switzerland
Tel. +41 61 683 77 34
Fax +41 61 302 89 18
www.mdpi.com

Polymers Editorial Office
E-mail: polymers@mdpi.com
www.mdpi.com/journal/polymers

www.ingramcontent.com/pod-product-compliance
Lightning Source LLC
Chambersburg PA
CBHW061445220926
51850CB00004B/14
9783036572710